Conversion Factors for SI and U.S. Units*

To convert from	To	Multiply by	
Length, area, volume			
foot (ft)	meter (m)	0.30480	3.2808
inch (in.)	m	0.025400	39.370
statute mile (mi)	m	1609.3	6.2137×10^{-4}
foot^2 (ft^2)	meter^2 (m^2)	0.092903	10.764
inch^2 (in.^2)	m^2	6.4516×10^{-4}	1550.0
foot^3 (ft^3)	meter^3 (m^3)	0.028317	35.315
inch^3 (in.^3)	m^3	1.6387×10^{-5}	61024
Velocity			
feet/second (ft/sec)	meter/second (m/s)	0.30480	3.2808
feet/minute (ft/min)	m/s	0.0050800	196.85
knot (nautical mi/hr)	m/s	0.51444	1.9438
mile/hour (mi/hr)	m/s	0.44704	2.2369
mile/hour (mi/hr)	kilometer/hour (km/h)	1.6093	0.62137
Acceleration			
feet/second^2 (ft/sec^2)	meter/second^2 (m/s^2)	0.30480	3.2808
inch/second^2 (in./sec^2)	m/s^2	0.025400	39.370
Mass			
slug ($\text{lb-sec}^2\text{/ft}$)	kg	14.594	0.068522
Force			
pound (lb) or pound-force (lbf)	newton (N)	4.4482	0.22481
Density			
slug/foot^3 (slug/ft^3)	kg/m^3	515.38	0.0019403
Energy, work, or moment of force			
foot-pound or pound-foot (ft-lb) (lb-ft)	joule (J) or newton · meter (N · m)	1.3558	0.73757
Power			
foot-pound/second (ft-lb/sec)	watt (W)	1.3558	0.73756
horsepower (hp) (550 ft-lb/sec)	W	745.70	0.0013410
Stress, pressure			
pound/inch^2 (lb/in.^2 or psi)	N/m^2 (or Pa)	6894.8	1.4504×10^{-4}
pound/foot^2 (lb/ft^2)	N/m^2 (or Pa)	47.880	0.020886
Mass moment of inertia			
slug-foot^2 (slug-ft^2 or lb-ft-sec^2)	$\text{kg} \cdot \text{m}^2$	1.3558	0.73756
Momentum (or linear momentum)			
slug-foot/second (slug-ft/sec)	kg · m/s	4.4482	0.22481
Impulse (or linear impulse)			
pound-second (lb-sec)	N · s (or kg · m/s)	4.4482	0.22481
Moment of momentum (or angular momentum)			
$\text{slug-foot}^2\text{/second}$ ($\text{slug-ft}^2\text{/sec}$)	$\text{kg} \cdot \text{m}^2\text{/s}$	1.3558	0.73756
Angular impulse			
pound-foot-second (lb-ft-sec)	$\text{N} \cdot \text{m} \cdot \text{s}$ (or $\text{kg} \cdot \text{m}^2\text{/s}$)	1.3558	0.73756

*Rounded to the five digits cited. Note, for example, that 1 ft = 0.30480 m, so that

$$(\text{Number of feet}) \times \left(\frac{0.30480\ \text{m}}{1\ \text{ft}}\right) = \text{Number of meters}$$

ENGINEERING MECHANICS
AN INTRODUCTION TO
DYNAMICS

PWS SERIES IN ENGINEERING

Third Edition

ENGINEERING MECHANICS
AN INTRODUCTION TO
DYNAMICS

DAVID J. MCGILL AND WILTON W. KING
Georgia Institute of Technology

PWS Publishing Company
Boston

ITP An International Thomson Publishing Company

Boston • Albany • Bonn • Cincinnati • Detroit • London • Madrid
Melbourne • Mexico City • New York • Paris • San Francisco
Singapore • Tokyo • Toronto • Washington

PWS PUBLISHING COMPANY
20 Park Plaza, Boston, Massachusetts 02116-4324

This book is printed on recycled, acid-free paper

I(T)P™
International Thomson Publishing
The trademark ITP is used under license.

For more information, contact:

PWS Publishing Company
20 Park Plaza
Boston, MA 02116

International Thomson Publishing Europe
Berkshire House I68-I73
High Holborn
London WC1V 7AA
England

Thomas Nelson Australia
102 Dodds Street
South Melbourne, 3205
Victoria, Australia

Nelson Canada
1120 Birchmount Road
Scarborough, Ontario
Canada M1K 5G4

International Thomson Editores
Campos Eliseos 385, Piso 7
Col. Polanco
11560 Mexico D.F., Mexico

International Thomson Publishing GmbH
Konigswinterer Strasse 418
53227 Bonn, Germany

International Thomson Publishing Asia
221 Henderson Road
#05-10 Henderson Building
Singapore 0315

International Thomson Publishing Japan
Hirakawacho Kyowa Building, 31
2-2-1 Hirakawacho
Chiyoda-ku, Tokyo 102
Japan

Library of Congress Cataloging-in-Publication Data

McGill, David J., 1939–
Engineering Mechanics, an introduction to dynamics / David J. McGill and Wilton W. King.
— 3rd ed.
p. cm.
At head of title: Engineering mechanics.
Includes index.
ISBN 0-534-93399-8
1. Dynamics. I. King, Wilton W., . II. Title.
III. Title: Engineering mechanics.
TA352.M385 1994
620.1'04 — dc20 94-33846
CIP

Printed and bound in the United States of America
95 96 97 98 99 — 10 9 8 7 6 5 4 3 2

Sponsoring Editor: Jonathan Plant
Editorial Assistant: Cynthia Harris
Developmental Editor: Mary Thomas
Production Coordinator: Kirby Lozyniak
Marketing Manager: Nathan Wilbur
Manufacturing Coordinator: Marcia Locke
Production: York Production Services

Cover Designer: Julie Gecha
Interior Designer: York Production Services
Compositor: Progressive Information Technologies
Cover Photo: © by Walter Bibikow, courtesy of Image Bank
Cover Printer: John P. Pow Company, Inc.
Text Printer: Quebecor Printing, Hawkins

TO OUR WIVES, CAROLYN AND KAY

CONTENTS

PREFACE

An Introduction to Dynamics is the second of two volumes covering basic topics of mechanics. The first two-thirds of the book contains most of the topics traditionally taught in a first course in dynamics at most colleges of engineering.

In the writing of this text we have followed one basic guideline — to write the book the same way we teach the course. To this end, we have written many explanatory footnotes and included frequent questions interspersed throughout the chapters. These questions are the same kind as the ones we ask in class; to make the most of them, treat them as serious homework as you read, and look at the answers only after you have your own answer in mind. The questions are intended to encourage thinking about tricky points and to emphasize the basic principles of the subject.

In addition to the text questions, a set of approximately one dozen review questions and answers are included at the end of each chapter. These true-false questions are designed for both classroom discussion and for student review. Homework problems of varying degrees of difficulty appear at the end of every major section. There are over 1,100 of these exercises, and the answers to the odd-numbered ones constitute Appendix D in the back of the book.

There are a number of reasons (besides carelessness) why it may be difficult to get the correct answer to a homework problem on the first try. The problem may require an unusual amount of thinking and insight; it may contain tedious calculations; or it may challenge the student's advanced mathematics skills. We have placed an asterisk beside especially difficult problems falling into one or more of these categories.

Some examples and problems are presented in SI (Système International) metric units, whereas others use traditional United States engineering system units. Whereas the United States is slowly and painfully converting to SI units, our consulting activities make it clear that much engineering work is still being performed using traditional units. Most United States engineers still tend to think in pounds instead of newtons and in feet instead of meters. We believe students will become much better engineers, scientists, and scholars if they are thoroughly familiar with both systems.

Dynamics is a subject rich in its varied applications; therefore, it is important that the student develop a feel for realistically modeling an engineering situation. Consequently, we have included a large number of actual engineering problems among the examples and exercises. Being aware of the assumptions and accompanying limitations of the model and the solution method is a valuable skill that can only be developed by sweating over many problems outside the classroom. Only in this way can a student develop the insight and creativity that must be brought to bear on engineering problems.

Kinematics of the particle, or of a material point of a body, is covered in Chapter 1. The associated kinetics of particles and mass centers of bodies follows logically in Chapter 2. Here it will be seen that we have not dwelt at length upon the "point-mass" model of a body. Since the engineering student will be dealing with bodies of finite dimensions, we believe that it is important to present equations of motion valid for such bodies as quickly as possible. Thus Euler's laws have been introduced relatively early; this provides for a compact presentation of general principles without, in our opinion and experience, any loss of understanding on the student's part. This is not meant to deprecate the point-mass model, which surely plays an important role in classical physics and can be utilized in a number of engineering problems. As we shall see in Chapter 2, however, these problems may be attacked directly through the equation of motion of the mass center of a body without detracting from the view that the body has finite dimensions. Trajectory problems, sometimes placed with particle kinematics, will be found in this kinetics chapter also, since a law of motion is essential to their formulation.

The rigid body in plane motion is treated in detail in the center of the book — the kinematics in Chapter 3 and the kinetics in Chapters 4 and 5. In Chapter 3 the topic of rolling is discussed only after both the velocity and acceleration equations relating two points of a rigid body have been covered. Further, we treat the equations of velocity and acceleration of a point moving relative to two frames of reference ("moving frames") in Chapters 3 (plane kinematics) and 6 (three-dimensional, or general, kinematics), after the student has been properly introduced to the angular velocity vector in these chapters.

Chapter 4 approaches plane kinetics from the equations of motion, written with the aid of a free-body diagram of the body being studied — that is, a sketch of the body depicting all the external forces and couples but excluding any vectors expressing acceleration. Thus the free-body diagram means the same thing in dynamics as it does in statics, facilitating the student's transition to the more difficult subject. Moments and products of inertia are covered right where they appear in the development of kinetics. This presentation gives students an appreciation of these concepts, as well as a sense of history, as they encounter them along the same paths that were traveled by the old masters.

Chapter 5 is dedicated to solving plane kinetics problems of rigid bodies with certain special yet general solutions (or integrals) of the equations of Chapters 2 and 4. These are known as the principles of work and kinetic energy, impulse and momentum, and angular impulse and angular momentum.

Chapters 6 and 7 deal comprehensively with the kinematics and kinetics, respectively, of rigid bodies in three dimensions. There is no natural linear extension from plane to general motion, and the culprit is the angular velocity vector $\boldsymbol{\omega}$, which depends in a much more complicated way than "$\dot{\theta}\hat{\mathbf{k}}$" on the angles used to orient the body in three dimensions. We have found that if students understand the angular velocity vector $\boldsymbol{\omega}$, they will have little trouble with the general motion of rigid bodies. Thus we begin Chapter 6 with a study of $\boldsymbol{\omega}$ and its properties. In three dimensions, the definition of angular velocity is motivationally developed through the relationship between derivatives of a vector in two different frames of reference. While this point of view is often associated with more advanced texts, we have found that college students at the junior level are quite capable of appreciating and exploiting the power of this approach. In particular, it allows the student to attack, in an orderly way, intimidating problems such as motions of gear systems and those of universal joints connecting noncollinear shafts.

Chapter 8 is an introduction to three special topics in the area of dynamics: vibrations, mass redistribution problems, and central force motion.

We have received a number of helpful suggestions from those who taught from earlier editions of the text, and we are especially grateful to Lawrence Malvern of the *University of Florida* and to our colleagues at *Georgia Institute of Technology*, in particular Don Berghaus, Mike Bernard, Al Ferri, Satya Hanagud, Dewey Hodges, Larry Jacobs, John Papastavridis, Jianmin Qu, George Rentzepis, Virgil Smith, Charles Ueng, Ray Vito, James Wang, Gerry Wempner and Wan-Lee Yin.

We also wish to thank our reviewers this time around:

William Bickford
Arizona State University

Donald E. Carlson
University of Illinois at Urbana

Robert L. Collins
University of Louisville

John Dickerson
University of South Carolina

John F. Ely
North Carolina State University

Laurence Jacobs
Georgia Institute of Technology

Seymour Lampert
University of Southern California

Vincent WoSang Lee
University of Southern California

Joseph Longuski
Purdue University

Robert G. Oakberg
Montana State University

Joseph E. Panarelli
University of Nebraska

Mario P. Rivera
Union College

Wallace S. Venable
West Virginia University

Carl Vilmann
Michigan Technological University

We are grateful to the following professors, who each responded to a questionnaire we personally sent out in 1991: Don Carlson, *University of Illinois;* Patrick MacDonald and John Ely, *North Carolina State University;* Vincent Lee, *University of Southern California;* Charles Krousgrill, *Purdue University;* Samuel Sutcliffe, *Tufts University;* Larry Malvern and Martin Eisenberg, *University of Florida;* John Dickerson, *University of South Carolina;* Bill Bickford, *Arizona State University;* James Wilson, *Duke University;* Mario Rivera, *Union College;* and Larry Jacobs, *Georgia Institute of Technology.* Their comments were also invaluable.

Special appreciation is expressed to our insightful editor, Jonathan Plant, our typist, Meghan Root, and to the following individuals involved in the smooth production of this third edition: Mary Thomas and Kirby Lozyniak of *PWS Publishing Company,* and Tamra Winters of *York Production Services.*

David J. McGill
Wilton W. King

We are pleased to introduce to this edition a new set of model-based problems. These problems, presented in a full-color insert bound into the book, introduce students to the process of building three-dimensional models from commonly found objects in order to observe as well as calculate mechanical behavior. Many students beginning their engineering education lack a hands-on, intuitive feel for this behavior, and these specially designed problems can help build confidence in their observational and analytical abilities.

We wish to acknowledge the following contributors to the model-based problems insert:

David J. McGill and Wilton W. King for their initial conception and presentation of the model-based problem idea in *Dynamics Model Problems,* written to accompany *Engineering Mechanics: An Introduction to Dynamics,* Third Edition.

David Barnett, *Stanford University,* Mario P. Rivera, *Union College;* Robert G. Oakberg, *Montana State University;* John F. Ely, *North Carolina State University;* Carl Vilmann, *Michigan Technological University;* Robert L. Collins, *University of Louisville;* Nicholas P. Jones, *Johns Hopkins University* and William B. Bickford, *Arizona State University* for their evaluations of McGill and King's *Dynamics Model Problems.*

We thank Mario P. Rivera, Robert G. Oakberg, and John F. Ely, for developing additional model problems for the insert. And a very special thanks to Michael K. Wells, *Montana State University,* for developing and editing the final text of the insert, and for providing an introduction and additional problems.

PWS Publishing Company

ENGINEERING MECHANICS

AN INTRODUCTION TO

DYNAMICS

1 Kinematics of Material Points or Particles

1.1 Introduction

Dynamics is the general name given to the study of the motions of bodies and the forces that accompany or cause those motions. The branch of the subject that deals only with considerations of space and time is called **kinematics.** The branch that deals with the relationships between forces and motions is called **kinetics,** but since the force-motion relationships involve kinematic considerations, it is necessary to study kinematics first.

In this chapter we present some fundamentals of the kinematics of a material point or, equivalently, an infinitesimal element of material. We shall use the term **particle** for such an element, but we shall also use this term in a broader sense to denote a piece of material sufficiently small that the locations of its different material points need not be distinguished. The vagueness of this definition correctly suggests that, for *some* purposes, a truck or a space vehicle or even a planet might be modeled adequately as a particle.

The key elements of the kinematics of a point (or particle) are its position, velocity and acceleration. Velocity is rate-of-change of position, and acceleration is rate-of-change of velocity. It is the acceleration in Newton's Second Law and the position-velocity-acceleration relationships that allow us to deduce two things: the forces that must act for a particle to achieve a certain motion; or the evolution of a particle's position under the action of a set of prescribed forces.

Position, velocity, and acceleration vectors are defined in Section 1.3. These definitions are independent of the choice of any particular coordinate system. However, solution of a practical problem almost always involves the use of some specific coordinate system, the most common being rectangular (Cartesian), cylindrical and spherical. The rectangular and the cylindrical coordinate systems are treated in great detail in this chapter (Sections 1.4–1.6) because we judge them to be of greatest practical use, particularly for problems of motion confined to a plane. These developments are sufficient for establishing the procedures to follow should the reader later find it desirable to develop counterpart relationships using some other system.

If we focus our attention on the path being traversed by a point (or particle), we find that the velocity of the point is tangent to the path and that the acceleration has parts normal and tangent to the path with special significances. These characteristics are developed and exploited in Section 1.7.

We now begin our study of particle kinematics with a preliminary section devoted to the calculus of vectors that depend upon scalars. In particular, we need to understand how to take derivatives of vectors with respect to time and to acknowledge the crucial role of the frame of reference.

1.2 Reference Frames and Vector Derivatives

In the next section and throughout the book, we are going to be differentiating vectors; the derivative of the position vector of a point will be its velocity, for example. Thus in this preliminary section it seems wise to examine the concept of the derivative of a vector **A**, which is a function of time t. The definition of $d\mathbf{A}/dt$, which is also commonly written as $\dot{\mathbf{A}}$, is deceptively simple:

$$\frac{d\mathbf{A}}{dt} \equiv \lim_{\Delta t \to 0} \left[\frac{\mathbf{A}(t + \Delta t) - \mathbf{A}(t)}{\Delta t} \right] \tag{1.1}$$

This definition closely parallels the definition of the derivative of a scalar, such as dy/dx, as found in any calculus text. But what we must realize about a vector is that it can change with time to *two* ways—in direction as well as in magnitude. This means that $\dot{\mathbf{A}}$ is intrinsically tied to the frame of reference in which the derivative is taken.

To illustrate this idea, consider the two points P and Q on the surface of the phonograph record in Figure 1.1. The record rests on a turntable that revolves in the indicated direction at, say, $33\frac{1}{3}$ rpm.

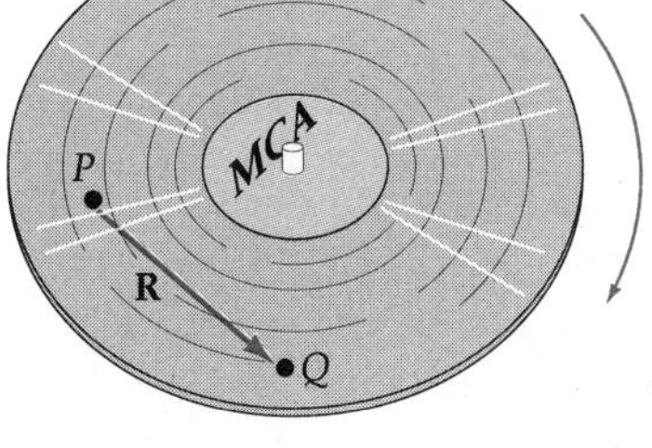

Figure 1.1

Suppose we call **R** the vector that is the directed line segment from P to Q, and inquire about the rate at which **R** changes with time as the turntable rotates. Even though we perceive the record and turntable to behave as a rigid body so that the distance between P and Q (that is, the magnitude of **R**) is constant, most of us would judge $d\mathbf{R}/dt$ to be nonzero, owing to the varying direction of **R**. This conclusion follows from our automatically having adopted the building (or earth) as our frame of reference. If we were to ride on the turntable and blind ourselves to the surroundings, however, our perception would be that **R** is a constant vector and consequently has a vanishing derivative. Thus **R**, relative to the turntable, is a constant vector and, relative to the building, is a constant-magnitude but varying-direction vector. It is therefore seen that $d\mathbf{R}/dt$ cannot be evaluated except by specific association with a **frame of reference,** which is nothing more nor less than a **rigid body.** We shall discuss the frame of reference concept further in the next section and again in Chapter 3.

We shall be needing several vector derivative relationships that have analogs in the calculus of scalars; these relationships follow directly from the definition (Equation 1.1). If α is a scalar and **A** and **B** are vector functions of t, then

$$\frac{d}{dt}(\alpha\mathbf{A}) = \left(\frac{d\alpha}{dt}\right)\mathbf{A} + \alpha\left(\frac{d\mathbf{A}}{dt}\right) \tag{1.2}$$

$$\frac{d}{dt}(\mathbf{A} + \mathbf{B}) = \frac{d\mathbf{A}}{dt} + \frac{d\mathbf{B}}{dt} \tag{1.3}$$

$$\frac{d}{dt}(\mathbf{A} \cdot \mathbf{B}) = \left(\frac{d\mathbf{A}}{dt}\right) \cdot \mathbf{B} + \mathbf{A} \cdot \left(\frac{d\mathbf{B}}{dt}\right) \tag{1.4}$$

$$\frac{d}{dt}(\mathbf{A}\times\mathbf{B}) = \left(\frac{d\mathbf{A}}{dt}\right)\times\mathbf{B} + \mathbf{A}\times\left(\frac{d\mathbf{B}}{dt}\right) \tag{1.5}$$

The first and second of these equations allow us to be more specific about the manner in which differentiation is linked to a frame of reference. Suppose that $\hat{\mathbf{e}}_1, \hat{\mathbf{e}}_2, \hat{\mathbf{e}}_3$ are mutually perpendicular unit vectors* and A_1, A_2, A_3 are the corresponding scalar components of a vector $\mathbf{A}$ so that

$$\mathbf{A} = A_1\hat{\mathbf{e}}_1 + A_2\hat{\mathbf{e}}_2 + A_3\hat{\mathbf{e}}_3 \tag{1.6}$$

If $\mathcal{I}$ is the frame of reference** and we denote the derivative of $\mathbf{A}$ relative to $\mathcal{I}$ by ${}^{\mathcal{I}}d\mathbf{A}/dt$, then

$$\begin{aligned}\frac{{}^{\mathcal{I}}d\mathbf{A}}{dt} = \left(\frac{dA_1}{dt}\right)^{\dagger}\hat{\mathbf{e}}_1 + A_1\,{}^{\mathcal{I}}\!\left(\frac{d\hat{\mathbf{e}}_1}{dt}\right) + \left(\frac{dA_2}{dt}\right)\hat{\mathbf{e}}_2 \\ + A_2\,{}^{\mathcal{I}}\!\left(\frac{d\hat{\mathbf{e}}_2}{dt}\right) + \left(\frac{dA_3}{dt}\right)\hat{\mathbf{e}}_3 + A_3\,{}^{\mathcal{I}}\!\left(\frac{d\hat{\mathbf{e}}_3}{dt}\right)\end{aligned} \tag{1.7}$$

Now if we choose $\hat{\mathbf{e}}_1$, $\hat{\mathbf{e}}_2$, $\hat{\mathbf{e}}_3$ to have fixed directions in $\mathcal{I}$, they are each constant there and

$$\frac{{}^{\mathcal{I}}d\mathbf{A}}{dt} = \frac{dA_1}{dt}\hat{\mathbf{e}}_1 + \frac{dA_2}{dt}\hat{\mathbf{e}}_2 + \frac{dA_3}{dt}\hat{\mathbf{e}}_3 \tag{1.8}$$

which is the most straightforward way to express the derivative of a vector and its intrinsic association with a frame of reference. We now give one example of the use of Equation (1.8) and, assuming the reader to be familiar with Equations (1.1) to (1.5), then move on to Section 1.3 and the task of describing the motion of a point (particle) P.

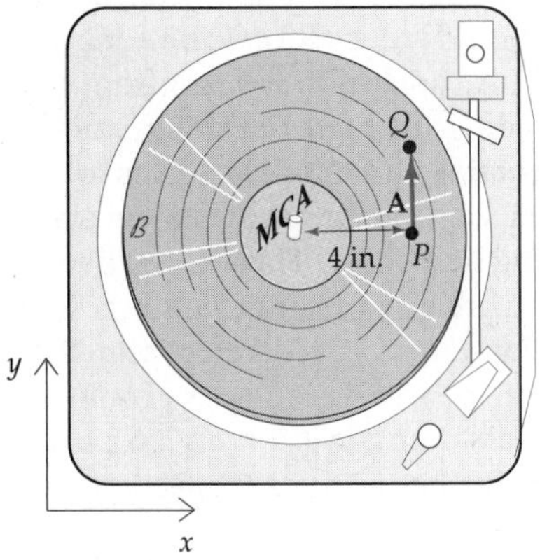

Figure E1.1a

EXAMPLE 1.1

If the distance from P to Q on the $33\frac{1}{3}$ rpm record in Figure E1.1a is 3 in. and if $\mathbf{A}$ is the vector from P to Q, find ${}^{\mathcal{I}}\dot{\mathbf{A}}$, where the frame $\mathcal{I}$ is the cabinet of the stereo in which the axes (x, y, z) are embedded. It is also given that the line PQ is in the indicated position (parallel to y) when $t = 0$.

Solution

At a later time t (in seconds), the vector $\mathbf{A}$ is seen in Figure E1.1b to make an angle $\theta(t)$ with y of

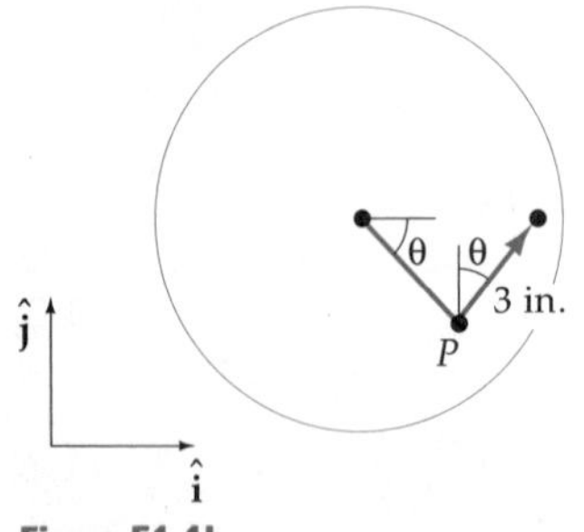

Figure E1.1b

---

* Note that we could use any set of base vectors (that is, linearly independent reference vectors) here, in which case A_1, A_2, A_3 are not necessarily orthogonal components of $\mathbf{A}$. Equation (1.6) simply illustrates the most common choice of scalars and base vectors. When this is the case, the magnitude of $\mathbf{A}$, written $|\mathbf{A}|$ or sometimes A, is $\sqrt{A_1^2 + A_2^2 + A_3^2}$.

** Throughout the book, frames (rigid bodies) are denoted by capital script letters. These are intended simply to be the capital cursive letters we use in writing; thus we *write* the names of bodies and *print* the names of points. We do this because, as we shall see in Chapter 3, points and rigid bodies have very different motion properties.

† Note that the derivatives of the scalar components of $\mathbf{A}$, such as dA_1/dt, need not be "tagged" since they are the same in any frame.

$$\theta = (33\tfrac{1}{3})\left(\frac{2\pi}{60}\right)t = 3.49t \text{ rad}$$

The vector **A**, expressed in terms of the unit vectors $\hat{\mathbf{i}}$ and $\hat{\mathbf{j}}$ in the respective directions of x and y, then has the following form:

$$\mathbf{A} = 3(\sin\theta\hat{\mathbf{i}} + \cos\theta\hat{\mathbf{j}}) \text{ in.}$$

Noting that the unit vectors do not change in direction in $\mathcal{I}$, we obtain, using Equation (1.8),

$$\frac{{}^{\mathcal{I}}d\mathbf{A}}{dt} = {}^{\mathcal{I}}\dot{\mathbf{A}} = 3\cos\theta\frac{d\theta}{dt}\hat{\mathbf{i}} - 3\sin\theta\frac{d\theta}{dt}\hat{\mathbf{j}}$$

$$= 3\cos(3.49t)(3.49)\hat{\mathbf{i}} - 3\sin(3.49t)(3.49)\hat{\mathbf{j}}$$

$$= 10.5(\cos\theta\hat{\mathbf{i}} - \sin\theta\hat{\mathbf{j}}) \text{ in./sec}$$

We see from this result, for example, that:

1. At $\theta = 0$, ${}^{\mathcal{I}}\dot{\mathbf{A}}$ is in the x direction.
2. At $\theta = \pi/2$, ${}^{\mathcal{I}}\dot{\mathbf{A}}$ is in the $-y$ direction.
3. At $\theta = \pi$, ${}^{\mathcal{I}}\dot{\mathbf{A}}$ is in the $-x$ direction.
4. At $\theta = 3\pi/2$, ${}^{\mathcal{I}}\dot{\mathbf{A}}$ is in the y direction.

In all four cases, and at all intermediate angles as well, the derivative of **A** in $\mathcal{I}$ is seen to be that of the cross product:

$$[\dot{\theta}(-\hat{\mathbf{k}})] \times \mathbf{A}$$

The bracketed vector represents what will come to be called *the angular velocity* of the record ($\mathcal{B}$) in the reference frame (stereo cabinet) $\mathcal{I}$. In later chapters we shall see that it is precisely this cross product that must be added to ${}^{\mathcal{B}}\dot{\mathbf{A}}$ to obtain ${}^{\mathcal{I}}\dot{\mathbf{A}}$. Here, of course, **A** is constant relative to the turntable $\mathcal{B}$ so that its derivative in $\mathcal{B}$ (that is, ${}^{\mathcal{B}}\dot{\mathbf{A}}$) vanishes.

PROBLEMS ▶ Section 1.2

In Problems 1.1–1.8, $\hat{\mathbf{i}}$, $\hat{\mathbf{j}}$, $\hat{\mathbf{k}}$ are mutually perpendicular unit vectors having directions fixed in the frame of reference. In each case t is time measured in seconds ("s" in SI units). Determine at $t = 3$ s the rate of change (with respect to time) of vector **L**.

1.1 $\mathbf{L} = 2\hat{\mathbf{i}} + 3t^2\hat{\mathbf{j}} - 8t\hat{\mathbf{k}}$ kg-m/s

1.2 $\mathbf{L} = 20\sin(\pi t/4)\hat{\mathbf{i}} - 20\cos(\pi t/4)\hat{\mathbf{k}}$ slug-ft/sec

1.3 $\mathbf{L} = -100e^{-t/2}\hat{\mathbf{i}} + 20t\hat{\mathbf{j}} - 5t^2\hat{\mathbf{k}}$ slug-ft/sec

1.4 $\mathbf{L} = 20\cosh(\pi t/4)\hat{\mathbf{i}} + 20\sinh(\pi t/4)\hat{\mathbf{k}}$ kg-m/s

1.5 $\mathbf{L} = 5t\hat{\mathbf{i}} - \dfrac{12}{t^2}\hat{\mathbf{j}} + \dfrac{6}{t^3}\hat{\mathbf{k}}$ slug-ft/sec

1.6 $\mathbf{L} = e^{-6t}(5\hat{\mathbf{i}} + 8t\hat{\mathbf{j}})$ slug-ft/sec

1.7 $\mathbf{L} = -2te^{-t^2}\hat{\mathbf{i}} + 3^t\hat{\mathbf{j}}$ kg-m/s

1.8 $\mathbf{L} = 50\hat{\mathbf{i}} + 60\ln t\hat{\mathbf{j}}$ kg-m/s

1.9–1.16 If the vectors enumerated in Problems 1.1–1.8 represent various forces **F**, find the integral of each force over the time interval from $t = 2$ through 5 sec. Let the metric units become newtons and the U.S. units become pounds.

1.3 Position, Velocity, and Acceleration

In this short but important section, we present the definitions of the position, velocity, and acceleration vectors of a material point P as it moves relative to a frame of reference $\mathcal{I}$. It is important to mention that while a frame of reference is usually identified by the material constituting the reference body (for example, the earth, the moon, or the body of a truck), the frame is actually composed of all those material points *plus* the points generated by a rigid extension of the body to all of space. Thus, for example, we refer to a point on the centerline of a straight pipe as a point in (or of) the pipe.

We now consider a point P as it moves along a path as shown in Figure 1.2. The **path** is the locus of points of $\mathcal{I}$ that P occupies as time passes. If we select a point O of $\mathcal{I}$ to be our reference point (or origin), then the depicted vector from O to P is called a **position vector** for P in $\mathcal{I}$ and is written $\mathbf{r}_{OP}$.

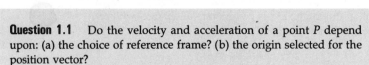

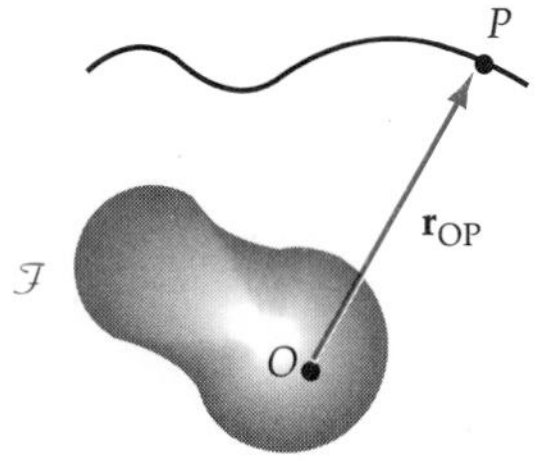

Figure 1.2 Position vector for P in $\mathcal{I}$.

The first and second derivatives (with respect to time) of the position vector are respectively called the **velocity** ($\mathbf{v}_P$) and **acceleration** ($\mathbf{a}_P$) of point P in $\mathcal{I}$:

$$\mathbf{v}_P = \frac{d\mathbf{r}_{OP}}{dt} = \dot{\mathbf{r}}_{OP} \qquad \text{(The } \textit{magnitude} \text{ of } \mathbf{v}_P \text{ is called the } \textbf{speed} \text{ of } P.) \tag{1.9}$$

$$\mathbf{a}_P = \frac{d^2\mathbf{r}_{OP}}{dt^2} = \ddot{\mathbf{r}}_{OP} = \frac{d\mathbf{v}_P}{dt} = \dot{\mathbf{v}}_P \tag{1.10}$$

The derivatives in Equations (1.9) and (1.10) are calculated in frame $\mathcal{I}$, the only frame under consideration here. Later, however, we shall sometimes find it necessary to specify the frame in which derivatives, velocities, and accelerations are to be computed; we shall then tag the derivatives as in Equation (1.8) and write

$$\mathbf{v}_{P/\mathcal{I}} = {}^{\mathcal{I}}\dot{\mathbf{r}}_{OP} \tag{1.11}$$

Whenever there is just one frame involved, we shall omit the $\mathcal{I}$ on both sides and write an equation such as (1.11) in the form of (1.9).

Throughout the text we have inserted questions for the reader to think about. (The answer is always on the same page as the question.) Here is the first question:

Question 1.1 Do the velocity and acceleration of a point P depend upon: (a) the choice of reference frame? (b) the origin selected for the position vector?

Answer 1.1 (a) Yes; we could simply define a frame in which P is fixed, and it would then have $\mathbf{v}_P = \mathbf{0} = \mathbf{a}_P$. (b) No; letting O' be a second origin in $\mathcal{I}$ and differentiating the relationship $\mathbf{r}_{OP} = \mathbf{r}_{OO'} + \mathbf{r}_{O'P}$ in $\mathcal{I}$ shows that: $\mathbf{v}_P$ (with origin O) $= \mathbf{v}_P$ (with origin O'). The derivative of $\mathbf{r}_{OO'}$ in $\mathcal{I}$ is, of course, zero! See Problem 1.17.

At this point it is reasonable to wonder why we have not chosen to introduce time derivatives of the position vector higher than the second. The reason is that the relationships between forces and motions do not involve those higher derivatives. As we shall see later when we study kinetics, if we know the accelerations of the particles making up a body, the force-motion laws will yield the external forces; conversely, for rigid bodies, if we know the external forces, we can calculate the accelerations and then, by integrating twice, the position vectors. The force-motion laws turn out to be valid only in certain frames of reference; for that reason writers sometimes refer to motion relative to such a frame as *absolute motion*. We have not used the word *absolute* here because we wish to emphasize that kinematics inherently expresses relationships of geometry and time, independent of any laws linking forces and motions. *Thus in kinematics all frames of reference are of the same importance.*

Finally, we note that positions (or locations) of points are normally established through the use of a coordinate system. The ways in which positions, velocities, and accelerations are expressed in two of the most common systems, rectangular and cylindrical, are presented in the next three sections.

PROBLEMS ▶ Section 1.3

1.17 Show that the velocity (and therefore the acceleration also) of a point P in a frame $\mathcal{I}$ does not depend on the choice of the origin. *Hint:* Differentiate the following relationship in $\mathcal{I}$ (see Figure P1.17):

$$\mathbf{r}_{OP} = \mathbf{r}_{OO'} + \mathbf{r}_{O'P}$$

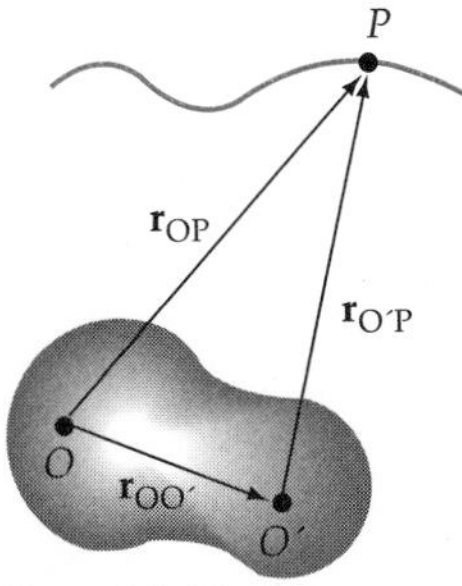

Figure P1.17

In Problems 1.18–1.22, $\hat{\mathbf{i}}, \hat{\mathbf{j}}, \hat{\mathbf{k}}$ are mutually perpendicular unit vectors having directions fixed in the frame of reference; $\mathbf{v}_P$ is the velocity of a point P moving in the frame; t is time measured in seconds. Determine at $t = 2$ s the acceleration of the point for the velocity given.

1.18 $\mathbf{v}_P = 12t\hat{\mathbf{i}} + \dfrac{36}{t^2}\hat{\mathbf{j}} - 3t^2\hat{\mathbf{k}}$ km/s $\;\Rightarrow a = \langle 12, -\frac{72}{t^3}, -6t \rangle$

$a(2) = \langle 12, -9, -12 \rangle$

1.19 $\mathbf{v}_P = 20e^{-0.1t}\left(\sin\dfrac{\pi t}{4}\hat{\mathbf{i}} - \cos\dfrac{\pi t}{4}\hat{\mathbf{j}}\right)$ ft/sec

1.20 $\mathbf{v}_P = 20\sin\dfrac{\pi t}{4}\hat{\mathbf{i}} - 20\cos\dfrac{\pi t}{4}\hat{\mathbf{j}}$ m/s

1.21 $\mathbf{v}_P = t\left(\sin\dfrac{\pi t}{4}\hat{\mathbf{i}} + \cos\dfrac{\pi t}{4}\hat{\mathbf{j}}\right)$ ft/sec

1.22 $\mathbf{v}_P = 5e^{-0.1t}\hat{\mathbf{i}} - 4e^{-0.4t}\hat{\mathbf{k}}$ m/s

1.23–1.27 The **displacement** of a point over a time interval t_1 to t_2 is defined to be the difference of the position vectors—that is, $\mathbf{r}(t_2) - \mathbf{r}(t_1)$. For the cases enumerated in Problems 1.18–1.22, find the displacement and the magnitude of the displacement over the interval $t = 4$ s to $t = 6$ s.

(23) $r(t) = \int_4^6 v_P\, dt = \int_4^6 \left(12t + \frac{36}{t^2} - 3t^2\right) dt$

$r(t) = 6t^2 - \dfrac{72}{t} - t^3 \Big|_4^6$

$r(6) = 216\hat{i} - 6\hat{j} - 216\hat{k}$

$-r(4) = 96\hat{i} - 9\hat{j} - 64\hat{k}$

⟹ $r(6) - r(4) = 120\hat{i} + 3\hat{j} - 152\hat{k}$

$|r| = \sqrt{120^2 + 3^2 + 15^2} = 193.78$

1.4 Kinematics of a Point in Rectilinear Motion

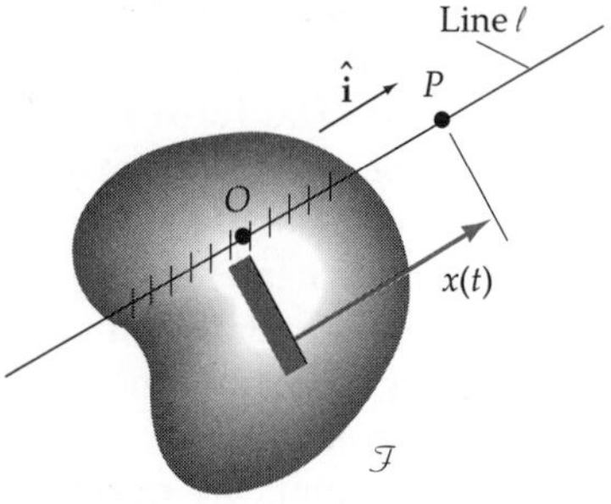

Figure 1.3

In this section we study problems in which point P moves along a straight line in the reference frame $\mathcal{I}$. This situation is called **rectilinear motion**, and the position of P may be expressed with a single coordinate x measured along the fixed line on which P moves (see Figure 1.3).

A position vector for P is simply

$$\mathbf{r}_{OP} = x\hat{\mathbf{i}} \tag{1.12}$$

in which the unit vector $\hat{\mathbf{i}}$ is parallel to the line as shown in Figure 1.3 and hence does not change in either magnitude or direction in $\mathcal{I}$. Therefore P has the following very simple velocity and acceleration expressions:

$$\mathbf{v}_P = \dot{\mathbf{r}}_{OP} = \dot{x}\hat{\mathbf{i}} \tag{1.13}$$

$$\mathbf{a}_P = \ddot{\mathbf{r}}_{OP} = \ddot{x}\hat{\mathbf{i}} \tag{1.14}$$

In rectilinear motion, there are three interesting cases worthy of special note:

1. Acceleration is a known function of time, $f(t)$.
2. Acceleration is a known function of velocity, $g(v)$, where $v = \dot{x}$.
3. Acceleration is a known function of position, $h(x)$.

In each case, we can go far with general integrations. We shall consider each case in turn and give an example.

First, if $\ddot{x} = f(t)$, then

$$\ddot{x} = f(t) \Rightarrow \dot{x} = \int f(t)\,dt + C_1 \Rightarrow x = \int\int f(t)\,dt + C_1 t + C_2 \tag{1.15}$$

in which C_1 and C_2 are to be determined by the initial conditions on velocity and position, respectively, once the problem (and thus $f(t)$) is stated and the indefinite integrals are performed. Alternatively, we might know the values of x at two times, instead of one position and one velocity. In any case, we need two constants.

EXAMPLE 1.2

The acceleration of a point P in rectilinear motion is given by the equation $\ddot{x} = 5t^2$ m/s², with initial conditions $\dot{x}(0) = 2$ m/s and $x(0) = -7$ m. Find $x(t)$.

Solution

We note that this is the problem of a point moving with a quadratically varying acceleration magnitude and with the initial conditions being the position and velocity of P at $t = 0$ as shown in Figure E1.2. Integrating as above,

$$\dot{x} = \int 5t^2\,dt + C_1 = \tfrac{5}{3}t^3 + C_1 \text{ m/s}$$

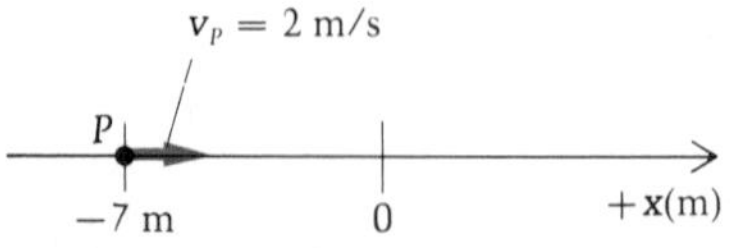

Figure E1.2

And integrating once more,

$$x = \tfrac{5}{12}t^4 + C_1 t + C_2 \text{ m}$$

The constants are found from the initial conditions to be $C_1 = 2$ m/s and $C_2 = -7$ m, as follows:

$$\dot{x}(0) = 2 = (\tfrac{5}{3})(0)^3 + C_1 \Rightarrow C_1 = 2 \text{ m/s}$$
$$x(0) = -7 = (\tfrac{5}{12})(0)^4 + 2(0) + C_2 \Rightarrow C_2 = -7 \text{ m}$$

Thus the motion of the point P is given by the integrated function of time:

$$x = \tfrac{5}{12}t^4 + 2t - 7 \text{ m}$$

If $\ddot{x} = g(v)$, then

$$\ddot{x} = \frac{dv}{dt} = g(v)$$

Suppose $v(t)$ can be inverted to give $t(v)$; then

$$\frac{dt}{dv} = \frac{1}{g(v)}$$

and

$$t + C_3 = \int \frac{dv}{g(v)}$$

If the integral can be found as a function $p(v)$, then we may be able to solve the equation $p(v) = t + C_3$ for the velocity:

$$v = q(t)$$

If so, then

$$v = \frac{dx}{dt} = q(t)$$

so that

$$x = \int q(t)\, dt + C_4 \tag{1.16}$$

This procedure should become clearer with the following example.

Example 1.3

Suppose that the acceleration of a point P in one-dimensional motion is proportional to velocity according to $\ddot{x} = -2v$ m/s^2 with the same initial conditions as in the previous example. Solve for the motion $x(t)$.

Solution

$$\ddot{x} = \frac{dv}{dt} = -2v \Rightarrow \int \frac{dv}{-2v} = t + C_3$$

so that, integrating,* we get

$$t + C_3 = \frac{-\ln v}{2} \Rightarrow v = e^{-2t - 2C_3}$$

Since $v = 2$ when $t = 0$, then $C_3 = (-\ln 2)/2$ and

$$\frac{dx}{dt} = v = e^{-2t + \ln 2} = 2e^{-2t} \text{ m/s}$$

Therefore

$$x = \int 2e^{-2t}\,dt + C_4 = -e^{-2t} + C_4 \text{ m}$$

But $x = -7$ m when $t = 0$ s gives $C_4 = -6$ m, and so we obtain our solution:

$$x = -6 - e^{-2t} \text{ m}$$

When acceleration is a function of position, $a = \ddot{x} = h(x)$, we may combine $a = \dot{v}$ and $v = \dot{x}$ to obtain the useful relation

$$a\frac{dx}{dt} = v\frac{dv}{dt} \tag{1.17}$$

Then, if a function $r(x)$ exists such that $h(x) = \dfrac{d\,r(x)}{dx}$, we obtain, from Equation (1.17), the following:

$$\frac{dr}{dx}\frac{dx}{dt} = v\frac{dv}{dt}$$

$$\frac{dr}{dt} = v\frac{dv}{dt} \tag{1.18}$$

and integrating with respect to time,

$$r(x) = \frac{v^2}{2} + C_5 \tag{1.19}$$

Thus the square of the speed is

$$v^2 = 2r(x) - 2C_5$$

Equation (1.19) will be called an energy integral in Chapters 2 and 5.

EXAMPLE 1.4

Let $\ddot{x} = h(x) = -4x$ m/s². Find $v^2(x)$ if the initial conditions are the same as in Examples 1.2 and 1.3.

---

* This problem could also be solved by first integrating the linear differential equation $\dot{v} + 2v = 0$, observing that Ae^{-2t} is the general solution.

Solution

We are dealing with the equation

$$\ddot{x} + 4x = 0$$

Actually we know that the solution to this equation, by the theory of differential equations, is $x = A \sin 2t + B \cos 2t$—which, with $x(0) = -7$ m and $\dot{x}(0) = 2$ m/s, becomes $x = \sin 2t - 7 \cos 2t$ meters. But let us obtain the desired result by using the procedure described above, which applies even when $h(x)$ is *not* linear. Here $h(x) = -4x$, so with $r(x) = -2x^2$, Equation (1.19) gives

$$-2x^2 = \frac{v^2}{2} + C_5$$

or

$$v^2 = -4x^2 - 2C_5 = 200 - 4x^2$$

where C_5 has been found by using $v = 2$ m/s and $x = -7$ m at $t = 0$.

The *v-t* Diagram

In problems of rectilinear kinematics in which the acceleration is a known function of time (Case 1), we sometimes use what is called the v-t diagram. We shall give just one example of its use because it is a method somewhat limited in application. (We discuss this shortcoming at the end of the example.)

Example 1.5

A point P moves on a line, starting from rest at the origin with constant acceleration of 0.8 m/s^2 to the right. After 10 s, the acceleration of P is suddenly reversed to 0.2 m/s^2 to the left. Determine the total time elapsed when P is again passing through the origin.

Solution

If we graph the velocity versus time, the acceleration (dv/dt) will of course be the slope of the curve at every point. The v-t diagram for this problem is shown in Figure E1.5a.

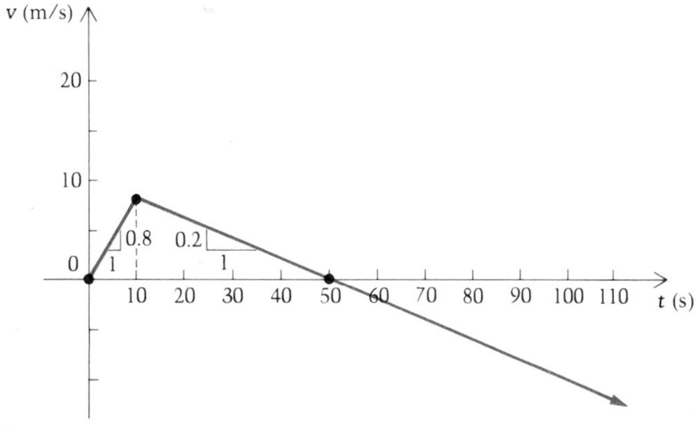

Figure E1.5a

We note not only that

$$a = \frac{dv}{dt} = \text{slope of diagram}$$

but also that

$$x = \int v\, dt + \text{constant}$$

Hence the change in the position x between any two times is nothing more than the area beneath the v-t diagram between those points. Thus four points, or times, are important in the diagram for this problem:

t_1 = starting time (in this case zero)

t_2 = time when acceleration changes (given to be 10 s)

t_3 = time when velocity has been reduced to zero (deceleration causes P to stop before moving in opposite direction)

t_4 = required total time elapsed before point P is again at origin

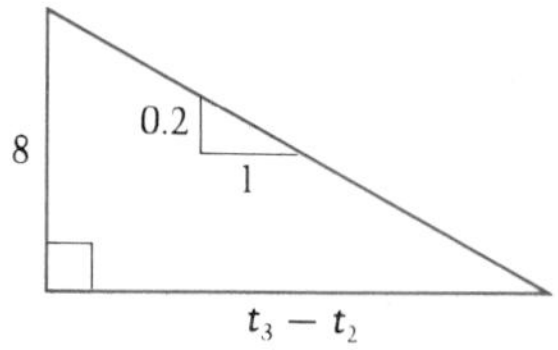

Figure E1.5b

The velocity at time t_2 is seen to be $0.8 \text{ m/s}^2 \times 10 \text{ s} = 8 \text{ m/s}$. To find the time interval $t_3 - t_2$, we use the similar triangles shown in Figure E1.5b.

$$\frac{0.2}{1} = \frac{8}{t_3 - t_2} \Rightarrow t_3 - t_2 = 40$$

$$t_3 = 50 \text{ s}$$

The total distance traveled before the point (momentarily) stops is thus

S_1 = area of 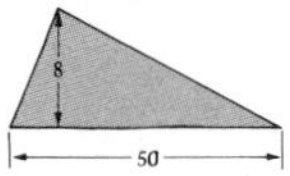

$= \frac{1}{2}(50 \text{ s})(8 \text{ m/s}) = 200 \text{ m}$

This is the distance traveled by the point in the positive direction (to the right).

The point will be back at $x = 0$ when the absolute value of the negative area *beneath* the t axis (the distance traveled back to the left) equals the 200 m traveled to the right (represented by the area *above* the axis):

$$\underbrace{\tfrac{1}{2}(t_4 - 50)}_{\text{base of triangle}}\underbrace{[0.2(t_4 - 50)]}_{\text{height of triangle}} = 200$$

which can be rewritten as

$$t_4^2 - 100t_4 + 500 = 0$$

The only root of this equation larger than 50 s is $t_4 = 94.7$ s, and this is the answer to the problem.

An alternative approach to the preceding v-t diagram solution is as follows. Integrating the acceleration during the interval $0 \le t < 10$ s, with x during this interval called x_1,

$$\ddot{x}_1 = 0.8 \Rightarrow \dot{x}_1 = 0.8t + C_1 = 0.8t \text{ m/s} \qquad (\text{since } \dot{x}_1 = 0 \text{ at } t = 0)$$

Integrating again (over the same interval), we get

$$x_1 = 0.4t^2 + C_2 = 0.4t^2 \text{ m} \qquad (\text{since } x_1 = 0 \text{ at } t = 0)$$

Thus at $t = 10$ s, by substitution,

$$x_1 = 40 \text{ m} \qquad \text{and} \qquad \dot{x}_1 = 8 \text{ m/s}$$

Next, after the deceleration starts, using x_2 in this interval,

$$\ddot{x}_2 = -0.2 \Rightarrow \dot{x}_2 = -0.2t + C_3 \qquad (\text{for } t \geq 10 \text{ s})$$

and since $\dot{x}_2 = 8$ m/s when $t = 10$ s, we obtain $C_3 = 10$. Therefore

$$\dot{x}_2 = -0.2t + 10 \text{ m/s}$$

Integrating again, we get

$$x_2 = -0.1t^2 + 10t + C_4 \text{ m}$$

And with $x_2 = 40$ m when $t = 10$ s, then $C_4 = -50$ m:

$$x_2 = -0.1t^2 + 10t - 50 \text{ m}$$

When $x_2 = 0$, we can solve for the time; the equation is the same as in the v-t diagram solution:

$$t_4^2 - 100t_4 + 500 = 0$$

Of the roots, $t_4 = 5.28$ and 94.7 s, only the latter is valid since 5.28 s occurs prior to the change of acceleration expressions.

Even though both approaches yield the correct answer of 94.7 s in the preceding example, we must recommend the latter approach of integrating the accelerations and matching velocities and positions between intervals. The reason is that when we are faced with *nonconstant* accelerations, the v-t diagram approach requires us to find areas under curves, the formulas for which are not ordinarily memorized.

It is interesting, in using the equations, to start a new time measurement t_2 at the beginning of the second interval:

$$\ddot{x}_2 = -0.2 \text{ m/s}^2 \Rightarrow \dot{x}_2 = -0.2t_2 + C_3 = -0.2t_2 + 8 \text{ m/s}$$

Integrating again, we get

$$x_2 = -0.1t_2^2 + 8t_2 + C_4 = -0.1t_2^2 + 8t_2 + 40 \text{ m}$$

Then $x_2 = 0$ yields the equation

$$t_2^2 - 80t_2 - 400 = 0$$

which has the positive root $t_2 = 84.7$—which, added to the 10-s duration of the first interval, gives again 94.7 s of total time elapsed. It is slightly easier to calculate the integration constants with this approach of "starting time over" than to use the same t throughout. The only price we pay for this convenience is that we must add the times at the end.

EXAMPLE 1.6

A point B starts from rest at the origin at $t = 0$ and accelerates at a constant rate k m/s^2 in rectilinear motion. After 6 s, the acceleration changes to the time-dependent function $0.006t_2^2$ m/s^2 in the *opposite* direction, where $t_2 = 0$ when

$t = 6$ s. If the point stops at $t = 26$ s (from the starting time) and reverses direction, find the acceleration k during the first interval and the distance traveled by B before it reverses direction. Then find the total time elapsed before B passes back through the origin.

Solution

We begin the solution by determining the motion ($x_1(t)$) during the first time interval; we integrate the acceleration to obtain the velocity and then again to get the position:

$$\ddot{x}_1 = k \text{ m/s}^2$$

$$\dot{x}_1 = kt_1 + c_1 = kt_1 \text{ m/s} \qquad (\text{since } \dot{x}_1 = 0 \text{ when } t_1 = 0)$$

$$x_1 = \frac{kt_1^2}{2} + c_2 = \frac{kt_1^2}{2} \text{ m} \qquad (\text{since } x_1 = 0 \text{ when } t_1 = 0)$$

At $t_1 = 6$ s, the acceleration changes to a negative value and point B "decelerates." At the beginning of this second interval, the speed and position of B are given by the "ending" values during the first interval. These values are $\dot{x}_1$ and x_1 at $t_1 = 6$:

$$x_2|_{t_2=0} = x_1|_{t_1=6} = \frac{k6^2}{2} = 18k \text{ m}$$

$$\dot{x}_2|_{t_2=0} = \dot{x}_1|_{t_1=6} = 6k \text{ m/s}$$

Note that we start time t_2 at the beginning of the second interval, during which we have

$$\ddot{x}_2 = -0.006t^2 \text{ m/s}^2$$

where we note that the minus sign is needed to express the *deceleration*. Integrating, we get

$$\dot{x}_2 = -0.002t_2^3 + c_3$$
$$= -0.002t_2^3 + 6k \text{ m/s}$$

since $\dot{x}_2 = 6k$ m/s when $t_2 = 0$. Integrating a second time, we get

$$x_2 = -0.0005t_2^4 + 6kt_2 + c_4$$
$$= -0.0005t_2^4 + 6kt_2 + 18k \text{ m}$$

where c_4 was computed by using the initial condition that $x_2 = 18k$ meters when $t_2 = 0$.

Now we use the fact that $\dot{x}_2$ is zero at time $t_2 = 26 - 6 = 20$ s; this strategy will allow us to determine k:

$$0 = -0.002(20^3) + 6k$$
$$k = 2.67 \text{ m/s}^2$$

Substituting k into the x_2 expression at $t_2 = 20$ s gives us the position of B at the "turnaround":

$$x_{2\text{STOP}} = -0.0005(20^4) + 6(2.67)(20) + 18(2.67)$$
$$= 288 \text{ m}$$

Finally, to obtain the time $t_{2\text{END}}$ when B is passing back through the origin we set

$$x_2 = 0 = -0.0005t_{2\text{END}}^4 + 6(2.67)t_{2\text{END}} + 18(2.67)$$

Rewriting, we get

$$t_{2\text{END}}^4 - 32{,}000 t_{2\text{END}} - 96{,}100 = 0 \qquad (1)$$

On a calculator, the only positive root to this equation* is found in a matter of minutes to be (to three significant figures):

$$t_{2\text{END}} = 32.7 \text{ s}$$

The total time is $t_{2\text{END}}$ plus the duration of the first interval, or 38.7 s.

Before we leave this example we wish to note that during the first time interval, *while the acceleration is constant,*

$$x = \frac{kt^2}{2} + v_0 t + x_0 \text{ m} \qquad (1.20)$$

where

$$x_0 = x(0) \text{ m}$$
$$v_0 = \dot{x}(0) \text{ m/s}$$

Letting $v = \dot{x}$, we note further that

$$v = kt + v_0 \text{ m/s} \qquad (1.21)$$

and eliminating t we obtain

$$v^2 = v_0^2 + 2k(x - x_0) \text{ m}^2/\text{s}^2 \qquad (1.22)$$

This expression gives us the magnitude of the velocity in terms of displacement. Most students have used this relationship in high school or perhaps elementary college physics. There is sometimes a tendency, however, to forget the conditions under which it is valid; it holds only for *rectilinear* motion with *constant acceleration.* Thus it could not be used during the second interval of the preceding example, nor could the equations for x and v from which it was derived.

EXAMPLE 1.7

Two cars in a demolition derby are approaching a common point (the origin in Figure E1.7), each at 55 mph in a straight line as indicated. Car $\mathcal{C}_1$ does not speed up or slow down; the driver of $\mathcal{C}_2$ applies the brakes. Find the smallest rate of deceleration of $\mathcal{C}_2$ that will allow $\mathcal{C}_1$ to precede it through the intersection, if:

a. $d_2 = 200$ ft
b. $d_2 = 100$ ft

---

* Descartes' rule of signs tells us that the maximum number of positive real roots to Equation (1) is one (the number of changes in sign on the left-hand side). And there will be exactly one because the left side is negative at $t_{2\text{END}} = 0$ and positive for large values of $t_{2\text{END}}$.

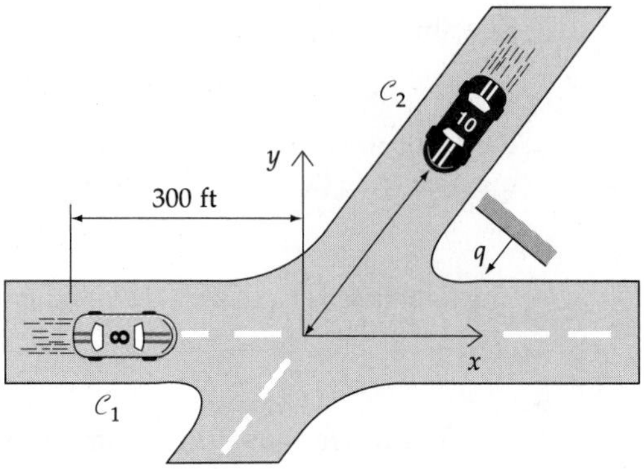

Figure E1.7

Solution

Placing the origin at the point of intersection of the two cars' paths, we have the following for $\mathcal{C}_1$:

$$\dot{x}_1 = 55 \text{ mph}\left(\frac{88 \text{ ft/sec}}{60 \text{ mph}}\right) = 80.7 \text{ ft/sec}$$

$$x_1 = 80.7t + C_1 \text{ ft}$$

Using the initial condition that $x_1 = -300$ ft when $t = 0$, we get

$$x_1 = 80.7t - 300 \text{ ft}$$

The back of car $\mathcal{C}_1$ will be at the origin (point of possible collision) when $x = 0$:

$$0 = 80.7t_0 - 300$$

$$t_0 = 3.72 \text{ sec}$$

Now let us study the motion of car $\mathcal{C}_2$. We use the coordinate q as shown for this car. Calling the unknown deceleration K, we obtain

$$\ddot{q} = -K \text{ ft/sec}^2$$

so that

$$\dot{q} = -Kt + C_2 = -Kt + 80.7 \text{ ft/sec}$$

and

$$q = \frac{-Kt^2}{2} + 80.7t + C_3 \text{ ft}$$

But $C_3 = 0$, since $q = 0$ at $t = 0$.

Next we see that at 3.72 sec the position of $\mathcal{C}_2$ is

$$q = \frac{-K(3.72^2)}{2} + 80.7(3.72)$$

$$= -6.92K + 300 \text{ ft}$$

Finally, car $\mathcal{C}_2$ just passes the rear of $\mathcal{C}_1$ if q is d_2 at this time:

$$d_2 = -6.92K + 300 \text{ ft}$$

Hence:

a. If $d_2 = 200$ ft, then $K = 14.5$ ft/sec^2.
b. If $d_2 = 100$ ft, then $K = 28.9$ ft/sec^2.

Note also that if $d_2 = 300$ ft, then $K = 0$; this is because *no* deceleration is needed for the same distances at the same speeds. Further, if $d_2 > 300$, then K is negative, meaning that car $\mathcal{C}_2$ would have to *accelerate* to arrive at the intersection at the same time as car $\mathcal{C}_1$.

Sometimes there are special conditions in a problem that require ingenuity in expressing the kinematics. If there is an inextensible rope, string, cable, or cord present, for example, we may have to express the constancy of length mathematically. This is the case in the following example.

Example 1.8

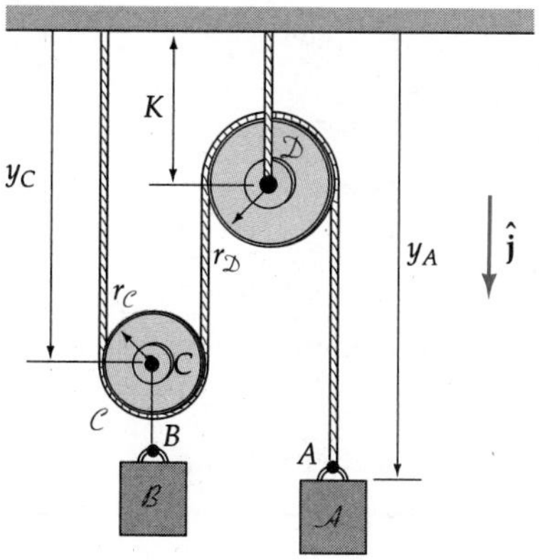

Figure E1.8

Block $\mathcal{A}$ travels downward with $\mathbf{v}_A = 3t^2 \downarrow$ m/s. Find the velocity of block $\mathcal{B}$ when $t = 4$ s.

Solution

The length L of the rope that passes around both small pulleys is a constant. This is a constraint equation that must be used in the solution. The procedure is as follows (see Figure E1.8):

$$L = y_C + \pi r_{\mathcal{C}} + (y_C - K) + \pi r_{\mathcal{D}} + (y_A - K) \text{ m}$$

Differentiating and noting that L, π, $r_{\mathcal{C}}$, $r_{\mathcal{D}}$, and K are constants, we get

$$0 = 2\dot{y}_C + \dot{y}_A \Rightarrow \dot{y}_C = -\frac{\dot{y}_A}{2} = -\frac{3}{2}t^2$$

The velocities of C and B are equal since both points move on the same path with a constant length separating them. Hence

$$\mathbf{v}_B = -\frac{3}{2}t^2\hat{\mathbf{j}} \text{ m/s} \qquad \text{(Note that C moves upward since } \hat{\mathbf{j}} \text{ is downward!)}$$

Therefore

$$\mathbf{v}_B\big|_{t=4} = -24\hat{\mathbf{j}} \text{ m/s} \qquad \text{or} \qquad 24 \uparrow \text{ m/s}$$

Example 1.9

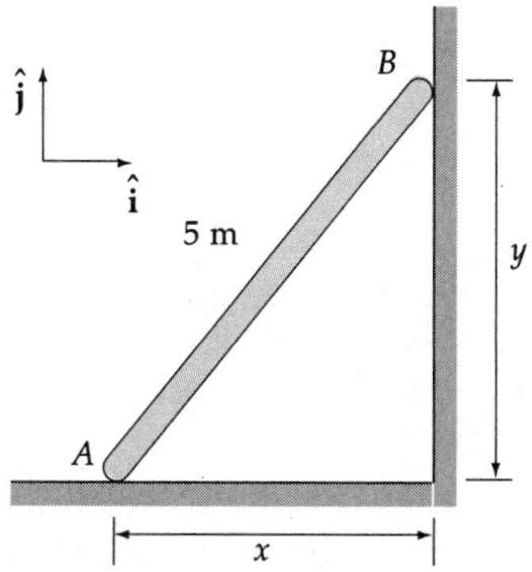

Figure E1.9

The ends A and B of the rigid bar in Figure E1.9 are to move along the horizontal and vertical guides as shown. End A moves to the right at a constant speed of 8 m/s. Find the velocity and acceleration of B at the instant when A is 3 m from the corner.

Solution

In terms of the parameters and unit vectors shown in Figure E1.9,

$$\mathbf{v}_A = -\dot{x}\hat{\mathbf{i}} = 8\hat{\mathbf{i}}\ \text{m/s}, \qquad \mathbf{v}_B = \dot{y}\hat{\mathbf{j}}$$

$$\mathbf{a}_A = -\ddot{x}\hat{\mathbf{i}} = 0 \qquad \mathbf{a}_B = \ddot{y}\hat{\mathbf{j}}$$

Using the fact that the distance from A to B is a constant 5 m,

$$x^2 + y^2 = 25$$

so that

$$2\,x\dot{x} + 2y\dot{y} = 0$$

or

$$x\dot{x} + y\dot{y} = 0$$

Thus when $x = 3$ m, $y = 4$ m and

$$3(-8) + 4\dot{y} = 0$$

or

$$\dot{y} = 6\ \text{m/s}$$

so that

$$\mathbf{v}_B = 6\hat{\mathbf{j}}\ \text{m/s}$$

Differentiating again,

$$\dot{x}\dot{x} + x\ddot{x} + \dot{y}\dot{y} + y\ddot{y} = 0$$

Therefore at the instant of interest

$$(-8)(-8) + (3)(0) + (6)(6) + 4\ddot{y} = 0$$

or

$$\ddot{y} = -25\ \text{m/s}^2$$

and

$$\mathbf{a}_B = -25\hat{\mathbf{j}}\ \text{m/s}^2$$

Our last example illustrates a different kind of constraint, that of a point on a body maintaining contact with a surface (or line) on another moving body.

EXAMPLE 1.10

The curve AB on block $\mathcal{B}$ (see Figure E1.10a) is a parabola whose vertex is at A. Its equation is $x^2 = (64/3)y$. The block $\mathcal{B}$ is pushed to the left with a constant velocity of 10 ft/sec. The rod $\mathcal{R}$ slides on the parabola so that the plate $\mathcal{P}$ is forced upward. Find the acceleration of the plate.

Solution

We first note that plate $\mathcal{P}$ and rod $\mathcal{R}$ together constitute a single body, each of whose points has one-dimensional (y) motion. The velocities and accelerations of

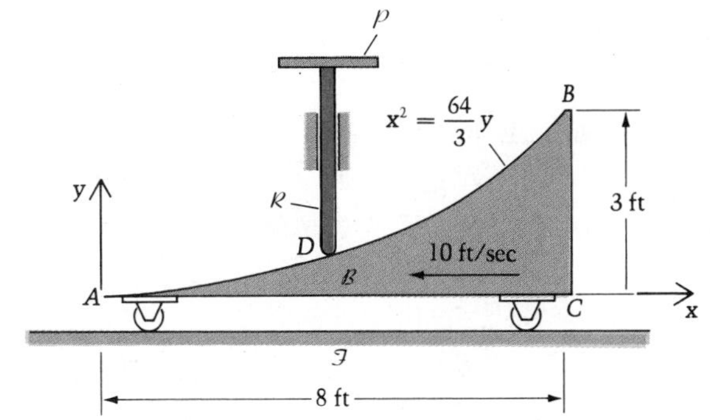

Figure E1.10a

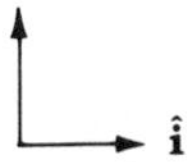

all these points are therefore the same. We shall then focus on point D, the lowest point of $\mathcal{R}$, which is in contact with $\mathcal{B}$.

Defining the ground to be the reference frame $\mathcal{I}$, we establish its origin at O as shown in Figure E1.10b.

$$\mathbf{r}_{OA} = -x\hat{\mathbf{i}} \qquad \text{and} \qquad \mathbf{r}_{OD} = y\hat{\mathbf{j}} \text{ ft} \tag{1}$$

But because D always rests on the parabolic surface of $\mathcal{B}$, $y = (3/64)x^2$ so that

$$\mathbf{r}_{OD} = \tfrac{3}{64}x^2\hat{\mathbf{j}} \text{ ft} \tag{2}$$

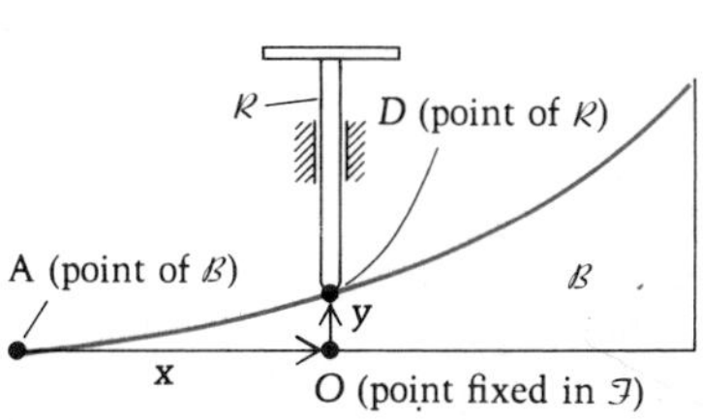

Figure E1.10b

To get the acceleration of D, we first find its velocity:

$$\mathbf{v}_D = \tfrac{3}{32}x\dot{x}\hat{\mathbf{j}} \text{ ft/sec} \tag{3}$$

To obtain $\dot{x}$, we differentiate $\mathbf{r}_{OA}$ from Equation (1):

$$\dot{\mathbf{r}}_{OA} = \mathbf{v}_A = -\dot{x}\hat{\mathbf{i}} = -10\hat{\mathbf{i}} \text{ ft/sec} \tag{4}$$

since it is given that all points of the body $\mathcal{B}$ have the constant velocity 10 ft/sec to the left.

Substitution of $\dot{x} = 10$ into Equation (3) then gives

$$\mathbf{v}_D = \tfrac{3}{32}x(10)\hat{\mathbf{j}} = \tfrac{15}{16}x\hat{\mathbf{j}} \text{ ft/sec} \tag{5}$$

and we see that the velocity of D depends upon x. Differentiating $\mathbf{v}_D$ will give us the acceleration of D:

$$\mathbf{a}_D = \frac{15}{16}\dot{x}\hat{\mathbf{j}} = \frac{150}{16}\hat{\mathbf{j}} = 9.38\hat{\mathbf{j}} \text{ ft/sec}^2 \tag{6}$$

Equation (6) gives the acceleration of all the points of the plate. Note that the acceleration of D is a constant.

Question 1.2 Would $\mathbf{a}_D$ be a constant if instead of being quadratically shaped, the inclined surface were (a) flat or (b) cubic?

Answer 1.2 If the surface is flat, then $\mathbf{a}_D$ vanishes. If it is cubic, then $\mathbf{a}_D$ is linear in x.

PROBLEMS ▶ Section 1.4

1.28 A slider block moves rectilinearly in a slot (see Figure P1.28) with an acceleration given by

$$\mathbf{a}_s = \ddot{x}\hat{\mathbf{i}} = -\pi^2 \sin \pi t \hat{\mathbf{i}} \text{ m/s}^2$$

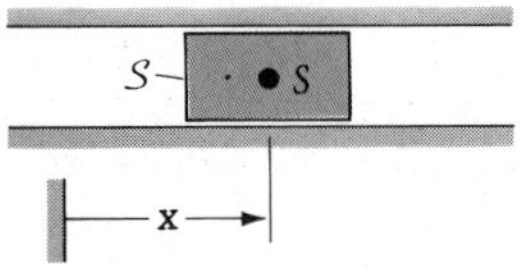

Figure P1.28

Find the motion $x(t)$ of the slider block if at $t = 0$:

a. It is passing through the origin, and
b. It has velocity $\dot{x}\hat{\mathbf{i}} = 2\pi\hat{\mathbf{i}}$ m/s.

1.29 Suppose an airplane touches down smoothly on a runway at 60 mph. If it then decelerates to a stop at the constant deceleration rate of 10 ft/sec², find the required length of runway.

1.30 A train is traveling at 60 km/hr. If its brakes give the train a constant deceleration of 0.5 m/s², find the distance from the station where the brakes should be applied so that the train will come to a stop at the station. How long will it take the train to stop?

1.31 A point P starts from rest and accelerates uniformly (meaning $\ddot{x}$ = constant) to a speed of 88 ft/sec after traveling 120 ft. Find the acceleration of P.

1.32 If in the preceding problem a braking deceleration of 2 ft/sec² is experienced beginning when P is at 120 ft, determine the time and distance required for stopping.

1.33 A car is traveling at 55 mph on a straight road. The driver applies her brakes for 6 sec, producing a constant deceleration of 5 ft/sec², and then immediately accelerates at 2 ft/sec². How long does it take for the car to return to its original velocity?

1.34 In the preceding problem, suppose the acceleration following the braking is not constant but is instead given by $\ddot{x} = 0.6t$ ft/sec². *Now* how long does it take to return to 55 mph?

In Problems 1.35–1.37, the graph depicts the velocity of a point P in rectilinear motion. Draw curves showing the position $x(t)$ and acceleration $a(t)$ of P if the point is at the indicated position x_0 at $t = 0$.

1.35 $x_0 = -1125$ m
Time interval: $0 \le t \le 20$ s (See Figure P1.35.)

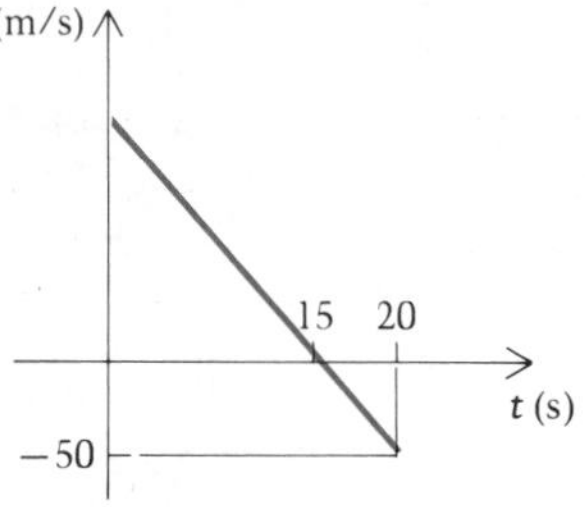

Figure P1.35

1.36 $x_0 = 10$ m
Time interval: $0 \le t \le 30$ s (See Figure P1.36.)

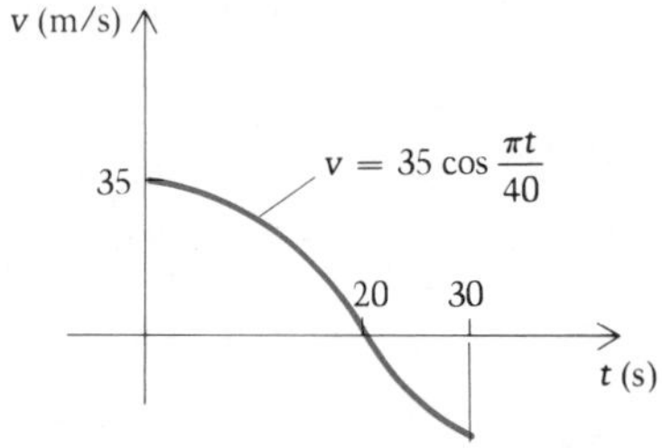

Figure P1.36

1.37 $x_0 = 10$ m
Time interval: $2 \le t \le 5$ s (See Figure P1.37.)

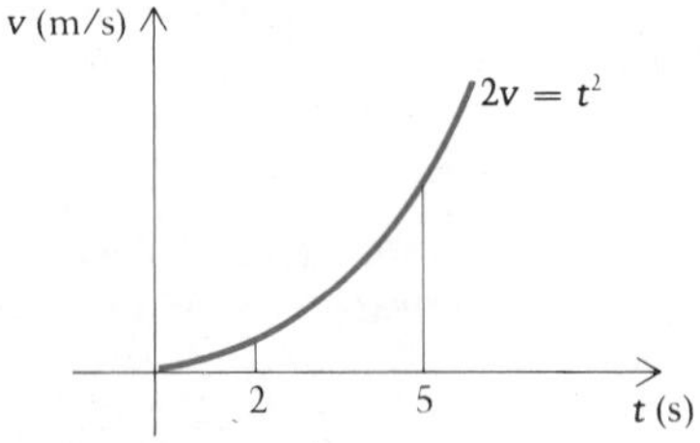

Figure P1.37

1.38 A hot rod enthusiast accelerates his dragster along a straight drag strip at a constant rate of acceleration from zero to 120 mph. Then he immediately decelerates at a constant rate to a stop. He finds that he has traveled a total distance of $\frac{1}{4}$ mi from start to stop. How much time passes from the instant he starts to the time he stops? *Hint*: Sketch a v-t diagram.

1.39 Ben Johnson set a world record of 9.83 seconds in the 100-meter dash on August 31, 1987. He had also set the record for the 60-meter dash of 6.40 seconds that same year. Assuming that in each race Johnson accelerated uniformly up to a certain speed v_o and then held that same maximum speed to the end of both races, find (a) the time t_o to reach v_o; (b) the value of v_o; and (c) the distance traveled while accelerating.

1.40 A train travels from one city to another which is 134 miles away. It accelerates from rest to a maximum speed of 100 mph in 4 min, averaging 65 mph during this time interval. It maintains maximum velocity until just before arrival, when it decelerates to rest at an average speed during the deceleration of 40 mph. If the total travel time was 110 min, find the deceleration interval.

1.41 A point Q in rectilinear motion passes through the origin at $t = 0$, and from then until 5 seconds have passed, the acceleration of Q is 6 ft/sec^2 to the right. Beginning at $t = 5$ seconds, the acceleration of Q is $12t$ ft/sec^2 to the left. If after 2 more seconds point Q is 13 feet to the right of the origin, what was the velocity of Q at $t = 0$?

1.42 A point begins at rest at $x = 0$ and experiences constant acceleration to the right for 10 s. It then continues at constant velocity for 8 more seconds. In the third phase of its motion, it decelerates at 5 m/s^2 and is observed to be passing again through the origin when the total time of travel equals 28 s. Determine the acceleration in the first 10 s.

1.43 An automobile passes a point P at a speed of 80 mph. At P it begins to decelerate at a rate proportional to time. If after 5 sec the car has slowed to 50 mph, what distance has it traveled?

1.44 Work the preceding problem, but suppose the deceleration is proportional to the square of time. The other information is the same.

1.45 A particle has a linearly varying rectilinear acceleration of $\mathbf{a} = \ddot{x}\hat{\mathbf{i}} = 12t\hat{\mathbf{i}}$ m/s^2. Two observations of the particle's motion are made: Its velocity at $t = 1$ s is $\dot{x}\hat{\mathbf{i}} = 2\hat{\mathbf{i}}$ m/s, and its position at $t = 2$ s is given by $x\hat{\mathbf{i}} = 3\hat{\mathbf{i}}$ meters.

a. Find the displacement of the particle at $t = 5$ s relative to where it was at $t = 0$.

b. Determine the distance traveled by the particle over the same time interval.

1.46 A point P moves on a line. The acceleration of P is given by $\mathbf{a}_P = \ddot{x}_P\hat{\mathbf{i}} = (3t^2 - 30t + 56)\hat{\mathbf{i}}$ m/s^2. The velocity of P at $t = 0$ is $-60\hat{\mathbf{i}}$ m/s, with the point at $x_P = 7$ m at that time. Find the distance traveled by P in the time interval $t = 0$ to $t = 13$ s.

* **1.47** The position of a point P on a line is given by the equation $x = t \sin(\pi t/2)$. The point starts moving at $t = 0$. Find the total distance traveled by P when it passes through the origin (counting the start as the first pass) for the third time.

1.48 A particle moving on a straight line is subject to an acceleration directly proportional to its distance from a fixed point P on the line and directed toward P. Initially the particle is 5 ft to the left of P and moving to the right with a velocity of 24 ft/sec. If the particle momentarily comes to rest 10 ft to the right of P, find its velocity as it passes through P.

1.49 A particle moving on the x axis has an acceleration always directed to the origin. The magnitude of the acceleration is nine times the distance from the origin. When the particle is 6 m to the left of the origin, it has a velocity of 3 m/s to the right. Find the time for the particle to get from this position to the origin.

1.50 A point P has an acceleration that is position-dependent according to the equation $\ddot{x} = -5x^2$ m/s^2. Determine the velocity of P as a function of its position x if P is at 0.3 m with $\dot{x} = 0.6$ m/s when $t = 0$.

1.51 Suppose initial conditions are the same as in the preceding problem but $\ddot{x} = -5\dot{x}^2$. Find $\dot{x}$ as a function of time.

1.52 The velocity of a particle moving along a horizontal path is proportional to its distance from a fixed point on the path. When $t = 0$, the particle is 1 ft to the right of the fixed point. When $v = 20$ ft/sec to the right, $a = 5$ ft/sec^2 to the right. Determine the position of the particle when $t = 4$ sec. (See Figure P1.52.)

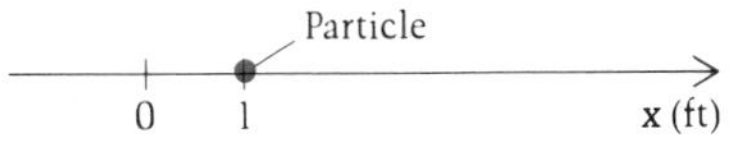

Figure P1.52

* Asterisks identify the more difficult problems.

1.53 A speeder zooms past a parked police car at a constant speed of 70 mph (Figure P1.53). Then, 3 sec later, the policewoman starts accelerating from rest at 10 ft/sec^2 until her velocity is 85 mph. How long does it take her to overtake the speeding car if it neither slows down nor speeds up?

Figure P1.53

1.54 In the preceding problem, suppose the speeder sees the policewoman 10 sec after she begins to move, and decelerates at 3 ft/sec^2. How long does it take the policewoman to *pass* the car if she is actually chasing a faster speeder ahead of it?

1.55 Two cars start from rest at the same location and at the same instant and race along a straight track. Car $\mathcal{A}$ accelerates at 6.6 ft/sec^2 to a speed of 90 mph and then runs at a constant speed. Car $\mathcal{B}$ accelerates at 4.4 ft/sec^2 to a speed of 96 mph and then runs at a constant speed.

- **a.** Which car will win the 3-mi race, and by what distance?
- **b.** What will be the maximum lead of $\mathcal{A}$ over $\mathcal{B}$?
- **c.** How far will the cars have traveled when $\mathcal{B}$ passes $\mathcal{A}$?

* **1.56** A car is 40 ft behind a truck; both are moving at 55 mph. (See Figure P1.56.) Suddenly the truck driver slams on his brakes after seeing an obstruction in the road ahead, and he decelerates at 10 ft/sec^2. Then, 2 sec later, the driver of the car reacts by slamming on her brakes, giving her car a deceleration a_C. Find the minimum value of a_C for which the car will not collide with the truck. *Hint*: Enforce $x_T > x_C$ for *all* time t before the vehicles are stopped.

Figure P1.56

1.57 Point B of block $\mathcal{B}$ has a constant acceleration of 10 m/s^2 upward. At the instant shown in Figure P1.57, it is 30 m below the level of point A of $\mathcal{A}$. At this time, $\mathbf{v}_A$ and $\mathbf{v}_B$ are zero. Determine the velocities of A and B as they pass each other.

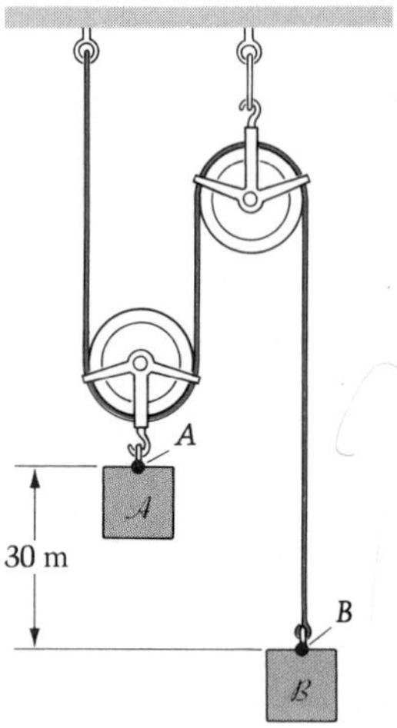

Figure P1.57

1.58 The accelerations of the translating blocks $\mathcal{A}$ and $\mathcal{B}$ are 2 m/s^2 ↓ and 4 m/s^2 ↑, respectively. (See Figure P1.58.)

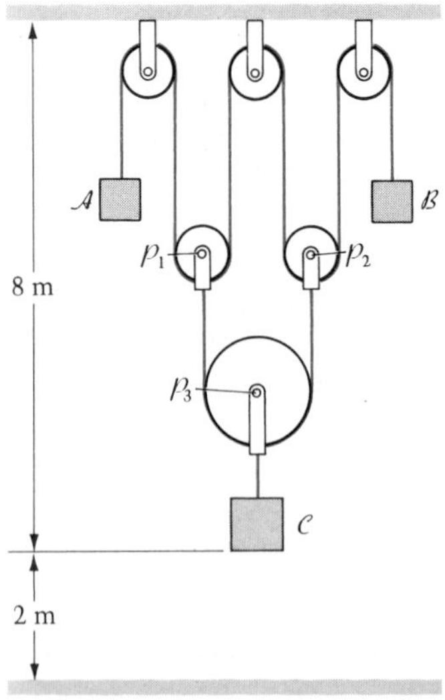

Figure P1.58

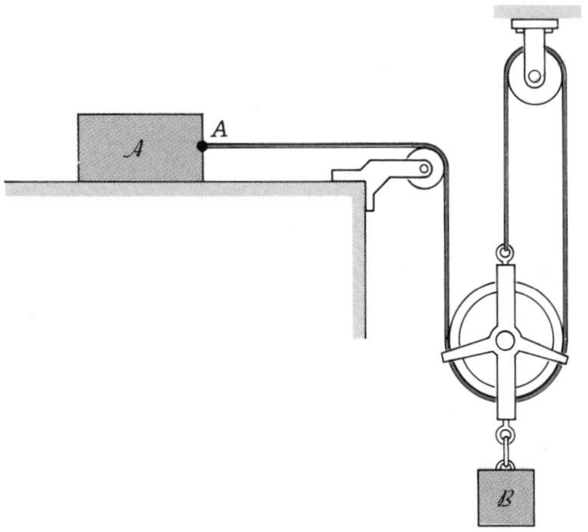

Figure P1.59

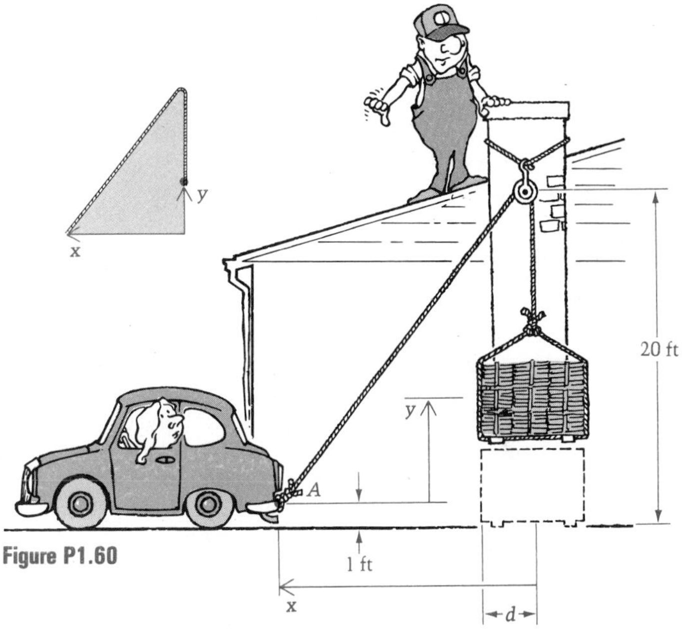

Figure P1.60

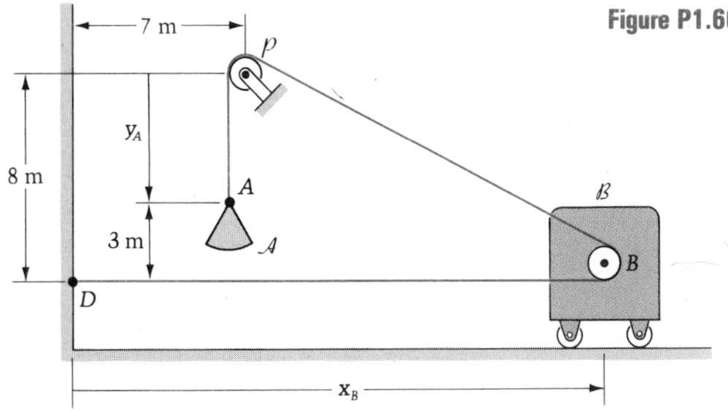

Figure P1.61

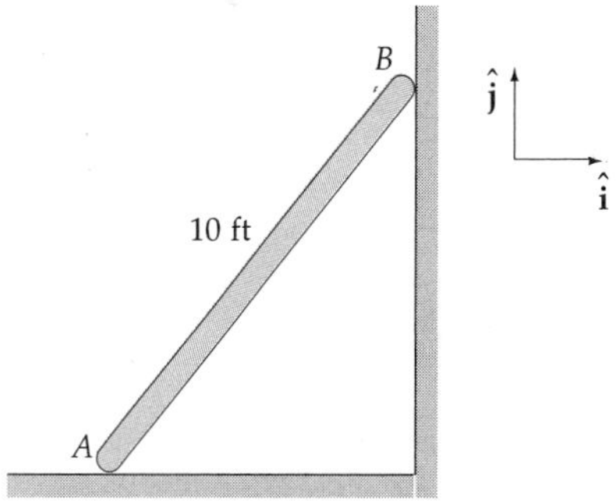

Figure P1.62

The entire system is at rest at the given instant. Find how long it will take for block C to hit the ground. (Do not assume that pulleys P_1 and P_2 remain at the same level!)

* **1.59** Block A has $\mathbf{v}_A = 10$ m/s to the right at $t = 0$ and a constant acceleration of 2 m/s^2 to the left. Find the distance traveled by block B during the interval $t = 0$ to 8 s. (See Figure P1.59.)

1.60 A man and his daughter have figured out an ingenious way to hoist 8000 lb of shingles onto their roof, several bundles at a time. They have rigged a pulley onto a frame around the chimney (Figure P1.60) and will use the car to raise the weights. When the bumper of the car is at $x = 0$ (neglect d), the pallet of shingles is on the ground with no slack in the rope. While the car is traveling to the left at a constant speed of $v_A = 2$ mph, find the velocity and acceleration of the shingles as a function of x. Do this by using the triangle to the left of the figure to express y as a function of x; then differentiate the result.

* **1.61** The cord shown in Figure P1.61, attached to the wall at D, passes around a small pulley fixed to B at B; it then passes around another small pulley P and ends at point A of body A. The cord is 44 m long, and the system is being held at rest in the given position. Suddenly point B is forced to move to the right with constant acceleration $a_B = 2$ m/s^2. Determine the velocity of A just before it reaches the pulley.

1.62 The ends of the rigid bar in Figure P1.62 move while maintaining contact with floor and wall. End A moves toward the wall at the constant rate of 2 ft/sec. What is the acceleration of B at the instant when A is 6 feet from the wall?

1.63 The velocity of point A in Figure P1.63 is a constant 2 m/s to the right. Find the velocity of B when $x = 10$ m.

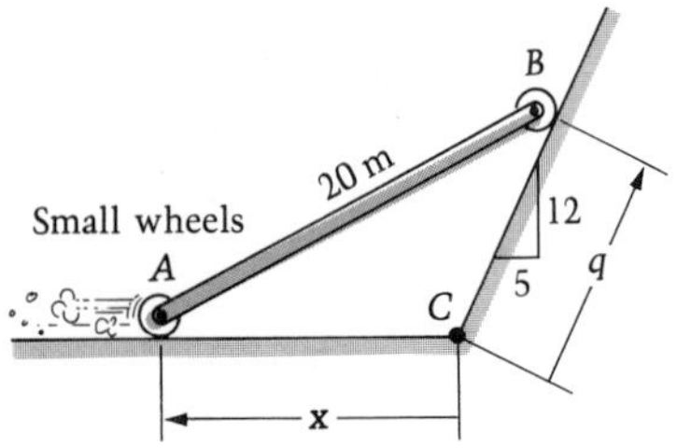

Figure P1.63

1.64 The collars in Figure P1.64 are attached at C_1 and C_2 to the rod by ball and socket joints. Point C_2 has a velocity of $-3\hat{\mathbf{i}}$ m/s and no acceleration at the instant shown. Find the velocity and acceleration of C_1 at this instant.

1.65 The wedge-shaped cam in Figure P1.65 is moving to the left with constant acceleration a_0. Find the acceleration of the follower $\mathcal{F}$.

1.66 In Example 1.10, let the equation of the incline be given by $x^3 = (512/3)y$. If the motion starts when $x = y = 0$, find the acceleration of the plate $\mathcal{P}$ when $y = 2$ ft.

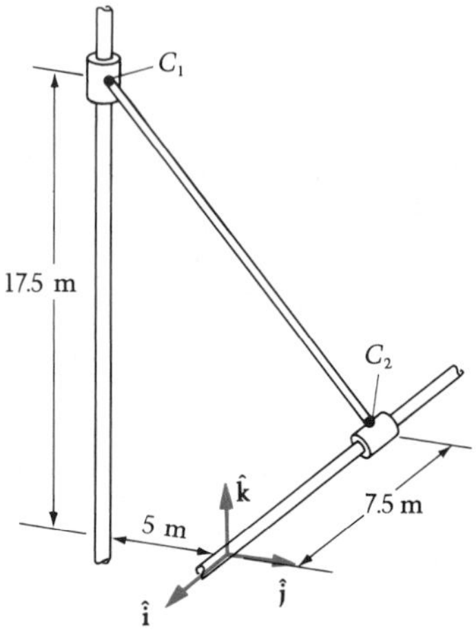

Figure P1.64

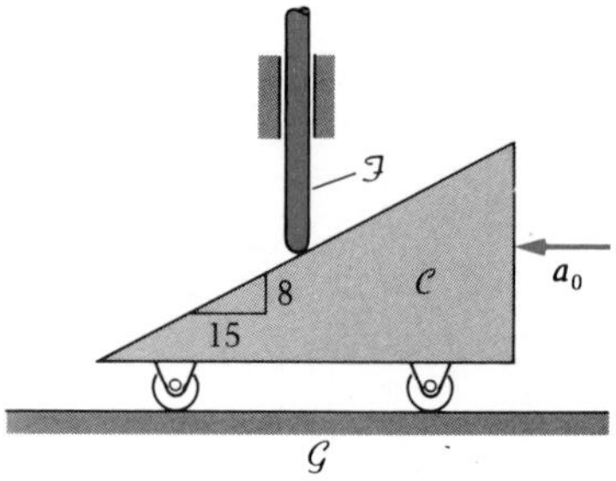

Figure P1.65

1.5 Rectangular Cartesian Coordinates

In this section we merely add the y and/or z components of position to the rectilinear (x) component studied in the preceding section. This step allows the point P to move on a curve in two- or three-dimensional space instead of being constrained to movement along a straight line in the reference frame $\mathcal{I}$.

Suppose that P is in a state of general (three-dimensional) motion in frame $\mathcal{I}$. We may study this motion by embedding a set of orthogonal axes in $\mathcal{I}$ as shown in Figure 1.4 on the next page. The position vector of point P may then be expressed as

$$\mathbf{r}_{OP} = x\hat{\mathbf{i}} + y\hat{\mathbf{j}} + z\hat{\mathbf{k}} \tag{1.23}$$

in which (x, y, z) are **rectangular Cartesian coordinates** of P measured along the embedded axes and $(\hat{\mathbf{i}}, \hat{\mathbf{j}}, \hat{\mathbf{k}})$ are unit vectors respectively parallel to these axes (Figure 1.4). Using the basic definitions (Equations 1.9 and 1.10), we may differentiate $\mathbf{r}_{OP}$ and obtain expressions for velocity and acceleration in rectangular Cartesian coordinates:

$$\mathbf{v}_P = \dot{x}\hat{\mathbf{i}} + \dot{y}\hat{\mathbf{j}} + \dot{z}\hat{\mathbf{k}} \tag{1.24}$$

$$\mathbf{a}_P = \ddot{x}\hat{\mathbf{i}} + \ddot{y}\hat{\mathbf{j}} + \ddot{z}\hat{\mathbf{k}} \tag{1.25}$$

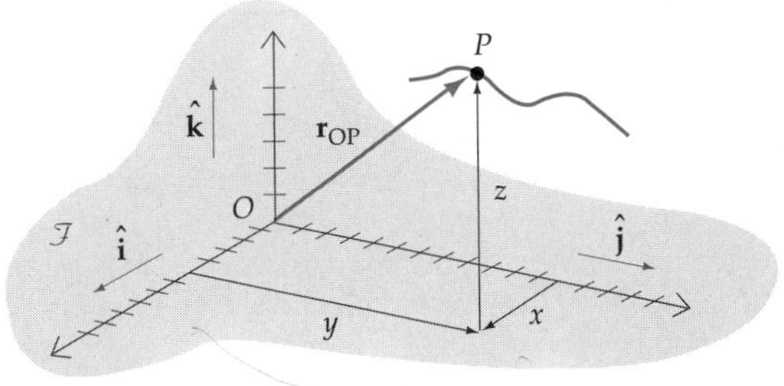

Figure 1.4 **Rectangular Cartesian coordinates of *P*.**

We shall now consider examples in which points move in two and three dimensions.

EXAMPLE 1.11

The position vector of a point P is given as

$$\mathbf{r}_{OP} = 2t\hat{\mathbf{i}} + t^3\hat{\mathbf{j}} + 3t^2\hat{\mathbf{k}} \text{ ft}$$

Find the velocity and acceleration of P at $t = 1$ sec.

Solution

Differentiating the position vector, we obtain the velocity vector of P:

$$\mathbf{v}_P = 2\hat{\mathbf{i}} + 3t^2\hat{\mathbf{j}} + 6t\hat{\mathbf{k}} \text{ ft/sec}$$

Another derivative yields the acceleration of P:

$$\mathbf{a}_P = 6t\hat{\mathbf{j}} + 6\hat{\mathbf{k}} \text{ ft/sec}^2$$

Therefore, at $t = 1$ sec, the velocity and acceleration of P are

$$\mathbf{v}_P|_{t=1} = 2\hat{\mathbf{i}} + 3\hat{\mathbf{j}} + 6\hat{\mathbf{k}} = 7\left(\frac{2\hat{\mathbf{i}} + 3\hat{\mathbf{j}} + 6\hat{\mathbf{k}}}{\sqrt{2^2 + 3^2 + 6^2}}\right) \text{ ft/sec}$$

$$\mathbf{a}_P|_{t=1} = 6\hat{\mathbf{j}} + 6\hat{\mathbf{k}} = 6\sqrt{2}\left(\frac{\hat{\mathbf{j}} + \hat{\mathbf{k}}}{\sqrt{2}}\right) \text{ ft/sec}^2$$

Note that the speed (magnitude of velocity) of P at $t = 1$ is $|\mathbf{v}_P| = 7$ ft/sec and the magnitude of the acceleration at $t = 1$ is $6\sqrt{2}$ ft/sec². We shall return to this example in Section 1.7.

We see from the previous example that if the position vector of P is known as a function of time, it is a very simple matter to obtain the velocity and acceleration of the point. In the following example we are given the acceleration of P and asked for its *position.* Since this problem requires integration instead of differentiation, initial conditions enter the picture. These conditions allow us to compute the constants of integration, just as they did for rectilinear motion in the preceding section.

EXAMPLE 1.12

A point Q has the acceleration vector

$$\mathbf{a}_Q = 4\hat{\mathbf{i}} - 6t\hat{\mathbf{j}} + \sin 0.2t\hat{\mathbf{k}} \text{ m/s}^2$$

At $t = 0$, the point Q is located at $(x, y, z) = (1, 3, -5)$ m and has a velocity vector of $2\hat{\mathbf{i}} - 7\hat{\mathbf{j}} + 3.4\hat{\mathbf{k}}$ m/s. When $t = 3$ s, find the speed of Q and its distance from the starting point.

Solution

Integrating, we get

$$\mathbf{v}_Q = 4t\hat{\mathbf{i}} - 3t^2\hat{\mathbf{j}} - 5 \cos 0.2t\hat{\mathbf{k}} + \mathbf{c} \text{ m/s}$$

in which $\mathbf{c}$ is a vector constant. Using the initial condition for velocity at $t = 0$, we obtain

$$\mathbf{v}_Q|_{t=0} = 0\hat{\mathbf{i}} - 0\hat{\mathbf{j}} - 5\hat{\mathbf{k}} + \mathbf{c} = 2\hat{\mathbf{i}} - 7\hat{\mathbf{j}} + 3.4\hat{\mathbf{k}} \text{ m/s}$$

so that

$$\mathbf{c} = 2\hat{\mathbf{i}} - 7\hat{\mathbf{j}} + 8.4\hat{\mathbf{k}} \text{ m/s}$$

Therefore

$$\mathbf{v}_Q = (4t + 2)\hat{\mathbf{i}} - (3t^2 + 7)\hat{\mathbf{j}} + (8.4 - 5 \cos 0.2t)\hat{\mathbf{k}} \text{ m/s}$$

Integrating again, we get

$$\mathbf{r}_{OQ} = (2t^2 + 2t)\hat{\mathbf{i}} - (t^3 + 7t)\hat{\mathbf{j}} + (8.4t - 25 \sin 0.2t)\hat{\mathbf{k}} + \mathbf{c}' \text{ m}$$

where $\mathbf{c}'$ is another vector constant, evaluated below from the initial condition for the *position* of Q at $t = 0$:

$$\mathbf{r}_{OQ}|_{t=0} = 0\hat{\mathbf{i}} - 0\hat{\mathbf{j}} + 0\hat{\mathbf{k}} + \mathbf{c}' = \hat{\mathbf{i}} + 3\hat{\mathbf{j}} - 5\hat{\mathbf{k}} \text{ m}$$

so that

$$\mathbf{c}' = \hat{\mathbf{i}} + 3\hat{\mathbf{j}} - 5\hat{\mathbf{k}} \text{ m}$$

and thus

$$\begin{aligned}\mathbf{r}_{OQ} &= (2t^2 + 2t + 1)\hat{\mathbf{i}} - (t^3 + 7t - 3)\hat{\mathbf{j}} \\ &\quad + (8.4t - 25 \sin 0.2t - 5)\hat{\mathbf{k}} \text{ m}\end{aligned}$$

Substituting the required time, $t = 3$ s, into the expressions for $\mathbf{v}_Q$ and $\mathbf{r}_{OQ}$ will give the answers:

$$\begin{aligned}\mathbf{v}_Q|_{t=3} &= 14\hat{\mathbf{i}} - 34\hat{\mathbf{j}} + (8.4 - 5 \cos 0.6)\hat{\mathbf{k}} \\ &= 14\hat{\mathbf{i}} - 34\hat{\mathbf{j}} + 4.27\hat{\mathbf{k}} \text{ m/s}\end{aligned}$$

Thus the speed of Q is given by

$$v_Q|_{t=3} = \sqrt{14^2 + (-34)^2 + 4.27^2} = 37.0 \text{ m/s}$$

Continuing, we have

$$\begin{aligned}\mathbf{r}_{OQ}|_{t=3} &= 25\hat{\mathbf{i}} - 45\hat{\mathbf{j}} + (20.2 - 25 \sin 0.6)\hat{\mathbf{k}} \\ &= 25\hat{\mathbf{i}} - 45\hat{\mathbf{j}} + 6.08\hat{\mathbf{k}} \text{ m}\end{aligned}$$

The distance d between Q and its starting point is therefore given by

$$\begin{aligned} d &= |\mathbf{r}_{OQ}(3) - \mathbf{r}_{OQ}(0)| \\ &= \sqrt{(25-1)^2 + (-45-3)^2 + [6.08-(-5)]^2} \\ &= 54.8 \text{ m} \end{aligned}$$

Example 1.13

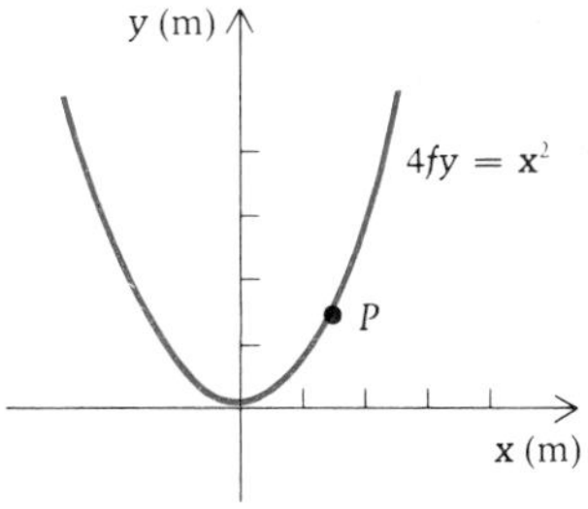

Figure E1.13

The point P in Figure E1.13 travels on the parabola (with focal distance $f = \frac{1}{2}$ m) at the constant speed of 0.2 m/s. Determine the acceleration of P: (a) as a function of x and (b) at $x = 2$ m.

Solution

We may obtain the velocity components by differentiating:

$$2y = x^2$$

$$2\dot{y} = 2x\dot{x} \Rightarrow \dot{y} = x\dot{x}$$

Thus

$$\mathbf{v}_P = \dot{x}\hat{\mathbf{i}} + \dot{y}\hat{\mathbf{j}} = \dot{x}\hat{\mathbf{i}} + x\dot{x}\hat{\mathbf{j}} \text{ m/s} \qquad (1)$$

Similarly the acceleration of P is

$$\mathbf{a}_P = \ddot{x}\hat{\mathbf{i}} + \ddot{y}\hat{\mathbf{j}} = \ddot{x}\hat{\mathbf{i}} + (\dot{x}^2 + x\ddot{x})\hat{\mathbf{j}} \text{ m/s}^2 \qquad (2)$$

Since $|\mathbf{v}_P|$, or v_P, is constant, we have

$$v_P = 0.2 = \sqrt{\dot{x}^2 + (x\dot{x})^2} \text{ m/s}$$

$$\dot{x} = \frac{0.2}{\sqrt{1+x^2}} \text{ m/s} \qquad (3)$$

We also see from (2) that we need $\ddot{x}$; differentiating (3), we get

$$\ddot{x} = \frac{-0.2\,x\dot{x}}{(1+x^2)^{3/2}} = \frac{-0.2\left(\dfrac{0.2}{\sqrt{1+x^2}}\right)x}{(1+x^2)^{3/2}} \text{ m/s}^2$$

or

$$\ddot{x} = \frac{-0.04x}{(1+x^2)^2} \text{ m/s}^2 \qquad (4)$$

Substituting (3) and (4) into (2), we get

$$\mathbf{a}_P = \frac{-0.04x}{(1+x^2)^2}\hat{\mathbf{i}} + \left[\frac{0.04}{1+x^2} - \frac{0.04x^2}{(1+x^2)^2}\right]\hat{\mathbf{j}} \text{ m/s}^2$$

When $x = 2$ m,

$$\begin{aligned} \mathbf{a}_P &= \frac{-0.04(2)}{5^2}\hat{\mathbf{i}} + \left[\frac{0.04}{5} - \frac{0.04(2^2)}{5^2}\right]\hat{\mathbf{j}} \\ &= -0.0032\hat{\mathbf{i}} + 0.0016\hat{\mathbf{j}} \text{ m/s}^2 \end{aligned}$$

In closing this section, we remark that the simple forms of Equations (1.24) and (1.25) are due to the fact that the unit vectors $\hat{\mathbf{i}}$, $\hat{\mathbf{j}}$, $\hat{\mathbf{k}}$ remain constant in both magnitude and direction when the axes are fixed in the frame of reference. For *planar* applications (Chapter 3), we shall set the z component of velocity identically to zero, obtaining the following for a point in plane motion (moving only in a plane parallel to the xy plane):

$$\mathbf{r}_{OP} = x\hat{\mathbf{i}} + y\hat{\mathbf{j}} + z\hat{\mathbf{k}} \quad \text{(where } z \text{ is constant)} \tag{1.26}$$

$$\mathbf{v}_P = \dot{x}\hat{\mathbf{i}} + \dot{y}\hat{\mathbf{j}} \tag{1.27}$$

$$\mathbf{a}_P = \ddot{x}\hat{\mathbf{i}} + \ddot{y}\hat{\mathbf{j}} \tag{1.28}$$

PROBLEMS ▶ Section 1.5

1.67 The moving pin P of a rotating crank has a location defined by

$$x = 20 \cos \pi t \text{ m}$$

$$y = 20 \sin \pi t \text{ m}$$

Find the velocity of P when $t = 0, \frac{1}{2}, \frac{3}{2}$, and 2 s.

1.68 A bar of length $2L$ moves with its ends in contact with the guides shown in Figure P1.68. Find the velocity and acceleration of point C in terms of θ and its derivatives.

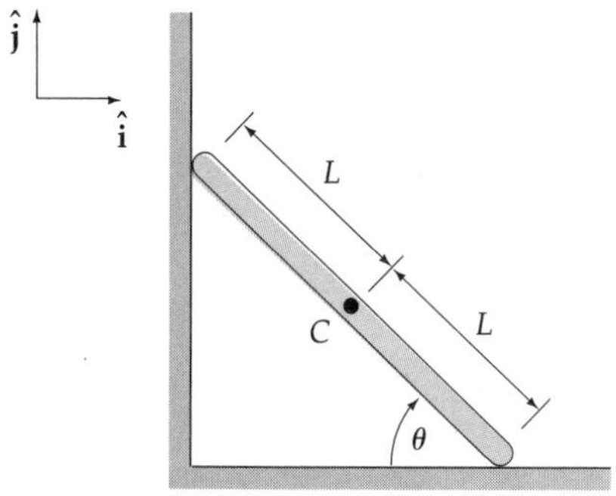

Figure P1.68

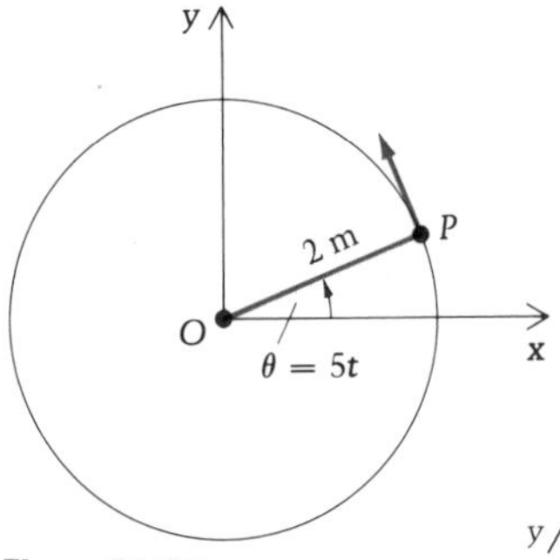

Figure P1.69

1.69 A point P moves on a circle in the direction shown in Figure P1.69. Express $\mathbf{r}_{OP}$ in (x, y, z) coordinates and differentiate to obtain $\mathbf{v}_P$ and $\mathbf{a}_P$. (Angle θ is in radians.)

1.70 Repeat the preceding problem. In this case, however, the angle θ increases quadratically, instead of linearly, with time according to $\theta = 2t^2$ rad.

1.71 A point P starts at the origin and moves along the parabola shown in Figure P1.71 with a constant x-component of velocity, $\dot{x} = 3$ ft/sec. Find the velocity and acceleration of P at the point $(x, y) = (1, 1)$.

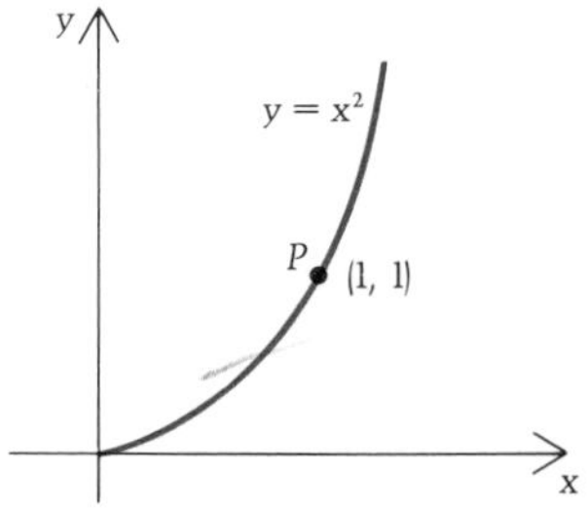

Figure P1.71

1.72 Point P is constrained to move in the two slots shown: one cut in the body $\mathcal{A}$, the other cut in the reference frame $\mathcal{R}$. The constant acceleration of $\mathcal{A}$ is given to be 4 cm/s² to the left. If point P reaches the bottom of the slot (in $\mathcal{A}$) 2 sec after the instant shown in Figure P1.72,

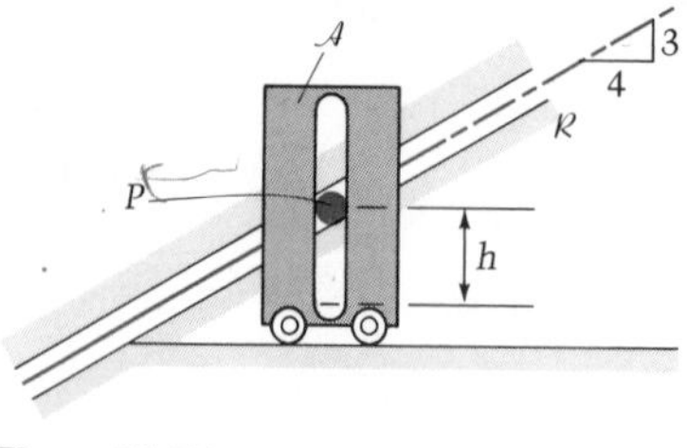

Figure P1.72

when $\mathcal{A}$ is at rest,

a. Through what height h did the marble move?
b. What distance did the marble travel?

1.73 The pin P shown in Figure P1.73 moves in a parabolic slot cut in the reference frame $\mathcal{I}$ and is guided by the vertical slot in body $\mathcal{B}$. For body $\mathcal{B}$, $x = 0.05t^3$ m locates the centerline of its slot.

a. Find the acceleration of P at $t = 5$ s.
b. Find the time(s) when the x and y components of $\mathbf{a}_P$ are equal.

1.74 A pin P moves in a slot that is cut in the shape of a hyperbolic sine as shown in Figure P1.74. It is guided along by the vertical slotted body $\mathcal{B}$, all the points of which have velocity 0.08 m/s to the right. Find $\mathbf{v}_P$ and $\mathbf{a}_P$ when $x = 0.2$ m.

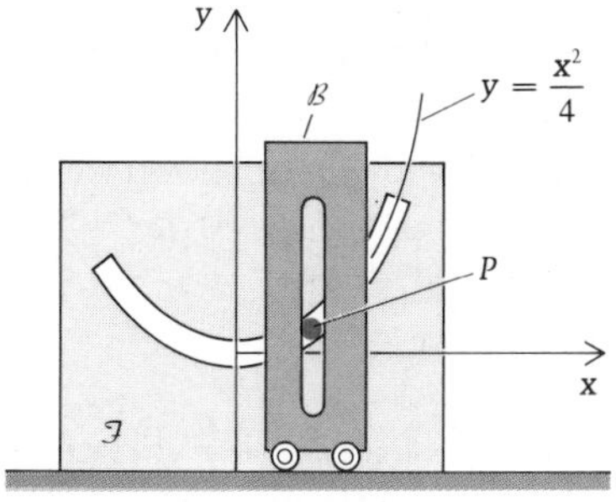

Figure P1.73

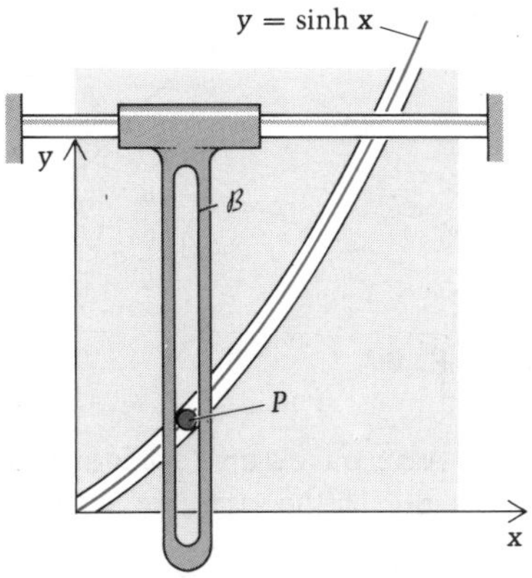

Figure P1.74

1.75 A point P travels on a path and has the following coordinates as functions of time t (in seconds):

$$x = 12\cos\frac{\pi t}{2}\text{ m} \qquad y = 8\sin\frac{\pi t}{2}\text{ m}$$

$$z = 0$$

a. Find the velocity $\mathbf{v}_P(t)$ and acceleration $\mathbf{a}_P(t)$ of P.
b. Find the position, velocity, and acceleration of P when $t = 4$ s.
c. Eliminate the time t from the x and y expressions and obtain the equation of the path of P.

1.76 The motion of a particle P is given by $x = C\cosh kt$ and $y = C\sinh kt$, where C and k are constants. Find the equation of the path of P by eliminating time t.

1.77 In the preceding problem, find the speed of P as a function of the distance $r (= \sqrt{x^2 + y^2})$ from the origin to P.

1.78 Describe precisely the path of a particle's motion if its xy coordinates are given by $(2.5t^2 + 7, 6t^2 + 9)$ meters when t is in seconds.

1.79 A particle P moves in the xy plane. The motion of P is given by

$$x = 30t + 6 \text{ ft}$$
$$y = 20t - 7 \text{ ft}$$

Find the equation of the path of P in the form $y = f(x)$.

1.80 Repeat Problem 1.79 if:

$$x = 5t \text{ m}$$
$$y = -250t^3 \text{ m}$$

1.81 Repeat Problem 1.79 if:

$$x = 2 + 3\sin t \text{ ft}$$
$$y = 4\cos t \text{ ft}$$

1.82 A cycloid is the curve traced out by a point (such as P) on the rim of a rolling wheel. In terms of the parameter θ (the angle in Figure P1.82), the equations of the cycloid are:

$$x = a(\theta - \sin\theta)$$
$$y = a(1 - \cos\theta)$$

Noting that θ changes with time, find the speed of P at $\theta = 0, \pi/2, \pi$, and $3\pi/2$ radians, in terms of a and $\dot{\theta}$.

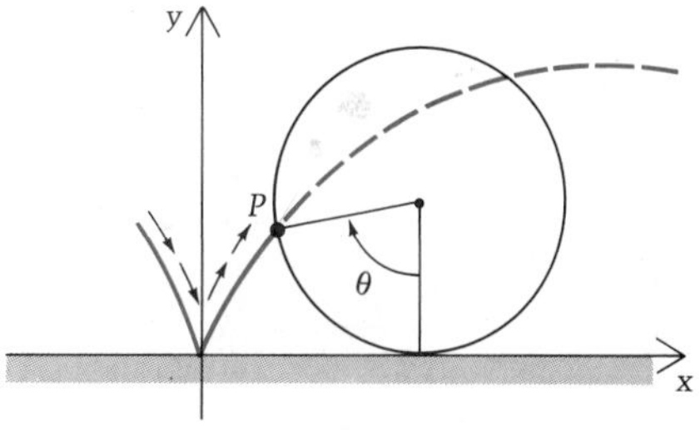

Figure P1.82

In Problems 1.83–1.86 (see Figures P1.83–P1.86), a point P travels on the curve with a constant x component of velocity, $\dot{x} = 3$ in./sec. Each starts on the curve at $x = 1$ when $t = 0$. Find the velocity vector of P when $t = 10$ sec in each case.

1.83 Logarithmic curve

1.84 Exponential curve

1.85 First-quadrant branch of rectangular hyperbola

1.86 First-quadrant branch of semicubical parabola

1.87–1.90 Find the respective acceleration vectors at $t = 10$ s of the points whose motions are described in Problems 1.83–1.86.

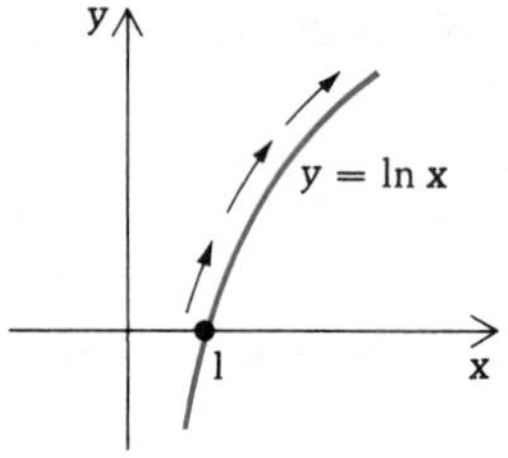

Figure P1.83

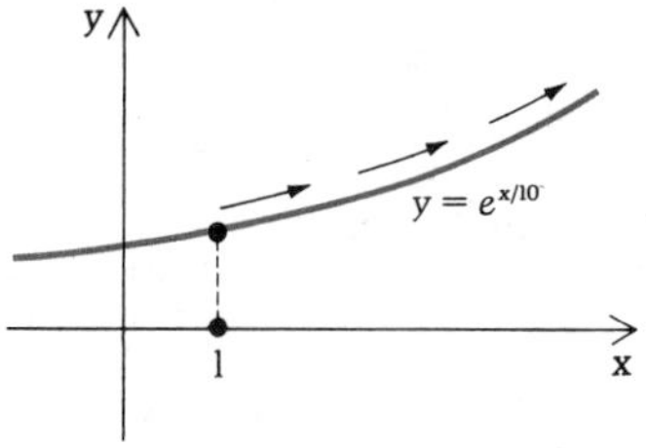

Figure P1.84

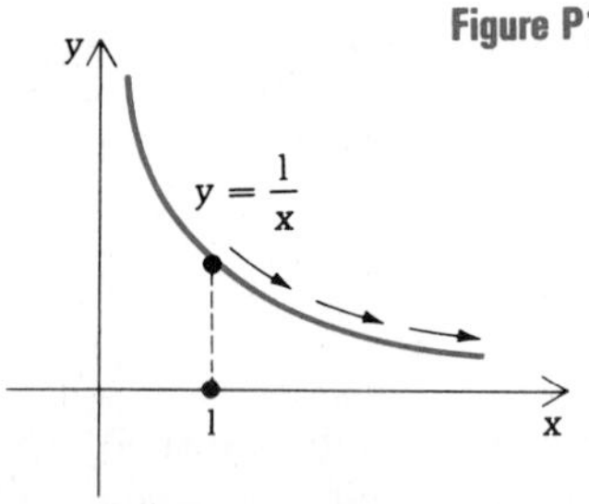

Figure P1.85

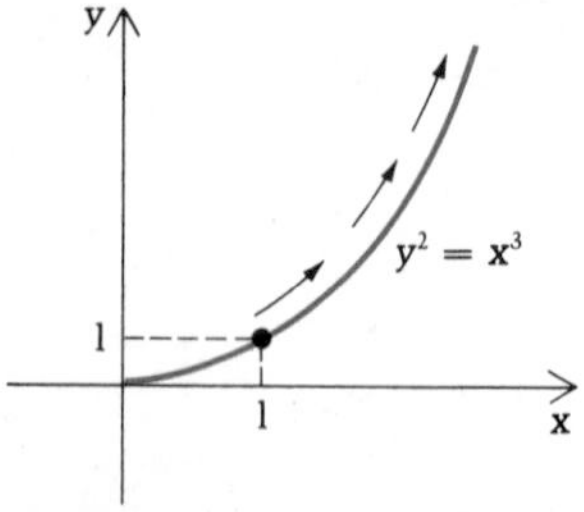

Figure P1.86

1.91 Two points P and Q have position vectors in a reference frame that are given by $\mathbf{r}_{OP} = 50t\hat{\mathbf{i}}$ meters and $\mathbf{r}_{OQ} = 40\hat{\mathbf{i}} - 20t\hat{\mathbf{j}}$ meters. Find the minimum distance between P and Q and the time at which this occurs.

1.92 Describe the path of a point P that has the following rectangular Cartesian coordinates as functions of time: $x = a\cos\omega t$, $y = a\sin\omega t$, and $z = bt$, where a, b, and ω are constants. Identify the meanings of the three constants.

1.93 For the following values of the constants, find the velocity of P at $t = 5$ s in the preceding problem: $a = 2$ m, $b = 0.5$ m/s, and $\omega = 1.2$ rad/s.

1.94 The acceleration of a point is given by

$$\mathbf{a}_P = 6t\hat{\mathbf{i}} + 12t^2\hat{\mathbf{j}} - 4\hat{\mathbf{k}}\ \text{m/s}^2$$

At $t = 0$, the initial conditions are that $\mathbf{v}_P = 2\hat{\mathbf{i}}$ m/s and $\mathbf{r}_{OP} = \hat{\mathbf{i}} + 3\hat{\mathbf{j}} + 9\hat{\mathbf{k}}$ meters. Find the position vector of P at $t = 3$ s, and determine how far P then is from its position at $t = 0$.

1.95 A point moves on a path, with a position vector as a function of time given by $\mathbf{r}_{OP} = \sin 2t\hat{\mathbf{i}} + 3t\hat{\mathbf{j}} + e^{6t}\hat{\mathbf{k}}$, in units of meters when t is in seconds. Find:

- **a.** The speed of the point at $t = 0$.
- **b.** Its acceleration at $t = \pi/2$ s.
- **c.** The component of the velocity vector, at $t = 0$, which is parallel to the line ℓ in the xy plane given by $y = \frac{5}{12}x - 6$ shown in Figure P1.95.

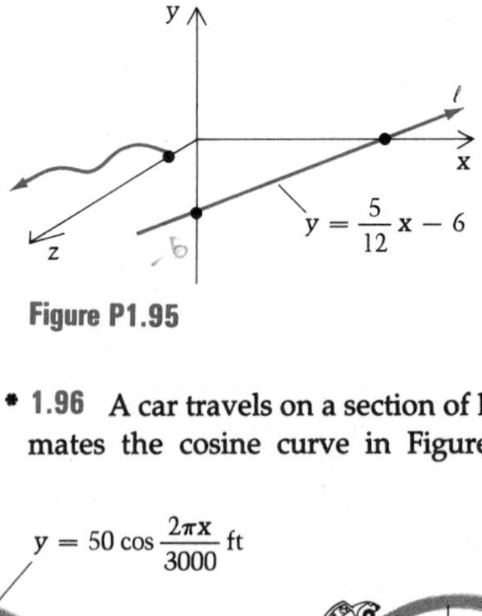

Figure P1.95

• **1.96** A car travels on a section of highway that approximates the cosine curve in Figure P1.96. If the driver

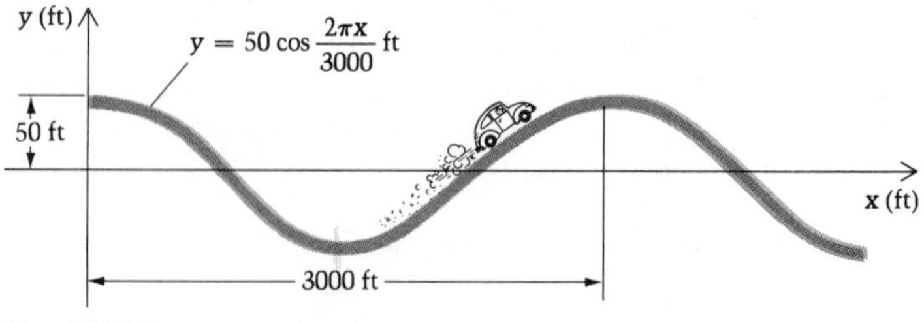

Figure P1.96

maintains a constant speed of 55 mph, determine his x and y components of velocity when $x = 2500$ ft.

* **1.97** A car travels along the highway of the preceding problem with a constant x-component of velocity of 54.9 mph. Over what sections of the highway does the driver exceed the speed limit of 55 mph?

1.98 Determine the minimum magnitude of acceleration of the car in Problem 1.96. Where on the curve is this acceleration experienced?

1.99 Find the maximum magnitude of acceleration of the car in Problem 1.97. Where does it occur on the curve?

1.6 Cylindrical Coordinates

If a point P is moving in such a way that its projection into the xy plane is more easily described with polar (r and θ) coordinates than with x and y, then we may use **cylindrical coordinates** to advantage. These coordinates are nothing more than the **polar coordinates** r and θ together with an "axial" coordinate z. Thus r and θ locate the projection point of P in a plane, while z gives the distance of P *from* the plane.

Embedding the same set of rectangular axes (x, y, z) in the reference frame $\mathcal{I}$ as we did in the preceding section, we now show the coordinates r and θ as well (Figure 1.5). Note that P' is the projection of P into the plane xy. From Figure 1.5 we see that the unit vectors $\hat{\mathbf{e}}_r$ and $\hat{\mathbf{e}}_\theta$ are drawn in the xy plane and that:

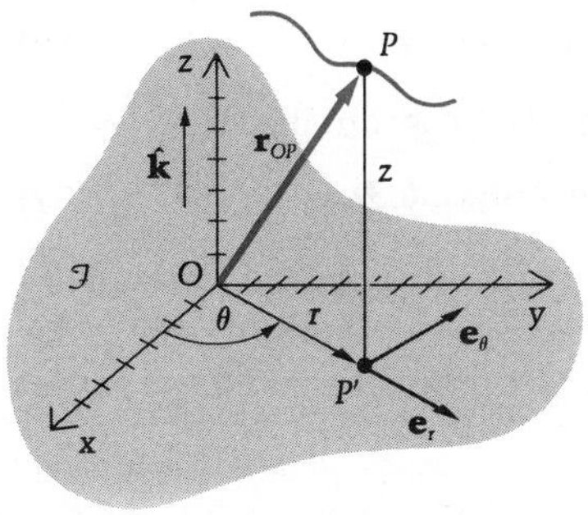

Figure 1.5 Cylindrical coordinates of P.

1. The direction of $\hat{\mathbf{e}}_r$ is that of OP'.
2. $\hat{\mathbf{e}}_\theta$ is normal to $\hat{\mathbf{e}}_r$ in the direction of increasing θ.

It will be helpful later in the section to note carefully at this point that $\hat{\mathbf{e}}_r$ and $\hat{\mathbf{e}}_\theta$ change (in direction) with changes in θ, but not with r or z. Thus if the point P moves along either a radial line or a vertical line, the two unit vectors remain the same. But if P moves in such a way as to alter θ, then the directions of $\hat{\mathbf{e}}_r$ and $\hat{\mathbf{e}}_\theta$ will vary.

The rectangular and cylindrical coordinates (both having z in common) are related through

$$x = r\cos\theta$$
$$y = r\sin\theta \tag{1.29}$$

which can be differentiated to produce, by Equations (1.24) and (1.25), formulas for velocity and acceleration in terms of the cylindrical coordinates (and their derivatives) and the unit vectors $\hat{\mathbf{i}}$, $\hat{\mathbf{j}}$, and $\hat{\mathbf{k}}$. It is usually more desirable, however, to express the velocity and acceleration in terms of the set of unit vectors ($\hat{\mathbf{e}}_r$, $\hat{\mathbf{e}}_\theta$, $\hat{\mathbf{k}}$), which are naturally associated with cylindrical coordinates. Thus it is useful to express a position vector $\mathbf{r}_{OP}$ as

$$\mathbf{r}_{OP} = r\hat{\mathbf{e}}_r + z\hat{\mathbf{k}} \tag{1.30}$$

Question 1.3 Why is there no $\hat{\mathbf{e}}_\theta$ term in Equation (1.30)?

Answer 1.3 From Figure 1.5, we see that $\hat{\mathbf{e}}_\theta$ is perpendicular to $\mathbf{r}_{OP}$. Note, however, that implicit in the writing and use of Equation (1.30) is the polar angle θ.

Differentiating Equation (1.30), we obtain the velocity of P:

$$\mathbf{v}_P = \dot{r}\hat{\mathbf{e}}_r + r\dot{\hat{\mathbf{e}}}_r + \dot{z}\hat{\mathbf{k}} \tag{1.31}$$

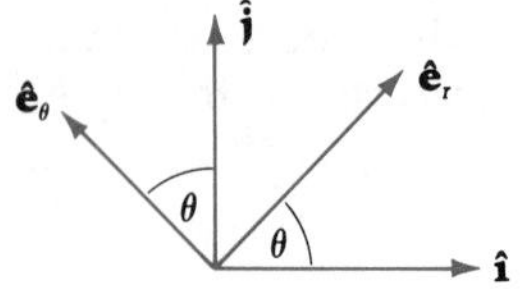

Figure 1.6

To evaluate $\dot{\hat{\mathbf{e}}}_r$ we note from Figure 1.6 that

$$\hat{\mathbf{e}}_r = \cos\theta\hat{\mathbf{i}} + \sin\theta\hat{\mathbf{j}} \tag{1.32a}$$

$$\hat{\mathbf{e}}_\theta = -\sin\theta\hat{\mathbf{i}} + \cos\theta\hat{\mathbf{j}} \tag{1.32b}$$

Hence

$$\dot{\hat{\mathbf{e}}}_r = \dot{\theta}(-\sin\theta\hat{\mathbf{i}} + \cos\theta\hat{\mathbf{j}}) = \dot{\theta}\hat{\mathbf{e}}_\theta \tag{1.33}$$

and thus the velocity in cylindrical coordinates takes the form

$$\mathbf{v}_P = \dot{r}\hat{\mathbf{e}}_r + r\dot{\theta}\hat{\mathbf{e}}_\theta + \dot{z}\hat{\mathbf{k}} \tag{1.34}$$

Differentiating again, we get

$$\mathbf{a}_P = \ddot{r}\hat{\mathbf{e}}_r + \dot{r}\dot{\hat{\mathbf{e}}}_r + \dot{r}\dot{\theta}\hat{\mathbf{e}}_\theta + r\ddot{\theta}\hat{\mathbf{e}}_\theta + r\dot{\theta}\dot{\hat{\mathbf{e}}}_\theta + \ddot{z}\hat{\mathbf{k}} \tag{1.35}$$

Using Equations (1.32), we find that

$$\dot{\hat{\mathbf{e}}}_\theta = \dot{\theta}(-\cos\theta\hat{\mathbf{i}} - \sin\theta\hat{\mathbf{j}}) = -\dot{\theta}\hat{\mathbf{e}}_r \tag{1.36}$$

Thus the acceleration expression in cylindrical coordinates is

$$\mathbf{a}_P = (\ddot{r} - r\dot{\theta}^2)\hat{\mathbf{e}}_r + (r\ddot{\theta} + 2\dot{r}\dot{\theta})\hat{\mathbf{e}}_\theta + \ddot{z}\hat{\mathbf{k}} \tag{1.37}$$

In the special case for which the motion is in a plane defined by $z =$ constant, we have $\dot{z} = \ddot{z} = 0$. In this case we need only the polar coordinates r and θ, and the directions of $\hat{\mathbf{e}}_r$ and $\hat{\mathbf{e}}_\theta$ are said to be **radial** and **transverse**, respectively.

Question 1.4 If a point P moves with $z \equiv 0$, then $\mathbf{r}_{OP} = r\hat{\mathbf{e}}_r$; thus $|\mathbf{r}_{OP}| = r$. Why then isn't $|\mathbf{v}_P| = \dot{r}$?

Before turning to the examples in this section, we return briefly to calculation of the derivatives of the unit vectors $\hat{\mathbf{e}}_r$ and $\hat{\mathbf{e}}_\theta$. We note that each derivative turns out to be perpendicular to the vector being differentiated. To understand why this is the case, we consider an alternative derivation of the formula for $\dot{\hat{\mathbf{e}}}_r$. The time dependence of $\hat{\mathbf{e}}_r$ is due to the time dependence of the coordinate θ on which $\hat{\mathbf{e}}_r$ depends explicitly; thus we can write

$$\dot{\hat{\mathbf{e}}}_r = \frac{d\hat{\mathbf{e}}_r}{d\theta}\dot{\theta}$$

Answer 1.4 For one thing, $\dot{r}$ can be negative. But $|\mathbf{v}_P| \neq |\dot{r}|$ either, because the magnitude of the derivative of a vector is not equal to the absolute value of the derivative of the magnitude of the vector. Differentiating $\mathbf{r}_{OP}$ produces a term $(r\dot{\theta}\hat{\mathbf{e}}_\theta)$ in $\mathbf{v}_P$ in addition to $\dot{r}\hat{\mathbf{e}}_r$.

Let us study the derivative $d\hat{\mathbf{e}}_r/d\theta$. By definition,

$$\frac{d\hat{\mathbf{e}}_r}{d\theta} = \lim_{\Delta\theta \to 0} \left[\frac{\hat{\mathbf{e}}_r(\theta + \Delta\theta) - \hat{\mathbf{e}}_r(\theta)}{\Delta\theta}\right] = \lim_{\Delta\theta \to 0} \left(\frac{\Delta\hat{\mathbf{e}}_r}{\Delta\theta}\right)$$

With the aid of Figure 1.7 we can see that:

1. The direction of $\Delta\hat{\mathbf{e}}_r$ (and hence $\Delta\hat{\mathbf{e}}_r/\Delta\theta$) approaches that of $\hat{\mathbf{e}}_\theta$ as $\Delta\theta$ approaches zero.
2. The magnitude of $\Delta\hat{\mathbf{e}}_r/\Delta\theta$ is $[2(1)\sin(\Delta\theta/2)]/\Delta\theta$, which approaches unity as $\Delta\theta$ approaches zero.

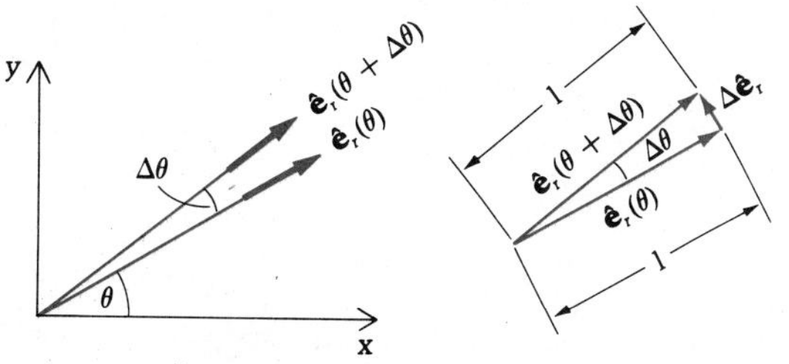

Figure 1.7 Change in $\hat{\mathbf{e}}_r$ as θ changes.

Thus $d\hat{\mathbf{e}}_r/d\theta = \hat{\mathbf{e}}_\theta$, and we obtain $\dot{\hat{\mathbf{e}}}_r = \dot{\theta}\hat{\mathbf{e}}_\theta$ in agreement with Equation (1.33). The reader may wish to sketch a similar geometric proof of Equation (1.36).

This mutual orthogonality of a vector and its derivative, incidentally, is not just restricted to unit vectors. It is in fact a property of all vectors of constant magnitude. We can show that this is the case by noting that if $\mathbf{A}$ is such a vector, then

$$\mathbf{A} \cdot \mathbf{A} = |\mathbf{A}|^2 = \text{constant}$$

and thus

$$\frac{d}{dt}(\mathbf{A} \cdot \mathbf{A}) = 0$$

or

$$\frac{d\mathbf{A}}{dt} \cdot \mathbf{A} + \mathbf{A} \cdot \frac{d\mathbf{A}}{dt} = 0$$

or

$$2\mathbf{A} \cdot \frac{d\mathbf{A}}{dt} = 0 \tag{1.38}$$

Hence, provided that neither the vector nor its derivative vanishes, the vector and the derivative are mutually perpendicular. This is a result we shall make use of frequently throughout the book. We now proceed to some examples of velocity and acceleration in cylindrical coordinates.

EXAMPLE 1.14

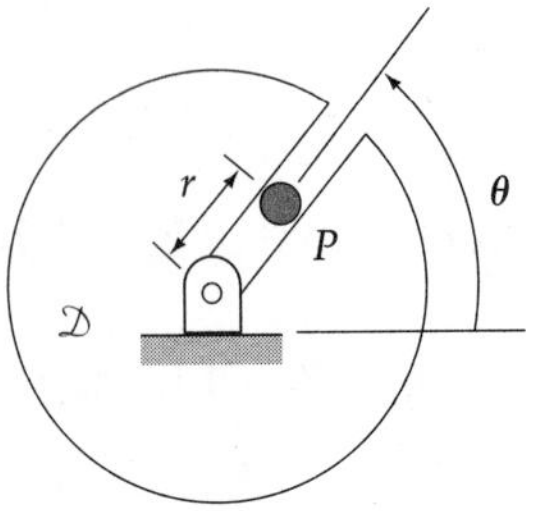

Figure E1.14a

The pin P in Figure E1.14a moves outward with respect to a horizontal circular disk, and its radial coordinate r is given as a function of time by $r = 3t^2/2$ meters. The disk $\mathscr{D}$ turns with the time-dependent angle $\theta = 4t^2/3$ rad. Find the velocity and acceleration of P at $t = 1$ s.

Solution

From Equation (1.34) we have

$$\begin{aligned}\mathbf{v}_P &= \dot{r}\hat{\mathbf{e}}_r + r\dot{\theta}\hat{\mathbf{e}}_\theta + \dot{z}\hat{\mathbf{k}}\\ &= 3t\hat{\mathbf{e}}_r + \left(\frac{3}{2}t^2\right)\left(\frac{8}{3}t\right)\hat{\mathbf{e}}_\theta + 0\\ &= 3t\hat{\mathbf{e}}_r + 4t^3\hat{\mathbf{e}}_\theta \text{ m/s}\end{aligned}$$

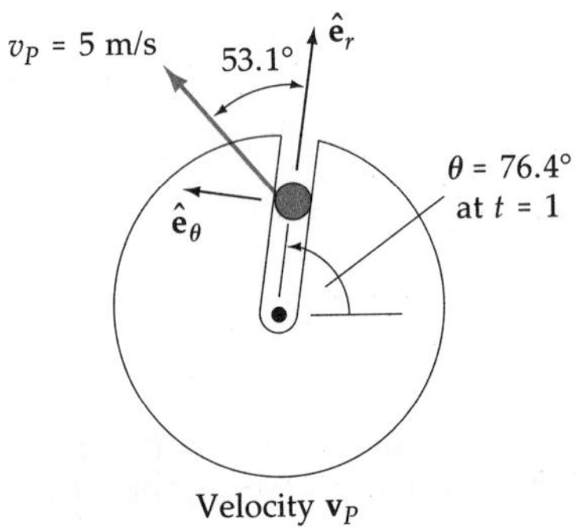

Velocity $\mathbf{v}_P$

Figure E1.14b

Thus

$$\mathbf{v}_P|_{t=1} = 3\hat{\mathbf{e}}_r + 4\hat{\mathbf{e}}_\theta \text{ m/s}$$

and we note that the speed of P at $t = 1$ s is 5 m/s.

Continuing, from Equation (1.37) we get

$$\begin{aligned}\mathbf{a}_P &= (\ddot{r} - r\dot{\theta}^2)\hat{\mathbf{e}}_r + (r\ddot{\theta} + 2\dot{r}\dot{\theta})\hat{\mathbf{e}}_\theta + \ddot{z}\hat{\mathbf{k}}\\ &= \left[3 - \left(\frac{3}{2}t^2\right)\left(\frac{8}{3}t\right)^2\right]\hat{\mathbf{e}}_r + \left[\left(\frac{3}{2}t^2\right)\frac{8}{3} + 2(3t)\left(\frac{8}{3}t\right)\right]\hat{\mathbf{e}}_\theta + 0\\ &= \left(3 - \frac{32}{3}t^4\right)\hat{\mathbf{e}}_r + 20t^2\hat{\mathbf{e}}_\theta\end{aligned}$$

Thus

$$\mathbf{a}_P|_{t=1} = \frac{-23}{3}\hat{\mathbf{e}}_r + 20\hat{\mathbf{e}}_\theta \text{ m/s}^2$$

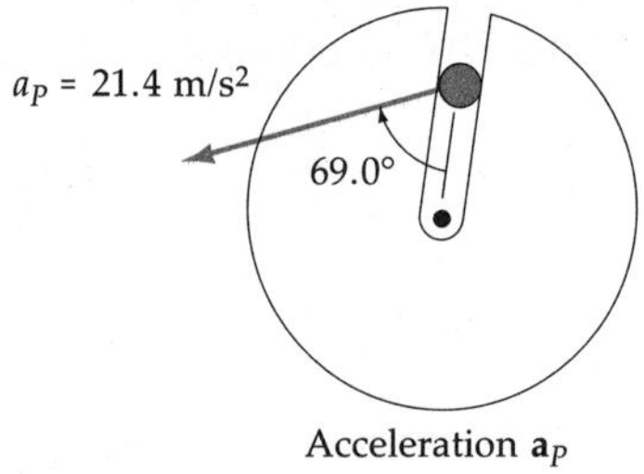

Acceleration $\mathbf{a}_P$

Figure E1.14c

Since at $t = 1$ we have $r = 3/2$ m and $\theta = 4/3$ rad, we show the preceding results pictorially in Figures E1.14b and c.

Note that there is a time, $t = \sqrt[4]{9/32}$ s, when the $\ddot{r}$ and $-r\dot{\theta}^2$ parts of the radial component of $\mathbf{a}_P$ cancel each other, making this component zero at that instant of time. The reader is urged to compute and sketch $\mathbf{v}_P$ and $\mathbf{a}_P$ at another time, say $t = 2$ s.

EXAMPLE 1.15

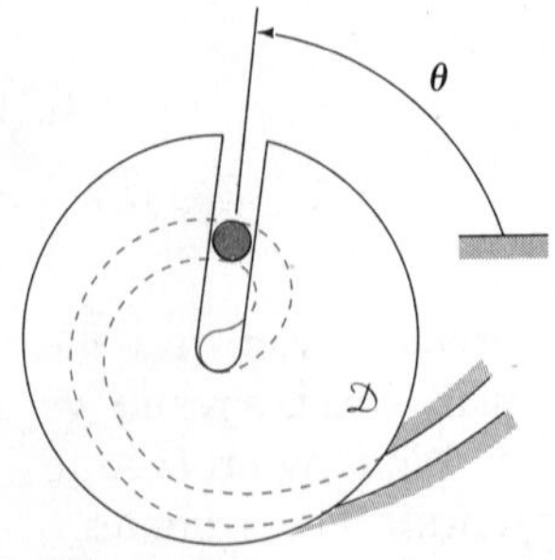

Figure E1.15a

In the preceding example, discard the given r and θ. Suppose instead that $\dot{\theta}_{\mathscr{D}}$ = constant = 0.3 rad/s and that the pin slides not only in the slot of disk $\mathscr{D}$ (see Figure E1.15a), but also in the spiral slot cut in the reference frame and defined by $r = 0.1\theta$ meters, with θ in radians. Find the velocity and acceleration of the pin when $\theta = \pi$ rad.

Solution

From $r = 0.1\theta$, we have $\dot{r} = 0.1\dot{\theta}$ and $\ddot{r} = 0.1\ddot{\theta}$, which is zero since $\dot{\theta}$ = constant. Therefore

$$\mathbf{v}_P = \dot{r}\hat{\mathbf{e}}_r + r\dot{\theta}\hat{\mathbf{e}}_\theta = 0.1\dot{\theta}\hat{\mathbf{e}}_r + 0.10\theta\dot{\theta}\hat{\mathbf{e}}_\theta$$

$.1(.2)\,\hat{e}_r + .1(\pi)(.3)\,\hat{e}_\theta \quad = 0.0300\hat{\mathbf{e}}_r + 0.0942\hat{\mathbf{e}}_\theta \text{ m/s}$

Now for the acceleration:

$$\mathbf{a}_P = (\ddot{r} - r\dot{\theta}^2)\hat{\mathbf{e}}_r + (r\ddot{\theta} + 2\dot{r}\dot{\theta})\hat{\mathbf{e}}_\theta$$

$$= (0 - 0.10\dot{\theta}^2)\hat{\mathbf{e}}_r + [0 + 2(0.1\dot{\theta})\dot{\theta}]\hat{\mathbf{e}}_\theta$$

$$= -0.0283\hat{\mathbf{e}}_r + 0.0180\hat{\mathbf{e}}_\theta \text{ m/s}^2$$

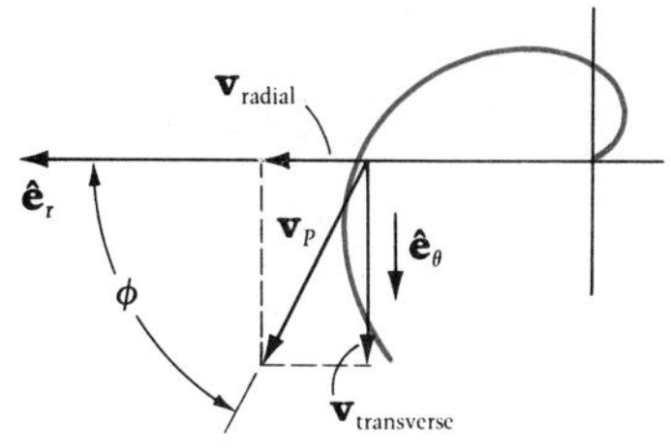

We shall learn in the next section that the velocity is always tangent to the path of the point. Thus the angle ϕ between the path and the $-x$ axis can be found from the velocity components, as shown in Figure E1.15b, as follows:

$$\phi = \tan^{-1}\left(\frac{0.0942}{0.0300}\right) = 72.3°$$

Figure E1.15b

In our next examples, there is motion in the z direction as well as the radial (r) and transverse (θ) of the previous examples.

EXAMPLE 1.16

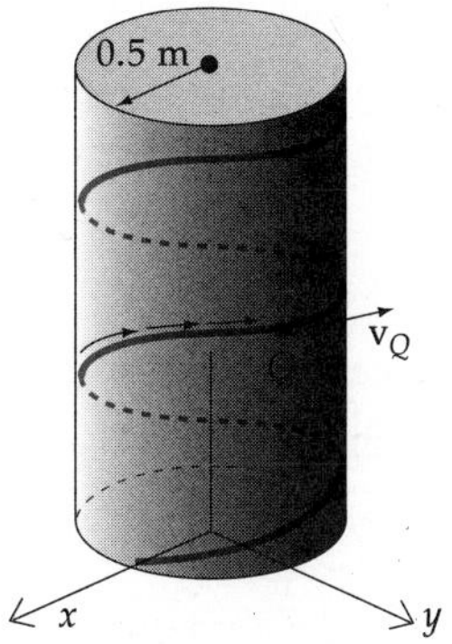

A point Q moves on a helix as shown in Figure E1.16a. The pitch, p, of the helix is 0.2 m, and the point travels at constant speed 20 m/s. Find the velocity of Q in terms of its cylindrical components.

Solution

The meaning of the *pitch* of a helix is the (constant) advance of Q in the z direction for each revolution in θ. Therefore

$$\frac{p\theta}{2\pi} = z \tag{1}$$

so that

$$\dot{\theta} = \frac{2\pi\dot{z}}{p} \tag{2}$$

or, for this problem,

$$\dot{\theta} = 31.42\dot{z} \tag{3}$$

Noting that $\dot{r} = 0$ since Q travels on a cylinder (with r therefore constant), Equation (1.34) then gives the following for the point's velocity:

$$\mathbf{v}_Q = r\dot{\theta}\hat{\mathbf{e}}_\theta + \dot{z}\hat{\mathbf{k}} \tag{4}$$

$$= 0.5\dot{\theta}\hat{\mathbf{e}}_\theta + \dot{z}\hat{\mathbf{k}} \tag{5}$$

or, using (3),

$$\mathbf{v}_Q = 15.71\dot{z}\hat{\mathbf{e}}_\theta + \dot{z}\hat{\mathbf{k}} \tag{6}$$

The speed of Q is constant at 20 m/s; thus

$$\sqrt{(15.71\dot{z})^2 + \dot{z}^2} = 20 \tag{7}$$

$$\dot{z} = 1.271 \text{ m/s} \tag{8}$$

Figure E1.16a

$r = .5m$

$\mathbf{v} = \dot{r}\,\hat{e}_r + r\dot{\theta}\,\hat{e}_\theta + \dot{z}\,\hat{k}$

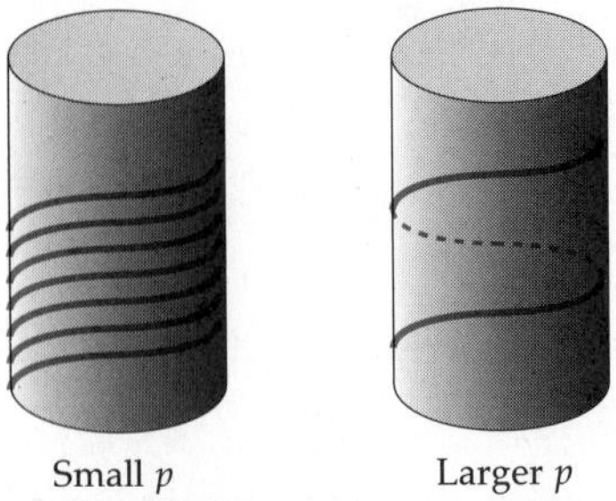

Figure E1.16b

From Equation (3), we then get

$$\dot{\theta} = 31.42(1.271) = 39.94 \text{ rad/s} \tag{9}$$

Hence the velocity vector of Q is (substituting (9) and (8) into (5))

$$\mathbf{v}_Q = 19.97\hat{\mathbf{e}}_\theta + 1.271\hat{\mathbf{k}} \text{ m/s} \tag{10}$$

Note that $|\mathbf{v}_Q| = 20.0$ m/s, as it must be. Note also that a larger pitch will spread out the helix (see Figure E1.16b). The equations of this example then show that the $\hat{\mathbf{k}}$ component will become larger in comparison to the $\hat{\mathbf{e}}_\theta$ component for larger p.

EXAMPLE 1.17

Find the acceleration of Q in the preceding example.

Solution

From Equation (1.37) we get

$$\mathbf{a}_Q = (\ddot{r} - r\dot{\theta}^2)\hat{\mathbf{e}}_r + (r\ddot{\theta} + 2\dot{r}\dot{\theta})\hat{\mathbf{e}}_\theta + \ddot{z}\hat{\mathbf{k}}$$

Because r is constant on the cylinder, this equation reduces to

$$\mathbf{a}_Q = -r\dot{\theta}^2\hat{\mathbf{e}}_r + r\ddot{\theta}\hat{\mathbf{e}}_\theta + \ddot{z}\hat{\mathbf{k}}$$

Furthermore, since $|\mathbf{v}_Q|$ is constant, Equations (8) and (3) of the previous example show that $\dot{z}$ and $\dot{\theta}$ are constants. Therefore there is only one non-vanishing acceleration component here:

$$\mathbf{a}_Q = -r\dot{\theta}^2\hat{\mathbf{e}}_r = -0.5(39.9^2)\hat{\mathbf{e}}_r = -796\hat{\mathbf{e}}_r \text{ m/s}^2$$

Note that even though point Q *never* has a radial component of velocity (see Figure E1.17), it has *only* a radial component of acceleration!

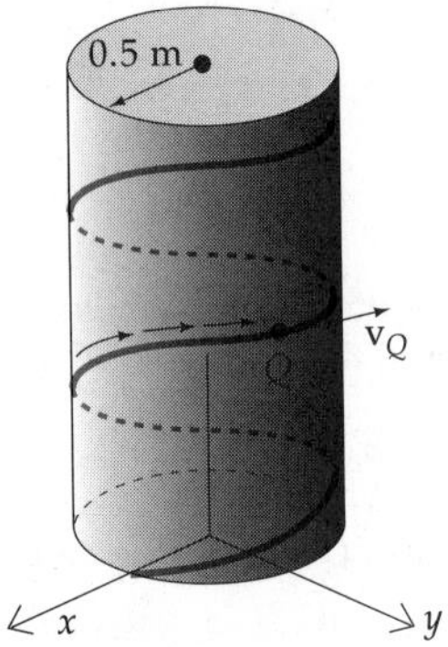

Figure E1.17

EXAMPLE 1.18

Find the velocity and acceleration vectors of point Q in Example 1.16 if, instead of the speed of Q being constant, we have its vertical position given as the function of time:

$$z = 0.08t^3 \text{ m}$$

$\dot{z} = .24t^2$

Solution

Referring to Example 1.16 (see Figure E1.18), we find

$\dot{\theta} = 31.42\dot{z} \Rightarrow$

$$\dot{\theta} = 31.4\dot{z} = 31.4(0.24t^2)$$
$$= 7.54t^2 \text{ rad/s}$$

Therefore

$$\mathbf{v}_Q = 0.5(7.54t^2)\hat{\mathbf{e}}_\theta + 0.24t^2\hat{\mathbf{k}}$$
$$= 3.77t^2\hat{\mathbf{e}}_\theta + 0.24t^2\hat{\mathbf{k}} \text{ m/s}$$

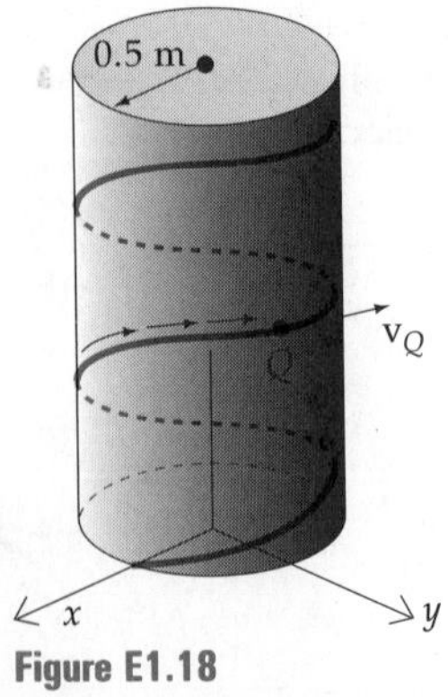

Figure E1.18

This time, however, the velocity is seen to depend on the time; for example, at $t = 10$ s,

$$\mathbf{v}_Q|_{t=10} = 377\hat{\mathbf{e}}_\theta + 24.0\hat{\mathbf{k}} \text{ m/s}$$

For the acceleration, we note that $\dot{r}$ is still zero, so that

$$\mathbf{a}_Q = -r\dot{\theta}^2\hat{\mathbf{e}}_r + r\ddot{\theta}\hat{\mathbf{e}}_\theta + \ddot{z}\hat{\mathbf{k}}$$

This time, all three terms are nonzero. We have

$$\dot{z} = 0.24t^2 \text{ m/s} \qquad \dot{\theta} = 7.54t^2 \text{ rad/s}$$

$$\ddot{z} = 0.48t \text{ m/s}^2 \qquad \ddot{\theta} = 15.1t \text{ rad/s}^2$$

Thus

$$\begin{aligned}\mathbf{a}_Q &= -0.5(7.54t^2)^2\hat{\mathbf{e}}_r + 0.5(15.1t)\hat{\mathbf{e}}_\theta + (0.48t)\hat{\mathbf{k}} \\ &= -28.4t^4\hat{\mathbf{e}}_r + 7.55t\hat{\mathbf{e}}_\theta + 0.48t\hat{\mathbf{k}} \text{ m/s}^2\end{aligned}$$

In the final example of this section, we consider the case in which, in addition to the changing θ and z of the preceding three examples, the radius varies.

EXAMPLE 1.19

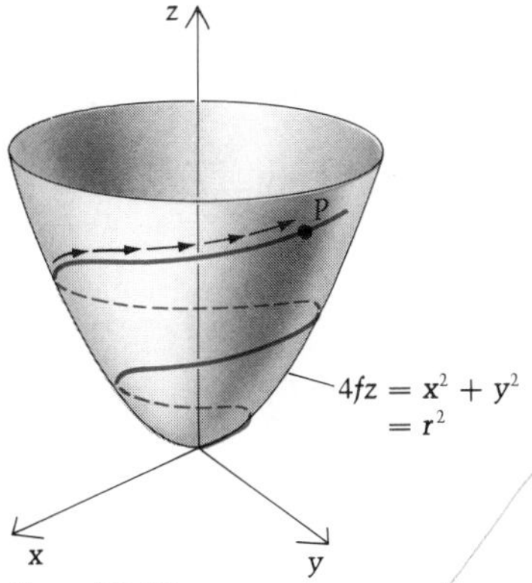

Figure E1.19

A point P moves on a spiraling path that winds around the paraboloid of revolution shown in Figure E1.19. The focal distance f is $\frac{1}{4}$ m, and the point P advances 4.0 m vertically with each revolution. If the speed of P is 0.7 m/s, a constant, determine the vertical component of the velocity vector of P as a function of r.

Solution

From $z = r^2$, we obtain

$$\dot{z} = 2r\dot{r} \Rightarrow \dot{r} = \frac{\dot{z}}{2r} \text{ m/s} \tag{1}$$

And from the pitch relationship $p\theta/2\pi = z$, we get

$$4.0\dot{\theta} = 2\pi\dot{z} \Rightarrow \dot{\theta} = \frac{\pi}{2}\dot{z} \text{ rad/s} \tag{2}$$

Therefore the speed of P may be expressed as

$$\begin{aligned}|\mathbf{v}_P| = 0.7 &= \sqrt{\dot{r}^2 + (r\dot{\theta})^2 + \dot{z}^2} \\ &= \sqrt{\left(\frac{\dot{z}}{2r}\right)^2 + \left(\frac{\pi}{2}r\dot{z}\right)^2 + \dot{z}^2} \text{ m/s}\end{aligned}$$

Thus the answer is

$$\dot{z} = \frac{1.4r}{\sqrt{1 + 4r^2 + \pi^2 r^4}} \text{ m/s}$$

Let us extend the preceding example slightly. We can see that $\dot{z}$ varies with the radius r (distance from the z axis to P) and that it is zero initially and approaches zero again for large r. Its maximum may be determined from calculus:

$$\frac{d\dot{z}}{dr} = 0$$

or

$$0 = \frac{\sqrt{1 + 4r^2 + \pi^2 r^4}\,\dfrac{d(1.4r)}{dr} - 1.4r\,\dfrac{d\sqrt{1 + 4r^2 + \pi^2 r^4}}{dr}}{(\sqrt{1 + 4r^2 + \pi^2 r^4})^2}$$

This yields the equation and result:

$$\pi^2 r^4 = 1 \Rightarrow r = 0.56 \text{ m}$$

at which

$$\dot{z}_{\max} = 0.44 \text{ m/s}$$

Note from Equation (2) in the example that at this value of $\dot{z}$,

$$\dot{\theta} = 0.69 \text{ rad/s}$$

From Equation (1), we see that at the same time

$$\dot{r} = \frac{\dot{z}}{2r} = 0.39 \text{ m/s}$$

and therefore, when $\dot{z}$ is maximum, the speed is

$$|\mathbf{v}_P| = \sqrt{\dot{r}^2 + (r\dot{\theta})^2 + \dot{z}^2}$$
$$= \sqrt{0.39^2 + (0.56 \times 0.69)^2 + 0.44^2} = 0.70 \text{ m/s}$$

as it should be, since it does not change with time.

Question 1.5 By inspection (with little or no writing), what is the maximum magnitude of the radial component of $\mathbf{v}_P$?

Answer 1.5 When $r = 0$, we have $\dot{z} = 0 = \dot{\theta}$. Thus $\dot{r}$, the radial component of $\mathbf{v}_P$, is maximum there at the value 0.7, which is the constant speed. (Note that $\dot{r}$ decreases continuously toward zero from there.)

PROBLEMS ▶ Section 1.6

1.100 The airplane in Figure P1.100 travels at constant speed at a constant altitude. The radar tracks the plane and computes the distance D, the angle θ, and the rate of change of θ ($\dot{\theta}$) at all times. In terms of θ, $\dot{\theta}$, and D, find the speed of the airplane.

1.101 A ball bearing is moving radially outward in a slotted horizontal disk that is rotating about the vertical z axis. At the instant shown in Figure P1.101, the ball bearing is 3 in. from the center of the disk. It is traveling radially outward at a velocity of 4 in./sec relative to the disk

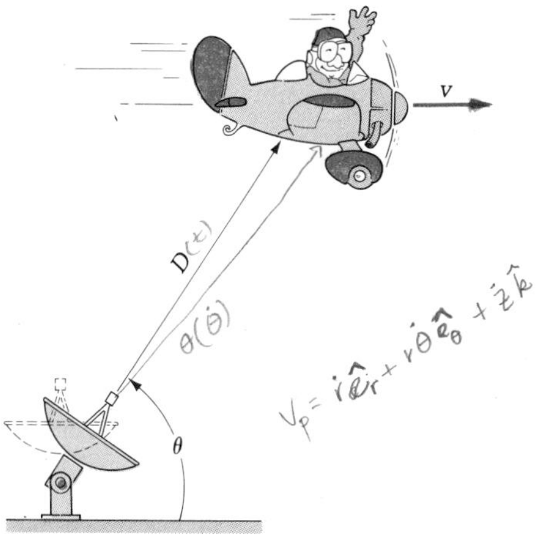

Figure P1.100

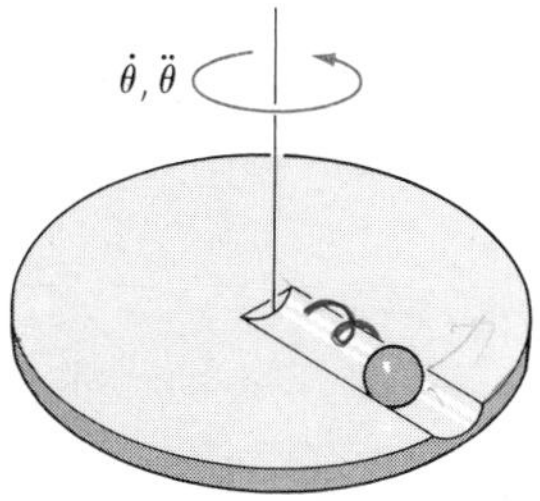

Figure P1.101

and has a radial acceleration with respect to the disk of 5 in./sec² outward. What would $\dot{\theta}$ and $\ddot{\theta}$ have to be at the instant shown for the ball bearing to have a total acceleration of zero?

1.102 The disk shown in Figure P1.102 is horizontal and turns so that $\dot{\theta} = ct$ about the vertical. Forces cause a marble to move in a slot such that its radial distance from the center equals kt^2. Note that c and k are constants.

a. Find the acceleration of the marble.

b. At what time does the radial acceleration vanish?

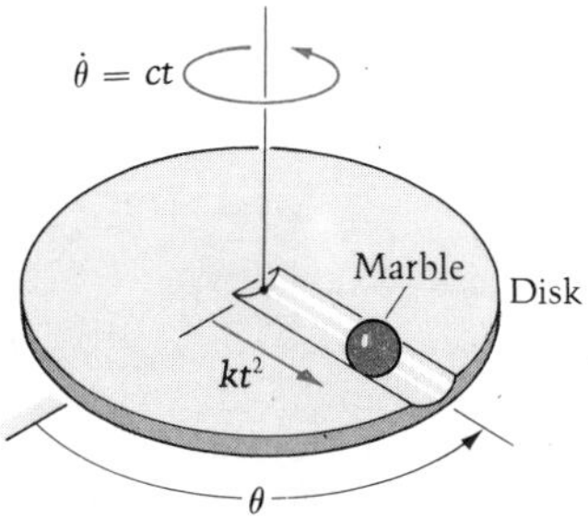

Figure P1.102

1.103 A particle moves on a curve called the "Lemniscate of Bernoulli," defined by $r^2 = 2 \cos 2\theta$ ft². It moves along the branch shown in Figure P1.103 with arrows, and passes through point P at $t = 0$. The angle θ increases with time according to $\theta = 3t^2 + 2t$ rad, with t measured in seconds. At the point P, find the velocity and acceleration of the particle.

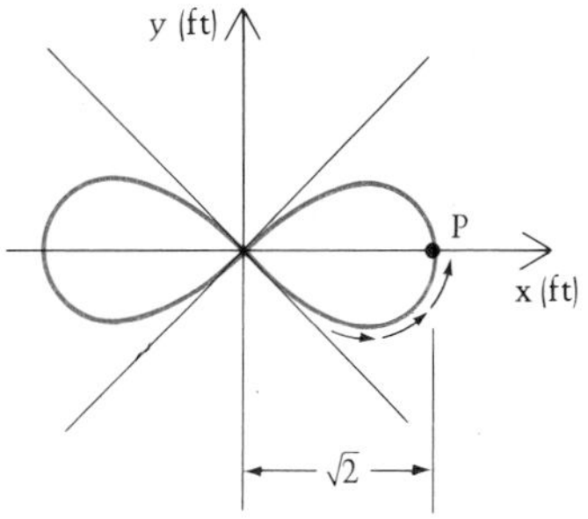

Figure P1.103

* **1.104** A point P moves on the "Spiral of Archimedes" at constant speed 2 m/s. (See Figure P1.104.) The equation of the spiral is $r = 3\theta$. Find the acceleration of P when $\theta = 180°$.

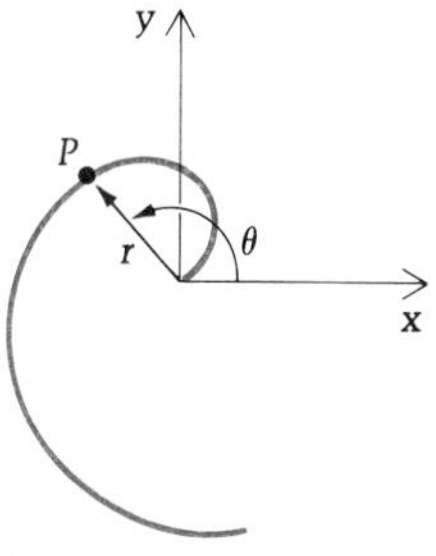

Figure P1.104

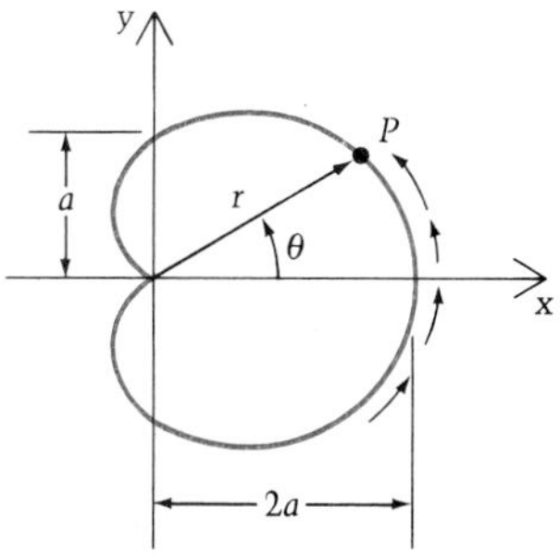

Figure P1.105

1.105 The cardioid in Figure P1.105 has the equation $r = a(1 + \cos \theta)$. Point P travels around this curve, in the direction indicated, in such a way that $\dot{\theta} = K =$ constant. In terms of K and the length a, find the velocity of P at the four points where its path intersects the coordinate axes. Express the result in terms of radial and transverse components, and then convert to rectangular components by expressing $\hat{\mathbf{e}}_r$ and $\hat{\mathbf{e}}_\theta$ in terms of $\hat{\mathbf{i}}$ and $\hat{\mathbf{j}}$ at each position.

1.106 In the preceding problem, find the acceleration of P at the same four points. Again, do the problem first in $(\hat{\mathbf{e}}_r, \hat{\mathbf{e}}_\theta)$ components and then convert the results to $(\hat{\mathbf{i}}, \hat{\mathbf{j}})$ components.

1.107 A point P starts at the origin and moves along the parabola shown in Figure P1.107 with a constant x-component of velocity, $\dot{x} = 3$ ft/sec. Using the following approach, find the radial and transverse components of the velocity and acceleration of P at the point $(x, y) = (1, 1)$: Find $\mathbf{v}_P$ and $\mathbf{a}_P$ in rectangular components (see Problem 1.71); then resolve these vectors along $\hat{\mathbf{e}}_r$ and $\hat{\mathbf{e}}_\theta$ to obtain their radial and transverse components.

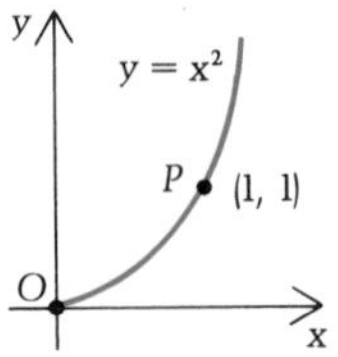

Figure P1.107

1.108 Solve the preceding problem by a different approach: Recall the polar coordinate relations $r = \sqrt{x^2 + y^2}$ and $\theta = \tan^{-1}(y/x)$, and differentiate to obtain $\dot{r}$, $\ddot{r}$, $\dot{\theta}$, and $\ddot{\theta}$ for entry into Equations (1.34) and (1.37).

1.109 The four-leaf rose in Figure P1.109 has the Equation $r = 3 \sin 2\theta$ ft. A particle P starts at the origin and travels on the indicated path with $\dot{\theta} = 1/6$ rad/sec $=$ constant. When P is at the highest point in the first quadrant, find:

a. the speed of P

b. the acceleration of P

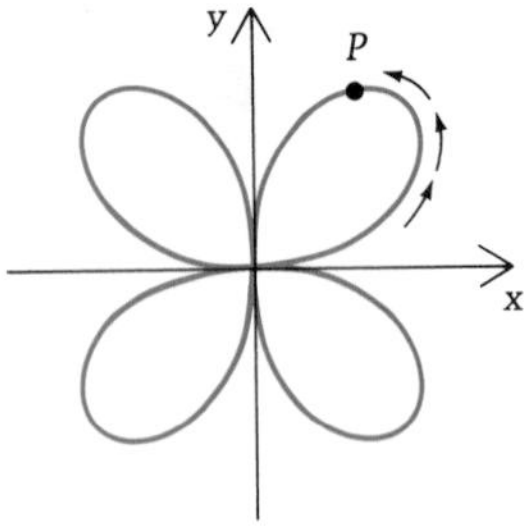

Figure P1.109

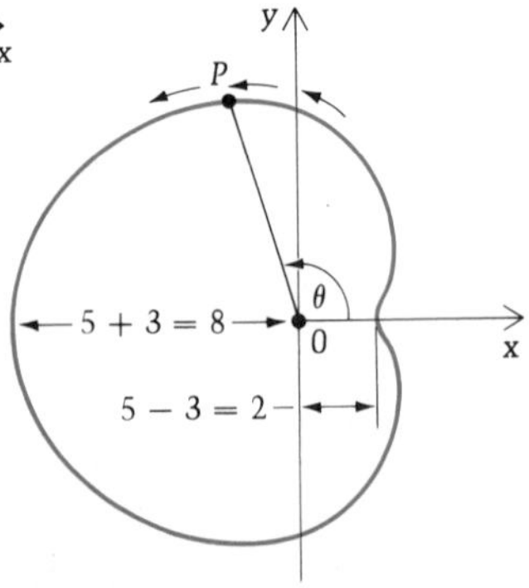

Figure P1.110

• **1.110** The point P in Figure P1.110 moves on the limacon defined in polar coordinates by

$$r = 5 - 3\cos\theta \text{ m}$$

If the polar angle is quadratic in time according to $\theta = 10t^2$ rad, find the velocity of P when it is at its highest point.

1.111 In the preceding problem, determine the acceleration of P at (a) the same highest point and (b) $\theta = \pi$ rad.

• **1.112** A point P moves on the figure eight in the indicated direction (Figure P1.112) at constant speed 2 m/s. Find the acceleration vector of P the next time its velocity is horizontal.

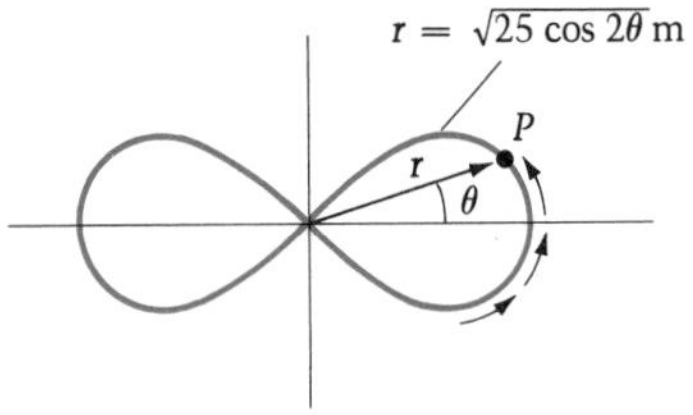

Figure P1.112

• **1.113** An insect is asleep on a $33\frac{1}{3}$ rpm record, 6 in. from the spindle. When the record is turned on, the insect wakes up and dizzily heads toward the center, in a straight line relative to the disk, at 1 in./sec (Figure P1.113). If the bug can withstand a maximum acceleration magnitude of 100 in./sec², does it make it to the spindle (a) if it starts after the record is up to speed? (b) if it starts as soon as the record is turned on? Assume that the turntable accelerates linearly (with time) up to speed in one revolution, and that $\ddot{r} = -\frac{1}{4}$ in./sec² until $\dot{r} = -1$ in./sec.

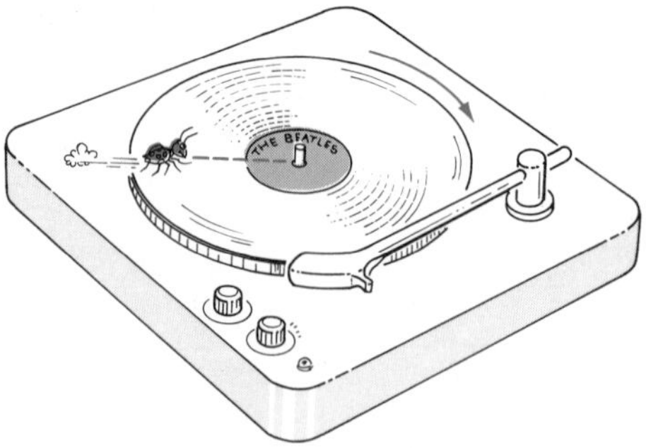

Figure P1.113

1.114 David throws a rock at Goliath with a sling. He whirls it around one revolution plus 135° more and releases it there, as shown in Figure P1.114, at 50 ft/sec.

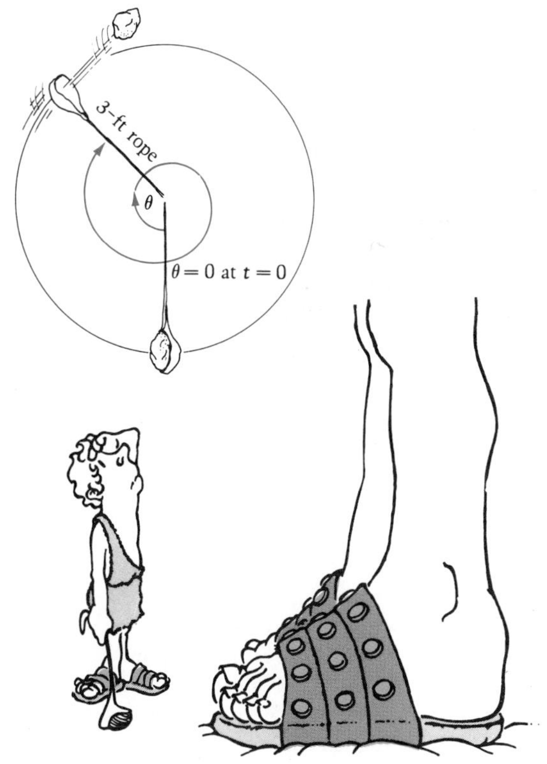

Figure P1.114

As he whirls the sling, the speed of the rock increases linearly with the time t; that is, $\dot{\theta} = kt$, where k is a constant. Find the acceleration of the rock just prior to release.

1.115 In Problem 1.60 show that the velocity of the shingles may also be obtained by simply taking the component of the velocity of the bumper attachment point A *along the rope.* Using the cylindrical coordinate expression for velocity, explain why this procedure works.

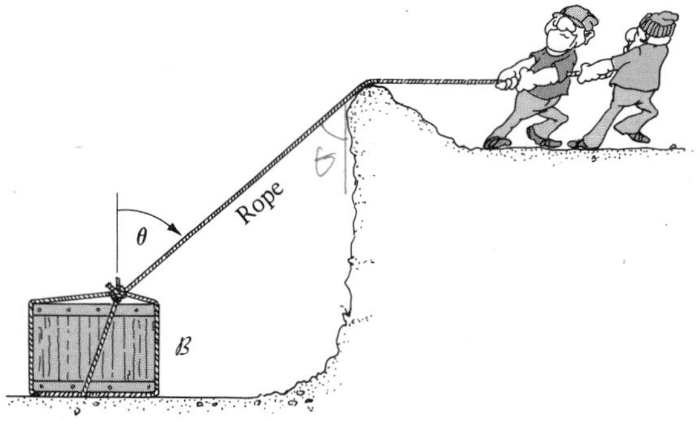

Figure P1.116

1.116 Two people moving at 2 ft/sec to the right are using a rope to drag the box $\mathcal{B}$ along the ground at the lower level (Figure P1.116). Determine the speed of $\mathcal{B}$ as a function of the angle θ between the rope and the vertical.

***1.117** The rigid rod $\mathcal{R}$ in Figure P1.117 moves so that its ends, A and B, remain in contact with the surfaces. If, at the instant shown, the velocity of A is 0.5 ft/sec to the right, find the velocity of B.

1.118 In Problem 1.117, find the acceleration of B at the instant in question if the acceleration of A is 2 ft/sec² to the left at that time.

1.119 An ant travels up the banister of a spiral staircase (Figure P1.119) according to

$$\mathbf{r}_{OA} = 2\cos\frac{t}{50}\,\hat{\mathbf{i}} + 2\sin\frac{t}{50}\,\hat{\mathbf{j}} + \frac{t}{50}\,\hat{\mathbf{k}}\text{ m}$$

Find the position and velocity of the ant when $t = 30$ s.

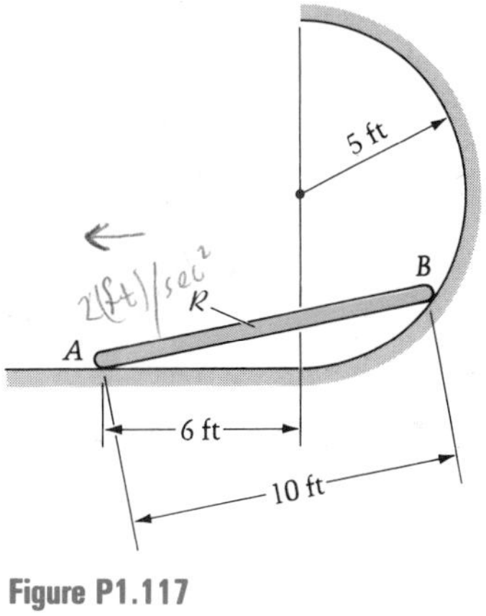

Figure P1.117

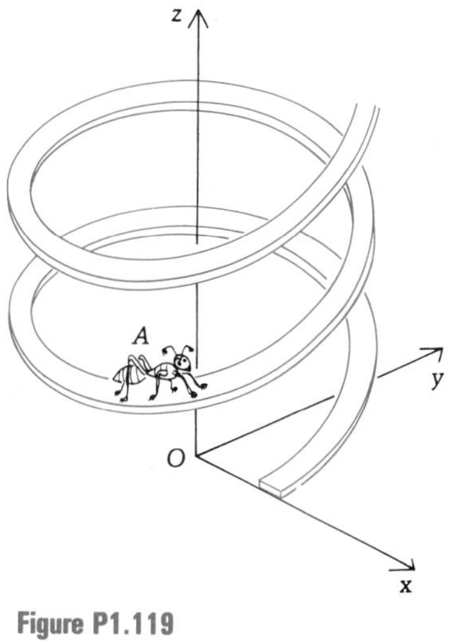

Figure P1.119

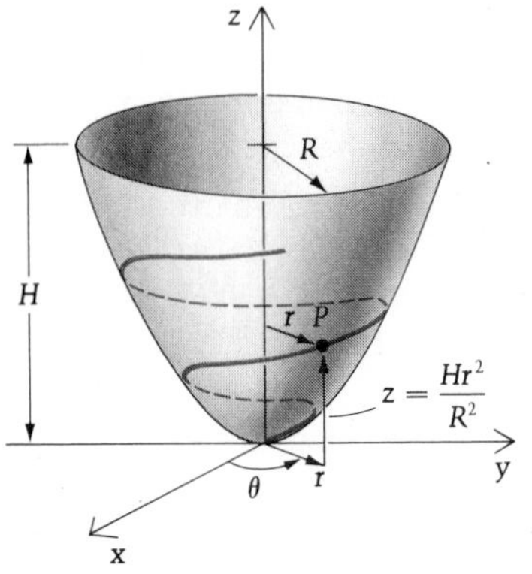

Figure P1.121

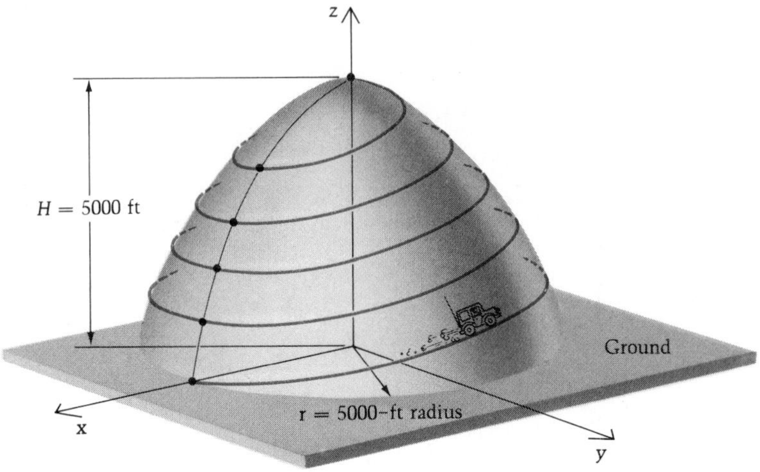

Figure P1.123

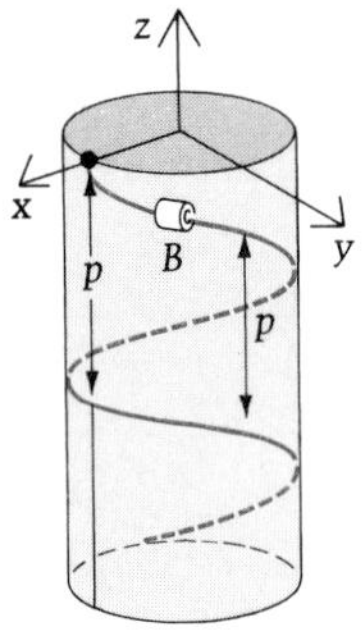

Figure P1.122

1.120 Find the acceleration of the ant (again at $t = 30$ s) in the preceding problem.

1.121 A point P starts at $t = 0$ at the origin and proceeds along a path on the paraboloid of revolution shown in Figure P1.121. The path is described (with time as a parameter) by

$$r = k_1 t$$

$$\theta = k_2 t^2$$

Find the position, velocity, and acceleration vectors of the point when it reaches the top edge of the paraboloid. (H, R, k_1, and k_2 are constants.)

1.122 A bead B slides down and around a cylindrical surface on a helical wire (Figure P1.122). The vertical drop of the bead as θ changes by 2π is called the pitch p of the helix; R is the radius of the helix.

- **a.** Noting that θ (and therefore also z) is a function of time, write the equations for $\mathbf{r}_{OB}$, $\mathbf{v}_B$, and $\mathbf{a}_B$ in cylindrical coordinates.
- **b.** For the values $R = 0.3$ m, $p = 0.2$ m, and $\theta = 0.6t$ rad/s, find and sketch the velocity and acceleration vectors of B when $t = 10$ s.

* **1.123** The mountain shown in Figure P1.123 is in the shape of the paraboloid of revolution $H - z = kr^2$, where H = height = 5000 ft, r is the radius at z, and k is a constant. The base radius is also 5000 ft. A car travels up the mountain on a spiraling path. Each time around, the car's altitude is 1000 ft higher. The car travels at the *constant speed* of 50 mph. Find the largest and smallest absolute values of the radial component of velocity on the journey, and tell where the car is at these two times.

* **1.124** In the preceding problem, find the locations of the car (r, θ, z) for which the following velocity components are equal.

- **a.** Radial ($\dot{r}$) and transverse ($r\dot{\theta}$)
- **b.** Radial and vertical ($\dot{z}$)
- **c.** Transverse and vertical

1.125 Show that the velocity of a point P in spherical coordinates (r, θ, ϕ) is given by

$$\mathbf{v}_P = \dot{r}\hat{\mathbf{e}}_r + r\dot{\theta}\hat{\mathbf{e}}_\theta + r\dot{\phi}\sin\theta\hat{\mathbf{e}}_\phi$$

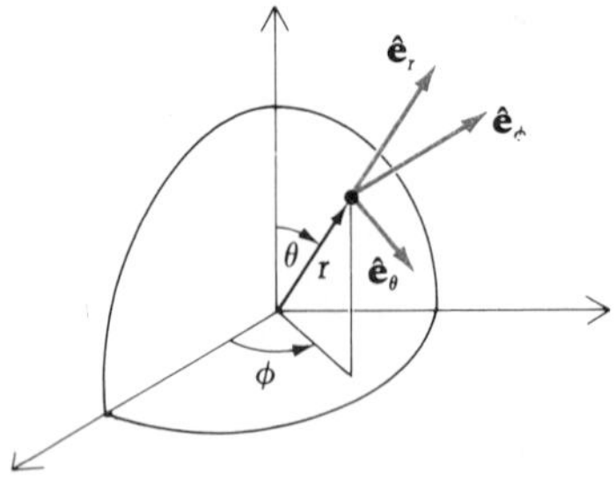

Figure P1.125

See Figure P1.125. *Hint*: As intermediate steps, obtain the results

$$\dot{\hat{\mathbf{e}}}_r = \dot{\theta}\hat{\mathbf{e}}_\theta + \dot{\phi}\sin\theta\hat{\mathbf{e}}_\phi$$
$$\dot{\hat{\mathbf{e}}}_\theta = -\dot{\theta}\hat{\mathbf{e}}_r + \dot{\phi}\cos\theta\hat{\mathbf{e}}_\phi$$
$$\dot{\hat{\mathbf{e}}}_\phi = -\dot{\phi}(\sin\theta\hat{\mathbf{e}}_r + \cos\theta\hat{\mathbf{e}}_\theta)$$

Then differentiate the simple position vector $\mathbf{r}_{OP} = r\hat{\mathbf{e}}_r$.

1.126 Show by differentiating $\mathbf{v}_P$ in the preceding problem that the corresponding expression for the acceleration in spherical coordinates is

$$\begin{aligned}\mathbf{a}_P = {} & (\ddot{r} - r\dot{\theta}^2 - r\dot{\phi}^2\sin^2\theta)\hat{\mathbf{e}}_r \\ & +(2\dot{r}\dot{\theta} + r\ddot{\theta} - r\dot{\phi}^2\sin\theta\cos\theta)\hat{\mathbf{e}}_\theta \\ & +(2\dot{r}\dot{\phi}\sin\theta + 2r\dot{\theta}\dot{\phi}\cos\theta + r\ddot{\phi}\sin\theta)\hat{\mathbf{e}}_\phi\end{aligned}$$

* **1.127** The velocity of a point P moving in a plane is the resultant of one part, $v\hat{\mathbf{e}}_r$, along the radius from a fixed point O to the point P, and another part, $u\hat{\mathbf{i}}$, which is always parallel to a fixed line. (See Figure P1.127.) Prove that the acceleration of P may be written as $a_v\hat{\mathbf{e}}_r + a_u\hat{\mathbf{i}}$, where

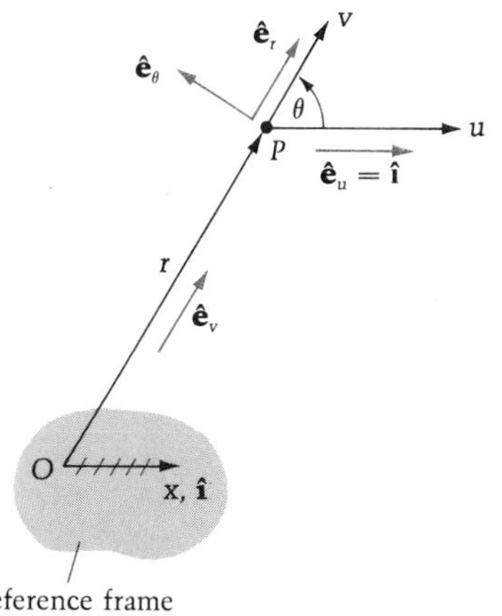

Figure P1.127

$$a_v = \frac{dv}{dt} - \frac{uv}{r}\cos\theta \quad \text{and} \quad a_u = \frac{du}{dt} + \frac{uv}{r}$$

where r is the length of the radius vector from O to P and θ is the angle it makes with the fixed direction.

1.7 Tangential and Normal Components

In this section we examine yet another means of expressing the velocity and acceleration of a point P. Instead of focusing on a specific coordinate system, this time we shall study the way in which the motion of P is related to its path. Consequently, the components of velocity and acceleration that result are sometimes called *intrinsic* or *natural.*

The path of point P, as mentioned in Section 1.3, is the locus of points of the reference frame $\mathcal{J}$ successively occupied by P as it moves. We begin, then, by defining some reference point on the path. From this arclength origin we then measure the arclength s along the path to the point P. Clearly, the arclength coordinate depends on the time; that is, $s = s(t)$.

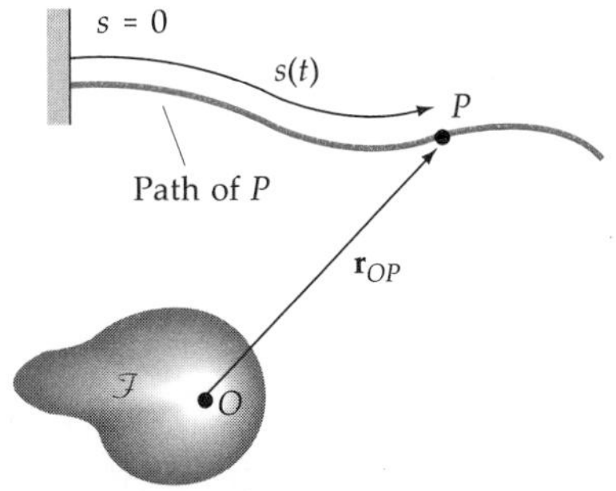

Figure 1.8 Arclength measurement of point *P* on its path.

In Figure 1.8 we see a position vector, $\mathbf{r}_{OP}$, for point P. This vector was seen in preceding sections to define the location of P, and thus it may be considered a function of the arclength s:

$$\mathbf{r}_{OP} = \mathbf{r}_{OP}(s) = \mathbf{r}_{OP}[s(t)] \tag{1.39}$$

Forming the velocity of P by differentiation (the definition is the same, regardless of how we choose to represent the vectors), we get

$$\mathbf{v}_P = \dot{\mathbf{r}}_{OP} = \frac{d\mathbf{r}_{OP}}{dt}$$

and, by the chain rule,

$$\begin{aligned}\mathbf{v}_P &= \frac{d\mathbf{r}_{OP}}{ds}\frac{ds}{dt} \\ &= \dot{s}\lim_{\Delta s \to 0}\left[\frac{\mathbf{r}_{OP}(s+\Delta s) - \mathbf{r}_{OP}(s)}{\Delta s}\right] \\ &= \dot{s}\lim_{\Delta s \to 0}\left(\frac{\Delta \mathbf{r}_{OP}}{\Delta s}\right)\end{aligned} \tag{1.40}$$

Figure 1.9 shows the quantities $\Delta\mathbf{r}_{OP}$ and Δs.

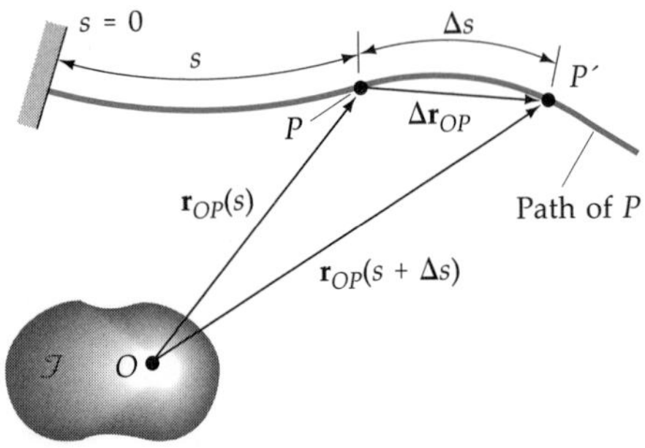

Figure 1.9 Changes in $\mathbf{r}_{OP}$ as s changes.

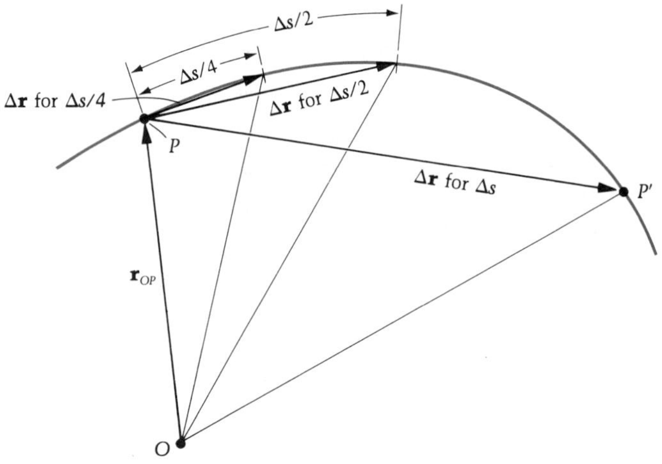

Figure 1.10 Shrinking Δs toward zero.

We suggest that the reader sketch an arc on a large sheet of paper and then use a straightedge to draw the triangle OPP' (Figure 1.10). Then the limit in Equation (1.40) can be taken by, let us say, dividing Δs in half each time. After just a few more divisions of Δs on the large sheet, it will become clear that as Δs approaches zero — that is, as P' backs up toward P — two interesting things happen:

1. $\Delta\mathbf{r}_{OP}$ becomes tangent to the path of P at arclength s.
2. The magnitude of $\Delta\mathbf{r}_{OP}/\Delta s$ approaches $\Delta s/\Delta s = 1$.

These two results, taken together, prove that $d\mathbf{r}_{OP}/ds$ is always a unit vector that is tangent to the path and pointing in the direction of increasing s. It is for these reasons that this vector is called $\hat{\mathbf{e}}_t$, the **unit tangent.** Equation (1.40) may then be rewritten as

$$\mathbf{v}_P = \dot{s}\hat{\mathbf{e}}_t \tag{1.41}$$

From Equation (1.41) we see that the velocity vector of point P is *always tangent to its path.* The absolute value $|\dot{s}|$ of the scalar part — which is the same as the magnitude $|\mathbf{v}_P|$ of the velocity vector $\mathbf{v}_P$ — is called the *speed* of P in $\mathcal{I}$, as we mentioned in Section 1.3.

Next we shall differentiate again in order to obtain the acceleration of P. Using Equation (1.41), we get

$$\mathbf{a}_P = \dot{\mathbf{v}}_P = \ddot{s}\hat{\mathbf{e}}_t + \dot{s}\frac{d\hat{\mathbf{e}}_t}{dt}$$

$$= \ddot{s}\hat{\mathbf{e}}_t + \dot{s}^2\frac{d\hat{\mathbf{e}}_t}{ds} \tag{1.42}$$

Since $\hat{\mathbf{e}}_t$ is a unit vector, $d\hat{\mathbf{e}}_t/ds$ is perpendicular to $\hat{\mathbf{e}}_t$ and hence perpendicular, or normal, to the path. Equation (1.42) shows an important separation of the acceleration into two parts, one tangent and the other normal to the path of P. The component tangent to the path, $\ddot{s}$, is (for $\dot{s} > 0$) the rate of change of the velocity *magnitude,* or speed, of P. The component normal to the path reflects the rate of change of the *direction* of the velocity vector.

Further examination of $d\hat{\mathbf{e}}_t/ds$ in Equation (1.42) is facilitated if we first restrict our attention to the case of a two-dimensional (plane) curve. To that end, let θ be the inclination of a tangent to the plane curve as shown in Figure 1.11. We can visualize that as s increases, $\hat{\mathbf{e}}_t$ turns in such a way that $d\hat{\mathbf{e}}_t/ds$ points toward the inside of the curve — that is, in the direction of $\hat{\mathbf{e}}_n$ shown in the figure. We can obtain this result analytically if we write

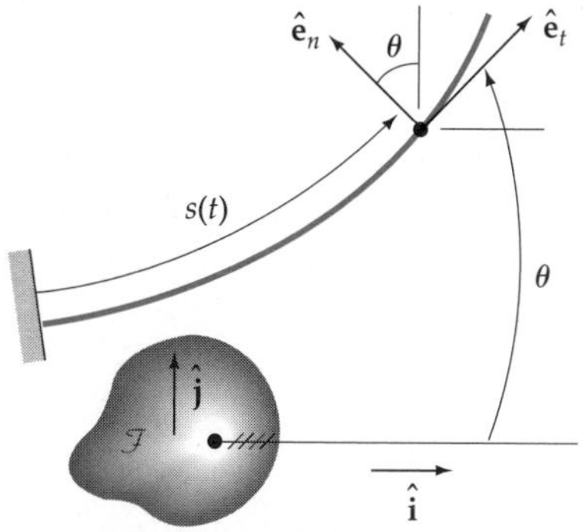

Figure 1.11 Tangent and normal to a plane curve.

$$\frac{d\hat{\mathbf{e}}_t}{ds} = \frac{d\theta}{ds}\frac{d\hat{\mathbf{e}}_t}{d\theta} \tag{1.43}$$

Noting from Figure 1.11 that

$$\hat{\mathbf{e}}_t = \cos\theta\hat{\mathbf{i}} + \sin\theta\hat{\mathbf{j}}$$

we may differentiate to obtain

$$\frac{d\hat{\mathbf{e}}_t}{d\theta} = -\sin\theta\hat{\mathbf{i}} + \cos\theta\hat{\mathbf{j}}$$

which is a unit vector normal to the curve. If $d\theta/ds$ is positive, as is illustrated in Figure 1.11, then $d\hat{\mathbf{e}}_t/d\theta = \hat{\mathbf{e}}_n$. If $d\theta/ds$ is negative (curve concave downward), then $d\hat{\mathbf{e}}_t/d\theta$ points toward the outside of the curve, as the reader may wish to confirm with a sketch, and $d\hat{\mathbf{e}}_t/ds$ again points toward the inside of the curve. Thus, in either case

$$\frac{d\hat{\mathbf{e}}_t}{ds} = \left|\frac{d\theta}{ds}\right|\hat{\mathbf{e}}_n \tag{1.44}$$

where $\hat{\mathbf{e}}_n$ is understood to point toward the inside of the curve.

From studies in calculus the reader probably recognizes $|d\theta/ds|$ as the **curvature** of a plane curve. The reciprocal of the curvature is the **radius of curvature** ρ. The radius of curvature is the radius of the circle that provides the best local approximation to an infinitesimal segment of the curve. Equation (1.43) may thus be written

$$\frac{d\hat{\mathbf{e}}_t}{ds} = \frac{1}{\rho}\hat{\mathbf{e}}_n \tag{1.45}$$

In three dimensions the situation is more difficult to visualize. We cannot use the preceding development because $\hat{\mathbf{e}}_t$ cannot be expressed as a function of a single angle such as θ. Consequently, in the general case we adopt a definition of curvature that, in two dimensions, reduces to what we have just established. That is, we simply define the curvature $1/\rho$ to be the magnitude of the vector $d\hat{\mathbf{e}}_t/ds$. Then the unit vector $\hat{\mathbf{e}}_n$ as defined by

$$\hat{\mathbf{e}}_n = \frac{1}{|d\hat{\mathbf{e}}_t/ds|}\frac{d\hat{\mathbf{e}}_t}{ds} = \rho\frac{d\hat{\mathbf{e}}_t}{ds} \tag{1.46}$$

is called the **principal unit normal** to the curve. Upon substituting into Equation (1.42), we then obtain

$$\mathbf{a}_P = \ddot{s}\hat{\mathbf{e}}_t + \frac{(\dot{s})^2}{\rho}\hat{\mathbf{e}}_n \tag{1.47}$$

An alternative form in which the arclength parameter s is not explicitly involved follows if we choose the measurement of s so that at the instant of interest $\dot{s} > 0$. Hence $\dot{s} = |\mathbf{v}_P|$ and

$$\mathbf{a}_P = \left(\frac{d}{dt}|\mathbf{v}_P|\right)\hat{\mathbf{e}}_t + \frac{|\mathbf{v}_P|^2}{\rho}\hat{\mathbf{e}}_n \tag{1.48}$$

This expression more vividly depicts the natural decomposition of acceleration into parts related to rate of change of magnitude of velocity and rate of change of direction of velocity. We now consider some examples of the use of tangential and normal components.

EXAMPLE 1.20

A car starts at rest at A and increases its speed around the track at 6 ft/sec², traveling counterclockwise (see Figure E1.20). Determine the position and the time at which the car's acceleration magnitude reaches 20 ft/sec².

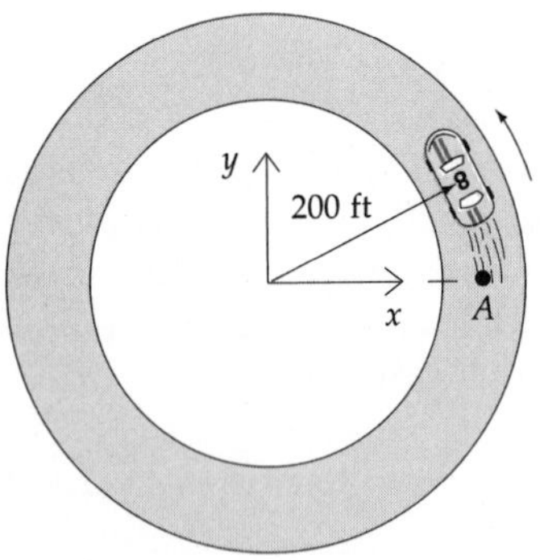

Figure E1.20

Solution

$$\ddot{s} = 6 \text{ ft/sec}^2$$

$$\dot{s} = 6t + \overset{0 \text{ ft/sec}}{C_1} \quad \text{(The constant is zero since } \dot{s} = 0 \text{ at } t = 0.)$$

$$s = 3t^2 + \overset{0 \text{ ft}}{C_2} \quad \text{(The constant is zero since } s = 0 \text{ at } t = 0.)$$

The acceleration magnitude of Q is

$$|\mathbf{a}_Q| = a_Q = \sqrt{\ddot{s}^2 + (\dot{s}^2/\rho)^2}$$
$$= \sqrt{36 + (36t^2/200)^2} \text{ ft/sec}^2$$

When $a_Q = 20$ ft/sec², we obtain the equation

$$20^2 = 36 + (0.18t^2)^2 = 36 + 0.0324t^4$$

$$t = \sqrt[4]{\frac{364}{0.0324}} = 10.3 \text{ sec}$$

At $t = 10.3$ sec, $s = 318$ ft, which represents $318/(2\pi r) = 0.253$ of a revolution, or $91.1°$ counterclockwise from the x axis.

Example 1.21

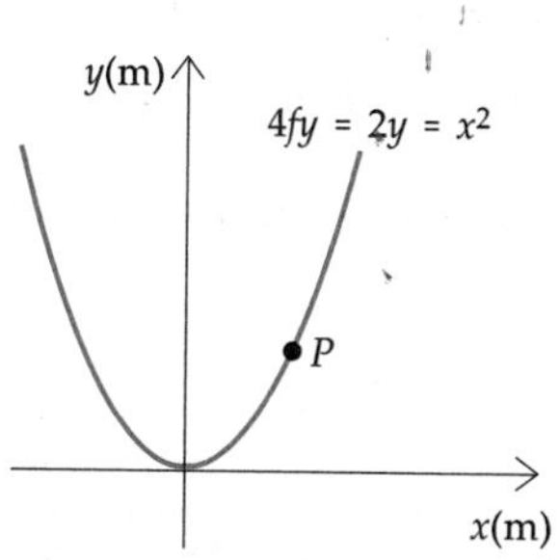

Figure E1.21a

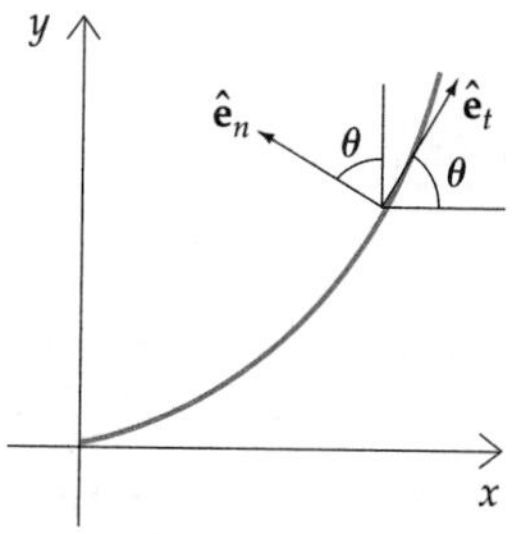

Figure E1.21b

Verify the results of Example 1.13, at $x = 2$ m, by using $\hat{\mathbf{e}}_t$ and $\hat{\mathbf{e}}_n$ components.

Solution

We are given that $\dot{s} = |\mathbf{v}_P| = 0.2$ m/s. Since $\mathbf{v}_P$ is tangent to the path of P, we can calculate $\hat{\mathbf{e}}_t$:

$$y = \frac{x^2}{4f} = \frac{x^2}{2} \text{ for } f = \tfrac{1}{2} \qquad \text{(see Figure E1.21a)}$$

$$\tan\theta = \frac{dy}{dx} = x = 2 \qquad \text{(at the given point)}$$

$$\theta = \tan^{-1}(2) = 63.4°$$

$$\hat{\mathbf{e}}_t = \cos\theta\hat{\mathbf{i}} + \sin\theta\hat{\mathbf{j}} \qquad \text{(see Figure E1.21b)}$$

$$= 0.448\hat{\mathbf{i}} + 0.894\hat{\mathbf{j}}$$

The radius of curvature comes from calculus:

$$\frac{1}{\rho} = \left|\frac{y''}{(1+y'^2)^{3/2}}\right| = \left|\frac{1}{(1+x^2)^{3/2}}\right| = \frac{1}{5^{3/2}}$$

or

$$\rho = 11.2 \text{ m}$$

Thus with $\hat{\mathbf{e}}_n$ seen to be $-\sin\theta\hat{\mathbf{i}} + \cos\theta\hat{\mathbf{j}}$, we have

$$\mathbf{a}_P = \ddot{s}\hat{\mathbf{e}}_t + \frac{\dot{s}^2}{\rho}\hat{\mathbf{e}}_n$$

$$= 0 + \frac{\dot{s}^2}{\rho}\hat{\mathbf{e}}_n \qquad \text{(since } \dot{s} = \text{constant)}$$

$$= \frac{0.04}{11.2}(-0.894\hat{\mathbf{i}} + 0.448\hat{\mathbf{j}})$$

$$= -0.0032\hat{\mathbf{i}} + 0.0016\hat{\mathbf{j}} \text{ m/s}^2 \qquad \text{(as before)}$$

Example 1.22

In Example 1.11 find the following for point P at $t = 1$ sec: tangential and normal components of acceleration, radius of curvature, and the principal unit normal.

Solution

We obtained

$$\mathbf{r}_{OP} = 2t\hat{\mathbf{i}} + t^3\hat{\mathbf{j}} + 3t^2\hat{\mathbf{k}} \text{ ft}$$

$$\mathbf{v}_P = 2\hat{\mathbf{i}} + 3t^2\hat{\mathbf{j}} + 6t\hat{\mathbf{k}} \text{ ft/sec}$$

$$\mathbf{a}_P = 6t\hat{\mathbf{j}} + 6\hat{\mathbf{k}} \text{ ft/sec}^2$$

If we write the velocity $\mathbf{v}_P$ as a magnitude times a unit vector, we can determine $\dot{s}$ and $\hat{\mathbf{e}}_t$ for P:

$$\mathbf{v}_P = \sqrt{4 + 9t^4 + 36t^2}\left(\frac{2\hat{\mathbf{i}} + 3t^2\hat{\mathbf{j}} + 6t\hat{\mathbf{k}}}{\sqrt{4 + 9t^4 + 36t^2}}\right) \text{ ft/sec}$$

$$= \dot{s}\hat{\mathbf{e}}_t$$

where we note that $\dot{s} > 0$ since we are choosing the direction of increasing s to be that of the velocity.

Let us find the tangential and normal components of the acceleration of P at $t = 1$ sec:

$$\mathbf{v}_P|_{t=1} = 7\left(\frac{2\hat{\mathbf{i}} + 3\hat{\mathbf{j}} + 6\hat{\mathbf{k}}}{7}\right) = |\mathbf{v}_P|\hat{\mathbf{e}}_t \text{ ft/sec}$$

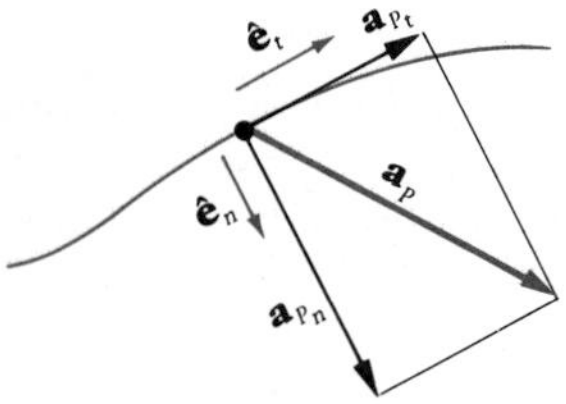

Figure E1.22

Now that we have $\hat{\mathbf{e}}_t$, we can use it to split the acceleration $\mathbf{a}_P = 6\hat{\mathbf{j}} + 6\hat{\mathbf{k}}$ ft/sec² (at $t = 1$) into its tangential and normal components (see Figure E1.22). The tangential component of $\mathbf{a}_P$ (that is, the component parallel to $\hat{\mathbf{e}}_t$) is seen from the figure to be the dot product of $\mathbf{a}_P$ with $\hat{\mathbf{e}}_t$:

$$a_{P_t}|_{t=1} = \mathbf{a}_P|_{t=1} \cdot \hat{\mathbf{e}}_t|_{t=1}$$

$$= 6\sqrt{2}\left(\frac{\hat{\mathbf{j}} + \hat{\mathbf{k}}}{\sqrt{2}}\right) \cdot \left(\frac{2\hat{\mathbf{i}} + 3\hat{\mathbf{j}} + 6\hat{\mathbf{k}}}{7}\right)$$

$$= \frac{6}{7}(3 + 6) = \frac{54}{7} \text{ ft/sec}^2$$

Next we obtain the normal acceleration component by vectorially subtracting the component $\mathbf{a}_{P_t}(=(54/7)\hat{\mathbf{e}}_t)$ from the total acceleration $\mathbf{a}_P$. That is, since

$$\mathbf{a}_P = \mathbf{a}_{P_t} + \mathbf{a}_{P_n}$$

we obtain

$$a_{P_n}|_{t=1} = |\mathbf{a}_{P_n}|\Big|_{t=1} = \left|\mathbf{a}_P|_{t=1} - \mathbf{a}_{P_t}|_{t=1}\right|$$

$$= \left|6\hat{\mathbf{j}} + 6\hat{\mathbf{k}} - \frac{54}{7}\left(\frac{2\hat{\mathbf{i}} + 3\hat{\mathbf{j}} + 6\hat{\mathbf{k}}}{7}\right)\right| \text{ ft/sec}^2$$

$$= \left|\frac{-108\hat{\mathbf{i}} + 132\hat{\mathbf{j}} - 30\hat{\mathbf{k}}}{49}\right| = 3.53 \text{ ft/sec}^2$$

And since $a_{P_n} = \dot{s}^2/\rho = |\mathbf{v}_P|^2/\rho$, we obtain the radius of curvature:

$$\rho = \frac{7^2}{3.53} = 13.9 \text{ ft}$$

The unit vector $\hat{\mathbf{e}}_n$ follows from

$$\hat{\mathbf{e}}_n = \frac{\mathbf{a}_{P_n}}{a_{P_n}} = \frac{\mathbf{a}_P - \mathbf{a}_t}{a_{P_n}} = \frac{-108\hat{\mathbf{i}} + 132\hat{\mathbf{j}} - 30\hat{\mathbf{k}}}{49(3.53)}$$

$$= -0.624\hat{\mathbf{i}} + 0.763\hat{\mathbf{j}} - 0.173\hat{\mathbf{k}}$$

It is instructive to make a direct calculation of $d|\mathbf{v}_P|/dt$ since we here know $|\mathbf{v}_P|$ as a function of time:

$$|\mathbf{v}_P| = \sqrt{4 + 9t^4 + 36t^2} \text{ ft/sec}$$

$$\frac{d|\mathbf{v}_P|}{dt} = \frac{\frac{1}{2}(36t^3 + 72t)}{\sqrt{4 + 9t^4 + 36t^2}} \text{ ft/sec}^2$$

Thus

$$\left.\frac{d|\mathbf{v}_P|}{dt}\right|_{t=1} = \frac{\frac{1}{2}(108)}{\sqrt{49}} = \frac{54}{7} \text{ ft/sec}^2$$

which is, of course, the result we have already obtained by investigating the components of the acceleration vector.

Question 1.6 How would you find the position vector from the origin O to the center of curvature at $t = 1$ sec?

In closing this section, we remark that tangential and normal components of velocity and acceleration will be very useful to us later when we happen to know the path of a point (the center C of a wheel rolling on a curved track, for instance). We can then use Equations (1.41) and (1.47) to express $\mathbf{v}_P$ and $\mathbf{a}_P$.

Answer 1.6 If we call the center of curvature C, then $\mathbf{r}_{OC} = \mathbf{r}_{OP} + \mathbf{r}_{PC} = \mathbf{r}_{OP} + \rho\hat{\mathbf{e}}_n$, with everything evaluated at the time of interest (in this case $t = 1$ sec).

PROBLEMS ▶ Section 1.7

1.128 Particle P moves on a circle (Figure P1.128) with an arclength given as a function of time as shown. Find the time(s) and the angle(s) θ when the tangential and normal acceleration components are equal.

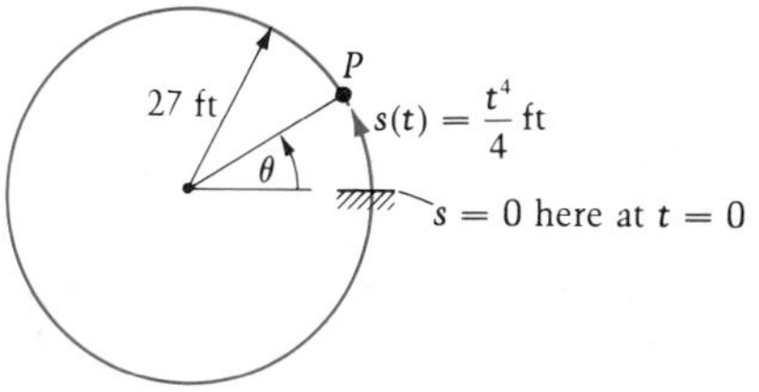

Figure P1.128

1.129 In Problem 1.78 find the arclength s as a function of time.

1.130 In Problem 1.67 determine the expression for $\dot{s}(t)$. Integrate, for a motion beginning at $t = 0$ at $(x, y) = (20, 0)$ m, and obtain $s(t)$. Evaluate the arclength at $t = 2$ s and show that the result, as it should be, is the circumference of the circle on which P travels.

1.131 A point P moves on a path with $s = ct^3$ where c = constant = 1 ft/sec³. At $t = 2$ sec, the magnitude of the acceleration is 15 ft/sec². At that time, find the radius of curvature of the path of P.

1.132 A point D moves along a curve in space with a speed given by $\dot{s} = 6t$ m/s, where t is measured from zero when D is at the arclength origin $s = 0$. If at a certain time t' the acceleration magnitude of D is 12 m/s² and the radius of curvature is 3 m, determine t'.

1.133 At a certain instant the velocity and acceleration of a point are as shown in Figure P1.133. At this instant find

a. $\dfrac{d|\mathbf{v}|}{dt}$

b. the radius of curvature of the path

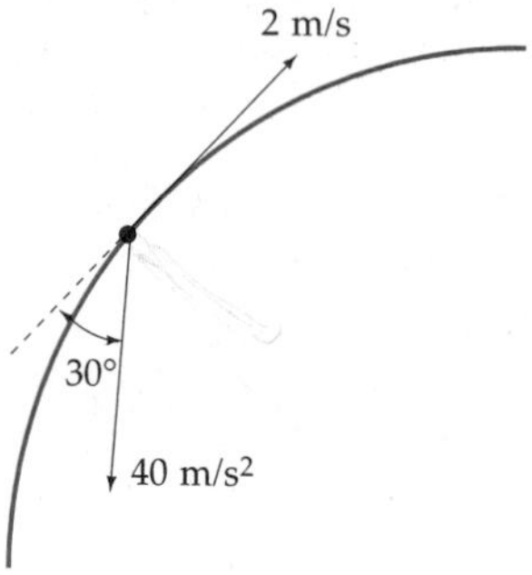

Figure P1.133

1.134 At a certain instant, the velocity and acceleration of a point are

$$\mathbf{v} = 3\hat{\mathbf{i}} + 4\hat{\mathbf{j}}\ \text{m/s}$$

$$\mathbf{a} = -10\hat{\mathbf{k}}\ \text{m/s}^2$$

At this instant find (a) $\dfrac{d|\mathbf{v}|}{dt}$, (b) the radius of curvature of the path, (c) the principal unit normal.

1.135 At an instant the velocity and acceleration of a point are

$$\mathbf{v} = -2\hat{\mathbf{i}}\ \text{m/s}$$

$$\mathbf{a} = -4\hat{\mathbf{i}} + 3\hat{\mathbf{j}} - 2\hat{\mathbf{k}}\ \text{m/s}^2$$

At this instant find:

a. $\dfrac{d}{dt}|\mathbf{v}|$

b. the radius of curvature of the path.

1.136 At a certain instant, the velocity and acceleration of a point are

$$\mathbf{v} = 4\hat{\mathbf{i}} - 3\hat{\mathbf{j}}\ \text{m/s}$$

$$\mathbf{a} = -10\hat{\mathbf{i}} + 20\hat{\mathbf{j}} + 12\hat{\mathbf{k}}\ \text{m/s}^2$$

Find:

a. $\dfrac{d}{dt}|\mathbf{v}|$

b. $\hat{\mathbf{e}}_n$

1.137 In Problem 1.103, find the radius of curvature of the path of P at the instant given. Note that $\hat{\mathbf{e}}_t = \hat{\mathbf{j}} = \hat{\mathbf{e}}_\theta$ and $\hat{\mathbf{e}}_n = -\hat{\mathbf{i}} = -\hat{\mathbf{e}}_r$.

1.138 Find the radius of curvature of the "Witch of Agnesi" curve at $x = 0$. (See Figure P1.138.)

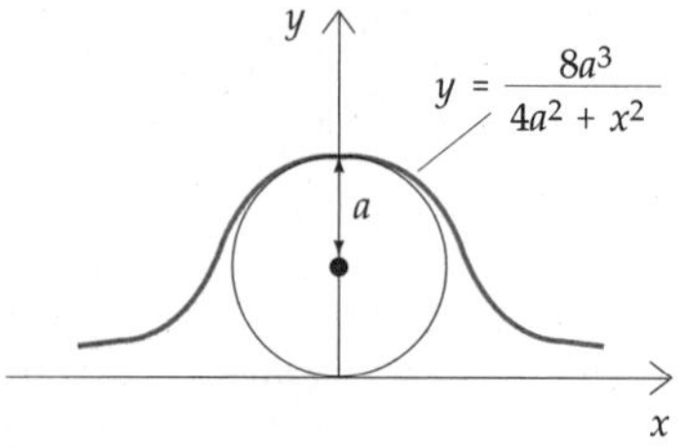

Figure P1.138

1.139 A point P moves from left to right along the curve defined in the preceding problem with a constant x component ($\dot{x}_0$) of velocity. Find the acceleration of P when it reaches the point $(x, y) = (0, 2a)$.

1.140 In Problem 1.105, at the same four points express $\mathbf{v}_P$ in terms of tangential and normal components.

1.141 In Problem 1.105, for the position $\theta = \pi/2$, express $\mathbf{a}_P$ in terms of tangential and normal components, and find the radius of curvature of the path of P at that point.

1.142 A point P starts at the origin and moves along the parabola shown in Figure P1.142 with a constant x-component of velocity, $\dot{x} = 3$ ft/sec. Find the tangential and normal components of the velocity and acceleration of P at the point $(x, y) = (1, 1)$.

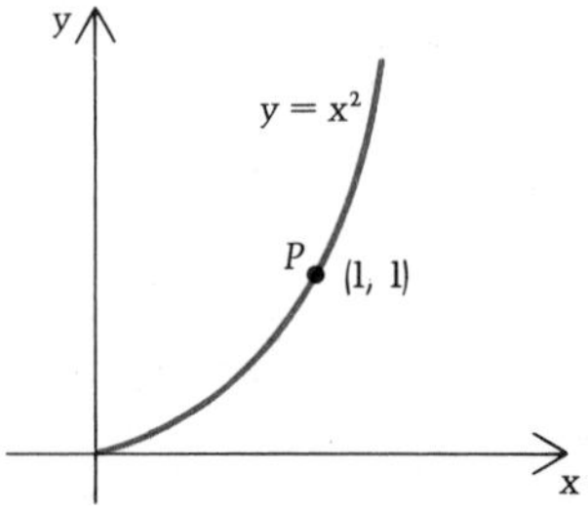

Figure P1.142

* **1.143** In Problem 1.104, find the center of curvature of the path of P when $\theta = 270°$.

1.144 At a particular instant a point has a velocity $3\hat{\mathbf{i}} + 4\hat{\mathbf{j}}$ in./sec and an acceleration $\hat{\mathbf{i}} + \hat{\mathbf{j}} - \hat{\mathbf{k}}$ in./sec². At this instant find: (a) the principal unit normal, (b) the curvature of the path, and (c) the time rate of change of the point's speed.

1.145 A point P has position vector $\mathbf{r}_{OP} = t^2\hat{\mathbf{i}} + t^3\hat{\mathbf{j}} - t^6\hat{\mathbf{k}}$ meters. Find the vector from the origin to the center of curvature of the path of P at $t = 1$ s. Find $d|\mathbf{v}_P|/dt$ at the same instant.

1.146 The position vector of a point is given as a function of time by $\mathbf{r}(t) = t^2\hat{\mathbf{i}} - t\hat{\mathbf{j}} + t^3\hat{\mathbf{k}}$ ft. Find the tangential and normal components of acceleration at $t = 1$ sec and determine the radius of curvature at that time.

1.147 A particle P has the x, y, and z coordinates $(3t, 0, 4 \ln t)$ meters as functions of time. What is the vector from the origin to the center of curvature of the path of P at $t = 1$ s?

1.148 Show by expressing the velocity and acceleration in tangential and normal components that

$$|\mathbf{v} \times \mathbf{a}| = va_n = \frac{|\dot{s}^3|}{\rho}$$

so that

$$\frac{1}{\rho} = \frac{|\mathbf{v} \times \mathbf{a}|}{|\mathbf{v}|^3} \quad \text{or} \quad \frac{|\mathbf{v} \times \mathbf{a}|}{v^3}$$

1.149 There is another formula for the radius of curvature ρ from the calculus; this one is in terms of a parameter such as time t, and for a plane curve:

$$\frac{1}{\rho} = \left|\frac{\dot{x}\ddot{y} - \dot{y}\ddot{x}}{(\dot{x}^2 + \dot{y}^2)^{3/2}}\right|$$

Derive this from the result of the preceding problem and use it to find the radius of curvature at 8 sec if

$$x = 6 \sinh 0.02t \text{ ft} \qquad y = t^3 - 10t^2 + 7.2 \text{ ft}$$

1.150 Find the difference between the velocities (and also the accelerations) of cars A and B in Figure P1.150 if, at the instant shown,

$$\dot{s}_A = 30 \text{ mph} \qquad \ddot{s}_A = 500 \text{ mph}^2$$

$$\dot{s}_B = 50 \text{ mph} \qquad \ddot{s}_B = -1000 \text{ mph}^2$$

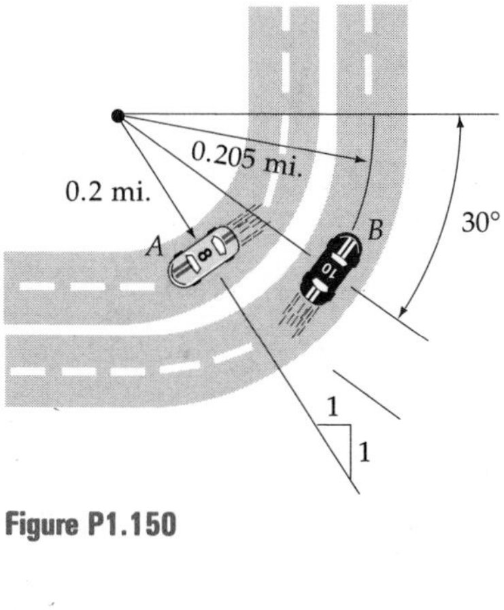

Figure P1.150

1.151 A particle moves on the curve $(x - a)^2 + y^2 = a^2$, where a is a constant distance in meters. The first and second time derivatives of the arclength s are related by

$$\dot{s} = Ka\ddot{s}$$

in which the constant K has the value 1 second/meter. The distance s is measured counterclockwise on the curve from the point $(2a, 0)$ meters. When $t = 0$, the speed of the particle is $\dot{s} = 1$ meter/second and $s = a$. Find the normal and tangential components of the acceleration at time $t = 0$. Show these components on a diagram.

1.152 Use Example 1.21 to show that the center of curvature does not have to be on the y axis for a curve symmetric about y. *Hint*: Use $f = 1$ and $x = 2$, find ρ, and compare with the distance from (2, 1) to the y axis along the normal to the tangent at this point.

* **1.153** In Problem 1.96 find the tangential and normal components of the car's acceleration when $x = 2500$ ft. Check your result by also computing $\ddot{x}$ and $\ddot{y}$ there and showing that $\sqrt{\ddot{x}^2 + \ddot{y}^2} = \sqrt{a_t^2 + a_n^2}$.

* **1.154** A particle P starts from rest at the origin and moves along the parabola shown in Figure P1.154. Its speed is given by $\dot{s} = 3s + 2$, where $\dot{s}$ is in meters per second when s is in meters. Determine the velocity of P when its x coordinate is 5 m. Also determine the elapsed time. *Hint*: $ds = \sqrt{dx^2 + dy^2} = \sqrt{1 + y'^2}\, dx$, so that $s = \int \sqrt{1 + y'^2}\, dx$. Substitute y' and use a table of integrals to get $s(x)$.

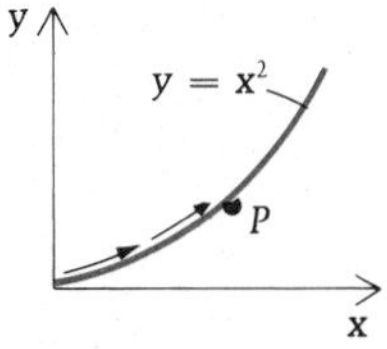

Figure P1.154

1.155 There is a third unit vector associated with the motion of a point on its path. It is called the binormal $\hat{\mathbf{e}}_b$ and forms an orthogonal moving trihedral with $\hat{\mathbf{e}}_t$ and $\hat{\mathbf{e}}_n$, defined by $\hat{\mathbf{e}}_b = \hat{\mathbf{e}}_t \times \hat{\mathbf{e}}_n$.

a. Differentiate $\hat{\mathbf{e}}_b \cdot \hat{\mathbf{e}}_t = 0$ with respect to s. Then, using $d\hat{\mathbf{e}}_t/ds = \hat{\mathbf{e}}_n/\rho$,† prove that $d\hat{\mathbf{e}}_b/ds \cdot \hat{\mathbf{e}}_t = 0$ and therefore that $d\hat{\mathbf{e}}_b/ds$ is parallel to $\hat{\mathbf{e}}_n$.

b. Using part (a), let $d\hat{\mathbf{e}}_b/ds = \tau\hat{\mathbf{e}}_n$.† ($\tau$ is called the torsion of the path or curve.) Then differentiate $\hat{\mathbf{e}}_n = \hat{\mathbf{e}}_b \times \hat{\mathbf{e}}_t$ with respect to s and prove that

$$\frac{d\hat{\mathbf{e}}_n}{ds} = -\left(\frac{1}{\rho}\hat{\mathbf{e}}_t + \tau\hat{\mathbf{e}}_b\right)^{\dagger}$$

The three equations marked with daggers give the derivatives of the three unit vectors associated with a space curve and are called the Serret-Frenet formulas.

1.156 The derivative of acceleration is called the *jerk* and is studied in the dynamics of vehicle impact and in the kinematics of mechanisms involving cams and followers. Show that the jerk of a point has the following form in terms of its intrinsic components:

$$\mathbf{J}_P = \dot{\mathbf{a}}_P = \left(\dddot{s} - \frac{\dot{s}^3}{\rho^2}\right)\hat{\mathbf{e}}_t + \left(\frac{3\dot{s}\ddot{s}}{\rho} - \frac{\dot{\rho}\dot{s}^2}{\rho^2}\right)\hat{\mathbf{e}}_n - \frac{\dot{s}^3}{\rho}\tau\hat{\mathbf{e}}_b$$

* **1.157** The following "pursuit" problem is very difficult, yet it illustrates exceptionally well the idea that the velocity vector is tangent to the path. Thus we include it along with a set of steps for the courageous student who wishes to "pursue" it. A dog begins at the point $(x, y) = (D, O)$ and runs toward his master at constant speed $2V_0$. (See Figure P1.157.) The dog's velocity direction is always toward his master, who starts at the same time at the origin and moves along the positive y direction at speed V_0. Find the man's position when his dog overtakes him, and determine how much time has elapsed. *Hints*: The man's y coordinate is y_M (which of course is $V_0 t$). Show that:

1. $\dfrac{-dy_D}{dx} = \dfrac{y_M - y_D}{x}$, where (x, y_D) represent the dog's coordinates at any time.
2. $2V_0 = \dfrac{ds_D}{dt} = \dfrac{\sqrt{dx^2 + dy_D^2}}{dt} = \dfrac{-dx\sqrt{1 + y'^2_D}}{dt}$.
3. $V_0 = dy_M/dt$.
4. From dividing and rearranging steps 2 and 3, we get $2y'_M = -\sqrt{1 + y'^2_D} \quad (y'_M = dy_M/dx)$.
5. From step 1, we have $y'_M = -xy''_D$.
6. From steps 4 and 5, we have $2xy''_D = \sqrt{1 + y'^2_D}$.
7. Letting $y'_D = p$, from step 6 we get $dp/\sqrt{1+p^2} = dx/2x$.
8. By integrating step 7 with a table of integrals, we get $$\ln(p + \sqrt{1 + p^2}) = \tfrac{1}{2}\ln(x) - \tfrac{1}{2}\ln(C_1) = \ln\left(\frac{x}{C_1}\right)^{1/2}.$$
9. From step 8, we get $p + \sqrt{1 + p^2} = (x/C_1)^{1/2}$.
10. From step 9, we get $p - \sqrt{x/C_1} = -\sqrt{1 + p^2}$.
11. From step 10, squaring both sides and solving for p, $$2p = \sqrt{\frac{x}{C_1}} - \sqrt{\frac{C_1}{x}} = 2y'_D$$
12. $y'_D = 0$ when $x = D$ (initial condition).
13. From steps 11 and 12 we have $C_1 = D$, so that $2y'_D = \sqrt{x/D} - \sqrt{D/x}$.
14. Integrating step 13, we get $$y_D = x^{3/2}/(3\sqrt{D}) - \sqrt{xD} + C_2.$$
15. $y_D = 0$ when $x = D$ (initial condition).
16. From steps 14 and 15 we have $C_2 = 2D/3$, so that $$y_D = \frac{x^{3/2}}{3\sqrt{D}} - \sqrt{xD} + \tfrac{2}{3}D$$
17. $y_M = V_0 t$.
18. Finally, write the conditions relating to y_M, y_D, and x when the dog overtakes his master, and wrap it up!

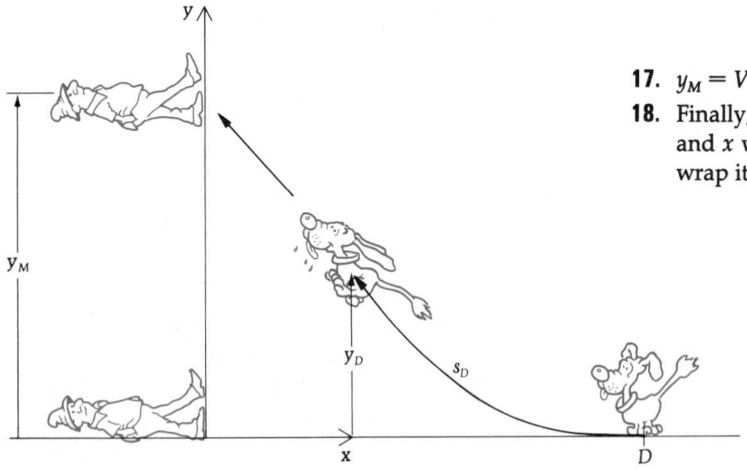

Figure P1.157

Computer Problem ▶ Chapter 1

* **1.158** A particle moves in the xy plane according to the equation $r = k\theta$, where k is a constant, and has the constant speed v_0. The particle passes through the origin with $r = \theta = 0$ at $t = 0$. (See Figure P1.158.)

a. Show that $k\dot{\theta}\sqrt{1+\theta^2} = v_0$.

b. With the trigonometric substitution $\theta = \tan\phi$, and then consulting integral tables, integrate the equation and obtain:

$$\frac{\theta\sqrt{1+\theta^2}}{2} + \frac{1}{2}\ln\left[\frac{\sqrt{1+\theta^2}+1+\theta}{\sqrt{1+\theta^2}+1-\theta}\right] = \frac{v_0 t}{k}$$

c. For the case $v_0/k = 1$, use the computer to plot θ versus time until θ has increased from 0 to 2π radians.

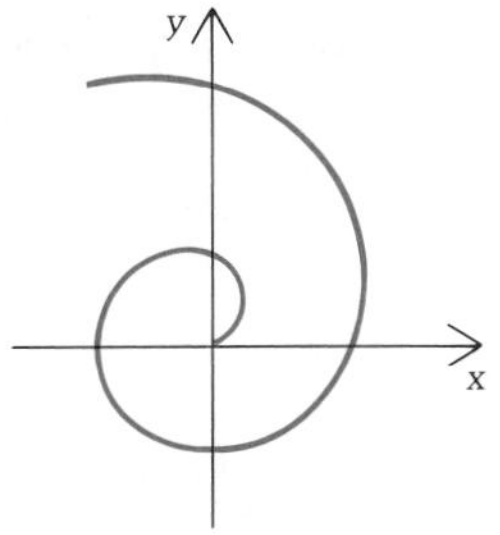

Figure P1.158

Summary ▶ Chapter 1

In this chapter we have studied the position, velocity and acceleration of a point (or particle). With O being a point fixed in the frame of reference and P denoting the moving point, then $\mathbf{r}_{OP}$ is a position vector for P and we defined

$$\text{Velocity:} \qquad \mathbf{v}_P = \frac{d\mathbf{r}_{OP}}{dt}$$

$$\text{Acceleration:} \qquad \mathbf{a}_P = \frac{d\mathbf{v}_P}{dt} = \frac{d^2\mathbf{r}_{OP}}{dt^2}$$

With rectangular coordinates and associated unit vectors, and with O chosen as the origin of the coordinate system,

$$\mathbf{r}_{OP} = x\hat{\mathbf{i}} + y\hat{\mathbf{j}} + z\hat{\mathbf{k}}$$
$$\mathbf{v}_P = \dot{x}\hat{\mathbf{i}} + \dot{y}\hat{\mathbf{j}} + \dot{z}\hat{\mathbf{k}}$$
$$\mathbf{a}_P = \ddot{x}\hat{\mathbf{i}} + \ddot{y}\hat{\mathbf{j}} + \ddot{z}\hat{\mathbf{k}}$$

In similar fashion for cylindrical coordinates,

$$\mathbf{r}_{OP} = r\hat{\mathbf{e}}_r + z\hat{\mathbf{k}}$$
$$\mathbf{v}_P = \dot{r}\hat{\mathbf{e}}_r + r\dot{\theta}\,\hat{\mathbf{e}}_\theta + \dot{z}\hat{\mathbf{k}}$$
$$\mathbf{a}_P = (\ddot{r} - r\dot{\theta}^2)\hat{\mathbf{e}}_r + (r\ddot{\theta} + 2\dot{r}\dot{\theta})\hat{\mathbf{e}}_\theta + \ddot{z}\hat{\mathbf{k}}$$

With a path-length parameter $s(t)$, describing the motion of P on a given curve (path) and with $\hat{\mathbf{e}}_t$ and $\hat{\mathbf{e}}_n$ being unit tangent and principal unit normal, respectively,

$$\mathbf{v}_P = \dot{s}\hat{\mathbf{e}}_t$$

$$\mathbf{a}_P = \ddot{s}\hat{\mathbf{e}}_t + \frac{(\dot{s})^2\hat{\mathbf{e}}_n}{\rho},$$

where ρ is the radius of curvature of the path at the point occupied by P at time t. Sometimes it is convenient to choose to measure $s(t)$ so that $\dot{s} > 0$ in an interval of time of interest. In this case, we have expressions that don't involve s explicitly:

$$\mathbf{v}_P = |\mathbf{v}_P|\,\hat{\mathbf{e}}_t$$

$$\mathbf{a}_P = \left(\frac{d}{dt}|\mathbf{v}_P|\right)\hat{\mathbf{e}}_t + \frac{|\mathbf{v}_P|^2}{\rho}\,\hat{\mathbf{e}}_n$$

REVIEW QUESTIONS ▶ Chapter 1

True or False?

1. The velocity $\mathbf{v}_P$ of a point P is always tangent to its path.
2. $\mathbf{v}_P$ depends on the reference frame chosen to express the position of P.
3. $\mathbf{v}_P$ depends on the origin chosen in the reference frame.
4. The magnitude and direction in space of $\mathbf{v}_P$ depend on the choice of coordinates used to locate the point relative to the reference frame.
5. $\mathbf{a}_P$ always has a nonvanishing component normal to the path.
6. For any point P, $|\mathbf{a}_P| = \sqrt{\ddot{s}^2 + \dot{s}^4/\rho^2}$.
7. A point can have $\ddot{r} = 0$ but still have a nonvanishing radial component of acceleration.
8. If a ball on a string is being whirled around in a horizontal circle at constant speed, the center of the ball has zero acceleration.
9. Studying the kinematics of a particle results in the same equations as studying the kinematics of a point.
10. In our study of the kinematics of a point, the following terms have not appeared in any of the equations: mass, force, moments, gravity, momentum, moment of momentum, inertia, or Newtonian (inertial) frames.

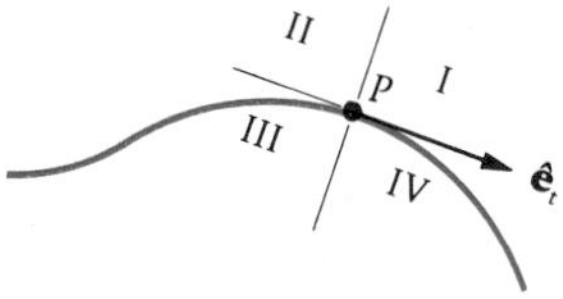

11. The acceleration vector of P, at the indicated point on the path shown in the figure, can lie in any of the four quadrants.
12. A particle moving in a plane, with constant values of $\dot{r}$ and $\dot{\theta}$, will at all times have zero acceleration.

Answers: 1. T 2. T 3. F 4. F 5. F 6. T 7. T 8. F 9. T 10. T 11. F 12. F

2 Kinetics of Particles and of Mass Centers of Bodies

2.1 Introduction

In this chapter we begin to consider the manner in which the motion of a body is related to external mechanical actions (forces and couples). Our kinematics notions of space and time must now be augmented by those of mass and force, which, like space and time, are **primitives** of the subject of mechanics. We simply have to agree in advance that some measures of **quantity of matter (mass)** and **mechanical action (force)** are basic ingredients in any attempt to analyze the motion of a body. We assume that the reader has a working knowledge, probably from a study of statics, of the characteristics of forces and moments and their vector descriptions. We use the term **body** to denote some material of fixed identity; we could think of a specific set of atoms, although the model we shall employ is based upon viewing material on a spatial scale such that mass is perceived to be distributed continuously. A body need not be rigid or even a solid, but, since our subject is classical dynamics (no relativistic effects), a body necessarily has constant mass.

In Section 2.2 we use Newton's laws for a particle and for interacting particles to deduce that the sum of the external forces on a body of any size is equal to the sum of the $m\mathbf{a}$'s of the body, alternatively expressed as the total mass multiplied by the acceleration of the mass center. This result is usually called Euler's first law. Applications of this are developed in Section 2.3 along with a review of the critically important concept of the free-body diagram.

The Principle of Work and Kinetic Energy for a particle is developed in Section 2.4, wherein are found expressions for the work done by several special types of forces. The concept of a conservative force is introduced, and the condition for which Work and Kinetic Energy becomes Conservation of Mechanical Energy is established. Finally, the implications of Work and Kinetic Energy for a system of particles are explored.

In Section 2.5 the impulse-momentum form of Euler's first law is developed, and conditions for conservation of momentum are demonstrated. Applications are made to problems of impact.

Euler's second law is the subject of Section 2.6. We return to Newton's laws so as to derive this important result which states that the sum of the moments of the external forces on a body (or system of particles) equals the sum of the moments of the body's $m\mathbf{a}$'s. Momentum forms of this are developed, primarily for later applications in Chapters 4, 5, and 7.

2.2 Newton's Laws and Euler's First Law

The usual starting point for relating the external forces on a body to its motion is Newton's laws. These were proposed in the Englishman Isaac Newton's famous work the *Principia,* published in 1687, and are commonly expressed today as:

1. If the resultant force **F** on a particle is zero, then the particle has constant velocity.
2. If $\mathbf{F} \neq \mathbf{0}$, then **F** is proportional to the time derivative of the particle's momentum $m\mathbf{v}$ (product of mass and velocity).
3. The interaction of two particles is through a pair of self-equilibrating forces. That is, they have the same magnitude, opposite directions, and a common line of action.

Clearly the first law may be regarded a special case of the second, and one must add an assumption about the frame of reference, since a point may have its velocity constant in one frame of reference and varying in another. Frames of reference in which these laws are valid are variously called Newtonian, Galilean, or **inertial.** Furthermore, the constant of proportionality in the second law can be made unity by appropriate choices of units so that the law becomes

$$\mathbf{F} = \frac{d}{dt}(m\mathbf{v}) = m\mathbf{a}$$

where **a** is the acceleration of the particle.

As we mentioned briefly in Chapter 1, a particle is a piece of material sufficiently small that we need not make distinctions among its material points with respect to locations (or to velocities or accelerations). We also noted that this definition allows, for *some* purposes, a truck or a space vehicle or even a planet to be adequately modeled as a particle. In the *Principia,* Newton used heavenly bodies as the particles in his examples and treated them as moving points subject only to universal gravitation and their own inertia.* Newton did not extend his work to problems for which it is necessary to account for the actual sizes of the bodies and how their masses are distributed. It was to be over 50 years before the Swiss mathematician Leonhard Euler presented the first of the two principles that have come to be called **Euler's laws.**

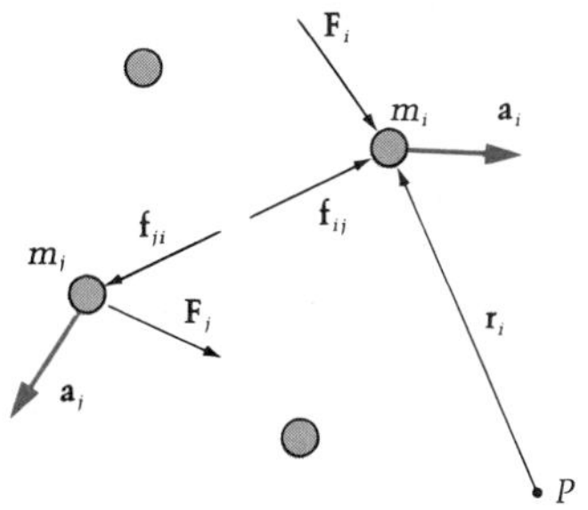

Figure 2.1

For a body composed of a set of N particles, we may deduce Euler's laws from Newton's laws. As suggested by Figure 2.1, we separate the forces acting on the ith particle into two groups: there are the N-1 forces exerted by other particles of the system, $\mathbf{f}_{ij}$ being that exerted by the jth particle; then there is $\mathbf{F}_i$, the net force exerted on the ith particle by things *external* to the system. Applying Newton's second law to the ith particle whose mass is m_i and whose acceleration is $\mathbf{a}_i$,

$$\mathbf{F}_i + \sum_{j=1}^{N} \mathbf{f}_{ij} = m_i\mathbf{a}_i \tag{2.1}$$

in which we understand

$$\mathbf{f}_{ii} = \mathbf{0}$$

* See C. Truesdell, *Essays in the History of Mechanics* (Berlin: Springer-Verlag, 1968).

Now we sum the N such equations to obtain

$$\sum_{i=1}^{N} \mathbf{F}_i + \sum_{i=1}^{N}\sum_{j=1}^{N} \mathbf{f}_{ij} = \sum_{i=1}^{N} m_i\mathbf{a}_i \tag{2.2}$$

But Newton's third law tells us that

$$\mathbf{f}_{ij} = -\mathbf{f}_{ji}$$

so that

$$\Sigma\Sigma\mathbf{f}_{ij} = \mathbf{0}$$

Thus we conclude that

$$\Sigma\mathbf{F}_i = \Sigma m_i\mathbf{a}_i \tag{2.3}$$

which is the particle-system form of Euler's first law and states that the sum of the *external forces* on the system equals the sum of the $m\mathbf{a}$'s of the particles making up the system.

For a body whose mass is continuously distributed, as depicted in Figure 2.2, the counterpart to equation (2.3) is

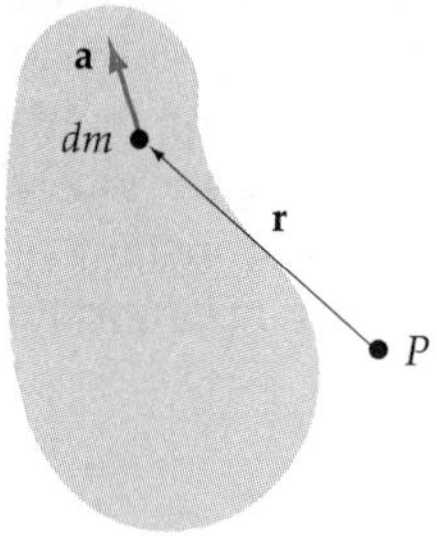

Figure 2.2

$$\Sigma\mathbf{F} = \int \mathbf{a}\, dm \tag{2.4}$$

where dm* is a differential element of mass, $\mathbf{a}$ is its acceleration, and $\Sigma\mathbf{F}$ is the sum of the external forces acting on the body.

Motion of the Mass Center

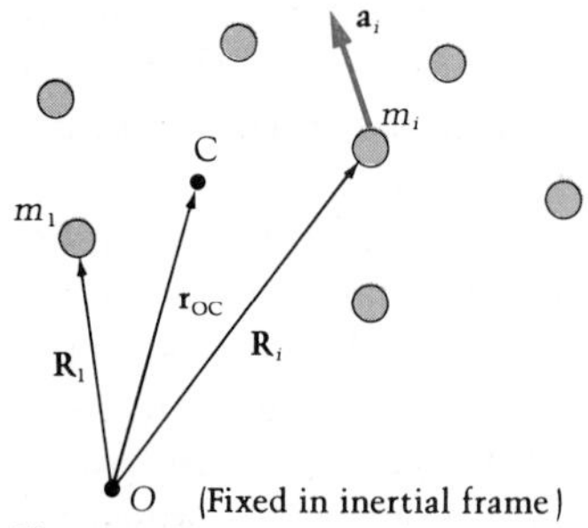

Figure 2.3a

We close this section by developing the relationship between the external forces acting on a body and the motion of its mass center. To do this we first construct position vectors for the particles of a system as shown in Figure 2.3a. Thus the acceleration of the ith particle may be written

$$\mathbf{a}_i = \frac{d^2\mathbf{R}_i}{dt^2} \tag{2.5}$$

Applying this to Equation (2.3),

$$\Sigma\mathbf{F}_i = \Sigma m_i \frac{d^2\mathbf{R}_i}{dt^2}$$

$$= \frac{d^2}{dt^2}(\Sigma m_i\mathbf{R}_i) \tag{2.6}$$

The location of the mass center, C, of a system of particles is defined by

$$\mathbf{r}_{OC} = \frac{\Sigma m_i\mathbf{R}_i}{m} \tag{2.7}$$

* $dm = \rho\, dV$ where ρ is mass density and dV is an infinitesimal element of volume. When using rectangular coordinates x, y, z, then $dV = dx\, dy\, dz$.

where m is the mass (Σm_i) of the system. Thus Euler's first law becomes

$$\Sigma \mathbf{F} = \frac{d^2}{dt^2}(m\mathbf{r}_{OC})$$

$$= m\frac{d^2\mathbf{r}_{OC}}{dt^2}$$

or

$$\Sigma \mathbf{F} = m\mathbf{a}_C \tag{2.8}$$

Question 2.1 What happened to the $\dot{m}$ and $\ddot{m}$ terms in going from $\frac{d^2}{dt^2}(m\mathbf{r}_{OC})$ to $m\frac{d^2\mathbf{r}_{OC}}{dt^2}$?

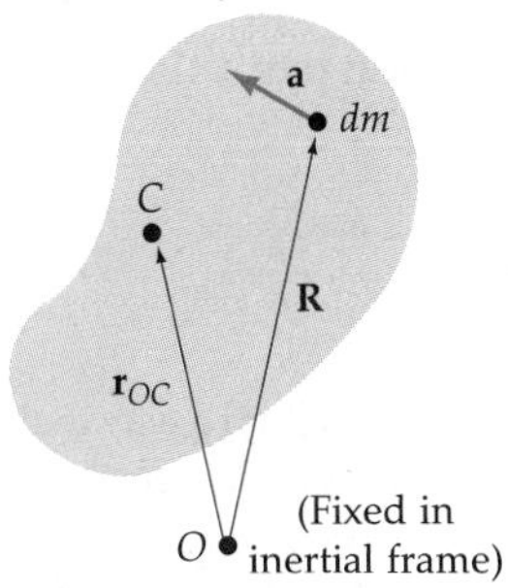

Figure 2.3b

For a continuous body (Figure 2.3b), the counterparts to Equations (2.5–8) are

$$\mathbf{a} = \frac{d^2\mathbf{R}}{dt^2}$$

$$\Sigma \mathbf{F} = \int \frac{d^2\mathbf{R}}{dt^2}\,dm = \frac{d^2}{dt^2}\int \mathbf{R}\,dm$$

and

$$m\mathbf{r}_{OC} = \int \mathbf{R}\,dm$$

resulting again in Equation (2.8). Thus we see that the resultant external force on the body is the product of the constant mass m of the body and the acceleration $\mathbf{a}_C$ of its mass center. Hence the motion of the mass center of a body is governed by an equation identical in form to Newton's second law for a particle. It is very important to realize that for a rigid body the mass center C coincides at every instant with a specific material point of the body or of its rigid extension (for example, the center of a hollow sphere). This is not the case for a deformable body.

Sometimes it is useful to subdivide a body into two parts, say of masses m_1 and m_2 with mass-center locations C_1 and C_2. Recalling a property of mass centers,

$$m\mathbf{r}_{OC} = m_1\mathbf{r}_{OC_1} + m_2\mathbf{r}_{OC_2}$$

so that after differentiating twice with respect to time,

$$m\mathbf{a}_C = m_1\mathbf{a}_{C_1} + m_2\mathbf{a}_{C_2}$$

They are zero since our definition of a body requires that its mass be constant.

We see that Equation (2.8) can also be written

$$\Sigma \mathbf{F} = m_1 \mathbf{a}_{C_1} + m_2 \mathbf{a}_{C_2} \tag{2.9}$$

The principal purpose of this section has been the derivation of Equation (2.8), and the next section is devoted wholly to applications of it. But the natural occurrence of the mass center between Equations (2.6) and (2.8) motivates a brief review of the calculation of its location. The example and problems that follow are designed to provide that review in instances for which the body comprises several parts, the mass centers of which are known.

EXAMPLE 2.1

A uniform prismatic rod of density ρ and length $2L$ is deformed in such a way that the right half is uniformly compressed to length $L/2$ with no change in cross-sectional area A. (See Figure E2.1.) The left half of the rod is not altered. Letting the x axis be the locus of cross-sectional centroids, find the coordinates of the mass center in the deformed configuration.

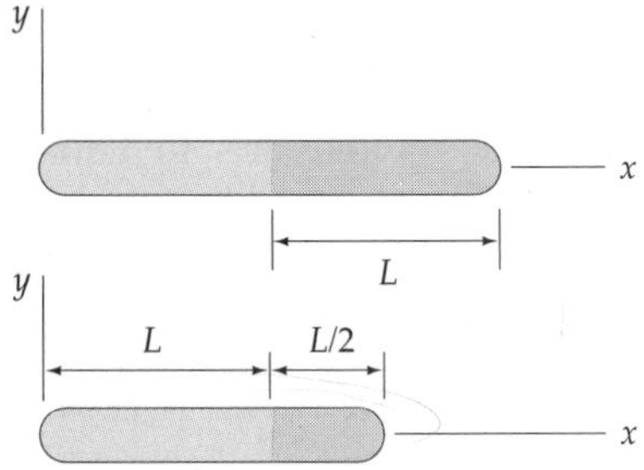

Figure E2.1

Solution

In the first configuration the center-of-mass coordinates are $(L, 0, 0)$; that is, the center of mass is at the interface of the two segments. In the second configuration, however,

$$2\rho AL\bar{x} = \rho AL\left(\frac{L}{2}\right) + \rho AL\left(\frac{5}{4}L\right)$$

$$\bar{x} = \frac{L}{4} + \frac{5}{8}L$$

$$= \frac{7}{8}L$$

Thus the mass center no longer lies in the interface. This example illustrates that the mass center of a deformable body does not in general coincide with the same material point in the body at different times.

PROBLEMS ▶ Section 2.2

2.1 Show that the mass center C of a body $\mathcal{B}$ is unique. *Hint*: Consider the two mass centers C_1 and C_2, respectively:

$$\mathbf{r}_{O_1C_1} = \frac{1}{m}\int \mathbf{R}_1\, dm$$

$$\mathbf{r}_{O_2C_2} = \frac{1}{m}\int \mathbf{R}_2\, dm$$

and relate $\mathbf{R}_1$ to $\mathbf{R}_2$. (See Figure P2.1.) Using this relation,

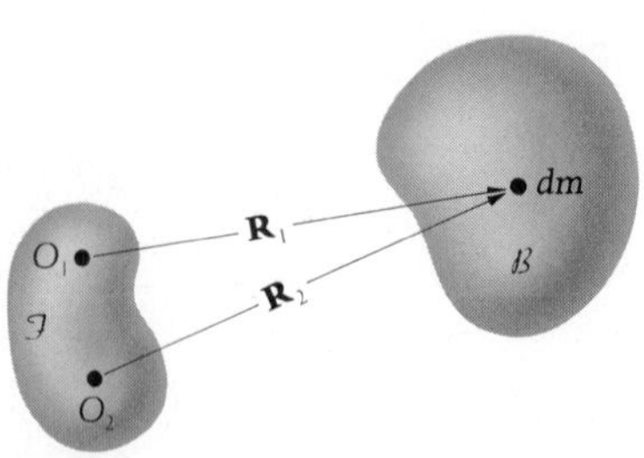

Figure P2.1

show that $\mathbf{r}_{O_1C_1} = \mathbf{r}_{O_1C_2}$, which means that C_1 and C_2 are the same point!

2.2 Find the mass center of the composite body shown in Figure P2.2. Note that the three parts are composed of different materials.

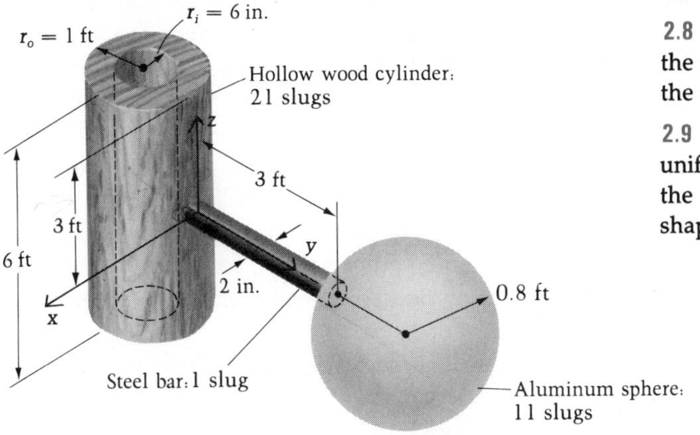

Figure P2.2

2.3 Find the center of mass of the body composed of two uniform slender bars and a uniform sphere in Figure 2.3.

2.4 Find the center of mass of the bent bar, each leg of which is parallel to a coordinate axis and as uniform density and mass m. (See Figure P2.4.)

2.5 Repeat Problem 2.4 if the four legs have uniform, but different, densities, so that the masses of $\mathcal{A}$, $\mathcal{B}$, $\mathcal{C}$, and $\mathcal{D}$ are, respectively, m, $2m$, $3m$, and $4m$.

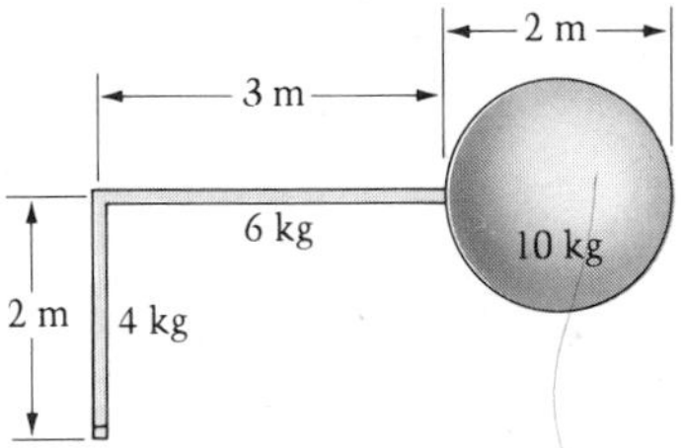

Figure P2.3

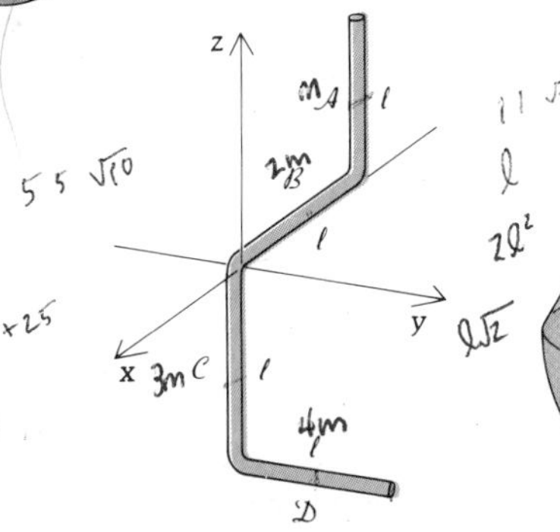

Figure P2.4

2.6 Consider a body that is a composite of a uniform sphere and a uniform cylinder, each of density ρ. Find the mass center of the body. (See Figure P2.6.)

2.7 Find the mass center of the body in Figure P2.7 which is a hemisphere glued to a solid cylinder of the same density, if $L = 2R$.

2.8 In the preceding problem, for what ratio of L to R is the mass center in the interface between the sphere and the cylinder?

2.9 In Figure P2.9 find the height H of the cone of uniform density (in terms of R) so that the mass center of the cone plus hemisphere is at the interface of the two shapes (i.e., $z = 0$).

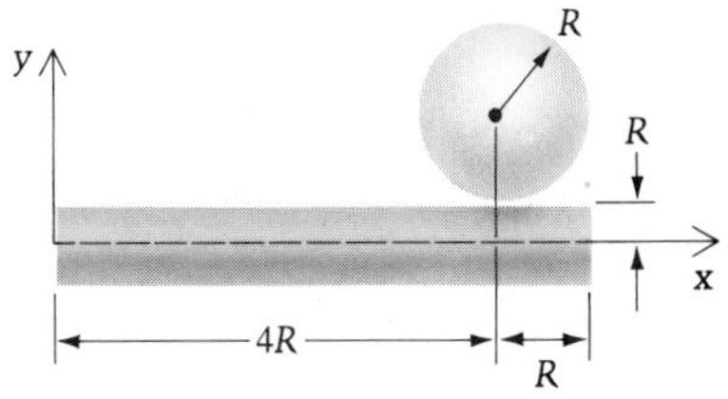

Figure P2.6

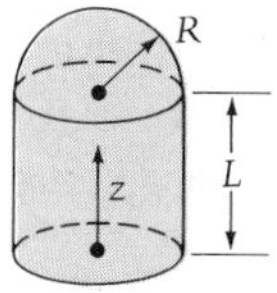

Figure P2.7

2.10 A thin wire is bent into the shape of an isosceles triangle (Figure P2.10). Find the mass center of the object, and show that it is at the same point as the mass center of a triangular plate of equal dimensions only if the triangle is equilateral. (Area of cross section = A and mass density = ρ, both constant.)

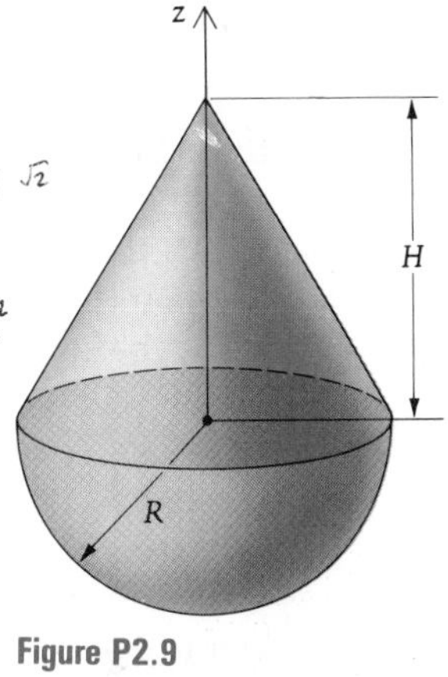

Figure P2.9

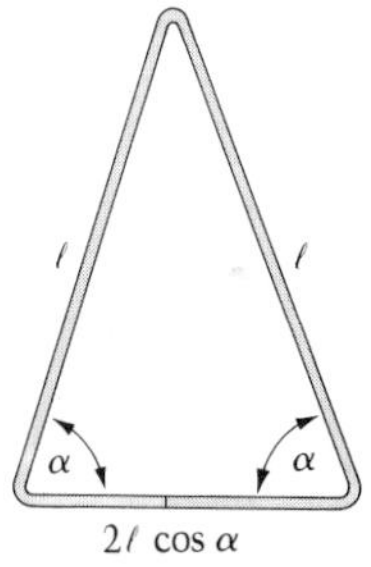

Figure P2.10

2.3 Motions of Particles and of Mass Centers of Bodies

Although the mass center of a body does not always coincide with a specific material point of the body, the mass center is nonetheless clearly an important point reflecting the distribution of the body's mass. Furthermore, there are a number of situations in which our objectives are satisfied if we can determine the motion of any material or characteristic point of the body. Clearly this is the case when we attempt to describe the orbits in which the planets move around the sun. Closer to home, a football coach is overjoyed if he finds a punter who can consistently kick the ball 60 yards in the air, regardless of whether the ball gets there end over end, spiraling, or floating like a "knuckleball." In such cases the material point upon which we focus our attention is unimportant. However, there is a strong computational advantage in focusing on the mass center: It is that the motion of that point is directly related to the external forces acting on the body.

We are more likely to think of the football as particle-like when exhibiting the knuckleball behavior than when it is rapidly spinning. Nonetheless, the *mass center's* motion in each case is governed by Euler's first law, although those motions might be quite different because of the different sets of external force induced by the differing interactions of the ball with the air.

If the external forces acting on the body are known functions of time, the mass center's motion can be calculated from Euler's first law:

$$\Sigma \mathbf{F} = m\mathbf{a}_C \tag{2.8}$$

or, alternatively,

$$\Sigma \mathbf{F} = m \frac{d^2\mathbf{r}_{OC}}{dt^2} \tag{2.10}$$

where $\mathbf{r}_{OC}$ is a position vector for the mass center. It is easily seen that two integrations of (2.10) with respect to time yield $\mathbf{r}_{OC}(t)$ provided that initial values of $\mathbf{r}_{OC}$ and $\dot{\mathbf{r}}_{OC}(= \mathbf{v}_C)$ are known.

The Free-Body Diagram

Only one thing remains to be done prior to studying several examples that make use of Euler's first law to analyze the motions of the mass centers of bodies. It is to review the concept of the **free-body diagram** which the reader should already have mastered in the study of statics. Without the ability to identify the external forces (and later the moments also), the student will not be able to write a correct set of equations of motion.

A free-body diagram is a sketch of a body in which all the external forces and couples acting upon it are carefully drawn with respect to location, direction, and magnitude. These forces might result from pushes or pulls, as the boy and girl are exerting on the crate and rope in Figure 2.4a. Or the forces might result from gravity, such as the weight of

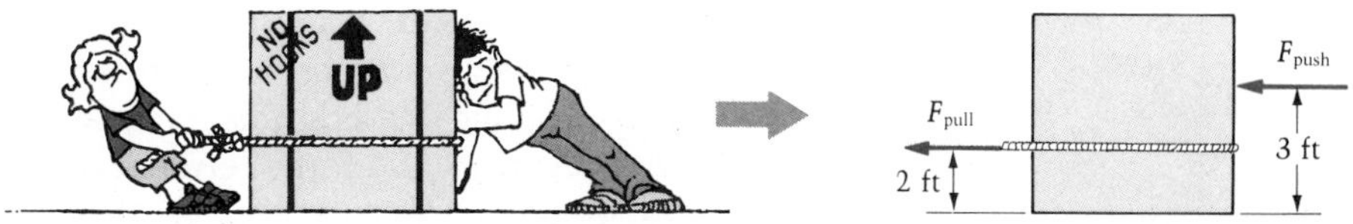

Figure 2.4a

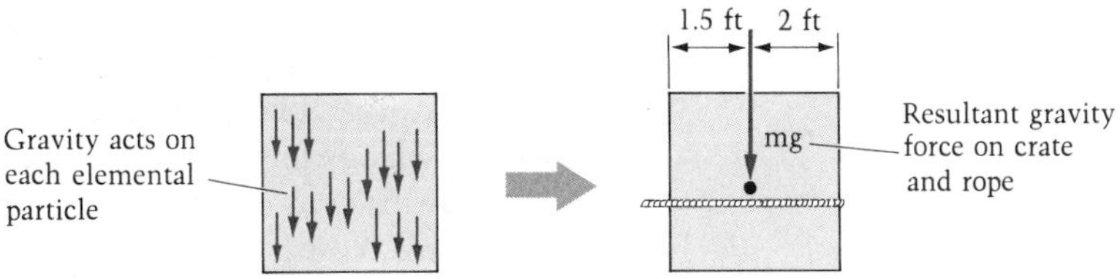

Figure 2.4b

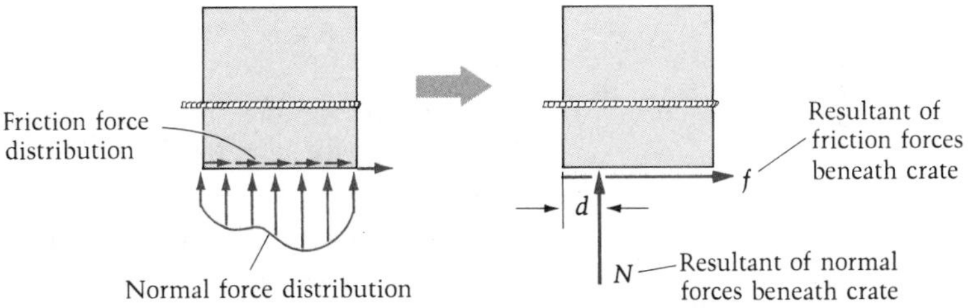

Figure 2.4c

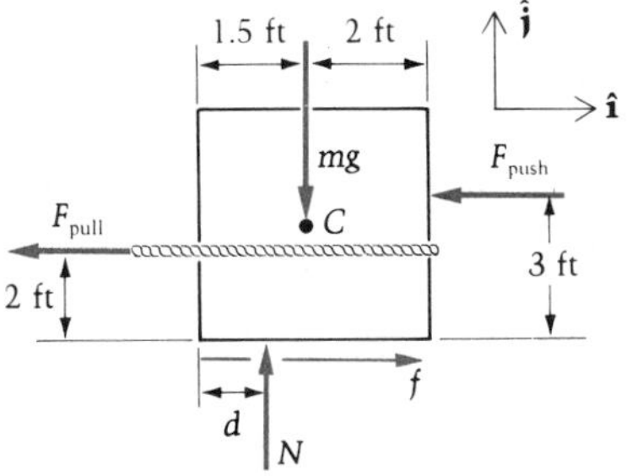

Figure 2.5

the crate in Figure 2.4b. (Note that the forces need not *touch* the body to be included in the free-body diagram; another such example is electromagnetic forces.) Or the forces might result from supports, such as the floor beneath the crate in Figure 2.4c. If the crate/rope body is acted upon simultaneously by all the forces in these figures, its complete free-body diagram is as shown in Figure 2.5.

It is important to recognize that the free-body diagram:

1. Clearly identifies the body whose motion is to be analyzed.
2. Provides a catalog of all the *external* forces (and couples) *on* the body.
3. Allows us to express, in a compact way, what we know or can easily conclude about the lines of action of known and unknown forces. For example, we know that the pressure (distributed normal force) exerted by the floor on the bottom of the box has a resultant that is a force with a vertical line of action. The symbol N along with the arrow is a code for communicating the fact that we have decided to express that unknown (vector) force as $N\hat{\mathbf{j}}$. The fact that we do not know the location of the line of action of that force is displayed by the presence of the unknown length d.

In dynamics, as in statics, the only characteristics of a force that are manifest in the equations of motion are the vector describing the force and the location of its line of action; that is, we must sum up all the

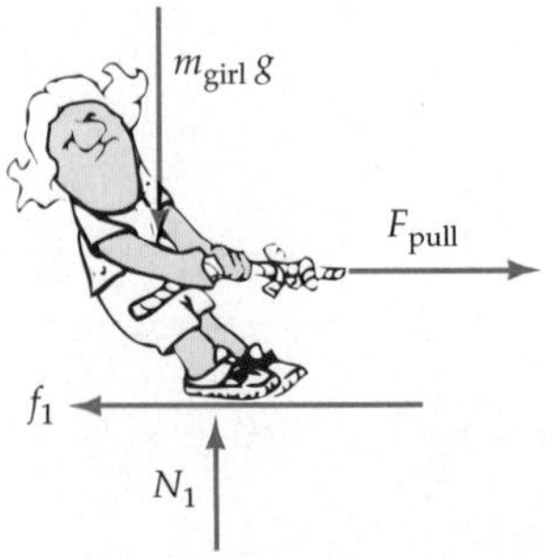

Figure 2.6

external forces, and we must also sum their moments about some point. Consequently, everything we need to know about the external forces is displayed on the free-body diagram, and we may readily check our work by glancing back and forth between our diagram and the equations we are writing.

When we focus individually on two or more interacting bodies, the free-body diagrams provide an economical way to satisfy—and show that we have satisfied—the principle of action and reaction. The free-body diagram of the girl in our example is shown in Figure 2.6. Since we have already established by Figure 2.5 that the force exerted by the girl on the rope will be $F_{pull}(-\hat{\mathbf{i}})$, then, by the action-reaction principle, the force exerted by the rope on the girl must be $F_{pull}(+\hat{\mathbf{i}})$ as shown in Figure 2.6. In other words, consistent forces of interaction are expressed through the single scalar F_{pull} and the arrow code.

EXAMPLE 2.2

Ignoring air resistance, find the trajectory of a golf ball hit off a tee at speed v_0 and angle θ with the horizontal.

Solution

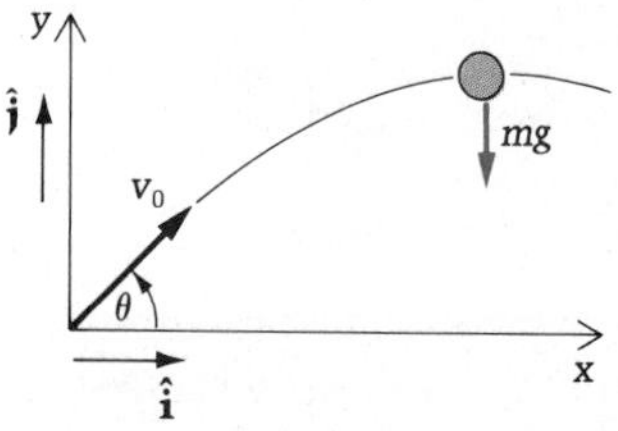

Figure E2.2

It is convenient here to set up a rectangular coordinate system as shown in Figure E2.2 and let time $t = 0$ be the instant at which the ball leaves the club. With x, y, and z as the coordinates of the mass center of the ball and since the only external force on the ball is its weight, $-mg\hat{\mathbf{j}}$, we have from Equation (2.10):

$$-mg\hat{\mathbf{j}} = m(\ddot{x}\hat{\mathbf{i}} + \ddot{y}\hat{\mathbf{j}} + \ddot{z}\hat{\mathbf{k}})$$

Thus, collecting the coefficients of $\hat{\mathbf{i}}$, $\hat{\mathbf{j}}$, and $\hat{\mathbf{k}}$, we obtain

$$\ddot{x} = 0 \qquad \ddot{y} = -g \qquad \ddot{z} = 0$$

Integrating, we get

$$\dot{x} = C_1 \qquad \dot{y} = -gt + C_2 \qquad \dot{z} = C_3$$

Because of the way we have aligned the x and z axes,

$$\dot{x}(0) = v_0 \cos\theta \qquad \dot{y}(0) = v_0 \sin\theta \qquad \dot{z}(0) = 0$$

Therefore

$$C_1 = v_0 \cos\theta \qquad C_2 = v_0 \sin\theta \qquad C_3 = 0$$

Integrating again, we get

$$x = v_0(\cos\theta)t + C_4$$

$$y = \frac{-gt^2}{2} + v_0(\sin\theta)t + C_5$$

$$z = C_6$$

Our location of the origin of the coordinate system at the "launch" site yields $x(0) = y(0) = z(0) = 0$, so that $C_4 = C_5 = C_6 = 0$ and the trajectory of the mass center of the ball is given by

$$x = v_0(\cos\theta)t$$

$$y = -\frac{gt^2}{2} + v_0(\sin\theta)t$$

$$z = 0$$

which describes a parabola in the xy plane—that is, in the vertical plane defined by the launch point and the direction of the launch velocity.

Letting the time of maximum elevation be t_1, we find that $\dot{y}(t_1) = 0$ yields

$$0 = -gt_1 + v_0 \sin\theta$$

so that $t_1 = (v_0/g)\sin\theta$ and the maximum elevation is

$$y(t_1) = -\frac{v_0^2}{2g}\sin^2\theta + \frac{v_0^2}{g}\sin^2\theta = \frac{v_0^2}{2g}\sin^2\theta$$

If t_2 is the time the ball strikes the fairway (assumed level), then

$$y(t_2) = 0 = -\frac{gt_2^2}{2} + v_0 t_2 \sin\theta$$

$$t_2 = \frac{2v_0}{g}\sin\theta$$

which is, not surprisingly, twice the time (t_1) to reach maximum elevation. The length of the drive is

$$x(t_2) = v_0(\cos\theta)\left(\frac{2v_0}{g}\sin\theta\right)$$

$$= \frac{2v_0^2}{g}\sin\theta\cos\theta$$

$$= \frac{v_0^2}{g}\sin 2\theta$$

which, with v_0 fixed, is maximized by $\theta = 45°$. That is, for a given launch speed we get maximum range when the launch angle is 45°.

The results of this analysis apply to the unpowered flight of any projectile as long as the path is sufficiently limited that the gravitational force is constant (magnitude and direction) and we can ignore the medium (air) through which the body moves. Interaction with the air is responsible not only for the drag (retarding of motion) on a golf ball but also for the fact that its path is usually not planar (slice or hook!). On one of the Apollo moon landings in the early 1970s, astronaut Alan Shepard drove a golf ball a "country mile" on the moon because of the absence of air resistance and, more important, because the gravitational acceleration at the moon's surface is only about one-sixth that at the surface of the earth.

EXAMPLE 2.3

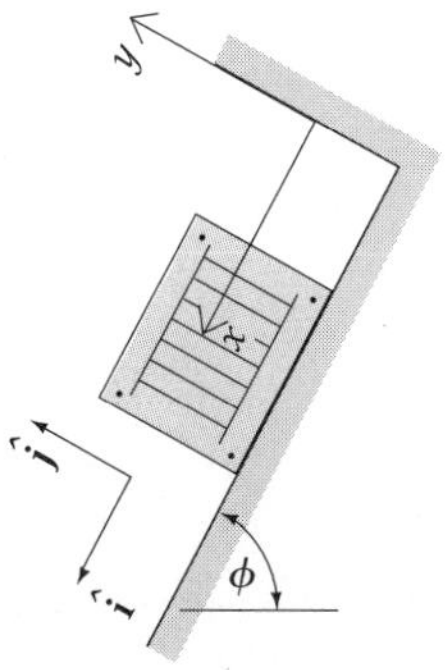

Figure E2.3a

If the 20-kg block shown in Figure E2.3a is released from rest, find its speed after it has descended a distance $d = 5$ m down the plane. The angle $\varphi = 60°$ and the (Coulomb) coefficients of friction are

$$\mu_s \text{ (static)} = 0.30$$
$$\mu_k \text{ (kinetic)} = 0.25$$

Solution

In the statement of the problem we are using some loose but common terminology in referring to the speed of the block. In fact we may only speak of the speed of a *point*, but here we are tacitly assuming that the block is *rigid* and *translating* so that every point in the block has the same velocity and the same acceleration. In contrast to the preceding example, note that here we do not know all the external forces on the body before we carry out the analysis, because the surface touching the block constrains its motion. That constraint is acknowledged by expressing the velocity of (the mass center of) the block by $\dot{x}\hat{\mathbf{i}}$ and its acceleration by $\ddot{x}\hat{\mathbf{i}}$.

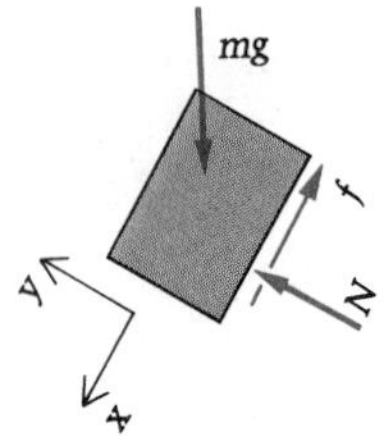

Figure E2.3b

Referring to the free-body diagram shown in Figure E2.3b,

$$\Sigma\mathbf{F} = m\mathbf{a}_C$$
$$N\hat{\mathbf{j}} + mg(\sin\varphi\hat{\mathbf{i}} - \cos\varphi\hat{\mathbf{j}}) - f\hat{\mathbf{i}} = m\ddot{x}\hat{\mathbf{i}}$$

or

$$N = mg\cos\varphi$$

and

$$mg\sin\varphi - f = m\ddot{x}$$

First we must determine if in fact the block will move. For equilibrium, $\ddot{x} = 0$ and f is limited by $0 \le f \le f_{\text{max}} = \mu_s N$. Hence

$$f = mg\sin\varphi \le \mu_s\, mg\cos\varphi = \mu_s N$$

or

$$\tan\varphi \le \mu_s$$

But

$$\tan 60° = 1.73 > 0.3 = \mu_s$$

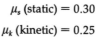

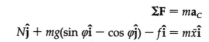

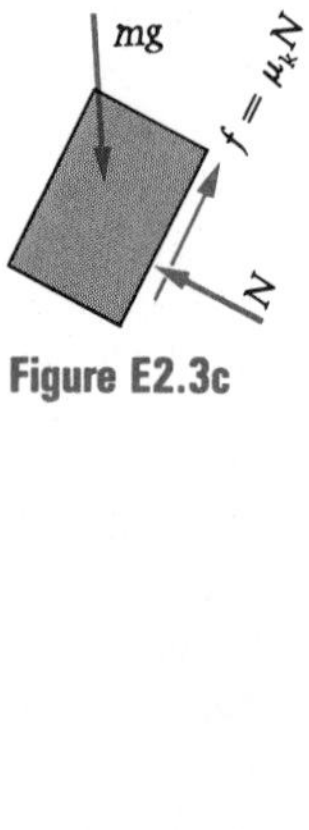

Figure E2.3c

Thus the block moves (and, as it does, is acted on by $\mu_k N$ up the plane as shown in Figure E2.3c). We note that $\tan^{-1}(\mu_s)$ is sometimes called the *angle of friction*. Here $\tan^{-1}(\mu_s)$ is 16.7°, and this of course is the angle for which $\tan\varphi = \mu_s$; it means that any angle $\varphi > 16.7°$ (like our 60°) will result in sliding, or a loss of equilibrium.

Having checked the statics and briefly reviewed friction, we now solve the equation of motion for $\ddot{x}$:

$$\begin{aligned} m\ddot{x} &= mg\sin\varphi - \mu_k N \\ &= mg\sin\varphi - \mu_k(mg\cos\varphi) \\ &= mg(\sin\varphi - \mu_k\cos\varphi) \end{aligned}$$

or

$$\ddot{x} = g(\sin\varphi - \mu_k\cos\varphi)$$

Thus

$$\dot{x} = g(\sin\varphi - \mu_k\cos\varphi)t + C_1$$

and $C_1 = 0$ since $\dot{x}(0) = 0$ if $t = 0$ is the instant at which the block is released. Hence

$$x = \frac{g}{2}(\sin\varphi - \mu_k \cos\varphi)t^2 + C_2$$

and $C_2 = 0$ if we choose the measurement of x so that $x(0) = 0$.

If we let t_1 be the time at which $x = d$, then

$$d = \frac{g}{2}(\sin\varphi - \mu_k \cos\varphi)t_1^2$$

For $\varphi = 60°$, $\mu_k = 0.25$, $d = 5$ m, and $g = 9.81$ m/s^2, we get

$$5 = \left(\frac{9.81}{2}\right)[0.866 - 0.25(0.5)]t_1^2$$

from which $t_1 = 1.17$ s.

Since the velocity is given by $\dot{x}\hat{\mathbf{i}}$, the speed at t_1 is merely the magnitude (or absolute value) of $\dot{x}(t_1)$ and

$$\begin{aligned}\dot{x}(t_1) &= (9.81)[0.866 - 0.25(0.5)](1.17)\\ &= 8.50 \text{ m/s}\end{aligned}$$

Finally we should note that the plausibility of our numerical results can be verified from the fact that, owing to the steep angle and moderate coefficient of friction, they should be of the same orders of magnitude as those arising from a free vertical drop (acceleration g) for which

$$t_1 = \sqrt{\frac{2d}{g}} = \sqrt{\frac{2(5)}{9.81}} = 1.01 \text{ s}$$

and

$$\text{Speed} = \sqrt{2gd} = \sqrt{2(9.81)(5)} = 9.90 \text{ m/s}$$

EXAMPLE 2.4

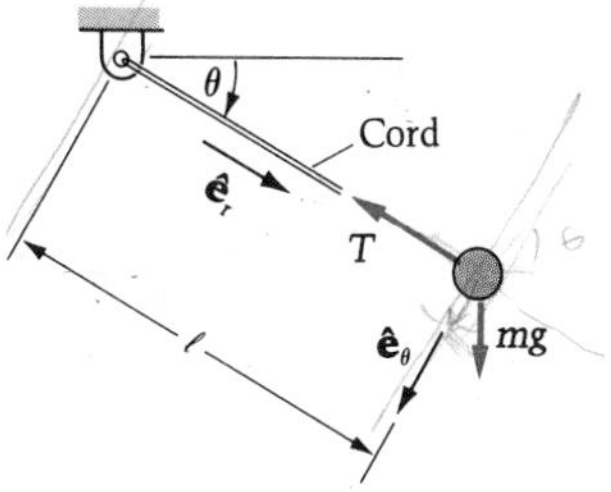

Figure E2.4

A ball of mass m (see Figure E2.4) is released from rest with the cord taut and $\theta = 30°$. Find the tension in the cord during the ensuing motion.

Solution

In this problem we make two basic assumptions:

1. The cord is inextensible.
2. The cord is attached to the ball at its mass center (or equivalently the ball is small enough to be treated as a particle). Either way the point whose motion is to be described has a path that is a circle. Thus the problem is similar to Example 2.3 in that the path of the mass center is known in advance (a circle here and a straight line there) and consequently among the external forces are unknowns caused by constraints (the tension in the cord here and the surface reaction in the preceding problem).

Using polar coordinates (Section 1.6), we may express the acceleration as

$$\mathbf{a} = (\ddot{r} - r\dot{\theta}^2)\hat{\mathbf{e}}_r + (r\ddot{\theta} + 2\dot{r}\dot{\theta})\hat{\mathbf{e}}_\theta$$

Since the polar coordinate r is the constant ℓ here, referring to the free-body diagram in Figure E2.4 we have

$$-T\hat{\mathbf{e}}_r + mg(\hat{\mathbf{e}}_\theta \cos\theta + \hat{\mathbf{e}}_r \sin\theta) = m(-\ell\dot{\theta}^2\hat{\mathbf{e}}_r + \ell\ddot{\theta}\hat{\mathbf{e}}_\theta)$$

so that

$$T = mg \sin\theta + m\ell\dot{\theta}^2 \tag{1}$$

and

$$m\ell\ddot{\theta} - mg\cos\theta = 0 \tag{2}$$

The first of these component equations (Equation (1)) yields the tension T if we know $\theta(t)$; the second (2) is the differential equation that we must integrate to obtain $\theta(t)$. In Example 2.3 the counterpart of Equation (2) is $\ddot{x}$ = constant, which of course was easily integrated.

Here not only do we have a nontrivial differential equation in that $\ddot{\theta}$ is a function of θ, but we have the substantial complication that Equation (2) is nonlinear because $\cos\theta$ is a nonlinear function of θ. However, a partial integration of Equation (2) can be accomplished; to this end we write the equation in the standard form

$$\ddot{\theta} - \frac{g}{\ell}\cos\theta = 0$$

and then multiply by $\dot{\theta}$ to obtain

$$\dot{\theta}\ddot{\theta} - \frac{g}{\ell}\dot{\theta}\cos\theta = 0$$

which we recognize to be

$$\frac{d}{dt}\left(\frac{\dot{\theta}^2}{2} - \frac{g}{\ell}\sin\theta\right) = 0$$

or

$$\frac{\dot{\theta}^2}{2} - \frac{g}{\ell}\sin\theta = C_1 \qquad \text{(a constant)} \tag{3}$$

Equation (3) is called an **energy integral** of Equation (2) and is closely related to the "work and kinetic energy" principle that is introduced in the next section.

For the problem at hand the constant C_1 may be obtained from the fact that when $\theta = 30°$, then $\dot{\theta} = 0$; thus

$$0 - \frac{g}{\ell}\sin 30° = C_1$$

or

$$C_1 = \frac{-g}{2\ell}$$

Thus from Equation (3) we get

$$\dot{\theta}^2 = \frac{g}{\ell}(2\sin\theta - 1) \tag{4}$$

which we may substitute in Equation (1) to obtain

$$T = mg \sin\theta + mg(2\sin\theta - 1)$$

or

$$T = mg(3\sin\theta - 1) \tag{5}$$

Even though we have not obtained the time dependence of the tension,* the energy integral has enabled us to find the way in which the tension depends on the position of the ball. As we would anticipate intuitively, the maximum tension occurs when $\theta = 90°$, at which time $T = [3(1) - 1]\,mg = 2\,mg$.

Example 2.5

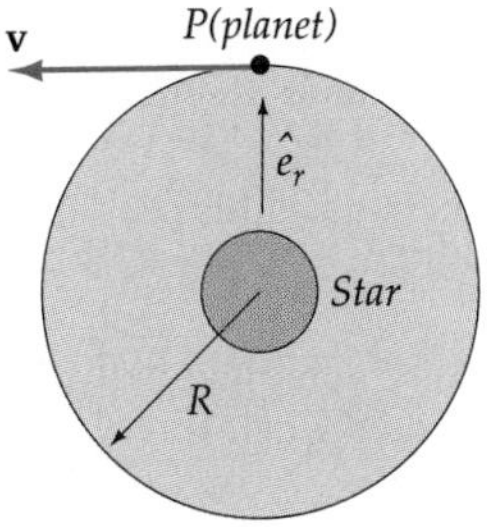

Figure E2.5

A planet P of mass m moves in a circular orbit around a star of mass M, far away from any other gravitational or other forces (see Figure E2.5). If the planet completes one orbit in τ units of time, find the orbit radius, using the fact that in a circular orbit the speed is constant.

Solution

Writing the component of the equation of motion for the planet P in the radial direction,

$$\Sigma F_r = m\mathbf{a}_{Cr}$$

The only external force on P is gravity, in the $-\hat{\mathbf{e}}_r$ direction (towards the star). Letting G be the universal gravitational constant, we substitute and obtain:

$$\frac{-GMm}{R^2} = m[\ddot{r} - r\dot{\theta}^2]$$

where from Equation (1.37), the radial acceleration component is $\ddot{r} - r\dot{\theta}^2$. Since $r = R =$ constant, we have, with $\dot{\theta} = \omega =$ orbital rate,

$$\frac{GM}{R^2} = R\omega^2$$

But $\omega = \dfrac{2\pi}{\tau}$, so that

$$R^3 = \frac{GM}{\left(\dfrac{2\pi}{\tau}\right)^2} = \frac{GM\tau^2}{4\pi^2}$$

or

$$R = \left(\frac{GM\tau^2}{4\pi^2}\right)^{1/3}$$

* This would require solving the differential equation (4) for $\theta(t)$ and substituting into Equation (5).

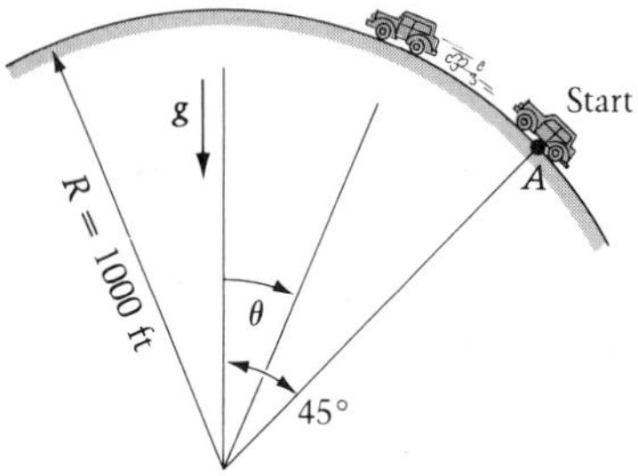

Figure E2.6a

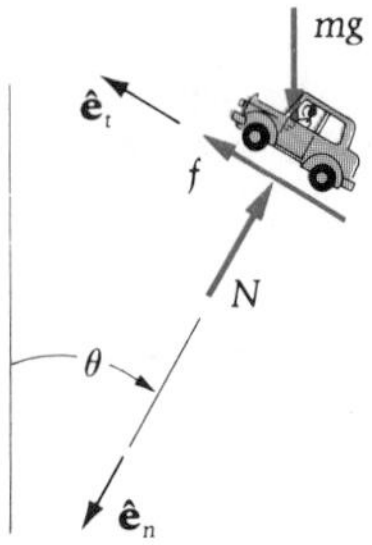

Figure E2.6b

EXAMPLE 2.6

A car accelerates from rest, increasing its speed at the constant rate of $K = 6$ ft / sec^2. (See Figure E2.6a.) It travels on a circular path starting at point A. Find the time and the position of the car when it first leaves the surface due to excessive speed.

Solution

Before the car (treated as a particle) leaves the surface, the free-body diagram is as shown in Figure E2.6b. We shall work this problem in general (without substituting numbers until the end). The purpose is to illustrate the concept of nondimensional parameters. The equation of motion in the tangential ($\hat{\mathbf{e}}_t$) direction is

$$\Sigma F_t = m\ddot{s}$$

$$f - mg\sin\theta = mK \tag{1}$$

Equation (1) shows that the friction exerted on the tires by the road is the external force which moves the car up the path. Note that after it passes the top of the circular hill, we have $\sin\theta < 0$ and then the gravity force *adds* to the friction in accelerating the car on the way down.

The following equation of motion is the one that will help us in this problem; it equates $\Sigma\mathbf{F}$ and $m\mathbf{a}_C$ in the normal direction.

Question 2.2 Are the components of $\Sigma\mathbf{F}$ and $m\mathbf{a}_C$ equal in *all* directions or just in coordinate directions?

$$\Sigma F_n = m\frac{\dot{s}^2}{\rho}$$

$$mg\cos\theta - N = \frac{m(Kt)^2}{R} \tag{2}$$

where $\dot{s} = \int \ddot{s}\,dt = Kt + C_1 = Kt$, since $\dot{s} = 0$ when $t = 0$.

We note that the car will lose contact with the road when N becomes zero. (The ground cannot pull down on the car for further increases of t, which would require $N < 0$!) Therefore, at the point of leaving the ground,

$$\not{m}g\cos\theta = \frac{\not{m}K^2t^2}{R} \tag{3}$$

Question 2.3 What is the *meaning* of the fact that m cancels in Equation (3)?

Now θ is related to s according to

$$s = R\left(\frac{\pi}{4} - \theta\right) \tag{4}$$

Answer 2.2 The components of the two sides of a vector equation are equal in *any* direction.

Answer 2.3 It means the answer does not depend on the mass of the car.

And from $\dot{s} = Kt$, we get another expression for s:

$$s = \frac{Kt^2}{2} + C_2 = \frac{Kt^2}{2} \qquad \text{(since } s = 0 \text{ when } t = 0) \tag{5}$$

Hence from equations (4) and (5) we get

$$\frac{Kt^2}{2} = R\left(\frac{\pi}{4} - \theta\right) \Rightarrow \theta = \frac{\pi}{4} - \frac{Kt^2}{2R} \tag{6}$$

Substituting for θ from (6) into (3), we have

$$g \cos\left(\frac{\pi}{4} - \frac{Kt^2}{2R}\right) = \frac{K^2t^2}{R}$$

or

$$\cos\left(\frac{\pi}{4} - \frac{Kt^2}{2R}\right) = \frac{Kt^2}{2R} \cdot \frac{2K}{g} \tag{7}$$

Equation (7) allows us to solve for the dimensionless parameter $q = (Kt^2/2R)$, once we have selected a value of the car's dimensionless acceleration K/g. In this problem, for example,

$$\cos\left(\frac{\pi}{4} - q\right) = 2q\left(\frac{6}{32.2}\right) = 0.373q \tag{8}$$

The following table shows how (with a calculator)* we can quickly arrive at the value of q that solves Equation (8):

q	$\cos\left(\frac{\pi}{4} - q\right)$	$0.373q$
0.1	0.7742	0.0373
0.5	0.9595	0.1865
07.854 (at the top)	1	0.2930
1.0	0.9771	0.3730
1.3	0.8705	0.4849
1.6	0.6862	0.5968
1.7	0.6101	0.6341
1.69	0.6180	0.6304
1.68	0.6258	0.6266

Thus at $q = Kt^2/2R \approx 1.68$, the car leaves the circular track due to excessive speed. Therefore for $K = 6$ ft/sec^2 and $R = 1000$ ft,

$$t = \sqrt{\frac{1.68 \times 2 \times 1000}{6}} = 23.7 \text{ sec}$$

---

* See Appendix B for a numerical solution to this problem using the Newton-Raphson method.

The angle at loss of contact is given by Equation (6):

$$\theta = \frac{\pi}{4} - \frac{Kt^2}{2R} = \frac{\pi}{4} - q$$

$$= \frac{\pi}{4} - 1.68$$

$$= -0.895 \text{ rad}$$

$$= -51.3^\circ$$

The speed at the point of loss of contact is $Kt = 6(23.7) = 142$ ft/sec $= 97.0$ mph. Note that for $K/g = 6/32.2$, the angle -51.3° is the angle of leaving for many combinations of t and R (so long as $Kt^2/2R = 1.68$).

EXAMPLE 2.7

In the system shown in Figure E2.7a, each of the blocks weighs 10 lb and the pulleys are very much lighter. Find the accelerations of the blocks, assuming the belt (or rope) to be inextensible and of negligible mass.

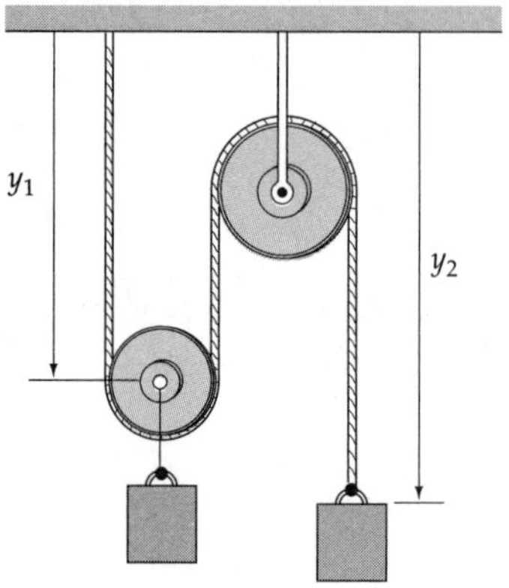

Figure E2.7a

Solution

If there is negligible friction in the bearings of the pulley, and the pulley is much lighter than other elements of the system, then the belt tension won't change from one side of the pulley to the other. So, referring to Figure E2.7b,

$$T_1 = T_2$$

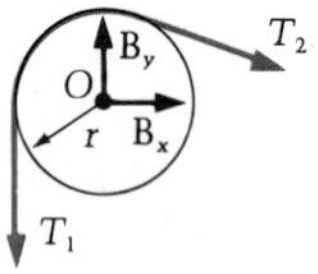

Figure E2.7b

Figure E2.7c

For the block in Figure E2.7c:

$$10 - T = \frac{10}{32.2}\ddot{y}_2 \tag{1}$$

And for the block and pulley in Figure E2.7d:

$$10 - 2T = \frac{10}{32.2}\ddot{y}_1 \tag{2}$$

We also have a kinematic constraint relationship between y_1 and y_2 because the belt is inextensible. For this problem it is that $\dot{y}_2 = -2\dot{y}_1$ (see Example 1.8) and consequently

$$\ddot{y}_2 = -2\ddot{y}_1 \tag{3}$$

Solving Equations (1), (2), and (3) simultaneously,

$$T = 6 \text{ lb}$$

$$\ddot{y}_1 = -6.44 \text{ ft/sec}^2$$

$$\ddot{y}_2 = 12.9 \text{ ft/sec}^2$$

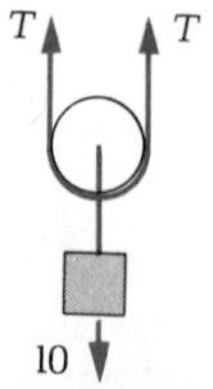

Figure E2.7d

so the left block accelerates upward at 6.44 ft/sec^2 and the right block accelerates downward at 12.9 ft/sec^2.

EXAMPLE 2.8

Find the accelerations of the blocks shown in Figure E2.8a when released from rest. Then repeat the problem with the friction coefficients reversed.

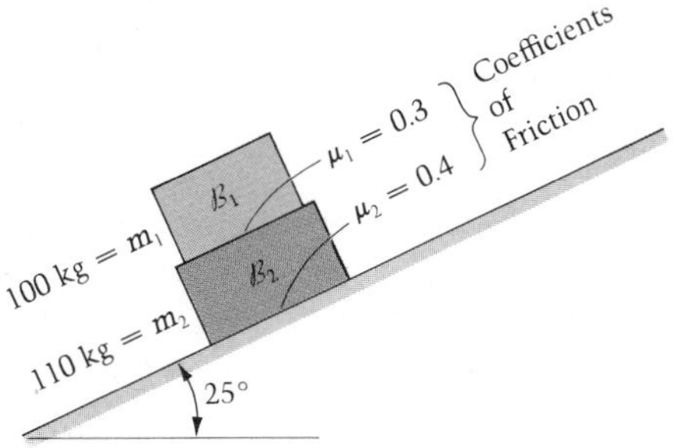

Figure E2.8a

Solution

We know from statics that *if* the two blocks move as a unit, their motion will occur when

$$\tan 25^\circ > \mu_2$$

that is, when

$$0.466 > 0.4$$

which is the case here. But before our solution is complete we must determine whether either block moves without the other. We consider the free-body diagrams of each translating block (see Figure E2.8b) and write the equations of motion:

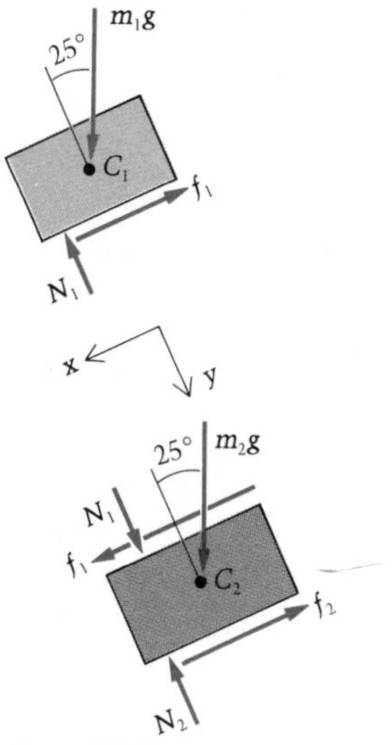

Figure E2.8b

$$\Sigma F_x = m_1\ddot{x}_{C_1} \Rightarrow 100(9.81)\sin 25^\circ - f_1 = 100\ddot{x}_{C_1} \tag{1}$$

$$\Sigma F_y = m_1\ddot{y}_{C_1} \Rightarrow 100(9.81)\cos 25^\circ - N_1 = 100)\overset{0}{\cancel{\ddot{y}_{C_1}}} \tag{2}$$

$$\Sigma F_x = m_2\ddot{x}_{C_2} \Rightarrow 110(9.81)\sin 25^\circ + f_1 - f_2 = 110\ddot{x}_{C_2} \tag{3}$$

$$\Sigma F_y = m_2\ddot{y}_{C_2} \Rightarrow 110(9.81)\cos 25^\circ + N_1 - N_2 = 110)\overset{0}{\cancel{\ddot{y}_{C_2}}} \tag{4}$$

We mention that the sum of Equations (1) and (3) gives the "x equation" of the overall system; the sum of (2) and (4) yields the "y equation" (C is the mass center of the combined blocks):

$$\begin{aligned} \Sigma F_x &= (100 + 110)9.81 \sin 25^\circ - f_2 \\ &= 100\ddot{x}_{C_1} + 110\ddot{x}_{C_2} \\ &= (100 + 110)\ddot{x}_C \end{aligned} \tag{5}$$

$$\begin{aligned} \Sigma F_y &= (100 + 110)9.81 \cos 25^\circ - N_2 \\ &= (100 + 110)\overset{0}{\cancel{\ddot{y}_C}} = 0 \end{aligned} \tag{6}$$

Note that f_1 and N_1 disappear in (5) and (6), as they become internal forces on the combined system.

Equation (6) tells us that $N_2 = 1870$ N, regardless of which motion takes place. The equation for the x motion (Equation 5) shows again that if $f_{2_{max}} < mg(\sin 25°)$, then one or both of the blocks must slide:

$$f_{2_{max}} = \mu_2 N_2 = 0.4(1870) = 748 < 210(9.81)(0.423) = 871 \text{ N}$$

and so $\ddot{x}_C$ cannot be zero. Assuming first that the blocks *both* move, then f_2 is at its maximum:

$$f_2 = f_{2_{max}} = 748 \text{ N}$$

If they move *together* as one body, then Equation (5) gives us

$$\begin{aligned} \ddot{x}_C (= \ddot{x}_{C_1} = \ddot{x}_{C_2}) &= (mg \sin 25° - f_{2_{max}})/m \\ &= (871 - 748)/210 \\ &= 0.586 \text{ m/s}^2 \end{aligned}$$

Substituting this acceleration into Equation (1), we can check to see if body $\mathcal{B}_1$ additionally slides relative to $\mathcal{B}_2$:

$$f_1 = -100(0.586) + 415 = 356 \text{ N}$$

But the maximum value that f_1 can have is given by

$$f_{1_{max}} = \mu_1 N_1 = 0.3(889) = 267 \text{ N}$$

Hence block $\mathcal{B}_1$ slides on $\mathcal{B}_2$ and the blocks do not move together; our assumption was incorrect. We then substitute $f_1 = \mu_1 N_1$ into Equation (1) and proceed:

$$\ddot{x}_{C_1} = (415 - 267) \div 100 = 1.48 \text{ m/s}^2$$

This is then the acceleration of the top block. Substituting f_1 into Equation (3) gives

$$\begin{aligned} 456 + 267 - f_2 &= 110\ddot{x}_{C_2} \\ 723 - f_2 &= 110\ddot{x}_{C_2} \end{aligned}$$

For no motion of the bottom block, $f_{2_{max}}$ clearly needs to be at least 723 N. Since it is in fact 748 N, the bottom block does *not* move for this combination of parameters, and $\ddot{x}_{C_2} = 0$.

If the friction coefficients are now swapped, nothing changes until we begin to analyze the six equations. We have

$$f_{2_{max}} = \mu_2 N_2 = 0.3(1870) = 561 < mg \sin 25° = 871 \text{ N} \qquad \text{(as before)}$$

Again, then, $\ddot{x}_C$ cannot be zero. Assuming again that the blocks both move, f_2 is its maximum and Equation (5) gives

$$\ddot{x}_C = (871 - 561) \div 210 = 1.48 \text{ m/s}^2$$

Substituting this acceleration into Equation (1), we get

$$f_1 = 415 - 100(1.48) = 267 < \mu_1 N_1 = 0.4(889) = 356 \text{ N}$$

This time we have more friction than we need in order to prevent $\mathcal{B}_1$ from slipping on $\mathcal{B}_2$. Thus both $\ddot{x}_{C_1}$ and $\ddot{x}_{C_2}$ are 1.48 m/s².

Problems ▶ Section 2.3

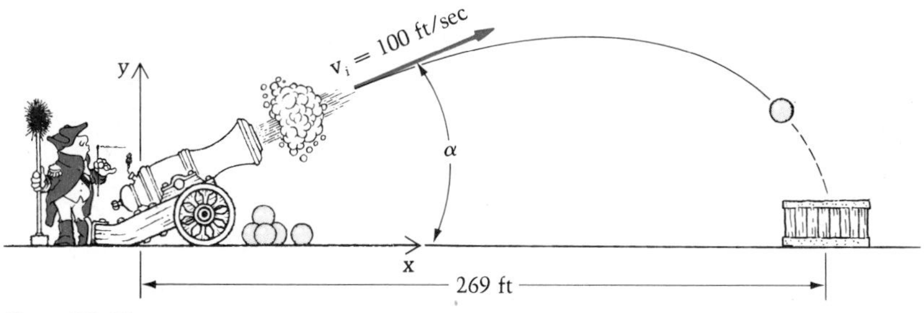

Figure P2.12

2.11 In Problem 1.114 what is the acceleration of the rock just *after* release?

2.12 A cannonball is fired as shown in Figure P2.12. Neglecting air resistance, find the angle α that will result in the cannonball landing in the box.

2.13 A baseball slugger connects with a pitch 4 ft above the ground. The ball heads toward the 10-ft-high center-field fence, 455 ft away. The ball leaves the bat with a velocity of 125 ft/sec and a slope of 3 vertical to 4 horizontal. Neglecting air resistance, determine whether the ball hits the fence (if it does, how high above the ground?) or whether it is a home run (if it is, by how much does it clear the fence?).

2.14 From a high vantage point in Yankee Stadium, a baseball fan observes a high-flying foul ball. Traveling vertically upward, the ball passes the level of the observer 1.5 sec after leaving the bat, and it passes this level again on its away down 4 sec after leaving the bat. Disregarding air friction, find the maximum height reached by the baseball and determine the ball's initial velocity as it leaves the bat (which is 3 ft above the ground at impact).

2.15 A soccer ball (Figure P2.15) is kicked toward the goal from 60 ft. It strikes the top of the goal at the highest point of its trajectory. Find the velocity and angle θ at which the ball was kicked, and determine the time of travel.

2.16 The motorcycle in Figure P2.16 is to be driven by a stunt man. Find the minimum takeoff velocity at A for which the motorcycle can clear the gap, and determine the corresponding angle θ for which the landing will be tangent to the road at B and hence smooth.

Figure P2.15

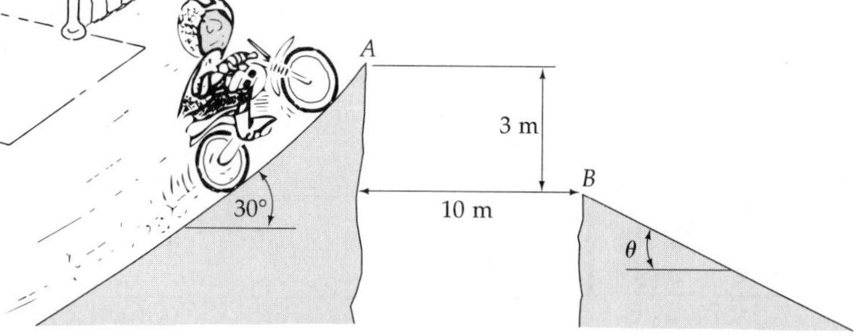

Figure P2.16

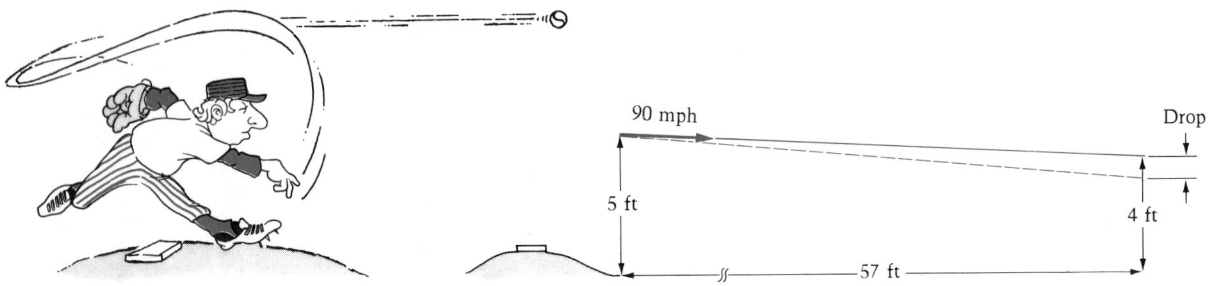

Figure P2.17

2.17 A baseball pitcher releases a 90-mph fastball 5 ft off the ground (Figure P2.17). If in the absence of gravity the ball would arrive at home plate 4 ft off the ground, find the drop in the actual path caused by gravity. Neglect air resistance.

2.18 In the preceding problem, find the radius of curvature of the path of the baseball's center at the instant it arrives at the plate.

2.19 In the preceding problem, the batter hits a pop-up that leaves the bat at a 45° angle with the ground. The shortstop loses the ball in the sun and it lands on second base, $90\sqrt{2}$ ft from home plate. What was the velocity of the baseball when it left the bat?

2.20 The pilot of an airplane flying at 300 km/hr wishes to release a package of mail at the right position so that it hits spot A. (See Figure P2.20.) What angle θ should his line of sight to the target make at the instant of release?

2.21 A darts player releases a dart at the position indicated in Figure P.2.21 with the initial velocity vector making a 10° angle with the horizontal. What must the dart's initial speed be if it scores a bull's-eye?

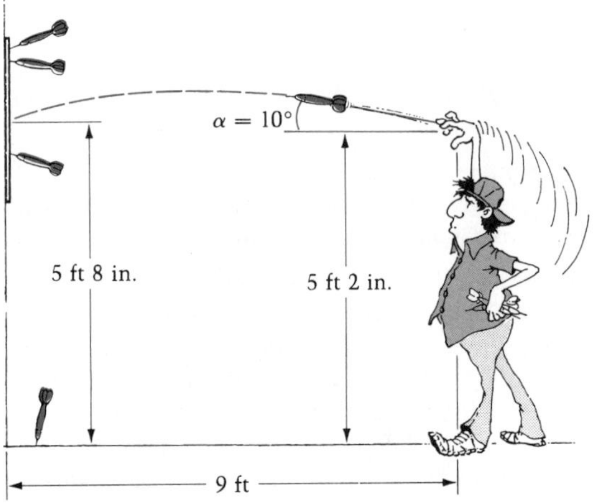

Figure P2.21

2.22 In the preceding problem, suppose the initial speed of the dart is 20 ft/sec. What must the angle α be if a bull's-eye is scored?

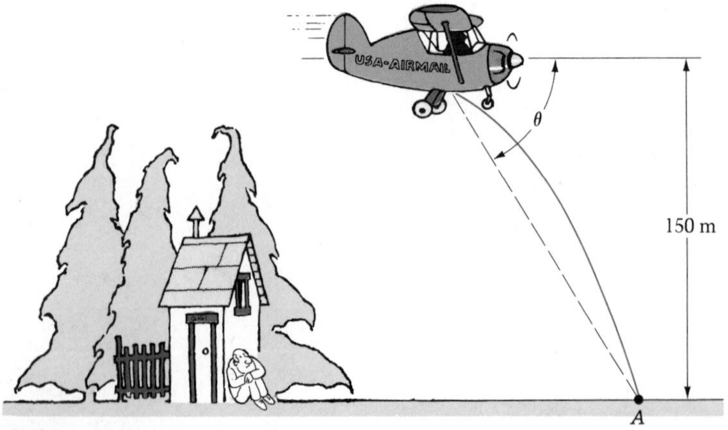

Figure P2.20

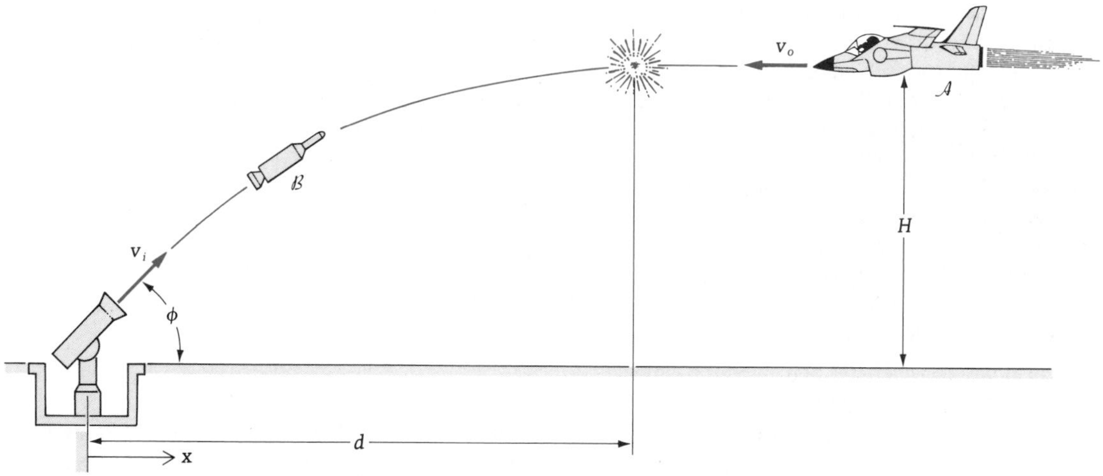

Figure P2.23

2.23 Find the angle ϕ, firing velocity v_i, and time t_f of intercept so that the ballistic missile shown in Figure P2.23 will intercept the bomber when $x = d$. The bomber, at $x = D$ when the missile is launched, travels horizontally at constant speed v_0 and altitude H. What has been neglected in your solution?

2.24 The garden hose shown in Figure P2.24 expels water at 13 m/s from a height of 1 m. Determine the maximum height H and horizontal distance D reached by the water.

* **2.25** In the preceding problem, use calculus to find the angle θ that will give maximum range D to the water.

* **2.26** Find the range R for a projectile fired onto the inclined plane shown in Figure P2.26. Determine the maximum value of R for a given muzzle velocity u. (Angle α = constant.)

2.27 If a baseball player can throw a ball 90 m on the fly on earth, how far can he throw it on the moon where the gravitational acceleration is about one-sixth that on earth? Neglect the height of the player and the air resistance on earth.

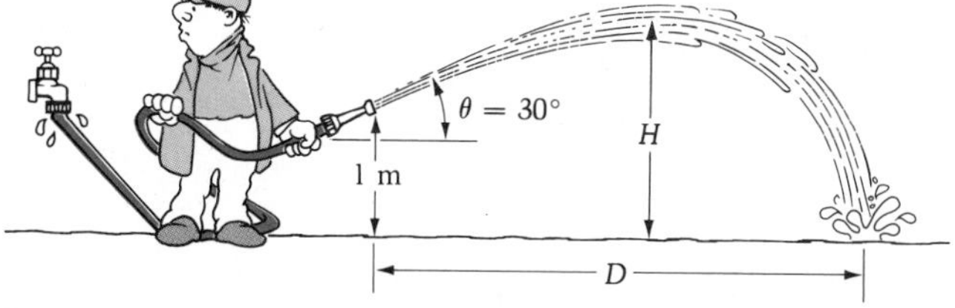

Figure P2.24

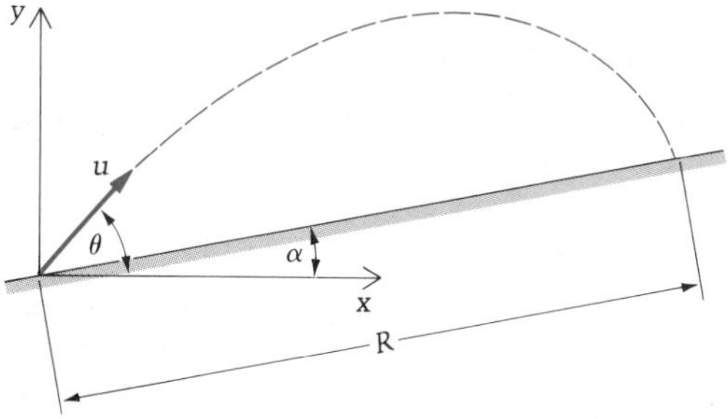

Figure P2.26

2.28 A child drops a rock into a well and hears it splash into the water at the bottom exactly 2 sec later. (See Figure P2.28.) If she is at a location where the speed of sound is $v_s = 1100$ ft/sec, determine the depth of the well with and without considering v_s. Compare the two results.

Figure P2.28

2.29 At liftoff the space shuttle is powered upward by two solid rocket boosters of 12.9×10^6 N each and by the three Orbiter main liquid-rocket engines with thrusts of 1.67×10^6 N each. At liftoff, the total weight of the shuttle (orbiter, tanks, payload, boosters) is about 19.8×10^6 N. Determine the acceleration experienced by the crew members at liftoff. (This differs from the initial acceleration on earlier manned flights; demonstrate this by comparing with the Apollo moon rocket, which weighed 6.26×10^6 lb at liftoff and was powered by five engines each with a thrust of 1.5×10^6 lb.) Neglect the change in mass between ignition and liftoff.

2.30 What is the apparent weight, as perceived through pressure on the feet, of a 200-lb passenger in an elevator accelerating at the rate of 10 ft / sec^2 upward (a) or downward (b)?

2.31 When a man stands on a scale at one of the poles of the earth, the scale indicates weight W. Assuming the earth to be spherical (4000-mile radius) and assuming the earth to be an inertial frame, what will the scale read when the man stands on it at the equator?

2.32 Assuming the earth's orbit around the sun to be circular and supposing that a frame containing the earth's center and poles and the center of the sun is inertial, repeat Problem 2.31. Neglect the earth's tilt.

2.33 In an emergency the driver of an automobile applies the brakes and locks all four wheels. Find the time and distance required to bring the car to rest in terms of the coefficient of sliding friction μ, the initial speed v, and the gravitational acceleration g.

2.34 A box is placed in the rear of a pickup truck. Find the maximum acceleration of the truck for which the block does not slide on the truck bed. The coefficient of friction between the box and truck bed is μ.

2.35 The truck in Figure P2.35 is traveling at 45 mph. Find the minimum stopping distance such that the 250-lb crate will not slide. Assume the crate cannot tip over.

Figure P2.35

2.36 The 200-lb block is at rest on the floor ($\mu = 0.2$) before the 50-lb force is applied as shown in Figure P2.36. What is the acceleration of the block immediately after application of the force? Assume the block is wide enough that it cannot tip over.

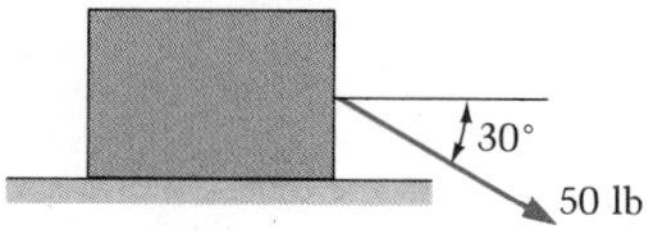

Figure P2.36

2.37 Repeat Problem 2.36 with $\mu = 0.1$.

2.38 Repeat Problem 2.36 for the case in which the 50-lb force is applied as shown in Figure P2.38.

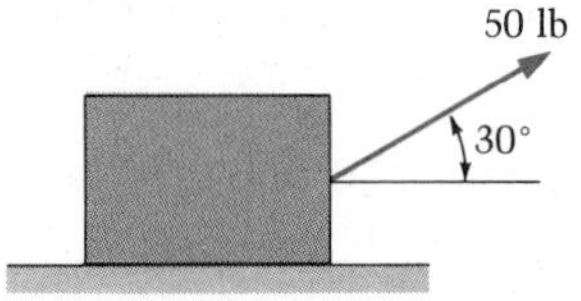

Figure P2.38

2.39 The two blocks in Figure P2.39 are at rest before the 100-Newton force is applied. If friction between $\mathcal{B}$ and the floor is negligible and if $\mu = 0.4$ between $\mathcal{A}$ and $\mathcal{B}$, find the magnitude and direction of the subsequent friction force exerted *on* $\mathcal{A}$ *by* $\mathcal{B}$.

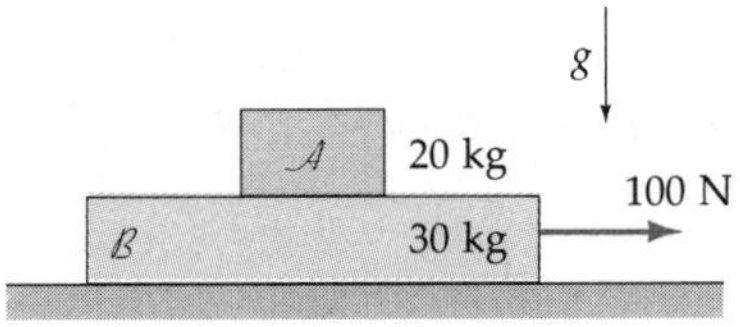

Figure P2.39

2.40 Find the largest force P for which $\mathcal{A}$ in Figure P2.40 will not slide on $\mathcal{B}$.

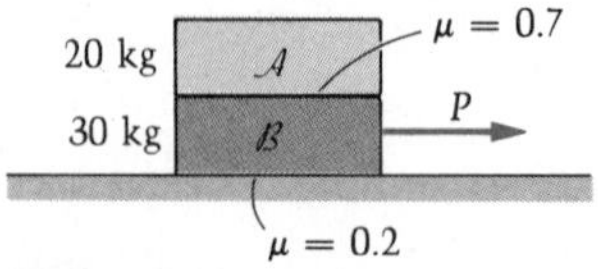

Figure P2.40

2.41 Work the preceding problem if P is applied to $\mathcal{A}$ instead of $\mathcal{B}$.

2.42 The blocks in Figure P2.42 are in contact as they slide down the inclined plane. The masses of the blocks are $m_{\mathcal{B}} = 25$ kg and $m_{\mathcal{A}} = 20$ kg, and the friction coefficients between the blocks and the plane are 0.5 for $\mathcal{A}$ and 0.1 for $\mathcal{B}$. Determine the force between the blocks and find their common acceleration.

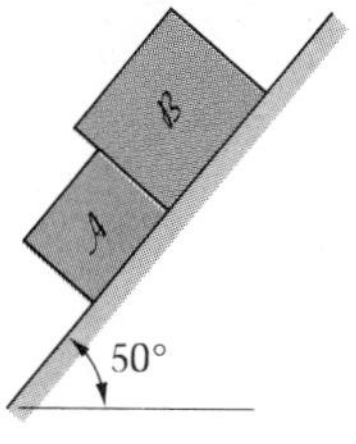

Figure P2.42

2.43 In the preceding problem, let μ be the coefficient of friction between $\mathcal{A}$ and the plane. Using the two motion equations of the blocks, find the range of values of μ for which the blocks will separate when released from rest.

2.44 If all surfaces are smooth for the setup of blocks and planes in Figure P2.44, find the force P that will give block $\mathcal{B}$ an acceleration of 4 ft / sec² up the incline.

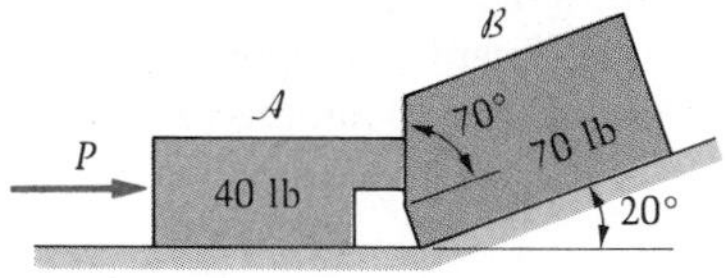

Figure P2.44

2.45 Work the preceding problem if the planes are still smooth but the friction coefficient between $\mathcal{A}$ and $\mathcal{B}$ is $\mu_s \approx \mu_k = 0.3$.

2.46 Work the preceding problem if the coefficient of friction is 0.3 for *all* contacting surfaces.

* **2.47** Generalizing Example 2.8, let the blocks, friction coefficients, and angle of the plane be as shown in Figure P2.47. Show that:

a. If $\tan\varphi > \mu_2$, motion will occur, and if so:

b. If $\mu_2 \le \mu_1$, the blocks move together

c. If $\mu_2 > \mu_1$, then $\mathcal{B}_1$ slides on $\mathcal{B}_2$. In this case, the lower block does not move if

$$\tan\varphi \le \mu_2 + (\mu_2 - \mu_1)\frac{m_1}{m_2}$$

d. If $\tan\varphi \le \mu_2$, then the lower block will not move. In this case, the upper block slides on it if and only if $\tan\varphi > \mu_1$.

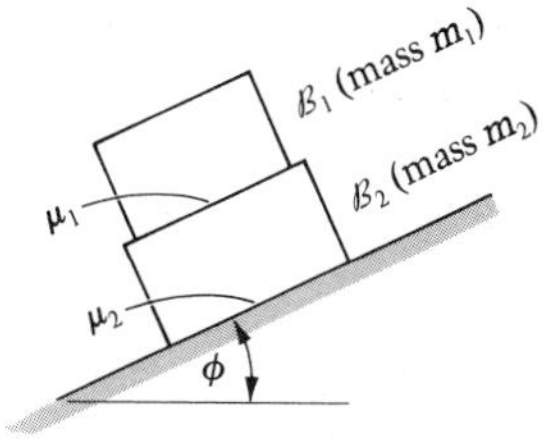

Figure P2.47

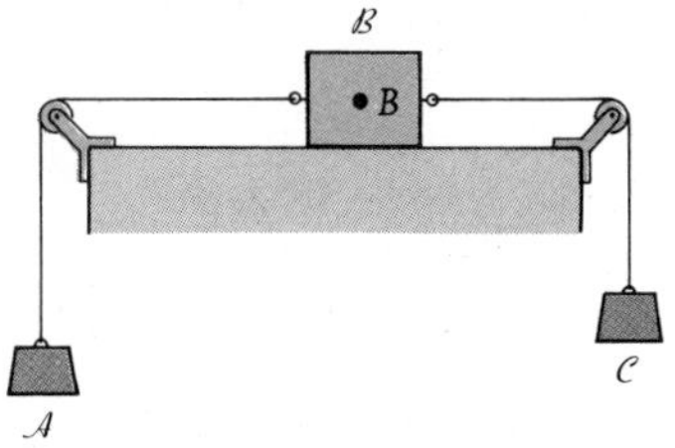

Figure P2.48

2.48 In Figure P2.48 the masses of $\mathcal{A}$, $\mathcal{B}$, and $\mathcal{C}$ are 10, 60, and 50 kg, respectively. The coefficient of friction between $\mathcal{B}$ and the plane is $\mu = 0.35$, and the pulleys have negligible mass and friction. Find the tensions in each cord, and the acceleration of B, upon release from rest.

2.49 If the system in Figure P2.49 is released from rest, how long does it take the 5-lb block to drop 2 ft? Neglect friction in the light pulley and assume the cord connecting the blocks to be inextensible.

2.50 The coefficient of friction $\mu = 0.1$ is the same between $\mathcal{A}$ and $\mathcal{B}$ as it is between $\mathcal{B}$ and the plane. (See Figure P2.50.) Find the tension in the cord at the instant the system is released from rest. Neglect friction in the light pulley.

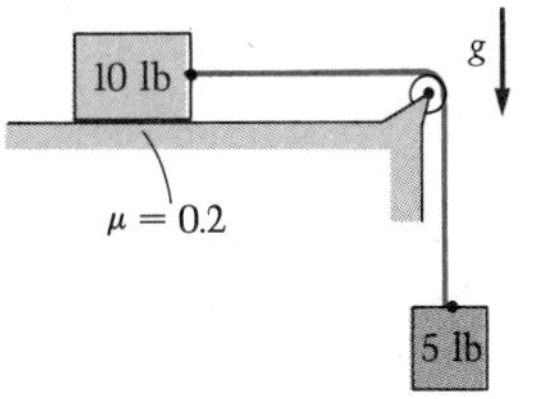

Figure P2.49

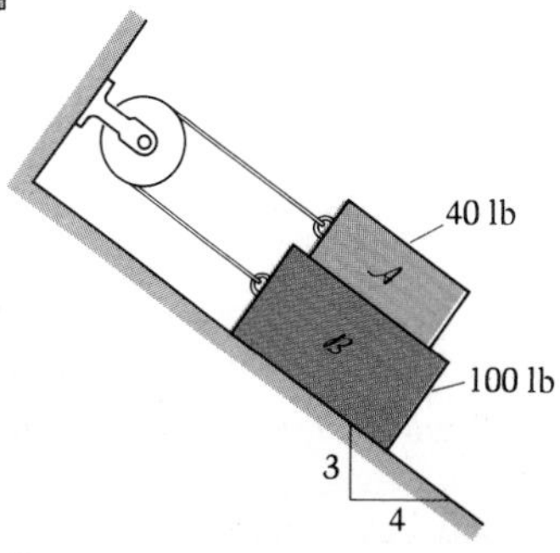

Figure P2.50

2.51 The system in Figure P2.51 is released from rest.

a. How far does block A move in 2 sec?

b. How would the solution be changed if the coefficient of friction between the floor and A were $\mu = 0.2$?

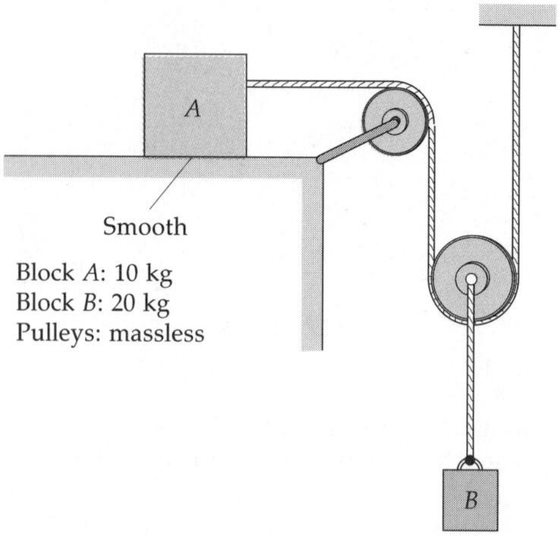

Figure P2.51

2.52 A child notices that sometimes the ball m does not slide down the inclined surface of toy $\mathcal{T}$ when she pushes it along the floor. (See Figure P2.52.) What is the minimum acceleration a_{min} of $\mathcal{T}$ to prevent this motion? Assume all surfaces are smooth.

Figure P2.52

2.53 In the preceding problem, suppose the acceleration of $\mathcal{T}$ is $2a_{min}$. What is the normal force between the vertical surface of $\mathcal{T}$ and the ball? The ball's weight is 0.06 lb.

2.54 Let the mass of $\mathcal{B}$ in Problem 1.57 be 20 kg. What then must be the mass of $\mathcal{A}$ to produce the prescribed motion? Neglect the masses of the pulleys.

* **2.55** Find the tension in the cord in Problem 1.61 at the onset of the motion if the mass of $\mathcal{A}$ is 10 kg.

2.56 For the cam-follower system of Problem 1.65 find the force that must be applied to the cam to produce the motion. Let the masses of cam and follower be m_1 and m_2 and neglect friction.

2.57 In Figure P2.57 the masses of blocks $\mathcal{A}$, $\mathcal{B}$, and $\mathcal{C}$ are 50, 20, and 30 kg, respectively. Find the accelerations of each if the table is removed. Which block will hit the floor first? How long will it take?

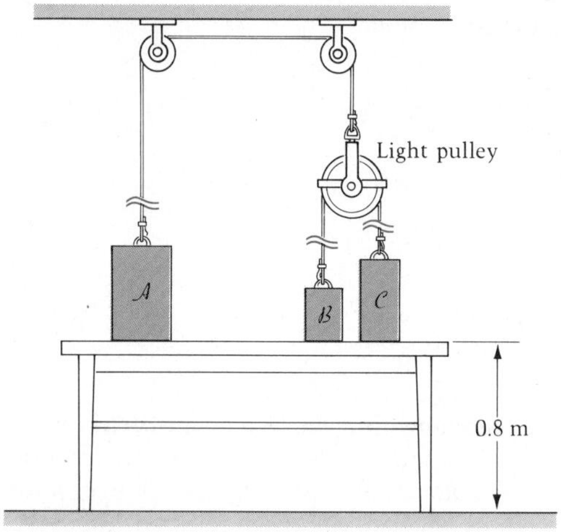

Figure P2.57

2.58 Body $\mathcal{A}$ in Figure P2.58 weighs 223 N and body $\mathcal{B}$ weighs 133 N. Neglect the weight of the rigid member connecting $\mathcal{A}$ and $\mathcal{B}$. The coefficient of friction is 0.3 between all surfaces. Determine the accelerations of $\mathcal{A}$ and $\mathcal{B}$ just after the cord is cut.

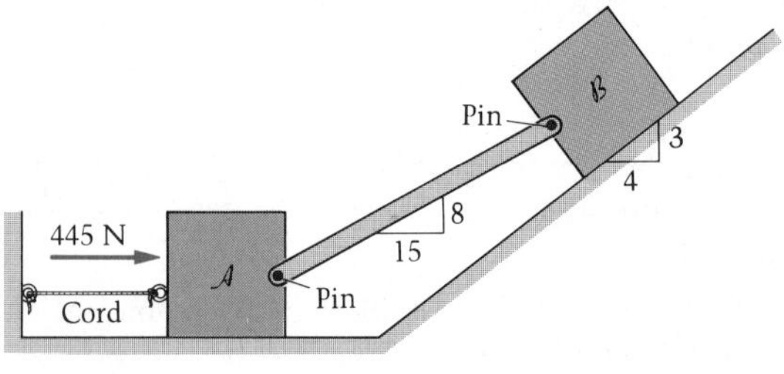

Figure P2.58

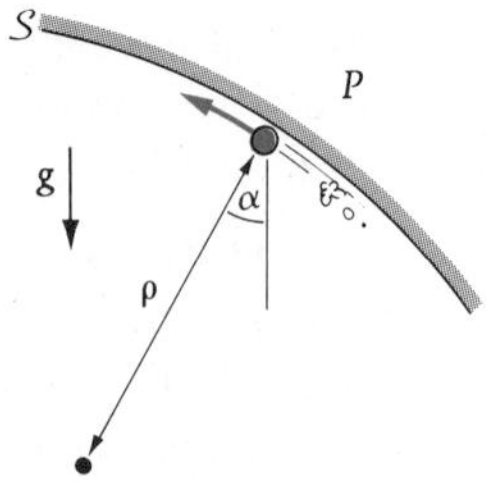

Figure P2.59

2.59 A particle P moves along a curved surface $\mathcal{S}$ as shown in Figure P2.59. Show that P will remain in contact with $\mathcal{S}$ provided that, at all times, $v \geq \sqrt{\rho g \cos \alpha}$.

2.60 Find the condition for retention of contact if P moves along the *outside* of a surface defined by the same curve as in the preceding problem. (See Figure P2.60.)

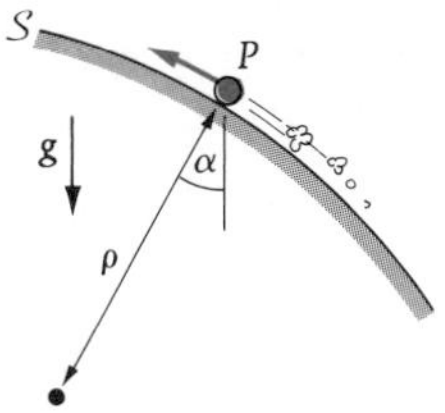

Figure P2.60

2.61 A ball of mass m on a string is swung at constant speed v_0 in a horizontal circle of radius R by a child. (See Figure P2.61.)

- **a.** What holds up the ball?
- **b.** What is the tension in the string?
- **c.** If the child increases the speed of the ball, what provides the force in the forward direction needed to produce the $\ddot{s}$? Explain.

Figure P2.61

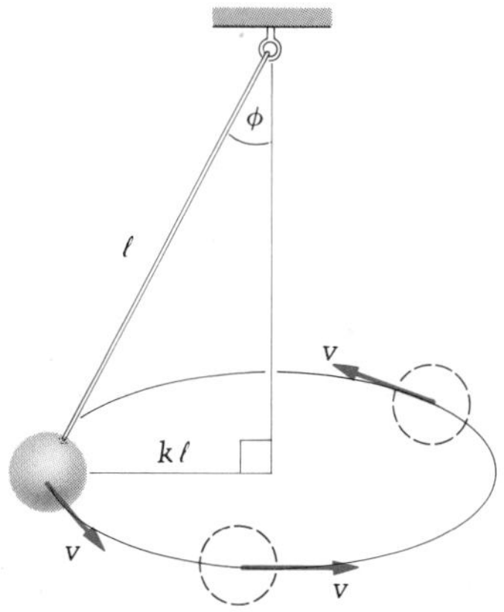

Figure P2.62

2.62 There is a speed, called the *conical speed*, at which a ball on a string, in the absence of all friction, moves on a specific horizontal circle (with the string sweeping out a conical surface) with no radial or vertical component of velocity (Figure P2.62). If ℓ is the length of the string and $k\ell$ is the radius of the circle on which the ball moves, find the conical speed in terms of k, ℓ, and the acceleration of gravity.

2.63 For an object at rest on the earth's surface, we can write $mg = (GMm)/R_e^2$, so that the unwieldy constant GM may be replaced by gR_e^2, which for the earth is approximately $32.2[3960(5280)]^2$ ft^3/sec^2. Use this, plus the result of Example 2.5, to solve for the distance above the earth of a satellite in a circular, 90-minute orbit. (Let the satellite replace the planet, and the earth replace the star, in the example.)

2.64 Communications satellites are placed in *geosynchronous orbit*, an orbit in which the satellites are always located in the same position in the sky (Figure P2.64).

- **a.** Give an argument why this orbit must lie in the equatorial plane. Why must it be circular?
- **b.** If the satellites are to remain in orbit without expending energy, find the important ratio of the orbit radius r_s to the earth's radius r_e. *Hint*: Use Newton's law of universal gravitation

$$F = \frac{Gm_s m_e}{r_s^2}$$

together with the law of motion in the radial direction, and note that if the satellite were sitting on the earth's surface, the force would be

$$F = m_s g = \frac{Gm_s m_e}{r_e^2}$$

so that the product Gm_e may be rewritten as gr_e^2, as in Problem 2.63. Use $r_e = 3960$ mi.

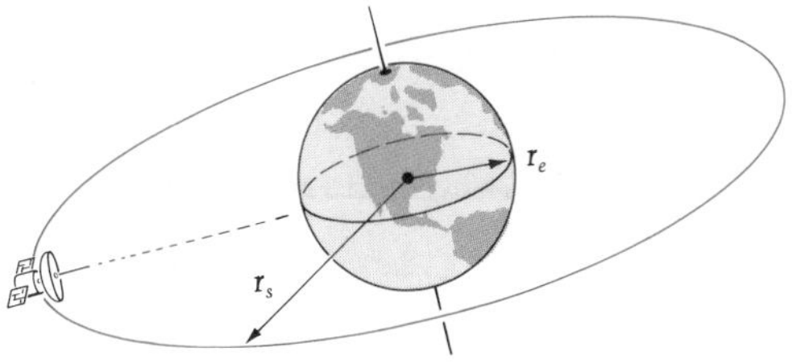

Figure P2.64

2.65 Using the result of the preceding problem, show that a minimum of three satellites in geosynchronous orbit are required for continuous communications coverage over the whole earth except for small regions near the poles.

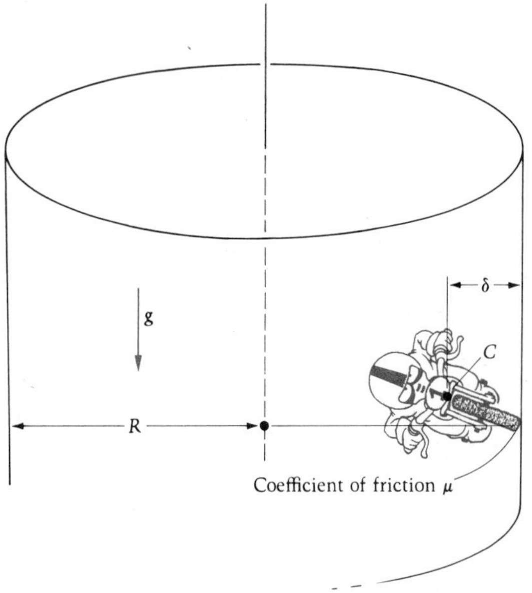

Figure P2.66

2.66 In terms of the parameters δ, R, μ, and g defined in Figure P2.66, find the minimum speed for which the motorcycle will not slip down the inside wall of the cylinder.

2.67 In the "spindle top" ride in an amusement park, people stand against a cylindrical wall and the cylinder is then spun up to a certain angular velocity ω_0. (See Figure P2.67.) The floor is then lowered, but the people remain against the wall at the same level. Use the equation $\Sigma F_n = ma_{Cn} = mv^2/R$ to explain the phenomenon. Noting that each person is "in equilibrium vertically," solve for the minimum ω_0 to prevent people from slipping if $R = 2$ m and the expected friction coefficient between the rough wall and the clothing is $\mu = 0.5$.

Figure P2.67

2.68 In preparation for Problem 2.69, for the ellipse shown in Figure P2.68, the equation is

$$\frac{x^2}{a^2} + \frac{y^2}{b^2} = 1$$

Show that the radius of curvature ρ of the ellipse, as a function of x, is

$$\rho = \frac{[a^2(a^2 - x^2) + b^2x^2]^{3/2}}{a^4 b}$$

Hint: Recall from calculus that if $y = y(x)$, then

$$\rho = \left| \frac{(1 + y'^2)^{3/2}}{y''} \right|$$

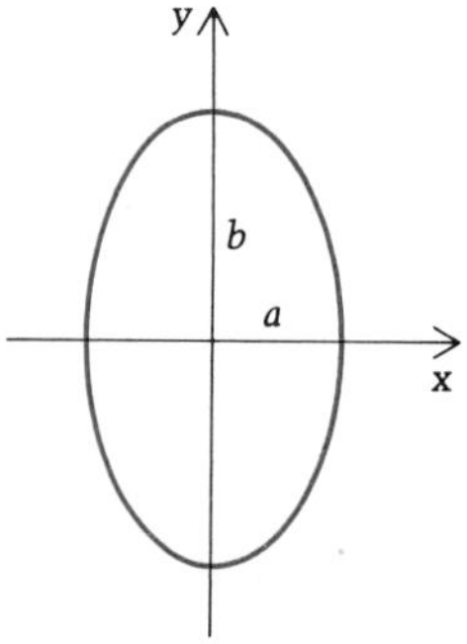

Figure P2.68

2.69 In a certain amusement park, the tallest loop in a somersaulting ride (Figure P2.69) is 100 ft high and shaped approximately like an ellipse with a width of 95 ft. The ride advertises "five times the earth's pull at over 50 mph." Use the result of the preceding exercise to compute the radius of curvature at the bottom of the loop. Assuming that the normal force resultant is 5 mg, determine whether or not the maximum speed is over 50 mph. Treat the cars as a single particle.

2.70 If bar $\mathcal{B}$ shown in Figure P2.70 were raised *slowly*, block $\mathcal{A}$ would start to slide at the angle $\theta = \tan^{-1}(\mu)$, which was seen in statics to be one way of determining the friction coefficient μ. Suppose now that the bar is suddenly rotated, starting from the position $\theta = 0$, at constant angular velocity ω_0 ↺. For $\mu = 0.5$ and $r\omega_0^2 = 0.1g$, compute the angle θ at which $\mathcal{A}$ slips downward on $\mathcal{B}_1$ and compare the result with $\tan^{-1}\mu = \tan^{-1}(0.5)$.

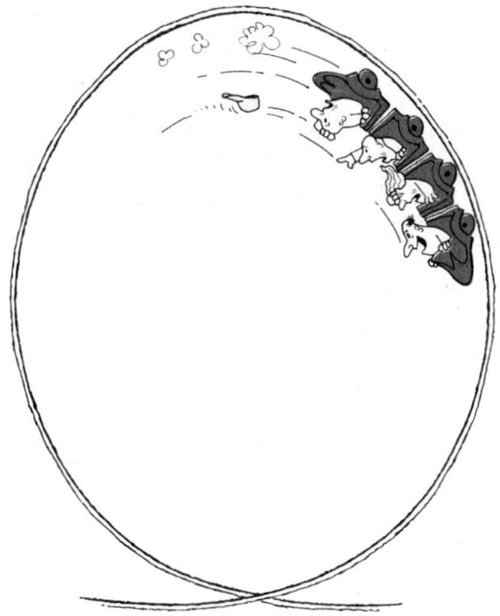

Figure P2.69

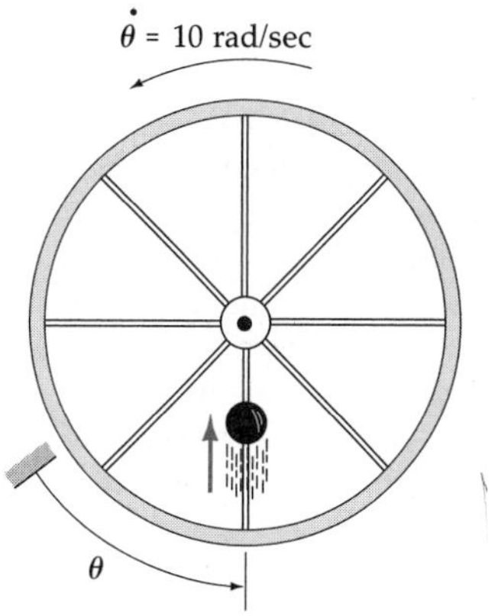

Figure P2.72

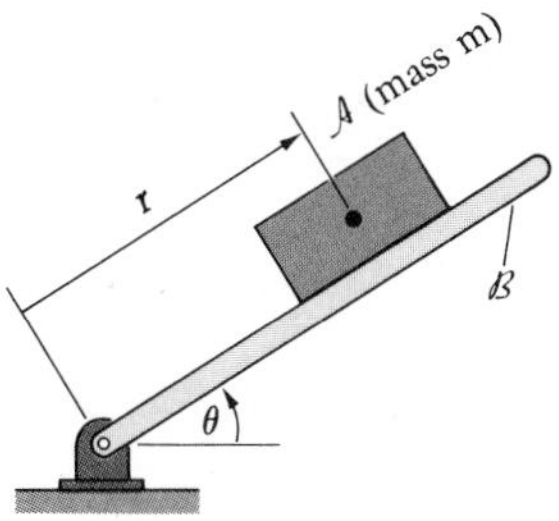

Figure P2.70

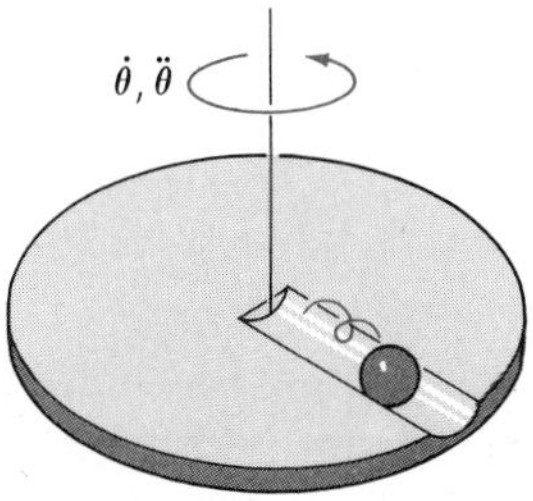

Figure P2.73

2.71 In the preceding problem, let μ remain at 0.5 but consider increasing the parameter $r\omega_0^2/g$. At what value of this parameter will $\mathcal{A}$ slide *outward* on $\mathcal{B}$? At what angle will this occur?

2.72 A horizontal wheel is rotating about its fixed axis at a rate of 10 rad/sec, and this angular speed is increasing at the given time at $\ddot{\theta} = 5$ rad/sec². (See Figure 2.72.) At this same instant, a bead is sliding inward relative to the spoke on which it moves at 5 ft/sec; this speed is slowing down at this time at 2 ft/sec². If the bead weighs 0.02 lb and is 1 ft from the center in the given configuration, find the external force exerted on the bead. Is it possible that this force can be exerted solely by the spoke and not in part by other external sources?

2.73 A ball bearing is moving radially outward in a slotted horizontal disk that is rotating about the vertical z axis. At the instant shown in Figure P2.73, the ball bearing is 3 in. from the center of the disk. It is traveling radially outward at a velocity of 4 in./sec relative to the disk. If $\dot{\theta} = 2$ rad/sec and is constant, find $\ddot{r}$ and the force exerted on the ball by the disk at this instant. Assume no friction and take the weight of the ball to be 0.05 lb.

2.74 A bead slides down a smooth circular hoop that, at a certain instant, has $\dot{\varphi} = 2$ rad/sec and $\ddot{\varphi} = 3$ rad/sec² in the direction shown in Figure P2.74. The angular speed of line OP at this time is $\dot{\theta} = 0.5$ rad/sec and $\theta = 135°$. Find the value of $\ddot{\theta}$ and the force exerted on the bead by the hoop at the given instant, if the mass of the bead is 0.1 kg and the radius of the hoop is 20 cm. *Hint*: Use spherical coordinates.

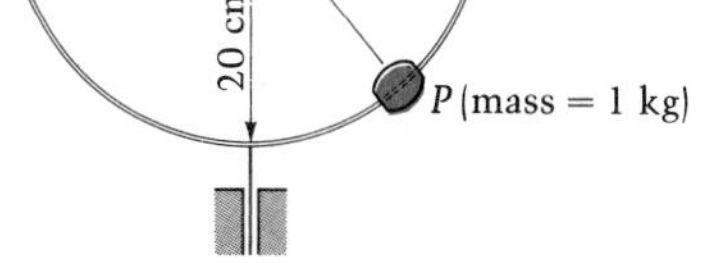

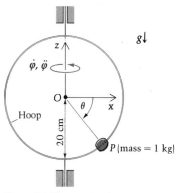

Figure P2.74

* **2.75** The four light rods are pinned at the origin and at each mass in such a way that as these seven bodies are spun up about the vertical, the masses m move outward and the mass M slides smoothly up along the vertical rod Oy. There is a relationship between ϕ, ω_0, ℓ, g, m, and M such that at the particular spin-speed ω_0, the bodies behave as one rigid body (meaning ϕ remains constant). Find the relationship. *Hint*: Use separate free-body diagrams of m and M, and write equations of motion for each. The unknowns are F_T (force in each top rod) and F_B (force in each bottom rod), ω_0 and ϕ. There will be three useful equations. See Figure P2.75.

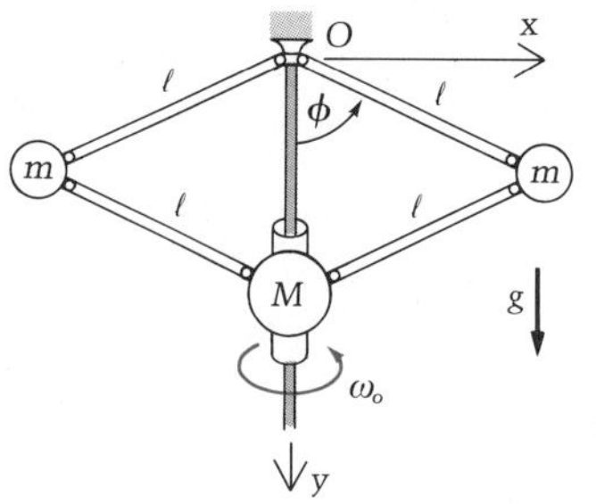

Figure P2.75

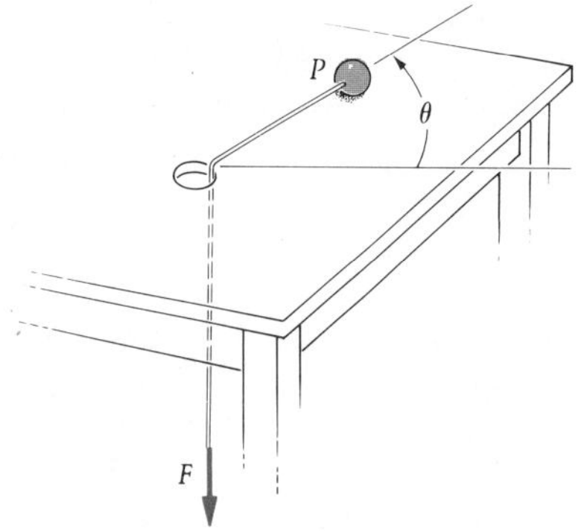

Figure P2.76

2.76 A particle P of mass m moves on a smooth, horizontal table and is attached to a light, inextensible cord that is being pulled downward by a force $F(t)$ as shown in Figure P2.76. Show that the differential equations of motion of P are

$$-F = m(\ddot{r} - r\dot{\theta}^2) \qquad (1)$$

$$0 = r\ddot{\theta} + 2\dot{r}\dot{\theta} \qquad (2)$$

Then show that Equation (2) implies that $r^2\dot{\theta} = $ constant.

2.77 In the preceding problem, let the particle be at $r = r_0$ at $t = 0$, and let the part of the cord beneath the table be descending at constant speed v_C. If the transverse component of velocity of P is $r\dot{\theta} = r_0\dot{\theta}_0$ at $t = 0$, find the tension in the cord as a function of time t.

2.78 A wintertime fisherman of mass 70 kg is in trouble—*he* is being reeled in by Jaws on a lake of frozen ice. At the instant shown in Figure P2.78, the man has a velocity component, perpendicular to the radius r, of $v_{1_\perp} = 0.3$ m/s at an instant when $r = R_1 = 5$ m. If Jaws pulls in the line with a force of 100 N, find the value of $v_{2_\perp}$ when the radius is $R_2 = 1$ m. *Hint*:

$$r\ddot{\theta} + 2\dot{r}\dot{\theta} = \frac{(d/dt)(r^2\dot{\theta})}{r} \qquad \text{and} \qquad \Sigma F_\theta = ?$$

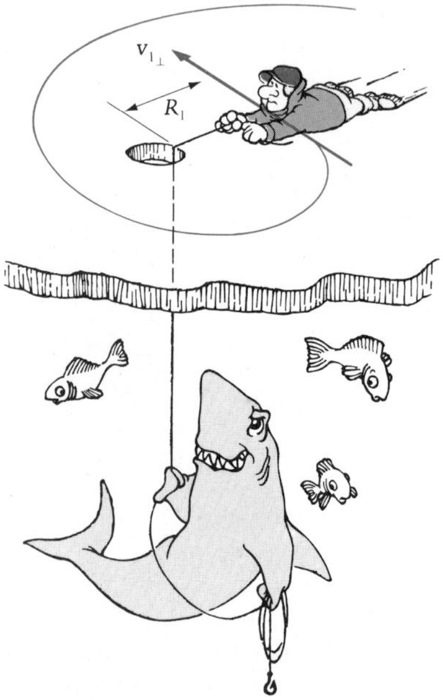

Figure P2.78

2.79 In the preceding problem, show that the differential equation of the man's radial motion is $\ddot{r} = (2.25/r^3) - (10/7)$. Use $\dot{r}\ddot{r} = d/dt\,(\dot{r}^2/2)$ to integrate this, and *if* $\dot{r} = 0$ when $r = 5$ m, show that the radial component ($\dot{r}$) of the man's velocity when $r = 1$ m is 3.04 m/s.

2.80 Particle P of mass m travels in a circle of radius a on the smooth table shown in Figure P2.80. Particle P is connected by an inextensible string to the stationary particle of mass M. Find the period of one revolution of P.

* **2.81** A weight of 100 lb hangs freely from a light rope (Figure P2.81). It is pulled up by a force that is 150 lb at $t = 0$ but diminishes uniformly in magnitude at 1 lb per foot pulled up. Find the time required to pull the weight up to the platform from rest, and determine its velocity upon reaching the top.

* **2.82** Rework Problem 2.81, but this time assume that the force *increases* by 1 lb per foot pulled up.

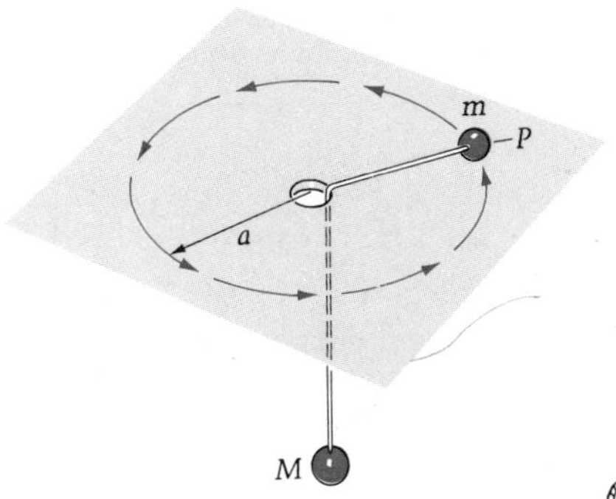

Figure P2.80

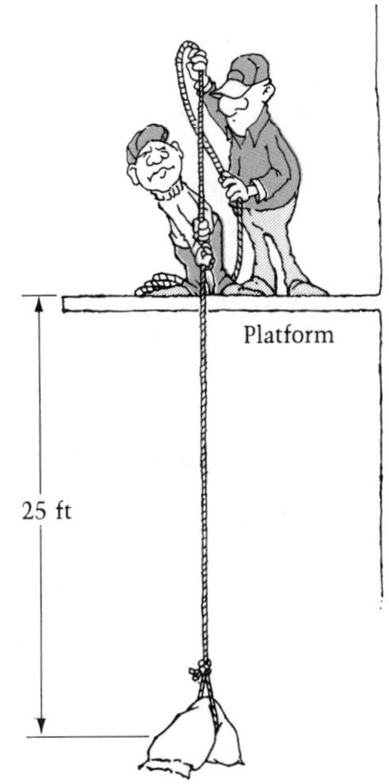

Figure P2.81

2.83 The acceleration of gravity varies with distance z above the earth's surface as

$$\ddot{z} = \frac{-gR^2}{(R + z)^2}$$

where g is the acceleration of gravity on the surface and R is the earth's radius. Find the minimum firing velocity v_i that a projectile must have in order to escape the earth if fired straight up (Figure P2.83). *Hint*: Not to return to earth requires the condition that $v \to 0$ as z gets large for the *minimum* possible v_i.

2.84 The mass m shown in Figure P2.84 is given an initial velocity of v_0 in the x direction. It moves in a medium that resists its motion with force proportional to its velocity, with proportionality constant K. By solving for $v(x)$, determine how far the mass travels before stopping. Then solve in a different manner for $v(t)$ if $v = v_0$ when $t = 0$.

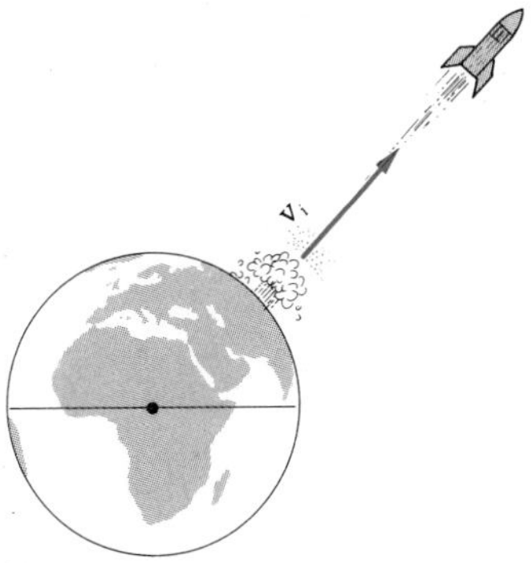

Figure P2.83

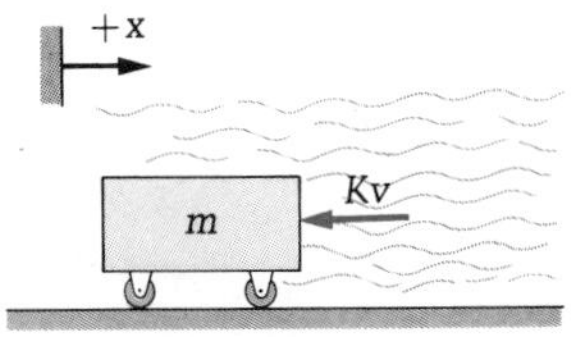

Figure P2.84

2.85 Using the result of the preceding problem and expressing v as dx/dt, solve for $x(t)$ if $x = 0$ when $t = 0$.

2.86 The identical plastic scottie dogs shown in Figure P2.86 are glued onto magnets and attract each other with a force $F = K/(2x)^2$, where K is a constant related to the strength of the magnets. Find the speeds at which the dogs collide if the magnets are initially separated by the distance S.

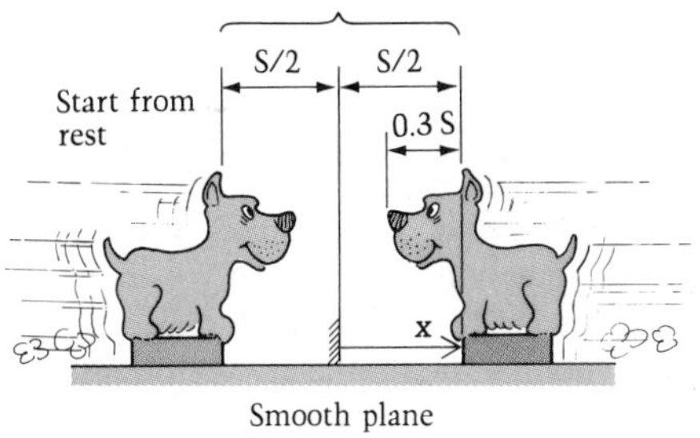

Figure P2.86

2.87 A ball is dropped from the top of a tall building. The motion is resisted by the air, which exerts a drag force given by Dv^2; D is a constant and v is the speed of the ball. Find the *terminal speed* (the limiting speed of fall) if there is no limit on the drop height. What is the drop height for which the ball will strike the ground at 95 percent of the terminal speed?

Figure P2.88

Figure P2.90

Figure P2.92

* **2.88** Over a certain range of velocities, the effect of air resistance on a projectile is proportional to the square of the object's speed. If the object can be regarded as a particle, the *drag force* is expressible as $F_D = \frac{1}{2}\rho A C_D v^2$, in which ρ is the density of the air, A is the projected area of the object onto a plane normal to the velocity vector, and C_D is a coefficient that depends on the object's shape. If $\rho A C_D = 0.0004$ lb-sec^2/ft^2 for the 76-lb cannonball of Figure P2.88, find the maximum height it reaches. Compare your result with the answer neglecting air resistance.

* **2.89** In the preceding problem, find the velocity of the cannonball just before it hits the ground; again compare with the case of no air resistance.

* **2.90** A 160-lb parachutist in the "free-fall spread-stable position" (Figure P2.90) reaches a velocity of 174 ft/sec in 12 sec after exiting a stationary blimp. Assuming velocity-squared air resistance, solve the differential equation of motion

$$\ddot{y} = g - \frac{k}{m}\dot{y}^2$$

and determine the constant k.

* **2.91** In the preceding problem, suppose the parachutist opens his chute at a height of 1000 ft. If the value of k then becomes 0.63 lb/(ft/sec)^2, find the velocity at which the parachutist strikes the ground, if $v_i = 174$ ft/sec.

2.92 The drag car of mass m shown in Figure P2.92, traveling at speed v_0, is to be initially slowed primarily by the deployment of a parachute. The parachute exerts a force F_d proportional to the square of the velocity of the car, $F_d = Cv^2$. Neglecting friction and the inertia of the wheels, determine the distance traveled by the car before its velocity is 40 percent of v_0. If the car and driver weigh 1000 lb and $C = 0.182$ lb-sec^2/ft^2, find the distance in feet.

2.93 In the preceding problem, suppose the drag car's speed at parachute release is 237 mph. Find the time it takes to reach 40 percent speed.

* **2.94** A 50-lb shell is fired from the cannon shown in Figure P2.94. The pressure of the expanding gases is inversely proportional to the volume behind the shell. Initially this pressure is 10 tons per square inch; just before exit, it is one-tenth this value. Find the exit velocity of the shell.

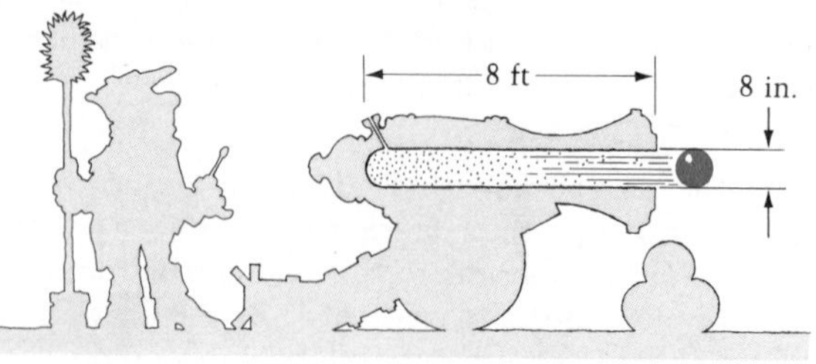

Figure P2.94

2.95 The block of mass m shown in Figure P2.95 is brought slowly down to the point of contact with the end of the spring, and then (at $t = 0$) the block is released. Write the differential equation governing the subsequent motion, clearly defining your choice of displacement parameter. What are the initial conditions? Find the maximum force induced in the spring and the first time at which it occurs.

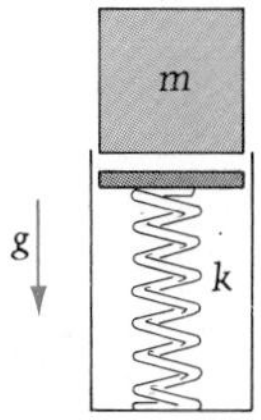

Figure P2.95

2.96 Show that for a particle P moving in a viscous medium in which the air resistance is proportional to velocity (Figure P2.96), the differential equations of motion are

$$m\ddot{x} = -k\dot{x}$$

$$m\ddot{y} = -k\dot{y} - mg$$

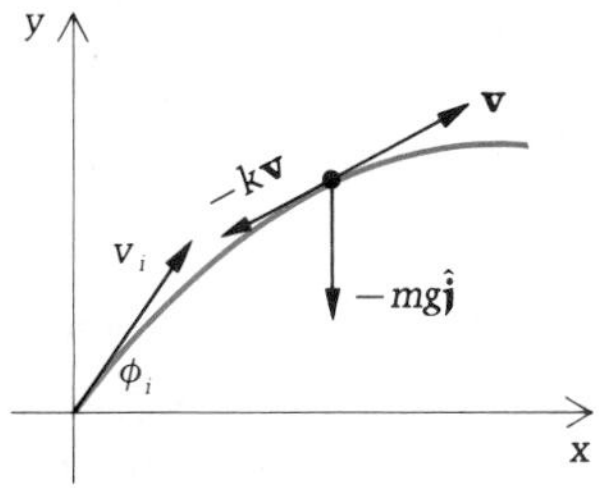

Figure P2.96

2.97 In the preceding problem, show by integration that the components of velocity of P are given by

$$\dot{x} = v_i \cos \phi_i \, e^{-kt/m}$$

$$\dot{y} = \frac{-mg}{k} + \left(v_i \sin \phi_i + \frac{mg}{k}\right) e^{-kt/m}$$

2.98 Continue the preceding exercise and show that the particle's position is given by the equations

$$x = \frac{mv_i \cos \phi_i}{k}(1 - e^{-kt/m})$$

$$y = \frac{m}{k}\left[\left(\frac{mg}{k} + v_i \sin \phi_i\right)(1 - e^{-kt/m}) - gt\right]$$

Show further that both x and $\dot{y}$ approach asymptotes as $t \to \infty$. (The limiting value of $\dot{y}$ is known as the terminal velocity of P, after which the weight is balanced by the viscous resistance so that the acceleration goes to zero.)

* **2.99** A particle moves on the inside of a fixed, smooth vertical hoop of radius a. It is projected from the lowest point A with velocity $\sqrt{7ga/2}$. Show that it will leave the hoop at a height $3a/2$ above A and meet the hoop again at A.

* **2.100** The two particles in Figure P2.100 are at rest on a smooth horizontal table and connected by an inextensible string that passes through a small, smooth ring fixed to the table. The lighter particle (mass m) is then projected at right angles to the string with velocity v_0. Prove that the other particle will strike the ring with velocity $v_0\sqrt{3}/(2\sqrt{n+1})$. *Hint*: Use polar coordinates and note that $r^2\dot{\theta}$ is constant for each particle.

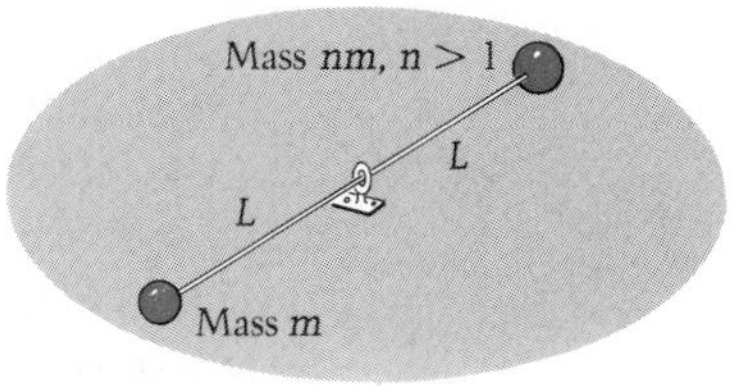

Figure P2.100

2.4 Work and Kinetic Energy for Particles

In Example 2.4 we were able to get useful information from an energy integral of the governing differential equation. The same result may be obtained in general by an integration of

$$\Sigma \mathbf{F} = m\mathbf{a}_C$$

Forming the dot product of each side with the velocity $\mathbf{v}_C$ of the mass center, we have

$$
\begin{aligned}
(\Sigma \mathbf{F}) \cdot \mathbf{v}_C &= m\mathbf{a}_C \cdot \mathbf{v}_C \\
&= m \frac{d\mathbf{v}_C}{dt} \cdot \mathbf{v}_C \\
&= \frac{m}{2} \frac{d}{dt} (\mathbf{v}_C \cdot \mathbf{v}_C) \\
&= \frac{d}{dt} \left(\frac{m}{2} \mathbf{v}_C \cdot \mathbf{v}_C \right) \\
&= \frac{d}{dt} \left(\frac{m}{2} |\mathbf{v}_C|^2 \right)
\end{aligned}
$$

Integrating,* we get

$$\int_{t_1}^{t_2} (\Sigma \mathbf{F}) \cdot \mathbf{v}_C \, dt = \frac{m}{2} [|\mathbf{v}_C(t_2)|^2 - |\mathbf{v}_C(t_1)|^2] \tag{2.11}$$

or, for a particle,

$$\int_{t_1}^{t_2} (\Sigma \mathbf{F}) \cdot \mathbf{v} \, dt = \frac{m}{2} [|\mathbf{v}(t_2)|^2 - |\mathbf{v}(t_1)|^2] \tag{2.12}$$

Work and Kinetic Energy for a Particle

For a particle, $\int_{t_1}^{t_2} (\Sigma \mathbf{F}) \cdot \mathbf{v} \, dt$ is called the work done on the particle by the resultant of external forces.† We note that if there are N forces acting on the particle, then the resultant $\Sigma \mathbf{F}$ is given by $\mathbf{F}_1 + \mathbf{F}_2 + \cdots + \mathbf{F}_N$ and

$$
\begin{aligned}
(\Sigma \mathbf{F}) \cdot \mathbf{v} &= (\mathbf{F}_1 + \mathbf{F}_2 + \cdots + \mathbf{F}_N) \cdot \mathbf{v} \\
&= \mathbf{F}_1 \cdot \mathbf{v} + \mathbf{F}_2 \cdot \mathbf{v} + \cdots + \mathbf{F}_N \cdot \mathbf{v}
\end{aligned}
$$

Each term of this equation represents the rate of work of one of the forces. Thus the left side of Equation (2.12) may be read as the sum of the works of the individual forces acting on the particle. These statements are all consistent with the presentation to come in Chapter 5 in which we define the rate of work done by a force $\mathbf{F}$ to be $\mathbf{F} \cdot \mathbf{v}$, where $\mathbf{v}$ is the velocity of the point of the body at which the force is applied. The left side of Equation (2.11) may then be interpreted as the work that *would* be done by the external forces *were* each to have a line of action through the mass center.

For a *particle*, $(m/2)|\mathbf{v}|^2$ is called the kinetic energy, usually written T. Thus for the particle, Equation (2.12) is the work and kinetic energy principle:

Work done on the particle = Change in the particle's kinetic energy

---

* Sometimes (t_i, t_f), referring to "initial" and "final," are used instead of (t_1, t_2).

† Thus the appropriate unit of work and of energy in SI is the joule (J), the joule being 1 N · m; in U.S. units the ft-lb is the unit of work and energy. The N · m and lb-ft are usually reserved for the moment of a force. Note that work, energy, and moment of force all have the same dimension.

or

$$W = \Delta T \tag{2.13}$$

For a *body*, the kinetic energy is defined to be the sum of the kinetic energies of the particles constituting the body. If all the points in a body $\mathcal{B}$ have the same velocity (which is then $\mathbf{v}_C$), then $(m/2)|\mathbf{v}_C|^2$ is the *total* kinetic energy of $\mathcal{B}$. In general, however, the body is turning or deforming (or both) and this is not the case; the body then has *additional* kinetic energy due to its changes in orientation (that is, due to its angular motion) or due to the deformation. For a body $\mathcal{B}$, we shall also see in Chapter 5 that, in general, the left side of Equation (2.11) does not constitute the total work done on $\mathcal{B}$ by the external forces and couples. This is because, for a body, the forces do not have to be concurrent as they are for a particle. Equation (2.13) still turns out to be true for a rigid body, however, with the two sides of Equation (2.11) representing *parts* of W and ΔT.

Finally, with no restrictions on the size of the body, the energy integral (Equation 2.11) states that the work that would be done if the external forces acted at the mass center equals the change in what would be the kinetic energy if every point in the body had the velocity of the mass center. We could call this result the "mass center work and kinetic energy principle."

Work Done by a Constant Force

Before attempting to apply the work-energy principle to a specific problem, it is helpful to determine the work done by two classes of forces. First, suppose $\mathbf{F}$ is a constant force and suppose we let $\mathbf{r}$ be a position vector for the particle. Then

$$\begin{aligned}\int_{t_1}^{t_2} \mathbf{F} \cdot \mathbf{v}\, dt &= \mathbf{F} \cdot \int_{t_1}^{t_2} \mathbf{v}\, dt \\ &= \mathbf{F} \cdot \int_{t_1}^{t_2} \frac{d\mathbf{r}}{dt}\, dt \\ &= \mathbf{F} \cdot [\mathbf{r}(t_2) - \mathbf{r}(t_1)] \end{aligned} \tag{2.14}$$

which states that the work done is the dot product of the force with the displacement of the particle. We recall that this dot product can be expressed as the product of the force magnitude and the component of displacement in the direction of the force *or* as the product of the displacement magnitude and the component of force in the direction of the displacement.

Work Done by a Central Force

The second case to which we give special attention is that of a central force. Such a force is defined to have a line of action always passing through the same fixed point in the frame of reference and a magnitude

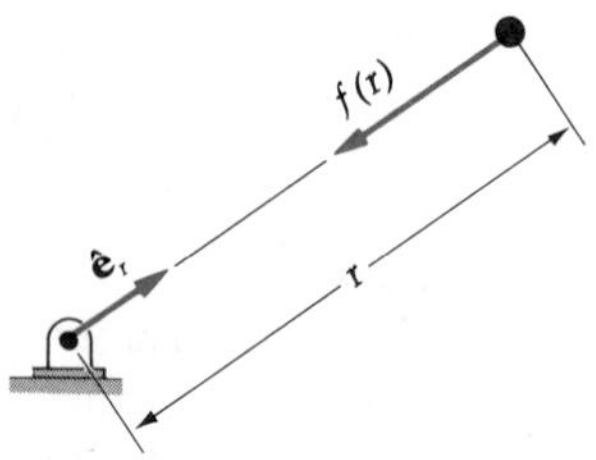

Figure 2.7

that depends only upon the distance r of the particle from that fixed point, as shown in Figure 2.7.

The velocity of the particle may be expressed as

$$\mathbf{v} = \frac{d}{dt}(r\hat{\mathbf{e}}_r) = \dot{r}\hat{\mathbf{e}}_r + r\dot{\hat{\mathbf{e}}}_r$$

$$\begin{aligned}\mathbf{F}\cdot\mathbf{v} &= [-f(r)\hat{\mathbf{e}}_r]\cdot[\dot{r}\hat{\mathbf{e}}_r + r\dot{\hat{\mathbf{e}}}_r]\\ &= -\dot{r}f(r)\end{aligned}$$

since by Equation (1.38) we know that $\hat{\mathbf{e}}_r \cdot \dot{\hat{\mathbf{e}}}_r = 0$. Thus the work done by $\mathbf{F} = -f(r)\hat{\mathbf{e}}_r$ is

$$\begin{aligned}W &= \int_{t_1}^{t_2} -f(r)\frac{dr}{dt}\,dt\\ &= -\int_{r(t_1)}^{r(t_2)} f(r)\,dr \end{aligned} \tag{2.15}$$

If φ is a function of r so that $f = d\varphi/dr$, then the work may be written as

$$\begin{aligned}W &= -\int_{r(t_1)}^{r(t_2)} \frac{d\varphi}{dr}\,dr\\ &= -\varphi[r(t_2)] + \varphi[r(t_1)]\end{aligned} \tag{2.16}$$

Work Done by a Linear Spring

A special central force is that exerted by a spring on a particle when the other end of the spring is fixed. In the case of a linear spring of instantaneous length r, we note that $f = k(r - L_0)$, where L_0 is its natural, or unstretched, length and k is called the spring modulus or stiffness. In this case, $\varphi = (k/2)(r - L_0)^2$ or, more simply, $\varphi = (k/2)\delta^2$, where δ is the spring stretch. Thus by equation (2.16),

$$W = -\frac{k}{2}[\delta^2(t_2) - \delta^2(t_1)] \tag{2.17}$$

Question 2.4 What assumption about the mass of the spring is to be understood in the force-stretch relationship?

Answer 2.4 $(k)\cdot$(stretch) gives the equal-in-magnitude but opposite-in-direction force acting at the ends of a spring in equilibrium. If particles in the spring are accelerating, as is generally the case in dynamics problems, there is no simple force-stretch law. If the spring is very light, however, so that its mass may be neglected (compared to the masses of other bodies in the problem), $\int_{\text{spring}} \mathbf{a}\,dm \approx \mathbf{0}$ and the forces on the spring are instantaneously related just as if the spring were in equilibrium. The hidden assumption is that the mass of the spring may be neglected.

Work Done by Gravity

A second special case of a central force is the gravitational force exerted on a body by the earth. By Newton's law of universal gravitation,

$$f = \frac{mgr_e^2}{r^2} \tag{2.18}$$

where r_e is the radius of the earth, m is the mass of the attracted body, and g is the gravitational strength (or acceleration) at the surface of the earth.* By Equation (2.15) we have

$$W = -\int_{r(t_1)}^{r(t_2)} \frac{mgr_e^2}{r^2}\,dr = mgr_e^2\left[\frac{1}{r(t_2)} - \frac{1}{r(t_1)}\right]$$

We note that for this case the function φ is given by

$$\varphi = \frac{-mgr_e^2}{r} \tag{2.19}$$

If the motion is sufficiently near the surface of the earth,

$$\frac{r_e^2}{r^2} \approx 1$$

and so $f \approx mg$, a constant. In this case,

$$\begin{aligned} W &= \frac{mgr_e^2[r(t_1) - r(t_2)]}{r(t_1)r(t_2)} \\ &\approx mg[r(t_1) - r(t_2)] \\ &= (\text{weight}) \times (\text{decrease in altitude of mass center of body}) \end{aligned} \tag{2.20}$$

and the φ function becomes (if z_C is positive upward)

$$\varphi = mgz_C \tag{2.21}$$

Conservative Forces

In each case we have considered, the work has depended only on the initial and final positions of the point where the force is applied. Such a force whose work is independent of the path traveled by the point on

* The force of gravity in fact results in infinitely many differential forces, each tugging on one of the body's particles. For nearly all applications on the planet earth, these forces may be thought of as equivalent to a single force through the mass center of the body. For applications in astronomy or in space vehicle dynamics, however, the gravity moment that accompanies the force at the mass center becomes important. In Skylab, for example, three huge control-moment gyros were present to "take out" the angular momentum built up by a gravity moment of only a few lb-ft. And the gravity moment exerted on the earth by the sun and moon's gravitation causes the earth's axis to precess in the heavens once every 25,800 yr. The gravity moment vanishes if the body is a uniform sphere (which the earth is not, being bulged at the equator and having varying density). A further discussion of this *luni-solar precession* is presented in Chapter 7.

which it acts is called **conservative.** Furthermore, the work may be expressed as the change in a scalar function of position; we saw this to be the case for the central force, and we may make the same statement for the constant force by defining φ to be $-\mathbf{F} \cdot \mathbf{r}$.

Question 2.5 Why the minus sign in front of $\mathbf{F} \cdot \mathbf{r}$?

Conservation of Energy

If all forces acting are conservative and if φ is the sum of all their φ functions, then the work and kinetic energy equation (2.13) becomes

$$\varphi(t_1) - \varphi(t_2) = T(t_2) - T(t_1)$$

or

$$T(t_2) + \varphi(t_2) = T(t_1) + \varphi(t_1)$$

or

$$T + \varphi = \text{constant} \tag{2.22}$$

We call φ the **potential energy** and $T + \varphi$ the **total (mechanical) energy.** Thus Equation (2.22) is a statement of conservation of mechanical energy when all the forces are conservative (path-independent).

Question 2.6 How would Equation (2.22) read if instead of φ we had chosen to construct the scalar function ψ so that the work done by a force is the *increase* in its ψ?

In closing it is important to realize that not all forces are conservative. An example is the force of friction acting on a block sliding on a fixed surface. That force does negative work regardless of the direction of the motion, and a potential function φ cannot be found for it.

Answer 2.5 It is needed so that the work equals the decrease in φ; that is, $\varphi[\mathbf{r}(t_1)] - \varphi[\mathbf{r}(t_2)]$.
Answer 2.6 $T - \psi =$ constant; we use the φ simply so that we may say that mechanical energy is the sum of its two parts.

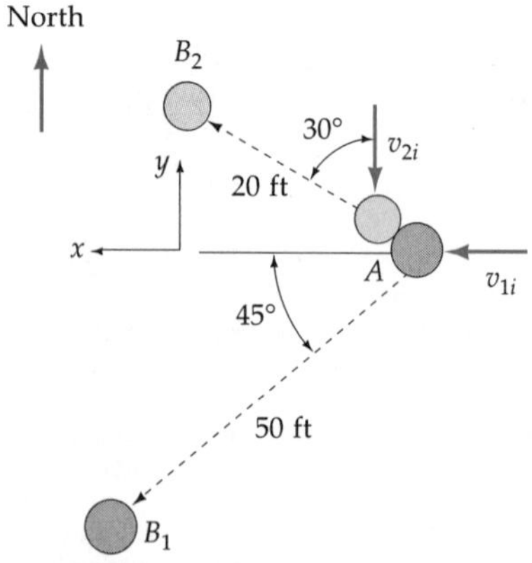

Figure E2.9

EXAMPLE 2.9

In an accident reconstruction, the following facts are known:

1. Identical cars 1 and 2, respectively headed west and south as indicated in Figure E2.9, collided at point A in an intersection.
2. With locked brakes indicated by skid marks, the cars skidded to the final positions B_1 and B_2 shown in the figure.

Assuming the cars are particles, determine their velocities immediately following the collision if the friction coefficient between the tires and road is 0.5.

Solution

After separation, each car is brought to rest by the friction force acting on its tires. For Car 1, we have

$$W = \Delta T = T_f - T_i$$

$$-\mu mg\,(50) = 0 - \frac{1}{2}\,mv_1^2$$

in which $T_f = 0$ since the cars end up at rest, and v_1 is the speed of Car 1 just after the cars separate. Thus

$$v_1 = \sqrt{2(50)0.5(32.2)} = 40.1 \text{ ft/sec}$$

Similarly for Car 2,

$$v_2 = \sqrt{2(50)0.5(32.2)} = 25.4 \text{ ft/sec}$$

In the next section, we will return to this example and use the principle of impulse and momentum to approximate the speeds of the cars *before* the impact.

Example 2.10

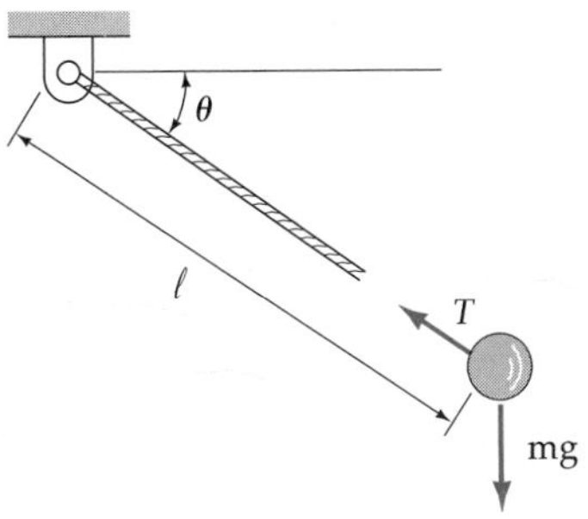

Figure E2.10

We repeat Example 2.4 (see Figure E2.10): For a ball of mass m released from rest with the cord taut and $\theta = 30°$, we wish to find the tension in the cord as a function of θ.

Solution

If, as before, we write the force-acceleration component equation in the radial direction, we have

$$T = mg \sin\theta + m\ell\dot{\theta}^2 \qquad (1)$$

Now we apply Equation (2.13) by letting t_1 be the initial time at which $\theta = 30°$, and letting t_2 be the time at which we are applying (1). We note that

$$|\mathbf{v}(t_1)| = 0$$

$$|\mathbf{v}(t_2)|^2 = \ell^2\dot{\theta}^2$$

and the work done by the cord tension T is zero since that force is always perpendicular to the velocity of (the center of mass of) the ball. By Equation (2.20) the work of the weight is $mg[\ell \sin\theta - \ell \sin 30°]$. Thus Equation (2.13) yields

$$mg[\ell \sin\theta - \ell \sin 30°] = \frac{1}{2}\,m(\ell^2\dot{\theta}^2) - 0$$

or

$$m\ell\dot{\theta}^2 = 2mg\left(\sin\theta - \frac{1}{2}\right)$$

Substituting in Equation (1) above, we get

$$T = mg \sin\theta + 2mg \sin\theta - mg$$

$$= mg(3 \sin\theta - 1)$$

which is precisely the result obtained previously.

EXAMPLE 2.11

The block shown in Figure E2.11a slides on an inclined surface for which the coefficient of friction is $\mu = 0.3$. Find the maximum force induced in the spring if the motion begins under the conditions shown.

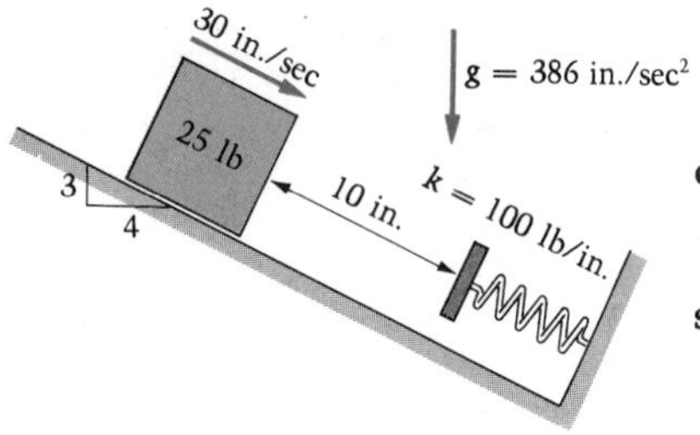

Figure E2.11a

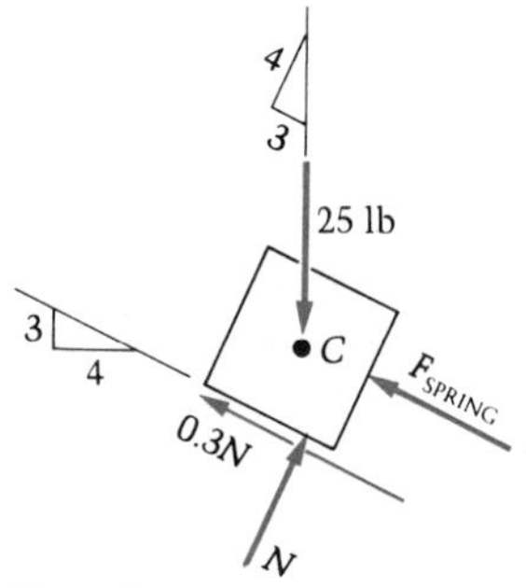

Figure E2.11b

Solution

We assume that the block can be treated as rigid; thus the end of the spring, once it contacts the block, will undergo the same displacements as the mass center (or for that matter any other point) of the block. To apply the mass-center work and kinetic energy principle we let t_1 be the initial time shown above and let t_2 be the time of maximum compression Δ of the spring. To catalog the external forces that do work, we consider a free-body diagram at some arbitrary instant between t_1 and t_2. (See Figure E2.11b.) Since the mass center has a path parallel to the inclined plane, there is no component of acceleration perpendicular to it and

$$N - \frac{4}{5}(25) = 0$$

or

$$N = 20 \text{ lb}$$

so that the friction force is $0.3(20) = 6$ lb.

Denoting the left side of Equation (2.11) by Work (t_1, t_2) we have

$$\text{Work}(t_1, t_2) = \frac{1}{2} m[|\mathbf{v}_C(t_2)|^2 - |\mathbf{v}_C(t_1)|^2]$$

where

$$|\mathbf{v}_C(t_2)| = 0$$

$$|\mathbf{v}_C(t_1)| = 30 \text{ in./sec}$$

and the various works are

1. For N, work$(t_1, t_2) = 0$ since the force is perpendicular to $\mathbf{v}_C$ at each instant.
2. For friction, work$(t_1, t_2) = -6(10 + \Delta)$ in.-lb
3. For the weight, work$(t_1, t_2) = [(\frac{3}{5})(25)](10 + \Delta)$ in.-lb
4. For the spring,

$$\text{Work}(t_1, t_2) = \text{work}(t_1, \text{contact}) + \text{work}(\text{contact}, t_2)$$

$$= 0 - \frac{100}{2}[(-\Delta)^2 - 0] \text{ in.-lb}$$

Thus, $W = \Delta T$ gives (with $g = 32.2 \times 12 = 386$ in./sec²)

$$-6(10 + \Delta) + 15(10 + \Delta) - 50\Delta^2 = 0 - \frac{1}{2}\left(\frac{25}{386}\right)(30)^2$$

or

$$50\Delta^2 - 9\Delta - 119 = 0$$

$$\Delta^2 - 0.180\Delta - 2.38 = 0$$

From the quadratic formula,

$$\Delta = 0.09 \pm \sqrt{0.0081 + 2.38} \text{ in.}$$

from which only the positive root is meaningful:

$$\Delta = 0.09 + 1.55 = 1.64 \text{ in.}$$

The corresponding force is $100(1.64) = 164$ lb.

EXAMPLE 2.12

In the preceding example, find the next position at which the block comes to rest.

Solution

At time t_2 the spring force (164 lb) exceeds the sum of the component of weight along the plane (15 lb) and the maximum frictional resistance (6 lb), so we know that the block is not in equilibrium and must then begin to move back up the plane, with the friction force now acting down the plane as shown in Figure E2.12. Suppose we let t_3 be the time at which the block next comes to rest and let d represent the corresponding compression of the spring. Then, since $|\mathbf{v}_C(t_2)| = |\mathbf{v}_C(t_3)| = 0$, we have

$$\text{Work}(t_2, t_3) = 0$$

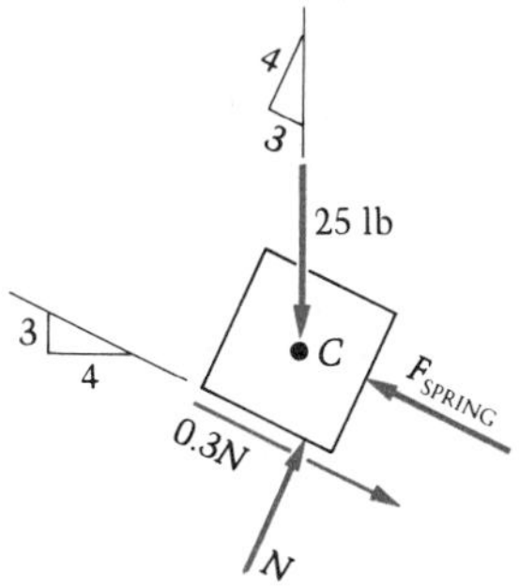

Figure E2.12

For the spring, the work is $(-100/2)[d^2 - (-\Delta)^2]$; for the friction force, the work is $-6(\Delta - d)$; for the weight, the work is $-(\frac{3}{5})(25)(\Delta - d)$. Thus

$$(\Delta - d)\left[\frac{100}{2}(\Delta + d) - 6 - 15\right] = 0$$

or

$$d = \frac{21}{50} - 1.64 = -1.22 \text{ in.}$$

The negative sign here tells us that the spring must be stretched 1.22 in. when the block again comes to rest. If, as intended here, the spring does not become permanently attached to the block on first contact (that is, contact is maintained only in compression), our analysis only tells us that contact is broken before the block comes to rest. We therefore need to modify the expression for the work done by the spring, which we now see should have been $-(100/2)(0 - \Delta^2)$. It is convenient to let e be the distance from the end of the spring to the block (measured up the plane). Then

$$-\frac{100}{2}[0 - \Delta^2] - 6(\Delta + e) - \frac{3}{5}(25)(\Delta + e) = 0$$

$$21(\Delta + e) = 50\Delta^2$$

$$e = \frac{50}{21}\Delta^2 - \Delta$$

$$= \frac{50}{21}(1.64)^2 - 1.64$$

$$= 4.76 \text{ in.}$$

It is instructive to obtain this result by using the work and kinetic energy principle over the interval t_1 to t_3, for which the net work done by the spring is zero. Noting then that the mass center of the block drops $\frac{3}{5}(10 - e)$ in. vertically and that the distance traveled by the block on the plane is $(10 + 1.64 + 1.64 + e)$ in., we have

$$\text{Work}(t_1, t_3) = \frac{1}{2}m|\mathbf{v}_C(t_3)|^2 - \frac{1}{2}m|\mathbf{v}_C(t_1)|^2$$

$$25\left[\frac{3}{5}(10 - e)\right] - 6(13.3 + e) = 0 - \frac{1}{2}\left(\frac{25}{386}\right)(30)^2$$

$$150 - 15e - 79.8 - 6e = -29.2$$

$$e = 4.73 \text{ in.}$$

which is the same result we obtained before except for the third significant figure — a consequence of rounding off at an intermediate step.

EXAMPLE 2.13

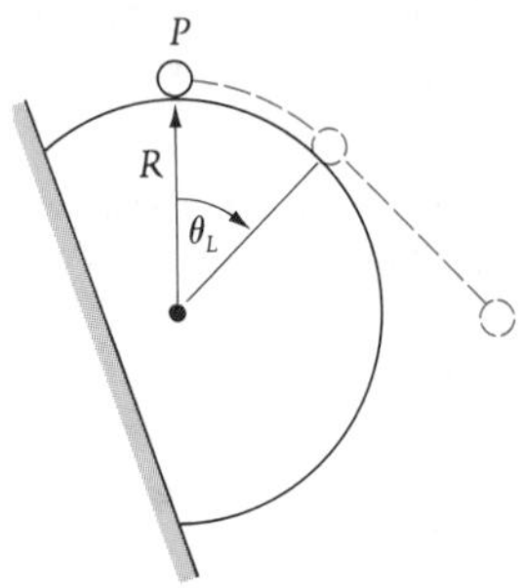

Figure E2.13a

A particle P of mass m rests atop a smooth spherical surface. (See Figure E2.13a.) A slight nudge starts it sliding downward in a vertical plane. Find the angle θ_L at which the particle leaves the surface.

Solution

Mechanical energy is conserved here because (1) there is no friction, (2) the normal force does not work since it is always normal to the velocity of P, and (3) the only other force is gravity. (See Figure E2.13b.) Therefore, using Equations (2.21) and (2.22),

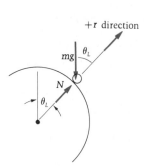

Figure E2.13b

$$T_1 + \varphi_1 = T_2 + \varphi_2$$

$$0 + mgR = \frac{1}{2}mv_2^2 + mgR\cos\theta_L$$

$$v_2^2 = 2gR(1 - \cos\theta_L) \tag{1}$$

Equation (1) contains two unknowns; to eliminate the velocity v_2, we use the equation of motion in the radial direction:

$$\Sigma F_r = ma_r$$

$$-mg\cos\theta_L + N = m\left(-\frac{v_2^2}{R}\right) \tag{2}$$

But N has just become zero when P is at the point of leaving. Therefore

$$v_2^2 = gR\cos\theta_L \tag{3}$$

Equating the right sides of Equations (1) and (3), we get

$$gR\cos\theta_L = 2gR(1 - \cos\theta_L)$$

$$3\cos\theta_L = 2$$

$$\theta_L = \cos^{-1}\left(\frac{2}{3}\right) = 48.2°$$

Work and Kinetic Energy for a System of Particles

As has been mentioned before, the kinetic energy of a system of particles is given by

$$T = \sum_{i=1}^{N} \frac{1}{2} m_i |\mathbf{v}_i|^2$$

or

$$T = \sum_{i=1}^{N} T_i$$

Now if we let W_i be the work of all the forces acting on the ith particle,

$$W_i = \Delta T_i$$

and, summing over all the particles,

$$\sum_{i=1}^{N} W_i = \Delta \sum_{i=1}^{N} T_i$$

or

$$W = \Delta T \tag{2.23}$$

for the system. The work, W, however, is the net work of all the forces external *and internal* that act on the particles. Sometimes Equation (2.23) can be used effectively because the net work of internal forces can be evaluated. For example, if two moving particles are joined by a linear spring, no simple formula can be written for the work of the spring force on *one* of the particles. However, the net work of the equal and opposite spring forces on the two particles is given by Equation (2.17) (see Problem 2.128). Particles that are rigidly connected interact through forces for which no net work is done. Thus, we shall find in Chapter 5 that when we use $W = \Delta T$ for a rigid body, the work only involves external forces.

PROBLEMS ► Section 2.4

2.101 A truck body weighing 4000 lb is carried by four light wheels that roll on the sloping surface. (See Figure P2.101.) The truck has a velocity of 5 ft/sec in the position shown. Determine the modulus of the spring if the truck is brought to rest by compressing the spring 6 in. *Note:* Light wheels with good bearings imply negligible friction.

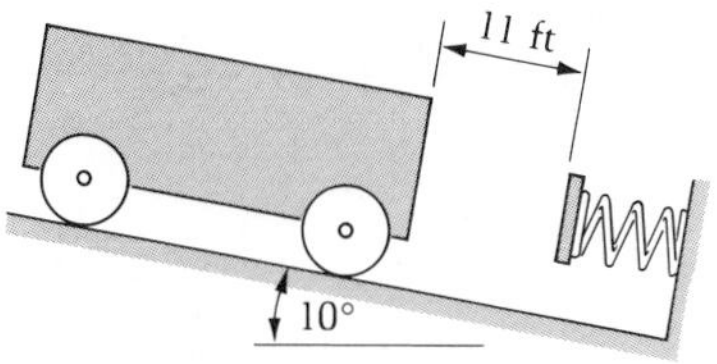

Figure P2.101

2.102 The block shown in Figure P2.102 weighs 100 lb and the spring's modulus is 10 lb/ft. The spring is unstretched when the block is released from rest. Find the minimum coefficient of friction μ such that the block will not start back up the plane after it stops.

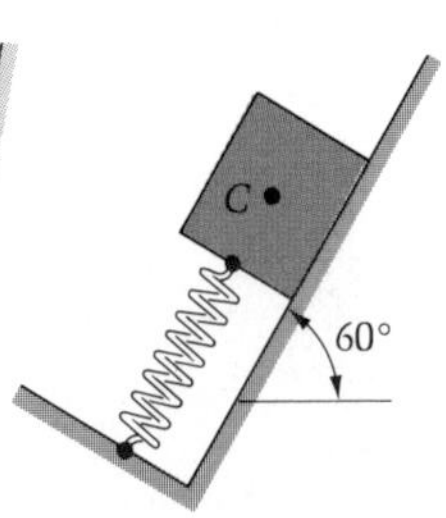

Figure P2.102

2.103 The block shown in Figure P2.103 is released from rest. What is its velocity when it first hits the spring?

2.104 How far does the block in the preceding problem rebound back up the plane after compressing the spring?

2.105 The 6-lb block shown in Figure P2.105 is released from rest when it just contacts the end of the unstretched spring. For the subsequent motion, find: (a) the maximum force in the spring; (b) the maximum speed of the block.

2.106 Block $\mathcal{A}$ weighs 16.1 lb and translates along a smooth horizontal plane with a speed of 36 ft/sec. (See Figure P2.106.) The coefficient of friction between $\mathcal{A}$ and the inclined surface is $\mu = 0.5$, and the spring constant is 100 lb/ft. Determine the distance that $\mathcal{A}$ moves up the incline before coming to rest.

2.107 The weight shown in Figure P2.107 is prevented from sliding down the inclined plane by a cable. An engineer wishes to lower the weight to the dashed position by inserting a spring and then cutting the cable. Find the modulus of a spring that will accomplish this task without allowing the block to move back up the incline after it stops. *Hint*: You are free to specify the initial stretch — try zero!

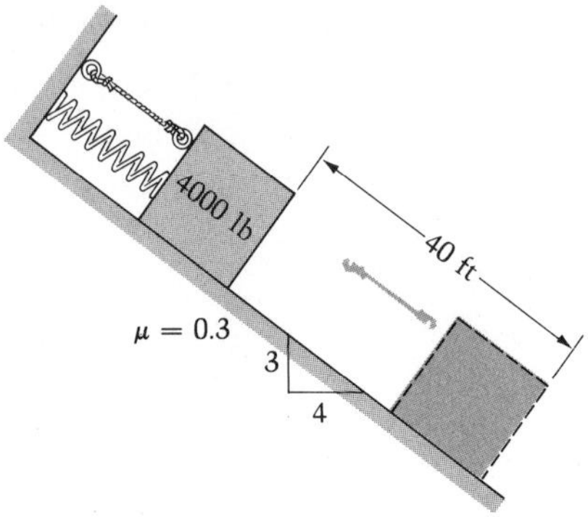

Figure P2.107

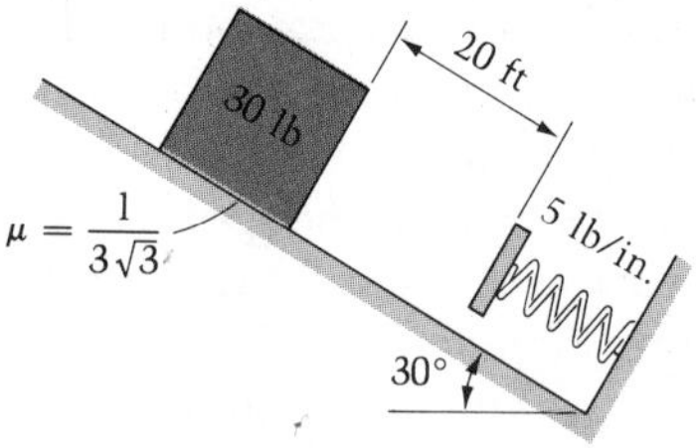

Figure P2.103

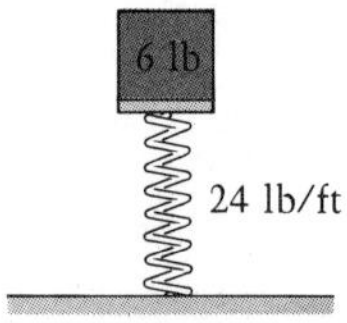

Figure P2.105

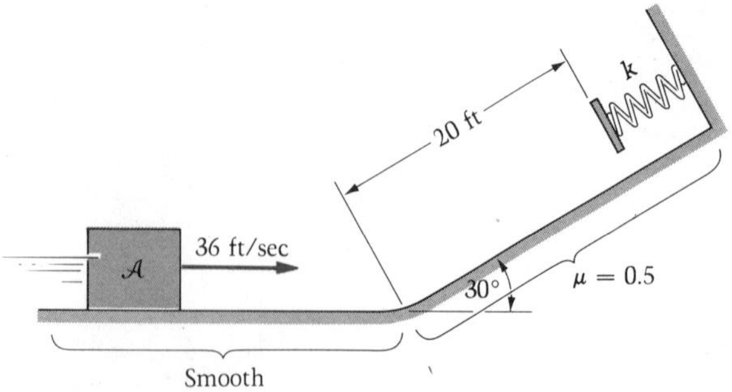

Figure P2.106

2.108 At the instant shown in Figure P2.108, the block is traveling to the left at 7 m/s and the spring is unstretched. Find the velocity of the block when it has moved 4 m to the left.

2.109 Repeat the preceding problem with $\mu = 0.2$. You will need

$$\int \frac{dx}{\sqrt{a^2 + x^2}} = \ln(x + \sqrt{a^2 + x^2}) + c$$

where a and c are constants.

2.110 The Bernoulli brothers posed and then solved the "brachistochrone problem." (See Figure P2.110.) The problem was to determine on which single-valued, continuous, smooth path a particle would arrive at B in minimum time under uniform gravity, after beginning at rest at a higher point A. Their solution, beyond the scope of this book, was that this path of "quickest descent" is a cycloid. Show that, regardless of the path, the *speed on arrival* is as if the particle had been dropped freely through the same height H.

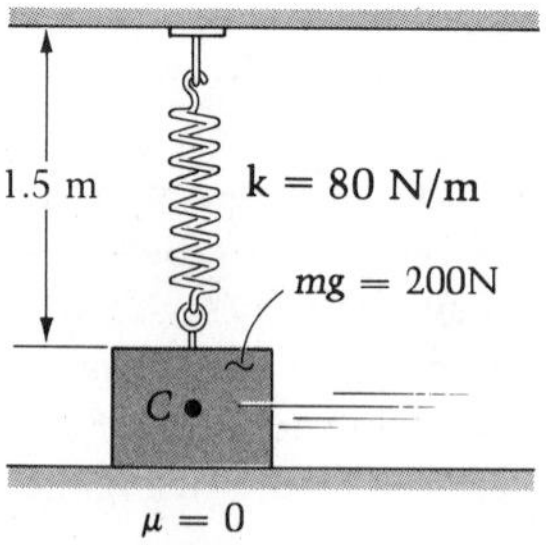

Figure P2.108

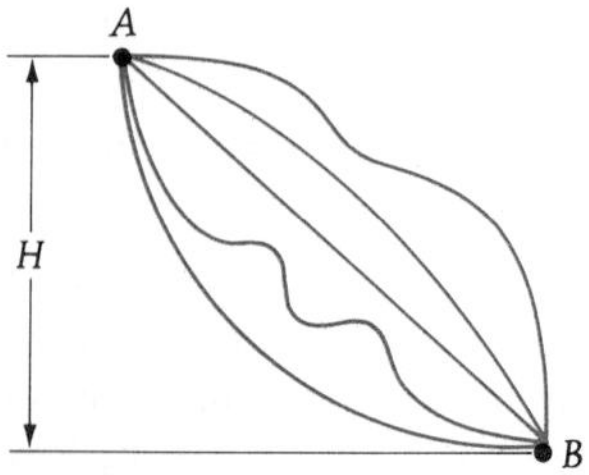

Figure P2.110

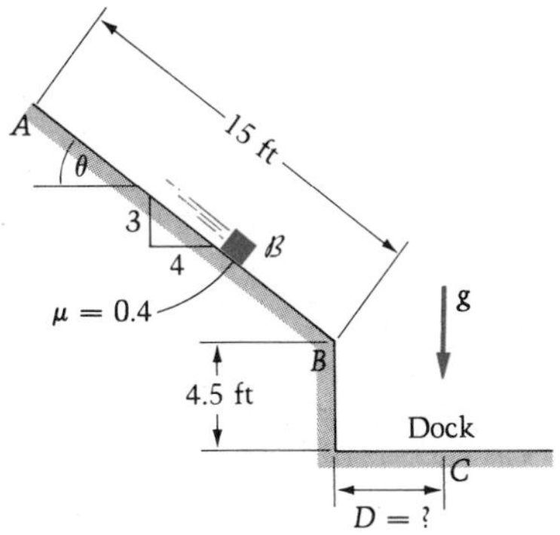

Figure P2.111

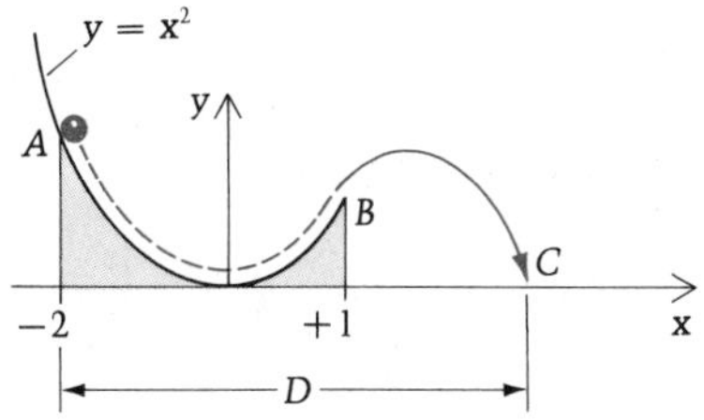

Figure P2.112

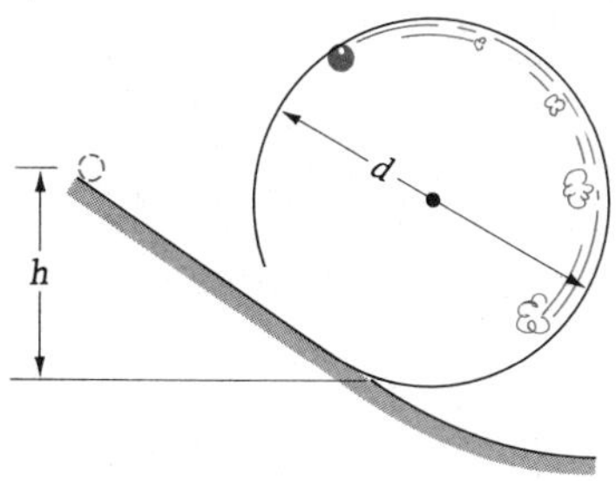

Figure P2.113

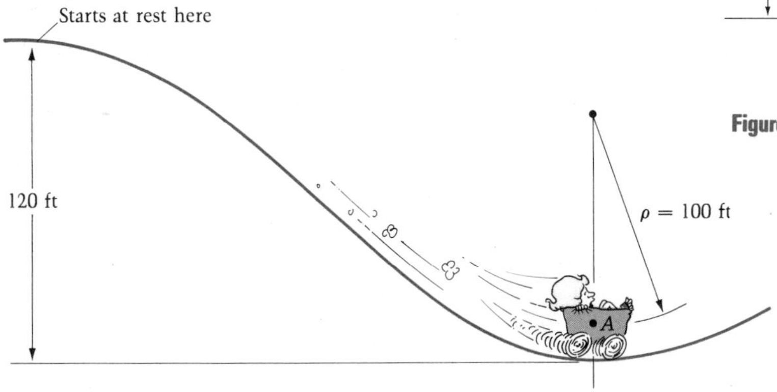

Figure P2.114

2.111 A small box $\mathcal{B}$ (see Figure P2.111) slides from rest down a rough inclined plane from A to B and then falls onto the loading dock. The coefficient of sliding friction between the box and plane is $\mu = 0.4$. Find the distance D to the point C where the box strikes the dock.

2.112 A particle is released at rest at A and slides on the smooth parabolic surface to B, where it flies off. (See Figure P2.112.) Find the total horizontal distance D that it travels before hitting the ground at C.

2.113 A particle of mass m slides down a frictionless chute and enters a circular loop of diameter d. (See Figure P2.113.) Find the minimum starting height h in order that the particle will make a complete circuit of the loop and exit normally (without having lost contact with the loop).

2.114 An 80-lb child rides a 10-lb wagon down an incline (Figure P2.114). Neglecting all frictional losses, find the "weight" of the child at A as indicated by a scale upon which she is sitting.

2.115 A skier descends the smooth slope, which may be approximated by the parabola $y = \frac{1}{20}x^2 - 5$ (see Figure P2.115). If she starts from rest at "A" and has a mass of 52 kg, determine the normal force she exerts on the ground the instant she arrives at "B", and her acceleration there. Note: treat the skier as a particle, and neglect friction.

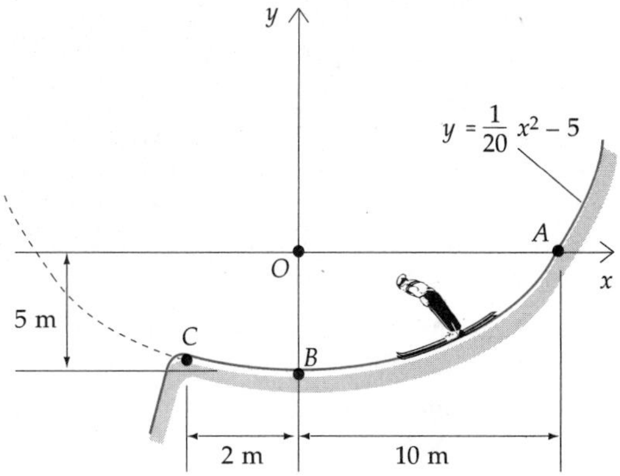

Figure P2.115

2.116 Find the speed sought in Problem 2.81 using work and kinetic energy.

2.117 Use work and kinetic energy to solve Problem 2.83.

2.118 Use work and kinetic energy to solve Problem 2.94.

2.119 Use work and kinetic energy to solve Problem 2.86.

2.120 Show that the equation $m\ell\dot{\theta}^2 = 2mg(\sin\theta - \frac{1}{2})$ can be obtained by conservation of mechanical energy ($T + \varphi =$ constant) in Example 2.10. Why can this principle not be used in Examples 2.11 and 2.12?

2.121 Check the solutions to Problems 2.78 and 2.79 by using the principle of work and kinetic energy.

2.122 Show that for central gravitational force $[\mathbf{F} = (-GMm/r^2)\hat{\mathbf{e}}_r]$, as distinct from the uniform gravity $(-mg\hat{\mathbf{k}})$ in the text, the potential is given by

$$\varphi = \frac{-GMm}{r}$$

where G is the universal gravitational constant and M and m are the masses of the two attracting bodies. Note that in view of Equation (2.19), one simply needs to show that $gr_e^2 = GM$.

2.123 Using the result of the preceding problem, calculate the work done by the earth's gravity on a satellite between the times of launch and insertion into a geosynchronous orbit with radius 6.61 times the radius of earth. (See Problem 2.64.)

2.124 For Problem 2.49, use $W = \Delta T$ to find the speeds of the blocks when the 5-lb block has droppped 2 ft.

2.125 Block $\mathcal{A}$ in Figure P2.125 is moving downward at 5 ft/sec at a certain time when the spring is compressed 6 in. The coefficient of friction between block $\mathcal{B}$ and the plane is 0.2, the pulley is light, and the weights of $\mathcal{A}$ and $\mathcal{B}$ are 161 and 193 lb, respectively.

a. Find the distance that $\mathcal{A}$ falls from its initial position before coming to zero speed.

b. Determine whether or not body $\mathcal{A}$ will start to move back upward.

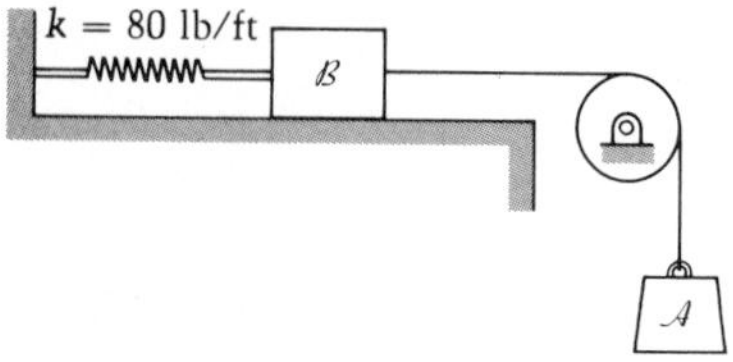

Figure P2.125

2.126 The system in Figure P2.126 consists of the 12-lb body $\mathcal{A}$, the light pulley $\mathcal{B}$, the 8-lb "rider" $\mathcal{C}$, and the 10-lb body $\mathcal{D}$. Everything is released from rest in the given position. Body $\mathcal{D}$ then falls through a hole in bracket $\mathcal{E}$, which stops body $\mathcal{C}$. Find how far $\mathcal{D}$ descends from its original position.

2.127 At the instant shown in Figure P2.127, block $\mathcal{B}$ is 30 m below the level of block $\mathcal{A}$. At this time, $\mathbf{v}_A$ and $\mathbf{v}_B$ are zero. Determine the velocities of $\mathcal{A}$ and $\mathcal{B}$ as they pass each other. $\mathcal{A}$ and $\mathcal{B}$ have masses of 15 kg and 5 kg, respectively. The pulleys are light.

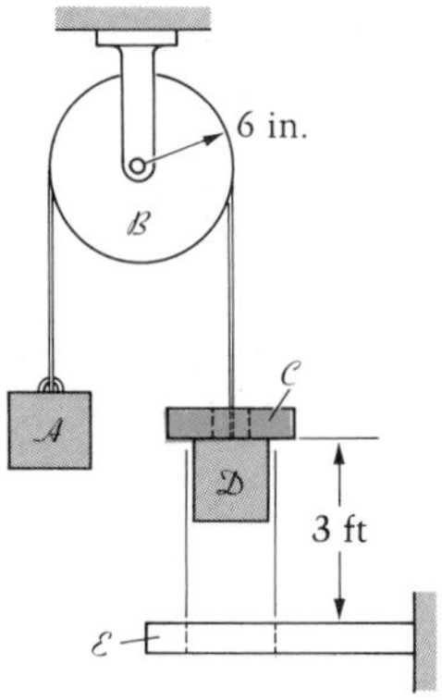

Figure P2.126

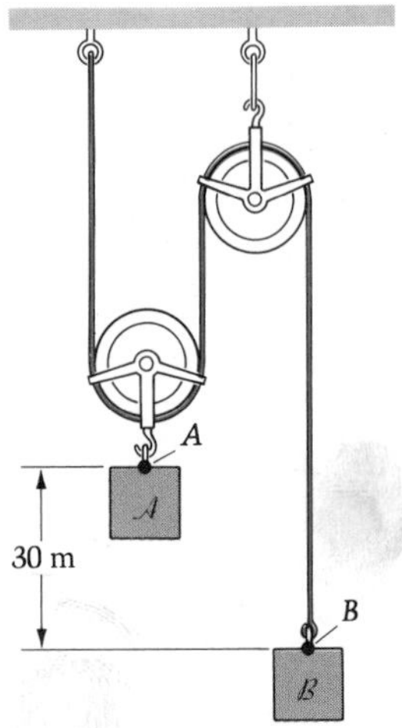

Figure P2.127

2.128 Suppose the ends of a spring are attached to "particles" of mass m_1 and m_2. Show that the sum of the works of the spring forces on the particles is given by Equation (2.17).

* **2.129** The blocks in Figure P2.129 are released from rest. Determine where they are when they stop permanently. What is the spring force then? *Hint:* Write the work-energy equation for each block, add the two equations, and use the result of Problem 2.125. Also, think about the motion of the mass center.

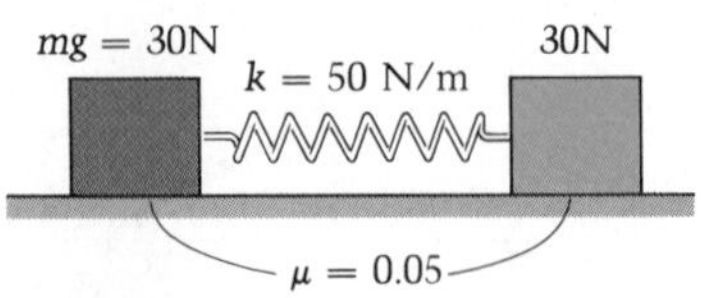

Figure P2.129

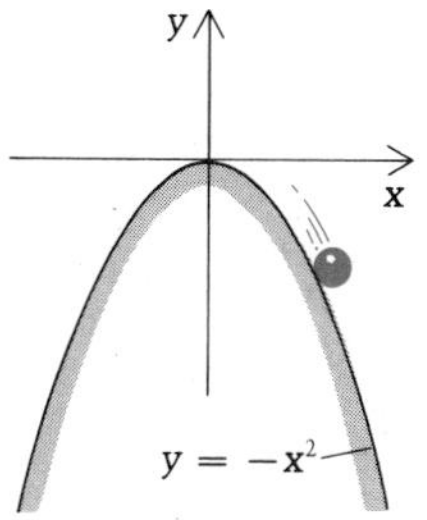

Figure P2.130

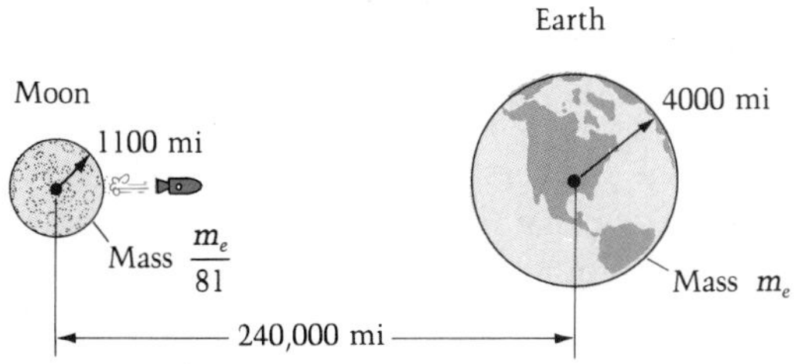

Figure P2.131

* **2.130** Show that if the surface in Example 2.13 is the parabola shown in Figure P2.130, the particle will *never* leave the surface. *Hint*: Show that

$$\rho = \left| \frac{(1 + y'^2)^{3/2}}{y''} \right| = \frac{(1 + 4x^2)^{3/2}}{2}$$

and use this in our equation:

$$\Sigma F_n = m \frac{\dot{s}^2}{\rho}$$

together with $W = \Delta T$.

* **2.131** Find the least velocity with which a particle could be projected from the moon and reach the earth. (See Figure P2.131.) For this problem assume that the centers of the moon and earth are both fixed in an inertial frame.

2.5 Momentum Form of Euler's First Law

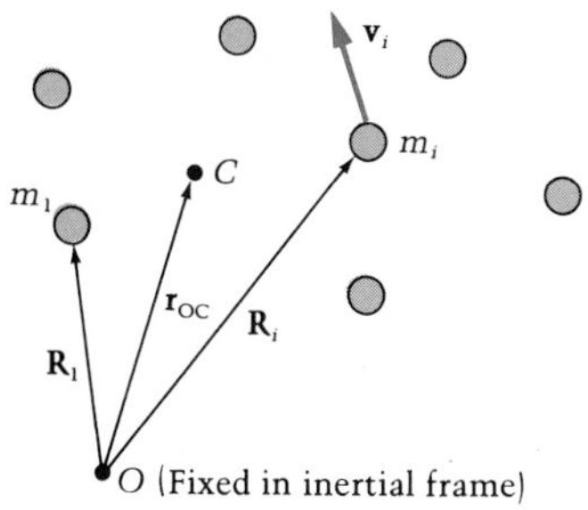

Figure 2.8a

The **momentum** of a particle is defined to be the product of its mass and its velocity. For a system of particles (Figure 2.8a) the momentum is defined to be the sum of the momenta of the particles in the system. Thus, if we denote the momentum of a system (or body) by **L**,* then

$$\mathbf{L} \equiv \sum_{i=1}^{N} m_i \mathbf{v}_i \tag{2.24}$$

or, for a body of continuously distributed mass (Figure 2.8b),

$$\mathbf{L} \equiv \int \mathbf{v}\, dm \tag{2.25}$$

If $\mathbf{R}_i$ is a position vector for the ith particle, then

$$\mathbf{v}_i = \frac{d\mathbf{R}_i}{dt}$$

and Equation (2.24) becomes

$$\mathbf{L} = \sum_{i=1}^{N} m_i \frac{d\mathbf{R}_i}{dt}$$

or

$$\mathbf{L} = \frac{d}{dt}(\Sigma m_i \mathbf{R}_i)$$

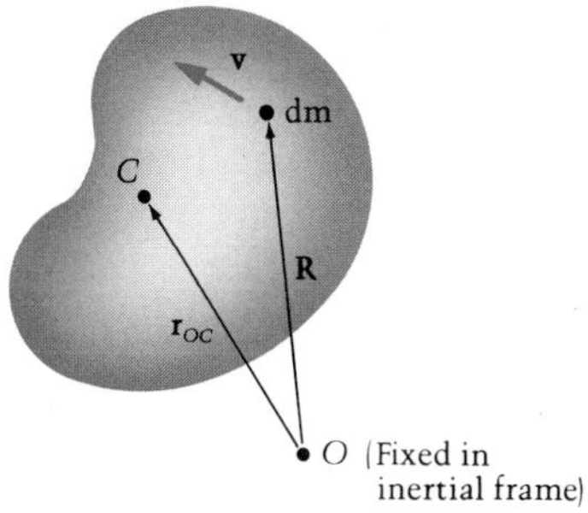

Figure 2.8b

* Momentum is sometimes called *linear momentum.*

Recalling [see Equation (2.7) and Figure 2.3a] that

$$\Sigma m_i \mathbf{R}_i = m\mathbf{r}_{OC}$$

then

$$\mathbf{L} = \frac{d}{dt}(m\mathbf{r}_{OC})$$

$$= m\frac{d\mathbf{r}_{OC}}{dt}$$

or

$$\mathbf{L} = m\mathbf{v}_C \tag{2.26}$$

where $\mathbf{v}_C$ is the velocity of the mass center of the system or body.

The connection between external forces and momentum now can be made easily by differentiating Equation (2.26) to obtain

$$\frac{d\mathbf{L}}{dt} = m\frac{d\mathbf{v}_C}{dt}$$

$$= m\mathbf{a}_C$$

But

$$\Sigma\mathbf{F} = m\mathbf{a}_C$$

so that

$$\Sigma\mathbf{F} = \frac{d\mathbf{L}}{dt} \tag{2.27}$$

which is the momentum form of Euler's first law.

Question 2.7 Should we expect Equation (2.27) to be valid for a system for which the mass is changing with time, such as a rocket with its varying-mass contents?

Impulse and Momentum; Conservation of Momentum

A straightforward integration of the first law of motion (Equation 2.27) yields

$$\int_{t_1}^{t_2} \Sigma\mathbf{F}\,dt = \mathbf{L}(t_2) - \mathbf{L}(t_1) = \text{change in momentum} = \Delta\mathbf{L} \tag{2.28}$$

where the integral is usually called the impulse imparted to the body by the external forces;* note that the impulse is intrinsically associated with

---

Answer 2.7 No, at several places in the development we have needed to require that the mass be constant.

* The impulse is sometimes called the *linear impulse.*

a specific time interval. If, during some time interval, the sum of the external forces vanishes, then $\dot{\mathbf{L}} = \mathbf{0}$ and hence the momentum is a constant, or is *conserved*, during that interval.

Since Equation (2.28) is a vector equation (unlike the scalar work and kinetic energy equation), we may use any or all of its component equations. For example:

$$\int_{t_1}^{t_2} \Sigma F_x \, dt = L_x(t_2) - L_x(t_1) = m\dot{x}_C(t_2) - m\dot{x}_C(t_1) \tag{2.29}$$

and similarly for y and z. We note that we may have a planar situation, for example, in which $\Sigma F_x = 0$ but $\Sigma F_y \neq 0$ over an interval. If this is the case, momentum is conserved in the x direction but not in the y direction.*

Impact

Sometimes it is possible, by conservation of momentum, to obtain limited quantitative information about the motions of colliding bodies. As a rule this can be done when the bodies interact for a relatively brief interval — *before* and *after* which it is reasonable to treat their motions as rigid. While the analysis is best discussed with examples, we make the observation here that generally it makes little sense to treat the bodies as rigid during the collision. If we wish to describe the motion that ensues when a bullet is fired into a wooden block, for example, the block clearly cannot be regarded as rigid during the penetration process. On the other hand, it may be quite plausible to assume that rigid motion of the block and embedded bullet occurs subsequent to permanent reorientation of material.

A key feature in the analysis of collision (or impact) problems is the fact that the momentum of a body made up of two parts is the sum of the momenta of the individual parts. This feature follows directly from the definition of a body's momentum as the integral

$$\begin{aligned} \mathbf{L} &= \int_{\mathcal{B}} \mathbf{v}\, dm = \int_{\mathcal{B}_1} \mathbf{v}\, dm + \int_{\mathcal{B}_2} \mathbf{v}\, dm \\ &= \mathbf{L}_1 + \mathbf{L}_2 \end{aligned} \tag{2.30}$$

where the subscripts (1 and 2) identify the two constituent parts of the body.

Example 2.14

A wooden block of mass m_1 is at rest on a smooth horizontal surface when it is struck by a bullet of mass m_2 traveling at a speed v as shown in Figure E2.14a. After the bullet becomes embedded in the block, the block slides to the right at speed V. Find the relationship between v and V.

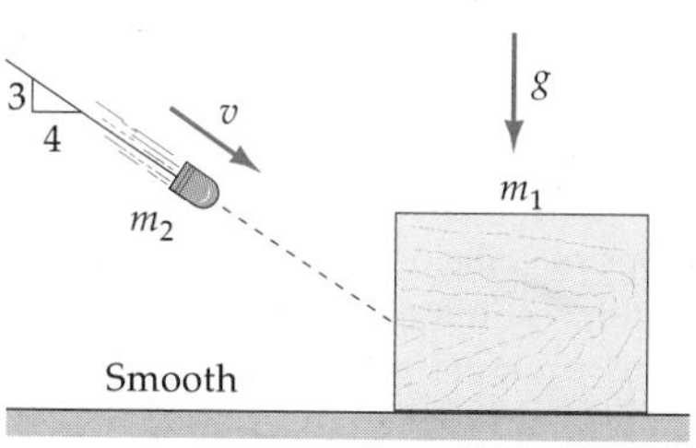

Figure E2.14a

* Ballistics problems are of this type if air resistance is neglected.

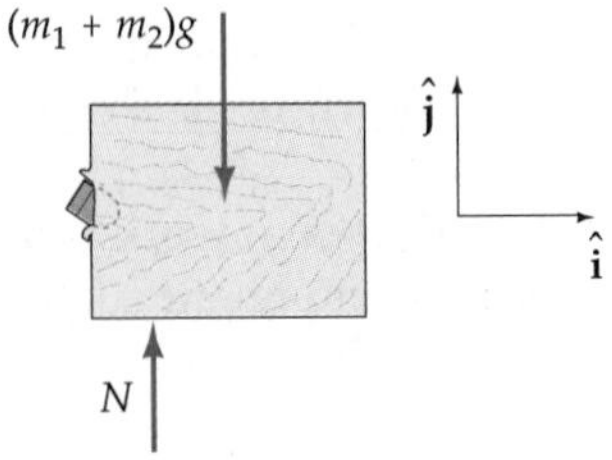

Figure E2.14b

Solution

Let t_1 be the time at which the bullet first contacts the block and let t_2 be the time after which the bullet / block composite behaves as a rigid body in translation. For $t_1 < t < t_2$, a complex process of deformation and redistribution of mass is occurring within the block. If we isolate the block / bullet system during this interval (see Figure E2.14b), Equation (2.28) yields

$$\int_{t_1}^{t_2} [N(t) - m_1 g - m_2 g]\hat{\mathbf{j}}\, dt = \mathbf{L}(t_2) - \mathbf{L}(t_1)$$

But

$$\mathbf{L}(t_2) = (m_1 + m_2)V\hat{\mathbf{i}}$$

and

$$\mathbf{L}(t_1) = m_2(0.8v\hat{\mathbf{i}} - 0.6v\hat{\mathbf{j}})$$

since v is the speed of the mass center of the bullet and the block has no momentum at t_1. Therefore, equating the $\hat{\mathbf{i}}$ coefficients, we get

$$(m_1 + m_2)V = 0.8m_2 v$$

or

$$V = \frac{0.8m_2 v}{m_1 + m_2}$$

We note that, in the absence of an external force with a horizontal component, the horizontal component of momentum is conserved.

While we cannot calculate the reaction N during the collision, we *can* calculate its impulse:

$$\int_{t_1}^{t_2} [N - (m_1 + m_2)g]\, dt = 0.6m_2 v$$

or

$$\int_{t_1}^{t_2} N\, dt = (m_1 + m_2)g(t_2 - t_1) + 0.6m_2 v$$

Similarly, the impulse of the force $\mathbf{F}$ exerted on the bullet by the block can be calculated if we apply Equation (2.28) to the bullet:

$$\int_{t_1}^{t_2} (\mathbf{F} - m_2 g\hat{\mathbf{j}})\, dt = m_2 V\hat{\mathbf{i}} - m_2(0.8v\hat{\mathbf{i}} - 0.6v\hat{\mathbf{j}})$$

$$\int_{t_1}^{t_2} \mathbf{F}\, dt = m_2(V - 0.8v)\hat{\mathbf{i}} + m_2[g(t_2 - t_1) + 0.6v]\hat{\mathbf{j}}$$

$$= m_2\left(\frac{0.8m_2 v}{m_1 + m_2} - 0.8v\right)\hat{\mathbf{i}} + m_2[g(t_2 - t_1) + 0.6v]\hat{\mathbf{j}}$$

$$= -\frac{0.8m_1 m_2}{m_1 + m_2} v\hat{\mathbf{i}} + m_2[g(t_2 - t_1) + 0.6v]\hat{\mathbf{j}}$$

The reader should note that with a high-speed collision occurring in a short period of time, the impulses can be accurately estimated by neglecting the impulses of the weights of the bodies. In this example we would have $0.6v \gg g(t_2 - t_1)$. Other examples of impact problems are treated in Chapter 5.

EXAMPLE 2.15

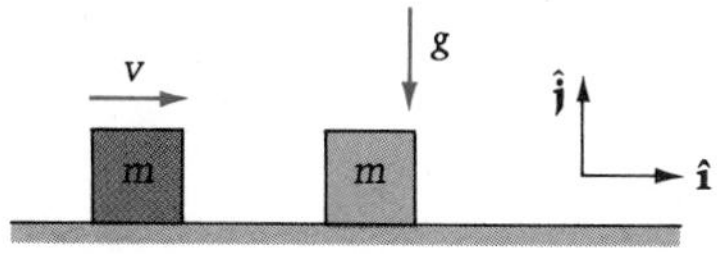

Figure E2.15a

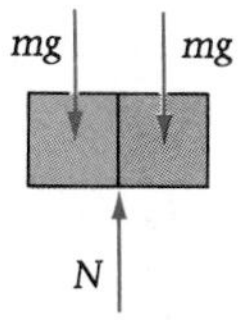

Figure E2.15b

A block is at rest on a smooth horizontal surface before being struck by an identical block sliding at speed v. (See Figure E2.15a.) Find the velocities of the two blocks after the collision assuming (1) that they stick together or (2) that the system experiences no loss in kinetic energy.

Solution

Let v_L and v_R be the speeds of the mass centers of the left and right blocks at the end of the collision; that is, $v_L\hat{\mathbf{i}}$ is the velocity of the left block. The free-body diagram of the system of two blocks during the collision (see Figure E2.15b) shows that there is no external force with a horizontal component. Thus the horizontal component (the only component not zero here) of momentum is conserved and

$$mv_L + mv_R = mv + m(0) \tag{1}$$

or

$$v_L + v_R = v \tag{2}$$

If the blocks remain attached after the collision is completed and they are behaving as rigid bodies, we have

$$v_R = v_L$$

so that

$$v_R = v_L = \frac{v}{2} \tag{3}$$

Question 2.8 What would be the common velocity if the right block had 100 times as much mass as the left block? If it had 10,000 times as much mass?

If, however, the blocks do not stick together, the conservation of momentum statement alone is not adequate to determine their subsequent velocities. What we need is some measure of their tendency to bounce off each other — or, to put it another way, a measure of how much energy is expended in permanent deformations or vibrations (or both) of the blocks. The parameter used to describe these effects (the coefficient of restitution) is discussed in the text that follows this example. At this point we simply note that when the blocks stick together the kinetic energy of the system is less after the collision than before. That loss is

$$\frac{1}{2}mv^2 - \left[\frac{1}{2}m\left(\frac{v}{2}\right)^2 + \frac{1}{2}m\left(\frac{v}{2}\right)^2\right] = \frac{1}{4}mv^2 \tag{4}$$

which is to say that one-half of the mechanical energy was dissipated in the collision in this case.

The other extreme case is that in which *no* mechanical energy is expended during the collision process. In this case

---

Answer 2.8 $v/101$; $v/10{,}001$.

$$\frac{1}{2}mv_L^2 + \frac{1}{2}mv_R^2 = \frac{1}{2}mv^2 \tag{5}$$

But since $v_R = v - v_L$ from Equation (2), we have

$$v_L^2 + (v - v_L)^2 = v^2$$

$$v_L^2 + v^2 - 2vv_L + v_L^2 = v^2 \tag{6}$$

or

$$2v_L(v_L - v) = 0 \tag{7}$$

Therefore either $v_L = 0$ and $v_R = v$, or $v_L = v$ and $v_R = 0$. The latter case must be rejected as physically meaningless since it would require the left block to pass through the stationary right block.

An extension of this result to the case of three blocks is shown in Figures E2.15c and E2.15d:

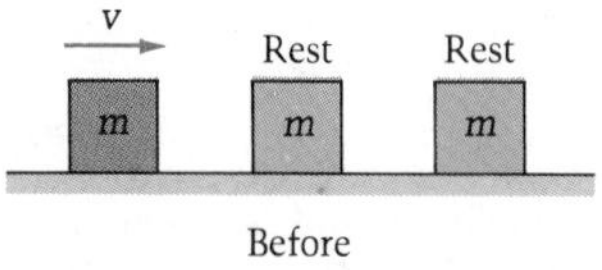

Figure E2.15c

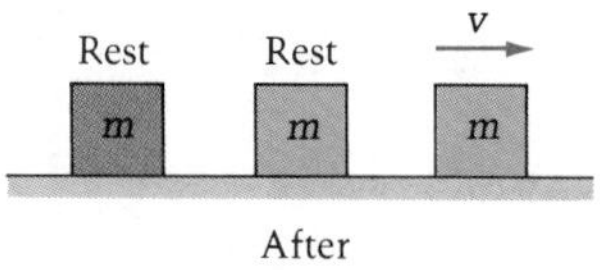

Figure E2.15d

If we let the spacing between the blocks initially at rest approach zero and add more of them, then we have the mechanism for a popular adult toy (see Figures E2.15e,f):

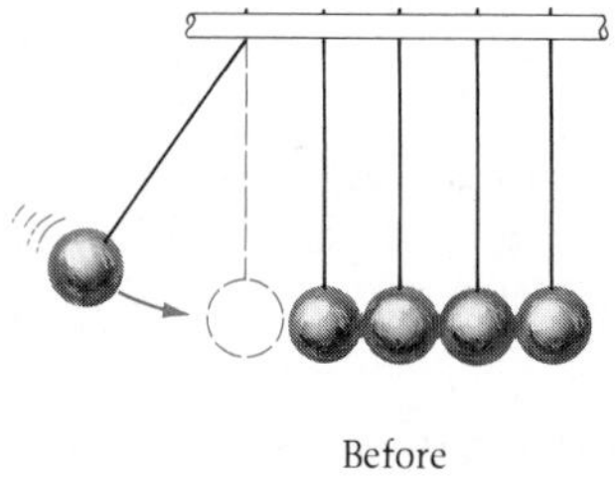

Figure E2.15e

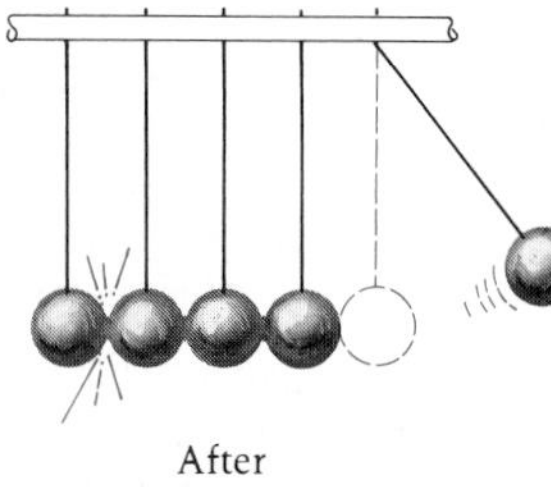

Figure E2.15f

Coefficient of Restitution

In the preceding example we noted the need for some measure of the capacity of colliding bodies to rebound off each other. The introduction of a parameter called the coefficient of restitution which provides this information is most easily accomplished through a simple example.

Suppose that, as depicted in Figure 2.9, two disks are sliding along a smooth floor. The paths are the same straight line and disk $\mathcal{A}$ is just about to overtake and contact disk $\mathcal{B}$ at time t_1. The centers of mass of the disks will approach each other until, at time t_2, they have the common velocity v_C. Then they will recede from each other until at time t_3 contact is broken. The equal and opposite forces of interaction $F(t)$ are shown on the disks in Figure 2.9. Applying the impulse-momentum principle dur-

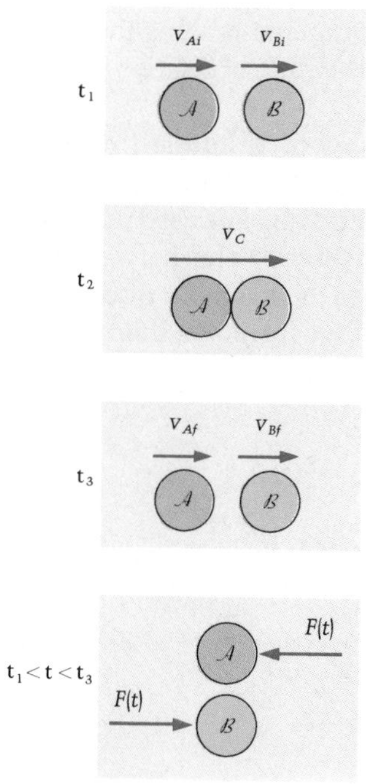

Figure 2.9

ing the intervals of approach and then separation of the centers of mass of the disks,

$$-\int_{t_1}^{t_2} F\,dt = m_{\mathcal{A}}(v_C - v_{Ai})$$

$$\int_{t_1}^{t_2} F\,dt = m_{\mathcal{B}}(v_C - v_{Bi})$$

$$-\int_{t_2}^{t_3} F\,dt = m_{\mathcal{A}}(v_{Af} - v_C) \tag{2.31}$$

and

$$\int_{t_2}^{t_3} F\,dt = m_{\mathcal{B}}(v_{Bf} - v_C)$$

Defining the coefficient of restitution, e, by

$$e \equiv \frac{\int_{t_2}^{t_3} F\,dt}{\int_{t_1}^{t_2} F\,dt} \tag{2.32}$$

we obtain, after using the impulse-momentum equations above,

$$e = \frac{v_{Af} - v_C}{v_C - v_{Ai}} = \frac{v_{Bf} - v_C}{v_C - v_{Bi}}$$

Eliminating v_C, there results

$$e = \frac{v_{Bf} - v_{Af}}{v_{Ai} - v_{Bi}} \tag{2.33}$$

which is seen to be the quotient of the "relative velocity of separation" and the "relative velocity of approach." The coefficient of restitution is inherently nonnegative, and the case $e = 0$ yields $v_{Af} = v_{Bf}$, which means that the disks stick together. In Example 2.15, for the case of no energy dissipation, we had

$$e = \frac{v - 0}{v - 0} = 1$$

Exercise Problem 2.147 provides an outline of proof that, under the conditions of our discussion here, $e \leqq 1$.

The impact just described is called *central,* because the line of action of the equal and opposite forces of interaction is the line joining the mass centers of the bodies. It is also called *direct* because the preimpact velocities are parallel to that line of action. Generalization to the case of indirect, but still central, impact is easily accomplished, assuming the disks are smooth and that the time of contact is so small that there are no significant changes in their positions during the collision. Then, the velocity components perpendicular to the line of action of the impulsive force (called the line of impact) are unchanged by the collision. Equations (2.31) and (2.33) now refer to velocity components along the line of impact. Thus, the coefficient of restitution is the ratio of relative velocity components along the line of impact.

Experiments* indicate that the coefficient of restitution depends upon just about everything involved in an impact: materials, geometry, and initial velocities. Therefore, numerical values must be used with care. Nonetheless, the fact that the coefficient must have a value between zero and unity is valuable information in bounding the behavior of colliding bodies. The use of the coefficient of restitution for other than central impact is discussed in Chapter 5.

EXAMPLE 2.16

Two identical hockey pucks collide, coming into contact in the positions shown. Their velocities before the collision are also shown in Figure E2.16a.

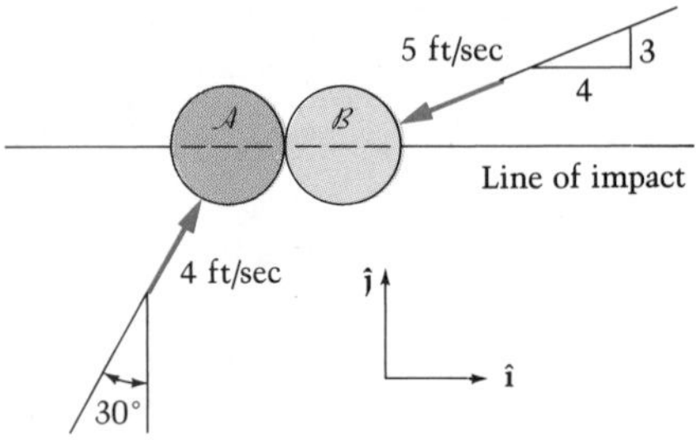

Figure E2.16a

If the coefficient of restitution is 0.8, find the velocities of the pucks after the collision. Then find the impulse of the force of interaction.

Solution

Neglecting friction and assuming insignificant deformation, the forces of interaction will act along the line of impact shown in Figure E2.16a. It is convenient to choose $\hat{\mathbf{i}}$ and $\hat{\mathbf{j}}$ parallel and perpendicular to this line as shown. Let m be the mass of each puck and let $\mathbf{v}_{Af}$ and $\mathbf{v}_{Ai}$ be the final and initial velocities of puck $\mathcal{A}$ and similarly for puck $\mathcal{B}$.

$$\mathbf{v}_{Ai} = 4(\sin 30°\hat{\mathbf{i}} + \cos 30°\hat{\mathbf{j}}) = 2\hat{\mathbf{i}} + 3.46\hat{\mathbf{j}} \text{ ft/sec}$$

Thus

$$\mathbf{v}_{Af} = \mathbf{v}_1\hat{\mathbf{i}} + 3.46\hat{\mathbf{j}}$$

where v_1 is the unknown component along the line of impact. Also

$$\mathbf{v}_{Bi} = 5\left(-\frac{4}{5}\hat{\mathbf{i}} - \frac{3}{5}\hat{\mathbf{j}}\right) = -4\hat{\mathbf{i}} - 3\hat{\mathbf{j}} \text{ ft/sec}$$

and

$$\mathbf{v}_{Bf} = v_2\hat{\mathbf{i}} - 3\hat{\mathbf{j}}$$

* See W. Goldsmith, *Impact* (London: Edward Arnold Publishers, Ltd., 1960).

Since there are no external forces on the system of two pucks, momentum is conserved:

$$m\mathbf{v}_{Af} + m\mathbf{v}_{Bf} = m\mathbf{v}_{Ai} + m\mathbf{v}_{Bi}$$

The component equation for the directions perpendicular to the line of impact is automatically satisfied, and for the $\hat{\mathbf{i}}$-direction

$$mv_1 + mv_2 = m(2) + m(-4)$$

or

$$v_1 + v_2 = -2 \tag{1}$$

By the definition of the coefficient of restitution

$$e = 0.8 = \frac{v_2 - v_1}{2 - (-4)}$$

or

$$v_2 - v_1 = 4.80 \tag{2}$$

Solving (1) and (2),

$$v_1 = -3.40$$

$$v_2 = 1.40$$

so that

$$\mathbf{v}_{Af} = -3.40\hat{\mathbf{i}} + 3.46\hat{\mathbf{j}} \text{ ft/sec}$$

and

$$\mathbf{v}_{Bf} = 1.40\hat{\mathbf{i}} - 3\hat{\mathbf{j}} \text{ ft/sec}$$

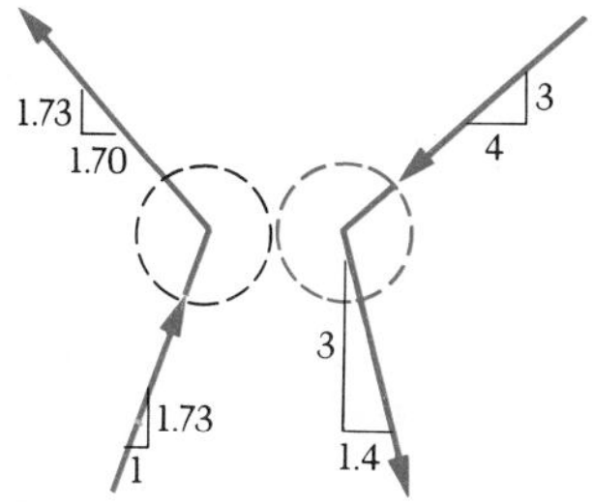

Figure E2.16b

Thus the paths of the pucks are as shown in Figure E2.16b.

To compute the impulse of the force of interaction we apply the impulse-momentum principle to puck $\mathcal{B}$, noting that a hockey puck weighs about 6 ounces. Thus with $m = (6/16)/32.2 = 0.0116$ slug,

$$\begin{aligned} \int F\,dt &= 0.0116[1.40 - (-4)] \\ &= 0.0626 \text{ lb-sec} \end{aligned}$$

Example 2.17

In Example 2.9 we found the velocities of the identical Cars 1 and 2, just after they collided in the given position, to be 40.1 and 25.4 ft/sec, respectively, as shown in Figure E2.17. Now, using the principle of impulse and momentum, find the velocities of the cars *prior* to impact. Remember that Cars 1 and 2 were heading west and south, respectively. Assume the collision is instantaneous.

Solution

If the impact occurs over a vanishingly small time Δt, then the impulse from the road (due to the friction force on the tires) during the impact is negligible, so that the linear momentum of the system of two cars may be assumed to be conserved

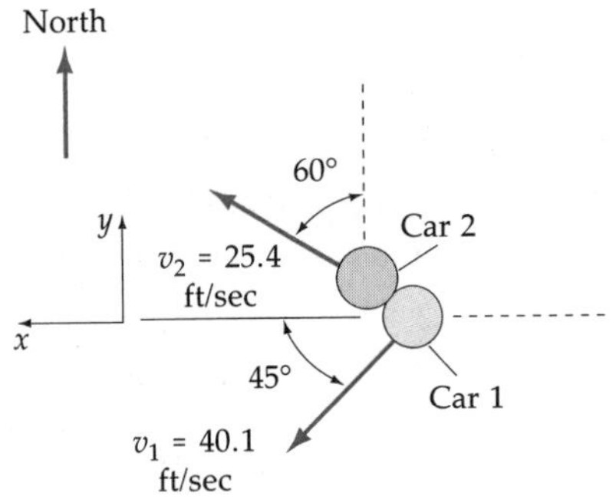

Figure E2.17

during Δt. In expressing this conservation, we shall use "i" for initial (before impact) and "f" for final (after impact):

$$mv_{1i}\hat{\mathbf{i}} + mv_{2i}(-\hat{\mathbf{j}}) = m(v_{1fx}\hat{\mathbf{i}} + v_{1fy}\hat{\mathbf{j}}) + m(v_{2fx}\hat{\mathbf{i}} + v_{2fy}\hat{\mathbf{j}})$$

or

$$v_{1i} = v_{1fx} + v_{2fx} \tag{1}$$

and

$$-v_{2i} = v_{1fy} + v_{2fy} \tag{2}$$

Using the results of Example 2.9 and the angles in the figure above, we obtain the following post-collision velocity components:

$$v_{1fx} = 40.1 \cos 45° = 28.4 \text{ ft/sec}$$
$$v_{1fy} = -40.1 \sin 45° = -28.4 \text{ ft/sec}$$
$$v_{2fx} = 25.4 \sin 60° = 22.0 \text{ ft/sec}$$
$$v_{2fy} = 25.4 \cos 60° = 12.7 \text{ ft/sec}$$

Using Equations (1,2), we find

$$\begin{aligned} v_{1i} &= v_{1fx} + v_{2fx} \\ &= 28.4 + 22.0 = 50.4 \text{ ft/sec} \end{aligned}$$

$$\begin{aligned} v_{2i} &= -v_{1fy} - v_{2fy} \\ &= +28.4 - 12.7 = 15.7 \text{ ft/sec} \end{aligned}$$

We remark that the energy lost during the collision may now be calculated:

$$\begin{aligned} E_{\text{lost}} = KE_i - KE_f &= [\tfrac{1}{2}m(50.4)^2 + \tfrac{1}{2}m(15.7)^2] - [\tfrac{1}{2}m(40.1)^2 + \tfrac{1}{2}m(25.4)^2] \\ &= 1390m - 1130m = 260m \end{aligned}$$

Note that the work done by the road friction was (see Example 2.9) equal to $0.5(32.2)(50 + 20)m = 1130m$, and that this energy change plus the $260m$ lost in the collision (to deformation, sound, vibration, etc.) gives the total original kinetic energy, 1390 m.

PROBLEMS ▶ Section 2.5

2.132 Figure P2.132 presents data pertaining to a system of two particles. At the instant shown find the:

a. Position of the mass center
b. Kinetic energy of the system
c. Linear momentum of the system
d. Velocity of the mass center
e. Acceleration of the mass center

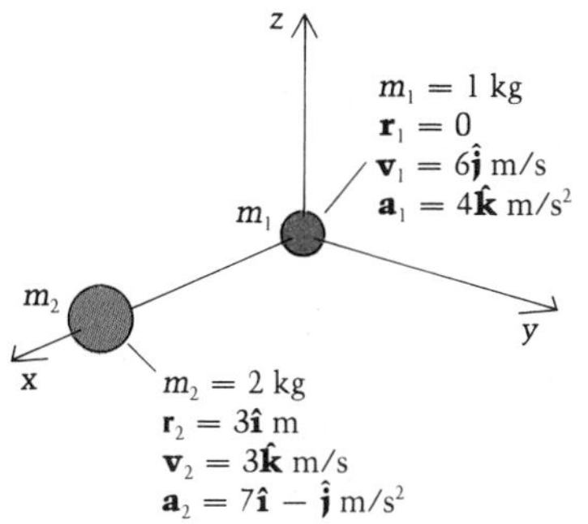

Figure P2.132

Figure P2.133

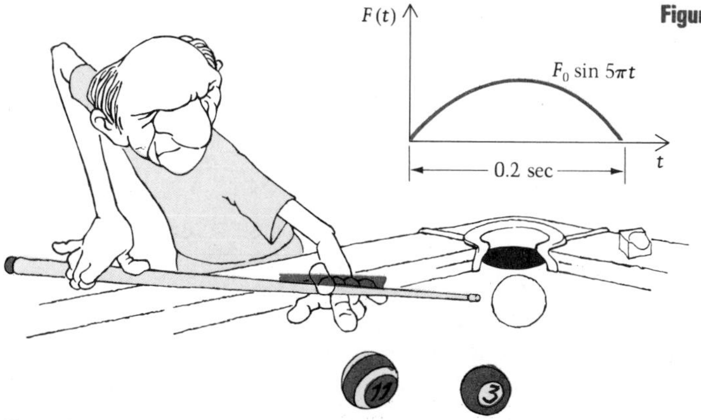

Figure P2.134

2.133 The astronaut in Figure P2.133 is finding it difficult to stop his forward momentum while jogging on the moon. Using a friction coefficient of $\mu = 0.3$ and a gravitational acceleration one-sixth that of earth's, illustrate the difficulty of stopping a forward momentum of mv = (5 slugs)(12 ft / sec). Specifically, use the principle of impulse and momentum to find the time it takes to stop on earth versus on the moon.

2.134 A horizontal force $F(t)$ is applied for 0.2 sec to a cue ball (weighing 0.55 lb) by a cue stick; the form of the force is as shown in Figure P2.134. If the velocity of the center of the ball is 8 ft / sec after contact with the stick is broken, find the peak magnitude F_0 of the force. Neglect friction. Force F is measured in pounds.

2.135 The 50-lb box shown in Figure P2.135 is at rest before the force $F(t) = 5 + 2t$ pounds is applied at $t = 0$. Assume the box to be wide enough not to tip over and suppose the coefficient of friction between box and floor to be 0.2. Find the velocity of (the mass center of) the box at $t = 10$ sec.

2.136 Repeat the preceding problem for the case in which the force $F(t)$ has a vertical component as shown in Figure P2.136.

2.137 A force P applied to $\mathcal{C}$ at $t = 0$ varies with time according to $P = 25 \sin(\pi t / 60)$ lb, where t is in seconds. (See Figure P2.137.) How long will it take for $\mathcal{C}$ to begin sliding? What will be its velocity at $t = 30$ sec?

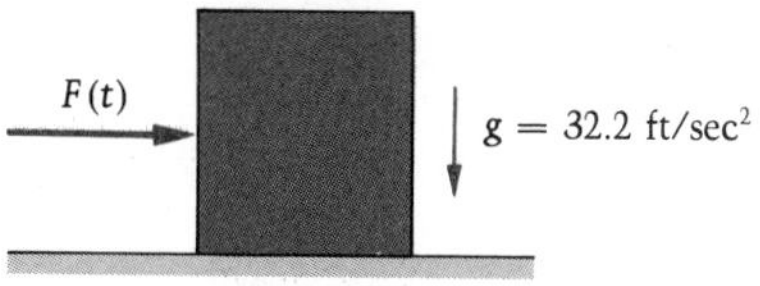

Figure P2.135

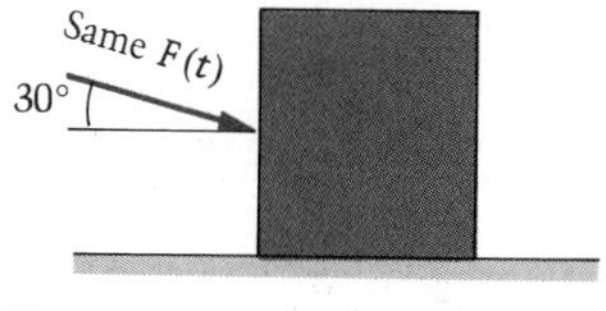

Figure P2.136

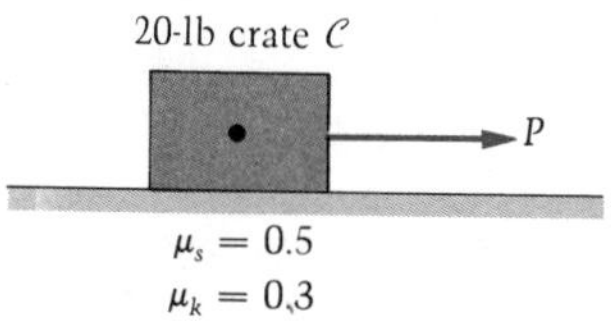

Figure P2.137

2.138 An unattached 2.2-lb roofing shingle slides downward and strikes a gutter. (See Figure P2.138.) The angle at which the shingle would be just on the verge of slipping is 20°. Determine the impulse imparted to the shingle by the gutter if there is no rebound. If the interval of impact is 0.1 sec, find the average force imparted to the gutter by the shingle.

2.139 Two railroad cars are coupled by a collision occurring just after the instant shown in Figure P2.139. Neglecting the impulse caused by friction from the tracks, determine the final velocity of the two cars as they move together.

2.140 In the preceding problem, find the average impulsive force between the cars if the coupling requires 0.6 sec of contact.

2.141 In a rail yard a freight car moving at speed v strikes two identical cars at rest. (See Figure P2.141.) Neglecting any resistance to rolling, find the common velocity of the three-car system after the coupling has been completed and any associated vibrations have died out.

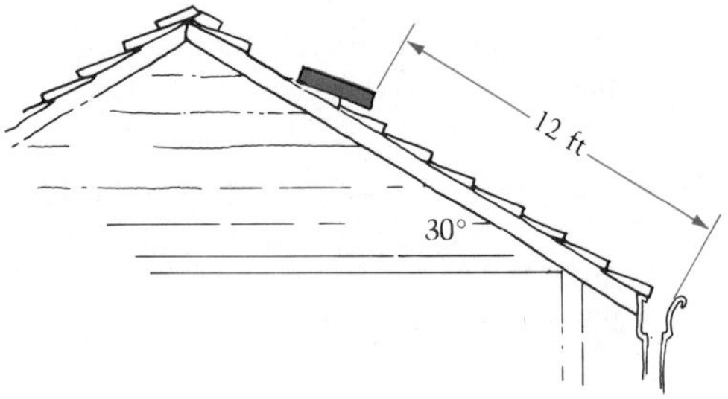

Figure P2.138

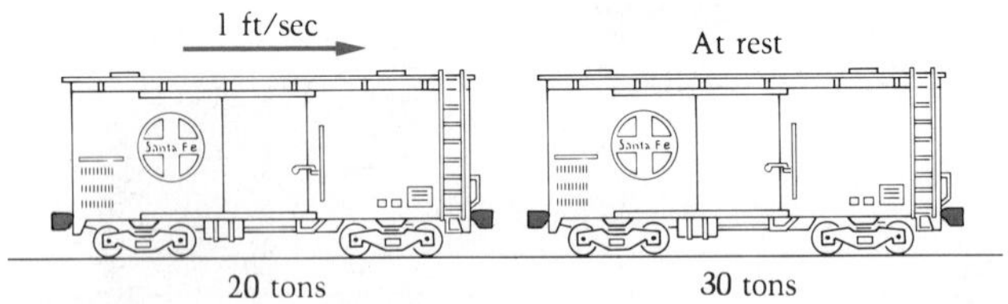

Figure P2.139

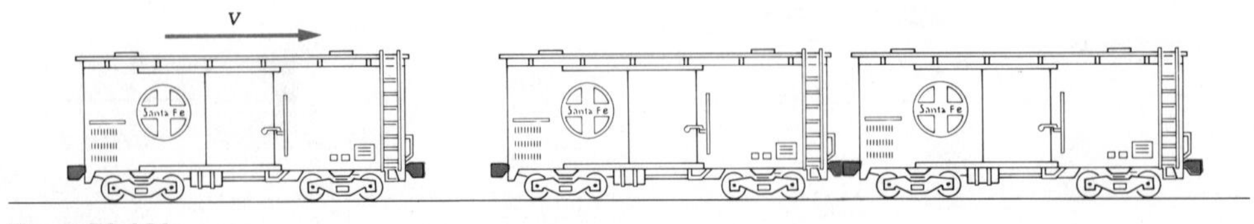

Figure P2.141

Figure P2.142

2.142 A man of mass m and a boat of mass M are at rest as shown in Figure P2.142. If the man walks to the front of the boat, show that his distance from the pier is then $L\mathcal{M}/(1 + \mathcal{M})$, where $\mathcal{M} = m/M$ is the ratio of the masses of man and boat. Explain the answer in the limiting cases in which $m \ll M$ and $M \ll m$. Neglect the resistance of the water to the boat's motion.

* **2.143** Two men each of mass m stand on the ends of a flatcar of mass M. The car is free to move on frictionless level tracks. All is at rest initially. One man runs to the right end of the car and jumps off horizontally, parallel to the tracks with a velocity U relative to the car. Then the other man runs to the left end of the car and jumps off horizontally, parallel to the tracks also with a velocity U relative to the car. Find the final velocity of the car and indicate clearly the direction of its motion.

* **2.144** In Figure P2.144 the man of mass m stands at end A of a 20-ft plank of mass $3m$ that is held at rest on the smooth inclined plane by the cord. The man cuts the cord and runs down to end B of the plank. When he gets there, end B is in the same position on the plane as it was originally. Find the time it takes the man to run down from A to B.

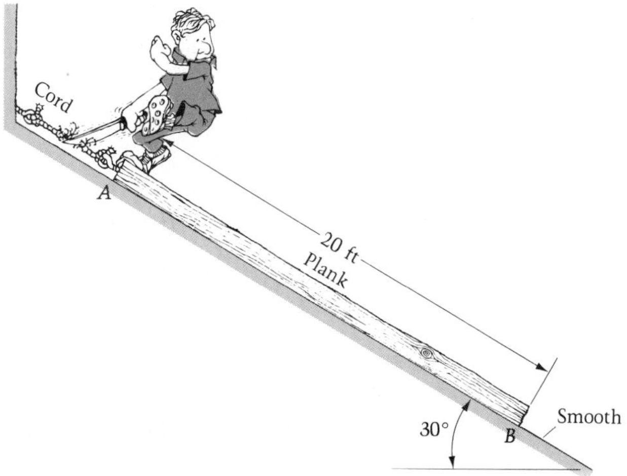

Figure P2.144

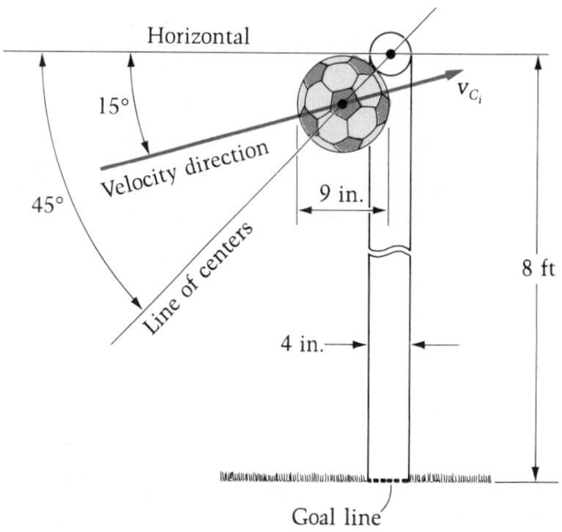

Figure P2.149

2.145 A ball is dropped from a height H and bounces. (See Figure P2.145.) If the coefficient of restitution is e, find the height to which the ball rises after the second bounce.

2.146 Two identical elastic balls $\mathcal{A}$ and $\mathcal{B}$ move toward each other. Find the approach velocity ratio v_{A_1}/v_{B_1} that will result in $\mathcal{A}$ coming to rest following the collision. The coefficient of restitution is e. (See Figure P2.146.)

2.147 Use the two equations

$$m_{\mathcal{A}} v_{A_i} + m_{\mathcal{B}} v_{B_i} = m_{\mathcal{A}} v_{A_f} + m_{\mathcal{B}} v_{B_f}$$

and

$$e = \frac{v_{B_f} - v_{A_f}}{v_{A_i} - v_{B_i}} \qquad (= \text{coefficient of restitution})$$

to prove that the loss in kinetic energy as the bodies $\mathcal{A}$ and $\mathcal{B}$ collide (Figure P2.147) is

$$\Delta T = \frac{m_{\mathcal{A}} m_{\mathcal{B}} (1 - e^2)(v_{A_i} - v_{B_i})^2}{2(m_{\mathcal{A}} + m_{\mathcal{B}})}$$

Deduce from this result that $e \le 1$!

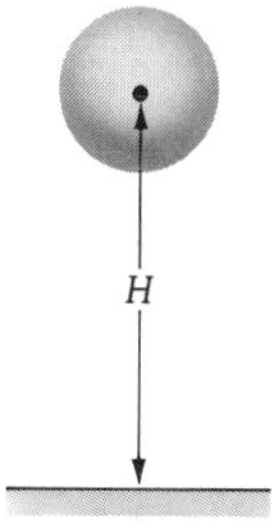

Figure P2.145

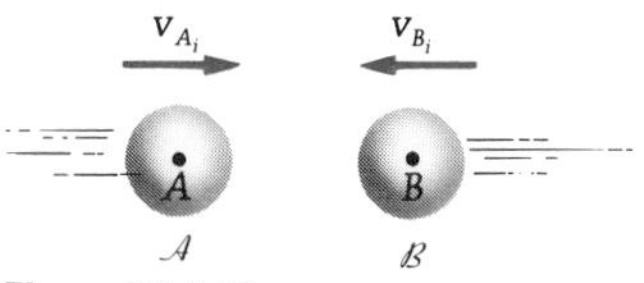

Figure P2.146

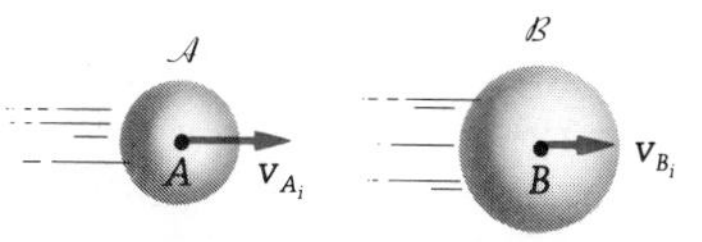

Figure P2.147

2.148 Use the result of the preceding problem to show that for a head-on collision at equal speeds v and equal masses m,

$$\Delta T = 2\,\frac{mv^2}{2}(1 - e^2)$$

so that if $e = 0$ then *all* of the initial T is lost and if $e = 1$ then *none* of T is lost. Is this true for differing speeds and masses?

2.149 In soccer, a goal is scored only when the *entire* ball is over the *entirety* of the 4-in.-wide goal line. (See Figure P2.149.) Neglecting friction between ball and post, determine the maximum coefficient of restitution for which a goal will be scored before the ball hits the ground. The velocity of the ball's center C makes an angle with the horizontal of 15°. Neglect the deviation caused by gravity on the trajectory between post and ground.

2.150 Repeat Example 2.16 for the line of impact shown in Figure P2.150.

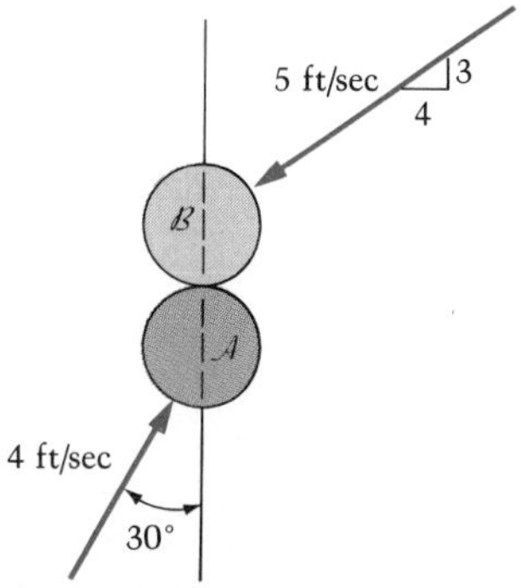

Figure P2.150

2.151 Let disk $\mathcal{A}$ in Problem 2.150 weigh 8 oz (and $\mathcal{B}$ weigh 6 oz as before), and then repeat the problem.

2.152 Repeat Example 2.16 for the case where the line of impact is parallel to the before-collision velocity of $\mathcal{B}$.

2.153 Let disk $\mathcal{A}$ in Problem 2.152 weigh 9 oz (and $\mathcal{B}$ weigh 6 oz as before), and then repeat the problem.

2.154 A 10-kg block swings down as shown in Figure P2.154 and strikes an identical block. Assume that the 6 m rope breaks during impact and the blocks stick together after colliding. How long will it be before they come to rest? How far will they have traveled?

2.155 Using the angle α that will land the cannonball of Problem 2.12 in the cart, find the maximum deflection of the spring. (See Figure P2.155.)

2.156 Find the total time after firing for the cannonball and box to either stop or strike the wall, whichever comes first. (See Figure P2.156.)

2.157 A cannonball is fired as shown in Figure P2.157 with an initial speed of 1600 ft/sec at 60°. Just after the cannon fires, it begins to recoil, and strikes a plate attached to a spring. Find the maximum spring deflection if the plane is smooth and the spring modulus is 500 lb/ft.

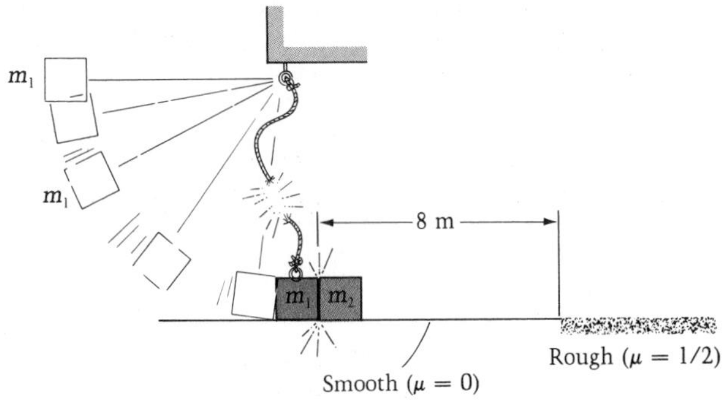

Figure P2.154

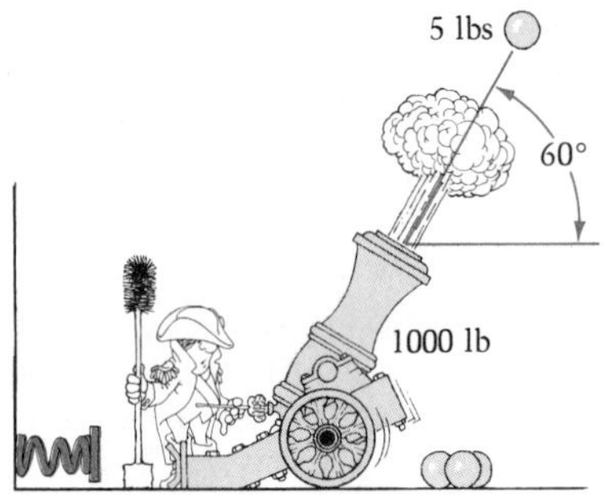

Figure P2.157

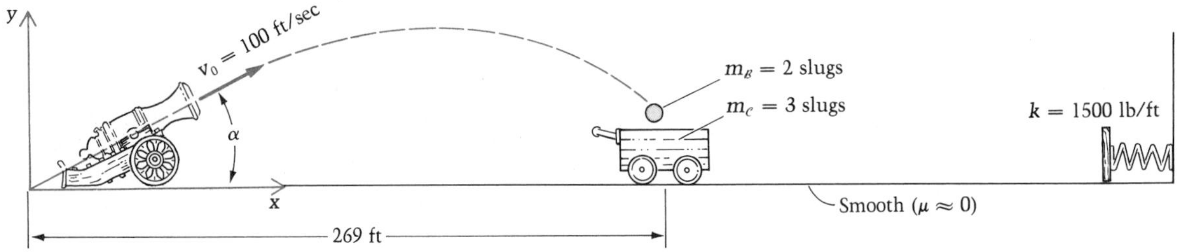

Figure P2.155

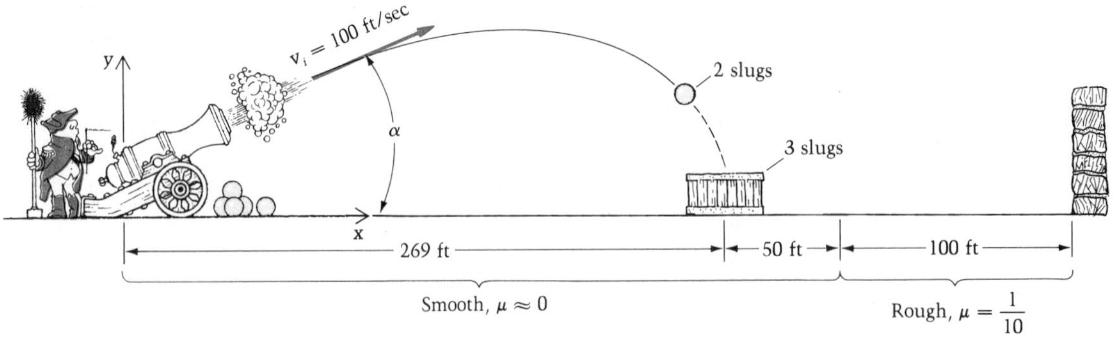

Figure P2.156

2.158 A $\frac{3}{4}$-oz bullet is fired with a speed of 1800 ft/sec into a 10-lb block. (See Figure P2.158.) If the coefficient of friction between block and plane is 0.3, find, neglecting the impulse of friction during the collision:

- **a.** The distance through which the block will slide
- **b.** The percentage of the bullet's loss of initial kinetic energy caused by sliding friction, and the percentage caused by the collision
- **c.** How long it takes block and bullet to come to rest after the impact.

2.159 Weight W_1 falls from rest through a distance H; it lands on another weight W_2, which was in equilibrium atop a spring of modulus k. (See Figure P2.159.) If the coefficient of restitution is zero, find the spring compression when the weights are at their lowest point.

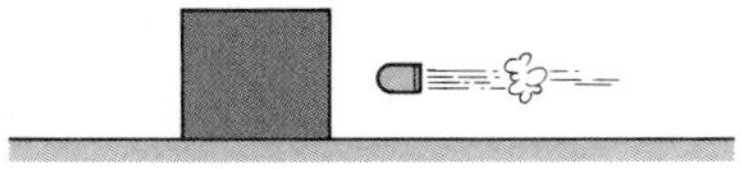

Figure P2.158

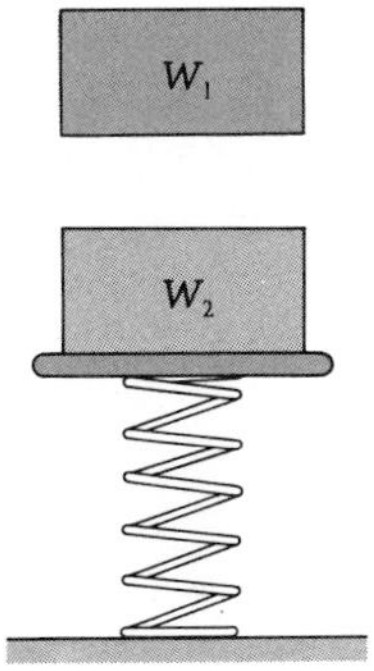

Figure P2.159

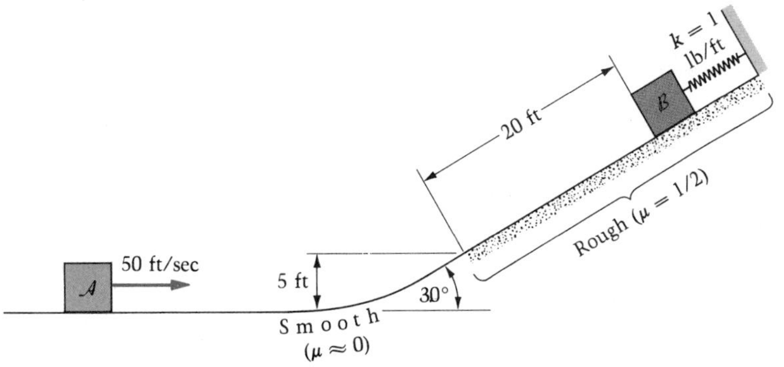

Figure P2.160

2.160 Block $\mathcal{A}$ in Figure P2.160 weighs 16.1 lb and is traveling to the right on the smooth plane at 50 ft/sec. Block $\mathcal{B}$ weighs 8.05 lb and is in equilibrium with the spring barely preventing it from sliding down the rough section of plane. Body $\mathcal{A}$ impacts $\mathcal{B}$; the coefficient of restitution $e = \frac{1}{2}$. Find the maximum spring deflection.

2.161 The 16-kg body $\mathcal{A}$ and the 32-kg body $\mathcal{B}$ shown in Figure P2.161 are connected by a light spring of modulus 12,000 N/m. The unstretched length of the spring is 0.15 m. The blocks are pulled apart on the smooth horizontal plane until the distance between them is 0.3 m and then released from rest. Determine the velocity of each block when the distance between them has decreased to 0.22 m. *Hint*: As in Problem 2.129, form the sum of the work-energy equations for the two blocks.

2.162 The cart and block in Figure P2.162 are initially at rest, when the bullet slams into the block at speed $100\sqrt{gL}$ and sticks inside it. The combined body then starts sliding on the cart.

Find:

- **a.** the speed of the block just after impact;
- **b.** the energy lost during impact;
- *** c.** the time when the block leaves the cart.

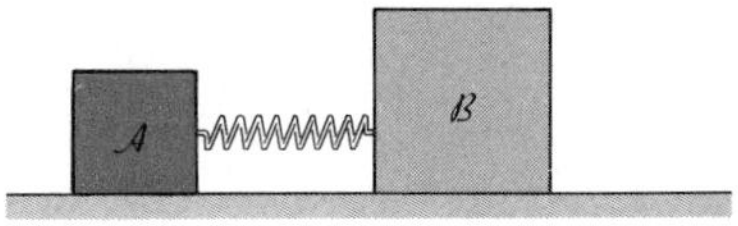

Figure P2.161

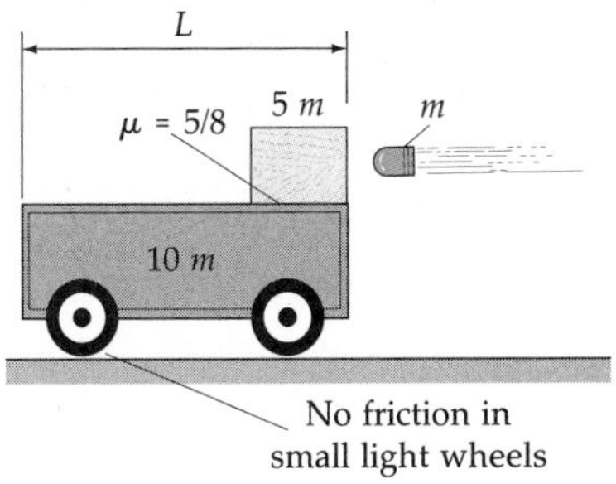

Figure P2.162

2.163 A chain of length L and mass per unit length β is held vertically above the platform scale shown in Figure P2.163 and is released from rest with its lower end just touching the platform. Assume that the links quickly come to rest as they stack up on the platform and that they do not interfere with the links still in free fall above the platform. Draw a free-body diagram of the entire chain and express the momentum as a function of the distance through which the upper end has fallen. Then determine the force read on the scale in terms of this distance.

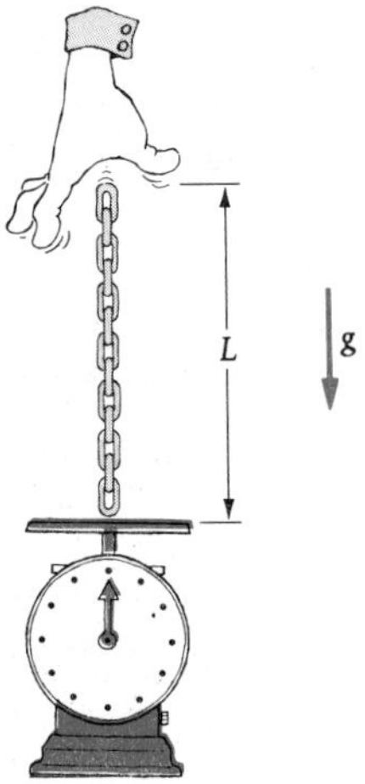

Figure P2.163

• **2.164** A block of mass mL, which can move on a smooth horizontal table, is attached to one end of a uniform chain of mass m per unit length. Initially the block and the chain are at rest, and the chain is completely coiled on the table. A constant horizontal force mLf is then applied to the block so that the chain begins to uncoil. Show that the length x uncoiled after time t is given by

$$(L + x)^2 = Lft^2 + L^2$$

until the chain is completely uncoiled. If the length of the chain is very large compared with L, show that the velocity of the block is approximately equal to $(Lf)^{1/2}$ at the moment when the chain is completely uncoiled.

• **2.165** An important problem in the dynamics of deformable solids is that of describing the motion which ensues when pressure is rapidly applied to the end of a slender, uniform, elastic bar. A useful approximate theory yields the one-dimensional wave equation as the governing equation of motion. This theory predicts that a pressure applied at one end of the bar creates a disturbance (wave) that propagates into the bar at a constant speed c. To be specific, suppose the bar shown in Figure P2.165a is at rest for $t < 0$ and is subjected to the uniform pressure (over the end of area A) shown in Figure P2.165b. If the disturbance has not reached the right end, that is if $t < L/c$, then for $t > t_0$ the particle velocities $\dot{u}\hat{\mathbf{i}}$ and accelerations $\ddot{u}\hat{\mathbf{i}}$, which vary only with x and t, are as shown in Figures P2.165c and d, where ρ is the density of the bar.

The first part of this problem is to evaluate the integral

$$\int_{\mathcal{B}} \mathbf{a}\, dm = \hat{\mathbf{i}} \int_0^L \int_A \rho \ddot{u}\, dA\, dx$$

The value that should be obtained is $p_0 A\hat{\mathbf{i}}$, and, since this equals the external force on the bar, Equation (2.4) is thus confirmed for this case. It is important to recognize that only the interval from $x = ct - ct_0$ to $x = ct$ contributes to the value of the integral; that is, only the particles in that region are accelerating.

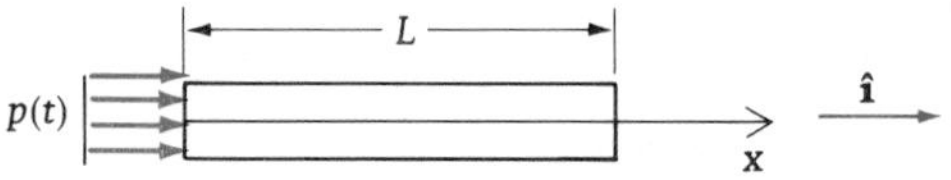

Figure P2.165a

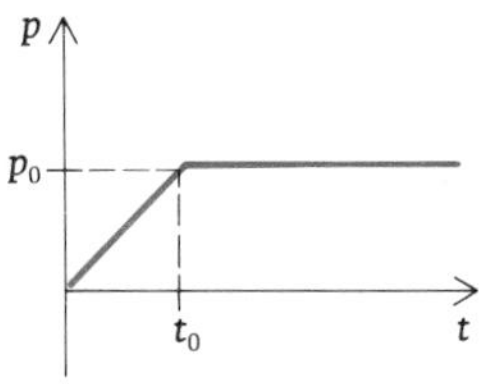

Figure P2.165b

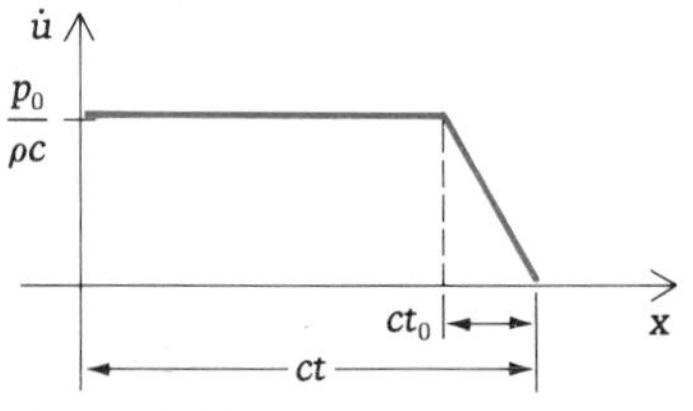

Figure P2.165c

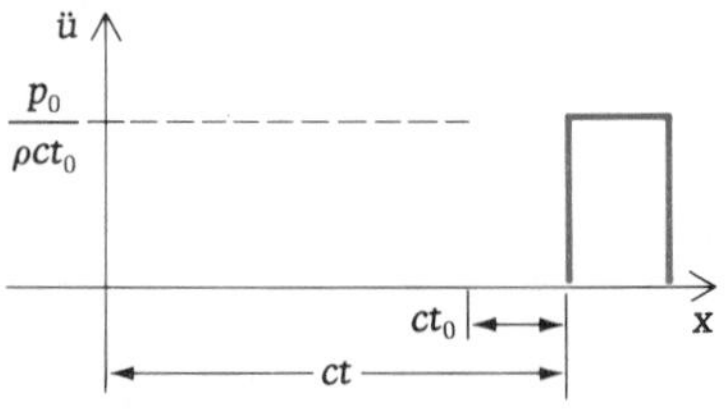

Figure P2.165d

The second element of this problem is to evaluate the momentum

$$\mathbf{L} = \int_{\mathscr{B}} \mathbf{v}\, dm = \hat{\mathbf{i}} \int_0^L \int_A \rho \dot{u}\, dA\, dx$$

The result will be

$$\mathbf{L} = p_0 A \left[(t - t_0) + \frac{t_0}{2} \right] \hat{\mathbf{i}}$$

The second term in the brackets, a constant, is the contribution from integrating over the interval ct_0 where the particles are accelerating. The time dependence of $\mathbf{L}$ appears through the increasing number of particles having velocity $(p_0/\rho c)\hat{\mathbf{i}}$. As expected, we see that $\dot{\mathbf{L}} = p_0 A \hat{\mathbf{i}}$.

In effect we have confirmed Euler's law, $\Sigma\mathbf{F} = \dot{\mathbf{L}}$, in two forms. In the first,

$$\dot{\mathbf{L}} = \int \frac{d\mathbf{v}}{dt}\, dm = \int \mathbf{a}\, dm$$

In the second,

$$\dot{\mathbf{L}} = \frac{d}{dt} \int \mathbf{v}\, dm$$

For the case at hand there is no reason to express a preference for the order of differentiating with respect to time and integrating over the body. If the pressure were suddenly applied at full strength ($t_0 = 0$), however, there would be a discontinuity in particle velocity (shock wave) and a consequent undefined acceleration at the wavefront $x = ct$. Because of this undefined (or infinite) acceleration, $\int \mathbf{a}\, dm$ becomes meaningless and no longer provides $\dot{\mathbf{L}}$. There is no difficulty involved in evaluating $\mathbf{L}$, however, since the particle velocities are $p_0/\rho c$ for $x < ct$ and zero for $x > ct$. Thus

$$\mathbf{L} = \frac{p_0}{\rho c} (\rho A c t)\hat{\mathbf{i}} = p_0 A t \hat{\mathbf{i}}$$

and $\dot{\mathbf{L}} = p_0 A \hat{\mathbf{i}}$.

2.6 Euler's Second Law (The Moment Equation)

A second relationship between the external forces on a particle system or a body is obtained if, referring to Figure 2.10, we take the cross product of $\mathbf{r}_i$ with both sides of Equation (2.1):

$$\mathbf{r}_i \times \mathbf{F}_i + \sum_{j=1}^{N} \mathbf{r}_i \times \mathbf{f}_{ij} = \mathbf{r}_i \times m_i \mathbf{a}_i \tag{2.31}$$

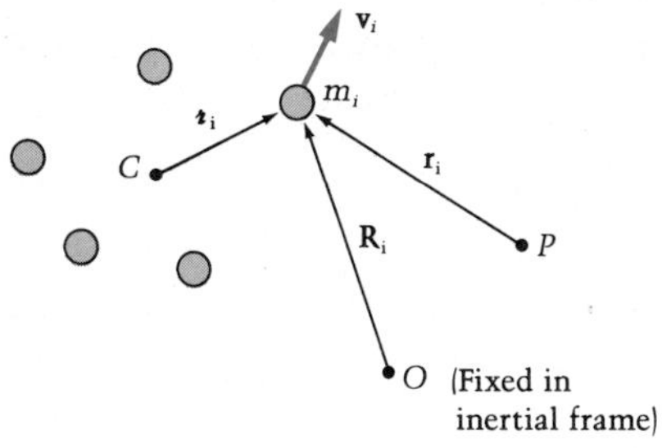

Figure 2.10

The first term on the left is recognized as the moment about point P of the external force $\mathbf{F}_i$. The cross product $\mathbf{r}_i \times \mathbf{f}_{ij}$ is the moment with respect to P of the force exerted on the ith particle by the jth particle. As before, we now sum the N equations typified by Equation (2.31) to obtain

$$\sum_{i=1}^{N} \mathbf{r}_i \times \mathbf{F}_i + \sum_{i=1}^{N} \sum_{j=1}^{N} \mathbf{r}_i \times \mathbf{f}_{ij} = \sum_{i=1}^{N} \mathbf{r}_i \times m_i \mathbf{a}_i \tag{2.32}$$

Terms in the double sum occur in pairs, such as

$$\mathbf{r}_1 \times \mathbf{f}_{12} + \mathbf{r}_2 \times \mathbf{f}_{21}$$

But $\mathbf{r}_2 \times \mathbf{f}_{21} = \mathbf{r}_1 \times \mathbf{f}_{21}$ since $\mathbf{r}_2$ and $\mathbf{r}_1$ both terminate on the line of action of $\mathbf{f}_{21}$. Moreover, $\mathbf{f}_{21} = -\mathbf{f}_{12}$ so that

$$\begin{aligned} \mathbf{r}_1 \times \mathbf{f}_{12} + \mathbf{r}_2 \times \mathbf{f}_{21} &= \mathbf{r}_1 \times (\mathbf{f}_{12} + \mathbf{f}_{21}) \\ &= \mathbf{0} \end{aligned}$$

and similarly for other such pairs. That is, the moments of the internal forces of interaction sum to zero. Thus

$$\Sigma \mathbf{r}_i \times \mathbf{F}_i = \Sigma \mathbf{r}_i \times m_i \mathbf{a}_i$$

or

$$\Sigma \mathbf{M}_P = \Sigma \mathbf{r}_i \times m_i \mathbf{a}_i \tag{2.33}$$

which is the particle-system form of Euler's second law and states that the sum of the moments of the external forces about a point equals the sum of the moments of the $m\mathbf{a}$'s about that point.

For a body whose mass is continuously distributed, the counterpart to Equation (2.33) is

$$\Sigma \mathbf{M}_P = \int \mathbf{r} \times \mathbf{a}\, dm \tag{2.34}$$

Question 2.9 In Equation 2.33 (or 2.34) must point P be fixed in the inertial frame of reference?

Equations (2.4) and (2.34) play the same roles in dynamics as do the equations of equilibrium in statics. And in fact we obtain those equations, $\Sigma \mathbf{F} = \mathbf{0}$ and $\Sigma \mathbf{M} = \mathbf{0}$, from (2.4) and (2.34), if we set to zero the accelerations of all points of a body.

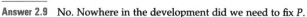

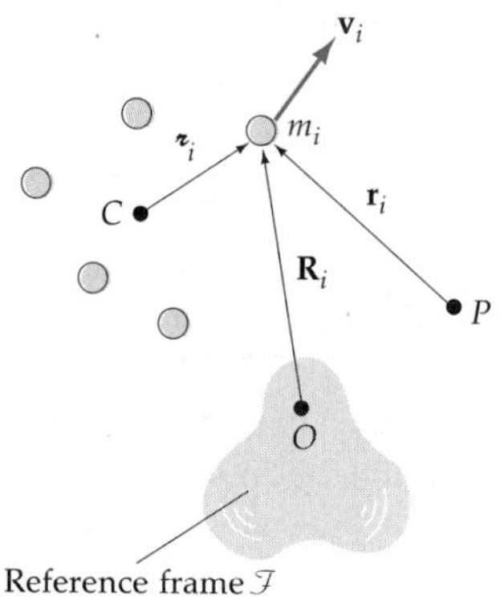

Figure 2.11

Moment of Momentum

Just as Euler's first law can be expressed in terms of the time derivative of momentum of a body, so Euler's second law can be expressed in terms of the time derivative of a quantity called **moment of momentum,** or **angular momentum.*** The moment of momentum with respect to a point P is designated $\mathbf{H}_P$ and is defined to be the sum of the moments (with respect to P) of the momenta of the individual particles making up the body. Referring to Figure 2.11, where $\mathbf{v}_i$ is the velocity in reference frame $\mathcal{I}$ of the ith particle, we have

$$\mathbf{H}_P = \Sigma \mathbf{r}_i \times m_i \mathbf{v}_i \tag{2.35}$$

Before proceeding to the development of several of the forms of Euler's second law, we shall develop a very useful relationship between moments of momentum. Noting from the definition, Equation (2.35), and from Figure 2.11 that

$$\mathbf{H}_C = \Sigma\, \boldsymbol{\imath}_i \times m_i \mathbf{v}_i$$

and that

$$\mathbf{r}_i = \mathbf{r}_{PC} + \boldsymbol{\imath}_i$$

then for any point P,

$$\begin{aligned}\mathbf{H}_P &= \Sigma(\boldsymbol{\imath}_i + \mathbf{r}_{PC}) \times \mathbf{m}_i \mathbf{v}_i = \Sigma\, \boldsymbol{\imath}_i \times m_i \mathbf{v}_i + \mathbf{r}_{PC} \times \Sigma m_i \mathbf{v}_i \\ &= \mathbf{H}_C + \mathbf{r}_{PC} \times \Sigma m_i \mathbf{v}_i\end{aligned}$$

Answer 2.9 No. Nowhere in the development did we need to fix P.

* The term angular momentum stems from the fact that the moment of momentum of a rigid body is related to the angular velocity of the body.

But from Section 2.5, $\Sigma m_i \mathbf{v}_i$ is the momentum $\mathbf{L}$, also expressed as

$$\mathbf{L} = m\mathbf{v}_C$$

so that

$$\mathbf{H}_P = \mathbf{H}_C + \mathbf{r}_{PC} \times \mathbf{L} \tag{2.36}$$

Thus the moment of the momentum of a body about *any* point P (not necessarily fixed in the reference frame) is the sum of the moment of momentum about its mass center, C, and the moment of its (linear) momentum $\mathbf{L}$ about P, where $\mathbf{L}$ is given a "line of action" through C.

Now we can return to the definition of moment of momentum, Equation (2.35), and apply it for the case of a point, O, fixed in the frame of reference. Thus, using definition (2.35) for the third time,

$$\mathbf{H}_O = \Sigma \mathbf{R}_i \times m_i \mathbf{v}_i$$

Now, differentiating with respect to time in $\mathcal{I}$,

$$\frac{d\mathbf{H}_O}{dt} = \Sigma(\dot{\mathbf{R}}_i \times m_i \mathbf{v}_i + \mathbf{R}_i \times m_i \mathbf{a}_i) \tag{2.37}$$

But because O is fixed in $\mathcal{I}$,

$$\dot{\mathbf{R}}_i = \mathbf{v}_i$$

so that

$$\dot{\mathbf{R}}_i \times \mathbf{v}_i = 0$$

Therefore

$$\frac{d\mathbf{H}_O}{dt} = \Sigma \mathbf{R}_i \times m_i \mathbf{a}_i \tag{2.38}$$

Momentum Forms of Euler's Second Law

Now the fundamental form of Euler's second law, Equation (2.33), tells us that if the frame $\mathcal{I}$ in which the $\mathbf{a}_i$ are calculated is an inertial frame, then

$$\Sigma \mathbf{M}_O = \Sigma \mathbf{R}_i \times m_i \mathbf{a}_i \tag{2.39}$$

so that, from the last two equations, we see that

$$\Sigma \mathbf{M}_O = \frac{d\mathbf{H}_O}{dt} \tag{2.40}$$

Another similar form of Euler's second law can be deduced if we first use Equation (2.36) in the case of a fixed point O:

$$\mathbf{H}_O = \mathbf{H}_C + \mathbf{r}_{OC} \times \mathbf{L}$$

Differentiating with respect to time in the reference frame $\mathcal{I}$,

$$\frac{d\mathbf{H}_O}{dt} = \frac{d\mathbf{H}_C}{dt} + \mathbf{v}_C \times \mathbf{L} + \mathbf{r}_{OC} \times \frac{d\mathbf{L}}{dt} \tag{2.41}$$

where we have used the fact that

$$\frac{d\mathbf{r}_{OC}}{dt} = \mathbf{v}_C$$

But we again recall that

$$\mathbf{L} = m\mathbf{v}_C \qquad (\text{so that } \mathbf{v}_C \times \mathbf{L} = \mathbf{0})$$

and therefore

$$\frac{d\mathbf{H}_O}{dt} = \frac{d\mathbf{H}_C}{dt} + \mathbf{r}_{OC} \times \frac{d\mathbf{L}}{dt}$$

Now we know from our study of equipollent force systems in statics that the external forces on the body must produce moments about O and C that are related by

$$\Sigma\mathbf{M}_O = \Sigma\mathbf{M}_C + \mathbf{r}_{OC} \times (\Sigma\mathbf{F}) \tag{2.42}$$

This law of resultants has nothing to do with whether or not the body is in equilibrium. And since $\mathcal{I}$ is an inertial frame, then we also know

$$\Sigma\mathbf{F} = \frac{d\mathbf{L}}{dt} \qquad \text{and} \qquad \Sigma\mathbf{M}_O = \frac{d\mathbf{H}_O}{dt}$$

and thus we may subtract the two Equations (2.41) and (2.42) to obtain

$$\Sigma\mathbf{M}_C = \frac{d\mathbf{H}_C}{dt} \tag{2.43}$$

Neither of the above equations $\Sigma\mathbf{M}_C = \dot{\mathbf{H}}_C$ or $\Sigma\mathbf{M}_O = \dot{\mathbf{H}}_O$ is any more basic or special than the other, as each one can be derived from the other. They are therefore equivalent forms. However, the equation does *not* hold for any arbitrary point P, i.e., in general $\Sigma\mathbf{M}_P \neq \dot{\mathbf{H}}_P$.

Conservation of Moment of Momentum

We next note that — as was the case with linear momentum — there are situations in which a moment of momentum is conserved. In particular, if for an interval of time $\Sigma\mathbf{M}_O = \mathbf{0}$, then during that interval $\dot{\mathbf{H}}_O = \mathbf{0}$, and thus $\mathbf{H}_O$ is constant. For example, let the body of interest be a single spherical planet in its motion around its star. The gravitational force exerted on the planet by the star always passes through the star's mass center O, so $\Sigma\mathbf{M}_O = \mathbf{0}$ and thus $\mathbf{H}_O$ of the planet is a constant. This result is shown in Section 8.4 to lead to the elliptical orbit of the earth around the sun.

For an arbitrary point P, there is a form of Euler's second law that is of particular value in analyzing the motions of rigid bodies, although it remains valid for non-rigid bodies as well. To derive it we again use our knowledge about force systems to write

$$\Sigma\mathbf{M}_P = \Sigma\mathbf{M}_C + \mathbf{r}_{PC} \times (\Sigma\mathbf{F}) \tag{2.44}$$

Thus, using Euler's laws (Equations (2.27 and (2.43))

$$\Sigma \mathbf{M}_P = \frac{d\mathbf{H}_C}{dt} + \mathbf{r}_{PC} \times \frac{d\mathbf{L}}{dt}$$

or

$$\Sigma \mathbf{M}_P = \dot{\mathbf{H}}_C + \mathbf{r}_{PC} \times m\mathbf{a}_C \tag{2.45}$$

which we shall use later.

Question 2.10 Must point P be fixed in the inertial frame of reference, (a) for Equation (2.36) to be true? (b) for Equation (2.45) to be true?

Finally, we remind the reader that all of the relationships of this chapter pertain only to a specific collection of material — that is, a system (or body) of constant mass. However, the momentum forms of Euler's laws provide the natural starting point for developing relationships appropriate to "variable mass" systems such as rockets. If desired, the reader now has the proper background to study that special topic which is found in Section 8.3.

Answer 2.10 (a) No; (b) No. Point P was unrestricted in both derivations.

EXAMPLE 2.18

Figure E2.18a

Two gymnasts of equal weight (see Figure E2.18a) are hanging in equilibrium at the ends of a rope passing over a relatively light pulley for which the bearing friction can be neglected. Then the gymnast on the right begins to climb the rope, while the gymnast on the left simply holds on. When the right gymnast has raised himself through height h (relative to the floor), what has been the change in position of the left gymnast?

Solution

Constructing a free-body diagram (Figure E2.18b) of the pulley-rope-gymnasts system in which we neglect the weights of the pulley and the rope, we see that

$$\begin{aligned}\dot{\mathbf{H}}_O &= \Sigma \mathbf{M}_O \\ &= mgr\hat{\mathbf{k}} + mgr(-\hat{\mathbf{k}}) \\ &= \mathbf{0}\end{aligned}$$

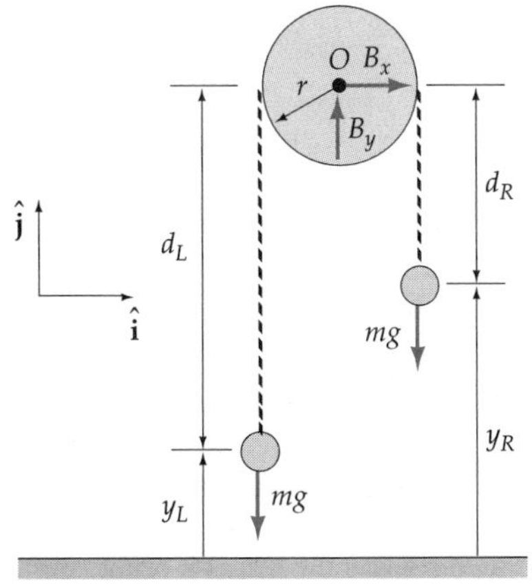

Figure E2.18b

Therefore, $\mathbf{H}_O$ is constant during the motion, and since everything starts from rest,

$$\mathbf{H}_O = \mathbf{0}$$

Treating the gymnasts as particles and neglecting the moment of momentum of the pulley,

$$\mathbf{H}_O = (r\hat{\mathbf{i}} - d_R\hat{\mathbf{j}}) \times (m\dot{y}_R\hat{\mathbf{j}}) + (-r\hat{\mathbf{i}} - d_L\hat{\mathbf{j}}) \times (m\dot{y}_L\hat{\mathbf{j}})$$

or

$$\mathbf{H}_O = rm\dot{y}_R\hat{\mathbf{k}} + rm\dot{y}_L(-\hat{\mathbf{k}})$$

But

$$\mathbf{H}_O = \mathbf{0}$$

so that

$$\dot{y}_R = \dot{y}_L$$

and the left gymnast, "going along for the ride," rises at the same rate as the right one. Thus when the right gymnast has pulled himself up height h, the left one has been pulled up the same height h. Note that if the rope is inextensible, the right gymnast therefore would have climbed $2h$ relative to the rope!

EXAMPLE 2.19

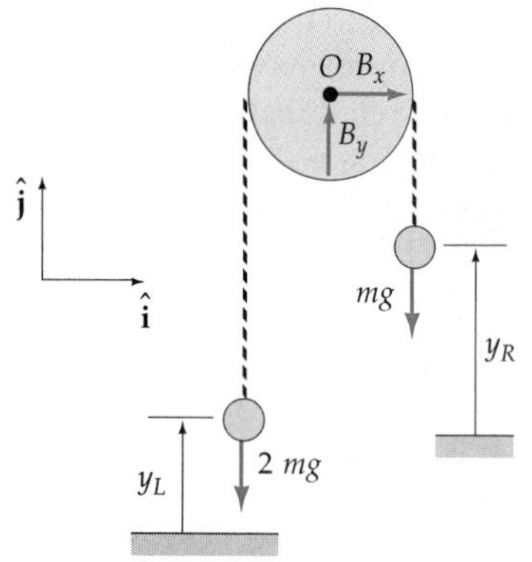

Figure E2.19

Suppose the "counterweight" gymnast on the left in the preceding example were to weigh twice that of the climbing gymnast, as suggested by Figure E2.19. What then would be the relationship between their elevation changes?

Solution

From the free-body diagram

$$\Sigma \mathbf{M}_O = 2mgr\hat{\mathbf{k}} - mgr\hat{\mathbf{k}}$$
$$= mgr\hat{\mathbf{k}}$$

Since $mgr\hat{\mathbf{k}} \neq \mathbf{0}$, the moment of momentum is not conserved this time. Integrating,

$$\int_0^t \Sigma \mathbf{M}_O d\tau = \mathbf{H}_O(t) - \cancelto{0}{\mathbf{H}_O(0)}$$

or

$$mgrt\hat{\mathbf{k}} = rm\dot{y}_R\hat{\mathbf{k}} + r(2m)\dot{y}_L(-\hat{\mathbf{k}})$$

so that

$$\dot{y}_L = \frac{1}{2}\dot{y}_R - \frac{gt}{2}$$

If we define y_R and y_L so that $y_R = y_L = 0$ at $t = 0$, then

$$y_L = \frac{1}{2}y_R - \frac{gt^2}{4}$$

Thus we see that it's possible for the lighter gymnast to raise the heavier gymnast by climbing rapidly enough.

For an inextensible rope, the right gymnast climbs, as before, at a rate of $\dot{y}_R + \dot{y}_L$ relative to the rope.

PROBLEMS ▶ Section 2.6

2.166 The uniform rigid bar $\mathcal{B}$ in Figure P2.166 weighs 60 lb and is pinned at A (and fastened by the cable DB) to the frame $\mathcal{F}$. If the frame is given an acceleration $a = 32.2$ ft/sec^2 as shown, determine the tension T in the cable and the force exerted by the pin at A on the bar.

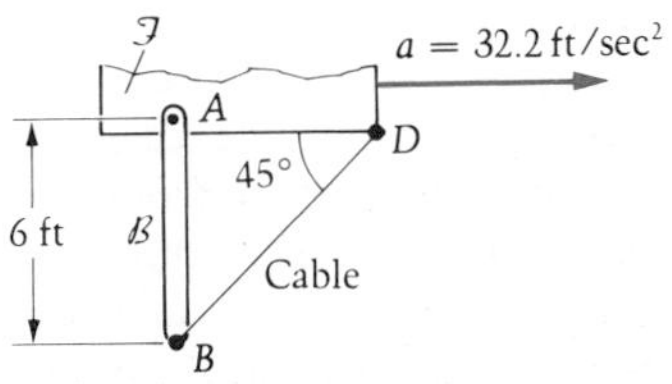

Figure P2.166

2.167 A force F causes the carriage to move with rectilinear horizontal motion defined by a constant acceleration of 20 ft/sec² (see Figure P2.167). A rigid, slender, homogeneous rod of weight 32.2 lb and length 6 ft is welded to the carriage at B and projects vertically upward. Find, in magnitude and direction, the bending moment that the carriage exerts on the rod at B.

2.168 A uniform slender bar of density ρ, cross-sectional area A, and length L undergoes small-amplitude, free transverse vibrations according to $y(x, t) = Y \sin(\pi x/L) \sin \omega t$, where y is the displacement perpendicular to the axis (x) of the bar. (See Figure P2.168.) Neglecting other components of displacement (and hence acceleration), calculate the maximum force generated at one of the supports during the motion.

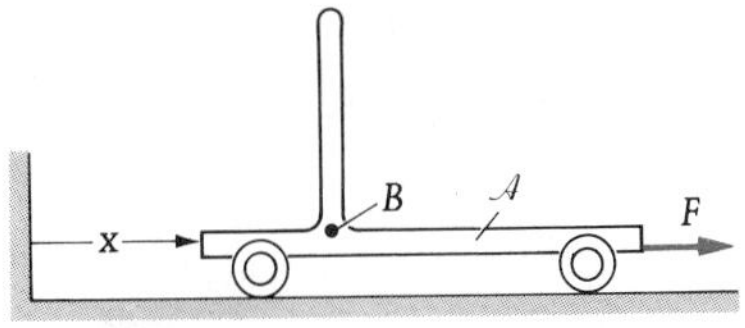

Figure P2.167

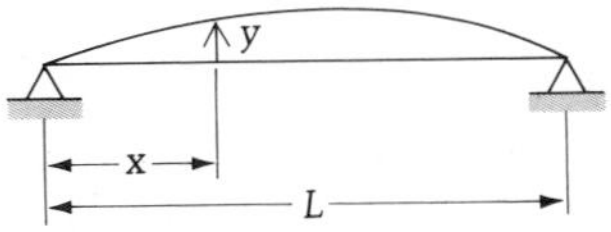

Figure P2.168

2.169 Show that in Equation (2.36), point C need not be the mass center, i.e., if Q is another arbitrary point like P, then show that $\mathbf{H}_P = \mathbf{H}_Q + \mathbf{r}_{PQ} \times \mathbf{L}$ (where $\mathbf{L}$ is of course still $\Sigma m_i \mathbf{v}_i = m\mathbf{v}_C$).

2.170 Let S be a set of vectors $\mathbf{Q}_1, \mathbf{Q}_2, \ldots, \mathbf{Q}_i, \ldots, \mathbf{Q}_N$ of equal dimension. Define the resultant of S as $\mathbf{R} = \Sigma \mathbf{Q}_i$, and place each $\mathbf{Q}_i$ at a point P_i. Define the moment of the set of vectors about a point A by

$$\Sigma \mathbf{M}_A = \Sigma(\mathbf{r}_{AP_i} \times \mathbf{Q}_i)$$

and show that

$$\Sigma \mathbf{M}_P = \Sigma \mathbf{M}_A + \mathbf{r}_{PA} \times \mathbf{R}$$

Now note that (a) if the $\mathbf{Q}_i$ are a set of forces $\mathbf{F}_i$, then Equation (2.42) results with $\mathbf{R}$ being $\Sigma \mathbf{F}_i$; and (b) if the $\mathbf{Q}_i$ are a set of momenta $m_i \mathbf{v}_i$ of a group of particles, then Equation (2.36) results with $\mathbf{R}$ being $\Sigma m_i \mathbf{v}_i$ (or $\mathbf{L}$ or $m\mathbf{v}_c$). Thus we may conclude that both equations, (2.36) and (2.42), are practical examples of the very same law of resultants!

2.171 The angular momentum about point Q is defined as

$$\mathbf{H}_Q = \Sigma(\mathbf{r}_{QP_i} \times m_i \mathbf{v}_i)$$

Differentiate this expression in the inertial reference frame $\mathcal{I}$, and show by the result and that of the preceding problem that, in general, (i.e., not just at an isolated instant of time):

$$\Sigma \mathbf{M}_Q = \dot{\mathbf{H}}_Q$$

only if (a) Q is fixed in $\mathcal{I}$, or (b) Q is the mass center C; or (c) $\mathbf{v}_Q$ is parallel to $\mathbf{v}_C$.

2.172 In Problem 2.132 find: (a) the angular momentum of the system with respect to the origin; (b) the angular momentum of the system with respect to the mass center.

2.173 A massless rope hanging over a massless, frictionless pulley supports two monkeys (one of mass M, the other of mass $2M$). The system is released at rest at $t = 0$, as shown in Figure P2.173. During the following 2 sec, monkey B travels down 15 ft of rope to obtain a massless peanut at end P. Monkey A holds tightly to the rope during these 2 sec. Find the displacement of A during the time interval.

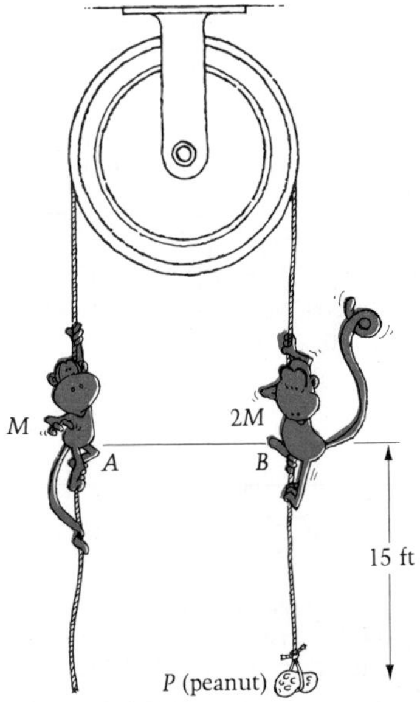

Figure P2.173

2.174 A starving monkey of mass m spies a bunch of delicious bananas of the same mass. (See Figure P2.174.) He climbs at a varying speed relative to the (light) rope. Determine whether the monkey reaches the bananas before they sail over the pulley if the pulley's mass is negligible ($\ll m$).

Figure P2.174

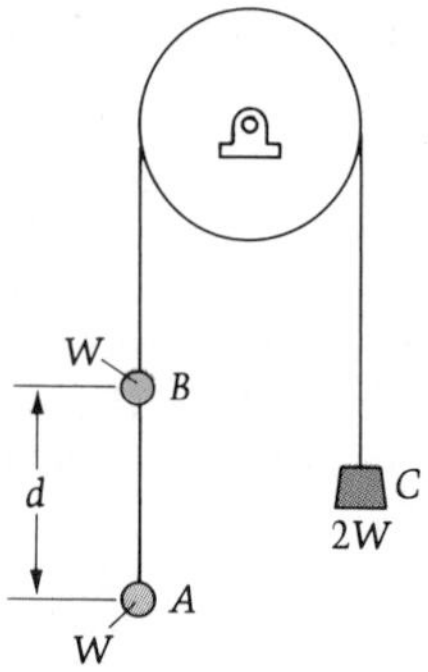

Figure P2.175

2.175 Two gymnasts A and B, each of weight W, hold onto the left side of a rope that passes over a light pulley to a counterweight C of weight $2W$. (See Figure P2.175.) Initially the gymnast A is at depth d below B. He climbs the rope to join gymnast B. Determine the displacement of the counterweight C at the end of the climb.

2.176 Define the angular momentum of a particle about a fixed axis and state the conditions under which the angular momentum remains constant. A man (to be regarded as a particle) stands on a swing. His distance from the smooth horizontal axis of the swing is L when he crouches and $L-H$ when he stands. As the swing falls he crouches; as it rises he stands — the changeover is assumed instantaneous. If the swing falls through an angle α and then rises through an angle β, show that

$$\sin\frac{\beta}{2} = \left(\frac{L}{L-H}\right)^{3/2} \sin\frac{\alpha}{2}$$

The *relative angular momentum* of a body $\mathcal{B}$ with respect to a point P is defined to be

$$\mathbf{H}_{P_{rel}} = \int \mathbf{R} \times (\mathbf{v} - \mathbf{v}_P)\, dm$$

(See Figure P2.177, where the velocity $\mathbf{v}$ of dm is the derivative of $\mathbf{r}$ in inertial frame $\mathcal{I}$.) Note that what makes it "relative" is that the velocity in the integral is the difference between $\mathbf{v}$ (of dm) and $\mathbf{v}_P$. Now solve the following problems.

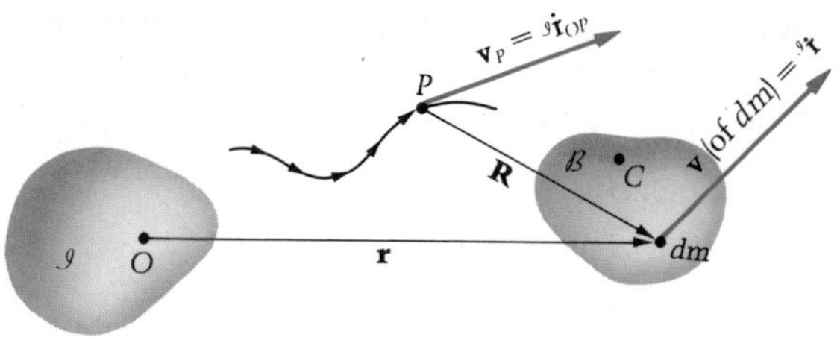

Figure P2.177

2.177 Show that $\mathbf{H}_P = \mathbf{H}_{P_{rel}} + m\mathbf{r}_{PC} \times \mathbf{v}_P$. (Thus $\mathbf{H}_C = \mathbf{H}_{C_{rel}}$ always!)

2.178 Show that $\Sigma\mathbf{M}_P = \dot{\mathbf{H}}_P + \mathbf{v}_P \times m\mathbf{v}_C$. (Thus $\Sigma\mathbf{M}_P$ is not generally equal to $\dot{\mathbf{H}}_P$!)

2.179 Show that $\Sigma\mathbf{M}_P = \dot{\mathbf{H}}_{P_{rel}} + \mathbf{r}_{PC} \times m\mathbf{a}_P$.

2.180 Show that $\dot{\mathbf{H}}_P = \dot{\mathbf{H}}_{P_{rel}}$ if and only if $\mathbf{v}_P \times \mathbf{v}_C = \mathbf{r}_{PC} \times \mathbf{a}_P$.

COMPUTER PROBLEMS ► Chapter 2

2.181 In the table on the next page are data of the mass center velocity versus time for a 30-lb crate that was lifted approximately straight up by two people. The "velocity 1" column represents a taller person than the "velocity 2" column. Use the computer to integrate numerically the velocity from $t = 0$ to 4.4 sec, thereby obtaining and comparing the heights to which the crate was lifted by the two people.

Time (sec)	Velocity 1 (in./sec)	Velocity 2 (in./sec)	Time (sec)	Velocity 1 (in./sec)	Velocity 2 (in./sec)
0.0	0.0	0.0	2.3	25.0 (peak)	23.5
0.2	0.0	0.0	2.4	24.0	25.5
0.4	1.5	0.0	2.6	22.0	32.0
0.6	3.5	0.0	2.7	20.5	37.5 (peak)
0.8	5.5	0.0	2.8	19.0	35.0
1.0	8.5	2.0	3.0	16.5	25.0
1.2	12.0	4.0	3.2	14.5	21.5
1.4	16.5	6.5	3.4	13.0	17.0
1.6	19.0	10.0	3.6	12.0	15.0
1.8	21.0	14.0	3.8	10.0	13.0
2.0	22.5	17.0	4.0	8.5	9.5
2.2	24.5	21.0	4.2	7.5	8.5
			4.4	6.5	8.0

SUMMARY ▶ Chapter 2

In this chapter we have set out the fundamental relationships between forces on a body and its motion, and we have illustrated their use for the solution of a variety of problems, many of which are closely associated with our everyday experience.

The starting point here was Newton's second law for a particle,

$$\Sigma \mathbf{F} = m\mathbf{a}$$

where $\Sigma\mathbf{F}$ is the sum of all the forces acting on the particle, m is the mass of the particle and $\mathbf{a}$ is its acceleration relative to an inertial frame of reference. Extending to a system of particles, the ith having mass m_i and acceleration $\mathbf{a}_i$,

$$\Sigma \mathbf{F} = \Sigma m_i \mathbf{a}_i$$

where $\Sigma\mathbf{F}$ is the sum of the *external* forces on the system. Another form of particular value is

$$\Sigma \mathbf{F} = m\mathbf{a}_C$$

where $m = \Sigma m_i$ is the mass of the system of particles, or the body comprising them, and $\mathbf{a}_C$ is the acceleration of the mass center C. In addition, it is sometimes useful to decompose a body into two (or more) parts with masses m_1 and m_2 and mass centers C_1 and C_2, and then we may use:

$$\Sigma \mathbf{F} = m_1\mathbf{a}_{C_1} + m_2\mathbf{a}_{C_2}$$

The preceding equations, for a body of finite size, are forms of what is often called Euler's first law and are counterparts in dynamics to the equilibrium equation, $\Sigma\mathbf{F} = \mathbf{0}$, studied in statics. We have used them to solve a variety of problems such as finding accelerations and constraining forces when some forces and paths were prescribed and also integrating to find the motion of a particle (or mass center of a body) when external forces were prescribed. Central to the problem-solving process was the free-body diagram, the importance of which cannot be overstated.

The Principle of Work and Kinetic Energy is very useful in solving problems in which the speeds of a particle at different locations in space are of interest. The Kinetic Energy, T, of a particle is defined to be

$$T \equiv \tfrac{1}{2}m\,|\mathbf{v}|^2$$

The principle states that the work done by all the forces acting over an interval of time is equal to the change in kinetic energy. Or, in symbols,

$$W = T_2 - T_1$$

This is a derived result following from integrating and assigning the term "work of a force on a particle" to $\Sigma\mathbf{F} = m\mathbf{a}$

$$W = \int_{t_1}^{t_2} \mathbf{F}\cdot\mathbf{v}dt = \int_{\mathbf{r}(t_1)}^{\mathbf{r}(t_2)} \mathbf{F}\cdot d\mathbf{r}$$

Two special forces arise frequently enough in problems to evaluate the work and express it in symbols:

a. Constant force: $W = \mathbf{F}\cdot(\mathbf{r}_2 - \mathbf{r}_1)$, or in words: (magnitude of force) * (magnitude of displacement) * (cosine of angle between force and displacement). For weight (force exerted by gravity near the earth's surface) this means (weight) * (decrease in altitude of mass center).

b. Force exerted by a linear spring:

$$W = -\frac{k}{2}(\delta_2^2 - \delta_1^2)$$

where δ is spring stretch and k is the spring modulus.

A force whose work does not depend upon the path of the point of application is called conservative and a potential energy, φ, is associated with it so that the work done is the negative of the change in that potential energy,

$$W = -[\varphi_2 - \varphi_1]$$

Combining this with $W = \Delta T$, assuming all forces to be conservative,

$$T_2 + \varphi_2 = T_1 + \varphi_1$$

or

$$T + \varphi = \text{constant}$$

which means that in this case, kinetic plus potential energy is conserved. A potential energy for a linear spring is

$$\varphi = \tfrac{1}{2}k\delta^2$$

and for weight (with z being elevation)

$$\varphi = mgz$$

For a system of particles,

$$T = \Sigma\tfrac{1}{2}m_i|\mathbf{v}_i|^2$$

and the Principle of Work and Kinetic Energy applies so long as one considers the work of all *internal* forces as well as the work of external forces; that is,

$$W_{\text{external}} + W_{\text{internal}} = \Delta T$$

This is of practical value only in special situations where it's possible to readily evaluate the work of internal forces. One example is a rigid body or system of rigidly connected particles, for then $W_{\text{internal}} = 0$. Another case is that of a pair of particles joined by a linear spring, for which the net work of the equal and opposite internal forces is

$$W_{\text{internal}} = -\frac{k}{2}[\delta_2^2 - \delta_1^2].$$

The concept of momentum is particularly useful in problems of impact or collision in which very intense forces of interaction may act for a very brief interval. Momentum is defined to be

$$\mathbf{L} = m\mathbf{v} \qquad \text{(particle)}$$

$$\mathbf{L} = \Sigma m_i \mathbf{v}_i \qquad \text{(system of particles)}$$

from which, for a body in general,

$$\mathbf{L} = m\mathbf{v}_C$$

Euler's first law can be written

$$\Sigma\mathbf{F} = \frac{d\mathbf{L}}{dt} \quad \text{or} \quad \dot{\mathbf{L}}$$

which when integrated yields the impulse-momentum principle

$$\int_{t_1}^{t_2} \Sigma\mathbf{F}\, dt = \mathbf{L}(t_2) - \mathbf{L}(t_1)$$

So if external forces do not act during the interval,

$$\mathbf{L}(t_2) = \mathbf{L}(t_1),$$

and so momentum is conserved. This is quite often (approximately) the case in problems of collision.

Finally in this chapter we have developed the counterpart in dynamics to the second equilibrium equation, $\Sigma\mathbf{M} = \mathbf{0}$, in statics. In dynamics this is, for a system of particles,

$$\Sigma\mathbf{M}_P = \Sigma\mathbf{r}_i \times m_i\mathbf{a}_i$$

when $\Sigma\mathbf{M}_P$ refers, as it did in statics, to the moments of *external* forces. This is often called Euler's second law. There are several different forms in which this law can be expressed, among them expressions involving the moment of momentum, defined as

$$\mathbf{H}_P = \Sigma\mathbf{r}_i \times m_i\mathbf{v}_i$$

for a system of particles. A useful relationship is

$$\mathbf{H}_P = \mathbf{H}_C + \mathbf{r}_{PC} \times \mathbf{L},$$

but the key expressions are the forms that Euler's second law can take,

$$\Sigma \mathbf{M}_C = \dot{\mathbf{H}}_C$$

and

$$\Sigma \mathbf{M}_O = \dot{\mathbf{H}}_O$$

where "O" is a point fixed in the inertial frame of reference.

REVIEW QUESTIONS ▶ Chapter 2

True or False?

1. At a given time, the mass center of a deformable body can be shown to be a unique point.
2. The momentum of any body (or system of bodies) in a frame $\mathcal{I}$ can be shown to be equal to the total mass times the velocity of the mass center in $\mathcal{I}$, even if $\mathcal{I}$ is not an inertial frame.
3. Euler's first law ($\Sigma \mathbf{F} = \dot{\mathbf{L}}$) applies to deformable bodies whether solid, liquid, or gaseous, as well as to rigid bodies and particles.
4. Neither the laws of motion nor the inertial frame is of any value without the other.
5. The mass center of a body $\mathcal{B}$ has to be a physical, or material, point of $\mathcal{B}$.
6. The work done by a linear spring depends on the paths traversed by its endpoints between the initial and final positions.
7. The work done by the friction force upon a block sliding on a fixed plane depends on the path taken by the block.
8. The work done by gravity on a body $\mathcal{B}$ depends on the lateral as well as the vertical displacement of the mass center of $\mathcal{B}$.
9. Since no external work was done on the two bodies of Example 2.14 *during the impact,* their total kinetic energy is the same after the collision as it was before.
10. For all bodies of constant density, the centroid of volume and the center of mass coincide.
11. In studying the motion of the earth around the sun, it is acceptable to treat the earth as a particle; in studying the daily rotation of the earth on its axis, however, it would not make sense to consider the earth as a particle.
12. The external forces acting on a body $\mathcal{B}$, which together form the resultant $\mathbf{F}_r$, must each have a line of action passing through the mass center of $\mathcal{B}$ in order for Euler's first law to apply.
13. Euler's second law can take the form $\Sigma \mathbf{M}_P = \dot{\mathbf{H}}_P$ regardless of the motion of point P in the inertial frame.

Answers: 1. T 2. T 3. T 4. T 5. F 6. F 7. T 8. F 9. F 10. T 11. T 12. F 13. F

3 Kinematics of Plane Motion of a Rigid Body

3.1 Introduction

In this chapter our goals are to develop the relationships between velocities, accelerations, angular velocity, and angular acceleration when a rigid body $\mathcal{B}$ moves in plane motion in a reference frame $\mathcal{I}$. Before doing so, however, we shall first explain precisely what we mean by such terms as *rigid body, plane motion, rigid extension, reference plane,* and several other concepts we shall be needing in this chapter and those to follow.

A **rigid body** is taken to be a body in which the distance between each and every pair of its points remains the same throughout the motion.* There is, of course, no such thing as a truly rigid body (since all bodies do *some* deforming); however, the deformations of many bodies are sufficiently small during their motions to allow the bodies to be treated as though they were rigid with good results.

The significance of the rigid-body model is that velocities of different points will be found to differ by something proportional to the rate at which the body turns, what we shall come to call its angular velocity. And accelerations will be found to be related through the angular velocity and its rate of change which we know as the angular acceleration. Thus a very small amount of information will characterize all the accelerations in the body. There are a number of ways in which this is important, but foremost is the fact that the right-hand side of our moment equation in Chapter 2 will in Chapter 4 be seen to take on a compact form involving the angular velocity and the angular acceleration.

Plane motion is treated in this book as motion in the xy plane (fixed in $\mathcal{I}$) or in planes parallel to it. Let a point P be located originally at coordinates (x_p, y_p, z_p). To say that P has plane motion simply means that it stays in the plane $z = z_p$ throughout its motion. Extending this definition, we say that rigid body $\mathcal{B}$ has plane motion whenever *all* its points remain in the same planes (parallel to xy) in which they started.

Question 3.1 How few points of a rigid body must be in plane motion to ensure that they *all* are?

A third concept we need to understand in rigid-body kinematics is that of the **body extended,** also called a **rigid extension** of the body. This idea, briefly mentioned in Chapter 1, says that we sometimes need to imagine points (which are not physical or material points of $\mathcal{B}$) moving with $\mathcal{B}$ as though they were in fact attached to it. An example would be the points on the axis of a pipe that are in the space inside it but of course move rigidly with it. We shall imagine a "rigid extension" of the body to pick up such points whenever it is useful to do so. Note that *any* point Q

* We have already encountered this concept in Chapter 1, where it was seen to be synonymous with the concept of *frame.*

Answer 3.1 Three noncollinear points are needed.

may be considered a point of *any* body $\mathcal{B}$ extended, provided that Q moves with $\mathcal{B}$ as if it were rigidly attached to it.

With these three concepts in mind, we are now prepared to define the **reference plane.** First we must realize that we are faced with the problem of determining where all the points of a body are as functions of time t. These points' locations $(x(t), y(t))$ would take forever to find if we had to do so for each of the infinitely many points of $\mathcal{B}$. Fortunately, for a rigid body in plane motion, if we know the locations of all its points in any one plane of $\mathcal{I}$ (which we shall call the reference plane), then we automatically know the locations of all its *other* points in all *other* planes. The reason for this is as follows. For each point B of $\mathcal{B}$ that does not lie in the reference plane, there is a "companion point" of $\mathcal{B}$ in the reference plane (suggested by A in Figure 3.1) that has the same (x, y) coordinates at the beginning of the motion. It then follows that the (x, y) coordinates of A and B *always* match *throughout* the motion of $\mathcal{B}$!

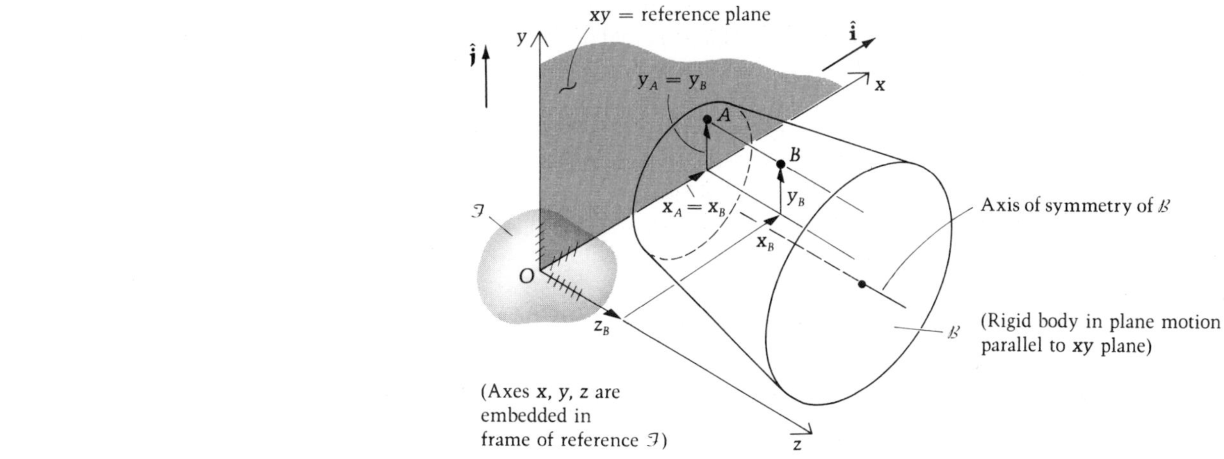

Figure 3.1

Question 3.2 Why is $x_A \equiv x_B$ and $y_A \equiv y_B$ as time passes?

Body $\mathcal{B}$ in Figure 3.1 is a cone pulley, which turns about its axis of symmetry. Note from its varying cross-sectional diameter that a body need not have constant cross section to be in plane motion.

Having laid the necessary groundwork, we now let xy be our reference plane. From Figure 3.1 we see that

$$\mathbf{r}_{OA} = x_A\hat{\mathbf{i}} + y_A\hat{\mathbf{j}} \tag{3.1}$$

Answer 3.2 If ever $x_A \neq x_B$ or $y_A \neq y_B$ (or both), either the rigid body or the plane motion assumptions (or both) will have been violated.

and we may differentiate this equation to obtain

$$\mathbf{v}_A = \dot{x}_A\hat{\mathbf{i}} + \dot{y}_A\hat{\mathbf{j}} = \dot{x}_B\hat{\mathbf{i}} + \dot{y}_B\hat{\mathbf{j}} = \mathbf{v}_B \tag{3.2}$$

$$\mathbf{a}_A = \ddot{x}_A\hat{\mathbf{i}} + \ddot{y}_A\hat{\mathbf{j}} = \ddot{x}_B\hat{\mathbf{i}} + \ddot{y}_B\hat{\mathbf{j}} = \mathbf{a}_B \tag{3.3}$$

In these equations we have used the facts that $x_B \equiv x_A$, $y_B \equiv y_A$, and z_B = constant. Equations (3.2) and (3.3) show clearly that if we completely describe the velocities and accelerations in one reference plane, we then know them for *all* the points of the body. This allows us to focus on one plane of the body throughout this chapter and most of the next two as well.

The reference plane is thus a very important concept, for it allows us to study the motion of an entire body by concerning ourselves only with those of its points that lie in this plane. We say we "know the motion" of $\mathcal{B}$ when we know where all its points are at all times. We have already reduced this task to knowing the locations of the points in the reference plane. (The rest of the body "goes along for the ride.") But in fact if we know the location of just *two* points (say P_1 and P_2) of the reference plane, then we know the whereabouts of *all* points of this plane and thus of the whole body! This is because each point of the reference plane must maintain the same position relative to the points P_1 and P_2. This idea is illustrated in Figure 3.2. Note in the figure that if P_1 and P_2 are correctly located with respect to the reference frame $\mathcal{I}$, all other points of $\mathcal{B}$ are necessarily in their correct positions.

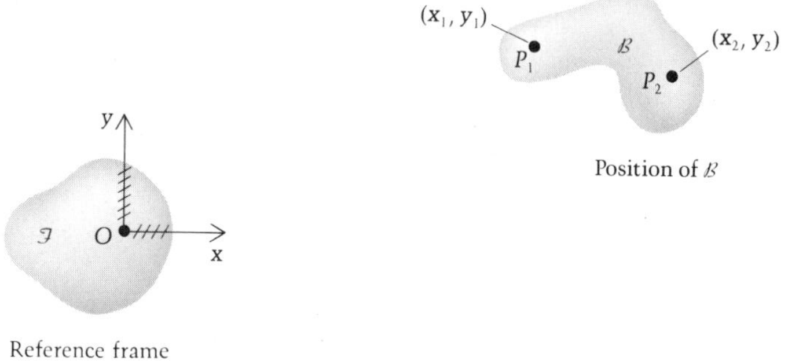

Figure 3.2

> **Question 3.3** Is knowledge of the locations of two points sufficient for us to know the motion of a body in general (three-dimensional) motion?

Instead of knowing the locations of two points of the body — (x_1, y_1) of P_1 and (x_2, y_2) of P_2 — we may alternatively locate the body if we know where just *one* point, P, is located *plus* the value of the orientation angle θ (about an axis through P_1 and parallel to z); see Figure 3.3.

Answer 3.3 No; the body could rotate around the line joining the two points. In three dimensions it takes three points, not all on the same line!

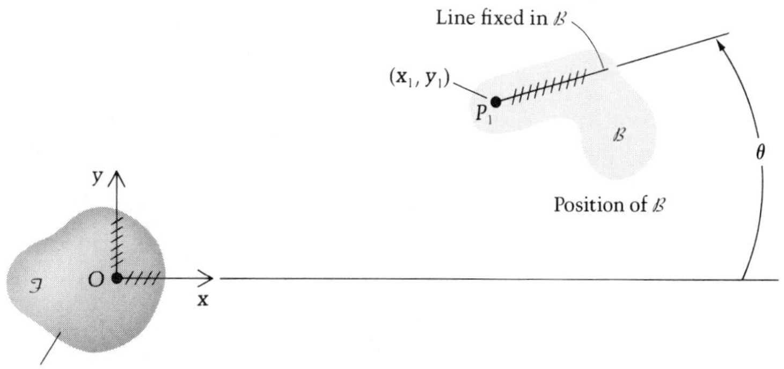

Figure 3.3

Question 3.4 Knowing the locations of two points requires four variables (x_1, y_1, x_2, y_2), whereas one point plus the angle takes but three (x_1, y_1, θ). Why do these numbers of variables differ?

The foregoing is intended to suggest quite correctly that the velocities of different points in the reference plane will be linked together because of the rigidity of the body, and similarly for accelerations. In the next section we will turn to the development of the relationship between the velocities of points such as P_1 and P_2 in Figure 3.2.

Answer 3.4 The $2 \times 2 = 4$ coordinates of P_1 and P_2 are not independent. The distance between the points is a constant, so

$$\sqrt{(x_1 - x_2)^2 + (y_1 - y_2)^2} = \text{constant}$$

can be used to find any one of x_1, y_1, x_2, y_2 in terms of the other three.

PROBLEMS ▸ Section 3.1

3.1 Which of the bodies $\mathscr{B}$ shown in Figure P3.1(a–f) are in plane motion in frame $\mathscr{I}$?

(a) A turkey being barbecued by slowly turning on a rotisserie.

(b) A cone rolling on a tabletop.

(c) A spinning coin if the base is fixed.

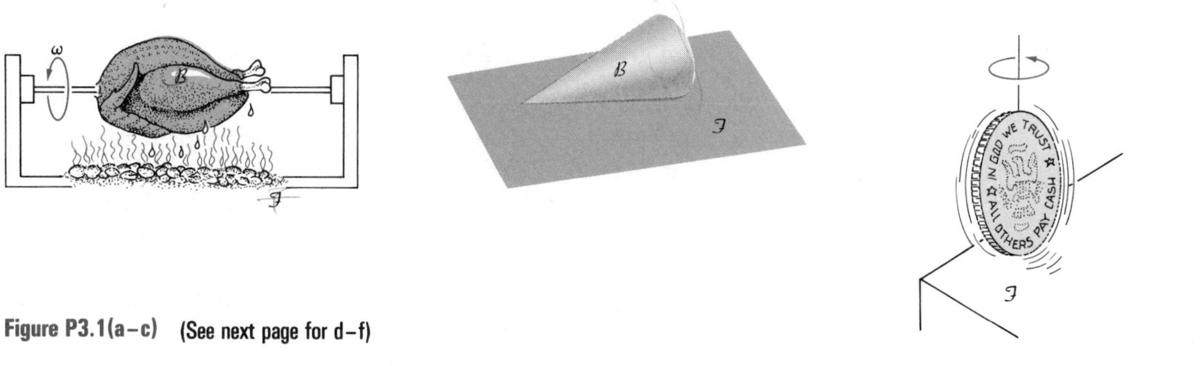

Figure P3.1(a–c) (See next page for d–f)

(d) A can rolling down an inclined plane.

(e) The bevel gear $\mathcal{B}$, which meshes with another bevel gear $\mathcal{A}$.

(f) The (shaded) crosspiece of a universal joint.

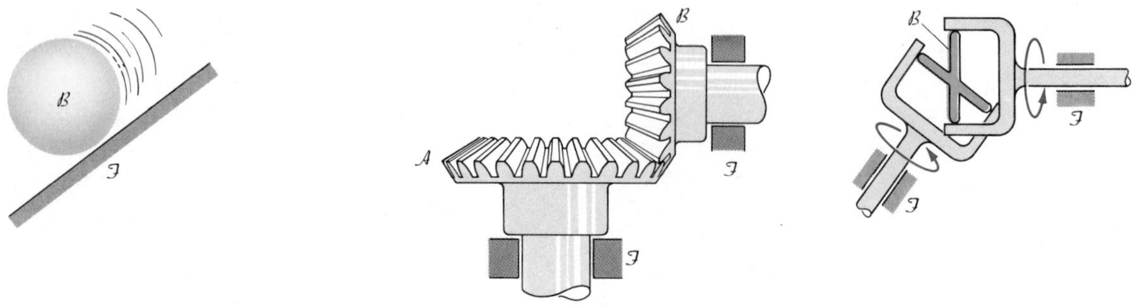

Figure P3.1(d–f)

3.2 Give three examples of plane motion besides those in the previous problem. Then give three examples of motion that is not planar.

3.2 Velocity and Angular Velocity Relationship for Two Points of the Same Rigid Body

In this section we derive a very useful relationship between the velocities in $\mathcal{F}$ of any two points in the reference plane of a rigid body $\mathcal{B}$ in plane motion and the angular velocity vector of $\mathcal{B}$ in $\mathcal{F}$. Let P and Q denote these two points of $\mathcal{B}$, and let us embed the axes (x, y, z) in reference frame $\mathcal{F}$ as shown in Figure 3.4.

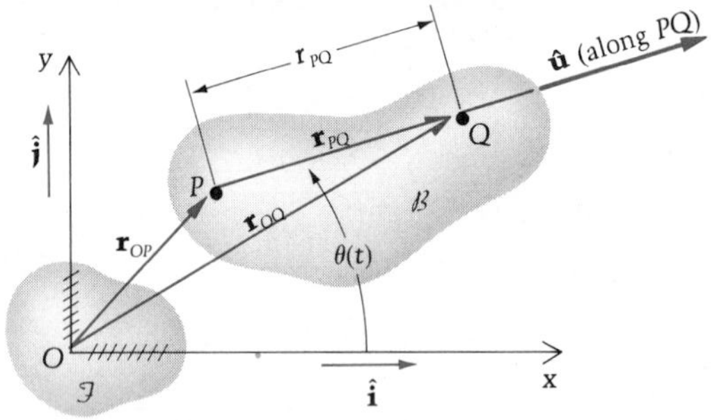

Figure 3.4

We are saying that even though $\mathcal{B}$ may move with respect to the reference frame $\mathcal{F}$, the xy plane of $\mathcal{F}$ always contains the points of interest P and Q of $\mathcal{B}$. A good example is found in the classroom; let the body $\mathcal{B}$ be a blackboard eraser. Letting the blackboard itself be the reference frame $\mathcal{F}$ (so that x and y are fixed in the plane of the blackboard), the eraser undergoes plane motion whenever the professor erases the board. Our points P and Q are any two points of the erasing surface of the eraser. Note how each point of the eraser remains the same z distance from the blackboard (where $z = 0$) during the erasing. The eraser is no longer in plane motion, however, once it *leaves* the surface of the board and its points move with z components of velocity.

Notice from Figure 3.4 that $\hat{\mathbf{u}}$ is a unit vector always directed from P toward Q so that $\mathbf{r}_{PQ} = r_{PQ}\hat{\mathbf{u}}$, where r_{PQ} is the distance PQ (that is, the magnitude of the vector $\mathbf{r}_{PQ}$). Note further that the orientation (angular rotation) of the body is described by the angle θ, measured between any line fixed in the reference frame $\mathcal{I}$ (we shall here use the x axis) and any line fixed in the body (for the moment, we shall use the line segment from P to Q).

Development of the Velocity and Angular Velocity Relationship

We are now ready to develop the velocity and angular velocity relationship for rigid bodies. From Figure 3.4 we see that

$$\mathbf{r}_{OQ} = \mathbf{r}_{OP} + \mathbf{r}_{PQ} \tag{3.4}$$

so that we have, upon differentiation in frame $\mathcal{I}$,

$$\dot{\mathbf{r}}_{OQ} = \dot{\mathbf{r}}_{OP} + \dot{\mathbf{r}}_{PQ}$$

Recognizing the first two vectors as the definitions of the velocities of P and Q (in $\mathcal{I}$, where O is fixed), we may write

$$\mathbf{v}_Q = \mathbf{v}_P + \dot{\mathbf{r}}_{PQ} \tag{3.5}$$

In obtaining Equation (3.5), all derivatives were taken in $\mathcal{I}$, so that, for example, there is no need to write $\mathbf{v}_{Q/\mathcal{I}}$.

In order to write $\dot{\mathbf{r}}_{PQ}$ as a vector we can use, we express $\mathbf{r}_{PQ}$ as a magnitude times a unit vector. With the help of Figure 3.4 we get

$$\mathbf{r}_{PQ} = r_{PQ}\hat{\mathbf{u}} = r_{PQ}(\cos\theta\hat{\mathbf{i}} + \sin\theta\hat{\mathbf{j}}) \tag{3.6}$$

Differentiating this expression as in Section 1.6, we have

$$\begin{aligned}\dot{\mathbf{r}}_{PQ} &= r_{PQ}\dot{\theta}(-\sin\theta\hat{\mathbf{i}} + \cos\theta\hat{\mathbf{j}}) = r_{PQ}\dot{\theta}\hat{\mathbf{k}} \times (\cos\theta\hat{\mathbf{i}} + \sin\theta\hat{\mathbf{j}})^{*} \\ &= r_{PQ}\dot{\theta}\hat{\mathbf{k}} \times \hat{\mathbf{u}} = \dot{\theta}\hat{\mathbf{k}} \times r_{PQ}\hat{\mathbf{u}}\end{aligned}$$

Therefore we have derived a useful expression for $\dot{\mathbf{r}}_{PQ}$:

$$\dot{\mathbf{r}}_{PQ} = \dot{\theta}\hat{\mathbf{k}} \times \mathbf{r}_{PQ} \tag{3.7}$$

Question 3.5 In the preceding development, why is $\dot{r}_{PQ} = 0$?

Substituting Equation (3.7) into (3.5) yields

$$\mathbf{v}_Q = \mathbf{v}_P + \dot{\theta}\hat{\mathbf{k}} \times \mathbf{r}_{PQ} \tag{3.8}$$

---

* Throughout this book $\hat{\mathbf{i}}$, $\hat{\mathbf{j}}$, and $\hat{\mathbf{k}}$ constitute a right-handed system so that $\hat{\mathbf{i}} \times \hat{\mathbf{j}} = \hat{\mathbf{k}}$.

Answer 3.5 Because $\mathcal{B}$ is rigid, r_{PQ} is the *constant* distance between points P and Q.

which relates the velocities of the points P and Q and introduces the **angular velocity** of $\mathcal{B}$ in reference frame $\mathcal{I}$. The Greek letter omega is usually used to denote this vector:

$$\boldsymbol{\omega}_{\mathcal{B}/\mathcal{I}} = \dot{\theta}\hat{\mathbf{k}} \tag{3.9}$$

When there is no confusion about the body and reference frame involved, we may drop the subscripts and write $\boldsymbol{\omega}_{\mathcal{B}/\mathcal{I}}$ as simply $\boldsymbol{\omega}$. Also, some prefer to write $\boldsymbol{\omega}$ as $\omega\hat{\mathbf{k}}$ rather than $\dot{\theta}\hat{\mathbf{k}}$; we shall use both forms, feeling that the latter is a nice reminder that in plane motion, angular velocity is proportional to the time rate of change of an angle.*

The magnitude $|\dot{\theta}|$ (or $|\omega|$ or $|\boldsymbol{\omega}|$) of the angular velocity is called the **angular speed** of $\mathcal{B}$ in frame $\mathcal{I}$. Note that $\dot{\theta}$ (or ω) itself can be negative. Note further that neither the angular velocity vector nor Equation (3.8) depends on which body-fixed line segment (such as PQ above) is chosen to measure θ. The proof of the preceding statement is not difficult and will be given later as an exercise.

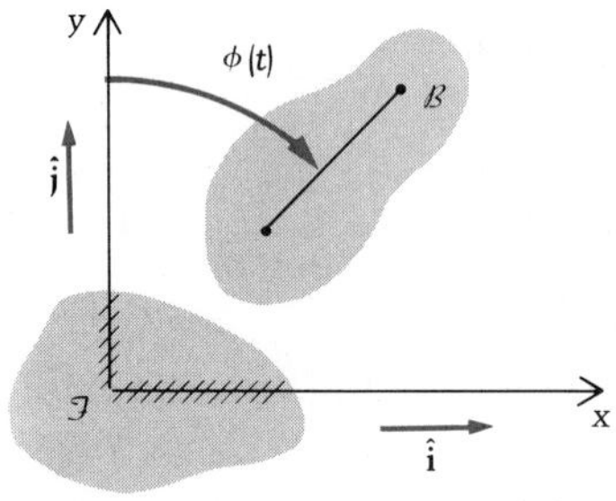

Figure 3.5

Note that if our angle of orientation were chosen as shown in Figure 3.5, then the angular velocity would be given by

$$\boldsymbol{\omega} = \dot{\varphi}(-\hat{\mathbf{k}}) = -\dot{\varphi}\hat{\mathbf{k}}$$

The angular velocity vector is always in the direction given by the right-hand rule when we turn our fingers in the direction of rotation of the body. Referring to both Figures 3.4 and 3.5, $\dot{\varphi} = -\dot{\theta}$, and

$$\boldsymbol{\omega} = \dot{\theta}\hat{\mathbf{k}} = -\dot{\varphi}\hat{\mathbf{k}}$$

and $\boldsymbol{\omega}$ is directed out of the page if the body is turning counterclockwise, and into the page if it's turning clockwise.

Important Things to Remember About Equation (3.8)

We now have the result that the angular velocity vector is a property of the overall body $\mathcal{B}$, and *not* a property of its individual *points*. This idea cannot be overemphasized. Remember:

1. A *point* has position, velocity, and acceleration.
2. A *body* has orientation, angular velocity, and angular acceleration.†

Remember too that a *point* does not have orientation, $\boldsymbol{\omega}$, or $\boldsymbol{\alpha}$,† and a finite-sized *body* does not have a unique $\mathbf{r}$, $\mathbf{v}$, and $\mathbf{a}$.**

We shall emphasize these property differences between points and bodies in the following way: throughout this book, points are denoted by capital italic printed letters while bodies are denoted by ordinary capital

---

* We remark that in general (three-dimensional) motion, such a simple relationship as $\dot{\theta}\hat{\mathbf{k}}$ between angular velocity and body orientation does *not* exist.

† Angular acceleration ($\boldsymbol{\alpha}$), the derivative of angular velocity, is discussed in Section 3.5.

** Of course the *particle*, being treated as small enough that we need not distinguish between the locations of its points, must be considered as having an $\mathbf{r}$, $\mathbf{v}$, and $\mathbf{a}$ — and *not* an $\boldsymbol{\omega}$ or $\boldsymbol{\alpha}$.

cursive letters. Hence, for example, P, A, and B denote points, while $\mathcal{P}$, $\mathcal{A}$, and $\mathcal{B}$ denote bodies. Therefore, we shall simply print the names of points, and write the names of bodies in cursive script.

We now move toward a number of examples of the use of our new Equation (3.8); in each application of this equation, the following three rules must be followed *without exception:*

3. This vector extends *from* the point (P) on *this* (right) side of the equation *to* the point (Q) on the *other* (left) side.

$$\mathbf{v}_Q = \mathbf{v}_P + \dot{\theta}\hat{\mathbf{k}} \times \mathbf{r}_{PQ} \tag{3.10}$$

1. These two points are on the *same* rigid body $\mathcal{B}$.

2. This is the angular velocity vector of $\mathcal{B}$.

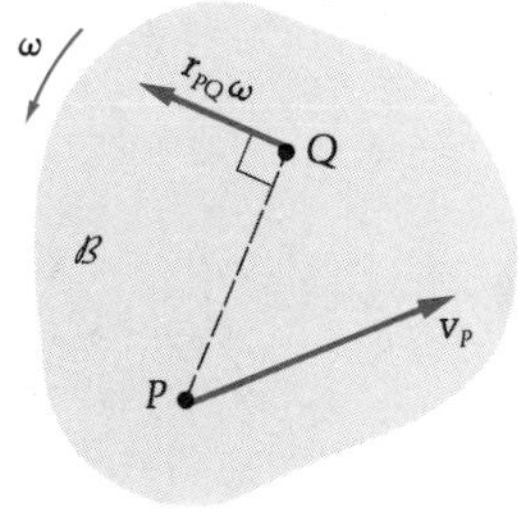

Figure 3.6

Also helpful in using Equation (3.8) is the kinematic diagram presented in Figure 3.6. The velocity of Q, from Equation (3.8), is the sum of the two vectors in Figure 3.6 (see Figure 3.7). Note that depending on the relative sizes of v_P and $r_{PQ}\omega$, the velocity $\mathbf{v}_Q$ could lie on either side, or even along, line PQ. Note further that the difference between the velocities of Q and P — that is, $\mathbf{v}_Q - \mathbf{v}_P$ — is simply $\dot{\theta}\hat{\mathbf{k}} \times \mathbf{r}_{PQ}$. This means that the only way in which the velocities of two points of a rigid body $\mathcal{B}$, in motion in frame $\mathcal{I}$, can differ is by the $r\omega$ term normal to the line joining them. We shall return to this idea following the first three examples of this section.

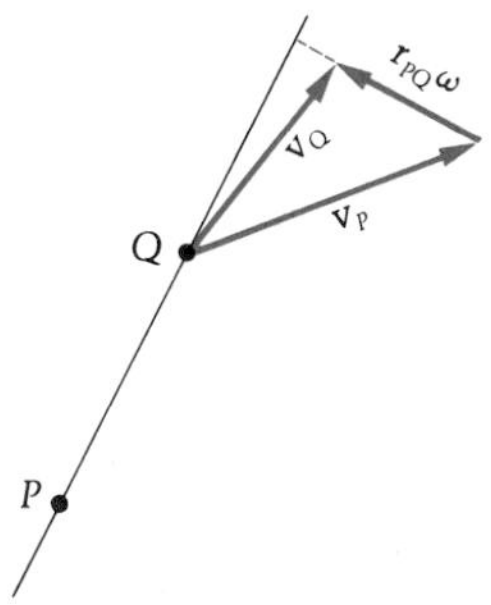

Figure 3.7

Incidentally, some books describe $\mathbf{v}_Q - \mathbf{v}_P$ as "the velocity of point Q relative to point P." We mention this only by way of explanation; our definition of $\mathbf{v}_P$ in Section 1.3 shows that points have velocities relative to frames, *not* relative to other points. If one uses the phrase "the velocity of point Q relative to point P," one means the velocity of Q in a reference frame in which P is fixed and which translates relative to $\mathcal{I}$.*

In each of the examples that follow, note the importance of selecting *and depicting* the unit vectors to be used in the solution. Also, in each of the first three examples, pay careful attention to the way the velocity of a point (say, B) is expressed if the tangent to the path of B is known; using what we learned in Section 1.7, $\mathbf{v}_B$ is expressed as a single unknown scalar (whose absolute value is the speed of B) times a unit vector along the known tangent.

Example 3.1

A 30-ft ladder is slipping down in a warehouse with the upper contact point T moving downward on the wall at a speed of 2 ft/sec in the position shown in Figure E3.1. Find the velocity of point B, which is sliding on the floor.

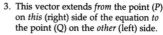

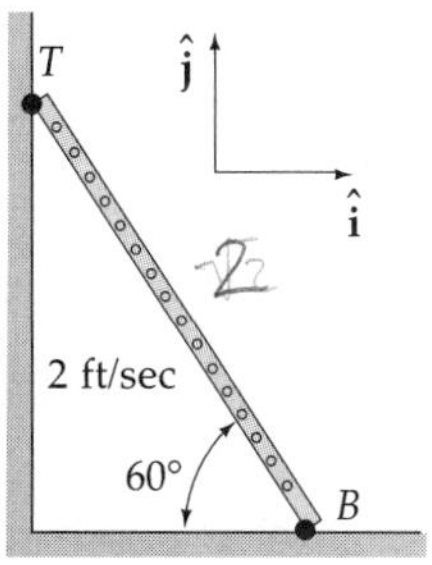

Figure E3.1

* "Translates" means that the frame moves in $\mathcal{I}$ without rotating. Translation is discussed in more detail in Section 3.3.

Solution

We relate $\mathbf{v}_T$ and $\mathbf{v}_B$ by using Equation (3.8):

$$\mathbf{v}_B = \mathbf{v}_T + \dot{\theta}\hat{\mathbf{k}} \times \mathbf{r}_{TB}$$

Noting that $\mathbf{v}_B$ has no $\hat{\mathbf{j}}$-component and that $\mathbf{v}_T$ has no $\hat{\mathbf{i}}$-component, we write:

$$v_B\hat{\mathbf{i}} = -2\hat{\mathbf{j}} + \dot{\theta}\hat{\mathbf{k}} \times 30\left(\frac{1}{2}\hat{\mathbf{i}} - \frac{\sqrt{3}}{2}\hat{\mathbf{j}}\right)$$

$$= \left(30\frac{\sqrt{3}}{2}\dot{\theta}\right)\hat{\mathbf{i}} + (-2 + 15\dot{\theta})\hat{\mathbf{j}} \text{ ft/sec}$$

Matching the $\hat{\mathbf{j}}$ coefficients, we have

$$0 = -2 + 15\dot{\theta} \Rightarrow \dot{\theta} \text{ (or } \omega) = \frac{2}{15} = 0.133 \text{ rad/sec}$$

Matching the $\hat{\mathbf{i}}$ coefficients, we have

$$v_B = 15\sqrt{3}\dot{\theta} = 15\sqrt{3}\left(\frac{2}{15}\right) = 3.46 \text{ ft/sec}$$

Thus the velocity of B is $\mathbf{v}_B = 3.46\hat{\mathbf{i}}$ ft/sec (or 3.46 → ft/sec).

Note that a direction indicator must be attached to $\dot{\theta}$ in order to specify correctly the angular velocity vector of the ladder:

$$\boldsymbol{\omega} = \dot{\theta}\hat{\mathbf{k}} = 0.133\hat{\mathbf{k}} \text{ rad/sec}$$

or, alternatively,

$$\omega = 0.133 \circlearrowleft \text{ rad/sec}$$

Note that the directions of $\mathbf{v}_B(\rightarrow)$ and ω ($\circlearrowleft$) make sense. Such visual checks on solutions should be made whenever possible.

EXAMPLE 3.2

At the instant shown in Figure E3.2, the velocity of point A is 0.2 m/s to the right. Find the angular velocity of rod $\mathcal{B}$, and determine the velocity of its other end (point B), which is constrained to move in the circular slot.

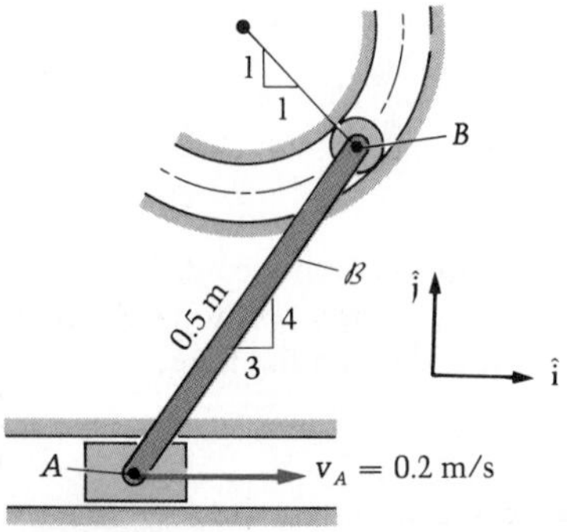

Figure E3.2

Solution

We shall use Equation (3.8), featuring the points A and B of the rod:

$$\mathbf{v}_B = \mathbf{v}_A + \dot{\theta}\hat{\mathbf{k}} \times \mathbf{r}_{AB}$$

Noting that the velocity of B has a known direction (tangent to its path), we write $\mathbf{v}_B$ as an unknown scalar times a unit vector in this direction:

$$v_B\left(\frac{\hat{\mathbf{i}} + \hat{\mathbf{j}}}{\sqrt{2}}\right) = 0.2\hat{\mathbf{i}} + \dot{\theta}\hat{\mathbf{k}} \times (0.3\hat{\mathbf{i}} + 0.4\hat{\mathbf{j}})$$

The component equations are:

$$\hat{\mathbf{i}} \text{ coefficients:} \quad \left(\frac{1}{\sqrt{2}}\right)v_B = 0.2 - 0.4\dot{\theta} \qquad (1)$$

$$\hat{\mathbf{j}} \text{ coefficients:} \qquad \left(\frac{1}{\sqrt{2}}\right) v_B = 0.3\dot{\theta} \tag{2}$$

Solving Equations (1) and (2) gives

$$v_B = 0.121 \text{ m/s} \qquad \dot{\theta} = 0.286 \text{ rad/s}$$

and therefore the answers (*vectors* are what are asked for!) are

$$\mathbf{v}_B = 0.121 \;\measuredangle_{1}^{1}\; \text{m/s} \qquad \text{and} \qquad \omega = 0.286 \circlearrowleft \text{rad/s}$$

or, equivalently,

$$\mathbf{v}_B = 0.0856\hat{\mathbf{i}} + 0.0856\hat{\mathbf{j}} \text{ m/s} \qquad \text{and} \qquad \dot{\theta}\hat{\mathbf{k}} = \boldsymbol{\omega} = 0.286\hat{\mathbf{k}} \text{ rad/s}$$

In the next example, *two* bodies have angular velocity; thus we shall have to subscript the ω's (or $\dot{\theta}$'s). We shall simply denote by $\boldsymbol{\omega}_1$ (or $\dot{\theta}_1\hat{\mathbf{k}}$) the angular velocity of $\mathcal{B}_1$ and $\boldsymbol{\omega}_2$ (or $\dot{\theta}_2\hat{\mathbf{k}}$) the angular velocity of $\mathcal{B}_2$.

Example 3.3

The crank arm $\mathcal{B}_1$ shown in Figure E3.3a turns about a horizontal z axis, through its pinned end O, with an angular velocity of 10 rad/sec clockwise at the given instant. Find the velocity of the piston pin B.

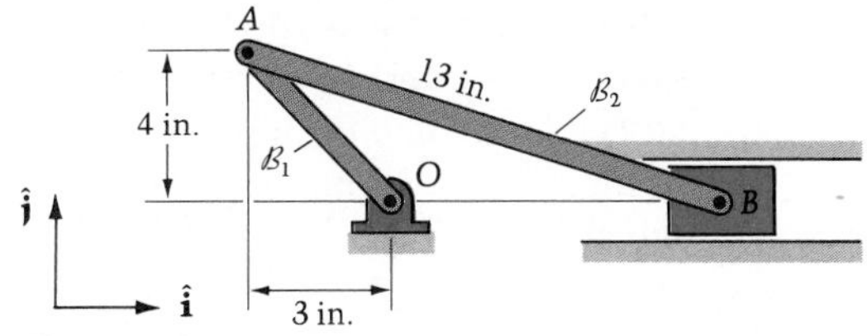

Figure E3.3a

Solution

We apply Equation (3.8) first to relate the velocities of A and O on body $\mathcal{B}_1$ and then to relate $\mathbf{v}_B$ to $\mathbf{v}_A$ on rod $\mathcal{B}_2$. Note that A is a "linking point" of both $\mathcal{B}_1$ and $\mathcal{B}_2$, since it belongs to *both* bodies. On body $\mathcal{B}_1$:

$$\begin{aligned} \mathbf{v}_A &= \mathbf{v}_O + \omega_1\hat{\mathbf{k}} \times \mathbf{r}_{OA} \\ &= 0 + (-10\hat{\mathbf{k}}) \times (-3\hat{\mathbf{i}} + 4\hat{\mathbf{j}}) \\ &= 40\hat{\mathbf{i}} + 30\hat{\mathbf{j}} \text{ in./sec} \end{aligned}$$

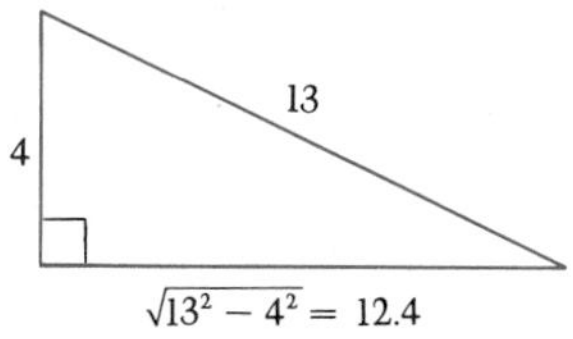

Figure E3.3b

On body $\mathcal{B}_2$:

$$\mathbf{v}_B = \mathbf{v}_A + \omega_2\hat{\mathbf{k}} \times \mathbf{r}_{AB}$$

and using the Pythagorean theorem (see Figure E3.3b),

$$\mathbf{v}_B = (40\hat{\mathbf{i}} + 30\hat{\mathbf{j}}) + \omega_2\hat{\mathbf{k}} \times (12.4\hat{\mathbf{i}} - 4\hat{\mathbf{j}})$$

Now point B is constrained to move only horizontally. Therefore,

$$v_B\hat{\mathbf{i}} = (40 + 4\omega_2)\hat{\mathbf{i}} + (30 + 12.4\omega_2)\hat{\mathbf{j}}$$

Equating the $\hat{\mathbf{i}}$ coefficients:

$$v_B = 40 + 4\omega_2 \tag{1}$$

Equating the $\hat{\mathbf{j}}$ coefficients:

$$0 = 30 + 12.4\omega_2$$

$$\omega_2 = -2.42 \text{ rad/sec}$$

Therefore,

$$\omega_2 = 2.42 \circlearrowright \text{ rad/sec} \tag{2}$$

Substituting ω_2 into (1), we have $\mathbf{v}_B = 30.3$ in./sec and $\mathbf{v}_B = 30.3\hat{\mathbf{i}}$ in./sec.

In all three of the preceding examples, we re-emphasize that it is absolutely essential to correctly incorporate the kinematic constraints imposed by slots, walls, floors, and so forth.

It is often helpful in studying the kinematics of rigid bodies to make use of the following result, which is a corollary of Equation (3.8):

> Corollary: If P and Q are two points of a rigid body, their velocity components along the line joining them must be equal.

Intuitively, we see that the difference between these components is the rate of stretching of the line PQ, and this has to vanish. Also, we have seen that $\mathbf{v}_Q$ and $\mathbf{v}_P$ differ only by the term $\dot{\theta}\hat{\mathbf{k}} \times \mathbf{r}_{PQ}$, which is clearly *normal* to the line PQ joining the points. Mathematically, we can see this immediately by dotting Equation (3.8) with the unit vector parallel to $\mathbf{r}_{PQ}$, which is $\mathbf{r}_{PQ}/r_{PQ}$:

$$\underbrace{\frac{\mathbf{r}_{PQ}}{r_{PQ}} \cdot \mathbf{v}_Q}_{\substack{\text{component of} \\ \mathbf{v}_Q \text{ along } PQ}} = \underbrace{\frac{\mathbf{r}_{PQ}}{r_{PQ}} \cdot \mathbf{v}_P}_{\substack{\text{component of} \\ \mathbf{v}_P \text{ along } PQ}} + \underbrace{(\dot{\theta}\hat{\mathbf{k}} \times \mathbf{r}_{PQ}) \cdot \frac{\mathbf{r}_{PQ}}{r_{PQ}}}_{\substack{\text{zero (since } \dot{\theta}\hat{\mathbf{k}} \times \mathbf{r}_{PQ} \\ \text{is } \perp \mathbf{r}_{PQ})}} \tag{3.11}$$

Thus, if we know the velocity of one point of the body, we can find any other without involving the angular velocity by using Equation (3.11). For instance, in Example 3.2, the unit vector $\mathbf{r}_{AB}/r_{AB}$ is simply $(3\hat{\mathbf{i}} + 4\hat{\mathbf{j}})/5$, and dotting this with the equation

$$v_B \frac{\hat{\mathbf{i}} + \hat{\mathbf{j}}}{\sqrt{2}} = 0.2\hat{\mathbf{i}} + \omega\hat{\mathbf{k}} \times (0.3\hat{\mathbf{i}} + 0.4\hat{\mathbf{j}})$$

from that example gives

$$v_B\left(\frac{3(1) + 4(1)}{5\sqrt{2}}\right) = \frac{3}{5}\,(0.2) = 0.120$$

or

$$v_B = 0.121 \Rightarrow \mathbf{v}_B = 0.121 \ {\angle}^{1}_{1} \text{ m/s} \qquad \text{(as before)}$$

The algebra is seen to be simpler; we have worked with one equation in one unknown rather than two in two.

We now return to the vector formulation (Equation 3.8) for two final examples in this section.

Example 3.4

In the linkage shown in Figure E3.4, the velocities of A and C are given to be

$$\mathbf{v}_A = 2 \leftarrow \text{m/s}$$
$$\mathbf{v}_C = 3 \uparrow \text{m/s}$$

at the instant given. Find the velocity of point B at the same instant.

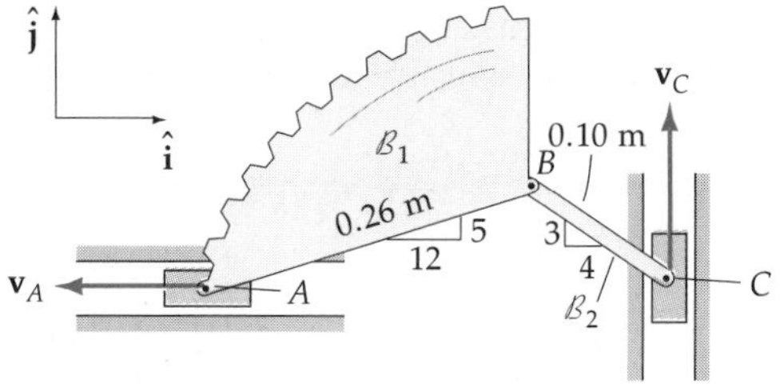

Figure E3.4

Solution

On body $\mathcal{B}_1$:

$$\begin{aligned}\mathbf{v}_B &= \mathbf{v}_A + \omega_1\hat{\mathbf{k}} \times \mathbf{r}_{AB}\\ &= -2\hat{\mathbf{i}} + \omega_1\hat{\mathbf{k}} \times (0.24\hat{\mathbf{i}} + 0.10\hat{\mathbf{j}})\\ &= (-2 - 0.1\omega_1)\hat{\mathbf{i}} + (0.24\omega_1)\hat{\mathbf{j}} \text{ m/s} \qquad (1)\end{aligned}$$

On bar $\mathcal{B}_2$:

$$\begin{aligned}\mathbf{v}_B &= \mathbf{v}_C + \omega_2\hat{\mathbf{k}} \times \mathbf{r}_{CB}\\ &= 3\hat{\mathbf{j}} + \omega_2\hat{\mathbf{k}} \times (-0.08\hat{\mathbf{i}} + 0.06\hat{\mathbf{j}})\\ &= (-0.06\omega_2)\hat{\mathbf{i}} + (3 - 0.08\omega_2)\hat{\mathbf{j}} \text{ m/s} \qquad (2)\end{aligned}$$

Equating the two vector expressions for $\mathbf{v}_B$, we get

$$\hat{\mathbf{i}} \text{ coefficients:} \quad -2 - 0.1\omega_1 = -0.06\omega_2$$
$$\hat{\mathbf{j}} \text{ coefficients:} \quad 0.24\omega_1 = 3 - 0.08\omega_2$$

Solving these two equations,

$$\omega_1 = 0.893 \quad \text{and} \quad \omega_2 = 34.8 \text{ rad/s}$$

From Equation (1), it follows that

$$\mathbf{v}_B = -2.09\hat{\mathbf{i}} + 0.214\hat{\mathbf{j}} \text{ m/s}$$

and the same result follows from (2), as a check.

Question 3.6 If the velocities of A and C were given to be $2 \leftarrow$ m/s and $3 \uparrow$ m/s for an *interval* of time, and not just at the instant shown, would the solution be any different at (a) the same instant? (b) some other instant?

Answer 3.6 (a) No. (b) Yes, because the geometry would be different.

EXAMPLE 3.5

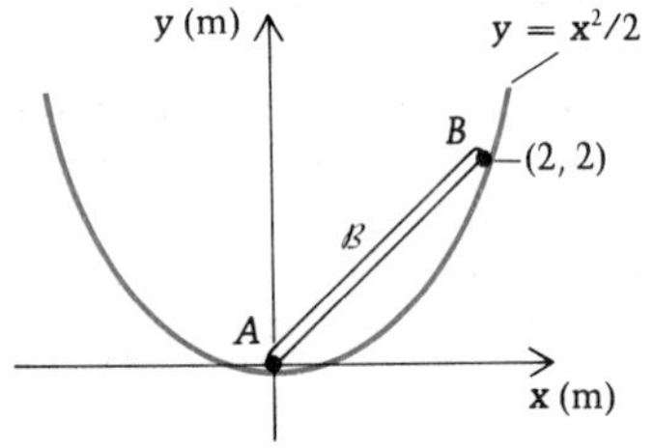

Figure E3.5a

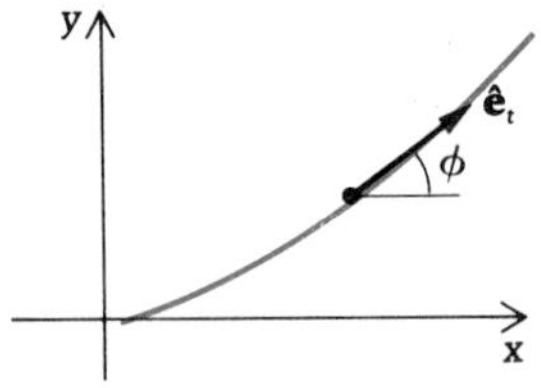

Figure E3.5b

The end B of rod $\mathcal{B}$ travels up the right half of the parabolic incline in Figure E3.5a at the constant speed of 0.3 m/s. Find the angular velocity of $\mathcal{B}$ and the velocity of point A, which is at the origin at the given instant.

Solution

We shall relate $\mathbf{v}_B$ to $\mathbf{v}_A$ using Equation (3.8):

$$\mathbf{v}_B = \mathbf{v}_A + \boldsymbol{\omega} \times \mathbf{r}_{AB} \tag{1}$$

Next we use Equation (1.41) to express $\mathbf{v}_B$:

$$\mathbf{v}_B = 0.3\hat{\mathbf{e}}_t$$

To get the unit tangent $\hat{\mathbf{e}}_t$ for point B, we use Figure E3.5b, noting that $\hat{\mathbf{e}}_t$ is tangent to the parabola at all times:

$$\phi = \tan^{-1}\left(\frac{dy}{dx}\right)$$

$$= \tan^{-1} 2 = 63.4°$$

Therefore, for point B,

$$\hat{\mathbf{e}}_t = \cos\phi\hat{\mathbf{i}} + \sin\phi\hat{\mathbf{j}}$$

$$= 0.447\hat{\mathbf{i}} + 0.894\hat{\mathbf{j}}$$

And thus

$$\mathbf{v}_B = 0.3\hat{\mathbf{e}}_t = 0.134\hat{\mathbf{i}} + 0.268\hat{\mathbf{j}} \text{ m/s}$$

Since point A likewise has a velocity tangent to *its* path, we may write

$$\mathbf{v}_A = v_A\hat{\mathbf{i}}$$

and so Equation (1) gives

$$0.134\hat{\mathbf{i}} + 0.268\hat{\mathbf{j}} = v_A\hat{\mathbf{i}} + \dot{\theta}\hat{\mathbf{k}} \times (2\hat{\mathbf{i}} + 2\hat{\mathbf{j}})$$

Collecting the coefficients of $\hat{\mathbf{i}}$ and $\hat{\mathbf{j}}$, we have

$$\hat{\mathbf{j}} \text{ coefficients:} \quad 0.268 = 2\dot{\theta} \Rightarrow \dot{\theta} = 0.134 \text{ rad/s}$$

so that

$$\boldsymbol{\omega} = \dot{\theta}\hat{\mathbf{k}} = 0.134\hat{\mathbf{k}} \text{ rad/s or } 0.134 \circlearrowleft \text{rad/s}$$

$$\hat{\mathbf{i}} \text{ coefficients:} \quad 0.134 = v_A - 2\dot{\theta}$$

Substituting for $\dot{\theta}$ and solving,

$$v_A = 0.402 \text{ m/s}$$

so that

$$\mathbf{v}_A = 0.402\hat{\mathbf{i}} \text{ m/s}$$

Applications of Equation (3.8) to rolling bodies are presented in Section 3.6 after we have examined that topic in detail.

PROBLEMS ▶ Section 3.2

3.3 The angular velocity of the bent bar is indicated in Figure P3.3. Find the velocity of the endpoint B in this position.

3.4 The velocities of the two endpoints A and B of a rigid bar in plane motion are shown in Figure P3.4. Find the velocity of the midpoint of the bar in the given position.

3.5 If $\mathbf{v}_A = 80\hat{\mathbf{i}}$ in./sec, find ω_2 and ω_3. See Figure P3.5.

3.6 At a certain instant, the coordinates of two points A and B of a rigid body $\mathcal{B}$ in plane motion are given in Figure P3.6. Point A has $\mathbf{v}_A = 2\hat{\mathbf{i}}$ m/s, and the velocity of B is vertical. Find $\mathbf{v}_B$ and the angular velocity of $\mathcal{B}$.

3.7–3.11 In the following five problems involving a "four-bar linkage" (the fourth bar in each case is the rigid ground length between fixed pins!), the angular velocity of one of the bars is indicated. Find the angular velocities of the other two bars.

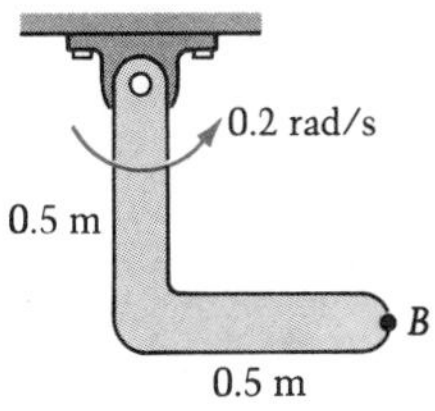

Figure P3.3

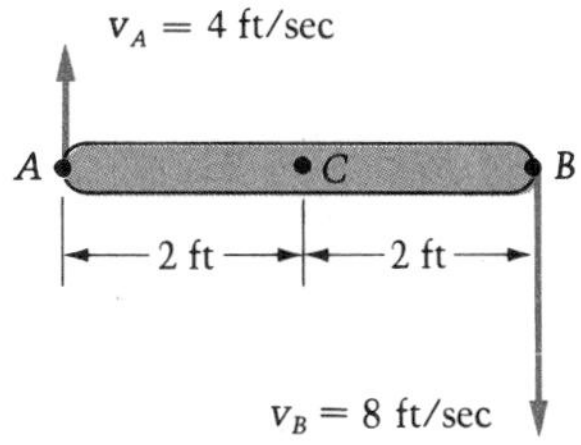

Figure P3.4

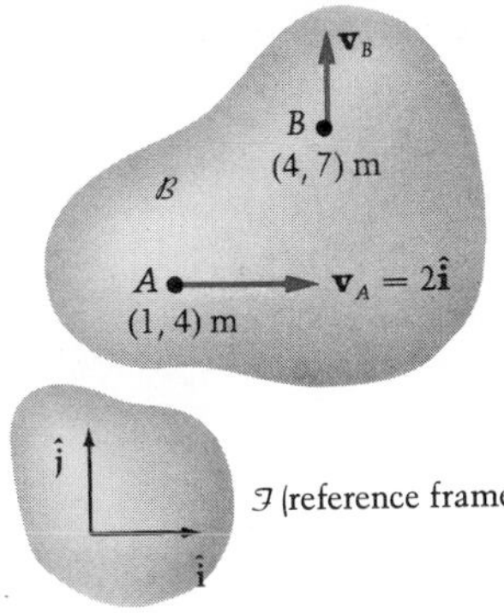

Figure P3.6

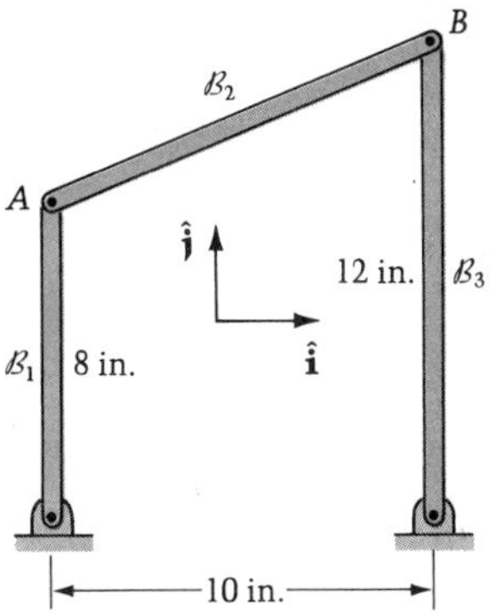

Figure P3.5

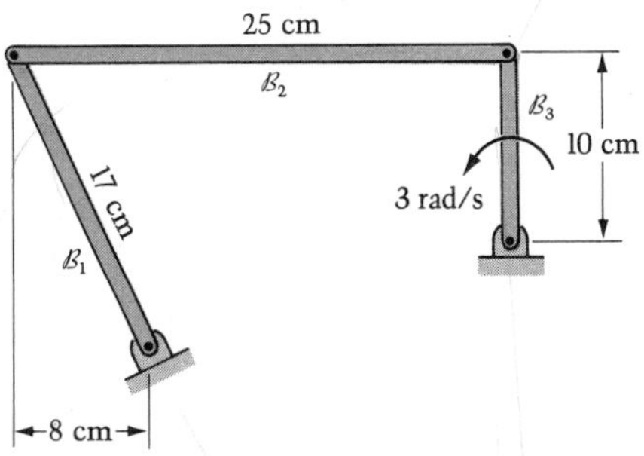

Figure P3.7

3.12 The equilateral triangular plate $\mathcal{B}_1$ shown in Figure P3.12 has three sides of length 0.3 m each. The bar $\mathcal{B}_2$ has an angular velocity $\omega_2 = 2$ rad/s counterclockwise and is pinned to $\mathcal{B}_1$ at A. Body $\mathcal{B}_1$ is also pinned to a block at B, which moves in the indicated slot. At the given time, find the angular velocity of $\mathcal{B}_1$.

3.13 Crank arm $\mathcal{B}_1$ shown in Figure P3.13 turns counterclockwise at a constant rate of 1 rad/s. Rod $\mathcal{B}_2$ is pinned to $\mathcal{B}_1$ at A and to a roller at B that slides in a circular slot. Determine the velocity of B and the angular velocity of $\mathcal{B}_2$ at the given instant.

3.14 The wheel shown in Figure P3.14 turns and slips in such a manner that its angular velocity is 2 rad/s ↻ while the velocity of the center C is 0.3 m/s to the left. Determine the velocity of point A.

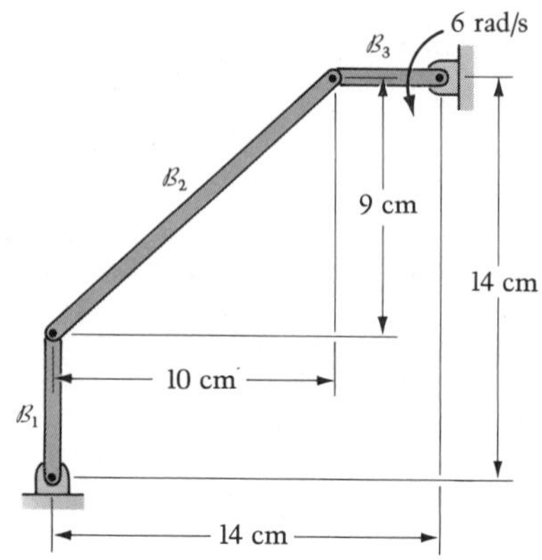

Figure P3.8

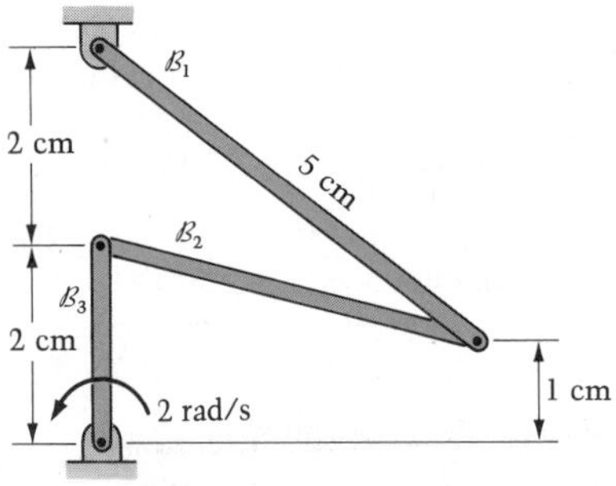

Figure P3.9

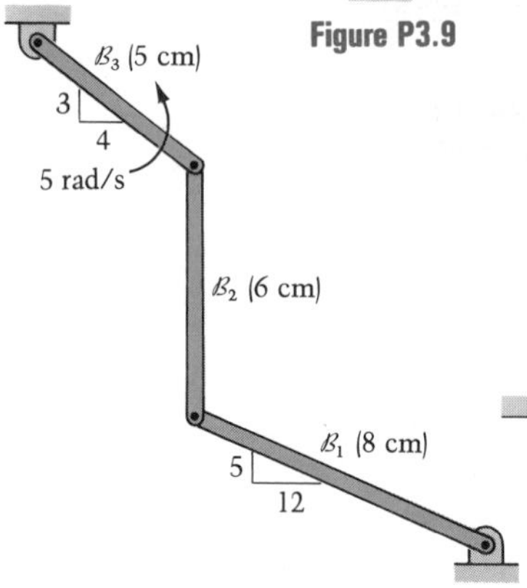

Figure P3.10

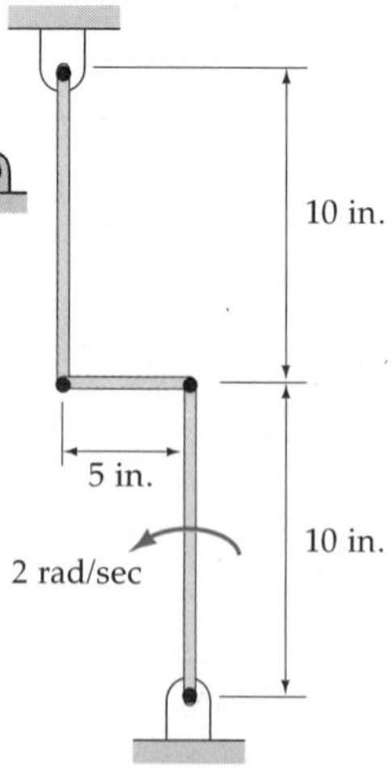

Figure P3.11

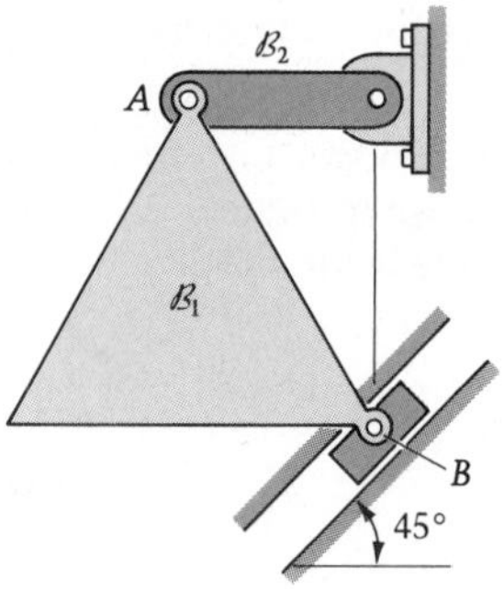

Figure P3.12

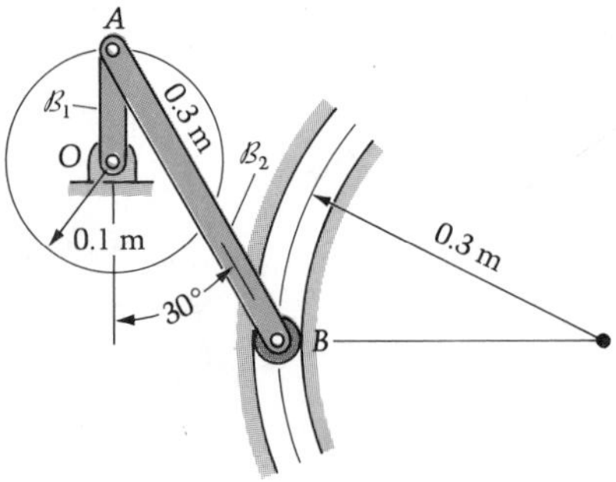

Figure P3.13

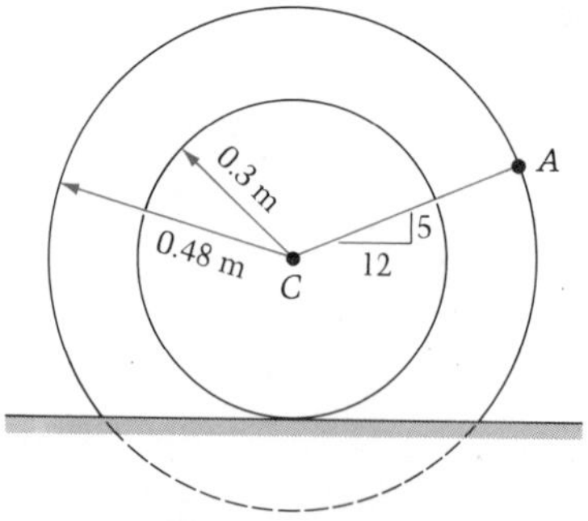

Figure P3.14

3.15 For the configuration shown in Figure P3.15, find the velocity of point P of the disk $\mathcal{B}_3$.

3.16 The speed of block $\mathcal{B}_1$ in Figure P3.16 has the value shown. Find the angular velocity of rod $\mathcal{B}_2$ and determine the velocity of pin A of block $\mathcal{B}_3$, when $\theta = 60°$.

3.17 Wheel $\mathcal{B}_1$ (Figure P3.17) turns and slips in such a way that its angular velocity is 2↻ rad/s while the velocity of C is 0.4 m/s to the left. Determine the velocity of point B, which slides on the plane. Bar $\mathcal{B}_2$ is pinned to $\mathcal{B}_1$ at D.

3.18 Point A of the rod slides along an inclined plane as in Figure P3.18, while the other end, B, slides on the horizontal plane. In the indicated position, $\boldsymbol{\omega} = 1.5\hat{\mathbf{k}}$ rad/sec. Find the velocity of the midpoint of the rod at this instant.

3.19 Wheel $\mathcal{B}_1$ in Figure P3.19 has a counterclockwise angular velocity of 6 rad/s. What is the velocity of point B at the instant shown?

3.20 Block $\mathcal{B}_1$ in Figure P3.20, which slides in a vertical slot, is pinned to bars $\mathcal{B}_2$ and $\mathcal{B}_3$ at A. The other ends of $\mathcal{B}_2$ and $\mathcal{B}_3$ are pinned to blocks that slide in horizontal slots. Block $\mathcal{B}_4$ translates to the left at constant speed 0.2 m/s. Find the velocity of B: (a) at the given instant; (b) when C is at point D; (c) when C is at point E.

3.21 The four links shown in Figure P3.21 each have length 0.4 m, and two of their angular velocities are indicated. Find the velocity of point C and determine the angular velocities of $\mathcal{B}_2$ and $\mathcal{B}_3$ at the indicated instant.

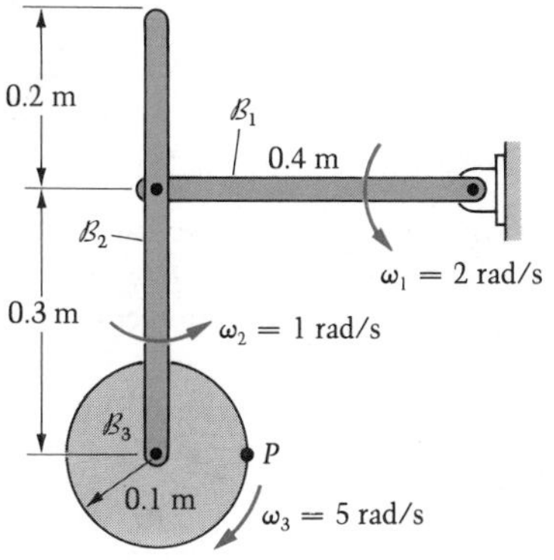

Figure P3.15

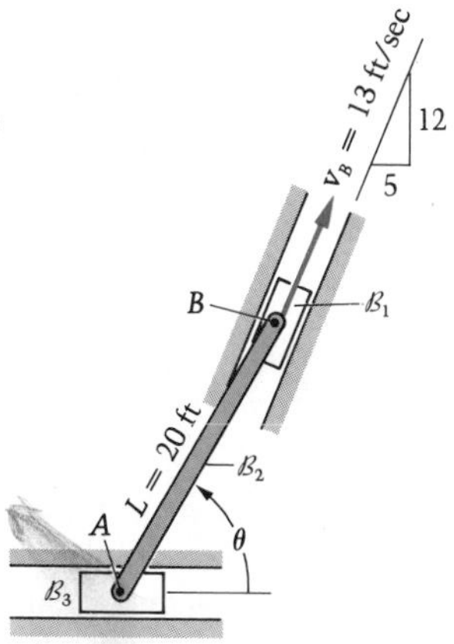

Figure P3.16

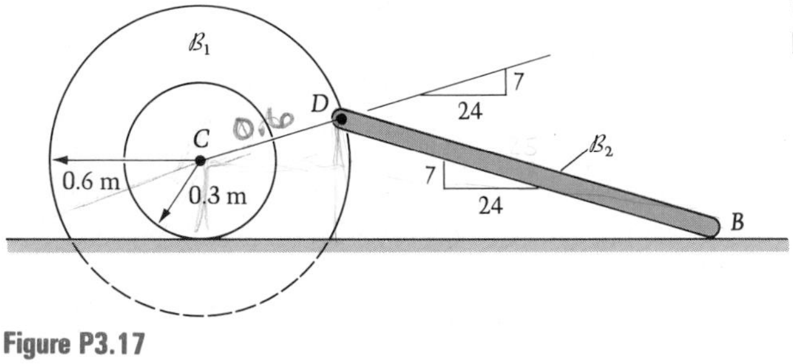

Figure P3.17

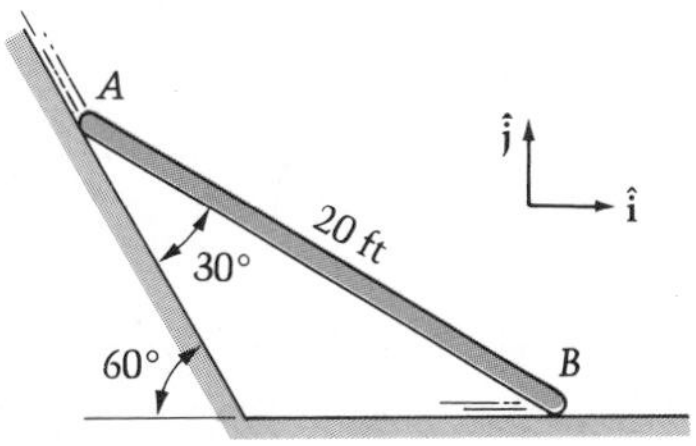

Figure P3.18

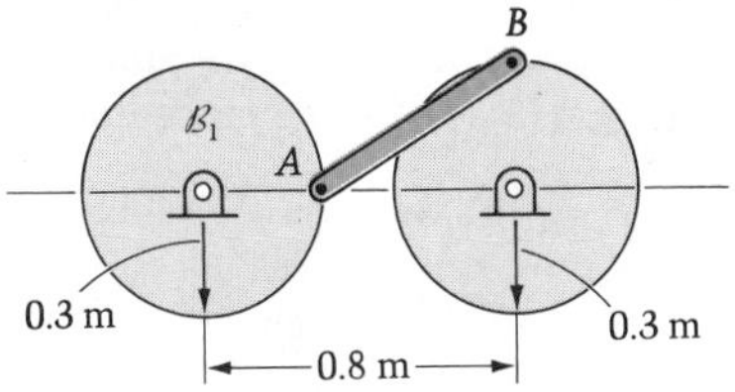

Figure P3.19

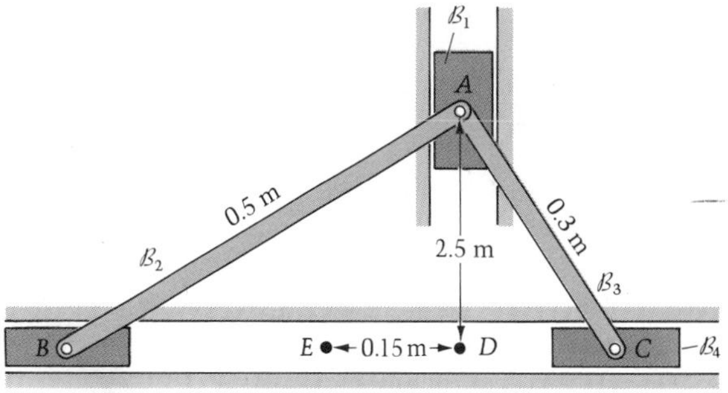

Figure P3.20

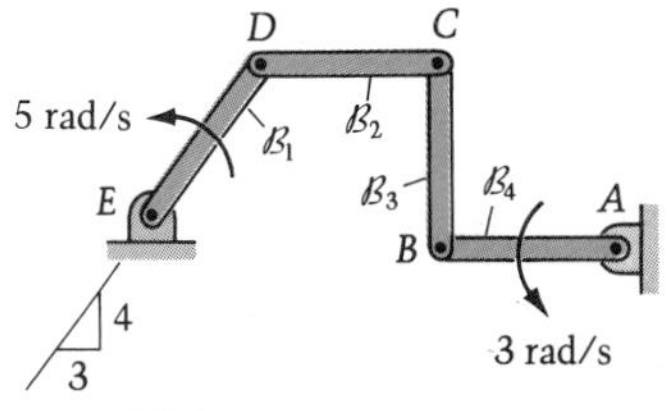

Figure P3.21

3.22 In the mechanism shown in Figure P3.22, the sleeve $\mathcal{B}_1$ is connected to the pivoted bar $\mathcal{B}_2$ by the 15-cm link $\mathcal{B}_3$. Over a certain range of motion of $\mathcal{B}_2$, the angle θ varies according to $\theta = 0.02t^2$ rad, starting at $t = 0$ with $\mathcal{B}_2$ and $\mathcal{B}_3$ horizontal. Find the velocity of pin S and the angular velocities of $\mathcal{B}_2$ and $\mathcal{B}_3$ when $\theta = 30°$. Time t is measured in seconds.

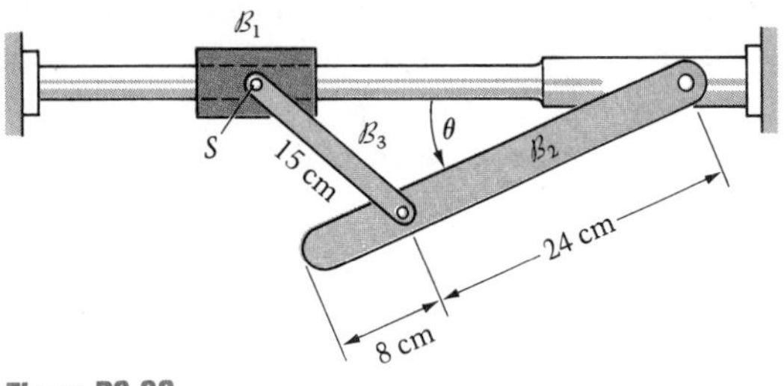

Figure P3.22

• **3.23** Find the velocity of point B of the rod if end A has constant velocity 2 m/s to the right as shown in Figure P3.23. The rollers are small. Compare the use of Equation (3.8) with the procedure used to solve Problem 1.63.

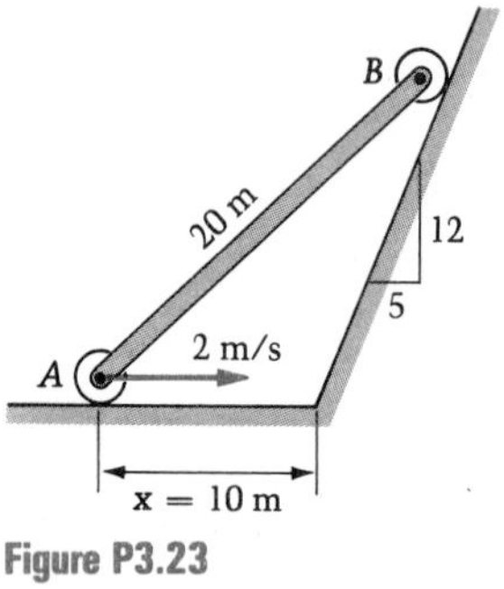

Figure P3.23

3.24 Find the velocity of the guided block at the instant shown in Figure P3.24.

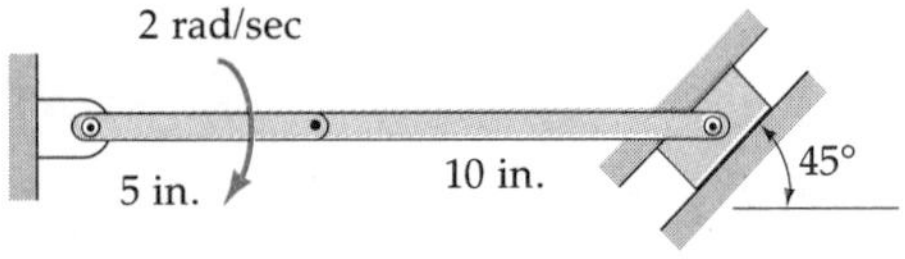

Figure P3.24

• **3.25** Block $\mathcal{B}$ has a controlled position in the slot given by $y = \sqrt{120}\,\sin(\pi t/10)$ in. for $0 \le t \le 10$ sec. (See Figure P3.25.) The time is $t = 0$ sec in the indicated position. Find the angular velocities of the rod and the wheel at (a) $t = 0$ sec and (b) $t = 5$ sec.

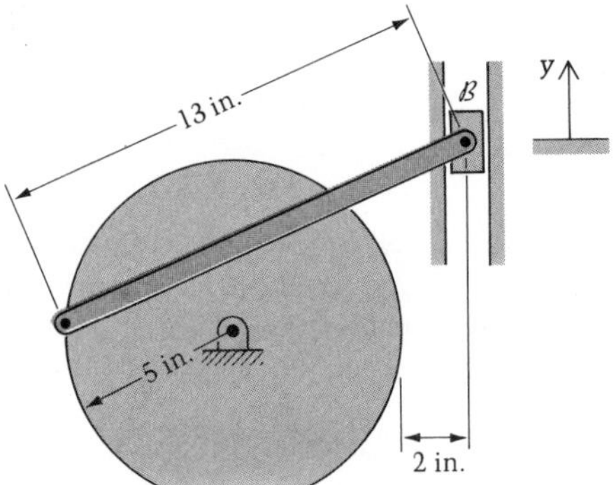

Figure P3.25

3.26 Crank $\mathcal{B}_1$ of the slider-crank mechanism shown in Figure P3.26 has a constant angular speed $\dot{\theta}$. Find the equation for the angular velocity $\dot{\phi}$ of the connecting rod $\mathcal{B}_2$ as a function of r, ℓ, θ, and $\dot{\theta}$.

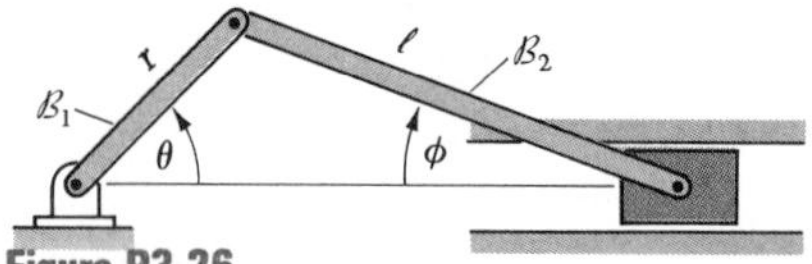

Figure P3.26

3.27 In the preceding problem, plot $\dot{\phi}/\dot{\theta}$ as a function of θ, from $\theta = 0$ to 2π, for $\ell/r = 1$, 2, and 5.

3.28 Referring to Section 3.2, show that neither the angular velocity vector nor Equation (3.8) depends on which body-fixed line segment (such as PQ in the text) is chosen to measure θ. Use two other points P' and Q' and their angle ϕ as suggested in Figure P3.28 for your proof.

• **3.29** Rod $\mathcal{B}_1$ begins moving at $\theta = 0$ (see Figure P3.29) and is made to turn at the constant angular rate $\dot{\theta} = 0.2$ rad/s. The cord is attached to the end of $\mathcal{B}_1$ and passes around a pulley. The other end of the cord is tied to weight $\mathcal{B}_2$ at point B. Observe that $\mathcal{B}_2$ moves downward until $\theta = 90°$, when it reverses direction. Write an equation that gives the velocity of point B as a function of θ for $\pi \ge \theta \ge \pi/2$. *Hint*: Using trigonometry, write y as a function of θ and the length L of the cord. Then differentiate.

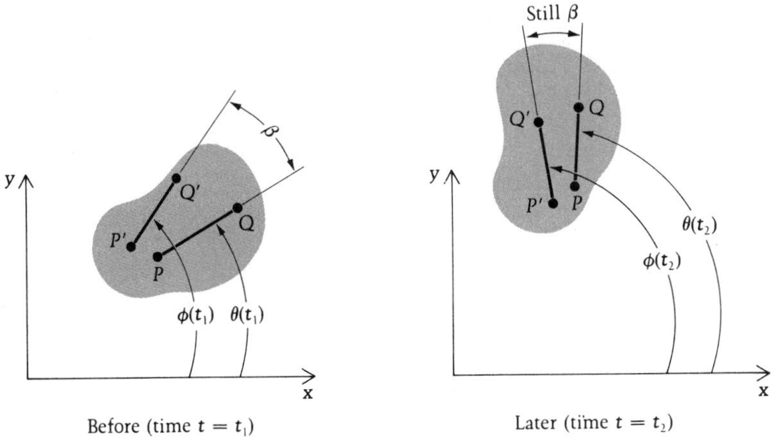

Figure P3.28

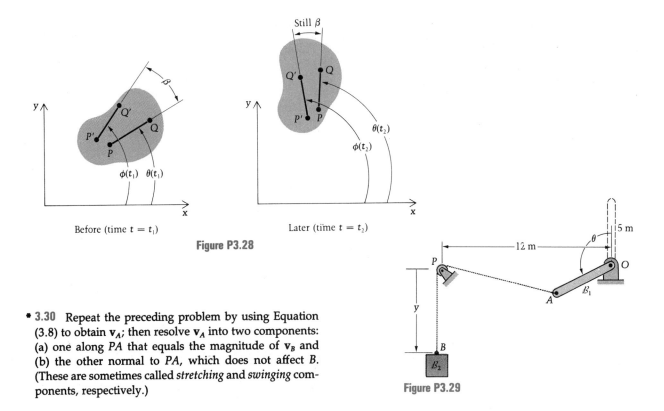

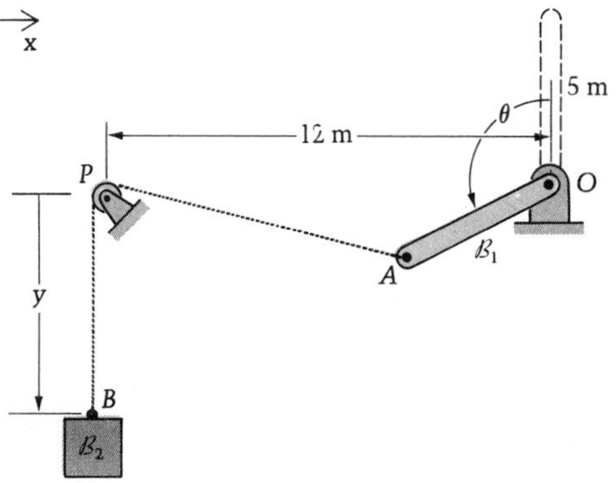

Figure P3.29

• **3.30** Repeat the preceding problem by using Equation (3.8) to obtain $\mathbf{v}_A$; then resolve $\mathbf{v}_A$ into two components: (a) one along PA that equals the magnitude of $\mathbf{v}_B$ and (b) the other normal to PA, which does not affect B. (These are sometimes called *stretching* and *swinging* components, respectively.)

3.3 Translation

When a rigid body $\mathcal{B}$ moves during a certain time interval in such a way that its angular velocity vector remains identically zero, then the body is said to be **translating,** or to be in a state of **translational motion** during that interval. From Equation (3.8) we thus see that for translation

$$\mathbf{v}_Q = \mathbf{v}_P \tag{3.12}$$

That is, all points of the body have the same velocity vector. By differentiating Equation (3.12), we see that the accelerations of all points of $\mathcal{B}$ are also equal for translation. Note that if $\dot{\theta} = 0$ only at an *instant* (that is, at a single value of time rather than over an interval), then all points of the body have equal velocities at that instant but *need not have equal accelerations.*

Question 3.7 Why is this the case?

Answer 3.7 The derivative of $\dot{\theta}\hat{\mathbf{k}} \times \mathbf{r}_{PQ}$ is not zero merely because $\dot{\theta}$ happens to be zero at one instant of time. To be able to differentiate $\mathbf{v}_Q = \mathbf{v}_P$, this equation must be valid for *all* values of t and not just one!

Translation can be either:

1. *Rectilinear:* Each point of $\mathcal{B}$ moves along a straight line in $\mathcal{I}$.
2. *Curvilinear:* Each point of $\mathcal{B}$ moves on a curved path in $\mathcal{I}$.

Examples of translation are shown in Figure 3.8. Part (a) shows an example of rectilinear translation: Body $\mathcal{B}$ is constrained to move in a straight slot. Part (b) shows an example of curvilinear translation: Body $\mathcal{B}$ is constrained by the identical links.

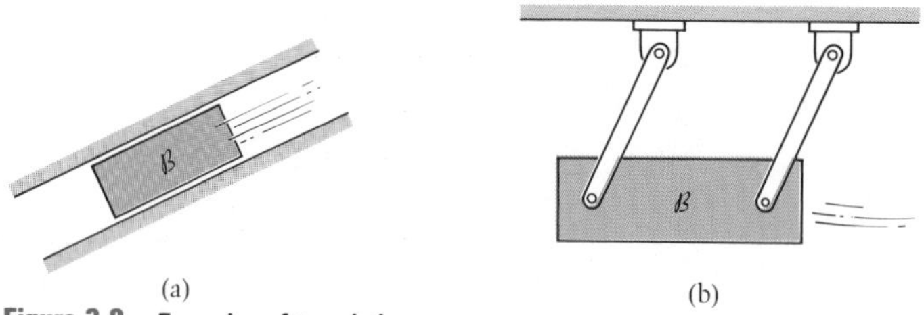

Figure 3.8 Examples of translation.

Perhaps an even better pair of examples is the blackboard eraser (Figure 3.9), which we used earlier to explain plane motion in Section 3.2. In part (a), the professor moves the eraser so that each of its points stays on a straight line; it is therefore in a state of rectilinear translation. In part (b), the professor moves the eraser on a curve; but if the word *eraser* is always horizontal during the erasing, then $\dot{\theta} \equiv 0$ and the eraser is in a state of curvilinear translation. Even though each of its points moves on a curve, all the velocities (and accelerations) are equal at all times. There is one notable exception to our earlier statement that "points, not bodies, have velocities and accelerations." In this present case of translation, since all the points have the *same* **v**'s and **a**'s, one could loosely refer to "the velocity of the eraser" without ambiguity.

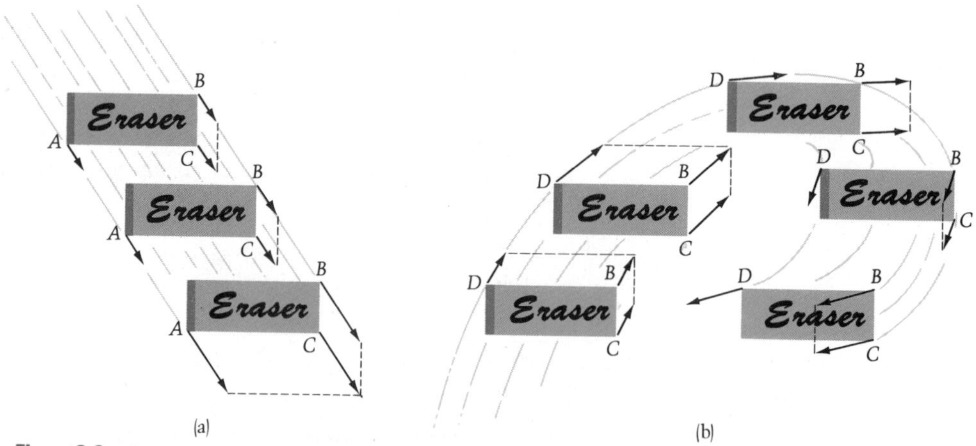

Figure 3.9 Another example of translation.

There are no examples or problems in this section because translation problems of rigid bodies require no new theory beyond what was developed in Chapter 1.

Summarizing, when a body is translating (either rectilinearly or curvilinearly), its angular velocity $\dot{\theta}\hat{\mathbf{k}}$ is identically zero, and all its points have equal velocities (and accelerations). If $\dot{\theta} = 0$ only at an instant, then all the points of the body have the same velocity at that instant but need not have equal accelerations.

3.4 Instantaneous Center of Zero Velocity

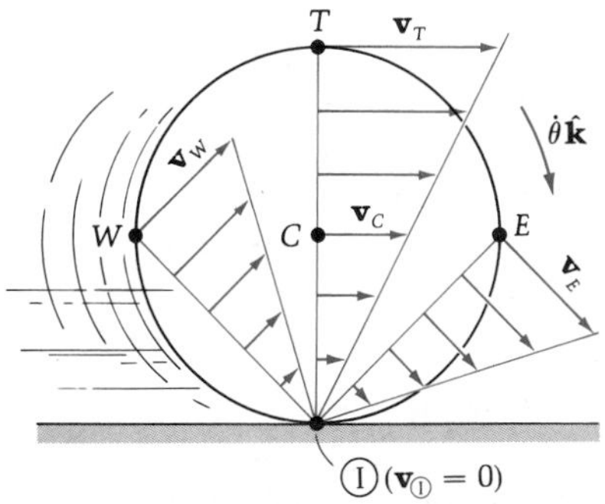

Figure 3.10 Instantaneous center of a rolling wheel.

If P is a point in the reference plane having zero velocity at some instant, then the velocity field of $\mathcal{B}$ is the same as if the body were constrained at that instant to rotate about an axis through P normal to the reference plane. This axis is called the **instantaneous axis of rotation**, and point P is called the **instantaneous center** (abbreviated Ⓘ) of **zero velocity*** of $\mathcal{B}$. Thus if Q is any *other* point of $\mathcal{B}$, then we have

$$\mathbf{v}_Q = \overset{0}{\cancel{\mathbf{v}_{\text{Ⓘ}}}} + \dot{\theta}\hat{\mathbf{k}} \times \mathbf{r}_{\text{Ⓘ}Q} = \dot{\theta}\hat{\mathbf{k}} \times \mathbf{r}_{\text{Ⓘ}Q} \quad \text{or} \quad \boldsymbol{\omega} \times \mathbf{r}_{\text{Ⓘ}Q} \qquad (3.13)$$

and since $\mathbf{v}_Q$ is then normal to both of the vectors $\boldsymbol{\omega}$ and $\mathbf{r}_{\text{Ⓘ}Q}$, we see that *each point moves with its velocity perpendicular to the line joining it to* Ⓘ. This concept is illustrated in Figure 3.10 for a rolling† wheel, in which Ⓘ is the contact point.

Proof of the Existence of the Instantaneous Center

We can show that if a body $\mathcal{B}$ has $\dot{\theta} \neq 0$ at a given instant, then it has an instantaneous center.

Question 3.8 Why can there be no point Ⓘ whenever $\dot{\theta}$ is zero?

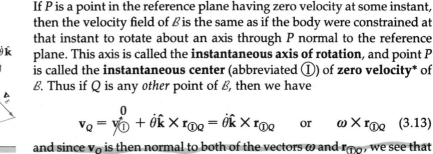

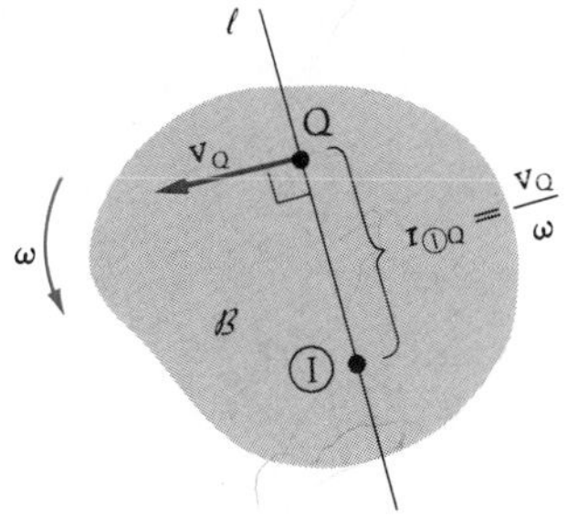

Figure 3.11

To demonstrate the existence of Ⓘ, we shall use Equation (3.13) in conjunction with Figure 3.11. As we have noted above, the vector $\mathbf{v}_Q$, being equal to $\boldsymbol{\omega} \times \mathbf{r}_{\text{Ⓘ}Q}$ for the point Ⓘ having $\mathbf{v}_{\text{Ⓘ}} = \mathbf{0}$, is normal to both $\boldsymbol{\omega}$ and to $\mathbf{r}_{\text{Ⓘ}Q}$. Hence we have these results:

1. The vector $\mathbf{r}_{\text{Ⓘ}Q}$ lies in the reference plane and is normal to $\mathbf{v}_Q$. It thus lies along the line ℓ in Figure 3.11.

---

* The phrase is admittedly redundant, but it is in common usage. "Instantaneous center of velocity" would perhaps be more concise, and "center of velocity" even more so. "Instantaneous center," however, is inadequate because of the possibility of confusion with points of zero acceleration.

† Rolling means no slipping, according to the definition we adopt in this book (see Section 3.6).

Answer 3.8 If $\dot{\theta} = 0$, Equation (3.8) says that $\mathbf{v}_Q = \mathbf{v}_P$; that is, all points of $\mathcal{B}$ have the *same* velocity vector. This common velocity vector is then zero only if the body is at rest. Incidentally, some think of $\mathcal{B}$ as having an instantaneous center Ⓘ at infinity when $\dot{\theta} = 0$.

2. The point Ⓘ exists (and is unique) because $|\mathbf{r}_{ⒾQ}|$ is therefore seen to be $|\mathbf{v}_Q/\omega|$ in order that $\mathbf{v}_Q = \boldsymbol{\omega} \times \mathbf{r}_{ⒾQ}$.

> **Question 3.9** Why is Ⓘ *below* Q in Figure 3.11 instead of being the same distance *above* Q?

We have thus verified the existence of the instantaneous center (unless $\dot{\theta} = 0$), because we know how to get to it from any arbitrary point Q of the body $\mathcal{B}$ whenever the angular velocity $\omega\hat{\mathbf{k}}$ of $\mathcal{B}$ and the velocity $\mathbf{v}_Q$ of the point Q are known. We also note again that the velocity magnitude of every point of $\mathcal{B}$ (in the reference plane!) equals the product of $|\omega|$ and the distance to the point from Ⓘ.

Sometimes because of a constraint on the motion we know the location of Ⓘ at the outset. This is the case for the rolling wheel of Figure 3.10 in which the bottom point grips the ground and is held at rest (thereby becoming Ⓘ) for the instant of its contact. If the radius of the wheel is 15″ and the velocity of its center C is 88 ft/sec →, then from the above discussion, the angular speed of the wheel is

$$\omega = \frac{v_C}{r_{ⒾC}} = \frac{88}{15/12} = 70.4 \text{ rad/sec}$$

Note that when viewed from Ⓘ the senses of the velocity direction of any point and the angular velocity direction of the body must always agree. For example, these are possible situations:

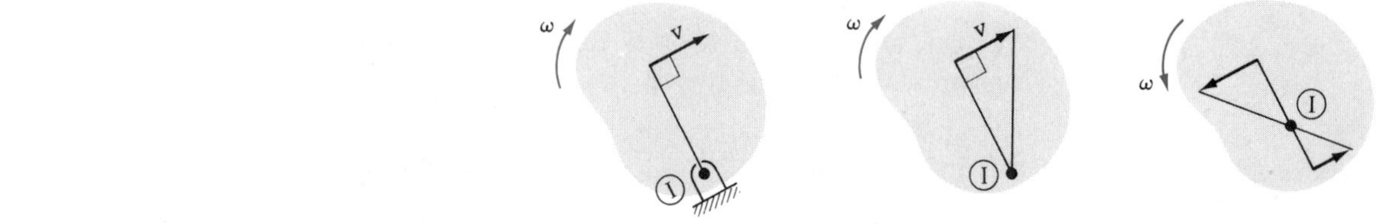

These are not:

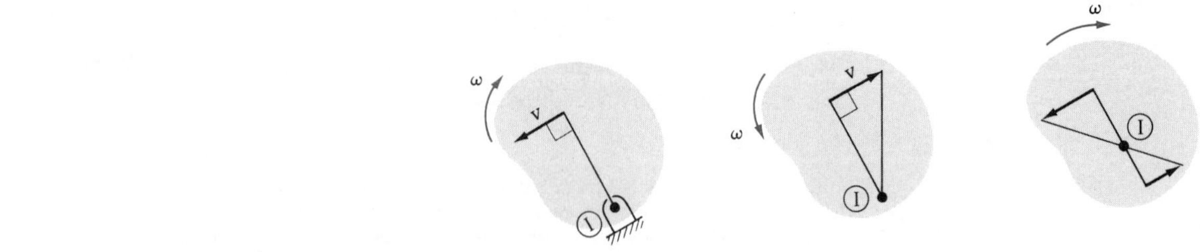

Answer 3.9 If point I were above Q, then $\boldsymbol{\omega} \times \mathbf{r}_{ⒾQ}$ would give an incorrect direction for the velocity $\mathbf{v}_Q$—it would be opposite to the actual direction. The senses of $\mathbf{v}_Q$ and $\boldsymbol{\omega}$ when viewed from Ⓘ must agree, as we point out later in this section.

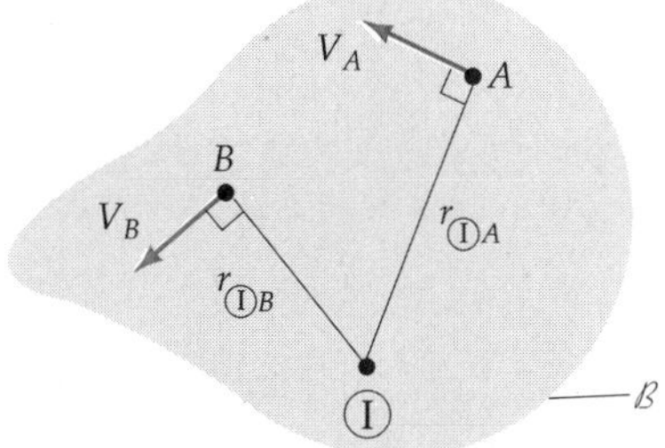

Figure 3.12

Therefore the direction of ω of the wheel in Figure 3.10 is clockwise (↻) in order that the known velocity direction of point C (→) and the angular velocity of the body be in agreement as the body rotates about Ⓘ at the instant shown.

Even if the angular velocity of $\mathcal{B}$ is not known, we can still easily find the instantaneous center of zero velocity Ⓘ if we know the velocities — or really, just the directions of the velocities — of *two* points A and B of $\mathcal{B}$.

Constructing perpendicular lines to the velocities of A (at A) and of B (at B) as shown in Figure 3.12, we immediately recognize Ⓘ as the intersection point of the two lines.

Question 3.10 Why?

From Figure 3.12 and the discussion above, we know that:

$$\frac{v_A}{r_{ⒾA}} = \omega = \frac{v_B}{r_{ⒾB}} \qquad \text{(Recall that there is only one } \omega \text{ for the body.)}$$

where we are abbreviating $|\mathbf{v}_A|$ by v_A, $|\mathbf{v}_B|$ by v_B, $|\mathbf{r}_{ⒾA}|$ by $r_{ⒾA}$, $|\mathbf{r}_{ⒾB}|$ by $r_{ⒾB}$, and $|\omega|$ by ω.

We now present three examples of the use of the above procedure for locating Ⓘ when two velocity directions are known in advance.

Answer 3.10 The point Ⓘ is unique. Since there is only one common point on the lines drawn perpendicular to the velocities (to $\mathbf{v}_B$ at B and to $\mathbf{v}_A$ at A), that point is the instantaneous center.

EXAMPLE 3.6

Ladders commonly carry a warning that for safe placement, the distance B in Figure E3.6a should be $\frac{1}{4}$ of the length L (i.e., of $\sqrt{B^2 + H^2}$). Let us suppose that a careless painter temporarily set a ladder against a wall in a dangerous position with $B/L = 0.5$, and went off to get his paint and brushes. Suppose further that the ladder began to slip, with the top of the ladder, point P, sliding down the wall and the bottom, point Q, slipping along the ground as shown. When B is 15 ft, find the instantaneous center of zero velocity Ⓘ of the ladder, and discuss the path of Ⓘ in space as the ladder falls further.

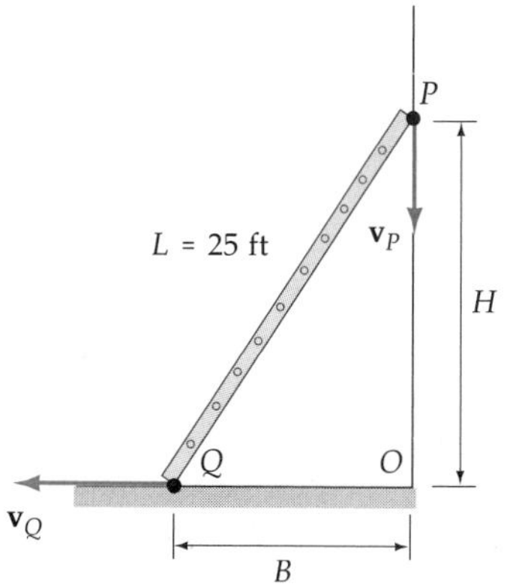

Figure E3.6a

Solution

When $B = 15$ ft, the normals to $\mathbf{v}_P$ at P and to $\mathbf{v}_Q$ at Q intersect at point Ⓘ as shown in Figure E3.6b on the next page.

If we imagine a rigid sheet of very light plastic glued to the ladder as in Figure E3.6c, then the ladder has been "rigidly extended." Note that *only for this instant,* we can think of the extended body as rotating about an imaginary pin at the intersection point Ⓘ of the normals to two velocities as shown. Note further from Figure E3.6c that the velocities of *all* points of the rigid sheet are perpendicular to lines drawn to them from Ⓘ. The velocity magnitude of each point is proportional to the distance from that point to Ⓘ, with the proportionality

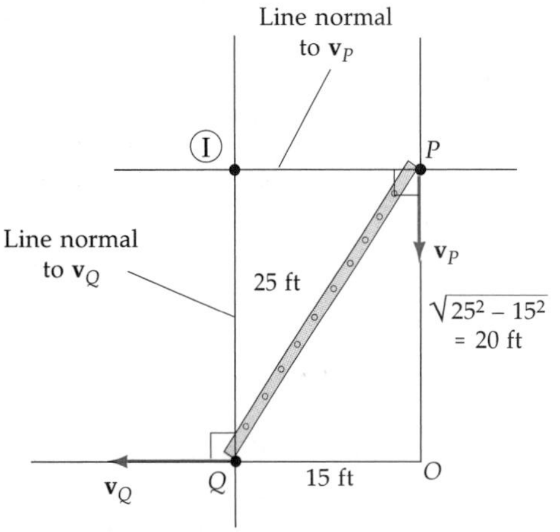

Figure E3.6b

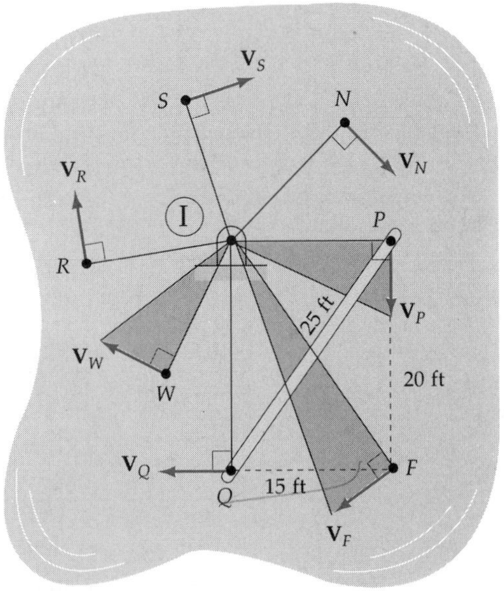

Figure E3.6c

constant being ω of the body at the instant. Hence all triangles like the three shaded in Figure E3.6c are similar.

As the ladder falls, the location of the point Ⓘ changes on the imagined rigid extension (sheet) as time passes, because the perpendiculars to $\mathbf{v}_A$ and $\mathbf{v}_B$ intersect at different points of the sheet, as seen below in Figure E3.6d:

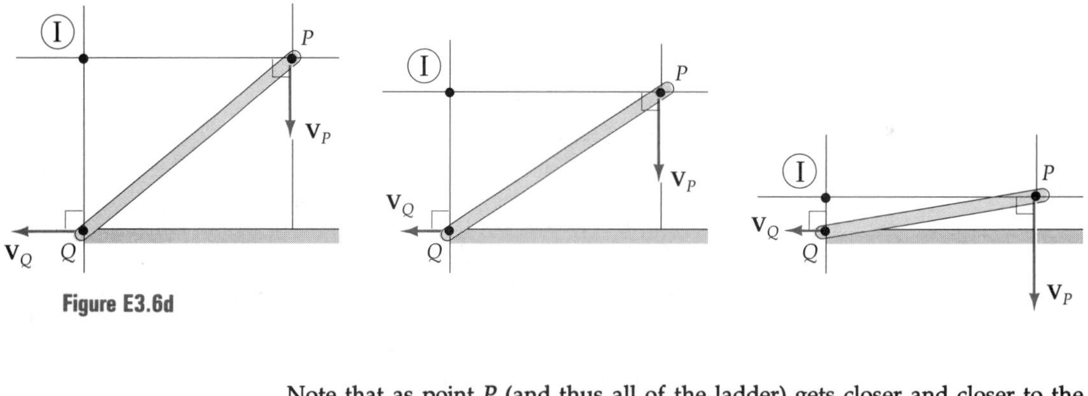

Figure E3.6d

Note that as point P (and thus all of the ladder) gets closer and closer to the ground, point Ⓘ gets closer and closer to point Q. Thus even though ω increases, $r_{ⒾQ}\, \omega$ gets smaller and smaller, until, in the limit (as P contacts the ground) Q *becomes* Ⓘ (the intersection of the perpendiculars) and V_Q is then zero.

When the body $\mathcal{B}$ has a **pivot** (a point that never moves throughout the body's motion, such as a pin), it is clearly *always* Ⓘ; in this case the motion is called *pure rotation.* But otherwise, the point Ⓘ is *not* the same point of $\mathcal{B}$ throughout the motion, as we have already seen with the wheel and ladder. In each of the next two examples, one of the bodies has a pivot.

EXAMPLE 3.7

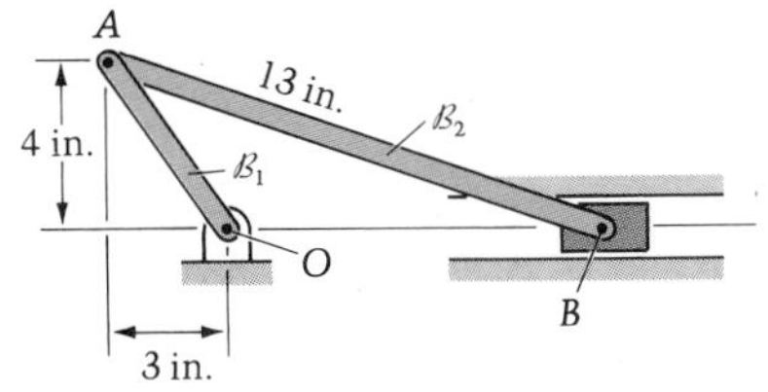

Figure E3.7a

Rework Example 3.3 using instantaneous centers. (See Figure E3.7a) The crank arm $\mathcal{B}_1$ turns about a horizontal z axis, through its pinned end O, with an angular velocity of 10 rad/sec clockwise at the given instant. Find the velocity of the piston pin B.

Solution

Since O is the point Ⓘ for body $\mathcal{B}_1$, we have

$$\mathbf{v}_A = r\omega \measuredangle^{3}_{4} = 5(10) \measuredangle^{3}_{4}$$

$$= 50 \measuredangle^{3}_{4} \text{ in./sec}$$

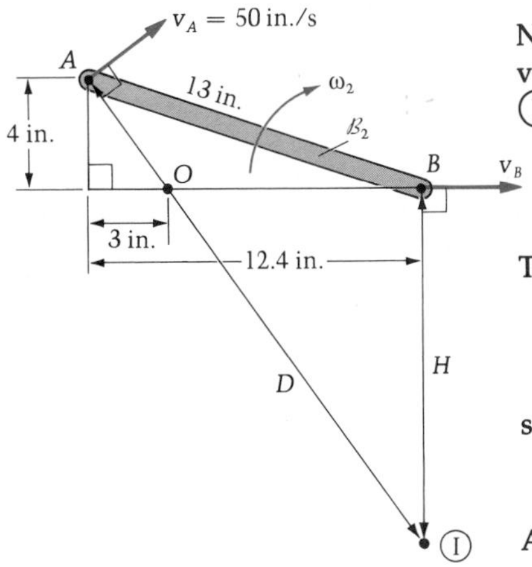

Figure E3.7b

Next we find the Ⓘ of $\mathcal{B}_2$, using the fact that it is on lines perpendicular to the velocities of A and B as shown in Figure 3.7b. If we next find the distance D from Ⓘ to A, then ω_2 will be $v_A/D = 50/D$. By similar triangles:

$$\frac{D}{12.4} = \frac{5}{3} \Rightarrow D = 20.7 \text{ in.}$$

Then

$$\omega_2 = \frac{v_A}{D} = \frac{50}{20.7} = 2.42$$

so

$$\omega_2 = 2.42 \circlearrowright \text{ rad/sec}$$

Again by similar triangles:

$$\frac{H+4}{12.4} = \frac{4}{3} \Rightarrow H = 12.5 \text{ in.}$$

so that $v_B = H\omega_2 = 12.5(2.42) = 30.3$ in./sec to the right, as we have seen before.

EXAMPLE 3.8

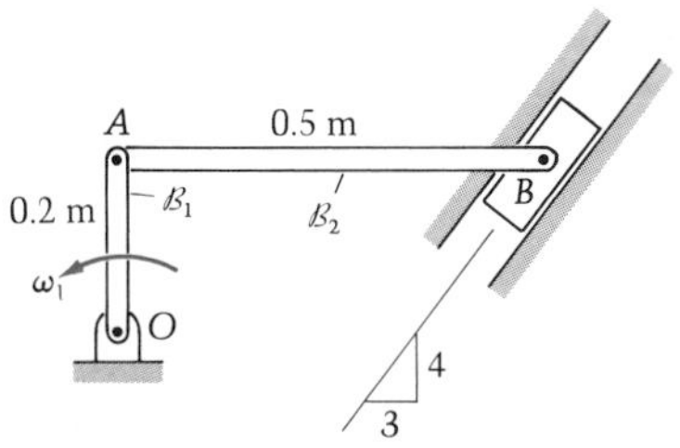

Figure E3.8a

At the instant shown in Figure E3.8a, the angular velocity of bar $\mathcal{B}_1$ is $\omega_1 = 5 \circlearrowleft$ rad/sec. Find the velocity of pin B connecting bar $\mathcal{B}_2$ to the slider block, constrained to slide in the slot as shown.

Solution

As seen in Figure E3.8b on the next page, the point Ⓘ for $\mathcal{B}_1$ is O, since it is pinned to the reference frame. The velocity of A is perpendicular to the line from Ⓘ to A (that is, from O to A) and has a direction in agreement with the angular velocity of $\mathcal{B}_1$ as the body turns about O. Its value is $r_{ⒾA}\omega_1$, or 1 m/s $\leftarrow$.

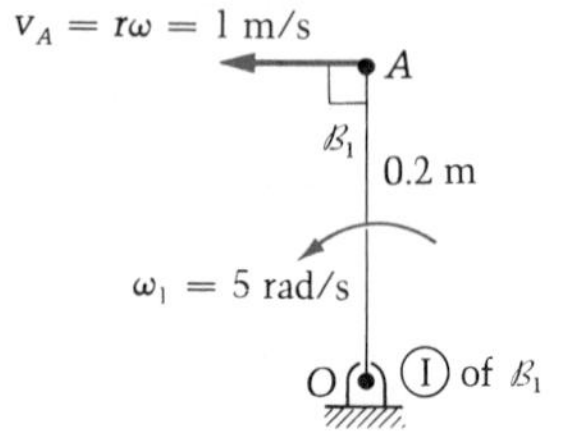

Figure E3.8b

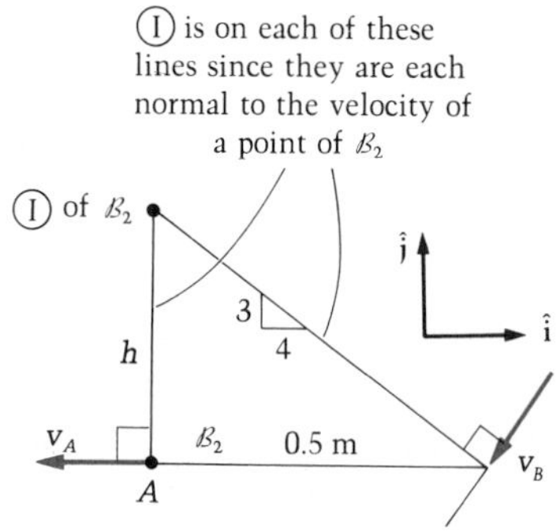

Figure E3.8c

Next we sketch body $\mathcal{B}_2$ and note the position of Ⓘ for $\mathcal{B}_2$, as explained in Figure E3.8c. Similar triangles yield the height H of Ⓘ above A:

$$\frac{H}{0.5} = \frac{3}{4}$$

$$H = 0.375 \text{ m}$$

Question 3.11 Why does $\mathbf{v}_B$ have to be "southwest" along the slot and not "northeast"?

We may now use Ⓘ of $\mathcal{B}_2$ to get the angular velocity of $\mathcal{B}_2$; using vectors this time,

$$\mathbf{v}_A = \omega_2 \hat{\mathbf{k}} \times \mathbf{r}_{ⒾA}$$

Substituting, we get

$$-1\hat{\mathbf{i}} = \omega_2 \hat{\mathbf{k}} \times (-0.375\hat{\mathbf{j}})$$

Solving gives

$$\omega_2 = -2.67 \Rightarrow \boldsymbol{\omega}_2 = -2.67\hat{\mathbf{k}} \text{ or } 2.67 \circlearrowright \text{ rad/s}$$

Note that when we write $\boldsymbol{\omega}_2 = \omega_2 \hat{\mathbf{k}}$, we are saying that ω_2 is counterclockwise in accordance with the sign convention adopted for the problem in the figure if its value turns out positive. Thus when its value is now found to be negative, we know that $\mathcal{B}_2$ is turning clockwise at the given instant. Of course, as we have seen, we do not have to use vectors on such a simple problem; we can use what we know about the instantaneous center in scalar form to get a quick solution:

$$v_A = H|\omega_2| \Rightarrow |\omega_2| = \frac{v_A}{H} = \frac{1}{0.375} = 2.67 \Rightarrow \boldsymbol{\omega}_2 = 2.67 \circlearrowright \text{ rad/s}$$

where we assign the direction in accordance with the known velocity direction of A and the position of Ⓘ.

Next we use Ⓘ of $\mathcal{B}_2$ to obtain $\mathbf{v}_B$:

$$\begin{aligned} v_B &= r_{ⒾB}|\omega_2| \\ &= \sqrt{0.375^2 + 0.5^2}|\omega_2| \text{ m/s} \\ &= 0.625(2.67) = 1.67 \text{ m/s} \end{aligned}$$

The velocity of B is thus $\mathbf{v}_B = 1.67 \underset{3}{\overset{4}{\swarrow}}$ m/s.

Note that the arrow in this sketch is just as descriptive of the direction of the vector velocity of B as is the unit vector $-0.6\hat{\mathbf{i}} - 0.8\hat{\mathbf{j}}$.

Answer 3.11 The known velocity direction of A dictates that $\mathcal{B}_2$ is turning clockwise around Ⓘ, so $\mathbf{v}_B$ has to be "southwest" for this to be the case.

The Special Case in Which the Normals Do Not Intersect

Two things can go wrong with the procedure we have been following of intersecting the normals to two points' velocity vectors to find Ⓘ. The first of these is that the two perpendicular lines may be parallel and hence not intersect, as suggested by Figure 3.13 below for the points A and B of bar $\mathcal{B}$. Note that A and B are each at the top of the vertical plane circles on which they move, and since their velocity vectors are tangent to their paths, each is horizontal at this instant.

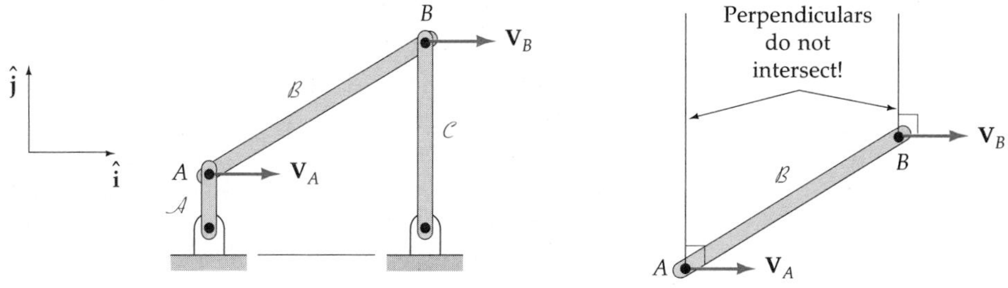
ĵ
î
B
$\mathbf{v}_B$
$\mathcal{B}$
$\mathcal{C}$
A
$\mathbf{v}_A$
$\mathcal{A}$
Perpendiculars do not intersect!
$\mathbf{v}_B$
B
$\mathcal{B}$
A
$\mathbf{v}_A$

Figure 3.13

Let us examine Equation 3.8 for this case:

$$\mathbf{v}_B = \mathbf{v}_A + \boldsymbol{\omega} \times \mathbf{r}_{AB}$$

Since $\mathbf{v}_B$ and $\mathbf{v}_A$ have only $\hat{\mathbf{i}}$-components while $\boldsymbol{\omega} \times \mathbf{r}_{AB}$ has both $\hat{\mathbf{i}}$ *and* $\hat{\mathbf{j}}$, then $\boldsymbol{\omega}$ must be zero at this instant. This does not mean the body is translating; that occurs when $\boldsymbol{\omega}$ is zero *all the time*. Rather, in this case the body is just stopped for one instant in its angular motion, as its angular velocity is changing from clockwise to counterclockwise (see Figure 3.14).

The equation above also shows that at such an instant when $\boldsymbol{\omega} = \mathbf{0}$, all points of the body have identical velocities. So in Figure 3.13, $\mathbf{v}_A = \mathbf{v}_B = \mathbf{v}_{\textit{any point of } \mathcal{B}}$ at that instant. Conversely, any time two points of a body have equal velocities in plane motion, the body's angular velocity vanishes at that instant.

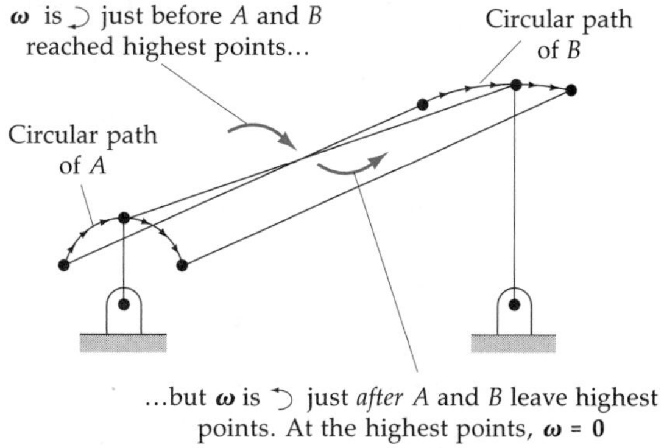
ω is ↻ just before A and B reached highest points...
Circular path of B
Circular path of A
...but ω is ↺ just *after* A and B leave highest points. At the highest points, ω = 0

Figure 3.14

The Special Case in Which the Normals are Coincident

The second exceptional case occurs when the perpendiculars to two velocities are one and the same line (see Figures 3.15(a,b)):

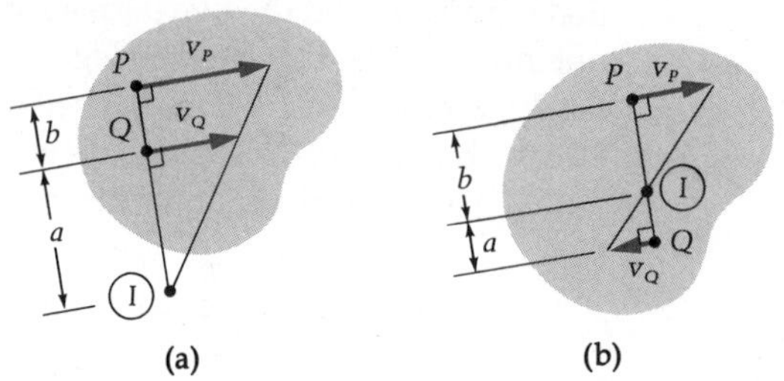

Figure 3.15

In this case, we can find the instantaneous center using similar triangles as shown. This simple procedure works, because, for the case shown in Figure 3.15a,

$$v_Q = a\omega \qquad \text{and} \qquad v_P = (a+b)\omega$$

so that

$$\frac{v_P}{a+b} = \frac{v_Q}{a}$$

If we should get coincident normals with the directions of $\mathbf{v}_P$ and $\mathbf{v}_Q$ *opposite*, as in Figure 3.15b, then Ⓘ lies *between* P and Q, and it may again be found by similar triangles. This time,

$$v_P = b\omega \qquad \text{and} \qquad v_Q = a\omega$$

so that

$$\frac{v_P}{b} = \frac{v_Q}{a}$$

In the examples we have presented in this section, note that use of the instantaneous center may be made with or without vector algebra. Its advantage is in finding and using points of zero velocity in order to simplify the resulting mathematics. Instantaneous centers never *have* to be used to effect a solution. Sometimes they are helpful, but at other times, they may be more trouble to locate than they are worth!

Examples of both the above special cases occur in our last example in this section. It involves four different positions of the same system:

EXAMPLE 3.9

Figure E3.9a shows a rolling wheel $\mathcal{B}_1$ of a large vehicle that travels at a constant speed of 60 mph →. Find the velocity of the piston when θ equals: (a) 0°, (b) 90°, (c) 180°, (d) 270°.

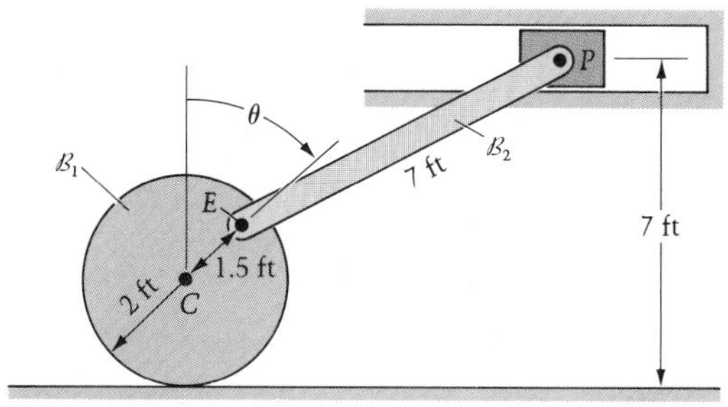

Figure E3.9a

Solution

First we solve for the velocities by using several approaches. In each case, the speed of the wheel's center is the same as the speed of the vehicle: 60 mph or 88 ft/sec. And since the piston translates, all its points have equal velocities and equal accelerations at every instant.

Case (a): As we shall see in detail in Section 3.6, the instantaneous center of a wheel rolling on a fixed track is at the point of contact. Since velocities increase linearly with distance from this point Ⓘ, we have for the point E of $\mathcal{B}_1$:

$$v_E = \frac{2 + 1.5}{2}(88) = 154 \text{ ft/sec}$$

If we draw lines at E and P perpendicular to $\mathbf{v}_E$ and $\mathbf{v}_P$ (P is constrained to move horizontally), they are parallel and thus will not intersect (see Figure E3.9b). Therefore $\omega_2 = 0$ at that instant. Thus $\mathbf{v}_P = \mathbf{v}_E = 154 \rightarrow$ ft/sec.

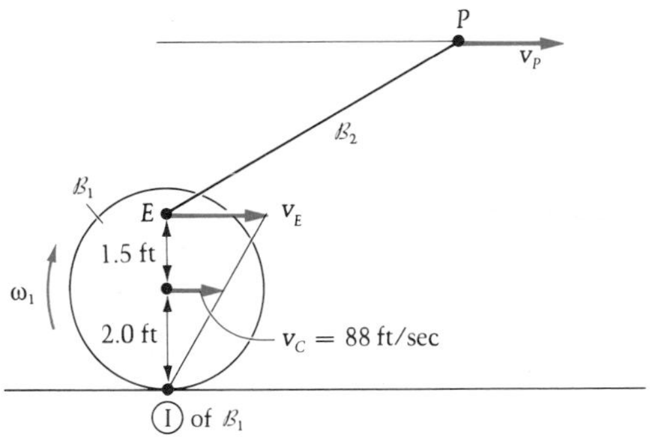

Figure E3.9b

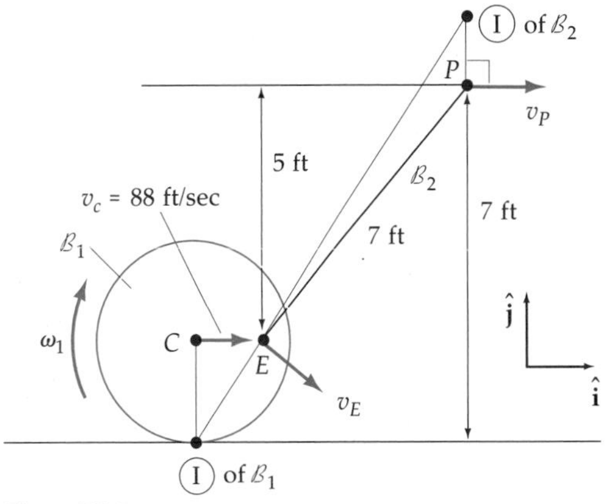

Figure E3.9c

Case (b): This time we shall use vectors; on the wheel (see Figure E3.9c),

$$\mathbf{v}_C = \cancelto{0}{\mathbf{v}_{Ⓘ}} + \omega_1\hat{\mathbf{k}} \times \cancelto{2\hat{\mathbf{j}}}{\mathbf{r}_{ⒾC}}$$

$$88\hat{\mathbf{i}} = -2.0\omega_1\hat{\mathbf{i}} \Rightarrow \omega_1 = -44 \Rightarrow \boldsymbol{\omega}_1 = -44\hat{\mathbf{k}}$$

or

$$\omega_1 = 44.0 \circlearrowright \text{ rad/sec}$$
$$\mathbf{v}_E = -44.0\hat{\mathbf{k}} \times (1.5\hat{\mathbf{i}} + 2\hat{\mathbf{j}}) = 88.0\hat{\mathbf{i}} - 66.0\hat{\mathbf{j}}$$

or

$$\mathbf{v}_E = 110 \overset{4}{\searrow}_3 \text{ ft/sec}$$

On $\mathcal{B}_2$ now (after noting the trigonometry results in Figure E3.9d):

$$\mathbf{v}_P = v_P\hat{\mathbf{i}} = \mathbf{v}_E + \omega_2\hat{\mathbf{k}} \times \mathbf{r}_{EP}$$
$$= 88.0\hat{\mathbf{i}} - 66.0\hat{\mathbf{j}} + \omega_2\hat{\mathbf{k}} \times (4.90\hat{\mathbf{i}} + 5.00\hat{\mathbf{j}})$$

Equating coefficients of $\hat{\mathbf{i}}$ and then of $\hat{\mathbf{j}}$, we obtain:

$$\hat{\mathbf{i}}: \quad v_P = 88.0 - 5.00\omega_2$$
$$\hat{\mathbf{j}}: \quad 0 = -66.0 + 4.90\omega_2 \Rightarrow \omega_2 = +13.5 \text{ rad/sec}$$

Therefore

$$v_P = 88.0 - 5.00(+13.5) = 20.5 \text{ ft/sec}$$

so that

$$\mathbf{v}_P = 20.5\hat{\mathbf{i}} \text{ ft/sec}$$

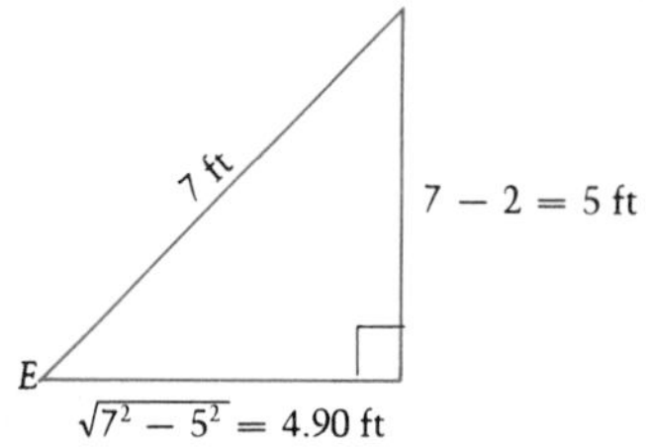

Figure E3.9d

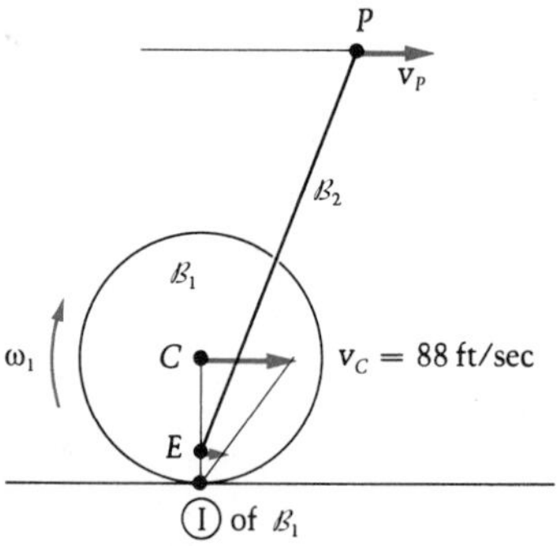

Figure E3.9e

Case (c): In all four cases, $\omega_1 = 44.0 \circlearrowright$ rad/sec. This time, then (see Figure E3.9e),

$$v_E = (2.0 - 1.5)\omega_1 = 22.0 \text{ ft/sec}$$
$$\mathbf{v}_E = 22.0 \rightarrow \text{ft/sec}$$

Again, as in Case (a), body $\mathcal{B}_2$ has $\omega_2 = \mathbf{0}$ so that all points of $\mathcal{B}_2$ have the same velocity. Thus

$$\mathbf{v}_P = 22.0\hat{\mathbf{i}} \text{ ft/sec}$$

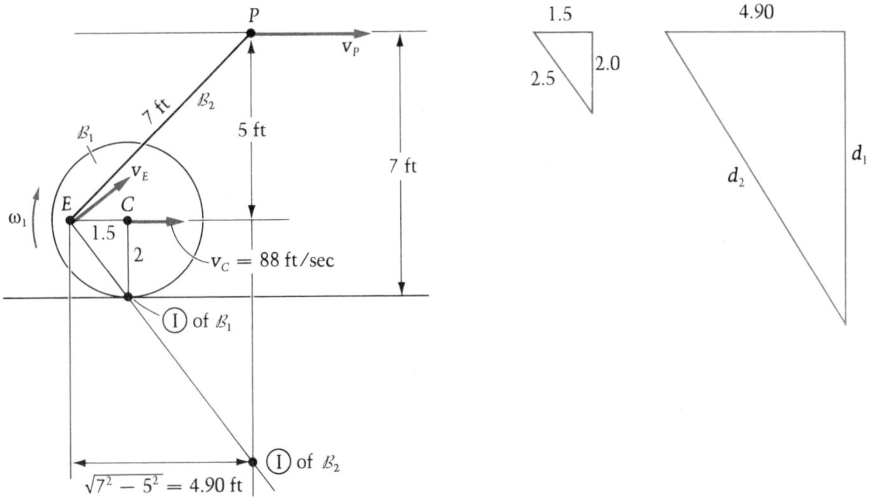

Figure E3.9f

Case (d): Using Figure E3.9f, we see that, on body $\mathcal{B}_1$,

$$\mathbf{v}_E = \underbrace{88\hat{\mathbf{i}}}_{\mathbf{v}_C} + \underbrace{(-44.0\hat{\mathbf{k}})}_{\omega_1\mathbf{k}} \times \underbrace{(-1.5\hat{\mathbf{i}})}_{\mathbf{r}_{CE}}$$

$$= 88.0\hat{\mathbf{i}} + 66.0\hat{\mathbf{j}}$$

We shall now use the instantaneous center of $\mathcal{B}_2$. From the similar triangles in the above figure,

$$d_1 = \frac{2.0}{1.5}(4.90) = 6.53 \text{ ft}$$

$$d_2 = \frac{2.5}{1.5}(4.90) = 8.17 \text{ ft}$$

Therefore, on body $\mathcal{B}_2$,

$$\omega_2 = \frac{v_E}{8.17} = \frac{\sqrt{88.0^2 + 66.0^2}}{8.17} = \frac{110}{8.17} = 13.5 \text{ rad/sec}$$

$$\omega_2 = 13.5 \circlearrowright \text{ rad/sec}$$

and

$$v_P = (6.53 + 5)\omega_2 = 156 \text{ ft/sec} \Rightarrow \mathbf{v}_P = 156 \rightarrow \text{ft/sec}$$

Note from the four answers in the above example that the piston is moving faster during the parts of the wheel's revolution in which $\mathbf{v}_E$ makes small angles with rod $\mathcal{B}_2$; conversely, it moves slower when $\mathbf{v}_E$ is making a large angle with $\mathcal{B}_2$. This is because the components of $\mathbf{v}_E$ and $\mathbf{v}_P$ along $\mathcal{B}_2$ must always be the same, as we saw earlier.

PROBLEMS ▸ Section 3.4

3.31 The angular velocity of $\mathcal{B}_1$ in Figure P3.31 is $3 \circlearrowleft$ rad/s = constant. Trace the five sketches and then show on parts (a) to (d) the position of Ⓘ for the rod $\mathcal{B}_2$. In part (e), using the proper length of $\mathcal{B}_2$, draw the positions of $\mathcal{B}_2$ at the two times when $\mathbf{v}_B = \mathbf{0}$. Rod $\mathcal{B}_2$ has length 0.9 m.

3.32 Solve Problem 3.16 by using instantaneous centers.

3.33 Solve Problem 3.7 by using instantaneous centers.

3.34 Solve Problem 3.8 by using instantaneous centers.

3.35 Solve Problem 3.6 by using instantaneous centers.

3.36 Solve Problem 3.24 by using instantaneous centers.

3.37 Solve Problem 3.11 by using instantaneous centers.

3.38 In Figure P3.38 the crank arm $\mathcal{B}_1$ is 4 in. long and has constant angular velocity $\omega_1 = 2 \circlearrowleft$ rad/sec. It is pinned to the triangular plate $\mathcal{B}_2$, which is also pinned to the block in the slot at D. Find the velocity of D at the instant shown.

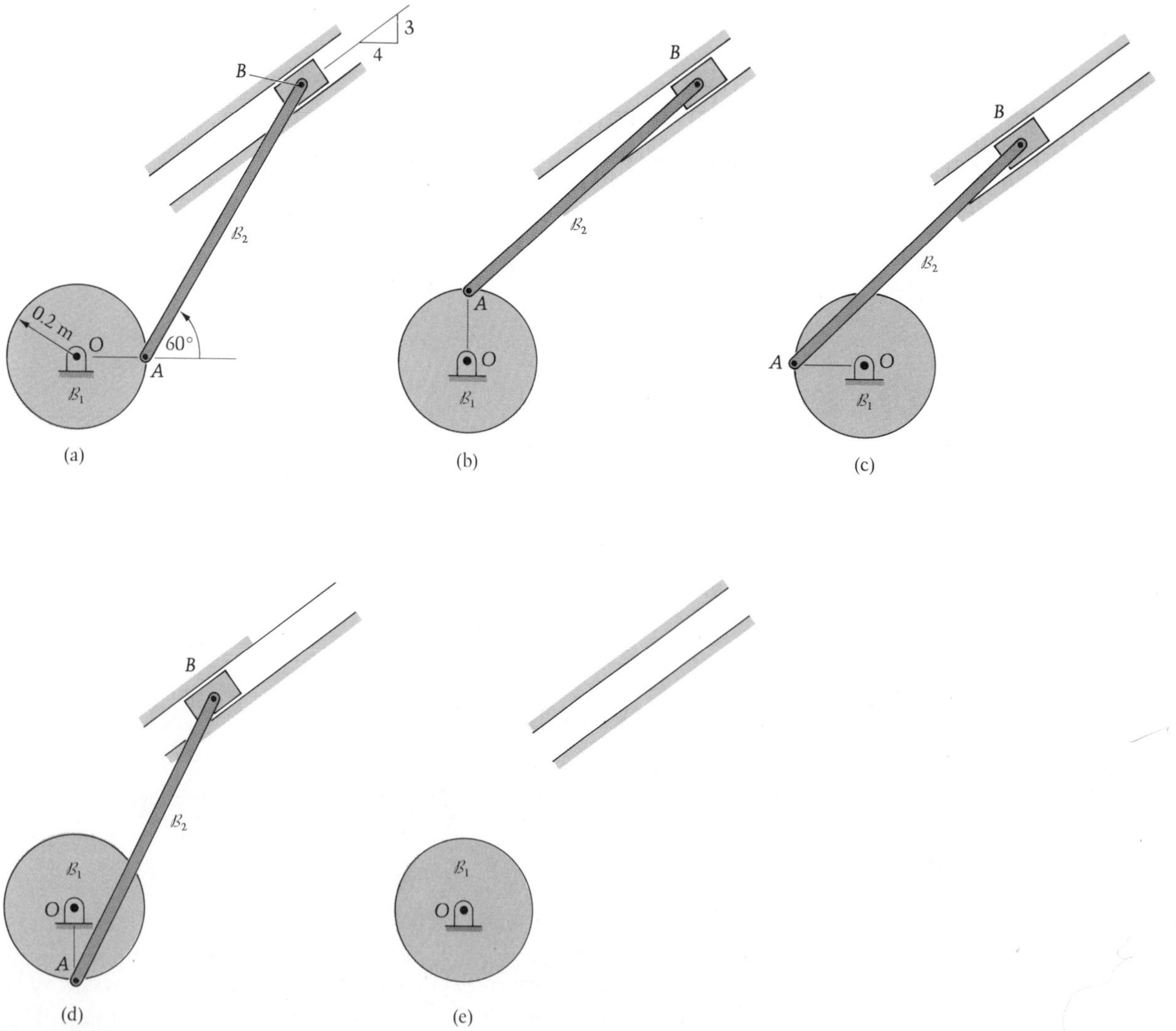

Figure P3.31

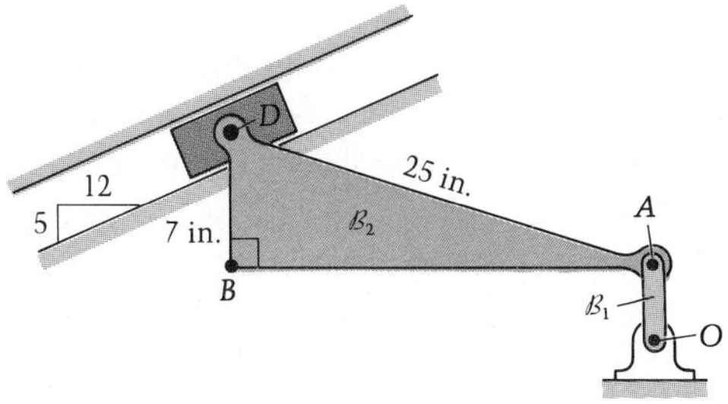

Figure P3.38

3.39 See Figure P3.39. The angular velocity of rod $\mathcal{B}_1$ is a constant: $\boldsymbol{\omega}_1 = 0.3\hat{\mathbf{k}}$ rad/s. Determine the angular velocities of plate $\mathcal{B}_2$ and bar $\mathcal{B}_3$ in the indicated position.

3.40 The pin at B (Figure P3.40) has a constant speed of 51 cm/s and moves in a circle in the clockwise direction. Find the angular velocities of bars $\mathcal{B}_1$ and $\mathcal{B}_2$ in the given position.

3.41 Solve Example 3.5 by using the instantaneous center of the rod.

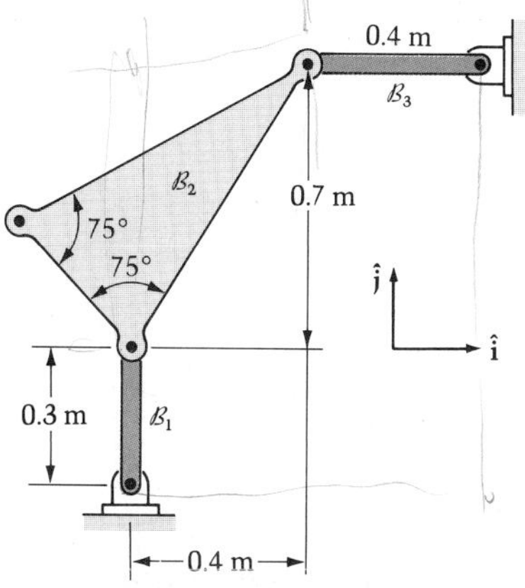

Figure P3.39

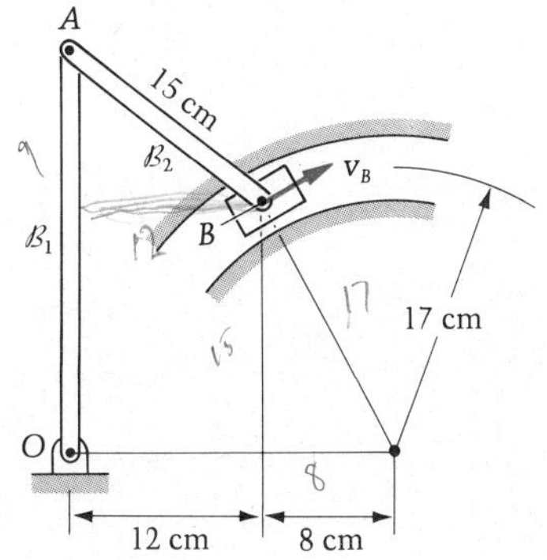

Figure P3.40

3.42 Rods $\mathcal{B}_1$ and $\mathcal{B}_2$ are pinned at B and move in a vertical plane with the constant angular velocities shown in Figure P3.42. Locate the instantaneous center of $\mathcal{B}_2$ for the given position, and use it to find the velocity of point C. Then check by calculating $\mathbf{v}_C$ by relating it (on $\mathcal{B}_2$) to the velocity of B. Note that sometimes Ⓘ is more trouble to locate than it is worth!

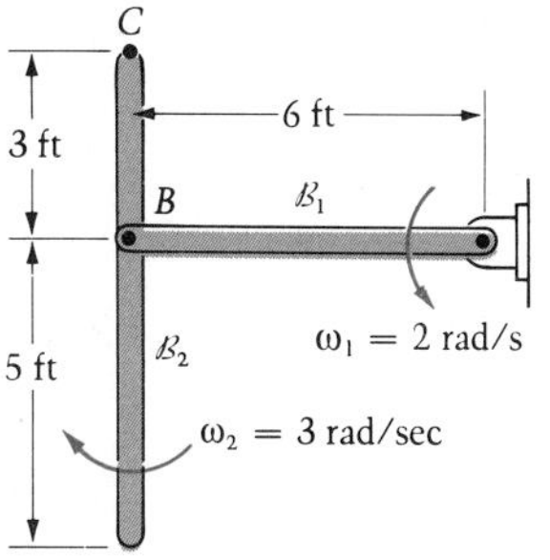

Figure P3.42

3.43 The linkage shown in Figure P3.43 is made up of rods $\mathcal{B}_1$, $\mathcal{B}_2$, and $\mathcal{B}_3$. Rod $\mathcal{B}_1$ has constant angular velocity $\boldsymbol{\omega}_1 = 0.6 \circlearrowleft$ rad/sec. Determine the angular velocities of $\mathcal{B}_2$ and $\mathcal{B}_3$ when the angle θ is equal to 90° as shown.

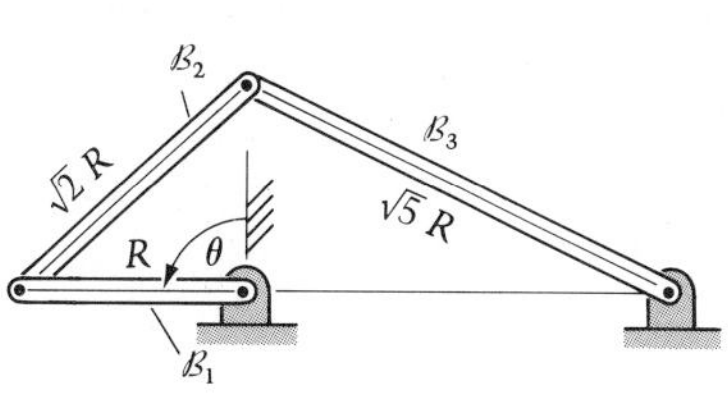

Figure P3.43

3.44 A bar of length $2L$ moves with its ends in contact with the planes shown in Figure P3.44. Find the velocity and acceleration of point C, in terms of θ and its derivatives, by writing and then differentiating the position vector of C. Then check the velocity solution by using the instantaneous center.

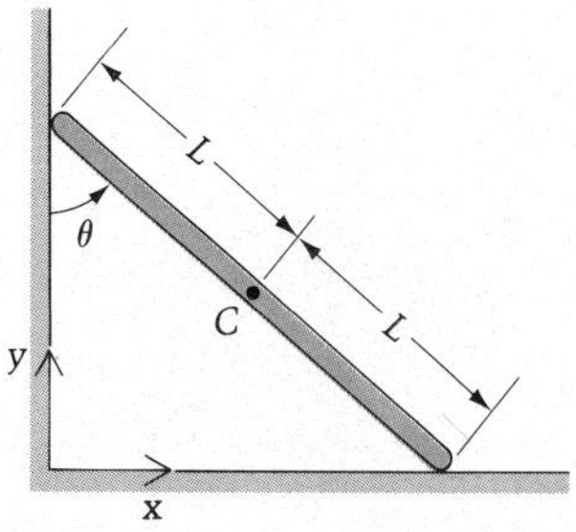

Figure P3.44

3.45 The piston rod of the hydraulic cylinder shown in Figure P3.45 moves outward at the constant speed of 0.13 m/s. Find the angular velocity of $\mathcal{B}_1$ at the instant shown.

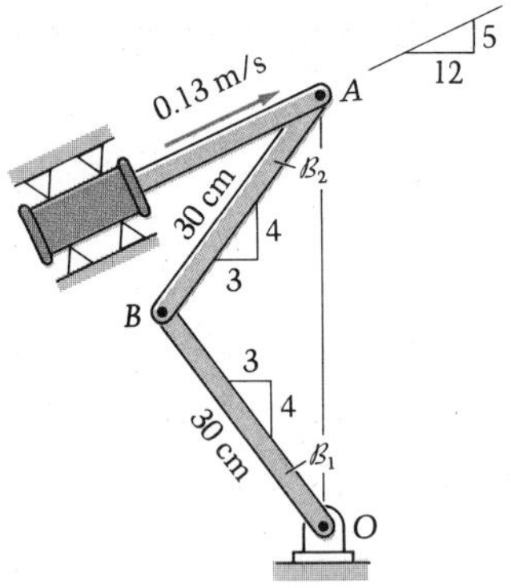

Figure P3.45

3.46 Using the method of instantaneous centers, find the velocity of point B in Figure P3.46, which is constrained to move in the slot as shown. The angular velocity of $\mathcal{B}_1$ is 0.3 ↺ rad/s at the indicated instant.

3.47 The center of block $\mathcal{B}_1$ in Figure P3.47 travels at a constant speed of 30 mph to the right. Disk $\mathcal{B}_2$ is pinned to $\mathcal{B}_1$ at A and spins at 100 rpm counterclockwise. Find: (a) the velocity of P; (b) the instantaneous center Ⓘ of $\mathcal{B}_2$; and (c), using Ⓘ, the velocities of Q, S, and R.

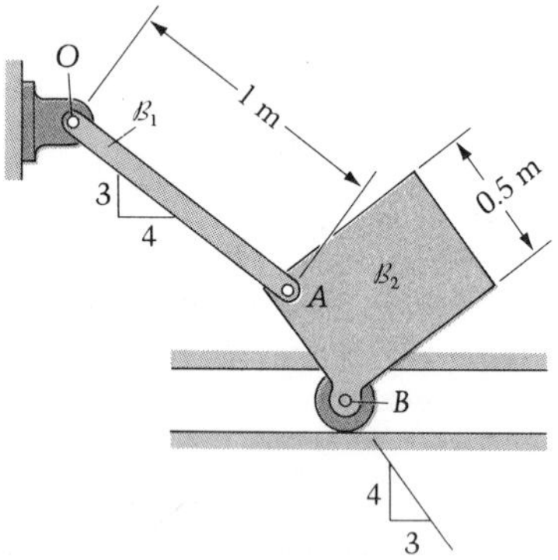

Figure P3.46

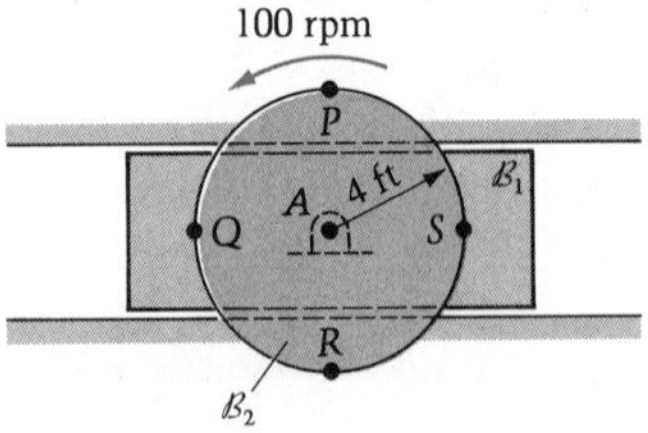

Figure P3.47

3.48 The roller at B, which moves in the parabolic slot, is pinned to bar $\mathcal{B}_1$ as shown in Figure P3.48. Bar $\mathcal{B}_1$ is pinned to $\mathcal{B}_2$ at A. The angular velocity of $\mathcal{B}_2$ at this instant is shown. Find the angular velocity of $\mathcal{B}_1$ at this time.

3.49 Bars $\mathcal{B}_1$ and $\mathcal{B}_2$ (see Figure P3.49) are pinned together at A. Find the angular velocity of bar $\mathcal{B}_1$ and the velocity of point B when the bars are next collinear. *Hint*: To find this configuration, draw a series of rough sketches of $\mathcal{B}_1$ and $\mathcal{B}_2$ as $\mathcal{B}_2$ turns counterclockwise from the position shown, and you will see $\mathcal{B}_1$ and $\mathcal{B}_2$ coming into alignment.

3.50 Repeat the preceding problem at the *second* instant of time when the bars are collinear. Follow the same hint, but this time start just past the first collinear position, found in Problem 3.49.

3.51 The constant angular velocity of wheel $\mathcal{B}_1$ is $\omega_1 = 100$ ↻ rad/sec. It is in rolling (i.e., "no slip") contact with $\mathcal{B}_2$, which means the contacting points have the same velocity. Find the angular velocity of the bar $\mathcal{B}_4$ at the instant shown in Figure P3.51.

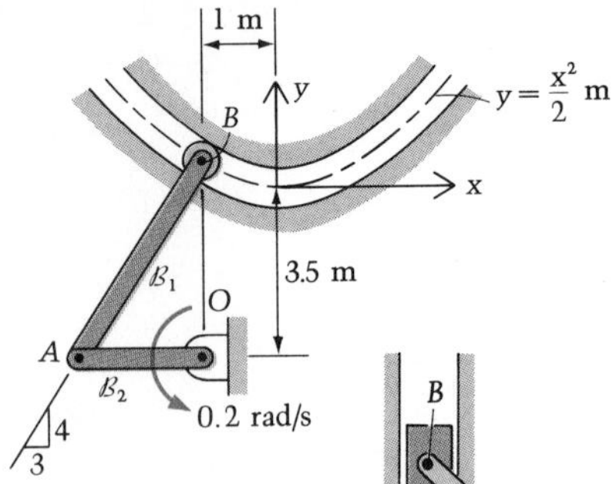

Figure P3.48

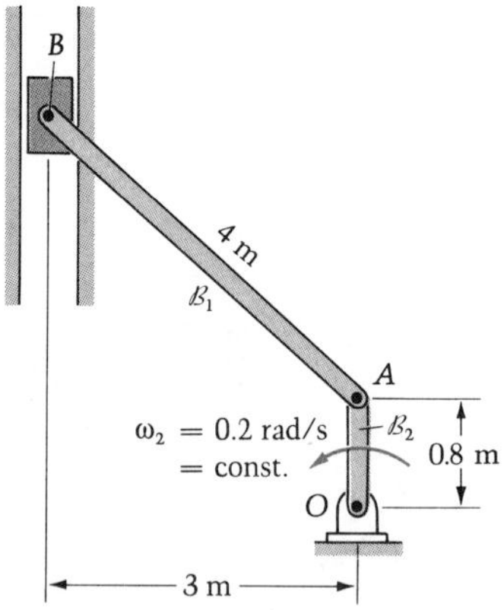

Figure P3.49

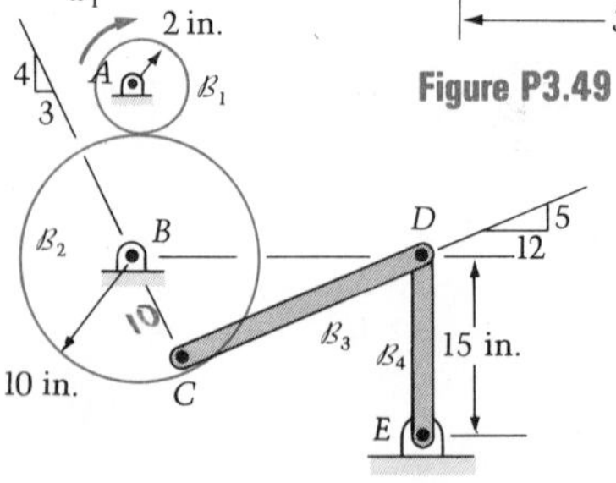

Figure P3.51

3.5 Acceleration and Angular Acceleration Relationship for Two Points of the Same Rigid Body

The **angular acceleration** vector of a rigid body $\mathcal{B}$ in plane motion in a frame $\mathcal{I}$ is defined as the derivative in $\mathcal{I}$ of the angular velocity and is called $\boldsymbol{\alpha}$:

$$\boldsymbol{\alpha}_{\mathcal{B}/\mathcal{I}} = \boldsymbol{\alpha} = \frac{d\boldsymbol{\omega}_{\mathcal{B}/\mathcal{I}}}{dt} = \dot{\boldsymbol{\omega}} = \ddot{\theta}\hat{\mathbf{k}} \tag{3.14}$$

or

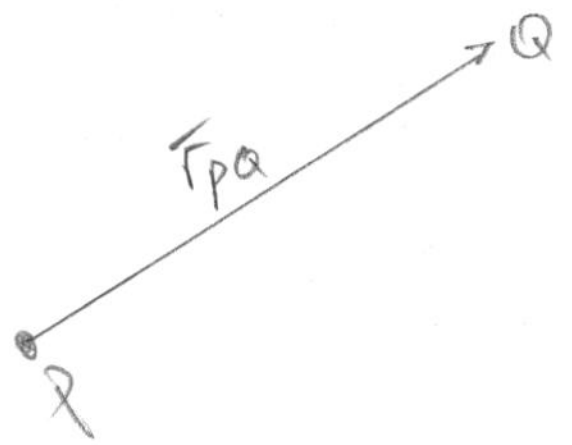

$$\boldsymbol{\alpha} = \ddot{\theta}\hat{\mathbf{k}} \quad \text{or} \quad \alpha\hat{\mathbf{k}} \tag{3.15}$$

where $\hat{\mathbf{k}}$ is a constant vector in both $\mathcal{B}$ and $\mathcal{I}$. Note that, as with $\boldsymbol{\omega}$, we delete the subscript when there is no confusion about the frame of reference being used.

Development of the Acceleration and Angular Acceleration Relationship

We now develop the relationship between the accelerations of two points P and Q of a rigid body $\mathcal{B}$. Differentiating both sides of Equation (3.8) yields

$$\dot{\mathbf{v}}_Q = \mathbf{a}_Q = \mathbf{a}_P + \ddot{\theta}\hat{\mathbf{k}} \times \mathbf{r}_{PQ} + \dot{\theta}\hat{\mathbf{k}} \times \dot{\mathbf{r}}_{PQ} \tag{3.16}$$

Now using Equation (3.7),* we may rewrite the last term as

$$\begin{aligned} \dot{\theta}\hat{\mathbf{k}} \times \dot{\mathbf{r}}_{PQ} &= \dot{\theta}\hat{\mathbf{k}} \times (\dot{\theta}\hat{\mathbf{k}} \times \mathbf{r}_{PQ}) \\ &= \dot{\theta}\hat{\mathbf{k}}(\dot{\theta}\hat{\mathbf{k}} \cdot \mathbf{r}_{PQ}) - \mathbf{r}_{PQ}(\dot{\theta}\hat{\mathbf{k}} \cdot \dot{\theta}\hat{\mathbf{k}}) \end{aligned} \tag{3.17}$$

or

$$\dot{\theta}\hat{\mathbf{k}} \times \dot{\mathbf{r}}_{PQ} = -\dot{\theta}^2\mathbf{r}_{PQ} \tag{3.18}$$

Question 3.12 Why is the dot product $\dot{\theta}\hat{\mathbf{k}} \cdot \mathbf{r}_{PQ}$ in Equation (3.17) zero?

Equations (3.16) and (3.18) then yield the desired relation between the accelerations of P and Q:

$$\mathbf{a}_Q = \mathbf{a}_P + \ddot{\theta}\hat{\mathbf{k}} \times \mathbf{r}_{PQ} - \dot{\theta}^2\mathbf{r}_{PQ} \tag{3.19}$$

We note that the same three rules spelled out in Equation (3.10) also hold for the use of Equation (3.19). Unlike velocities, however, the accelerations of P and Q do not generally have equal components along the line PQ joining them; these components differ by $r\dot{\theta}^2 = r\omega^2$. Likewise, the components *perpendicular* to PQ differ by $r\ddot{\theta} = r\alpha$ (in the same way as the velocity components normal to PQ differ by $r\dot{\theta} = r\omega$).

Answer 3.12 Because $\mathbf{r}_{PQ}$ lies in the (xy) reference plane and is therefore perpendicular to $\hat{\mathbf{k}}$.

* And the vector identity $\mathbf{A} \times (\mathbf{B} \times \mathbf{C}) = \mathbf{B}(\mathbf{A} \cdot \mathbf{C}) - \mathbf{C}(\mathbf{A} \cdot \mathbf{B})$.

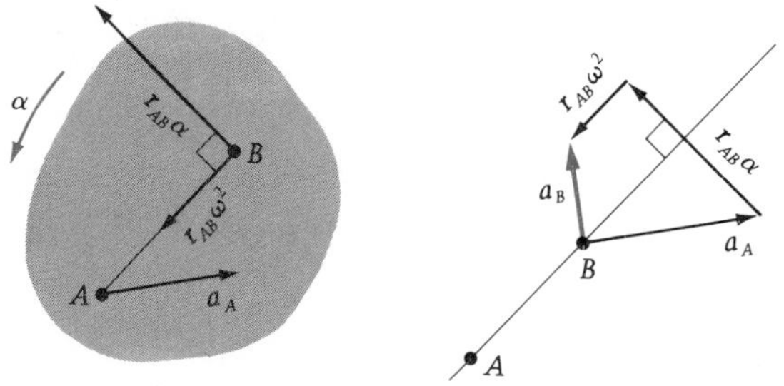

Figure 3.16

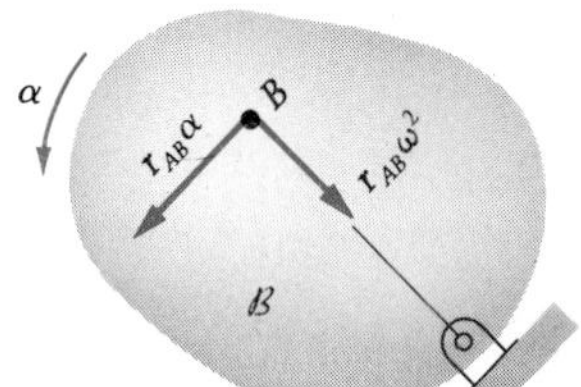

Figure 3.17

If the acceleration of point A of a body $\mathcal{B}$ is $\mathbf{a}_A$, for example, then the acceleration of any point B is the sum of the three vectors shown in Figure 3.16. If A is a pinned point, or pivot,* then $\mathbf{a}_B$ has two components: one along the line from B toward A and the other perpendicular to it and tangent to the circle (on which it necessarily travels when there is a pivot at A). Note that the direction of the tangential component depends on the direction of $\boldsymbol{\alpha}$, but the "radial" component is *always* inward toward the pivot. This case is shown in Figure 3.17. We now illustrate the use of Equation (3.19) with several examples.

EXAMPLE 3.10

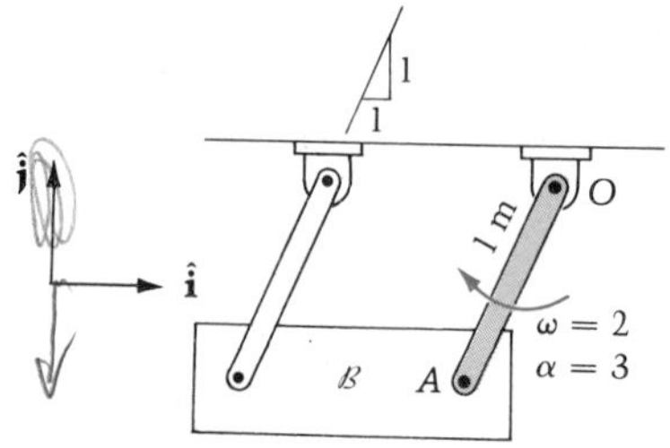

Figure E3.10

In Figure E3.10, let the links have length 1 m and let them each have $\omega = 2$ ↺ rad/s and $\alpha = 3$ ↺ rad/s² at a time when they make an angle of 45° with the ceiling. Find the acceleration of block $\mathcal{B}$. (That is, find the acceleration of any of its points — they are all the same since $\mathcal{B}$ is translating.)

Solution

All points of $\mathcal{B}$ have the same $\mathbf{v}$ and $\mathbf{a}$ as point A. Using Equation (3.19) for the link OA, we get

$$\begin{aligned}\mathbf{a}_A &= \overset{0}{\mathbf{a}_O} + \boldsymbol{\alpha} \times \mathbf{r}_{OA} - \omega^2 \mathbf{r}_{OA} \\ &= 3\hat{\mathbf{k}} \times (-0.707\hat{\mathbf{i}} + 0.707\hat{\mathbf{j}}) - 2^2(-0.707\hat{\mathbf{i}} + 0.707\hat{\mathbf{j}}) \\ &= (-2.12 + 2.83)\hat{\mathbf{i}} + (-2.12 - 2.83)\hat{\mathbf{j}} \\ &= 0.71\hat{\mathbf{i}} - 4.95\hat{\mathbf{j}} \text{ m/s}^2\end{aligned}$$

EXAMPLE 3.11

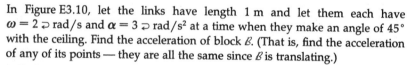

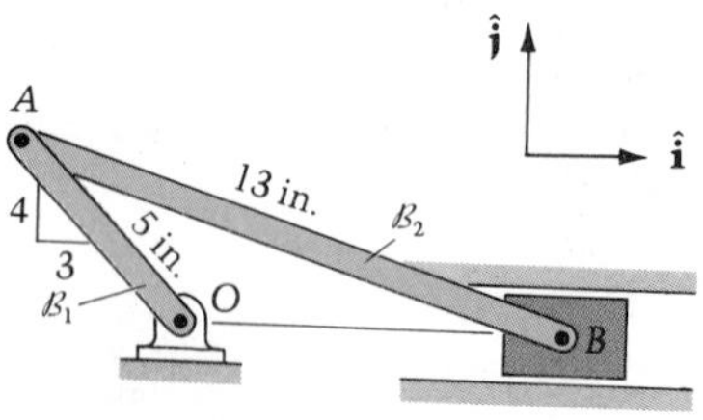

Figure E3.11

In Example 3.3 find the acceleration of the translating piston at the given instant if $\omega_1 = 10$ ↺ rad/sec and $\alpha_1 = 5$ ↻ rad/sec². See Figure E3.11.

* Again, *pivot* means a point of $\mathcal{B}$ that does not move throughout a motion. It includes, but is not limited to, a hinge.

Question 3.13 What does it mean when α is in the opposite direction from that of ω?

Solution

Relating O and A on body $\mathcal{B}_1$ by Equation (3.19), we have

$$\begin{aligned}\mathbf{a}_A &= \overset{0\text{ (pinned)}}{\cancel{\mathbf{a}_O}} + \alpha_1\hat{\mathbf{k}} \times \mathbf{r}_{OA} - \omega_1^2\mathbf{r}_{OA} \\ &= 5\hat{\mathbf{k}} \times (-3\hat{\mathbf{\imath}} + 4\hat{\mathbf{\jmath}}) - 10^2(-3\hat{\mathbf{\imath}} + 4\hat{\mathbf{\jmath}}) \\ &= 280\hat{\mathbf{\imath}} - 415\hat{\mathbf{\jmath}} \text{ in./sec}^2\end{aligned}$$

And relating A and B on body $\mathcal{B}_2$, we have

$$\mathbf{a}_B = \mathbf{a}_A + \alpha_2\hat{\mathbf{k}} \times \mathbf{r}_{AB} - \omega_2^2\mathbf{r}_{AB}$$

In the previous example we found

$$\mathbf{r}_{AB} = 12.4\hat{\mathbf{\imath}} - 4\hat{\mathbf{\jmath}} \text{ in.} \qquad \text{and} \qquad \omega_2 = 2.42 \circlearrowright \text{ rad/sec}$$

Noting that the piston translates horizontally, we get

$$\begin{aligned}\mathbf{a}_B = a_B\hat{\mathbf{\imath}} &= 280\hat{\mathbf{\imath}} - 415\hat{\mathbf{\jmath}} + 12.4\alpha_2\hat{\mathbf{\jmath}} + 4\alpha_2\hat{\mathbf{\imath}} - 2.42^2(12.4\hat{\mathbf{\imath}} - 4\hat{\mathbf{\jmath}}) \\ &= (207 + 4\alpha_2)\hat{\mathbf{\imath}} + (-392 + 12.4\alpha_2)\hat{\mathbf{\jmath}}\end{aligned}$$

The coefficients of $\hat{\mathbf{\jmath}}$ yield

$$\alpha_2 = \frac{392}{12.4} = 31.6 \text{ rad/sec}^2$$

The coefficients of $\hat{\mathbf{\imath}}$ then give our answer:

$$\mathbf{a}_B = a_B\hat{\mathbf{\imath}} = [207 + 4(31.6)]\hat{\mathbf{\imath}} = 333\hat{\mathbf{\imath}} \text{ in./sec}^2$$

Let us find the acceleration of the instantaneous center of zero velocity of the rod $\mathcal{B}_2$ in Examples 3.7 and 3.11. Using those examples and Equation (3.19), we have

$$\begin{aligned}\mathbf{a}_{\text{Ⓘ}} &= \mathbf{a}_B + \alpha_2\hat{\mathbf{k}} \times \mathbf{r}_{B\text{Ⓘ}} - \omega_2^2\mathbf{r}_{B\text{Ⓘ}} \\ &= 333\hat{\mathbf{\imath}} + 31.6\hat{\mathbf{k}} \times (-12.5\hat{\mathbf{\jmath}}) - (2.42)^2(-12.5\hat{\mathbf{\jmath}}) \\ &= 728\hat{\mathbf{\imath}} + 73.2\hat{\mathbf{\jmath}} \text{ in./sec}^2\end{aligned}$$

We see from this result that the instantaneous center of zero *velocity* does not generally have zero acceleration. Unless the point Ⓘ is a pivot (i.e., a permanently fixed point, such as a pin connecting the point to the reference frame), it should *never* be assumed that $\mathbf{a}_{\text{Ⓘ}}$ is zero.*

Answer 3.13 It means that the angular speed of the body is decreasing.

* Actually, in nontranslational cases there is a point of zero acceleration, but unless it is a pivot point, or a point of rolling contact at an instant when $\omega = 0$, it is more trouble to find than it is worth. See Problem 3.83.

EXAMPLE 3.12

In Example 3.5, find the angular acceleration of the rod and the acceleration of its endpoint A. See Figure E3.12a.

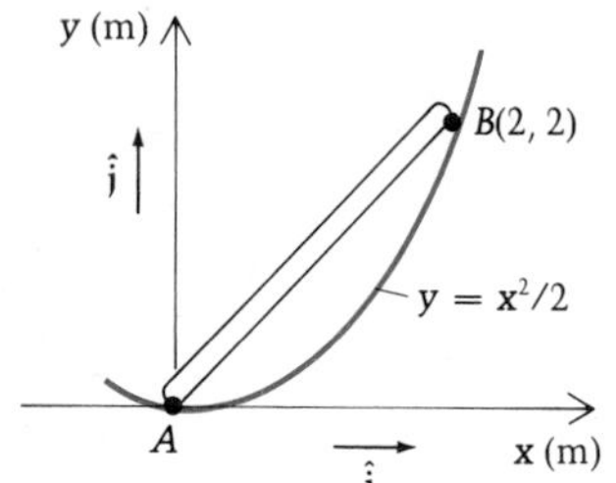

Figure E3.12a

Solution

Relating the accelerations of B and A with Equation (3.19),

$$\mathbf{a}_B = \mathbf{a}_A + \boldsymbol{\alpha} \times \mathbf{r}_{AB} - \omega^2 \mathbf{r}_{AB} \tag{1}$$

The acceleration of point B is

$$\mathbf{a}_B = \ddot{s}\hat{\mathbf{e}}_t + \frac{\dot{s}^2}{\rho}\hat{\mathbf{e}}_n = \frac{\dot{s}^2}{\rho}\hat{\mathbf{e}}_n$$

where $\ddot{s}_B = 0$ since $\dot{s}_B = \text{constant} = 0.3$ m/s. The curvature formula from calculus gives us the radius of curvature at point B:

$$\frac{1}{\rho} = \left| \frac{y''}{(1 + y'^2)^{3/2}} \right|$$

$$= \frac{1}{(1 + x^2)^{3/2}}$$

Therefore

$$\rho_B = (1 + 2^2)^{3/2}$$

$$= 5^{3/2}$$

$$= 11.2 \text{ m}$$

Therefore,

$$\mathbf{a}_B = \frac{(0.3)^2}{11.2}\hat{\mathbf{e}}_n = 0.00804\hat{\mathbf{e}}_n \text{ m/s}^2$$

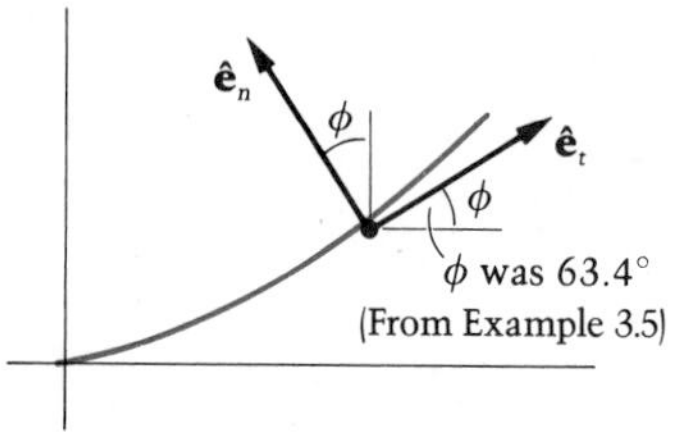

Figure E3.12b

The unit vector $\hat{\mathbf{e}}_n$ is seen in Figure E3.12b to be

$$\hat{\mathbf{e}}_n = -\sin\phi\hat{\mathbf{i}} + \cos\phi\hat{\mathbf{j}}$$

$$= -0.894\hat{\mathbf{i}} + 0.448\hat{\mathbf{j}}$$

Therefore

$$\mathbf{a}_B = -0.00804\hat{\mathbf{e}}_n$$

$$= -0.00719\hat{\mathbf{i}} + 0.00360\hat{\mathbf{j}} \text{ m/s}^2$$

Substituting a_B into Equation (1) gives

$$-0.00719\hat{\mathbf{i}} + 0.00360\hat{\mathbf{j}} = \ddot{s}_A\hat{\mathbf{e}}_t + \frac{\dot{s}_A^2}{\rho_A}\hat{\mathbf{e}}_n$$

$$+ \alpha\hat{\mathbf{k}} \times (2\hat{\mathbf{i}} + 2\hat{\mathbf{j}}) - \omega^2(2\hat{\mathbf{i}} + 2\hat{\mathbf{j}})$$

where $\mathbf{r}_{AB} = 2\hat{\mathbf{i}} + 2\hat{\mathbf{j}}$ has also been substituted.

Next, the radius of curvature of the path of A at the instant of interest is

$$\rho_A = (1 + x^2)^{3/2} = (1 + 0^2)^{3/2} = 1 \text{ m}$$

Substituting $\rho_A = 1$, $\dot{s}_A = 0.403$ and $\omega = 0.134$ from Example 3.5 (along with $\hat{\mathbf{e}}_t = \hat{\mathbf{i}}$ and $\hat{\mathbf{e}}_n = \hat{\mathbf{j}}$ at point A), we obtain the vector equation:

$$-0.00719\hat{\mathbf{i}} + 0.00360\hat{\mathbf{j}} = \ddot{s}_A\hat{\mathbf{i}} + \frac{(0.403)^2}{1}\hat{\mathbf{j}} - 2\alpha\hat{\mathbf{i}} + 2\alpha\hat{\mathbf{j}} - 2(0.134)^2(\hat{\mathbf{i}} + \hat{\mathbf{j}})$$

Writing the component equations, we have

$$\hat{\mathbf{i}} \text{ coefficients:} \qquad -0.00719 = \ddot{s}_A - 2\alpha - 2(0.134)^2$$

$$\hat{\mathbf{j}} \text{ coefficients:} \qquad 0.00360 = \frac{(0.403)^2}{1} + 2\alpha - 2(0.134)^2$$

The $\hat{\mathbf{j}}$ equation gives

$$\alpha = -0.0612 \Rightarrow \boldsymbol{\alpha} = -0.0612\hat{\mathbf{k}} \text{ rad/s}^2$$

And the $\hat{\mathbf{i}}$ equation then yields

$$\ddot{s}_A = -0.0936 \text{ m/s}^2$$

Therefore

$$\mathbf{a}_A = -0.0936\hat{\mathbf{i}} + \frac{(0.403)^2}{1}\hat{\mathbf{j}}$$

$$= -0.0936\hat{\mathbf{i}} + 0.162\hat{\mathbf{j}} \text{ m/s}^2$$

Many more examples of the use of Equation (3.19) will be found in the next section, after we have discussed the topic of rolling.

PROBLEMS ▸ Section 3.5

3.52 In Figure P3.52 the angular velocity of the bent bar is 0.2 rad/s counterclockwise at an instant when its angular acceleration is 0.3 rad/s² clockwise. Find the acceleration of the endpoint B in the indicated position.

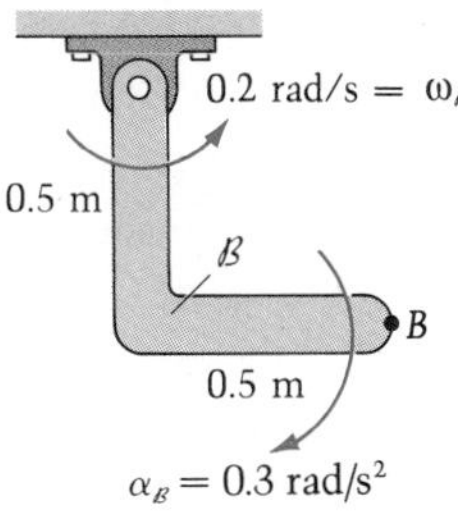

Figure P3.52

3.53 The acceleration of pin B in Figure P3.53 is 9.9 ft/sec² down and to the left, and its velocity is 4 ft/sec up and to the right, when $\mathcal{L}$ passes the horizontal. At this instant, find the angular acceleration of $\mathcal{L}$.

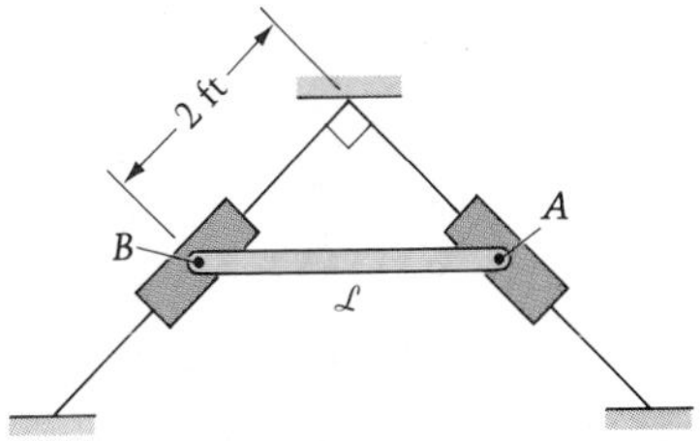

Figure P3.53

3.54 End B of the rod shown in Figure P3.54 has a constant velocity of 10 ft/sec down the plane. For the position shown (rod horizontal) determine the velocity and acceleration of end A of the rod.

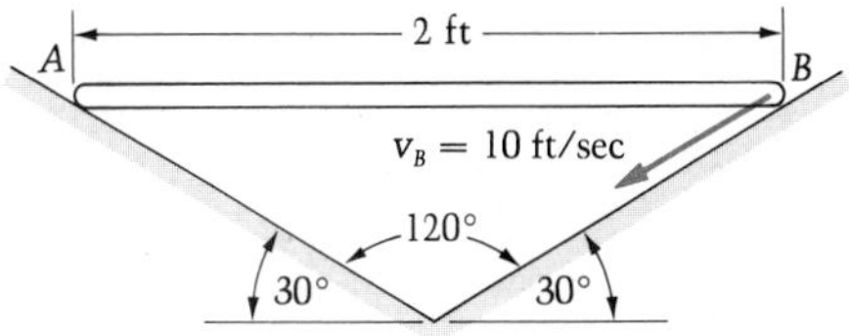

Figure P3.54

3.55 The velocities and accelerations of the two endpoints A and B of a rigid bar in plane motion are as shown in Figure P3.55. Find the acceleration of the midpoint of the bar in the given position.

3.56 In Problem 3.42 find the acceleration of C in the position given in the figure.

3.57 At the instant given, the angular velocity and angular acceleration of bar $\mathcal{B}_1$ are $0.2 \circlearrowleft$ rad/sec and $0.1 \circlearrowright$ rad/sec². (See Figure P3.57.) Find the angular accelerations of $\mathcal{B}_2$ and $\mathcal{B}_3$ at this instant.

3.58 Bar $\mathcal{B}_1$ rotates with a constant angular velocity of $2\hat{\mathbf{k}}$ rad/sec. Find the angular velocities and angular accelerations of $\mathcal{B}_2$ and $\mathcal{B}_3$ at the instant shown in Figure P3.58.

3.59 In the position indicated in Figure P3.59, the slider block $\mathcal{B}$ has the indicated velocity and acceleration. Find the angular acceleration of the wheel at this instant.

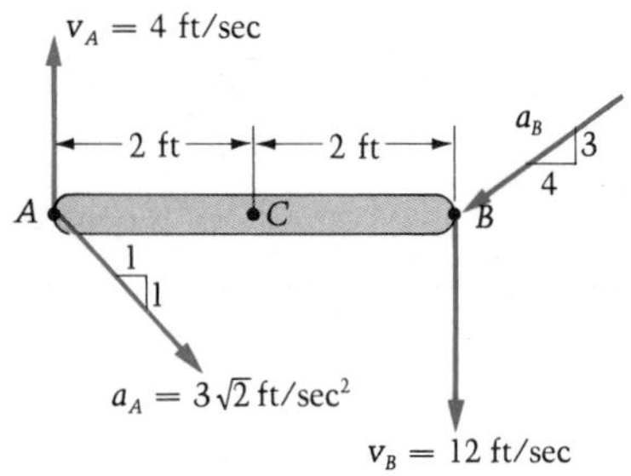

Figure P3.55

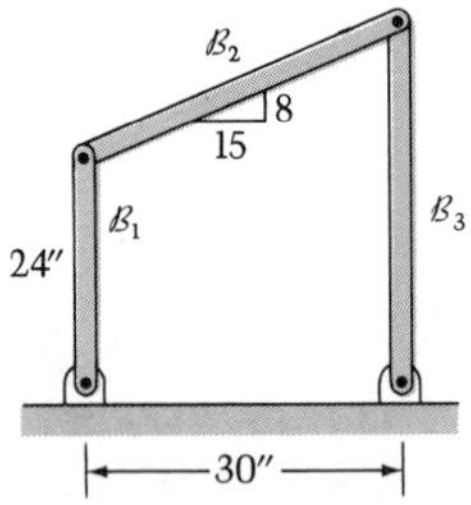

Figure P3.57

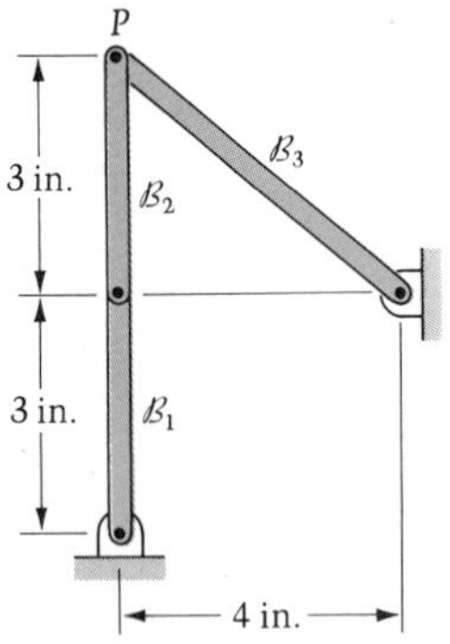

Figure P3.58

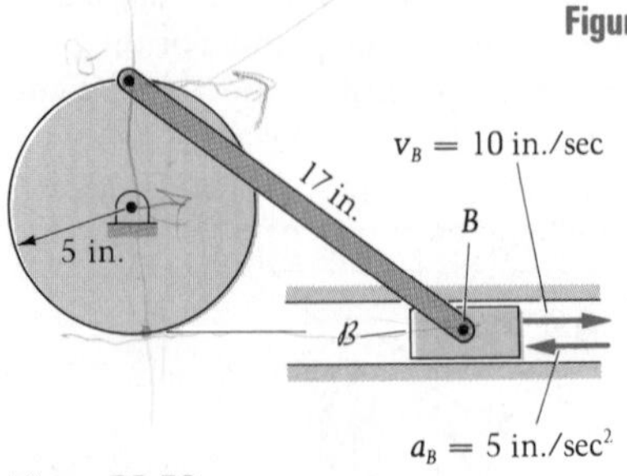

Figure P3.59

3.60 If in Problem 3.11 the bar whose angular velocity is given to be 2 rad/sec has an angular acceleration of zero at that instant, find the angular acceleration of the 5-inch (horizontal) bar.

* **3.61** The angular velocity of $\mathcal{B}_1$ in Figure P3.61 is a constant 3 rad/sec clockwise. Find the velocity and acceleration of point C in the given configuration, and determine the acceleration of point C when $\mathbf{v}_C = \mathbf{0}$.

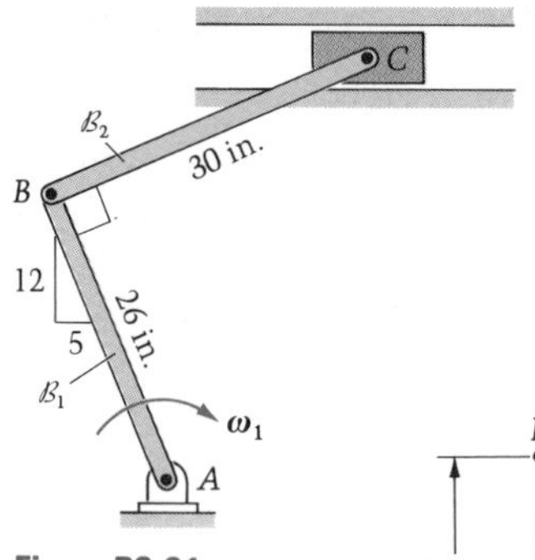

Figure P3.61

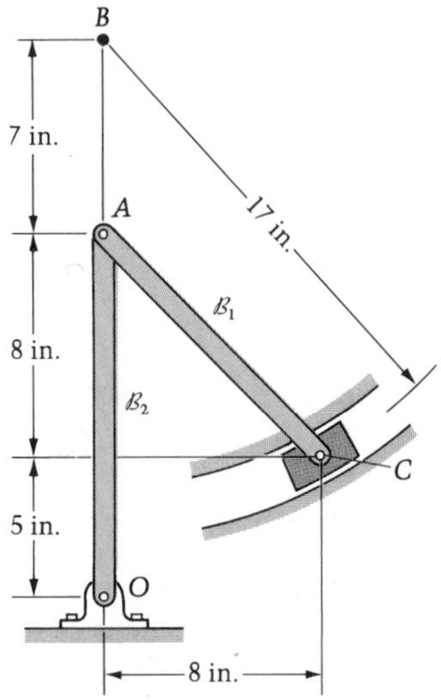

Figure P3.64

3.62 Refer to the preceding problem. At the instant of time when $\omega_2 = 0$, find the acceleration of point C.

3.63 If in Problem 3.24 the angular velocity of the 5-inch bar is constant throughout an interval which includes the instant shown, find at that instant the acceleration of the guided block.

3.64 At the instant of time shown in Figure P3.64, the angular velocity and angular acceleration of rod $\mathcal{B}_1$ are $3 \circlearrowleft$ rad/sec and $2 \circlearrowright$ rad/sec². At the same time, find the angular acceleration of bar $\mathcal{B}_2$.

3.65 In Problem 3.26 find the equation for the angular acceleration $\ddot{\phi}$ of the connecting rod $\mathcal{B}_2$ a a function of r, ℓ, θ, and $\dot{\theta}$.

3.66 In the preceding problem, plot $\ddot{\phi}/\dot{\theta}^2$ versus θ $(0 \le \theta \le 2\pi)$ for $\ell/r = 1, 2,$ and 5.

3.67 Crank $\mathcal{B}_1$ in Figure P3.67 is pinned to rod $\mathcal{B}_2$; the other end of $\mathcal{B}_2$ slides on a parabolic incline and is at the origin in the position shown. The angular velocity of $\mathcal{B}_1$ is $3 \circlearrowleft$ rad/s = constant. Determine the acceleration of A and the angular acceleration of $\mathcal{B}_2$ at the given instant. *Hint*: The radius of curvature ρ of a plane curve $y = y(x)$ can be calculated from

$$\frac{1}{\rho} = \left| \frac{y''}{(1 + y'^2)^{3/2}} \right|$$

Use this result in computing the normal component of $\mathbf{a}_A$ as in Example 3.11.

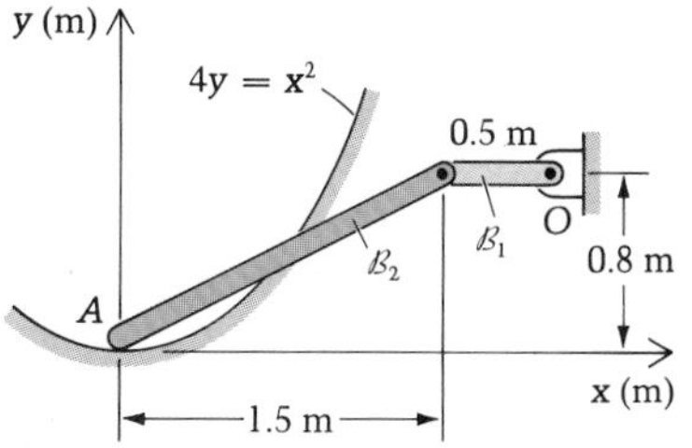

Figure P3.67

3.68 The 10-ft bar in Figure P3.68 is sliding down the 13-ft-radius circle as shown. For the position shown, the bar has an angular velocity of 2 rad/sec and an angular acceleration of 3 rad/sec², both clockwise. Find the x and y components of acceleration of point B for this position.

3.69 In Problem 3.23, find the acceleration of B when $x = 10$ m.

3.70 In Problem 3.31 find the acceleration of the upper end B of rod $\mathcal{B}_2$ in position (a).

3.71 Find the acceleration of P in Problem 3.58 if at the same instant the body $\mathcal{B}_1$ has angular acceleration $\boldsymbol{\alpha}_1 = 3\hat{\mathbf{k}}$ rad/sec² instead of zero.

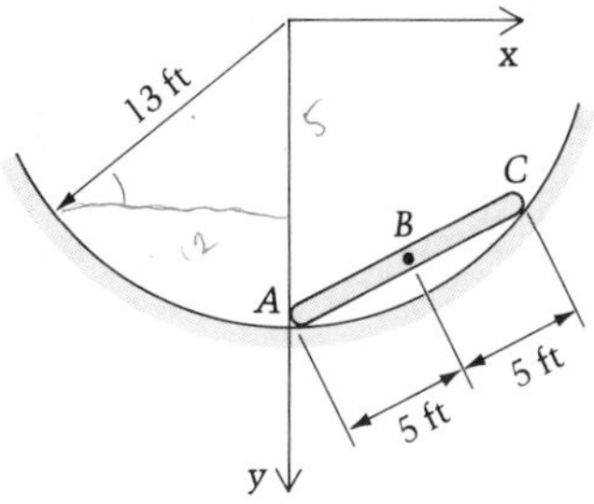

Figure P3.68

3.72 In Problem 3.49 find the acceleration of point B and the angular acceleration of body $\mathcal{B}_1$ at the described instant.

3.73 In Problem 3.50 find the acceleration of point B and the angular acceleration of body $\mathcal{B}_1$ at the described instant.

3.74 In Problem 3.22 find the acceleration of S and the angular accelerations of $\mathcal{B}_2$ and $\mathcal{B}_3$ at the instant when $\theta = 30°$.

3.75 In Problem 3.38 determine the angular acceleration of the plate and the acceleration of pin D in the indicated position.

3.76 In Problem 3.45 determine the angular accelerations of $\mathcal{B}_1$ and $\mathcal{B}_2$ in the indicated position.

3.77 The motion of a rotating element in a mechanism is controlled so that the rate of change of angular speed ω with angular displacement θ is a constant K. If the angular speed is ω_0 when both θ and the time t are zero, determine θ, ω, and the angular acceleration α as *functions of time*.

* **3.78** Rod $\mathcal{B}_1$ in Figure P3.78 is pinned to disk $\mathcal{B}_2$ at A and B. Disk $\mathcal{B}_2$ rotates about a fixed axis through O. The rod makes an angle θ radians with the line $\ell\ell$ as shown, where $\theta = \sin t$. The time is given by t in seconds. Determine the horizontal and vertical components of acceleration of the midpoint P of segment AB when $\theta = -\pi/6$ rad.

3.79 A rod is pinned at A and B to the centers of two small rollers. (See Figure P3.79.) The speed of A is kept constant at v_O even after B encounters the parabolic surface. Find the acceleration of B just after its roller begins to move on the parabola.

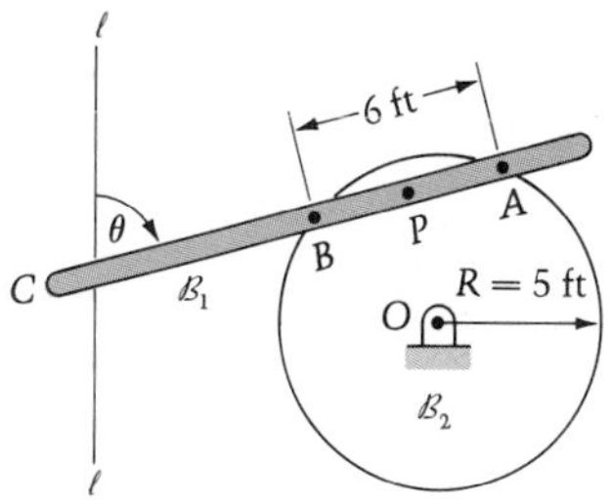

Figure P3.78

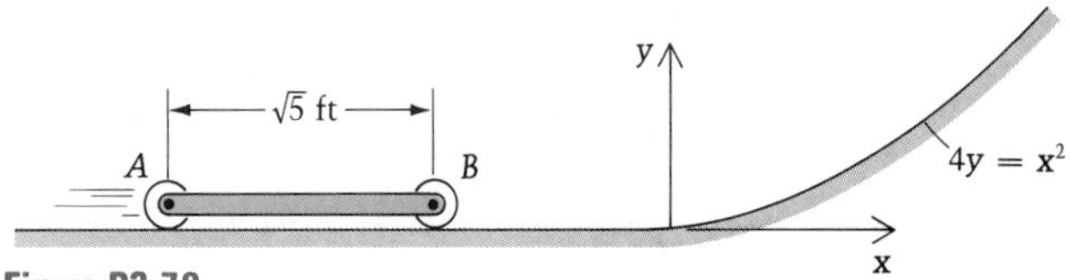

Figure P3.79

* **3.80** In the preceding problem, find $\mathbf{a}_B$ just after the *left* roller has begun to travel on the parabola.

* **3.81** The bent bar shown in Figure P3.81 slides on the vertical and horizontal surfaces. For the position shown, A has an acceleration of 4 ft/sec² to the left, while the bar has an angular velocity of 2 rad/sec clockwise and an angular acceleration of $\boldsymbol{\alpha}$.

a. Determine $\boldsymbol{\alpha}$ for the position shown.

b. Find, for the position shown, the angle θ and the distance PA such that point P has zero acceleration.

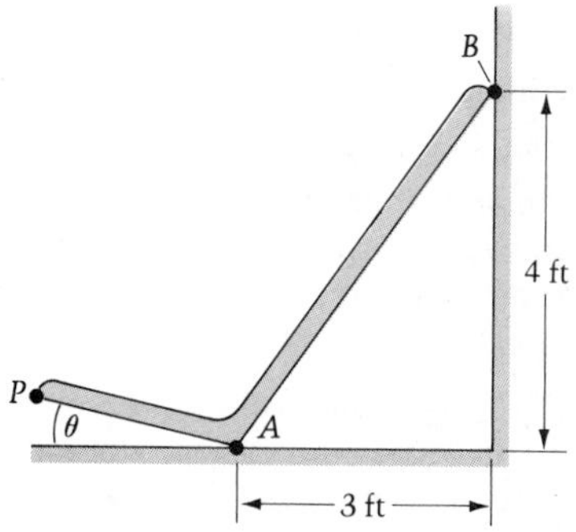

Figure P3.81

* **3.82** The right end P of bar $\mathcal{B}$ is constrained to move to the right on the sine wave shown in Figure P3.82 at the constant speed $\sqrt{10}$ in./sec. The left end A of $\mathcal{B}$ is constrained to slide along the x-axis. At the instant when $x = \pi$ in., find (a) $\omega_{\mathcal{B}}$; (b) $\mathbf{a}_P$.

* **3.83** Show that for a rigid body $\mathcal{B}$ in plane motion, as long as ω and α are not *both* zero there is a point of $\mathcal{B}$ having zero acceleration. *Hint*: Let P be a reference point with acceleration $\mathbf{a}_P = a_{P_x}\hat{\mathbf{i}} + a_{P_y}\hat{\mathbf{j}}$. See if you can find a vector $\mathbf{r}_{PT} = x\hat{\mathbf{i}} + y\hat{\mathbf{j}}$ from P to a point T of zero acceleration. That is, solve

$$\mathbf{a}_T = \mathbf{0} = \mathbf{a}_P + \alpha\hat{\mathbf{k}} \times \mathbf{r}_{PT} - \omega^2\mathbf{r}_{PT}$$

for x and y.

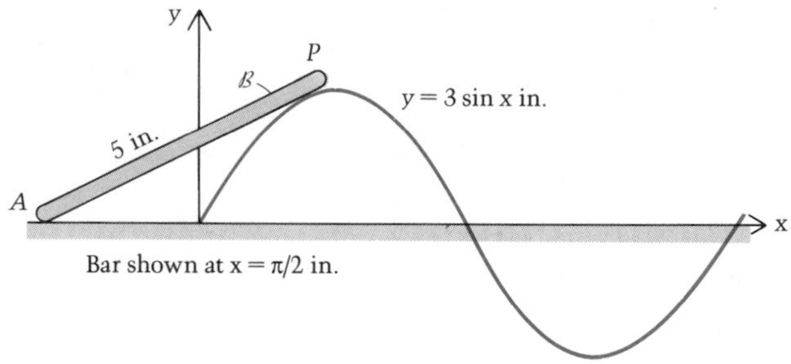

Figure P3.82

3.6 Rolling

Let $\mathcal{B}_1$ and $\mathcal{B}_2$ be two rigid bodies in motion. We define **rolling** to exist between $\mathcal{B}_1$ and $\mathcal{B}_2$ if during their motion:

1. A continuous sequence of points on the surface of $\mathcal{B}_1$ comes into one-to-one contact with a continuous sequence of points on the surface of $\mathcal{B}_2$.
2. At each instant during the interval of the motion, the contacting points have the same velocity vector.

Note that according to this definition there can be no slipping or sliding between the surfaces of $\mathcal{B}_1$ and $\mathcal{B}_2$ if rolling exists. Many authors, however, use the phrase "rolling without slipping" to describe the motion defined here. In their context, "rolling and slipping" would denote turning without the contact points having equal velocities; in our context, rolling *means* no slipping, so we shall say "turning and slipping" in cases of unequal contact-point velocities.

In this section we consider three classes of problems involving rolling contact:

1. Rolling of a wheel on a fixed straight line
2. Rolling of a wheel on a fixed plane curve
3. Gears

Rolling of a Wheel on a Fixed Straight Line

If the wheel ($\mathcal{B}_1$) shown in Figure 3.18 is rolling on the ground ($\mathcal{B}_2$, the reference frame in this case), then the continuous (shaded) sequences of points of $\mathcal{B}_1$ and $\mathcal{B}_2$ are in contact, one pair of points at a time. Since the points of $\mathcal{B}_2$ are all at rest, each point P_1 on the rim of $\mathcal{B}_1$ comes instantaneously to rest as it contacts a point P_2 of $\mathcal{B}_2$ (and is gripped for an instant by the ground). In this case, the velocities of P_1 and P_2 are each zero, although they need not vanish in general for rolling; all that is required is that $\mathbf{v}_{P_1} = \mathbf{v}_{P_2}$.*

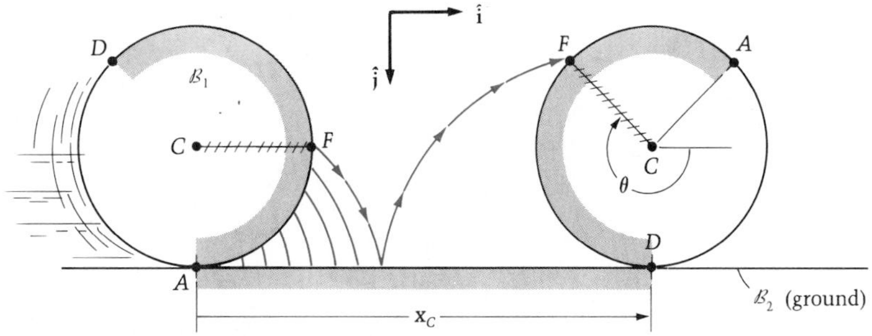

Figure 3.18 Rolling wheel, illustrating contacting sequence of points.

Using Equation (3.8) and noting that the center point C moves only horizontally and that the contact point is always the instantaneous center of $\mathcal{B}_1$,

$$\mathbf{v}_C = \mathbf{v}_{\text{①}} + \dot{\theta}\hat{\mathbf{k}} \times \mathbf{r}_{\text{①}C} \tag{3.20}$$

$$\dot{x}_C\hat{\mathbf{i}} = \mathbf{0} + \dot{\theta}\hat{\mathbf{k}} \times (-r\hat{\mathbf{j}}) \tag{3.21}$$

$$\dot{x}_C\hat{\mathbf{i}} = r\dot{\theta}\hat{\mathbf{i}} \tag{3.22}$$

For the rolling wheel, therefore, the velocity of its center C and the body's angular velocity are related quite simply by

$$\dot{x}_C = r\dot{\theta} \quad \text{or} \quad r\omega \tag{3.23}$$

Question 3.14 How would this expression be different were θ to have been chosen so that it would increase with *counterclockwise* turning of the body?

The relation between the displacement of C and the rotation of $\mathcal{B}_1$ follows from integrating Equation (3.23):

$$x_C = r\theta + C_1 \tag{3.24}$$

where the integration constant is zero if we choose $x_C = 0$ when $\theta = 0$.

---

* If $\mathcal{B}_2$ is the flatbed of a truck, for example, itself in motion with respect to a ground reference frame $\mathcal{B}_3$, then $\mathbf{v}_{P_1} = \mathbf{v}_{P_2} \neq \mathbf{0}$, but $\mathcal{B}_1$ still rolls on $\mathcal{B}_2$.

Answer 3.14 Then we would have $\dot{x}_C = -r\dot{\theta}$.

Another approach to rolling is to begin with a small displacement Δx_C while the base point $\textcircled{I}$ grips the ground. If the angle of rotation produced is $\Delta\theta$, then

$$\Delta x_C = r\,\Delta\theta$$

if we envision the body turning about its instantaneous center. Dividing by a small time increment Δt and taking the limit, we get

$$\lim_{\Delta t \to 0}\left(\frac{\Delta x_C}{\Delta t}\right) = \dot{x}_C = \lim_{\Delta t \to 0}\left(\frac{r\,\Delta\theta}{\Delta t}\right) = r\dot{\theta} = r\omega$$

as was obtained in Equation (3.23).

Next we consider accelerations. From Equation (3.22):

$$\mathbf{a}_C = \dot{\mathbf{v}}_C = \ddot{x}_C\hat{\mathbf{i}} = r\ddot{\theta}\hat{\mathbf{i}} = r\alpha\hat{\mathbf{i}} \tag{3.25}$$

and the center point C accelerates parallel to the plane, as it must since it (alone among all points of the wheel) has rectilinear motion. Now let us compute the acceleration of the instantaneous center $\textcircled{I}$ of the wheel. Using Equation (3.19), we obtain

$$\begin{aligned}\mathbf{a}_{\textcircled{I}} &= \mathbf{a}_C + \ddot{\theta}\hat{\mathbf{k}} \times \mathbf{r}_{C\textcircled{I}} - \dot{\theta}^2\mathbf{r}_{C\textcircled{I}}\\ &= r\ddot{\theta}\hat{\mathbf{i}} + \ddot{\theta}\hat{\mathbf{k}} \times r\hat{\mathbf{j}} - \dot{\theta}^2 r\hat{\mathbf{j}}\\ &= -r\dot{\theta}^2\hat{\mathbf{j}} \quad \text{or} \quad -r\omega^2\hat{\mathbf{j}} \quad \text{or} \quad r\omega^2\uparrow \end{aligned} \tag{3.26}$$

Thus the contact point of a wheel rolling on a flat, fixed plane is accelerated toward its center with a magnitude $r\omega^2$. We see once again that a point of zero velocity need *not* be a point of zero acceleration, although of course it will be such if it is pinned to the reference frame. The point $\textcircled{I}$ in the example is at rest instantanously, but it has an acceleration — which is why its velocity changes from zero as soon as it moves and a new $\textcircled{I}$ takes its place in the rolling. We now take up several examples,* each of which deals with a round object rolling on a flat surface.

EXAMPLE 3.13

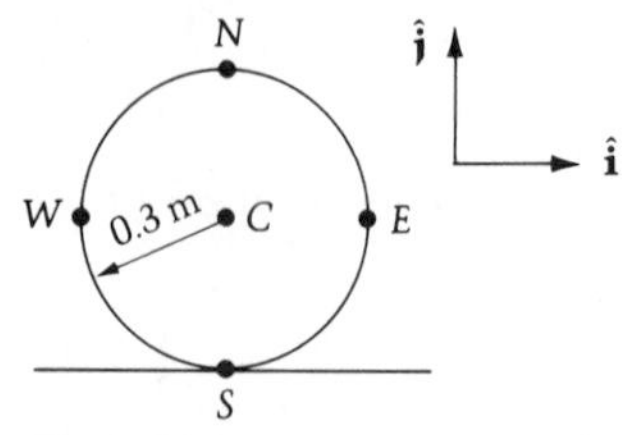

Figure E3.13

At a given instant, the rolling cylinder in Figure E3.13 has $\omega = 2 \circlearrowright$ rad/s and $\alpha = 1.5 \circlearrowleft$ rad/s^2. Find the velocity and acceleration of points N and E.

Solution

We shall relate the velocities and accelerations of N and E to those of point C. Calculating these, we have, because of the rolling (with S being $\textcircled{I}$),

$$\mathbf{v}_C = \dot{x}_C\hat{\mathbf{i}} = \boldsymbol{\omega} \times \mathbf{r}_{SC} = -2\hat{\mathbf{k}} \times 0.3\hat{\mathbf{j}} = 0.3(2)\hat{\mathbf{i}} = 0.6\hat{\mathbf{i}}\ \text{m/s}$$

* The examples of this section make continued use of Equations (3.8), (3.13), and (3.19), with the added feature that a rolling body is involved in each problem.

and

$$\mathbf{a}_C = \ddot{x}_C\hat{\mathbf{i}} = \mathbf{a}_S + \boldsymbol{\alpha} \times \mathbf{r}_{SC} - \omega^2\mathbf{r}_{SC} = 0.3(2)^2\hat{\mathbf{j}} + \alpha\hat{\mathbf{k}} \times \mathbf{r}_{SC} - (2)^2 0.3\hat{\mathbf{j}}$$
$$= \alpha\hat{\mathbf{k}} \times \mathbf{r}_{SC}$$
$$= 1.5\hat{\mathbf{k}} \times 0.3\hat{\mathbf{j}}$$
$$= -0.45\hat{\mathbf{i}} \text{ m/s}^2$$

Therefore

$$\mathbf{v}_N = \mathbf{v}_C + \boldsymbol{\omega} \times \mathbf{r}_{CN}$$
$$= 0.6\hat{\mathbf{i}} + (-2\hat{\mathbf{k}}) \times 0.3\hat{\mathbf{j}}$$
$$= 1.2\hat{\mathbf{i}} \text{ m/s}$$

Note the agreement of this result with

$$\mathbf{v}_N = \boldsymbol{\omega} \times \mathbf{r}_{Ⓘ N}$$
$$= (-2\hat{\mathbf{k}}) \times 0.6\hat{\mathbf{j}}$$
$$= 1.2\hat{\mathbf{i}} \text{ m/s}$$

Continuing, we get

$$\mathbf{a}_N = \mathbf{a}_C + \boldsymbol{\alpha} \times \mathbf{r}_{CN} - \omega^2\mathbf{r}_{CN}$$
$$= -0.45\hat{\mathbf{i}} + 1.5\hat{\mathbf{k}} \times 0.3\hat{\mathbf{j}} - 2^2(0.3\hat{\mathbf{j}})$$
$$= -0.90\hat{\mathbf{i}} - 1.2\hat{\mathbf{j}} \text{ m/s}^2$$

For point E,

$$\mathbf{v}_E = \mathbf{v}_C + \boldsymbol{\omega} \times \mathbf{r}_{CE}$$
$$= 0.6\hat{\mathbf{i}} + (-2\hat{\mathbf{k}}) \times 0.3\hat{\mathbf{i}}$$
$$= 0.6\hat{\mathbf{i}} - 0.6\hat{\mathbf{j}} \text{ m/s}$$

and

$$\mathbf{a}_E = \mathbf{a}_C + \boldsymbol{\alpha} \times \mathbf{r}_{CE} - \omega^2\mathbf{r}_{CE}$$
$$= -0.45\hat{\mathbf{i}} + 1.5\hat{\mathbf{k}} \times 0.3\hat{\mathbf{i}} - 2^2(0.3\hat{\mathbf{i}})$$
$$= -1.65\hat{\mathbf{i}} + 0.45\hat{\mathbf{j}} \text{ m/s}^2$$

The reader may wish to obtain the results for $\mathbf{a}_N$ and $\mathbf{a}_E$ by relating them instead to $\mathbf{a}_{Ⓘ}$, which as we have seen is $r\omega^2\uparrow$, *not* zero.

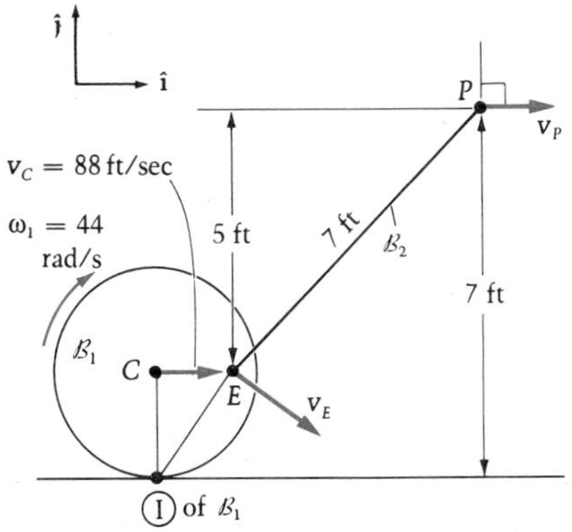

Figure E3.14

EXAMPLE 3.14

In Example 3.9 find the piston acceleration when $\theta = 90°$. See Figure E3.14.

Solution

In *Case (b)* of Example 3.9 we found $\omega_1 = 44.0 \circlearrowleft$ rad/sec, $\mathbf{v}_E = 88.0\hat{\mathbf{i}} + 66.0\hat{\mathbf{j}}$ ft/sec, $\omega_2 = 13.5 \circlearrowleft$ rad/sec, and $\mathbf{v}_P = 20.5\hat{\mathbf{i}}$ ft/sec. On body $\mathcal{B}_1$, the velocity of C is constant ($88\hat{\mathbf{i}}$), so that $a_C = 0$. Also, $a_C = r\alpha_1$ so that $\alpha_1 = 0$.

Therefore

$$\begin{aligned}\mathbf{a}_E &= \mathbf{a}_C + \alpha_1\hat{\mathbf{k}} \times \mathbf{r}_{CE} - \omega_1^2\mathbf{r}_{CE}\\ &= \mathbf{0} + \mathbf{0} - 44.0^2(1.5\hat{\mathbf{i}})\\ &= -2900\hat{\mathbf{i}}\ \text{ft/sec}^2\end{aligned}$$

On $\mathcal{B}_2$, we now relate $\mathbf{a}_E$ to the desired $\mathbf{a}_P$:

$$\underbrace{\mathbf{a}_P = a_P\hat{\mathbf{i}}}_{\substack{\text{kinematic}\\ \text{constraint}}} = \mathbf{a}_E + \alpha_2\hat{\mathbf{k}} \times \mathbf{r}_{EP} - \omega_2^2\mathbf{r}_{EP}$$

$$= -2900\hat{\mathbf{i}} + \underbrace{\alpha_2\hat{\mathbf{k}} \times (4.90\hat{\mathbf{i}} + 5.00\hat{\mathbf{j}})}_{\text{Note that this is a cross product, but}} - \underbrace{13.5^2(4.90\hat{\mathbf{i}} + 5.00\hat{\mathbf{j}})}_{\dots\ \textit{this}\ \text{is not!}}$$

$$a_P\hat{\mathbf{i}} = (-2900 - 5.00\alpha_2 - 893)\hat{\mathbf{i}} + (4.90\alpha_2 - 911)\hat{\mathbf{j}}$$

Equating the $\hat{\mathbf{j}}$ coefficients first eliminates the unknown $\mathbf{a}_P$:

$$0 = 4.90\alpha_2 - 911 \Rightarrow \alpha_2 = 186\ \text{rad/sec}^2 \qquad \text{or} \qquad \alpha_2 = 186 \circlearrowleft \text{rad/sec}^2$$

Then the coefficients of $\hat{\mathbf{i}}$ yield our answer:

$$a_P = -4720\ \text{ft/sec}^2 \qquad \text{or} \qquad \mathbf{a}_P = 4720 \leftarrow \text{ft/sec}^2$$

EXAMPLE 3.15

The wheel $\mathcal{B}_1$ in Figure E3.15 rolls to the right on plane $\mathcal{P}$. At the instant shown in the figure, $\mathcal{B}_1$ has angular velocity $\omega_1 = 1.2 \circlearrowright$ rad/s. Rod $\mathcal{B}_2$ is pinned to $\mathcal{B}_1$ at A, and the other end B of $\mathcal{B}_2$ slides along a plane Q parallel to $\mathcal{P}$. Determine the velocity of B and the angular velocity of $\mathcal{B}_2$ at the given instant.

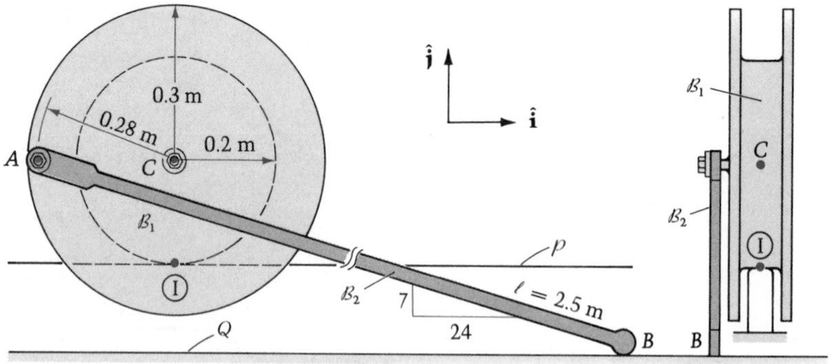

Figure E3.15

Solution

We shall use Equation (3.8) together with the results we have derived for rolling. We first seek the velocity of A; when we have $\mathbf{v}_A$ we shall then relate it to $\mathbf{v}_B$ on rod $\mathcal{B}_2$. We may find $\mathbf{v}_A$ in either of two ways, each on wheel $\mathcal{B}_1$:

$$\mathbf{v}_A = \overset{0}{\cancel{\mathbf{v}_{\text{Ⓘ}}}} + \boldsymbol{\omega}_1 \times \mathbf{r}_{\text{Ⓘ}A}$$
$$= (-1.2\hat{\mathbf{k}}) \times (-0.28\hat{\mathbf{i}} + 0.2\hat{\mathbf{j}})$$
$$= 0.24\hat{\mathbf{i}} + 0.336\hat{\mathbf{j}} \text{ m/s}$$

$$\mathbf{v}_A = \mathbf{v}_C + \boldsymbol{\omega}_1 \times \mathbf{r}_{CA}$$
$$= 0.2(1.2)\hat{\mathbf{i}} + (-1.2\hat{\mathbf{k}}) \times (-0.28\hat{\mathbf{i}})$$
$$= 0.24\hat{\mathbf{i}} + 0.336\hat{\mathbf{j}} \text{ m/s}$$

We note for interest that when point A is beneath Ⓘ, its velocity is to the left — that is, it is going backwards!

Next, on $\mathcal{B}_2$,

$$\mathbf{v}_B = \mathbf{v}_A + \boldsymbol{\omega}_2 \times \mathbf{r}_{AB}$$

Point B is constrained to move horizontally; therefore

$$v_B\hat{\mathbf{i}} = 0.24\hat{\mathbf{i}} + 0.336\hat{\mathbf{j}} + \omega_2\hat{\mathbf{k}} \times (2.4\hat{\mathbf{i}} - 0.7\hat{\mathbf{j}})$$

or

$$v_B\hat{\mathbf{i}} = \hat{\mathbf{i}}(0.24 + 0.7\omega_2) + \hat{\mathbf{j}}(0.336 + 2.4\omega_2)$$

$\hat{\mathbf{j}}$ coefficients: $\quad 0 = 0.336 + 2.4\omega_2 \Rightarrow \omega_2 = -0.140$

so that

$$\boldsymbol{\omega}_2 = -0.140\hat{\mathbf{k}} \text{ rad/s or } 0.140 \circlearrowright \text{ rad/s}$$

$\hat{\mathbf{i}}$ coefficients: $\quad v_B = 0.24 + 0.7(-0.140) = 0.142$

so that

$$\mathbf{v}_B = 0.142\hat{\mathbf{i}} \text{ m/s}$$

The reader is encouraged to mentally locate the Ⓘ point for $\mathcal{B}_2$ and from it deduce that the directions of $\boldsymbol{\omega}_2$ and $\mathbf{v}_B$ are correct.

Example 3.16

In the preceding example, at the same instant, $\alpha_1 = 0.8 \circlearrowright \text{ rad/s}^2$. Find the acceleration of point B and the angular acceleration of rod $\mathcal{B}_2$.

Solution

As we did with the velocity of A, we can relate $\mathbf{a}_A$ to the acceleration of either Ⓘ or C:

$$\mathbf{a}_A = \mathbf{a}_{\text{Ⓘ}} + \boldsymbol{\alpha}_1 \times \mathbf{r}_{\text{Ⓘ}A} - \omega_1^2\mathbf{r}_{\text{Ⓘ}A}$$

The terms are

$$\mathbf{a}_{\text{Ⓘ}} = r\omega_1^2\uparrow = 0.2(1.2^2)\hat{\mathbf{j}}$$
$$= 0.288\hat{\mathbf{j}} \qquad \text{(note not zero!)}$$
$$\boldsymbol{\alpha}_1 \times \mathbf{r}_{\text{Ⓘ}A} = -0.8\hat{\mathbf{k}} \times (-0.28\hat{\mathbf{i}} + 0.2\hat{\mathbf{j}})$$
$$= 0.16\hat{\mathbf{i}} + 0.224\hat{\mathbf{j}}$$
$$-\omega_1^2\mathbf{r}_{\text{Ⓘ}A} = -1.2^2(-0.28\hat{\mathbf{i}} + 0.2\hat{\mathbf{j}})$$
$$= 0.403\hat{\mathbf{i}} - 0.288\hat{\mathbf{j}}$$

Adding the terms, we get

$$\mathbf{a}_A = 0.563\hat{\mathbf{i}} + 0.224\hat{\mathbf{j}} \text{ m/s}^2$$

$$\mathbf{a}_A = \mathbf{a}_C + \boldsymbol{\alpha}_1 \times \mathbf{r}_{CA} - \omega_1^2\mathbf{r}_{CA}$$

The terms are

$$\mathbf{a}_C = r\alpha_1\hat{\mathbf{i}} = 0.2(0.8)\hat{\mathbf{i}}$$
$$= 0.16\hat{\mathbf{i}}$$
$$\boldsymbol{\alpha}_1 \times \mathbf{r}_{CA} = (-0.8\hat{\mathbf{k}}) \times (-0.28\hat{\mathbf{i}})$$
$$= 0.224\hat{\mathbf{j}}$$
$$-\omega_1^2\mathbf{r}_{CA} = -1.2^2(-0.28\hat{\mathbf{i}})$$
$$= 0.403\hat{\mathbf{i}}$$

Again adding the terms, we get

$$\mathbf{a}_A = 0.563\hat{\mathbf{i}} + 0.224\hat{\mathbf{j}} \text{ m/s}^2$$

Now we relate $\mathbf{a}_A$ to $\mathbf{a}_B$ on the rod; note that the acceleration of B is constrained by the plane to be horizontal:

$$\begin{aligned}\mathbf{a}_B = a_B\hat{\mathbf{i}} &= \mathbf{a}_A + \boldsymbol{\alpha}_2 \times \mathbf{r}_{AB} - \omega_2^2\mathbf{r}_{AB} \\ &= 0.563\hat{\mathbf{i}} + 0.224\hat{\mathbf{j}} + \alpha_2\hat{\mathbf{k}} \times (2.4\hat{\mathbf{i}} - 0.7\hat{\mathbf{j}}) \\ &\quad - (-0.140)^2(2.4\hat{\mathbf{i}} - 0.7\hat{\mathbf{j}})\end{aligned}$$

$\hat{\mathbf{j}}$ coefficients: $\quad 0 = 0.224 + 2.4\alpha_2 + 0.0137 \Rightarrow \alpha_2 = -0.0991 \text{ rad/s}^2$

so that

$$\boldsymbol{\alpha}_2 = -0.0991\ \hat{\mathbf{k}} \text{ rad/s}^2$$

$\hat{\mathbf{i}}$ coefficients: $\quad a_B = 0.563 + 0.7(-0.0991) - 0.0470 = 0.447$

so that

$$\mathbf{a}_B = 0.447\hat{\mathbf{i}} \text{ m/s}^2$$

Rolling of a Wheel ($\mathcal{B}_1$) on a Fixed Plane Curve ($\mathcal{B}_2$)

In the second class of rolling problems to be considered in this section, the contact surface is curved. Let $\hat{\mathbf{e}}_t$ and $\hat{\mathbf{e}}_n$ be the principal unit tangent and normal vectors for the center point C of the wheel. (See Figure 3.19.) Then, since we are again defining a motion such that the contact point has zero velocity, Equation (3.8) yields

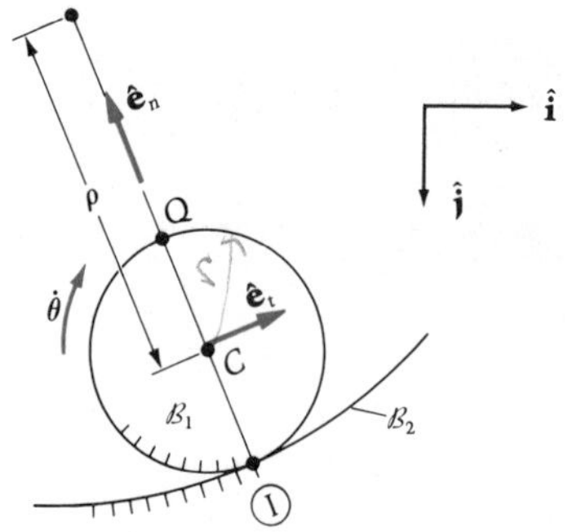

Figure 3.19 Wheel rolling on curved path concave upward.

$$\begin{aligned}\mathbf{v}_C &= \mathbf{v}_{Ⓘ} + \dot{\theta}\hat{\mathbf{k}} \times \mathbf{r}_{ⒾC} \\ &= \mathbf{0} + \dot{\theta}\hat{\mathbf{k}} \times r\hat{\mathbf{e}}_n \\ &= r\dot{\theta}\hat{\mathbf{e}}_t \end{aligned} \tag{3.27}$$

which gives us the velocity of C. Differentiating Equation (3.27), we get

$$\begin{aligned}\mathbf{a}_C &= r\ddot{\theta}\hat{\mathbf{e}}_t + r\dot{\theta}\dot{\hat{\mathbf{e}}}_t \\ &= r\ddot{\theta}\hat{\mathbf{e}}_t + r\dot{\theta}\frac{\dot{s}}{\rho}\hat{\mathbf{e}}_n \end{aligned} \tag{3.28}$$

in which we have used Equations (1.42) and (1.45), where ρ is the instantaneous radius of curvature of the path on which C moves. If this path is a circle, as is often the case, then ρ = constant = radius of the circle.

Using $\dot{s} = v_C = r\dot{\theta}$ in Equation (3.28), we have

$$\mathbf{a}_C = r\ddot{\theta}\hat{\mathbf{e}}_t + \frac{(r\dot{\theta})^2}{\rho}\hat{\mathbf{e}}_n \tag{3.29}$$

The acceleration of Ⓘ is interesting, and it follows from Equation (3.19):

$$\mathbf{a}_{Ⓘ} = \mathbf{a}_C + \ddot{\theta}\hat{\mathbf{k}} \times \mathbf{r}_{CⒾ} - \dot{\theta}^2\mathbf{r}_{CⒾ} \tag{3.30}$$

or

$$\mathbf{a}_{Ⓘ} = r\ddot{\theta}\hat{\mathbf{e}}_t + \frac{(r\dot{\theta})^2}{\rho}\hat{\mathbf{e}}_n - r\ddot{\theta}\hat{\mathbf{e}}_t + r\dot{\theta}^2\hat{\mathbf{e}}_n = \left(1 + \frac{r}{\rho}\right) r\dot{\theta}^2\hat{\mathbf{e}}_n \tag{3.31}$$

Comparing the accelerations of the contact points Ⓘ of Figures 3.18 and 3.19, we observe from our results (Equations 3.26 and 3.31) that the point contacting the curved track has a greater acceleration than the one touching the flat track, due to the r/ρ term of Equation (3.31). This term represents the normal component of acceleration of C, which was zero on the flat track.

It is also interesting to examine the acceleration of point Q at the top of the wheel in Figure 3.19:

$$\begin{aligned}\mathbf{a}_Q &= \mathbf{a}_C + \ddot{\theta}\hat{\mathbf{k}} \times \mathbf{r}_{CQ} - \dot{\theta}^2\mathbf{r}_{CQ} \\ &= \mathbf{r}\ddot{\theta}\hat{\mathbf{e}}_t + \frac{(r\dot{\theta})^2}{\rho}\hat{\mathbf{e}}_n + r\ddot{\theta}\hat{\mathbf{e}}_t - r\dot{\theta}^2\hat{\mathbf{e}}_n \\ &= 2r\ddot{\theta}\hat{\mathbf{e}}_t - r\dot{\theta}^2\left(1 - \frac{r}{\rho}\right)\hat{\mathbf{e}}_n \qquad (3.32)\end{aligned}$$

Note that it is possible for the $\hat{\mathbf{e}}_n$ (normal) component of $\mathbf{a}_Q$ to be either away from or toward C (or even zero), depending on whether $r > \rho$ or $r < \rho$ (or $r = \rho$), respectively (see Figure 3.20).

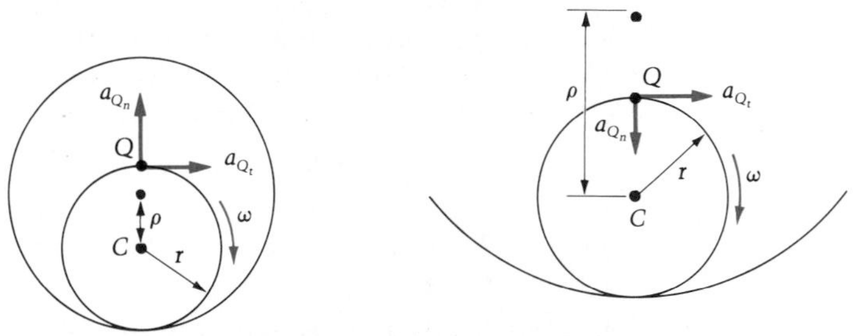

$r > \rho$; a_{Q_n} is ↑ $\qquad$ $r < \rho$; a_{Q_n} is ↓

Figure 3.20

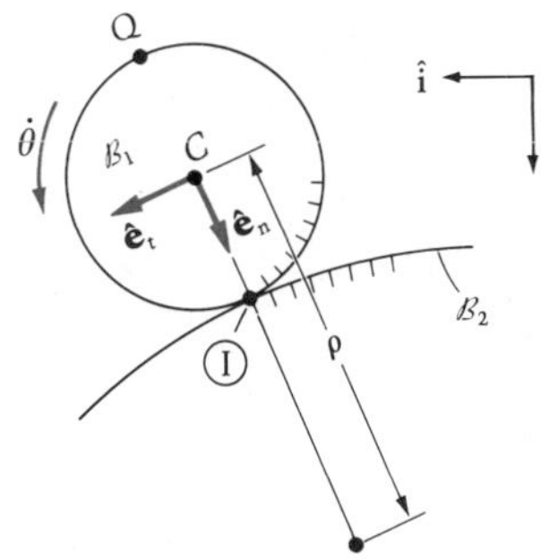

Figure 3.21

If the track is concave *downward* as shown in Figure 3.21, similar results may be obtained for the accelerations of C, Ⓘ, and Q. From Equation (3.29), which still holds:

$$\mathbf{a}_C = r\ddot{\theta}\hat{\mathbf{e}}_t + \frac{(r\dot{\theta})^2}{\rho}\hat{\mathbf{e}}_n \qquad (3.33)$$

After relating $\mathbf{a}_{Ⓘ}$ and $\mathbf{a}_C$ on $\mathcal{B}_1$, we obtain:

$$\mathbf{a}_{Ⓘ} = \left(-1 + \frac{r}{\rho}\right) r\dot{\theta}^2\hat{\mathbf{e}}_n \qquad (3.34)$$

And relating Q to either C or Ⓘ gives

$$\mathbf{a}_Q = 2r\ddot{\theta}\hat{\mathbf{e}}_t + \left(1 + \frac{r}{\rho}\right) r\dot{\theta}^2\hat{\mathbf{e}}_n \qquad (3.35)$$

Note that this time the radius r cannot exceed ρ, so $\mathbf{a}_{Ⓘ}$ is always outward; the normal component of the acceleration of Q is seen to be always inward in this case.

The reader may find it useful to remember the following form that doesn't depend upon any particular choice of coordinate system:

$$\mathbf{v}_C = \boldsymbol{\omega} \times \mathbf{r}_{\text{Ⓘ}C}$$

$$\mathbf{a}_C = \boldsymbol{\alpha} \times \mathbf{r}_{\text{Ⓘ}C} + \text{(normal part)}$$

If we knew either $\boldsymbol{\alpha}$ or the tangential component of $\mathbf{a}_C$, for example, we could get the other without concern for expressing the normal component of $\mathbf{a}_C$. We now consider several examples that feature a body rolling on a curved surface.

EXAMPLE 3.17

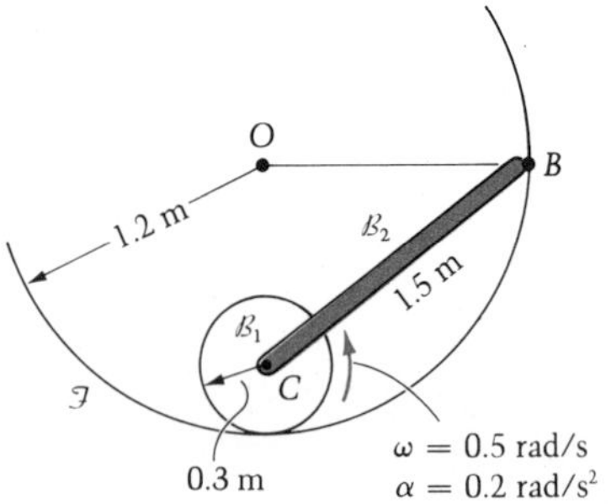

Figure E3.17a

The cylinder $\mathcal{B}_1$ shown in Figure E3.17a is rolling on the fixed, circular track with the indicated angular velocity and acceleration when $\mathcal{B}_1$ is at the bottom of the track. Rod $\mathcal{B}_2$ is pinned to the center C of $\mathcal{B}_1$, and its other end, B, slides on track $\mathcal{J}$. Find the velocity and acceleration of B.

Solution

As we have seen in Section 3.5, problems like this one have two parts. The "velocity part" must be solved before the "acceleration part" because the velocities as well as angular velocities are needed in the expressions for acceleration. We shall use instantaneous centers to get $\mathbf{v}_B$; the steps are:

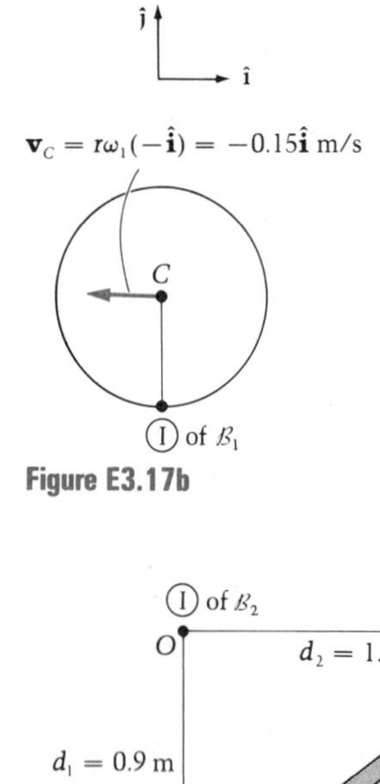

Figure E3.17b

1. The contact point of $\mathcal{B}_1$ is its Ⓘ point, since the body is rolling (see Figure E3.17b).
2. $\mathbf{v}_C$ is determined as in the diagram from $r_{\text{Ⓘ}C}\,\omega_1$.
3. The velocity of B is vertical (tangent to the path of the point).
4. Thus Ⓘ of $\mathcal{B}_2$ is at the intersection of the normals to $\mathbf{v}_C$ and $\mathbf{v}_B$, namely at point O (see Figure E3.17c).
5. Then $\omega_2 = v_C/d_1 = 0.15/0.9 = 0.167$ or $\boldsymbol{\omega}_2 = 0.167 \circlearrowleft$ rad/s
6. Finally, $v_B = d_2\omega_2 = 1.2(0.167) = 0.200$ or $\mathbf{v}_B = 0.200 \downarrow$ m/s

Next, to find $\mathbf{a}_B$ we shall relate it to the acceleration of C, which is, from Equation (3.29),

$$\mathbf{a}_C = r\alpha(-\hat{\mathbf{i}}) + \frac{(r\omega)^2}{\rho}\hat{\mathbf{j}}$$

$$= 0.3(0.2)(-\hat{\mathbf{i}}) + \frac{(0.3 \times 0.5)^2}{0.9}\hat{\mathbf{j}}$$

$$= -0.06\hat{\mathbf{i}} + 0.0250\hat{\mathbf{j}} \text{ m/s}^2$$

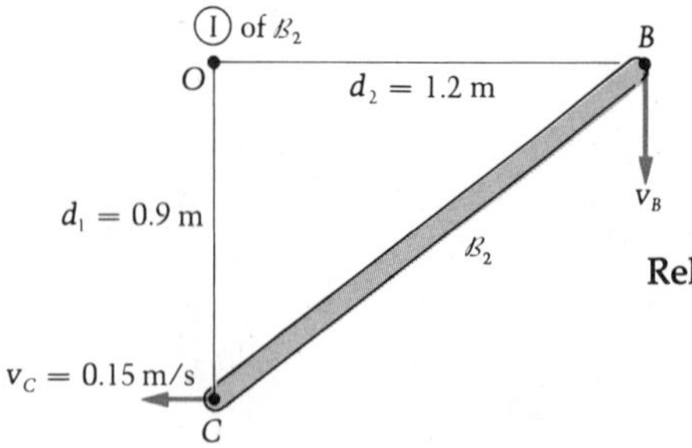

Figure E3.17c

Relating $\mathbf{a}_C$ to $\mathbf{a}_B$, we get, using Figure E3.17c:

$$\mathbf{a}_B = \mathbf{a}_C + \alpha_2\hat{\mathbf{k}} \times \mathbf{r}_{CB} - \omega_2^2\mathbf{r}_{CB}$$

$$\ddot{s}_B(-\hat{\mathbf{j}}) + \frac{\dot{s}_B^2}{\rho_B}(-\hat{\mathbf{i}}) = -0.06\hat{\mathbf{i}} + 0.0250\hat{\mathbf{j}} + \alpha_2\hat{\mathbf{k}} \times (1.2\hat{\mathbf{i}} + 0.9\hat{\mathbf{j}})$$

$$-(0.167)^2(1.2\hat{\mathbf{i}} + 0.9\hat{\mathbf{j}})$$

In substituting for $\mathbf{a}_B$ we have used the facts that for B we have $\hat{\mathbf{e}}_t = -\hat{\mathbf{j}}$ (direction vector of $\mathbf{v}_B$) and $\hat{\mathbf{e}}_n = -\hat{\mathbf{i}}$ (always toward the center of curvature of the path of the point). Substituting $\dot{s}_B = |\mathbf{v}_B| = 0.2$ m/s and $\rho_B = 1.2$ m, we then have two scalar equations in the two unknowns $\ddot{s}_B$ and α_2:

$\hat{\mathbf{i}}$ coefficients:

$$-0.0333 = -0.06 - 0.9\alpha_2 - 0.0333$$
$$\alpha_2 = -0.0667 \text{ rad/s}^2$$

$\hat{\mathbf{j}}$ coefficients:

$$-\ddot{s}_B = 0.0250 + 1.2\alpha_2 - 0.0250 = -0.0800$$
$$\ddot{s}_B = 0.0800 \text{ m/s}^2$$

Therefore

$$\mathbf{a}_B = -0.0333\hat{\mathbf{i}} - 0.0800\hat{\mathbf{j}} \text{ m/s}^2$$

We wish to make a very important point regarding the preceding example. You may have noticed that the values of α_2 and $\ddot{s}_B$ could have been obtained more quickly from

$$|\alpha_2| = \frac{a_C}{r_{OC}} = \frac{0.06}{0.9} = 0.0667 \text{ rad/s}^2$$

and

$$\ddot{s}_B = r_{OB}|\alpha_2| = 1.2(0.0667) = 0.08 \text{ m/s}^2$$

This shortcut is very dangerous because it is not always valid! It is essential to understand when and why this procedure works. (It is *not* because O is the instantaneous center of zero velocity of $\mathcal{B}_2$!) For a counterexample, consider the results of Examples 3.3, 3.7, and 3.11:

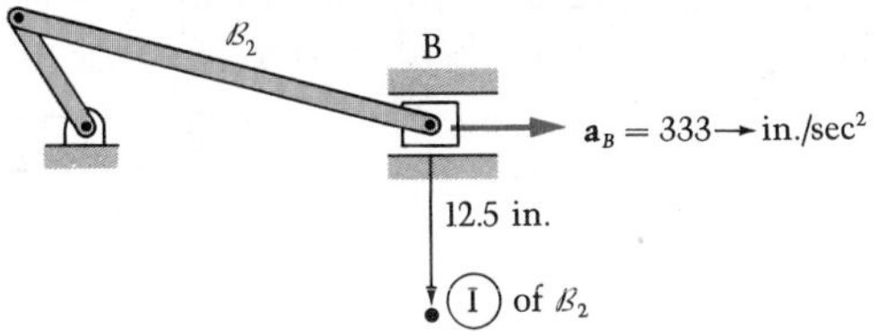

Figure 3.22

If we were to divide $|\mathbf{a}_B|$ by $|\mathbf{r}_{\text{(I)}B}|$ (see Figure 3.22), we would erroneously obtain 26.6 rad/sec² with a direction indicator ↻. But $\boldsymbol{\alpha}_2$ is 31.6 ↺ rad/sec²! The answer to the question of when the procedure is legitimate is covered by the following text question:

Question 3.15 When can we use $r_{\text{(I)}B}\alpha$ to obtain the correct acceleration component of B normal to line (I)B?

Answer 3.15 From Figure 3.13 we can see exactly when the component of $\mathbf{a}_B$ normal to line AB is given by $r_{AB}\alpha$: It is when $\mathbf{a}_A$ has no component normal to AB. (This result does not require that A be the point (I), incidentally.)

EXAMPLE 3.18

The velocity magnitude of G in Figure E3.18a is $v_G = t^2$ ft/sec, and G moves on the 8-ft circle in a clockwise direction. The position shown is at $t = 2$ sec. Find the acceleration of B at this instant.

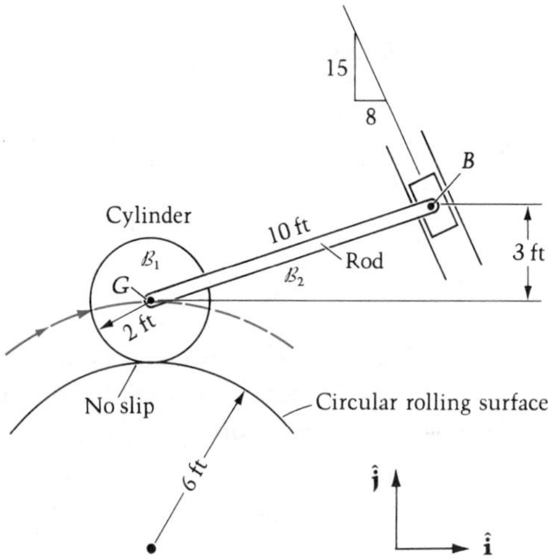

Figure E3.18a

Solution

We shall relate $\mathbf{a}_B$ to $\mathbf{a}_G$ on bar $\mathcal{B}_2$:

$$\mathbf{a}_B = \mathbf{a}_G + \alpha_2\hat{\mathbf{k}} \times \mathbf{r}_{GB} - \omega_2^2\mathbf{r}_{GB} \qquad (1)$$

First, we determine $\mathbf{a}_G$. Since $v_G = \dot{s}_G = t^2$ in this problem, we have $\ddot{s}_G = 2t = 2(2) = 4$ ft/sec². Therefore

$$\mathbf{a}_G = \ddot{s}_G\hat{\mathbf{e}}_t + \frac{\dot{s}_G^2}{\rho_G}\hat{\mathbf{e}}_n = 2t\hat{\mathbf{e}}_t + \frac{t^4}{8}\hat{\mathbf{e}}_n$$

$$= 4\hat{\mathbf{i}} - 2\hat{\mathbf{j}} \text{ ft/sec}^2 \qquad (\text{at } t = 2 \text{ sec})$$

From Equation (1)

$$\mathbf{a}_B = a_B\left(\frac{8\hat{\mathbf{i}} - 15\hat{\mathbf{j}}}{17}\right) = \overset{(4\hat{\mathbf{i}} - 2\hat{\mathbf{j}})}{\mathbf{a}_G} + \alpha_2\hat{\mathbf{k}} \times \overset{(\sqrt{91}\,\hat{\mathbf{i}} + 3\hat{\mathbf{j}})}{\mathbf{r}_{GB}} - \omega_2^2\mathbf{r}_{GB}$$

in which the acceleration of B is an unknown magnitude in a known direction as signified by the unit vector $(8\hat{\mathbf{i}} - 15\hat{\mathbf{j}})/17$ down the slot.

> **Question 3.16** If the direction of $\mathbf{a}_B$ turns out to be up the slot, will the solution be valid?

Answer 3.16 Yes; this will be manifested by a_B turning out negative. Note that (negative a_B) $\cdot$ $[(8\hat{\mathbf{i}} - 15\hat{\mathbf{j}})/17]$ is the same as (positive a_B) $\cdot$ $[(-8\hat{\mathbf{i}} + 15\hat{\mathbf{j}})/17]$.

We have two equations ($\hat{\mathbf{i}}$ and $\hat{\mathbf{j}}$ coefficients) in the three unknowns a_B, α_2, and ω_2, but the angular speed ω_2 may be found from the instantaneous center of $\mathcal{B}_2$. From the geometry and similar triangles, we use Figure E3.18b to obtain:

$$\frac{\sqrt{91}}{H} = \frac{15}{8} \Rightarrow H = 5.09 \text{ ft}$$

$$D = H - 3 = 2.09 \text{ ft}$$

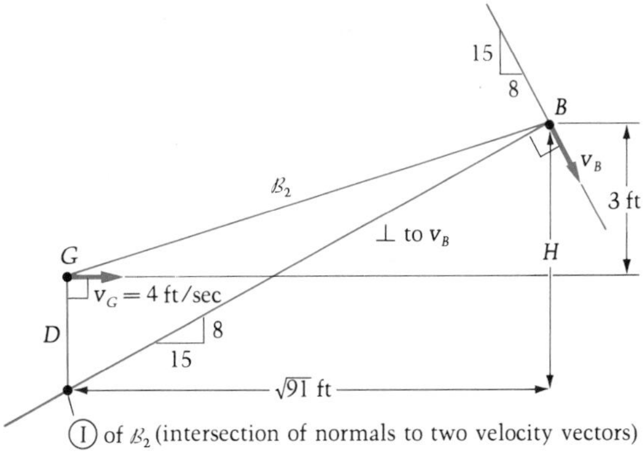

Figure E3.18b

Thus

$$\omega_2 = \frac{v_G}{D} = \frac{4}{2.09} = 1.91 \text{ rad/sec}$$

$$\omega_2 = 1.91 \circlearrowright \text{ rad/sec}$$

Note that until Ⓘ of $\mathcal{B}_2$ is located, we do not know whether $\mathbf{v}_B$ is up or down the slot; however, the *normal* to $\mathbf{v}_B$ is the same line in either case. Once Ⓘ is established, $\mathbf{v}_G$ to the right gives ω_2 clockwise — and then we know that $\mathbf{v}_B$ is *down* the slot. Substituting, we get

$$a_B\left(\frac{8\hat{\mathbf{i}} - 15\hat{\mathbf{j}}}{17}\right) = 4\hat{\mathbf{i}} - 2\hat{\mathbf{j}} + \sqrt{91}\alpha_2\hat{\mathbf{j}} - 3\alpha_2\hat{\mathbf{i}} - 1.91^2(\sqrt{91}\,\hat{\mathbf{i}} + 3\hat{\mathbf{j}})$$

$\hat{\mathbf{i}}$ coefficients: $\quad \frac{8}{17}a_B = 4 - 3\alpha_2 - 1.91^2\sqrt{91}$

$\hat{\mathbf{j}}$ coefficients: $\quad \frac{-15}{17}a_B = -2 + \sqrt{91}\,\alpha_2 - 1.91^2(3)$

Eliminating α_2 gives $a_B = -181$, so that

$$\mathbf{a}_B = 181 \text{ (slope 15/8)} \quad \text{or } (-85.2\hat{\mathbf{i}} + 160\hat{\mathbf{j}}) \text{ ft/sec}^2$$

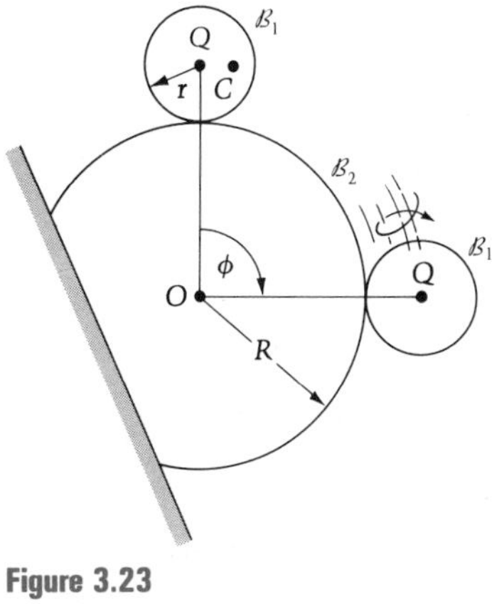

Figure 3.23

The final example in this part of Section 3.6 will be very helpful to us later in situations such as the one shown in Figure 3.23. Suppose we need to know the location of point C of cylinder $\mathcal{B}_1$ after $\mathcal{B}_1$ has rolled on $\mathcal{B}_2$ to the lower position.* The problem is to find the angle θ turned through by $\mathcal{B}_1$ for a given rotation ϕ of the line OQ. The procedure to be followed in solving this problem is illustrated by the next example.

EXAMPLE 3.19

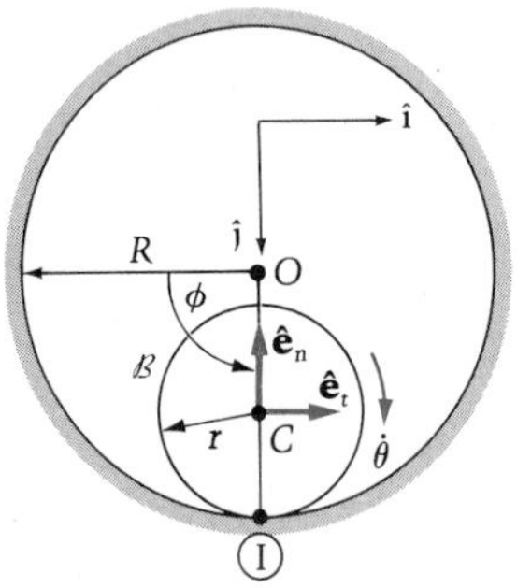

Figure E3.19

Find the relationship between the angle ϕ (locating the line OC in Figure E3.19) and the angle of rotation θ of the rolling cylinder $\mathcal{B}$.

Solution

Treating C as a point whose path is a known circle, we get

$$\mathbf{v}_C = \dot{s}\hat{\mathbf{e}}_t = (R - r)\dot{\phi}\hat{\mathbf{i}}$$

Alternatively, we may also treat C as a point on the cylinder with instantaneous center at (I):

$$\mathbf{v}_C = \omega\hat{\mathbf{k}} \times \mathbf{r}_{\text{(I)}C} = \dot{\theta}\hat{\mathbf{k}} \times (-r\hat{\mathbf{j}}) = r\dot{\theta}\hat{\mathbf{i}}$$

Thus, equating the two expressions for $\mathbf{v}_C$, we have

$$(R - r)\dot{\phi} = r\dot{\theta}$$

Integrating, and letting $\theta = 0$ when $\phi = 0$, we get

$$(R - r)\phi = r\theta + \overset{0}{\cancel{C_1}}$$

or

$$\theta = \left(\frac{R - r}{r}\right)\phi \tag{3.36}$$

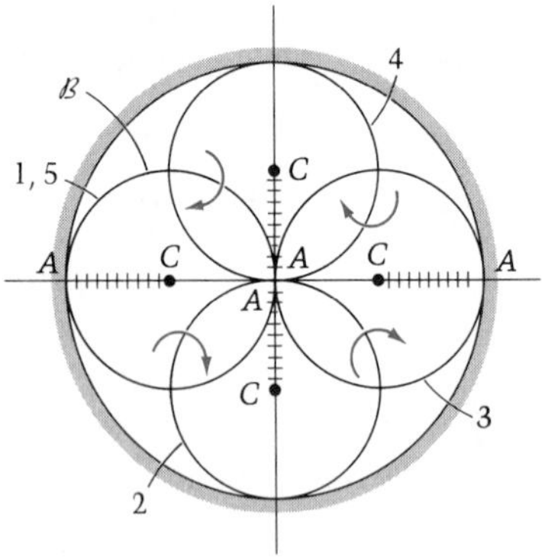

Figure 3.24

From Example 3.19, if we let $R = 2r$, then we see (Figure 3.24) that $\theta \equiv \phi$. Even though the circumferences of cylinder and track are $2\pi r$ and $4\pi r (= 2\pi R)$,† the curvature forces the angular velocities of the line OC and the cylinder to be the same. If the outer track were *straight* and of length $4\pi r$, the cylinder would turn *two* revolutions in space instead of just one in traversing it.

It is seen that the line AC of $\mathcal{B}$ (and hence $\mathcal{B}$ itself) revolves once as C completes its circular path for the case $R = 2r$. When $R > 2r$, then $\theta > \phi$. Such a case is shown in Figure 3.25. If we now let $R = 7r$, then Equation (3.36) gives $\theta = 6\phi$ and $\mathcal{B}$ now revolves once in space for each 60° of turn of line OC.

If the track were *convex*, then for $R = 2r$:

$$\mathbf{v}_C = 3r\dot{\phi}\hat{\mathbf{e}}_t \qquad \text{and} \qquad \mathbf{v}_C = r\dot{\theta}\hat{\mathbf{e}}_t \Rightarrow 3\phi = \theta$$

* This need will arise, for example, in kinetics problems in which we seek the work done by gravity on $\mathcal{B}_1$ if its mass center is offset from its geometric center.

† Which at first glance might lead one to believe that the cylinder would revolve twice per revolution of OC.

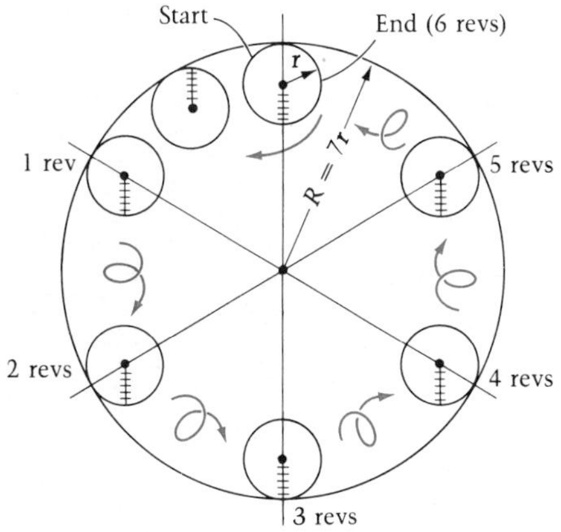

Figure 3.25

and the wheel would turn in space *three times* as fast and as far as line OC (see Figure 3.26).

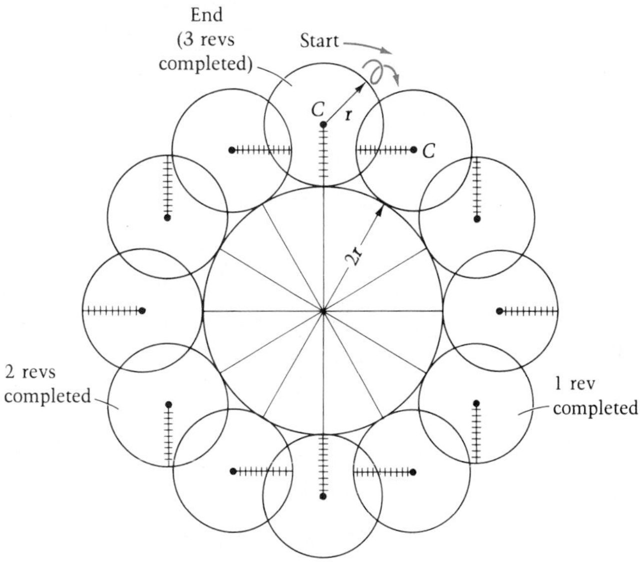

Figure 3.26

Gears

The final class of rolling problems is concerned with gears. Gears are used to transmit power. The teeth of the gears are cut so that they will give constant speed to the driven gear when the driving gear is itself turning at constant angular speed.

However, gears violate the rolling condition; there is necessarily some sliding since the contacting points do not have equal velocities (except at $\theta = 0$), as can be seen in Figure 3.27 for spur gears; nonetheless, the teeth are cut so that we may correctly treat the gears for dynamic

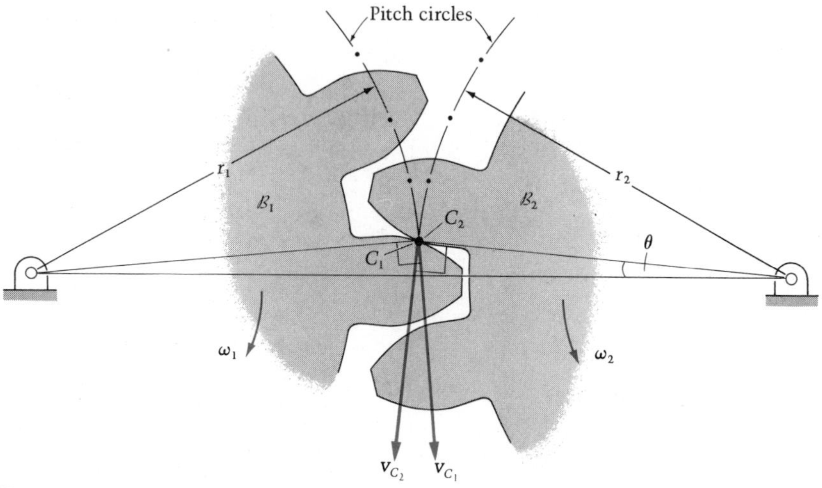

Figure 3.27

purposes as if they were two cylinders rolling on each other at the pitch circles. Thus when the centers are pinned as in Figure 3.27, we may use the relation

$$r_1\omega_1 = r_2\omega_2 \tag{3.37}$$

where r_1 and r_2 are the respective pitch radii of the gears $\mathcal{B}_1$ and $\mathcal{B}_2$. We may also use the derivative of Equation 3.37 (since it is valid for all t):

$$r_1\alpha_1 = r_2\alpha_2 \tag{3.38}$$

We note that the radius ratio is inversely proportional to the ratio of angular speeds (and directly proportional to the ratio of numbers of gear teeth, since the shape and spacing of teeth must match). We now consider several examples.

EXAMPLE 3.20

Find the angular speed of the front sprocket (rigidly fixed to the pedal crank) of the bicycle in Figure E3.20a if the man is traveling at 10 mph. There are 26 teeth on the front sprocket and 9 on the rear sprocket (which turns rigidly with the rear wheel). The wheel diameters are 26 in.

Figure E3.20a

Solution

The velocities of A and B, the two ends of the straight upper length of chain, are equal. (As shown in Figure E3.20b, A is just leaving the rear sprocket $\mathcal{B}_1$; B is just about to enter the front sprocket $\mathcal{B}_2$.) To prove this, we note that the translating section AB of chain is behaving as if rigid, so that, calling this "body" $\mathcal{B}_3$,

$$\mathbf{v}_B = \mathbf{v}_A + \boldsymbol{\omega}_3 \times \mathbf{r}_{AB}$$

But $\boldsymbol{\omega}_3 = \mathbf{0}$, so that

$$\mathbf{v}_B = \mathbf{v}_A$$

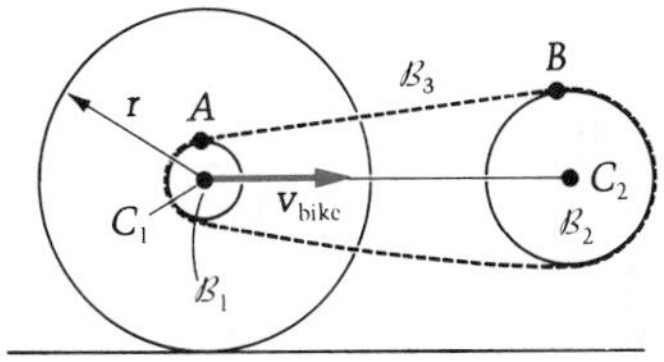

Figure E3.20b

Next we relate the equal velocities of A and B to the respective centers of their sprockets $\mathcal{B}_1$ and $\mathcal{B}_2$:

$$\mathbf{v}_{C_1} + \boldsymbol{\omega}_1 \times \mathbf{r}_{C_1A} = \mathbf{v}_{C_2} + \boldsymbol{\omega}_2 \times \mathbf{r}_{C_2B}$$

Now the velocities of C_1 and C_2 are each equal to the "velocity of the bike," meaning the common velocity of all the points on the translating part of the bike, such as points of the frame and seat. Therefore $\mathbf{v}_{C_1}$ and $\mathbf{v}_{C_2}$ cancel, leaving

$$\boldsymbol{\omega}_1 \times \mathbf{r}_{C_1A} = \boldsymbol{\omega}_2 \times \mathbf{r}_{C_2B}$$

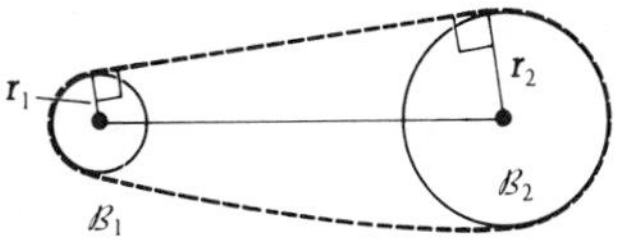

Figure E3.20c

This says simply that (see Figure E3.20c):

$$r_1\omega_1 = r_2\omega_2$$

Now if the speed of the bike is to be 10 mph, we have

$$v_{C_1} = \overset{13\text{ in.}}{\cancel{r}}\omega_1 = 10\text{ mph}\left(\frac{88\text{ ft/sec}}{60\text{ mph}}\right)\frac{12\text{ in.}}{1\text{ ft}} = 176\text{ in./sec}$$

or

$$\omega_1 = 13.5 \circlearrowright \text{rad/sec}$$

Thus

$$\omega_2 = \frac{r_1}{r_2}\,\omega_1 = \frac{9}{26}\,(13.5) \qquad \text{(Radii are proportional to number of teeth!)}$$

$$= 4.67\text{ rad/sec}$$

So the rider must turn the pedal crank at $4.67/2\pi = 0.743$ rev/sec.

Example 3.21

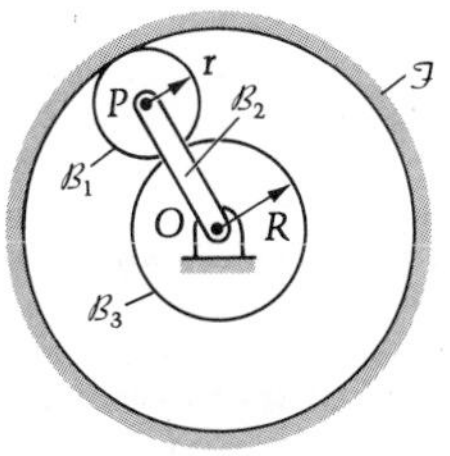

Figure E3.21a

Frame $\mathcal{I}$ is a fixed ring gear with internal teeth (not shown in Figure E3.21a) that mesh with those of the planetary gear $\mathcal{B}_1$. The teeth of $\mathcal{B}_1$ also mesh with those of the sun gear $\mathcal{B}_3$, which is pinned at its center point O to frame $\mathcal{I}$. The crank arm $\mathcal{B}_2$ is pinned at its ends to O and to the center point P of $\mathcal{B}_1$. The arm $\mathcal{B}_2$ has angular speed $\omega_2(t)$ counterclockwise. Find the angular velocity of $\mathcal{B}_3$ in terms of R, r, and ω_2.

Solution

We take $\mathcal{I}$ to be our reference frame, to which all motions are referred. We work first with the crank $\mathcal{B}_2$, since we know its angular velocity and the velocity of one of its points ($\mathbf{v}_O = \mathbf{0}$). From the sketch of $\mathcal{B}_2$ (Figure E3.21b), we see that we can write

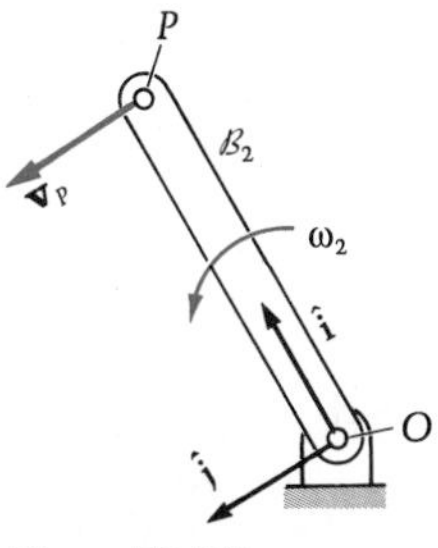

Figure E3.21b

$$\mathbf{v}_P = \overset{0}{\cancel{\mathbf{v}_O}} + \omega_2\hat{\mathbf{k}} \times (R + r)\hat{\mathbf{i}}$$
$$= (R + r)\omega_2\hat{\mathbf{j}} \qquad (1)$$

(Note that we align $\hat{\mathbf{i}}$ parallel to OP for convenience; we need not always draw it to the right.)

Next, the points of $\mathcal{B}_1$ and $\mathcal{B}_2$ that are pinned together at P have the same velocity at all times. Furthermore, the points of $\mathcal{I}$ and $\mathcal{B}_1$ at D (see Figure E3.21c)

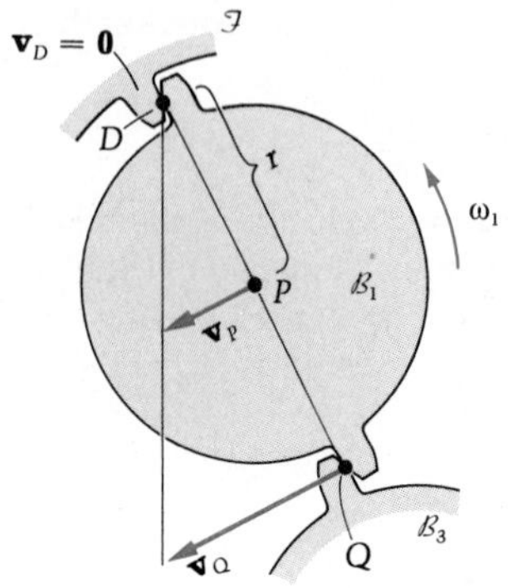

Figure E3.21c

are in contact and each has zero velocity* since D is fixed in $\mathcal{I}$. Thus

$$\mathbf{v}_P = \overset{0}{\cancel{\mathbf{v}_D}} + \omega_1 \hat{\mathbf{k}} \times (-r\hat{\mathbf{i}})$$

or

$$(R + r)\omega_2 \hat{\mathbf{j}} = -r\omega_1 \hat{\mathbf{j}}$$

Solving for ω_1 gives

$$\omega_1 = \frac{-(R + r)\omega_2}{r} \quad \text{or} \quad \omega_1 = \frac{R + r}{r}\omega_2 \circlearrowright \tag{2}$$

We are now in a position to obtain the velocity of point Q of $\mathcal{B}_1$, the point in contact with the tooth of $\mathcal{B}_3$:

$$\mathbf{v}_Q = \overset{0}{\cancel{\mathbf{v}_D}} + \omega_1 \hat{\mathbf{k}} \times (-2r\hat{\mathbf{i}})$$

Substituting for ω_1 in terms of ω_2 from Equation (2), we get

$$\mathbf{v}_Q = 2(R + r)\omega_2 \hat{\mathbf{j}} \tag{3}$$

Note that Q has twice the speed of P since it is twice as far from the instantaneous center D of $\mathcal{B}_1$ as is P.

Finally, we come to the body $\mathcal{B}_3$ of interest (see Figure E3.21d). Knowing that the points Q and Q' (the respective tooth points in contact on $\mathcal{B}_1$ and $\mathcal{B}_3$) have equal velocities as they move together tangent to the pitch circle, we obtain

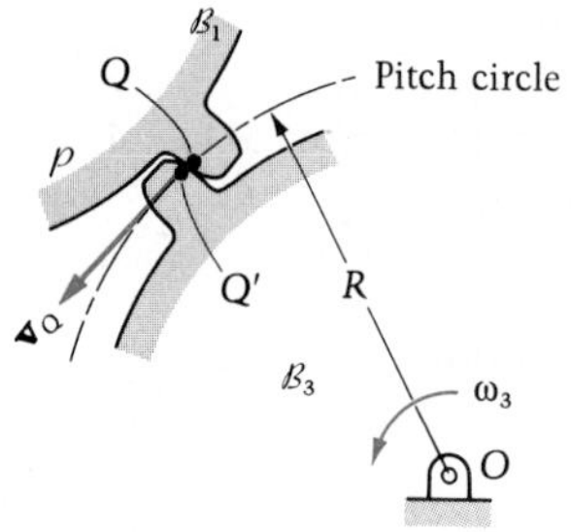

Figure E3.21d

$$\mathbf{v}_{Q'} = \mathbf{v}_O + \omega_3 \hat{\mathbf{k}} \times R\hat{\mathbf{i}}$$

$$2(R + r)\omega_2 \hat{\mathbf{j}} = \mathbf{0} + R\omega_3 \hat{\mathbf{j}}$$

Thus the angular speed of $\mathcal{B}_3$ in $\mathcal{I}$ is

$$\omega_3 = \frac{2(R + r)\omega_2}{R} \tag{4}$$

and the angular velocity of $\mathcal{B}_3$ in $\mathcal{I}$ is

$$\omega_3 \hat{\mathbf{k}} = \frac{2(R + r)\omega_2}{R} \hat{\mathbf{k}}$$

Note in the previous example that the point of $\mathcal{B}_2$ passing over point Q has velocity $R\omega_2$, which is of course less than the velocity $\mathbf{v}_Q$ of the gear teeth in contact at Q. This velocity is $R\omega_3 = 2(R + r)\omega_2$, which is more than twice as fast as $R\omega_2$.

We also remark that since the answers

$$\omega_1(t) = \frac{R + r}{r}\omega_2(t) \quad \text{and} \quad \omega_3(t) = \frac{2(R + r)}{R}\omega_2(t)$$

---

* As we have pointed out, the contact points of gear teeth necessarily slide relative to each other. The points used in the analysis are actually not tooth points, however, but imaginary points on the pitch circles of the gears. Furthermore, the radii given in the examples and problems are the radii of these circles.

are completely general functions of time, the angular accelerations are obtainable immediately by differentiation, with $\alpha_2 = \dot{\omega}_2$:

$$\alpha_1(t) = \frac{R + r}{r}\,\alpha_2(t) \qquad \text{and} \qquad \alpha_3(t) = \frac{2(R + r)}{R}\,\alpha_2(t)$$

Rather than differentiating, however, we shall obtain these two results in the following example in another manner: by repeated use of Equation (3.19). The purpose is to gain insight into its use in gearing situations involving several bodies. The procedure in the next example would have to be followed if the previous example had been worked using instantaneous values instead of generally (with symbols).

Example 3.22

Find the angular accelerations α_1 and α_3 of the planetary and sun gears in the previous example in terms of R, r, and ω_2 and α_2, which are given functions of the time t.

Solution

Relating the accelerations of P and O on body $\mathcal{B}_2$ gives (see the bar in Figure E3.22a):

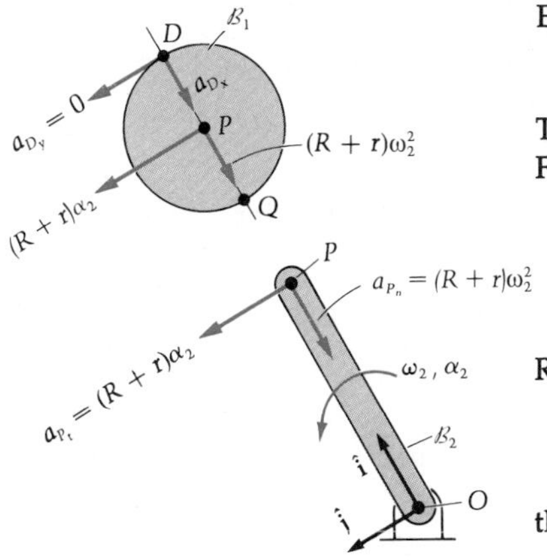

Figure E3.22a

$$\mathbf{a}_P = -(R + r)\omega_2^2\hat{\mathbf{i}} + (R + r)\alpha_2\hat{\mathbf{j}}$$

This acceleration is then carried over to the coincident point P on $\mathcal{B}_1$ (again see Figure E3.22a). Relating D and P on the planetary gear $\mathcal{B}_1$, we have

$$\begin{aligned}\mathbf{a}_D = a_D\hat{\mathbf{i}} &= \mathbf{a}_P + \alpha_1\hat{\mathbf{k}} \times \overset{r\hat{\mathbf{i}}}{\mathbf{r}_{PD}} - \omega_1^2\,\overset{r\hat{\mathbf{i}}}{\mathbf{r}_{PD}} \\ &= -(R + r)\omega_2^2\hat{\mathbf{i}} + (R + r)\alpha_2\hat{\mathbf{j}} + \alpha_1\hat{\mathbf{k}} \times r\hat{\mathbf{i}} - \omega_1^2 r\hat{\mathbf{i}}\end{aligned}$$

Recalling that

$$\omega_1 = \frac{R + r}{r}\,\omega_2$$

the coefficients of $\hat{\mathbf{i}}$ then give

$$a_D = \frac{-(R + 2r)(R + r)}{r}\,\omega_2^2$$

and the $\hat{\mathbf{j}}$ coefficients yield

$$\alpha_1 = \frac{-(R + r)\alpha_2}{r} \qquad \text{or} \qquad \boldsymbol{\alpha}_1 = \frac{R + r}{r}\,\alpha_2 \circlearrowleft$$

We now need $\mathbf{a}_Q$, where Q is again the tooth point of $\mathcal{B}_1$ in contact with the sun gear $\mathcal{B}_3$:

$$\begin{aligned}\mathbf{a}_Q &= \mathbf{a}_D + \alpha_1\hat{\mathbf{k}} \times \overset{-2r\hat{\mathbf{i}}}{\mathbf{r}_{DQ}} - \omega_1^2\mathbf{r}_{DQ} \\ &= \frac{-(R + 2r)(R + r)}{r}\,\omega_2^2\hat{\mathbf{i}} + 2(R + r)\alpha_2\hat{\mathbf{j}} + \frac{2r(R + r)^2}{r^2}\,\omega_2^2\hat{\mathbf{i}} \\ &= \frac{(R + r)R}{r}\,\omega_2^2\hat{\mathbf{i}} + 2(R + r)\alpha_2\hat{\mathbf{j}}\end{aligned}$$

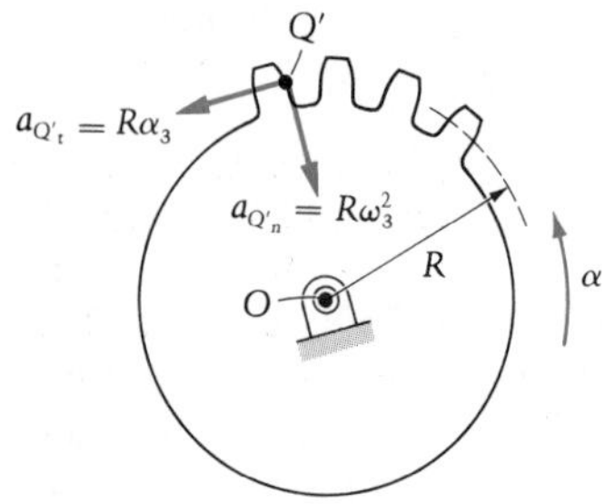

Figure E3.22b

We now go to body $\mathcal{B}_3$ to complete the solution. Relating the tooth point Q' to O on $\mathcal{B}_3$ gives the components of $a_{Q'}$. (see Figure E3.22b). The tangential acceleration components of Q and Q' are equal* as the teeth contact and move together:

$$a_{Q'_t} = a_{Q'_y} = 2(R + r)\alpha_2 = R\alpha_3$$

Thus

$$\alpha_3 = \frac{2(R + r)}{R}\alpha_2 \circlearrowleft$$

Note that we may express the normal acceleration component of Q' in terms of ω_2 by using the result for ω_3 from Example 3.21:

$$a_{Q'_n} = -a_{Q'_x} = +R\omega_3^2 = +R\left[\frac{2(R + r)}{R}\omega_2\right]^2$$
$$= \frac{4(R + r)^2\omega_2^2}{R}$$

We also note that we could have alternatively obtained the accelerations of D and Q as points on rims of wheels rolling on curved tracks by using Equations (3.31) and (3.32). Noting that ρ for P is $(R + r)$, we present these partial checks on our solution:

$$\mathbf{a}_D = \left(1 + \frac{r}{\rho}\right) r\dot{\theta}^2\hat{\mathbf{e}}_n = \left(1 + \frac{r}{R + r}\right) r\omega_1^2(-\hat{\mathbf{i}})$$
$$= -\left(\frac{R + 2r}{R + r}\right) r\omega_1^2\hat{\mathbf{i}} = -\left(\frac{R + 2r}{R + r}\right) r\left(\frac{R + r}{r}\omega_2\right)^2\hat{\mathbf{i}}$$
$$= \frac{-(R + 2r)(R + r)}{r}\omega_2^2\hat{\mathbf{i}} \qquad \text{(as above)}$$

$$\mathbf{a}_Q = 2r\ddot{\theta}\hat{\mathbf{e}}_t - r\dot{\theta}^2\left(1 - \frac{r}{\rho}\right)\hat{\mathbf{e}}_n$$
$$= 2r(-\alpha_1)(+\hat{\mathbf{j}}) - r\omega_1^2\left(1 - \frac{r}{R + r}\right)(-\hat{\mathbf{i}})$$
$$= \frac{Rr}{R + r}\omega_1^2\hat{\mathbf{i}} - 2r\alpha_1\hat{\mathbf{j}}$$
$$= \frac{Rr}{R + r}\left(\frac{R + r}{r}\omega_2\right)^2\hat{\mathbf{i}} - 2r\left[\frac{-(R + r)}{r}\alpha_2\right]\hat{\mathbf{j}}$$
$$= \frac{(R + r)R}{r}\omega_2^2\hat{\mathbf{i}} + 2(R + r)\alpha_2\hat{\mathbf{j}} \qquad \text{(as above)}$$

* This is in fact true even when *neither* body's center is fixed and the geometry is irregular. As long as there is rolling, the acceleration components of the contacting points *in the plane tangent to the two bodies* are equal in plane motion at all times. See "Contact Point Accelerations in Rolling Problems," D. J. McGill, *Mechanics Research Communications,* 7(3), 175–179, 1980.

PROBLEMS ▶ Section 3.6

3.84 The wheel in Figure P3.84 rolls on the plane with constant angular velocity 1⟳ rad/sec. Find the velocity of point Q by using the instantaneous center Ⓘ of zero velocity. Then check by using Equation (3.8) to relate $\mathbf{v}_Q$ to $\mathbf{v}_P$.

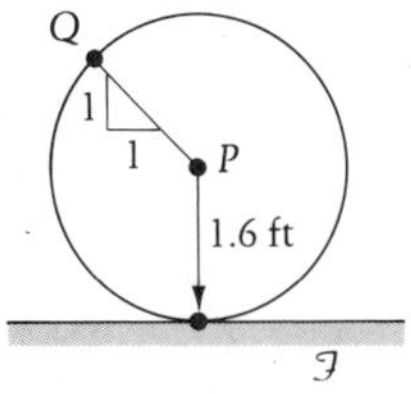

Figure P3.84

3.85 In the preceding problem, suppose that the plane ℐ on which the wheel rolls is not fixed to the reference frame but instead translates on it (this time the reference frame is 𝒢) at constant velocity 3 ft/sec to the left. (See Figure P3.85.)

a. Find the instantaneous center of zero velocity Ⓘ

b. Find $\mathbf{v}_Q$ again.

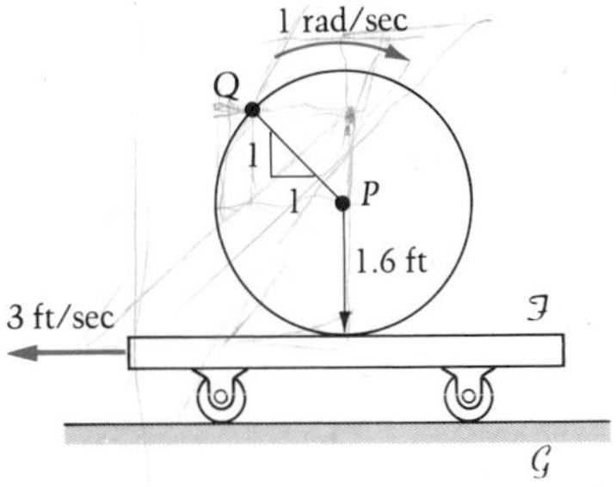

Figure P3.85

3.86 The wheel in Figure P3.86 rolls on the bar. If at a certain instant the bar has a velocity of 2 m/s to the right and the wheel has counterclockwise angular velocity of 0.5 rad/s, determine the velocity of (a) the center of the wheel and (b) point P.

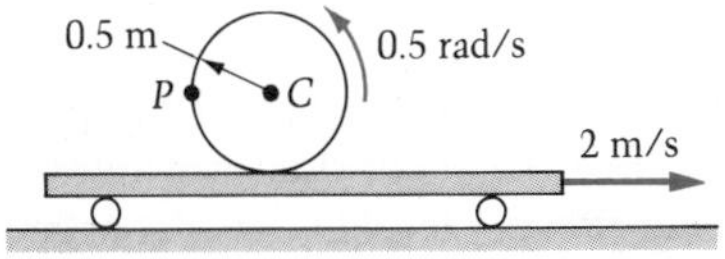

Figure P3.86

3.87 Figure P3.87(a) shows the manner in which a train wheel rests on the track. If the train travels at a constant speed of 80 mph and does not slip on the track, determine the velocities of points A, B, D, and E on the vertical line through the center C in Figure P3.87(b). Which point is traveling backward? Why?

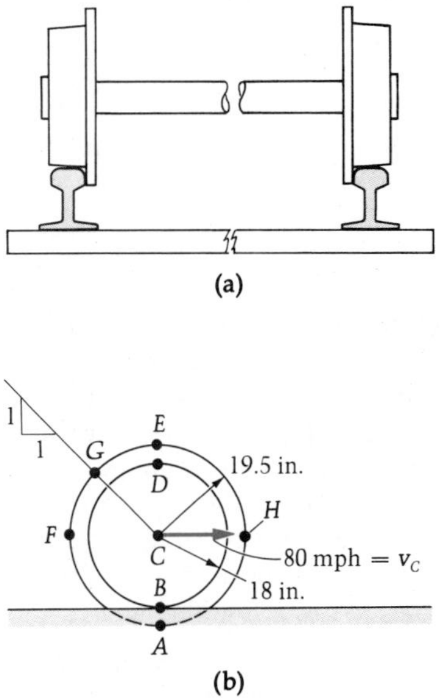

Figure P3.87

3.88 In the preceding problem find the velocities of points F, G, and H.

3.89 Find the velocities of points B and C in Figure P3.89 if the cylinder does not slip on the translating bodies $\mathcal{B}_1$ and $\mathcal{B}_2$.

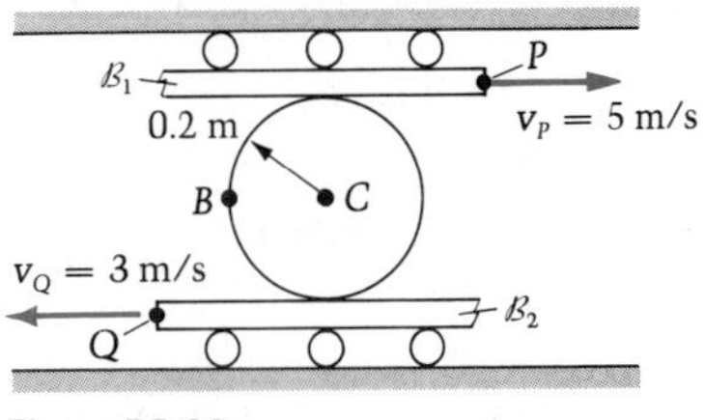

Figure P3.89

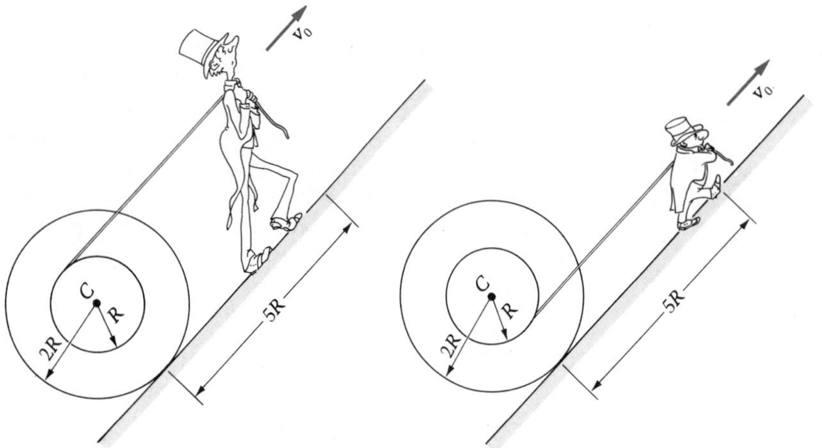

Figure P3.90

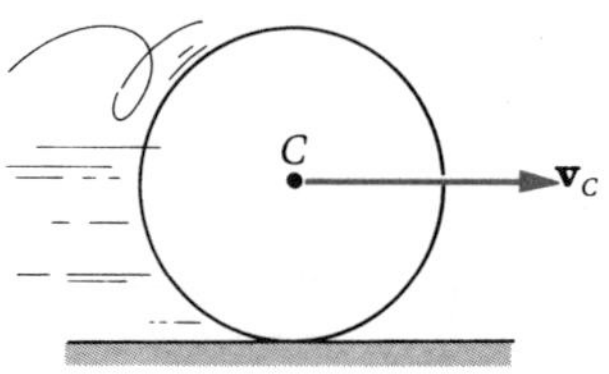

Figure P3.92

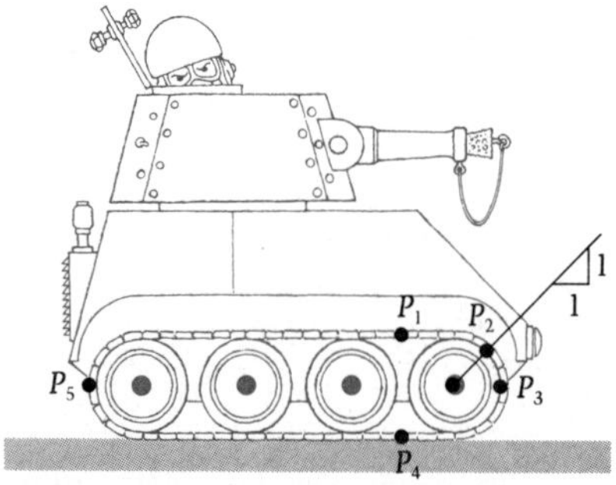

Figure P3.93

3.90 Two men, a tall one and a short one, travel up identical inclines, pulling identical spools by means of ropes wrapped around the hubs. (See Figure P3.90.) The men travel at the same constant speed v_0, and the ropes are wrapped in the opposite directions indicated. If the spools do not slip on the plane, one of the men will be run over by his own spool. Prove which one it is, and show how long it will take, from the instant depicted, for the spool to roll over him.

3.91 A cylinder $\mathcal{B}$ of radius r rolls over a circular arc of constant radius of curvature R (see Figure P3.91). What is the ratio of the angular speed of the cylinder to $\dot{\phi}$?

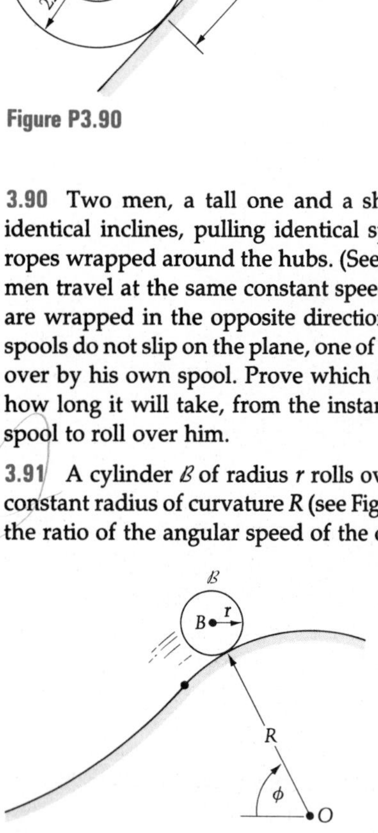

Figure P3.91

3.92 A disk with diameter 1.2 m rolls along the plane as indicated in Figure P3.92. Its center point C has velocity

$$\mathbf{v}_C = (t^2 + 3t + 4)\hat{\mathbf{i}} \text{ m/s}$$

where t is the time in seconds. Find the velocity of the point that lies 0.3 m directly below C when (a) $t = 2$ s and (b) $t = 5$ s.

3.93 The tank shown in Figure P3.93 is translating to the right, and at a certain instant it has velocity $v_0\hat{\mathbf{i}}$ and acceleration $a_0\hat{\mathbf{i}}$. (These values are $\mathbf{v}$ and $\mathbf{a}$ for all points in the body of the tank and for the centers of its wheels.) Find the velocities of the five points P_1, P_2, P_3, P_4, and P_5 if there is no slipping. The wheels have radius R.

3.94 The wheel rolls on both $\mathcal{B}_1$ and $\mathcal{B}_2$ (See Figure P3.94.) The constant angular velocity of the wheel in frame $\mathcal{I}$ is shown in the figure. Find:

a. The velocity of the points of $\mathcal{B}_1$ relative to $\mathcal{B}_2$

b. The constant velocity of C in $\mathcal{I}$ for which the velocities of T (on $\mathcal{B}_1$) and B (on $\mathcal{B}_2$) in $\mathcal{I}$ are equal in magnitude and opposite in direction.

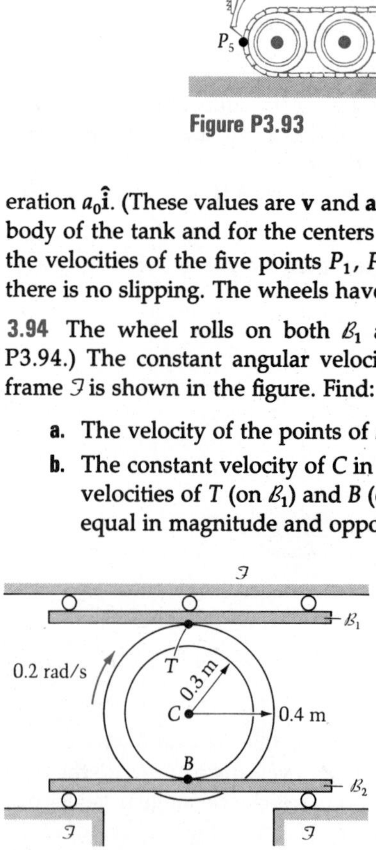

Figure P3.94

3.95 An inextensible string is wrapped around the cylinder in Figure P.3.95, fitting in a small slot near the rim. The center C is moving down the plane at a constant speed of 0.1 m/s. Find the velocities of points A, B, D, and

E. *Hint*: The cylinder is not rolling on the plane, but it *is* rolling on __?__.

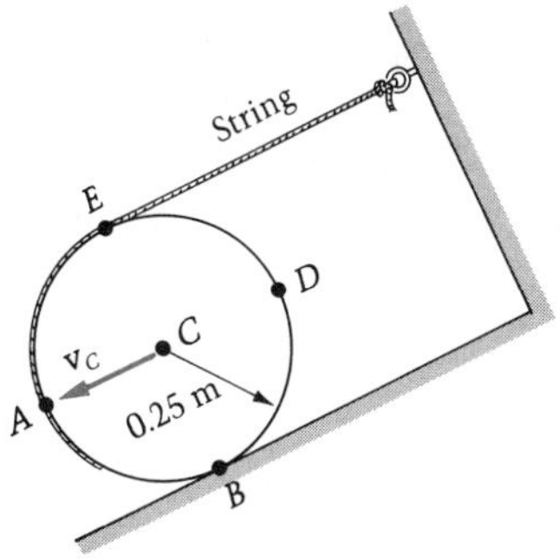

Figure P3.95

3.96 At the instant shown in Figure P3.96, point B of the block (to which rod $\mathcal{B}$ is pinned) has $\mathbf{v}_B = 0.5 \downarrow$ ft/sec. Find the angular velocity of the rolling cylinder.

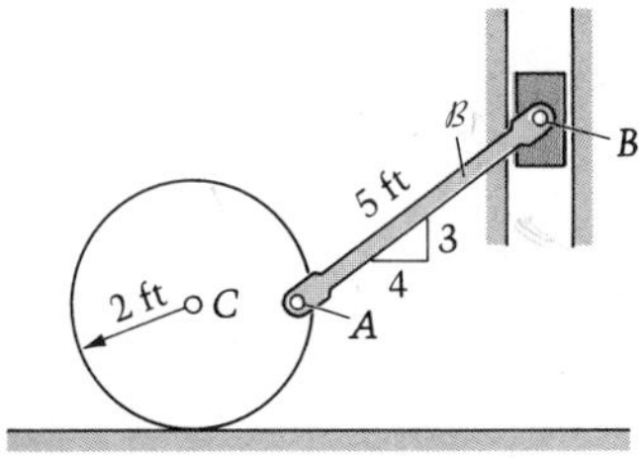

Figure P3.96

3.97 Figure P3.97 shows a circular cam $\mathcal{B}_1$ and an oscillating roller follower consisting of the roller $\mathcal{B}_2$ (which rolls on $\mathcal{B}_1$) and the follower bar $\mathcal{B}_3$. If the cam turns at the constant angular velocity 0.3 ⮌ rad/s, find the angular velocity of the follower bar and of the roller at the given instant.

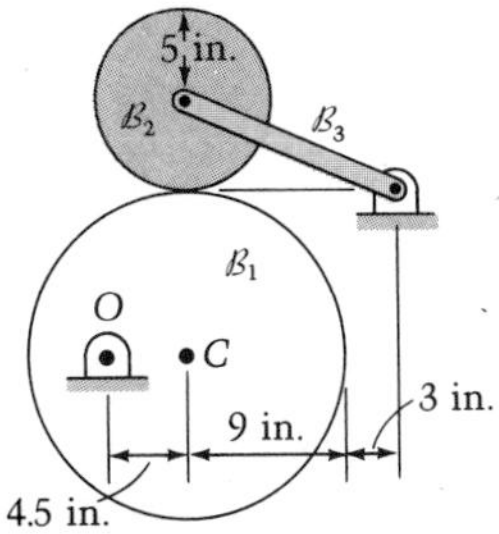

Figure P3.97

3.98 The cylinder in Figure P3.98 is rolling to the left with constant center speed v_C. A stick is pinned to the cylinder at B, and its other end A slides on the plane. Find the velocity of A when $\theta = 0°, 90°, 180°$, and $270°$.

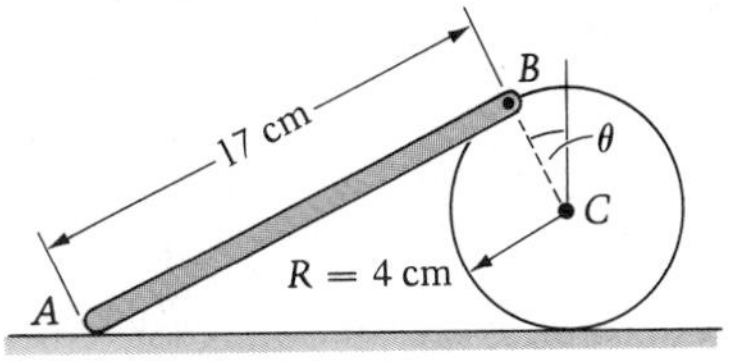

Figure P3.98

3.99 Wheel $\mathcal{B}_1$ in Figure P3.99 has angular velocity 3 ⮏ rad/sec. Find the angular velocities of wheel $\mathcal{B}_2$ and the bent bar $\mathcal{B}_3$.

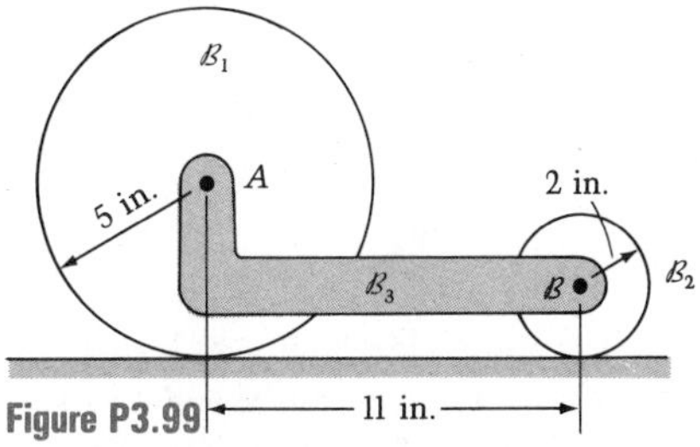

Figure P3.99

* **3.100** The cart $\mathcal{B}_1$ in Figure P3.100 travels from left to right, with its rear wheels $\mathcal{B}_2$ rolling at constant angular velocity 0.2 ⮏ rad/sec. The front wheels $\mathcal{B}_3$ are rolling up the parabolic surface shown. The wheels have radius 0.4 m, and are pinned to the cart. Find the angular velocity of cart $\mathcal{B}_1$ at the given instant. *Hint*:

$$\frac{dy}{dx} = \tan\phi = 2$$

$$\phi = 63.43°$$

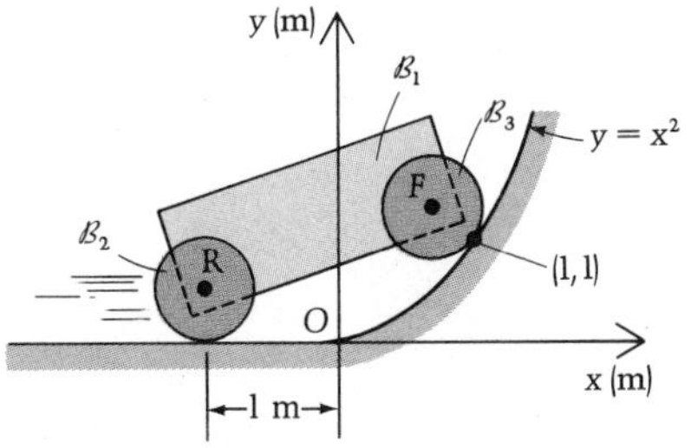

Figure P3.100

3.101 The constant angular velocities of the ring gear $\mathcal{B}_1$ and the spider arm $\mathcal{B}_2$ shown in Figure P3.101 are 2 ↺ rad/s and 10 ↻ rad/s, respectively. Determine the angular velocity of gear $\mathcal{B}_3$ and the velocity of the point of $\mathcal{B}_4$ having maximum speed in the given position. The centers of $\mathcal{B}_1$ and $\mathcal{B}_2$ are pinned to the reference frame $\mathcal{I}$.

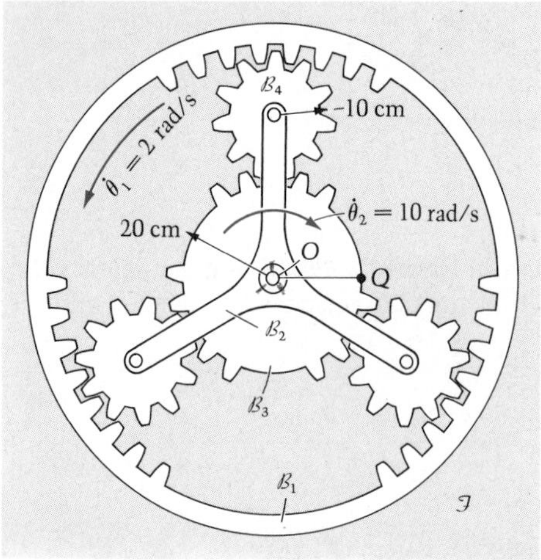

Figure P3.101

3.102 The rod $\mathcal{R}$, which is pinned to cylinder $\mathcal{B}_2$, translates upward in the y-direction at the constant speed 4 ft/sec (see Figure P3.102). The rod $\mathcal{B}_1$ is pinned to the reference frame at O and rests against the rim of $\mathcal{B}_2$ as shown in the figure. There is rolling contact between $\mathcal{B}_1$ and $\mathcal{B}_2$. Find the angular velocities of $\mathcal{B}_1$ and $\mathcal{B}_2$ at the instant shown.

3.103 If the given velocities of P and Q in Problem 3.89 are constant, find the accelerations of C and B.

3.104 If, in Problem 3.89, the respective accelerations of P and Q are 1 ← m/^{2} and 2 → m/s^2, find the angular acceleration of the cylinder.

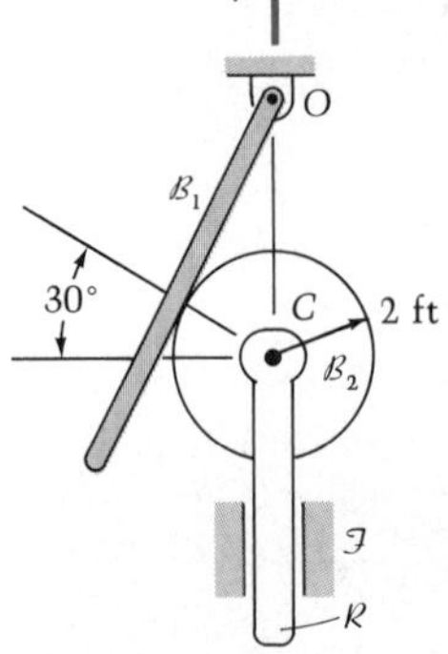

Figure P3.102

3.105 The shaded arcs on $\mathcal{B}_1$ and $\mathcal{B}_2$ (Figure P3.105(a)) are always equal if the two bodies are in rolling contact; however, the converse is not necessarily true. Just because the contacting arclengths are equal does not mean that $\mathcal{B}_1$ rolls on $\mathcal{B}_2$. For the wheel $\mathcal{B}_1$ on the plane $\mathcal{B}_2$ shown in Figure P3.105(b), give constant values $\dot{x}_C$ and $\dot{\theta}$ for which the arclengths of contact are equal but the velocities of the contact points are not. *Hint*: Look at the shaded arcs on $\mathcal{B}_1$ and $\mathcal{B}_2$ in Figure P3.105(b).

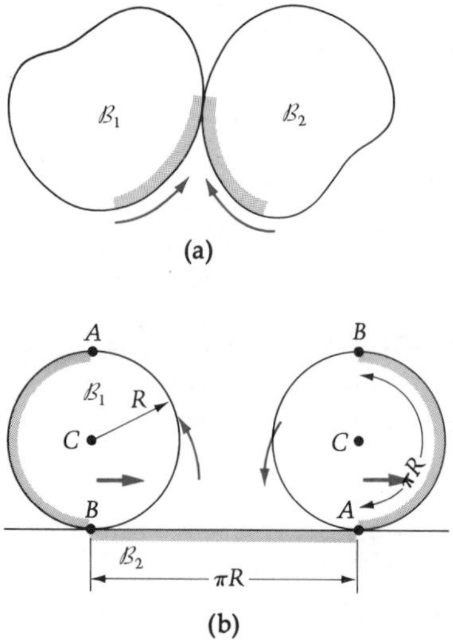

Figure P3.105

3.106 If in Problem 3.99 the angular acceleration of $\mathcal{B}_1$ is $\alpha_1 = 1$ ↺ rad/sec^2 at the given instant, find α_2 and α_3 at that time.

3.107 In Problem 3.93 find the accelerations of the same five points P_1, P_2, P_3, P_4, and P_5. (See Figure P3.107.)

3.108 During startup of the two friction wheels (see Figure P3.108), the angular velocity of $\mathcal{B}_1$ is $\omega_1 = 5t^2$ ↺ rad/sec. Assuming rolling contact, compute the acceleration of the point T, which is at the top of $\mathcal{B}_2$ when $t = 3$ sec.

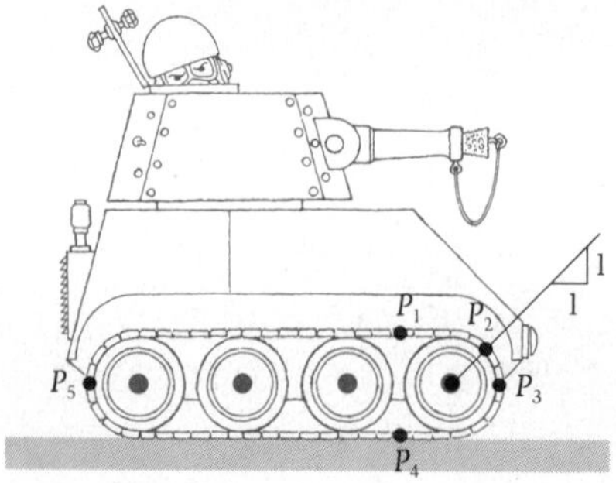

Figure P3.107

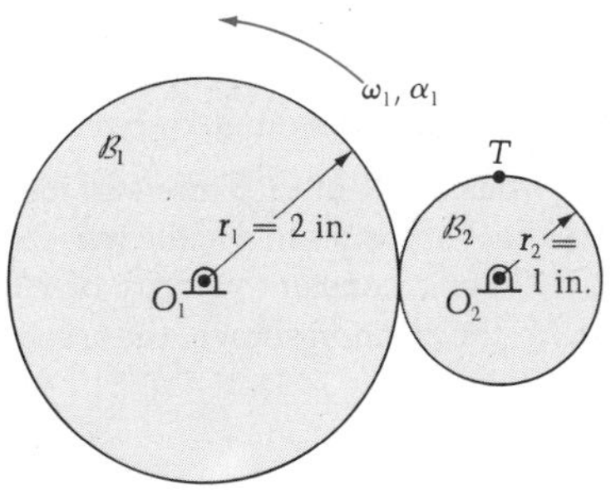

Figure P3.108

3.109 The wheel in Figure P3.109 rolls, its center having a constant velocity of 10 ft/sec to the right. Find $d|\mathbf{v}_A|/dt$ at the instant shown.

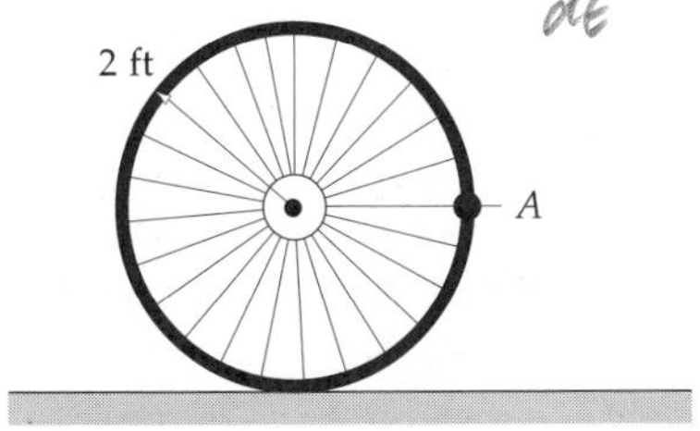

Figure P3.109

3.110 A wheel rolls on a 10-cm-radius track. (See Figure P3.110.) At the instant shown the wheel has an angular velocity of $4\hat{\mathbf{k}}$ rad/s and an angular acceleration of $-3\hat{\mathbf{k}}$ rad/s^2. At the instant shown, find:

- **a.** The velocity and acceleration of C
- **b.** The velocity and acceleration of A
- **c.** $(d/dt)|\mathbf{v}_A|$
- **d.** The center of curvature of the path of point T.

3.111 In Problem 3.87 find the accelerations of points A, B, C, D, and E.

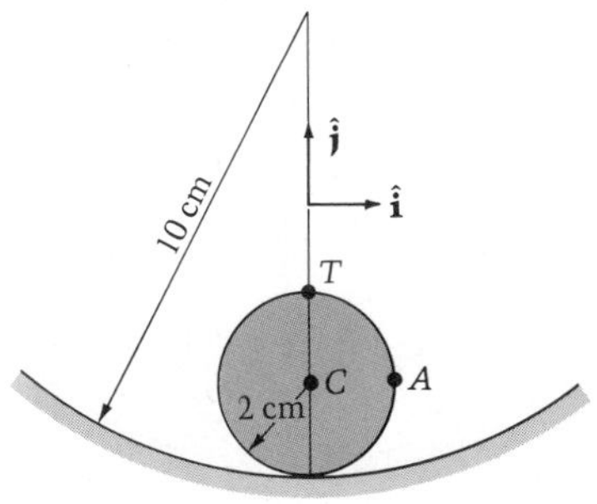

Figure P3.110

3.112 In Problem 3.88 find the accelerations of points F, G, and H.

3.113 A moment applied to gear $\mathcal{B}_1$ in Figure P3.113 results in a constant angular acceleration $\boldsymbol{\alpha}_1 = 2 \circlearrowright$ rad/sec^2. The other gear, $\mathcal{B}_2$, is fixed in the reference frame. Determine:

- **a.** The time required for C to return to its starting point after one revolution around $\mathcal{B}_2$ from rest
- **b.** The number of revolutions turned through in space by $\mathcal{B}_1$ during the complete revolution.

3.114 The center of the rolling wheel in Figure P3.114 moves to the right at a constant speed of 10 in./sec. The bar is pinned to the wheel at A, and end B always stays in contact with the ground. Find the acceleration of B at the instant shown.

3.115 Calculate the acceleration of the instantaneous center of rod $\mathcal{B}_2$ in Example 3.15.

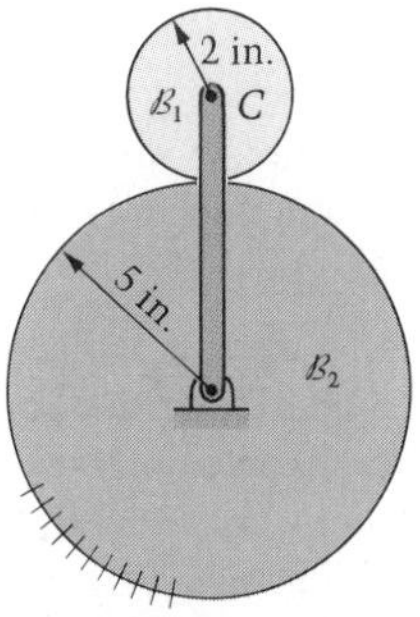

Figure P3.113

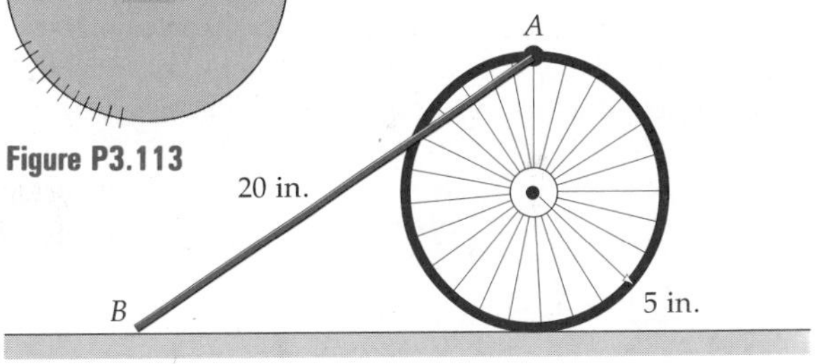

Figure P3.114

3.116 Gear $\mathcal{B}_1$ and crank $\mathcal{B}_2$ have angular speeds ω_0 and angular acceleration magnitudes α_0 at the instant shown in Figure P3.116 in the indicated directions. Find the angular velocities and angular accelerations of gears $\mathcal{B}_3$ and $\mathcal{B}_4$ at the same time, if $\mathcal{B}_2$ is pinned to $\mathcal{B}_1$, $\mathcal{B}_3$ and $\mathcal{B}_4$.

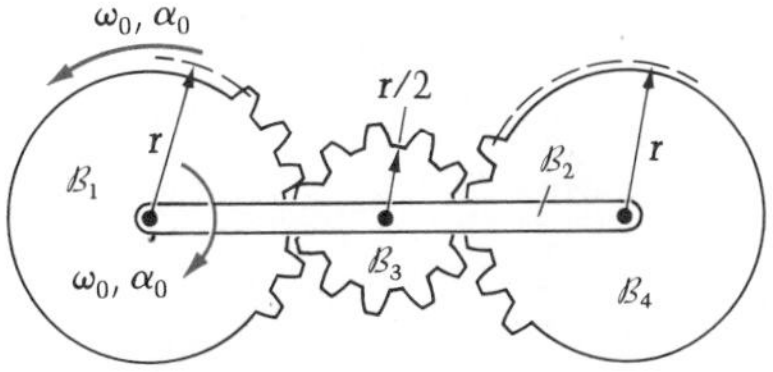

Figure P3.116

• **3.117** The center C of the small cylinder $\mathcal{B}_1$ in Figure P3.117 has a speed of $0.1t^2$ m/s as it moves clockwise on a circle. Body $\mathcal{B}_1$ rolls on the large cylinder $\mathcal{B}_2$. In the position given in the figure, $t = 10$ s. Find the acceleration of point B of the stick $\mathcal{B}_3$ that is in contact with $\mathcal{B}_2$ at the given instant.

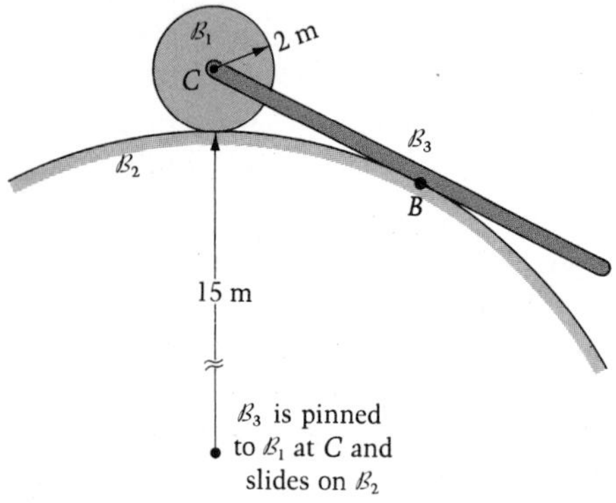

Figure P3.117

3.118 In Problem 3.89 let $\mathbf{a}_P = 0.2 \rightarrow$ m/s² and $\mathbf{a}_Q = 0.1 \leftarrow$ m/s² at the instant shown. Again assuming no slipping, find the accelerations of B and C.

3.119 The two identical cylinders $\mathcal{B}_1$ and $\mathcal{B}_2$ (Figure P3.119) are connected by bar $\mathcal{B}_3$ (which is pinned to their centers), and they roll on the surface as shown. If the angular velocity of $\mathcal{B}_1$ is $\boldsymbol{\omega}_1 = \omega_0 \circlearrowright =$ constant, find the angular accelerations of both $\mathcal{B}_2$ and $\mathcal{B}_3$ at the given instant.

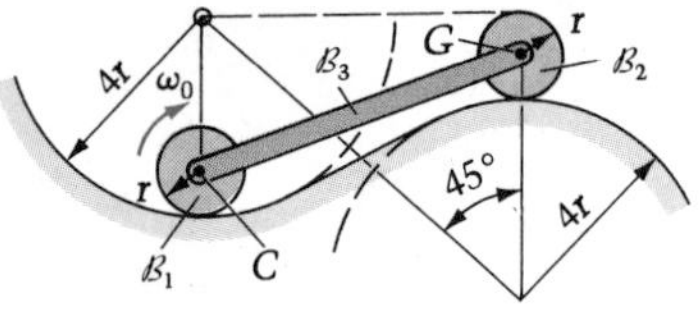

Figure P3.119

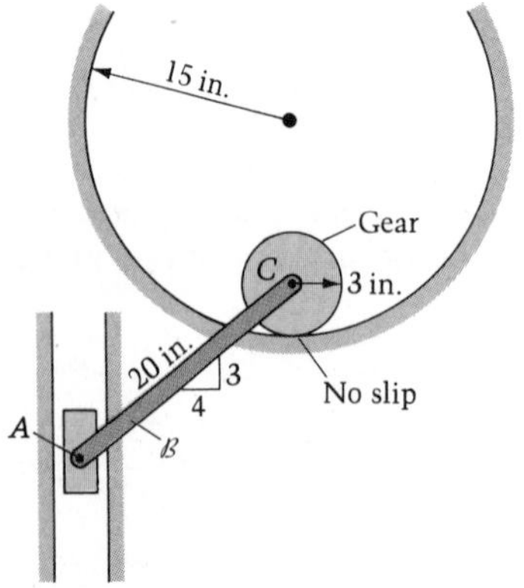

Figure P3.120

3.120 Point A of the slider block has, at the instant shown in Figure P3.120, $\mathbf{v}_A = 12 \uparrow$ in./sec and $\mathbf{a}_A = 6 \downarrow$ in./sec². Find the angular acceleration of bar $\mathcal{B}$.

3.121 Two 5-in.-radius wheels roll on a plane surface. (See Figure P3.121.) A 13-in. bar $\mathcal{B}$ is pinned to the wheels at A and B as shown. If C has a constant velocity of 20 ft/sec to the right, find, for the position shown, the acceleration of A.

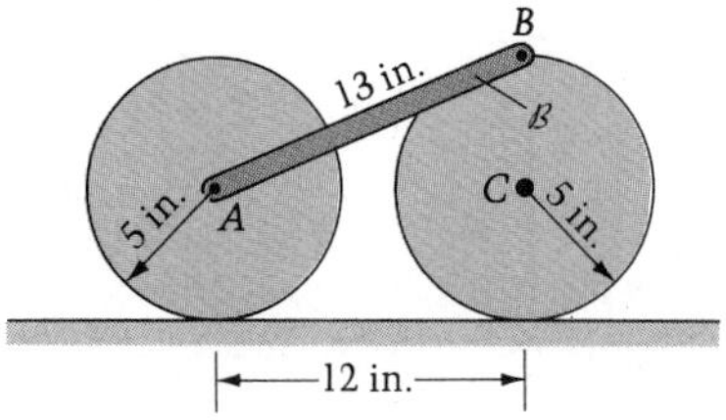

Figure P3.121

3.122 See Figure P3.122. The velocity of the pin in block $\mathcal{B}$ is $0.2 \downarrow$ m/s, and its acceleration is $0.5 \uparrow$ m/s² in the given position. Find at this instant the angular velocity and angular acceleration of the cylinder if there is sufficient friction to prevent it from slipping.

3.123 A wheel rolls along a curved surface. In the position shown in Figure P3.123, its angular velocity and angular acceleration are $\omega = 3 \circlearrowright$ rad/s and $\alpha = 5 \circlearrowright$ rad/s². Determine at this instant the angular acceleration of bar $\mathcal{B}$ and the acceleration of pin B of the slider block.

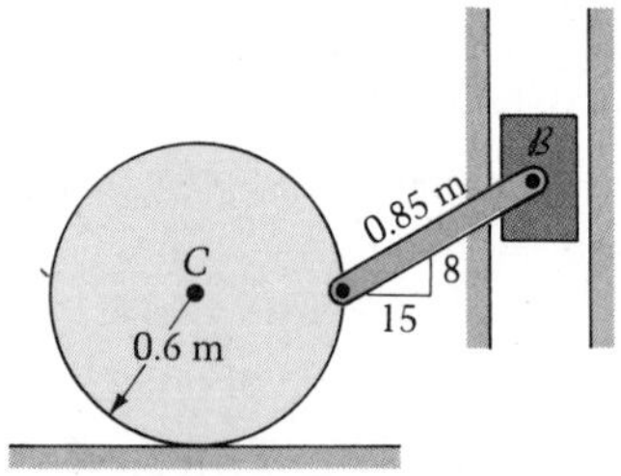

Figure P3.122

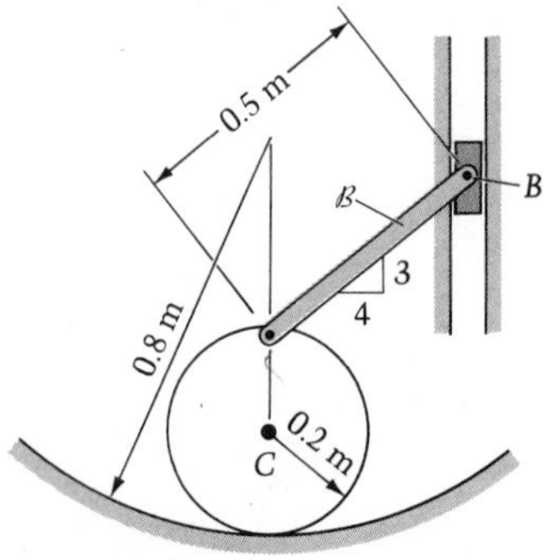

Figure P3.123

3.124 In Example 3.14 find the piston acceleration when $\theta = 0°$, $180°$, and $270°$.

3.125 The center point C of gear $\mathcal{B}_1$ in Figure P3.125 moves in a horizontal plane at constant speed v_0. The ring gear $\mathcal{I}$ is fixed in the reference frame, and the constant angular velocity of $\mathcal{B}_1$ is clockwise. Find the acceleration of point Q of $\mathcal{B}_1$.

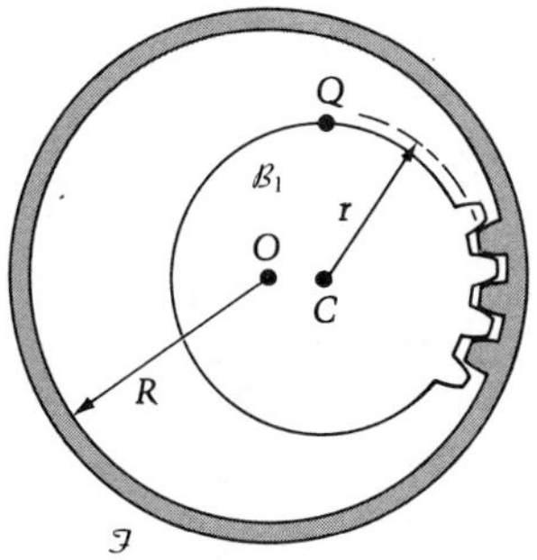

Figure P3.125

3.126 If the ball in a ball bearing assembly (Figure P3.126) neither slips on the shaft nor on the fixed housing, find the velocity and acceleration of the center of the ball in terms of the angular velocity and angular acceleration of the shaft (ω and $\dot{\omega}$).

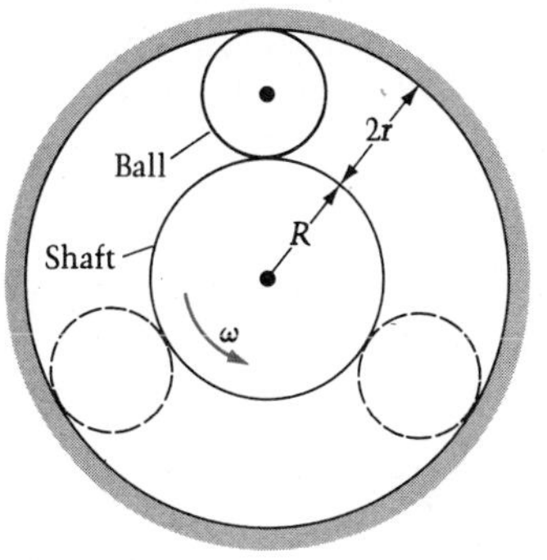

Figure P3.126

3.127 The ball in Figure P3.127 rolls on the fixed surface and at the instant shown has angular velocity $\boldsymbol{\omega} = 3\hat{\mathbf{k}}$ rad/s and angular acceleration $\boldsymbol{\alpha} = -2\hat{\mathbf{k}}$ rad/s². At this instant find:

- **a.** The velocities of A and B
- **b.** The accelerations of A and B
- **c.** $(d/dt)|\mathbf{v}_B|$

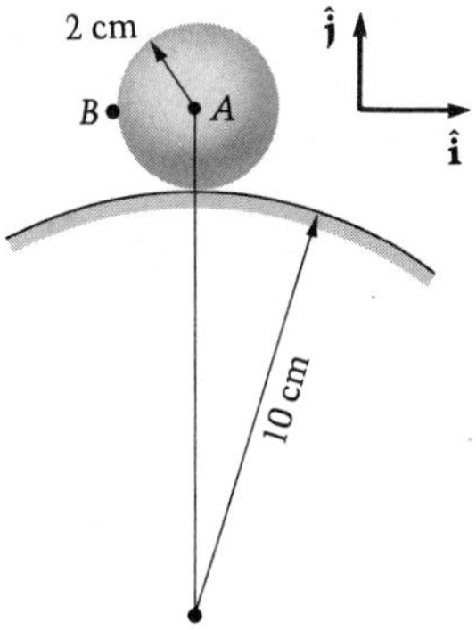

Figure P3.127

3.128 The angular velocity of crank $\mathcal{B}_1$ in Figure P3.128 is a constant 3 ⟲ rad/s. In the given position, find the velocity of the center C of wheel $\mathcal{B}_2$ and also determine the angular acceleration of $\mathcal{B}_2$, which rolls on the circular track.

3.129 Cylinders $\mathcal{B}_1$ and $\mathcal{B}_2$ in Figure P3.129 have a radius of 10 in. each and roll on the respective planes. Bar $\mathcal{B}_3$ has length 48 in. and is pinned to the centers of the cylinders. The center G of $\mathcal{B}_1$ has velocity $\mathbf{v}_G = -10t \rightarrow$ in./sec. If the time at the instant shown is $t = 5$ sec, find $\boldsymbol{\alpha}_2$ and $\boldsymbol{\alpha}_3$ at that instant.

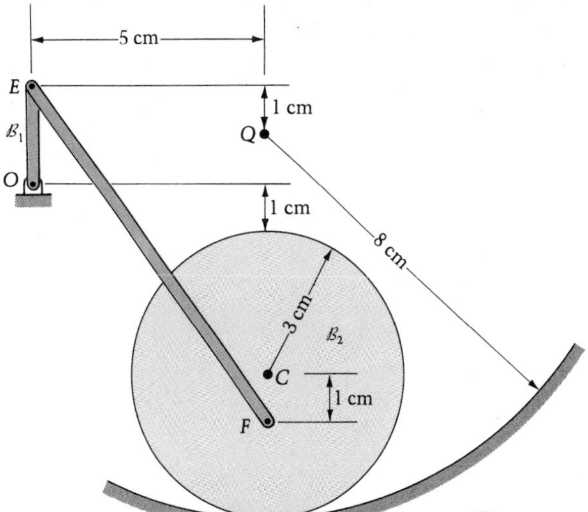

Figure P3.128

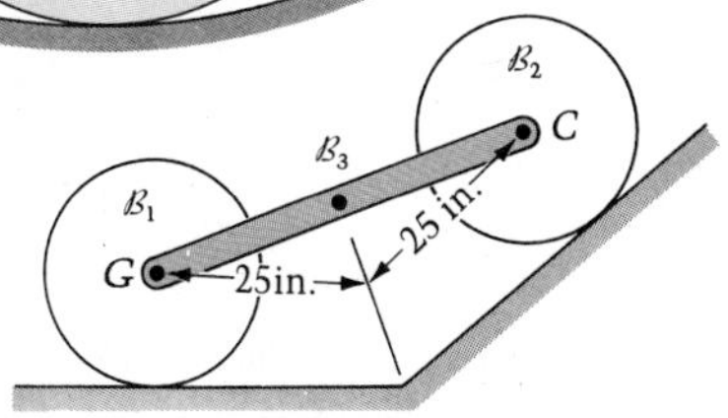

Figure P3.129

3.130 Bar $\mathcal{B}$ in Figure P3.130 is 25 cm long and is pinned to the rolling cylinder at B. The other end of $\mathcal{B}$ is pinned to the roller at A as shown. The center of the cylinder has $v_C = 11.2$ cm/s and $a_C = 16.8$ cm/s^2 down the plane at the given instant; at this time line BⒾ is vertical and AⒾ is horizontal, and BC is parallel to the plane beneath it. Find the acceleration of point A and the angular acceleration of body $\mathcal{B}$ at the given instant.

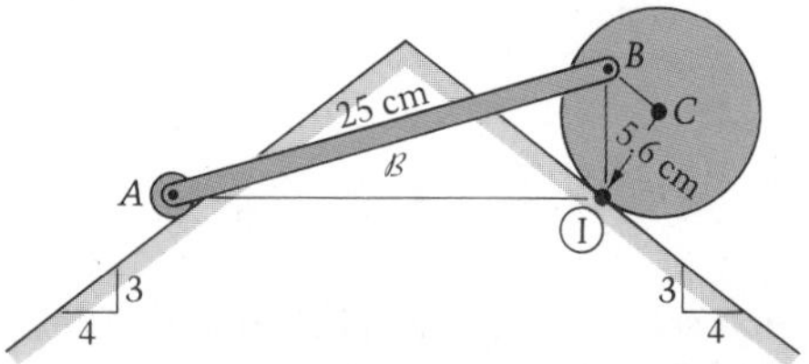

Figure P3.130

3.131 Figure P3.131 shows a 10-ft-radius disk that rolls on a plane surface. It has, at the instant shown, an angular velocity of 2 rad/sec and an angular acceleration of 3 rad/sec^2, both counterclockwise. Find a point on the disk or the disk extended that has zero acceleration at this instant.

3.132 Find the acceleration of point B, the pin connecting rod $\mathcal{B}$ to the block in Figure P3.132, at $t = 1$ sec. The rod is horizontal at $t = 0$, and the velocity of C is $\mathbf{v}_C = 2t^2 \rightarrow$ ft/sec.

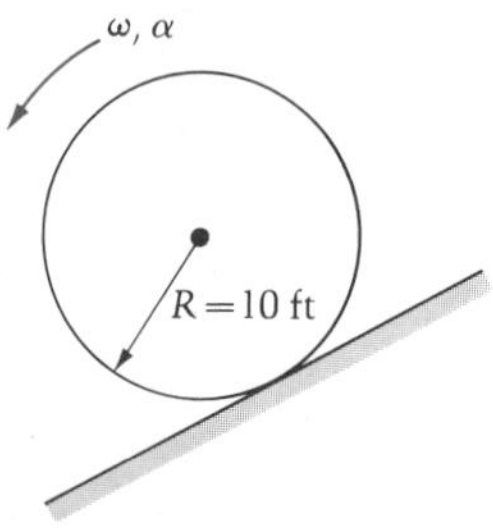

Figure P3.131

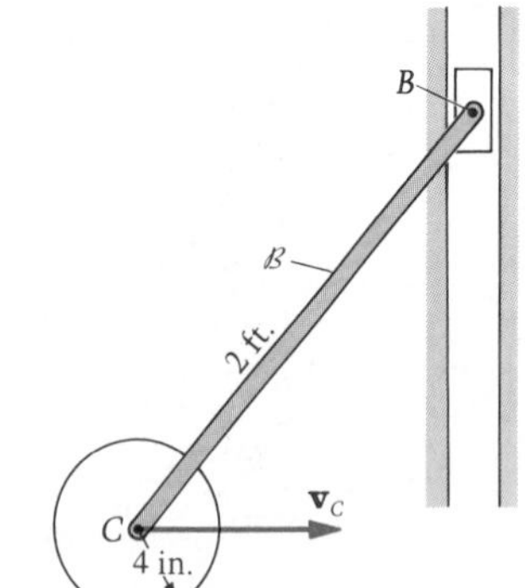

Figure P3.132

3.133 Referring to Problem 3.121, if at the instant shown $\mathbf{v}_C = 20 \rightarrow$ in./sec and $\mathbf{a}_C = 5 \leftarrow$ in./sec^2, find $\mathbf{a}_A$ at this time.

* **3.134** Gears $\mathcal{B}_1$ and $\mathcal{B}_2$ in Figure P3.134 have 25 and 50 teeth, respectively. Rod $\mathcal{B}_3$ is 2 ft long, and the radius of $\mathcal{B}_2$ is 1 ft. Determine the acceleration of point A when $t = 0$ if $x_B = 0.2 \sin \pi t$ ft (positive to the left) with $\theta = 90°$ at $t = 0$.

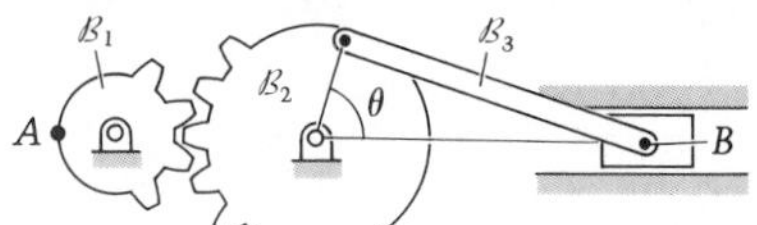

Figure P3.134

* **3.135** At the instant shown in Figure P3.135, bar $\mathcal{B}_1$ has $\omega_1 = \pi$ ↺ rad/sec and $\alpha_1 = (\pi/3)$ ↻ rad/sec^2; and the gear $\mathcal{B}_2$ has $\omega_2 = 2\pi$ rad/sec ↻ and $\alpha_2 = \pi/2$ ↺ rad/sec^2. At this instant, determine the accelerations of each of the two gear tooth contacting points.

* **3.136** The outer gear $\mathcal{I}$ in Figure P3.136 is stationary. Crank $\mathcal{B}_1$ turns at the constant angular velocity of 10 rad/sec counterclockwise and is pinned at its ends to the centers of the sun gear $\mathcal{B}_2$ (at S) and the planetary gear $\mathcal{B}_3$ (at P). Find the accelerations of the points of $\mathcal{B}_2$ and $\mathcal{B}_3$ that are in contact with each other, if the radii of $\mathcal{B}_2$ and $\mathcal{B}_3$ are, respectively, 3 in. and 10 in.

* **3.137** Point O is pinned to the reference frame $\mathcal{I}$. (See Figure P3.137.) The pitch radii of gears $\mathcal{B}_1$ and $\mathcal{B}_2$ are each 0.2 m. The angular velocities of $\mathcal{B}_3$ and $\mathcal{B}_4$ are 2 rad/s, clockwise for $\mathcal{B}_3$ and counterclockwise for $\mathcal{B}_4$, and both constant. Find the maximum acceleration magnitude experienced by any point of $\mathcal{B}_1$.

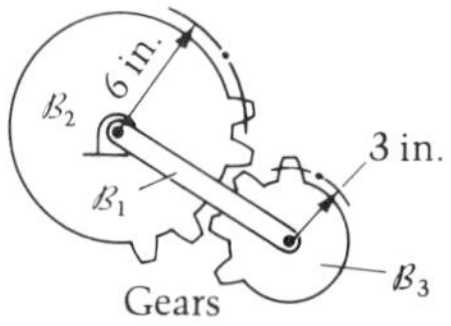

Figure P3.135

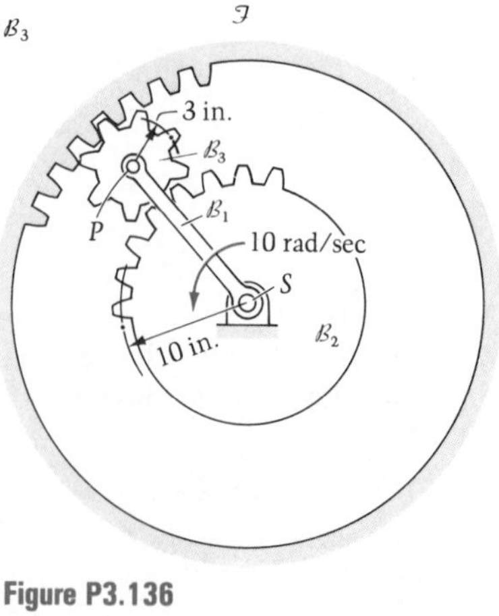

Figure P3.136

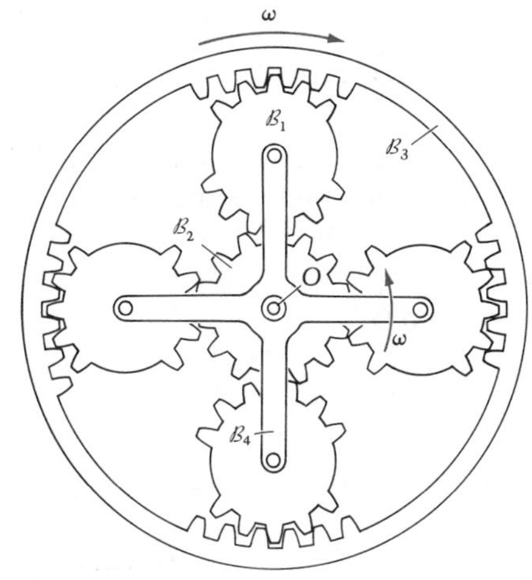

Figure P3.137

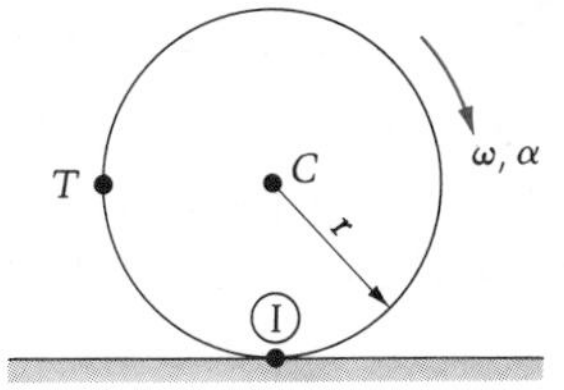

Figure P3.138

* **3.138** The wheel in Figure P3.138 rolls on the plane. Find the radius of curvature and the center of curvature of the path of point T at the given time in terms of r.

* **3.139** A *cycloid* is the curve traced out by a point on the rim of a rolling wheel. The equations for the rectangular coordinates of a point on the cycloid, in terms of the parameter φ (the angle shown in Figure P3.139), are

$$x = a(\varphi - \sin\varphi)$$

$$y = a(1 - \cos\varphi)$$

where a is the wheel's radius. Recall from calculus that the curvature of a plane curve is

$$\frac{1}{\rho} = \left| \frac{y''}{(1 + y'^2)^{3/2}} \right|$$

where ρ is the radius of curvature.

a. Use the chain rule

$$y' = \frac{dy}{dx} = \frac{dy}{d\varphi}\frac{d\varphi}{dx}$$

and show that, for the cycloid,

$$\rho = -2^{3/2}\, a\sqrt{1 - \cos\varphi}$$

b. Explain what the minus sign in the expression for ρ means.

c. Observe that P is at its highest point when $\varphi = \pi$ and that $|\rho| = 4a$ there. In this configuration, show that the following two expressions for the acceleration of P agree:

$$\mathbf{a}_P = \underbrace{a\alpha\hat{\mathbf{i}}}_{\mathbf{a}_C} + \alpha\hat{\mathbf{k}} \times \mathbf{r}_{CP} - \omega^2\mathbf{r}_{CP}$$

$$\mathbf{a}_P = \ddot{s}\hat{\mathbf{e}}_t + \frac{\dot{s}^2}{\rho}\hat{\mathbf{e}}_n$$

* **3.140** Show that for a rigid body $\mathcal{B}$ in the plane motion, as long as $\alpha \neq 0$ there is a circle of points P of $\mathcal{B}$ whose accelerations pass through any point C of $\mathcal{B}$. *Hint*: Write $\mathbf{a}_P = \mathbf{a}_C + \alpha\hat{\mathbf{k}} \times \mathbf{r}_{CP} - \omega^2\mathbf{r}_{CP}$ and dot both sides with the vector $\hat{\mathbf{k}} \times \mathbf{r}_{CP}$. Assume that $\mathbf{a}_P \parallel \mathbf{r}_{CP}$ and see if you can exhibit $\mathbf{r}_{CP}$. For the rolling wheel, show that the points are as shown in Figure P3.140.

* **3.141** Show that for the rolling uniform cylinder in Figure P3.141 there is a point of zero acceleration at the indicated position J if $\boldsymbol{\omega}$ and $\boldsymbol{\alpha}$ are in the given directions and are not both zero. You should find that the coordinates of J are $(x_J, y_J) = [r\alpha\omega^2/(\omega^4 + \alpha^2), r\alpha^2/(\omega^4 + \alpha^2)]$.

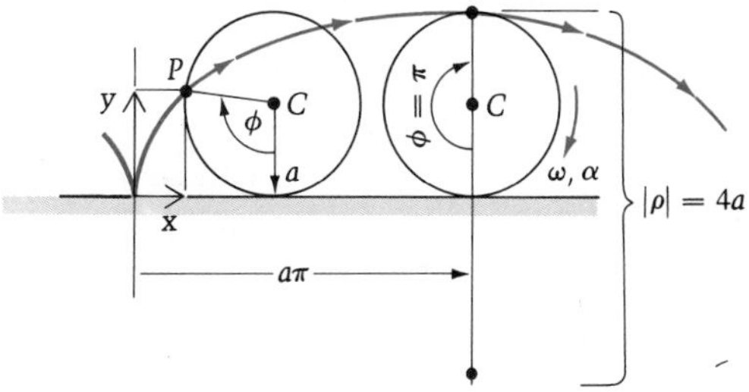

Figure P3.139

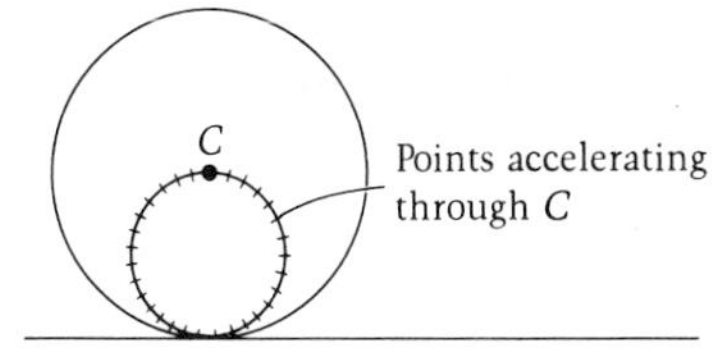

Figure P3.140

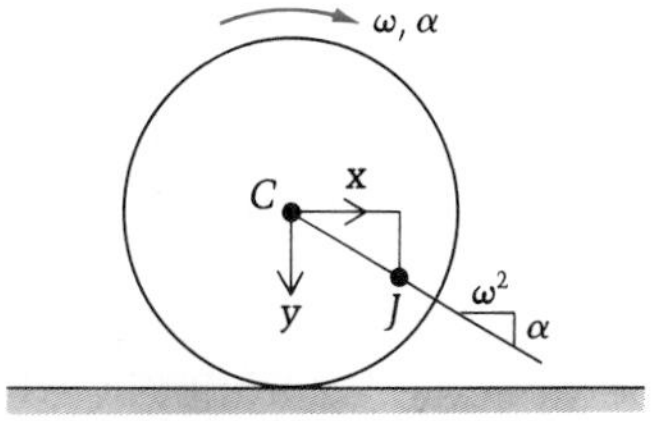

Figure P3.141

3.7 Relationship Between the Velocities of a Point with Respect to Two Different Frames of Reference

While Equation (3.8) gives us the relationship between the velocities in $\mathcal{I}$ of two points of the *same* rigid body, we often need another equation relating the velocities of the *same point* relative to two *different* frames or bodies. This relationship (together with a companion equation for accelerations to be developed in the next section) will be essential in solving kinematics problems involving bodies moving in special ways relative to others (such as a pin of one body sliding in a slot of another).

Relationship Between the Derivatives of a Vector in Two Frames

To develop this equation, we must first find the relationship between the derivatives of an arbitrary vector **A** in two frames $\mathcal{I}$ and $\mathcal{B}$. To do this, we begin by embedding axes X and Y in $\mathcal{I}$, and x and y in $\mathcal{B}$, as suggested by the hatch marks in Figure 3.28. Further, we let $(\hat{\mathbf{I}}, \hat{\mathbf{J}})$ and $(\hat{\mathbf{i}}, \hat{\mathbf{j}})$ be pairs of unit vectors, always respectively parallel to (X, Y) and to (x, y).

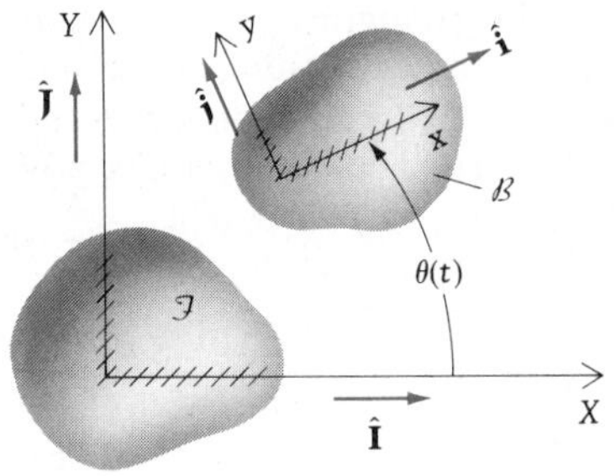

Figure 3.28

We note that if $\theta(t)$ again locates the x axis relative to X as shown, then

$$\hat{\mathbf{i}} = \cos\theta\hat{\mathbf{I}} + \sin\theta\hat{\mathbf{J}} \tag{3.39}$$

and

$$\hat{\mathbf{j}} = -\sin\theta\hat{\mathbf{I}} + \cos\theta\hat{\mathbf{J}} \tag{3.40}$$

Differentiating in frame $\mathcal{I}$ and noting that $\hat{\mathbf{I}}$ and $\hat{\mathbf{J}}$ are constants there, we obtain

$$\frac{d\hat{\mathbf{i}}}{dt} = \dot{\hat{\mathbf{i}}} = (-\sin\theta\hat{\mathbf{I}} + \cos\theta\hat{\mathbf{J}})\dot{\theta} = \dot{\theta}\hat{\mathbf{j}} \tag{3.41}$$

$$\frac{d\hat{\mathbf{j}}}{dt} = \dot{\hat{\mathbf{j}}} = -(\cos\theta\hat{\mathbf{I}} + \sin\theta\hat{\mathbf{J}})\dot{\theta} = -\dot{\theta}\hat{\mathbf{i}} \tag{3.42}$$

Now let the arbitrary vector **A** be written in frame $\mathcal{B}$ (meaning that **A** is expressed in terms of its components there—i.e., in terms of unit vectors fixed in direction in $\mathcal{B}$):

$$\mathbf{A} = A_x\hat{\mathbf{i}} + A_y\hat{\mathbf{j}}$$

Differentiating this vector in $\mathcal{I}$, we get

$${}^{\mathcal{I}}\dot{\mathbf{A}} = \dot{A}_x\hat{\mathbf{i}} + \dot{A}_y\hat{\mathbf{j}} + A_x{}^{\mathcal{I}}\dot{\hat{\mathbf{i}}} + A_y{}^{\mathcal{I}}\dot{\hat{\mathbf{j}}} \tag{3.43}$$

We now note that the first two terms on the right side of Equation (3.43) add up to the derivative of vector **A** in $\mathcal{B}$, because $\hat{\mathbf{i}}$ and $\hat{\mathbf{j}}$ do not change in magnitude *or* direction with time there. Thus

$${}^{\mathcal{I}}\dot{\mathbf{A}} = {}^{\mathcal{B}}\dot{\mathbf{A}} + A_x{}^{\mathcal{I}}\dot{\hat{\mathbf{i}}} + A_y{}^{\mathcal{I}}\dot{\hat{\mathbf{j}}}$$

Substituting the derivatives of $\hat{\mathbf{i}}$ and $\hat{\mathbf{j}}$ in $\mathcal{I}$ from Equations (3.41) and (3.42) yields

$$\begin{aligned} {}^{\mathcal{I}}\dot{\mathbf{A}} &= {}^{\mathcal{B}}\dot{\mathbf{A}} + \dot{\theta}(A_x\hat{\mathbf{j}} - A_y\hat{\mathbf{i}}) \\ &= {}^{\mathcal{B}}\dot{\mathbf{A}} + \dot{\theta}\hat{\mathbf{k}} \times (A_x\hat{\mathbf{i}} + A_y\hat{\mathbf{j}}) \end{aligned}$$

or

$${}^{\mathcal{I}}\dot{\mathbf{A}} = {}^{\mathcal{B}}\dot{\mathbf{A}} + \dot{\theta}\hat{\mathbf{k}} \times \mathbf{A} \tag{3.44}$$

The angular velocity $\dot{\theta}\hat{\mathbf{k}}$ has reappeared, and Equation (3.44) shows us that this vector has a more general purpose than merely relating velocities in kinematics. It is in fact the link that allows us to relate the derivatives of any vector in two different frames. (This same result is in fact true in general three-dimensional motion, with three-dimensional vectors and a more general expression for angular velocity substituted, as will be seen in Chapter 6.)

Velocity Relationship in Two Frames

We now use Equation (3.44) to relate the velocities of a point P in two frames $\mathcal{B}$ and $\mathcal{I}$. From Figure 3.29, the position vectors of P in these two frames are related by

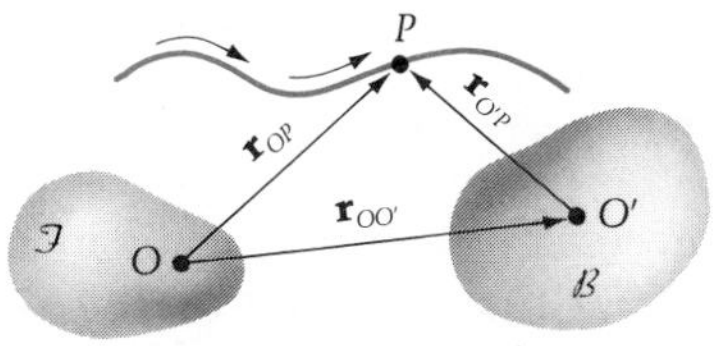

Figure 3.29

$$\mathbf{r}_{OP} = \mathbf{r}_{OO'} + \mathbf{r}_{O'P} \tag{3.45}$$

Differentiating this equation in $\mathcal{I}$, we have

$${}^{\mathcal{I}}\dot{\mathbf{r}}_{OP} = {}^{\mathcal{I}}\dot{\mathbf{r}}_{OO'} + {}^{\mathcal{I}}\dot{\mathbf{r}}_{O'P} \tag{3.46}$$

The first two vectors in Equation (3.46) are the velocities of P and O' in $\mathcal{I}$ by definition:

$$\mathbf{v}_{P/\mathcal{I}} = \mathbf{v}_{O'/\mathcal{I}} + {}^{\mathcal{I}}\dot{\mathbf{r}}_{O'P} \tag{3.47}$$

Question 3.17 (a) Why is the last vector in Equation (3.47) not the velocity of P in $\mathcal{I}$? (b) Why is it not the velocity of P in $\mathcal{B}$?

To replace ${}^{\mathcal{I}}\dot{\mathbf{r}}_{O'P}$ by a vector that we can operate with, we use Equation (3.44), with $\mathbf{r}_{O'P}$ becoming the vector $\mathbf{A}$:

$${}^{\mathcal{I}}\dot{\mathbf{r}}_{O'P} = {}^{\mathcal{B}}\dot{\mathbf{r}}_{O'P} + \dot{\theta}\hat{\mathbf{k}} \times \mathbf{r}_{O'P} \tag{3.48}$$

Therefore, recognizing that ${}^{\mathcal{B}}\dot{\mathbf{r}}_{O'P}$ is $\mathbf{v}_{P/\mathcal{B}}$ and substituting Equation (3.48) into (3.47), we obtain

$$\mathbf{v}_{P/\mathcal{I}} = \mathbf{v}_{P/\mathcal{B}} + \mathbf{v}_{O'/\mathcal{I}} + \dot{\theta}\hat{\mathbf{k}} \times \mathbf{r}_{O'P} \tag{3.49}$$

Another way of expressing Equation (3.49) is to think of frame $\mathcal{B}$ as a "moving frame" with respect to a "fixed" reference frame $\mathcal{I}$. Then the velocities of P can be written as simply $\mathbf{v}_P$ when the reference is $\mathcal{I}$ (thus $\mathbf{v}_P = \mathbf{v}_{P/\mathcal{I}}$) and as $\mathbf{v}_{\text{rel}}$ when the reference is the "moving frame" $\mathcal{B}$ (thus

---

Answer 3.17 (a) It is not $\mathbf{v}_{P/\mathcal{I}}$ because the origin of the position vector is not fixed in $\mathcal{I}$. (b) And it is not $\mathbf{v}_{P/\mathcal{B}}$ because the derivative is not taken in $\mathcal{B}$.

$\mathbf{v}_{\text{rel}} = \mathbf{v}_{P/\mathcal{B}}$). Hence we can write Equation (3.49) in abbreviated notation as

$$\mathbf{v}_P = \mathbf{v}_{\text{rel}} + \mathbf{v}_{O'} + \boldsymbol{\omega} \times \mathbf{r} \tag{3.50}$$

where $\mathbf{r} = \mathbf{r}_{O'P}$, the position of P in the moving frame, and $\boldsymbol{\omega}$ is the angular velocity of $\mathcal{B}$ relative to frame $\mathcal{I}$. The reader may find this form of Equation (3.49) easier to use when there is just one "moving frame."

Now let us denote by P' the point of $\mathcal{B}$ (or $\mathcal{B}$ extended) that is coincident with P. Then $\mathbf{r}_{O'P} = \mathbf{r}_{O'P'}$ and the last two terms of Equation (3.49) or (3.50) are seen (by Equation 3.8) to be the velocity of P' in $\mathcal{I}$:

$$\mathbf{v}_{P/\mathcal{I}} = \mathbf{v}_{P/\mathcal{B}} + \mathbf{v}_{P'/\mathcal{I}} \tag{3.51}$$

In words, Equation (3.51) says that at any time, the velocity of P in $\mathcal{I}$ is the sum of the velocity of P in $\mathcal{B}$ plus the velocity in $\mathcal{I}$ of the point of $\mathcal{B}$ coincident with P.*

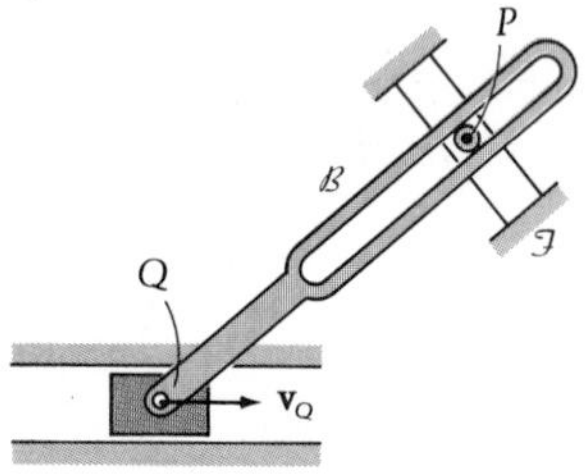

Figure 3.30

As a preliminary example, consider Figure 3.30, in which pin Q is moving to the right. Let P be the center of the other pin, which is attached to frame $\mathcal{I}$. Then we may write

$$\mathbf{v}_{P/\mathcal{I}} = \mathbf{0} = \mathbf{v}_{P/\mathcal{B}} + \mathbf{v}_{P'/\mathcal{I}} \Rightarrow \mathbf{v}_{P'/\mathcal{I}} = -\mathbf{v}_{P/\mathcal{B}}$$

Now the center of the pin's motion in $\mathcal{B}$ is necessarily along a straight line within the slot of $\mathcal{B}$. Therefore the velocity of P' (the point of $\mathcal{B}$ extended coincident with P) is seen to be *also* parallel to the slot and in a direction ($\nearrow$) opposite to that of $\mathbf{v}_{P/\mathcal{B}}$ ($\swarrow$). We now consider several detailed examples of the use of Equation (3.49).

EXAMPLE 3.23

A yellowjacket walks radially outward at a constant 2 in./sec in a straight line relative to a record turning at $33\frac{1}{3}$ rpm. (See Figure E3.23.) Find the velocity of the yellowjacket in frame $\mathcal{I}$, which is the cabinet on which the stereo rests.

Solution

We shall treat the yellowjacket as a point Y, and we note that the unit vectors in Figure E3.23 are fixed in our "moving frame" $\mathcal{B}$. Also, the "moving" (O') and "fixed" (O) origins are coincident. Then we have, using Equation (3.49):

Figure E3.23

$$\begin{aligned}
\mathbf{v}_{Y/\mathcal{I}} &= \mathbf{v}_{Y/\mathcal{B}} + \mathbf{v}_{O'/\mathcal{I}} + \dot{\theta}\hat{\mathbf{k}} \times \mathbf{r}_{O'Y} \\
&= 2\hat{\mathbf{i}} + \mathbf{0} + \left(33\frac{1}{3}\right)\left(\frac{2\pi}{60}\right)\hat{\mathbf{k}} \times r\hat{\mathbf{i}} \\
&= 2\hat{\mathbf{i}} + \frac{10\pi}{9} r\hat{\mathbf{j}} \text{ in./sec}
\end{aligned}$$

Note how the second term grows linearly with the radius.

* This latter term is sometimes called the *vehicle velocity* of P.

In the next example, there is more than one "moving frame." To avoid three levels of subscripts, we shall name the bodies $\mathcal{B}$, $\mathcal{R}$, and $\mathcal{C}$ rather than the usual $\mathcal{B}_1$, $\mathcal{B}_2$, $\mathcal{B}_3$. Thus $\boldsymbol{\omega}_1$ becomes $\boldsymbol{\omega}_{\mathcal{B}}$, and so on.

EXAMPLE 3.24

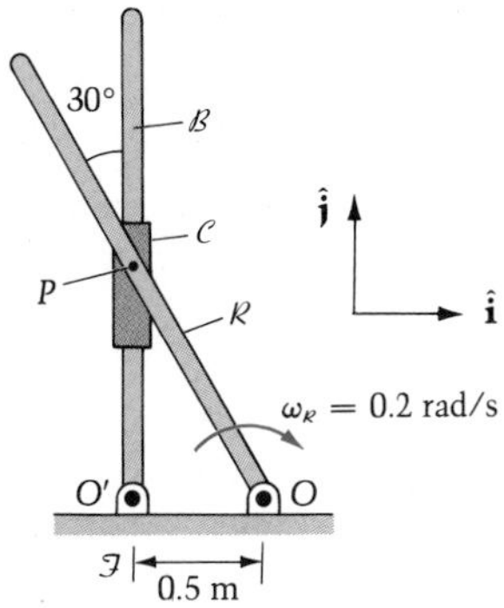

Figure E3.24

Collar $\mathcal{C}$ in Figure E3.24 is pinned to rod $\mathcal{R}$ at P and is free to slide along rod $\mathcal{B}$. The angular velocity of $\mathcal{R}$ is 0.2↻ rad/s at the instant shown. Find the angular velocity of $\mathcal{B}$ at this time, and determine the velocity of P relative to $\mathcal{B}$.

Solution

We relate the velocities of P in $\mathcal{I}$ and in $\mathcal{B}$:

$$\mathbf{v}_{P/\mathcal{I}} = \mathbf{v}_{P/\mathcal{B}} + \mathbf{v}_{O'/\mathcal{I}} + \boldsymbol{\omega}_{\mathcal{B}} \times \mathbf{r}_{O'P}$$

$$\overbrace{\underbrace{\mathbf{v}_{O/\mathcal{I}}}_{\text{zero}} + \boldsymbol{\omega}_{\mathcal{R}} \times \mathbf{r}_{OP}}^{\text{on } \mathcal{R}} = v_{P/\mathcal{B}}\hat{\mathbf{j}} + (\underbrace{\mathbf{v}_{O'/\mathcal{I}}}_{\text{zero}} + \boldsymbol{\omega}_{\mathcal{B}} \times \mathbf{r}_{O'P})$$

$$-0.2\hat{\mathbf{k}} \times (-0.5\hat{\mathbf{i}} + 0.866\hat{\mathbf{j}}) = v_{P/\mathcal{B}}\hat{\mathbf{j}} + \omega_{\mathcal{B}}\hat{\mathbf{k}} \times 0.866\hat{\mathbf{j}}$$

$\hat{\mathbf{i}}$ coefficients: $\quad 0.173 = -0.866\omega_{\mathcal{B}}$

$$\omega_{\mathcal{B}} = -0.2 \Rightarrow \boldsymbol{\omega}_{\mathcal{B}} = -0.2\hat{\mathbf{k}} \text{ rad/s} \quad \text{or} \quad 0.2 \circlearrowright \text{ rad/s}$$

$\hat{\mathbf{j}}$ coefficients: $\quad 0.1 = v_{P/\mathcal{B}}$

$$\mathbf{v}_{P/\mathcal{B}} = 0.1\hat{\mathbf{j}} \text{ m/s}$$

Thus the pin is moving outward on $\mathcal{B}$, which is turning clockwise.

> **Question 3.18** Will $\mathcal{B}$ *always* have the same $\boldsymbol{\omega}$ as does $\mathcal{R}$?

Note that in the preceding example, we did not need to know the angular velocity of $\mathcal{C}$. Nonetheless, it is important for the reader to realize that $\boldsymbol{\omega}_{\mathcal{C}} \equiv \boldsymbol{\omega}_{\mathcal{B}}$. The reason is that $\mathcal{B}$ and $\mathcal{C}$ can only *translate* relative to each other; thus lines fixed in each will turn at the same time rates in $\mathcal{I}$. This observation will sometimes be needed (as in Problems 3.147 and 3.149).

EXAMPLE 3.25

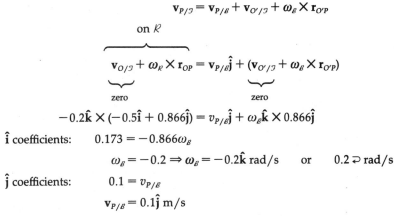

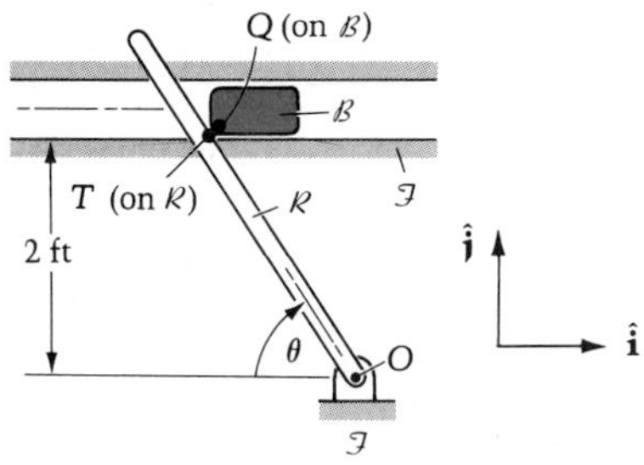

Figure E3.25a

Block $\mathcal{B}$ translates in a horizontal slot (see Figure E3.25a) and is pushed along by a bar $\mathcal{R}$ that turns at angular velocity $\omega_{\mathcal{R}} = 10$↻ rad/sec about the pin at point O. Find $\mathbf{v}_Q$, the velocity of the contact point of $\mathcal{B}$, when $\theta = 60°$.

Answer 3.18 Definitely not! This happened because of the geometry at the given instant.

Solution

Let the ground be the reference frame $\mathcal{J}$, and note that T is the point of $\mathcal{R}$ coincident with Q at the given instant (the point we have been calling P' in the theory). Using Equation (3.51), we obtain

$$\mathbf{v}_{Q/\mathcal{J}} = \mathbf{v}_{Q/\mathcal{R}} + \mathbf{v}_{T/\mathcal{J}}$$

Therefore

$$\mathbf{v}_{Q/\mathcal{J}} = v_{Q/\mathcal{R}}\left(\frac{\hat{\mathbf{i}}}{2} - \frac{\sqrt{3}}{2}\hat{\mathbf{j}}\right) + (-10\hat{\mathbf{k}}) \times \left(\frac{-2}{\sqrt{3}}\hat{\mathbf{i}} + 2\hat{\mathbf{j}}\right) \tag{1}$$

Note that the velocity of Q relative to bar $\mathcal{R}$ has an unknown magnitude $v_{Q/\mathcal{R}}$ but a *known* direction (along $\mathcal{R}$). Now we also know the direction of $\mathbf{v}_{Q/\mathcal{J}}$, so that

$$\mathbf{v}_{Q/\mathcal{J}} = v_{Q/\mathcal{J}}\hat{\mathbf{i}} = \frac{20}{\sqrt{3}}\hat{\mathbf{j}} + 20\hat{\mathbf{i}} + \frac{v_{Q/\mathcal{R}}}{2}\hat{\mathbf{i}} - v_{Q/\mathcal{R}}\frac{\sqrt{3}}{2}\hat{\mathbf{j}} \tag{2}$$

Equating the x components of Equation (2), we get

$$v_{Q/\mathcal{J}} = 20 + \frac{v_{Q/\mathcal{R}}}{2} \tag{3}$$

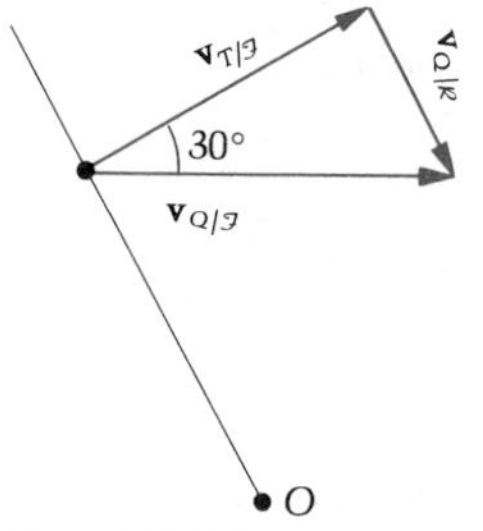

Figure E3.25b

And equating the y components:

$$0 = \frac{20}{\sqrt{3}} - v_{Q/\mathcal{R}}\frac{\sqrt{3}}{2} \tag{4}$$

Equation (4) gives $v_{Q/\mathcal{R}} = 13.3$ ft/sec, from which Equation (3) then yields

$$v_{Q/\mathcal{J}} = v_{Q/\mathcal{J}}\hat{\mathbf{i}} = 26.7\hat{\mathbf{i}} \text{ ft/sec}$$

The correct triangle relating the velocities of Q and T is shown in Figure E3.25b. As a check, $v_Q \cos 30° = (26.7)\sqrt{3}/2 = 23.1$ and $v_r = \sqrt{(20/\sqrt{3})^2 + 20^2} = 23.1$ ft/sec.

Question 3.19 Can the preceding example also be worked by using $\mathbf{v}_{T/\mathcal{J}} = \mathbf{v}_{T/\mathcal{B}} + \mathbf{v}_{Q/\mathcal{J}}$?

Answer 3.19 Yes, provided we recognize that the direction of $\mathbf{v}_{T/\mathcal{B}}$ is along the axis of the rod (see Problem 3.146). That is, $\mathbf{v}_{T/\mathcal{B}}$ must be tangent to the surface at which $\mathcal{R}$ and $\mathcal{B}$ touch. It is important to realize, however, that while the path of Q in $\mathcal{R}$ is a straight line, the path of T in $\mathcal{B}$ is not. Thus $\mathbf{a}_{T/\mathcal{B}}$ would *not* be in the direction of the axis of $\mathcal{R}$.

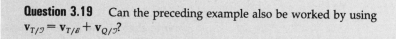

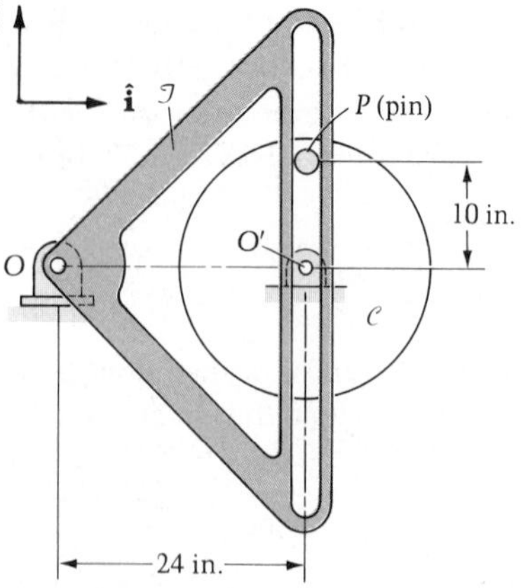

Figure E3.26a

EXAMPLE 3.26

Disk $\mathcal{C}$ of Figure E3.26a, with its attached pin P, has limited angular motion. After a 45° clockwise rotation from the original position (see Figure E3.26b), disk $\mathcal{C}$ has angular velocity $\omega_{\mathcal{C}} = 2$ ↻ rad/sec. At this time, find $\omega_{\mathcal{J}}$ of the slotted triangular body $\mathcal{J}$ and determine the velocity of pin P relative to $\mathcal{J}$.

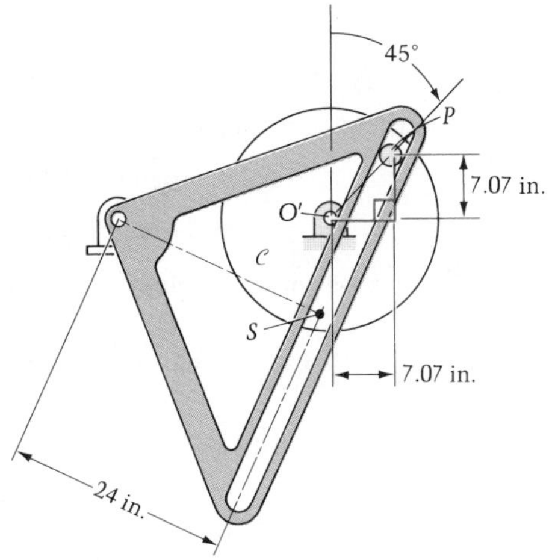

Figure E3.26b

Solution

Calling the ground frame $\mathcal{I}$, we relate the velocities of pin P in $\mathcal{J}$ and in $\mathcal{I}$:

$$\mathbf{v}_{P/\mathcal{I}} = \mathbf{v}_{P/\mathcal{J}} + \mathbf{v}_{P'/\mathcal{I}} \tag{1}$$

where P' is the point of $\mathcal{J}$ coincident with P.

We may find $\mathbf{v}_{P/\mathcal{I}}$ by relating it on body $\mathcal{C}$ to $\mathbf{v}_{O'/\mathcal{I}}$ (which vanishes):

$$\begin{aligned}\mathbf{v}_{P/\mathcal{I}} &= \omega_{\mathcal{C}}\hat{\mathbf{k}} \times \mathbf{r}_{O'P} = -2\hat{\mathbf{k}} \times (7.07\hat{\mathbf{i}} + 7.07\hat{\mathbf{j}})\\ &= 1.41\hat{\mathbf{i}} - 1.41\hat{\mathbf{j}} \text{ in./sec}\end{aligned}$$

To find $\mathbf{v}_{P/\mathcal{J}}$, we need the orientation of the slot. At 45° ↻, the configuration is as shown in Figure E3.26c. Note that the slot center S moves on a circle about O and that a tangent to this circle at S must pass through P at all times, since P must stay within the slot. From the diagram at the left we get, from geometry and trigonometry,

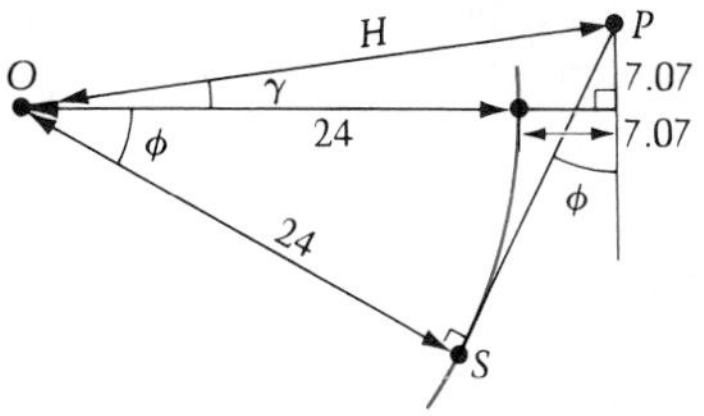

Figure E3.26c

$$\gamma = \tan^{-1}\left(\frac{7.07}{31.07}\right) = 12.8°$$

$$H = \frac{7.07}{\sin\gamma} = 31.9 \text{ in.}$$

$$\varphi + \gamma = \cos^{-1}\left(\frac{24}{31.9}\right) = 41.2°$$

$$\varphi = 28.4°$$

Using φ to form $\mathbf{v}_{P/\mathcal{J}}$ in Equation (1), we have:

$$1.41\hat{\mathbf{i}} - 1.41\hat{\mathbf{j}} = v_{P/\mathcal{J}}(\sin 28.4°\hat{\mathbf{i}} + \cos 28.4°\hat{\mathbf{j}}) + \mathbf{v}_{P'/\mathcal{I}}$$

in which we have used the fact that we know the direction but not the magnitude of $\mathbf{v}_{P/\mathcal{J}}$. (It moves in the slot at the angle calculated earlier.) Further, relating the velocities of points P' and O on $\mathcal{J}$, we have

$$\mathbf{v}_{P'/\mathcal{I}} = \overset{0}{\cancel{\mathbf{v}_{O/\mathcal{I}}}} + \omega_{\mathcal{J}}\hat{\mathbf{k}} \times \mathbf{r}_{OP'} = \omega_{\mathcal{J}}\hat{\mathbf{k}} \times (31.1\hat{\mathbf{i}} + 7.07\hat{\mathbf{j}})$$

or

$$\mathbf{v}_{P'/\mathcal{J}} = -7.07\omega_{\mathcal{J}}\hat{\mathbf{i}} + 31.1\omega_{\mathcal{J}}\hat{\mathbf{j}}$$

Substituting, and equating the coefficients of $\hat{\mathbf{i}}$ and of $\hat{\mathbf{j}}$, we get

$$\hat{\mathbf{i}} \text{ coefficients:} \quad 0.476v_{P/\mathcal{J}} - 7.07\omega_{\mathcal{J}} = 1.41$$

$$\hat{\mathbf{j}} \text{ coefficients:} \quad 0.880v_{P/\mathcal{J}} + 31.1\omega_{\mathcal{J}} = -1.41$$

Solving these equations gives $v_{P/\mathcal{J}} = 1.61$, so that:

$$\mathbf{v}_{P/\mathcal{J}} = 1.61 \angle 28.4° \text{ in./sec}$$

$$\omega_{\mathcal{J}} = 0.0909 \text{ rad/sec or } \boldsymbol{\omega}_{\mathcal{J}} = 0.0909 \circlearrowright \text{rad/sec}$$

In working out the following problems, the student is urged to begin by thinking carefully about the selection of a point whose velocities in two frames are to be related with Equation (3.49).

PROBLEMS ▶ Section 3.7

3.142 Boat $\mathcal{B}$ in Figure P3.142 departs from A and is supposed to arrive at point B some 100 ft downstream and on the other side of a river with a current of 5 ft/sec. If $\mathcal{B}$ can move at 10 ft/sec relative to the water, and if it travels on a straight line from A toward B, how long will it take?

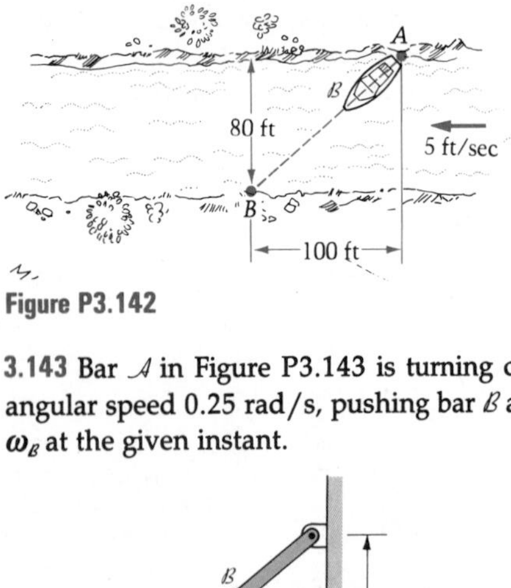

Figure P3.142

3.143 Bar $\mathcal{A}$ in Figure P3.143 is turning clockwise with angular speed 0.25 rad/s, pushing bar $\mathcal{B}$ as it goes. Find $\omega_{\mathcal{B}}$ at the given instant.

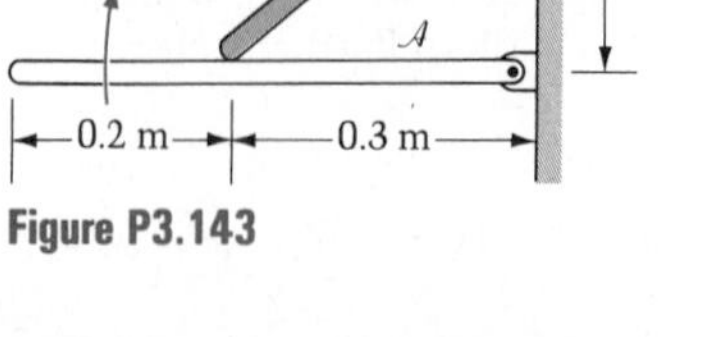

Figure P3.143

3.144 Cylinder $\mathcal{C}$ in Figure P3.144 rolls on a circular surface. When it is at the lowest point of the circle, its angular velocity and acceleration are $\omega_{\mathcal{C}} = 0.2 \circlearrowleft$ rad/s and $\alpha_{\mathcal{C}} = 0.02 \circlearrowleft$ rad/s^2. Rod $\mathcal{B}$ is pinned to $\mathcal{C}$ at E and is also pinned to a block at P that slides in the slot of $\mathcal{S}$. The constant angular velocity of $\mathcal{S}$ is $0.3 \circlearrowleft$ rad/s. Find the velocity of P in $\mathcal{S}$ and the angular velocity of $\mathcal{B}$ at the given instant.

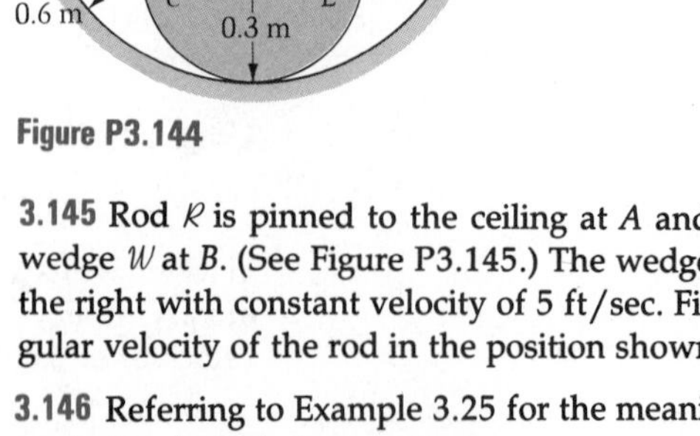

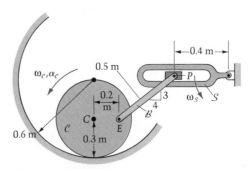

Figure P3.144

3.145 Rod $\mathcal{R}$ is pinned to the ceiling at A and slides on wedge $\mathcal{W}$ at B. (See Figure P3.145.) The wedge moves to the right with constant velocity of 5 ft/sec. Find the angular velocity of the rod in the position shown.

3.146 Referring to Example 3.25 for the meanings of the symbols, show that $\mathbf{v}_{Q/\mathcal{R}} = -\mathbf{v}_{T/\mathcal{B}}$.

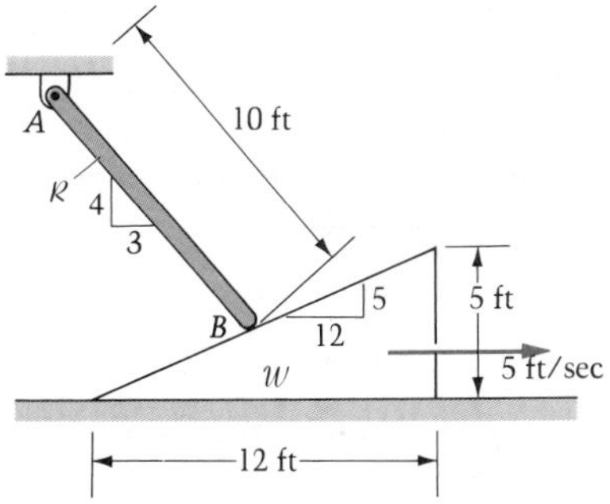

Figure P3.145

3.147 Bar $\mathcal{B}$ slides through a collar in body $\mathcal{C}$ (see Figure P3.147) and is pinned at P to a second bar $\mathcal{R}$. Both $\mathcal{R}$ and $\mathcal{C}$ are pinned to the reference frame as shown, and $\mathcal{R}$ rotates with limited motion at constant angular velocity $\dot{\theta}_{\mathcal{R}} = 1$ rad/s counterclockwise. Find the angular velocity of $\mathcal{C}$ when point P is at the top of the circle on which it travels.

3.148 Plank $\mathcal{P}$ slides on the floor at A and on block $\mathcal{D}$ at Q. Block $\mathcal{D}$ moves to the right with a constant velocity of 6 ft/sec while end A moves to the left with a constant velocity of 4 ft/sec. For the position shown in Figure P3.148, find the angular velocity of the plank.

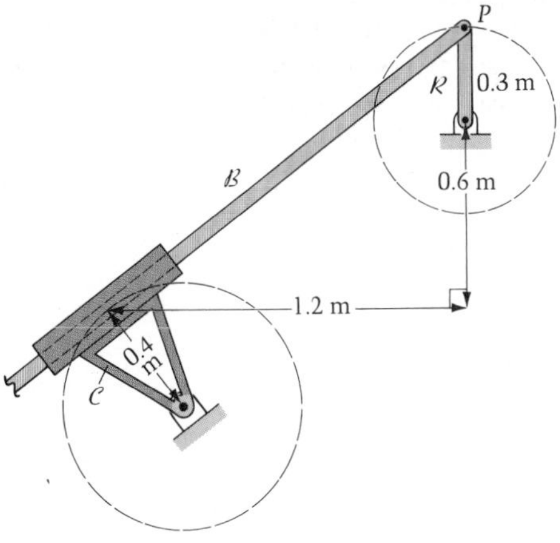

Figure P3.147

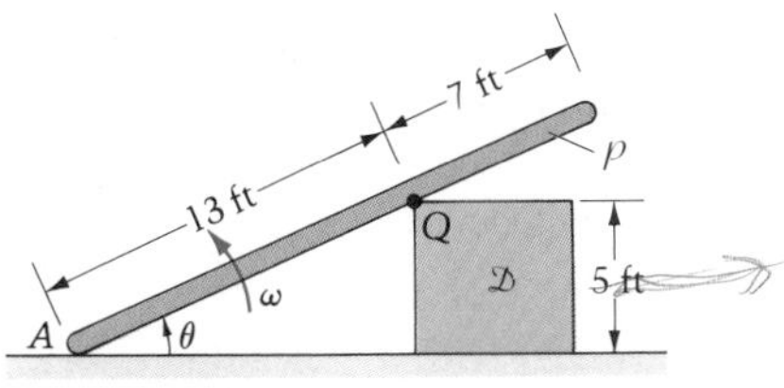

Figure P3.148

3.149 Rod $\mathcal{B}$ in Figure P3.149 has angular velocity 5 $\circlearrowleft$ rad/s. It is pinned to another rod $\mathcal{R}$, which passes through a slot in $\mathcal{A}$ as shown. At the given instant, find the angular velocity of body $\mathcal{A}$ and the velocity of any point of $\mathcal{R}$ relative to $\mathcal{A}$. *Hint*: All points of $\mathcal{R}$ translate in $\mathcal{A}$—what does this mean about the angular velocities of $\mathcal{R}$ and $\mathcal{A}$?

3.150 A mechanism consists of crank $\mathcal{C}$ pinned to O, rocker $\mathcal{R}$ pinned to O', and a small body $\mathcal{B}$ that is pinned to $\mathcal{C}$ and slides in the slot of $\mathcal{R}$. (See Figure P3.150.) The length of $\mathcal{C}$ is $\ell = D\sqrt{2}$, where D is the distance between O and O'. If $\mathcal{C}$ has constant angular velocity $\omega_{\mathcal{C}} \circlearrowright$ over a range of its motion, find $\omega_{\mathcal{R}}$ when: (a) $\theta_{\mathcal{R}} = \tan^{-1}(1/2)$; (b) $\theta_{\mathcal{R}} = 90°$

3.151 Collars $\mathcal{A}$ and $\mathcal{C}$ in Figure P3.151 are pinned together at C, and they slide on rods $\mathcal{B}$ and $\mathcal{S}$, respectively. Rod $\mathcal{B}$ has a constant angular velocity $\dot{\theta}\hat{\mathbf{k}}$ for $10° \le \theta \le 45°$. Find the velocity of point C relative to $\mathcal{B}$, as a function of D, θ, and $\dot{\theta}$ in this range of angles.

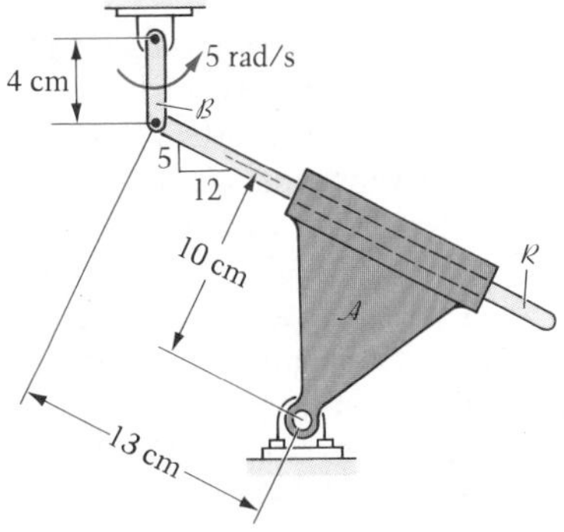

Figure P3.149

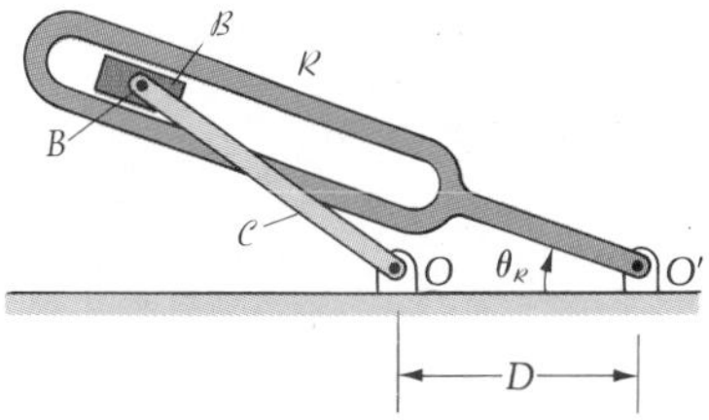

Figure P3.150

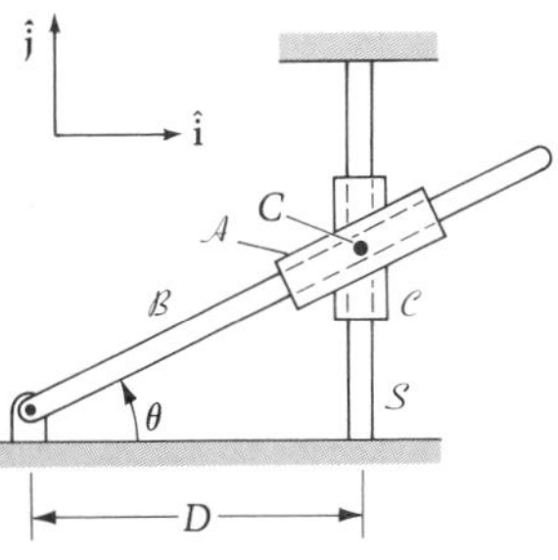

Figure P3.151

3.152 Figure P3.152 shows a circular cam $\mathcal{B}_1$ and a flat-face translating follower $\mathcal{B}_2$. If $\mathcal{B}_1$ rotates with constant angular velocity ω_0 ⟲, find the maximum velocity in reference frame $\mathcal{I}$ of any point of $\mathcal{B}_2$, in terms of ω_0 and the offset distance δ.

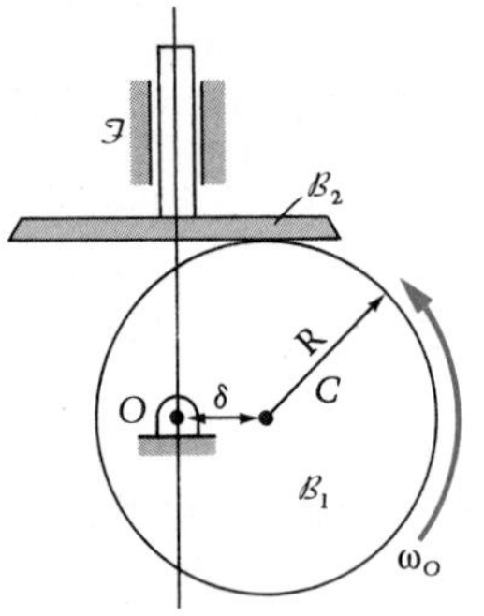

Figure P3.152

* **3.153** Figure P3.153 illustrates a "Geneva mechanism," in which disk $\mathcal{A}$ is driven with a constant counterclockwise angular speed and produces an intermittent (starting and stopping, waiting, then repeating) rotational motion of the slotted disk $\mathcal{B}$. Pin P is fixed to $\mathcal{A}$ and drives disk $\mathcal{B}$ by pressing on the surfaces of the slots. Show with the use of Equation (3.49) that disk $\mathcal{B}$ will have zero angular speed in the two positions shown, a varying angular speed in between these positions, and zero angular speed while P is returning to the $\theta = 135°$ position. Note that the operation of the mechanism requires that the distance between O and O' be $\sqrt{2}R$.

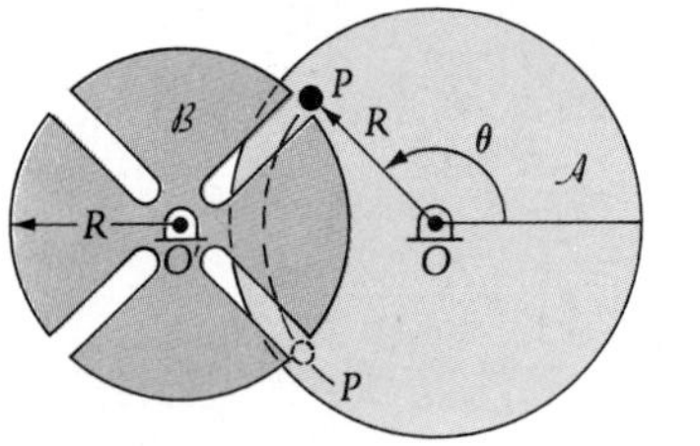

Figure P3.153

* **3.154** In the preceding problem let $R = 0.1$ m and $\dot{\theta}_{\mathcal{A}} = 5$ rad/s = constant. Find the angular velocity of the slotted body $\mathcal{B}$ at the instant when $\theta = 160°$.

* **3.155** Rods $\mathcal{R}$ and $\mathcal{L}$ in Figure P3.155 are pinned at O and O' to a reference frame $\mathcal{I}$. Rod $\mathcal{L}$ is also pinned to the slotted body $\mathcal{B}$ at B. The upper end of $\mathcal{R}$ is pinned at P to a roller that moves freely in the slot of $\mathcal{B}$. The angular velocities of rod $\mathcal{R}$ and link $\mathcal{L}$ are constants:

$$\omega_{\mathcal{R}} = 0.2 \circlearrowright \text{ rad/s}$$

$$\omega_{\mathcal{L}} = 0.4 \circlearrowright \text{ rad/s}$$

Determine the velocity of P in $\mathcal{B}$ and the angular velocity of $\mathcal{B}$ at the given instant.

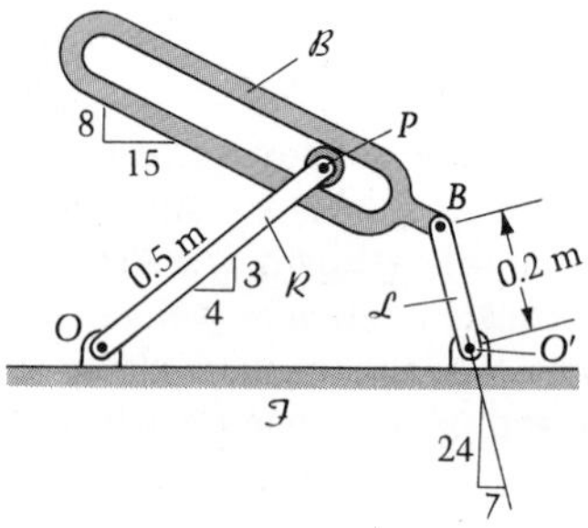

Figure P3.155

* **3.156** In Figure P3.156 collar $\mathcal{C}$ is fixed to arm $\mathcal{A}$ and slides along rod $\mathcal{R}$. Arm $\mathcal{A}$ is pinned to a second collar $\mathcal{K}$ at A; this collar slides on rod $\mathcal{L}$. At the given instant, $\omega_{\mathcal{R}} = 0.2 \circlearrowright$ rad/s, $\omega_{\mathcal{L}} = 0.1 \circlearrowright$ rad/s, and the velocity of all points of $\mathcal{C}$ relative to rod $\mathcal{R}$ is 0.3 m/s outward (along OC). Give the angular velocity of $\mathcal{A}$ by inspection, and find the velocity of the points of $\mathcal{K}$ relative to $\mathcal{L}$.

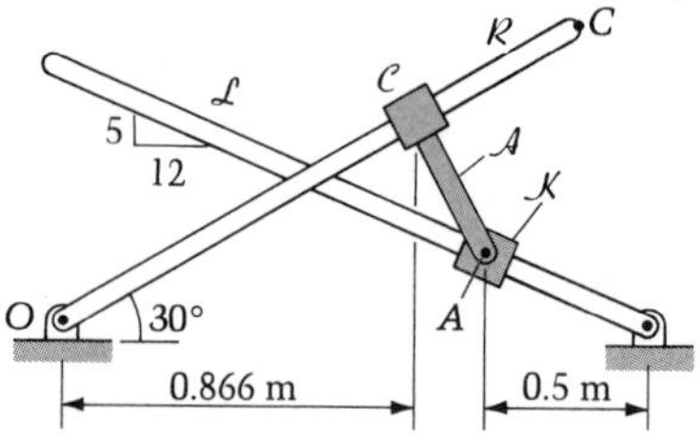

Figure P3.156

3.157 Show that the rigid-body velocity equation (3.8) can be derived from Equation (3.49). *Hint*: Fix P to $\mathcal{B}$ and the equation will relate the velocities in $\mathcal{I}$ of the two points P and O' of $\mathcal{B}$.

3.8 Relationship Between the Accelerations of a Point with Respect to Two Different Frames of Reference

We shall now derive the companion equation to (3.49), this one relating the accelerations of P in $\mathcal{I}$ and $\mathcal{B}$. Differentiating Equation (3.49) in $\mathcal{I}$, we get

$$\mathbf{a}_{P/\mathcal{I}} = {}^{\mathcal{I}}\dot{\mathbf{v}}_{P/\mathcal{B}} + \mathbf{a}_{O'/\mathcal{I}} + \ddot{\theta}\hat{\mathbf{k}} \times \mathbf{r}_{O'P} + \dot{\theta}\hat{\mathbf{k}} \times {}^{\mathcal{I}}\dot{\mathbf{r}}_{O'P} \tag{3.52}$$

Using Equation (3.44) to "move the derivative" in the first and last terms on the right side of Equation (3.52) gives

$$\begin{aligned}\mathbf{a}_{P/\mathcal{I}} = ({}^{\mathcal{B}}\dot{\mathbf{v}}_{P/\mathcal{B}} + \dot{\theta}\hat{\mathbf{k}} \times \mathbf{v}_{P/\mathcal{B}}) + \mathbf{a}_{O'/\mathcal{I}} + \ddot{\theta}\hat{\mathbf{k}} \times \mathbf{r}_{O'P} \\ + \dot{\theta}\hat{\mathbf{k}} \times ({}^{\mathcal{B}}\dot{\mathbf{r}}_{O'P} + \dot{\theta}\hat{\mathbf{k}} \times \mathbf{r}_{O'P})\end{aligned} \tag{3.53}$$

Recognizing that ${}^{\mathcal{B}}\dot{\mathbf{r}}_{O'P} = \mathbf{v}_{P/\mathcal{B}}$ and that ${}^{\mathcal{B}}\dot{\mathbf{v}}_{P/\mathcal{B}} = \mathbf{a}_{P/\mathcal{B}}$, and rearranging terms, from Equation (3.53) we obtain

$$\mathbf{a}_{P/\mathcal{I}} = \mathbf{a}_{P/\mathcal{B}} + (\mathbf{a}_{O'/\mathcal{I}} + \ddot{\theta}\hat{\mathbf{k}} \times \mathbf{r}_{O'P} - \dot{\theta}^2\mathbf{r}_{O'P}) + 2\dot{\theta}\hat{\mathbf{k}} \times \mathbf{v}_{P/\mathcal{B}} \tag{3.54}$$

in which $\dot{\theta}\hat{\mathbf{k}} \times (\dot{\theta}\hat{\mathbf{k}} \times \mathbf{r}_{O'P}) = -\dot{\theta}^2\mathbf{r}_{O'P}$ as we have already seen in Section 3.5.

The parenthesized term in Equation (3.54) is seen from Equation (3.19) to be $\mathbf{a}_{P'/\mathcal{I}}$, where, as before, P' is the point of $\mathcal{B}$ (or $\mathcal{B}$ extended) that is coincident with P. Therefore we have the following for our result:

$$\mathbf{a}_{P/\mathcal{I}} = \mathbf{a}_{P/\mathcal{B}} + \mathbf{a}_{P'/\mathcal{I}} + 2\dot{\theta}\hat{\mathbf{k}} \times \mathbf{v}_{P/\mathcal{B}} \tag{3.55}$$

In words: The acceleration of P in $\mathcal{I}$ equals its acceleration in $\mathcal{B}$, plus the acceleration in $\mathcal{I}$ of the point of $\mathcal{B}$ coincident with P, *plus* the **Coriolis acceleration**, $2\dot{\theta}\hat{\mathbf{k}} \times \mathbf{v}_{P/\mathcal{B}}$. The Coriolis acceleration is seen to provide an unexpected but essential difference between the forms of Equations (3.51) and (3.55).

Question 3.20 If we differentiate Equation (3.51) instead of (3.49), we might (*erroneously!*) obtain

$$\begin{aligned}\mathbf{a}_{P/\mathcal{I}} &= {}^{\mathcal{I}}\dot{\mathbf{v}}_{P/\mathcal{B}} + \mathbf{a}_{P'/\mathcal{I}} \\ &= {}^{\mathcal{B}}\dot{\mathbf{v}}_{P/\mathcal{B}} + \boldsymbol{\omega}_{\mathcal{B}/\mathcal{I}} \times \mathbf{v}_{P/\mathcal{B}} + \mathbf{a}_{P'/\mathcal{I}} \\ &= \mathbf{a}_{P/\mathcal{B}} + \mathbf{a}_{P'/\mathcal{I}} + \boldsymbol{\omega}_{\mathcal{B}/\mathcal{I}} \times \mathbf{v}_{P/\mathcal{B}}\end{aligned}$$

and we come up (incorrectly) short by half on the Coriolis term. What is wrong with this approach?

In the same procedure we used for velocities, we can simplify the notation of Equation (3.54) if there is but one "moving frame" ($\mathcal{B}$) involved, which is in motion relative to the reference frame ($\mathcal{I}$):

$$\mathbf{a}_P = \mathbf{a}_{\text{rel}} + \mathbf{a}_{O'} + \boldsymbol{\alpha} \times \mathbf{r} - \omega^2\mathbf{r} + 2\boldsymbol{\omega} \times \mathbf{v}_{\text{rel}} \tag{3.56}$$

Answer 3.20 The error is that the derivative ${}^{\mathcal{I}}\dot{\mathbf{v}}_{P'/\mathcal{I}}$ is not equal to $\mathbf{a}_{P'/\mathcal{I}}$, because P' denotes a *succession of points* of $\mathcal{B}$ which are at each instant coincident with P.

In this equation, $\mathbf{a}_P$ and $\mathbf{a}_{\text{rel}}$ are the respective accelerations of P in $\mathcal{I}$ and in $\mathcal{B}$. The vectors $\boldsymbol{\omega}$ and $\boldsymbol{\alpha}$ are the angular velocity and angular acceleration of $\mathcal{B}$ in $\mathcal{I}$ (equal to $\dot{\theta}\hat{\mathbf{k}}$ and $\ddot{\theta}\hat{\mathbf{k}}$) and $\mathbf{r} = \mathbf{r}_{O'P}$ (the position vector of P in the "moving frame"). We now consider several examples showing the use of Equations (3.54) and (3.55).

EXAMPLE 3.27

Find the acceleration in frame $\mathcal{I}$ of the yellowjacket of Example 3.23.

Solution

Using Equation 3.54, we obtain:

$$\mathbf{a}_{Y/\mathcal{I}} = \mathbf{a}_{Y/\mathcal{B}} + (\mathbf{a}_{O'/\mathcal{I}} + \ddot{\theta}\hat{\mathbf{k}} \times \mathbf{r}_{O'Y} - \dot{\theta}^2\mathbf{r}_{O'Y}) + 2\dot{\theta}\hat{\mathbf{k}} \times \mathbf{v}_{Y/\mathcal{B}}$$

$$= 0 + 0 + 0 - \left(\frac{10\pi}{9}\right)^2 (r\hat{\mathbf{i}}) + 2\left(\frac{10\pi}{9}\hat{\mathbf{k}}\right) \times 2\hat{\mathbf{i}}$$

$$= -\frac{100\pi^2}{81} r\hat{\mathbf{i}} + \frac{40\pi}{9}\hat{\mathbf{j}} \text{ in./sec}^2$$

The "$\hat{\mathbf{j}}$-term" in this example is the Coriolis acceleration. Note that the yellowjacket has two nonzero acceleration components, even though both $\ddot{r}$ and $\ddot{\theta}$ are zero in this example.

EXAMPLE 3.28

If in Example 3.24 we have the additional data that $\boldsymbol{\alpha}_{\mathcal{R}} = 0.25\hat{\mathbf{k}}$ rad/s² at the given time, find $\boldsymbol{\alpha}_{\mathcal{B}}$ and $\mathbf{a}_{P/\mathcal{B}}$. (See Figure E3.28.)

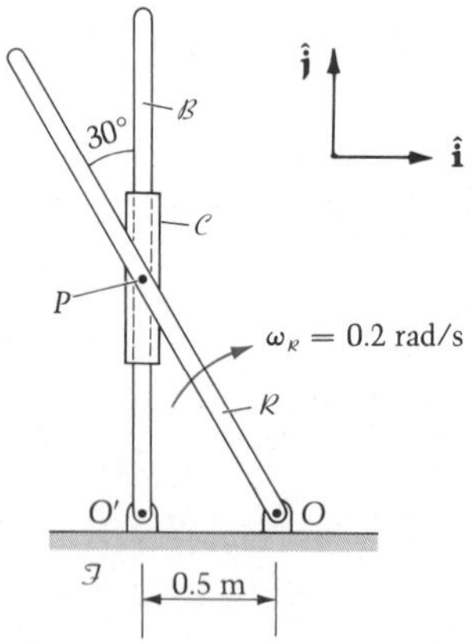

Figure E3.28

Solution

Relating the accelerations of P in $\mathcal{B}$ and $\mathcal{I}$, we have

$$\mathbf{a}_{P/\mathcal{I}} = \mathbf{a}_{P/\mathcal{B}} + \mathbf{a}_{O'/\mathcal{I}} + \boldsymbol{\alpha}_{\mathcal{B}} \times \mathbf{r}_{O'P} - \omega_{\mathcal{B}}^2\mathbf{r}_{O'P} + 2\boldsymbol{\omega}_{\mathcal{B}/\mathcal{I}} \times \mathbf{v}_{P/\mathcal{B}}$$

$$\overbrace{\underbrace{\mathbf{a}_{O/\mathcal{I}}}_{\text{zero}} + \boldsymbol{\alpha}_{\mathcal{R}} \times \mathbf{r}_{OP} - \omega_{\mathcal{R}}^2\mathbf{r}_{OP}}^{\text{on } \mathcal{R}} = a_{P/\mathcal{B}}\hat{\mathbf{j}} + \underbrace{\mathbf{a}_{O'/\mathcal{I}}}_{\text{zero}} + \boldsymbol{\alpha}_{\mathcal{B}} \times \mathbf{r}_{O'P} - \omega_{\mathcal{B}}^2\mathbf{r}_{O'P} + 2(-0.2\hat{\mathbf{k}}) \times (0.1\hat{\mathbf{j}})$$

in which the answers to Example 3.24 are used in the Coriolis term. Filling in the variables, we get

$$0.25\hat{\mathbf{k}} \times (-0.5\hat{\mathbf{i}} + 0.866\hat{\mathbf{j}}) - 0.04(-0.5\hat{\mathbf{i}} + 0.866\hat{\mathbf{j}})$$
$$= a_{P/\mathcal{B}}\hat{\mathbf{j}} + \alpha_{\mathcal{B}}\hat{\mathbf{k}} \times 0.866\hat{\mathbf{j}} - 0.04(0.866\hat{\mathbf{j}}) + 0.04\hat{\mathbf{i}}$$

$\hat{\mathbf{i}}$ coefficients: $\quad -0.217 + 0.02 = -0.866\alpha_{\mathcal{B}} + 0.04$

$$\alpha_{\mathcal{B}} = 0.274 \Rightarrow \boldsymbol{\alpha}_{\mathcal{B}} = 0.274\hat{\mathbf{k}} \text{ rad/s}^2$$

$\hat{\mathbf{j}}$ coefficients: $\quad -0.125 - 0.0346 = a_{P/\mathcal{B}} - 0.0346$

$$\mathbf{a}_{P/\mathcal{B}} = -0.125\hat{\mathbf{j}} \text{ m/s}^2$$

Thus point P is slowing down as it moves outward along $\mathcal{B}$, and $\mathcal{B}$ is slowing down as it rotates clockwise.

The reader should note in the preceding example that $\boldsymbol{\alpha}_{\mathcal{C}}$ and $\boldsymbol{\alpha}_{\mathcal{B}}$ are identical. The sleeve forces the two bodies $\mathcal{B}$ and $\mathcal{C}$ to translate relative to each other, so that, as was pointed out at the end of Example 3.23, $\boldsymbol{\omega}_{\mathcal{B}} \equiv \boldsymbol{\omega}_{\mathcal{C}}$. By differentiating this equation, we see that the angular accelerations are also always equal.

Example 3.29

In Example 3.25, suppose that at the given instant ($\theta = 60°$) all the data are the same and in addition $\boldsymbol{\alpha}_{\mathcal{R}} = 30 \circlearrowleft$ rad/sec². Find the acceleration of block $\mathcal{B}$ (see Figure E3.29).

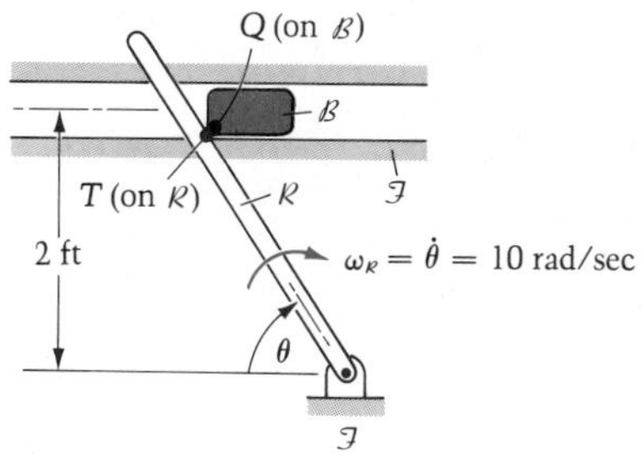

Figure E3.29

Solution

Note that if we again use Q as our point (of the block $\mathcal{B}$) that is moving relative to two bodies ($\mathcal{R}$ and $\mathcal{I}$), we know the direction of $\mathbf{a}_{Q/\mathcal{R}}$. (It is *along* $\mathcal{R}$ since point Q moves on a *straight line* in $\mathcal{R}$.)

Equation (3.55) thus gives

$$\mathbf{a}_{Q/\mathcal{I}} = \mathbf{a}_{Q/\mathcal{R}} + \mathbf{a}_{Q'/\mathcal{I}} + 2\omega_{\mathcal{R}}\hat{\mathbf{k}} \times \mathbf{v}_{Q/\mathcal{R}}$$

Noting that point Q' is the point T, we obtain

$$a_{Q/\mathcal{I}}\hat{\mathbf{i}} = a_{Q/\mathcal{R}}\left(-\frac{1}{2}\hat{\mathbf{i}} + \frac{\sqrt{3}}{2}\hat{\mathbf{j}}\right) + (\overset{0}{\mathbf{a}_O} + \overset{30}{\alpha_{\mathcal{R}}}\hat{\mathbf{k}} \times \overbrace{\mathbf{r}_{OP}}^{\left(\frac{-2}{\sqrt{3}}\hat{\mathbf{i}} + 2\hat{\mathbf{j}}\right)} - \overset{10^2}{\omega_{\mathcal{R}}^2}\mathbf{r}_{OP})$$

$$+ 2(-10\hat{\mathbf{k}}) \times \left(\frac{40}{3}\right)\left(-\frac{1}{2}\hat{\mathbf{i}} + \frac{\sqrt{3}}{2}\hat{\mathbf{j}}\right)$$

This result yields the following scalar component equations:

$\hat{\mathbf{j}}$ coefficients:
$$0 = a_{Q/\mathcal{R}}\frac{\sqrt{3}}{2} - \frac{60}{\sqrt{3}} - 200 + \frac{800}{6}$$

$$a_{Q/\mathcal{R}} = 117 \text{ in./sec}^2$$

$\hat{\mathbf{i}}$ coefficients:
$$a_{Q/\mathcal{I}} = 117\left(-\frac{1}{2}\right) - 60 + \frac{200}{\sqrt{3}} + \frac{800}{6}\sqrt{3}$$

Thus the acceleration of the translating block is:

$$\mathbf{a}_{Q/\mathcal{I}} = a_{Q/\mathcal{I}}\hat{\mathbf{i}} = 228\hat{\mathbf{i}} \text{ in./sec}^2$$

EXAMPLE 3.30

If in Example 3.26 we add $\boldsymbol{\alpha}_C = 10 \circlearrowleft$ rad/sec² to the data, find the angular acceleration of $\mathcal{I}$ and the acceleration of P relative to $\mathcal{I}$. (See Figures E3.30a,b.)

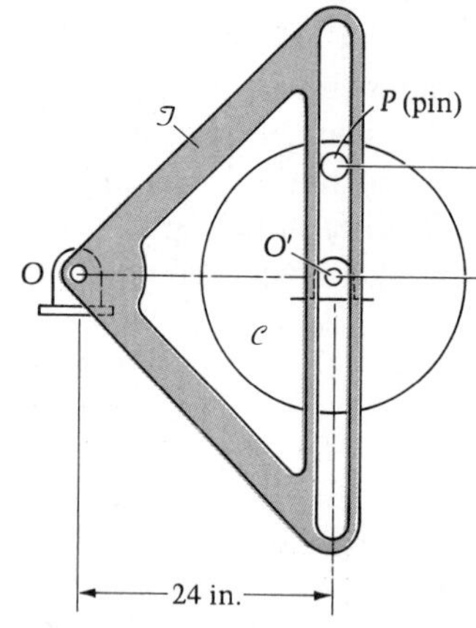

Figure E3.30a

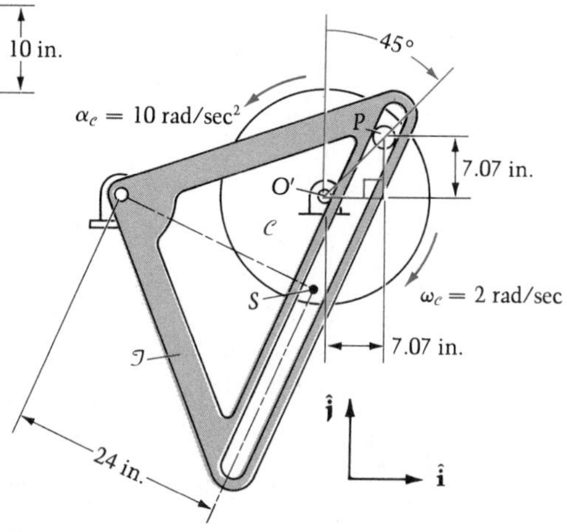

Figure E3.30b

Solution

Again we apply Equation (3.55):

$$\mathbf{a}_{P/\mathcal{I}} = \mathbf{a}_{P/\mathcal{I}} + \mathbf{a}_{P'/\mathcal{I}} + 2\omega_{\mathcal{I}}\hat{\mathbf{k}} \times \mathbf{v}_{P/\mathcal{I}}$$

$$\overset{0}{\cancel{\mathbf{a}}_{O'/\mathcal{I}}} + \overset{10}{\cancel{\alpha}_{\mathcal{C}}}\hat{\mathbf{k}} \times \overset{(7.07\hat{\mathbf{i}} + 7.07\hat{\mathbf{j}})}{\cancel{\mathbf{r}}_{O'P}} - \overset{2^2}{\cancel{\omega^2_{\mathcal{C}}}}\mathbf{r}_{O'P} = a_{P/\mathcal{I}}(\sin 28.4°\hat{\mathbf{i}} + \cos 28.4°\hat{\mathbf{j}}) + \overset{0}{\cancel{\mathbf{a}}_{O'/\mathcal{I}}}$$

$$+ \alpha_{\mathcal{I}}\hat{\mathbf{k}} \times (31.1\hat{\mathbf{i}} + 7.07\hat{\mathbf{j}}) - \overset{0.091^2}{\cancel{\omega^2_{\mathcal{I}}}}(31.1\hat{\mathbf{i}} + 7.07\hat{\mathbf{j}})$$

$$+ 2(-0.091\hat{\mathbf{k}}) \times 1.62(\sin 28.4°\hat{\mathbf{i}} + \cos 28.4°\hat{\mathbf{j}})$$

and again we equate the $\hat{\mathbf{i}}$ coefficients, and then the $\hat{\mathbf{j}}$ coefficients, to obtain the scalar component equations:

$\hat{\mathbf{i}}$ coefficients: $\quad -99.0 = 0.476a_{P/\mathcal{I}} - 7.07\alpha_{\mathcal{I}} - 0.258 + 0.259$

$\hat{\mathbf{j}}$ coefficients: $\quad 42.4 = a_{P/\mathcal{I}}(0.880) + 31.1\alpha_{\mathcal{I}} - 0.0586 - 0.140$

Solving these equations results in $\alpha_{\mathcal{I}} = 5.12$ rad/sec², so that:

$\boldsymbol{\alpha}_{\mathcal{I}} = 5.12 \circlearrowleft$ rad/sec²

$a_{P/\mathcal{I}} = -132$ in./sec² $\quad$ or $\quad \mathbf{a}_{P/\mathcal{I}} = 132$ (direction arrow) 28.4° in./sec²

Example 3.31

Pin P in Figure E.3.31 is attached to cart $\mathcal{B}$ and slides in the smooth slot cut in wheel $\mathcal{C}$. The wheel rolls on the rough plane $\mathcal{I}$. The cart's position is given by $x_B = 0.3t^2$, with x_B in meters when t is in seconds. Find $\omega_{\mathcal{C}}$ and $\alpha_{\mathcal{C}}$ at the given instant (which is at $t = 3$ s), and determine the acceleration of P in the slot at this time.

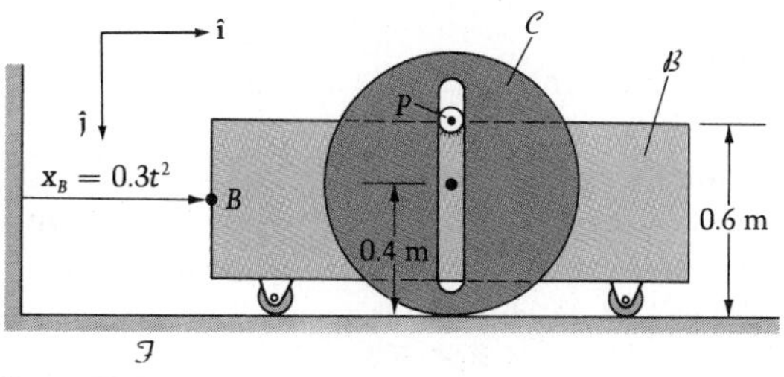

Figure E3.31

Solution

Since $x_B = 0.3t^2$, we have $\mathbf{v}_B = 0.6t\hat{\mathbf{i}}$ and $\mathbf{a}_B = 0.6\hat{\mathbf{i}}$; these are the velocity and acceleration vectors of all points of the cart $\mathcal{B}$, in particular of P. At $t = 3$, we have $\mathbf{v}_B = 1.8\hat{\mathbf{i}}$ m/s and $\mathbf{a}_B = 0.6\hat{\mathbf{i}}$ m/s^2.

Relating the velocities of P in frames $\mathcal{C}$ and $\mathcal{I}$, we obtain

$$\mathbf{v}_{P/\mathcal{I}} = \mathbf{v}_{P/\mathcal{C}} + \mathbf{v}_{P'/\mathcal{I}}$$

$$v_B\hat{\mathbf{i}} = v_{P/\mathcal{C}}\hat{\mathbf{j}} + 0.6\omega_{\mathcal{C}}\hat{\mathbf{i}}$$

$\hat{\mathbf{i}}$ coefficients: $\quad v_B = 1.8 = 0.6\omega_{\mathcal{C}} \Rightarrow \omega_{\mathcal{C}} = 3 \circlearrowright$ rad/s

$\hat{\mathbf{j}}$ coefficients: $\quad v_{P/\mathcal{C}} = 0$

Relating the accelerations using Equation (3.54), we get

$$\mathbf{a}_{P/\mathcal{I}} = \mathbf{a}_{P/\mathcal{C}} + \underbrace{\mathbf{a}_{P'/\mathcal{I}}} + 2\boldsymbol{\omega}_{\mathcal{C}/\mathcal{I}} \times \mathbf{v}_{P/\mathcal{C}}$$

$$a_B\hat{\mathbf{i}} = a_{P/\mathcal{C}}\hat{\mathbf{j}} + \overbrace{0.6\alpha_{\mathcal{C}}\hat{\mathbf{i}} + 0.2\omega_{\mathcal{C}}^2\hat{\mathbf{j}}} + 0$$

Note that we have related P' (the point of $\mathcal{C}$-extended coincident with P) and the center of $\mathcal{C}$ (call it C) to get

$$\mathbf{a}_{P'/\mathcal{I}} = \mathbf{a}_C + \alpha_{\mathcal{C}}\hat{\mathbf{k}} \times \overbrace{\mathbf{r}_{CP'}}^{-0.2\hat{\mathbf{j}}} - \omega_{\mathcal{C}}^2\mathbf{r}_{CP'}$$

$$= \overbrace{R}^{0.4}\alpha_{\mathcal{C}}\hat{\mathbf{i}} + 0.2\alpha_{\mathcal{C}}\hat{\mathbf{i}} + 0.2\omega_{\mathcal{C}}^2\hat{\mathbf{j}}$$

$$= 0.6\alpha_{\mathcal{C}}\hat{\mathbf{i}} + 0.2\omega_{\mathcal{C}}^2\hat{\mathbf{j}}$$

Solving, we obtain

$\hat{\mathbf{i}}$ coefficients: $\quad a_B = 0.6 = 0.6\alpha_{\mathcal{C}} \Rightarrow \alpha_{\mathcal{C}} = 1 \circlearrowright$ rad/s^2

$\hat{\mathbf{j}}$ coefficients: $\quad a_{P/\mathcal{C}} = -0.2\omega_{\mathcal{C}}^2 = -0.2(3^2) = -1.8$ m/s^2

Thus the acceleration of P in the slot (which is its acceleration in $\mathcal{C}$) is $1.8 \uparrow \text{m/s}^2$; it is upward because we assumed it to be in the positive y direction ($\downarrow$) and got a negative answer. Note that although P is momentarily stopped in the slot ($v_{P/\mathcal{C}} = 0$), it has to be "getting ready" to move outward since $\mathcal{B}$ is translating to the right and $\mathcal{C}$ is rolling that way. This is what is indicated by $\mathbf{a}_{P/\mathcal{C}} = 1.8 \uparrow \text{m/s}^2$.

PROBLEMS ▶ Section 3.8

3.158 A bug $\mathcal{B}$ is crawling outward at a uniform speed relative to the rotating arm of 3 ft/ sec. In the position shown in Figure P3.158, for the arm $\omega = 2$ rad/sec and $\alpha = 4$ rad/sec^2, both counterclockwise. What is the acceleration of the bug? Indicate the direction in a sketch.

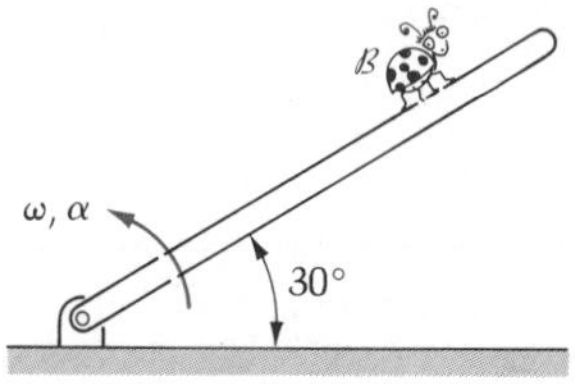

Figure P3.158

3.159 The mechanism shown in Figure P3.159 is used to raise and lower hammer $\mathcal{H}$. The 26-cm crank $\mathcal{C}$ turns clockwise at the constant rate of 30 rpm. It is pinned to block $\mathcal{B}$, which slides in a slot in $\mathcal{H}$. If at $t = 0$ point A is directly above O, find the velocity and acceleration of $\mathcal{H}$ as a function of time. (The block and hammer are slightly offset from $\mathcal{C}$ so they do not interfere with the pin at O.)

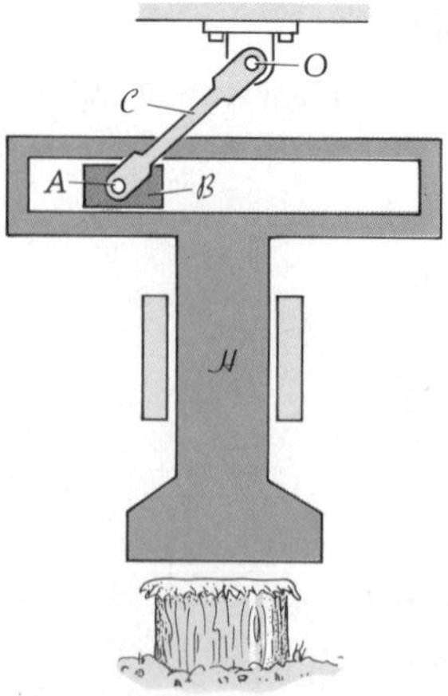

Figure P3.159

3.160 Bar $\mathcal{A}$ in Figure P3.160 has angular velocity 0.25 ↻ rad/s and angular acceleration 0.15 ↺ rad/s^2 at the given instant. Find the angular acceleration of $\mathcal{B}$ at this time. (See Problem 3.143.)

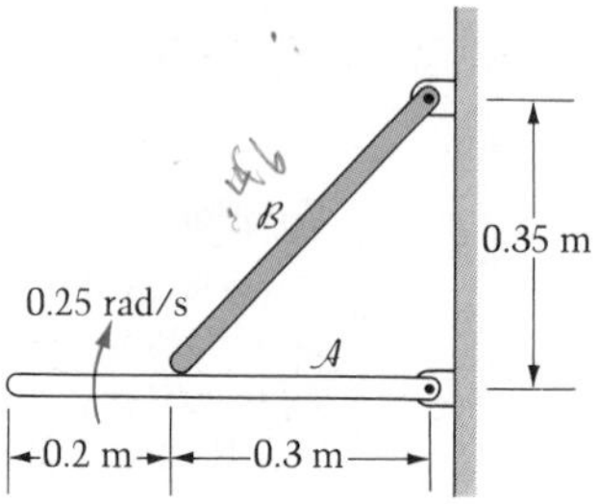

Figure P3.160

3.161 In Problem 3.144 find the acceleration of P in $\mathcal{S}$ and the angular acceleration of $\mathcal{B}$.

3.162 In Problem 3.155 find, at the same instant of time, the acceleration of P in $\mathcal{B}$ and the angular acceleration of $\mathcal{B}$.

3.163 In Figure P3.163,

$$\mathbf{v}_A = 4 \rightarrow \text{in./sec} \qquad \text{and} \qquad \frac{d}{dt}|\mathbf{v}_A| = 3 \text{ in./sec}^2$$

If the bar stays in contact with both the step and circular

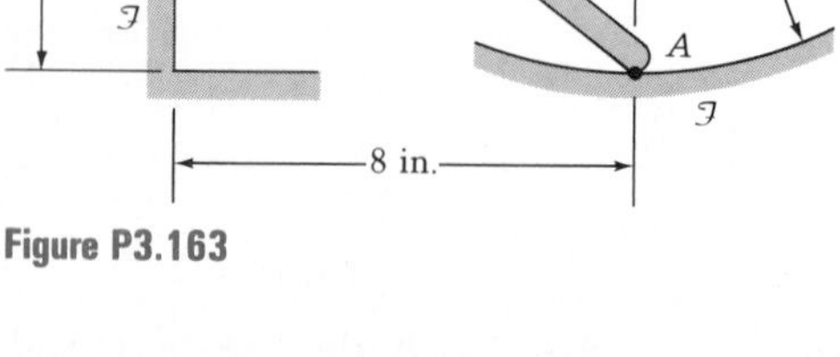

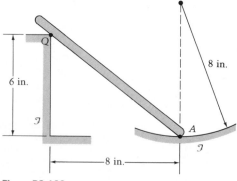

Figure P3.163

trough, find its angular acceleration. *Hint*: Treat point Q (fixed to the step) as the "moving point," and note that Q moves on a straight line relative to the bar.

3.164 In Problem 3.151 determine the acceleration of C relative to $\mathcal{B}$ as a function of D, θ, and $\dot{\theta}$.

3.165 Rods $\mathcal{A}$ and $\mathcal{B}$ (see Figure P3.165) pass smoothly through the short collars, which can turn relative to each other by virtue of the ball-and-socket connection. If bars $\mathcal{A}$ and $\mathcal{B}$ turn with constant angular velocities 0.4 ⟲ rad/sec and 0.2 ⟳ rad/sec, respectively, find the velocity and acceleration of the ball-and-socket connection with respect to $\mathcal{A}$ and to $\mathcal{B}$ in the indicated position.

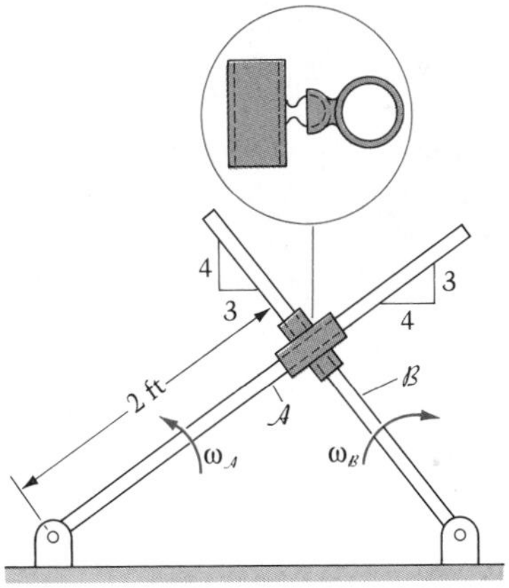

Figure P3.165

* **3.166** Wheel $\mathcal{W}$ in Figure P3.166 has a constant clockwise angular velocity of 2 rad/sec. It is connected by link $\mathcal{L}$ to block $\mathcal{E}$. End B of rod $\mathcal{R}$ slides in a vertical slot in block $\mathcal{E}$. For the position shown, find the angular velocity and angular acceleration of rod $\mathcal{R}$ if block $\mathcal{E}$ translates.

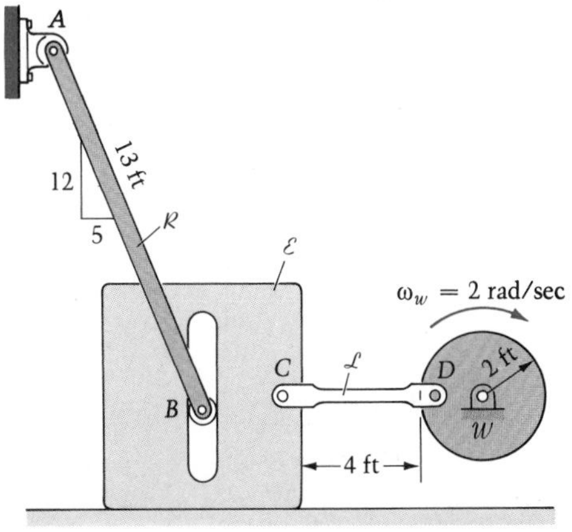

Figure P3.166

* **3.167** Extending Problem 3.149, suppose that the angular velocity of rod $\mathcal{B}$, 5 ⟲ rad/s, is constant in time. (See Figure P3.167.) Find, at the given instant, the angular acceleration of $\mathcal{A}$ and the acceleration of any point of $\mathcal{R}$ relative to $\mathcal{A}$.

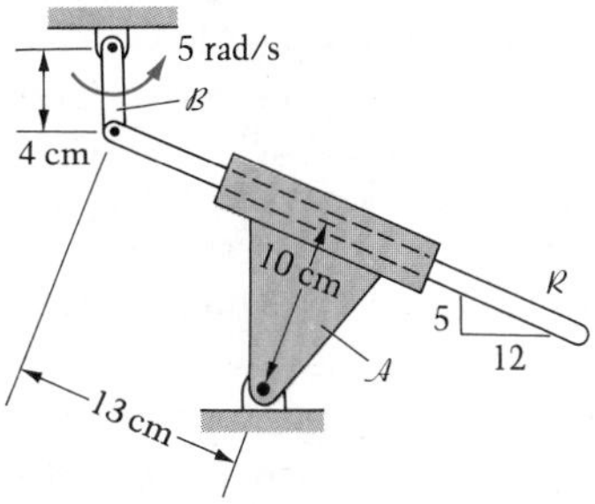

Figure P3.167

3.168 Referring to Example 3.29 for the meanings of the symbols, show that $\mathbf{a}_{Q/\mathcal{R}} \neq -\mathbf{a}_{T/\mathcal{B}}$.

* **3.169** A circular turntable $\mathcal{T}$ (see Figure P3.169) rotates about a vertical axis through O (normal to the plane of the paper) with θ changing at a constant rate ω_0. A block $\mathcal{B}$ rests in a groove cut in the turntable. If the cable is reeled in at a constant velocity v relative to $\mathcal{T}$, find expressions for the radial and transverse components of the block's acceleration. Check your answers using the expression for acceleration in cylindrical coordinates.

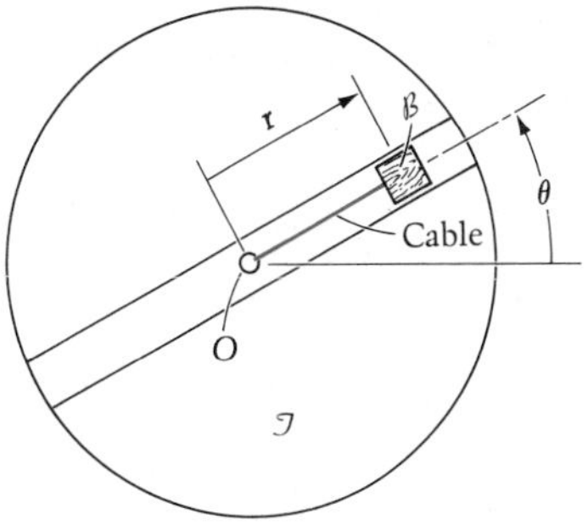

Figure P3.169

3.170 Show that the rigid-body acceleration equation (3.19) can be *derived* from Equation (3.54). *Hint*: If you fix P to $\mathcal{B}$, the equation will then relate the accelerations in $\mathcal{T}$ of the two points P and O' of $\mathcal{B}$.

* **3.171** In Problem 3.153 let $R = 0.1$ m and $\ddot{\theta}_{\mathcal{A}} = 5$ rad/s as in Problem 3.154. This time, again at $\theta = 160°$, find: (a) the angular acceleration of $\mathcal{B}$; (b) the acceleration of P in $\mathcal{B}$ $(= \mathbf{a}_{P/\mathcal{B}})$.

* **3.172** The 26-ft rod $\mathcal{R}$ in Figure P3.172 slides on a plane surface at A and on the fixed half-cylinder at Q as shown below. If end A is moved at a constant velocity of 4 ft/sec to the right along the plane, determine the vertical component of the acceleration of B for the position shown.

* **3.173** Pin P in Figure P3.173 moves along a curved path and is controlled by the motions of the slotted links $\mathcal{A}$ and $\mathcal{B}$. At the instant shown, each point of $\mathcal{A}$ has a velocity of 5 ft/sec and an acceleration of 20 ft/sec², both to the right, while each point of $\mathcal{B}$ has a velocity of 3 ft/sec and an acceleration of 30 ft/sec² in the direction shown in the figure. Find the radius of curvature of the path of P in this position.

* **3.174** Considering the instantaneous center Ⓘ of the bar $\mathcal{B}$ in Figure P3.174 as a point of $\mathcal{B}$ extended, find the acceleration of Ⓘ at the instant shown, if $\mathbf{v}_A = 2\hat{\mathbf{i}}$ in./sec = constant. Also find the acceleration of the point B' of $\mathcal{B}$ passing over the pin, and note that it is not along the slot.

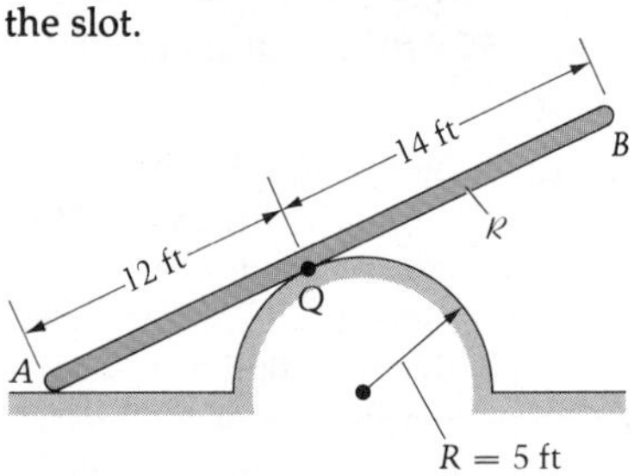

Figure P3.172

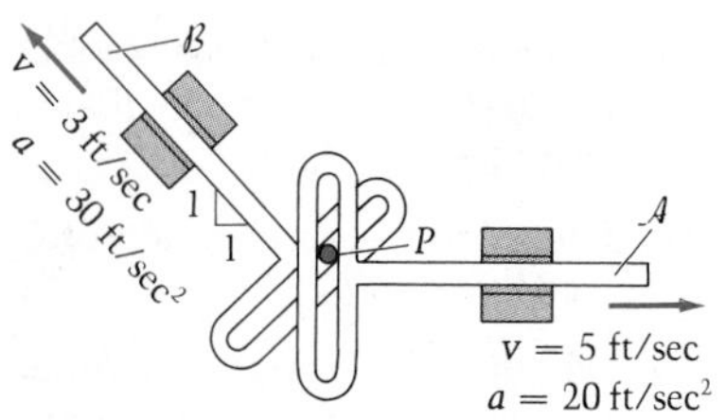

Figure P3.173

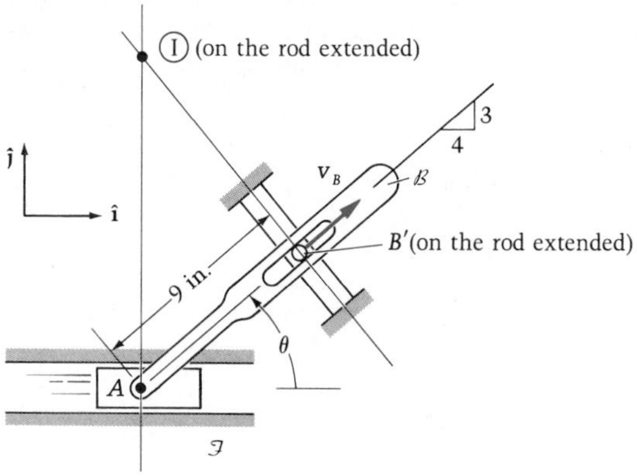

Figure P3.174

COMPUTER PROBLEMS ▶ Chapter 3

* **3.175** Crank $\mathcal{B}_1$ in Figure P3.175 is driven at a constant angular speed ω clockwise. Show that the speed of the piston is maximum when θ satisfies the equation

$$\cos\theta + \frac{\cos 2\theta}{\sqrt{(\ell_2/\ell_1)^2 - \sin^2\theta}} + \frac{\sin^2 2\theta}{4[(\ell_2/\ell_1)^2 - \sin^2\theta]^{3/2}} = 0$$

Solve with a computer for the first root of this equation when the lengths of $\mathcal{B}_1$ and $\mathcal{B}_2$ are 8 cm and 20 cm, respectively. Note that the answer is independent of the value of ω. You may wish to read Appendix B and make use of the Newton-Raphson method described there.

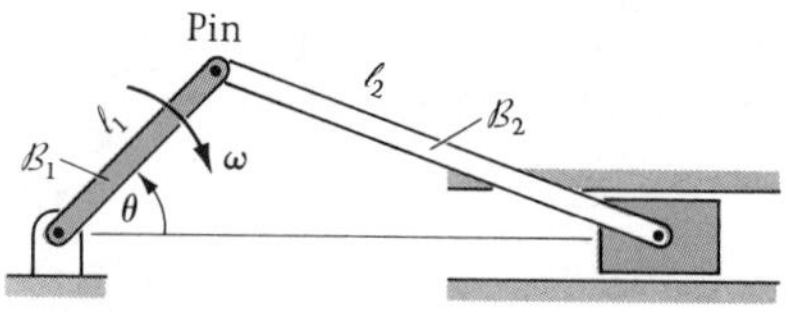

Figure P3.175

* **3.176** Crank $\mathcal{B}_1$ in Figure P3.176 rotates at constant angular velocity $\dot{\theta}\hat{\mathbf{k}}$. Use a computer to generate data for a plot of the following two quantities as functions of θ for the case in which $D = 2\ell$:

   a. The angle φ that locates slider $\mathcal{B}_2$
   b. The ratio $\dot{\varphi}/\dot{\theta}$ of the angular speeds of $\mathcal{B}_2$ and $\mathcal{B}_1$.

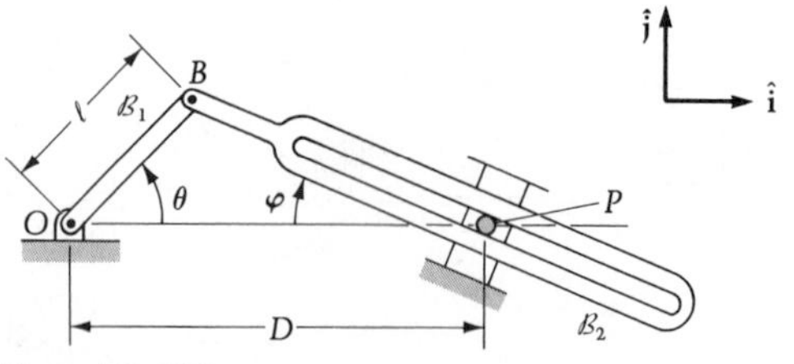

Figure P3.176

SUMMARY ▶ Chapter 3

This chapter has been devoted to presentation of the velocity and acceleration relationships that pertain to a rigid body in plane motion.

If A and B are two points of the body lying in the same plane of motion, then their velocities are linked through the angular velocity $\omega\hat{\mathbf{k}}$ ($\hat{\mathbf{k}}$ is perpendicular to the plane of motion) by

$$\begin{aligned}\mathbf{v}_B &= \mathbf{v}_A + \omega\hat{\mathbf{k}} \times \mathbf{r}_{AB}\\ &= \mathbf{v}_A + r\omega\hat{\mathbf{e}}\end{aligned}$$

If a point instantaneously has zero velocity, call it Ⓘ, then:

$$\begin{aligned}\mathbf{v}_P &= \overset{\mathbf{0}}{\mathbf{v}_{Ⓘ}} + \omega\hat{\mathbf{k}} \times \mathbf{r}_{ⒾP}\\ &= r\omega\hat{\mathbf{e}}\end{aligned}$$

which shows that the speed of any point is the product of the angular speed and the distance between the point and the instantaneous center. For accelerations, where $\alpha\hat{\mathbf{k}}$ is the angular acceleration, we have

$$\mathbf{a}_B = \mathbf{a}_A + \alpha\hat{\mathbf{k}} \times \mathbf{r}_{AB} - \omega^2\,\mathbf{r}_{AB}$$

When one body rolls on another, the points of contact have the same velocity. Rolling of a wheel on a fixed surface means that the point on the wheel in contact with the surface is the instantaneous center of velocity, and with C denoting the center of the wheel,

$$\mathbf{v}_C = \omega\hat{\mathbf{k}} \times \mathbf{r}_{ⒾC}$$

and

$$\mathbf{a}_C = \alpha\hat{\mathbf{k}} \times \mathbf{r}_{ⒾC} + (\text{a part normal to the path of } C)$$

Finally, we have investigated the relationships between velocities and then accelerations of a single point P relative to two reference frames. The results can be expressed compactly if we think of the underlying basic frame of reference as "fixed" and a second body as moving. Then with

$\mathbf{v}_P$ = velocity of P relative to the fixed frame,

$\mathbf{v}_{\text{rel}}$ = velocity of P relative to the moving body,

$\omega\hat{\mathbf{k}}$ = angular velocity of the moving body,

and similarly for accelerations and with O' being a point fixed in the moving body, we have:

$$\mathbf{v}_P = \mathbf{v}_{\text{rel}} + \mathbf{v}_{O'} + \omega\hat{\mathbf{k}} \times \mathbf{r}_{O'P}$$

and

$$\mathbf{a}_P = \mathbf{a}_{\text{rel}} + \mathbf{a}_{O'} + \alpha\hat{\mathbf{k}} \times \mathbf{r}_{O'P} - \omega^2\mathbf{r}_{O'P} + 2\omega\hat{\mathbf{k}} \times \mathbf{v}_{\text{rel}}$$

The last term in the second equation is called the Coriolis acceleration.

REVIEW QUESTIONS ▶ Chapter 3

True or False?

In these questions, P and Q are points in the reference plane of a rigid body $\mathcal{B}$ in plane motion.

1. For a rigid body in plane motion with angular velocity $\dot{\theta}\hat{\mathbf{k}}$, $\dot{\theta}$ depends on a certain choice of points whose velocities are to be related.
2. For a rigid body in plane motion, there is always an instantaneous center $\textcircled{I}$ of zero velocity located a finite distance from the body.
3. $\mathbf{a}_{\textcircled{I}}$ is not necessarily zero.
4. $\mathbf{v}_P = \dot{\theta}\hat{\mathbf{k}} \times \mathbf{r}_{\textcircled{I}P}$ if the angular velocity of $\mathcal{B}$ is not zero.
5. For a rigid body $\mathcal{B}$ in rectilinear translation, the velocities of all points of $\mathcal{B}$ are equal, and so are the accelerations.
6. For a rigid body $\mathcal{B}$ in curvilinear translation, the velocities of all points of $\mathcal{B}$ are equal, but the accelerations are not.
7. At a given instant, any point can be considered to lie on a rigid extension of any rigid body.
8. The smallest number of scalar parameters required to locate a rigid body in plane motion is four.
9. For a rigid body $\mathcal{B}$ in plane motion,

$$\mathbf{v}_P = \mathbf{v}_Q + \dot{\theta}\hat{\mathbf{k}} \times \mathbf{r}_{QP}$$

10. A point P can have an angular velocity.
11. The components of $\mathbf{v}_P$ and $\mathbf{v}_Q$ along the line PQ are not always equal.
12. The components of $\mathbf{a}_P$ and $\mathbf{a}_Q$ along the line PQ are not always equal.

Answers: 1. F 2. F 3. T 4. T 5. T 6. F 7. T 8. F 9. T 10. F 11. F 12. T

4 Kinetics of a Rigid Body in Plane Motion/ Development and Solution of the Differential Equations Governing the Motion

4.1 Introduction

In this chapter we apply Euler's laws to the plane motions of rigid bodies. Motion of the mass center of any body, rigid or not, is governed by Euler's first law as discussed in Chapter 2. The *rotational* motion of a rigid body is governed by Euler's second law. We saw in Chapter 2 that this law can be expressed in terms of a moment of momentum for any body, rigid or not. However, the moment of momentum for a rigid body can be expressed in a particularly compact way that involves moments and products of inertia of the body and its angular velocity; because of this, the term *angular momentum* is used synonymously with *moment of momentum.*

There is, however, one type of plane motion of a rigid body $\mathcal{B}$ that can be immediately studied, prior to the introduction of angular momentum. This class of motions, called *translation,* is characterized by the angular velocity of $\mathcal{B}$ being always zero. The translation problems treated in the next section (4.2) differ from the particle/mass-center motion problems of Chapter 2 in that a moment equation is required for their solution in addition to the mass-center equation $\Sigma\mathbf{F} = m\mathbf{a}_c$.

In Section 4.3, when we develop the expression for the angular momentum of a rigid body, the moments and products of inertia suddenly appear — in the same way the mass center did, back in Chapter 2. Thus we spend some time in Section 4.4 studying these inertia properties before moving on.

In Sections 4.5 and 4.6 we deduce several especially useful forms of Euler's second law in terms of the inertia properties. After each form of the equation, we illustrate its use with a set of examples. Some of these examples might be termed "snapshot" problems; in these we investigate the relationships between external forces on a body and its accelerations at a single instant. These problems are rather natural extensions of those the student has encountered in statics — that is, we know the geometrical configuration and seek information about forces on the body. Other problems might be called "movie" problems; in one class of these the geometry is of sufficient simplicity that Euler's laws produce differential equations which we can readily integrate so as to predict the motion of the body during some interval of time.

In Section 4.7, we take up the very special problem of rotation of an unbalanced body about a fixed axis. Here we establish the criteria for the technologically important problem of balancing.

In the next chapter, we will continue our study of plane-motion kinetics of rigid bodies by investigating the use of three special solutions (which can be obtained in general) to the differential equations of motion. These special integrals are known as the principles of work and kinetic energy; linear impulse and momentum; and angular impulse and angular momentum.

Finally, we mention to the reader that it is possible to obtain all the results of this chapter on plane motion of rigid bodies from the general three-dimensional results developed in Chapter 7. It is not necessary to travel this complex route in order to learn plane motion, however, and in this chapter we take a simpler path. It is worth noting that while the planar case covers a restricted class of motions, it does in fact contain a large number of problems with important engineering applications.

4.2 Rigid Bodies in Translation

In Chapter 2 we presented the equation

$$\Sigma \mathbf{M}_P = \int \mathbf{r} \times \mathbf{a}\, dm \tag{2.34}$$

which is valid both for *any* body and *any* point P. In particular, if the body is translating, then by definition all its points have the same acceleration — including its mass center — and if we label that common acceleration $\mathbf{a}$, then it may be factored, leaving:

$$\begin{aligned} \Sigma \mathbf{M}_P &= \left(\int \mathbf{r}\, dm\right) \times \mathbf{a} \\ &= m\mathbf{r}_{PC} \times \mathbf{a} \qquad \text{(using the definition of the mass center)} \\ &= \mathbf{r}_{PC} \times (m\mathbf{a}) \end{aligned} \tag{4.1a}$$

and therefore

$$\Sigma \mathbf{M}_C = \mathbf{0} \tag{4.1b}$$

In this section, we shall apply this simple equation to several examples of translating rigid bodies in plane motion. The reader should note, however, that the two forms of Equation (4.1) apply whether the motion is plane or not. It also doesn't even require the body to be "physically rigid," although if it is translating, it is necessarily behaving like a rigid body during that motion.

Question 4.1 If point P is arbitrary in Equation (4.1a), where does the inertial frame come into the equation?

Before wading into the examples, we wish to note what is new here. In Chapter 2, we were concerned with the *mass center motions* of bodies, whether they were rigid or not. In those sections there was no need to sum moments, and *that* is what will distinguish this chapter from those preceding sections. Now the simplest problems by far in which a moment equation is sometimes needed are those involving translation. Translation is simple because it can be studied prior to the introduction of angular momentum forms for rigid bodies with their accompanying inertia properties and angular velocities.

We now examine three examples of translation, and in each, the reader is urged to note two things: (1) how the problem could not be

---

Answer 4.1 The acceleration $\mathbf{a}$ is the second derivative, taken in an inertial frame, of the position vector from an origin in that frame to any point of the body.

solved without the use of Equation (4.1); and (2) how for translation, the moments generally do *not* sum to zero. They do sum to zero at the mass center, and also at points lying on the line of **a** drawn through C (for then $\mathbf{r}_{PC}$ is parallel to **a**)*, but *not otherwise.* (Equations for translation at constant velocity are trivial and identical to the equilibrium equations; these were studied in statics and are not considered in this book.)

EXAMPLE 4.1

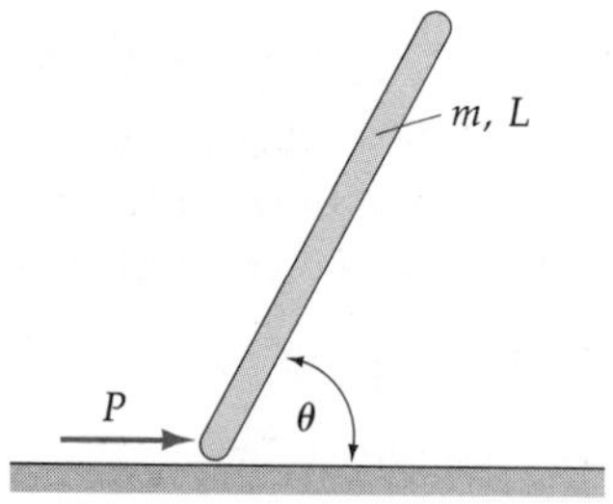

Figure E4.1a

Find the angle θ for which the bar in Figure E4.1a will translate to the right at the given constant acceleration "a." Then find the force P required to produce this motion.

Solution

We sum moments at the contact point A, using the FBD in Figure E4.1b and Equation (4.1):

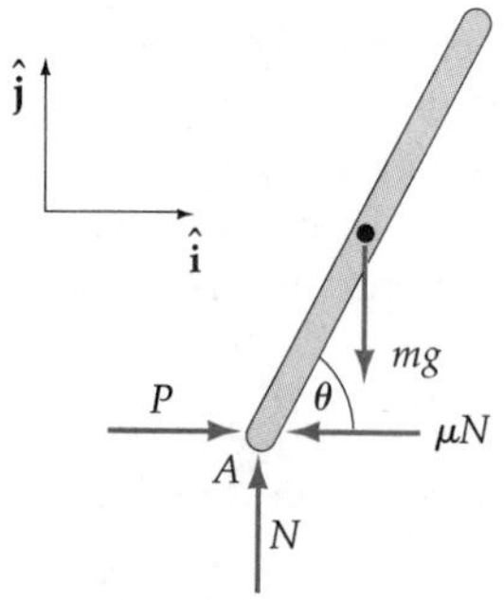

Figure E4.1b

$$\Sigma \mathbf{M}_A = \mathbf{r}_{AC} \times m\mathbf{a}$$

$$-mg\frac{L}{2}\cos\theta\,\hat{\mathbf{k}} = \frac{L}{2}(\cos\theta\hat{\mathbf{i}} + \sin\theta\hat{\mathbf{j}}) \times ma\hat{\mathbf{i}}$$

$$-mg\frac{L}{2}\cos\theta\,\hat{\mathbf{k}} = -\frac{Lma}{2}\sin\theta\,\hat{\mathbf{k}}$$

Therefore θ in terms of a is given by:

$$\tan\theta = \frac{g}{a} \Rightarrow \theta = \tan^{-1}\left(\frac{g}{a}\right)$$

To determine P, we write the mass-center equations:

$$\Sigma F_x = P - \mu N = ma \qquad (1)$$

and

$$\Sigma F_y = N - mg = m\ddot{y}_c = 0 \Rightarrow N = mg$$

Substituting this value of N into Equation (1) gives

$$P - \mu mg = ma$$

or

$$P = m(\mu g + a)$$

Note that part of P balances the friction and the rest, the unbalanced force in the x-direction, produces the "ma."

---

* and of course at times when the equal accelerations of all points, **a**, are zero.

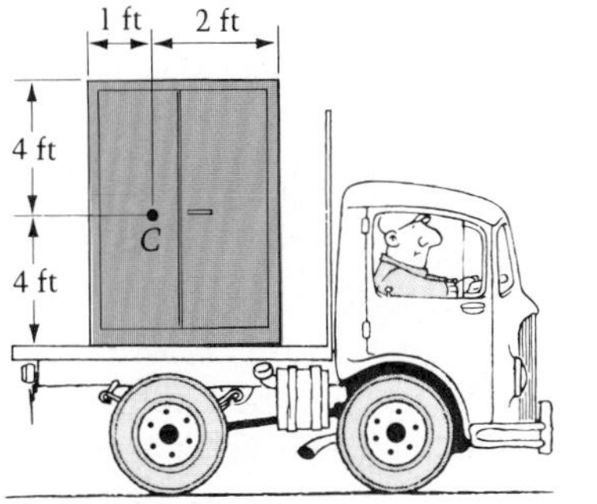

Figure E4.2a

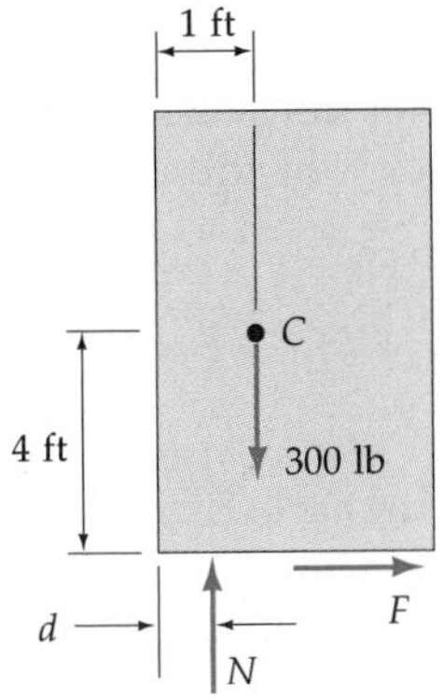

Figure E4.2b

Example 4.2

A 300-lb cabinet is to be transported on a truck as shown in Figure E4.2a. Assuming sufficient friction so that the cabinet will not slide on the truck, for forward acceleration find

a. the forces exerted by the truck bed on the cabinet for 5 ft/sec^2 acceleration;

b. the maximum acceleration for which the cabinet will not tip over.

Solution

a. The acceleration is 5 ft/sec^2 $\rightarrow$, so using the free body diagram (Figure E4.2b) and putting $\mathbf{\Sigma F} = m\mathbf{a}$ into component form,

$$\overset{+}{\longrightarrow} \qquad \Sigma F_x = ma_x$$

$$F = \left(\frac{300}{32.2}\right) 5 = 46.6 \text{ lb}$$

$$+\!\uparrow \qquad \Sigma F_y = ma_y$$

$$N - 300 = 0$$

$$N = 300 \text{ lb}$$

To locate the line of action of N, that is, to find the distance d, we need the moment equation:

$$\Sigma M = 0$$

$$4F - (1 - d)N = 0$$

$$4(46.6) - (1 - d)(300) = 0$$

$$1 - d = 0.621$$

$$d = 0.379 \text{ ft}$$

We might note at this point that the minimum coefficient of friction for which this motion is possible is

$$\mu_{min} = \frac{46.6}{300} = 0.155$$

b. Assuming we still have adequate friction to prevent slip, but treating the acceleration magnitude, a, as an unknown, we can retrace our steps as in part (a) to obtain

$$F = \frac{300}{32.2} a$$

$$N = 300$$

and

$$4F - (1 - d)N = 0$$

When the cabinet is on the verge of tipping, the line of action of N is at the left corner, $d = 0$, so for that condition:

$$4F - (1 - 0)(300) = 0$$

$$F = 75 \text{ lb}$$

and then from

$$F = \frac{300}{32.2}a$$

we find

$$a = \frac{32.2}{300}(75) = 8.05 \text{ ft/sec}^2$$

The reader should note that this tendency to tip over "backwards" is a phenomen uniquely of dynamics; there's nothing quite like it in statics. In addition we sometimes tend to think of friction in oversimplified terms, as perhaps "always opposing motion"; of course it is precisely the friction that here *provides* the motive force to cause the cabinet to accelerate.

In the preceding example we saw how the state of translation could be jeopardized by the tendency of a body to "rock." In the next we *additionally* explore a tendency to slip.

Figure E4.3a

EXAMPLE 4.3

The coefficient of friction at both ends of the uniform slender bar in Figure E4.3a is 0.5. Find the maximum forward acceleration that the truck may have without the bar moving relative to the truck.

Solution

One possibility is that the upper end separates from the truck body. The free-body diagram in Figure E4.3b shows the situation when the end is barely about to break away. Because the bar is in translation, we may write:

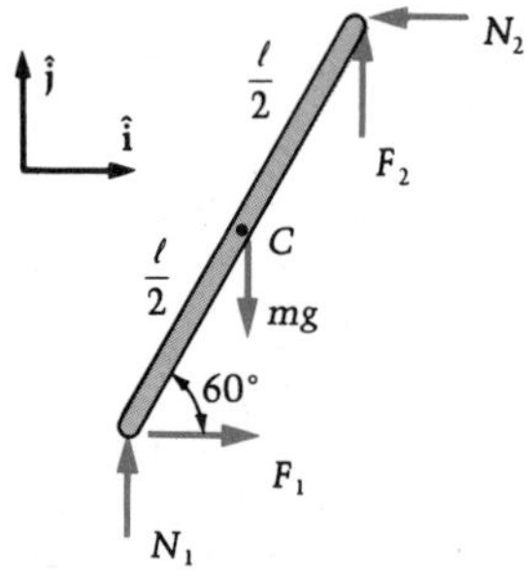

Figure E4.3b

$$\Sigma M_C = 0$$

$$F_1 \frac{\ell}{2} \sin 60° - N_1 \frac{\ell}{2} \cos 60° = 0$$

$$\frac{\sqrt{3}}{2} F_1 - \frac{1}{2} N_1 = 0$$

$$F_1 = N_1/\sqrt{3} = 0.577 N_1$$

But this much friction cannot be generated because $\mu = 0.5 < 0.577$. Therefore motion of the rod relative to the truck will not be initiated by this mechanism.

The other possibility, as shown in the second free-body diagram, Figure E4.3c, is that the bar is on the verge of slipping where it contacts the truck; this must occur at the two surfaces simultaneously. The equations of motion are (with $\mathbf{a} = \ddot{x}\hat{\mathbf{i}}$)

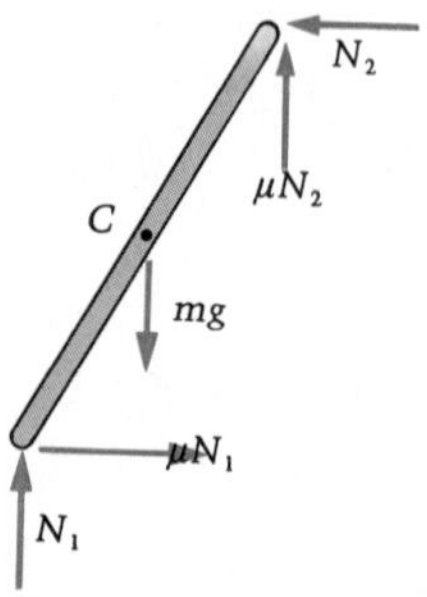

Figure E4.3c

$$\Sigma F_x = m\ddot{x}$$

$$\mu N_1 - N_2 = m\ddot{x} \tag{1}$$

$$\Sigma F_y = m\ddot{y} = 0$$

$$N_1 + \mu N_2 - mg = 0 \tag{2}$$

$$\Sigma M_C = 0$$

$$\frac{\sqrt{3}}{2}\left(\frac{\ell}{2}\right)\mu N_1 - \frac{1}{2}\left(\frac{\ell}{2}\right)N_1 + \frac{1}{2}\left(\frac{\ell}{2}\right)\mu N_2 + \frac{\sqrt{3}}{2}\left(\frac{\ell}{2}\right)N_2 = 0 \qquad (3)$$

With $\mu = 0.5$, Equation (3) yields

$$N_1 = 16.7 N_2$$

and from (2)

$$N_2 = 0.0581\ mg$$

so that

$$N_1 = 16.7 N_2 = 0.971\ mg$$

Equation (1) then gives

$$0.5(0.971\ mg) - 0.0581\ mg = m\ddot{x}$$

$$\ddot{x} = 0.427\ g$$

which is 13.8 ft/sec² (for $g = 32.2$ ft/sec²) or 4.19 m/s² (for $g = 9.81$ m/s²).

The reader is encouraged to rework the previous example using Equation (4.1a) to sum moments about the bottom point of the translating bar.

PROBLEMS ▶ Section 4.2

4.1 For what force P is it possible for the uniform slender bar (Figure P4.1) to translate across the smooth floor in the position shown? The bar has mass m and length ℓ.

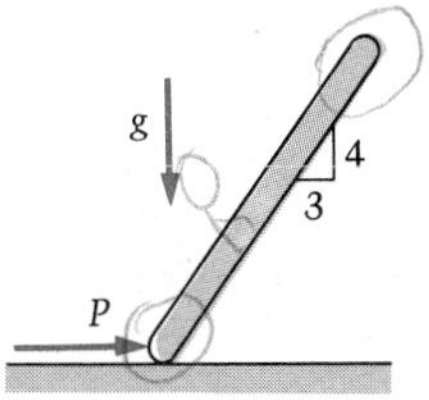

Figure P4.1

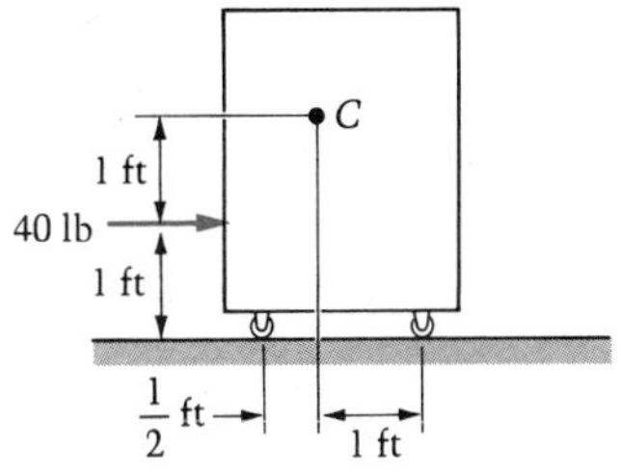

Figure P4.2

4.2 A 100-lb cabinet, rolling on small wheels, is subjected to a 40-lb force as shown in Figure P4.2. Neglecting friction, find (a) the acceleration of the cabinet; (b) the reactions of the floor on the wheels.

4.3 Repeat the preceding problem for the case where the 40-lb force is applied 1 ft *above* C.

4.4 Find the value of F for which one of the wheels of the door in Figure P4.4 lifts out of its track. Which one? Assume negligible friction.

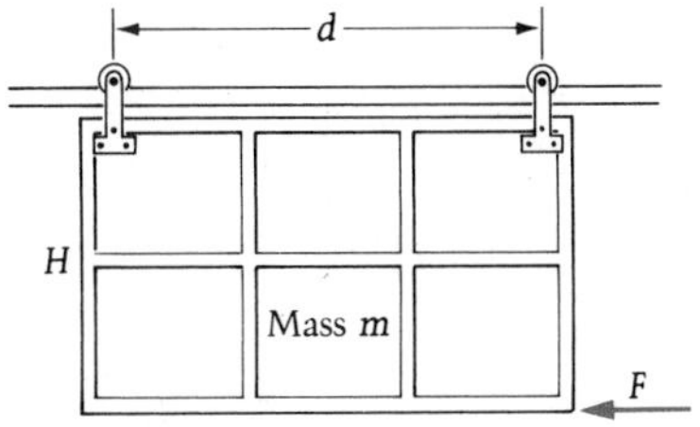

Figure P4.4

4.5 The force P causes the uniform rectangular box of weight W in Figure P4.5 to slide. Find the range of values of H for which the box will not tip about either the front or rear lower corner as it slides on the smooth floor, if $P = W$.

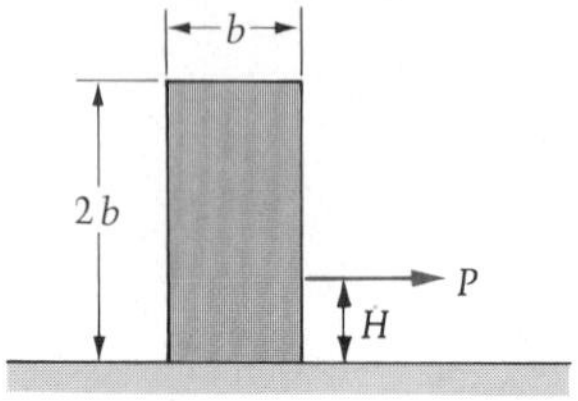

Figure P4.5

4.6 Repeat the preceding problem for a coefficient of sliding friction of 0.2.

4.7 A 400-lb cabinet is to be transported on a truck as shown in Figure P4.7. Assuming sufficient friction so that the cabinet will not slide on the truck, what is the maximum forward acceleration for which the cabinet will not tip over?

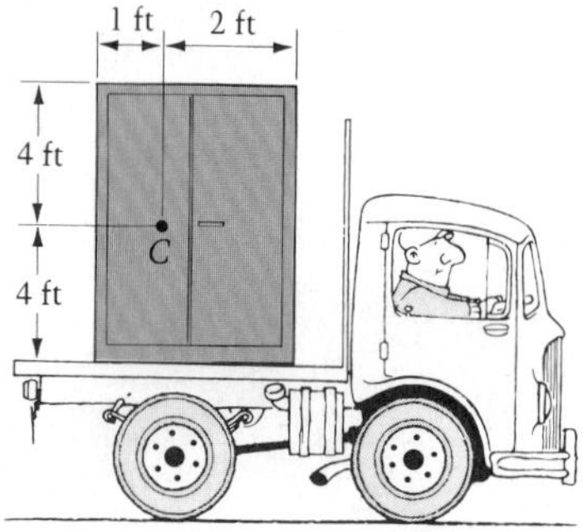

Figure P4.7

4.8 The uniform bar $\mathcal{B}$ in Figure P4.8 weighs 60 lb and is pinned at A (and fastened by the cable DB) to the frame $\mathcal{F}$. If the frame is given an acceleration $a = 32.2$ ft/sec² as shown, determine the tension T in the cable and the force exerted by the pin at A on the bar.

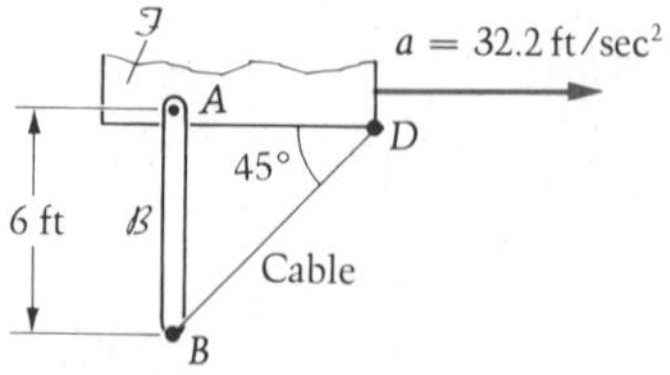

Figure P4.8

4.9 The cords in Figure P4.9 have a tensile strength of 12 N. Cart $\mathcal{C}$ has a mass of 35 kg exclusive of the 10-kg and 1.2-m vertical rod $\mathcal{R}$, which is pinned to it at A. Find the maximum value of P that can be exerted without breaking either cord if: (a) P acts to the right as shown; (b) P acts to the left. Neglect friction, and assume negligible tension in each cord when the cart is at rest.

4.10 A force F, alternating in direction, causes the carriage to move with rectilinear horizontal motion defined by the equation $x = 2 \sin \pi t$ ft, where x is the displacement in feet and t is the time in seconds. (See Figure P4.10.) A rigid, slender, homogeneous rod of weight 32.2 lb and length 6 ft is welded to the carriage at B and projects vertically upward. Find, in magnitude and direction, the bending moment that the carriage exerts on the rod at B when $t = \frac{1}{2}$ sec.

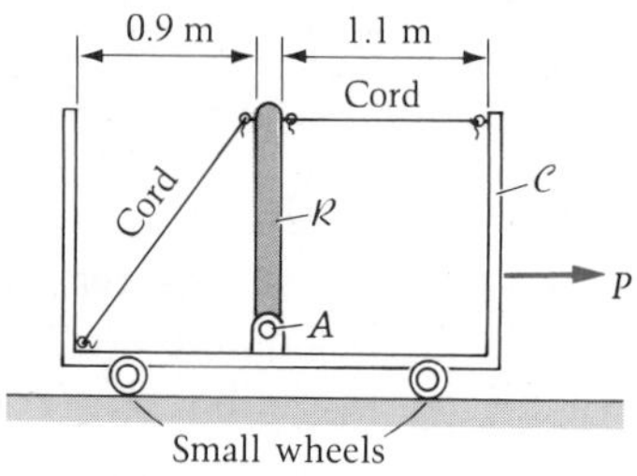

Figure P4.9

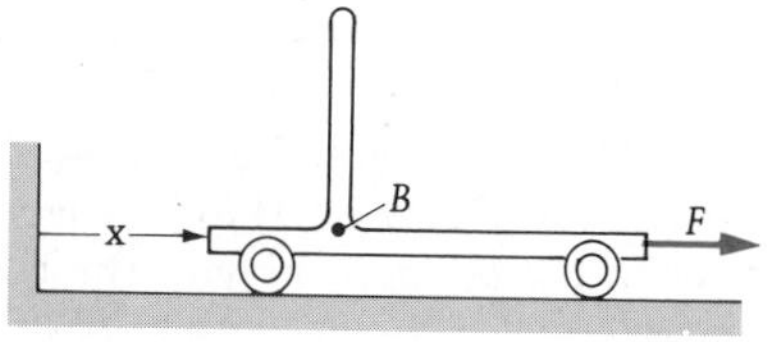

Figure P4.10

4.11 A child notices that sometimes the ball $\mathcal{M}$ does not roll down the inclined surface of toy $\mathcal{T}$ when she pushes it along the floor. (See Figure P4.11.) What is the minimum acceleration a_{min} of $\mathcal{T}$ to prevent this rolling?

Figure P4.11

4.12 In the preceding problem, suppose the acceleration of $\mathcal{T}$ is $2a_{min}$. What is the normal force between the smooth vertical surface of $\mathcal{T}$ and the ball? The ball's weight is 0.06 lb.

4.13 A can $\mathcal{C}$ that may be considered a uniform solid cylinder (see Figure P4.13) is pushed along a surface $\mathcal{B}$ by a moving arm $\mathcal{A}$. If it is observed that $\mathcal{C}$ *translates* to the right with $\ddot{x}_C = g/10$, what must be the minimum coefficient of friction between $\mathcal{A}$ and $\mathcal{C}$? The coefficient of friction between $\mathcal{C}$ and $\mathcal{B}$ is μ.

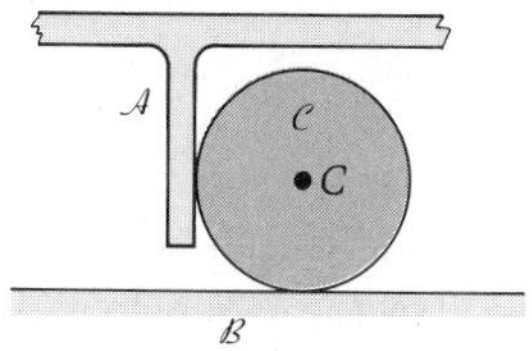

Figure P4.13

4.14 The force P is applied to cart $\mathcal{C}$, and increases slowly from zero, always acting to the right. (See Figure P4.14.) Point C is the mass center of $\mathcal{B}$. At what value of P will the bodies $\mathcal{B}$ and $\mathcal{C}$ no longer move as one?

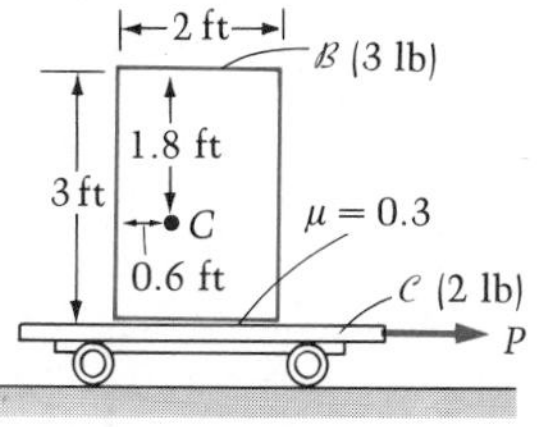

Figure P4.14

4.15 A nonuniform block rests on a flatcar as shown in Figure P4.15. If the coefficient of friction between car and block is 0.40, for what range of accelerations of the car will the block neither tip nor slide?

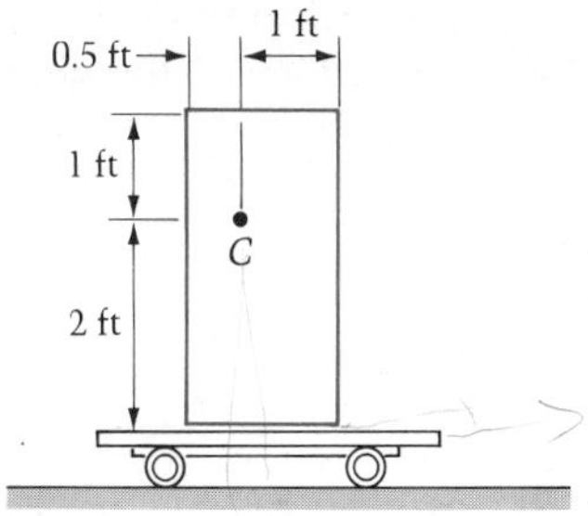

Figure P4.15

4.16 The truck in Figure P4.16 is traveling at 45 mph. Find the minimum stopping distance such that the 250-lb crate will neither slide nor tip over.

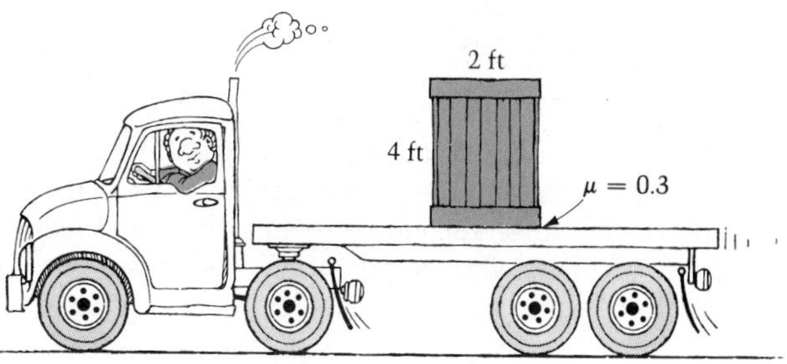

Figure P4.16

4.17 The block $\mathcal{B}$ of mass m in Figure P4.17 is resting on the cart $\mathcal{C}$ of mass M. Force P is applied to the cart, starting it in motion to the right. The wheels are small and frictionless, and the coefficient of friction between $\mathcal{B}$ and $\mathcal{C}$ is $\mu = \frac{1}{4}$. Find the largest value of P for which $\mathcal{B}$ and $\mathcal{C}$ will move together, considering all cases.

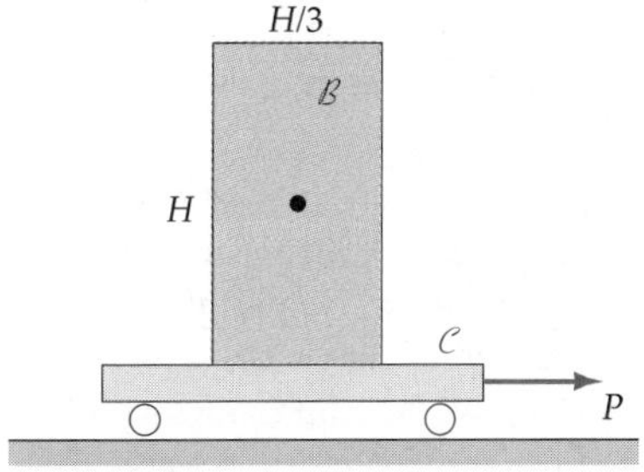

Figure P4.17

4.18 The 25-lb triangular plate is smoothly pinned at vertex A to a small, light wheel (see Figure P4.18). Find the value of force P so that the plate, in theory, will translate along the incline. Also find the acceleration.

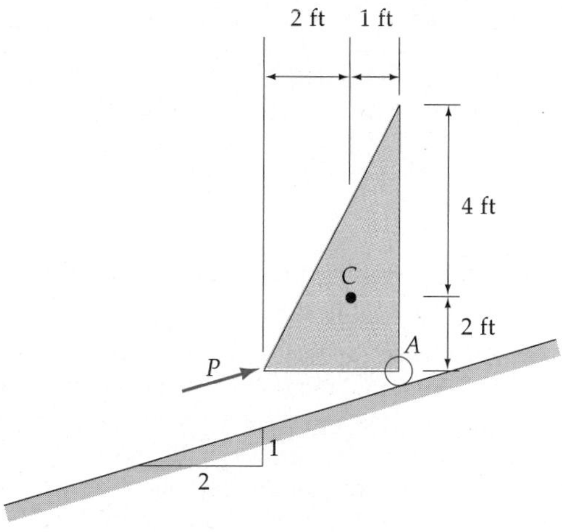

Figure P4.18

4.19 Repeat the preceding problem if the angle of the plane is changed from $\frac{1}{2}$ to $\frac{3}{4}$.

4.20 The monorail car in Figure P4.20 is driven through its front wheel and moves forward from left to right. If the coefficient of friction between wheels and track is $\mu = 0.55$, determine the maximum acceleration possible for the car.

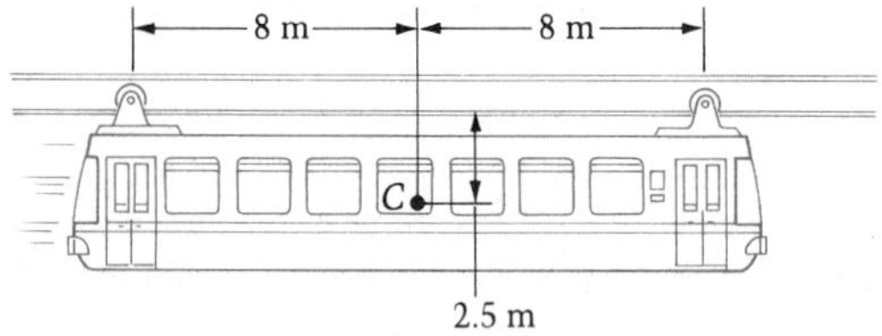

Figure P4.20

4.21 A dragster is all set for the annual neighborhood race. (See Figure P4.21.)

a. In terms of the dimensions b, H, and d and the coefficient of friction μ, find the maximum possible acceleration of the car. Neglect the rotational inertia of the wheels.

b. How would you adjust the four parameters b, H, d, and μ to further increase the driver's acceleration?

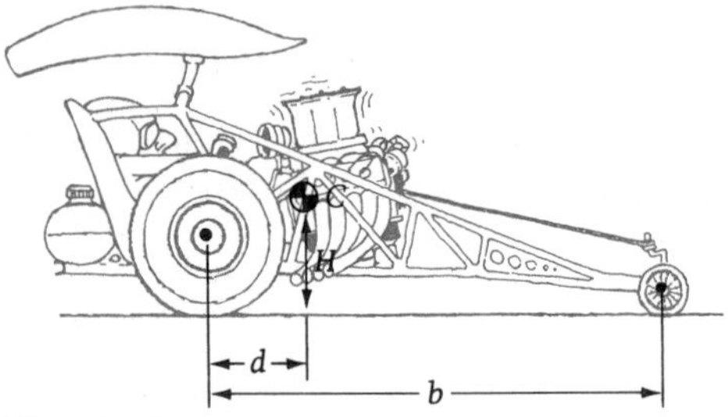

Figure P4.21

4.22 Rework the preceding problem for a car with (a) front-wheel drive and (b) four-wheel drive.

4.23 In an emergency the driver of an automobile applies his brakes; the front brakes fail and the rear wheels are locked. Find the time and distance required to bring the car to rest. Neglect the masses of the wheels, and express the results in terms of the coefficient of sliding friction μ, the initial speed v, the gravitational acceleration g, and the dimensions shown in Figure P4.23.

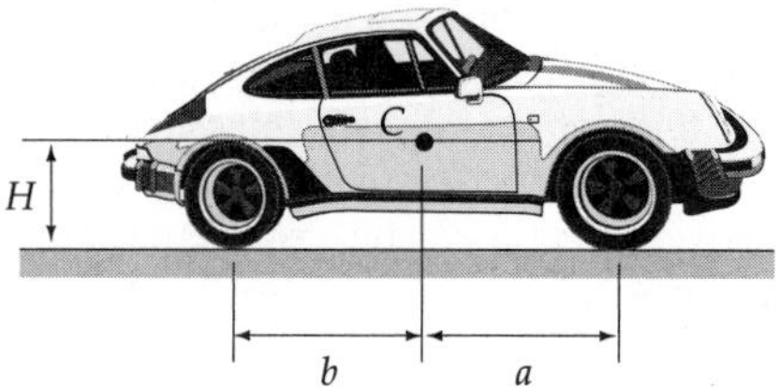

Figure P4.23

4.24 A uniform rod $\mathcal{A}$ of length L and weight W is connected to smooth hinges at E and D by the light members B and $\mathcal{C}$, each of length L. In the position shown in Figure P4.24, $\mathcal{B}$ has an angular velocity of ω rad/sec clockwise. Find the forces in members $\mathcal{B}$ and $\mathcal{C}$, and determine the acceleration of center C of rod $\mathcal{A}$ in terms of the given variables.

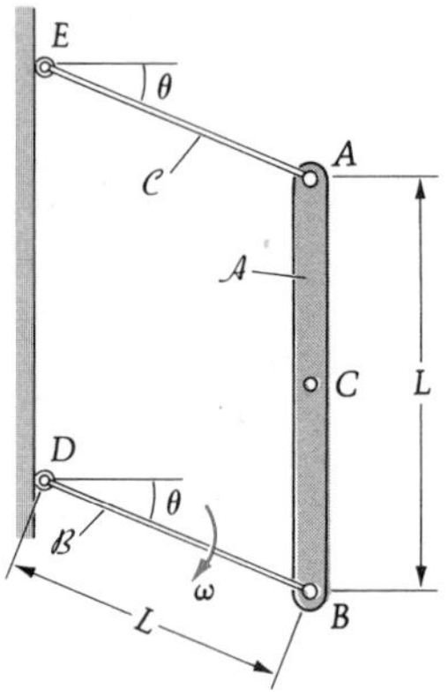

Figure P4.24

* **4.25** Find the range of accelerations of $\mathcal{W}$ that F can produce without $\mathcal{B}$ moving in *any* manner relative to $\mathcal{W}$. (See Figure P4.25.) Note carefully the position of the mass center of $\mathcal{B}$.

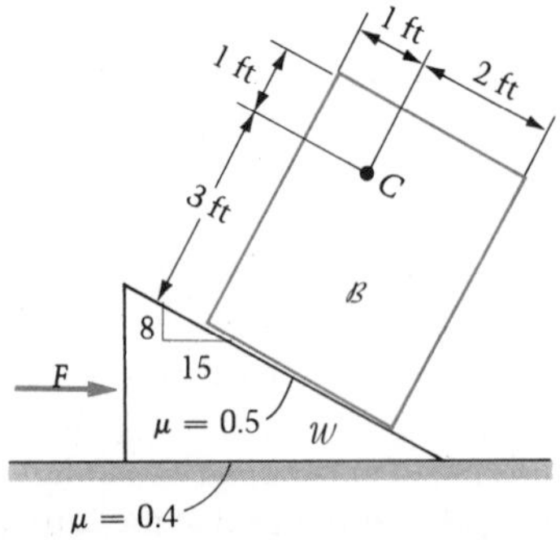

Figure P4.25

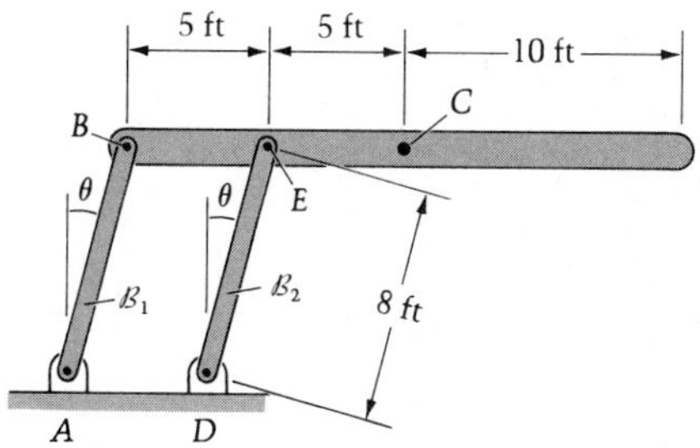

Figure P4.26

* **4.26** A slender homogeneous rod weighing 64.4 lb and 20 ft long is supported as shown in Figure P4.26. Bars $\mathcal{B}_1$ and $\mathcal{B}_2$ are of negligible mass and have frictionless pins at each end. The system is released from rest with $\theta = 0$.

   a. Derive expressions for the angular velocity and acceleration of bars $\mathcal{B}_1$ and $\mathcal{B}_2$ as functions of θ.

   b. Derive expressions for the axial force in bars $\mathcal{B}_1$ and $\mathcal{B}_2$ as a function of θ.

4.3 Moment of Momentum (Angular Momentum)

We recall from Chapter 2 that the moment of momentum of any body $\mathcal{B}$ with respect to a point P (not necessarily fixed in either the body or in the reference frame) was defined by Equation (2.35):

$$\mathbf{H}_P = \int_{\mathcal{B}} \mathbf{R} \times \mathbf{v}\, dm$$

where in this section $\mathbf{R}$ is the vector from P to the element of mass dm of the body, and $\mathbf{v}$ is the velocity of dm in the reference frame $\mathcal{I}$. (See Figure 4.1.) We note that $\mathbf{H}_P$ depends on the location of P as well as the distributions of mass and velocities in the body; it is seen to be the sum of the moments of momenta of all the mass elements of $\mathcal{B}$.

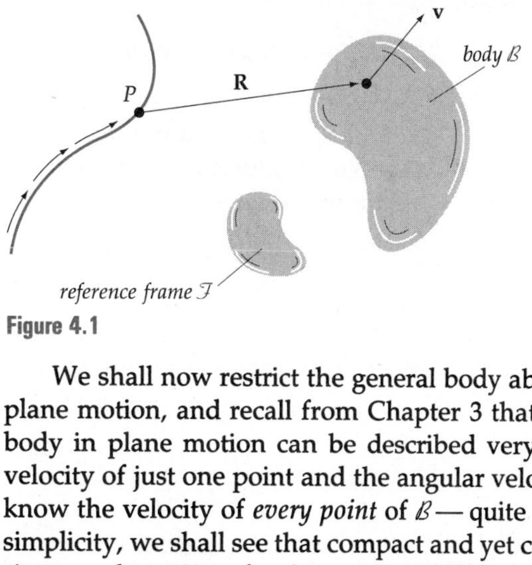

Figure 4.1

We shall now restrict the general body above to be rigid, place it in plane motion, and recall from Chapter 3 that the kinematics of a rigid body in plane motion can be described very simply. If we know the velocity of just one point and the angular velocity of the body, then we know the velocity of *every point* of $\mathcal{B}$—quite a bargain. Because of this simplicity, we shall see that compact and yet completely general expressions can be written for the moment of momentum (or angular momentum, as it is often called for rigid bodies). We shall further restrict the generic point P in the foregoing to be a point of $\mathcal{B}$.

The rectangular axes (x, y, z) have origin at P as shown in Figure 4.2, and the xy-plane is the reference plane, or plane of motion, as described in Sections 3.1, 2.

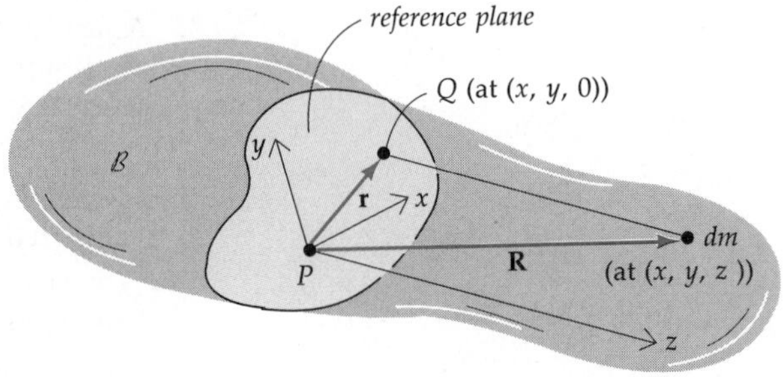

Figure 4.2

Recall also from Section 3.1 that each point of the rigid body has a "companion point" in the reference plane which always has the same x and y as the point, and hence also has the same velocity and acceleration. For the point at "dm" in Figure 4.2, the companion point is Q. Thus, using Equation (3.8),

$$\mathbf{v} = \mathbf{v}_Q = \mathbf{v}_P + \omega\hat{\mathbf{k}} \times \mathbf{r}_{PQ}$$

Substituting this expression for $\mathbf{v}$ and the coordinates of dm into R, $\mathbf{H}_P$ becomes:

$$\mathbf{H}_P = \int \underbrace{(x\hat{\mathbf{i}} + y\hat{\mathbf{j}} + z\hat{\mathbf{k}})}_{\mathbf{R}} \times [\mathbf{v}_P + \omega\hat{\mathbf{k}} \times \underbrace{(x\hat{\mathbf{i}} + y\hat{\mathbf{j}})}_{\mathbf{r}_{PQ}}]\, dm$$

or

$$\mathbf{H}_P = \left(\int \mathbf{R}\, dm\right) \times \mathbf{v}_P + \int (x\hat{\mathbf{i}} + y\hat{\mathbf{j}} + z\hat{\mathbf{k}}) \times [\omega\hat{\mathbf{k}} \times (x\hat{\mathbf{i}} + y\hat{\mathbf{j}})]\, dm$$

The integral $\int \mathbf{R}\, dm$ in the first term above is equal to $m\mathbf{r}_{PC}$ by the definition of the mass center. Making this substitution, and also carrying out the cross products in the second term, gives the moment of momentum vector in terms of $\mathbf{v}_P$, ω, and certain mass distribution integrals:

$$\mathbf{H}_P = \mathbf{r}_{PC} \times m\mathbf{v}_P + \hat{\mathbf{k}}\omega\int (x^2 + y^2)\, dm - \hat{\mathbf{i}}\omega\int xz\, dm - \hat{\mathbf{j}}\omega\int yz\, dm$$

Inertia Properties

We call the integrals in this equation inertia properties. Specifically:

$$\int (x^2 + y^2)\, dm = I_{zz}^P = \text{moment of inertia of mass of } \mathcal{B} \text{ about } z \text{ axis through } P \tag{4.2a}$$

$$-\int xz\, dm = I_{xz}^P = \text{product of inertia of mass of } \mathcal{B} \text{ with respect to } x \text{ and } z \text{ axes through } P^* \tag{4.2b}$$

* If the products of inertia are defined with the minus sign as above, then and only then will the inertia properties transform as a tensor — a topic beyond the scope of this book, however.

$$-\int yz\, dm = I_{yz}^P = \text{product of inertia of mass of } \mathcal{B} \text{ with respect to } y \text{ and } z \text{ axes through } P \qquad (4.2c)$$

Thus,

$$\mathbf{H}_P = \mathbf{r}_{PC} \times m\mathbf{v}_P + I_{xz}^P \omega \hat{\mathbf{i}} + I_{yz}^P \omega \hat{\mathbf{j}} + I_{zz}^P \omega \hat{\mathbf{k}} \qquad (4.3)$$

We are now at the point in our development where the inertia properties, like the mass center in Chapter 2, have arisen naturally. We shall spend the next section studying the moments and products of inertia; readers already familiar with inertia properties may wish to skip Section 4.4.

Before leaving Equation (4.3), we note for future reference that its first term, $\mathbf{r}_{PC} \times m\mathbf{v}_P$, vanishes if P is the mass center or has zero velocity* In both these cases, which will prove valuable to us, $\mathbf{H}_P$ takes the form

$$\mathbf{H}_P = I_{xz}^P \omega \hat{\mathbf{i}} + I_{yz}^P \omega \hat{\mathbf{j}} + I_{zz}^P \omega \hat{\mathbf{k}} \qquad (\text{if } P \text{ is } C, \text{ or if } \mathbf{v}_P = \mathbf{0}) \qquad (4.4)$$

Thus in these cases the moment of momentum can be expressed in terms of the angular velocity of $\mathcal{B}$ (hence its other name: **angular momentum**), along with three measures of its mass distribution.

Question 4.2 In Equation (4.3), does the mass center have to lie in the reference plane with P? How about in Equation (4.4) when P has zero velocity?

4.4 Moments and Products of Inertia / The Parallel-Axis Theorems

Examples of Moments of Inertia

From the definition of moment of inertia,

$$I_{zz}^P = \int (x^2 + y^2)\, dm$$

we see that I_{zz}^P is a measure of "how much mass is located how far" from the z axis through P. In cylindrical coordinates we have $I_{zz}^P = \int r^2\, dm$, and thus I_{zz}^P measures the sum total of mass times distance squared over the body's volume. The quantity I_{zz}^P is thus seen to be always positive. We now compute the mass-center moments of inertia of a number of common shapes. In Examples 4.4–4.12, we are seeking I_{zz}^C.

* Or if $\mathbf{r}_{PC}$ is parallel to $\mathbf{v}_P$, a case we need not consider here.

Answer 4.2 No. No.

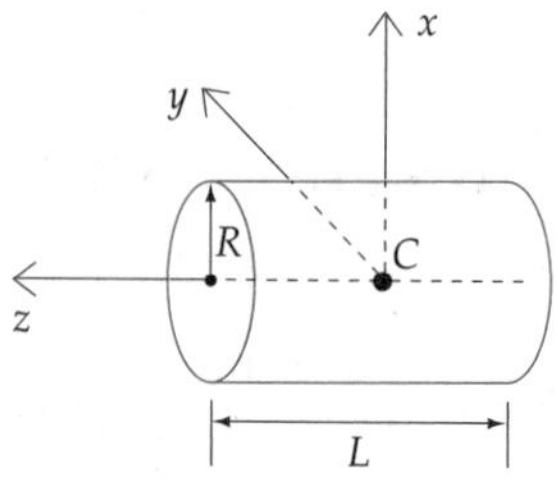

Figure E4.4

EXAMPLE 4.4

Homogeneous solid cylinder if z is its axis (see Figure E4.4).

Solution

Noting that $dm = \rho\, dV$, where ρ = mass density,

$$I_{zz}^{C} = \int_{\text{vol}} (x^2 + y^2)\overbrace{\rho\, dV}^{dm} = \int_{z=-L/2}^{L/2}\int_{\theta=0}^{2\pi}\int_{r=0}^{R} r^2\rho(r\, dr\, d\theta\, dz)$$

$$I_{zz}^{C} = \rho \frac{r^4}{4}\bigg]_0^R \theta\bigg]_0^{2\pi} z\bigg]_{-L/2}^{L/2} = (\rho\pi R^2 L)\frac{R^2}{2} = \boxed{\frac{mR^2}{2}}$$

EXAMPLE 4.5

Homogeneous solid cylinder if z is an axis *normal* to the axis of the cylinder (see Figure E4.5a).

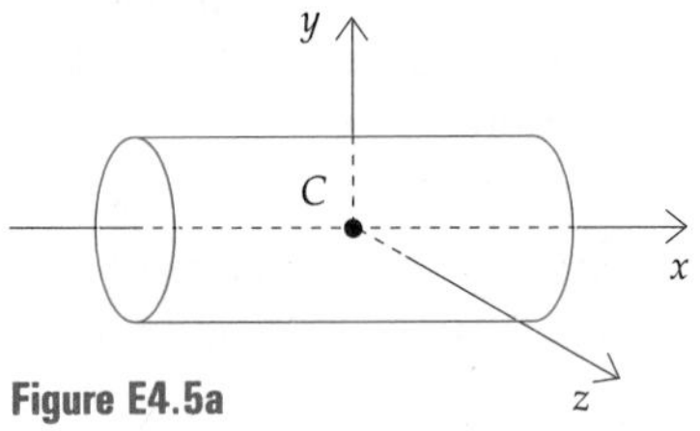

Figure E4.5a

Solution

$$I_{zz}^{C} = \int_{\text{vol}} (x^2 + y^2)\rho\, dV = \int_{x=-L/2}^{L/2}\int_{\theta=0}^{2\pi}\int_{r=0}^{R} [\underbrace{(r\sin\theta)^2}_{y = r\sin\theta} + x^2]\rho r\, dr\, d\theta\, dx$$

$$= \int_{x=-L/2}^{L/2}\int_{\theta=0}^{2\pi} \left(\frac{r^4}{4}\underbrace{\sin^2\theta}_{\frac{1-\cos 2\theta}{2}} + \frac{x^2r^2}{2}\right)\bigg]_0^R \rho\, d\theta\, dx$$

$$= \int_{-L/2}^{L/2}\left[\frac{R^4}{4}\left(\frac{\theta}{2} - \frac{\sin 2\theta}{4}\right)\bigg|_0^{2\pi} + \frac{x^2R^2\theta}{2}\bigg|_0^{2\pi}\right]\rho\, dx$$

$$= \rho\frac{\pi R^4}{4}x\bigg|_{-L/2}^{L/2} + \frac{\rho R^2\pi x^3}{3}\bigg|_{-L/2}^{L/2} = (\rho\pi R^2 L)\left(\frac{R^2}{4} + \frac{L^2}{12}\right)$$

$$= \boxed{\frac{mR^2}{4} + \frac{mL^2}{12}}$$

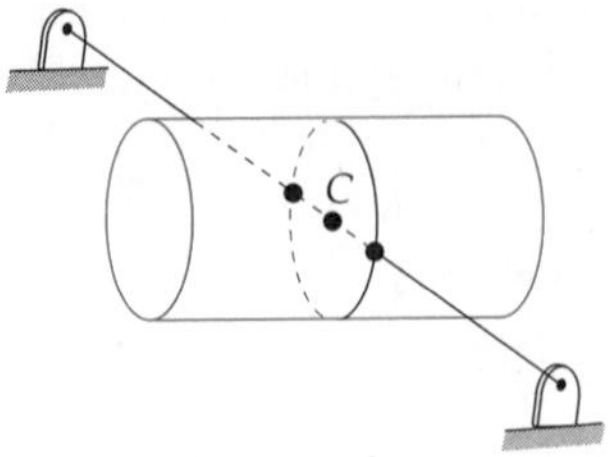

Figure E4.5b

It may be confusing at first to know which moment of inertia to use for a cylinder in a plane kinetics problem; the answer is that it is always the value associated with the axis normal to the xy plane of the motion. If the problem is a rolling cylinder, $I_{zz}^{C} = mR^2/2$. If we have a cylinder turning around a diametral axis (see Figure E4.5b), then $I_{zz}^{C} = mR^2/4 + mL^2/12$.

EXAMPLE 4.6

Two special cases of Example 4.5: Slender rods and disks (see Figures E4.6a, b).

Solution

1. If the body is "pencil-like" — that is, $L \gg R$ — the moment of inertia for a lateral axis through C is approximated as $mL^2/12$. This is also a correct result even if the cross section is not circular but has a maximum dimension within the cross section much less than L. Such a body is called a *slender bar* or *rod.*
2. If the body is a disk, however, we have $R \gg L$ and the moment of inertia is approximately $mR^2/4$.

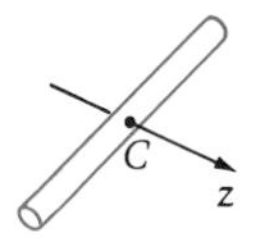

Figure E4.6a

For a cylinder with the dimensions of a pencil, for example, with $R = \frac{1}{8}$ in. and $L = 7$ in., we see that (see Figure E4.6a)

$$I_{zz}^C = \frac{mL^2}{12} + \frac{mR^2}{4} = \frac{mL^2}{12}\left[1 + 3\left(\frac{R}{L}\right)^2\right]$$

$$= \frac{mL^2}{12}\left[1 + 3\left(\frac{1/8}{7}\right)^2\right]$$

$$\approx \boxed{\frac{mL^2}{12}}$$

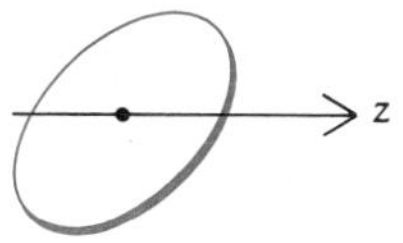

Figure E4.6b

where the second term is less than 0.1 of 1 percent of the retained $mL^2/12$ term. For a typical coin, on the other hand, with $R \approx \frac{15}{16}$ in. and $L \approx \frac{1}{16}$ in., we obtain (see Figure E4.6b).

$$I_{zz}^C = \frac{mR^2}{4} + \frac{mL^2}{12} = \frac{mR^2}{4}\left[1 + \frac{1}{3}\left(\frac{L}{R}\right)^2\right]$$

$$= \frac{mR^2}{4}\left[1 + \frac{(1/16)^2}{3(15/16)^2}\right]$$

$$\approx \frac{mR^2}{4}$$

This time it is the $mL^2/12$ term that is negligible; it is less than 0.15 of 1 percent of the $mR^2/4$ term. We emphasize, however, that with respect to the *axis* of any solid homogeneous cylinder (disk, rod, or anything in between), the moment of inertia is $mR^2/2$.

EXAMPLE 4.7

A uniform rectangular solid (see Figure E4.7).

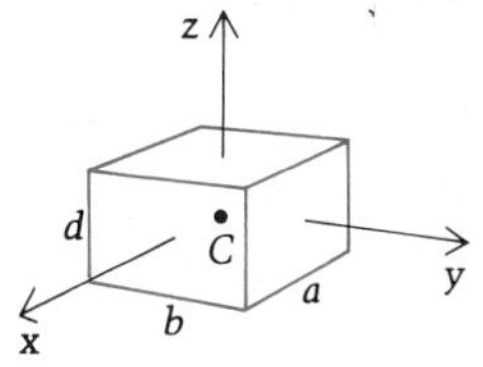

Figure E4.7

Solution

$$I_{zz}^C = \int_{-d/2}^{d/2}\int_{-b/2}^{b/2}\int_{-a/2}^{a/2} (x^2 + y^2)\rho \, dx \, dy \, dz$$

This integration yields

$$I_{zz}^{C} = (\rho abd)\,\frac{a^2+b^2}{12}$$

$$= \boxed{\frac{m}{12}\,(a^2+b^2)}$$

EXAMPLE 4.8

Special case of Example 4.7: A rectangular plate.

Solution

If the rectangular solid is a plate — that is, it has one edge much smaller than the other two dimensions — then, referring to Figure E4.8a–c, we have:

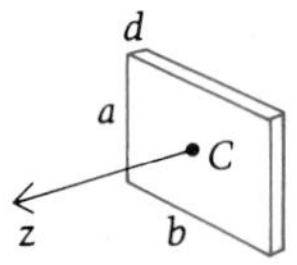

Figure E4.8a

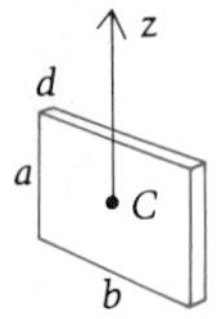

Figure E4.8b

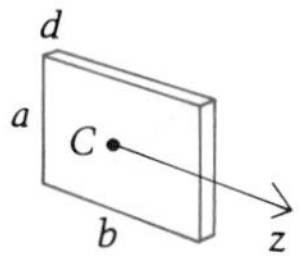

Figure E4.8c

$$I_{zz}^{C} = \boxed{\frac{m}{12}\,(a^2+b^2)} \qquad I_{zz}^{C} = \frac{m}{12}\,(b^2+d^2) \approx \boxed{\frac{mb^2}{12}} \qquad I_{zz}^{C} = \frac{m}{12}\,(a^2+d^2) \approx \boxed{\frac{ma^2}{12}}$$

Again it depends on how the body's plane motion is set up as to which axis is z (normal to the plane of the motion) and hence which formula to use.

EXAMPLE 4.9

Solid, homogeneous, right circular cone about its axis (see Figure E4.9).

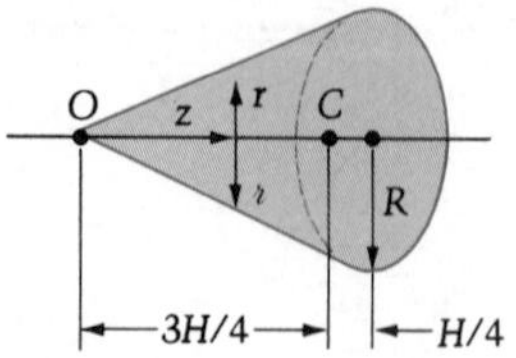

Figure E4.9

Solution

Here we encounter a variable limit, since $r = r(z)$. Noting that C and O are on the same z axis, then (see Figure E4.9):

$$I_{zz}^{C} = I_{zz}^{O} = \int (x^2+y^2)\,dm = \int_{z=0}^{H}\int_{\theta=0}^{2\pi}\int_{r=0}^{r(z)} r^2 \rho r\,dr\,d\theta\,dz$$

From similar triangles,

$$\frac{r}{z} = \frac{R}{H} \Rightarrow r = \frac{Rz}{H} = r(z)$$

which gives the varying radius in terms of z. Then

$$I_{zz}^C = \int_{z=0}^{H} \int_{\theta=0}^{2\pi} \frac{\rho r^4}{4}\Bigg]_0^{Rz/H} \rho\, d\theta\, dz = \int_{z=0}^{H} \frac{\rho R^4 z^4}{4H^4}\theta\Bigg]_0^{2\pi} dz$$

$$= \frac{\rho\pi R^4}{2H^4}\frac{z^5}{5}\Bigg]_0^H = \left(\frac{\rho\pi R^2 H}{3}\right)\frac{3R^2}{10} = \boxed{\frac{3}{10}mR^2}$$

EXAMPLE 4.10

Hollow, homogeneous cylinder about its axis (see Figure E4.10).

Solution

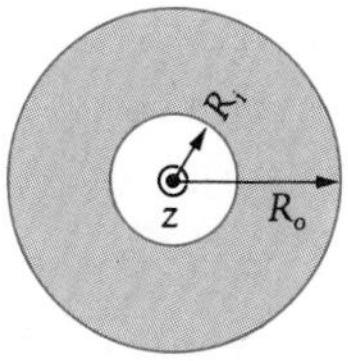

Figure E4.10

$$I_{zz}^C = \int_{-L/2}^{L/2} \int_0^{2\pi} \int_{R_i}^{R_o} r^2 \rho r\, dr\, d\theta\, dz = \frac{\rho\pi(R_o^4 - R_i^4)L}{2}$$

$$= \underbrace{[\rho\pi(R_o^2 - R_i^2)L]}\frac{(R_o^2 + R_i^2)}{2} = \boxed{\frac{m(R_o^2 + R_i^2)}{2}}$$

The same result can be obtained by substracting the moment of inertia of the "hole" (H) from that of the "whole" (W). The basis for this procedure is that we may integrate over *more* than the required region provided we subtract away the integral over the part that is not to be included:

$$I_{zz}^C = \frac{m_W R_o^2}{2} - \frac{m_H R_i^2}{2} = \frac{\rho\pi R_o^2 L R_o^2}{2} - \frac{\rho\pi R_i^2 L R_i^2}{2}$$

$$= \underbrace{\rho\pi(R_o^2 - R_i^2)L}_{m}\frac{(R_o^2 + R_i^2)}{2} = \frac{m(R_o^2 + R_i^2)}{2}$$

Note that if the wall thickness is small, we have a cylindrical shell (or a hoop if the length is small) for which $R_o \approx R_i$ and

$$I_{zz}^C \approx \frac{m(2R^2)}{2} = \boxed{mR^2}$$

(It is obvious that if all the mass is the same distance R from the axis z we should indeed get mR^2.)

EXAMPLE 4.11

A uniform solid sphere about any diameter.

Solution

$$I_{zz}^C = \int (x^2 + y^2)\, dm$$

Also:

$$I_{xx}^C = \int (y^2 + z^2)\, dm$$

and

$$I_{yy}^C = \int (z^2 + x^2)\, dm$$

Adding:

$$\underbrace{I_{xx}^C + I_{yy}^C + I_{zz}^C}_{\substack{= 3I \text{ since they are all} \\ \text{equal by symmetry}}} = \int 2\underbrace{(x^2 + y^2 + z^2)}_{r^2}\, dm$$

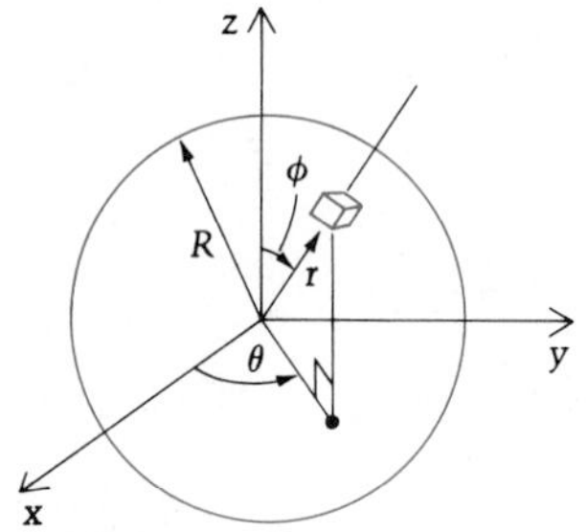

Figure E4.11

Thus (see Figure E4.11 for the spherical coordinates):

$$3I = 2\int r^2\rho_0\, dV = 2\rho_0 \int_{\theta=0}^{2\pi}\int_{\phi=0}^{\pi}\int_{r=0}^{R} r^2\underbrace{(r^2 \sin\phi\, dr\, d\phi\, d\theta)}_{dV \text{ in spherical coordinates}}$$

$$I = \frac{2}{3}\rho_0 \frac{r^5}{5}\Bigg]_0^R (-\cos\phi)\Bigg]_0^\pi \theta\Bigg]_0^{2\pi} = \frac{8\pi\rho_0 R^5}{15}$$

$$= \underbrace{\left(\frac{4}{3}\pi R^3\rho_0\right)}_{m}\frac{2}{5}R^2 = \boxed{\frac{2}{5}mR^2}$$

A less tricky way to do the sphere is to use spherical coordinates directly; the integral is

$$I_{zz}^C = \int_{\theta=0}^{2\pi}\int_{\phi=0}^{\pi}\int_{r=0}^{R} \rho_0(r \sin\phi)^2\, dr(r\, d\phi)(r \sin\phi\, d\theta)$$

which yields the same result of $(2/5)mR^2$, as the reader may show by carrying out the integration.

EXAMPLE 4.12

An example in which the density is not constant. Sometimes a body's density varies; if it does, it must stay inside the integral when we calculate the inertia properties. An example is the earth; we now know that the density of the solid central core of the earth is about four times that of the outermost part of its crust and, moreover, that this central density is nearly twice that of steel!

Let us imagine a sphere with the same mass and radius as in the preceding example but with a density that varies linearly and is twice as high at $r = 0$ as at $r = R$. We shall find I about any diameter. (See Figure E4.12(a).)

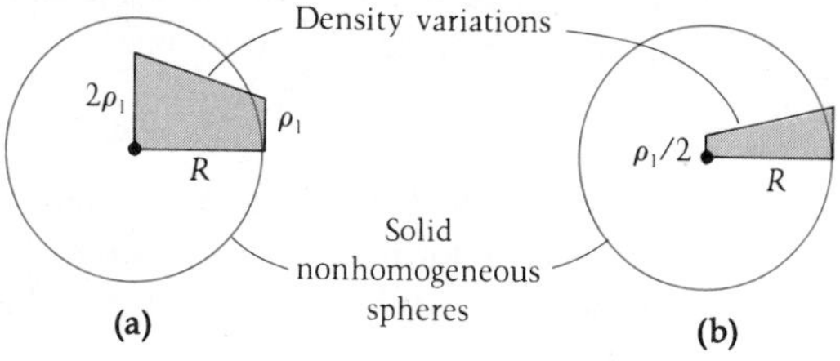

Figure E4.12

Solution

The mass of the body is

$$m = \int \rho\, dV = \int_0^{2\pi} \int_0^{\pi} \int_0^{R} \rho r^2 \sin\phi\, dr\, d\phi\, d\theta$$

Letting the density at R be ρ_1, then

$$\rho = \frac{-\rho_1}{R} r + 2\rho_1$$

Substituting and integrating with the same limits as before, and then equating the new mass to the old, gives

$$\rho_1 = \tfrac{4}{5}\rho_0$$

Thus to have the same mass as the uniform sphere, the density varies from $(8/5)\rho_0$ to $(4/5)\rho_0$ going outward from the center. Then, integrating to find I,

$$3I = 2\int_0^{2\pi} \int_0^{\pi} \int_0^{R} r^2 \rho(r)(r^2 \sin\phi\, dr\, d\phi\, d\theta)$$

which gives

$$I = \frac{28}{75} mR^2 = 0.373mR^2 \qquad \text{(slightly less than } \tfrac{2}{5}mR^2\text{)}$$

Alternatively, if the density varies linearly but is only half as much $(\rho_1/2)$ at the center as at R, then the results, if m and R are again the same as in the uniform case, are (see Figure E4.12(b)):

$$\rho_1 = \frac{8}{7}\rho_0 \qquad \text{and} \qquad I = \frac{44}{105} mR^2 = 0.419mR^2 \qquad \text{(slightly more than } \tfrac{2}{5}mR^2\text{)}$$

The Parallel-Axis Theorem for Moments of Inertia

In many applications the body consists of a number of different smaller bodies of familiar shapes. In such cases, there is fortunately no need to integrate in order to find the inertia of each part with respect to a common axis of interest, thanks to what is called the **parallel-axis theorem** or **transfer theorem**. If we know the moment of inertia about an axis

two constrains
1) The known Moment of inertia has to be through the mass center

2) The desiered axis has to be Parrallel to the known axis.

through the mass center C of any body $\mathcal{B}$, we can then easily find it about any axis parallel to C by a simple calculation. The theorem states that the moment of inertia of the mass of $\mathcal{B}$ about *any* line is the moment of inertia about a parallel line through C plus the mass of $\mathcal{B}$ times the square of the distance between the two axes:

$$I_{zz}^P = I_{zz}^C + md^2$$

To prove this theorem, we let (x, y, z) and (x_1, y_1, z_1) be rectangular cartesian coordinate axes through P and C with the corresponding axes respectively parallel, as shown in Figure 4.3. Then we have, by definition,

$$I_{zz}^C = \int (x_1^2 + y_1^2)\, dm \qquad \text{and} \qquad I_{zz}^P = \int (x^2 + y^2)\, dm$$

$$d = \sqrt{\bar{x}^2 + \bar{y}^2}$$

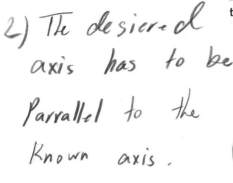

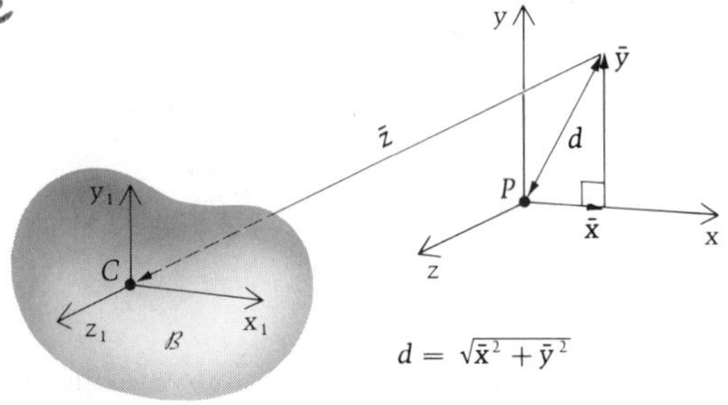

Figure 4.3

Since it is seen that

$$x = x_1 + \bar{x} \qquad \text{and} \qquad y = y_1 + \bar{y},$$

substitution gives

$$\begin{aligned}
I_{zz}^P &= \int [(x_1 + \bar{x})^2 + (y_1 + \bar{y})^2]\, dm \\
&= \int (x_1^2 + y_1^2)\, dm + (\bar{x}^2 + \bar{y}^2)\int dm \\
&\quad + 2\bar{x}\int x_1\, dm + 2\bar{y}\int y_1\, dm
\end{aligned} \tag{4.5}$$

or

$$I_{zz}^P = I_{zz}^C + md^2 \tag{4.6}$$

in which $\bar{x}^2 + \bar{y}^2 = d^2$, the square of the distance between z axes through C and P.

Question 4.3   Why are the last two integrals in Equation (4.5) zero?

Equation (4.6) is the parallel-axis theorem for moments of inertia. But

---

Answer 4.3   Any integral such as $\int_{\mathcal{B}} x\, dm$, where x is measured from an origin at, say, Q is equal to $m\bar{x}$; this *is* the definition of the mass center. So if the origin is C, then $\bar{x} = 0$ and $\int_{\mathcal{B}} x\, dm = 0$.

note: *We can only transfer from the mass center C and not from any other point A about which we may happen to know* I_{zz}^A.

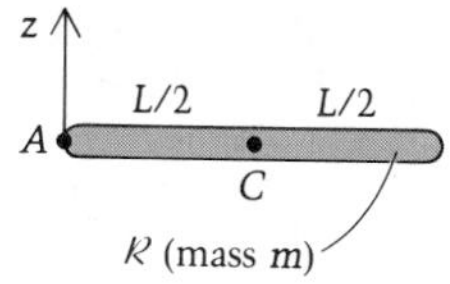

Figure E4.13

Example 4.13

For the uniform slender rod $\mathcal{R}$ shown in Figure E4.13 find I_{zz}^A, the moment of inertia of the mass of $\mathcal{R}$ with respect to a lateral axis through one end. (This exercise will be useful in pendulumlike applications in which a rod is pinned at one end.)

Solution

$$I_{zz}^A = I_{zz}^C + md^2 = \frac{mL^2}{12} + m\left(\frac{L}{2}\right)^2 = \boxed{\frac{mL^2}{3}}$$

We next consider an example of the buildup of the moment of inertia for a composite body.

Example 4.14

Find I_{zz}^O for the body shown in Figure E4.14. The mass densities each $= \rho$ = constant, so that the respective masses are:

$$m_{\text{sph}} = m_1 = \rho \frac{4}{3} \pi R^3$$

$$m_{\text{bar}} = m_2 = \rho \frac{\pi\, d^2 L}{4}$$

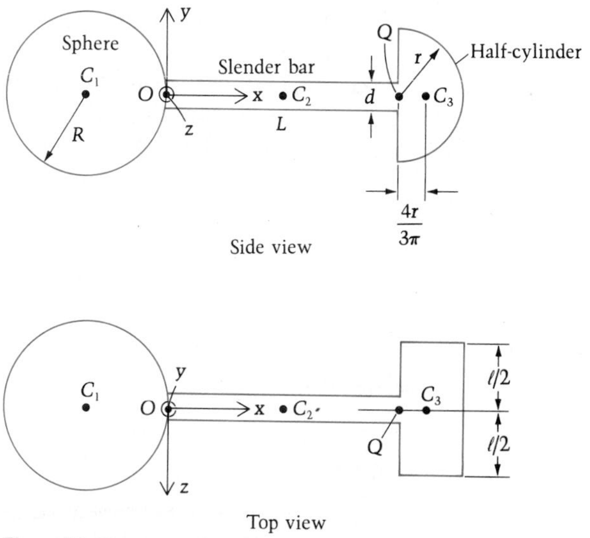

Figure E4.14

and

$$m_{\text{cyl}} = m_3 = \rho \frac{\pi r^2 \ell}{2}$$

Solution

First we observe that $I_{zz}^{O} = I_{zz}^{\text{sph}} = I_{zz}^{\text{bar}} + I_{zz}^{\text{cyl}}$ since the inertia integral may be carried out over the bodies separately, so long as we cover all the elemental masses of the total body. Filling in the separate integrals, we get

$$I_{zz}^{O} = \left(\frac{2}{5} m_1 R^2 + m_1 R^2\right) + \underbrace{\left[\frac{m_2 L^2}{12} + m_2 \left(\frac{L}{2}\right)^2\right]}_{\text{or } m_2 L^2/3}$$

$$+ \left\{\left[\frac{m_3 r^2}{2} - m_3 \left(\frac{4r}{3\pi}\right)^2\right] + m_3 \left(L + \frac{4r}{3\pi}\right)^2\right\}$$

in which, for the half-cylinder,

$$I_{zz}^{C_3} = I_{zz}^{Q} - m_3 \left(\frac{4r}{3\pi}\right)^2 = \frac{m_3 r^2}{2} - m_3 \left(\frac{4r}{3\pi}\right)^2$$

Note that we cannot correctly transfer the inertia of the semicylinder from Q to O; it must be done from the mass center C_3. So first we go "through the back door" to *find* $I_{zz}^{C_3}$ (since we know the moment of inertia already with respect to Q) and only *then* may we transfer to O.

EXAMPLE 4.15

A closed, empty wooden box is 5 ft × 3 ft × 2 ft and weights 124 lb (see Figure E4.15).

a. Find its moment of inertia about an axis through C parallel to the 2-ft dimension.

b. If the box is then filled with homogeneous material weighing 240 lb (excluding the box), how much does the moment of inertia about the axis of part (a) increase?

Figure E4.15

Solution

a. The masses of the various sides of the box are proportional to their areas (thickness and density assumed constant):

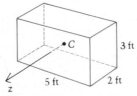

1. 3×5's: $2 \times 15 = 30 \text{ ft}^2$
2. 2×3's: $2 \times 6 = 12 \text{ ft}^2$
3. 2×5's: $2 \times 10 = 20 \text{ ft}^2$

total area $= 62 \text{ ft}^2$

$$W_1 = \frac{30}{62}(124) = 60 \text{ lb for two } 3 \times 5\text{'s}$$

$$W_2 = \frac{12}{62}(124) = 24 \text{ lb for two } 2 \times 3\text{'s}$$

$$W_3 = \frac{20}{62}(124) = 40 \text{ lb for two } 2 \times 5\text{'s}$$

Therefore, taking the contributions from the three pairs of sides,

$$I_{zz}^C = I_{zz}^{C_1} + I_{zz}^{C_2} + I_{zz}^{C_3}$$

$$= \frac{1}{32.2}\left[\frac{2(30)(3^2+5^2)}{12} + 2(12)\left(\frac{3^2}{12} + 2.5^2\right) + 2(20)\left(\frac{5^2}{12} + 1.5^2\right)\right]$$

$$= 15.9 \text{ slug-ft}^2 \qquad (\text{or lb-ft-sec}^2)$$

b. $$I_{zz_{\text{contents}}}^C = \frac{240(5^2+3^2)}{32.2(12)} = 21.1 \text{ slug-ft}^2$$

Even though the box weighs only about half as much as the contents, the position of its mass makes its moment of inertia over three-fourths that of the contents. The *total* moment of inertia is 37.0 slug-ft².

The Radius of Gyration

There is a distance called the **radius of gyration** that is often used in connection with moments of inertia. The radius of gyration of the mass of a body about a line z (through a point P) is called k_{zP}, or just k_P if the axis is understood to be z, and is defined by the equation

$$I_{zz}^P = mk_P^2 \tag{4.7}$$

If one insists on a physical interpretation of k_P, it may be thought of as the distance from P, in any direction perpendicular to z, at which a point mass, with the same mass as the body, would have the same resulting moment of inertia that the body itself has about axis z. For example, a solid homogeneous cylinder has a radius of gyration with respect to its axis of $R/\sqrt{2}$, since $I_{zz}^C = mk_C^2 = \frac{1}{2}mR^2$. The usefulness of k_C is seen here, since regardless of the mass of a cylinder (and hence of its density) k_C will be the same for all homogeneous cylinders of equal radii.

Note further that (using the parallel-axis theorem)

$$mk_P^2 = I_{zz}^P = I_{zz}^C + md^2 = m(k_C^2 + d^2)$$

Thus

$$k_P^2 = k_C^2 + d^2 \tag{4.8}$$

and we see from Equation (4.8) that the radius of gyration, like the moment of inertia itself, is a minimum at C.

Products of Inertia

We now turn to the other two measures of mass distribution that have arisen in our study of plane motion of a rigid body — namely I_{yz}^P and I_{xz}^P, taken here to be with respect to axes (x, y, z) through any point P.*

Our first step is to gain insight into the meaning of products of inertia as we show that they in fact vanish for two large classes of commonly occurring symmetry. These classes are defined by the two conditions (with ρ constant in both): (1) z is an axis of symmetry and (2) xy is a plane of symmetry. Let us examine why the two products of inertia I_{xz}^P and I_{yz}^P are zero in these cases. We recall that their definitions are

$$I_{xz}^P = -\int xz\, dm \qquad I_{yz}^P = -\int yz\, dm \tag{4.9}$$

Class 1: z Is an Axis of Symmetry. For each dV at (x, y, z) there is a corresponding dV at (−x, −y, z). Thus the contributions of these two elements cancel in both the I_{xz}^P and I_{yz}^P integrals. Since *each* point of $\mathcal{B}$ has a "canceling point" reflected through the z axis, I_{xz}^P and I_{yz}^P are each zero for this class of bodies. (See Figure 4.4.)

Class 2: xy Is a Plane of Symmetry. In this case each differential volume dV at (x, y, z) necessarily has a mirror image at (x, y, −z). Thus the contributions of these two elements cancel in both integrals, and taken over the whole of $\mathcal{B}$ we see again that I_{xz}^P and I_{yz}^P are each zero. (See Figure 4.5.)

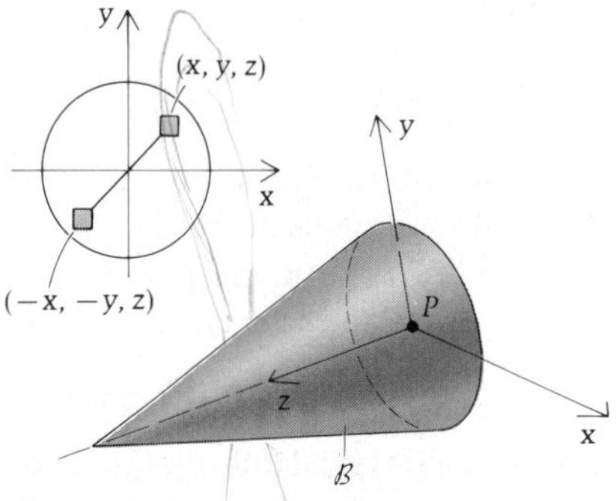

Figure 4.4

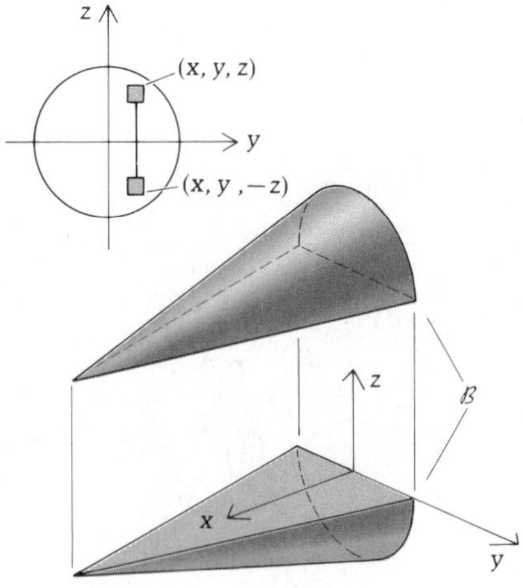

Figure 4.5

* In general (three-dimensional) motion, there arise six distinct inertia properties: three moments of inertia and three products of inertia.

Just because a body does not belong to either of these two classes does not mean it cannot have zero products of inertia. However, these are simply common cases worthy of note.

Question 4.4 Think of a rigid body for which both products of inertia are zero, but which does not fall into either of the two classes.

Transfer Theorem for Products of Inertia

There is a transfer theorem for products of inertia, just as there is one for moments of inertia. To derive it, we write from Figure 4.3:

$$I_{xz}^{P} = -\int xz\,dm = -\int (x_1 + \bar{x})(z_1 + \bar{z})\,dm$$
$$= -\int x_1 z_1\,dm - \bar{x}\bar{z}\int dm - \bar{x}\int z_1\,dm - \bar{z}\int x_1\,dm$$

The last two terms vanish by virtue of the definition of the mass center (for example, $\int z_1\,dm = m$ times the z distance from C to C, which is zero). Thus

$$I_{xz}^{P} = I_{xz}^{C} - m\bar{x}\bar{z} \qquad (4.10a)$$

We note that the factor of m in Equation (4.10a) is alternatively the product of the x and z coordinates of P in an axis system with origin at C. Similarly, we have

$$I_{yz}^{P} = I_{yz}^{C} - m\bar{y}\bar{z} \qquad (4.10b)$$

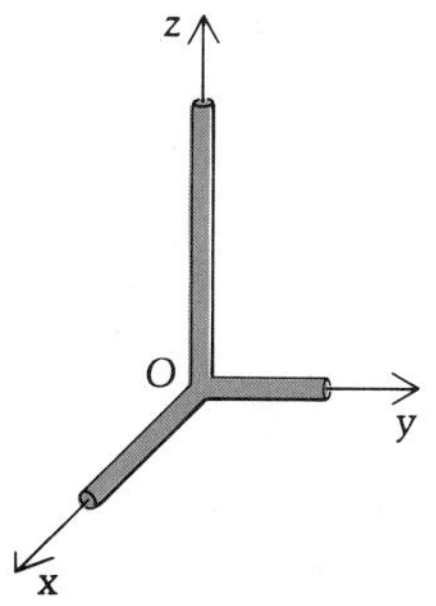

Answer 4.4 Both products of inertia are zero for the body shown in the diagram at the left. Three cylindrical bars of any lengths lying along the (x, y, z) axes are joined at the origin to form a rigid body. Neither the z axis nor the xy plane is one of symmetry, yet I_{xz}^{O} and I_{yz}^{O} are zero.

Example 4.16

Find I_{xz}^{P} and I_{yz}^{P} for the body shown in Figure E4.16; it is composed of eight identical uniform slender rods, each of mass m and length ℓ.

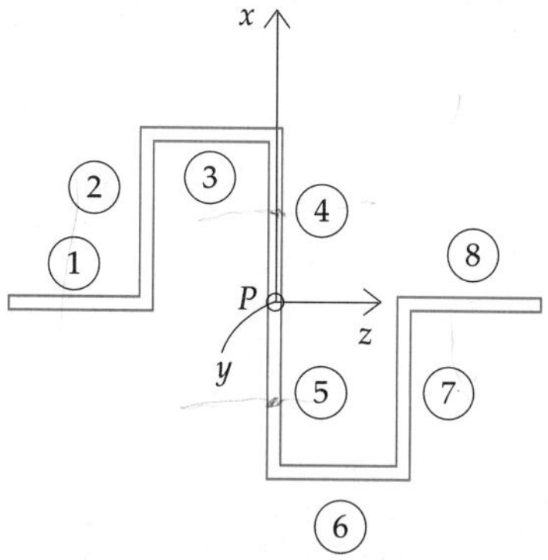

Figure E4.16

Solution

We have $I_{yz}^P = 0$ since xz is a plane of symmetry. Recall that when this happens, the two products of inertia containing (as a subscript) the coordinate normal to the plane are zero.* With superscripts identifying the various rods, we then have the following for the other product of inertia:

$$I_{xz}^P = I_{xz_P}^{①} + I_{xz_P}^{②} + I_{xz_P}^{③} + I_{xz_P}^{④} + I_{xz_P}^{⑤} + I_{xz_P}^{⑥} + I_{xz_P}^{⑦} + I_{xz_P}^{⑧}$$

Note that by symmetry each rod has zero I_{xz} about axes through its *own* center of mass parallel to x and z. Therefore the eight terms listed above will consist only of transfer terms in this problem.

Furthermore, since $\bar{z}$ is zero for rods 4 and 5, and since $\bar{x}$ is zero for rods 1 and 8, only four rods contribute to the overall I_{xz}^P:

$$I_{xz}^P = I_{xz_P}^{②} + I_{xz_P}^{③} + I_{xz_P}^{⑥} + I_{xz_P}^{⑦}$$

$$= -m\left(\frac{\ell}{2}\right)(-\ell) - m(\ell)\left(-\frac{\ell}{2}\right) - m(-\ell)\left(\frac{\ell}{2}\right) - m\left(-\frac{\ell}{2}\right)\ell$$

$$= 4\left(\frac{m\ell^2}{2}\right) = 2m\ell^2$$

Note that the "unbalanced" masses lie in the second and fourth quadrants in this example; hence the sign of I_{xz}^P is positive since its definition carries a minus sign outside the integral. We shall return to this example later in the chapter and examine the reactions caused by the nonzero value of I_{xz}^P when the body is spun up in bearings about the z axis.

PROBLEMS ▶ Section 4.4

4.27 An ellipsoid of revolution is formed by rotating the ellipse about the x axis as in Figure P4.27. Find the moment of inertia of this solid body of density 15 slug/ft^3 about the x axis.

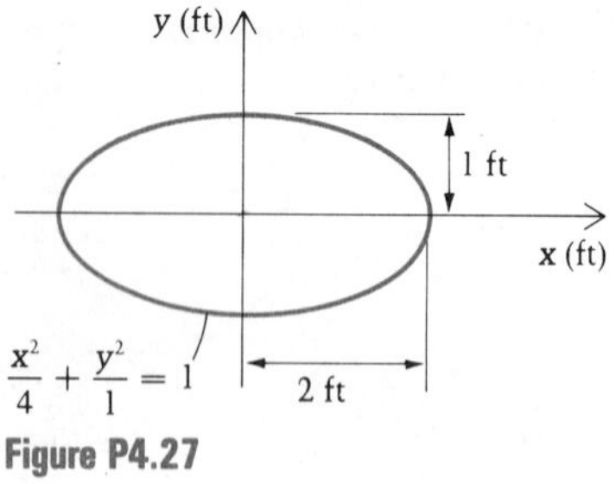

Figure P4.27

4.28 In Figure P4.28, the area bounded by the x and y axes and the parabola $y^2 = 1 - x$ is rotated about the x axis to form a solid of revolution. The density is $\rho = 1000 \cdot (1 - x)^8$ kg/m^3. Find the moment of inertia of the solid mass about the x axis.

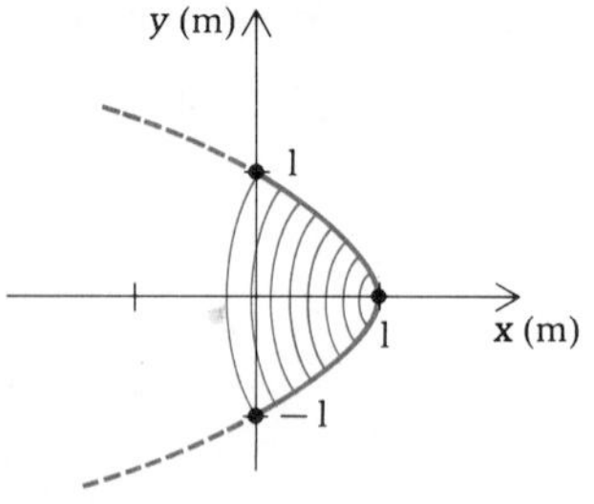

Figure P4.28

* Thus I_{xy}^P is also zero in this example, but I_{xy}^P does not appear anywhere in the plane-motion equations, as we have seen.

4.29 The slender rod in Figure P4.29 has a mass density given by

$$\rho = \rho_1 \left(\frac{x}{L}\right)^2 + \rho_0$$

in which ρ_0 and ρ_1 are constants. The rod has length L. Find its moment of inertia about the line defined by $x = 0$, $y = L/2$.

* **4.30** Find I_{zz}^C for the semielliptical prism shown in Figure P4.30 (density ρ, length normal to plane of paper $= L$).

4.31 The midplane of a uniform triangular plate is shown in Figure P4.31. Find, by integration:

a. I_{xx}^O
b. I_{yy}^O
c. I_{zz}^O
d. I_{xy}^O
e. I_{xz}^O and I_{yz}^O.

What would be good approximations were the plate thin?

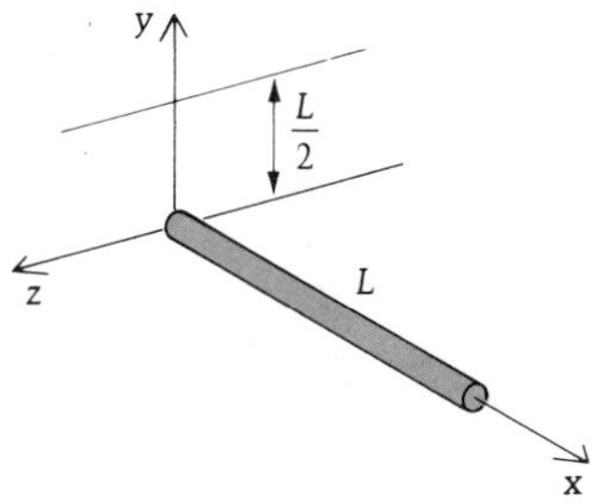

Figure P4.29

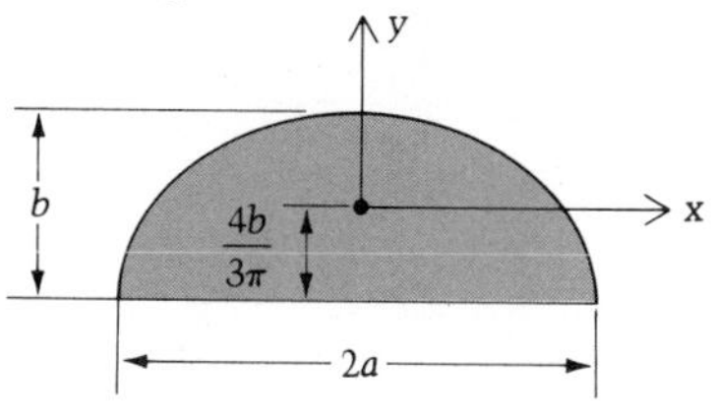

Figure P4.30

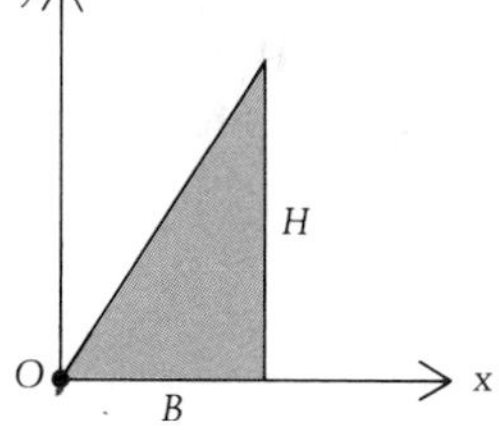

Density $= \rho$, thickness $= t$

Figure P4.31

4.32 Use the parallel-axis theorems and the results of the preceding problem to find, for that plate, the moments and products of inertia at the center of mass.

4.33 For a uniform thin plate with xy axes (and origin O) in the midplane, show that

$$I_{zz}^O \approx I_{xx}^O + I_{yy}^O$$

Confirm this statement with the results of Problem 4.31 for the case when that plate is thin.

4.34 Find I_{xx}^C for a uniform thin plate in the form of a pie-shaped circular sector as shown in Figure P4.34.

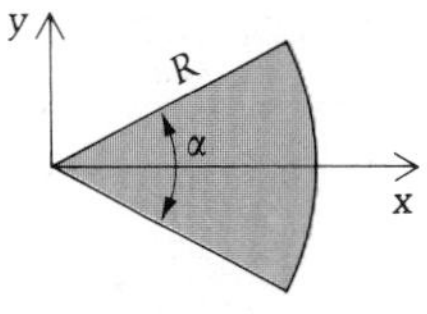

Figure P4.34

4.35 Find I_{yy}^C for the plate in the preceding problem.

4.36 The surface area of a solid of revolution is formed by rotating the curve $y = x^2$ (for $0 \leq x \leq 2$ m) about the x axis. (See Figure P4.36.) The density of the material varies according to the equation $\rho = 20x$, where ρ is in kg/m^3 when x is in meters. Find I_{xx}^O and tell why your answer is also I_{xx}^C.

4.37 In the preceding problem, find I_{yy}^O and I_{yy}^C.

* **4.38** See Figure P4.38. (a) Show that the moment of inertia (I_{xx}^O) for a solid homogeneous cone about a lateral axis through the base is $I_{xx}^O = (m/20)(3R^2 + 2H^2)$. (b) Using the transfer theorem, find the expression for I_{xx}^C.

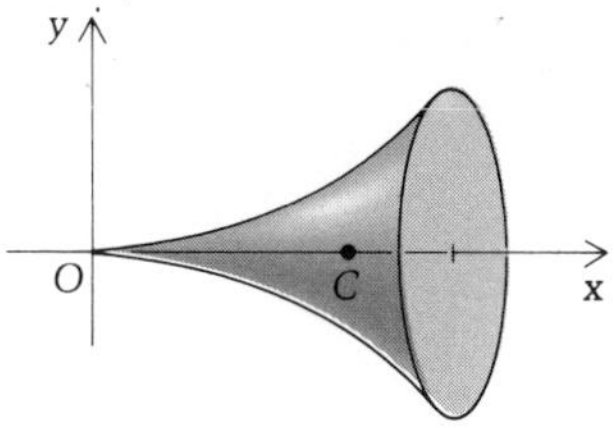

Figure P4.36

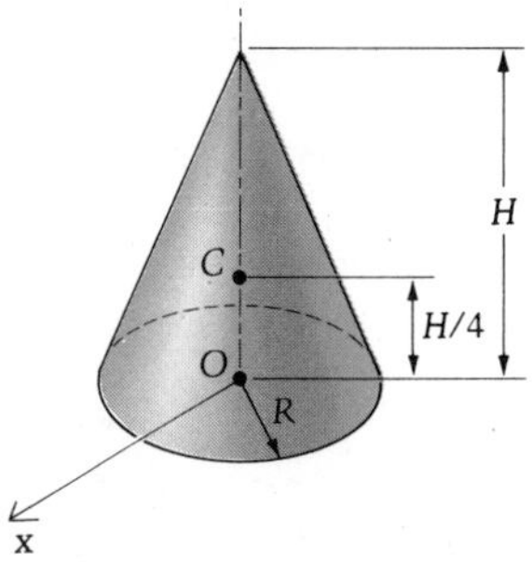

Figure P4.38

* Asterisks identify the more difficult problems.

*4.39 In the previous problem, find I_{xx}^{O} for the body in the figure if the part above $z = H/2$ is cut away to form a truncated cone.

4.40 The body shown in Figure P4.40 is composed of a slender uniform bar ($m = 4$ slugs) and a uniform sphere ($m = 5$ slugs). Find I_{zz}^{C} for the body, where z is normal to the figure.

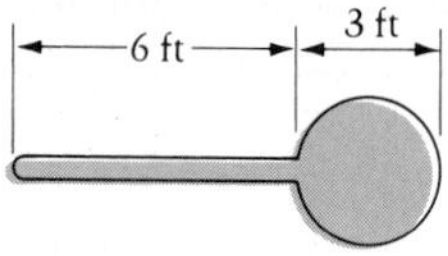

Figure P4.40

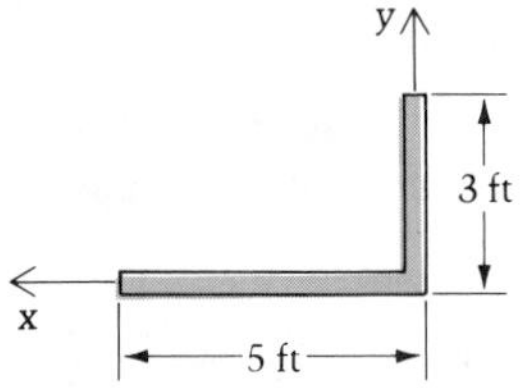

Figure P4.41

4.41 Two bars, each weighing 5 lb per foot, are welded together as shown in Figure P4.41. (a) Locate the center of mass of the body. (b) Find I_{zz}^{C}.

4.42 Find I_{xx} in Problem 4.41.

4.43 Use the result of the preceding problem, together with the transfer (parallel axis) theorem, to find I_{xx}^{C}.

4.44 Find I_{yy} in Problem 4.41.

4.45 Use the result of the preceding problem, together with the transfer theorem, to determine I_{yy}^{C}.

4.46 Find the moment of inertia of the mass of $\mathcal{B}$ about axis z_B if $I_{zz}^{A} = 40\ \text{kg} \cdot \text{m}^2$. (See Figure P4.46.)

4.47 Find the moment of inertia of a uniform hemispherical solid about the lateral axis x_C through its mass center. (See Figure P4.47.)

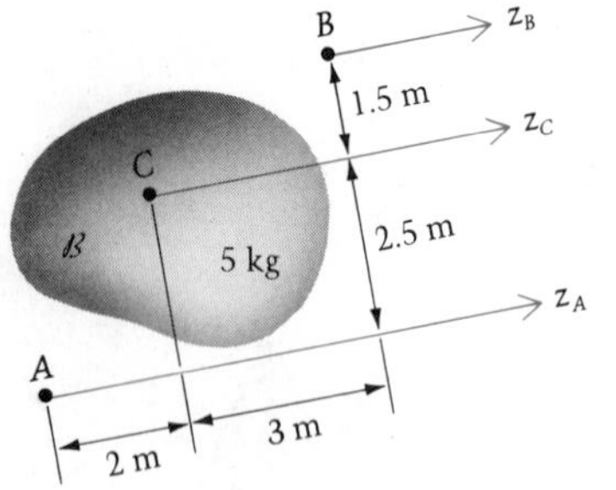

Figure P4.46

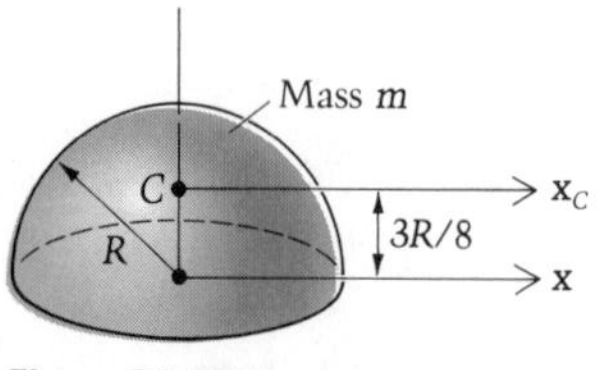

Figure P4.47

4.48 Find I_{zz}^{C} for the semicircular ring $\mathcal{B}$ in Figure P4.48. *Hint*: If the dashed portion were present, I_{zz}^{O} would be $(2m)R^2$; by symmetry, our semicircular ring contributes half of this, so that for $\mathcal{B}$ we have

$$I_{zz}^{O} = mR^2$$

Now use the transfer theorem to complete the solution without integration.

4.49 A rod $\mathcal{R}$ of length 1 m is welded on its ends to a disk $\mathcal{D}$ and a sphere $\mathcal{S}$. (See Figure P4.49.) The uniform bodies have masses $m_{\mathcal{R}} = 10$ kg, $m_{\mathcal{D}} = 5$ kg, and $m_{\mathcal{S}} = 15$ kg. The radii of $\mathcal{D}$ and $\mathcal{S}$ are 0.3 m and 0.1 m, respectively. Find I_{zz}.

4.50 The three bodies shown in Figure P4.50, welded together to form a single body $\mathcal{B}$, have masses of 64 (rectangular plate), 56 (rod), and 48 (disk), all in kilograms. Find the moment of inertia of $\mathcal{B}$ with respect to the z_Q axis.

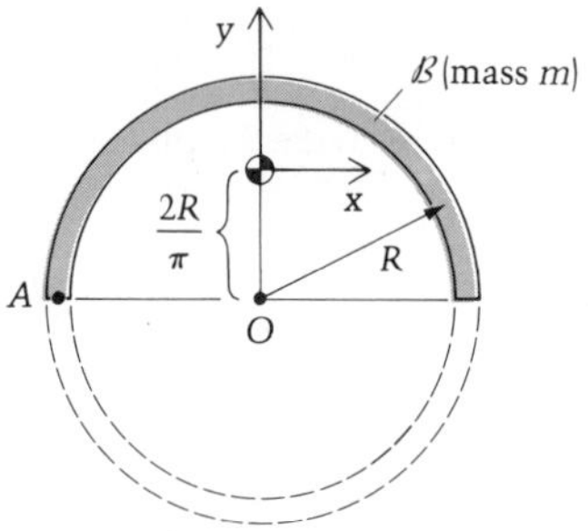

Figure P4.48

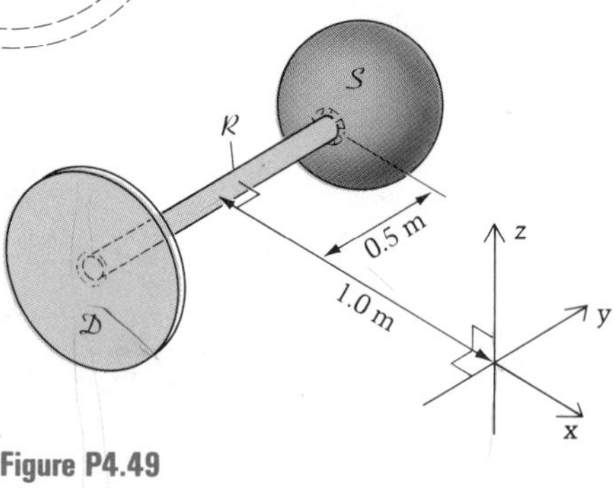

Figure P4.49

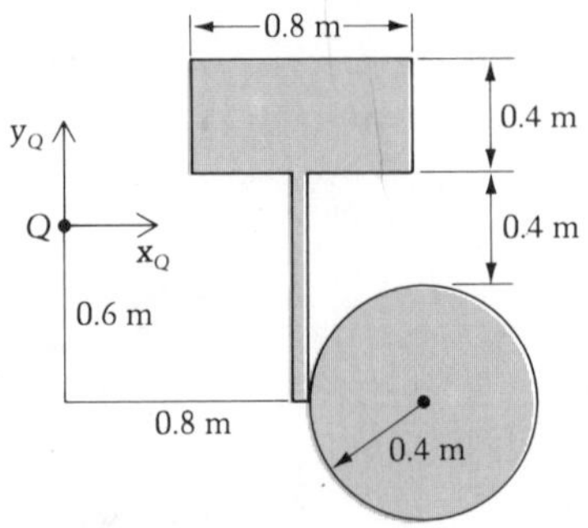

Figure P4.50

4.51 In Figure P4.51, $\mathcal{S}$ is a solid sphere, $\mathcal{C}$ is a solid cylinder, and $\mathcal{R}_1$ and $\mathcal{R}_2$ are slender rods. The center lines of $\mathcal{R}_1$ and $\mathcal{R}_2$ pass through the mass centers of $\mathcal{S}$ and $\mathcal{C}$, respectively. Find I_{zz}^O for the system of four bodies.

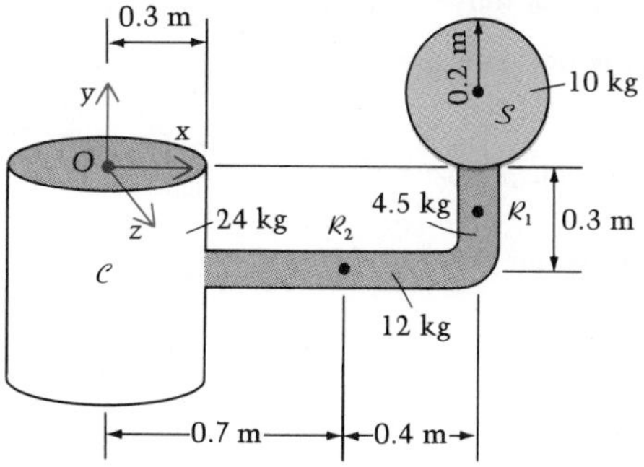

Figure P4.51

4.52 Find I_{xx}^C (which equals I_{yy}^C) for a thin plate (density ρ, thickness t) in the shape of a quarter-circle. (See Figure P4.52.)

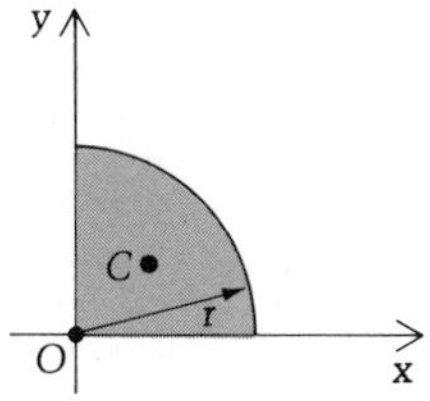

Figure P4.52

4.53 Find the moments of inertia of the pendulum about axes x, y, and z. (See Figure P4.53.) Axes x and y are in the plane of the pendulum, $\mathcal{R}$ is a slender rod, and $\mathcal{D}$ is a semicircular disk, each of constant density.

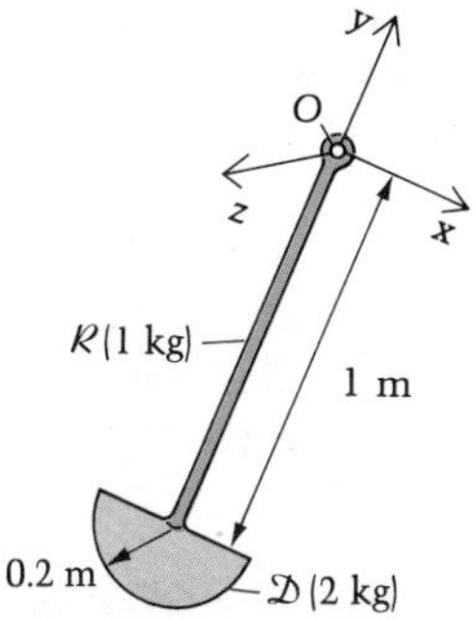

Figure P4.53

4.54 The bent slender rod in Figure P4.54 is located with the axes of the rods parallel to the x- and z-axes, as shown. Find the value of I_{zz}^O.

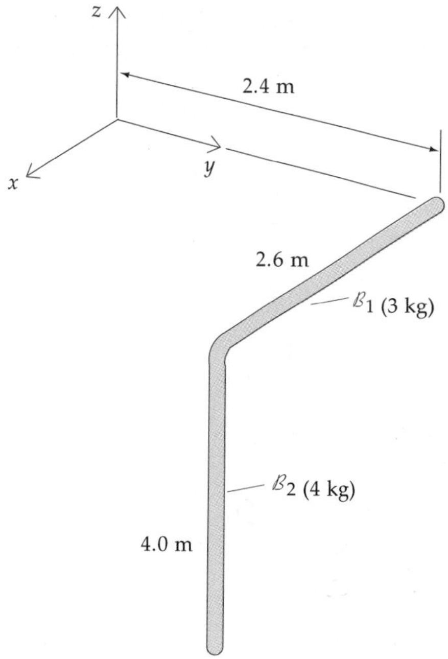

Figure P4.54

4.55 The cylinder $\mathcal{C}$ in Figure P4.55 has a mass of 6 kg and a radius of 0.4 m. Show that the moment of inertia about an axis z_C normal to the page is 0.48 kg · m², and that the corresponding radius of gyration is $k_C = 0.283$ m.

4.56 In the preceding problem, show that it is possible to drill a hole through $\mathcal{C}$ that is below the geometric center Q and satisfy all of the following:

- **a.** the remaining mass is 5.5 kg;
- **b.** the distance between Q and the *new* mass center C (see Figure P4.56) is 0.02 m.

Find the radius r of the hole, the center distance d, and the new value of k_C.

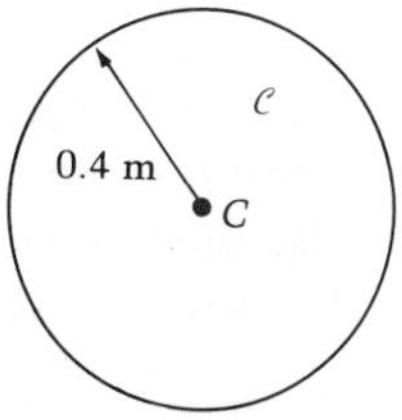

Figure P4.55

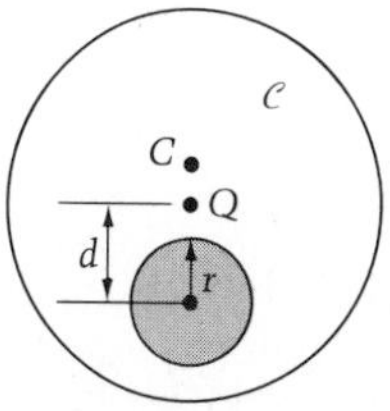

Figure P4.56

4.57 The antenna $\mathcal{A}$ in Figure P4.57 has a moment of inertia about z_C of I_C, and the counterweights $\mathcal{C}$ have a collective moment of inertia about z_G of I_G. Points C and G are the respective mass centers of the antenna and its counterweights. The purpose of the counterweights is to place the combined mass center at O to reduce stresses. Thus $MD = md$, where we neglect the mass of the connecting frame for this problem. Compute the values of M and D that will minimize the total moment of inertia I_O (I of $\mathcal{C}$ about O plus I of $\mathcal{A}$ about O). Use $I_G = Mk_G^2$, where k_G is a constant.

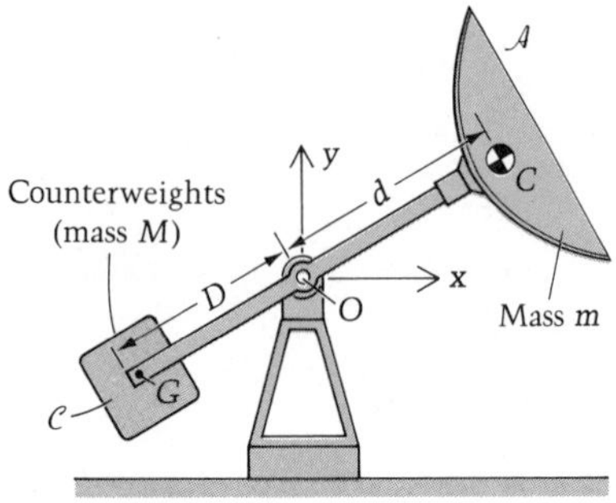

Figure P4.57

4.58 Find the product of inertia I_{xy}^A for the ring of Problem 4.48.

4.59 Find I_{xy}^Q for the welded body $\mathcal{B}$ of Problem 4.50.

4.60 Determine I_{xy} in Problem 4.49. The xy plane contains the centers of $\mathcal{D}$, $\mathcal{R}$, and $\mathcal{S}$.

4.61 Show in the following three ways that the moment of inertia of a uniform, thin spherical shell, about any line through its mass center, is $(2/3)\,mr^2$. (See Figure P4.61.) Which of the three approaches would work if the object were hollow but not thin — that is, if $t \gg R$?

a. Use spherical coordinates:

$$I_{zz}^C = \int_{\theta=0}^{2\pi}\int_{\phi=0}^{\pi} (r\sin\phi)^2 \rho t r^2 \sin\phi\, d\phi\, d\theta$$

where r is the average radius, $R - \dfrac{t}{2}$.

b. Use the idea:

$$I_{zz}^C = \frac{2}{5} m_{\text{WHOLE}} R^2 - \frac{2}{5} m_{\text{HOLE}}(R-t)^2$$

c. Use $I_{zz_{\text{sphere}}}^C = \dfrac{2}{5} mR^2 = \dfrac{2}{5}\rho\left(\dfrac{4}{3}\pi R^5\right) = \dfrac{8}{15}\rho\pi R^5$.
Let R increase by ΔR, and compute the change, ΔI_{zz}^C, by using differential calculus with $\Delta R \ll R$. Note that this change *is* the moment of inertia of the shell!

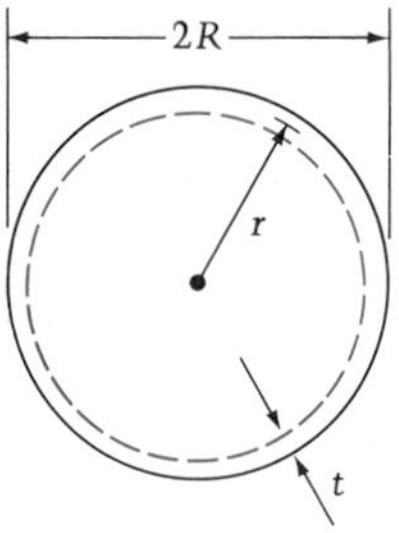

Figure P4.61

4.5 The Mass-Center Form of the Moment Equation of Motion

Development of the Equations of Plane Motion

In Chapter 2 we developed several different forms of Euler's second law. In this section we will continue to study the mass-center form, Equation (2.43):

$$\Sigma \mathbf{M}_C = \dot{\mathbf{H}}_C \tag{2.43}$$

where C is the mass center of an arbitrary body $\mathcal{B}$. If we now restrict the body $\mathcal{B}$ to be rigid and the reference frame to be an inertial frame in Equation (4.4), then we may substitute from that equation for $\mathbf{H}_C$ into Equation (2.43) above and obtain:

$$\Sigma \mathbf{M}_C = \frac{d}{dt}\mathbf{H}_C = \frac{d}{dt}(I_{xz}^C\omega\hat{\mathbf{i}} + I_{yz}^C\omega\hat{\mathbf{j}} + I_{zz}^C\omega\hat{\mathbf{k}}) \tag{4.11}$$

At this stage we must make a decision with regard to how the x and y axes of Equation (4.11), which have their origin at C, will be allowed to change relative to the inertial frame of reference where the derivative in Equation (4.11) is to be taken, and to which $\boldsymbol{\omega}$ of the body is referred. Note that the direction of the z axis has already been fixed perpendicular to the reference plane. If we fix the directions of x and y relative to the inertial frame, then $\hat{\mathbf{i}}$ and $\hat{\mathbf{j}}$ (as well as $\hat{\mathbf{k}}$) are constant relative to that frame but I_{xz}^C and I_{yz}^C are in general time-dependent. This choice is very difficult to deal with.

Question 4.5 However, I_{zz}^C will not change in this case. Why not?

A much more convenient choice is to let the axes x, y, and z all be fixed in the body so that the moments and products of inertia are all constant. Now $\hat{\mathbf{i}}$ and $\hat{\mathbf{j}}$ are time-dependent relative to the inertial frame $\mathcal{I}$. Figure 4.6 shows the unit vectors $\hat{\mathbf{i}}$ and $\hat{\mathbf{j}}$ (fixed to $\mathcal{B}$) expressed in terms of $\hat{\mathbf{I}}$ and $\hat{\mathbf{J}}$ (which are fixed in the inertial reference frame $\mathcal{I}$):

$$\hat{\mathbf{i}} = 1(\cos\theta\hat{\mathbf{I}} + \sin\theta\hat{\mathbf{J}})$$
$$\hat{\mathbf{j}} = 1(-\sin\theta\hat{\mathbf{I}} + \cos\theta\hat{\mathbf{J}})$$

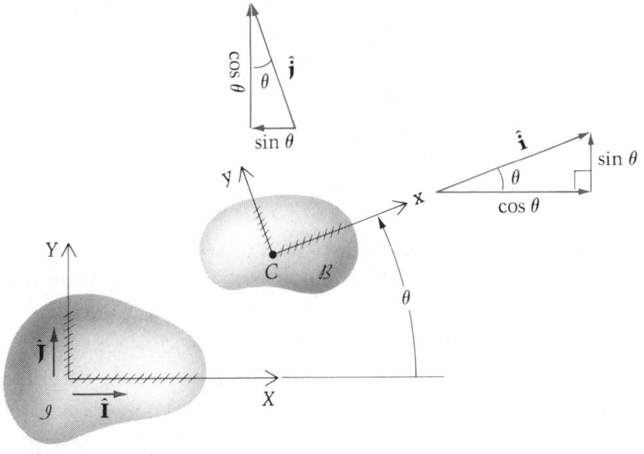

Figure 4.6

The derivatives of these unit vectors, obtained earlier as Equations (3.41) and (3.42), are:

$$\frac{d\hat{\mathbf{i}}}{dt} = (-\sin\theta\hat{\mathbf{I}} + \cos\theta\hat{\mathbf{J}})\dot{\theta} = \dot{\theta}\hat{\mathbf{j}} \text{ or } \omega\hat{\mathbf{j}} \tag{4.12a}$$

Answer 4.5 Let (x, y, z) be fixed in the reference frame $\mathcal{I}$. As the body rotates with respect to frame $\mathcal{I}$, its mass is then distributed differently, at different times, with respect to x and y. But in plane motion the z axis is fixed in direction in *both* the body and in space. Thus since the square of the distance from the z axis is always the same r^2 regardless of the orientation of x and y, we see that I_{zz}^C does not change as the body turns.

$$\frac{d\hat{\mathbf{j}}}{dt} = -(\cos\theta\hat{\mathbf{I}} + \sin\theta\hat{\mathbf{J}})\dot{\theta} = -\dot{\theta}\hat{\mathbf{i}} \text{ or } -\omega\hat{\mathbf{i}} \tag{4.12b}$$

Therefore, carrying out the differentiations in Equation (4.11), we have

$$\Sigma\mathbf{M}_C = I_{xz}^C\dot{\omega}\hat{\mathbf{i}} + I_{yz}^C\dot{\omega}\hat{\mathbf{j}} + I_{zz}^C\dot{\omega}\hat{\mathbf{k}} + I_{xz}^C\omega(\omega\hat{\mathbf{j}}) + I_{yz}^C\omega(-\omega\hat{\mathbf{i}})$$

or

$$\Sigma\mathbf{M}_C = (I_{xz}^C\alpha - I_{yz}^C\omega^2)\hat{\mathbf{i}} + (I_{yz}^C\alpha + I_{xz}^C\omega^2)\hat{\mathbf{j}} + I_{zz}^C\alpha\hat{\mathbf{k}} \tag{4.13}$$

This expression represents three scalar equations:

$$\Sigma M_{Cx} = I_{xz}^C\alpha - I_{yz}^C\omega^2 \tag{4.14a}$$

$$\Sigma M_{Cy} = I_{yz}^C\alpha + I_{xz}^C\omega^2 \tag{4.14b}$$

$$\Sigma M_{Cz} = I_{zz}^C\alpha \tag{4.14c}$$

Equations (4.14a,b), along with $\Sigma F_z = 0$, tell us about the nature of reactions necessary to maintain the plane motion. If I_{xz}^C and I_{yz}^C are both zero,* then $\Sigma M_{Cx} = 0 = \Sigma M_{Cy}$ and the system of external forces (loads plus reactions) has a planar resultant, with this plane containing the mass center. Thus, with a coplanar system of external *loads*, the resultant *reactions* must have a resultant in that same plane. This is the basis for "two-dimensionalizing" the analysis of problems for which the two products of inertia above vanish; we shall first work with symmetrical bodies for which this is the case. Then later in this chapter we shall examine some problems in which I_{xz}^C and I_{yz}^C are *not* both zero. In any case, however, the rotational motion of the body is governed by the simple kinetics equation

$$\Sigma M_{Cz} = I_{zz}^C\alpha \qquad (=I_{zz}^C\ddot{\theta}) \tag{4.14d}$$

We see that while force produces acceleration with the "resistance" being the mass, it is also true that moment produces angular acceleration with the "resistance" being the body's moment of inertia. Note also that the moment of the external forces about the z axis through the mass center is $\Sigma\mathbf{M}_C \cdot \hat{\mathbf{k}}$, and we have $\Sigma\mathbf{M}_C \cdot \hat{\mathbf{k}} = I_{zz}^C\alpha$. Thus the resultant moment about this axis equals the moment of inertia about the axis multiplied by the angular acceleration of the body, regardless of whether or not the products of inertia vanish.

We will now restrict our attention in the remainder of this section to a special class of problems of plane, rigid-body motions. This class is defined by the following pair of conditions:

1. $I_{xz}^C = I_{yz}^C = 0$ (usually because the body is either symmetric about the plane of motion of the mass center or else has an axis of physical symmetry which remains normal to the reference plane of motion); and

2. The external *loads* have a planar resultant with the plane containing the mass center.

---

* When I_{xz}^C or I_{yz}^C is *not* zero, there has to be a nonzero ΣM_{Cx} or ΣM_{Cy} present to maintain the motion; these are usually formed by lateral forces (such as bearing reactions) at different positions along the z-axis.

These two conditions necessitate external *reactions* likewise equipollent to a coplanar (in the same plane as the loads) force system. Euler's laws are then effectively reduced to the following three scalar equations:

$$\Sigma F_x = m\ddot{x}_C \tag{4.15a}$$

$$\Sigma F_y = m\ddot{y}_C \tag{4.15b}$$

$$\Sigma M_{Cz} = I_{zz}^C\ddot{\theta} \tag{4.15c}$$

where x_C and y_C are coordinates of the mass center in a rectangular coordinate system fixed in the inertial frame.* Equations (4.15a,b) are of course the x and y components of the mass center equation of motion valid for any body, rigid or not. Since there will be no confusion about the axis in question, we shall often write Equation (4.15c) as simply

$$\Sigma M_C = I_C\ddot{\theta}$$

or

$$\Sigma M_C = I_C\alpha \tag{4.16}$$

Helpful Steps to Follow in Generating and Solving the Equations of Motion

In some instances Equations (4.15a–c) will yield a differential equation(s) that can be readily integrated so that we predict the ongoing motion of a body. More commonly we shall be dealing with what we might call "snapshot" problems where at a specific instant we calculate forces and accelerations. These problems are rather natural extensions of statics, all of the operations in the analysis being algebraic. For problems of either class the following steps are recommended:

1. Draw a free-body diagram (FBD) of each body in the problem.

2. Define a set of unit vectors, or, equivalently, a coordinate system, in terms of which unknown forces and accelerations may be expressed.

3. Substitute into the three equations of motion; some may prefer to do this in vector algebraic form, explicitly displaying previously defined unit vectors, while others may prefer to initiate the analysis using the three component equations (4.15a,b and 4.16).

4. Often the number of scalar unknowns will exceed the number of independent equations (maximum of three), and we must look for supplementary information. This might mean nothing more than applying the Coulomb law of friction, but often the supplementary information will be in the form of a kinematic constraint. For example, for a wheel having its mass center at the geometric center, rolling would imply (with appropriate definitions of variables) $\ddot{x}_C = r\ddot{\theta}$. If some point A has its motion constrained, then the restriction on its acceleration along with $\mathbf{a}_C = \mathbf{a}_A + \alpha\hat{\mathbf{k}} \times \mathbf{r}_{AC} - \omega^2\mathbf{r}_{AC}$ can be used to relate acceleration of

* Which component equations are used and the specific forms they take depend, of course, on what type of coordinate system is used to describe the motion of C. A rectangular coordinate system is the natural choice for most of the problems taken up in this chapter. A polar coordinate system is the natural choice, however, for problems of orbital mechanics (see Section 8.4).

the mass center and angular acceleration of the body (presuming velocities, and thus ω, are known). (Of course, $\mathbf{a}_C$ and α are the kinematical variables which naturally appear in the equations of motion, as we have seen.)

5. Solve for unknown forces and/or accelerations. All problems appearing in this text are "rigid-body dynamically determinate," so this will always be possible. In the case of "movie" problems, integrate accelerations to get velocities and then velocities to get positions as functions of time.

6. Be sure to check on dimensional consistency of results and, in the case of numerical results, check to see that units are correct (especially important in these days of transition from U.S. to SI units). Also, see if your answer seems to make sense.

We now proceed toward a set of examples which will make use of the equations (4.15a,b and 4.16), which we have developed in this section. These examples are designed to illustrate the kinds of problems the student should learn to solve. The first one is a "movie" problem in which one resultant force component and the moment are zero, while the other force component is not.

EXAMPLE 4.17

A horseshoe pitcher releases a horseshoe with $\omega_i = 4.71 \circlearrowleft$ rad/sec in the position shown in Figure E4.17a. If the horseshoe turns exactly once in plane motion and scores a ringer, find the initial velocity of the horseshoe's mass center.

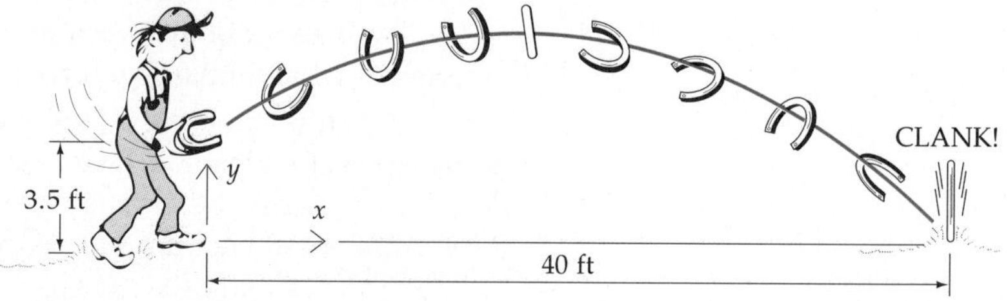

Figure E4.17a

A free-body diagram of the horseshoe in flight is shown in Figure E4.17b. We write and then integrate Equations (4.17a–c):

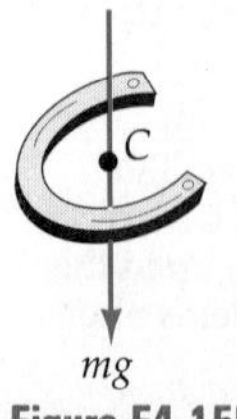

Figure E4.15b

$$\Sigma F_x = 0 = m\ddot{x}_C$$

$$\ddot{x}_C = 0$$

$$\dot{x}_C = \dot{x}_i$$

(the initial velocity of C in x-direction)

$$x_C = \dot{x}_i t + \cancel{x_i}^{\,0}$$

since $x_C = 0$ at $t = 0$

$$\Sigma F_y = -mg = m\ddot{y}_C$$

$$\ddot{y}_C = -g$$

$$\dot{y}_C = -gt + \dot{y}_i$$

(the initial velocity of C in y-direction)

$$y_C = \frac{-gt^2}{2} + \dot{y}_i t + \cancel{y_i}^{\,3.5}$$

since horseshoe is released at $y_C = 3.5$ ft

$$\Sigma M_C = 0 = I_C\ddot{\theta}$$

$$\ddot{\theta} = 0$$

$$\dot{\theta} = \omega_i = 4.71$$

$$\theta = 4.71t + \cancel{\theta_i}^{\,0}$$

Now when the horseshoe lands, $\theta = 2\pi$, so

$$2\pi = 4.71t_f \Rightarrow t_f = 1.33 \text{ sec}$$

Thus

$$x_C = 40 = \dot{x}_i(1.33) \Rightarrow \dot{x}_i = 30.1 \text{ ft/sec}$$

and

$$y_C = 0 = -\frac{32.2}{2}(1.33)^2 + \dot{y}_i(1.33) + 3.5 \Rightarrow \dot{y}_i = 18.8 \text{ ft/sec}$$

Hence the initial velocity of the horseshoe's mass center is

$$\mathbf{v}_i = 30.1\hat{\mathbf{i}} + 18.8\hat{\mathbf{j}} \text{ ft/sec}$$

The next example involves rolling, and is a "movie" (ongoing time) problem:

EXAMPLE 4.18

The cylinder (mass m, radius r) is released from rest on the inclined plane shown in Figure E4.18a. The coefficient of friction between cylinder and plane is μ. Determine the motion of C, assuming that μ is large enough to prevent slipping. (How large must it be?)

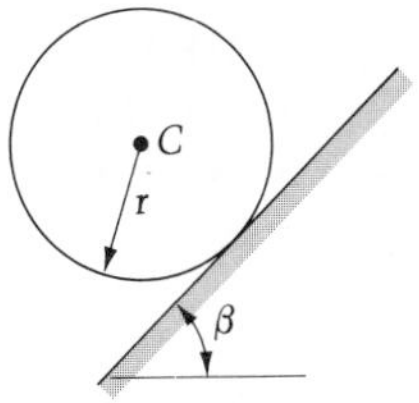

Figure E4.18a

Solution

As in statics, a good first step is to draw a free-body diagram (see Figure E4.18b). We are to assume that the cylinder rolls. In this case the friction force f is an unknown and has a value satisfying

$$0 \le f \le f_{max}$$

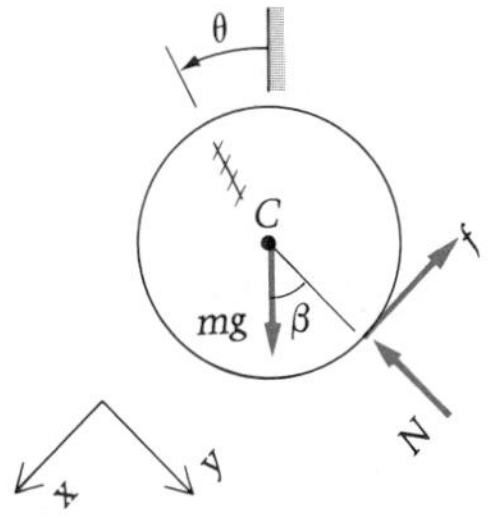

Figure E4.18b

where $f_{max} = \mu N$ from the study of Coulomb friction in statics. We shall also use a kinematic equation expressing the rolling. After solving for f, we shall then impose the condition that it be less than μN, since we know the cylinder is not slipping.

We choose x, y, and θ as shown, motivated by the fact that C will move down the plane and the cylinder will turn counterclockwise. The equations of motion are

$$\Sigma F_x = m\ddot{x}_C \Rightarrow mg \sin\beta - f = m\ddot{x}_C \tag{1}$$

$$\Sigma F_y = m\ddot{y}_C \Rightarrow mg \cos\beta - N = m\ddot{y}_C = 0 \tag{2}$$

(Note that, kinematically, y_C is constant so that $\ddot{y}_C$ vanishes.)

$$\Sigma M_C = I_C\alpha \Rightarrow fr = \tfrac{1}{2}mr^2\ddot{\theta} \tag{3}$$

We can solve (2) for N, getting $N = mg \cos\beta$. There remain two equations in the three unknowns f, $\ddot{x}_C$, and $\ddot{\theta}$. We must therefore supplement our equations of motion with the remaining kinematics result, which comes from the rolling condition:

$$\ddot{x}_C = r\ddot{\theta} \tag{4}$$

Solving Equations (1), (3), and (4) gives

$$\ddot{x}_C = \frac{2}{3} g \sin \beta \qquad \ddot{\theta} = \frac{2g \sin \beta}{3r} \qquad f = \frac{mg \sin \beta}{3}$$

Integrating twice, and noting that the integration constants vanish, we get

$$x_C = \frac{gt^2}{3} \sin \beta \qquad \theta = \frac{gt^2}{3r} \sin \beta$$

And since we are told that

$$f \leq f_{\max} = \mu N$$

that is,

$$\frac{mg \sin \beta}{3} \leq \mu mg \cos \beta$$

then we have

$$\mu \geq \frac{\tan \beta}{3}$$

for the rolling motion to occur.

In the preceding example, things would be quite different if the value of the friction force that satisfies the equations, namely $f = (mg \sin \beta)/3$, were larger than $f_{\max} = \mu N = \mu mg \cos \beta$. So if the solution were to yield

$$f > f_{\max}, \quad \text{i.e., if} \quad \mu < \frac{\tan \beta}{3},$$

then there would be insufficient friction to permit rolling. We would then need to abandon* the rolling condition, replacing it by the known maximum value of f, i.e.,

$$\ddot{x}_C \neq r\ddot{\theta} \text{ any longer,}$$

but now we know

$$f = \mu N = \mu mg \cos \beta$$

where N still equals $mg \cos \beta$. And since Equation (3) still holds,

$$fr = \mu mgr \cos \beta = \tfrac{1}{2} mr^2 \ddot{\theta}$$

or

$$\ddot{\theta} = \frac{2g\mu \cos \beta}{r}$$

Therefore, integrating twice, we get

$$\theta = \frac{\mu g t^2 \cos \beta}{r}$$

* When you assume something and later arrive at a contradiction of fact, then the logical conclusion is that the assumption was invalid.

where the integration constants are zero since $\theta = \dot{\theta} = 0$ at $t = 0$. Equation (1) now yields, with $f = \mu N$,

$$mg \sin \beta - \mu mg \cos \beta = m\ddot{x}_C$$

or

$$\ddot{x}_C = g(\sin \beta - \mu \cos \beta)$$

Thus

$$x_C = \frac{gt^2}{2}(\sin \beta - \mu \cos \beta)$$

and the solutions for the motion [$x_C(t)$ and $\theta(t)$] are indeed quite different when the cylinder turns and slips than they are when it rolls. If $\mu = 0.5$ and $\beta = 30°$, for example, then

$$\tan \beta = 0.577 \leq 3\mu = 1.5$$

and the cylinder rolls. But if $\mu = 0.2$ and $\beta = 60°$, then

$$\tan \beta = 1.73 \not\leq 3\mu = 0.6$$

and the cylinder slips. Note that in general (as one would expect) the cylinder rolls for larger μ and smaller β.

Finally, note that if we wish to distinguish between static and kinetic coefficients of friction (μ_s and μ_k), then the rolling assumption would be correct if $\tan \beta \leq 3\mu_s$. But if $\tan \beta > 3\mu_s$, we would then use $f = \mu_k N$ in the remainder of the solution, and the μ in the answers for x_C and θ would become μ_k.

Next we take up another rolling problem, only this time the mass center and geometric center are not the same point. As a result, the kinematics equations are more difficult. This example is a "snapshot" (occurring at one instant of time) problem.

Example 4.19

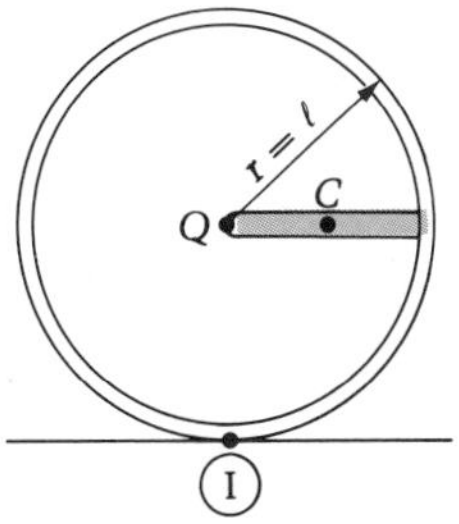

Figure E4.19a

The rigid body $\mathcal{B}$ in Figure E4.19a consists of a heavy bar of mass m welded to a light hoop; the radius of the hoop thus equals the length of the bar. Find the minimum coefficient of friction between the hoop and the ground for which the body will roll when released from rest in the given position.

Solution

The free-body diagram is shown in Figure E4.19b along with the base vectors adopted for the problem.

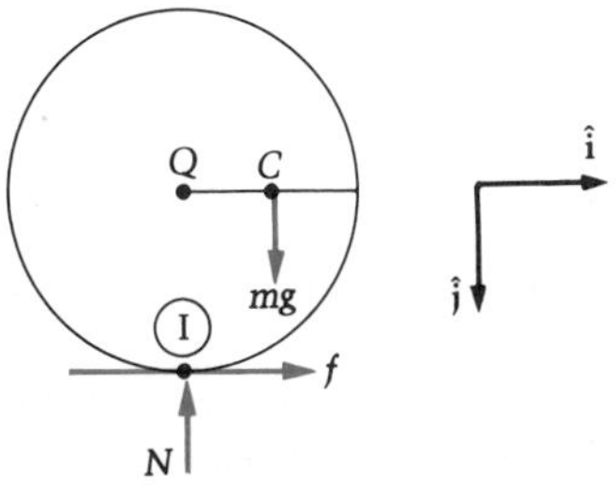

Figure E4.19b

> **Question 4.6** Why is this a good choice of base vectors to use in this problem?

Answer 4.6 Because the initial acceleration of the mass center C will be to the right and downward (↘) and the initial angular acceleration will be clockwise (↻).

Note that the gravity force resultant acts through the center of the bar since we are neglecting the weight of the hoop.

Next we write the three differential equations of motion, letting $\mathbf{a}_C = \ddot{x}_C\hat{\mathbf{i}} + \ddot{y}_C\hat{\mathbf{j}}$ and $\boldsymbol{\alpha} = \alpha\hat{\mathbf{k}}$.

$$\Sigma F_x = m\ddot{x}_C \Rightarrow f = m\ddot{x}_C \tag{1}$$

$$\Sigma F_y = m\ddot{y}_C \Rightarrow mg - N = m\ddot{y}_C \tag{2}$$

$$\Sigma M_C = I_C\alpha \Rightarrow \frac{N\ell}{2} - f\ell = \frac{m\ell^2}{12}\alpha \tag{3}$$

These equations contain the unknowns f, N, $\ddot{x}_C$, $\ddot{y}_C$, and α. We draw upon kinematics for two more equations. We know that the accleration of the geometric center Q of the hoop is $\mathbf{a}_Q = r\alpha\hat{\mathbf{i}}$, so that

$$\mathbf{a}_C = \underbrace{\mathbf{a}_Q}_{a_Q\hat{\mathbf{i}} = r\alpha\hat{\mathbf{i}}} + \alpha\hat{\mathbf{k}} \times \mathbf{r}_{QC} - \overset{0}{\cancel{\omega^2}}\,\mathbf{r}_{QC}$$

or

$$\ddot{x}_C\hat{\mathbf{i}} + \ddot{y}_C\hat{\mathbf{j}} = r\alpha\hat{\mathbf{i}} + \alpha\hat{\mathbf{k}} \times \left(\frac{\ell}{2}\hat{\mathbf{i}}\right)$$

Therefore, equating the $\hat{\mathbf{i}}$ coefficients and then the $\hat{\mathbf{j}}$ coefficients, we find

$$\ddot{x}_C = \ell\alpha \tag{4}$$

$$\ddot{y}_C = \frac{\ell\alpha}{2} \tag{5}$$

The student may wish to verify that (4) and (5) also result from relating $\mathbf{a}_C$ to $\mathbf{a}_{\textcircled{I}} = r\omega^2\uparrow$, true for any round body rolling on a flat, fixed plane. In this problem, $\mathbf{a}_{\textcircled{I}}$ is then zero at release because, until time passes, ω is still zero.

Solving Equations (1) to (5) for the five unknowns gives the results:

$$f = \frac{3}{8}mg \qquad N = \frac{13}{16}mg \qquad \ddot{x}_C = \frac{3g}{8} \qquad \ddot{y}_C = \frac{3g}{16} \qquad \alpha = \frac{3g}{8\ell}$$

To complete the solution, we must get the coefficient of friction μ into the picture. We know that for any friction force f,

$$0 \le f \le f_{\max} = \mu N$$

Therefore, in our problem,

$$\frac{3}{8}mg \le \mu\frac{13}{16}mg$$

so that

$$\mu \ge \frac{6}{13}$$

This means that for the body to roll, a friction coefficient of at least $\mu = 6/13$ is required; this is then the desired minimum.

We emphasize that students should always make "eyeball checks" of their answers — glancing over the results to be sure they make sense physically. In this problem, for instance, note that:

1. $N < mg$ as expected, for otherwise the mass center could not begin to move downward as the body rolls.
2. f is positive, and therefore in the correct direction to (since it is the only force in the x-direction) move the mass center to the right.
3. $\ddot{x}_C$, $\ddot{y}_C$, and α are all positive and therefore in the expected directions.

The next example features another "snapshot" problem — one which we are examining only at one instant. It also involves the interesting constraint of a taut string.

Example 4.20

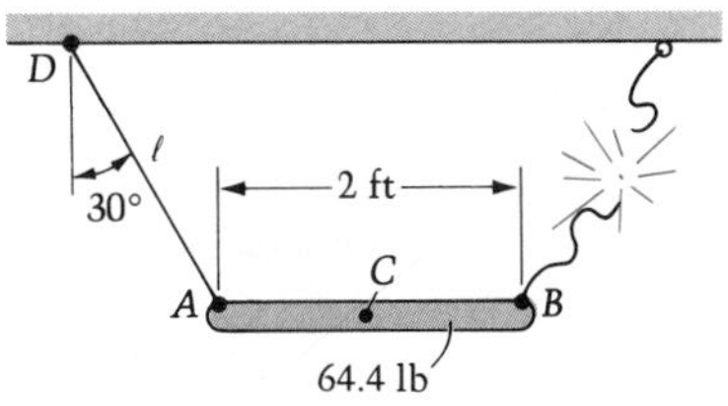

Figure E4.20a

A uniform rod is supported by two cords as shown in Figure E4.20a. If the right-hand cord suddenly breaks, determine the initial tension in the left cord AD. ("Initial" means before the rod has had time to move and before it has had time to generate any velocities.)

Solution

Using the free-body diagram in Figure E4.20b, the equations of motion are, with $\mathbf{a}_C = \ddot{x}_C\hat{\mathbf{i}} + \ddot{y}_C\hat{\mathbf{j}}$ and $\boldsymbol{\alpha} = \alpha\hat{\mathbf{k}}$:

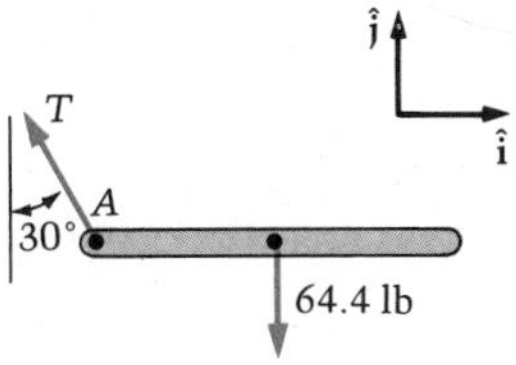

Figure E4.20b

$$\overset{+}{\longrightarrow} \quad \Sigma F_x = -T\left(\frac{1}{2}\right) = m\ddot{x}_C = 2\ddot{x}_C \tag{1}$$

$$+\!\uparrow \quad \Sigma F_y = -64.4 + T\left(\frac{\sqrt{3}}{2}\right) = m\ddot{y}_C = 2\ddot{y}_C \tag{2}$$

$$\circlearrowleft\!+ \quad \Sigma M_C = -T\left(\frac{\sqrt{3}}{2}\right)(1) = I_C\alpha = \frac{mL^2}{12}\alpha = \frac{2\alpha}{3} \tag{3}$$

Unfortunately, Equations (1–3) contain four unknowns (T, $\ddot{x}_C$, $\ddot{y}_C$, and α). Thus we seek an additional equation in these unknowns from kinematics. The point A is constrained to move (see Figure E4.20c) on a circle of radius ℓ about D. Thus point A has the tangential and normal components of acceleration shown (see Section 1.7). Furthermore, $v_A = 0$ at the instant of interest (nothing is moving yet!).

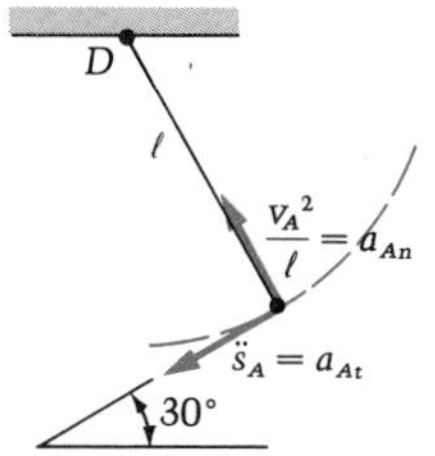

Figure E4.20c

We may relate this $\mathbf{a}_A$ to $\mathbf{a}_C$:

$$\mathbf{a}_C = \ddot{x}_C\hat{\mathbf{i}} + \ddot{y}_C\hat{\mathbf{j}} = \underbrace{\mathbf{a}_A}_{\ddot{s}_A \ \swarrow 30^\circ} + \underbrace{\alpha\hat{\mathbf{k}} \times \overset{1\hat{\mathbf{i}}}{\cancel{\mathbf{r}_{AC}}}}_{\alpha\hat{\mathbf{j}}} - \overset{\text{zero at } t=0}{\cancel{\omega^2}}\mathbf{r}_{AC} \tag{4}$$

Here now is a nice shortcut:
If we dot Equation (4) with a unit vector in the direction 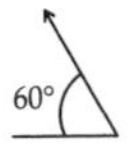, we will eliminate $\mathbf{a}_A$ (because it is perpendicular to that direction). This is easier than solving the two "$\hat{\mathbf{i}}$ and $\hat{\mathbf{j}}$ equations." Such a unit vector is

$$-\frac{1}{2}\hat{\mathbf{i}} + \frac{\sqrt{3}}{2}\hat{\mathbf{j}}$$

so that, doing the dotting,

$$\ddot{x}_C\left(-\frac{1}{2}\right) + \ddot{y}_C\left(\frac{\sqrt{3}}{2}\right) = 0 + \alpha\left(\frac{\sqrt{3}}{2}\right)$$

or

$$-\ddot{x}_C + \sqrt{3}\ddot{y}_C = \sqrt{3}\alpha \tag{5}$$

Equations (1), (2), and (3) yield:

$$\ddot{x}_C = -\frac{T}{4} \qquad \ddot{y}_C = -32.2 + T\left(\frac{\sqrt{3}}{4}\right) \qquad \alpha = \frac{-3\sqrt{3}T}{4}$$

Substituting these three results into Equation (5) results in:

$$+\frac{T}{4} + \sqrt{3}\left(-32.2 + \frac{\sqrt{3}T}{4}\right) = -\sqrt{3}\left(\frac{3\sqrt{3}T}{4}\right) = -\frac{9T}{4}$$

or

$$T = \frac{4}{13}(32.2)\sqrt{3} = 17.2 \text{ lb}$$

Note that *before* the right-hand string was cut, the tension, from statics, was:

$$\Sigma F_y = 0 = 2T_{\text{STATIC}}\frac{\sqrt{3}}{2} - 64.4 \Rightarrow T_{\text{STATIC}} = 37.2 \text{ lb}$$

Forces in *inextensible* strings (ropes, cables, cords) are capable of changing "instantaneously," and indeed we see that this is the case in this problem.

Question 4.7 Can spring forces change instantaneously in this way?

In the preceding example, back-substitution immediately yields

$$\ddot{x}_C = \frac{17.2}{4} = -4.30 \text{ ft/sec}^2$$

$$\ddot{y}_C = \frac{17.2\sqrt{3}}{4} - 32.2 = -24.8 \text{ ft/sec}^2$$

Thus the mass center will start to move off to the left and down.

$$\alpha = \frac{-17.2(3\sqrt{3})}{4} = -22.3 \text{ rad/sec}^2$$

Thus the body will start to turn clockwise.

The acceleration of A follows from Equation (4):

$$\begin{aligned}\mathbf{a}_A &= -4.30\hat{\imath} - 24.8\hat{\jmath} + 22.3\hat{\jmath} \\ &= -4.30\hat{\imath} - 2.5\hat{\jmath}\end{aligned}$$

and, as a check, the direction of $\mathbf{a}_A$, $\tan^{-1}\left(\frac{2.5}{4.3}\right)$, is 30°.

Answer 4.7 No, the length must change and that takes time.

The magnitude of $\mathbf{a}_A$ is $\sqrt{(-4.30)^2 + (-2.5)^2} = 4.97$ ft/sec², which is $\ddot{s}_A$ (see Section 1.7) at the initial instant. It is interesting to note that the initial angular acceleration of the *string DA* is $|\mathbf{a}_A|/\ell \circlearrowright$, or $4.97/\ell \circlearrowright$.

Finally, we close the section with an example containing two bodies in rolling contact. The plate is simply translating, but the pipe has a more complicated motion: it rolls on the pipe, but *not* on the inertial frame (ground).

Example 4.21

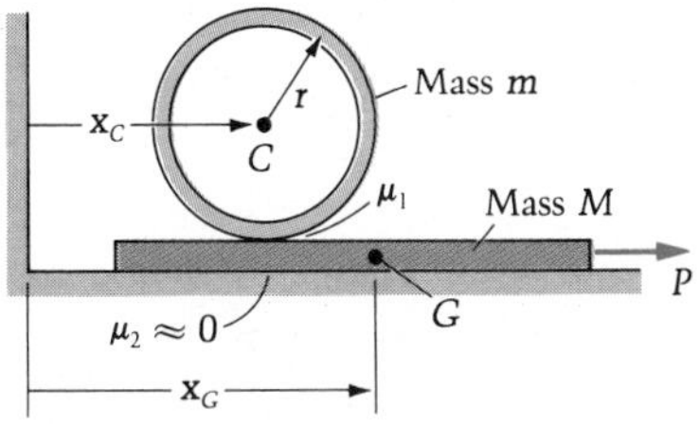

Figure E4.21a

Force P is applied to a plate that rests on a smooth surface. (See Figure E4.21a.) Find the largest force P for which the pipe will not slip on the plate.

Solution

For the pipe (Figure E4.21b), with $\boldsymbol{\alpha} = \alpha\hat{\mathbf{k}}$:

$$\xrightarrow{+} \quad \Sigma F_x = f = m\ddot{x}_C \tag{1}$$

$$+\uparrow \quad \Sigma F_y = N - mg = m\ddot{y}_C = 0 \Rightarrow N = mg \tag{2}$$

$$\circlearrowleft\!+ \quad \Sigma M_C = fr = I_C\alpha = mr^2\alpha \tag{3}$$

(Note that $I_C = (m/2)(r_o^2 + r_i^2) \approx mr^2$ if $r_o \approx r_i$. If the thickness $(r_o - r_i)$ is not given, assume it is small.)

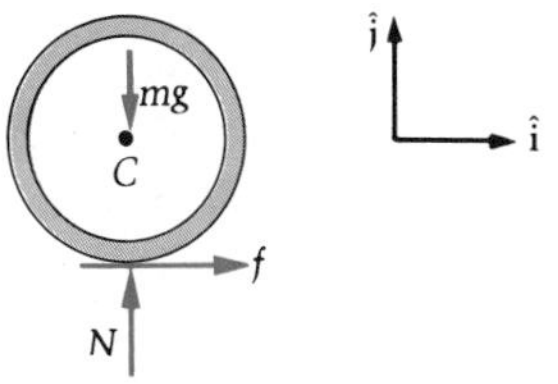

Figure E4.21b

For the plate (Figure E4.21c), we note that only the x equation of motion is of help; $\Sigma F_y = m\ddot{y}_G = 0$ gives $N_2 = N + Mg = (m + M)g$ as expected, and dimensions are not given so moments cannot be taken. (The moment equation would only give us the location of N_2, anyway.) Therefore

$$\Sigma F_x = P - f = M\ddot{x}_G{}^* \tag{4}$$

Eliminating f between (1) and (3) gives

$$m\ddot{x}_C = mr\alpha \tag{5}$$

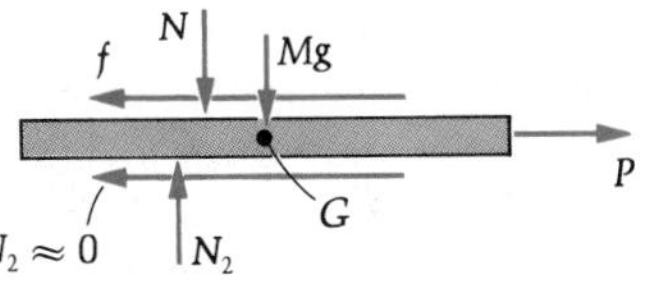

Figure E4.21c

And between (1) and (4) gives

$$P = m\ddot{x}_C + M\ddot{x}_G \tag{6}$$

We note that if $m_T = m + M$ and C_T is the mass center of pipe *plus* plate, then Equation (6) could have been written immediately from $\Sigma F_x = m_T\ddot{x}_{C_T}$ for the combined system. Here $\Sigma F_x = P$; the right side follows from two derivatives of the definition of the mass center ($m_T x_{C_T} = mx_C + Mx_G$).

The kinematics equation is tricky here. It is a rolling condition, but we must remember that x_G and x_C are necessarily measured relative to an inertial frame, here assumed to be fixed in the ground. Thus it is $\ddot{x}_C - \ddot{x}_G$ that is related to α.† For no slip, $\dot{x}_C + r\omega = \dot{x}_G$ which, when differentiated, yields

$$\ddot{x}_C - \ddot{x}_G = -r\alpha \tag{7}$$

Substituting α from (5) into (7) relates the accelerations of the two mass centers:

$$2\ddot{x}_C = \ddot{x}_G \tag{8}$$

---

* Sometimes G is used to designate a mass center.

† This difference is just the acceleration of C in the frame consisting of the translating plate.

Then (6) and (8) may be combined to give

$$P = (m + 2M)\ddot{x}_C \tag{9}$$

And combining (9) and (1) gives us the relationship between P and f:

$$P = \frac{m + 2M}{m} f \tag{10}$$

Since $f \le \mu_1 N$ for no slip, (10) gives:

$$\frac{m}{m + 2M} P \le \mu_1 mg \Rightarrow P \le (m + 2M)g\mu_1$$

Any larger P than $(m + 2M)g\mu_1$ will cause the pipe to slip on the plate.

PROBLEMS ▶ Section 4.5

4.62 A uniform sphere (radius r, mass m) rolls on the plane in Figure P4.62. If the sphere is released from rest at $t = 0$ when $x = L$, find $x(t)$.

4.63 A symmetric body $\mathcal{B}$ has mass m and radius R; a cord is wrapped around it as shown in Figure P4.63. Compute the downward acceleration of the center C if $\mathcal{B}$ is (a) a cylinder; (b) a sphere (with a small slot to accommodate the cord); (c) a thin ring. *Hint*: Work the problem just once with radius of gyration k_C; then substitute the three values $R/\sqrt{2}$, $\sqrt{\frac{2}{5}}R$, and $1R$ for k_C.

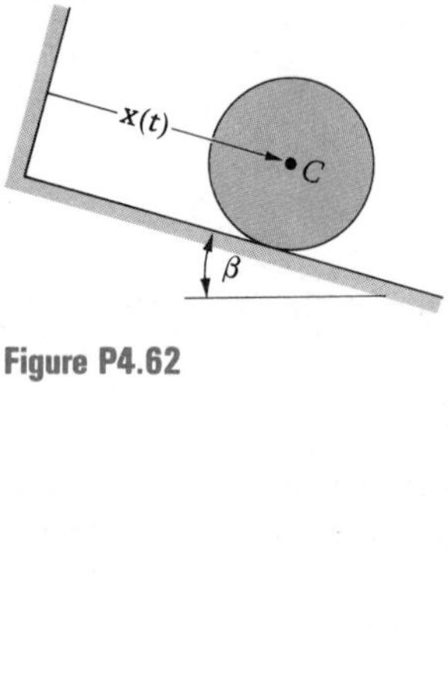

Figure P4.62

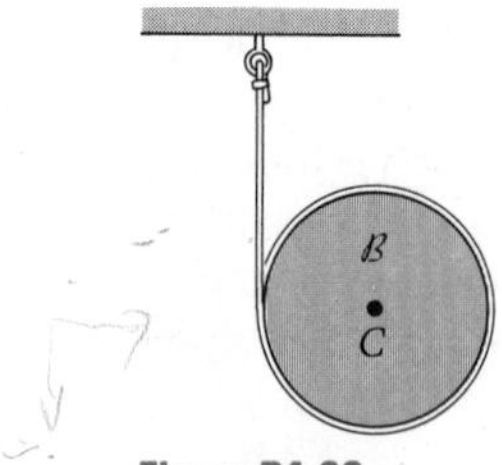

Figure P4.63

4.64 The cord in Figure P4.64 is wrapped around the cylinder, which is released from rest on the 60° incline shown. Find the velocity and position of C as a function of time t.

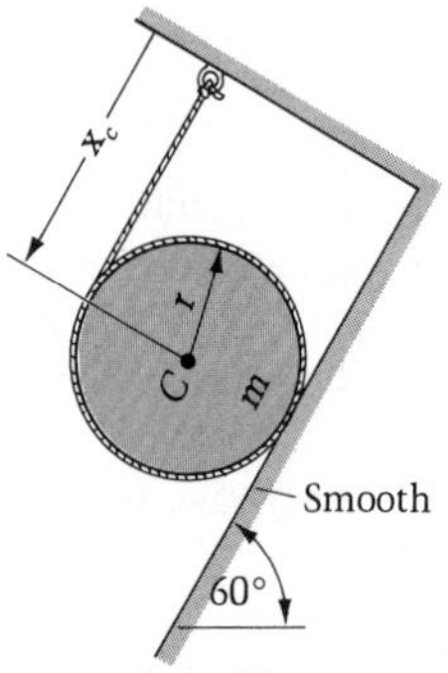

Figure P4.64

4.65 Sally Sphere, Carolyn Cylinder, Harry Hoop, and Wally Wheel each have mass m and radius R. Wally's spokes and rim are very light compared to his hub. (See Figure P4.65.) They are going to have a race by rolling down a rough plane. Give (a) the order in which they finish and (b) the times.

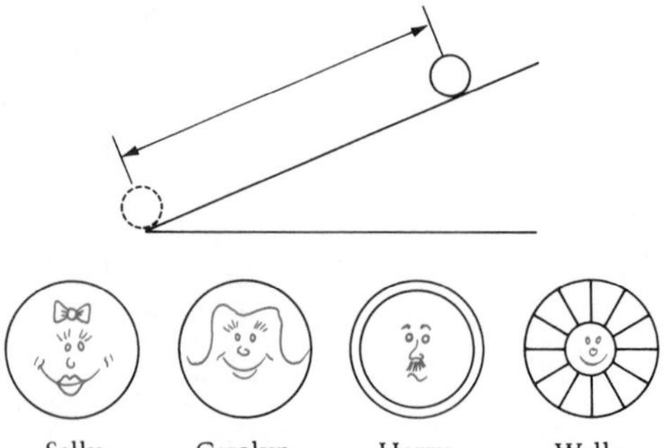

Figure P4.65

4.66 In the preceding problem, Wally and Carolyn are connected by a bar of negligible mass and released from rest on the same incline. (See Figure P4.66.) Determine the force in the bar.

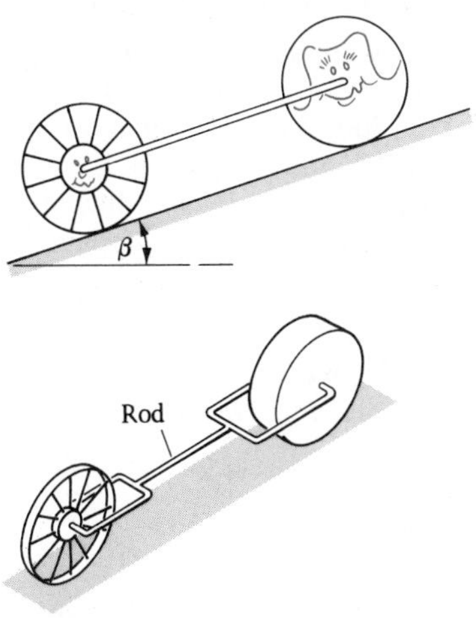

Figure P4.66

4.67 Repeat the preceding problem, but suppose that Wally and Carolyn switch places.

4.68 The two pulleys $\mathcal{P}_1$ and $\mathcal{P}_2$ and the block $\mathcal{B}$ in Figure P4.68 each have mass m and are connected by the cord.

a. Write a brief paragraph explaining in words why the system cannot be in equilibrium. Start with, "If the system were in equilibrium, the tension in the rope above $\mathcal{B}$ would equal mg." Then follow the rope around the pulleys until you reach a contradiction.

b. Find the acceleration of C_1.

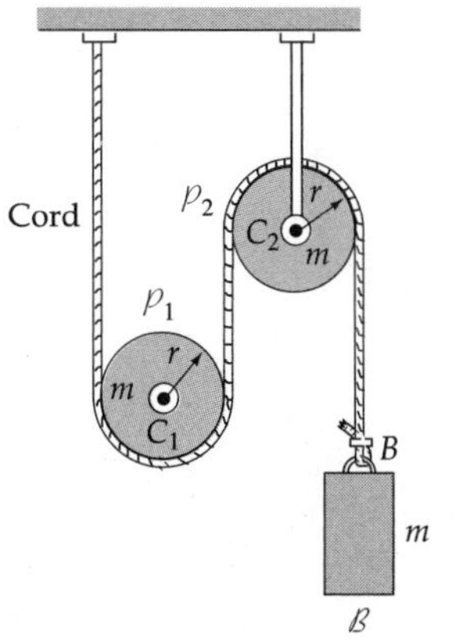

Figure P4.68

4.69 The uniform sphere (mass m, radius r) in Figure P4.69 is at rest before P is applied. If μ is the coefficient of friction between sphere and floor,

a. find the maximum P for there to be no slip;

b. for P twice that found in (a), find $\mathbf{a}_C$ and $\boldsymbol{\alpha}$. Note that $|\mathbf{a}_C|$ does not equal $r|\boldsymbol{\alpha}|$ as it would if there were no slipping, i.e., if the sphere were rolling.

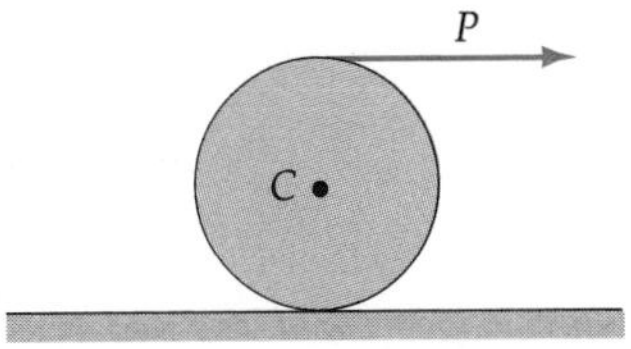

Figure P4.69

4.70 The uniform cylinder in Figure P4.70, of mass m and radius r, is at rest before it is subjected to a couple of moment M_o. The coefficient of friction between cylinder and floor is μ.

a. Find the largest value of M_o for which there is no slip.

b. For M_o twice the value found in (a), find $\mathbf{a}_C$ and $\boldsymbol{\alpha}$.

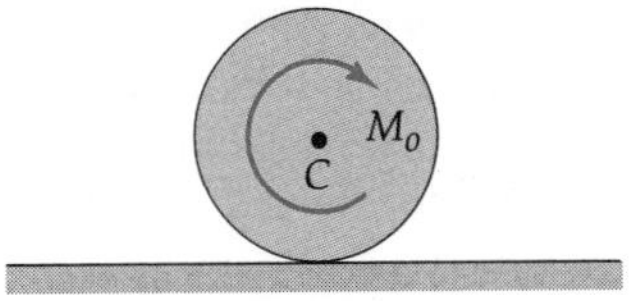

Figure P4.70

4.71 Find the ratio of r to R for which the force T in Figure P4.71 will cause the wheel to roll (no slip) no matter how small the friction. Treat the wheel as a uniform cylinder.

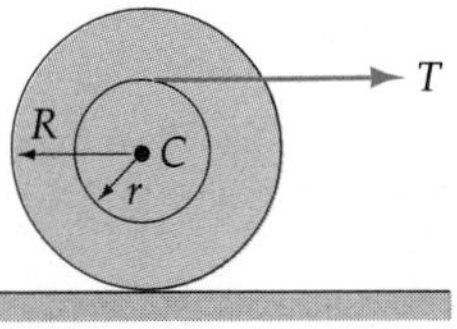

Figure P4.71

4.72 Force T is given to be small enough, and the friction coefficient large enough, that both wheels in Figure P4.72 will roll on the plane.

a. Give arguments why one wheel rolls left and the other right.

b. Find the ratio of r to R for which the accelerations of C are equal in magnitude.

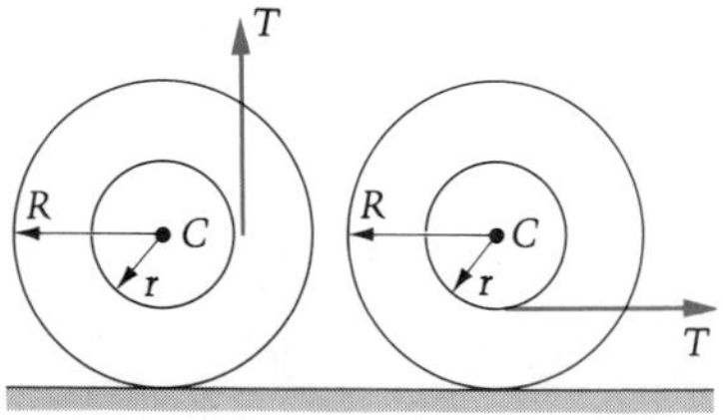

Figure P4.72

4.73 The uniform sphere in Figure P4.73, of mass m and radius r, is at rest when it is subjected to a couple of moment M_0. If there is no slip, find the acceleration of the center of the sphere.

4.74 A cylinder spinning at angular speed ω_0 rad/sec clockwise is placed on an inclined plane. (See Figure P4.74.) Show that the cylinder center will begin moving up the plane if $\mu > \tan\beta$. Why does this result have nothing to do with the size of ω_0?

4.75 The cylinder of weight W and radius r shown in Figure P4.75 has an angular velocity of 100 rad/s clockwise. It is lowered onto the rough incline. If its center C is observed to remain momentarily at rest, determine the coefficient of sliding friction. Find how long the center C remains at rest.

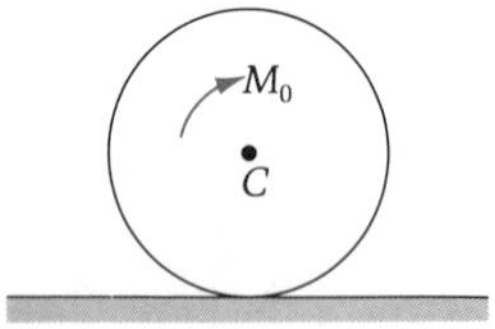

Figure P4.73

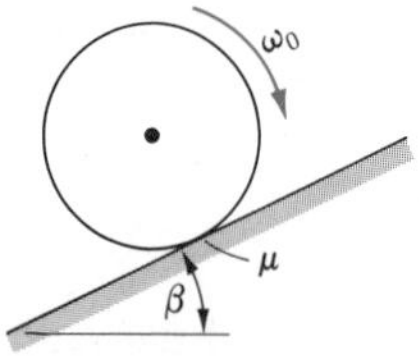

Figure P4.74

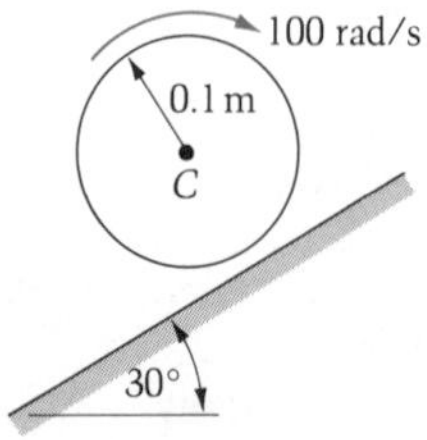

Figure P4.75

4.76 The bowling ball in Figure P4.76 is released with $v_C = 22$ ft/sec and $\omega = 0$ as it contacts the surface of the alley. Neglecting the effect of the three finger holes, and using a coefficient of friction of 0.3, find the distance traveled by the center of the ball before slipping stops.

* **4.77** The force $P = 60$ N is applied as shown in Figure P4.77 to the 10-kg cylinder $\mathcal{C}$, originally at rest beneath the mass center of the thin, 5-kg rectangular plate $\mathcal{P}$. The coefficient of friction between $\mathcal{C}$ and $\mathcal{P}$ is 0.5, and the plane beneath $\mathcal{C}$ is smooth. Determine: (a) the initial acceleration of C; (b) the value of x when $\mathcal{C}$ is slipping on both surfaces. The length of $\mathcal{P}$ is 2 m.

4.78 The constant force F_0 is applied to the cylinder, initially at rest, as shown in the two drawings constituting Figure P4.78. Show in the following two ways that the cylinder will slip provided that

$$\frac{F_0}{mg} > 3\mu$$

a. Assume rolling; then obtain the inequality from $f > \mu N$ after solving for f and N.

b. Assume slipping; then integrate $\ddot{x}_C$ and $\ddot{\theta}$ to obtain $\dot{x}_C$ and θ, and then find the velocity of the contact point B; if it is to the right (that is, positive), this is consistent with $f = \mu N$ to the left and we have slipping.

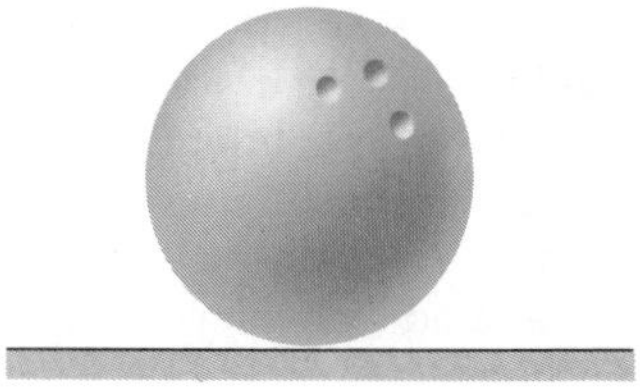

Figure P4.76

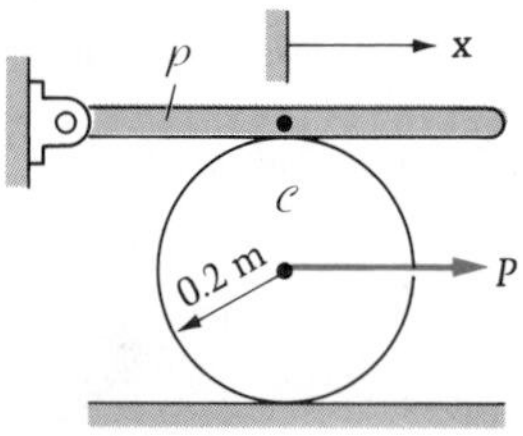

Figure P4.77

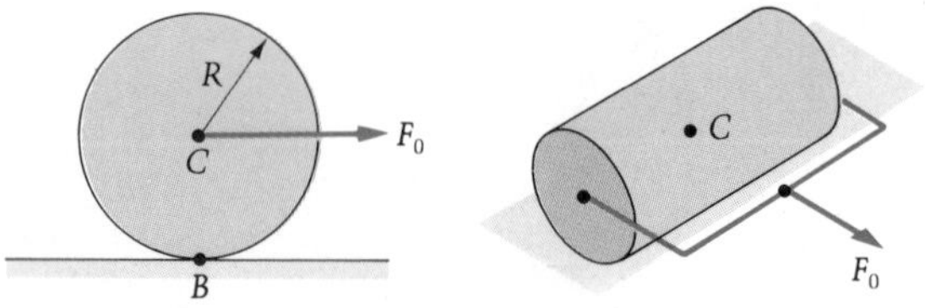

Figure P4.78

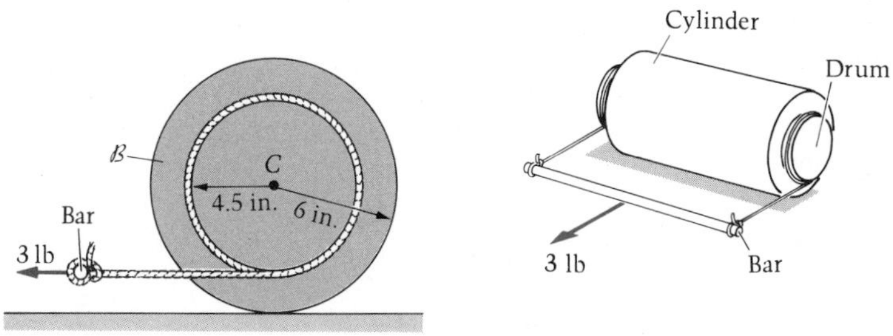

Figure P4.79

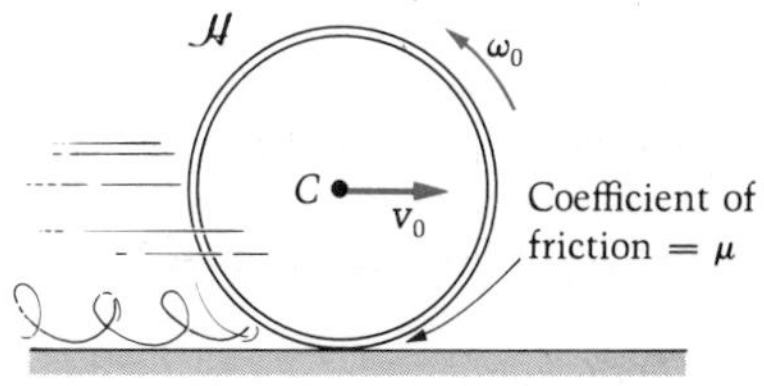

Figure P4.82

4.79 Two drums of radius 4.5 in. are mounted on each end of a cylinder of radius 6 in. to form a 40-lb rigid body $\mathcal{B}$ with radius of gyration $k_{zC} = 5$ in. (See Figure P4.79.) Ropes are wrapped around the drum and tied to a horizontal bar to which a 3-lb force is applied. As $\mathcal{B}$ rolls from rest, tell (a) the number of inches of rope wound or unwound (tell which) in three seconds and (b) the minimum friction coefficient needed for the rolling to take place.

* **4.80** Find the range of possible values of the couple M_0 for which the cylinder in Figure P4.80 will not slip in *either direction* when released from rest on the incline. The mass is 15 kg; the radius is 0.2 m; and the coefficient of friction is $\mu = 0.2$.

4.81 An airplane lands on a level strip at 200 mph. (See Figure P4.81.) Initially, just before the wheels touch the runway, the wheels are not turning. After they touch the runway they will skid for some distance and then roll free. If during this skidding the plane has a constant velocity of 200 mph and the normal force between the wheel and the runway is 10 times the wheel weight, find the length of the skid mark. (The coefficient of friction is $\frac{1}{2}$; the radius of gyration of the wheel is three-fourths of its radius.)

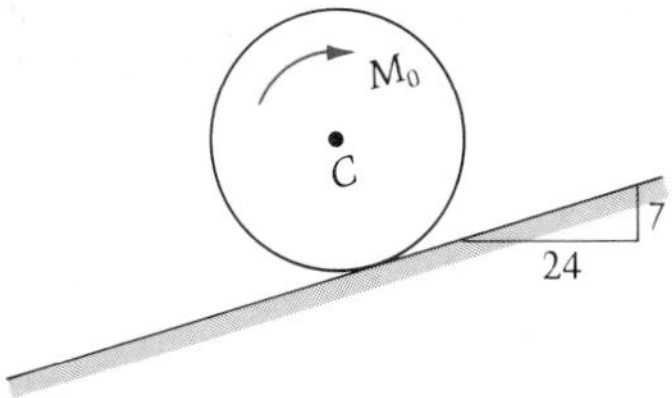

Figure P4.80

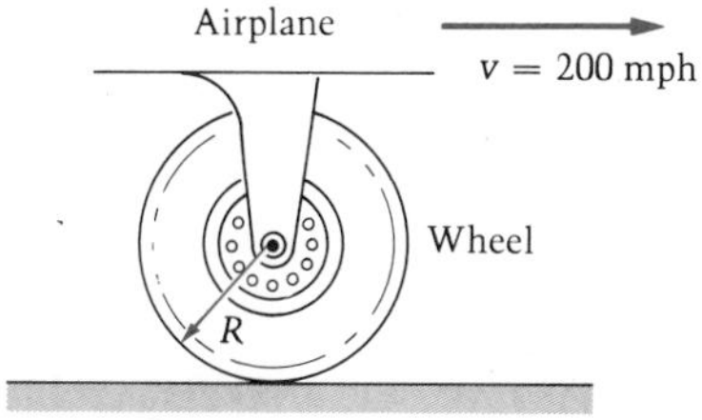

Figure P4.81

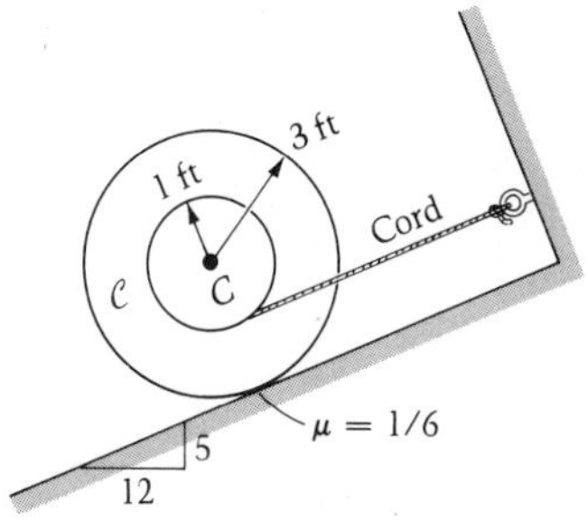

Figure P4.83

* **4.82** A hula hoop $\mathcal{H}$ (mass m, radius r) is thrown forward with backspin; $v_C = v_0$ to the right and $\omega = \omega_0$ counterclockwise. (See Figure P4.82.)

a. How long and how far does the mass center move before $\mathcal{H}$ stops slipping?

b. Find the relationship between v_0 and ω_0 such that when $\mathcal{H}$ stops slipping: (i) it rolls right; (ii) it rolls left; (iii) it stops.

4.83 The strong, flexible cable shown in Figure P4.83 is wrapped around a light hub attached to the 130-lb cylinder $\mathcal{C}$. Find the angular acceleration of $\mathcal{C}$ upon release from rest. Note that it is impossible for the wheel to roll down the plane (meaning without slipping); to do so the cord would have to break.

4.84 Repeat the previous problem for $\mu = 0.25$.

4.85 In Figure P4.85, find how far down the incline C travels in 5 s if the 20-kg cylinder $\mathcal{C}$ is released from rest.

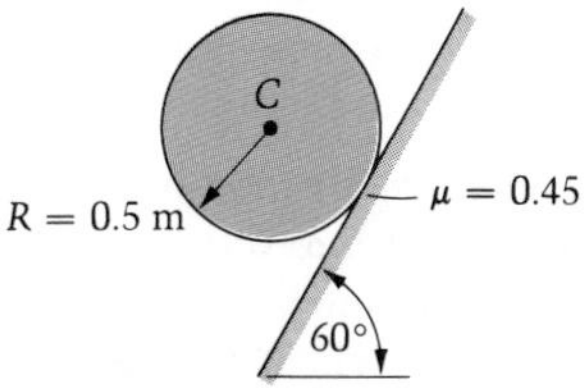

Figure P4.85

4.86 The 50-lb body $\mathcal{C}$ in Figure P4.86 may be treated as a solid cylinder of radius 2 ft. The coefficient of friction between $\mathcal{C}$ and the plane is $\mu = 0.2$, and a force $P = 10$ lb is applied vertically to a cord wrapped around the hub. Find the position of the center C, 10 sec after starting from rest.

4.87 Given that the slot (for the cord) in the cylinder in Figure P4.87 (mass 10 kg) has a negligible effect on I_C, find:

a. The largest θ for which no motion down the plane will occur

b. The time required for C to move 3 m down the incline if $\theta = 60°$.

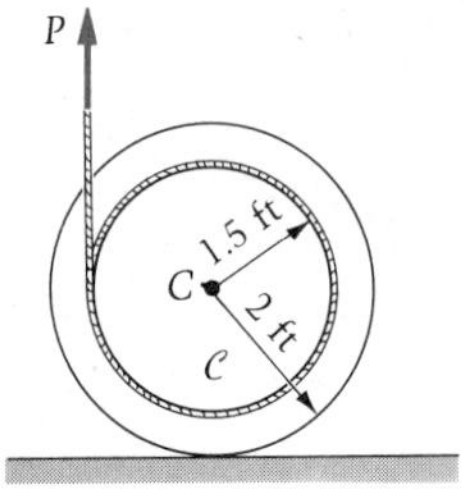

Figure P4.86

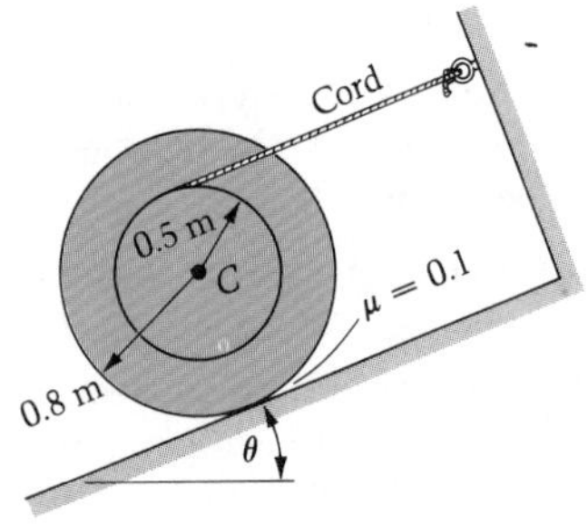

Figure P4.87

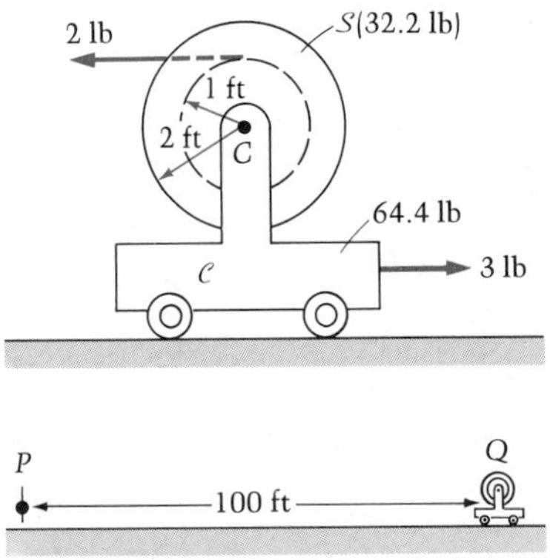

Figure P4.88

4.88 A light 100-ft cord is wrapped around the 32.2-lb spool $\mathcal{S}$, which is pinned at C to the cart $\mathcal{C}$ (see Figure P4.88). The radius of gyration of $\mathcal{S}$ with respect to an axis normal to the figure at C is 1.3 ft. The cart (without $\mathcal{S}$) has weight 64.4 lb. The wheels of $\mathcal{C}$ are small and light, so that friction beneath them is negligible. The 2- and 3-lb forces are applied to the system at rest. If upon complete unwrapping the cord is to end up between points P and Q in the lower figure, where should $\mathcal{C}$ be originally parked along PQ?

4.89 A child pulls on an old wheel with a force of 5 lb by means of a rope looped through the hub of the wheel. (See Figure P4.89.) The friction coefficient between wheel and ground is $\mu = 0.2$. Find I_C for the wheel, and use it to determine the location of C after 3 sec.

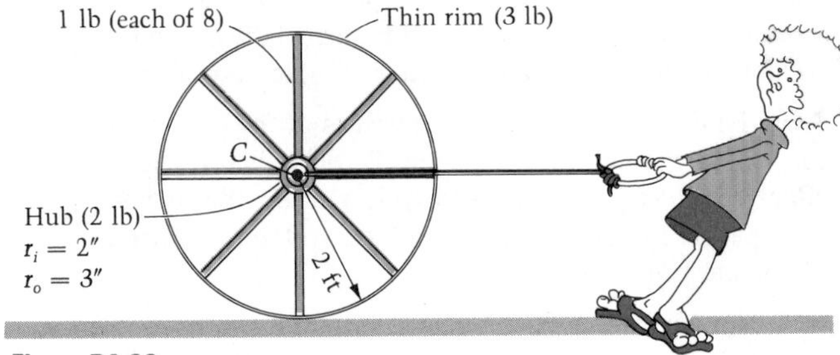

Figure P4.89

4.90 The wheel shown in Figure P4.90 has a mass of 10 kg, a radius of 0.4 m and a radius of gyration with respect to the z-axis through C of 0.3 m. Determine the angular acceleration of the wheel and how far the mass center C moves in 3 seconds if the wheel starts from rest.

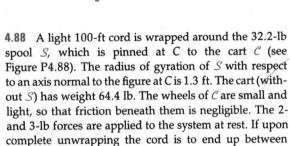

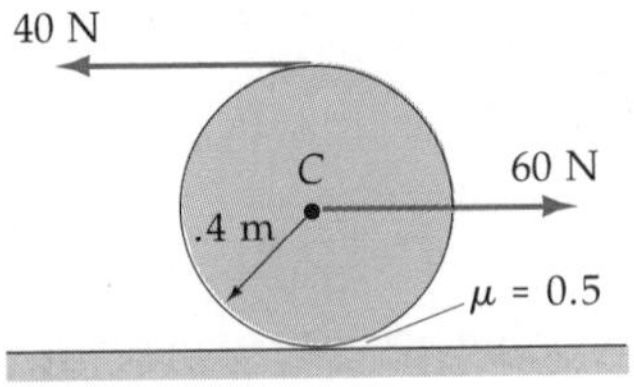

Figure P4.90

4.91 Rework the preceding problem if the friction coefficient is changed to $\mu = 0.2$.

4.92 Two cables are wrapped around the hub of the 10-kg spool shown in Figure P4.92, which has a radius of gyration of 500 mm with respect to its axis. A constant 40-N force is applied to the upper cable as shown. Find the mass center location 5 s after starting from rest if: (a) $\mu = 0.2$; (b) $\mu = 0.5$.

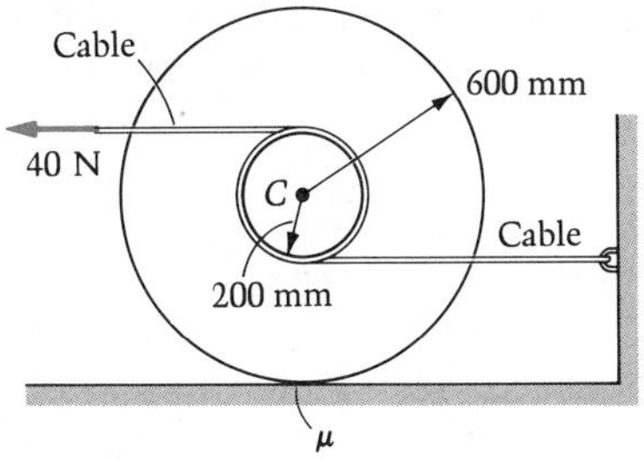

Figure P4.92

4.93 A sphere of radius $\frac{1}{2}$ ft and weight 16.1 lb is projected onto a horizontal plane (Figure P4.93). Its center has initial velocity v_0 at $t = 0$ and the sphere has initial angular velocity ω_0, defined as shown. If the coefficient of sliding friction between the sphere and the plane is 0.15, plot graphs of distance gone (x_C) against time t up to $t = 3$ sec for the following cases:

a. $v_0 = 10$ ft/sec; $\omega_0 = 100$ rad/sec
b. $v_0 = 10$ ft/sec; $\omega_0 = 50$ rad/sec
c. $v_0 = 10$ ft/sec; $\omega_0 = 30$ rad/sec

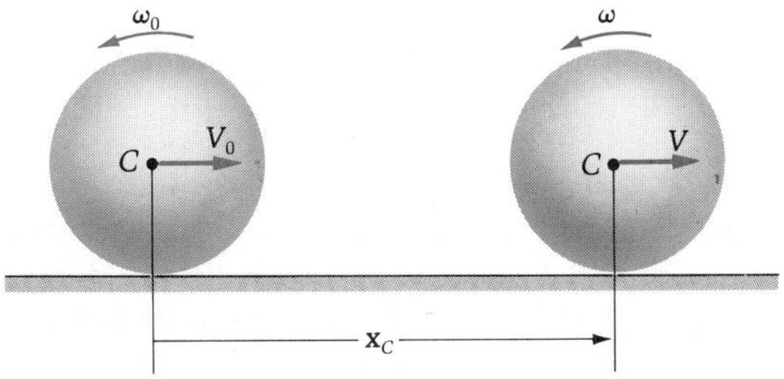

Figure P4.93

4.94 The cylinder $\mathcal{C}$ in Figure P4.94 has a thin slot cut in it which doesn't affect its moment of inertia appreciably. A cord is wrapped in the slot and connects to the cart $\mathcal{B}$, which rests on the plane on small, light wheels. The force of 10 lb is applied to $\mathcal{C}$ with the system initially at rest. Find the length of unwrapped cord after 4 seconds elapse. Assume enough friction to prevent slip of $\mathcal{C}$ on the plane.

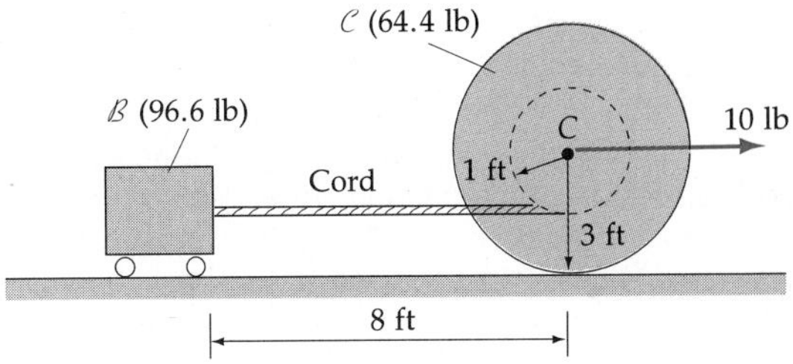

Figure P4.94

4.95 Assume that enough friction is available to prevent the cylinder in Figure P4.95 from slipping.

a. Show that
(i) $\mathcal{A}$ rolls to the right if $\theta < \cos^{-1}(r/R)$.
(ii) $\mathcal{A}$ rolls to the left if $\theta > \cos^{-1}(r/R)$.
(iii) $\mathcal{A}$ is in equilibrium if $\theta = \cos^{-1}(r/R)$ (and will translate if P increases enough to overcome friction).

b. Find $\ddot{x}_C$ and α if $r = 0.2$ m, $R = 0.4$ m, $P = 20$ N, $mg = 40$ N, and $\theta = 45°$.

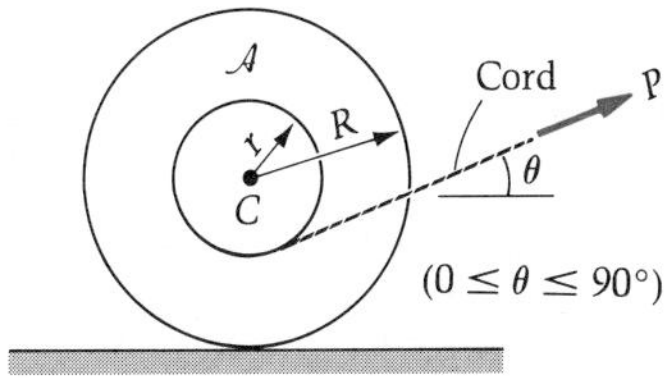

Figure P4.95

4.96 A homogeneous spool of weight W rolls on an inclined plane; a string tension of amount $4W$ acts up the plane as shown in Figure P4.96. With I_C given approximately by $WR^2/2g$, find the acceleration of C. Assume unlimited friction.

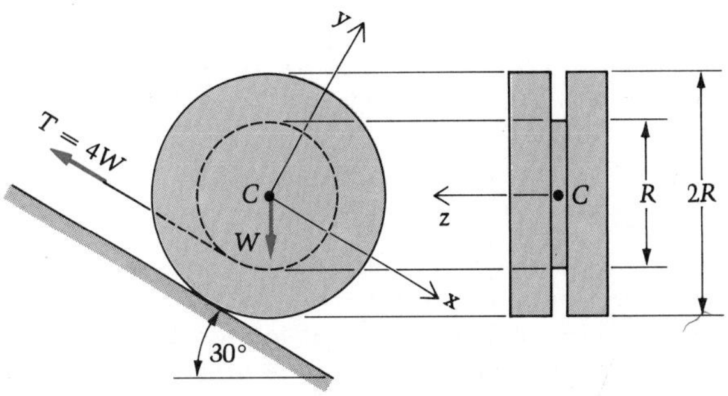

Figure P4.96

* **4.97** Pulley $\mathcal{B}_1$ in Figure P4.97 weighs 100 pounds and has a radius of gyration about the z-axis through O of $k_O = 7$ in. Pulley $\mathcal{B}_2$ weighs 20 lb and has $k_C = 3$ in. Find the angular acceleration of $\mathcal{B}_1$ just after the system is released from rest. Assume the rope doesn't slip on $\mathcal{B}_1$ but that there is *no friction* between $\mathcal{B}_2$ and the rope. Is this the angular acceleration for later times as well?

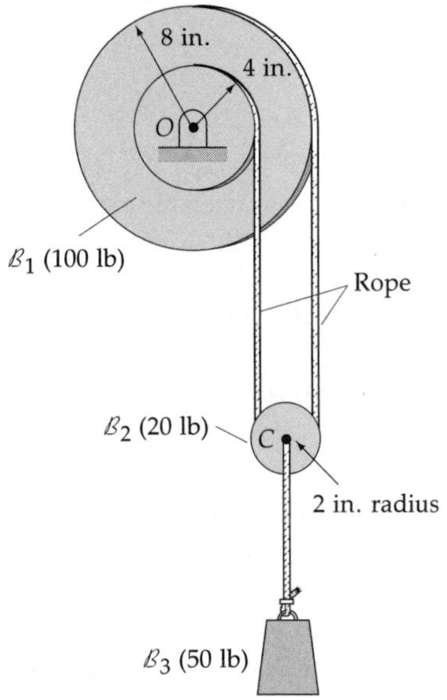

Figure P4.97

4.98 The 32.2-lb body $\mathcal{C}$ in Figure P4.98 is a spool having a radius of gyration $k_C = 6$ in. about its axis. Cords are wrapped around the peripheries; one is connected to a ceiling, the others to the 48.3-lb block $\mathcal{B}$. Find the accelerations of the centers C (of $\mathcal{C}$) and B (of $\mathcal{B}$).

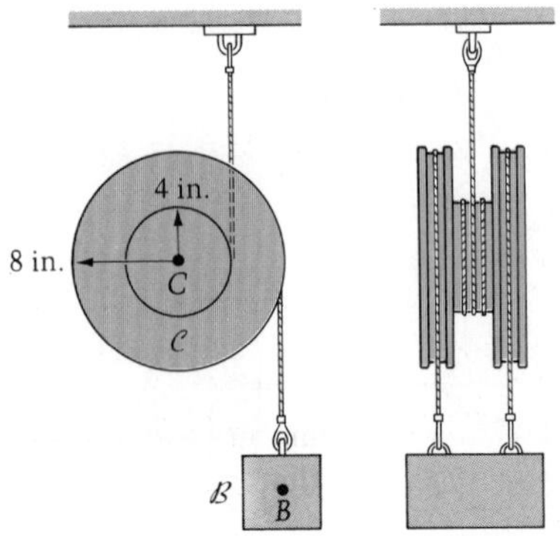

Figure P4.98

4.99 Wheel $\mathcal{C}$ is made up of the solid disk $\mathcal{A}$, rim $\mathcal{R}$, and four spokes $\mathcal{S}$. Masses and radii are given in Figure P4.99 and the table.

a. Compute I_{zz}^C for the wheel.

b. The coefficient of friction between $\mathcal{C}$ and the plane is $\mu = 0.3$. If a cord is wrapped around the disk and connected to the 50-kg body $\mathcal{B}$, determine the acceleration of the mass center C of $\mathcal{C}$.

Part	Mass (kg)
Disk $\mathcal{A}$	20
Spokes $\mathcal{S}$	5 (each)
Rim $\mathcal{R}$	10

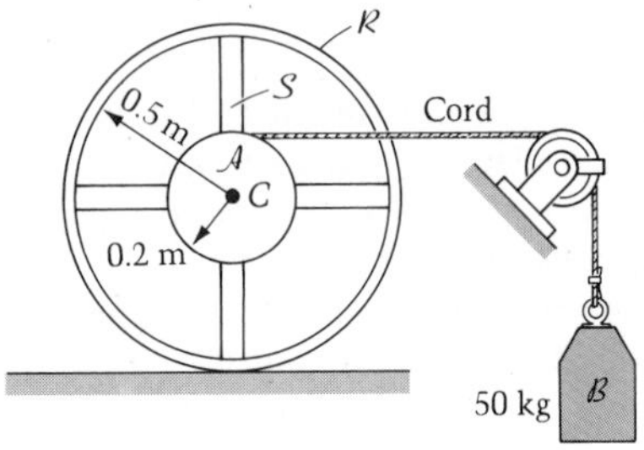

Figure P4.99

4.100 The radius of gyration of the 20-kg wheel in Figure P4.100 with respect to its axis is 0.3 m. Motion starts from rest. Find the acceleration of the mass center C, and determine how far C moves in 5 s.

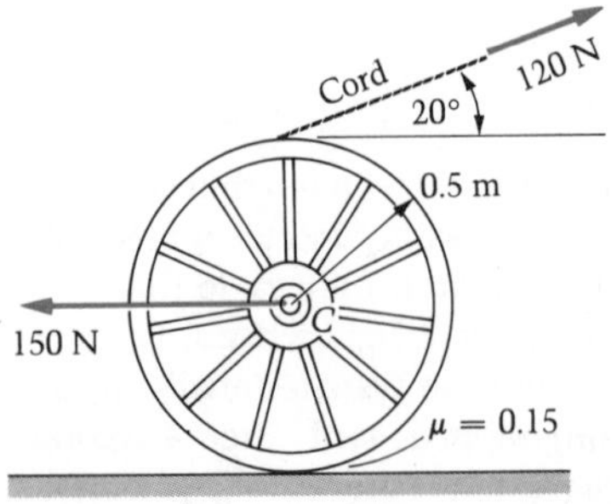

Figure P4.100

4.101 A string is wrapped around the hub of the spool shown in Figure P4.101. There are four indicated string directions. For the direction that will result in the largest displacement of C in 3 s, find this displacement. Assume sufficient friction to prevent slipping. The spool has a mass of 12 kg and a radius of gyration about z_C of 0.6 m. Each force equals 10 N, and the spool starts from rest.

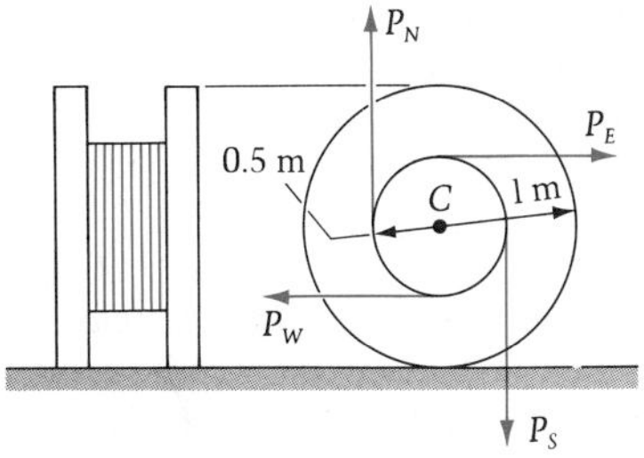

Figure P4.101

4.102 Cylinder $\mathcal{C}$ in Figure P4.102 has a mass of 4 slugs, and the effect of the hub on its moment of inertia is negligible. It is connected by means of a cord to the 1-slug block $\mathcal{B}$. The mass of the pulley is negligible. The coefficient of friction between $\mathcal{C}$ and the plane is $\mu = 0.5$, and the radii of $\mathcal{C}$ are given in the figure. If the system is released from rest, determine the time that will elapse before $\mathcal{B}$ hits the ground.

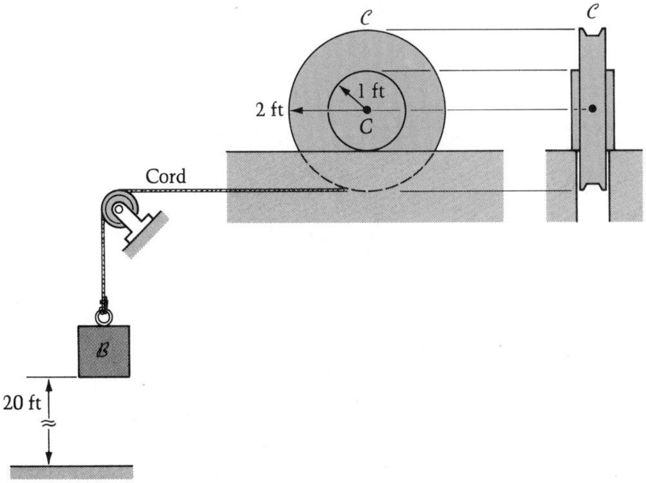

Figure P4.102

4.103 Find how long it takes for $\mathcal{A}$ to roll off the plane in Figure P4.103, assuming sufficient friction to prevent slipping. The system is released from rest.

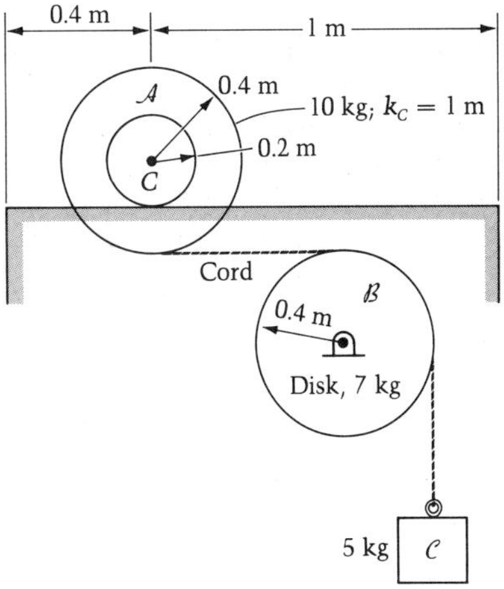

Figure P4.103

4.104 Disks $\mathcal{A}$ and $\mathcal{B}$ each weigh 64.4 lb and are rigidly attached to the light shaft $\mathcal{S}$ that joins their centers. (See Figure P4.104.) A 96.6-lb cylinder $\mathcal{C}$ has a hole drilled along its axis, through which $\mathcal{S}$ passes. A force of 20 lb is applied horizontally to an inextensible string wrapped around $\mathcal{C}$. If friction is negligible between $\mathcal{S}$ and $\mathcal{C}$, and if $\mathcal{A}$ and $\mathcal{B}$ roll on the plane, find:

- **a.** The angular acceleration of the cylinder
- **b.** The angular acceleration of the disks
- **c.** The minimum coefficient of friction between disks and plane for no slipping.

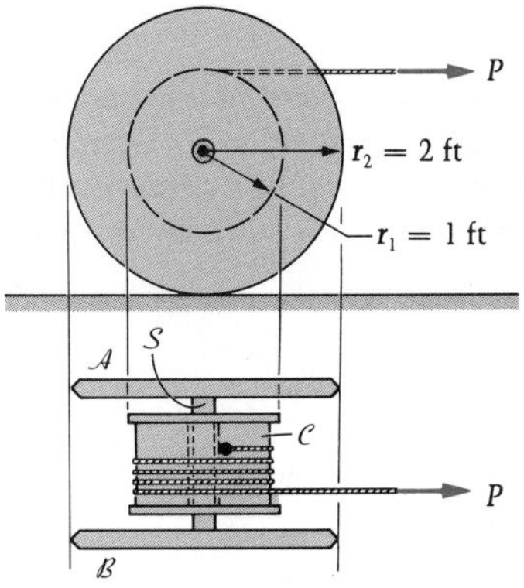

Figure P4.104

4.105 Rework the preceding problem, but this time assume that the string is wrapped so that it comes off the *bottom* of $\mathcal{C}$.

4.106 The two wheels are identical 16.1-lb cylinders with smooth axles at their centers. (See Figure P4.106.) The carriage weighs 32.2 lb and has its mass center at C. The cylinders do not slip on the inclined plane. Find the acceleration of point Q.

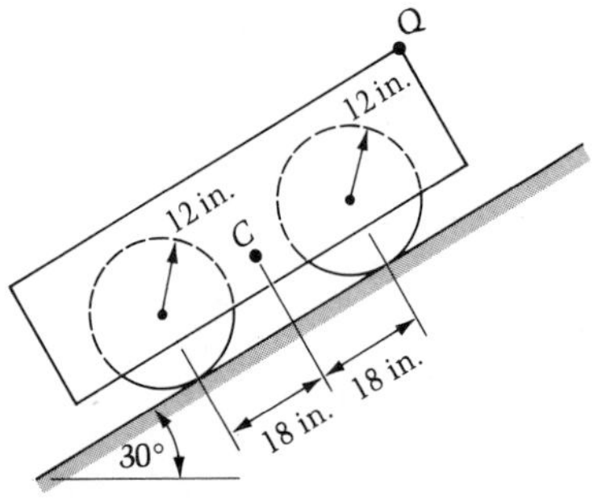

Figure P4.106

* **4.107** Cylinder $\mathcal{C}$ in Figure P4.107 weighs 100 lb; it is rolling on the plane and is pinned at its center C to the 10-lb rod $\mathcal{R}$. If v_C is initially 10 ft/sec to the left, and if the coefficient of kinetic friction between the plane and each body is $\mu = 0.4$, determine how long it will take the system to come to rest.

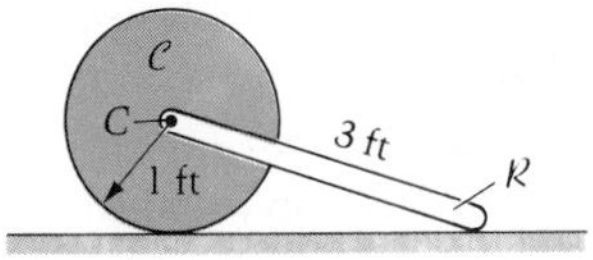

Figure P4.107

* **4.108** A uniform half-cylinder of radius r and mass m is held in the position shown in Figure P4.108 by the string tied to B. Find the reaction of the floor just after the string is cut. There is sufficient friction to prevent slipping.

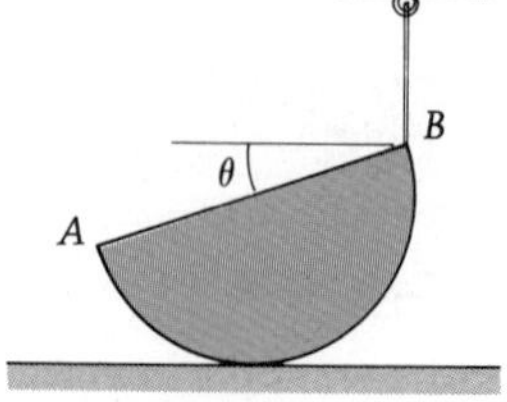

Figure P4.108

* **4.109** In Figure P4.109 the force P is applied to the cord at $t = 0$, when the 25-N cylinder is at rest. Find the position of the mass center when $t = 6$ s.

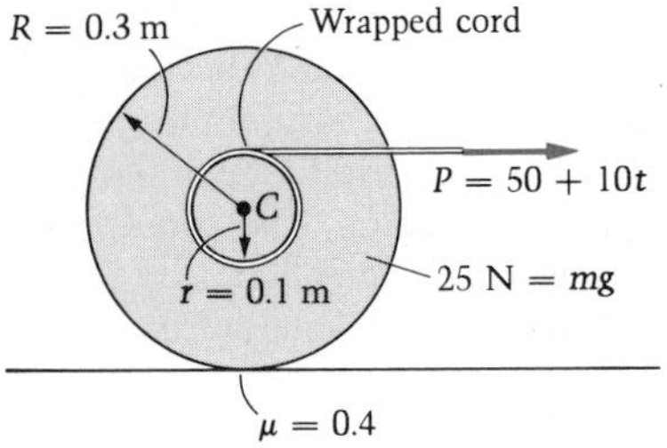

Figure P4.109

* **4.110** The semicylinder in Figure P4.110(a) is released from rest, and there is enough friction to prevent slipping throughout the ensuing motion (Figure P4.110(b)).

a. Find I_{zz}^C.

b. Write the three differential equations of motion of the body (good at any angle θ).

c. Find the two equations relating $\ddot{x}_C$ and $\ddot{y}_C$ to $\ddot{\theta}$, $\dot{\theta}$, and θ.

d. Eliminate f, N, $\ddot{x}_C$, and $\ddot{y}_C$ and obtain the single differential equation in the variable $\theta(t)$. Note the complexity of the equation!

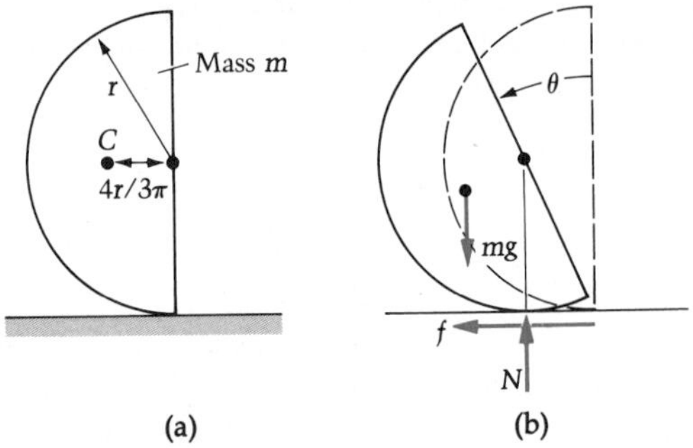

Figure P4.110

4.111 The 15-lb carriage shown in Figure P4.111 is supported by two uniform rollers each of weight 10 lb and radius 3 in. The rollers roll on the ground and on the carriage. Determine the acceleration of the carriage when the 5-lb force is applied to it.

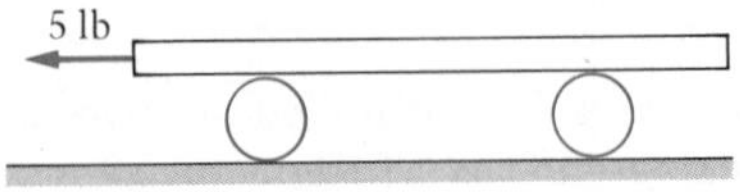

Figure P4.111

4.112 The 128.8-lb homogeneous plank shown in Figure P4.112 is placed on two homogeneous cylindrical rollers, each of weight 32.2 lb. The system is released from rest. Determine the initial acceleration of the plank if no slipping occurs. Is this the acceleration for later times as well?

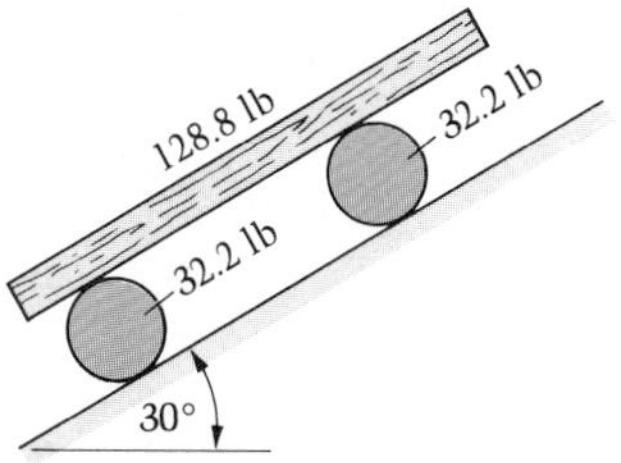

Figure P4.112

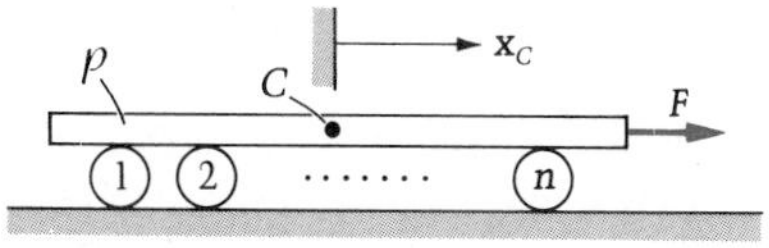

Figure P4.113

4.113 Body $\mathcal{P}$ in Figure P4.113 is a rigid plate of mass M, resting on a number n of cylinders each of mass m and radius R. Force F is constant and starts the system moving from the position shown. If there is no slipping at any surface, find: (a) the acceleration of the plate and (b) its position x_C as a function of M, m, F, n, and time t.

* **4.114** A 6-ft gymnast makes a somersault dive into a net by standing stiff and erect on the edge of a platform and allowing himself to overbalance. He loses foothold (without having slipped) when the platform's reaction on his feet becomes zero; he preserves his rigidity during his fall. Show that he falls flat on his back if the drop from the platform to the net is about 43 ft.

4.115 The homogeneous cylinder $\mathcal{C}$ in Figure P4.115 is at rest on the conveyor belt when the latter is started up with a constant acceleration of 3 ft/sec² to the right. If the cylinder rolls on the belt, find the elapsed time when the cylinder reaches the end A.

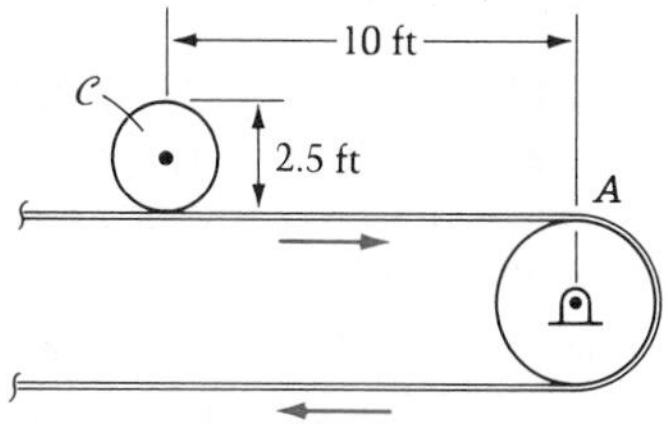

Figure P4.115

4.116 The pipe in Figure P4.116 has a mass of 500 kg and rests on the flatbed of the truck. The coefficient of friction between the pipe and truck bed is $\mu = 0.4$. The truck starts from rest with a constant acceleration a_0.

a. How large can a_0 be without the pipe slipping at any time?

b. For the value of a_0 in part (a), how far has the truck moved when the pipe rolls off the back?

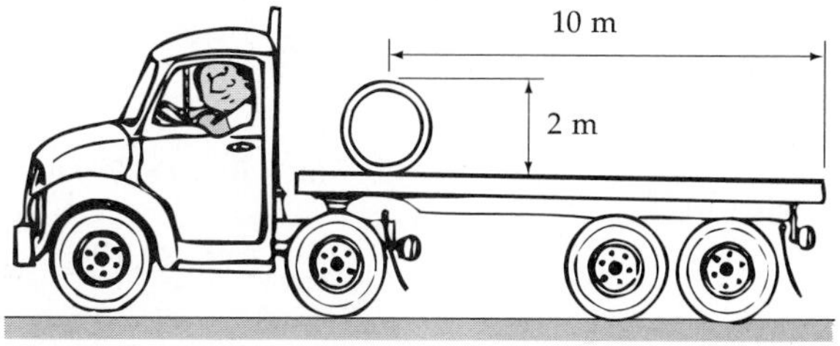

Figure P4.116

4.117 The uniform sphere (mass = 1 slug, radius = 1 ft) and the slab (mass = 2 slugs) shown in Figure P4.117 are at rest before the force $P = 24$ lb is suddenly applied to the slab. The coefficient of friction is 0.2 between the sphere and slab and between the slab and horizontal plane. (a) Does the sphere slip on the slab? (b) What is the acceleration of the center of the sphere?

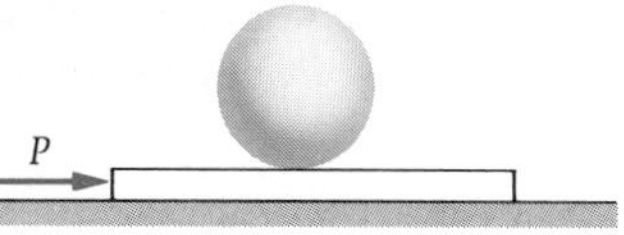

Figure P4.117

4.118 The homogeneous cylinder $\mathcal{C}$ in Figure P4.118 weighs 64.4 lb. The acceleration of the 96.6-lb cart $\mathcal{D}$ is 10 ft/sec² to the right.

a. Determine the acceleration of the center C of the cylinder and the friction force exerted on $\mathcal{C}$ by $\mathcal{D}$ if there is sufficient friction to prevent slipping.

b. How large does the friction coefficient μ have to be for this to occur?

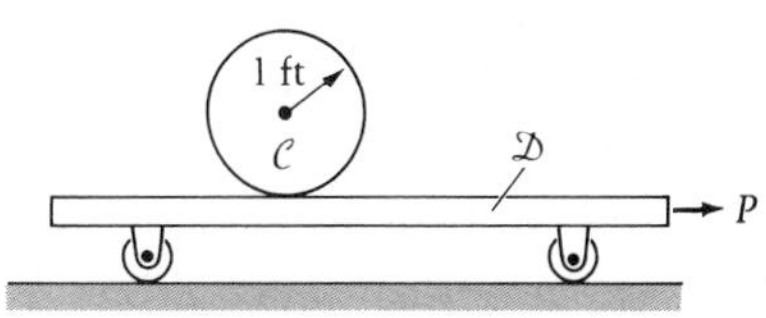

Figure P4.118

4.119 The homogeneous cylinder $\mathcal{A}$ in Figure P4.119 weighs 64.4 lb and rolls on the 96.6-lb truck $\mathcal{B}$. The mass of the truck rollers may be neglected. Find the force P such that C does not move relative to the plane.

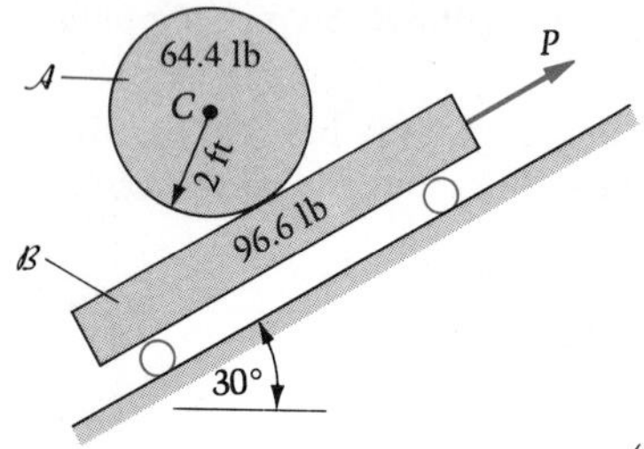

Figure P4.119

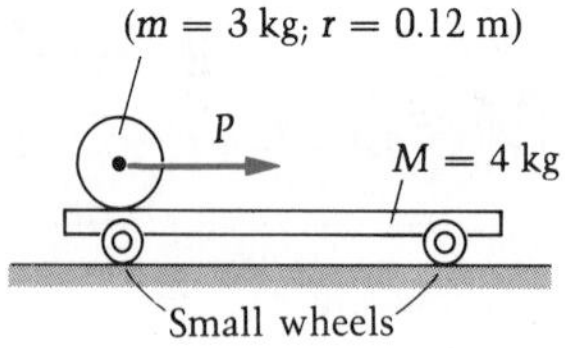

Figure P4.120

4.120 The system shown in Figure P4.120 is initially at rest. A force P is then applied that varies with time according to $P = 7t^2$, where P is in newtons and t in seconds. If the coefficient of friction between cylinder and cart is $\mu = 0.5$, find how much time elapses before the cylinder starts to slip on the cart.

4.121 In the previous problem, determine how much time passes (from $t = 0$) before the cylinder leaves the surface of the cart. Initially, the center of the cylinder is 2 m from the right end of the cart.

4.122 A slender homogeneous bar $\mathcal{B}$ weighing 193 lb has an angular velocity of 2 rad/sec clockwise and an angular acceleration of 8 rad/sec^2 clockwise when in the position shown in Figure P4.122. The wall at B is smooth; the coefficient of sliding friction at A is 0.10. Find the reactions at A and B on $\mathcal{B}$ in this position. *Hint*: The force P can be found.

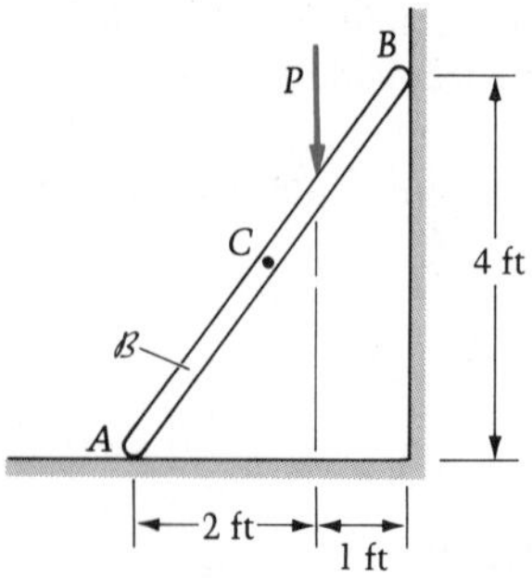

Figure P4.122

4.123 After release from a slightly displaced position, the rod in Figure P4.123 will remain in contact with the floor throughout its fall. Describe the path of C and find the reaction onto the floor just before the rod becomes horizontal.

4.124 The uniform slender bar of mass m and length L is released from rest in the position shown in Figure P4.124. Find the force exerted by the smooth floor at this instant.

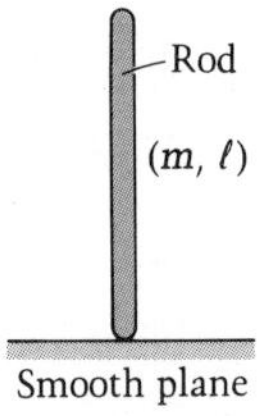

Figure P4.123

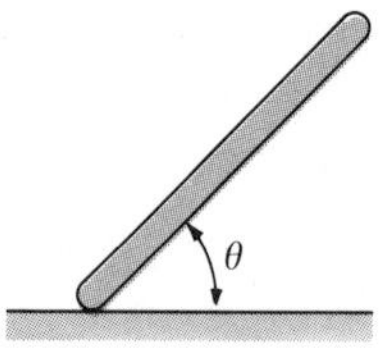

Figure P4.124

4.125 A thin rod AB of length ℓ and mass m is released from rest in the position shown in Figure P4.125. Point A of the rod is in contact with a surface whose coefficient of friction is μ.

a. Determine the minimum value of μ, say $\mu = \mu_{min}$, required to prevent end A from slipping upon release.
b. Find the acceleration of the mass center of the rod immediately after release for $\mu \geq \mu_{min}$ and for $\mu < \mu_{min}$.

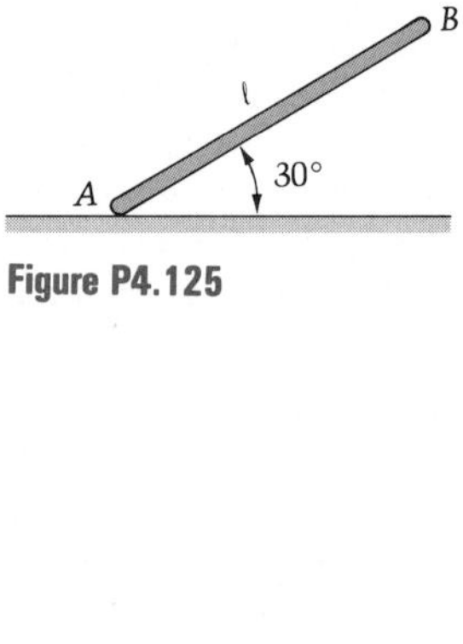

Figure P4.125

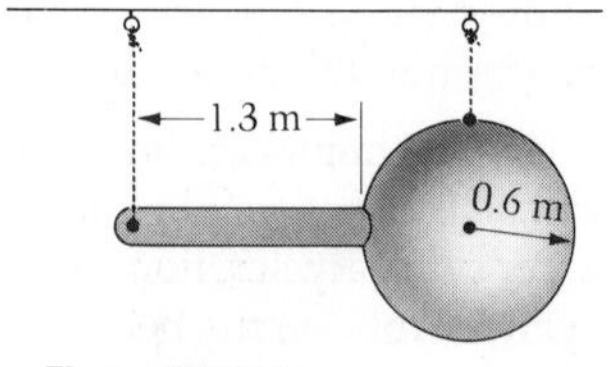

Figure P4.126

4.126 The 30-kg sphere and 15-kg rod in Figure P4.126 are welded together to form a single rigid body. Determine the angular acceleration of the body immediately after the right-hand string is cut.

4.127 If the right-hand string in Figure P4.127 is cut, find the initial tension in the left string. The slender rod has mass m and length L.

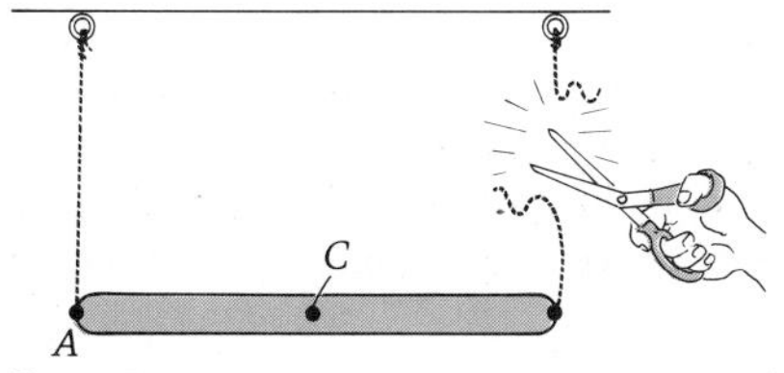

Figure P4.127

4.128 Repeat the preceding problem if the rod is replaced by a rectangular plate suspended from the two upper corners. The width (between strings) is B and the height is H.

4.129 A uniform slender rod, 10 ft long and weighing 90 lb, is supported by wires attached to its ends. (See Figure P4.129.) Find the tension in the right wire just after the left wire is cut. Assume the wires to be inextensible.

4.130 The left end of a slender uniform bar is attached to a light inextensible cable as shown in Figure P4.130. If the bar has mass m and length L and is released from rest in the position shown, find the angular acceleration of the bar at the instant after release.

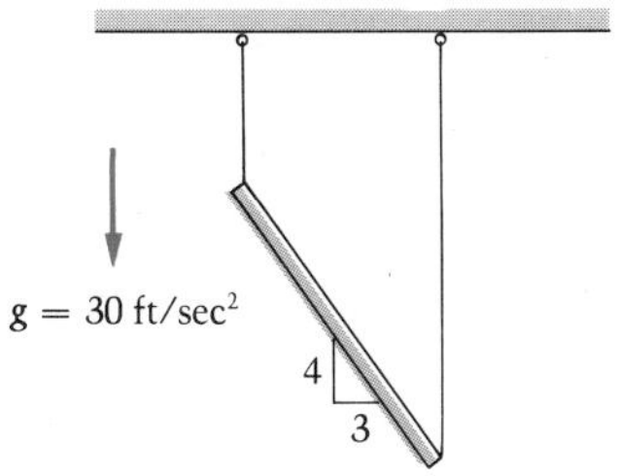

Figure P4.129

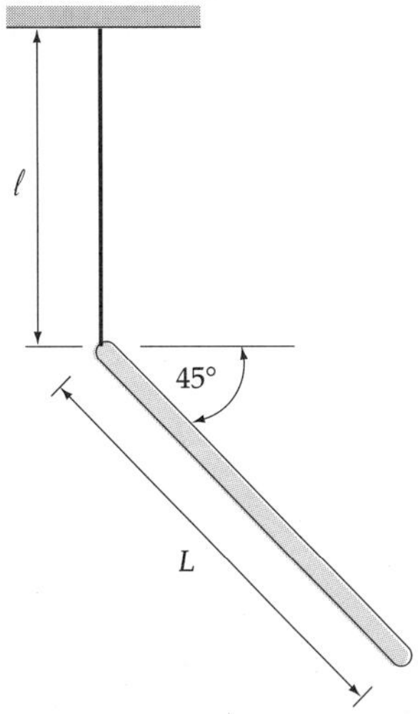

Figure P4.130

4.131 The uniform slender bar of mass m is released from rest in the position shown in Figure P4.131. Find $\mathbf{a}_A$ and the tension in the inextensible cord at the instant after release.

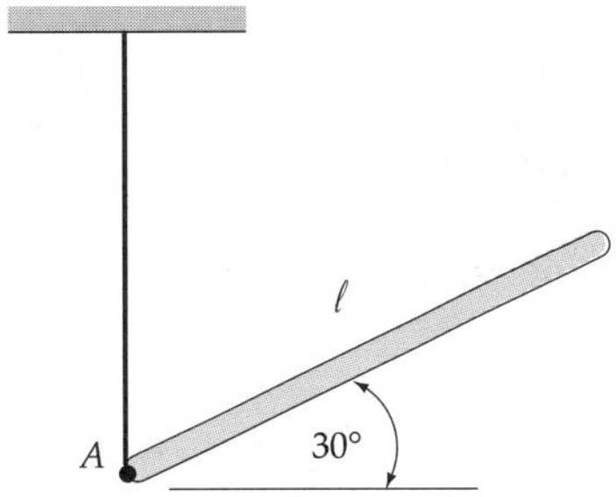

Figure P4.131

4.132 The uniform 20-lb bar is three feet long and has an angular velocity $\omega = 3$ ⤾ rad/sec with $\mathbf{v}_C = \mathbf{0}$ at the instant shown in Figure P4.132. Neglecting interaction with the air, what is the angular velocity of the bar after its center has dropped 10 feet?

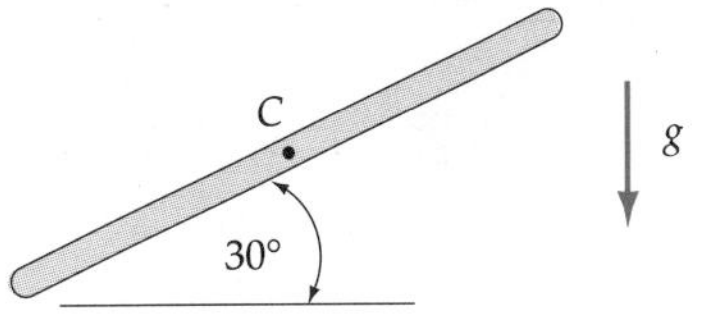

Figure P4.132

4.133 The disk shown in Figure P4.133 has mass m and radius r. Show that at the instant the right-hand string is cut, the tension in the other string changes to $\frac{2}{5}\,mg$, so that the acceleration of the mass center is $\frac{3}{5}\,g\downarrow$.

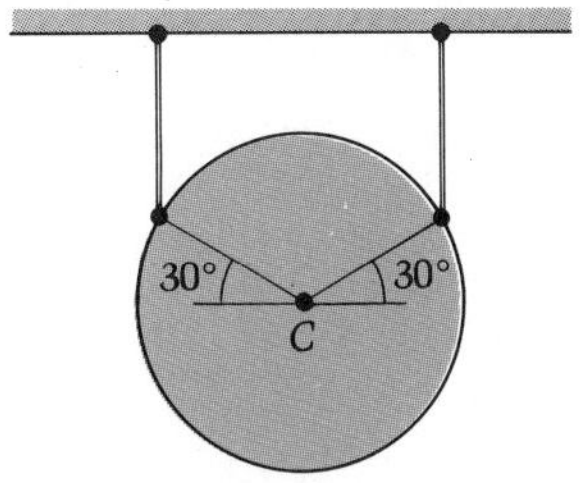

Figure P4.133

4.134 A uniform rod $\mathcal{R}$ is supported by two cords as shown in Figure P4.134. If the right-hand cord suddenly breaks, determine the initial tension in the left cord AD. ("Initial" means before the rod has had time to move and before it has had time to generate any velocities.)

4.135–4.140 The six equilateral triangular plates (Figures P4.135–4.140) are each supported on their rightmost corner B by a string; each has a different support condition at the left corner A. At the instant when the string at B is cut, find $\mathbf{a}_C$ and $\boldsymbol{\alpha}$ in each case. The length of each side is s.

4.141 The uniform 10-lb bar in Figure P4.141 is suspended by two inextensible cables. At the instant shown, when each point in the bar has a velocity of $10\hat{\mathbf{i}}$ ft/sec, the right cable breaks. Find the force in the left cable immediately after the break.

4.142 A slender bar $\mathcal{B}$ weighing 64.4 lb is attached by massless cables to a fixed pivot A as shown in Figure P4.142. The system is swinging about A as a pendulum. At $\theta = 0$ the angular velocity is 2 ⮌ rad/sec and cable AD breaks. Find the tension in cable AB just after the break.

* **4.143** A beam of length L and weight W per unit length is supported by two cables at A and B. (See Figure P4.143.) If the cable at B should break, find the shear force V and bending moment M at section xx just after the cable breaks. *Hint*: Euler's laws apply to every part of the body.

* **4.144** See Figure P4.144. Assuming that sufficient friction is present to prevent slipping between $\mathcal{C}$ and the plane, find the angular accelerations of $\mathcal{C}$ and $\mathcal{R}$ just after force P is applied to the bodies at rest. They are connected by a smooth pin.

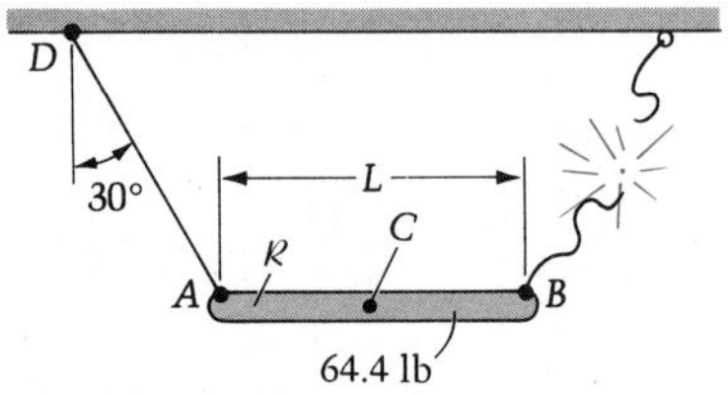

Figure P4.134

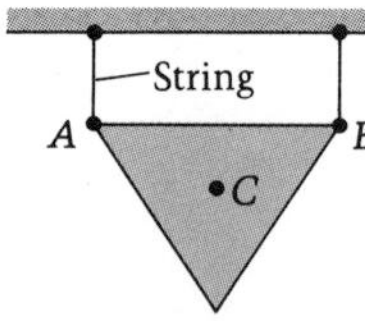

Figure P4.135

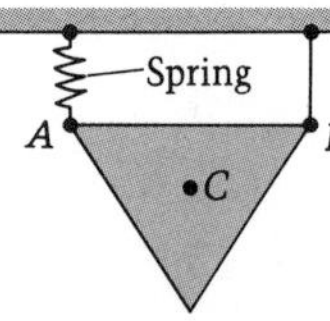

Figure P4.136

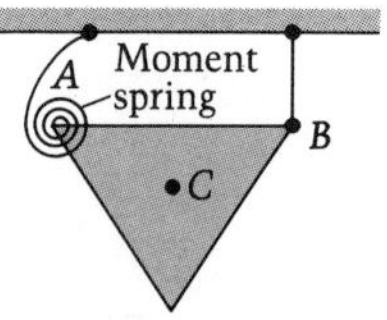

Figure P4.137

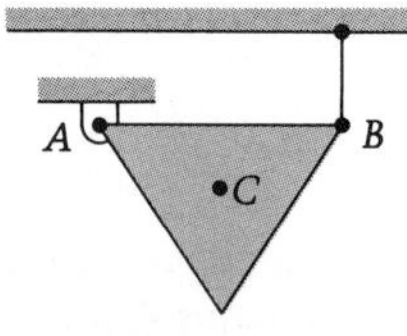

Figure P4.138

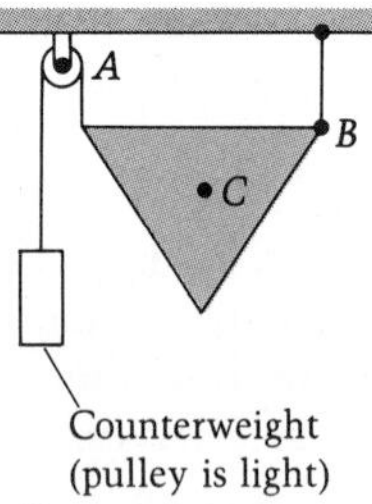

Figure P4.139

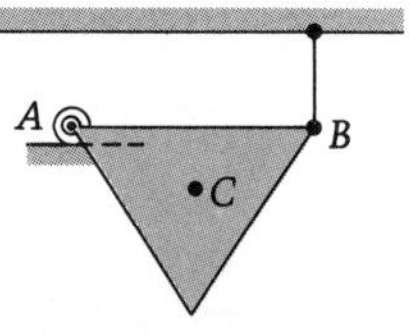

Figure P4.140

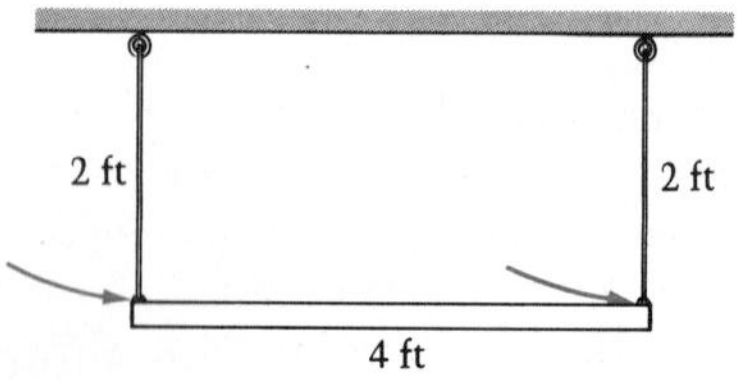

Figure P4.141

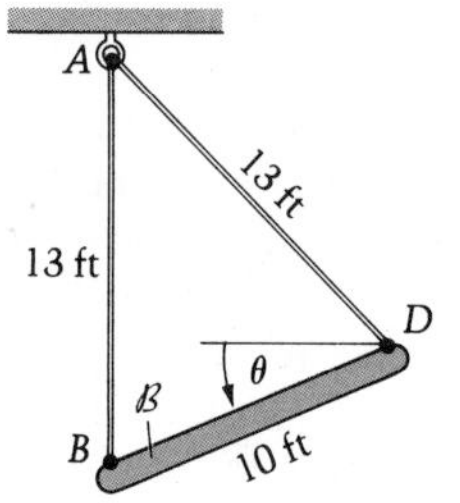

Figure P4.142

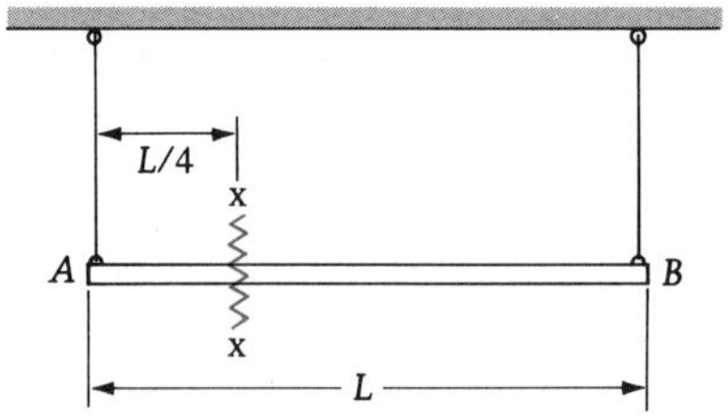

Figure P4.143

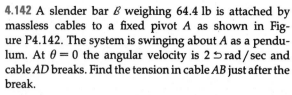

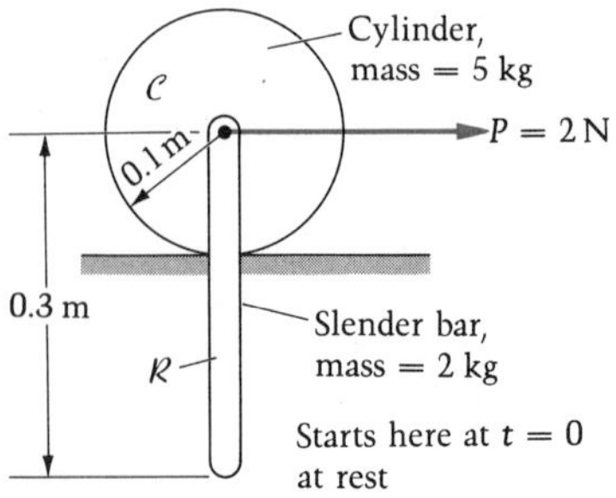

Figure P4.144

* **4.145** Rods $\mathcal{B}_1$ and $\mathcal{B}_2$ each have mass m. (See Figure P4.145.) Upon release from rest in the horizontal position indicated, find the reactions at O, and at A, onto $\mathcal{B}_1$.

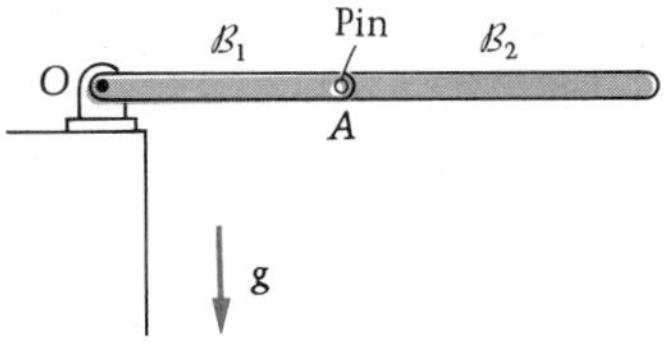

Figure P4.145

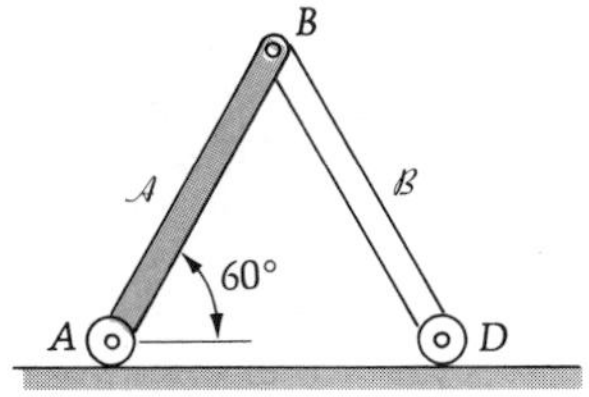

Figure P4.146

* **4.146** Two uniform bars $\mathcal{A}$ and $\mathcal{B}$ are released from rest in the position shown in Figure P4.146. Each bar is 2 ft long and weighs 10 lb. Determine the angular acceleration of each bar and the reactions at A and D immediately after release. The rollers are light and the pins smooth.

4.147 A constant torque T_0 ⮌ is applied to the crank arm $\mathcal{C}$ of the planetary mechanism shown in Figure P4.147. The axes of the identical gears $\mathcal{S}$ and $\mathcal{P}$ are vertical, and the ends of the crank are pinned to the centers of $\mathcal{S}$ and $\mathcal{P}$. Determine the angular acceleration of $\mathcal{C}$ if $\mathcal{S}$ is fixed in the inertial frame of reference. Treat the gears as uniform disks. The plane of the page is horizontal.

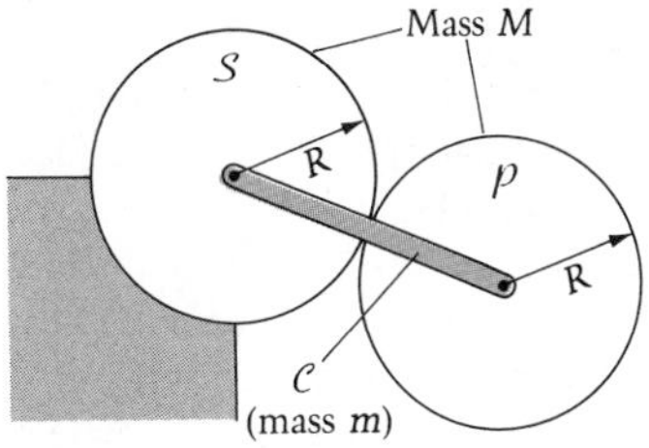

Figure P4.147

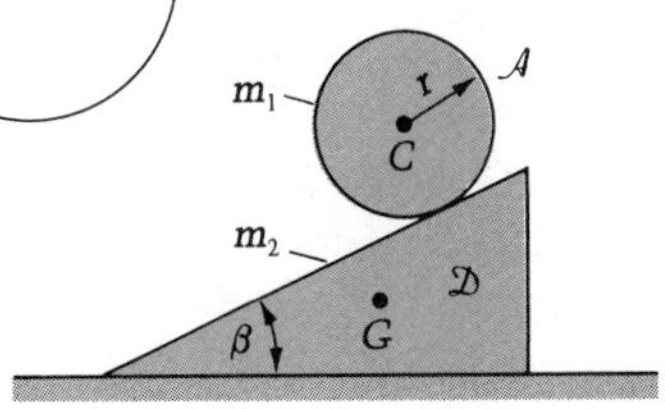

Figure P4.148

* **4.148** Cylinder $\mathcal{A}$ in Figure P4.148 rolls down a wedge that can slide without friction on a smooth floor. Show that the acceleration of wedge $\mathcal{D}$ is a constant given by the equation

$$a_G = \frac{m_1 g \sin 2\beta}{3(m_1 + m_2) - 2m_1 \cos^2 \beta}$$

Extended Problems

4.149 The stick of mass m, shown in Figure P4.149, originally at rest with $\theta = 0$, is disturbed slightly and begins to slide on a smooth wall and floor. Derive the differential equation of motion of the stick. Integrate the equation and find the angle θ at which contact with the wall is lost.

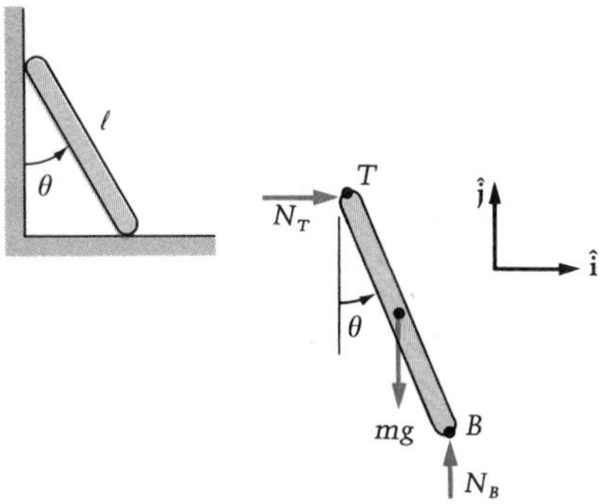

Figure P4.149

Hints: First verify the following equations of motion:

$$\Sigma F_x = N_T = m\ddot{x}_C \tag{1}$$

$$\Sigma F_y = N_B - mg = m\ddot{y}_C \tag{2}$$

$$\Sigma M_C = N_B \frac{\ell}{2} \sin\theta - N_T \frac{\ell}{2} \cos\theta = \frac{m\ell^2}{12}\ddot{\theta} \tag{3}$$

Then note that there are more unknowns than equations. Use kinematics to relate $\mathbf{a}_C$ to $\mathbf{a}_B$, and use the "$\hat{\mathbf{j}}$-component" of that equation to obtain a fourth equation, containing $\ddot{y}_C$. Then relate $\mathbf{a}_C$ to $\mathbf{a}_T$ and get a fifth equation, containing $\ddot{x}_C$. From your five equations, eliminate $\ddot{x}_C$, $\ddot{y}_C$, N_T, and N_B, obtaining the following differential equation governing θ:

$$\ddot{\theta} = \frac{3g}{2\ell} \sin\theta \tag{4}$$

Multiply (4) by $\dot{\theta}$ and integrate, using $\dot{\theta}\ddot{\theta} = d(\dot{\theta}^2)/dt$. Use the initial condition $\dot{\theta} = 0$ at $\theta = 0$ to evaluate the constant of integration. You will now know $\dot{\theta}$ as a function of θ. Then, with N_T expressed as a function of θ, set $N_T = 0$ in (1) to find the angle where contact is lost. Your answer should be $\cos^{-1}(\frac{2}{3})$.

4.150 Disks $\mathcal{B}_3$ and $\mathcal{B}_4$ each weigh 64.4 lb and are rigidly attached to the light shaft $\mathcal{B}_5$ that joins their centers. (See Figure P4.150.) A 96.6-lb cylinder $\mathcal{B}_1$ has a hole drilled along its axis, through which $\mathcal{B}_5$ passes. Let $\mathcal{B}_2$ represent the rigid body comprised of $\mathcal{B}_3$, $\mathcal{B}_4$, and $\mathcal{B}_5$.

While the body $\mathcal{B}_2$ is held fixed on the plane, the cylinder is spun up to an angular velocity $8 \circlearrowright$ rad/sec, and the system is then released. Assume that part of the reaction between the axle and the wall of the cylindrical hole in the cylinder is a friction couple proportional to the difference in angular velocities, with proportionality constant k. The friction couple acting on the axle will cause $\mathcal{B}_2$ to roll to the right; the opposite couple on $\mathcal{B}_1$ will slow its angular speed down. As time passes, the bodies $\mathcal{B}_1$ and $\mathcal{B}_2$ will approach the condition of moving as one. Show this, and find the common, limiting-case "terminal" angular velocity shared by $\mathcal{B}_1$ and $\mathcal{B}_2$. There is sufficient friction between $\mathcal{B}_2$ and the ground to prevent slipping there.

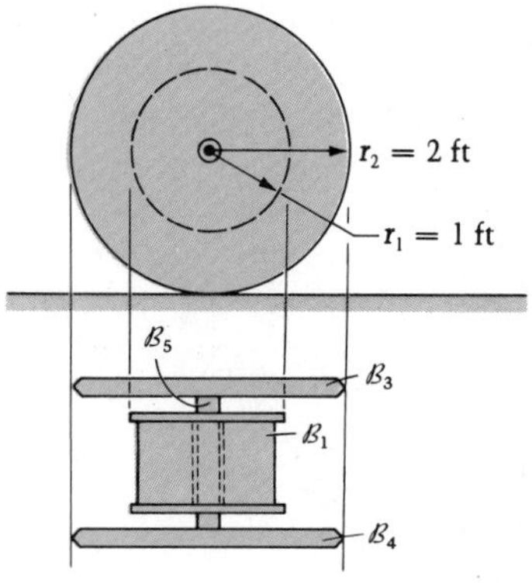

Figure P4.150

4.6 Other Useful Forms of the Moment Equation

For a rigid body in plane motion, there are several other forms of the moment equation of motion [besides the translation equation (4.1) and $\Sigma M_C = I_C\alpha$] which are worthy of special study. The idea is that it is often convenient and helpful to sum the moments about a point other than the mass center C. We will study three of these forms one-by-one and present examples of each as we go along.

Moment Equation in Terms of $\mathbf{a}_C$

To develop this form, we begin with Equation (2.45):

$$\Sigma \mathbf{M}_P = \dot{\mathbf{H}}_C + \mathbf{r}_{PC} \times m\mathbf{a}_C \tag{2.45}$$

where we recall that in this form there is *no restriction at all* on the location of point P, the type of body being studied, or the type of motion. Thus, specializing for a rigid body in plane motion and using the right-hand side of Equation (4.13) to replace $\dot{\mathbf{H}}_C$ for this case,

$$\begin{aligned}\Sigma \mathbf{M}_P = (I_{xz}^C\alpha - I_{yz}^C\omega^2)\hat{\mathbf{i}} + (I_{yz}^C\alpha + I_{xz}^C\omega^2)\hat{\mathbf{j}} \\ + I_{zz}^C\alpha\hat{\mathbf{k}} + \mathbf{r}_{PC} \times m\mathbf{a}_C\end{aligned} \tag{4.17}$$

Whenever the products of inertia in Equation (4.17) vanish, this equation takes the particularly simple form

$$\Sigma \mathbf{M}_P = I_{zz}^C\alpha\hat{\mathbf{k}} + \mathbf{r}_{PC} \times m\mathbf{a}_C \tag{4.18}$$

Note that the translation Equation (4.1) results from Equation (4.18) if $\alpha \equiv 0$.

Note further that if P and C are in the same (reference) plane of motion, then $\mathbf{r}_{PC} \times m\mathbf{a}_C$ is perpendicular to the plane of motion — that is, it is parallel to **k**. When this is the case, we will rewrite Equation (4.18) in scalar form*:

$$\Sigma M_P = I_{zz}^C \alpha + (\mathbf{r}_{PC} \times m\mathbf{a}_C)_z \tag{4.19}$$

in which $(\ )_z$ means the coefficient of the unit vector $\hat{\mathbf{k}}$ within the parenthesis.

We now present two examples of the use of Equation (4.18). In the first one, we eliminate reactions at a pin by summing moments there:

Example 4.22

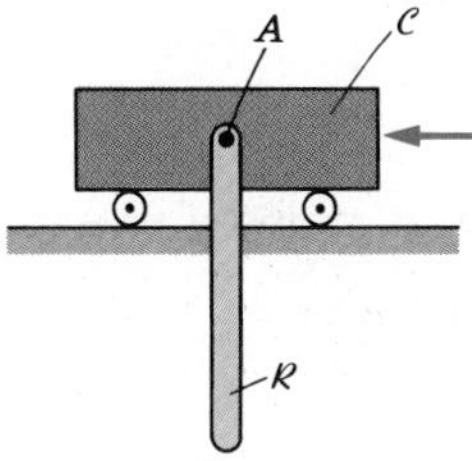

Figure E4.22a

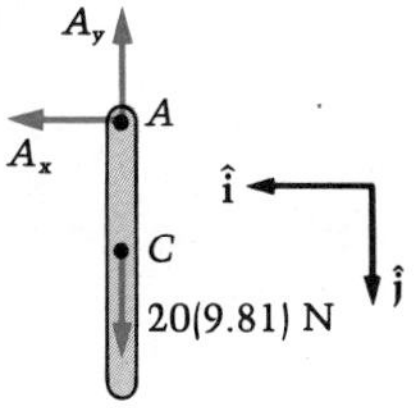

Figure E4.22b

The uniform rod $\mathcal{R}$ in Figure E4.22a (length 80 cm, mass 20 kg) is smoothly pinned to cart $\mathcal{C}$ (50 kg) at point A. Force P, applied to $\mathcal{C}$ with the system initially at rest, causes $\mathcal{C}$ to translate with the acceleration $3 \leftarrow \text{m/s}^2$. Find the initial angular acceleration of the rod.

Solution

By using Equation (4.19), we can sum moments about A of the forces on $\mathcal{R}$ and avoid having to use $\Sigma F_x = m\ddot{x}_C$ in this case. We obtain, using the free-body in Figure E4.22b,

$$\Sigma M_A = I_C \alpha + (\mathbf{r}_{AC} \times m\mathbf{a}_C)_z \tag{1}$$

We note that $I_C = m\ell^2/12 = 20(0.8)^2/12 = 1.07 \text{ kg} \cdot \text{m}^2$, and we get $\mathbf{a}_C$ from kinematics:

$$\mathbf{a}_C = \mathbf{a}_A + \alpha\hat{\mathbf{k}} \times (0.4\hat{\mathbf{j}}) - \overset{0 \text{ at } t=0}{\cancel{\omega^2}}(0.4\hat{\mathbf{j}})$$

$$\mathbf{a}_C = 3\hat{\mathbf{i}} - 0.4\alpha\hat{\mathbf{i}} = (3 - 0.4\alpha)\hat{\mathbf{i}}$$

so that, substituting into (1),

$$0 = 1.07\alpha + [0.4\hat{\mathbf{j}} \times 20(3 - 0.4\alpha)\hat{\mathbf{i}}]_z$$

$$0 = (1.07 + 3.2)\alpha - 24$$

$$\alpha = 5.62 \text{ rad/s}^2$$

Thus the rod starts off with $\alpha = 5.62 \circlearrowleft \text{rad/s}^2$.

Question 4.8 Why would the solution to the preceding example be much more complicated using the mass center form of the moment Equation (4.16)?

Next we use Equation (4.19) to rework an example from the previous section.

* When the products of inertia I_{xz}^C and I_{yz}^C both vanish, then by Equations (4.14a,b), $\Sigma M_{Cx} = 0 = \Sigma M_{Cy}$. This leads, when P and C are both in the reference plane, to $\Sigma M_{Px} = 0 = \Sigma M_{Py}$ since $\Sigma \mathbf{M}_P$ always equals $\Sigma \mathbf{M}_C + \mathbf{r}_{PC} \times \Sigma \mathbf{F}$ and $\Sigma \mathbf{F} = m\mathbf{a}_C$.

Answer 4.8 The pin reaction A_x would appear in ΣM_C. Thus we would need to also write Equation (4.15a) to eliminate A_x.

EXAMPLE 4.23

Find the starting angular acceleration for the body of Example 4.19, shown again in Figure E4.23a.

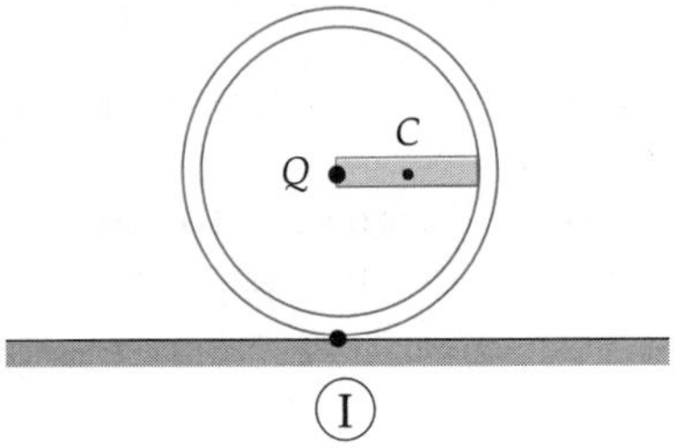

Figure E4.23a

Solution

We sum moments about point Ⓘ, with the help of Equation (4.19) and the free-body diagram in Figure E4.23b:

$$\Sigma M_{Ⓘ} = I_C\alpha + (\mathbf{r}_{ⒾC} \times m\mathbf{a}_C)_z$$

$$\circlearrowright \quad \frac{mg\ell}{2} = \frac{m\ell^2}{12}\alpha + \underbrace{\left[\left(\frac{\ell}{2}\hat{\mathbf{i}} - \ell\hat{\mathbf{j}}\right) \times m(\ddot{x}_C\hat{\mathbf{i}} + \ddot{y}_C\hat{\mathbf{j}})\right]_z}_{m\frac{\ell}{2}\ddot{y}_C + m\ell\ddot{x}_C}$$

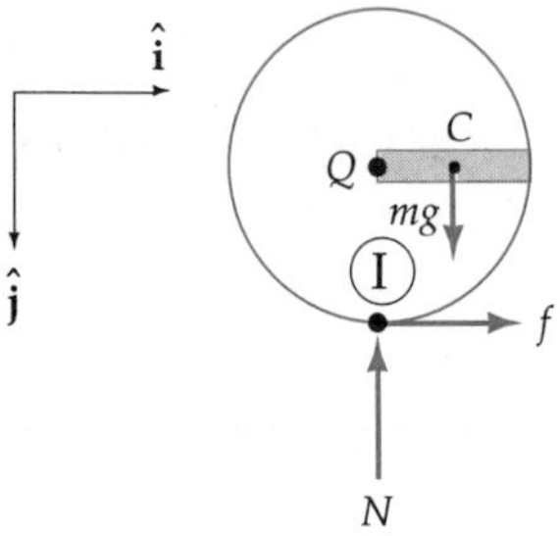

Figure E4.23b

From the kinematics in the earlier example, we know $\ddot{x}_C = \ell\alpha$ and $\ddot{y}_C = \frac{\ell}{2}\alpha$. Substituting these into the above equation after cancelling "m",

$$\frac{g\ell}{2} = \frac{\ell^2\alpha}{12} + \frac{\ell}{2}\left(\frac{\ell\alpha}{2}\right) + \ell(\ell\alpha)$$

Therefore,

$$\alpha = \frac{3}{8}\frac{g}{\ell}$$

as before.

Note that the "x" and "y" equations of motion, (4.15a,b), were not needed in this example. They would have been in Example 4.19, however, even if all we had been seeking was α.

Question 4.9 Why?

Moment Equation in Terms of $\mathbf{a}_P$

Another form of the moment equation of motion that is sometimes useful involves the acceleration and inertia properties at some point other than the mass center. Recalling that if P is a point of a rigid body in plane motion, then by Equation (4.3),

$$\mathbf{H}_P = \mathbf{r}_{PC} \times m\mathbf{v}_P + I_{xz}^P\omega\hat{\mathbf{i}} + I_{yz}^P\omega\hat{\mathbf{j}} + I_{zz}^P\omega\hat{\mathbf{k}} \tag{4.3}$$

And for *any* point P, we know from Equation (2.36) that:

$$\mathbf{H}_P = \mathbf{H}_C + \mathbf{r}_{PC} \times m\mathbf{v}_C \tag{2.36}$$

Equating these two expressions for $\mathbf{H}_P$ and then differentiating with respect to time,

Answer 4.9 ΣM_C would have included moments of both f and N.

$$\dot{\mathbf{H}}_C + \mathbf{r}_{PC} \times m\mathbf{a}_C + \dot{\mathbf{r}}_{PC} \times m\mathbf{v}_C$$

$$= \dot{\mathbf{r}}_{PC} \times m\mathbf{v}_P + \mathbf{r}_{PC} \times m\mathbf{a}_P + \frac{d}{dt}(I_{xz}^P\omega\hat{\mathbf{i}} + I_{yz}^P\omega\hat{\mathbf{j}} + I_{zz}^P\omega\hat{\mathbf{k}}) \quad (4.20)$$

By Equations (2.8) and (2.43), we know that $m\mathbf{a}_C = \Sigma\mathbf{F}$ and $\dot{\mathbf{H}}_C = \Sigma\mathbf{M}_C$. Using these results and Equation (2.42), we may replace by $\Sigma\mathbf{M}_P$ the first two terms on the left-hand side of this equation and obtain

$$\Sigma\mathbf{M}_P = \dot{\mathbf{r}}_{PC} \times m(\mathbf{v}_P - \mathbf{v}_C) + \mathbf{r}_{PC} \times m\mathbf{a}_P$$
$$+ \frac{d}{dt}(I_{xz}^P\omega\hat{\mathbf{i}} + I_{yz}^P\omega\hat{\mathbf{j}} + I_{zz}^P\omega\hat{\mathbf{k}}) \quad (4.21)$$

On the right side of Equation (4.21) the first term vanishes since $\mathbf{v}_P - \mathbf{v}_C = -\dot{\mathbf{r}}_{PC}$, and the third term is of the same form as $\dot{\mathbf{H}}_C$, except that here the inertia properties are with respect to axes with origin at P. Thus retracing the steps between Equations (4.11) and (4.13), we obtain

$$\Sigma\mathbf{M}_P = (I_{xz}^P\alpha - I_{yz}^P\omega^2)\hat{\mathbf{i}} + (I_{yz}^P\alpha + I_{xz}^P\omega^2)\hat{\mathbf{j}}$$
$$+ I_{zz}^P\alpha\hat{\mathbf{k}} + \mathbf{r}_{PC} \times m\mathbf{a}_P \quad (4.22)$$

Question 4.10 Why can we say, as was done above, that $\mathbf{v}_P - \mathbf{v}_C = -\dot{\mathbf{r}}_{PC}$ and why does that cause the first term on the right side of Equation (4.16) to vanish?

When the products of inertia I_{xz}^P and I_{yz}^P vanish, Equation (4.22) simplifies to

$$\Sigma\mathbf{M}_P = I_{zz}^P\alpha\hat{\mathbf{k}} + \mathbf{r}_{PC} \times m\mathbf{a}_P \quad (4.23)$$

As in Section 4.5, when P and C are in the same (reference) plane the $\hat{\mathbf{i}}$ and $\hat{\mathbf{j}}$ components of this equation vanish. Thus the scalar form of the equation is

$$\Sigma M_P = I_{zz}^P\alpha + (\mathbf{r}_{PC} \times m\mathbf{a}_P)_z \quad (4.24)$$

where $(\ )_z$ again means the coefficient of $\hat{\mathbf{k}}$ within the parentheses.

This equation is very useful if we happen to know the acceleration of a point other than the mass center, as in the following examples:

Answer 4.10 Let $\mathbf{r}_{OP}$ and $\mathbf{r}_{OC}$ be position vectors for P and C; differentiating $\mathbf{r}_{OC} = \mathbf{r}_{OP} + \mathbf{r}_{PC}$ gives $\mathbf{v}_C = \mathbf{v}_P + \dot{\mathbf{r}}_{PC}$. The cross product of parallel vectors always vanishes.

Example 4.24

Solve the problem of Example 4.22 by using Equation 4.24.

Solution

Equation (4.24) makes a problem such as this even simpler than did Equation (4.19):

$$\circlearrowleft^{+} \quad \Sigma M_A = I_A\alpha + (\mathbf{r}_{AC} \times m\mathbf{a}_A)_z$$

$$0 = \frac{m\ell^2}{3}\alpha - \frac{\ell}{2}ma_A = \frac{20(0.8)^2}{3}\alpha - 0.4(20)3$$

$\alpha = 5.62$ rad/sec^2, as before.

To determine other information in the preceding example, for instance the starting value of force P, we can write the x-component of the mass center equation of motion for the cart and then for the bar (see Figure 4.7):

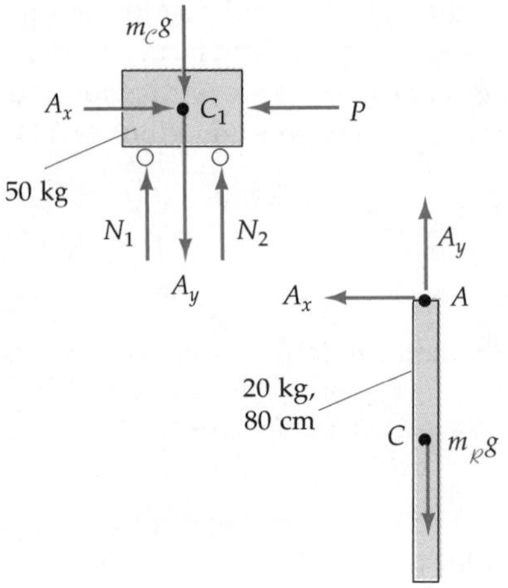

Figure 4.7

For $\mathcal{C}$:

$$\overset{+}{\longleftarrow} \quad \Sigma F_x = P - A_x = m_{\mathcal{C}}\ddot{x}_{C_1} = 50\,\ddot{x}_{C_1}$$

For $\mathcal{R}$:

$$\overset{+}{\longleftarrow} \quad \Sigma F_x = A_x = m_{\mathcal{R}}\ddot{x}_C = 20\,\ddot{x}_C$$

where

$$\mathbf{a}_C = \mathbf{a}_A + \boldsymbol{\alpha} \times \mathbf{r}_{AC} - \overset{0 \text{ at } t=0}{\cancel{\omega^2}}\,\mathbf{r}_{AC}$$

$$= \ddot{x}_{C_1}\hat{\mathbf{i}} + \alpha\hat{\mathbf{k}} \times \frac{\ell}{2}\hat{\mathbf{j}}$$

$$\mathbf{a}_C = (\ddot{x}_{C_1}\hat{\mathbf{i}} - \frac{\ell}{2}\alpha)\hat{\mathbf{i}}$$

or

$$\ddot{x}_C = \ddot{x}_{C_1} - \frac{\ell}{2}\alpha$$

Thus

$$A_x = 20\left(3 - \frac{0.8}{2}5.62\right) = 15.0 \text{ N}$$

and

$$P - A_x = P - 15.0 = 50(3)$$
$$P = 165 \text{ N}$$

EXAMPLE 4.25

Repeat Examples 4.19,23, using Equation (4.24) this time to find the starting angular acceleration.

Solution

We again sum moments about point Ⓘ:

$$\Sigma M_{Ⓘ} = I_{Ⓘ}\alpha + (\mathbf{r}_{ⒾC} \times m\mathbf{a}_{Ⓘ})_z$$

$$\circlearrowright^{+} \quad \frac{mg\ell}{2} = \left[\frac{m\ell^2}{12} + m\left(\left(\frac{\ell}{2}\right)^2 + \ell^2\right)\right]\alpha + \left[\left(\frac{\ell}{2}\hat{\mathbf{i}} - \ell\hat{\mathbf{j}}\right) \times m(-\ell\underbrace{\omega^2}_{0 \text{ initially}}\hat{\mathbf{j}})\right]_z$$

$$\frac{mg\ell}{2} = \frac{4}{3}m\ell^2\alpha$$

$$\alpha = \frac{3}{8}\frac{g}{\ell}$$

once again. Notice that the cross-product term is simpler this time, but the moment of inertia at Ⓘ has to be calculated by the parallel axis theorem. In Problems 4.153, 154, this problem is to be reworked one last time using the geometric center of the hoop as point P in Equations 4.19 and 4.24, respectively.

In the preceding example, it *accidentally* happened that $\Sigma M_{Ⓘ} = I_{Ⓘ}\alpha$. *This is not generally true.* For example, an instant later ω would not be zero and the cross-product term would *not* be zero.

Moment Equation for Fixed-Axis Rotation (The "Pivot" Equation)

Consider now the case in which the acceleration of a point P of the rigid body in plane motion is identically zero. Equation (4.22) tells us that for such a point,

$$\Sigma \mathbf{M}_P = (I_{xz}^P\alpha - I_{yz}^P\omega^2)\hat{\mathbf{i}} + (I_{yz}^P\alpha + I_{xz}^P\omega^2)\hat{\mathbf{j}} + I_{zz}^P\alpha\hat{\mathbf{k}} \tag{4.25}$$

If we take the dot product of this equation with $\hat{\mathbf{k}}$, we see that whether or not the products of inertia I_{xz}^P and I_{yz}^P vanish, we always have:

$$\Sigma \mathbf{M}_P \cdot \hat{\mathbf{k}} = (\Sigma \mathbf{M}_P)_z = I_{zz}^P\alpha \tag{4.26}$$

When the point P is fixed in the inertial frame as well as in the body, we will then usually label the point as O for emphasis. In this case, the only way the body can move is to rotate about the z-axis through O, a motion which we call *fixed-axis rotation.* The point O, which does not move during the motion of interest, is called a *pivot,* and we abbreviate Equation (4.26) as

$$(\Sigma M_O)_z = I_O\alpha \tag{4.27}$$

or as

$$\Sigma M_{axis} = I_{axis}\alpha \tag{4.28}$$

When the products of inertia vanish at a pivot O, as they will for all problems in this section, then of course $(\Sigma M_O)_z$ can be further abbreviated to simply ΣM_O, for then only the z-component of $\Sigma \mathbf{M}_O$ is non-zero. This is *not* always the case, as we shall later see in the final section 4.7 of this chapter. When it is, however, we simply write

$$\Sigma M_O = I_O\alpha \tag{4.29}$$

Question 4.11 Since Equation (4.4) applies for any point P having zero velocity, why can we not use equations such as (4.29) for the instantaneous center Ⓘ of $\mathcal{B}$ when Ⓘ is *not* a pivot?

Because of the importance of fixed-axis rotation in engineering, we present more examples of it than we did for the earlier forms in this section. In the first example, we examine a composite body rotating about a pivot:

Answer 4.11 Although $\mathbf{H}_O$ may always be written as $I^O_{xz}\omega\hat{\mathbf{i}} + I^O_{yz}\omega\hat{\mathbf{j}} + I^O_{zz}\omega\hat{\mathbf{k}}$ whenever $\mathbf{v}_O$ is zero, its derivative is only equal to $\Sigma\mathbf{M}_O$ when $\mathbf{v}_O$ is *identically* zero — in other words, *zero all the time.*

EXAMPLE 4.26

The rod $\mathcal{R}$ and sphere $\mathcal{S}$ in Figure E4.26a are welded together to form a combined rigid body which is attached to the ground at O by means of a smooth pin. Find the force exerted by the pin onto the body, upon release of the system from rest.

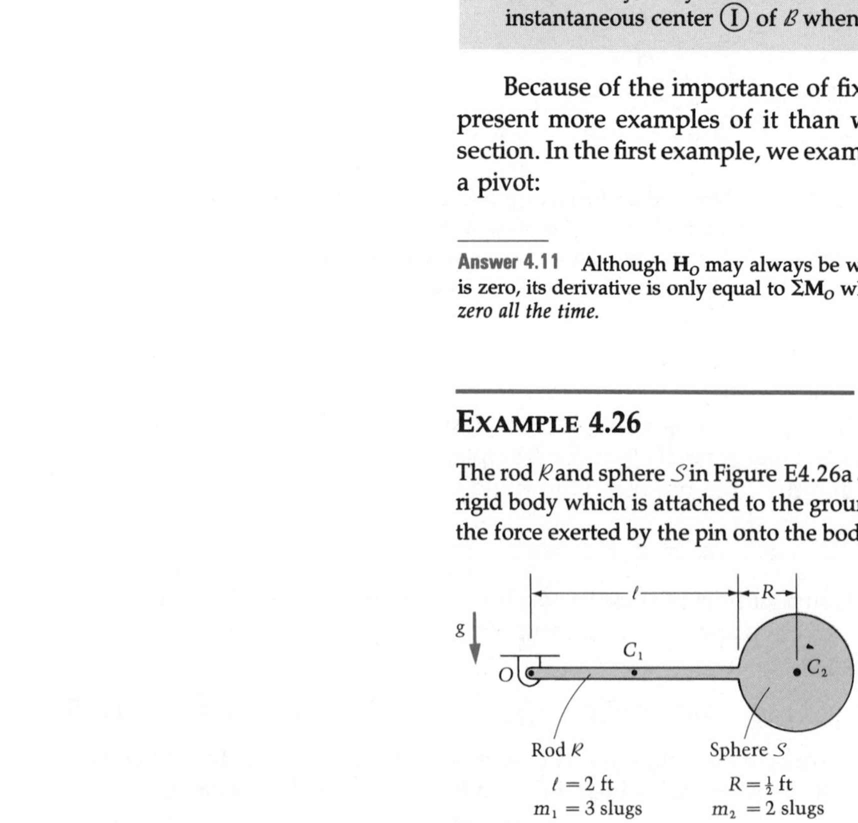

Figure E4.26a

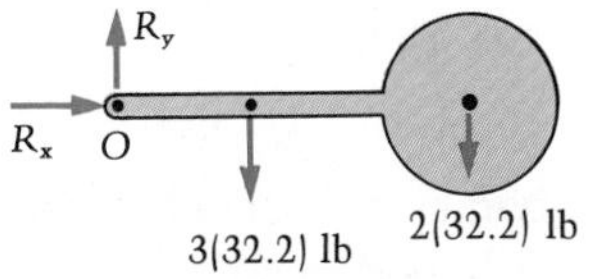

Figure E4.26b

Solution

Because O is a pivot of the combined body, we use Equation 4.29:

$$\Sigma \mathbf{M}_O = I_O \alpha \hat{\mathbf{k}}$$

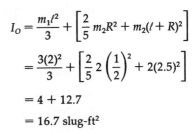

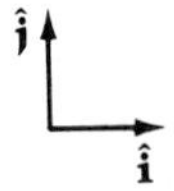

where

$$\begin{aligned} I_O &= \frac{m_1 \ell^2}{3} + \left[\frac{2}{5} m_2 R^2 + m_2(\ell + R)^2\right] \\ &= \frac{3(2)^2}{3} + \left[\frac{2}{5} 2 \left(\frac{1}{2}\right)^2 + 2(2.5)^2\right] \\ &= 4 + 12.7 \\ &= 16.7 \text{ slug-ft}^2 \end{aligned}$$

so that, at release, using the FBD in Figure E4.26b,

$$-3(32.2)(1) - 2(32.2)(2.5) = 16.7\alpha$$

or

$$\alpha = -15.4 \text{ rad/sec}^2$$

To calculate the pin reaction we shall use $\Sigma \mathbf{F} = m\mathbf{a}_C$. We first locate the mass center, C, of the body. The distance from O to C is

$$d = \frac{3(1) + 2(2.5)}{3 + 2} = \frac{8}{5} = 1.6 \text{ ft}$$

At this instant we have

$$\begin{aligned} \mathbf{a}_C &= \cancelto{0}{\mathbf{a}_O} + \boldsymbol{\alpha} \times \mathbf{r}_{OC} - \cancelto{0}{\omega^2} \mathbf{r}_{OC} \\ \mathbf{a}_C &= (-15.4\hat{\mathbf{k}}) \times (1.6\hat{\mathbf{i}}) \\ &= -24.6\hat{\mathbf{j}} \text{ ft/sec}^2 \end{aligned}$$

Thus

$$(R_x\hat{\mathbf{i}} + R_y\hat{\mathbf{j}}) - 96.6\hat{\mathbf{j}} - 64.4\hat{\mathbf{j}} = 5(-24.6\hat{\mathbf{j}})$$

So

$$\begin{aligned} R_x &= 0 \\ R_y &= 161 - 123 = 38 \text{ lb} \end{aligned}$$

An alternate approach to the mass center calculation in the preceding example would *not* require that we explicitly locate C, because

$$\Sigma \mathbf{F} = m\mathbf{a}_C = m_1 \mathbf{a}_{C_1} + m_2 \mathbf{a}_{C_2}$$

or

$$(R_x\hat{\mathbf{i}} + R_y\hat{\mathbf{j}}) - 96.6\hat{\mathbf{j}} - 64.4\hat{\mathbf{j}} = 3[(-15.4\hat{\mathbf{k}}) \times \hat{\mathbf{i}}] + 2[(-15.4\hat{\mathbf{k}}) \times 2.5\hat{\mathbf{i}}]$$

from which

$$R_x = 0$$

and

$$R_y = 161 - 46.2 - 77 = 38 \text{ lb}$$

as above.

In our second example, we feature distinct bodies connected by an unwinding rope. One body has a pivot and the others don't.

EXAMPLE 4.27

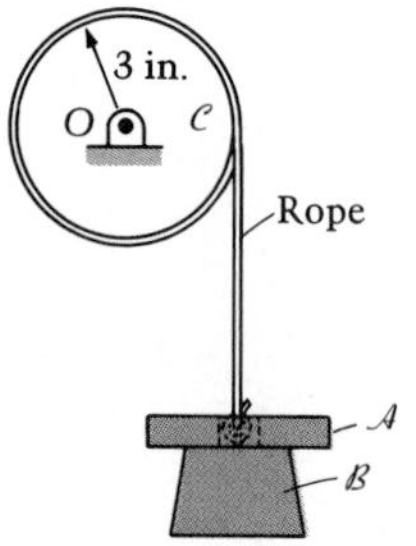

Figure E4.27a

A rope is wrapped around the 10-lb cylinder $\mathcal{C}$ as indicated in Figure E4.27a. The rope passes through a hole in the 5-lb annular disk $\mathcal{A}$ and is then tied to the 15-lb block $\mathcal{B}$. When the system is let go from rest with the rope just taut, what is the reaction exerted on $\mathcal{A}$ by $\mathcal{B}$?

Solution

Using the free-body diagrams in Figures E4.27b, 4.27c, and 4.27d, we write the following equations of motion of the respective bodies. For $\mathcal{C}$, using the "pivot equation" (4.29):

$$\circlearrowleft{+} \quad \Sigma M_O = I_O\alpha$$

$$T\left(\frac{3}{12}\right) = \left[\frac{1}{2}\left(\frac{10}{32.2}\right)\left(\frac{3}{12}\right)^2\right]\alpha$$

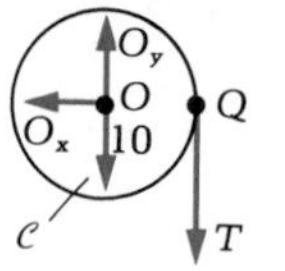

Figure E4.27b

which gives us the tension T in terms of α:

$$T = 0.0388\alpha \tag{1}$$

For the block $\mathcal{B}$ by itself,

$$\downarrow{+} \quad \Sigma F_y = m\ddot{y}_B$$

$$15 - T + R = \frac{15}{32.2}\ddot{y}_B \tag{2}$$

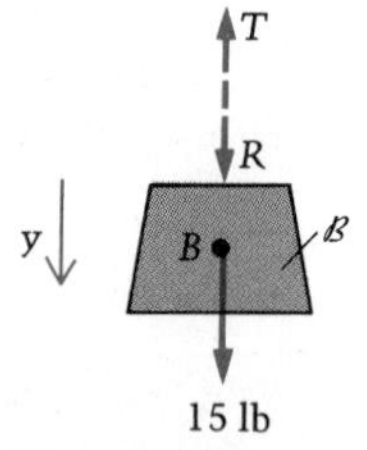

Figure E4.27c

where R is the force exerted by $\mathcal{A}$ onto $\mathcal{B}$. Now, on the disk $\mathcal{A}$,

$$\downarrow{+} \quad \Sigma F_y = m\ddot{y}_A$$

$$5 - R = \frac{5}{32.2}\ddot{y}_A \tag{3}$$

At this point we have three equations in the five unknowns T, α, R, $\ddot{y}_B$, and $\ddot{y}_A$.

One constraint is that the vertical component of $\mathbf{a}_Q$ (see Figure E4.27b) is the same as the acceleration of the points of the straight portion of rope, and these accelerations are each $\ddot{y}_B$:

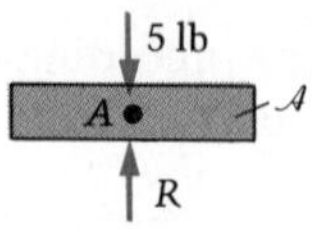

Figure E4.27d

$$\ddot{y}_B = (\mathbf{a}_Q)_y = \frac{3}{12}\alpha \tag{4}$$

Also, the accelerations of $\mathcal{B}$ and $\mathcal{A}$ are equal. Without any rope tension, they would each be $g\downarrow$; with this tension, the acceleration of $\mathcal{B}$ is slowed, guaranteeing continuing contact of the two bodies. Therefore:

$$\ddot{y}_A = \ddot{y}_B \tag{5}$$

Adding Equations (1) and (2) and using (4),

$$15 + R = \left[0.0388(4) + \frac{15}{32.2}\right] \ddot{y}_B$$

or

$$0.621\ddot{y}_B - R = 15 \tag{6}$$

Equations (5) and (3) give:

$$0.155\ddot{y}_B + R = 5 \tag{7}$$

Adding Equations (6) and (7),

$$\ddot{y}_B = \frac{20}{0.776} = 25.8 \text{ ft/sec}^2$$

so that, by (6),

$$R = -15 + 0.621(25.8) = 1.02 \text{ lb}$$

Note that the acceleration of $\mathcal{B}$ is less than "g" (32.2 ft/sec²), as it must be, and that R is less than the weight of $\mathcal{A}$, thereby allowing it to fall, but not freely.

In our third example, we again have two bodies, but this time *both* have pivots and one is in fact in equilibrium. The contact between these bodies involves sliding friction:

EXAMPLE 4.28

Just after the brake arm in Figure E4.28a contacts the top of the cylinder $\mathcal{A}$, the cylinder is turning at 1000 rpm ↻. The coefficient of kinetic friction between $\mathcal{A}$ and $\mathcal{B}$ is $\mu = 0.3$. Find (a) how long it takes for $\mathcal{A}$ to come to rest under the constant force $P = 40$ N; and (b) the pin reactions exerted onto $\mathcal{A}$ at the pin O.

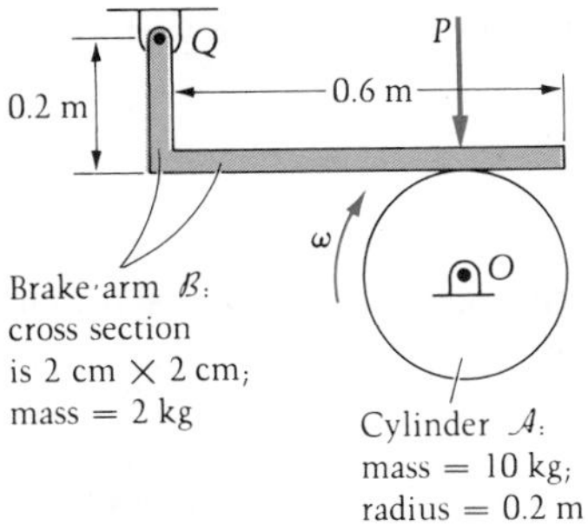

Figure E4.28a

Solution

Since body $\mathcal{B}$ is in equilibrium, we may find the normal force between it and the cylinder by statics. The bar's weight is proportioned between its horizontal and vertical parts as shown in Figure E4.28b. Note that equilibrium requires that $f = Q_x$, where $f = f_{max} = \mu N$ since slipping is taking place.

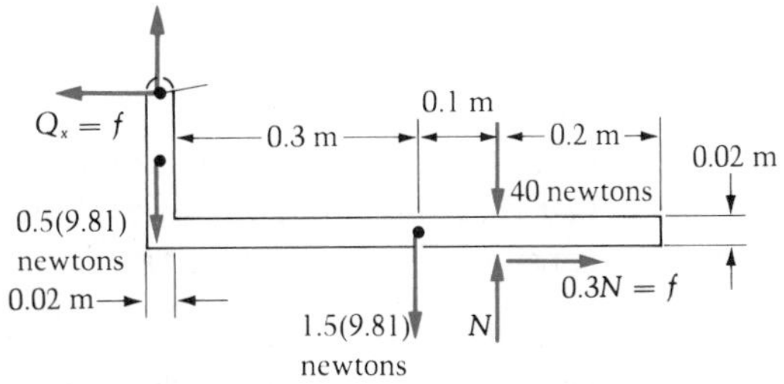

Figure E4.28b

Summing moments about Q, we have

$$0.401N + 0.2(0.3N) - 40(0.401) - 1.5(9.81)(0.301) = 0$$

from which we get

$$N = \frac{20.5}{0.461} = 44.6 \text{ newtons}$$

and

$$\mu N = 0.3(44.6) = 13.4 \text{ newtons}$$

Question 4.12 Would the solution for the normal force N be any different if ω were counterclockwise?

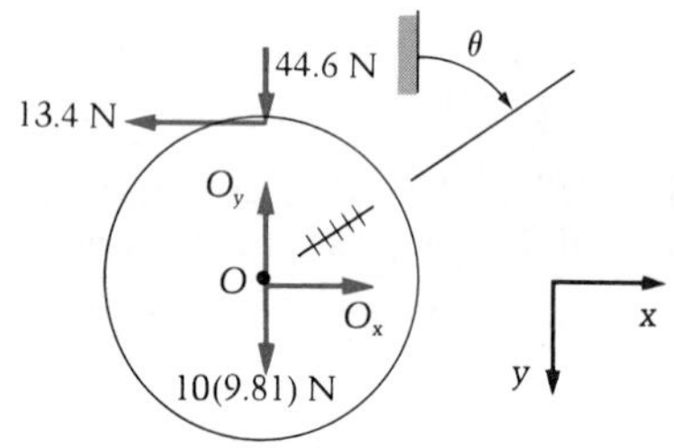

Figure E4.28c

The motion of body $\mathcal{A}$ is one of pure rotation. Its free-body diagram is shown in Figure E4.28c. We use $\ddot{\theta}$ for α since we are ultimately interested in equating $\dot{\theta}$ to zero.

$$\circlearrowleft^{+} \quad \Sigma M_O = I_O\ddot{\theta}$$

$$-13.4(0.2) = \left[\frac{1}{2}\,10(0.2)^2\right]\ddot{\theta}$$

$$\ddot{\theta} = \frac{-2.68}{0.20} = -13.4 \text{ rad/s}^2$$

Integrating, we get

$$\dot{\theta} = -13.4t + C_1$$

$$= -13.4t + 1000\left(\frac{2\pi}{60}\right)$$

where the initial condition on ω allows us to calculate the integration constant C_1.

Body $\mathcal{A}$ stops when $\dot{\theta} = 0$ at a time t_s that we are now in a position to calculate:

$$0 = -13.4t_s + 105$$

$$t_s = 7.84 \text{ s}$$

Note that the mass center $O = C$ of $\mathcal{A}$ is fixed in the inertial frame, so that the pin reactions follow from the mass center equations:

$$\xrightarrow{+} \quad \Sigma F_x = -13.4 + O_x = m\ddot{x}_C = 0 \Rightarrow O_x = 13.4 \text{ newtons}$$

$$+\!\downarrow \quad \Sigma F_y = +44.6 + 98.1 - O_y = m\ddot{y}_C = 0 \Rightarrow O_y = 143 \text{ newtons}$$

In the fourth example, we are concerned with *internal* forces and with the fact that the equations of motion can be applied to a *part* of a body, considered as a body in itself:

Answer 4.12 Yes, for then the friction force would be in the opposite direction and ΣM_Q would give a larger N. The normal force N would then have to balance the moments about Q of all three of P, f, and the weight.

Example 4.29

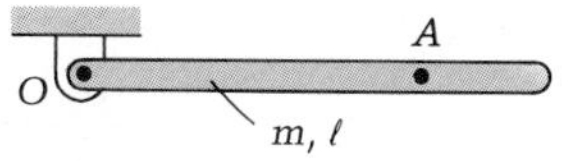

Figure E4.29a

The slender bar, pinned smoothly at O, is released from rest in the position shown in Figure E4.29a. After a rotation of 45°↻, its angular speed is $\omega = \sqrt{(3g)/(\sqrt{2}\ell)}$. Find at that instant the axial force, shear force, and bending moment in the bar at point A, which is one-fourth the length of the bar from its free end.

Solution

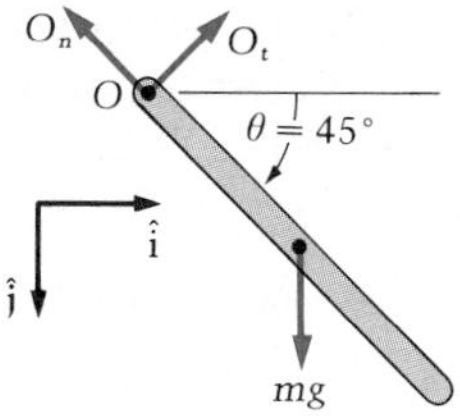

Figure E4.29b

First, we find the angular acceleration, making use of the FBD in Figure E4.29b:

$$\Sigma M_O = I_O \alpha$$

$$\frac{mg}{\sqrt{2}}\frac{\ell}{2} = \left(\frac{1}{3}m\ell^2\right)\alpha$$

$$\alpha = \frac{3g}{2\sqrt{2}\ell}$$

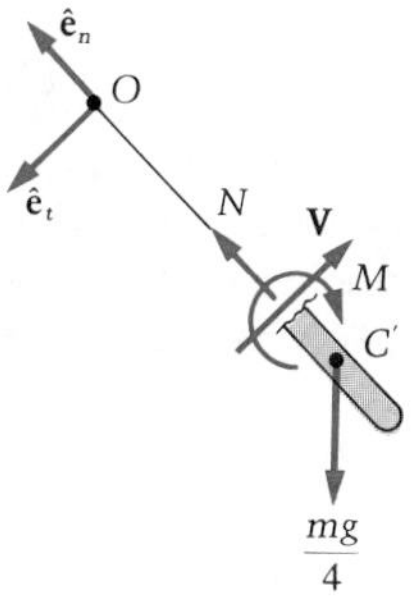

Figure E4.29c

Next we expose the desired forces and moment by drawing a free-body diagram (Figure E4.29c) of the lower fourth of the bar, and writing the equations of motion for just that body (C' is its mass center):

$$\Sigma F_t = m a_{C't}$$

$$\frac{mg}{4}\frac{1}{\sqrt{2}} - V = \frac{m}{4}\left(\frac{7}{8}\ell\alpha\right) = \frac{7}{32}m\ell\left(\frac{3g}{2\sqrt{2}\ell}\right)$$

so that

$$V = \frac{-5}{64\sqrt{2}}mg$$

where we note that ω and α are the same for this "sub-body" as they were for the whole bar, and that O is a pivot of the sub-body extended. The other mass center equation is:

$$\Sigma F_n = m a_{C'n}$$

$$N - \frac{mg}{4}\frac{1}{\sqrt{2}} = \frac{m}{4}\left(\frac{7}{8}\ell\omega^2\right) = \frac{7\,m\ell}{32}\frac{3g}{\sqrt{2}\ell} = \frac{21mg}{32\sqrt{2}}$$

$$N = \frac{29mg}{32\sqrt{2}}$$

Finally, the "moment equation of motion," written for the sub-body this time, is

$$\Sigma M_{C'} = I_{C'}\alpha$$

$$M + V\left(\frac{\ell}{8}\right) = \left(\frac{\frac{m}{4}\left(\frac{\ell}{4}\right)^2}{12}\right)\left(\frac{3g}{2\sqrt{2}\ell}\right)$$

$$M = \frac{3}{256\sqrt{2}}mg\ell$$

The final three results are summarized pictorially on the cut section in Figure E4.29d.

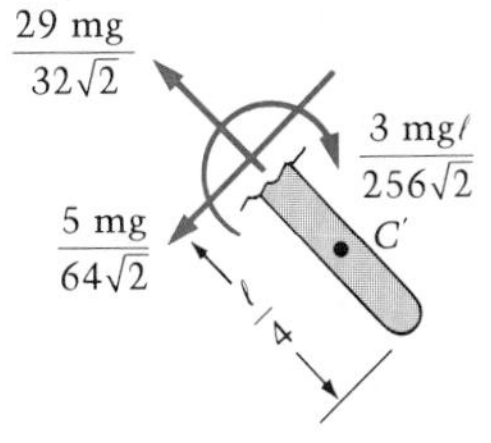

Figure E4.29d

We now examine a pivot problem in which the mass center of the body is offset from the fixed axis of rotation:

EXAMPLE 4.30

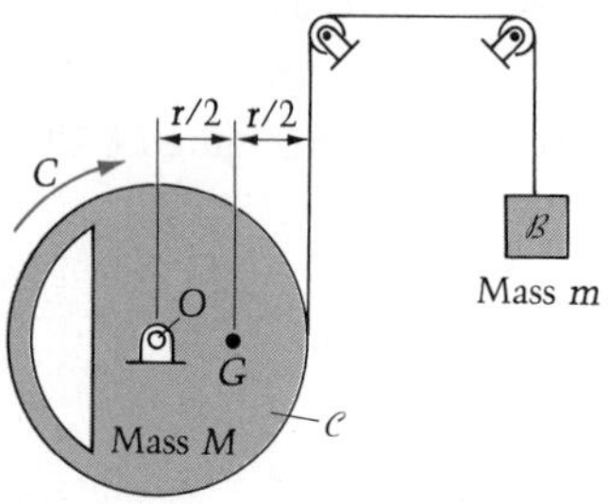

Figure E4.30a

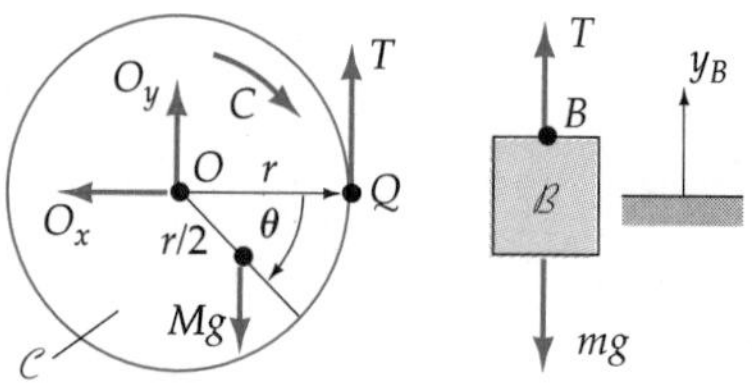

Figure E4.30b

How must the applied couple C in Figure E4.30a vary with time in order to turn the unbalanced (but round) wheel $\mathcal{C}$ at constant angular velocity ω_O↻? What is the initial angular acceleration if the couple is absent and the wheel is released in the position shown in the figure? The moment of inertia of the mass of $\mathcal{C}$ with respect to its axis of rotation is I_O, and the mass center of $\mathcal{C}$ is located at G.

Solution

Free-body diagrams of $\mathcal{C}$ and $\mathcal{B}$ are shown in Figure E4.30b. Since $\Sigma M_O = I_O\alpha$ for body $\mathcal{C}$, we have:

$$\circlearrowright\!+ \qquad -Tr + C + Mg\frac{r}{2}\cos\theta = I_O\ddot{\theta} \tag{1}$$

And for the translating block $\mathcal{B}$ we may write*

$$\uparrow + \qquad T - mg = m\ddot{y}_B \tag{2}$$

The point Q of $\mathcal{C}$ located where the rope leaves the rim has the same velocity ($r\dot{\theta}\downarrow$) and tangential acceleration component ($r\ddot{\theta}\downarrow$) as does the rope itself at that point. Since the rope is assumed inextensible, this acceleration has the same magnitude as $\ddot{y}_B$. Therefore our kinematics gives us the following additional equation for the translating block:

$$\ddot{y}_B = r\ddot{\theta} \tag{3}$$

Substituting (3) into (2) gives

$$T = mg + mr\ddot{\theta}$$

And substituting T into the moment equation (1) for $\mathcal{C}$ then yields

$$C = mgr - Mg\frac{r}{2}\cos\theta + (I_O + mr^2)\ddot{\theta}$$

Since $\dot{\theta} = \omega_0 =$ constant, we have $\theta = \omega_0 t$ and $\ddot{\theta} = 0$, so that

$$C = mgr - \frac{Mgr}{2}\cos\omega_0 t$$

and the required couple varies harmonically.

If there is *no* couple C and the system is released from rest in the position shown, then the equations are still valid and, with couple $C = 0$,

$$(I_O + mr^2)\ddot{\theta} = Mg\frac{r}{2}\cos\theta - mgr$$

The initial angular acceleration of $\mathcal{C}$ (with $\theta = 0$) is thus seen to be

$$\ddot{\theta}_0 = \frac{(Mgr/2) - mgr}{I_O + mr^2}$$

which is positive (↻) if $M > 2m$.

* Note that since $\mathcal{B}$ translates, the acceleration of its mass center is $\ddot{y}_B\downarrow$.

Question 4.13 What happens if $M < 2m$? If $M = 2m$?

In the preceding example, it is interesting to write the equations of motion in the following form for the wheel:

$$\overset{+}{\leftarrow} \quad \Sigma F_x = O_x = M\ddot{x}_G \qquad (4)$$

$$+\uparrow \quad \Sigma F_y = O_y + T - Mg = M\ddot{y}_G \qquad (5)$$

$$\circlearrowleft + \quad \Sigma M_G = O_y \frac{r}{2}\cos\theta - O_x \frac{r}{2}\sin\theta - T\left(r - \frac{r}{2}\cos\theta\right) = I_G\ddot{\theta} \qquad (6)$$

Equations (4) and (5) are useful if the pin reactions are desired,* but Equation (6) is nowhere near as handy to use as $\Sigma M_O = I_O\,\alpha$, which we have used earlier in the example since the body has a pin. The student may wish to eliminate O_x, O_y and T from (6) by using (4), (5), and the previous equation for $\mathcal{B}$ ($T = mg + mr\ddot{\theta}$) and to show that the same result is obtained (after a good deal more work than in the example) for $\ddot{\theta}$. (Kinematics must also be used to relate $\ddot{x}_G$ and $\ddot{y}_G$ to θ, $\dot{\theta}$ and $\ddot{\theta}$!)

In our last "pivot-equation" example, we take up a much longer problem, involving two bodies neither of which is translating. Only one of the bodies has a pivot, and so we shall have to use both the pivot equation (4.29) *and* the mass center form of the moment equation (4.16) before finally getting the problem solved:

Answer 4.13 If $M = 2m$, then (when the couple is not present) $T = mg$ and $\ddot{\theta} = 0$ are solutions to the problem and there is *no* motion. If $M < 2m$, the block moves *downward* and $\ddot{\theta}$ is negative (↺).

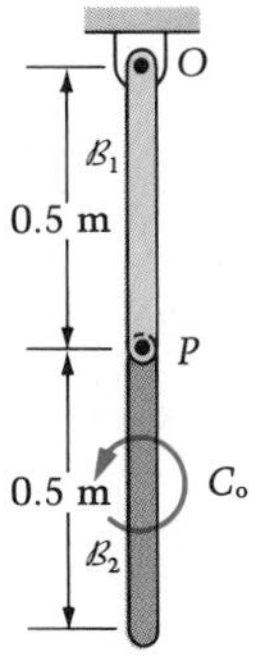

Figure E4.31a

Example 4.31

The two uniform, slender rods $\mathcal{B}_1$ and $\mathcal{B}_2$ in Figure E4.31a, each of mass 2 kg, are pinned together at P, and then $\mathcal{B}_1$ is suspended from a pin at O. (This arrangement is called a *double pendulum.*) The counterclockwise couple C_O, having moment 150 N · m, is applied to $\mathcal{B}_2$ beginning at $t = 0$. Find the angular accelerations of $\mathcal{B}_1$ and of $\mathcal{B}_2$ upon application of the couple, and the force exerted on $\mathcal{B}_2$ at P.

Solution

The equations of motion for $\mathcal{B}_1$ and $\mathcal{B}_2$, using the respective free-body diagrams in Figures E4.31b and 4.31c, are:

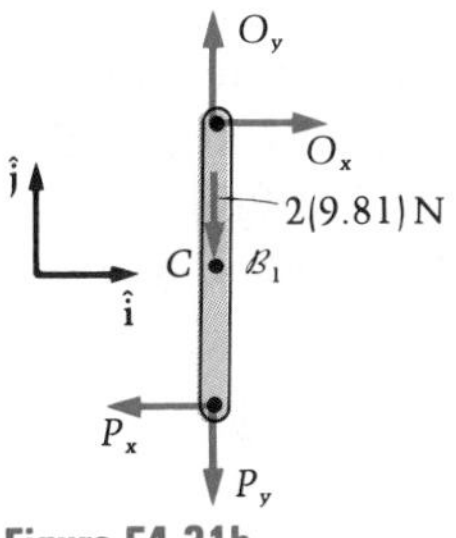

Figure E4.31b

$$\Sigma F_x = O_x - P_x = 2\ddot{x}_C \qquad (1)$$

$$\Sigma F_y = O_y - P_y - 19.6 = 2\ddot{y}_C \qquad (2)$$

$$\Sigma M_O = -0.5P_x = \frac{2(0.5)^2}{3}\,\alpha_1 \qquad (3)$$

* For instance, the pins must be designed strong enough to take the forces caused by the accelerations.

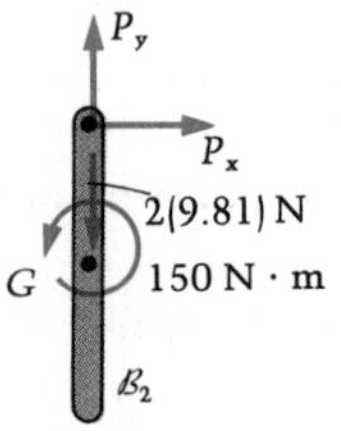

Figure E4.31c

(Note in Equation (3) that O is a pivot of $\mathcal{B}_1$.)

$$\Sigma F_x = P_x = 2\ddot{x}_G \tag{4}$$

$$\Sigma F_y = P_y - 19.6 = 2\ddot{y}_G \tag{5}$$

$$\Sigma M_G = -P_x(0.25) + 150 = \frac{2(0.5)^2}{12}\alpha_2 \tag{6}$$

Thus far we have six equations in the *ten* unknowns O_x, O_y, P_x, P_y, $\ddot{x}_C$, $\ddot{y}_C$, α_1, $\ddot{x}_G$, $\ddot{y}_G$, and α_2. Kinematics gives the four additional equations we will need:

$$\underbrace{\ddot{x}_C\hat{\mathbf{i}} + \ddot{y}_C\hat{\mathbf{j}}}_{\mathbf{a}_C} = \overset{0}{\cancel{\mathbf{a}_O}} + \alpha_1\hat{\mathbf{k}} \times \overset{-0.25\hat{\mathbf{j}}}{\cancel{\mathbf{r}_{OC}}} - \overset{0 \text{ at } t=0}{\cancel{\omega_1^2}}\mathbf{r}_{OC}$$

$$\hat{\mathbf{i}}\text{-coefficients} \Rightarrow \ddot{x}_C = 0.25\alpha_1 \tag{7}$$

$$\hat{\mathbf{j}}\text{-coefficients} \Rightarrow \ddot{y}_C = 0 \tag{8}$$

Also,

$$\overbrace{\ddot{x}_G\hat{\mathbf{i}} + \ddot{y}_G\hat{\mathbf{j}}}^{\mathbf{a}_G} = \mathbf{a}_P + \alpha_2\hat{\mathbf{k}} \times \overset{-0.25\hat{\mathbf{j}}}{\cancel{\mathbf{r}_{PG}}} - \overset{0 \text{ at } t=0}{\cancel{\omega_2^2}}\mathbf{r}_{PG}$$

But

$$\mathbf{a}_P = \overset{0}{\cancel{\mathbf{a}_O}} + \alpha_1\hat{\mathbf{k}} \times (-0.5\hat{\mathbf{j}}) - \overset{0}{\cancel{\omega_1^2}}(-0.5\hat{\mathbf{j}})$$
$$= 0.5\alpha\hat{\mathbf{i}}$$

Thus

$$\ddot{x}_G = 0.5\alpha_1 + 0.25\alpha_2 \tag{9}$$

$$\ddot{y}_G = 0 \tag{10}$$

Solving these equations gives:

$\boldsymbol{\alpha}_1 = -771\hat{\mathbf{k}}$ rad/s^2
$\boldsymbol{\alpha}_2 = 2060\hat{\mathbf{k}}$ rad/s^2 and
Force on $\mathcal{B}_2$ at $P = P_x\hat{\mathbf{i}} + P_y\hat{\mathbf{j}} = 257\hat{\mathbf{i}} + 19.6\hat{\mathbf{j}}$ N

As a point of interest, the values of the other four variables (besides $\ddot{y}_C = \ddot{y}_G = 0$) are: $O_x = -129$ N, $O_y = 39.2$ N, $\ddot{x}_C = -193$ m/s^2, and $\ddot{x}_G = 129$ m/s^2.

Finally, we remark that any of the alternative moment equations discussed in the preceding two sections may be used. However, the student is cautioned to realize that, just as in statics, once the vector "force equation" and "moment equation" have been written, no new (independent) information will arise from summing moments at a different point.

PROBLEMS ▶ Section 4.6

4.151 The rod is pinned to the light roller, which moves in the smooth slot, and with the system at rest as shown in Figure P4.151, the 6-N force is applied. Find the angular acceleration of the rod at the given instant, by using Equation (4.19) together with $\Sigma F_x = m\ddot{x}_C$. Then check your solution using only $\Sigma M_C = I_C\alpha$.

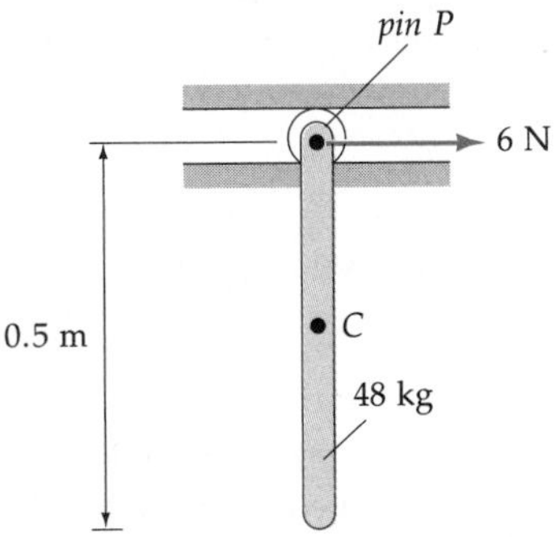

Figure P4.151

4.152 If in the preceding problem we replace the 6-N force by a force F which produces a constant acceleration of pin P of 0.5 m/s², again find the initial value of α, this time using only Equation (4.24). Explain why the answers are the same. (*Hint:* Solve for F at the given instant.)

4.153 Solve the problem of Examples 4.19, 23, 25 using Equation (4.19), with the point P (about which moments are taken) being the geometric center Q of the hoop.

4.154 Solve the problem of Examples 4.19, 23, 25 using Equation (4.24), with the point P again (see the preceding problem) being the geometric center Q of the hoop.

* **4.155** The cylinder shown in Figure P4.155 is made of two halves of different densities. The left half is steel, with mass density $\rho_1 = 15.2$ slug/ft³; the right half is wood with $\rho_2 = 1.31$ slug/ft³. Recalling that the mass center of each half is located $4r/3\pi$ from the geometric center Q, find the acceleration of Q when the cylinder is released from rest. Assume enough friction to prevent slipping. *Hint*: Use Equation (4.17) with Q as point P.

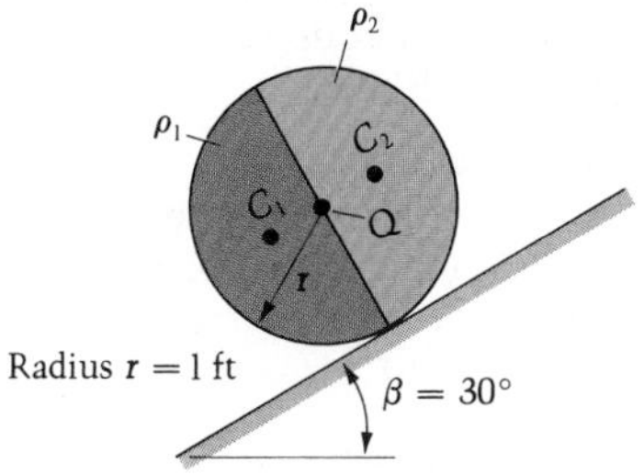

Figure P4.155

4.156 (a) Use Equation (4.24) to categorize the restrictions on point P for which we may correctly write $\Sigma M_P = I_P\alpha$. Show that there are only three cases, and that the mass center form is one of them, while the "fixed-axis-of-rotation form" is but a special case of one of the other two. (b) Note that the instantaneous center of zero velocity Ⓘ is *not* a point P for which, in general, $\Sigma M_P = I_P\alpha$. (c) Finally, determine in which of the problems in Figure P4.156 (a–e) it is true that $\Sigma M_Q = I_Q\alpha$.

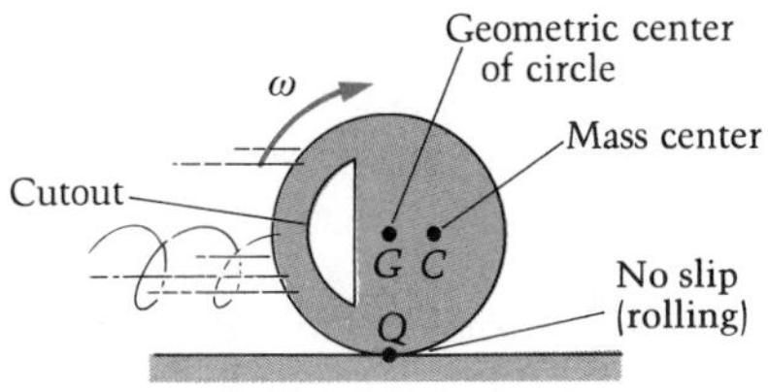

Figure P4.156(a)

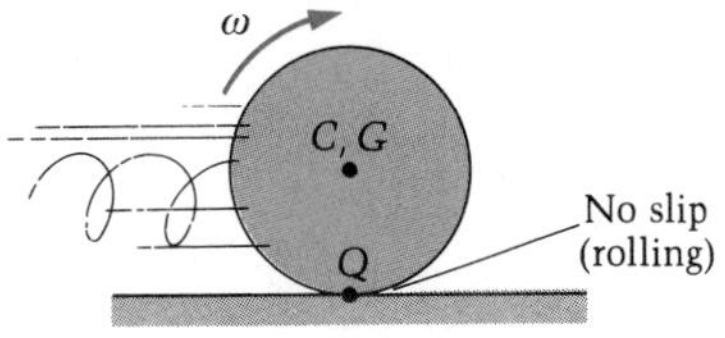

Figure P4.156(b)

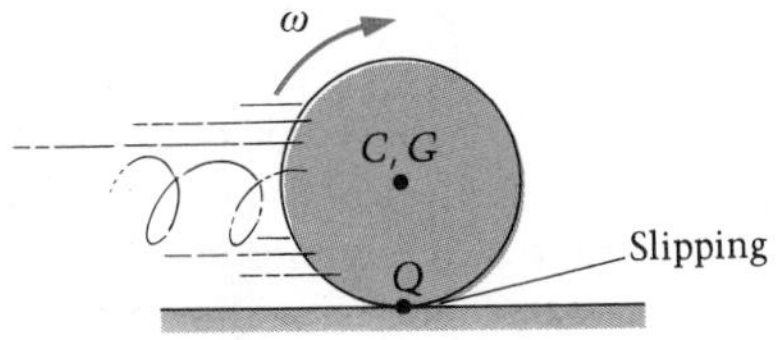

Figure P4.156(c)

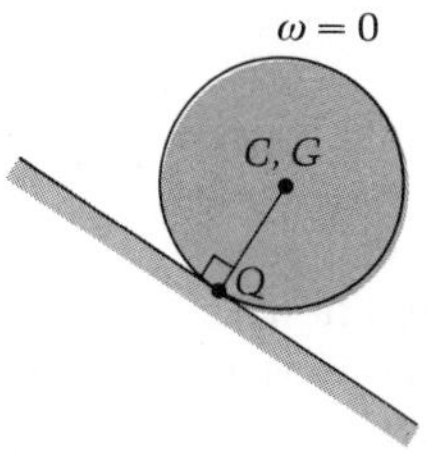

Figure P4.156(d)

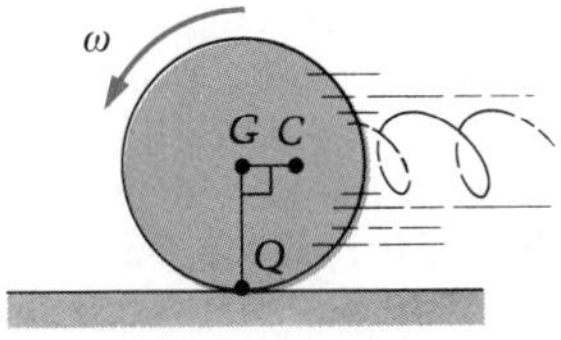

Figure P4.156(e)

4.157 Let $\mathcal{B}$ be a rigid body in plane motion, in constant contact with a surface $\mathcal{S}$. Let Q be the point of $\mathcal{B}$ in contact with $\mathcal{S}$ (Q can be different points of $\mathcal{B}$ at different times). Use the result of Problem 4.156 to prove that if the following four conditions hold, then $\Sigma M_Q = I_Q\alpha$ at all times: (1) $\mathcal{S}$ is fixed in an inertial frame; (2) $\mathcal{B}$ is rolling on $\mathcal{S}$; (3) $\mathcal{B}$ is round; and (4) the mass center of $\mathcal{B}$ is at its geometric center.

* **4.158** In the preceding problem, suppose that conditions (1) and (2) hold, and that at a certain instant, ω of $\mathcal{B}$ is zero. Using the result of Problem 4.156, show that, *at that instant*, $\Sigma M_Q = I_Q\alpha$ regardless of whether conditions (3) or (4) hold.

4.159 The thin-walled hollow sphere of Figure P4.159 (moment of inertia about any diameter $= \frac{2}{3}mr^2$) has average radius $r = 0.5$ m and mass 50 kg. It is pinned at the bottom of the cart. The force F applied to the cart at rest produces a constant acceleration of all points of the cart of $a\hat{\mathbf{i}}$.

a. Find the maximum value of a if the sphere is to translate and the breaking strength of each of the cords is 100 N.

b. Suppose a is twice the result of (a), so that one cord breaks at $t = 0$. Find the angular acceleration of the sphere at the instant it has turned through 90°, using Equation (4.24). Does the answer to (b) depend on the value of a?

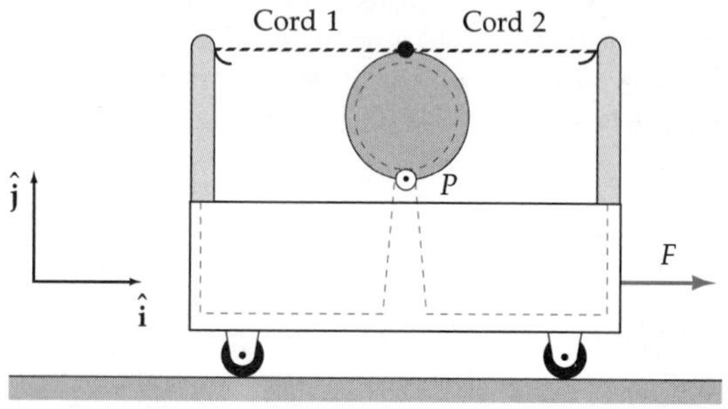

Figure P4.159

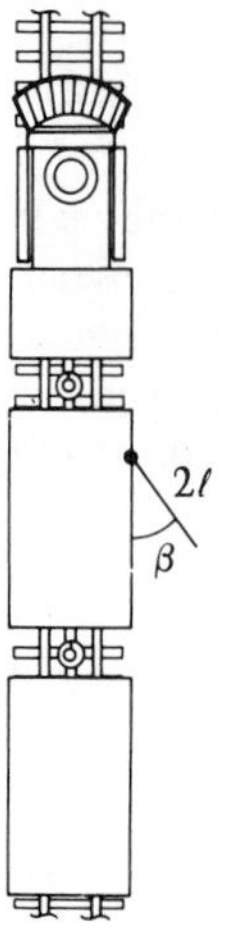

Figure P4.160

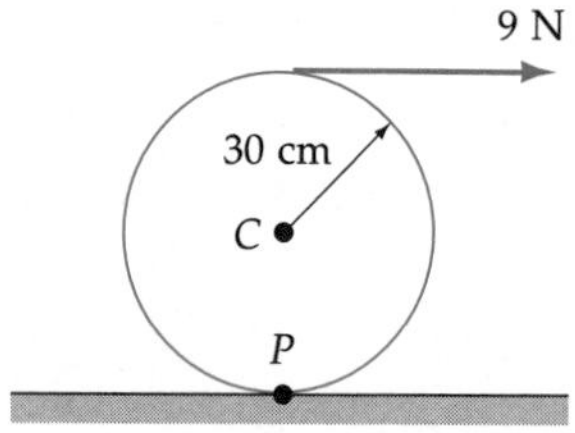

Figure P4.161

4.160 The rectangular door of a railroad car has mass m (Figure P4.160); it is of uniform width 2ℓ and has its hinges on the side of the doorway closest to the engine. Initially the door makes an angle $\mathcal{B}$ with the train, which begins to move forward from rest at constant acceleration a_o. Find the initial resultant horizontal reaction component that the hinges exert on the door.

4.161 The cylinder in Figure P4.161 has mass 10 kg. The 9-N force is applied to a string wrapped around a thin slot near the surface of the cylinder. Find the angular acceleration of the cylinder, assuming enough friction to prevent slip, using (a) Equation (4.19); (b) Equation (4.24). Observe how both right-hand-sides add up to $\frac{3}{2}mr^2\alpha$ even though the individual terms are different.

4.162 In Problem 4.152 at a later time, let θ be the angle between the vertical and the rod (see Figure P4.162). Using Equation (4.24), find the angular acceleration of the rod as a function of θ.

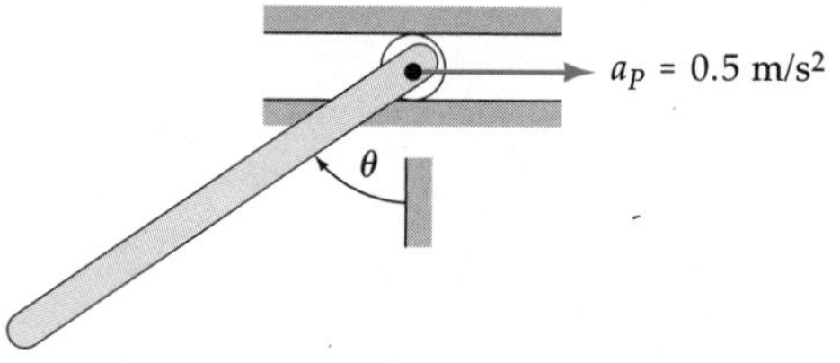

Figure P4.162

4.163 The crank arm OP is turned by the couple M_o at constant angular velocity 1.0 rad/s ⤼. In the position shown in Figure P4.163, determine the reaction of the smooth plane onto the 20-kg slender bar PQ. *Hints*: First solve for $\mathbf{a}_P$ and for $\boldsymbol{\alpha}$ of PQ using kinematics, then use Equation (4.24).

4.164 A body $\mathcal{B}$ weighing 805 lb with radius of gyration 0.8 ft about its z_C axis (see Figure P4.164) is pinned at its mass center. A clockwise couple of magnitude e^t lb-ft is applied to $\mathcal{B}$ starting at $t = 0$. Find the angle through which $\mathcal{B}$ has turned during the interval $0 \le t \le 3$ sec.

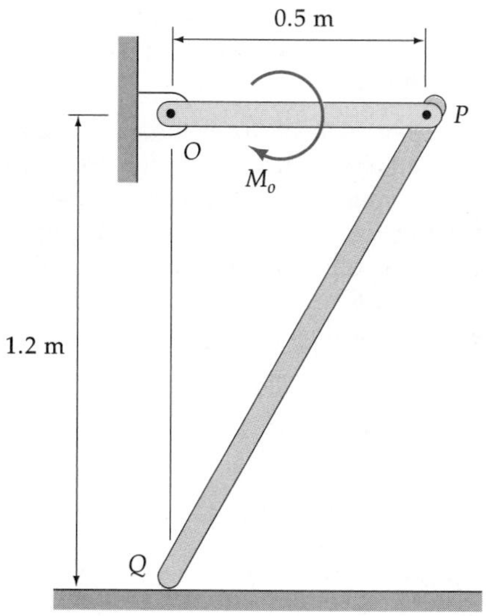

Figure P4.163

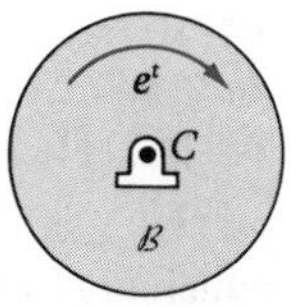

Figure P4.164

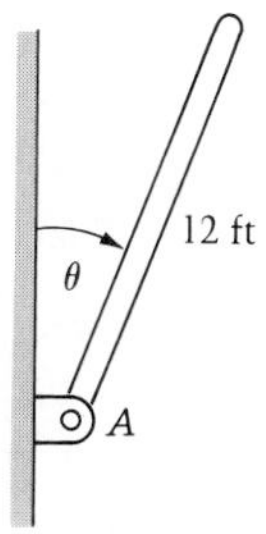

Figure P4.165

4.165 The slender homogeneous rod in Figure P4.165 is 12 ft long and weighs 5 lb; it is connected by a rusty hinge to a support at A. Because of friction in the hinge, the hinge exerts a couple of 9 lb-ft on the rod when it rotates. If the rod is released from rest with $\theta = 30°$, find: (a) the angular acceleration of the rod when $\theta = 30°$, $60°$ and $90°$; (b) the angle θ at which the angular acceleration of the rod is zero.

4.166 A wagon wheel spinning counterclockwise is placed in a corner and contacts the wall and floor. (See Figure P4.166.)

- **a.** Show with a free-body diagram that the wheel cannot climb the wall.
- **b.** Show with a free-body diagram that the wheel cannot move to the right along the floor either.
- **c.** Therefore the wheel stays in the corner. Treat it as a ring of mass m and radius R, with friction coefficient μ at both surfaces of contact. Determine how long it takes for the wheel to stop, and find how many radians it has turned through since first contacting the surfaces.

4.167 The cylinder in Figure P4.167 has a mass of 30 kg and rotates about an axis normal to the clevis at O. At the instant shown, $\omega = 5 \circlearrowright$ rad/s and $\dot{\omega} = 10 \circlearrowleft$ rad/s^2. Find the force P that acts on the cylinder, and determine the reactions exerted by the pin onto the clevis at O, all at the given instant.

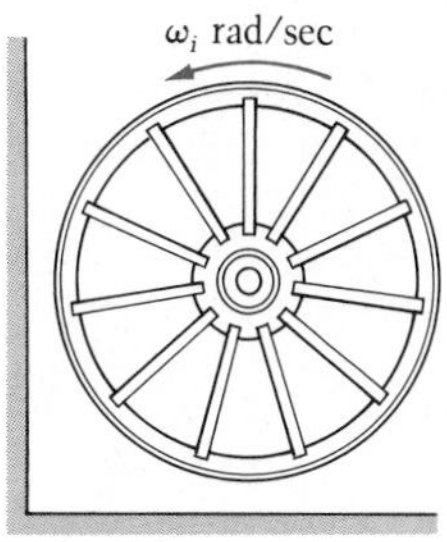

Figure P4.166

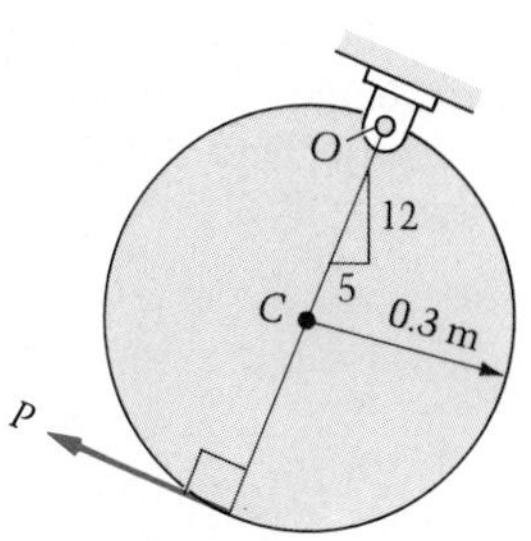

Figure P4.167

4.168 Figure P4.168 shows a scene from Edgar Allen Poe's "The Pit and the Pendulum." Find the reaction of the pin onto the bar if the pendulum is instantaneously at rest in a horizontal position.

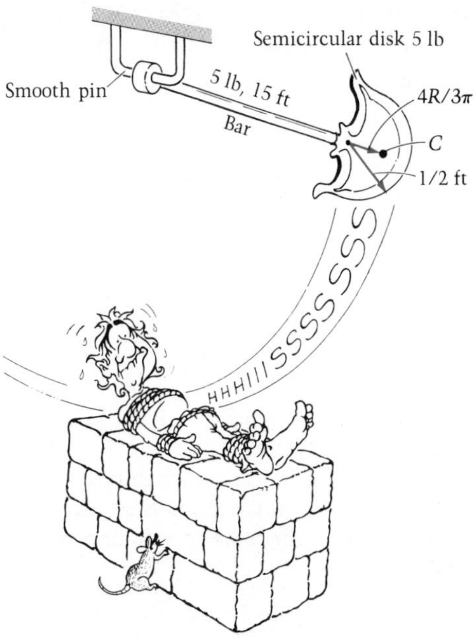

Figure P4.168

4.169 The uniform slender bar of mass m is released from rest in the position shown in Figure P4.169. Find the angular acceleration when the bar has turned through 45°

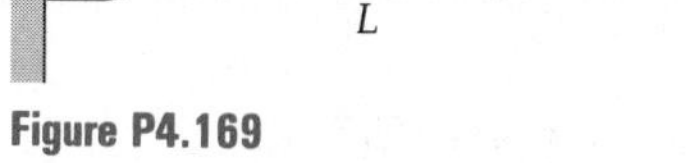

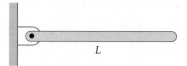

Figure P4.169

* **4.170** The uniform slender bar in Figure P4.170, of mass m and length ℓ, is released from rest at $\theta =$ zero plus a tiny increment. Find the magnitude of the bearing reaction when $\theta = \pi/2$.

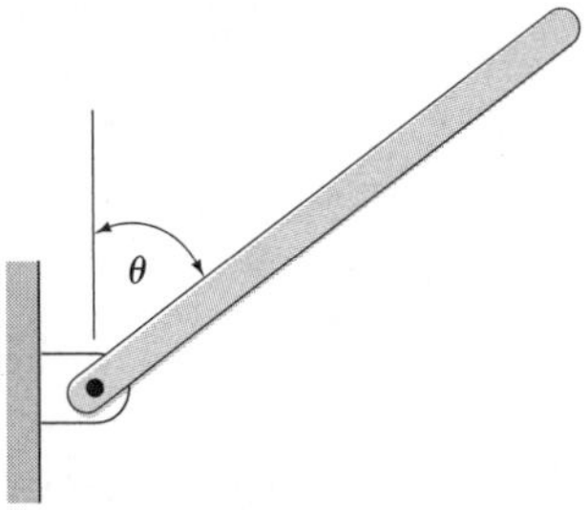

Figure P4.170

4.171 Body $\mathcal{B}$ is a slender bar bent into the shape of a quarter-circle (Figure P4.171). Find the tensions in strings OA and OB when the system is released from rest.

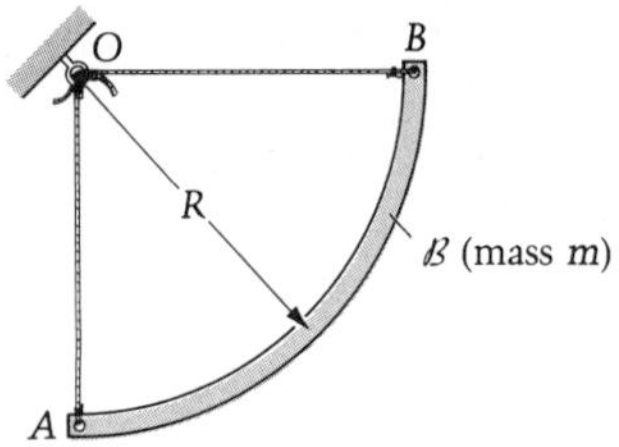

Figure P4.171

4.172 Two weights W_1 and W_2 in Figure P4.172 are connected by an inextensible cord that passes over a pulley. The pulley weighs W and its mass is concentrated at the rim of radius R. Show that if the system is released and the cord does not slip on the pulley, the acceleration magnitude of W_1 and W_2 is

$$\left| \frac{W_2 - W_1}{W_1 + W_2 + W} g \right|$$

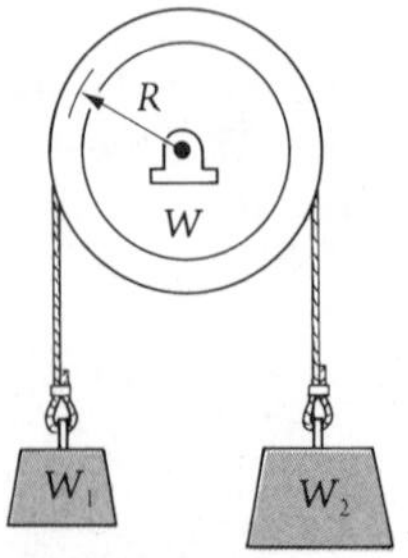

Figure P4.172

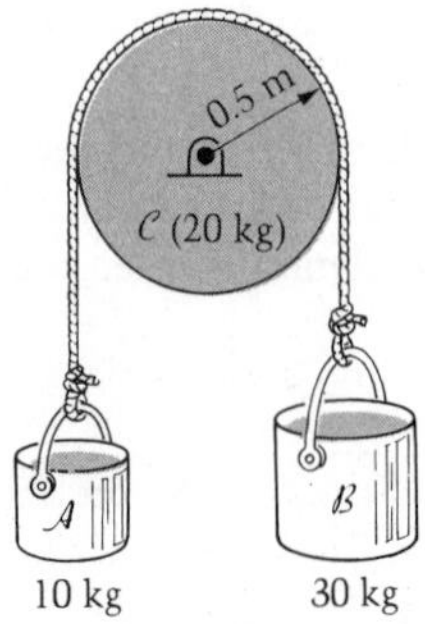

Figure P4.173

4.173 Find the angular acceleration of cylinder $\mathcal{C}$ in Figure P4.173. The rope passes over it without slipping and ties to $\mathcal{A}$ and $\mathcal{B}$ as shown.

4.174 Body $\mathcal{C}$ is a pulley made of the cylinders $\mathcal{D}$ and $\mathcal{E}$, which are butted together and rigidly attached. (See Figure P4.174.) The combined body $\mathcal{C}$ is smoothly pinned to the ground through its axis of symmetry (which passes through its mass center). Ropes wrapped around $\mathcal{E}$ and $\mathcal{D}$ are tied to bodies $\mathcal{A}$ and $\mathcal{B}$, respectively. If the system is released from rest, what will be the angular acceleration of $\mathcal{C}$?

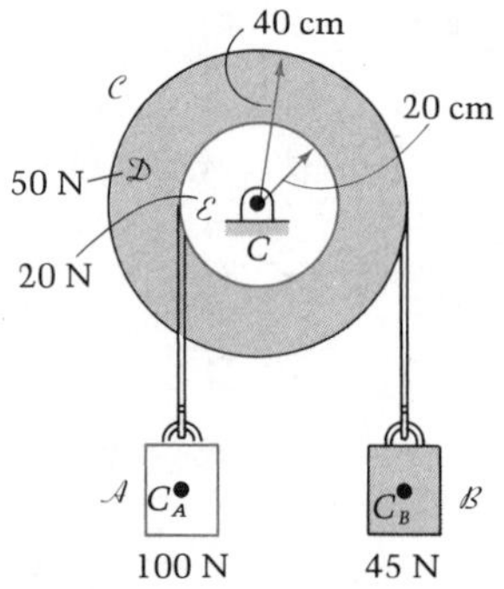

Figure P4.174

4.175 The 32.2-lb particle P rests on the 128.8-lb plank as shown in Figure P4.175. If the cord at B suddenly breaks, find the initial acceleration of the particle, and the force exerted on it by the bar.

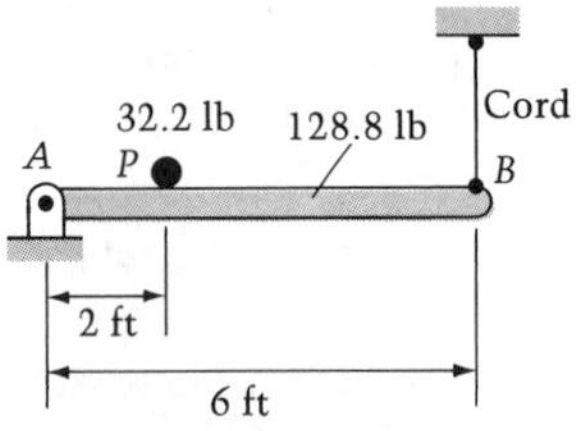

Figure P4.175

4.176 In Example 4.29, note that pivot point O is also a point of the "subbody" used in that example. Thus for that body, $\Sigma M_O = I_O\alpha$. Separately compute ΣM_O and $I_O\alpha$ for the subbody and show that their values are the same.

* **4.177** The uniform slender bar of weight W and length L in Figure P4.177 is released from rest at $\theta = 0$ and pivots on its square end about corner O.

   a. If the bar is observed to slip at $\theta = 30°$, find the coefficient of limiting static friction μ_s.

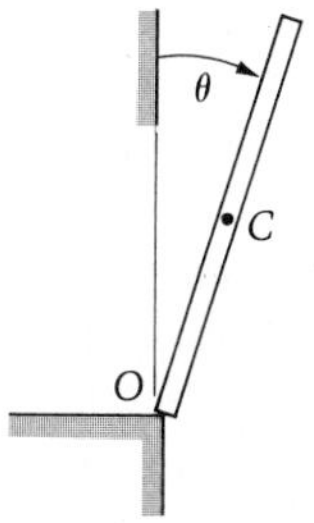

Figure P4.177

b. If the end of the bar is notched so that it cannot slip, find the angle θ at which contact between bar and corner ceases. *Hint*: Write the moment equation of motion about the pivot O, multiply it by $\dot{\theta}$, and integrate, obtaining $\dot{\theta}$ as a function of θ. Use this relation together with the component equation of $\Sigma \mathbf{F} = m\mathbf{a}_C$ in the $\hat{\mathbf{e}}_n$ direction.

4.178 The chain drive in Figure P4.178 may be considered as two disks with equal density and thickness. The larger sprocket has a mass of 2 kg and a radius of 0.2 m. If the couple is applied starting from rest at $t = 0$, find the angular speed of the smaller sprocket at $t = 10$ s. *Hint*: What does a dentist look at?

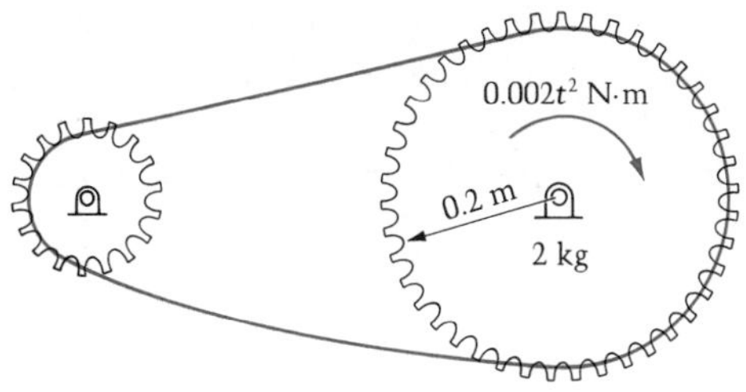

Figure P4.178

* **4.179** Cylinder $\mathcal{C}$ in Figure P4.179 with four cutouts is rotating at 200 rpm initially. A uniform 100-lb cylinder $\mathcal{D}$ is placed in the position shown, and the friction produces a braking moment that will stop $\mathcal{C}$. The friction coefficient is $\mu = \frac{1}{3}$, and before the four holes were drilled the uniform body $\mathcal{C}$ weighed 200 lb. For whichever rotation direction of $\mathcal{C}$ results in a quicker stop, find the stopping time.

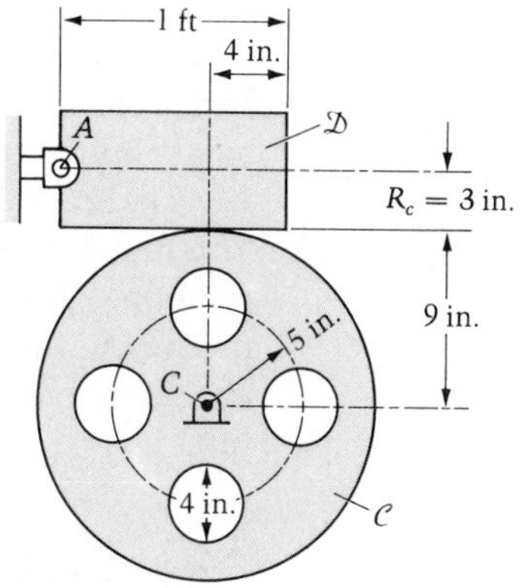

Figure P4.179

* **4.180** The slender, homogeneous rod in Figure P4.180 is supported by a cord at A and a horizontal pin at B. The cord is cut. Determine, at that instant, the location of pin B that will result in the maximum initial angular acceleration of the rod.

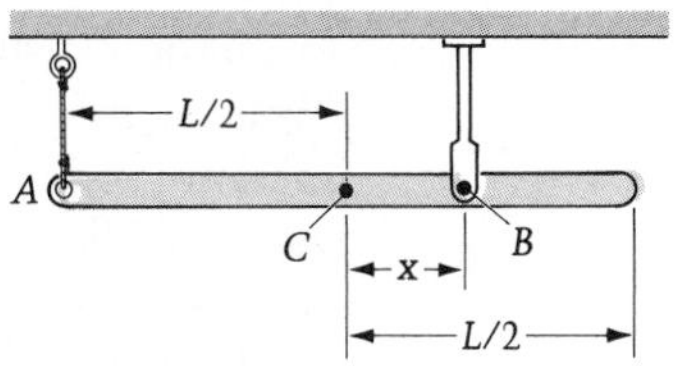

Figure P4.180

4.181 The uniform rod of mass m is released from rest in the horizontal position indicated in Figure P4.181. Consider the force exerted by the smooth pin.

a. How does the magnitude of the force vary with the angle θ through which it has turned?

b. What is the maximum value of this magnitude?

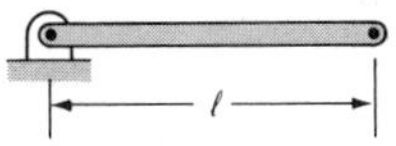

Figure P4.181

4.7 Rotation of Unbalanced Bodies

When a rigid body is mounted in bearings and made to rotate by means of a moment about the axis of the bearings, it is said to be balanced (for rotation about that axis) if the external reactions exerted by the bearings onto the body are only what are required to support the weight of the body. The bearing reactions accompanying *im*balance result in vibration and wear of rotating machinery and are the reason, for example, for balancing automobile tires.

There are two distinct causes for a rotating body to be out of balance. The first is if the mass center is located (a distance "d") off the axis of rotation. Then as the body turns, there will be forces at the bearings producing and equaling $m\mathbf{a}_C$. Clearly, these forces will be constantly changing in direction (relative to the inertial frame) if not also in magnitude.

Question 4.14 What would cause them to change in magnitude?

Moving the mass center onto the axis of rotation by addition or deletion of mass is called static balancing. It carries that name because only then will the body remain in equilibrium when turned to any position and released, this being true regardless of the orientation of the axis.

The second cause of imbalance is nonzero products of inertia I_{xz}^P and/or I_{yz}^P, where z is the axis of rotation and P is a point on that axis. In the same way that an off-axis mass center causes bearing forces which produce $m\mathbf{a}_C$, these products of inertia likewise cause bearing forces to exist; for a system which has been statically balanced, they produce the "bearing moments" (see Equations 4.14 a,b) ΣM_{Cx} and ΣM_{Cy}. And similar to removing the offset "d," we can also add or delete material to force the values of I_{xz}^P and I_{yz}^P to be zero. When this is done, in addition to having ensured that C lies on the axis, the body is then said to be dynamically (and of course also statically) balanced. We shall develop the equations to accomplish this in what follows.

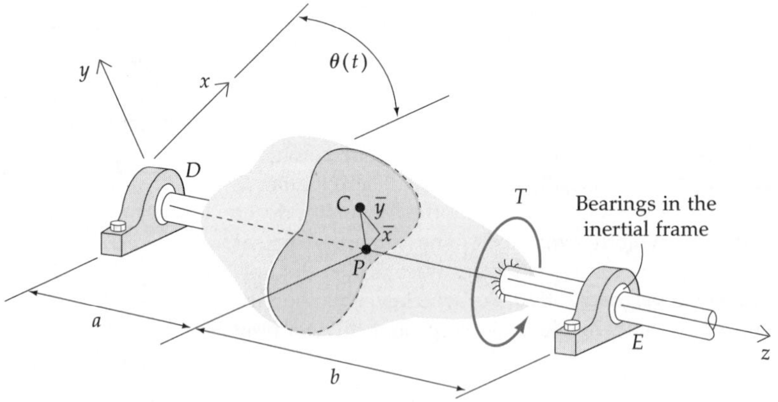

Figure 4.8

Answer 4.14 Changes in rotational speed.

We show body $\mathcal{B}$ in Figure 4.8, set in ball bearings at D and E. T is an externally applied torque about z, the axis of rotation. Let us say that T is the driving torque less any frictional resistance moments from the air or the bearings. Finally, note that the x and y axes are also fixed in $\mathcal{B}$, and $(\bar{x}, \bar{y}, a)$ are the coordinates of C in this system.

For an unbalanced body rotating about a horizontal axis, the bearing reactions required to support the weight of the body (when it is not rotating) may be simply added to the dynamic reactions that would be generated were there no gravity. Therefore, for the sake of simplicity we shall ignore the effects of gravity in our discussion here.

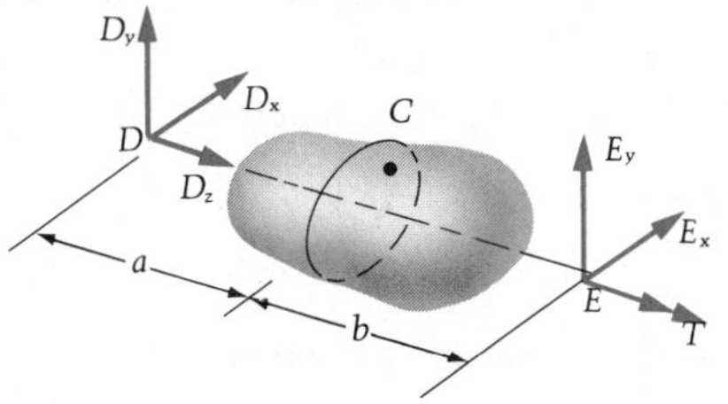

Figure 4.9

Now consider the free-body diagram, Figure 4.9, and let the bearing-reaction components be referred to the body-fixed axes (x, y, z).

Then $\Sigma\mathbf{F} = m\mathbf{a}_C$ yields the component equations

$$D_x + E_x = m(-\omega^2\bar{x} - \alpha\bar{y}) \quad (4.24a)$$

$$D_y + E_y = m(-\omega^2\bar{y} + \alpha\bar{x}) \quad (4.24b)$$

$$D_z = 0 \quad (4.24c)$$

Using Euler's second law in the form

$$\begin{aligned}\Sigma\mathbf{M}_D = &(I_{xz}^D\alpha - I_{yz}^D\omega^2)\hat{\mathbf{i}} \\ &+ (I_{yz}^D\alpha + I_{xz}^D\omega^2)\hat{\mathbf{j}} \\ &+ I_{zz}^D\alpha\hat{\mathbf{k}}\end{aligned} \quad (4.20)$$

we obtain the following component equations:

$$-(a + b)E_y = I_{xz}^D\alpha - I_{yz}^D\omega^2 \quad (4.25a)$$

$$(a + b)E_x = I_{yz}^D\alpha + I_{xz}^D\omega^2 \quad (4.25b)$$

$$T = I_{zz}^D\alpha \quad (4.25c)$$

We observe, then, that if we know ω, α, and the geometric and inertia properties of the body, we can solve Equations (4.24a,b) and (4.25a,b) for the bearing reactions D_x, D_y, E_x, and E_y. We now illustrate such a calculation with an example before taking up the issue of how to balance a body.

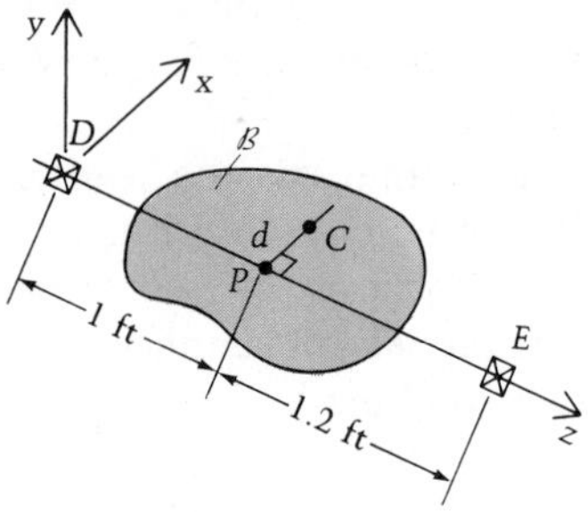

Figure E4.32

EXAMPLE 4.32

The body in Figure E4.32 has mass $m = 2$ slugs, and its mass center is off-axis by the amount $d = 1/64$ in. in the x-z plane so that $\bar{x} = 0.0156$ in. and $\bar{y} = 0$. Its products of inertia are $I_{xz}^C = I_{yz}^C = 0.000380$ slug-ft². If the body is spun up to a constant angular speed of 3000 rpm, what then are the dynamic reactions at bearings D and E?

Solution

By the parallel-axis theorem

$$I_{xz}^D = 0.000380 - 2\left(\frac{.0156}{12}\right)(1) = -0.00222 \text{ slug-ft}^2$$

$$I_{yz}^D = 0.000380 - 2(0)(1) = 0.000380 \text{ slug-ft}^2$$

Also,

$$\alpha = 0 \qquad \text{and} \qquad \omega = 3000(2\pi)/60 = 100\pi \text{ rad/sec}$$

Thus by Equations (4.24) and (4.25) we obtain

$$E_y = \frac{0.000380}{2.2}(100\pi)^2 = 17.0 \text{ lb}$$

$$D_y = 0 - E_y = -17.0 \text{ lb}$$

$$E_x = -\frac{0.00222(100\pi)^2}{2.2} = -99.6 \text{ lb}$$

and

$$D_x = -2\left(\frac{.0156}{12}\right)(100\pi)^2 + 99.6 = -157 \text{ lb}$$

Let us now assume that we have a rotating body (ω and α not both zero) mounted in bearings. For such a body, we can show that the bearing reactions vanish if and only if $\bar{x} = 0 = \bar{y}$ and $I_{xz}^D = 0 = I_{yz}^D$. The "if" proof is simple, for if $I_{xz}^D = 0 = I_{yz}^D$, Equations (4.25 a,b) yield $E_x = 0 = E_y$, and substituting these zero values along with $\bar{x} = 0 = \bar{y}$ into Equations (4.24 a,b), we find $D_x = 0 = D_y$.

For the "only if" part of the proof, if the bearing reactions are all zero, Kramer's rule applied to Equations (4.24 a,b) gives $\bar{x} = 0 = \bar{y}$; and to Equations (4.25 a,b) yields $I_{xz}^D = 0 = I_{yz}^D$. When these two products of inertia are zero at a point D, then z is called a *principal axis of inertia* at D; this concept is discussed in considerable detail in Chapter 7.

In summary, then, we can say that the bearing reactions vanish, and hence the body is balanced, if and only if the axis of rotation is a principal axis of inertia containing the mass center of the body.

Now let us see what can be done about *correcting* imbalance. Suppose values of m, $\bar{x}$, $\bar{y}$, I_{xz}^P and I_{yz}^P of a body are known, where P, lying on the axis of rotation, is the origin of the coordinates. We can, for example, determine the coordinates (x_A, y_A) and (x_B, y_B) and masses (m_A and m_B) of a pair of weights which, when placed in two "correction planes" A (at $z = z_A$) and B (at $z = z_B$), will ensure that the shaft is dynamically balanced. All we have to do is (a) force the mass center C^* of the combined system (m plus m_A and m_B) to lie on the axis of the shaft, and (b) force the products of inertia of the combined system to vanish:

$$x\text{-coordinate of } C^* = 0: \qquad m_A x_A + m_B x_B + m\bar{x} = 0 \quad (4.26a)$$

$$y\text{-coordinate of } C^* = 0: \qquad m_A y_A + m_B y_B + m\bar{y} = 0 \quad (4.26b)$$

$$^*I_{xz}^P = 0: \qquad -m_A x_A z_A - m_B x_B z_B + I_{xz}^P = 0 \quad (4.26c)$$

$$^*I_{yz}^P = 0: \qquad -m_A y_A z_A - m_B y_B z_B + I_{yz}^P = 0 \quad (4.26d)$$

Note that we assume that the "balance weights" are small enough to be treated as particles.

These four equations (4.26) may be solved for the four quantities $m_A x_A$, $m_B x_B$, $m_A y_A$, and $m_B y_B$. Thus there is some freedom to select two of the six quantities m_A, m_B, x_A, x_B, y_A, and y_B, provided there is no other condition linking them; for an example of such a constraint, the weights might have to be placed on a circle of given radius (such as when tires are balanced and weights are clamped to a rim). In this case we would additionally have, for example,

$$x_A^2 + y_A^2 = R_A^2$$

and

$$x_B^2 + y_B^2 = R_B^2$$

and now there are six equations in six unknowns. Let us illustrate the use of these equations in the following example.

EXAMPLE 4.33

In Example 4.32, suppose that we are to balance the body by adding weights in two correction planes midway between C and the two bearings. Furthermore, the weights are each to be placed on a circle of radius ½ ft. Find the masses and coordinates of the weights.

Solution

We had $m = 2$ slugs, $\bar{x} = 1/64$ in., $\bar{y} = 0$, and $I_{xz}^C = I_{yz}^C = 0.000380$ slug-ft². If we choose P to have the same axial position as C, then $z_A = -0.5$ ft, $z_B = 0.6$ ft, and $\bar{z} = 0$. Also, $I_{xz}^P = I_{xz}^C - 0 = 0.000380$ slug-ft² and $I_{yz}^P = I_{yz}^C - 0 = 0.000380$ slug-ft², and:

$$(4.26a) \Rightarrow m_A x_A + m_B x_B = -2\left(\frac{1}{64(12)}\right) = -0.00260$$

$$(4.26c) \Rightarrow m_A x_A(-0.5) - m_B x_B(0.6) = -0.000380$$

Solving these we get

$$m_B x_B = -0.000836$$

$$m_A x_A = -0.00176$$

Similarly,

$$(4.26b) \Rightarrow m_A y_A + m_B y_B = 0$$

$$(4.26d) \Rightarrow -m_A y_A(-0.5) - m_B y_B(0.6) = -0.000380$$

from which we obtain

$$m_B y_B = 0.000345$$

$$m_A y_A = -0.000345$$

Squaring and adding,

$$m_B^2 x_B^2 + m_B^2 y_B^2 = 0.818 \times 10^{-6}$$

$$m_B^2 \underbrace{(x_B^2 + y_B^2)}_{\left(\frac{1}{2}\right)^2} = 0.818 \times 10^{-6}$$

$$m_B = 1.81 \times 10^{-3} \text{ slug}$$

So the weight of B is $W_B = 1.81 \times 10^{-3}(32.2) = 0.0582$ lb, or 0.932 oz. For the coordinates,

$$x_B = \frac{-0.000836}{1.81 \times 10^{-3}} = -0.462 \text{ ft}$$

and

$$y_B = \frac{0.000345}{1.81 \times 10^{-3}} = 0.191 \text{ ft}$$

(These add vectorially to $\sqrt{(0.462)^2 + (0.191)^2} = 0.500$ ft, as a check.)

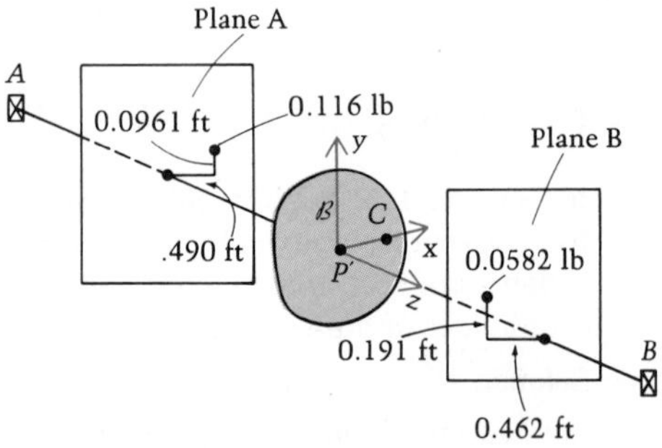

Figure E4.33

Now, for the mass and coordinates of m_A in plane A, we have, by again squaring and adding,

$$m_A^2 \underbrace{(x_A^2 + y_A^2)}_{\left(\frac{1}{2}\right)^2} = 3.22 \times 10^{-6}$$

$$m_A = 3.59 \times 10^{-3} \text{ slug}$$

Thus the weight of A is $W_A = (3.59 \times 10^{-3})\ 32.3 = 0.116$ lb, or 1.85 oz. The coordinates are

$$x_A = \frac{-0.00176}{3.59 \times 10^{-3}} = 0.490 \text{ ft}$$

$$y_A = \frac{-0.000345}{3.59 \times 10^{-3}} = 0.0961 \text{ ft}$$

Again checking,

$$\sqrt{x_A^2 + y_A^2} = 0.499 \text{ ft}$$

The above results are all shown in Figure E.4.33.

In our last example, we shall add mass in the form of two rods to balance the body in Example 4.16.

Example 4.34

For the body of Example 4.16, find the length L of the pair of rods, each of mass $3m$, that will dynamically balance the shaft when attached to it as shown in Figure E4.34.

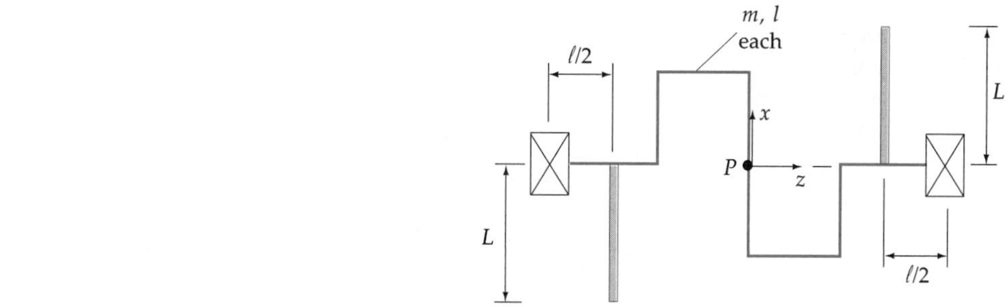

Figure E4.34

Solution

In the previous example, I_{xz}^P was computed to be $2m\ell^2$. Note that the mass center of the original and modified systems is at P, so the system is already statically balanced. Thus since $I_{yz}^P = 0$ (all the mass is still in the xz-plane), all we need for dynamic balance is:

$$(I_{xz}^P)_{\text{TOTAL}} = (I_{xz}^P)_{\text{RODS}} + 2m\ell^2 = 0$$

$$2\left[0 - 3m\left(-\frac{L}{2}\right)\left(-\frac{3}{2}\ell\right)\right] = -2m\ell^2$$

$$\frac{9}{2}mL\ell = 2m\ell^2$$

$$L = \frac{4}{9}\ell$$

PROBLEMS ▶ Section 4.7

4.182 Explain why the uniform plate in Figure P4.182 is dynamically balanced.

4.183 A light rod of length ℓ, with a concentrated end mass M, is welded to a vertical shaft turning at constant ω. (See Figure P4.183.) Find the force and moment exerted by the rod onto the shaft. Include the effect of gravity.

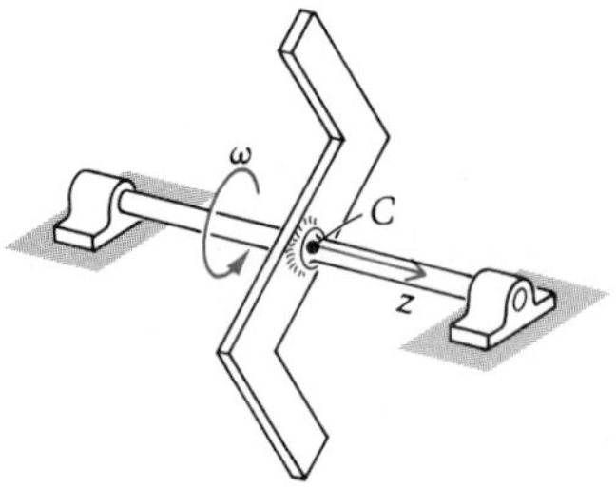

Figure P4.182

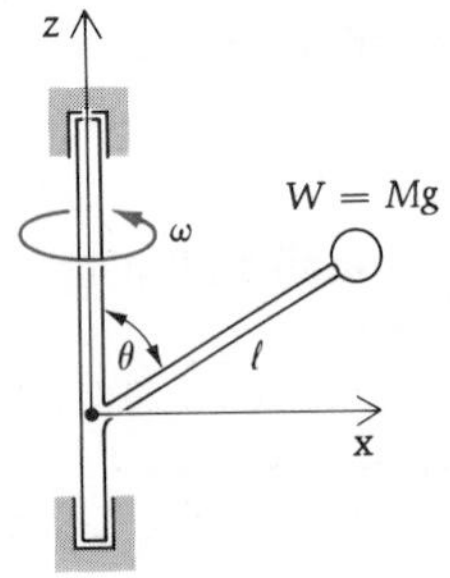

Figure P4.183

4.184 The shaft in Figure P4.184 turns at constant angular velocity 10 rad/sec. If the bars are light compared with the two weights, determine the bending moment exerted on $\mathcal{S}_2$ (length 2ℓ) by $\mathcal{S}_1$ at the point where they are welded together. Sketch the way the shaft will deform in reality under the action of this couple. Ignore gravity.

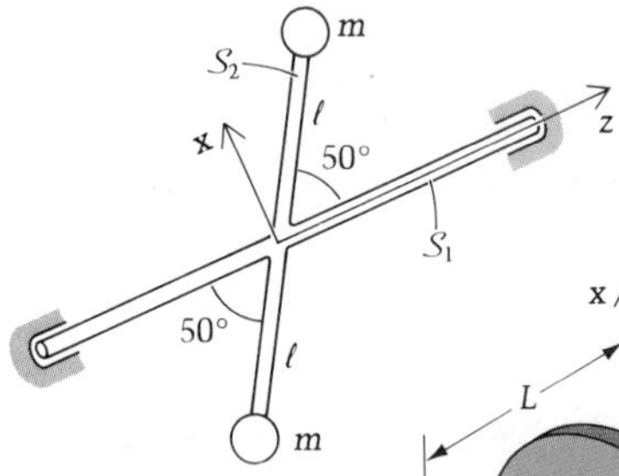

Figure P4.184

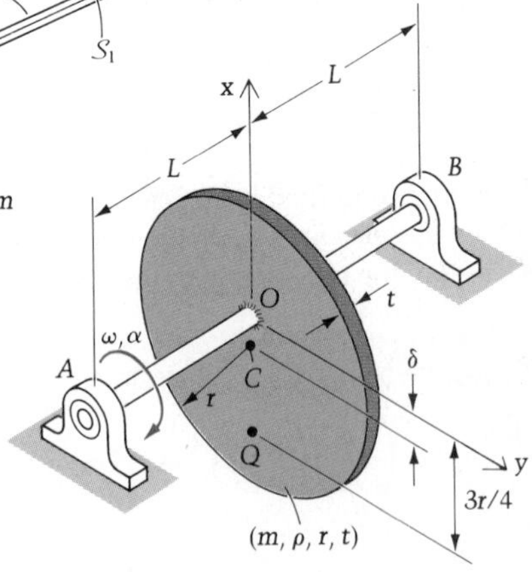

Figure P4.185

4.185 The circular disk in Figure P4.185 is offset by the amount δ from the shaft to which it is attached.

a. Find the dynamic bearing reactions at A and B in terms of the system parameters shown in the figure.

b. If $\delta = r/20$, find the radius of a hole (in terms of r) at Q that will eliminate these bearing reactions.

4.186 Two thin disks are mounted on a shaft, each midway between the center and one of the bearings, as indicated in Figure P4.186. The disks are each mounted off center by the amount $\delta = 0.05$ in. as shown. Determine the x and y locations of two small 4-oz magnetic weights (one for each disk), which when stuck to the disks will balance the shaft. Neglect the thicknesses of the disks, and treat the weights as particles.

4.187 In Figure P4.187, $\mathcal{A}$ is the axle of a bicycle, mounted in bearings $2d$ apart. The cranks $\mathcal{C}$ are rigidly connected to the axle and also to the pedal shafts $\mathcal{P}_1$ and $\mathcal{P}_2$. If the rigid body consisting of axle $\mathcal{A}$, the cranks $\mathcal{C}$, and the pedal shafts is turning freely about axis z_C at constant angular speed ω, find the forces exerted on the bearings in the given configuration.

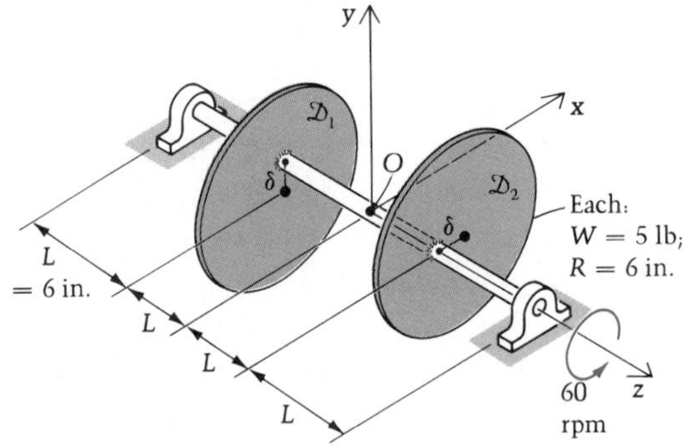

Figure P4.186

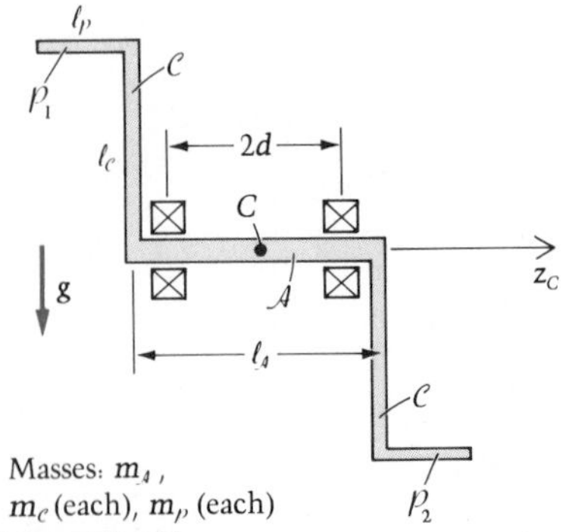

Figure P4.187

4.188 A solid cylinder (of mass m, radius r, and length $\ell = 4r$) and a light rod are welded together at angle ψ as shown in Figure P4.188. The rigid assembly is spun up to ω_0 rad/sec and then maintained at that speed. Find the dynamic bearing reactions after $\omega = \omega_0$.

$$I_{xz} = (-6.5 \sin 2\psi)(mr^2/12)$$

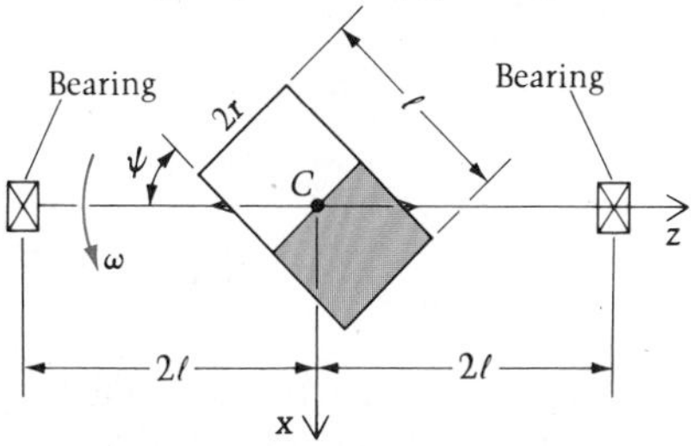

Figure P4.188

4.189 Repeat the preceding problem if the lower half (shaded) of the cylinder is missing. (The mass is now $m/2$.)

4.190 The S-shaped shaft in Figure P4.190(a) is made of two half-rings, each of radius R and mass $m/2$. Find the dynamic bearing reactions for the instant given. *Hint*: For a half-ring, the mass center is located as shown in Figure P4.190(b).

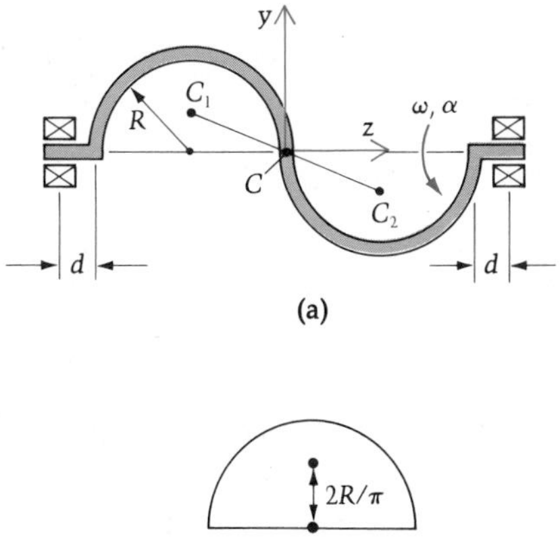

Figure P4.190

4.191 Show that, if a body mounted on a shaft is statically balanced, and if I_{xz}^D and I_{yz}^D are zero for any point D on the shaft, it follows that I_{xz}^Q and I_{yz}^Q are zero for any *other* point Q on the shaft.

4.192 The shaft in Figure P4.192 supports the eccentrically located weights W_1 (0.1 lb) and W_2 (0.2 lb) as shown. It is desired to add a 0.3-lb weight in plane A and a 0.4-lb weight in plane B to balance the shaft dynamically. Determine the x and y coordinates of the added weights.

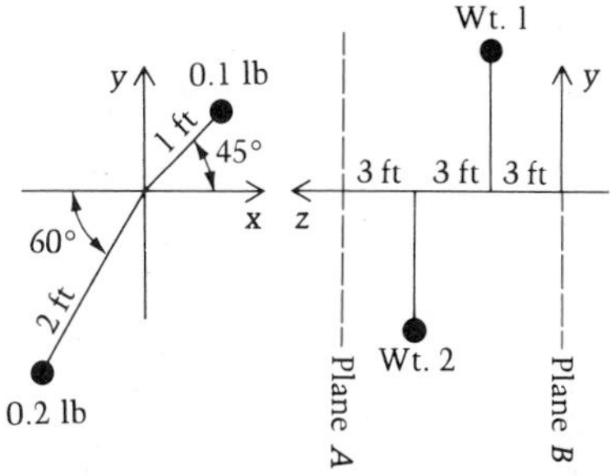

Figure P4.192

4.193 Rotor $\mathcal{R}$ in Figure P4.193 has a mass of 2 slugs, and its mass center C is at a 5-in. offset from its shaft as shown ($x = 3$ in., $y = 4$ in., and $z = 10$ in. in the coordinate system fixed in the shaft at the point B). The products of inertia of $\mathcal{R}$ with respect to the center of mass axes (x_C, y_C, z_C) are $I_{xz}^C = -\frac{5}{3}$ lb-in.-sec^2 and $I_{yz}^C = -\frac{5}{6}$ lb-in.-sec^2. By adding a $\frac{1}{7}$-slug mass in each of the two correction planes $\mathcal{A}$ and $\mathcal{B}$, balance the rotor. That is, determine the x and y coordinates of each of the two added masses by ensuring that the mass center of the final system is on the shaft and that the products of inertia vanish.

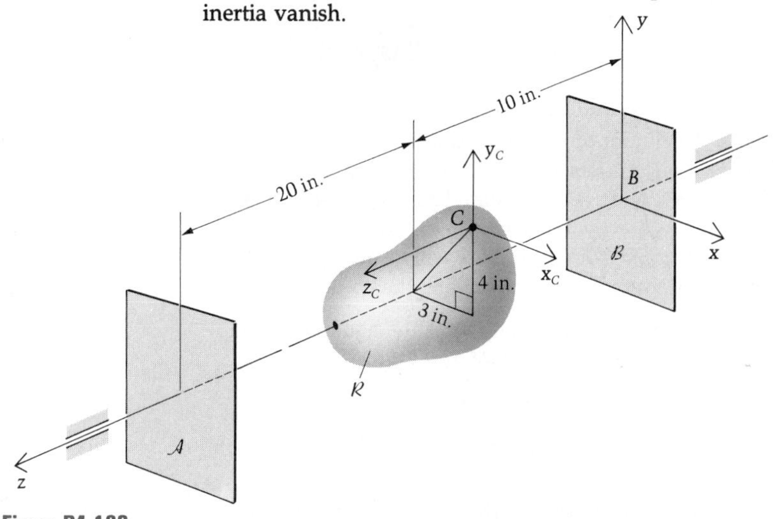

Figure P4.193

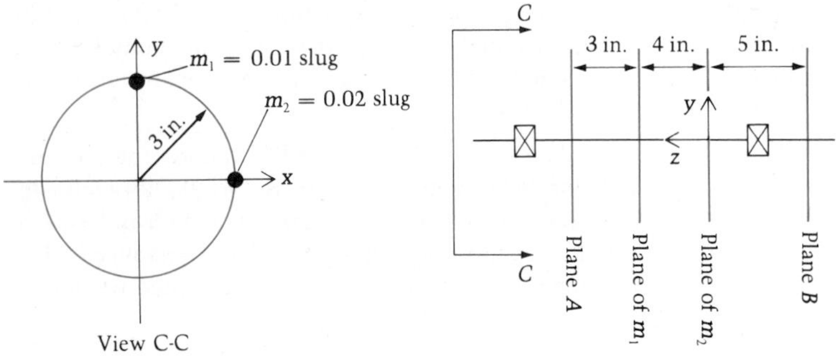

Figure P4.194

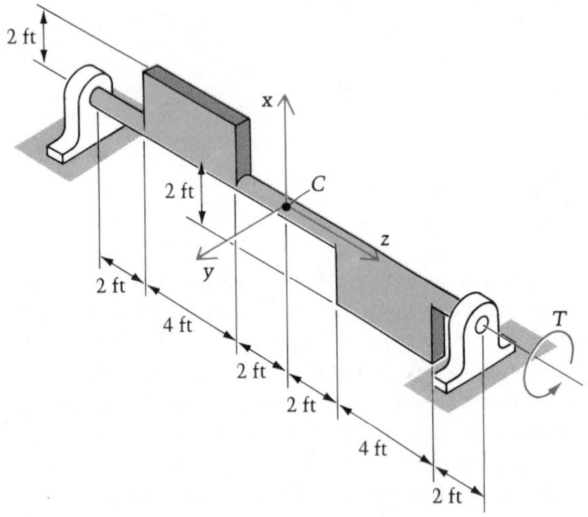

Figure P4.195

4.194 Balance the shaft in Figure P4.194 by adding a mass of 0.003 slug in plane A and a mass of 0.004 slug in plane B.

* **4.195** Two plates, each weighing 32.2 lb, are welded to a light shaft as shown in Figure P4.195. A torque T of 10 lb-ft is applied about the z axis until the assembly is turning at angular speed ω_0, then T is removed. If the bearings can hold a force perpendicular to the shaft of no more than 320 lb, find the maximum value that ω_0 can be without failure. Note that xz is the plane of the plates and (x, y, z) are fixed to the assembly.

COMPUTER PROBLEM ▶ Chapter 4

* **4.196** A cylinder of mass m and radius R is rolling to the left and encounters a pothole of length s, as shown in Figure P4.196(a). The angular velocity when the mass center C is directly above O is ω_i↺. We are interested in the condition(s) for which there will be no slip at O while the cylinder pivots prior to striking the corner at A.

a. Show that for no slip at O, the equations of motion are (see Figure P4.196(b)):

1. $mR\alpha = mg \sin\theta - f$

2. $mR\omega^2 = mg\cos\theta - N$

3. $mgR\sin\theta = \dfrac{3}{2} mR^2\alpha$

where $a_{Ct} = R\alpha$ and $a_{Cn} = R\omega^2$ have been substituted.

b. Multiply Equation (3) by $\dot{\theta}$ and integrate, obtaining

4. $mgR\,(1 - \cos\theta) = \dfrac{3}{4} mR^2\omega^2 - \dfrac{3}{4} mR^2\omega_i^2$

Solve these equations for f and N, and show that the no-slip condition $f \leqq \mu N$ requires that

$$\sin\theta \leqq \mu(7\cos\theta - 4 - 3r\omega_i^2/g)$$

Note that for very low ω_i, this is easily satisfied if μ is not too small and s (and therefore θ) is not too large. But, for example, if $\omega_i^2 = g/R$, then the cyl-

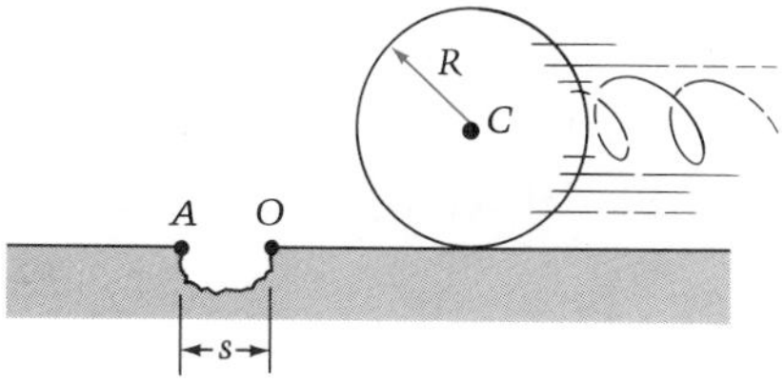

Figure P4.196(a)

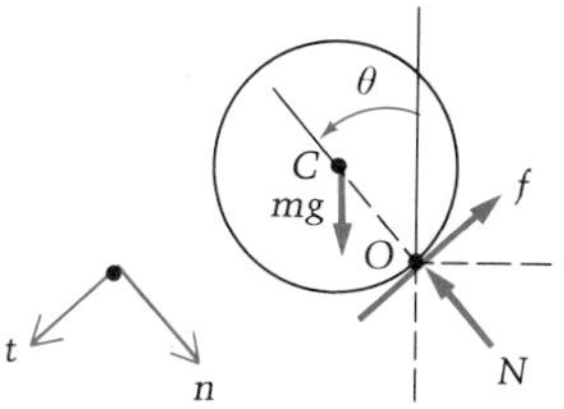

Figure P4.196(b)

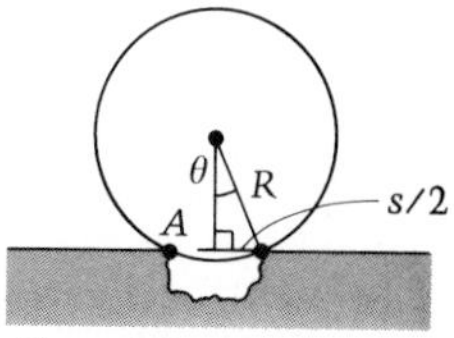

Figure P4.196(c)

inder will slip regardless of the value of μ because then the inequality cannot hold. Note too that if the cylinder is not slipping just prior to impact, it has not slipped at all.

c. Next, use Figure P4.196(c) to compute the angle θ for which the cylinder will strike the left corner A of the depression, and show that no slip will have occurred at any time during the pivoting if

$$\frac{s}{2R} \leqq \mu\left(7\sqrt{1-\left(\frac{s}{2R}\right)^2} - 4 - 3R\omega_i^2/g\right)$$

Finally, use the computer to create data for plots of the minimum μ required for no slip at O versus $R\omega_i^2/g$ for three values of $s/2R$: 0.1, 0.2, 0.5. Draw the three curves on the same graph.

Extra Credit Project Problem ▶ Chapter 4

4.197 Construct a round object with a challenging I_{zz}^C to calculate (as an example, see Figure P4.197(a)). On an inclined plane (see Figure P4.197(b)), roll your object down a 5-ft, 15° grade and with a stop-watch measure the descent time. Do this twice and average the times. Then explain the experiment, calculate the expected time, compare with the actual time, and give possible reasons for the difference in a brief report. (Note: It is fun to do all the students' experiments in the same session.)

Figure P4.197(a)

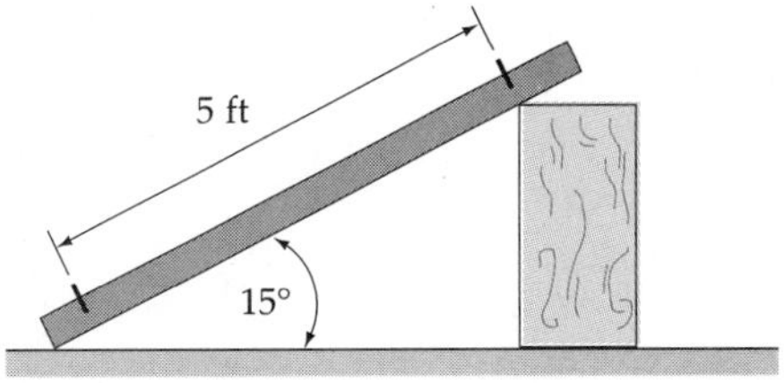

Figure P4.197(b)

Summary ▶ Chapter 4

In this chapter we have developed compact forms for the right-hand sides of moment equations for plane motion of a rigid body. The most general forms we have studied are, for an arbitrary point P,

$$\Sigma \mathbf{M}_P = (I_{xz}^C\,\alpha - I_{yz}^C\,\omega^2)\hat{\mathbf{i}} + (I_{yz}^C\,\alpha + I_{xz}^C\,\omega^2)\hat{\mathbf{j}} + I_{zz}^C\,\alpha\hat{\mathbf{k}} + \mathbf{r}_{PC} \times m\mathbf{a}_C$$

and

$$\Sigma \mathbf{M}_P = (I_{xz}^P\,\alpha - I_{yz}^P\,\omega^2)\hat{\mathbf{i}} + (I_{yz}^P\,\alpha + I_{xz}^P\,\omega^2)\hat{\mathbf{j}} + I_{zz}^P\,\alpha\hat{\mathbf{k}} + \mathbf{r}_{PC} \times m\mathbf{a}_P$$

where moments and products of inertia are defined by

$$I_{zz} = \int (x^2 + y^2)\, dm$$

$$I_{xz} = -\int xz\, dm \qquad \text{and} \qquad I_{yz} = -\int yz\, dm$$

and for which there are the very useful parallel-axis theorems

$$I_{zz}^P = I_{zz}^C + m(\bar{x}^2 + \bar{y}^2)$$

and

$$I_{xz}^P = I_{xz}^C - m\bar{x}\bar{z} \qquad \text{and} \qquad I_{yz}^P = I_{yz}^C - m\bar{y}\bar{z}$$

Except for the topic of balancing of rotating bodies (Section 4.7) we have restricted our attention to situations in which the products of inertia vanish, usually because of the body having an xy plane of symmetry. In those cases,

$$\Sigma \mathbf{M}_P = I_{zz}^C\,\alpha\,\hat{\mathbf{k}} + \mathbf{r}_{PC} \times m\mathbf{a}_C$$

and

$$\Sigma \mathbf{M}_P = I_{zz}^P\,\alpha\,\hat{\mathbf{k}} + \mathbf{r}_{PC} \times m\mathbf{a}_P$$

Important special cases are

a. Translation (in which every point has the same acceleration, **a**, and of course $\alpha = 0$), for which:

$$\Sigma \mathbf{M}_P = \mathbf{r}_{PC} \times m\mathbf{a}$$

and so

$$\Sigma \mathbf{M}_C = \mathbf{0}$$

b. Summing moments at the mass center C:

$$\Sigma \mathbf{M}_C = I_{zz}^C\alpha\hat{\mathbf{k}}$$

or, more simply,

$$\Sigma M_C = I_C\alpha$$

c. P is a pivot (body rotates about a fixed axis), so that $\mathbf{a}_P = \mathbf{0}$:

$$\Sigma \mathbf{M}_P = I_{zz}^P\alpha\hat{\mathbf{k}}$$

or, more simply,

$$\Sigma M_P = I_P\alpha$$

It is more important to realize that while we have a number of options as to form for the moment equation, one moment equation plus the force equation, $\Sigma \mathbf{F} = m\mathbf{a}_C$, are all we can bring to bear independently for a given (free) body. That is, the situation is the same as in statics: we may sum moments wherever we like, but the two vector equations — one force and one moment — give us all the independent relationships involving external forces on the body. Many practical problems are solved by augmenting these equations with kinematic constraint conditions that can be invoked to generate relationships between $\mathbf{a}_C$ and α.

A body rotating about a fixed axis is said to be statically balanced if the mass center is located on the axis. It is said to be dynamically balanced (no bearing reactions induced by the rotation) if in addition the products of inertia associated with the rotation axis all vanish. Industrial equipment and automobile tires are modified by the addition of "balance weights" so as to ensure these conditions.

REVIEW QUESTIONS ► Chapter 4

True or False?

These questions all refer to rigid bodies in plane motion.

1. Euler's second law enables us to study the rotational motion of rigid bodies.
2. The moment of inertia is always positive, whereas the products of inertia can have either sign.
3. The formula $m\ell^2/12$ gives the exact value of the moment of inertia of a slender rod about a lateral axis through its mass center.
4. Euler's second law, $\Sigma\mathbf{M}_O = \dot{\mathbf{H}}_O$, is valid only in an inertial frame (meaning that the position vectors and velocities inherent in $\mathbf{H}_O$, the origin O, and the time derivative are all taken in an inertial frame).
5. In $I_{zz}^P = I_{zz}^C + md^2$, the quantity d is the distance between the points P and C. (C is in the reference plane, whereas P is *any* point of the body.)
6. Euler's second law, $\Sigma\mathbf{M}_O = \dot{\mathbf{H}}_O$, applies to deformable bodies, liquids, and gases, as well as to rigid bodies.
7. If Ⓘ represents the instantaneous center of zero velocity, then $\Sigma M_{Ⓘ} \neq I_{Ⓘ}\alpha$ in general.
8. Products of inertia are not found in the equations of plane motion.
9. $\Sigma\mathbf{M}_C = \dot{\mathbf{H}}_C$ is just as general as $\Sigma\mathbf{M}_O = \dot{\mathbf{H}}_O$, where O is fixed in an inertial frame.
10. In translation problems, the moments of external forces and couples taken about any point add to zero.
11. Suppose you buy a new set of automobile tires and a dynamic balance is performed on each wheel by adding weights in two planes (inner and outer rims). The products of inertia I_{xz} and I_{yz} have thus been eliminated, which otherwise would have caused bearing reactions and vibration.
12. $\Sigma M_C = I_C\alpha$ applies to deformable as well as to rigid bodies, as long as they are in plane motion.
13. For two bodies $\mathcal{B}_1$ and $\mathcal{B}_2$, the sum of the equations $\Sigma\mathbf{F} = m_i\mathbf{a}_{C_i}$ written for each will be $\Sigma\mathbf{F} = m\mathbf{a}_C$ for the combined body.
14. If the bodies of Question (13) are turning relative to each other, it makes no sense to talk about a combined $\Sigma M_C = I_C\alpha$ equation.

Answers: 1. T 2. T 3. F 4. T 5. F 6. T 7. T 8. F 9. T 10. F 11. T 12. F 13. T 14. T

5

Special Integrals of the Equations of Plane Motion of Rigid Bodies: Work-Energy and Impulse-Momentum Methods

5.1 Introduction

Just as in Chapter 4, the framework here is rigid bodies in plane motion. But we shall focus our attention now on problems which most efficiently can be attacked by using work-and-kinetic-energy and/or impulse-and-momentum principles. We shall employ these principles, rather broadly stated in Chapter 2, taking advantage of the simple forms that kinetic energy and angular momentum take when the body is rigid and constrained to plane motion.

In Chapter 2 we defined the kinetic energy of a body to be the sum of the kinetic energies of the particles making up the body; that is, $T = \frac{1}{2}\Sigma m_i v_i^2$. Because we have found in Chapter 3 that velocities of different points in a rigid body are related through the body's angular velocity, the reader should not be surprised to find kinetic energy for such a body to be expressible in terms of the velocity of one point and the angular velocity. Moreover, in Chapter 2 we observed that, for a body in general, change in kinetic energy equals work of external *and* internal forces. But for a rigid body the net work of internal forces vanishes, so that the work W in $W = \Delta T$ is the work only of external forces. We shall derive this work-and-kinetic-energy relationship directly from the force and moment equations (Euler's laws) as studied in Chapter 4, but it is helpful to recall the discussion of Chapter 2 and note the consistency of that material with the result we shall develop here.

The relationship between angular impulse and angular momentum developed in Section 5.3 takes on a quite useful form for a rigid body in plane motion, owing to the fact, as shown in Chapter 4, that the angular momentum can be expressed then in terms of inertia properties and angular velocity. Thus we shall find ourselves in a position to evaluate sudden changes in rates of turning for colliding bodies and to study quantitatively the relationship between the spin rate and the arm-trunk configuration of a skater.

It is very important for the reader to always keep in mind that the principle of work and kinetic energy and the principles of impulse and momentum do not stand as principles somehow separate from Newton's laws or their extensions to bodies of finite size, Euler's laws. Rather, here it will be seen, as was observed before in Chapter 2, that these relationships, which involve velocities, are really just special first integrals of the more fundamental second-order expressions relating forces and accelerations. Thus the principles of this chapter allow us to begin our solutions halfway between accelerations and positions. They therefore involve velocities but not accelerations.

5.2 The Principle(s) of Work and Kinetic Energy

Kinetic Energy of a Rigid Body in Plane Motion

There is a principle, derived from the equations of motion, that will help us to solve for unknowns of interest in kinetics problems. In this section we shall see that this principle arises from first deriving and then differentiating the kinetic energy of the body.

Kinetic energy, which we have examined in Chapter 2, is usually denoted by the letter T; for any body or system of bodies, it is defined as the summation of $\frac{1}{2}(dm)v^2$ over all its elements of mass:

$$T = \frac{1}{2} \int (\mathbf{v} \cdot \mathbf{v})\, dm \tag{5.1}$$

In this section we need to specialize Definition (5.1) for a rigid body $\mathcal{B}$ in plane motion. To this end we kinematically relate the velocity $\mathbf{v}$ of the differential mass to the velocity of the mass center C. Using the fact that $\mathbf{v}$ is at all times equal to the velocity of its companion point in the reference plane containing C (see Figure 5.1), we may write

$$\mathbf{v} = \mathbf{v}_C + \omega\hat{\mathbf{k}} \times (x\hat{\mathbf{i}} + y\hat{\mathbf{j}}) = \mathbf{v}_C + \omega(-y\hat{\mathbf{i}} + x\hat{\mathbf{j}})$$

y
C
x
z
dm
y
x
Companion point of dm in plane of mass center C
Rigid body $\mathcal{B}$ in plane motion

Figure 5.1

We note that the x and y axes are fixed in $\mathcal{B}$ with their origin at C. Forming v^2, that is, $\mathbf{v} \cdot \mathbf{v}$, we have

$$\mathbf{v} \cdot \mathbf{v} = \mathbf{v}_C \cdot \mathbf{v}_C + \omega^2(x^2 + y^2) + 2\omega\mathbf{v}_C \cdot (x\hat{\mathbf{j}} - y\hat{\mathbf{i}}) \tag{5.2}$$

Thus the kinetic energy becomes, substituting (5.2) into (5.1),

$$T = \frac{1}{2}\mathbf{v}_C \cdot \mathbf{v}_C \int dm + \frac{\omega^2}{2} \int (x^2 + y^2)\, dm$$
$$+ \omega(\mathbf{v}_C \cdot \hat{\mathbf{j}}) \int x\, dm - \omega(\mathbf{v}_C \cdot \hat{\mathbf{i}}) \int y\, dm$$

Recognizing the moment of inertia integral in the second term, we obtain

$$T = \frac{1}{2} m v_C^2 + \frac{1}{2} I_{zz}^C \omega^{2*} \tag{5.3}$$

in which $\mathbf{v}_C \cdot \mathbf{v}_C = |\mathbf{v}_C|^2 = v_C^2$, the square of the magnitude of the velocity of C.

Question 5.1 Why do $\int x\, dm$ and $\int y\, dm$ both vanish?

We note that the (scalar) kinetic energy has two identifiable *parts (not components!)*: one relating to the motion of the mass center C ($T_v = \frac{1}{2}mv_C^2$)

---

* Henceforth, we shall use the abbreviation I_C for I_{zz}^C throughout this chapter.

Answer 5.1 By the definition of the mass center.

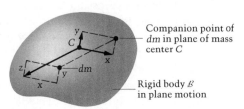

and the other to the motion of the body relative to C ($T_\omega = \frac{1}{2} I_C \omega^2$). This clear division of T even exists in general motion of rigid bodies (that is, in three dimensions), though there are more terms in T_ω then.

Example 5.1

Calculate the kinetic energy of the round rolling body $\mathcal{B}$ in Figure E5.1, which has mass m, radius R, and radius of gyration k_C with respect to the z_C axis. The mass center C (see Figure E5.1) lies at the geometric center.

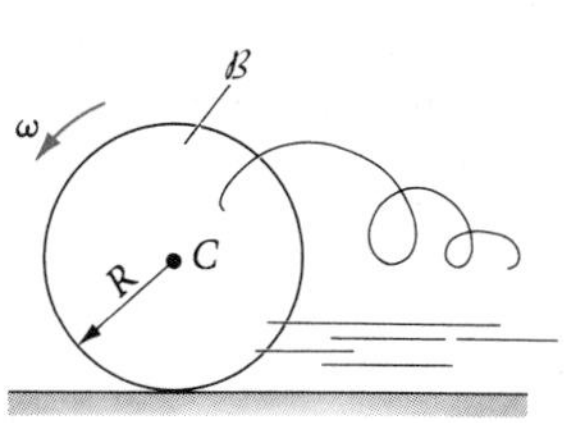

Figure E5.1

Solution

$$T = \frac{1}{2} m v_C^2 + \frac{1}{2} \overbrace{I_C}^{mk_C^2} \omega^2$$

$$= \frac{1}{2} m[(R\omega)^2 + k_C^2 \omega^2]$$

$$= \frac{mR^2\omega^2}{2}\left[1 + \left(\frac{k_C}{R}\right)^2\right]$$

Note that if $\mathcal{B}$ is a solid cylinder, then $k_C = R/\sqrt{2}$ and

$$T = \frac{mR^2\omega^2}{2}\left(1 + \frac{1}{2}\right)$$

In this case, two-thirds of the kinetic energy rests in the translation term of T ($\frac{1}{2} m v_C^2$).

If $\mathcal{B}$ is a ring (or hoop), however, $I_C = mR^2$ so that $k_C = R$ and

$$T = \frac{mR^2\omega^2}{2}(1 + 1)$$

and this time half the kinetic energy is in each of the translational and rotational terms.

Example 5.2

Work Example 5.1 for the case when the mass center C is offset by a distance r from the geometric center Q of a round rolling body $\mathcal{B}$. (See Figure E5.2.)

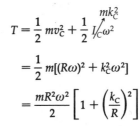

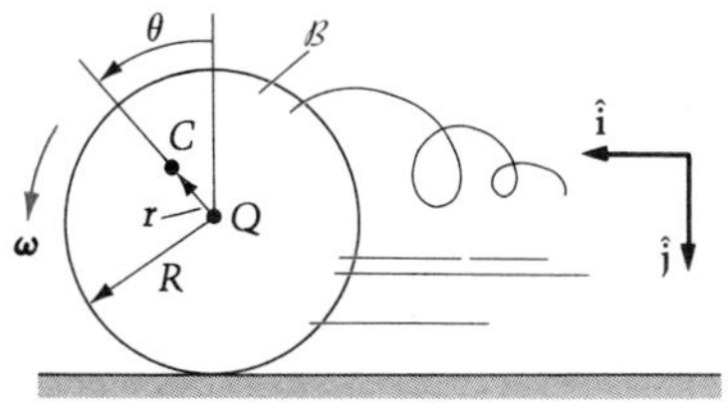

Figure E5.2

Solution

In order to use our equation for kinetic energy,

$$T = \frac{1}{2} m v_C^2 + \frac{1}{2} I_C \omega^2$$

we must first calculate v_C^2:

$$\mathbf{v}_C = \mathbf{v}_Q + \dot{\theta}\hat{\mathbf{k}} \times \overbrace{\mathbf{r}_{QC}}^{r \sin\theta\hat{\mathbf{i}} - r\cos\theta\hat{\mathbf{j}}}$$

$$= (R\dot{\theta} + r\dot{\theta}\cos\theta)\hat{\mathbf{i}} + r\dot{\theta}\sin\theta\hat{\mathbf{j}}$$

Therefore

$$v_C^2 = R^2\dot{\theta}^2 + r^2\dot{\theta}^2 + 2Rr\dot{\theta}^2\cos\theta$$

Substituting, we get

$$T = \frac{mR^2\dot{\theta}^2}{2}\left[1 + \frac{2r}{R}\cos\theta + \left(\frac{r}{R}\right)^2 + \left(\frac{k_C}{R}\right)^2\right]$$

Note that if $r = 0$, the answer agrees as it should with Example 5.1.

An Alternative Form for Kinetic Energy

There is an alternative means of writing the kinetic energy T of a rigid body in plane motion by making use of the instantaneous center of zero velocity Ⓘ (see Figure 5.2):

$$\begin{aligned} T &= \overbrace{\frac{1}{2}mv_C^2}^{T_v} + \overbrace{\frac{1}{2}I_C\omega^2}^{T_\omega} \\ &= \frac{1}{2}m(d\omega)^2 + \frac{1}{2}I_C\omega^2 \\ &= \frac{1}{2}(I_C + md^2)\omega^2 \end{aligned}$$

Thus by using the parallel-axis theorem we obtain

$$T = \frac{1}{2}I_{Ⓘ}\omega^2 \tag{5.4}$$

The translational (T_v) and rotational (T_ω) terms composing the scalar T are thus seen to collapse into the one term $\frac{1}{2}I_{Ⓘ}\omega^2$ if we choose to work with Ⓘ instead of C.

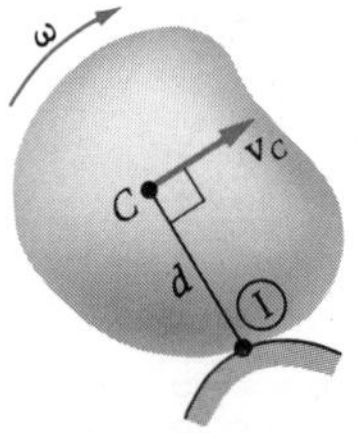

Figure 5.2

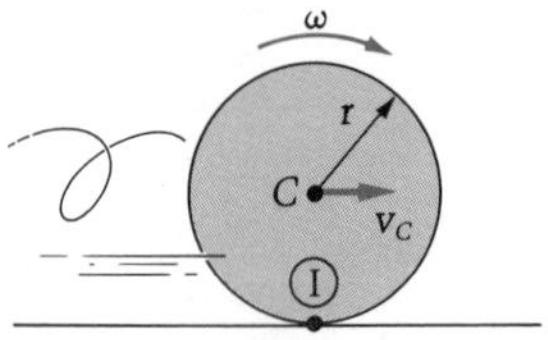

Figure 5.3

As an example, we consider a rolling cylinder again (see Figure 5.3):

$$\begin{aligned} T = T_v + T_\omega &= \frac{1}{2}mv_C^2 + \frac{1}{2}I_C\omega^2 \\ &= \frac{1}{2}m(r\omega)^2 + \frac{1}{2}\left(\frac{1}{2}mr^2\right)\omega^2 \\ &= \frac{1}{2}mr^2\omega^2\left(1 + \frac{1}{2}\right) = \frac{3}{4}mr^2\omega^2 \end{aligned}$$

We have noted that two-thirds of the cylinder's kinetic energy is associated with the "translational part" of T and one-third with the "rotational part." If we now use Ⓘ, we get *all* of T at once:

$$T = \frac{1}{2} I_{Ⓘ}\omega^2 = \frac{1}{2}\Big[\underbrace{\underbrace{\frac{1}{2}mr^2}_{I_C} + \underbrace{mr^2}_{\text{transfer term}}}_{I_{Ⓘ}}\Big]\omega^2 = \frac{3}{4}mr^2\omega^2 \qquad \text{(as above)}$$

As a second example of the use of Equation (5.4), consider the slender rod $\mathcal{B}$ swinging about a pivot at A as shown in Figure 5.4. The kinetic energy of $\mathcal{B}$ may be found in either of two ways:

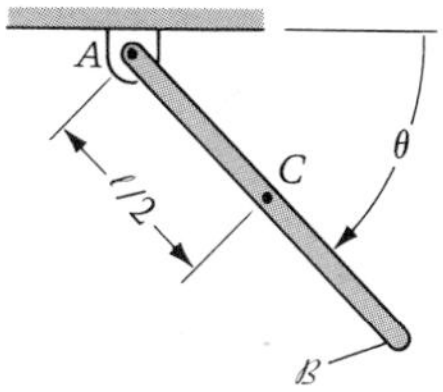

Figure 5.4

$$T = \frac{1}{2} I_{Ⓘ}\omega^2 = \frac{1}{2}\left[\frac{1}{3}m\ell^2\right]\omega^2 = \frac{m\ell^2\omega^2}{6}$$

$$T = \frac{1}{2}mv_C^2 + \frac{1}{2}I_C\omega^2 = \frac{1}{2}m\left(\frac{\ell}{2}\omega\right)^2 + \frac{1}{2}\left(\frac{1}{12}m\ell^2\right)\omega^2 = m\ell^2\omega^2\left(\frac{1}{8}+\frac{1}{24}\right) = \frac{m\ell^2\omega^2}{6}$$

Derivation of the Principle $W = \Delta T$; Power and Work of Systems of Forces and Couples

Returning now to the derivation of our principle, we next compute the rate of change of kinetic energy:

$$\frac{dT}{dt} = \frac{d}{dt}\left(\frac{1}{2}m\mathbf{v}_C \cdot \mathbf{v}_C + \frac{1}{2}I_C\omega\hat{\mathbf{k}} \cdot \omega\hat{\mathbf{k}}\right)$$

$$= \frac{1}{2}m\underbrace{(\mathbf{a}_C \cdot \mathbf{v}_C + \mathbf{v}_C \cdot \mathbf{a}_C)}_{2\mathbf{a}_C \cdot \mathbf{v}_C} + \frac{1}{2}I_C\underbrace{(\alpha\hat{\mathbf{k}} \cdot \omega\hat{\mathbf{k}} + \omega\hat{\mathbf{k}} \cdot \alpha\hat{\mathbf{k}})}_{2\alpha\hat{\mathbf{k}} \cdot \omega\hat{\mathbf{k}}}$$

Therefore

$$\frac{dT}{dt} = m\mathbf{a}_C \cdot \mathbf{v}_C + (I_C\alpha\hat{\mathbf{k}}) \cdot \omega\hat{\mathbf{k}} \tag{5.5}$$

Recalling that $\Sigma\mathbf{F} = m\mathbf{a}_C$ and that the z component of $\Sigma\mathbf{M}_C$ is $I_C\alpha$ for rigid bodies in plane motion, we may write

$$\frac{dT}{dt} = \Sigma\mathbf{F} \cdot \mathbf{v}_C + \Sigma\mathbf{M}_C \cdot \omega\hat{\mathbf{k}} \tag{5.6}$$

Question 5.2 Since $\Sigma\mathbf{M}_C$ can contain x and y components (see Equation 4.13) why may we substitute the *total* vector $\Sigma\mathbf{M}_C$ for just the z component $I_C\alpha\hat{\mathbf{k}}$ in Equation 5.5?

Our next goal is to get the individual external forces and couples acting on the body $\mathcal{B}$ into the equation. (See Figure 5.5.) Note the abbreviations $\mathbf{r}_{CP_1} = \mathbf{r}_1$, $\mathbf{r}_{CP_2} = \mathbf{r}_2$, and so on, of the vectors to the points of application of $\mathbf{F}_1$, $\mathbf{F}_2$, and so on. We assume that the external mechanical actions on the body arise from a system of forces ($\mathbf{F}_1, \mathbf{F}_2 \ldots$) and couples with moment vectors ($\mathbf{C}_1, \mathbf{C}_2, \ldots$), as shown in Figure 5.5. Further, we let ($\mathbf{v}_1, \mathbf{v}_2, \ldots$) be the velocities of the material points ($P_1, P_2, \ldots$) on which the forces act instantaneously.

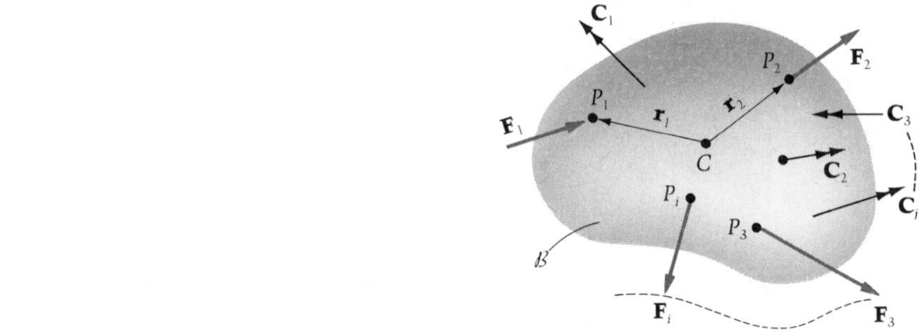

Figure 5.5

Clearly, then, the resultant of the external forces is

$$\Sigma\mathbf{F} = \mathbf{F}_1 + \mathbf{F}_2 + \cdots \tag{5.7}$$

and the moment of the $\mathbf{F}_i$'s and $\mathbf{C}_j$'s about C is

$$\Sigma\mathbf{M}_C = \mathbf{r}_1 \times \mathbf{F}_1 + \mathbf{r}_2 \times \mathbf{F}_2 + \cdots + \mathbf{C}_1 + \mathbf{C}_2 + \cdots \tag{5.8}$$

Substituting Equations (5.7) and (5.8) into (5.6) gives

$$\begin{aligned}\frac{dT}{dt} &= (\mathbf{F}_1 + \mathbf{F}_2 + \cdots) \cdot \mathbf{v}_C + (\mathbf{r}_1 \times \mathbf{F}_1 + \mathbf{r}_2 \times \mathbf{F}_2 + \cdots) \cdot \omega\hat{\mathbf{k}} \\ &\quad + (\mathbf{C}_1 + \mathbf{C}_2 + \cdots) \cdot \omega\hat{\mathbf{k}} \\ &= (\mathbf{F}_1 + \mathbf{F}_2 + \cdots) \cdot \mathbf{v}_C + \omega\hat{\mathbf{k}} \cdot (\mathbf{r}_1 \times \mathbf{F}_1 + \mathbf{r}_2 \times \mathbf{F}_2 + \cdots) \\ &\quad + (\mathbf{C}_1 + \mathbf{C}_2 + \cdots) \cdot \omega\hat{\mathbf{k}}\end{aligned} \tag{5.9}$$

Now since the dot and cross may be interchanged without altering the value of a scalar triple product,

$$\begin{aligned}\frac{dT}{dt} &= (\mathbf{F}_1 + \mathbf{F}_2 + \cdots) \cdot \mathbf{v}_C + (\omega\hat{\mathbf{k}} \times \mathbf{r}_1) \cdot \mathbf{F}_1 + (\omega\hat{\mathbf{k}} \times \mathbf{r}_2) \cdot \mathbf{F}_2 \\ &\quad + \cdots + (\mathbf{C}_1 + \mathbf{C}_2 + \cdots) \cdot \omega\hat{\mathbf{k}}\end{aligned} \tag{5.10}$$

Answer 5.2 Because $(\Sigma M_{Cx}\hat{\mathbf{i}} + \Sigma M_{Cy}\hat{\mathbf{j}}) \cdot \omega\hat{\mathbf{k}}$ is zero!

But the velocities of P_1 and C are related:

$$\mathbf{v}_{P_1} = \mathbf{v}_1 = \mathbf{v}_C + \omega\hat{\mathbf{k}} \times \mathbf{r}_1$$

so that

$$\frac{dT}{dt} = \mathbf{F}_1 \cdot \mathbf{v}_1 + \mathbf{F}_2 \cdot \mathbf{v}_2 + \cdots + (\mathbf{C}_1 + \mathbf{C}_2 + \cdots) \cdot \omega\hat{\mathbf{k}} \quad (5.11)$$

The right-hand side of Equation (5.11) is called the **power**, or **rate of work**, of the external system of forces and couples acting on the body. The power of a force is its dot product with the velocity of the point on which it acts; the power of a couple is its dot product with the angular velocity of the body on which it acts:

$$\text{Rate of work of force } \mathbf{F}_1 = \mathbf{F}_1 \cdot \mathbf{v}_1 = \text{power of } \mathbf{F}_1 \quad (5.12)$$

$$\text{Rate of work of couple } \mathbf{C}_1 = \mathbf{C}_1 \cdot \omega\hat{\mathbf{k}} = \text{power of } \mathbf{C}_1 \quad (5.13)$$

Hence one form of the principle of this section is

$$\text{Power} = \frac{dT}{dt} \quad (5.14)$$

or

$$P = \dot{T}$$

Integrating, we obtain another principle.*

$$\int_{t_1}^{t_2} P\,dt = T(t_2) - T(t_1) = T_2 - T_1$$

or

$$W = \Delta T = \left(\frac{1}{2}\,mv_C^2 + \frac{1}{2}\,I_C\omega^2\right)\Big]_{t_1}^{t_2} \quad (5.15)$$

where the integral of the power is called the **work** W of the external forces and couples. It is the work done by the $\mathbf{F}_i$'s and $\mathbf{C}_i$'s on the body between the two times t_1 and t_2. Hence we have a principle that can be stated in words:

Work done by external forces and couples on $\mathcal{B}$	=	Change of kinetic energy of $\mathcal{B}$

Restriction of $W = \Delta T$ to a Rigid Body; A Notable Exception

It is essential to recognize that our derivation of the principle of work and kinetic energy *depends crucially on the body being rigid.* In fact the work of external forces on a deformable body is *not* in general equal to the change in its kinetic energy. That is the case even when the "deformable" body is

* Sometimes (t_i, t_f) is used to denote the time interval, rather than (t_1, t_2); the subscripts stand for "initial" and "final" values.

composed of several individually rigid parts. However, there are a number of special circumstances, usually easy to recognize, for which the principle is valid for such a system of rigid bodies. To give an example for which this is true, suppose we have two rigid bodies, $\mathcal{B}_1$ and $\mathcal{B}_2$, making up the system, and suppose the bodies are connected by a pin (or hinge) with negligible friction as shown in Figure 5.6.

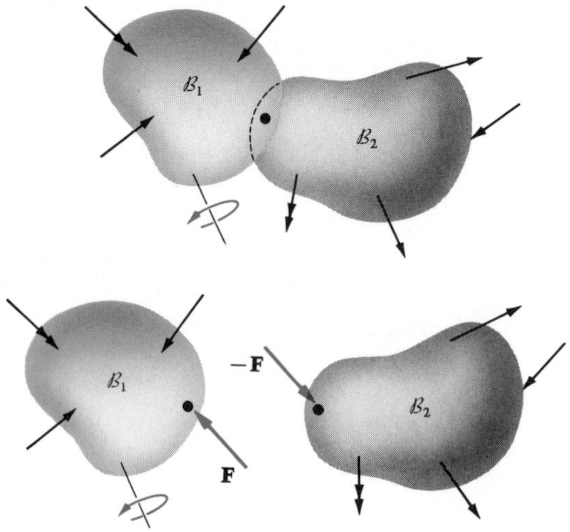

Figure 5.6

Let $\mathbf{F}$ be the force exerted by $\mathcal{B}_1$ on $\mathcal{B}_2$ at the pin, and consequently $-\mathbf{F}$ is the force exerted by $\mathcal{B}_2$ on $\mathcal{B}_1$. Furthermore, let

$\mathbf{v}$ = common velocity of attachment points in the two bodies

P_{E_1} = power (rate of work) of forces acting on $\mathcal{B}_1$ that are also external to system

P_{E_2} = power of forces acting on $\mathcal{B}_2$ that are also external to system

$T_{\mathcal{B}_1}$ = kinetic energy of $\mathcal{B}_1$

$T_{\mathcal{B}_2}$ = kinetic energy of $\mathcal{B}_2$

Now if we apply Equations (5.11) and (5.14) to each of the bodies, we obtain

$$P_{E_1} + (-\mathbf{F}) \cdot \mathbf{v} = \frac{dT_{\mathcal{B}_1}}{dt}$$

and

$$P_{E_2} + \mathbf{F} \cdot \mathbf{v} = \frac{dT_{\mathcal{B}_2}}{dt}$$

which may be added to yield

$$P_{E_1} + P_{E_2} = \frac{d}{dt}(T_{\mathcal{B}_1} + T_{\mathcal{B}_2})$$

or

$$P = \frac{dT}{dt}$$

where P is the power of the external forces on the system and T is the kinetic energy of the system.

With friction in the pin, however, we also would have interactive couples $\mathbf{C}$ and $-\mathbf{C}$, and the sum of their work rates would be

$$\mathbf{C} \cdot (\omega_{\mathcal{B}_2} - \omega_{\mathcal{B}_1})\hat{\mathbf{k}}$$

which in general would *not* vanish.* This net rate of work of friction couples would be negative, reflecting the fact that the friction will reduce the kinetic energy of the system. We can expect the principle of work (of external forces) and kinetic energy to be valid for a system of rigid bodies whenever the interaction of the bodies leads neither to dissipation of mechanical energy by friction nor to a storing of energy as in a spring. When in doubt, follow the procedure we have just been through — that is, apply Equation (5.14) to each of the bodies, add the equations, and see whether the rates of work of interactive forces cancel out.

Computing the Work Done by Various Types of Forces and Moments

Before we can put Equation (5.15) to use, it is essential to demonstrate how to compute the work W done on $\mathcal{B}$ by a number of common types of forces and moments:

Type 1: $\mathbf{F}_1$ is constant. In this case, as in Chapter 2,

$$W = \int \mathbf{F}_1 \cdot \mathbf{v}_1 \, dt = \mathbf{F}_1 \cdot \int \mathbf{v}_1 \, dt \tag{5.16}$$

Type 2: $\mathbf{F}_1$ acts on the same point P_1 of $\mathcal{B}$ throughout its motion.† In this case,

$$W = \int_{t_i}^{t_f} \mathbf{F}_1 \cdot \mathbf{v}_1 \, dt = \int_{t_i}^{t_f} \mathbf{F}_1 \cdot \frac{d\mathbf{r}_1}{dt} \, dt = \int_{\mathbf{r}(t_i)}^{\mathbf{r}(t_f)} \mathbf{F}_1 \cdot d\mathbf{r}_1 \tag{5.17}$$

where $\mathbf{r}_{OP_1} = \mathbf{r}_1$ and i and f denote starting (initial) and ending (final) times and positions. It is true, of course, that the velocity $\mathbf{v}_1$, which combines with $\mathbf{F}_1$ to produce its power, is at each instant the derivative of *some* position vector. If the force acts on *different* material points of $\mathcal{B}$ at different times throughout a motion (such as friction from a brake), however, the path integral $\int \mathbf{F}_1 \cdot d\mathbf{r}_{OP_1}$ has no real functional utility and the general $\int \mathbf{F}_1 \cdot \mathbf{v}_1 \, dt$ must be used.

Type 3: $\mathbf{F}_1$ is due to gravity. This is an example of *both* Types 1 *and* 2. Thus, letting z be positive downward, we get

$$W = \int mg\hat{\mathbf{k}} \cdot d\mathbf{r}_{OC} = mg\hat{\mathbf{k}} \cdot \int d\mathbf{r}_{OC}\text{**}$$

---

* It would vanish, of course, if the friction were enough to prevent relative rotation so that $\omega_{\mathcal{B}_1} = \omega_{\mathcal{B}_2}$; then the system would behave as a single rigid body!

† Which was *necessarily* the case in Chapter 2!

** The work done by *any* constant force $\mathbf{F}$ always acting on the same point with position vector $\mathbf{r}$ is thus $\mathbf{F} \cdot [\mathbf{r}(t_f) - \mathbf{r}(t_i)]$.

Expressing the differential of the position vector in terms of rectangular cartesian coordinates, we get

$$d\mathbf{r}_{OC} = dx_C\hat{\mathbf{i}} + dy_C\hat{\mathbf{j}} + dz_C\hat{\mathbf{k}}$$

and substituting we obtain a simple result for the work of gravity:

$$W = mg \int_{z_{C_1}}^{z_{C_2}} dz_C = mg(z_{C_2} - z_{C_1}) = mgh \tag{5.18}$$

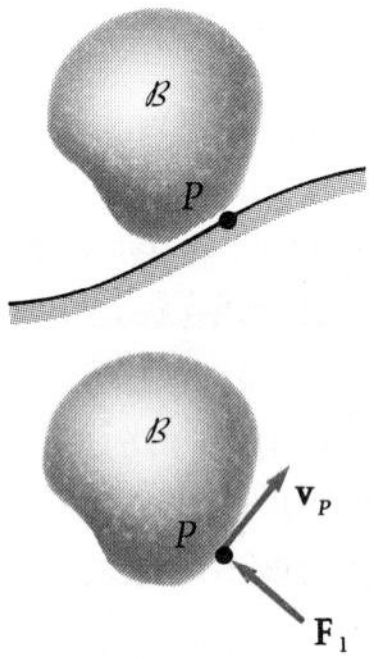

Figure 5.7

as we observed in Chapter 2. Note that gravity does positive work if the body moves downward. (Indeed, a good rule of thumb to remember is that a force does positive work if it "gets to move in the direction it wants to" — that is, it has a component in the direction of the motion of the point on which it acts. If it does not, it does negative work during the motion of that point.)

Type 4: $\mathbf{F}_1$ is the normal force exerted at the point of contact on a rigid body that is maintaining contact with a fixed surface, whether rolling or slipping. Note in the lower portion of Figure 5.7 that the normal force $\mathbf{F}_1$ is always perpendicular to the velocity of P. That is,

$$W = \int \mathbf{F}_1 \cdot \mathbf{v}_P \, dt = 0$$

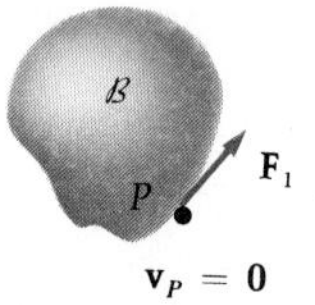

Figure 5.8

Type 5: $\mathbf{F}_1$ is the friction force exerted at the point of contact when a rigid body rolls on a fixed surface (Figure 5.8). This time, the force $\mathbf{F}_1$ (which may or may not be zero) does zero work because it always acts on a point of zero velocity:

$$W = \int \mathbf{F}_1 \cdot \cancelto{0}{\mathbf{v}_P} \, dt = 0$$

Type 6: $\mathbf{F}_1$ is the force in a linear spring connected to the same two points P and Q of bodies $\mathcal{B}$ and $\mathcal{R}$ during an interval of their motions. (See Figure 5.9.) We denote:

k = spring modulus (which when multiplied by the stretch yields the force in the linear spring)

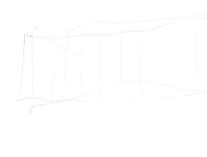

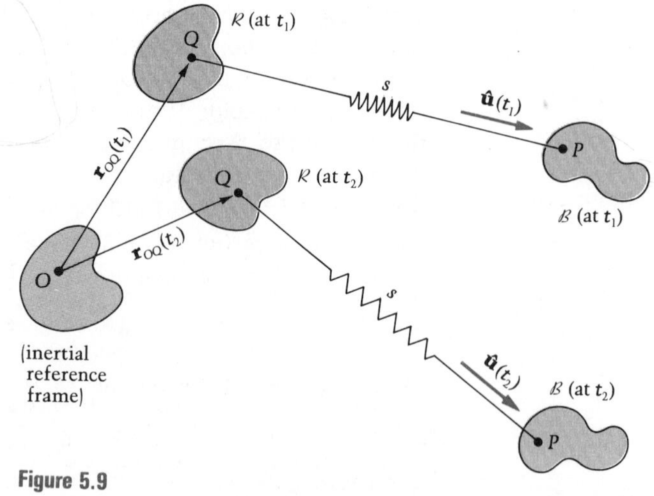

Figure 5.9

l_u = unstretched length

δ = stretch ($\delta < 0$ if compressed)

$\hat{\mathbf{u}}$ = unit vector along spring toward body $\mathcal{B}$

We first note that the work of spring S on body $\mathcal{B}$ is

$$W_{s \text{ on } \mathcal{B}} = \int_{t_1}^{t_2} \mathbf{F} \cdot \mathbf{v}_P \, dt = \int_{t_1}^{t_2} -k\delta \, \hat{\mathbf{u}} \cdot \mathbf{v}_P \, dt$$

Using

$$\mathbf{r}_{OP} = \mathbf{r}_{OQ} + (l_u + \delta)\hat{\mathbf{u}}$$

we may differentiate and obtain

$$\mathbf{v}_P = \mathbf{v}_Q + \dot{\delta}\hat{\mathbf{u}} + (l_u + \delta)\dot{\hat{\mathbf{u}}}$$

Therefore, substituting for $\mathbf{v}_P$, we get

$$W_{s \text{ on } \mathcal{B}} = \int_{t_1}^{t_2} -k \, \delta\hat{\mathbf{u}} \cdot [\mathbf{v}_Q + \dot{\delta}\hat{\mathbf{u}} + (l_u + \delta)\dot{\hat{\mathbf{u}}}] \, dt$$

$$= -\int_{t_1}^{t_2} k \, \delta\hat{\mathbf{u}} \cdot \mathbf{v}_Q \, dt - k \int_{t_1}^{t_2} \delta\dot{\delta} \, dt - k \int_{t_1}^{t_2} \delta(l_u + \delta)\hat{\mathbf{u}} \cdot \dot{\hat{\mathbf{u}}} \, dt$$

Since the derivative of a unit vector is perpendicular to the unit vector, the last integral vanishes and we obtain

$$W_{s \text{ on } \mathcal{B}} = -W_{s \text{ on } \mathcal{R}} - k \int_{\delta_1}^{\delta_2} \delta \, d\delta$$

Thus

$$W_{s \text{ on } \mathcal{B}} + W_{s \text{ on } \mathcal{R}} = W_{s \text{ on system of } (\mathcal{B} + \mathcal{R})} = \frac{k}{2}(\delta_1^2 - \delta_2^2) \qquad (5.19)$$

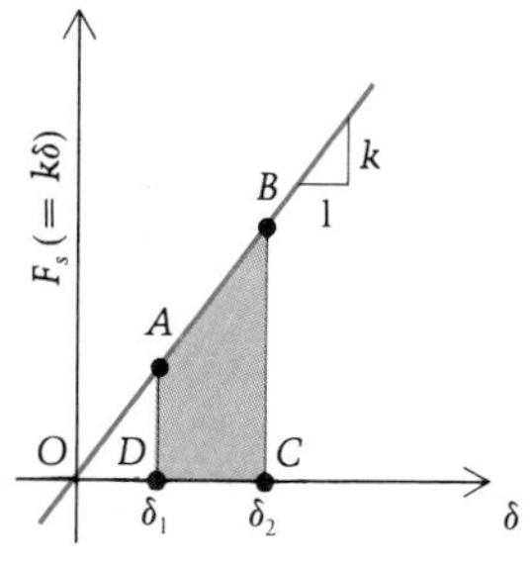

If Q is fixed in the inertial reference frame, the work of S on $\mathcal{B}$ alone is given by the right side of (5.19);* if Q moves, however, we can only say that the total work on *both* bodies by S is given by $(k/2)(\delta_1^2 - \delta_2^2)$.

We note from the spring's force-stretch diagram (Figure 5.10) that the work done by the spring is in fact the negative of the change in energy E stored in it; namely, in stretching from δ_1 to δ_2,

$$E = (\text{area of triangle } OCB) - (\text{are of } ODA)$$

$$= \frac{k}{2}(\delta_2^2 - \delta_1^2)$$

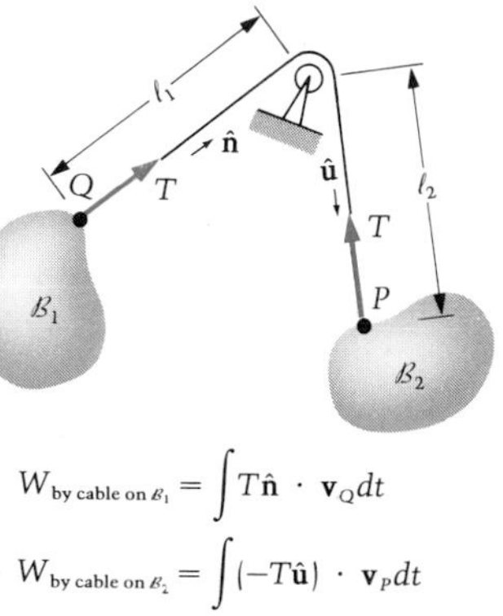

Type 7: We now consider the work done by the force in an inextensible cable (or rope, string, cord) connected to two points P and Q of bodies $\mathcal{B}_1$ and $\mathcal{B}_2$ during an interval of their motions (Figure 5.11). The cable under consideration may pass over one or more light, frictionless pulleys

Figure 5.10

$$W_{\text{by cable on } \mathcal{B}_1} = \int T\hat{\mathbf{n}} \cdot \mathbf{v}_Q dt$$

$$W_{\text{by cable on } \mathcal{B}_2} = \int (-T\hat{\mathbf{u}}) \cdot \mathbf{v}_P dt$$

Figure 5.11

* And its work on $\mathcal{R}$ is of course then zero.

between the bodies, but it is assumed to remain taut throughout the motion.

The work done by the cable tension on the system of $\mathcal{B}_1$ *plus* $\mathcal{B}_2$ is zero, which we proceed to prove as follows:

We write $\mathbf{v}_Q$ and $\mathbf{v}_P$ in terms of their components parallel and perpendicular to the cord:

$$\mathbf{v}_Q = \mathbf{v}_{Q\parallel} + \mathbf{v}_{Q\perp}$$
$$\mathbf{v}_P = \mathbf{v}_{P\parallel} + \mathbf{v}_{P\perp}$$

Noting that the perpendicular components $\mathbf{v}_{Q\perp}$ and $\mathbf{v}_{P\perp}$ have zero dot products with the unit vectors $\hat{\mathbf{n}}$ and $\hat{\mathbf{u}}$ (see Figure 5.11), we obtain for the works of the tensions:

$$W_{\text{by cable on } \mathcal{B}_1} = \int T\hat{\mathbf{n}} \cdot \mathbf{v}_{Q\parallel}\, dt = \int T\hat{\mathbf{n}} \cdot \frac{d\ell_1}{dt}(-\hat{\mathbf{n}})\, dt$$
$$= \int (-T)\, d\ell_1$$

and

$$W_{\text{by cable on } \mathcal{B}_2} = \int (-T\hat{\mathbf{u}}) \cdot \mathbf{v}_{P\parallel}\, dt = \int -T\hat{\mathbf{u}} \cdot \frac{d\ell_2}{dt}\hat{\mathbf{u}}\, dt$$
$$= \int (-T)\, d\ell_2$$

But by the cable's inextensibility, $d(\ell_1 + \ell_2) = 0$ so that $d\ell_2 = -d\ell_1$ and

$$W_{\text{by cable on } \mathcal{B}_2} = \int T\, d\ell_1 = -W_{\text{by cable on } \mathcal{B}_1}$$

so that

$$W_{\text{by cable on both bodies}} = 0$$

Type 8: We have a couple **C**. In this case, the work of the couple in plane motion is given by

$$W = \int_{t_1}^{t_2} \mathbf{C} \cdot \boldsymbol{\omega}\, dt = \int_{t_1}^{t_2} C\hat{\mathbf{k}} \cdot \dot{\theta}\hat{\mathbf{k}}\, dt$$
$$= \int_{t_1}^{t_2} C\dot{\theta}\, dt \quad \text{or} \quad \int_{\theta_1}^{\theta_2} C\, d\theta \tag{5.20}$$

Thus if C is constant, the work of the couple is given by

$$W = C(\theta_2 - \theta_1) \tag{5.21}$$

That is, the work of C is the strength of the couple times the angle through which the body turns. As with the work of forces, the couple's work is positive if it "gets to move" in the direction in which it acts (or turns, in this case).

Examples Solved by the Principle $W = \Delta T$

We are now in a position to solve some problems by using the principle of work and kinetic energy. A number of examples follow. In the first, work is done only by gravity, and $W = \Delta T$ is used to supplement the equations of motion.

EXAMPLE 5.3

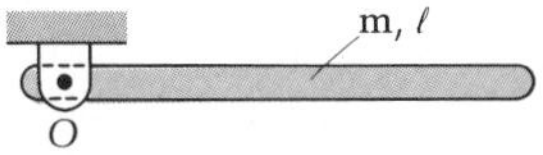

Figure E5.3a

Find the pin reaction at O when the uniform bar in Figure E5.3a has fallen through 45° from rest.

Solution

We first find the angular speed ω_2 in the final (45°) position by using the principle of work and kinetic energy $W = \Delta T$.

Letting T_2 be the kinetic energy in the final position, and noting that the work done by gravity $= mgh = mg[\ell/(2\sqrt{2})]$, we find:

$$mg\frac{\ell}{2\sqrt{2}} = \frac{1}{2}\overbrace{\left(\frac{1}{3}m\ell^2\right)}^{I_{①} = I_O}\omega_2^2 \Rightarrow \omega_2^2 = \frac{3g}{\sqrt{2}\ell} \tag{1}$$

We must now return to the differential equations to obtain equations in the desired reaction. (Note that $W = \Delta T$ alone can only give us the solution to one scalar unknown!) In the final position, we have:

$$\Sigma\mathbf{F} = m\mathbf{a}_C$$

Expressing this equation in its tangential and normal components with the help of the free-body diagram (Figure E5.3b),

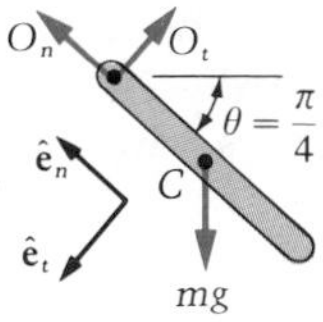

Figure E5.3b

$$\left(O_n - mg\frac{1}{\sqrt{2}}\right)\hat{\mathbf{e}}_n + \left(mg\frac{1}{\sqrt{2}} - O_t\right)\hat{\mathbf{e}}_t = m\mathbf{a}_C \tag{2}$$

But

$$\mathbf{a}_C = \overset{0}{\cancel{\mathbf{a}_O}} + \alpha\hat{\mathbf{k}} \times \mathbf{r}_{OC} - \omega^2\mathbf{r}_{OC}$$

and with $\hat{\mathbf{k}}$ defined as $\hat{\mathbf{e}}_t \times \hat{\mathbf{e}}_n$,

$$\mathbf{a}_C = \frac{\ell}{2}\alpha\hat{\mathbf{e}}_t + \frac{\ell}{2}\omega^2\hat{\mathbf{e}}_n$$

so that, from the $\hat{\mathbf{e}}_n$ component of Equation (2),

$$O_n - mg\frac{1}{\sqrt{2}} = m\frac{\ell}{2}\omega_2^2 = m\frac{\ell}{2}\frac{3g}{\sqrt{2}\ell}$$

where we have substituted ω_2^2 from Equation (1). Therefore, the normal component of the reaction is

$$O_n = \frac{5}{2\sqrt{2}}mg \tag{3}$$

Next, from the $\hat{\mathbf{e}}_t$ component of Equation (2),

$$mg\frac{1}{\sqrt{2}} - O_t = m\frac{\ell}{2}\alpha \tag{4}$$

Also, since point O is a pivot of the rod, we know:

$$\Sigma M_O = I_O\alpha$$

$$\frac{mg}{\sqrt{2}}\frac{\ell}{2} = \left(\frac{1}{3}m\ell^2\right)\alpha \Rightarrow \alpha = \frac{3g}{2\sqrt{2}\ell} \tag{5}$$

Substituting α from Equation (5) into (4) gives the tangential component of the reaction:

$$O_t = \frac{mg}{\sqrt{2}} - m\frac{\ell}{2}\frac{3g}{2\sqrt{2}\ell} = \frac{mg}{4\sqrt{2}}$$

Thus the pin reaction is

$$O_n\hat{\mathbf{e}}_n + O_t\hat{\mathbf{e}}_t = \frac{5mg}{2\sqrt{2}}\hat{\mathbf{e}}_n + \frac{mg}{4\sqrt{2}}\hat{\mathbf{e}}_t$$

$$= (1.77\hat{\mathbf{e}}_n + 0.177\hat{\mathbf{e}}_t)mg$$

In the second example, work is done only by a spring; however, calculating the final stretch is tricky.

EXAMPLE 5.4

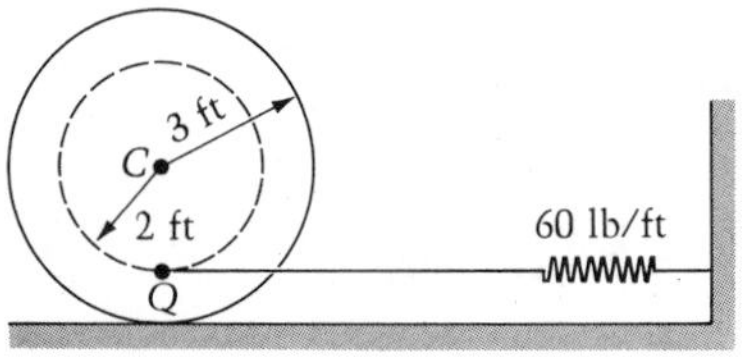

Figure E5.4a

One end of the linear spring in Figure E5.4a is attached to a thin inextensible cord that is lightly wrapped around a narrow groove in the wheel (mass = 1 slug, radius of gyration about center = 1.5 ft). If the wheel rolls, and starts from rest when the spring is stretched 1 ft, find the velocity of the center of the wheel when the center has moved 2 ft. The mass center of the wheel coincides with the geometric center.

Solution

We first note that the cord is not attached to a specific material point on the wheel. However, as time passes, the various "wrapping points" on the end of the straight portion of the cord (such as Q in Figure E5.4b) have, at every instant the cord is taut, the same velocity as the coincident point of the wheel at Q. Thus Equation (5.19) gives the work done on the wheel by the spring.

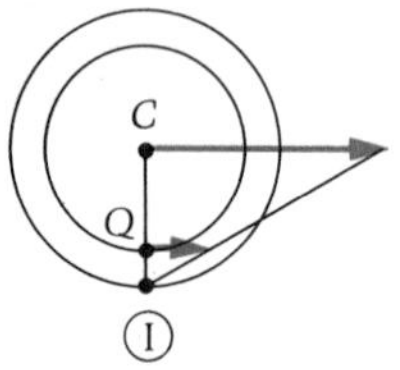

Figure E5.4b

We have at all times, by kinematics (see the figure),

$$\dot{x}_C = 3\omega \qquad \text{and} \qquad \dot{x}_Q = 1\omega$$

so that

$$\dot{x}_C = 3\dot{x}_Q$$

or

$$x_C = 3x_Q$$

Thus the net shortening of the spring when C has moved 2 ft to the right is $\frac{2}{3}$ ft. Another way to see this is to let C move to the right the amount x_C. This compresses the spring (if it were able to do so!) the same amount, x_C. Then turn the wheel clockwise about C through the angle $\theta = x_C/R = x_C/3$ radians, until the correct point is on the ground. ("Correct" means the point that would be on the ground had the wheel rolled normally over to the final position.) The rotation wraps $r\theta = 2\theta = 2x_C/3$ of string around the inner radius and "takes back" $\frac{2}{3}x_C$ of the compression. Thus $\frac{1}{3}x_C = \frac{1}{3}(2)$ is the reduction in the original 1 ft of stretch, leaving $\frac{1}{3}$ ft, as before. Hence

$$\delta_1 = 1 \text{ ft} \qquad \text{and} \qquad \delta_2 = 1 - \frac{2}{3} = \frac{1}{3} \text{ ft}$$

We then find, noting that gravity does no work here,

$$W = \Delta T = T_2 - \cancelto{0}{T_1}$$

$$\frac{60}{2}\left[(1)^2 - \left(\frac{1}{3}\right)^2\right] = \frac{1}{2}(1)v_C^2 + \frac{1}{2}[1(1.5)^2](\cancelto{v_C/3}{\omega})^2$$

$$v_C = \sqrt{42.7} = 6.53$$

$$\mathbf{v}_C = 6.53 \rightarrow \text{ft/sec}$$

We note that when the center has moved 3 ft, then $\frac{3}{3} = 1 = x_Q$ and all the stretch is gone. At this time, the spring would simply drop out of the problem.

The next example is an actual application of $W = \Delta T$ from industry.

Example 5.5

This example involves a practical application in the antenna industry of the work and kinetic energy principle. The antenna positioner in Figure E5.5 is equipped with a mechanical stop spring so that if the elevation drive overruns its lower limit, the antenna motion (a pure rotation about the horizontal elevation rotation axis) will be arrested before the reflector strikes another part and is damaged.

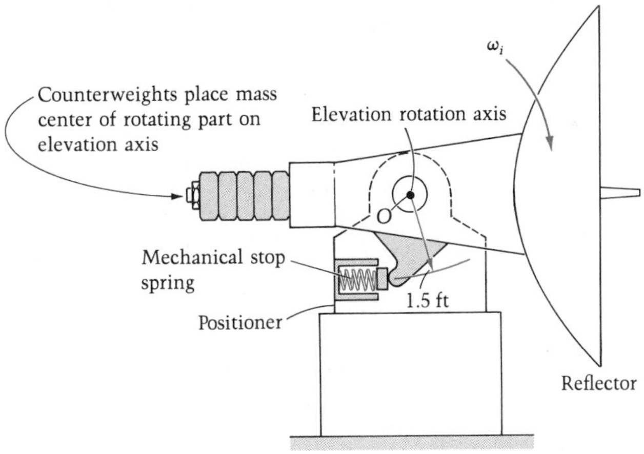

Figure E5.5

The elevation motor has an armature rotational mass moment of inertia of 0.01 lb-ft-sec^2 (or slug-ft^2) and drives the reflector through a gear reducer with a 700 : 1 gear ratio. The combined moment of inertia of the reflector, its counterweights, and the supporting structure is 12,000 slug-ft^2 = I_O.

It is desired to arrest a rotational speed of 30°/sec during a rotation from contact to full stop of 3°. The radius from the elevation rotation axis to the stop spring is 1.5 ft. The spring is unstretched at initial contact and may be assumed to have linear load-deflection behavior. It is further assumed that the motor is switched off but remains mechanically coupled while the rotation is being arrested. Find:

a. The required stiffness of the spring.

b. The maximum force induced in it.

c. The rotational position when it sustains its maximum force.

d. The angular accelerations of the reflector and motor armature at the position of maximum force. (Are these the maximum accelerations?)

Solution

Since the spring is linear, its greatest force is the spring stiffness times the maximum deflection. This is also the position for which motion is completely arrested. At this position the kinetic energy has been brought to zero with the stop spring storing the energy; the principle $W = \Delta T$ gives

$$W = \frac{1}{2}k(\overset{0}{\cancel{\delta_i^2}} - \delta_f^2) = \Delta T = \frac{1}{2}I_O\overset{0}{\cancel{\omega_f^2}} - \frac{1}{2}I_O\omega_i^2$$

Note that point O is Ⓘ for the rotating body and that gravity does no work between contact and stop.

Question 5.3 Why does gravity do no work?

The values of δ_f, I_O, and ω_i needed in the equation are calculated as follows:

$$\begin{aligned} I_O &= \text{total moment of inertia at axis of rotation} \\ &= I_{\text{motor}_O} + I_{(\text{reflector, counterweights, structure})_O} \\ &= 0.01 \times 700^2\,{*} + 12{,}000 = 16{,}900 \text{ slug-ft}^2 \end{aligned}$$

$$\omega_i = 30 \times \frac{\pi}{180} = 0.524 \text{ rad/sec}$$

$$\delta_f = 3 \times \frac{\pi}{180} \times 1.5 = 0.0785 \text{ ft}$$

(Note that over the very small angle of 3° the spring compression is approximately the arclength $R\theta$.)

Solving for the spring's stiffness, we get

$$k = \frac{I_O\omega_i^2}{\delta_f^2} = \frac{16{,}900 \times 0.524^2}{0.0785^2}$$

$$= 753{,}000 \text{ lb/ft}$$

The maximum spring force $= k\delta_f = 753{,}000 \times 0.0785 = 59{,}100$ lb. The rotational position is 3° beyond contact — that is, the position at full stop. The

---

Answer 5.3 Since the counterweights place the mass center on the elevation axis, the mass center does not move.

* As the reader may wish to prove, moments of inertia reflect through gear trains from input to output with the gear ratio squared as a factor; also, the torque increases (while the speed decreases) with the gear ratio as the multiplying factor.

angular acceleration of the reflector is

$$\alpha = \frac{\Sigma M_O}{I_O} = \frac{1.5 \times 59{,}100}{16{,}900} = 5.25 \text{ rad/sec}^2$$

and that of the motor armature is $5.25 \times 700^* = 3680$ rad/sec^2.

These are the maximum accelerations, since here the force (and torque) are greatest. In closing, we note that motor torque and friction, omitted in this problem for simplicity, limit the rebound in the actual case.

The next example illustrates work done by a force acting on different points of the body as time passes. A shortcut for calculating this work is presented.

Example 5.6

This example illustrates the work done by forces and couples belonging to Types 1, 2, and 8 on the preceding pages. The force **F** (52 lb) is applied to the uniform cylinder $\mathcal{C}$ at rest in Figure E5.6a at the left. (This type of force might be applied by a cord on a hub, as is suggested by Figure E5.6b.) If force **F** continues to act with the same magnitude and direction as the cylinder rolls, find:

a. The work done by **F** during transit to the dashed position
b. The velocity of C and the angular velocity of the cylinder in the dashed position

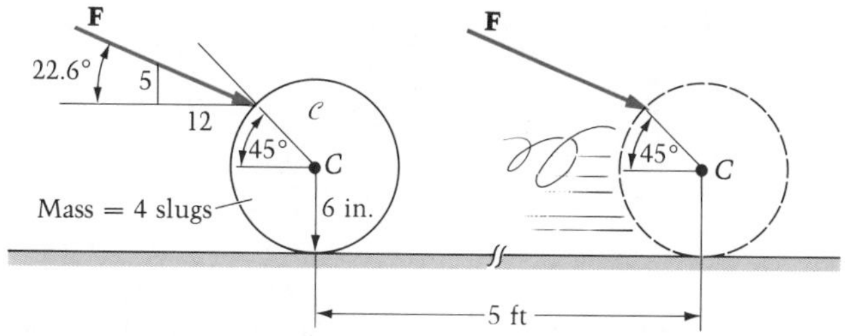

Figure E5.6a

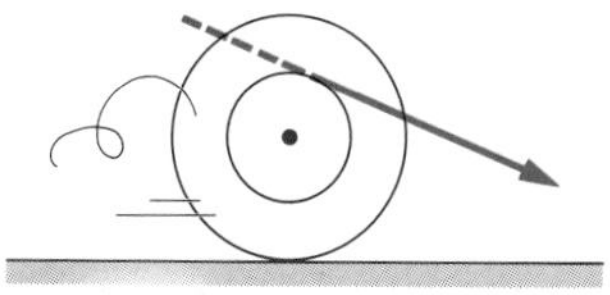

Figure E5.6b

Solution

We shall work part (a) in two ways. First, the definition of the work of **F** is

$$W = \int \mathbf{F} \cdot \mathbf{v}_Q \, dt$$

where Q is the point of $\mathcal{C}$ in contact with **F** at any time. The geometry in Figure E5.6c gives an angle of 45.1° between **F** and $\mathbf{v}_Q$, since

$$\alpha = \tan^{-1}\left(\frac{r/\sqrt{2}}{r + r/\sqrt{2}}\right) = 22.5°$$

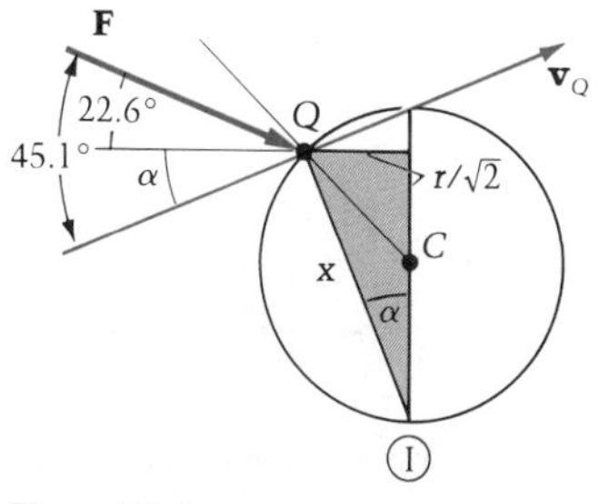

Figure E5.6c

* See preceding footnote.

Also,

$$r_{①Q} = \sqrt{\left(\frac{1}{\sqrt{2}}\right)^2 + \left(1 + \frac{1}{\sqrt{2}}\right)^2}\, r = 1.85r$$

Therefore

$$\mathbf{F} \cdot \mathbf{v}_Q = F(r_{①Q}\dot{\theta}) \cos 45.1°$$

and

$$W = \int (52 \cos 45.1°) 1.85r \frac{d\theta}{dt}\, dt$$

$$= (52 \cos 45.1°)(1.85) \frac{\theta}{2}\Big]_0^{5/0.5}$$

$$= 340 \text{ ft-lb}$$

A second and simpler approach is to note that **F** at Q may be moved to C, using the idea of resultants, as in Figure E5.6d. The force at Q is replaced by the force and couple at C which produce the same effect on the rigid body. Since, at C, force **F** always acts on the *same point of the body* (it didn't at Q!) we may write

$$W = \text{work of } \mathbf{F} \text{ at } Q = (\text{work of } \mathbf{F} \text{ at } C) + (\text{work of couple on } \mathcal{C})$$

$$= (F \cos 22.6°)x_C + (Fr \sin 22.4°)\theta = (0.923F)5 + (0.191F)10$$

$$= 6.53F = 340 \text{ ft-lb} \qquad \text{(as before)}$$

Note that the work of a constant couple in plane motion is simply the moment of the couple times the angle through which the body turns.

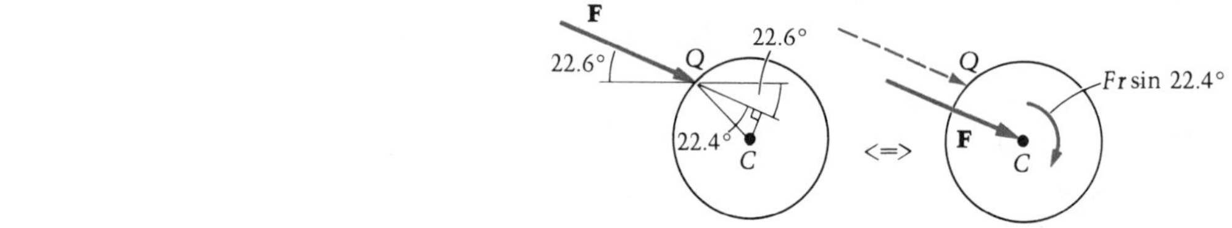

Figure E5.6d

For part (b) we equate the work to the change in the kinetic energy of $\mathcal{C}$:

$$W = \Delta T = \frac{1}{2} m v_{C_f}^2 + \frac{1}{2} I_C \omega_f^2 - 0 \qquad (\text{initial } T = 0)$$

$$340 = \frac{1}{2} 4 v_{C_f}^2 + \frac{1}{2}\left(\frac{1}{2} \cdot 4 \cdot \frac{1}{4}\right)\left(\frac{v_{C_f}}{1/2}\right)^2$$

$$v_{C_f} = \sqrt{\frac{340}{2+1}} = 10.6 \text{ ft/sec}$$

Note that the gravity, friction, and normal forces do no work in this problem, for the reasons given in Types 3, 4, and 5 of the text preceding the examples.

The hardest part of the next example is finding where the mass center is in the final position!

Example 5.7

The unstretched length of the spring in Figure E5.7a is $\ell_u = 0.3$ m. The initial angular velocity of body $\mathcal{A}$ in the top position is $\omega_i = 2.5 \circlearrowright$ rad/s. There is enough friction to prevent slipping of $\mathcal{A}$ on $\mathcal{B}$ at all times. Determine the modulus of the spring that will cause $\mathcal{A}$ to stop in the $\varphi = 90°$ position.

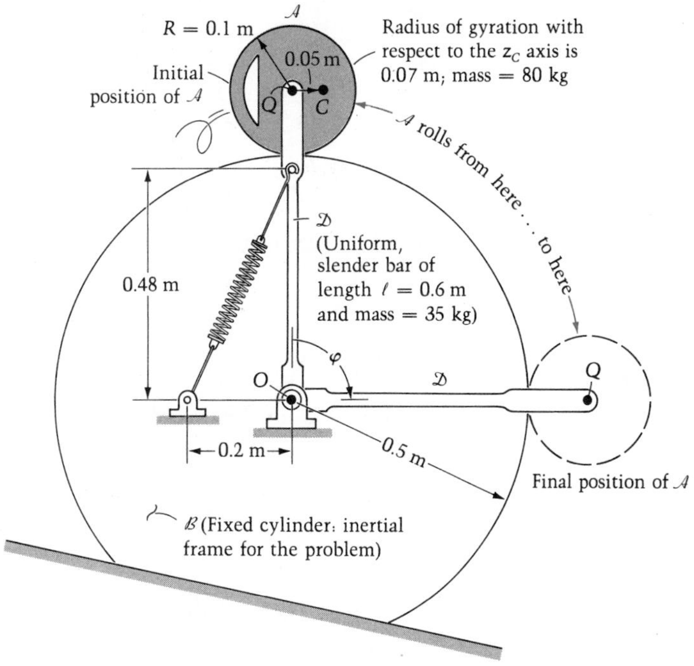

Figure E5.7a

Solution

Part of the work W in this problem is done by gravity. To express this work, we must determine where the mass center C is located when $\mathcal{A}$ reaches the final position. Reviewing the kinematics, we find that the velocity of the geometric center Q of $\mathcal{A}$ (see Figure E5.7b) is expressible in two ways:

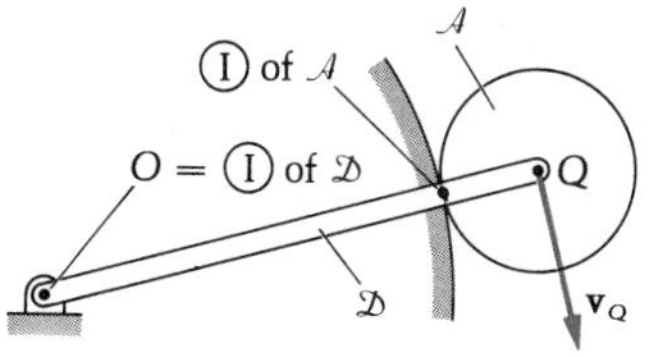

Figure E5.7b

1. As a point of $\mathcal{A}$, $\mathbf{v}_Q = \overset{0}{\mathbf{v}_{\text{Ⓘ}}} + R\dot{\theta}\hat{\mathbf{e}}_t$.

2. As a point of $\mathcal{D}$, $\mathbf{v}_Q = \overset{0}{\mathbf{v}_O} + \ell\dot{\varphi}\hat{\mathbf{e}}_t$.

Thus we see that $R\dot{\theta} = \ell\dot{\varphi}$. Integrating, we get

$$R\theta = \ell\varphi$$

in which the constant of integration is zero if we select $\theta = 0$ when $\varphi = 0$. Therefore when $\varphi = \pi/2$, we may find the orientation of body $\mathcal{A}$:

θ = angle that body $\mathcal{A}$ turns through in reference frame (angle seen by *stationary* observer in body $\mathcal{B}$)

$$= \frac{\ell}{R}\varphi = \frac{0.6}{0.1} \times \frac{\pi}{2} = 3\pi$$

And so the final position of C is to the left of Q (see Figure E5.7c). We can now write the work of gravity W_g because we now know the h moved through by C:

$$W_g = (mgh)_{\mathcal{D}} + (mgh)_{\mathcal{A}} = 35(9.81)(0.3) + 80(9.81)(0.6) = 574 \text{ J}$$

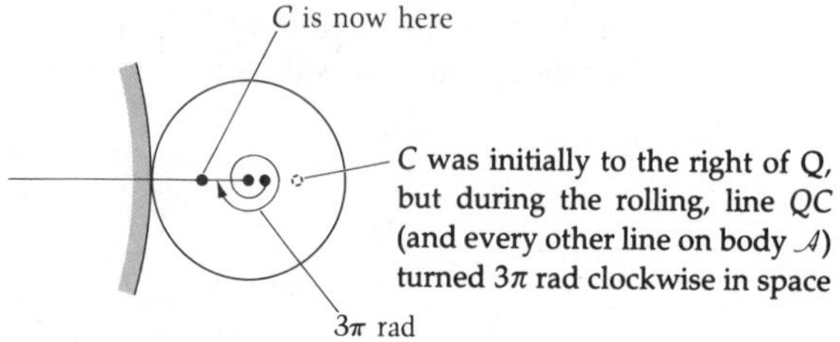

Figure E5.7c

The work done by the linear spring is *always* given by $(k/2)(\delta_i^2 - \delta_f^2)$:

$$W_s = \frac{k}{2}(\delta_i^2 - \delta_f^2) = \frac{k}{2}(0.220^2 - 0.380^2)$$

$$= -0.0480k \text{ J}$$

where k is our unknown and the initial and final stretches are computed as follows:

$$\ell_i = \ell_u + \delta_i \qquad \text{(unstretched length plus initial stretch = initial length of spring)}$$

and

$$\ell_f = \ell_u + \delta_f$$

so that the stretches are

$$\delta_i = \sqrt{0.2^2 + 0.48^2} - 0.3 = 0.520 - 0.300 = 0.220 \text{ m}$$

$$\delta_f = (0.48 + 0.2) - 0.3 = 0.680 - 0.300 = 0.380 \text{ m}$$

At $\varphi = 0°$:

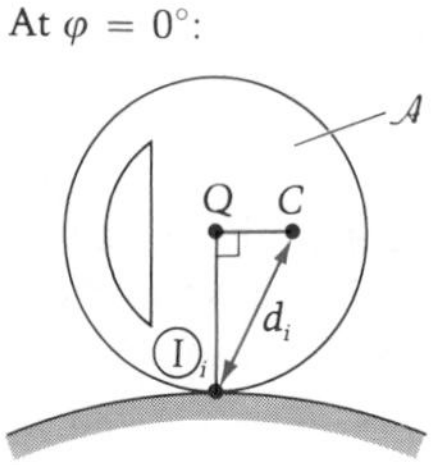

Figure E5.7d

For the kinetic energy side of $W = \Delta T$, we need the moments of inertia; first we consider body $\mathcal{A}$:

$$I_C = mk_C^2 = 80(0.07^2) = 0.392 \text{ kg-m}^2$$

We shall use the "short form" of T — namely $T = \frac{1}{2}I_{\text{Ⓘ}}\omega^2$ (always valid whenever $\omega \neq 0$ in plane motion). Thus we need $I_{\text{Ⓘ}i}$ and $I_{\text{Ⓘ}f}$.* Note that when Ⓘ is a different point of a body in the initial and final positions, the value of $I_{\text{Ⓘ}}$ is generally different in the two configurations, as is the case in this problem. Using Figures E5.7d and E5.7e, we find:

$$I_{\text{Ⓘ}i} = I_C + md_i^2 = 0.392 + 80(0.1^2 + 0.05^2) = 1.39 \text{ kg}\cdot\text{m}^2$$

$$I_{\text{Ⓘ}f} = I_C + md_f^2 = 0.392 + 80(0.1 - 0.05)^2 = 0.592 \text{ kg}\cdot\text{m}^2$$

At $\varphi = \pi/2$:

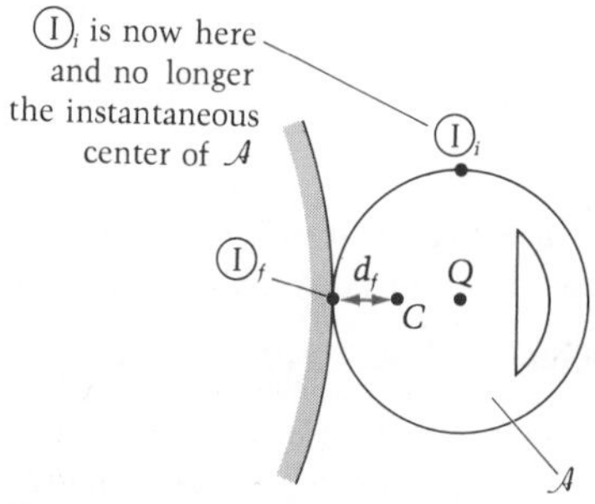

Figure E5.7e

* Since $\omega_f = 0$, we do not have to calculate $I_{\text{Ⓘ}f}$ here, but we do so to illustrate the procedure in general.

Therefore the kinetic energies of $\mathcal{A}$ we need are

$$T_i^{\mathcal{A}} = \frac{1}{2} I_{Ⓘi}\omega_i^2 = \frac{1}{2}(1.39)2.5^2 = 4.34 \text{ J}$$

$$T_f^{\mathcal{A}} = \frac{1}{2} I_{Ⓘf}\omega_f^2 = \frac{1}{2}(0.592)0^2 = 0 \text{ J}$$ (since the final angular speed is to be zero)

For the bar $\mathcal{D}$, $I_Ⓘ$ is the same in any position since Ⓘ is point O, which is pinned to the reference frame. Therefore, using $v_Q = R\omega_{\mathcal{A}}$, we have

$$T_i^{\mathcal{D}} = \frac{1}{2} I_{Ⓘi}\omega_i^2 = \frac{1}{2}\left(\frac{m\ell^2}{3}\right)\left(\frac{v_{Qi}}{r_{ⒾQ}}\right)^2 = \frac{1}{2}\left(\frac{35 \times 0.6^2}{3}\right)\left(\frac{0.1 \times 2.5}{0.6}\right)^2$$

$$= 0.365 \text{ J}$$

$$T_f^{\mathcal{D}} = \frac{1}{2} I_{Ⓘf}\omega_f^2 = \frac{1}{2}\left(\frac{m\ell^2}{3}\right)\left(\frac{v_{Qf}}{r_{ⒾQ}}\right)^2 = 0$$ (since $v_{Qf} = 0.1\omega_{\mathcal{A}f} = 0$)

Applying the work and kinetic energy principle, we get

$$W = \Delta T$$

$$W_g + W_s = T_f - T_i = -T_i$$

$$574 - 0.0480k = 0 - (4.34 + 0.365)$$

$$k = 12{,}100 \text{ N/m}$$

This is equivalent to 829 lb/ft of stiffness in the U.S. system of units, since 1 lb/ft is the same stiffness as 14.6 N/m.

Question 5.4 What happens if k is larger than the calculated value? What happens if it is smaller?

In the next example work and kinetic energy is used to help determine the point where rolling stops and slipping starts.

Answer 5.4 Larger: the rod will not reach the horizontal position; smaller: the system passes through this position without stopping.

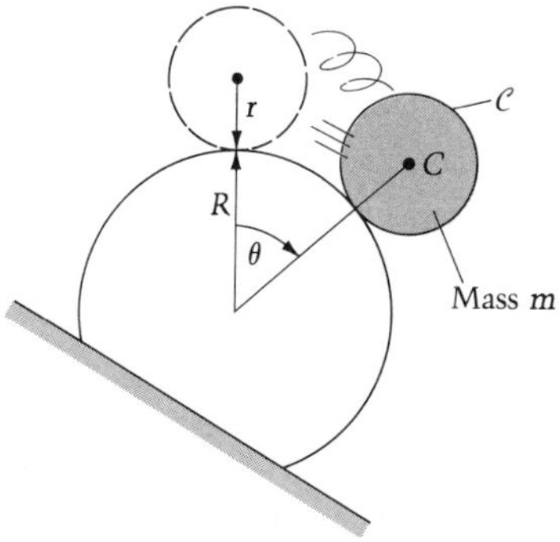

Figure E5.8a

EXAMPLE 5.8

As we found in Example 5.3, sometimes it is useful to combine the work and kinetic energy principle with one or more of the differential equations of motion in order to obtain a desired solution. This example involves such a combination. The small cylinder $\mathcal{C}$ starts from rest at $\theta = 0$ in the dotted position (see Figure E5.8a) and begins to roll down the large cylinder. Find the angle θ_s at which slipping starts, and show that the small cylinder will *always* slip before it leaves the surface for a finite coefficient of friction.

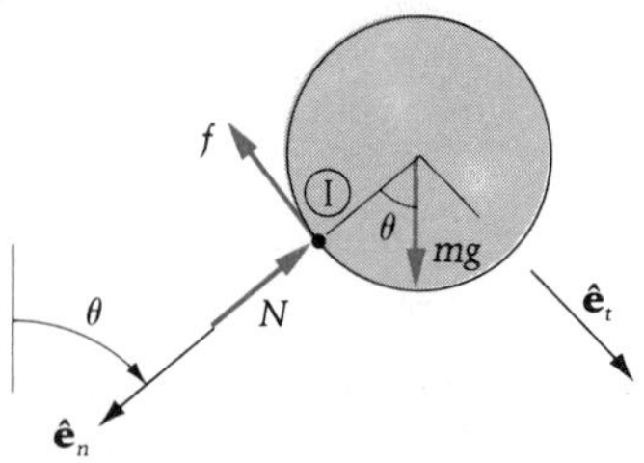

Figure E5.8b

Solution

Using the free-body diagram in Figure E5.8b, the equations of motion are

$$\Sigma F_n = mg\cos\theta - N = ma_{C_n} = \frac{mv_C^2}{R+r} = \frac{mr^2\omega^2}{R+r} \tag{1}$$

$$\Sigma F_t = mg\sin\theta - f = ma_{C_t} = m\ddot{s}_C = mr\alpha \tag{2}$$

$$\Sigma M_C = fr = I_C\alpha = \frac{mr^2}{2}\alpha \tag{3}$$

Just prior to slipping, the friction force $f \approx \mu N$ while a_C is still equal to $r\alpha$ and v_C is still equal to $r\omega$. Therefore the equations can be rewritten as

$$mg\cos\theta_s - N = \frac{mr^2\omega_s^2}{R+r} \tag{1a}$$

$$mg\sin\theta_s - \mu N = mr\alpha_s \tag{2a}$$

$$\mu N r = \frac{mr^2}{2}\alpha_s \tag{3a}$$

These equations may be supplemented with the work and kinetic energy equation for body $\mathcal{C}$, written between $\theta = 0$ and $\theta = \theta_s$:

$$W_g = mg(R+r)(1-\cos\theta_s) = \frac{1}{2}I_{①}\omega_s^2 = \frac{1}{2}\left(\frac{3}{2}mr^2\right)\omega_s^2 \tag{4}$$

Equations (1) to (4) may now be treated as four equations in the unknowns N, θ_s, ω_s^2, and α_s, where the last three quantities are the angle, angular velocity, and angular acceleration at slip. Solving them for θ_s yields the equation

$$7\mu\cos\theta_s - 4\mu = \sin\theta_s$$

Writing $\sqrt{1-\cos^2\theta_s}$ for $\sin\theta_s$, and then squaring and solving the resulting quadratic for $\cos\theta_s$, gives

$$\theta_s = \cos^{-1}\left(\frac{28\mu^2 + \sqrt{33\mu^2+1}}{1+49\mu^2}\right)$$

which plots as shown in Figure E5.8c.

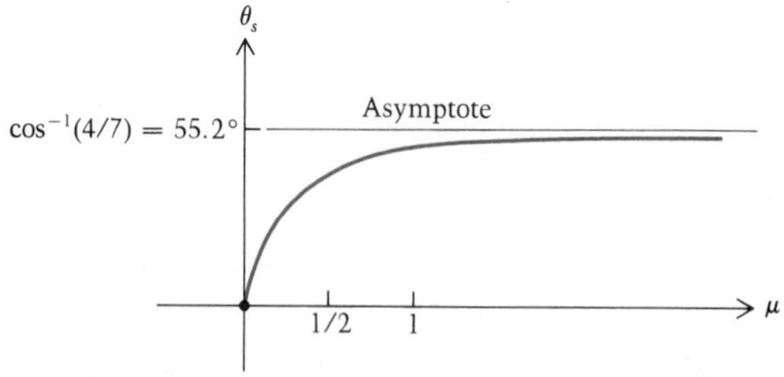

Figure E5.8c

The curve in the diagram gives the slipping angle as a function of the friction coefficient; this is *not* the angle at which body $\mathcal{C}$ leaves the surface. We note that if we were to look for the angle at which the small cylinder leaves the surface of the large cylinder, assuming no slipping has occurred, we would be trying to solve a problem with no solution if the friction coefficient is finite. The curve clearly shows that for $\mathcal{C}$ to reach the angle $\cos^{-1}(4/7)$, an infinite coefficient of friction is required. Since the solution to the "leaving without slip" problem is precisely $\cos^{-1}(4/7)$, shown below, then regardless of the friction coefficient (so long as it is finite) $\mathcal{C}$ will have to slip before it leaves.

Assuming now that the cylinder $\mathcal{C}$ leaves without having slipped, we obtain the (simpler) solution:

$$\text{Equation (1)} \Rightarrow mg\cos\theta_L - \overset{0 \text{ at leaving!}}{\cancel{N}} = \frac{mr^2\omega_L^2}{R+r}$$

$$\text{Equation (4)} \Rightarrow mg(R+r)(1-\cos\theta_L) = \frac{1}{2}\left(\frac{3}{2}mr^2\right)\omega_L^2$$

Eliminating ω_L gives

$$\theta_L = \cos^{-1}(4/7) = 55.2^\circ$$

As we have noted, this solution is valid only for an infinite coefficient of friction between the cylinders. If $\mathcal{C}$ were a *particle* (no rotational kinetic energy) with a smooth surface, we would obtain (Example 2.13) $\theta_L = \cos^{-1}(2/3) = 48.2^\circ$. Note the differences between these solutions.

Two Subcases of the Work and Kinetic Energy Principle

There is an important subcase of the principle of work and kinetic energy that we have already seen in Chapter 2. Using

$$\Sigma\mathbf{F} = \frac{d}{dt}(m\mathbf{v}_C)$$

we obtained

$$\int_{t_1}^{t_2} \Sigma\mathbf{F}\cdot\mathbf{v}_C\,dt = m\int_{t_1}^{t_2} \dot{\mathbf{v}}_C\cdot\mathbf{v}_C\,dt = \frac{1}{2}mv_C^2\Big]_{v_C(t_1)}^{v_C(t_2)} \tag{5.22}$$

This principle states again that the work done by the external force resultant, when considered to act on the mass center, equals the change in the translational part of the kinetic energy:

$$\int \Sigma\mathbf{F}\cdot\mathbf{v}_C\,dt = \Delta T_v \tag{5.23}$$

The integral of Equation (5.6) is

$$\int \Sigma\mathbf{F}\cdot\mathbf{v}_C\,dt + \int \Sigma\mathbf{M}_C\cdot\omega\hat{\mathbf{k}}\,dt = \int \frac{dT}{dt}\,dt = \Delta T = \Delta T_v + \Delta T_\omega \tag{5.24}$$

If we subtract (5.23) from (5.24), we obtain yet another result:

$$\int \Sigma\mathbf{M}_C\cdot\omega\hat{\mathbf{k}}\,dt = \Delta T_\omega = \Delta\left(\frac{1}{2}I_C\omega^2\right)$$

or

$$\int \Sigma M_C\,d\theta = \Delta T_\omega \tag{5.25}$$

This second subprinciple says that the work done by the external *moments* (about C) on the body, as it turns in the inertial frame, is equal to the change in the *rotational part* of the kinetic energy. We may use the "total" $W = \Delta T$ principle or either of its two "subparts" (Figure 5.12).

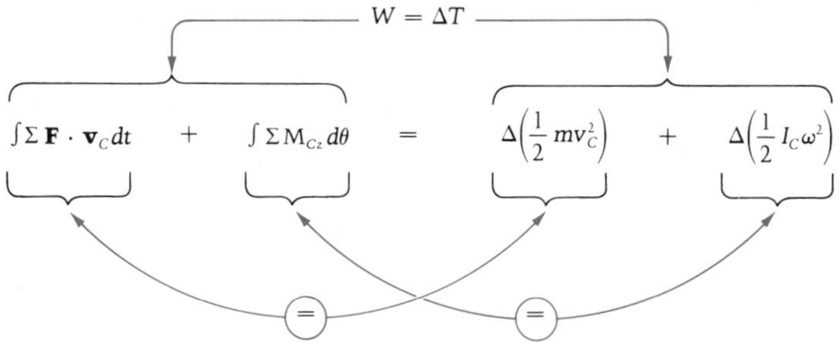

Figure 5.12

Let us now examine an example in which these subparts may be seen to add to the "total" $W = \Delta T$ equation:

EXAMPLE 5.9

As an illustration of the two subcases of the principle of work and kinetic energy, we consider the cylinder of mass m rolling down the inclined plane shown in Figure E5.9a. If the cylinder is released from rest, find the velocity v_C of its mass center as a function of the distance x_C traveled by C.

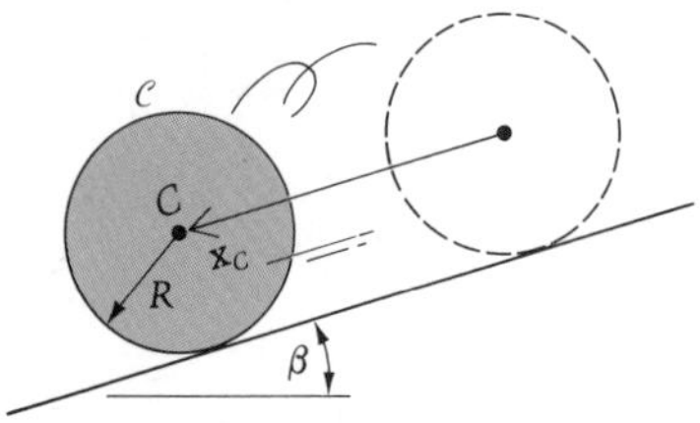

Figure E5.9a

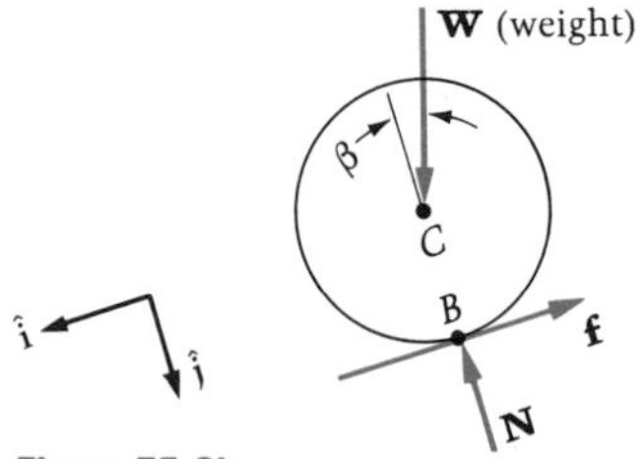

Figure E5.9b

Solution

Referring to the free-body diagram in Figure E5.9b, we see that the normal and friction forces do no work because, as the cylinder rolls on the incline, they always act on a point at rest. That is,

$$\int \mathbf{N} \cdot \mathbf{v}_B \, dt = 0 \qquad \text{and} \qquad \int \mathbf{f} \cdot \mathbf{v}_B \, dt = 0$$

Applying the principle that $W = \Delta T$, only the component of the gravity force **W** that acts parallel to the plane does any work:

$$\int (mg \sin \beta \hat{\mathbf{i}}) \cdot v_C \hat{\mathbf{i}} \; dt = \frac{1}{2} m v_C^2 + \frac{1}{2} I_C \omega^2 \tag{1}$$

Since **W** always acts on the same point (C) of $\mathcal{C}$, and since $dx_C/dt = v_C = R\omega$,

$$\int mg \sin \beta \, dx_C = \frac{1}{2} m v_C^2 + \frac{1}{2}\left(\frac{1}{2} m R^2\right) \frac{v_C^2}{R^2} \tag{2}$$

The mass center's velocity is therefore

$$\mathbf{v}_C = v_C\hat{\mathbf{i}} = \sqrt{\frac{4g \sin \beta \, x_C}{3}}\,\hat{\mathbf{i}} \tag{3}$$

Now suppose we apply Equation (5.23):

$$\int \Sigma\mathbf{F} \cdot \mathbf{v}_C \, dt = \Delta\left(\frac{1}{2} m v_C^2\right) \tag{4}$$

In this problem the resultant force acting on $\mathcal{C}$ is

$$\Sigma\mathbf{F} = \mathbf{W} + \mathbf{f} + \mathbf{N} = mg(\sin\beta\hat{\mathbf{i}} + \cos\beta\hat{\mathbf{j}}) - f\hat{\mathbf{i}} - N\hat{\mathbf{j}}$$

and Equation (4) becomes

$$\int mg \sin\beta \, dx_C - \int f v_C \, dt = \frac{1}{2} m v_C^2 \tag{5}$$

We see that, as expected, the friction force (though it does no *net* work) retards the motion of the mass center C while *turning* the cylinder, as can be seen from the *other* subcase of $W = \Delta T$:

$$\underbrace{\int \Sigma\mathbf{M}_C \cdot \boldsymbol{\omega} \, dt}_{\Sigma M_C \, d\theta} = \Delta \frac{1}{2} I_C \omega^2 \tag{6}$$

$$\int fR \, d\theta = \frac{1}{2}\left(\frac{1}{2} mR^2\right)\omega^2 \tag{7}$$

or

$$\int f\frac{d(R\theta)}{dt} dt = \int f v_C \, dt = \frac{1}{2}\left(\frac{1}{2} mR^2\right)\frac{v_C^2}{R^2} \tag{8}$$

And the sum of Equations (5) and (8) indeed gives Equation (2): the total $W = \Delta T$ equation!

Potential Energy, Conservative Forces, and Conservation of Mechanical Energy

In Section 2.4 we introduced the concept of **potential energy**, or the potential of a force. When the work done by a force on a body is independent of the path taken as the body moves from one configuration to another, the force is said to be **conservative** and the work is expressible as the decrease in a scalar function φ, the potential (energy). Thus as a body moves from a configuration at time t_1 to a second configuration at time t_2, the work done by an external conservative force is

$$W = \varphi(t_1) - \varphi(t_2)$$

or simply

$$W = \varphi_1 - \varphi_2$$

If all the external forces that do work on a rigid body are conservative and

φ is now the *sum* of the potentials of those forces, Equation (5.15) yields

$$\varphi_1 - \varphi_2 = W = \Delta T = T_2 - T_1$$

or

$$T_2 + \varphi_2 = T_1 + \varphi_1$$

or

$$T + \varphi = \text{constant}$$

which expresses the **conservation of mechanical energy**.

From Chapter 2 and earlier in this section we can easily identify two common conservative forces: (1) the constant force acting always on the same material point in the body and (2) the force exerted on a body by a linear spring attached at one end to the body and at the other to a point fixed in the inertial frame of reference.

In the case of the constant force, a potential is $\varphi = -\mathbf{F} \cdot \mathbf{r}$, where $\mathbf{r}$ is a position vector for the point of application. When the force is that exerted by gravity (weight) on a body near the surface of the earth,

$$\varphi = mgz$$

where h is the altitude of the mass center of the body.

For the linear spring, we recall that $\varphi = (k/2)\delta^2$, where k is the spring modulus, or stiffness, and δ is the stretch. It is important to recognize that when a spring is attached to, or between, two bodies that are both moving (relative to the inertial frame), then $(k/2)\delta^2$ is a potential for the two spring forces *taken together* (see Equation 5.19). That is, while neither of the forces acting on the bodies can be judged by itself to be conservative, the net work done on the two bodies by the two forces is expressible as a decrease in the potential, $\varphi = (k/2)\delta^2$. This is helpful in the analysis of problems in which we have two or more interacting rigid bodies. We have already noted earlier in this section that the work of the *external* forces on a system of rigid bodies is not in general equal to the change in kinetic energy of the system; this is because there may be net work done on the rigid bodies by the equal and opposite forces of interaction. Suppose now that our system is made up of two bodies joined by a spring, and suppose the spring forces are the *only* internal ones that produce net work on the system. We may then write $W = \Delta T$ for each rigid body. Upon adding these equations there results

(Work of forces external to system) + (Work of pair of spring forces)
= (Change in kinetic energy of system)

If the forces external to the system that do work are conservative, we may add the various potential energies associated with them to that for the pair of spring forces and conclude that

$$T + \varphi = \text{constant}$$

That is to say, in this case the mechanical energy of the *system* is conserved.

An example of a *nonconservative* force is sliding friction. A potential cannot be found for friction, since the work it does depends on the path taken by the body on which it acts. In this case, $W = \Delta T$ must be used, and it is seen to be more general than the principle of conservation of mechanical energy.

Example 5.10

Show that the same equation for the spring modulus in Example 5.7 is obtained by conservation of mechanical energy. (See Figure E5.10.)

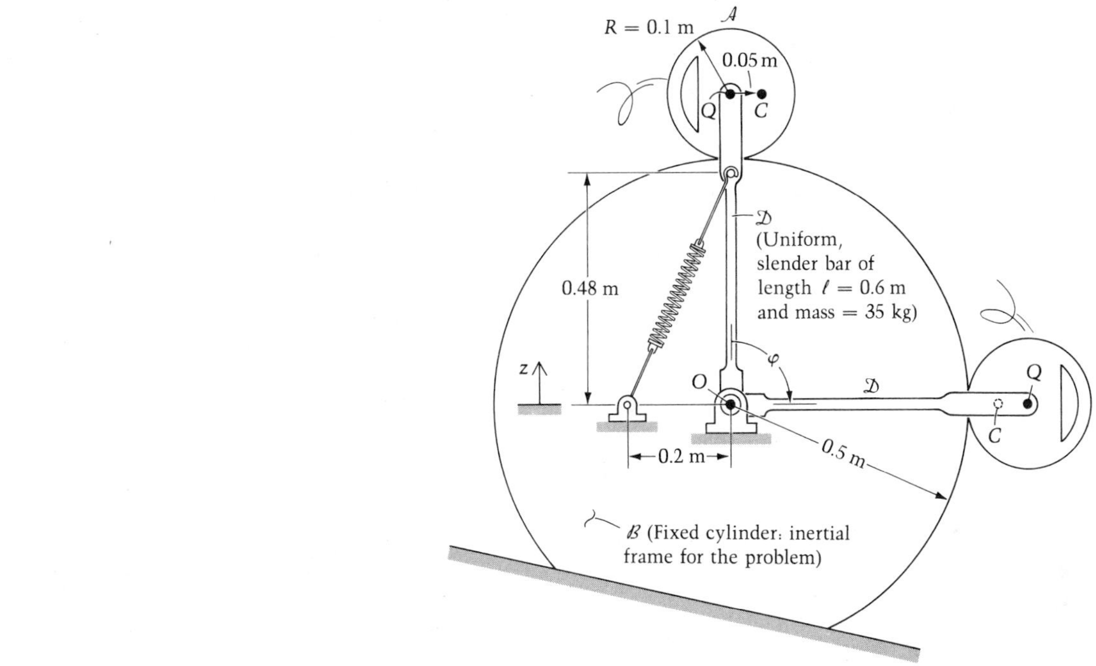

Figure E5.10

Solution

The potentials for gravity and for the spring are

$$\varphi_g = +m_{\mathcal{A}} g z_{C_{\mathcal{A}}} + m_{\mathcal{D}} g z_{C_{\mathcal{D}}} \qquad \varphi_{\text{spr}} = \frac{k\delta^2}{2}$$

Therefore, measuring z_C from O, we have

$$\begin{aligned}\varphi_{g_i} &= +80(9.81)(+0.6) + 35(9.81)(+0.3) \qquad & \varphi_{g_f} &= -mg(0)\\ &= +471 + 103 & & \\ &= +574 \text{ J} & &= 0\end{aligned}$$

For the spring, using i for the initial and f for the final configuration, we have

$$\varphi_{\text{spr}_i} = \frac{k(0.22)^2}{2} = 0.0242k \text{ J} \qquad \varphi_{\text{spr}_f} = \frac{k(0.38)^2}{2} = 0.0722k \text{ J}$$

Thus, adding the potentials ($\varphi = \varphi_g + \varphi_{spr}$), we get

$$\varphi_i = 574 + 0.0242k \text{ J} \qquad \varphi_f = 0 + 0.0722k \text{ J}$$

The kinetic energies were $T_i = 4.34 + 0.365 = 4.71$ J and $T_f = 0$. Therefore

$$\varphi_i + T_i = \varphi_f + T_f$$
$$574 + 0.0242k + 4.71 = 0.0722k + 0$$

or, rearranging,

$$574 - 0.0480k = -4.71$$

This is the same final equation that resulted from $W = \Delta T$ in the earlier Example 5.7.

PROBLEMS ▶ Section 5.2

5.1 Find the kinetic energy of the system of bodies $\mathcal{B}_1$, $\mathcal{B}_2$, and $\mathcal{B}_3$ at an instant when the speed of $\mathcal{B}_1$ is 5 ft/sec. (See Figure P5.1.)

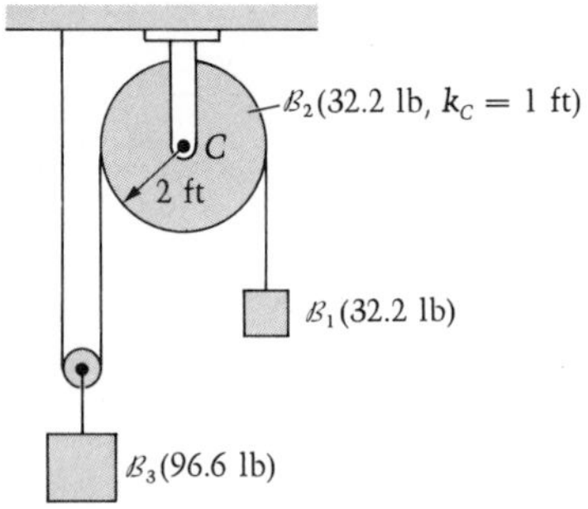

Figure P5.1

5.2 See Figure P5.2. (a) Explain why the friction force f does no work on the rolling cylinder $\mathcal{B}$ if the plane $\mathcal{I}$ is the reference frame. (b) If, however, $\mathcal{I}$ is the top surface of a moving block (dotted lines) and the reference frame is now the ground $\mathcal{G}$, does f then do work on $\mathcal{B}$? Why or why not?

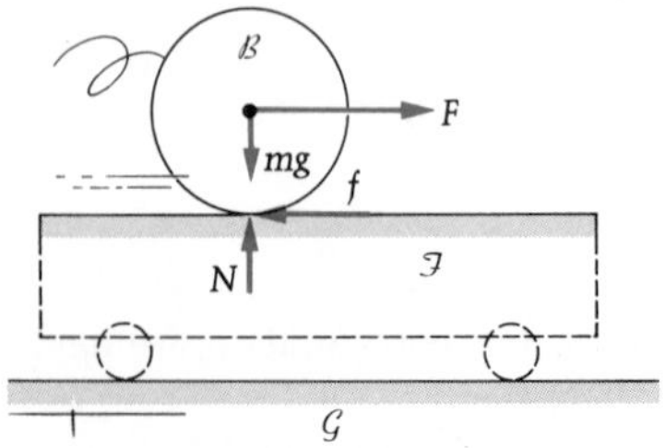

Figure P5.2

5.3 Upon application of the 10-N force F to the cord in Figure P5.3, the cylinder begins to roll to the right. After C has moved 5 m, how much work has been done by F?

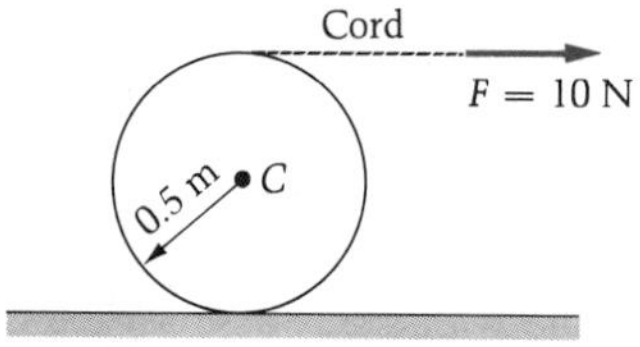

Figure P5.3

5.4 The suspended log shown in Figure P5.4 is to be used as a battering ram. At what angle θ should the ruffian release the log from rest so that it strikes the door at $\theta = 0$ with a velocity of 20 ft/sec?

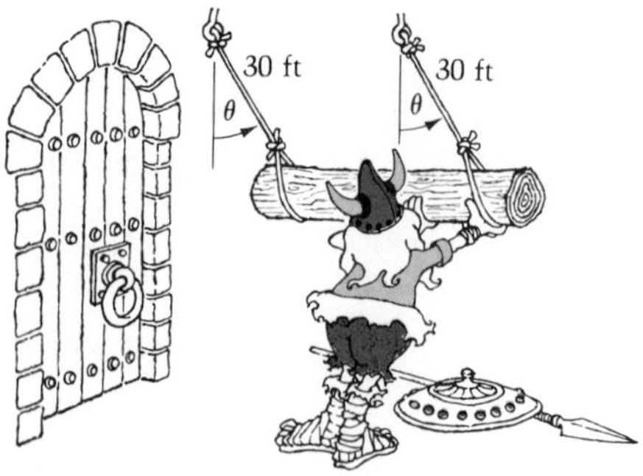

Figure P5.4

5.5 The 20-kg bar in Figure P5.5 has an angular velocity of 3 rad/s clockwise in the horizontal configuration shown. In that position the tensile force in the spring is

30 N. After a 90° clockwise rotation the angular velocity has increased to 4 rad/s. Determine the spring modulus k.

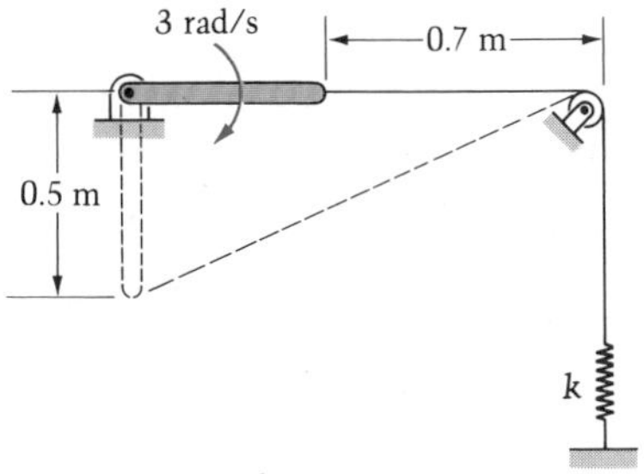

Figure P5.5

5.6 A uniform 40-lb sphere (radius = 1 ft) is released from rest in the position shown in Figure P5.6. If the sphere rolls (no slip), find its maximum angular speed.

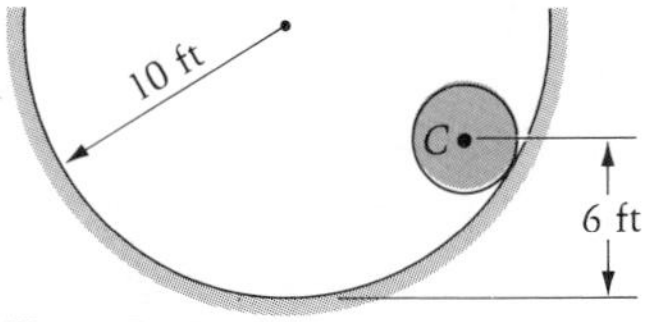

Figure P5.6

5.7 The unbalanced wheel of radius 2 ft and weight 64.4 lb shown in Figure P5.7 has a mass center moment of inertia of 6 slug-ft². In position 1, with C above O, the wheel has a clockwise angular velocity of 2 rad/sec. The wheel then rolls to position 2, where OC is horizontal. Determine the angular velocity of the wheel in position 2.

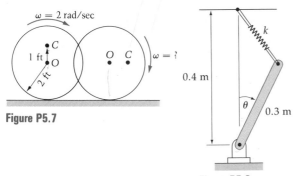

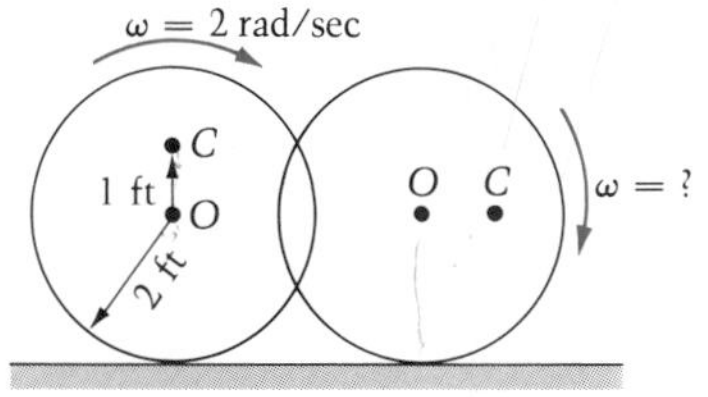

Figure P5.7

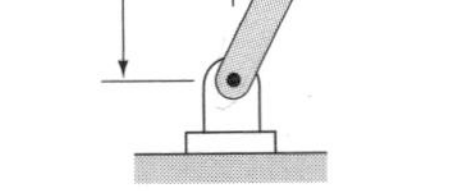

Figure P5.8

5.8 Determine the spring modulus that will allow the 2-kg bar in Figure P5.8 to arrive at the position $\theta = 90°$ at zero angular velocity if it passed through the vertical (where the spring is compressed 0.1 m) at 8 rad/s ↻.

5.9 Bar $\mathcal{B}_1$ is smoothly pinned to the support at A and smoothly pin-jointed to $\mathcal{B}_2$ at B. (See Figure P5.9.) End D slides on a smooth horizontal surface. If D starts from rest at $\theta = \theta_0$, determine the angular velocities of the rods just before they become horizontal.

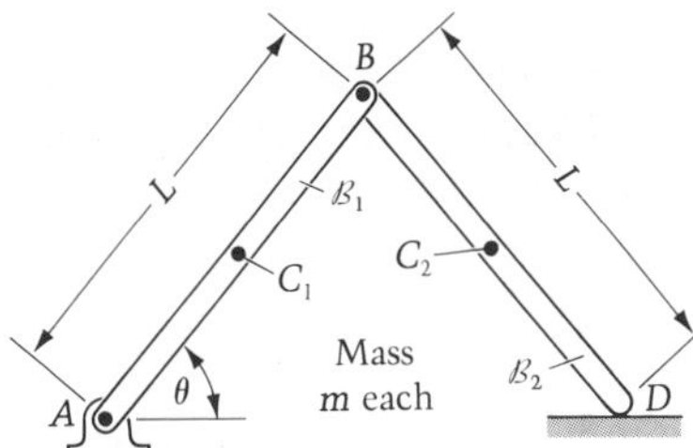

Figure P5.9

5.10 The prehistoric car shown in Figure P5.10 is powered by the falling rock m, connected to the main wheel (a cylinder of mass M) by a vine as shown. If the weights of the frame, pulley, and front wheel are small compared with Mg, find the velocity v_C of the car as a function of y if there is no slipping and it starts from rest with $y = 0$ at $t = 0$. Assume that m moves only vertically relative to the car's frame.

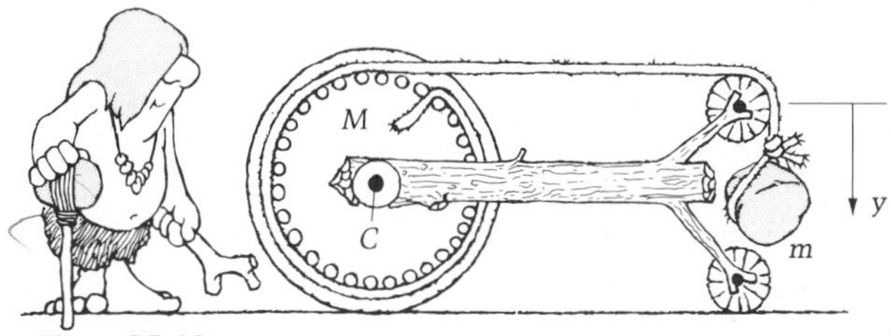

Figure P5.10

5.11 A truck body weighing 4000 lb is carried by four solid disk wheels that roll on the sloping surface. (See Figure P5.11.) Each wheel weighs 322 lb and is 3 ft in diameter. The truck has a velocity of 5 ft/sec in the position shown. Determine the modulus of the spring if the truck is brought to rest by compressing the spring 6 in.

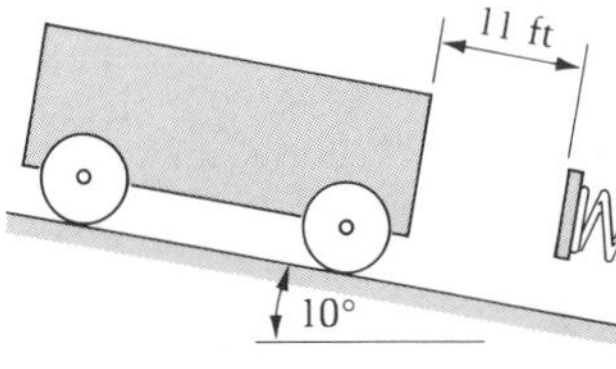

Figure P5.11

5.12 For the cylinder of Problem 4.70, assuming no slip and that the cylinder starts from rest, use work and kinetic energy to find the speed of its center in terms of the displacement of the center.

Ideally, the following five problems should be worked sequentially:

5.13 A cylinder with mass 6 kg has a 20-N force applied to it as shown in Figure P5.13. Find the angular velocity of the cylinder after it has rolled through 90° from rest.

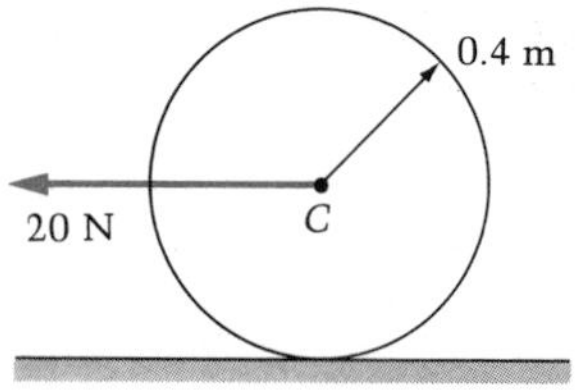

Figure P5.13

5.14 Rework Problem 5.13 if a slot is cut in the cylinder and a cord is wrapped around the slot, with the 20-N force now applied to the end of the cord as shown in Figure P5.14. Neglect the effect of the thin slot on the moment of inertia of the cylinder.

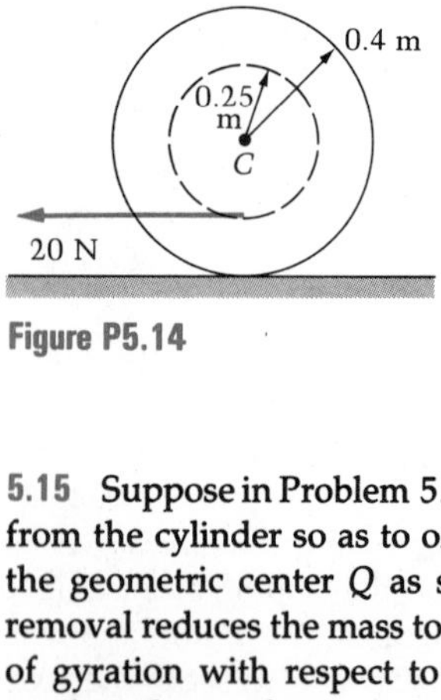

Figure P5.14

5.15 Suppose in Problem 5.14 we remove some material from the cylinder so as to offset the mass center C from the geometric center Q as shown in Figure P5.15. The removal reduces the mass to 5.5 kg and makes the radius of gyration with respect to the axis through C normal to the plane of the figure $k_C = 0.286$ m. Repeat the problem.

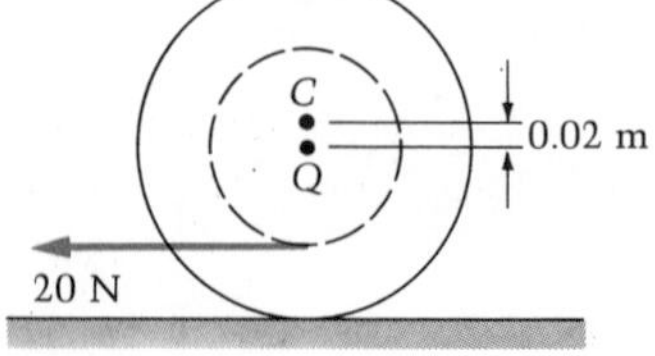

Figure P5.15

5.16 To the data of Problem 5.15 we add a constant counterclockwise couple of moment 2 N · m acting as shown in Figure P5.16. Repeat the problem.

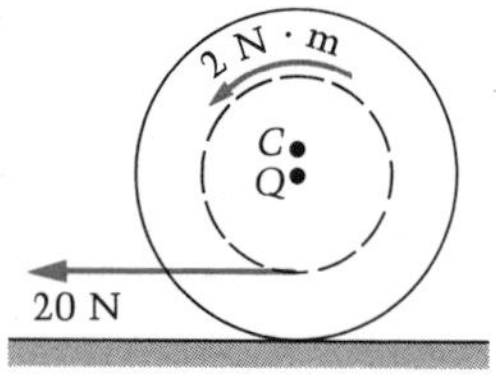

Figure P5.16

5.17 To the data of Problem 5.16, we add a spring, attached to a cord wrapped around a second slot in the cylinder near its outer rim as shown in Figure P5.17. The spring has modulus 6 N/m and is initially stretched 0.2 m. Repeat the problem.

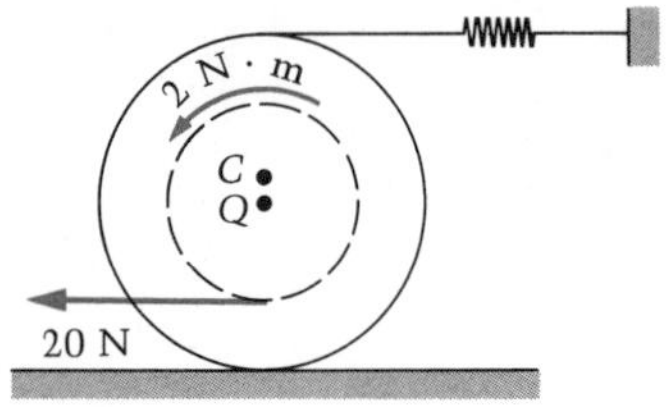

Figure P5.17

5.18 The 5-lb cylinder in Figure P5.18 rolls on the incline. If the velocity of the mass center C is 5 ft/sec down the plane in the upper (starting) position, find v_C in the bottom position.

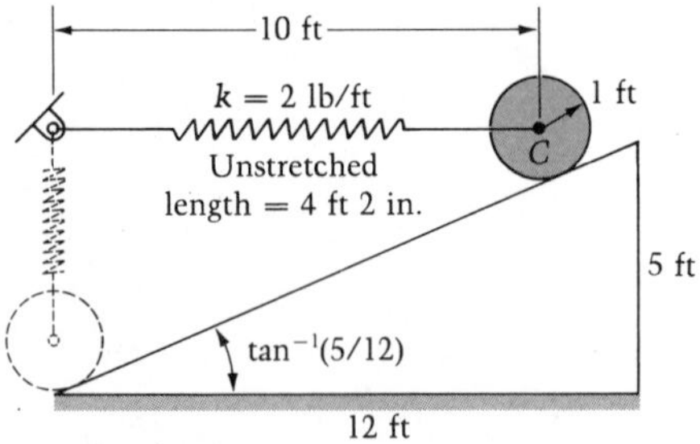

Figure P5.18

5.19 The spring in Figure P5.19 has an unstretched length of 0.8 m and a modulus of 60 N/m. The 20-kg wheel is released from rest in the upper position. Find its angular velocity when it passes through the lower (dashed) position if its radius of gyration is $k_C = 0.2$ m.

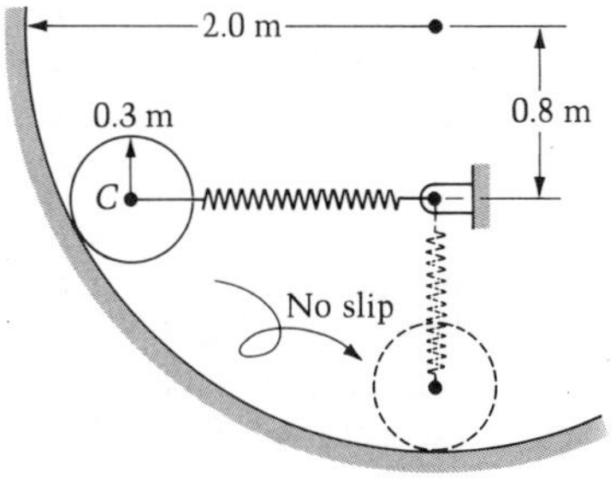

Figure P5.19

5.20 Bar $\mathcal{B}$ in Figure P5.20 is initially at rest in the vertical position, where the spring is unstretched. The wall and floor are smooth. Point B is then given a very slight displacement to the right, opening up a small angle $\Delta\theta$.

a. Draw a free-body diagram of the slightly displaced bar and use it to show that the bar will start to slide downward if $k < mg/2\ell$.

b. Find the angular velocity of $\mathcal{B}$ as a function of θ for such a spring.

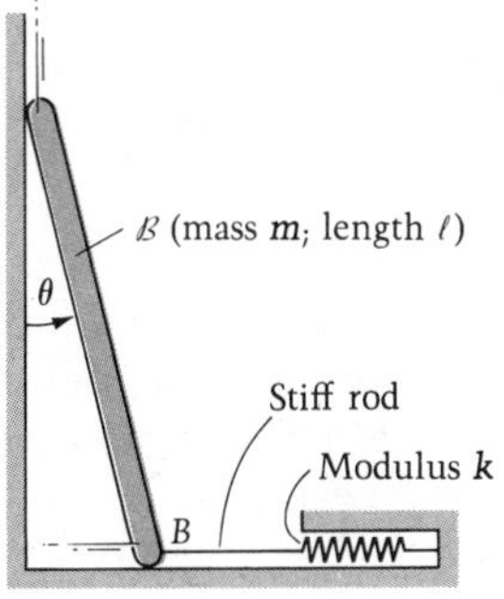

Figure P5.20

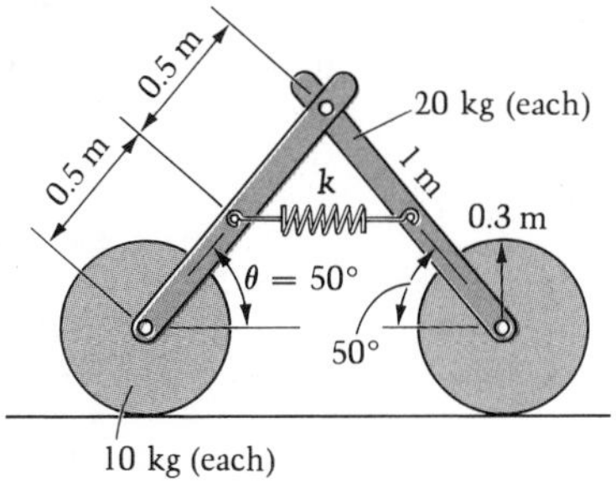

Figure P5.21

5.21 Find the spring modulus k that will result in the system momentarily stopping at $\theta = 35°$ after being released from rest at $\theta = 50°$ if the initial stretch δ_i in the spring is zero. (See Figure P5.21.) *Hint*: Use symmetry!

5.22 The wheel in Figure P5.22 has a mass of 5 slugs and a radius of gyration for the z axis through C of 0.7 ft. The spring has modulus 20 lb/ft and natural length 4 ft. The wheel is released from rest, and it rolls without slipping on the plane. Find how far down the plane the mass center C will move.

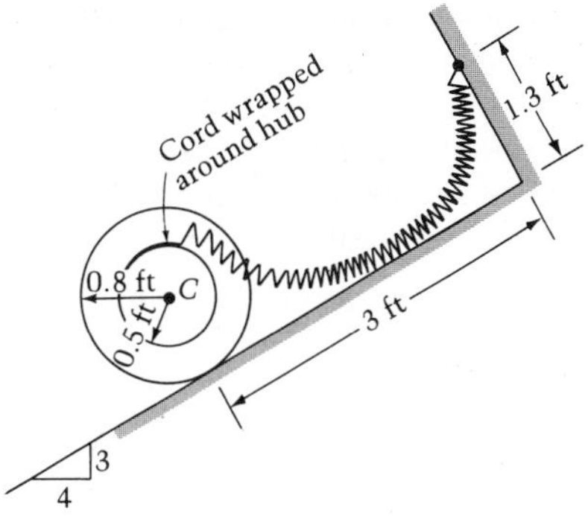

Figure P5.22

5.23 The wheel in Figure P5.23 weighs 200 N and has a radius of gyration 0.3 m with respect to the z_C axis. It is released from rest with the spring stretched $\frac{1}{2}$ m. If there is no slipping, find how far the cylinder center C moves

a. up, and

b. down the plane in the subsequent motion.

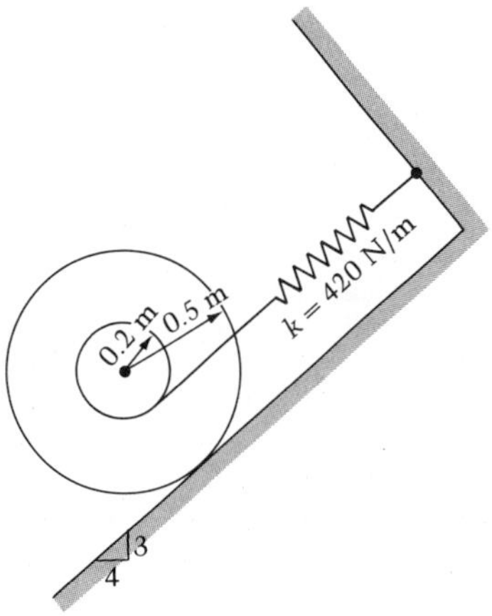

Figure P5.23

5.24 Show that if the rolling body in Example 5.8 is a sphere instead of a cylinder, it will slip at the angle θ_s satisfying the equation

$$\mu = \frac{2 \sin \theta_s}{17 \cos \theta_s - 10}$$

5.25 Use $W = \Delta T$ in Problem 4.98(b) to find the velocity of C when it has moved 3 m down the incline.

* **5.26** In Problem 4.134 use the principle of work and energy to obtain an upper bound on the rod's angular speed in its subsequent motion after the right-hand string is cut.

5.27 A thin disc of mass m and radius a is pinned smoothly at A to a thin rod of mass $m/2$ and length $3a$ (see Figure P5.27). The rod is then pinned at B. If the body is held in equilibrium in the configuration shown, then released from rest, find the velocity of point A as the system passes through the vertical.

5.28 Repeat the preceding problem if the pin at A is replaced by a weld.

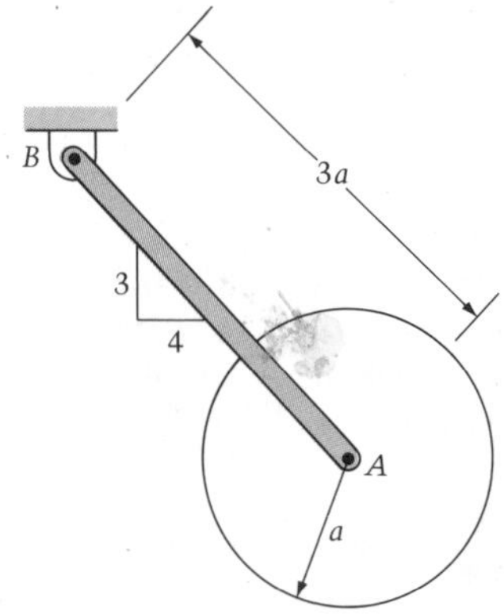

Figure P5.27

5.29 The 10-lb wheel shown in Figure P5.29 is attached at its center to a spring of modulus 20 lb/in. The radius of gyration of the wheel about the center is 2.5 in. The wheel rolls (no slip) after being released from rest with the spring stretched 1 in. Find: (a) the maximum magnitude of force in the spring; (b) the maximum speed of the center of the wheel during the ensuing motion.

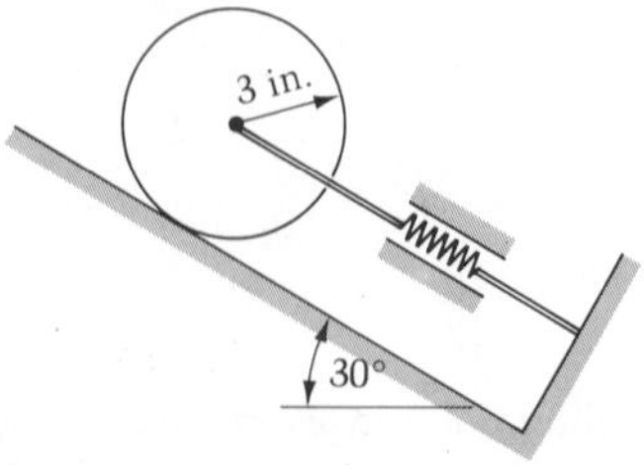

Figure P5.29

5.30 For the data of Problem 4.113 use $W = \Delta T$ to find the speed $\dot{x}_C$ of the plate as a function of the distance x_C it has traveled to the right. Use the $x_C = x_C(t)$ result to check your answer; differentiate and eliminate t to produce the same $\dot{x}_C = \dot{x}_C(x_C)$ result.

5.31 Figure P5.31 shows a fire door on the roof of a building. The door $\mathcal{B}_1$, 4 ft wide, 6 ft long, and 4 in. thick, is wooden (at 30 lb/ft³) and can rotate about a frictionless hinge at O. A cantilever arm $\mathcal{B}_2$ of negligible weight is rigidly fastened to the door and carries a 150-lb weight at its free end. During a fire the link $\mathcal{B}_3$ melts and the door swings open 45°. Find the angular velocity of the door just before the 150-lb weight hits the roof: (a) with no snow on the roof; (b) with snow at 1 lb/ft² on the roof.

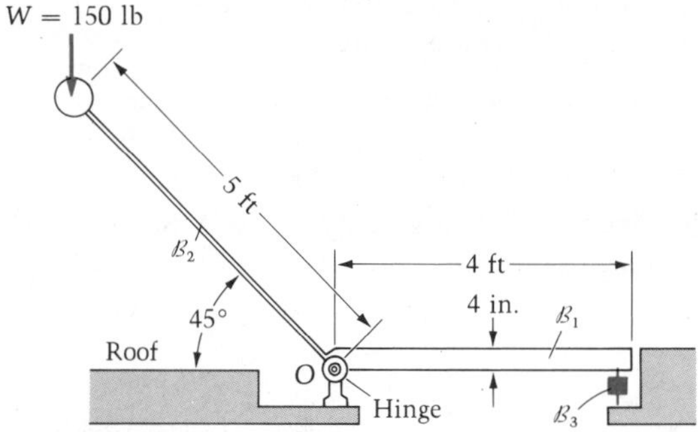

Figure P5.31

5.32 Block $\mathcal{B}_1$ in Figure P5.32 is moving downward at 5 ft/sec at a certain time when the spring is compressed 1 ft. The coefficient of friction between block $\mathcal{B}_2$ and the plane is 0.2, and the radius of cylinder $\mathcal{B}_3$ is 0.5 ft. Weights of $\mathcal{B}_1$, $\mathcal{B}_2$, and $\mathcal{B}_3$ are 161, 193, and 322 lb, respectively.

a. Find the distance that $\mathcal{B}_1$ falls from its initial position before coming to zero speed.

b. Determine whether or not body $\mathcal{B}_1$ will start to move back upward.

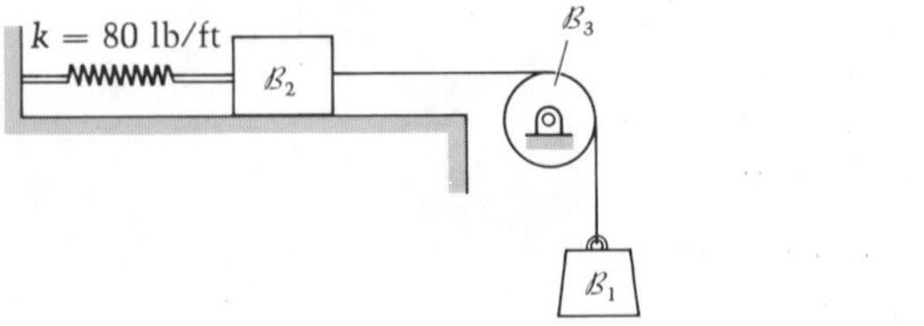

Figure P5.32

5.33 The system in Figure P5.33 consists of a cylinder $\mathcal{B}_1$ (100 kg) and (equilateral) triangular plate $\mathcal{B}_2$ (20 kg) pinned together at the mass center C_1 of the cylinder. The other two vertices of the plate are connected to springs,

* Asterisks identify the more difficult problems.

The study of classical mechanics is a profound experience. The deeper one delves into it, the more he appreciates the contributions of the great masters. Y. C. FUNG

INTRODUCTION

COMPREHENDING MECHANICS GOES beyond reading the textbook and working problems. Knowing dynamics means that you understand the physics embodied in the laws of mechanics, recognize their limitations and assumptions, and can correctly apply them to situations you encounter in practicing engineering. Knowing mechanics requires that you also develop a reasonable sense of the physical consequences of the fundamental principles along with the mathematical consequences.

GREAT MASTERS OF MECHANICS such as Galileo, Leonardo da Vinci, Hooke, Kepler, and Newton formulated the laws of statics and dynamics from the results of numerous observations and experiments. They devised simple experiments to test and clarify their ideas. Using empirical findings, they developed theories for predicting the behavior of mechanical systems. The principles they discovered and the mathematical expressions that describe these principles are the cornerstones of engineering mechanics.

STUDENTS (AND TEACHERS) of dynamics often overemphasize analysis and pay too little attention to the relationship between theory and the actual physical behavior of mechanical systems. Understanding both aspects of mechanics is essential. Engineers cannot successfully model the behavior of a mechanical system if they are unsure of the physics of the system. And the ability to predict successfully the behavior of physical systems is fundamental to the process of engineering design.

OVERVIEW

THIS SECTION PRESENTS experiments not unlike those used in early empirical studies of mechanics. These experiments demonstrate actual behaviors of simple mechanical systems and are intended to strengthen your understanding of the basic laws of dynamics. These exercises emphasize physical reality to help you develop qualitative intuitive skills that are essential in the practice of engineering. In addition, the demonstrations provide a way to check the soundness of certain mathematical models that are used to describe real-world problems.

THESE EXPERIMENTS PROVIDE only a starting point for your explorations and will raise additional questions as you conduct them. Do not leave those questions unanswered. To answer them you may need to modify a demonstration, design new experiments, or simply concentrate on interpreting a mathematical model. The important point is that you should pursue the answers. Along the way you will develop new insights into dynamics, you will become more proficient, and your ability to explain and predict the physical world will improve.

THE EXERCISES IN dynamics are keyed to specific sections and problems in the text. Be sure to review the text material before attempting these exercises. Each demonstration requires that you compare your observations with behavior predicted from a mathematical model. In most cases, experimental and theoretical results should be reasonably close. Remember, however, that models are only approximate and that experiments are never perfect. So, if your results disagree, find out why. To do so, verify your measurements and, if necessary, repeat or redesign the demonstration. Review the assumptions and limitations of the theory and check your analysis or computer program for mistakes. If you still find a disparity between the results, you may be applying the wrong principles or using incorrect equations.

MATERIALS

THE EXPERIMENTAL setups are simple and easy to construct. To conduct them, however, you will need some materials, all of which are readily available and can be obtained at little or no cost. These materials can be found at hardware stores, hobby shops, toy stores, and in the engineering shop at your college. We encourage the use of scrap materials and creative scrounging!

FOR MECHANICAL PARTS you should collect an assortment of cylinders, tubes, spheres, wheels, and rectangular blocks. The only requirement is that the parts be homogeneous and reasonably uniform. For example, if you need a cylindrical tube, select one that is straight and has a constant diameter and thickness. Manufactured tubes such as the following are excellent:

- Wood dowel
- PVC pipe, copper pipe, steel pipe
- Conduit tubing
- Aluminum rod, steel rod
- Empty coffee can with ends removed, tennis ball can
- Cardboard tube from roll of paper towels or toilet tissue
- Cardboard mailing tube
- Hockey puck
- Thread spool, metal adhesive tape spool

Be creative and resourceful when selecting materials.

IN ADDITION, you will need some laboratory supplies. They include string, duct tape, protractor, graph paper, stopwatch, tape measure or ruler, scissors, inexpensive calipers, and a scale or access to a scale for weighing parts.

Rolling Cylinders and Moments of Inertia

Lay out a length L on a flat board (or table), raise one end between 15° and 30°, and measure the angle. Select an assortment of cylinders, tubes, rods, wheels, and disks. Measure the time required for each object to roll distance L (see the figure below). Take several measurements for each object and obtain an average time. Work the theoretical problem of a round object starting from rest and rolling down an incline. Compare the results and explain any differences. Note that the results are independent of mass and radius. Repeat the procedure for different angles of inclination.

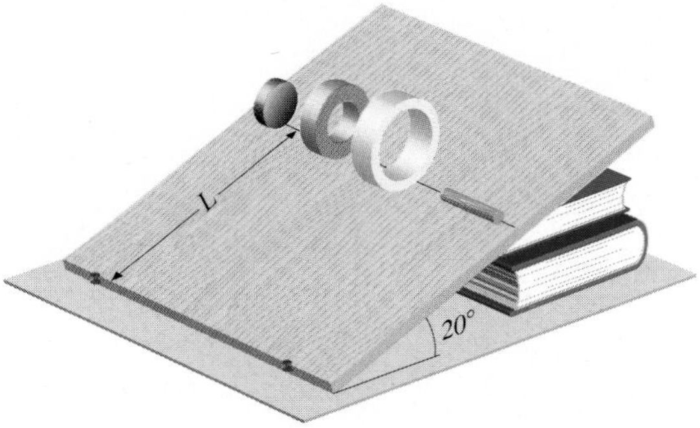

REFERENCE: *Dynamics* Sections 4.4-4.6

A "JUMPY" CYLINDER

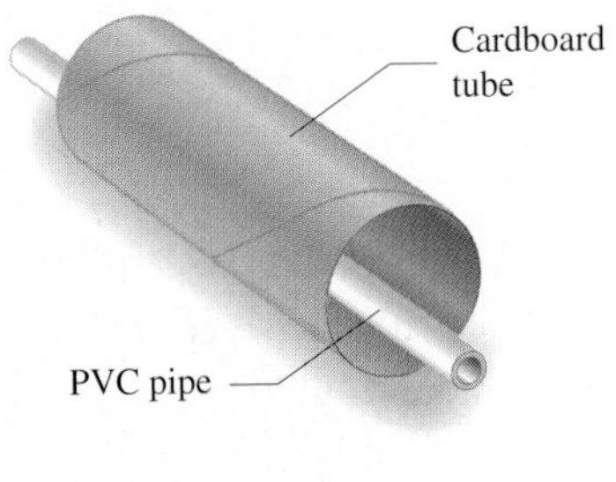

Build an unbalanced but round cylinder. One approach, shown in the figure to the right, is to tape or glue a rod or small cylinder to the inside of a cylindrical tube. (Be sure that the two axes are parallel.) A cardboard tube from a roll of paper towels and a length of 5/8" diameter PVC pipe work well for this demonstration. Place the cylinder on an inclined surface and release it from rest, as in the figure below. The slope of the surface should be great enough that the statically unbalanced cylinder rolls without stopping and does not slip. The "herky-jerky" motion of the cylinder should be obvious. Increase the inclination of the surface and repeat the test. The cylinder may begin to slip as it rolls. If so, increase the friction between the cylinder and the surface and continue your observations. You can increase the coefficient of friction by placing flat, wide rubber bands on the large cylinder to serve as treads or by wrapping the cylinder with several turns of duct tape. Experiment with different angles of inclination and different coefficients of friction until the composite body jumps up and loses contact with the incline. Did you anticipate this behavior? Can you explain it? The equations of motion predict this behavior and can help you interpret what you observed.

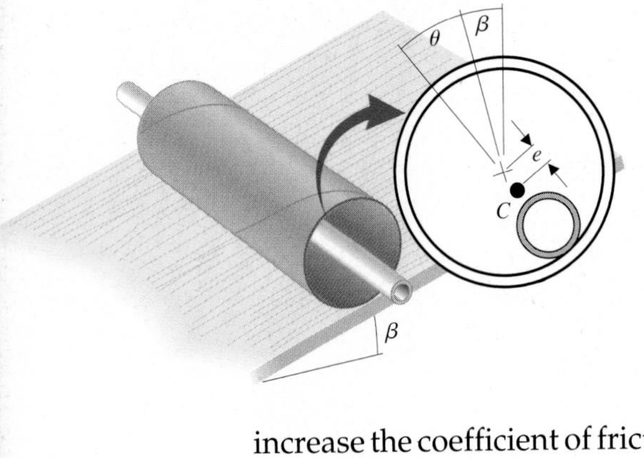

Derive the governing equations of motion for the unbalanced cylinder. Let m denote the mass of the body, e the eccentricity of the mass center C, I_G the body's moment of inertia about the central axis of the cylinder, R the cylinder radius, and g the acceleration of gravity. Assume that the cylinder rolls down the incline without slipping and show that

$$\ddot{\theta} = \frac{mg[e\cos\beta\sin\theta + (R + e\cos\theta)\sin\beta] + mRe\dot{\theta}^2\sin\theta}{I_G + mR^2 - 2m\,Re\cos\theta}$$

The equations also show that the normal force exerted on the body by the incline is

$$N = mg\cos\beta - me\dot{\theta}^2\cos\theta - me\ddot{\theta}\sin\theta$$

and that the upslope frictional force on the body is

$$F = mg\sin\beta + me\dot{\theta}^2\sin\theta - m(R + e\cos\theta)\ddot{\theta}$$

Note that N may vanish, which corresponds to the cylinder losing contact with the surface. Hence this equation predicts the possibility of the cylinder jumping up from the incline if certain conditions are met.

Develop the expression

$$\dot{\theta}^2 = \frac{2mgR\sin\beta(\theta - \theta_0) + 2mge[\cos(\theta_0 + \beta) - \cos(\theta + \beta)]}{I_G + mR^2 - 2m\,Re\cos\theta}$$

for the square of the body's angular speed if it starts from rest with $\theta = \theta_0$.

Construct a computer program that evaluates $\dot{\theta}$, $\ddot{\theta}$, N, F, and $|F|/N$ as functions of θ for a statically unbalanced cylinder that is released from rest and rolls down an incline without slipping. Your program should accept values of I_G, m, e, R, θ_0, and β as input and evaluate $\dot{\theta}$, $\ddot{\theta}$, N, F, and $|F|/N$ for equally spaced values of θ over several revolutions. A computer speadsheet works well for this analysis. Does the behavior predicted by your analysis match the actual motion of the unbalanced cylinder (at least qualitatively)?

REFERENCE: *Dynamics* Sections 4.4 - 4.6

ROLLING A WHEEL ON ITS AXLE

Assemble a single wheel and axle as shown in (a). You may use a pulley wheel (sheave) or a wheel with a grooved rim. The wheel should not slip around the axle. Tape a string to the wheel and wind it around the wheel at the center of the rim. The axle stubs need to ride on parallel rails that are high enough for the wheel to turn freely. The rails may be two boards on edge, two books of equal thickness, or two cylinders or tubes of equal diameters. Place the model on the rails, as in (b) and (c). Before pulling the string in the direction suggested in (c), decide which way the wheel will roll. Then pull the string and check your intuition. Solve the equation of motion for the angular acceleration α and confirm that the direction of α agrees with your observation. Repeat the experiment when the string force is vertical, as in (d), and when the force is horizontal and comes off the top of the wheel, as in (e). Verify your results using the equations of motion.

(a)

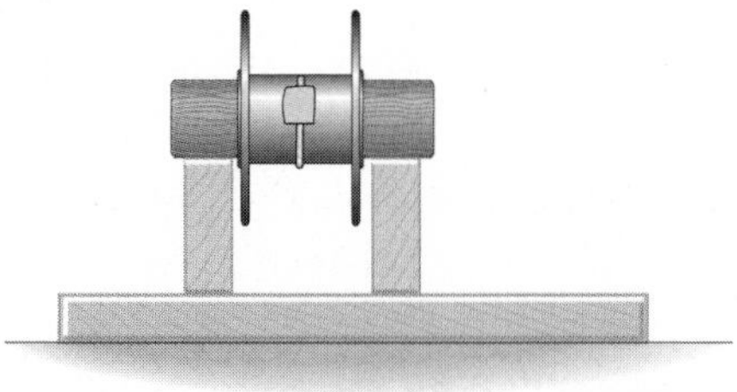

(b)

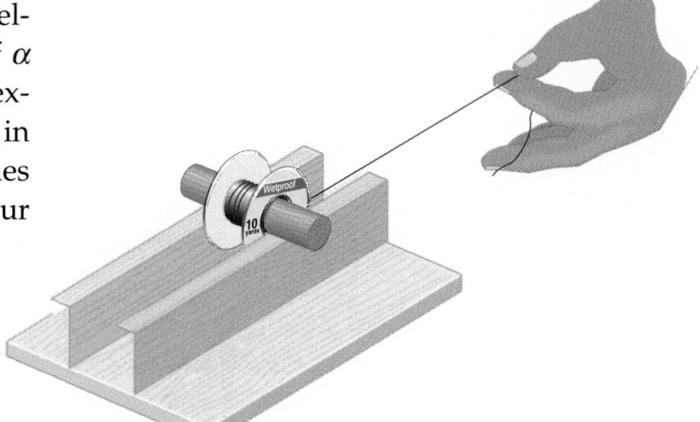

(c)

(d)

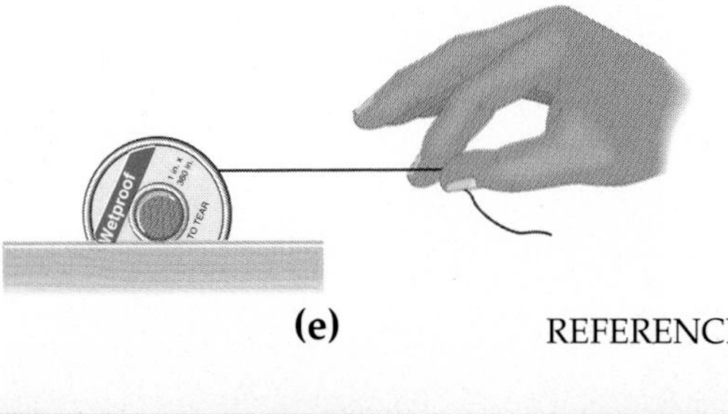

(e)

REFERENCE: *Dynamics* Sections 4.5 and 4.6

Unwinding a Spool on an Incline

Assemble a spool from two identical disks or cylinders and a single axle, as shown in the figure here. A metal adhesive tape spool or a yo-yo also works well for this demonstration. Tape a string to the center of the axle and wind it up. Bring the string from the bottom of the axle and hold it parallel to the surface of an inclined board, as shown in the top figure. Be sure that the string is attached midway between the disks. For small values of θ the spool remains in static equilibrium. Slowly increase the inclination of the board and observe when the spool begins to move. Note that the spool cannot roll down the plane (without slipping); to do so the string would break. The spool moves by slipping upward against the board at the point of contact while it actually rolls downward on the string. Derive the equations of motion for the mass center of the spool as it moves down the incline. Integrate twice and determine the time required for the center to move length L along the board. Mark off L on the incline and then measure the time for the center to traverse this distance when the spool is released from rest. Compare your predictions and observations. Repeat the experiment by pulling on the string from the top of the axle, as shown in the bottom figure.

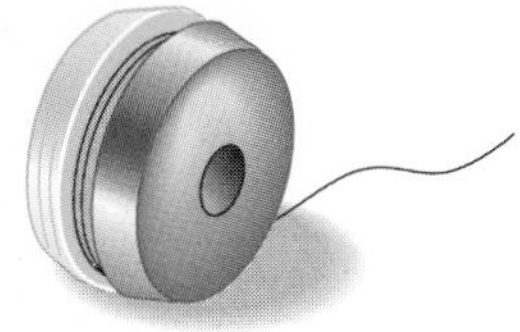

REFERENCE : *Dynamics* Sections 4.5 and 4.6

WALKING A YO-YO

Assemble a spool from two identical tubes or cylinders and a single axle, as shown in the top figure. Tape a string to the center of the axle and wind it up. Now you have a yo-yo! You may use a metal adhesive tape spool or an actual yo-yo for this demonstration. Be sure that the string is attached midway between the disks. If necessary, build up edges on both sides of the center of the axle with strips of tape or rubber bands. How will the spool move if you pull the string horizontally from the bottom of the axle, as shown in the middle figure? Vertically, as in the bottom figure? Gently pull the string from these positions and check your intuition. Use the equations of motion to compute α and compare its direction with your observations.

The interesting and somewhat unexpected behavior of the spool for these two situations should intrigue you and lead you to ask whether a pull angle between horizontal and vertical exists for which the spool will not roll either way. If so, what is that angle? Is the result the same if the string comes off the top of the axle?

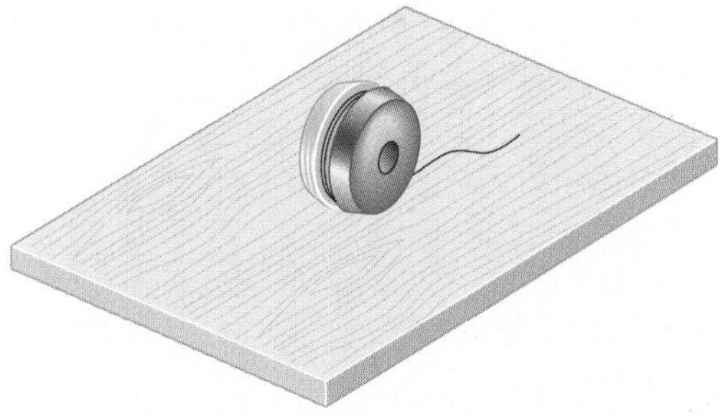

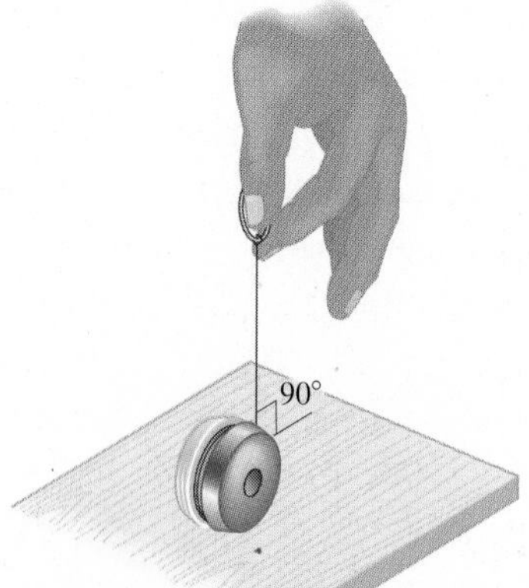

REFERENCE : *Dynamics* Sections 4.5 and 4.6

the left one of which (S_1) remains vertical in the slot. (Spring S_1 is shown only in its initial position.) The initial stretches of the two springs (in the position shown) are 0.2 m for S_1 and 0.04 m for S_2. The moduli are 40 N/m for S_1 and 10 N/m for S_2. If the system is released from rest in the given position, find the velocity of C_1 when vertex B reaches its lowest point in the slot. Assume sufficient friction to prevent $\mathcal{B}_1$ from slipping on the plane. The moment of inertia of an equilateral triangular plate of side s about its z_C axis is $ms^2/12$.

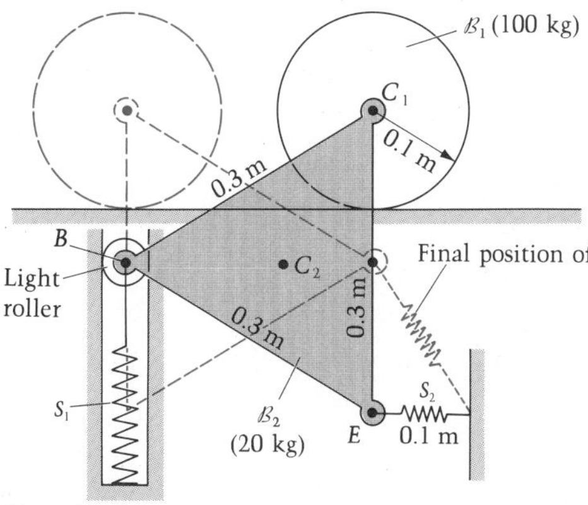

Figure P5.33

5.34 The mass center C of a rolling 2-kg wheel of radius $R = 15$ cm is located 5 cm from its geometric center Q. (See Figure P5.34.) The spring is attached at C and is not shown in position 2; its unstretched length is 0.3 m, and its modulus is 3 N/m. The radius of gyration is $k_C = 0.09$ m. Find the angular speed in position 2 (one-quarter turn from position 1).

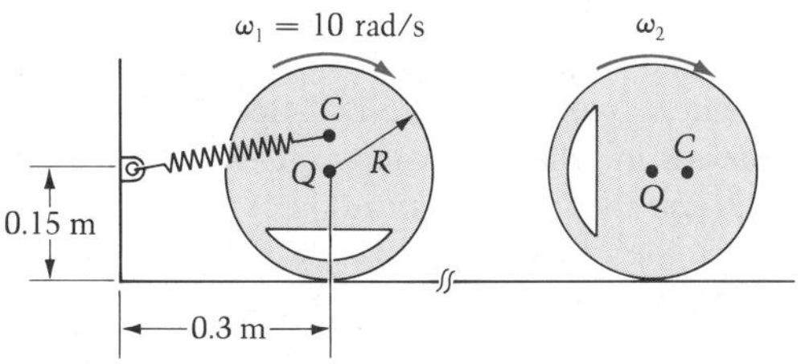

Figure P5.34

* **5.35** The three rods shown in Figure P5.35 are pinned together with one vertex also pinned to the ground. The length of the bar labeled $\mathcal{B}$ is given by $2b = 0.4$ m, and the density of the material of all bars is 7850 kg/m³. Their cross-sectional area is 0.002 m². Find the angular velocity of the combined body after it swings 90° from rest if: (a) $H = 2b$; (b) $H = \sqrt{3}\,b$; (c) $H = b$.

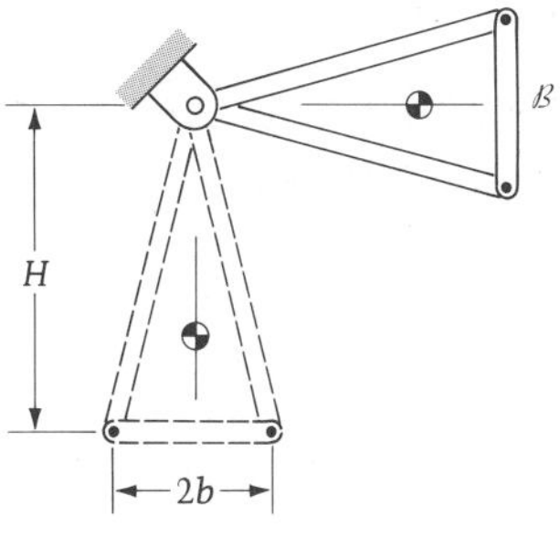

Figure P5.35

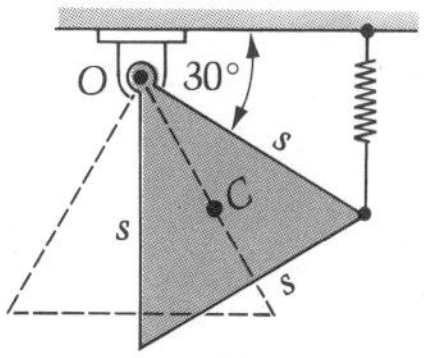

Figure P5.36

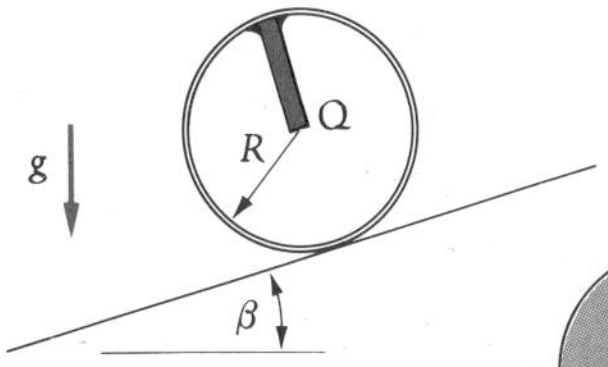

Figure P5.37

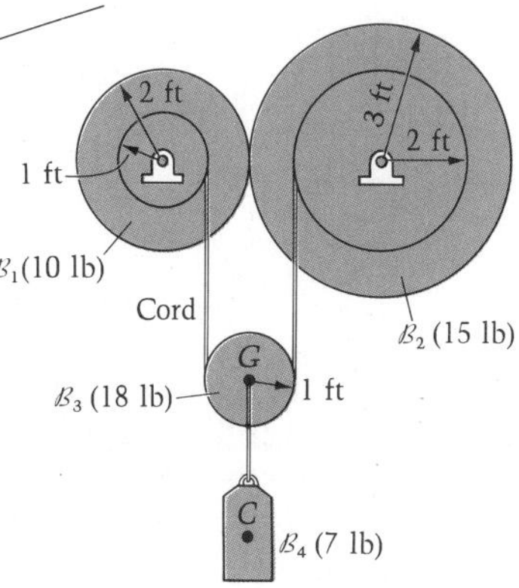

Figure P5.38

5.36 The center of mass of a uniform triangular plate is two-thirds of the distance from any vertex to the opposite side. The moment of inertia of an equilateral triangular plate is $ms^2/12$ with respect to the z axis through C. For the plate shown in Figure P5.36, with mass 30 kg and side 2 m, find its angular velocity when it reaches the dotted position where C is beneath O. The spring has unstretched length 0.5 m and modulus 20 N/m, and the plate is released from rest.

5.37 The bar in Figure P5.37 weighs the same (W) as the hoop to which it is welded. The combined body is released from rest on the incline in the position shown. If there is no slipping, determine the velocity of Q after one revolution of the hoop.

5.38 Cylinders $\mathcal{B}_1$ and $\mathcal{B}_2$ in Figure P5.38 are released from rest and turn without slip at the contact point. A cord is wrapped around an attached hub of each, which has negligible effect on the moment of inertia. There is enough friction to prevent the rope from slipping on the pulley. Find the velocity of the mass center of the pulley $\mathcal{B}_3$ after body $\mathcal{B}_4$ has fallen 20 ft.

5.39 A 5-lb cylinder is raised from rest by a force $P = 20$ lb. (See Figure P5.39.) Find the modulus of the spring that will cause the cylinder to stop after its center has been raised 2 ft. Will it then start back down? The spring is initially unstretched.

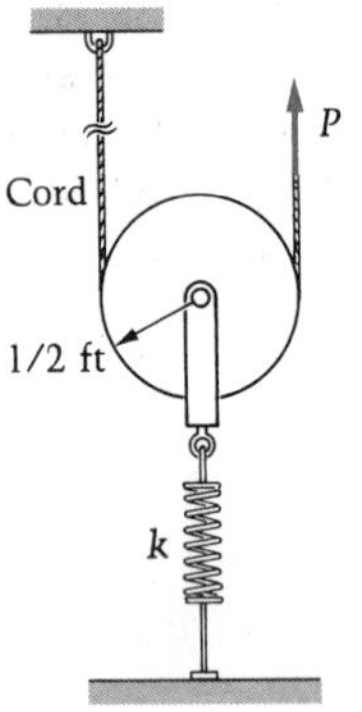

Figure P5.39

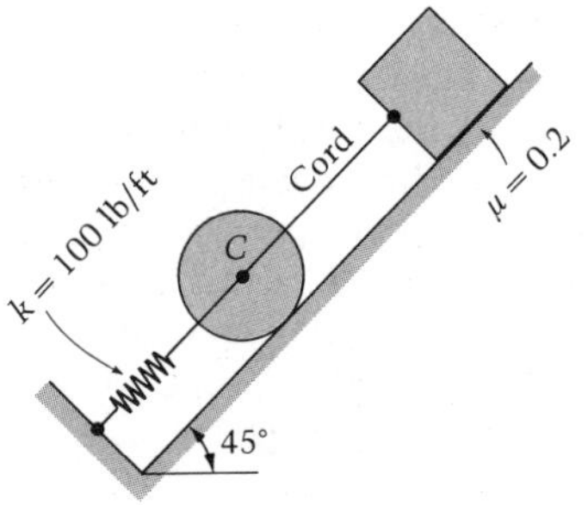

Figure P5.41

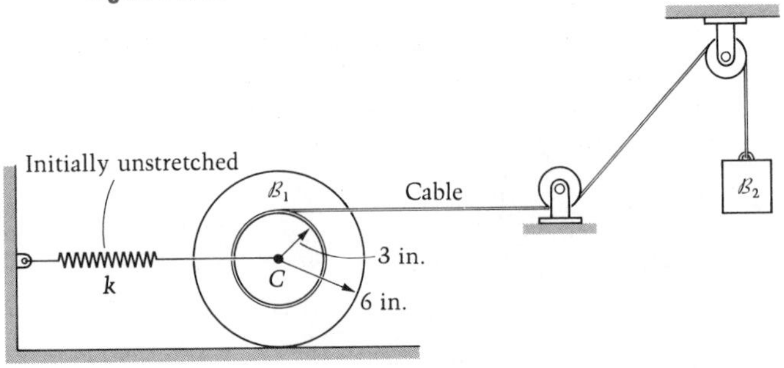

Figure P5.42

5.40 Body $\mathcal{B}_1$ in Figure P5.40 rolls to the right along the plane and has a radius of gyration with respect to its axis of symmetry of $k_C = 0.5$ m. The corresponding radius of gyration for $\mathcal{B}_2$ is 0.12 m. The spring is stretched 0.6 m at an instant when $\omega_1 = 5 \circlearrowleft$ rad/s. Find ω_1 after C has traveled 1 m to the right. (C_0 is an externally applied couple acting on $\mathcal{B}_1$.)

5.41 The cylinder and the block each weigh 100 lb. They are connected by a cord and released from rest on the inclined plane as shown in Figure P5.41. The spring, connected to the center C of the cylinder, is initially stretched 6 in. Find the velocity of the block at the instant the spring becomes unstretched, if there is sufficient friction between the plane and the cylinder to prevent it from slipping.

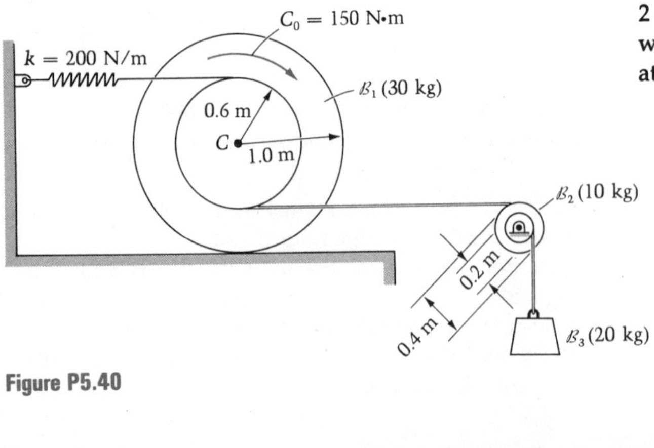

Figure P5.40

5.42 The 20-lb wheel $\mathcal{B}_1$ in Figure P5.42 has a radius of gyration of 4 in. with respect to its (z_C) axis. A cable wrapped around its inner radius passes under and over two small pulleys and is then tied to the 50-lb block $\mathcal{B}_2$. The spring has a modulus of 90 lb/ft and is constrained to remain horizontal. There is sufficient friction to prevent $\mathcal{B}_1$ from slipping on the plane. (a) If the system is released from rest, find the angular speed of $\mathcal{B}_1$ after the block then falls 1 ft. (b) Would the answer be different if block $\mathcal{B}_2$ were replaced by a device that keeps the cable force constant at 50 lb? Why or why not?

• **5.43** Rod $\mathcal{B}_1$ and disk $\mathcal{B}_2$ in Figure P5.43 have weights $W_1 = 5$ lb and $W_2 = 6$ lb. The rod's length is 8 in., the disk's radius is 4 in., the mass center offset of the disk is 2 in. from Q, and the radius of gyration of the mass of $\mathcal{B}_2$ with respect to the z axis through C is 3 in. It is desired to attach a spring between point Q and a fixed point so that

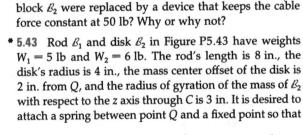

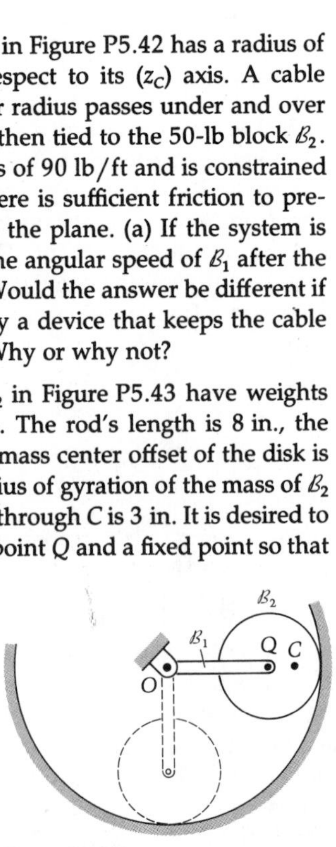

Figure P5.43

the disk and rod come to a stop (in the dotted position) after $\mathcal{B}_1$ turns 90° clockwise from rest. The spring has a modulus of 25.5 lb/ft and an unstretched length of 4 in.; it is to be unstretched initially. Find the final spring stretch, and from this result determine where to attach the fixed end of the spring. (There are two possible points!)

5.44 The bodies in Figure P5.44 have masses $m_1 = 0.3$ slug, $m_2 = 0.5$ slug, and m_3 negligible. A spring is attached to A that is stretched 25 in. in the dotted position when everything is at rest. Find the spring modulus if $\omega_{\mathcal{B}_1} = 2 \circlearrowleft$ rad/sec when $\mathcal{B}_1$ is horizontal.

5.45 A vertical rod is resting in unstable equilibrium when it begins to fall over. (See Figure P5.45.) End A slides along a smooth floor. Find the velocity of the mass center C as a function of L, g, and its height H above the floor.

5.46 Pulley $\mathcal{B}_1$ weighs 100 lb and has a centroidal radius of gyration $k_C = 7$ in. (See Figure P5.46.) The disk pulley $\mathcal{B}_2$ weighs 20 lb. Find the velocity of weight $\mathcal{B}_3$ (50 lb) after it falls 2 ft from rest. (Assume that the rope does not slip on the pulleys.)

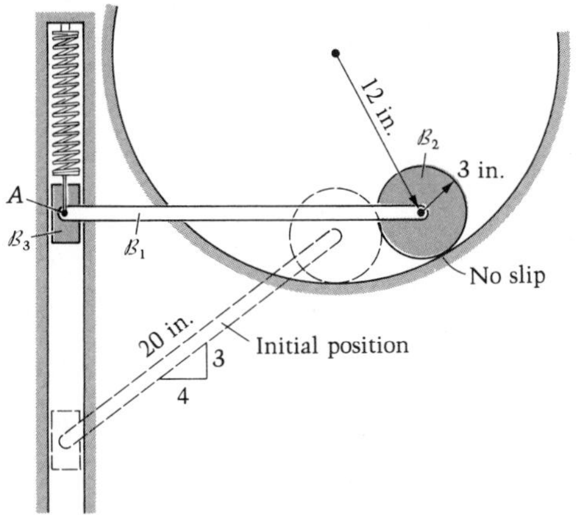

Figure P5.44

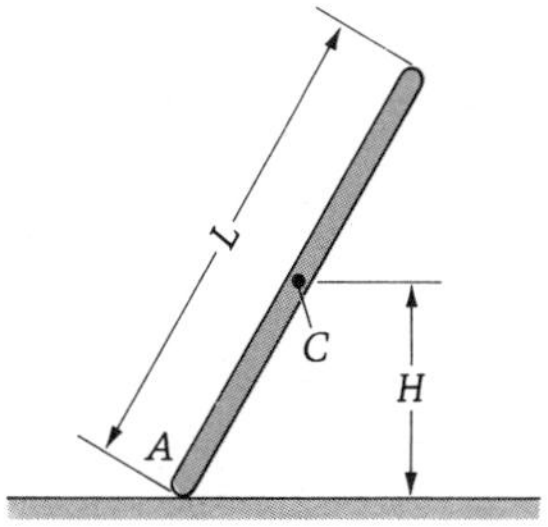

Figure P5.45

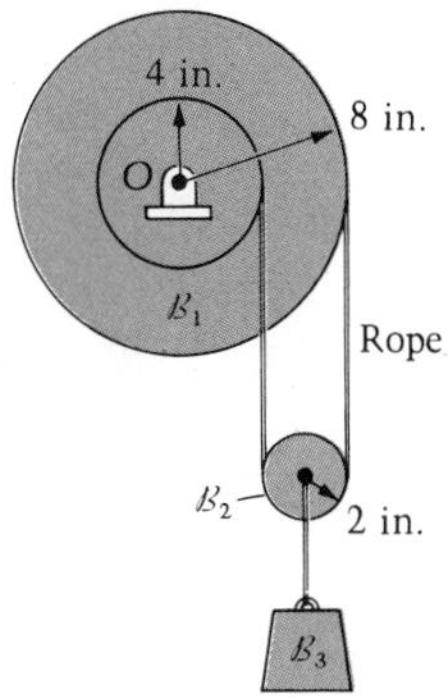

Figure P5.46

5.47 In Problem 4.108 determine the velocity of corner B of the half-cylinder when the diameter AB becomes horizontal for the first time.

5.48 Body $\mathcal{B}_1$ translates in the slot without friction. (See Figure P5.48.) Disk $\mathcal{B}_2$ (radius R) is pinned to block $\mathcal{B}_1$ through their mass centers at G. Body $\mathcal{B}_1$ and body $\mathcal{B}_2$ *each* has mass m; body $\mathcal{B}_3$ has mass $2m$. The system is released from rest a distance D above the floor. Find: (a) the starting accelerations of $\mathcal{B}_3$ and G; (b) the velocity of $\mathcal{B}_3$ when it hits the floor, using $W = \Delta T$.

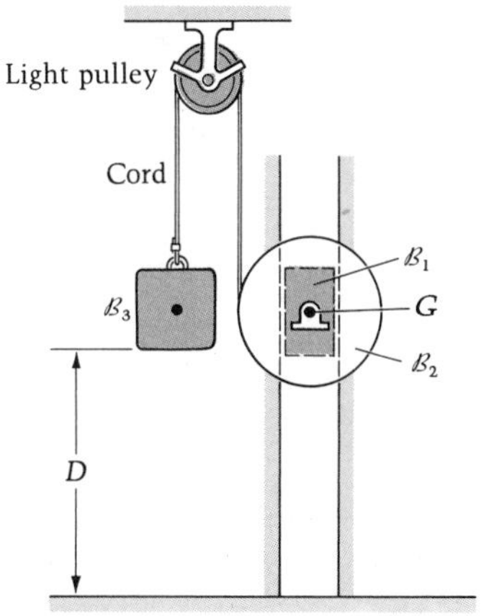

Figure P5.48

5.49 Link $\mathcal{B}_1$ weighs 10 lb and may be treated as a uniform slender rod (Figure P5.49). The 15-lb wheel is a circular disk with sufficient friction on the horizontal surface to prevent slipping. The spring is unstretched as shown. Link $\mathcal{B}_1$ is released from rest, and the light block $\mathcal{B}_2$ slides down the smooth slot. Neglecting friction in the pins, determine: (a) the angular velocity of the link as A strikes the spring with $\mathcal{B}_1$ horizontal; (b) the maximum deflection of the spring. (The modulus k of the spring is 10 lb/in.)

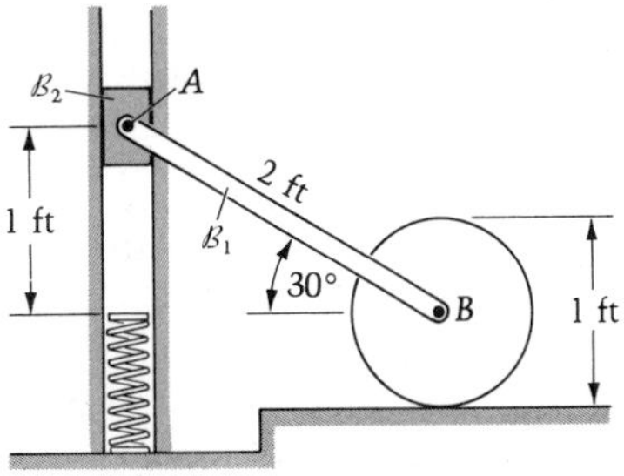

Figure P5.49

5.50 The masses of four bodies are shown in Figure P5.50. The radius of gyration of wheel $\mathcal{B}_4$ with respect to its axis is $k_C = 0.4$ m. Initially there is 0.6 m of slack in the cord between $\mathcal{B}_1$ and the linear spring. (Modulus $k = 1000$ N/m, and the spring is initially unstretched.) Determine how far downward body $\mathcal{B}_2$ will move.

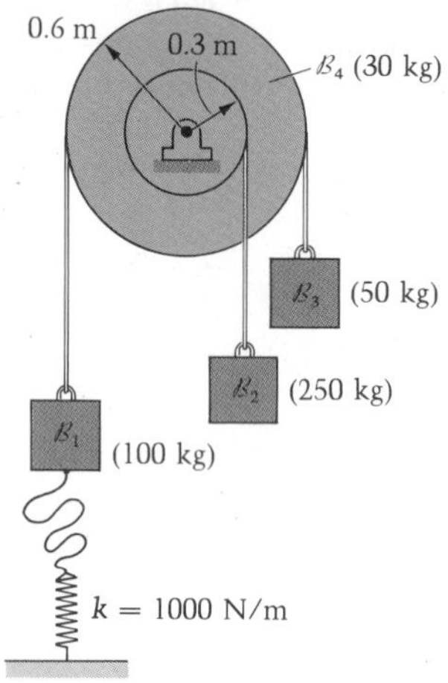

Figure P5.50

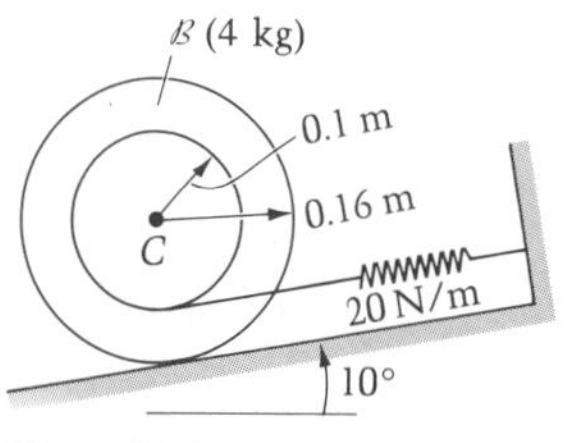

Figure P5.51

5.51 Cylinder $\mathcal{B}$ in Figure P5.51 is moving up the plane with $v_C = 0.3$ m/s at an initial instant when the spring is stretched 0.2 m. If $\mathcal{B}$ does not slip at any time, determine how far *down* the plane the point C will move in the subsequent motion. *Note*: The spring, connected to the cord, cannot be in compression.

5.52 Solve Problem 4.64 for v_C as a function of x_C using $W = \Delta T$.

5.53 In Problem 4.165, find the angular velocity of the rod when $\theta = 90°$.

5.54 Solve Problem 4.65 (a) by $W = \Delta T$.

5.55 For each of the wheels in Problem 4.72, solve for v_C as a function of x_C using $W = \Delta T$. The wheels start from rest.

* **5.56** Solve Problem 4.177 with the help of $W = \Delta T$. Ignore the hint.

5.57 The radius of gyration of the wheel and hub $\mathcal{B}$ in Figure P5.57, with respect to its axis of symmetry through C, is $k_C = 2.5$ m. The springs are unstretched at an initial position of rest, when the 50-N force is applied.

a. Find how far to the right the mass center moves in the ensuing motion, assuming sufficient friction to prevent slipping.

b. When $\mathcal{B}$ stops instantaneously at its farthest right point, what increase in the 50-N force and what minimum friction coefficient are needed to keep it there?

5.58 The slender nonuniform bar in Figure P5.58 (the mass is m and the radius of gyration with respect to the mass center C is $L/2$) is supported by two inextensible wires. If the bar is released from rest with $\theta = 0$, find the tension in each wire as a function of θ.

5.59 The system depicted in Figure P5.59 is released from rest with 2 ft of initial stretch in the spring. There is sufficient friction to prevent slipping at all times. Determine whether $\mathcal{B}_1$ will leave the horizontal surface during the subsequent motion. Note that the string $\mathcal{S}$ goes slack if the stretch tries to become negative.

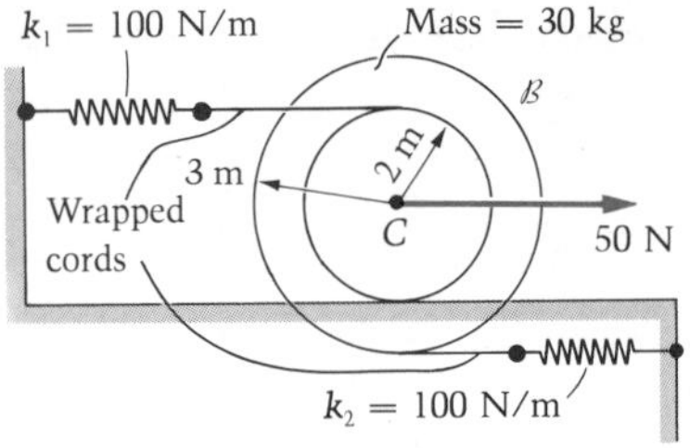

Figure P5.57

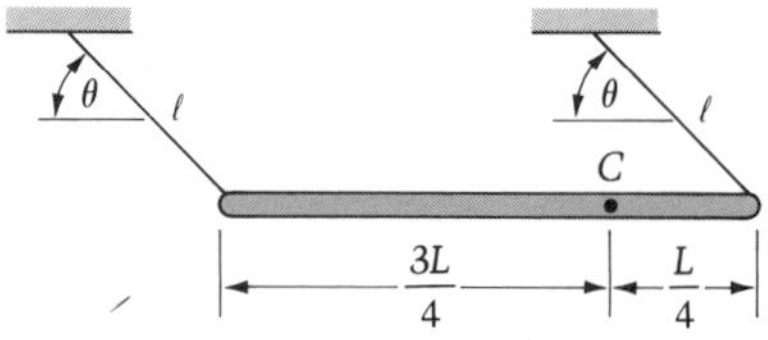

Figure P5.58

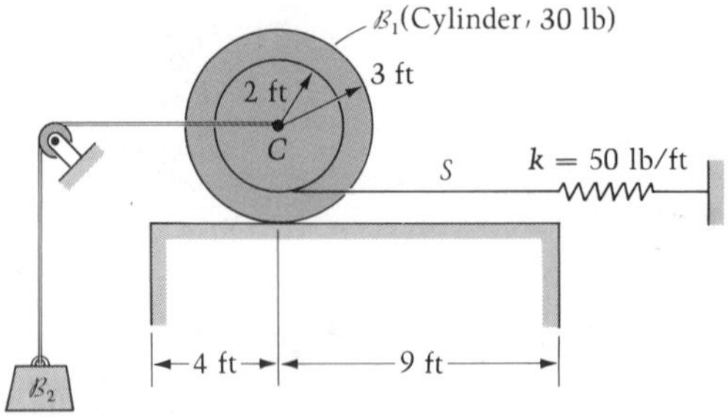

Figure P5.59

5.60 The system is released from rest in the position shown in Figure P5.60. Force P is constant, 60 lb, and the cord is wrapped around the inner radius of $\mathcal{B}_1$. Note the mass center of $\mathcal{B}_1$ is at C. Find the normal force exerted onto $\mathcal{B}_1$ by the plane (after using $W = \Delta T$ to get ω_1) at the instant when $\mathcal{B}_1$ has rotated 90° ↻. The spring is initially unstretched, and there is enough friction to prevent slipping.

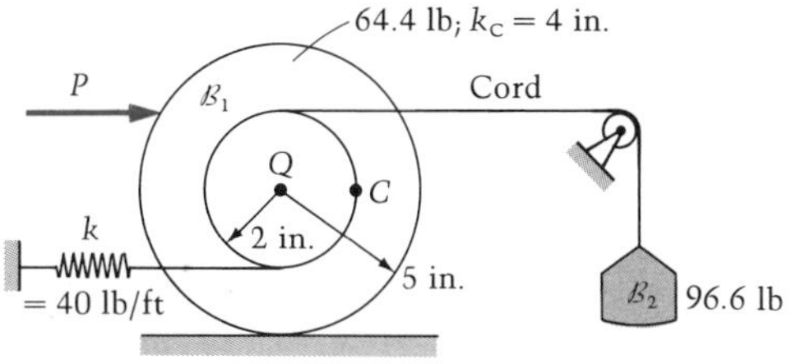

Figure P5.60

5.61 The two identical links $\mathcal{B}_1$ and $\mathcal{B}_2$ in Figure P5.61, each of mass m and length ℓ, are pinned together at A, and $\mathcal{B}_1$ is pinned to the ground at B. The end C of $\mathcal{B}_2$ slides in the vertical slot. Friction is negligible, and the system is released from rest. Find the velocity of point C just before point A reaches its lowest point.

5.62 The body of mass $2m$ in Figure P5.62 is composed of two identical uniform slender rods welded together. If friction in the bearing at O is neglected and the body is released from rest in the position shown, find the *magnitude* of the force exerted on the rod by the bearing after the body has rotated through 90°.

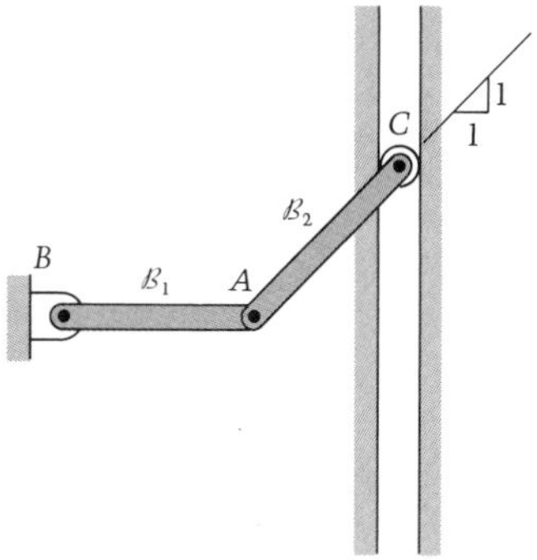

Figure P5.61

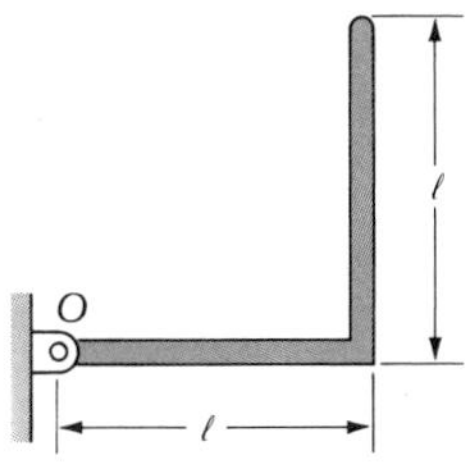

Figure P5.62

5.63 The cord connects the slotted cylinder $\mathcal{B}_1$ to the cylinder $\mathcal{B}_2$ as shown in Figure P5.63. Assume that neither body slips after the system is released from rest. The spring is initially unstretched, and is stiff and guided so it can take compression. Find the angular velocity of $\mathcal{B}_2$ after its center C has moved 1 m.

5.64 The 12-ft, 32.2-lb homogeneous rod $\mathcal{B}_1$ shown in Figure P5.64 is free to move on the smooth horizontal and vertical guides as shown. The modulus of the spring is 15 lb/ft and the spring is unstretched when in the position shown. Rod $\mathcal{B}_1$ is released from rest with $\theta = \pi/2$ and nudged to the right to begin motion. (a) Determine the angular velocity of the rod when it becomes horizontal. (b) What is the angular acceleration of the rod in this position ($\theta = 0$)?

5.65 The 50-kg wheel in Figure P5.65 is to be treated as a cylinder of radius $R = 0.2$ m. If it is rolling to the left with $v_C = 0.07$ m/s at an initial instant when the spring is unstretched, find: (a) the distance moved by C before v_C is instantaneously zero; (b) the minimum coefficient of friction that will prevent slip.

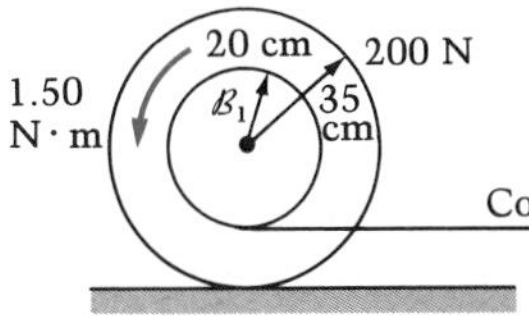

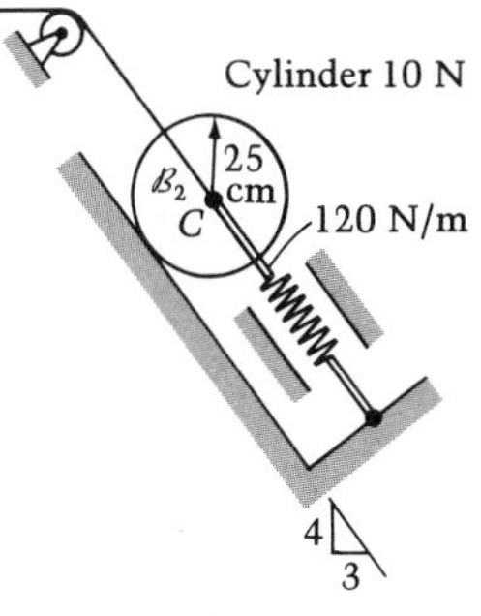

Figure P5.63

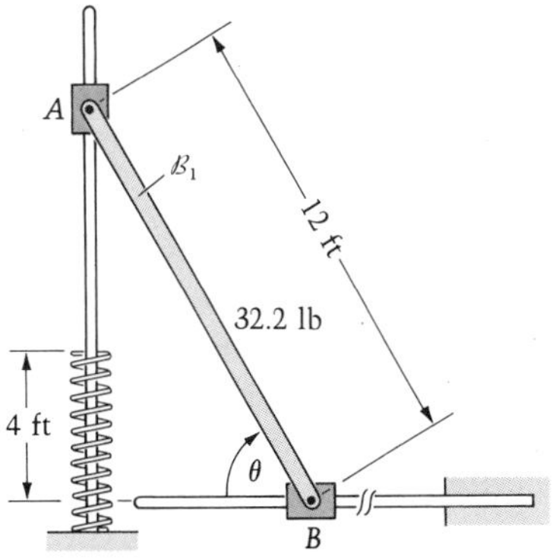

Figure P5.64

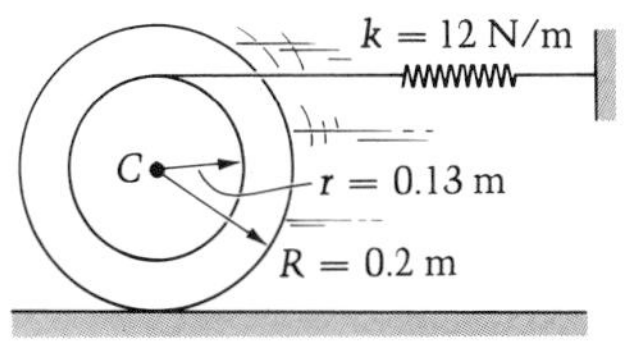

Figure P5.65

5.66 The cylinder in Figure P5.66 is rolling at $\omega = 2$ ↺ rad/sec in the initial (i) position, where the spring is unstretched. Other data are:

$$m = 2 \text{ slugs}$$

$$r = 3 \text{ ft}$$

$$k = 3 \text{ lb/ft}$$

$$\mu = 0.2$$

$$\ell_u = \text{unstretched spring length} = 9 \text{ ft}$$

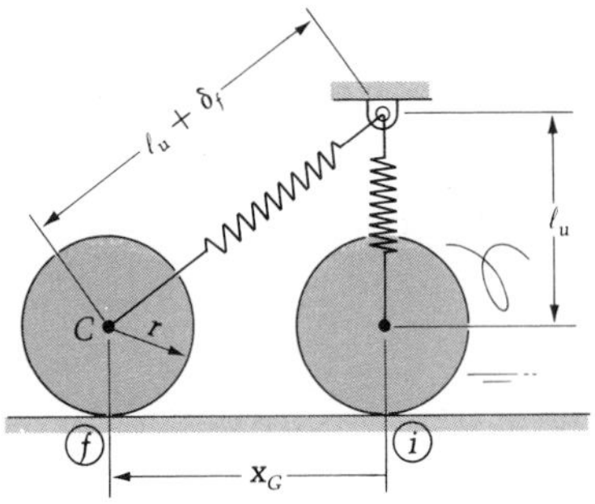

Figure P5.66

Find the final position of C (x_C) at which either the cylinder has stopped (for an instant) or started to slip, whichever comes first. *Hint*: Try one, check the other!

5.67 The uniform slender rod in Figure P5.67 (mass = 5 slugs, length = 10 ft) is released from rest in the position shown. Neglecting friction, find the force that the floor exerts on the lower end of the rod when the upper end is 6 ft above the floor. *Hint*: First use a free-body diagram and the equations of motion to deduce the path of the mass center.

5.68 In Figure P5.68, the ends of the bar are constrained to vertical and horizontal paths by the smooth rollers in the slots shown. The bar, originally vertical, is very gently nudged at its lower end to initiate motion. Find the reactions onto the bar at A and B just before the bar becomes horizontal.

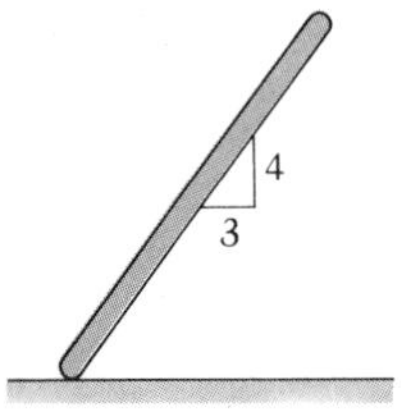

Figure P5.67

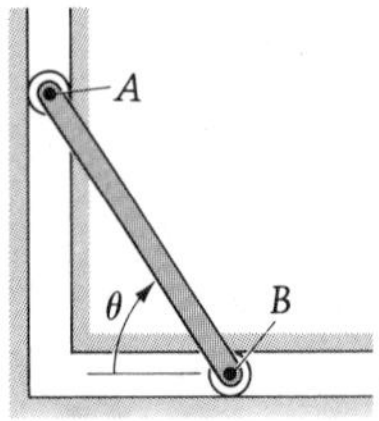

Figure P5.68

5.69 A slender uniform rod of weight W is smoothly hinged to a fixed support at A and rests on a block at B. (See Figure P5.69.) The block is suddenly removed. Find: (a) the initial angular acceleration and components of reaction at A; (b) the components of reaction at A when the rod becomes horizontal.

5.70 Two quarter-rings are pinned together at P and released from rest in the indicated position (Figure P5.70) on a smooth plane. Find the angular velocities of the rings when their mass centers are passing through their lowest points. *Hint*: By symmetry, point P always has only a vertical velocity component; this means that no work is done on either ring by the other, because (again by symmetry) the force between the rings has only a horizontal component normal to the velocity of P. More generally, as long as the pin is smooth, the work done by two pinned bodies in motion on each other will be the negative of each other because the velocities will be equal whereas the forces will be opposites.

* **5.71** A slender rod is placed on a table as shown in Figure P5.71. It will begin to pivot about the edge E and, at some angle θ_s, it will begin to slip. Find this angle, which will depend on the coefficient of friction μ and on k. *Hint*: Use all three equations of motion together with $W = \Delta T$. Eliminate α and ω^2, obtaining expressions for f and N. Setting $f = \mu N$ then permits a solution for θ_s. Solve the resulting equation when $\mu = 0.3$ and $k = 0.25$.

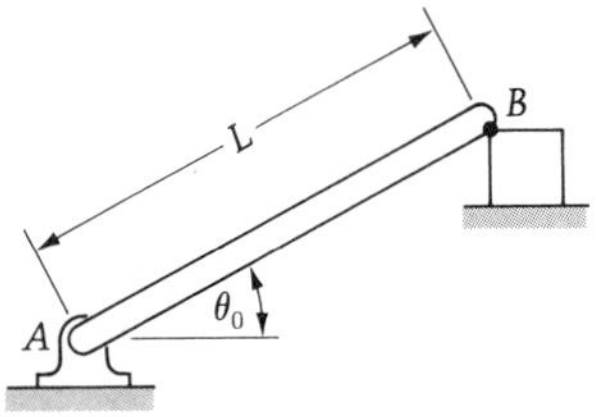

Figure P5.69

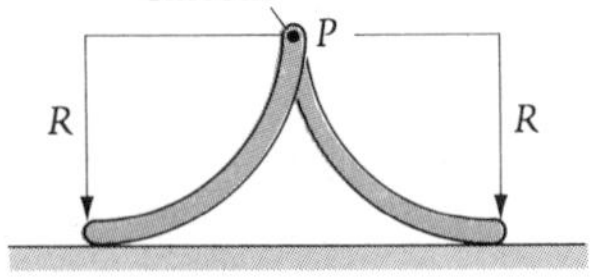

Figure P5.70

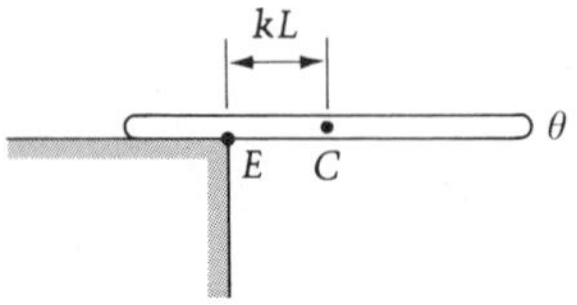

Figure P5.71

*** 5.72** The uniform equilateral triangular plate ABE in Figure P5.72 weighs W and is pinned to a fixed point at A and to a rope at E. The rope passes over a small, frictionless pulley at D and is then tied to the (equal) weight W which is constrained to move vertically. If the system is released in the given position with the angular velocity of ABE being 3 ↺ rad/sec, find the angular velocity of ABE in the dashed position (i.e., when side AE becomes horizontal).

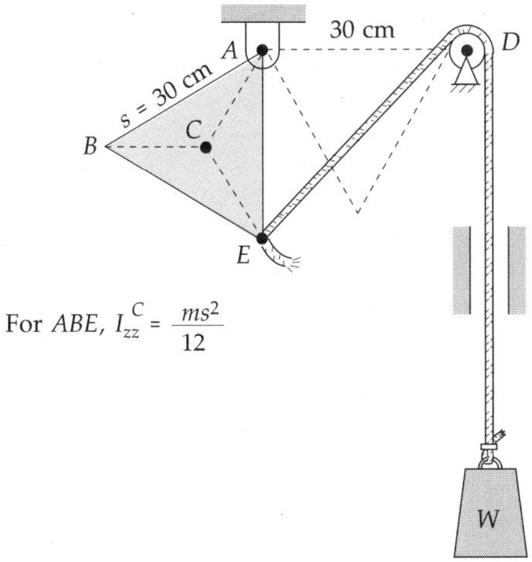

Figure P5.72

*** 5.73** This problem, continuing and using the results of Problem 4.150, is to verify that $W = \Delta T$ for that solution. First, note that only the friction couple does net work on the system of $\mathcal{B}_1$ plus $\mathcal{B}_2$, and verify:

$$\begin{aligned} W &= W_{\text{on } \mathcal{B}_1} + W_{\text{on } \mathcal{B}_2} \\ &= \int -k(\omega_1 - \omega_2)\omega_1 \, dt + \int k(\omega_1 - \omega_2)\omega_2 \, dt \\ &= -k \int (\omega_1 - \omega_2)^2 \, dt \\ &= -46.1 \text{ lb-ft} \end{aligned}$$

Next, show that an identical result is obtained (as it must be) for ΔT:

$$\Delta T = T_f - T_i = T_f^{\mathcal{B}_1} + T_f^{\mathcal{B}_2} - T_i^{\mathcal{B}_1} - \overset{0}{\cancel{T_i^{\mathcal{B}_2}}}$$

*** 5.74** The cylinder $\mathcal{B}_1$ in Figure P5.74 has a spring attached and is released from rest. Assume that there is sufficient friction between $\mathcal{B}_1$ and the plane to prevent slip throughout the motion, and that the slot around which the cord is wrapped has a negligible effect on the cylinder's moment of inertia. Find the velocity of B when the unwrapped length of rope is completely vertical (that is, when C has 0.3 m left to travel before it would be directly above E). Assume also that the weight $\mathcal{B}_2$ moves only vertically — that is, that the rope does not start to sway.

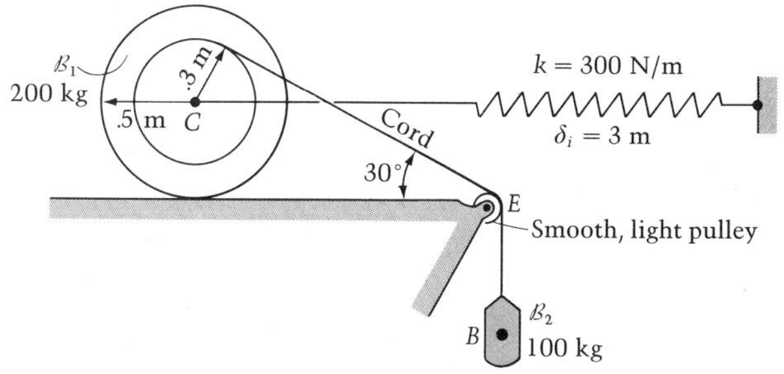

Figure P5.74

5.3 The Principles of Impulse and Momentum

The Equations of Impulse and Momentum for the Rigid Body in Plane Motion

The principle of work and kinetic energy is very helpful when the problem is posed in terms of positions and velocities. When time, rather than position, is the main concern, we often draw on a pair of principles concerned with impulse and momentum vectors. Just like $W = \Delta T$, these principles are obtained by general integrations of the equations of motion, but now the integration is directly with respect to time, without first dotting the equations with velocity. Thus they leave us with a set of vector equations instead of a single scalar result.

We have encountered one of the principles in Section 2.5 in our study of mass center motion: The impulse of the external forces imparted to any system equals its change of momentum over the same time interval. In Chapter 2 the system was general, so this principle holds for the rigid bodies we are now studying. From Equation (2.27), $\Sigma \mathbf{F} = \dot{\mathbf{L}}$ so that

$$\int \Sigma \mathbf{F}\, dt = \int \dot{\mathbf{L}}\, dt = \int d\mathbf{L} = \mathbf{L}\Big]_i^f = \mathbf{L}_f - \mathbf{L}_i = \Delta\mathbf{L} = m\mathbf{v}_{C_f} - m\mathbf{v}_{C_i} \tag{5.26}$$

Reviewing, we note that the integral $\int \Sigma\mathbf{F}\, dt$ is called the **impulse** (or **linear impulse**) imparted to the system by external forces. The vector $\Delta\mathbf{L}$ is the change in the system's momentum (or linear momentum) from the initial to the final time.

We need only the x and y components of Equation (5.26) for the rigid body in plane motion:

$$\int \Sigma F_x\, dt = \Delta L_x = \Delta(m\dot{x}_C) = m\dot{x}_{C_f} - m\dot{x}_{C_i} \tag{5.27}$$

$$\int \Sigma F_y\, dt = \Delta L_y = \Delta(m\dot{y}_C) = m\dot{y}_{C_f} - m\dot{y}_{C_i} \tag{5.28}$$

There is also a corresponding principle of *angular* impulse and momentum. From Equation (2.43),

$$\Sigma\mathbf{M}_C = \dot{\mathbf{H}}_C$$

so that

$$\int \Sigma\mathbf{M}_C\, dt = \int \dot{\mathbf{H}}_C\, dt = \int d\mathbf{H}_C = \mathbf{H}_C\Big]_i^f = \mathbf{H}_{C_f} - \mathbf{H}_{C_i} = \Delta\mathbf{H}_C \tag{5.29}$$

This equation may be put into a convenient form for rigid bodies in plane motion by recalling Equation (4.4) for the angular momentum:

$$\mathbf{H}_C = I_{xz}^C\omega\hat{\mathbf{i}} + I_{yz}^C\omega\hat{\mathbf{j}} + I_{zz}^C\omega\hat{\mathbf{k}}$$

For symmetric bodies in which the products of inertia vanish and $\Sigma\mathbf{M}_C = \Sigma M_C\hat{\mathbf{k}}$, this equation becomes

$$\mathbf{H}_C = I_{zz}^C\omega\hat{\mathbf{k}}$$

Therefore

$$\int \Sigma M_C\hat{\mathbf{k}}\, dt = \Delta\mathbf{H}_C = \Delta(I_{zz}^C\omega\hat{\mathbf{k}})$$

or

$$\int \Sigma M_C\, dt = \Delta(I_{zz}^C\omega) = (I_{zz}^C\omega)_f - (I_{zz}^C\omega)_i \tag{5.30}$$

The integral $\int \Sigma M_C\, dt$ is called the **angular impulse** imparted to the system by the external forces and couples, and the quantity $\Delta(I_{zz}^C\omega)$ is the change in angular momentum, both taken about C.

A subtle but important point regarding Equation (5.30) must be understood here. We note from Equation (5.29) that angular impulse equals the change in angular momentum for any body (deformable as well as rigid); therefore the use of Equation (5.30) only requires that the body of interest behave rigidly at the start (t_i) and end (t_f) of the time interval (t_i, t_f). At those times the moment of momentum is $\mathbf{H}_C = I_{zz}^C\omega\hat{\mathbf{k}}$, even though this simple expression for $\mathbf{H}_C$ may not apply *between* t_i and t_f. A good example is an ice skater drawing in her arms to increase angular speed, as we shall see later in Example 5.14.

In summary, for the rigid body in plane motion we have the following two principles at our disposal:

1. Linear impulse and momentum:

$$\int \Sigma \mathbf{F}\, dt = \Delta(m\mathbf{v}_C)$$

from which we get

$$\hat{\mathbf{i}} \text{ coefficients:} \quad \int_{t_i}^{t_f} \Sigma F_x\, dt = m(\dot{x}_{C_f} - \dot{x}_{C_i}) \tag{5.31}$$

$$\hat{\mathbf{j}} \text{ coefficients:} \quad \int_{t_i}^{t_f} \Sigma F_y\, dt = m(\dot{y}_{C_f} - \dot{y}_{C_i}) \tag{5.32}$$

where the directions of x, y, and z are fixed in the inertial frame.

2. Angular impulse and momentum:

$$\int_{t_i}^{t_f} \Sigma M_{Cz}\, dt = I_{zz_f}^C \omega_f - I_{zz_i}^C \omega_i \tag{5.33}$$

We note also that if the products of inertia are not zero, we have

$$\int \Sigma M_{Cx}\, dt = (I_{xz}^C \omega)_f - (I_{xz}^C \omega)_i \tag{5.34}$$

and

$$\int \Sigma M_{Cy}\, dt = (I_{yz}^C \omega)_f - (I_{yz}^C \omega)_i \tag{5.35}$$

Question 5.5 Are the coordinate axes associated with Equations (5.34) and (5.35) the same as those of (5.31) and (5.32)?

In the remainder of this section we treat only examples of symmetric bodies, for which Equations (5.31) to (5.33) are the impulse and momentum equations. The first example deals with both linear and angular impulse and momentum for a single body.

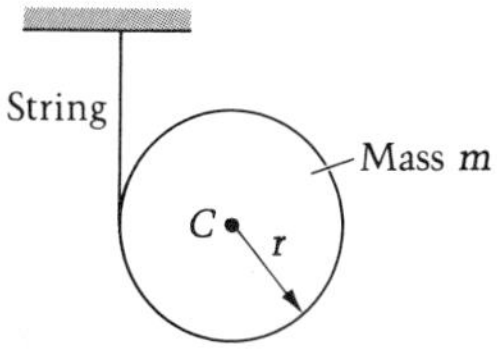

Figure E5.11a

Answer 5.5 Yes.

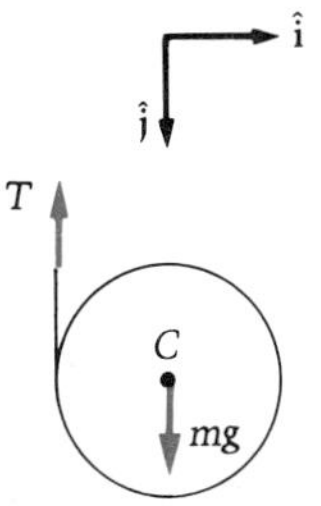

Figure E5.11b

Example 5.11

A cylinder has a string wrapped around it (see Figure E5.11a) and is released from rest. Determine the velocity of C as a function of time.

Solution

We choose the sign convention to be as shown in Figure E5.11b, since the cylinder turns clockwise as C moves downward. Applying the impulse and momentum equations in the y and θ directions (note that $\Sigma F_x = 0$ means $\ddot{x}_C = 0$ so that $\dot{x}_C = \text{constant} = 0$) gives

$$\int_0^t \Sigma F_y \, dt = m(\dot{y}_{C_f} - \dot{y}_{C_i})$$

$$\int_0^t (mg - T)\, dt = m(\dot{y}_C - 0)$$

$$mgt - \int_0^t T\, dt = m\dot{y}_C \tag{1}$$

and

$$\int_0^t \Sigma M_C \, dt = I_C(\omega_f - \omega_i)$$

$$\int_0^t Tr\, dt = \frac{1}{2} mr^2\omega$$

$$\int_0^t T\, dt = \frac{1}{2} mr\omega \tag{2}$$

We note that $\int T\,dt$ is itself an unknown and should be treated as such. (In this problem, use of the equations of motion would separately tell us that $T = mg/3$, so that the integral is in fact Tt. But sometimes T is time-dependent, in which case Tt would be incorrect for the value of the integral.)

Eliminating $\int_0^t T\,dt$ gives

$$mgt - m\dot{y}_C = \frac{1}{2} mr\omega$$

However, $\dot{y}_C = r\omega$ because the cylinder rolls on the rope, so that

$$gt = \left(1 + \frac{1}{2}\right)\dot{y}_C$$

which gives our result:

$$\dot{y}_C = \frac{2}{3} gt$$

The cylinder falls with "$\frac{2}{3}$ of a g" because of the retarding force of the rope.

In the next example, several bodies are involved, making the solution a little more difficult. Two limiting case checks are discussed at the end.

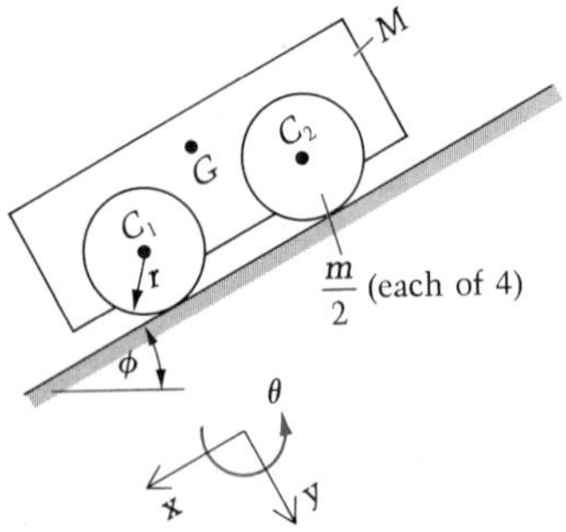

Figure E5.12a

EXAMPLE 5.12

The cart shown in Figure E5.12a has mass M exclusive of its four wheels, each of which is a disk of mass $m/2$. The front wheels and their axle are rigidly connected, and the same is true for the rear wheels. If the axles are smooth, find the velocity of G (the cart's mass center) as a function of time. The system starts from rest. Assume that there is enough friction to prevent the wheels from slipping.

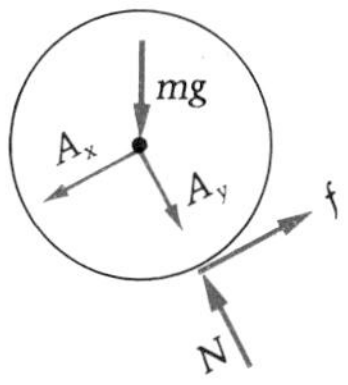

Figure E5.12b

Solution

We first consider the free-body diagram (Figure E5.12b) of a wheel pair (either front or back). Since the front and rear wheels are constrained to have identical angular velocities at all times, the fr's (front and back) must produce identical $I\alpha$'s; hence the friction force is the same for the rear wheels as for the front. And since the wheels' mass centers must always have identical velocities, the forces acting down the plane on each pair (front and back) must also be equal. These resultants are $A_x + mg \sin \phi - f$, so the reaction A_x is also the same on each pair of wheels.

Question 5.6 Are A_y and N also the same for front and rear wheels?

From the free-body diagram, we may write the following linear and angular equations of impulse and momentum:

$$\int_0^t \Sigma F_x \, dt = m(\dot{x}_{C_f} - \dot{x}_{C_i})$$

$$\int_0^t (A_x + mg \sin \phi - f) \, dt = m\dot{x}_{C_{1 \text{ or } 2}} \tag{1}$$

$$\int_0^t \Sigma M_C \, dt = I_C(\omega_f - \omega_i) = I_C(\omega - 0)$$

$$\int_0^t fr \, dt = \frac{1}{2} mr^2\omega \tag{2}$$

We may also isolate a free-body diagram of the translating cart (Figure E5.12c) and write its equation of impulse and momentum in the x direction down the plane:

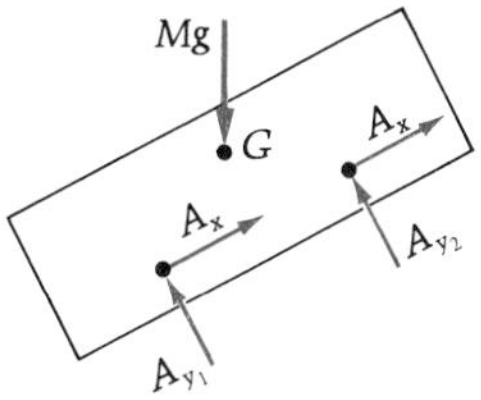

Figure E5.12c

$$\int_0^t \Sigma F_x \, dt = M(\dot{x}_{G_f} - \dot{x}_{G_i})$$

$$\int_0^t (-2A_x + Mg \sin \phi) \, dt = M(\dot{x}_G - 0) \tag{3}$$

Next we note that C_1 and C_2, the mass centers of the front and back wheel pairs, are also points of the cart; thus $\dot{x}_{C_1} \equiv \dot{x}_{C_2} \equiv \dot{x}_G$. Using this relation, and adding Equation (3) to twice Equation (1), eliminates the unknown impulse of the reaction A_x:

$$\int_0^t [(M + 2m)g \sin \phi - 2f] \, dt = (M + 2m)\dot{x}_G{}^* \tag{4}$$

Similarly, adding Equation (4) to twice Equation (2) results in an equation free of the unknown friction force:

$$\int_0^t (M + 2m)g \sin \phi \, dt = (M + 2m)\dot{x}_G + m \overbrace{(r\omega)}^{\dot{x}_G}$$

Answer 5.6 Not in general. They depend on the position of G relative to C_1 and C_2.

* This is the equation of linear impulse and momentum for the total (nonrigid) system of cart plus wheels.

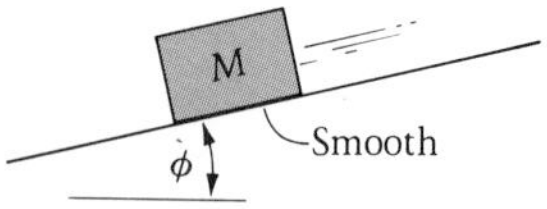

Figure E5.12d

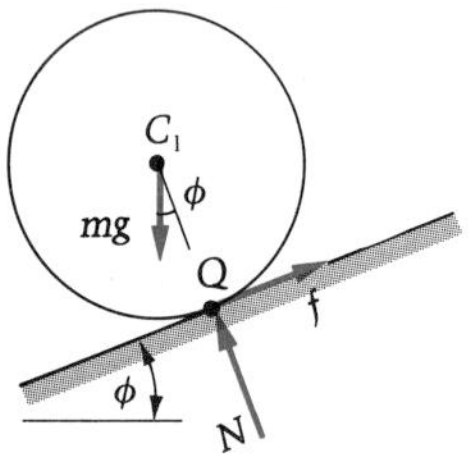

Figure E5.12e

Carrying out the integration and solving for $\dot{x}_G$, we get

$$\dot{x}_G = \frac{(M + 2m)\, gt \sin \phi}{M + 3m}$$

We note from this result that if the wheels are very light compared to the weight of the cart ($m \ll M$), then $\dot{x}_G = gt \sin \phi$, which is the answer for the problem of Figure E5.12d. Thus light wheels on smooth axles makes the cart move as if it were on a smooth plane, as expected.

The reader may wish to examine the other limiting case, that of the cart being light compared to heavy wheels ($M \ll m$). In this case the result, using the free-body diagram in Figure E5.12e, is $\dot{x}_G = \frac{2}{3} gt \sin \phi$.

In the next example, the two bodies — one rolling and the other translating — are connected by an inextensible cord.

EXAMPLE 5.13

Force P acts on the rolling cylinder $\mathcal{C}$ beginning at $t = 0$ with $\mathcal{C}$ at rest. (See Figure E5.13a.) Force P varies with the time t in seconds according to

$$P = 5 \sin \frac{\pi t}{10}\ \text{N} \qquad \text{(positive to the left as shown)}$$

Cylinder $\mathcal{C}$ and body $\mathcal{B}$ respectively weigh 100 and 40 N. Find the velocity of G (the mass center of $\mathcal{B}$) when $t = 10$ s. Neglect the effect of the hubs in Figure E5.13b (and the drilled hole to accommodate force P) on the moment of inertia of $\mathcal{C}$.

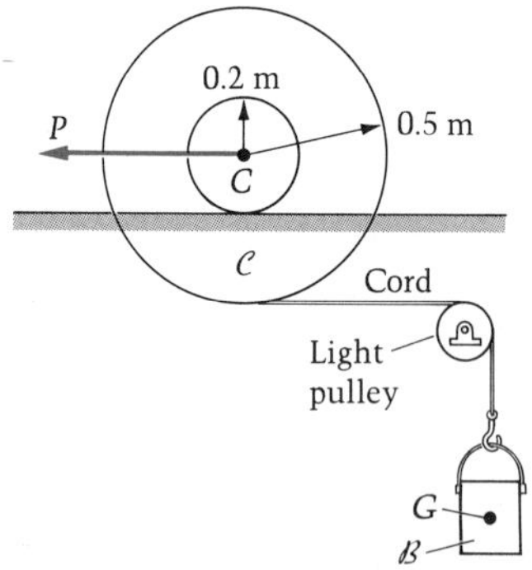

Figure E5.13a

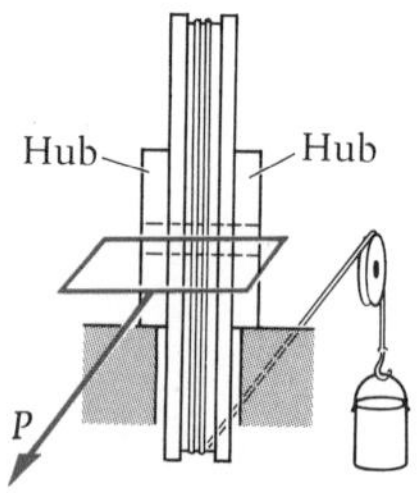

Figure E5.13b

Solution

Using the free-body diagrams (Figures E5.13c and E5.13d), we may write the equations of impulse and momentum. On $\mathcal{C}$, using Figure E5.13c,

$$\int \Sigma F_x \, dt = m\dot{x}_{C_f} - m\dot{x}_{C_i}^{\nearrow 0}$$

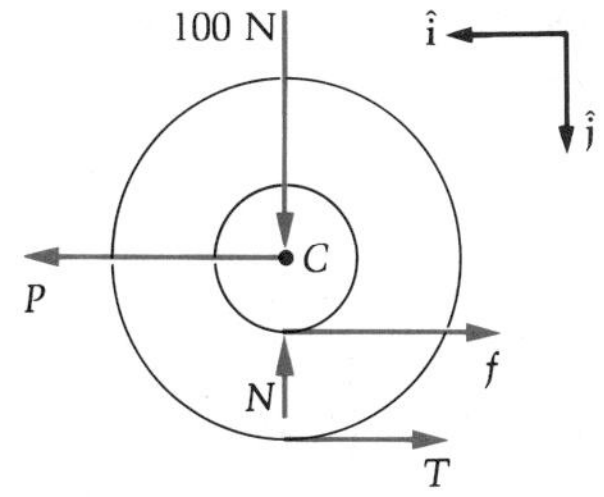

Figure E5.13c

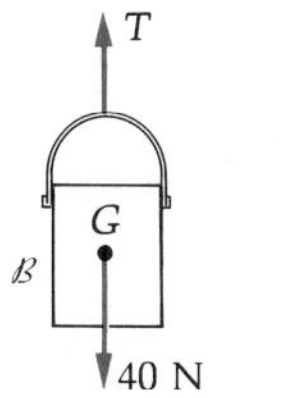

Figure E5.13d

or

$$\int_0^{10}\left(5\sin\frac{\pi t}{10} - f - T\right)dt = \frac{100}{9.81}\dot{x}_{C_f} = 10.2\dot{x}_{C_f} \tag{1}$$

Also on $\mathcal{C}$:

$$\int \Sigma M_C\, dt = I_C\omega_f - I_C\cancelto{0}{\omega_i}$$

or

$$\int_0^{10}(0.2f + 0.5T)\, dt = \left[\frac{1}{2}\frac{100}{9.81}(0.5)^2\right]\omega_f = 1.27\omega_f \tag{2}$$

On $\mathcal{B}$, using Figure E5.13d,

$$\int \Sigma F_y\, dt = m\dot{y}_{G_f} - m\cancelto{0}{\dot{y}_{G_i}}$$

or

$$\int_0^{10}(40 - T)\, dt = \frac{40}{9.81}\dot{y}_{G_f} = 4.08\dot{y}_{G_f} \tag{3}$$

Subtracting Equation (3) from (1), after integrating the sine function, we obtain

$$\frac{100}{\pi} - \int_0^{10} f\, dt - 40(10) = 10.2\dot{x}_{C_f} - 4.08\dot{y}_{G_f} \tag{4}$$

To obtain a second, independent equation that is also free of the integral of the unknown tension T, we add Equation (1) to twice Equation (2):

$$\frac{100}{\pi} - 0.6\int_0^{10} f\, dt = 10.2\dot{x}_{C_f} + 2.54\omega_f \tag{5}$$

Multiplying Equation (4) by 0.6 and subtracting from Equation (5) gives

$$\frac{40}{\pi} + 240 = 4.08\dot{x}_{C_f} + 2.54\omega_f + 2.45\dot{y}_{G_f} \tag{6}$$

Kinematics now relates $\dot{x}_{C_f}$, $\dot{y}_{G_f}$, and ω_f; the lowest point of $\mathcal{C}$ has the same velocity magnitude as does G because of the inextensibility of the cord (see Figure E5.13e):

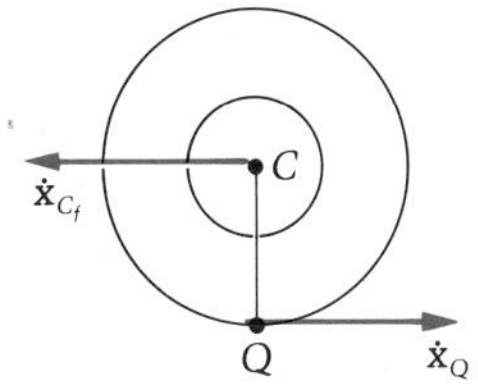

Figure E5.13e

$$\text{Kinematic conditions:}\quad \left.\begin{array}{l}\dot{x}_{C_f} = 0.2\omega_f\\ \dot{x}_Q = 0.3\omega_f\\ \dot{x}_Q = \dot{y}_{G_f}\end{array}\right\} \Rightarrow \left\{\begin{array}{l}\omega_f = 3.33\dot{y}_{G_f}\\ \dot{x}_{C_f} = 0.2\omega_f = 0.667\dot{y}_{G_f}\end{array}\right.$$

Substituting these expressions for $\dot{x}_{C_f}$ and ω_f into Equation (6) gives

$$253 = \dot{y}_{G_f}[4.08(0.667) + 2.54(3.33) + 2.45]$$

$$\dot{y}_{G_f} = \frac{253}{13.6} = 18.6 \text{ m/s}$$

Hence the velocity of the mass center of $\mathcal{B}$ at $t = 10$ s (when the force changes direction) is

$$\mathbf{v}_G = 18.6 \downarrow \text{m/s}$$

We emphasize that Equations (1), (2), and (3) in the preceding example are merely first integrals of the equations of motion studied in Chapter 4.

Conservation of Momentum

As we saw in Section 2.5, if the force in any direction (let us use x, for example) vanishes over a time interval, then the impulse in that direction vanishes also:

$$\int_{t_i}^{t_f} \Sigma F_x \, dt = 0$$

Since this impulse equals the change in momentum in the x direction, we have zero change when $\Sigma F_x \equiv 0$ and thus the momentum is *conserved* in that direction between t_i and t_f:

$$0 = m(\dot{x}_{C_f} - \dot{x}_{C_f})$$

or

$$m\dot{x}_{C_i} = m\dot{x}_{C_f} \tag{5.36}$$

We would of course also have **conservation of momentum** in the y (or any other) direction in which the force resultant vanished.

Finally, if the z component of $\Sigma \mathbf{M}_C$ is zero between t_i and t_f, then the *angular* impulse vanishes and we have **conservation of angular momentum:**

$$\int_{t_i}^{t_f} \Sigma M_{Cz} \, dt = 0 = H_{C_f} - H_{C_i}$$

or

$$H_{C_i} = H_{C_f} \tag{5.37}$$

For plane motion of symmetric bodies (I_{xz}^C and $I_{yz}^C = 0$) we have $\mathbf{H}_C = I_{zz}^C \omega \hat{\mathbf{k}}$; there is then no need for the z subscript, and Equation (5.37) may be rewritten

$$(I_C\omega)_i = (I_C\omega)_f \tag{5.38}$$

We now consider a well-known example of conservation of angular momentum.

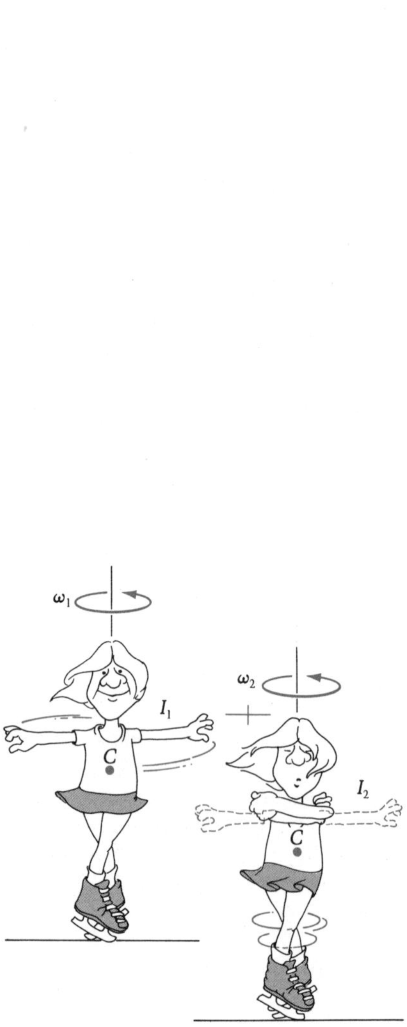

Figure E5.14

EXAMPLE 5.14

A skater spinning about a point on the ice (see Figure E5.14) draws in her arms and her angular speed increases.

a. Is angular momentum conserved?
b. Is kinetic energy conserved?
c. Account for any gains or losses if either answer is no.

Solution

We begin with part (a). *Before* the skater draws in her arms, we may treat her as a rigid body and thus $H_{Ci} = I_1\omega_1$. The same is true *after* the arms are drawn in, so that $H_{Cf} = I_2\omega_2$. If we neglect the small friction couple at the skates and the small drag moments caused by air resistance, then the answer to part (a) is yes because $\Sigma \mathbf{M}_C$ is then zero. Thus

$$I_1\omega_1 = I_2\omega_2$$

Therefore

$$\omega_2 = \frac{I_1}{I_2}\omega_1 > \omega_1 \qquad \text{(showing an angular speed increase since } I_1 > I_2\text{)}$$

For part (b) the kinetic energies are

$$T_1 = \frac{1}{2} I_1\omega_1^2$$

$$T_2 = \frac{1}{2} I_2\omega_2^2 = \frac{1}{2} I_1\omega_1^2\left(\frac{I_1}{I_2}\right) > T_1$$

Thus kinetic energy is *not* conserved.

For part (c) the change in kinetic energy is seen to be positive:

$$\Delta T = T_2 - T_1 = \frac{1}{2} I_1\omega_1^2 \underbrace{\left(\frac{I_1}{I_2} - 1\right)}_{> 0 \text{ since } I_1 > I_2}$$

Since there is no work done by the external forces and couples,* it is clear that this kinetic energy increase is accompanied by an internal energy decrease within the skater's body as her muscles do (nonexternal) work on her (nonrigid) arms in drawing them inward. Since *total* energy is always conserved (first law of thermodynamics), the skater has lost *internal* energy in the process.

The next example is similar to Examples 2.18 and 2.19 except that now the pulley has mass.

Example 5.15

Two identical twin gymnasts, L and R, of mass m are in equilibrium holding onto a stationary rope in the position shown in Figure E5.15a. The rope passes over the pulley $\mathcal{B}$, which has moment of inertia I with respect to the axis through O normal to the figure. The gymnasts then begin to move on the rope at speeds relative to it of $\dot{y}_{L\,\text{rel}}$ upward and $\dot{y}_{R\,\text{rel}}$ downward. When gymnast R reaches the end of the rope, he discovers he is in the same spot in space at which he began. How far up or down (tell which) has L moved (a) relative to the rope? (b) in space?

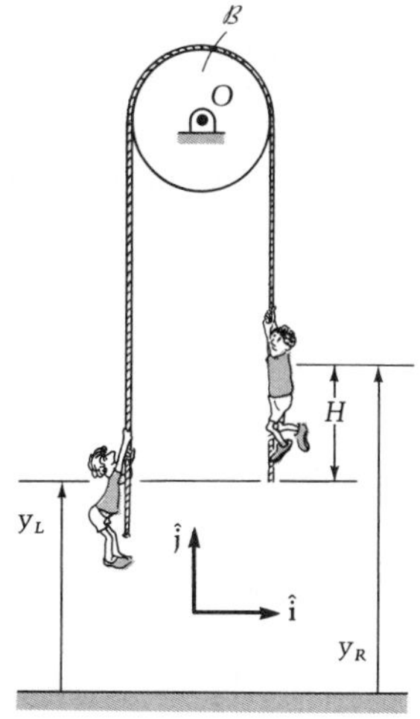

Figure E5.15a

* Assuming her arms are drawn in at the same level.

Solution

If we select our system to be "everything": L, R, the rope, and $\mathcal{B}$, then $\Sigma \mathbf{M}_O = \mathbf{0}$, and thus angular momentum about O is conserved. (Note that the gymnasts' gravity forces' moments about O cancel, and the pin reactions and weight of $\mathcal{B}$ pass through O.) Therefore,

$$\mathbf{H}_{O_f} = \mathbf{H}_{O_i} = \mathbf{0} \qquad \text{(since all bodies are at rest initially)} \tag{1}$$

After motion begins, the angular momentum about O of $\mathcal{B}$ is simply $I_O^{\mathcal{B}} \omega_{\mathcal{B}} \hat{\mathbf{k}}$, since O is a pivot of $\mathcal{B}$. For L and R, however, the situation is different and needs some discussion. The gymnasts, of course, are not rigid bodies, and we shall resort to Equation (2.38) to write their angular momenta with respect to O. For L, with velocity $\mathbf{v}_L$ in the inertial frame (the ground), and mass center C_L,

$$\mathbf{H}_O^L = \mathbf{H}_{C_L}^L + \mathbf{r}_{OC_L} \times m\mathbf{v}_{C_L}$$

We now assume that the gymnast is a particle; this is equivalent to neglecting $\mathbf{H}_{C_L}^L$, the gymnast's angular momentum about his mass center, in comparison with the "$\mathbf{r} \times m\mathbf{v}$" term. This is a good assumption because whatever body motions are not translatory are caused by parts (arms, mostly) in relative motion fairly close to the mass center. Thus,

$$\mathbf{H}_O^L \approx \mathbf{r}_{OC_L} \times m\mathbf{v}_L = \mathbf{r}_{OP} \times m\mathbf{v}_L$$

where P is shown in Figure E5.15b. Similarly,

$$\mathbf{H}_O^R \approx \mathbf{r}_{OC_R} \times m\mathbf{v}_R = \mathbf{r}_{OQ} \times m\mathbf{v}_R$$

Question 5.7 Why is $\mathbf{r}_{OC_R} \times m\mathbf{v}_R = \mathbf{r}_{OQ} \times m\mathbf{v}_R$?

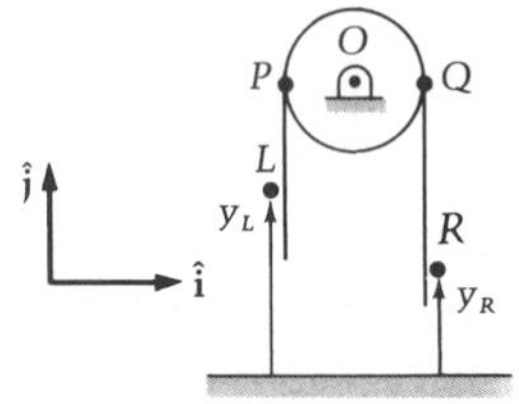

Figure E5.15b

Now, if we locate (see Figure E5.15b) the gymnasts' position *in space* with the coordinates y_L and y_R, then

$$\mathbf{H}_O^L = \mathbf{r}_{OP} \times m\mathbf{v}_L = -r\hat{\mathbf{i}} \times m\dot{y}_L\hat{\mathbf{j}} = -mr\dot{y}_L\hat{\mathbf{k}}$$

$$\mathbf{H}_O^R = \mathbf{r}_{OQ} \times m\mathbf{v}_R = r\hat{\mathbf{i}} \times m\dot{y}_R\hat{\mathbf{j}} = mr\dot{y}_R\hat{\mathbf{k}}$$

Note that if the gymnasts move in *opposite* vertical directions on opposite sides of the pulley, their angular momenta about O will be in the *same direction*. Note further that even though the "particles" L and R are in rectilinear motion, they still have angular momenta about points such as O that do not lie on their lines of motion.

Substituting the three angular momenta into Equation (1), we find:

$$\mathbf{H}_{O_f} = \mathbf{0}$$

$$\mathbf{H}_O^{\mathcal{B}} + \mathbf{H}_O^L + \mathbf{H}_O^R = 0$$

$$I\omega_{\mathcal{B}}\hat{\mathbf{k}} - mr\dot{y}_L\hat{\mathbf{k}} + mr\dot{y}_R\hat{\mathbf{k}} = 0 \tag{2}$$

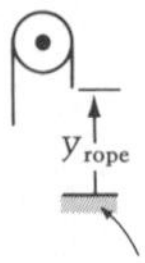

Starting postion of end of rope at $t = 0$

Figure E5.15c

Next, we know from the data that the rope moves counterclockwise around the pulley. Therefore, calling its speed $\dot{y}_{\text{rope}}$, we have (see Figure E5.15c)

$$\omega_B = \frac{\dot{y}_{\text{rope}}}{r}; \qquad \dot{y}_L = \dot{y}_{L\,\text{rel}} - \dot{y}_{\text{rope}}; \qquad \text{and} \qquad \dot{y}_R = -\dot{y}_{R\,\text{rel}} + \dot{y}_{\text{rope}}$$

Answer 5.7 Because $\mathbf{r}_{OC_R} = \mathbf{r}_{OQ} + \mathbf{r}_{QC_R}$, and $\mathbf{r}_{QC_R} \times m\mathbf{v}_{C_R} = \mathbf{0}$ since $\mathbf{r}_{QC_R}$ and $\mathbf{v}_{C_R}$ are parallel.

Substituting these into Equation (2), and simplifying the result, we obtain the equation

$$I\dot{y}_{\text{rope}} + mr^2(-\dot{y}_{L\text{ rel}} + \dot{y}_{\text{rope}} - \dot{y}_{R\text{ rel}} + \dot{y}_{\text{rope}}) = 0 \tag{3}$$

Integrating,

$$(I + 2mr^2)y_{\text{rope}} - mr^2 y_{L\text{ rel}} - mr^2 y_{R\text{ rel}} = C_1$$

But $C_1 = 0$ since $y_{\text{rope}} = y_{L\text{ rel}} = y_{R\text{ rel}} = 0$ at $t = 0$. Now at the time when R is at the end of the rope, $y_{R\text{ rel}} = H$, and $y_{\text{rope}} = H$ also, so that

$$mr^2 y_{L\text{ rel}} = (I + 2mr^2)H - mr^2 H = (I + mr^2)H \Rightarrow y_{L\text{ rel}} = \left(\frac{I + mr^2}{mr^2}\right)H$$

Therefore

$$y_L = y_{L\text{ rel}} - y_{\text{rope}} = \left(\frac{I + mr^2}{mr^2}\right)H - H = \frac{I}{mr^2}H$$

and gymnast L moves up $\left(\frac{I + mr^2}{mr^2}\right)H$ relative to the rope and $\frac{IH}{mr^2}$ in space from his original position.

We can check Example 2.18 by going back to Equation (2) of the preceding example, and letting $I \to 0$ (note that at this point nothing has been said about the relative motions):

$$\overset{0}{\cancel{I}}\,\omega_{\mathcal{B}} - mr\dot{y}_L + mr\dot{y}_R = 0$$

or

$$\dot{y}_L = \dot{y}_R$$

Therefore, regardless of relative motions (with respect to the rope), the two gymnasts rise in space equal amounts. The reader may wish to show that for the problem of Example 2.18 but with $I > 0$, the left gymnast is pulled up less than the height of climb in space of the right gymnast.

We now consider an example in which (unusually) both angular momentum *and* kinetic energy are conserved.

Example 5.16

The 2-kg collar $\mathcal{C}$ in Figures E5.16a,b turns along with the smooth rod $\mathcal{R}$ (see Figure E5.16a), which is 1 m long, has a mass of 3 kg, and is mounted in bearings with negligible friction. The angular speed is increased until the cord breaks (its tensile strength is 60 N), and at that instant the external moment is removed. Determine the angular velocity of $\mathcal{R}$ and the velocity of C (the mass center of $\mathcal{C}$) when the collar leaves the rod.

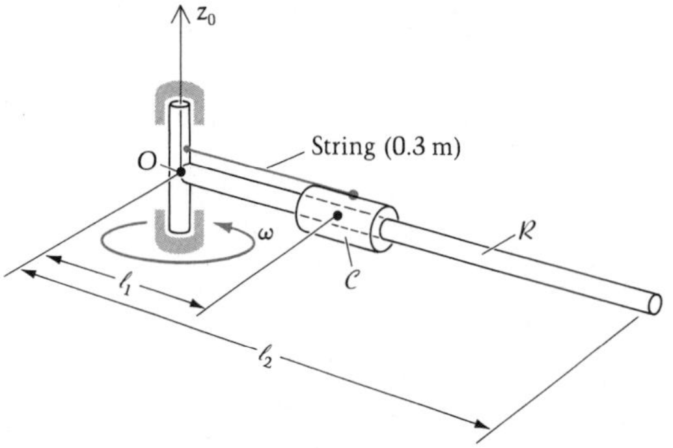

Figure E5.16a

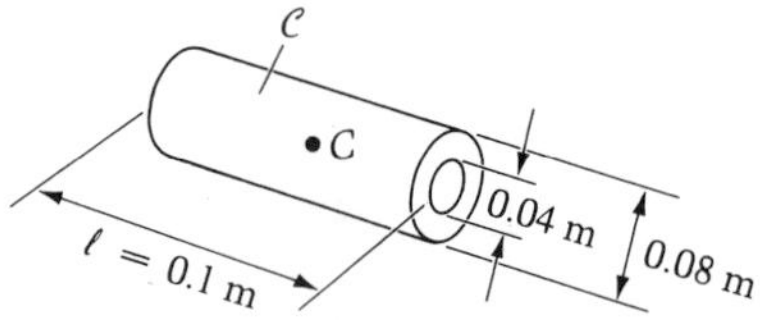

Figure E5.16b

Solution

The string provides the force causing the centripetal (inward) acceleration until it breaks. At that instant, we may solve for the angular velocity of $\mathcal{C}$:

$$\Sigma F_n = m a_{C_n}$$
$$T = m\ell_1\omega_1^2$$
$$60 = 2(0.3)\omega_1^2$$
$$\omega_1 = 10 \text{ rad/s}$$

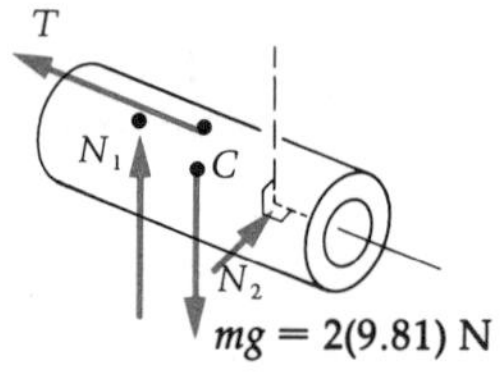

Figure E5.16c

In the accompanying free-body diagram, Figure E5.16c, N_1 and N_2 are the vertical and horizontal resultants of the pressures of the inside wall of $\mathcal{C}$ exerted by $\mathcal{R}$.

After the rope breaks at time t_1, collar $\mathcal{C}$ moves outward in addition to turning with $\mathcal{R}$; this is because there is no longer any inward force to keep it from "flying off on a tangent." Between times t_1 and t_2 (when it leaves $\mathcal{R}$), we have the following for the system ($\mathcal{C}$ plus $\mathcal{R}$):

1. Conservation of angular momentum H_O about z_O (because the external forces have no moment about z_O).
2. Conservation of kinetic energy T (since no net work is done on the system). Note that the normal forces between rod and collar, being equal in magnitude but opposite in direction, act on points with equal velocity components in the direction of either force; hence their net work vanishes.

Condition 1 gives

$$H_{O_i} = H_{O_f}$$

$$\underbrace{I_O^{\mathcal{C}}\omega_i + I_O^{\mathcal{R}}\omega_i}_{\text{Until the string breaks, } O \text{ is a point of both bodies!}} = I_O^{\mathcal{R}}\omega_f + \overbrace{I_C^{\mathcal{C}}\omega_f + \left[\left(\ell_2 + \frac{\ell}{2}\right)\hat{\mathbf{i}} \times m_{\mathcal{C}}\mathbf{v}_C\right]_z}^{H_{O_f}^{\mathcal{C}} = H_{C_f}^{\mathcal{C}} + (\mathbf{r}_{OC} \times m_{\mathcal{C}}\mathbf{v}_{C_f})_z}$$

Thus

$$\left[\underbrace{m_{\mathcal{C}}\left(\frac{r_i^2 + r_o^2}{4} + \frac{\ell^2}{12}\right)}_{I_C^{\mathcal{C}} = 0.00267} + m_{\mathcal{C}}\ell_1^2\right]10 + \frac{m_R\ell_2^2}{3}10 = \frac{m_R\ell_2^2}{3}\omega_f + I_C^{\mathcal{C}}\omega_f + \left(\ell_2 + \frac{\ell}{2}\right)^2\omega_f m_c$$

$$(0.00267 + 0.180)10 + \frac{3(1^2)}{3}10 = \omega_f + 0.00267\omega_f + (1.05)^2 2\omega_f$$

$$1.83 + 10 = \omega_f(1 + 0.00267 + 2.21)$$

$$\omega_f = 3.69 \text{ rad/s}$$

The component of $\mathbf{v}_C$ perpendicular to the rod $\mathcal{R}$ is thus $v_C = 1.05\omega_f = 3.87$ m/s. We can now obtain the radial component by conservation of T (condition 2):

$$\underbrace{\frac{1}{2} I_O \omega_i^2}_{O \text{ is } \textcircled{I} \text{ for both } \mathcal{C} \text{ and } \mathcal{R} \text{ initially}} = \frac{1}{2} I_O^{\mathcal{R}} \omega_f^2 + \frac{1}{2} m_{\mathcal{C}}(v_{C_\parallel}^2 + v_{C_\perp}^2) + \frac{1}{2} I_C^{\mathcal{C}} \omega_f^2$$

($v_{C_\parallel}$, $v_{C_\perp}$: components of $\mathbf{v}_C$ parallel and perpendicular to $\mathcal{R}$)

$$\tfrac{1}{2}(0.183 + 1)10^2 = \tfrac{1}{2}(1)3.69^2 + \tfrac{1}{2}(2)(v_{C_\parallel}^2 + 3.87^2) + \tfrac{1}{2}(0.00267)3.69^2$$

$$59.2 = 6.81 + v^2_{C_\parallel} + 15.0 + 0.0182$$

$$v_{C_\parallel} = 6.11 \text{ m/s}$$

Thus since the initial kinetic energy was 59.2 J and since

$$\frac{\frac{1}{2}(2)(6.11^2)}{59.2} = 0.631$$

we see that 63 percent of the original energy has gone into the outward motion of the collar.

Impact

We studied the impact of a pair of particles in Section 2.5. In this section we shall extend this study to two bodies colliding in plane motion.

The large forces occurring during an impact between two bodies $\mathcal{B}_1$ and $\mathcal{B}_2$ obviously deform the bodies. Because of vibrations and permanent deformations that are produced, some of the mechanical energy will be dissipated in the collision. However, it is often possible to treat a body as rigid *before*, and then again *after*, the impact in order to gain information of value. In impact problems we assume that:

1. Velocities and angular velocities may change greatly over the short impact interval Δt.
2. Positions of the bodies do not change appreciably.
3. Forces (and moments) that do not grow large over the interval Δt are neglected (such as gravity and spring forces). Such forces are called **nonimpulsive**; the large contact forces are called **impulsive**. It is the impulsive forces and moments that produce the sudden changes in velocities and angular velocities.

In Chapter 2 we introduced the coefficient of restitution as a measure of the capacity for colliding bodies to rebound off each other. We shall continue to use this parameter in this section, where now the relative velocities of separation and approach are of the impacting points of $\mathcal{B}_1$ and $\mathcal{B}_2$. Thus rigid-body kinematics will be needed to relate these velocities to those of the mass centers of the bodies. We emphasize again that the coefficient of restitution "e" is not the best of physical properties to measure; it depends upon the materials, geometry, and initial velocities. But as long as we take "e" with a grain of salt and remain aware of the limiting values $e = 0$ (bodies stick together) and $e = 1$ (no loss of energy), the definition of e does provide an approximate, much-needed equation that allows us to solve many problems of impact. We now consider two forms of the angular impulse and angular momentum equation that are applicable at the beginning and end of impacts involving the plane motion of bodies that may be regarded as rigid except during the collision phase of the motion.

If the body has a pivot O, we recall that

$$\mathbf{H}_O = I_{xz}^O\omega\hat{\mathbf{i}} + I_{yz}^O\omega\hat{\mathbf{j}} + I_{zz}^O\omega\hat{\mathbf{k}}$$

With O fixed in the inertial frame we have

$$\Sigma\mathbf{M}_O = \dot{\mathbf{H}}_O$$

Thus we may replace the C by an O in the angular impulse and momentum equation (5.33) for such pivot cases. The resulting equation about the axis of rotation is

$$\int_{t_i}^{t_f} \Sigma M_O\, dt = H_{Of} - H_{Oi} = I_{zz}^O(\omega_f - \omega_i)^* \tag{5.39}$$

This formula is of considerable value in impact problems because impulsive pivot reactions have no moment about O and thus do not appear in the equation. Note that if O is C, then Equation (5.39) is the same as our previous Equation (5.33) written about the mass center.

Another useful equation follows from

$$\Sigma\mathbf{M}_P = \dot{\mathbf{H}}_C + (\mathbf{r}_{PC} \times m\mathbf{a}_C) \tag{5.40}$$

In scalar form, for rigid bodies in plane motion this equation is

$$\Sigma M_P = I_{zz}^C\alpha + (\mathbf{r}_{PC} \times m\mathbf{a}_C)_z$$

in which P is an arbitrary point. If we state that P is now a fixed point O of the inertial frame $\mathcal{I}$, we may integrate this equation, getting

$$\int_{t_i}^{t_f} \Sigma M_O\, dt = (H_{Cf} - H_{Ci}) + (\mathbf{r}_{OC} \times m\mathbf{v}_C)_z\Big]_i^f$$

$$= \{I_{zz}^C\omega^* + (\mathbf{r}_{OC} \times m\mathbf{v}_C)_z\}\Big]_i^f \tag{5.41}$$

---

* We emphasize again that the angular momentum $\mathbf{H}_O$ (or $\mathbf{H}_C$) is not equal to its rigid body form $I_{zz}^O\omega\hat{\mathbf{k}}$ (or $I_{zz}^C\omega\hat{\mathbf{k}}$) *during* the impact, but these substitutions may be made at t_i before the collision and at t_f afterward.

Question 5.8 Why is the right-hand side not the integral of the right side of Equation (5.40) if O is moving?

Answer 5.8 If O is not fixed in $\mathcal{I}$, then $(d/dt)(\mathbf{r}_{OC} \times m\mathbf{v}_C) = \mathbf{r}_{OC} \times m\mathbf{a}_C - \mathbf{v}_O \times m\mathbf{v}_C$ and the second term is not zero then!

We shall now use these principles to solve a pair of example problems.

Example 5.17

An arrow of length L traveling with speed v_0 strikes a smooth hard wall obliquely as shown in the figure. End A does not penetrate but slides downward along the wall without friction or rebound. Find the angular velocity of the arrow after impact.

Solution

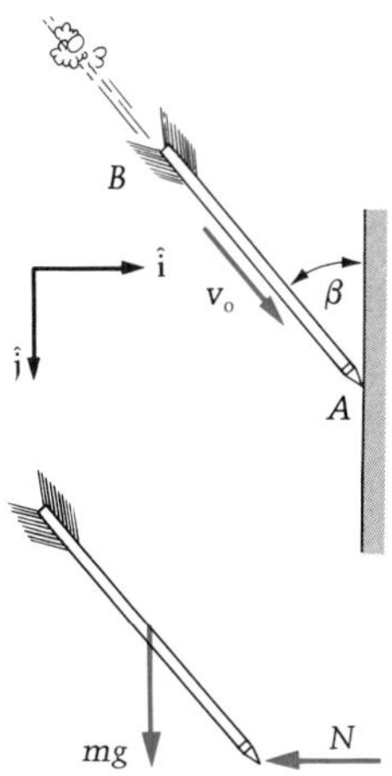

Figure E5.17

The only impulsive force acting on the arrow during its impact with the wall is the normal force N shown in the free-body diagram in Figure E5.17. We note that the gravity force over the short time interval is nonimpulsive:

$$\int_{t=0}^{\Delta t} mg\, dt\hat{\mathbf{j}} = mg\, \Delta t\hat{\mathbf{j}}$$

This is negligible in magnitude if Δt is very small, since mg does not grow large during impact. But for the normal force, that is not the case, as

$$\int_{t=0}^{\Delta t} N\, dt(-\hat{\mathbf{i}}) = -N_{\text{average}}\, \Delta t\hat{\mathbf{i}}$$

and this is non-negligible since N grows large "impulsively" during the short interval Δt, with average value N_{average}. In what follows, we shall delete the subscript and simply denote the impulse of $-N\hat{\mathbf{i}}$ as $-N\, \Delta t\hat{\mathbf{i}}$.

The impulse and momentum equation is then:

$$-N\, \Delta t\hat{\mathbf{i}} = m(v_{Cx_f} - v_0 \sin\beta)\hat{\mathbf{i}} + m(v_{Cy_f} - v_0 \cos\beta)\hat{\mathbf{j}}$$

or

$$\frac{-N\, \Delta t}{m} = v_{Cx_f} - v_0 \sin\beta \tag{1}$$

and

$$v_{Cy_f} = v_0 \cos\beta \tag{2}$$

where we note that momentum is conserved in the y-direction during impact.

The angular impulse and angular momentum principle yields:

$$N\, \Delta t \frac{L}{2} \cos\beta = \frac{mL^2}{12}\, \omega_f \tag{3}$$

"No rebound" means the x-component of $\mathbf{v}_{A_f}$ is zero; thus:

$$\mathbf{v}_C = v_{Cx_f}\hat{\mathbf{i}} + v_{Cy_f}\hat{\mathbf{j}} = v_{Ay_f}\hat{\mathbf{j}} + \omega_f\hat{\mathbf{k}} \times \mathbf{r}_{AC}$$

Using $\mathbf{r}_{AC} = L/2\,(-\sin\beta\hat{\mathbf{i}} - \cos\beta\hat{\mathbf{j}})$, we find that the x-component of this equation is:

$$v_{Cx_f} = \frac{L\omega_f}{2}\cos\beta \tag{4}$$

We have four equations in the unknowns v_{Cx_f}, v_{Cy_f}, ω_f, and $N\,\Delta t$. Solving, we find

$$\boldsymbol{\omega}_f = \omega_f\hat{\mathbf{k}} = \frac{6v_0\sin\beta\cos\beta}{L(1+3\cos^2\beta)}\hat{\mathbf{k}}$$

Note the obvious, that a head-on impact ($\beta = \pi/2$) brings the arrow to a dead stop with $\omega_f = 0$ and, from Equation (4), $v_{Cx_f} = 0$ also.

The next example, and the comments following it, constituted the solution to an actual engineering problem.

EXAMPLE 5.18

A 770-ton steel nuclear reactor vessel is being transported down a 6.5 percent grade using a specially designed suspended hauling platform together with crawler transporters. (See Figure E5.18a.) Determine the maximum velocity at which the reactor can be transported without tipping over if it should strike, and pivot about, a rigid obstacle at the front edge of the vessel's base ring.

Figure E5.18a *(Courtesy American Rigging Co.)*

Solution

The reactor vessel $\mathcal{B}$ will tip over if there is any kinetic energy left after it pivots about the front edge at O (see Figure E5.18b) and the mass center $\mathcal{C}$ reaches its highest point B, directly above O. Thus we solve for the velocity that will cause C to reach B; any higher velocity will cause overturning.

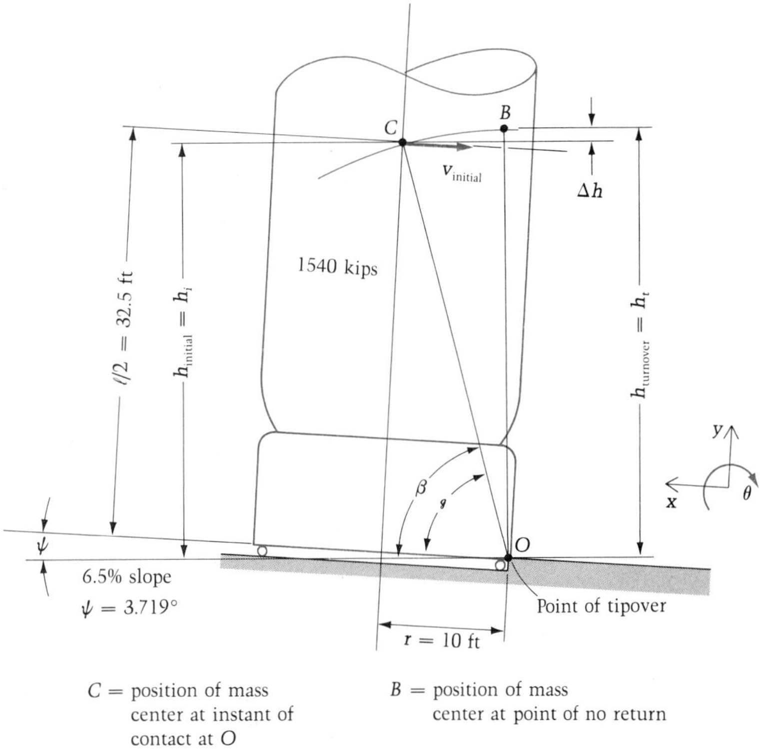

Figure E5.18b

We begin the solution with some preliminary geometric and trigonometric calculations based on the diagram. The 6.5 percent slope means $\psi = \tan^{-1}(0.065) = 3.719°$.* The turnover height h_t is given by the distance OC:

$$h_t = OC = OB = \sqrt{10^2 + 32.5^2} = 34.00 \text{ ft}$$

Also

$$\delta = \tan^{-1}\left(\frac{32.5}{10}\right) = 72.90°$$

and

$$\beta = \delta + \psi = 72.90° + 3.72° = 76.62°$$

The initial height h_i of C, above the horizontal line through O, is

$$h_i = (OC)\sin\beta$$
$$= (34.00)\sin 76.62° = 33.08 \text{ ft}$$

and thus the vertical distance through which C will move in reaching B is

$$\Delta h = h_t - h_i = 34.00 - 33.08$$
$$= 0.92 \text{ ft}$$

* We use four significant digits in this example.

(If we had adhered to three significant digits, the subtraction would have reduced us to just one good digit.)

There now remain two separate main parts to the solution of this problem. We first have to consider that mechanical energy is lost during the impact of $\mathcal{B}$ with the obstacle at O. Thus we are prevented from using the principle of work and kinetic energy over the short period of impact. What does apply, however, is conservation of angular momentum about O. This is because the impulsive forces (in both the x *and* y directions!) causing the sudden changes in the mass center velocity $\mathbf{v}_C$ and in the angular velocity ω are acting at O, so that $\Sigma\mathbf{M}_O = \mathbf{0}$. Therefore

$$\Sigma\mathbf{M}_O = \mathbf{0} = {}^{\mathcal{I}}\dot{\mathbf{H}}_O \Rightarrow \mathbf{H}_O = \text{constant vector}$$

or

$$\mathbf{H}_{Oi} = \mathbf{H}_{Of} \tag{1}$$

We shall use Equation (2.36) to express $\mathbf{H}_{Oi}$; this is the best formula for $\mathbf{H}_{Oi}$ for translation problems because $\mathbf{H}_{Ci} = \mathbf{0}$ in that case. For $\mathbf{H}_{Of}$, however, we have a nonzero ω. Thus we draw on the fact that since the vessel does not bounce at O, we may consider O a fixed point of both $\mathcal{B}$ and the inertial frame during and following the short period of impact. This in turn means that $\mathbf{H}_{Of}$ is simply $I_O\omega_f\hat{\mathbf{k}}$ after impact. Therefore Equation (1) becomes

$$\overset{0}{\cancel{\mathbf{H}_{Ci}}} + \mathbf{r}_{OC} \times m\mathbf{v}_{Ci} = I_O\omega_f\hat{\mathbf{k}} \tag{2}$$

Now since

$$\mathbf{r}_{OC} = +r\hat{\mathbf{i}} + \frac{\ell}{2}\hat{\mathbf{j}} \qquad \text{and} \qquad \mathbf{v}_{Ci} = -v_{Ci}\hat{\mathbf{i}}$$

the left side of Equation (2) is simply

$$\frac{\ell}{2}\, mv_{Ci}\hat{\mathbf{k}} \tag{3}$$

By the parallel-axis (transfer) theorem for moments of inertia we have

$$I_O = I_C + md^2 = m\left[k_C^2 + \left(\frac{\ell}{2}\right)^2 + r^2\right]$$

The vessel is essentially a thick shell; considering then that

$$I_C = mk_C^2 \approx \frac{mr^2}{2} + \frac{m\ell^2}{12} \tag{4}$$

we see that

$$I_O = \frac{m\ell^2}{3} + \frac{3}{2}mr^2 \tag{5}$$

By substituting Equations (3) and (5) into (2), we thus obtain

$$m\frac{\ell}{2}v_{Ci} = m\left(\frac{\ell^2}{3} + \frac{3}{2}r^2\right)\omega_f$$

and the angular velocity after impact is then

$$\omega_f = \frac{3\ell v_{Ci}}{2\ell^2 + 9r^2} \circlearrowleft \tag{6}$$

We are now ready to proceed to the second part of the solution.

Between the start of pivoting (immediately following impact) and the arrival at point B, the system is easily analyzed by work and kinetic energy:

$$W = \Delta T = T_f - T_i$$

To obtain the least possible value of v_{Ci} for no overturning, we set $T_f = 0$. The only work done in this phase of the vessel's motion is by gravity, so that

$$W = -mg\,\Delta h = 0 - T_i = -\frac{1}{2} I_O \omega_f^2$$

where ω_f is now an *initial* angular speed for this final stage of the problem and O is still a pivot point for $\mathcal{B}$. Thus

$$-mg\,\Delta h = \frac{1}{2} m \left(\frac{\ell^2}{3} + \frac{3}{2} r^2\right)\left(\frac{3\ell v_{C_i}}{2\ell^2 + 9r^2}\right)^2 \tag{7}$$

or

$$v_{Ci} = \sqrt{\frac{4g\,\Delta h(2\ell^2 + 9r^2)}{3\ell^2}} = \sqrt{\frac{4(32.17)(0.92)(2 \times 65^2 + 9 \times 10^2)}{3 \times 65^2}}$$

$$= 9.4 \text{ ft/sec}$$

There are several important follow-on remarks to be made about the preceding example. The first is that it can be shown (with a coefficient of restitution analysis) that more energy is lost with no bounce at O than if rebounding takes place. This energy loss for the $e = 0$ case just studied is

$$\Delta E = T_i - T_f$$

where i and f refer to the instants just before and after the impact. Substituting (for the case of no bounce), we get

$$\Delta E = \frac{1}{2} m v_{Ci}^2 - \frac{1}{2} I_O \omega_f^2$$

$$= \frac{1}{2} m\left[v_{Ci}^2 - \left(\frac{\ell^2}{3} + \frac{3}{2} r^2\right)\left(\frac{3\ell v_{Ci}}{2\ell^2 + 9r^2}\right)^2\right]$$

$$= \frac{1}{2} m v_{Ci}^2 \left(1 - \frac{9\ell^2/6}{2\ell^2 + 9r^2}\right)$$

For $\ell = 65$ ft and $r = 10$ ft, we obtain

$$\Delta E = \frac{1}{2} m v_{Ci}^2 (0.322)$$

Thus 32.2%, or nearly a third of the original mechanical energy, is lost during impact if the lower front corner of $\mathcal{B}$ sticks to, and pivots about, point O. Of great importance here is the fact that we do not *know* how much rebounding would actually occur in the physical situation and hence how much energy would be lost. This means that 9.4 ft/sec *may not be a conservative engineering answer* for the safe speed. If the plane is flat ($\psi = 0$), for example, it can be shown that:

1. The speed corresponding to pivoting as in this example is 11.9 ft/sec. (It has farther to pivot so it can be going faster prior to impact.)
2. At this speed initially, and with a no-energy-lost rebound, the vessel will easily overturn even though the striking corner backs up.

A conservative safe speed of the vessel in the inclined plane case can be obtained by assuming that no energy is dissipated during the impact and that all the vessel's initial kinetic energy goes into tilting it up about O. This approach gives

$$-mg\,\Delta h = -\frac{1}{2}\,mv_{Ci}^2$$

$$v_{Ci} = \sqrt{2g\,\Delta h} = 7.7 \text{ ft/sec}$$

In practice, the engineers in this case decided not to exceed 3 ft/sec, in view of the importance of the work and the danger involved.

The Center of Percussion

We now turn our attention to a new topic. Besides the mass center C (but of much lesser importance), there is another special point of interest associated with a rigid body $\mathcal{B}$ in plane motion, a point that differs from C in that it depends not only on the mass distribution of $\mathcal{B}$ but also on the motion of the body. This point lies along the resultant of the $m\mathbf{a}$ vectors of all the body's mass elements. The point is called the **center of percussion**, and it has value in certain applications such as impact testing.

Before getting into the theory behind the center of percussion, we first illustrate its existence and demonstrate its value by means of an example. If a youngster hits a baseball with a stick and does not translate his hands too much, we may model the situation as shown in Figure 5.13 and ask where the ball should hit the stick in order to eliminate the "sting" (transverse reaction R_y of the stick onto the boy's hands). If the boy hits the ball at just the right place, called the "sweet spot," he hits it a long way while hardly feeling it and is said to have gotten "good wood" on the ball. Assuming the stick to be rigid at the beginning ($t = 0$) and end ($t = \Delta t$) of the short impact interval, the two principles of impulse and momentum are used as follows.

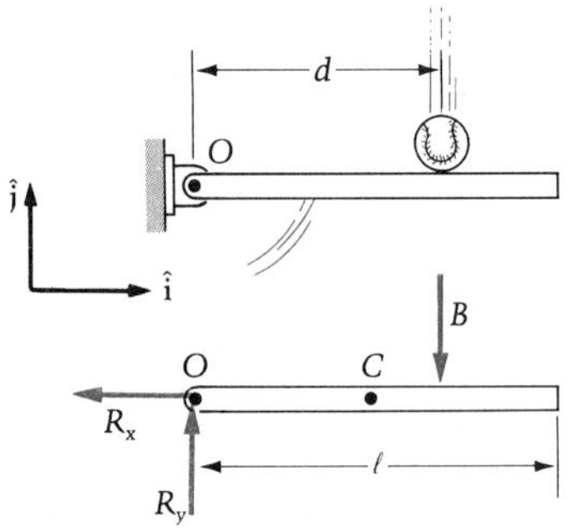

Figure 5.13

The impulse-momentum equation for the stick in the y direction is

$$\int_0^{\Delta t} (R_y - B)\,dt = m(\dot{y}_{C_f} - \dot{y}_{C_i}) \tag{5.42}$$

The angular impulse-angular momentum equation is

$$\int_0^{\Delta t} -Bd\,dt = I_O(\omega_f - \omega_i) = \frac{m\ell^2}{3}(\omega_f - \omega_i) \tag{5.43}$$

Using $R_y = 0$ for "no sting," setting $\dot{y}_C = (\ell/2)\omega$ by kinematics at $t = 0$ and Δt, and multiplying Equation (5.42) by d gives

$$-\int_0^{\Delta t} Bd\,dt = \frac{md\ell}{2}(\omega_f - \omega_i) \tag{5.44}$$

Dividing Equation (5.44) by (5.43) gives

$$1 = \frac{d}{2} \div \frac{\ell}{3} \Rightarrow d = \frac{2}{3}\ell$$

Thus if the ball is struck two-thirds of the way from O to the end of the stick, the transverse reaction R_y will be zero.

In this example the point at which the ball is struck ($d = 2\ell/3$) is the center of percussion of the stick. To show this, at least for the case when the bat is rigid, we first recall that in Chapter 2 we saw that for *any* point P (moving or not, fixed to $\mathcal{B}$ or not), we can always write

$$\Sigma \mathbf{M}_P = \int \mathbf{R} \times \mathbf{a}\, dm$$

in which $\mathbf{R}$ is the position vector from point P to a generic differential mass element. Therefore, since the integral of the vectors ($\mathbf{R} \times \mathbf{a}\, dm$) over the body in fact represents the resultant moment about P of the $m\mathbf{a}$ vectors over the mass of $\mathcal{B}$, this integral vanishes for all points P^* on the line along which the resultant of the $m\mathbf{a}$ vectors lies and hence $\Sigma \mathbf{M}_{P^*} = \mathbf{0}$ for these points.

Armed with the fact that $\Sigma \mathbf{M}_{P^*} = \mathbf{0}$ for the center of percussion, we can now derive the general equation for the distance from a pivot O to the center of percussion P^* in plane motion (Figure 5.14):

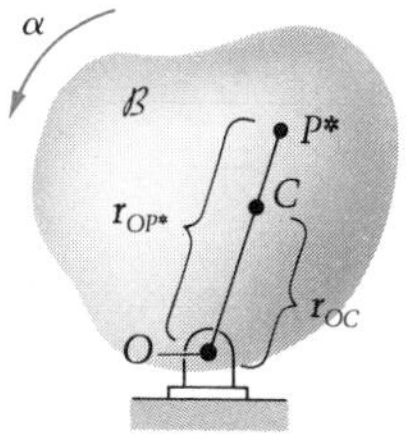

Figure 5.14

$$\Sigma M_O = \overset{0}{\cancel{\Sigma M_{P^*}}} + (\mathbf{r}_{OP^*} \times \overset{m\mathbf{a}_C}{\cancel{\Sigma \mathbf{F}}})_z$$

$$\alpha(I_C + mr_{OC}^2) = r_{OP^*} m (r_{OC}\alpha)$$

$$mk_C^2 + mr_{OC}^2 = mr_{OP^*} r_{OC}$$

Therefore

$$r_{OP^*} = r_{OC} + \frac{k_C^2}{r_{OC}} \tag{5.45}$$

and we see that, for the stick,

$$r_{OP^*} = \frac{\ell}{2} + \frac{\ell^2/12}{\ell/2} = \frac{2}{3}\ell \qquad \text{(as before)}$$

Note from Equation (5.45) that the center of percussion is always farther from the pivot than is the mass center. We make one final remark about the center of percussion. If we treat the "$\mathbf{a}\, dm$'s" of $\mathcal{B}$ as a collection of vectors, its resultant may be expressed (for a rigid body in plane motion) at the mass center (Figure 5.15), where

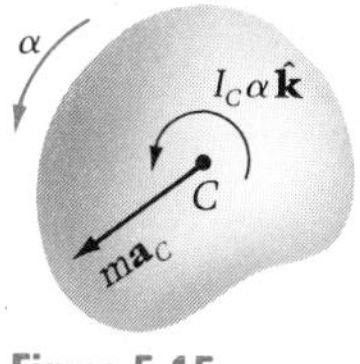

Figure 5.15

$$\int \mathbf{a}\, dm = m\mathbf{a}_C$$

$$\int \mathbf{r} \times \mathbf{a}\, dm = \Sigma \mathbf{M}_C = I_C \alpha \hat{\mathbf{k}}$$

In a manner identical to reducing a force and couple to its simplest form, we may reduce this resultant of the $m\mathbf{a}$ vectors as shown in Figure 5.16, where the distance D is $I_C\alpha/(ma_C)$. We note that there is in fact a *line* ℓ of points along the resultant of the $m\mathbf{a}$ vectors, making the location of a single point P^* ambiguous. However, the concept of the center of percussion is usually used in conjunction with problems in which the body has a pivot O (as in the previous example). In these problems P^* is the well-defined single point at the intersection of lines ℓ and OC as in Figure 5.16.

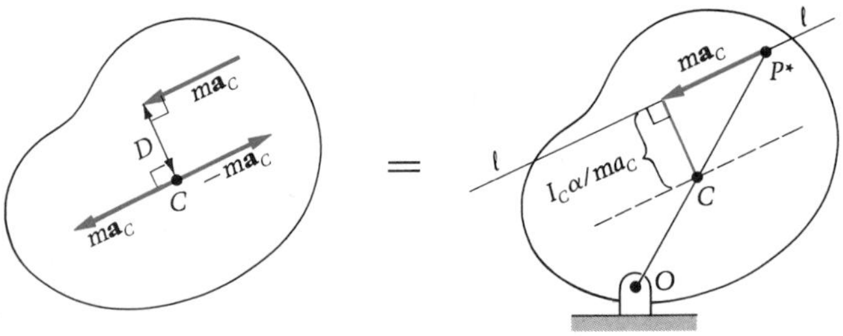

Figure 5.16

EXAMPLE 5.19

Find the center of percussion P^* for a pendulum $\mathcal{P}$ consisting of a rod plus disk, each of which has equal mass m. (See Figure E5.19.)

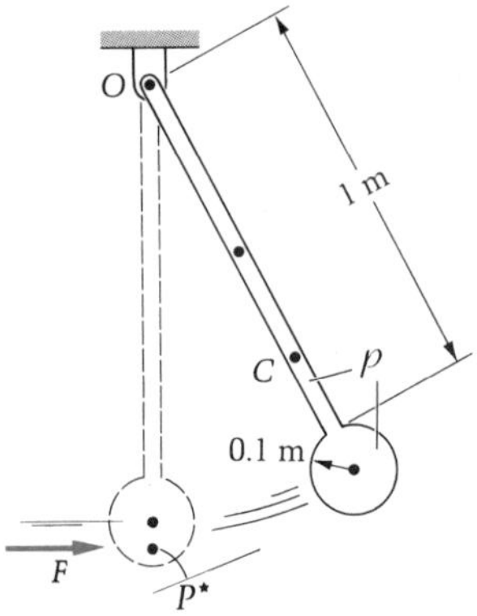

Figure E5.19

Solution

The mass center of $\mathcal{P}$ is located at a distance r_{OC} from O given by

$$m(0.5) + m(1.1) = 2mr_{OC}$$

$$r_{OC} = 0.8 \text{ m}$$

(Note that with equal masses C lies halfway between the mass centers of the rod and disk.) The radius of gyration with respect to C is calculated next:

$$I_{z_C} = \frac{m(1^2)}{12} + m(0.3^2) + \frac{m(0.1^2)}{2} + m(0.3^2)$$

$$= 0.268m = 2mk_C^2$$

Thus $k_C = 0.366$ m and Equation (5.45) then gives us the location of P^*:

$$r_{OP^*} = r_{OC} + \frac{k_C^2}{r_{OC}} = 0.8 + \frac{(0.366)^2}{0.8}$$

$$= 0.968 \text{ m}$$

Striking the pendulum at P^* eliminates the horizontal pin reaction at O, as we have seen.

PROBLEMS ▶ Section 5.3

5.75 Drum $\mathcal{B}_1$ has a radius of gyration of mass with respect to a horizontal axis through O of 1 m and a mass of 800 kg. Body $\mathcal{B}_2$ has a mass of 600 kg and a velocity of 20 m/s upward when in the position shown in Figure P5.75. Find the velocity of $\mathcal{B}_2$ 3 s later.

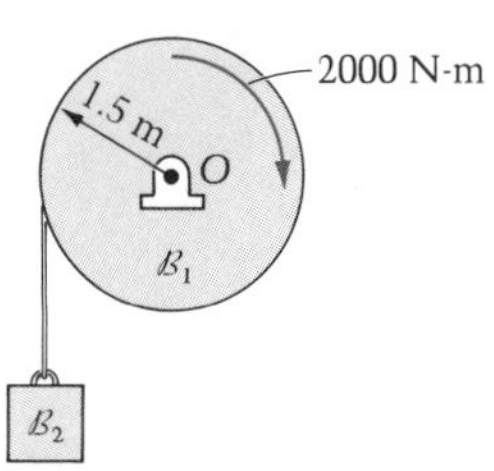

Figure P5.75

5.76 The hollow drum shown in Figure P5.76 weighs 161 lb and rotates about a fixed horizontal axis through O. The diameter of the drum is 2.4 ft, and the radius of gyration of the mass with respect to the axis through O is 0.8 ft. The angular speed changes from 30 rpm ⟳ to 90 rpm ⟲ during a certain time interval. Find the time interval.

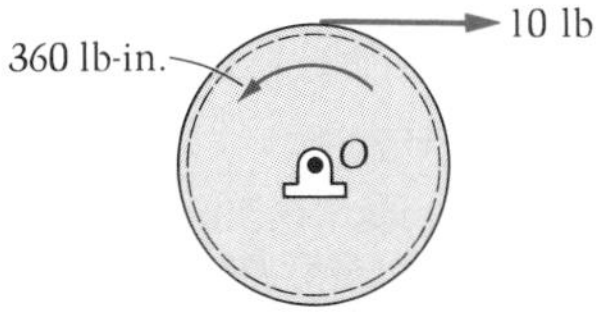

Figure P5.76

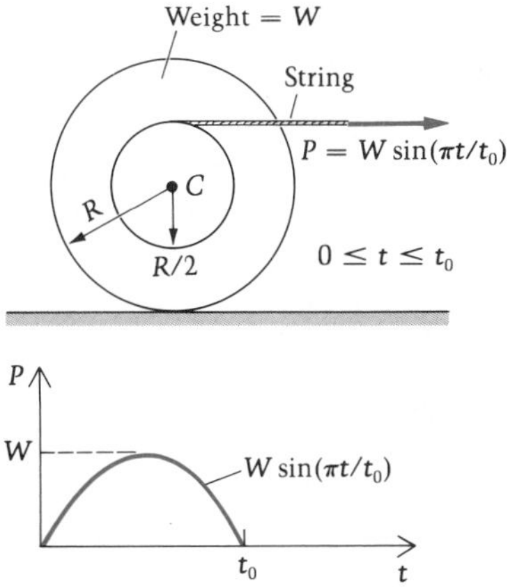

Figure P5.77

5.77 The sinusoidal force P is applied to the string in Figure P5.77 for a half-cycle. If the cylinder (initially at rest) does not slip, find its angular velocity at the end of the load application (at $t = t_0$).

5.78 A massless rope hanging over a frictionless pulley of mass M supports two monkeys (one of mass M, the other of mass $2M$). The system is released at rest at $t = 0$, as shown in Figure P5.78. During the following 2 sec, monkey B travels down 15 ft of rope to obtain a massless peanut at end P. Monkey A holds tightly to the rope during these 2 sec. Find the displacement of A during the time interval. Treat the pulley as a uniform cylinder of radius R.

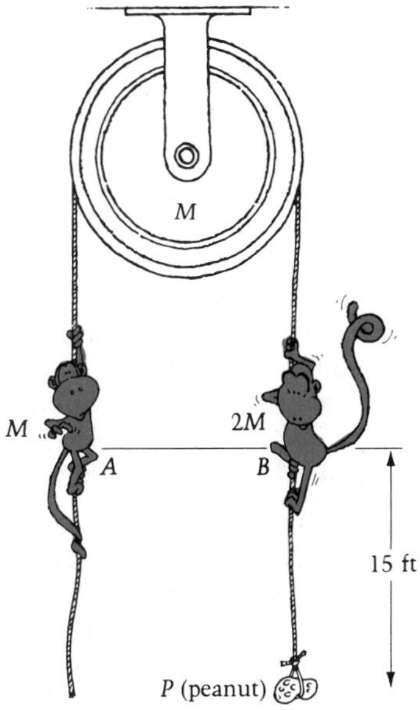

Figure P5.78

5.79 Force F in Figure P5.79 varies with time according to $F = 0.02t^2$ newtons, where t is measured in seconds. If there is enough friction to prevent slipping of the cylinder on the plane, find the velocity of C at: (a) $t = 3$ s; (b) $t = 10$ s. The cylinder starts from rest at $t = 0$.

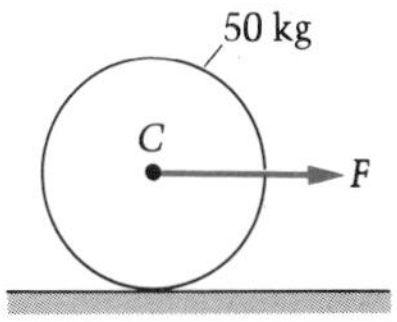

Figure P5.79

5.80 The cylinder in Figure P5.80 has mass $m = 3$ slugs and radius of gyration $k_C = 1.5$ ft with respect to C. There is sufficient friction to prevent slipping on the plane. A rope is wrapped around the inner radius, and a tension $T = 40$ lb is applied parallel to the plane as shown. Use impulse/momentum principles to find the velocity of C after 3 sec if motion starts from rest.

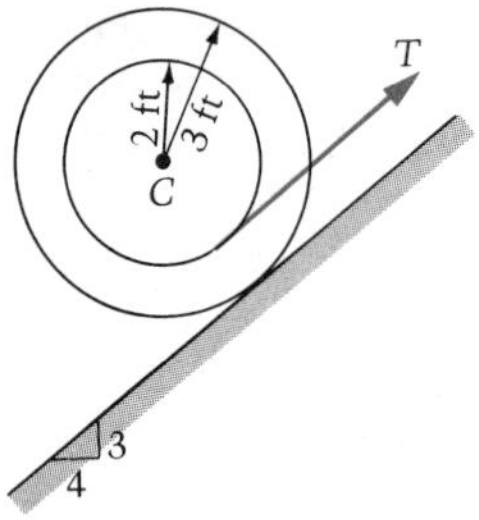

Figure P5.80

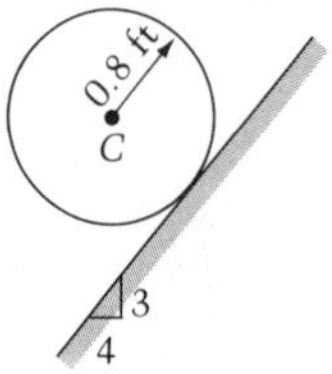

Figure P5.81

5.81 The 161-lb round body is rolling up the plane with $\omega = 5$ ⟲ rad/sec at the instant shown in Figure P5.81. The radius of gyration of the mass of the body with respect to the axis through the mass center C normal to the page is 0.7 ft. Find the time required for the mass center to reach its highest point.

5.82 The cylinder $\mathcal{B}_1$, turning at 200 rpm ⟲, is brought to rest by applying the 50-lb force to the light brake arm $\mathcal{B}_2$ as shown in Figure P5.82. Friction in the bearings at O produces a constant resistance torque of 7 lb-ft, and the coefficient of friction at the contact point A between $\mathcal{B}_1$ and $\mathcal{B}_2$ is $\mu = 0.3$. (a) Find the stopping time, and (b) find the number of revolutions turned by $\mathcal{B}_1$ during the braking.

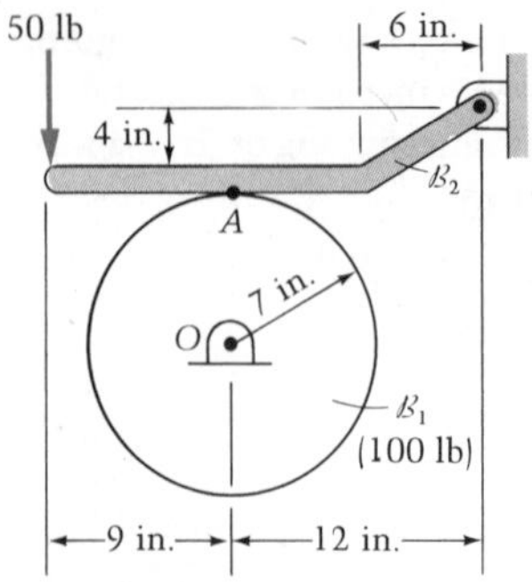

Figure P5.82

5.83 Acting on the gear is a couple C with a time-dependent strength given by $C = (6 + 0.8t)$ N-m, where t is measured in seconds. (See Figure P5.83.) If the system is released from rest at $t = 0$, find the velocity of block $\mathcal{B}$ when (a) $t = 3$ s; (b) $t = 10$ s. The centroidal radius of gyration of the gear is 0.25 m.

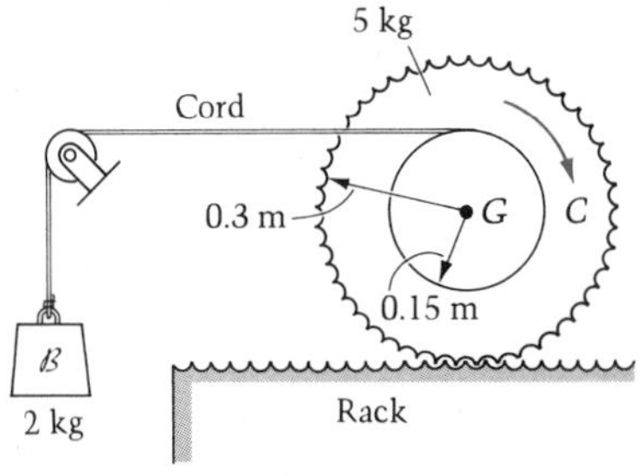

Figure P5.83

Solve the following problems by making use of the impulse and momentum, and/or angular impulse and angular momentum, methods.

5.84 Problem 4.107

5.85 Problem 4.166(c)

5.86 Problem 4.179

5.87 Problem 4.102

5.88 Problem 4.101

5.89 Problem 4.87(b)

5.90 A pipe rolls (from rest) down an incline (Figure P5.90). Using the *equations of motion*, find:

a. $\dot{x}_C$ at time t

b. $\dot{x}_C$ after C moves the distance x_C.

Then use work and energy to verify the answer to part (b) and impulse and momentum to verify part (a). Finally, give the minimum μ to prevent slipping.

5.91 A body $\mathcal{B}$ weighing 805 lb with radius of gyration 0.8 ft about its z_C axis (see Figure P5.91) is pinned at its mass center. A clockwise couple of magnitude e^t lb-ft is

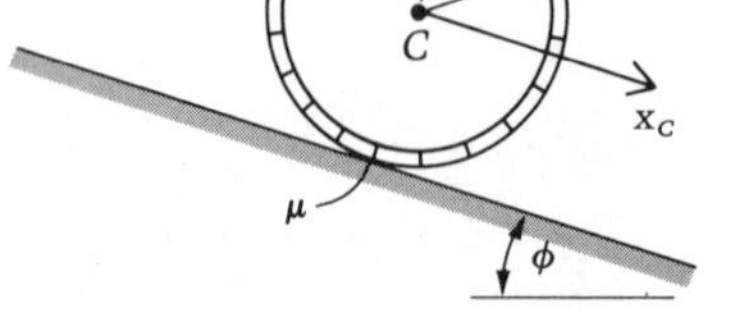

Figure P5.90

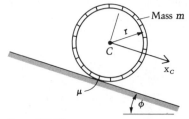

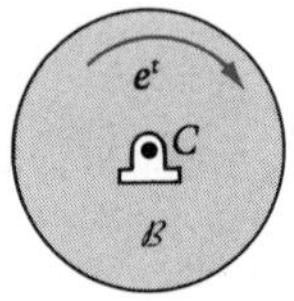

Figure P5.91

applied to $\mathcal{B}$ starting at $t = 0$. Find the angular velocity of $\mathcal{B}$ when $t = 3$ seconds.

5.92 Given that the slot (for the cord) in the cylinder in Figure P5.92 (mass 10 kg) has a negligible effect on I_C, find the velocity of the mass center C as a function of time, if $\theta = 60°$.

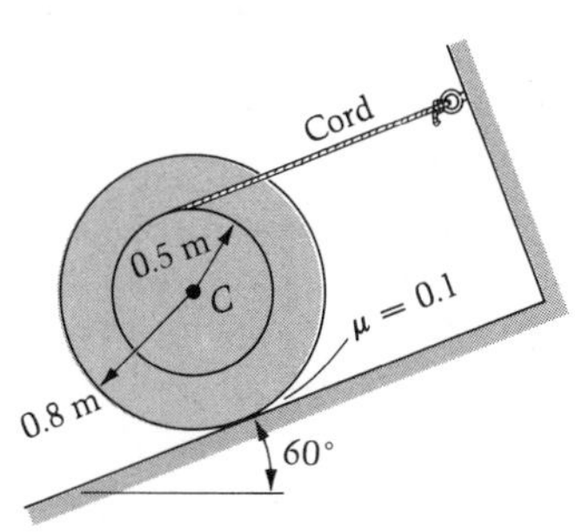

Figure P5.92

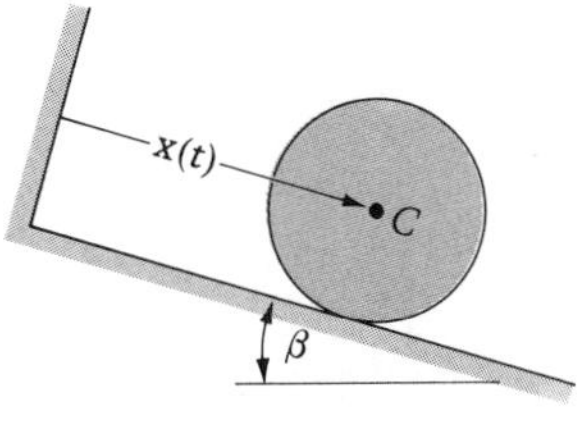

Figure P5.93

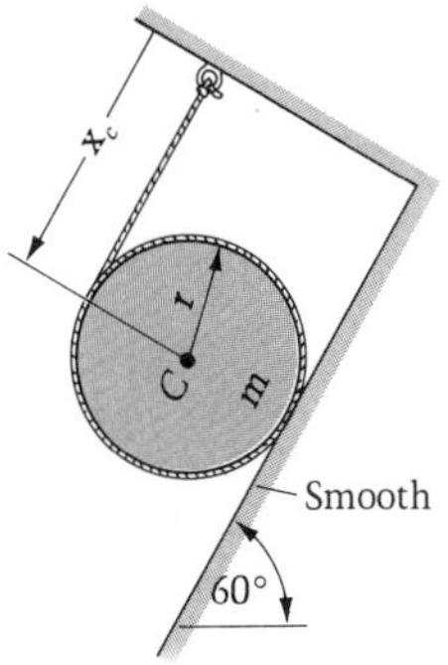

Figure P5.94

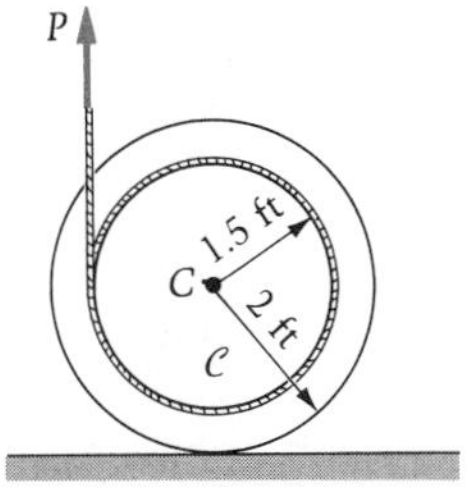

Figure P5.95

5.93 A uniform sphere (radius r, mass m) rolls on the plane in Figure P5.93. If the sphere is released from rest at $t = 0$ when $x = L$, find $\dot{x}(t)$.

5.94 The cord in Figure P5.94 is wrapped around the cylinder, which is released from rest on the 60° incline shown. Find the velocity of C as a function of time t.

* **5.95** The 50-lb body $\mathcal{C}$ in Figure P5.95 may be treated as a solid cylinder of radius 2 ft. The coefficient of friction between $\mathcal{C}$ and the plane is $\mu = 0.2$, and a force $P = 10$ lb is applied vertically to a cord wrapped around the hub. Find the velocity of the center C 10 sec after starting from rest.

* **5.96** A child pulls on an old wheel with a force of 5 lb by means of a rope looped through the hub of the wheel. (See Figure P5.96.) The friction coefficient between wheel and ground is $\mu = 0.2$. Find I_C for the wheel, and use it to determine the velocity of C 3 sec after starting from rest.

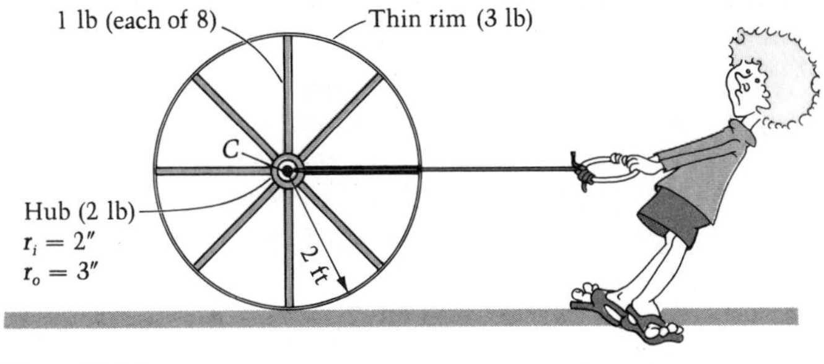

Figure P5.96

* **5.97** Two cables are wrapped around the hub of the 10-kg spool shown in Figure P5.97, which has a radius of gyration of 500 mm with respect to its axis. A constant 40-N force is applied to the upper cable as shown. Find the velocity of the mass center C 5 sec after starting from rest if: (a) $\mu = 0.2$; (b) $\mu = 0.5$.

* **5.98** The cart $\mathcal{B}_1$ is given an initial velocity v_i to the right at $t = 0$. The rod $\mathcal{B}_2$ is pinned to $\mathcal{B}_1$ at its mass center G, as shown in Figure P5.98(a). At $t = 0$, the mass center C of $\mathcal{B}_2$ is held fixed at the instant the cart starts off, then immediately released. At a later time (see Figure P5.98(b)), it is observed that $\mathcal{B}_2$ has $\omega_2 = 0$ at an instant when $\mathcal{B}_2$ has turned 90° clockwise. If $M = m$, find the velocity of G at that instant. Use $W = \Delta T$ *and* an impulse and momentum principle.

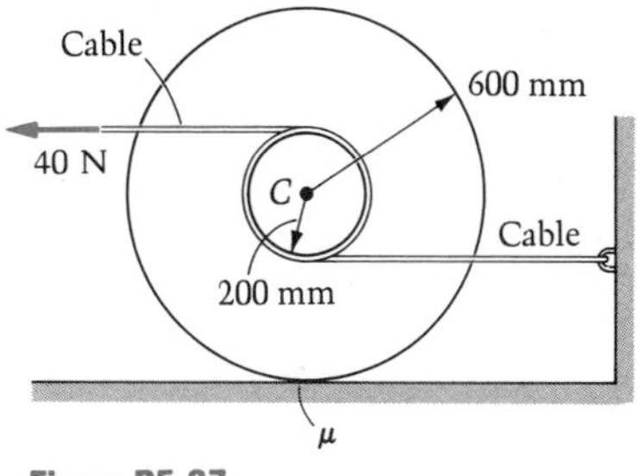

Figure P5.97

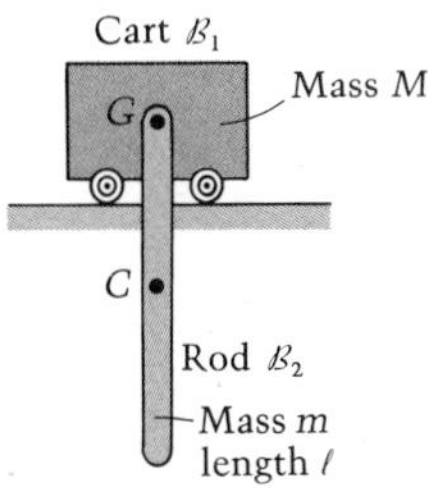

Figure P5.98(a)

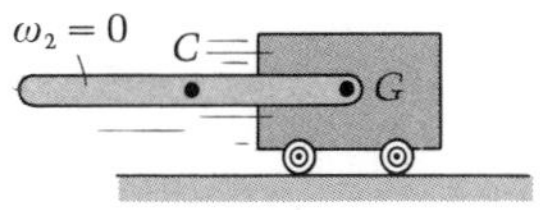

Figure P5.98(b)

5.99 Two gymnasts at A and B, each of weight W, hold onto the left side of a rope that passes over a cylindrical pulley (weight W, radius R) to a counterweight C of weight $2W$. (See Figure P5.99.) Initially the gymnast A is at depth d below B. He climbs the rope to join gymnast B. Determine the displacement of the counterweight C at the end of the climb.

5.100 Disk $\mathcal{B}_1$ and the light shaft in Figure P5.100 rotate freely at 40 rpm. Disk $\mathcal{B}_2$ (initially not turning) slides down the shaft and strikes $\mathcal{B}_1$; after a brief period of slipping, they move together. Find the average frictional moment exerted on $\mathcal{B}_1$ by $\mathcal{B}_2$ if the slipping lasts for 3 sec.

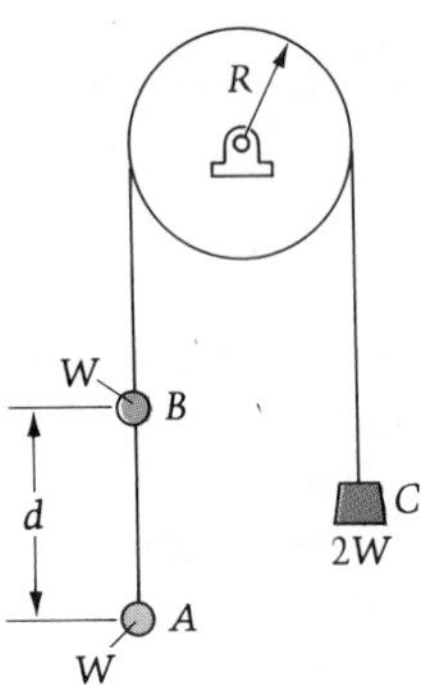

Figure P5.99

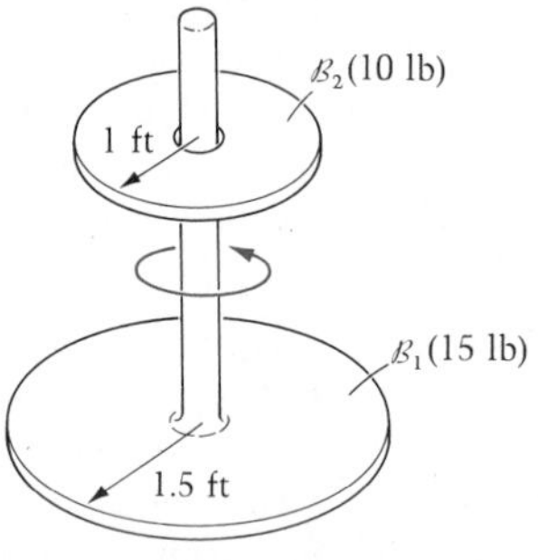

Figure P5.100

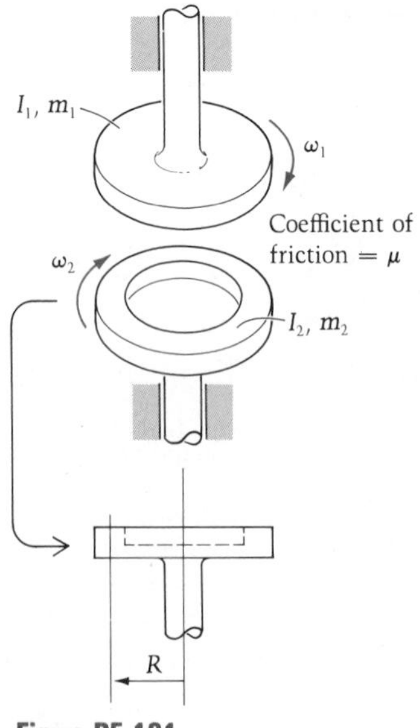

Figure P5.101

5.101 Two disks are spinning in the directions shown in Figure P5.101. The upper disk is lowered until it contacts the bottom disk (around the rim). Find how long it takes for the two disks to reach a common angular velocity, and determine its value. Finally, determine the energy lost. Show that if $I_1 = I_2$ and $\omega_1 = -\omega_2$, your solution predicts that 100 percent of the energy is lost (as it should). Determine which of the three answers (time, ω_f, energy loss) are the same if the two disks are instantaneously locked together instead of slipping.

5.102 Figure P5.102(a) shows a rough guess at a skater's mass distribution. Calculate the percentage increase in his angular speed about the vertical if he draws in his arms as shown in Figure P5.102(b). Assume that his arms are wrapped around the 6-in. radius circle of his upper body.

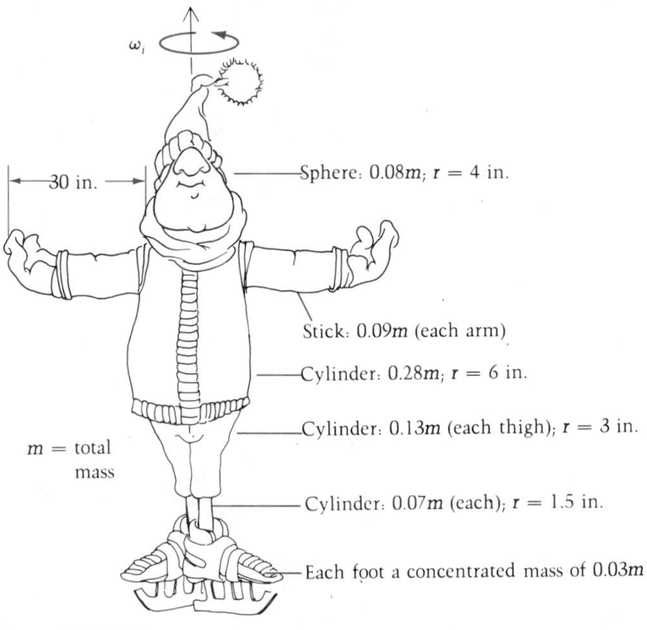

Figure P5.102(a)

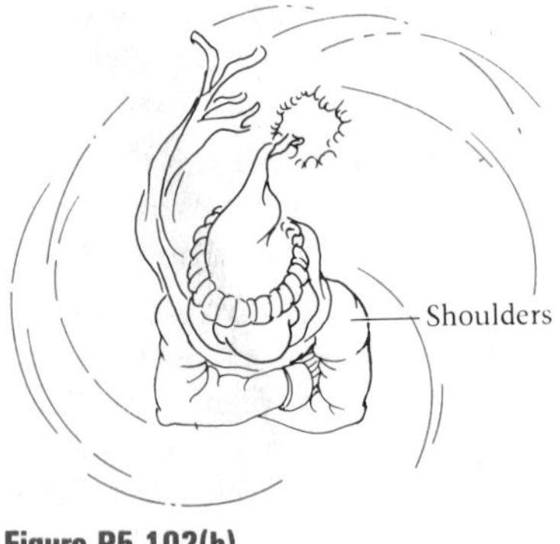

Figure P5.102(b)

* **5.103** A starving monkey of mass m spies a bunch of delicious bananas of the same mass. (See Figure P5.103.) He climbs at a varying speed relative to the (light) rope. Determine whether the monkey reaches the bananas before they sail over the pulley of radius R if:

  a. The pulley's mass is negligible ($\ll m$).

  b. The pulley's mass is fm, where $f > 0$ and the radius of gyration of the pulley with respect to its axis is k.

Figure P5.103

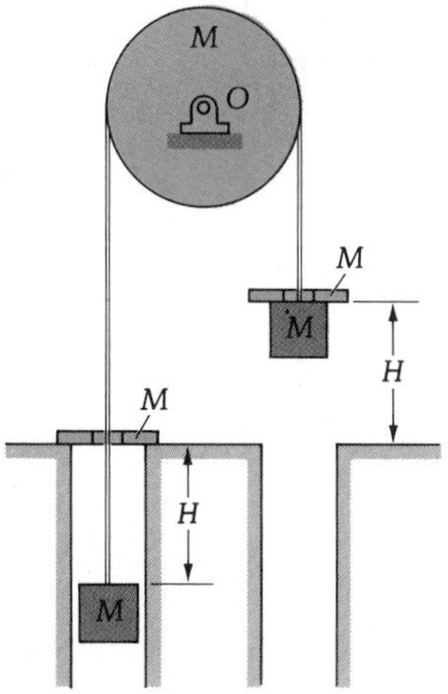

Figure P5.104

If either answer is yes, give the relationship between d and H in the figure for which overtaking the bananas is possible.

* **5.104** A circular disk of mass M rotates without friction about O. (See Figure P5.104.) A string passed over the disk (and not slipping on it) carries a mass M at each end. The system is released at rest as shown with the right-hand mass carrying a washer of mass M. As the system moves, the left-hand weight picks up a washer of mass M at the same instant the right mass deposits its washer. Find the velocity of the right-hand weight just after this exchange of washers.

* **5.105** A bird of mass m, flying horizontally at speed v_0 perpendicular to a stick, lands on the stick and holds fast to it. (See Figure P5.105.) The stick (mass M, length ℓ) is lying on a frozen pond. Find the angular velocity of the bird and stick as they move together. (Answer in terms of m, M, ℓ, and v_0.) Assume that the bird lands on the end of the stick.

Figure P5.105

* **5.106** A hemispherical block of mass M and radius a whose surfaces are smooth rests with its plane face in contact with a *smooth* horizontal table. A particle of mass m is placed at the highest point of the block and is slightly disturbed. Show that as long as the particle remains in contact with the block, the radius to the particle makes an angle θ with the upward vertical where

$$a\dot{\theta}^2(M + m\sin^2\theta) = 2g(M + m)(1 - \cos\theta)$$

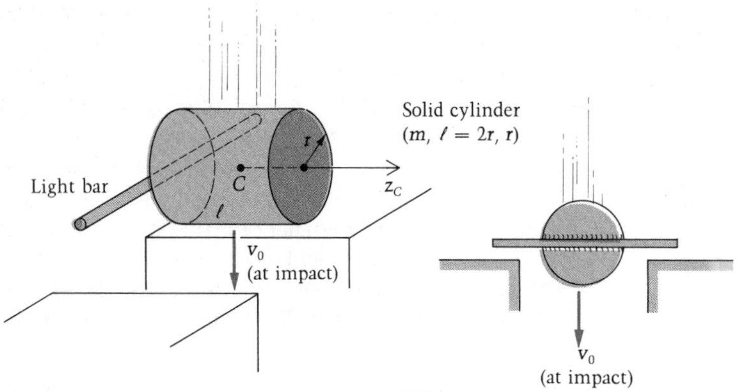

Figure P5.107

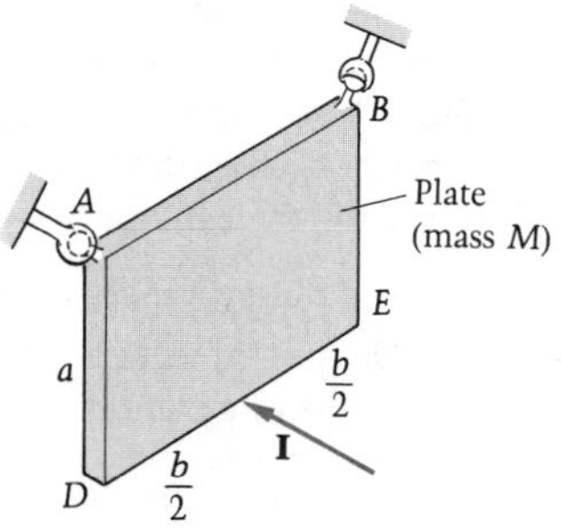

Figure P5.109

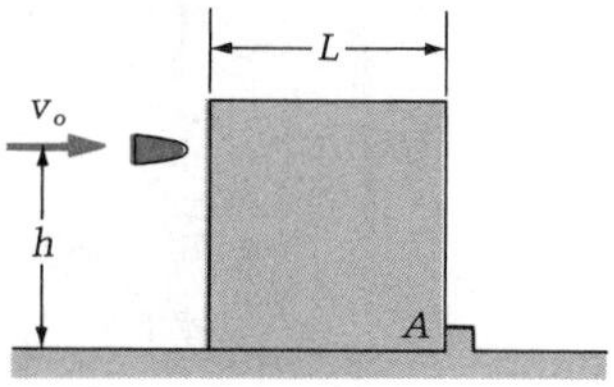

Figure P5.111

5.107 The bar in Figure P5.107 is welded to the end of the cylinder, which is traveling downward in translation. The bar strikes the tables at speed v_0, and the cylinder begins to rotate about the bar without rebound.

a. Find the angular velocity of the cylinder when C is at its lowest point.

b. Find the percentage of energy lost during the impact; that is

$$\left(1 - \frac{\text{new energy}}{\text{old energy}}\right) \times 100$$

5.108 A CARE package (Figure P5.108) consists of the box plus contents described in Example 4.12. At impact the crate has $\mathbf{v}_{G_i} = 50 \downarrow$ ft/sec and is translating. If there is no rebound, find the angular velocity of the box and the velocity of its mass center G just after the impact.

5.109 The plate in Figure P5.109, supported by ball joints at its top corners, is suddenly struck as shown with a force that produces the impulse I normal to the plate. Find the kinetic energy produced by the impact.

5.110 Work the preceding problem but assume that the plate is initially free. If you work both problems, show that the difference in the energies is $I^2/(2M)$.

5.111 A bullet (see Figure P5.111) of mass m_1 strikes a square homogeneous block of mass m_2, where $m_2 \gg m_1$. The bullet is traveling with initial velocity $\mathbf{v}_O$ and becomes embedded in the block. After impact, the block is observed to be pivoting about corner A. What is the maximum speed $|\mathbf{v}_O|$ of the bullet such that the block will not tip all the way over?

5.112 The cylinder $\mathcal{B}$ (radius 10 cm, length 40 cm) swings down from a position of rest where $\theta = 0$, and strikes the particle P of mass 5 kg. (See Figure P5.112.) The coeffi-

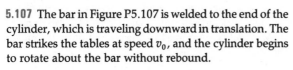

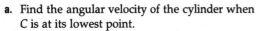

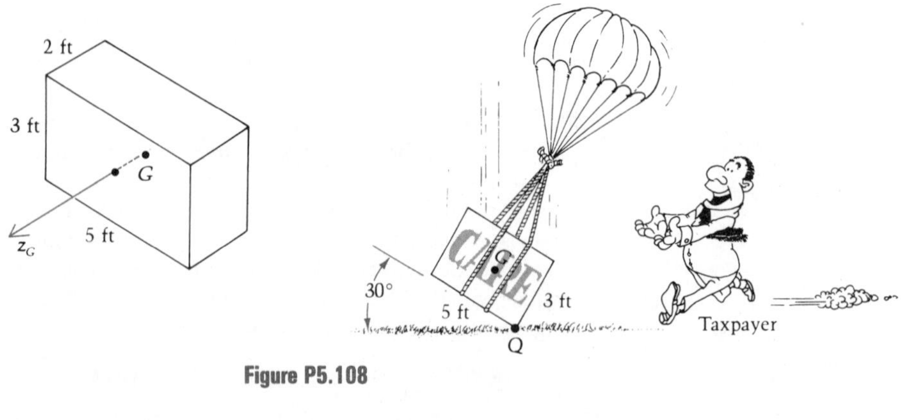

Figure P5.108

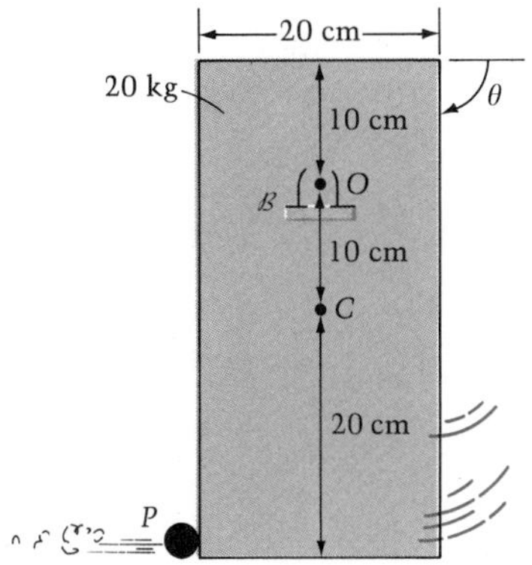

Figure P5.112

cient of restitution is $e = 0.5$, and at impact the particle has $\mathbf{v}_P = 2 \rightarrow$ m/s. Find the angle through which $\mathcal{B}$ will turn about its pivot O after impact.

5.113 An equilateral triangular plate of mass 2 slugs and side 2 ft is released in the upper position from rest (see Figure P5.113). It swings down and strikes the stationary cylinder. The coefficient of restitution for the impact is $e = 1/2$. Find elapsed time after impact until the cylinder no longer slips on the plane.

5.114 A block slides to the right and strikes a small obstruction at a speed of 20 ft/sec. (See Figure P5.114.)

- **a.** If the coefficient of restitution is zero, find the energy loss caused by the impact.
- **b.** What is the minimum striking velocity required to overturn the block after collision?

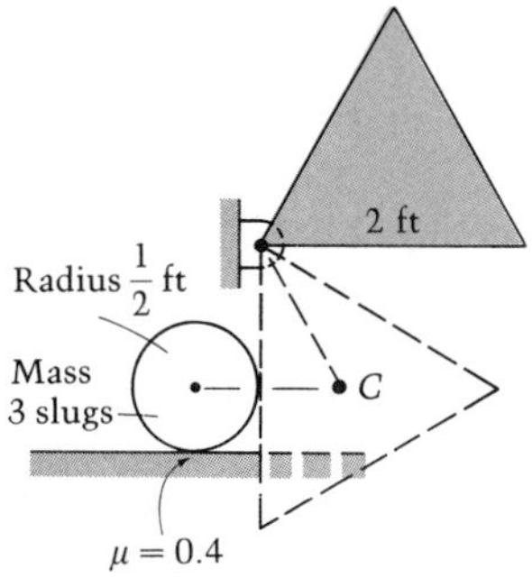

Figure P5.113

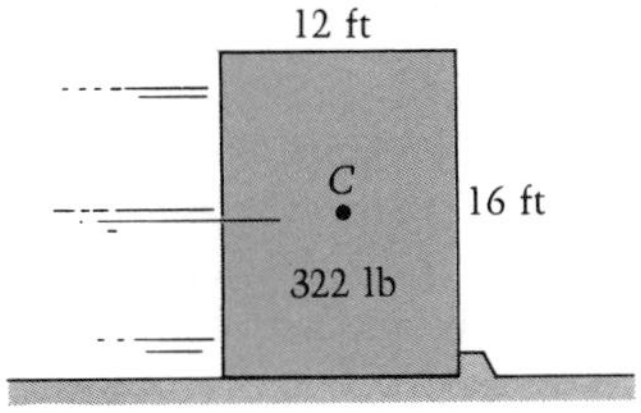

Figure P5.114

5.115 A homogeneous cube of side a and mass M slides on a level, frictionless table with velocity v_0. See Figure P5.115.) It strikes a small lip on the table at A of negligible height. Find the velocity of the center of mass just after impact if the coefficient of restitution is unity. (The centroidal moment of inertia of a cube about an axis parallel to an edge is $Ma^2/6$.)

5.116 There is only one height H above a pool table at which a cue ball may be struck by the stick without the ball slipping for a while after the impact. (See Figure P5.116.) Find this value of H, in terms of R, for which the ball immediately rolls.

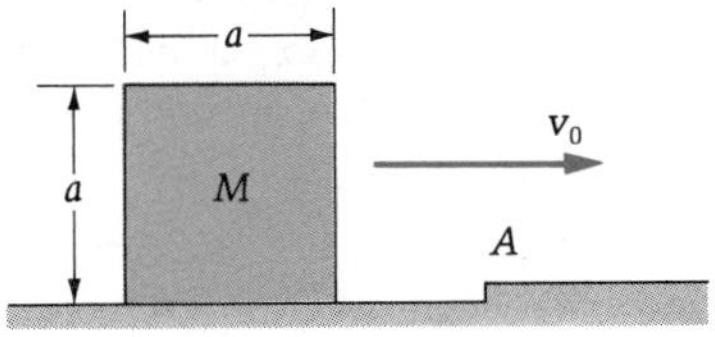

Figure P5.115

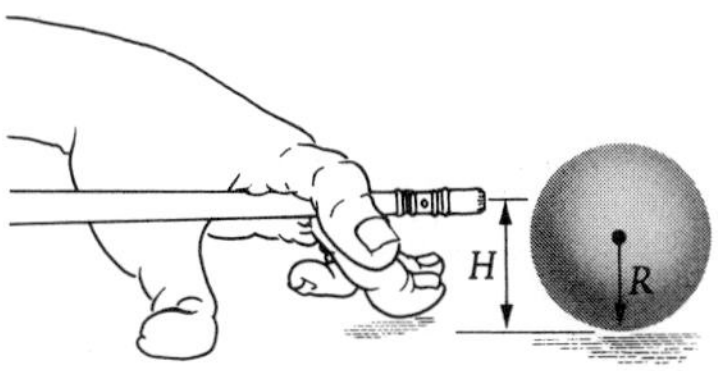

Figure P5.116

5.117 Compute the error in I_{zz}^C in Example 5.18 that was incurred in assuming the vessel to be a shell (so that I_{zz}^C was $m\ell^2/12 + mr^2/2$). Use the weight, height, outer radius, and density to compute the thickness of the vessel; then calculate a more accurate I_{zz}^C and compare.

5.118 A uniform rod of length L is dropped and translates downward at an angle θ with the vertical as shown in Figure P5.118. If end A does not rebound after striking the ground at speed v_0, find: (a) the energy lost during the impact of A with the ground; (b) the speed at which the other end B then hits the ground.

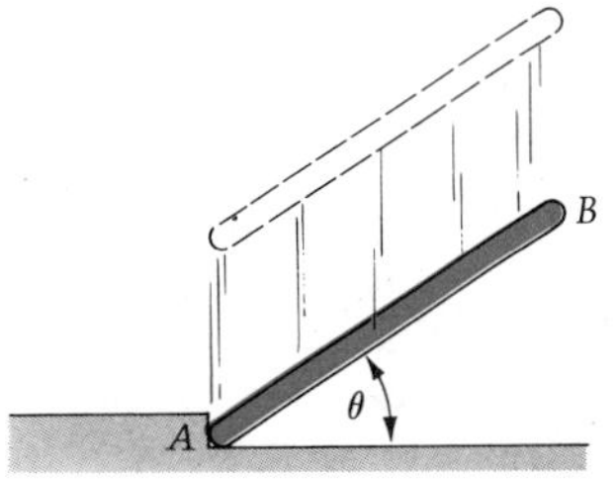

Figure P5.118

5.119 A uniform bar AB of length L and mass M is moving on a smooth horizontal plane with $\mathbf{v}_C = v_0\hat{\mathbf{i}}$ and $\boldsymbol{\omega} = \omega_0\hat{\mathbf{k}}$, when end B strikes a peg P (see Figure P5.119). If $\omega_0 = 2v_0/L$ and the coefficient of restitution $e = \frac{1}{6}$, find the loss of kinetic energy.

5.120 The 80-lb solid block hits a smooth, rigid wall (see Figure P5.120) and rebounds with a coefficient of restitution of $e = 0.2$. Prior to impact the block had: $\omega_i = 1.2 \circlearrowleft$ rad/sec and $\mathbf{v}_C = 0.8\hat{\mathbf{i}} + 0.6\hat{\mathbf{j}}$ ft/sec. Find the angular velocity of the block immediately following the impact.

5.121 The rod in Figure P5.121 is freely falling in a vertical plane. At a certain instant it is horizontal with its ends A and B having the velocities shown. If end A is suddenly fixed, prove that the rod will start to rise around end A provided that $v_1 < 2v_2$.

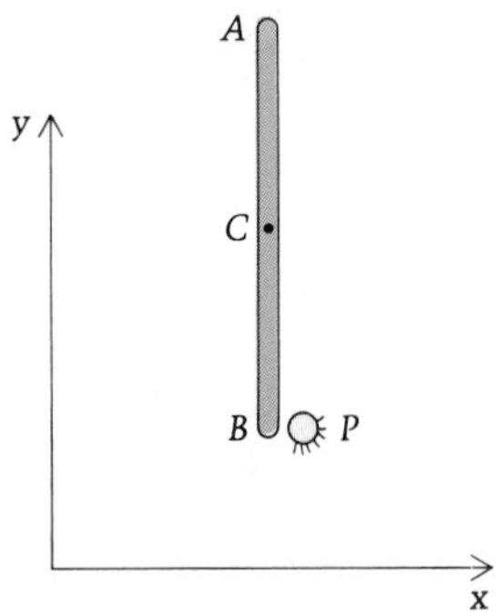

Figure P5.119

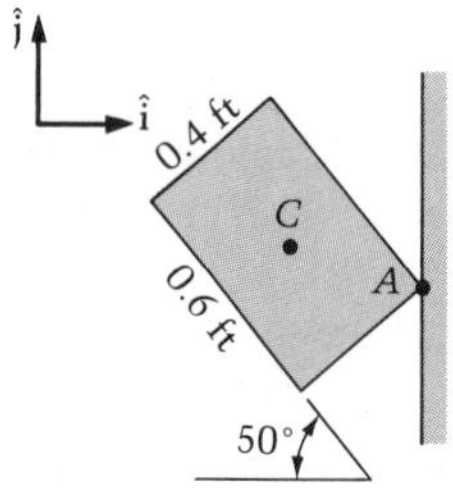

Figure P5.120

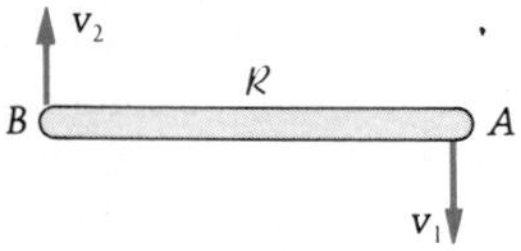

Figure P5.121

5.122 In the preceding problem show that the energy loss in instantaneously stopping point A is independent of v_2.

5.123 The bar in Figure P5.123 swings downward from the dotted horizontal position and strikes mass m. The bar has mass M and length ℓ. The collision takes place with a coefficient of restitution of zero. If the coefficient of friction between m and the plane is μ, find the distance moved by m before stopping. Treat m as a particle.

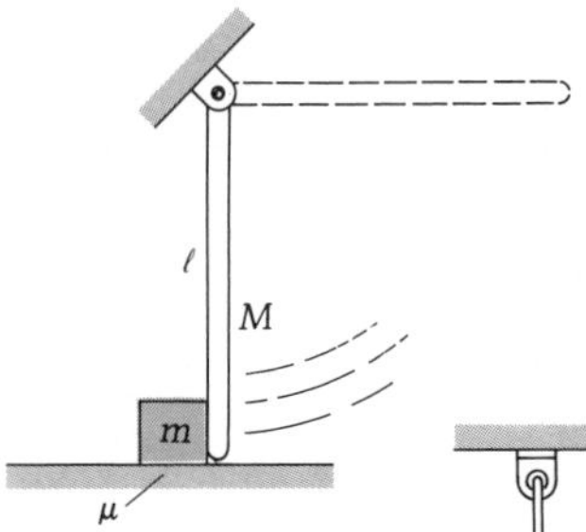

Figure P5.123

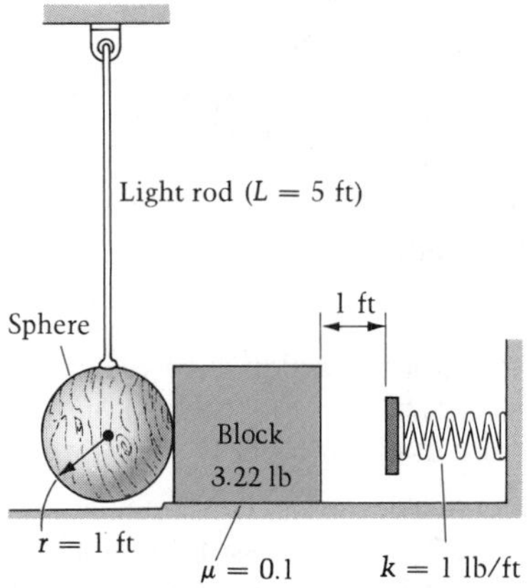

Figure P5.124

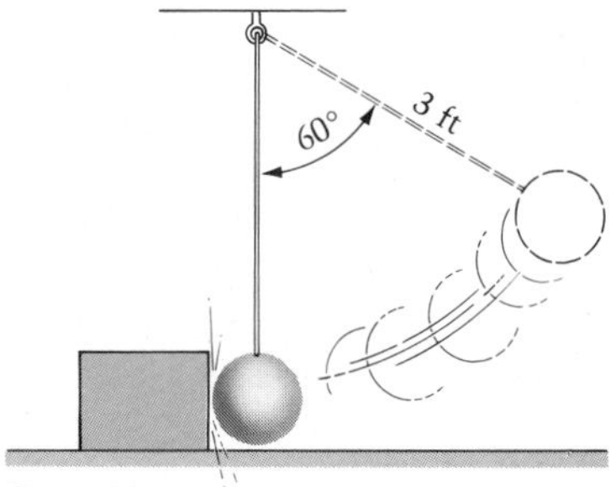

Figure P5.125

5.124 A wooden sphere weighing 0.644 lb swings down from a position where the rod is horizontal, and it impacts the block. (See Figure P5.124.) The coefficient of restitution is $e = 0.6$. The block weighs 3.22 lb and is initially at rest. Find the position of the block when it comes *permanently* to rest. (Assume that the sphere is removed from the problem after impact and that the spring cannot rebound past its original unstretched position.)

5.125 A 4-lb sphere is released from rest in the position shown in Figure P5.125, and two observations are then made: (1) The sphere comes immediately to rest after the impact; (2) the 5-lb block slides 3 ft before coming to rest. Using these observations, find the coefficients of restitution (between sphere and block) and friction (between block and floor).

5.126 The rod-sphere rigid body in Figure P5.126 is released from rest in the horizontal position. It swings down and at its lowest point strikes the box. Find how far the box slides before coming to rest if the coefficient of restitution is $e = 0.5$. The data are:

1. Rod: length = 1 m; mass = 3 kg
2. Sphere: radius = 0.2 m; mass = 10 kg
3. Block: $b = 0.3$ m; $H = 0.35$ m; mass = 5 kg
4. Coefficient of friction between block and plane = 0.3

Assume the sphere hits the block just once.

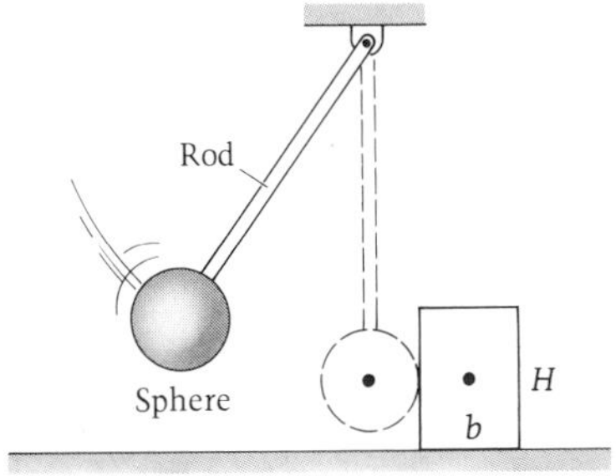

Figure P5.126

5.127 The assembly in Figure P5.127 is turning at $\omega_i = 2$ rad/sec when the collar is released. All surfaces are smooth, and the disk is fixed to the bar; the light vertical shaft ends in bearings and turns freely. The data are:

$$m_3 = \tfrac{1}{4} \text{ slug} \qquad \ell_2 = 2 \text{ ft}$$
$$m_1 = 1 \text{ slug} \qquad \ell_1 = 4 \text{ ft}$$
$$m_2 = \tfrac{1}{4} \text{ slug}$$

The collar moves outward and impacts the disk without rebound. Find: (a) the angular speed of the bar just before and just after impact; (b) the percentage of energy lost during impact. The radii of $\mathcal{B}_1$ and $\mathcal{B}_2$ are small compared to their lengths. Treat $\mathcal{B}_3$ as a particle.

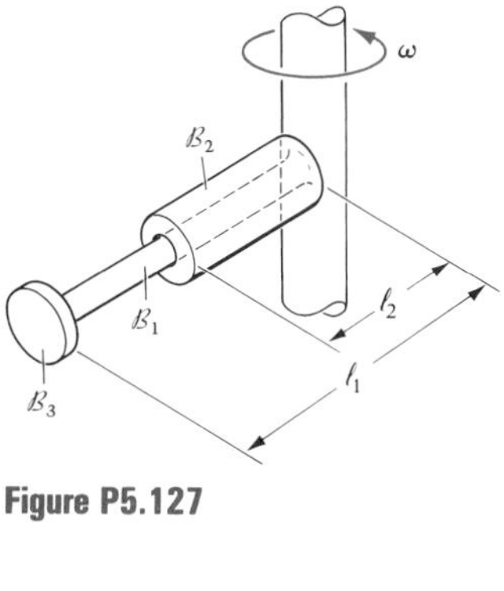

Figure P5.127

5.128 Two toothed gear wheels, which may be treated as uniform disks of radii a and b and masses M and m, respectively, are rotating in the same plane. They are not quite in contact and have angular velocities ω_1 and ω_2 about fixed axes through their centers. Their axes are then slightly moved so that the wheels engage. Prove that the loss of energy is

$$\frac{Mm}{4(M+m)}(a\omega_1 + b\omega_2)^2$$

5.129 Sphere $\mathcal{B}_1$ has mass m and radius r, and it rolls with mass center velocity $v_0 \rightarrow$ on a horizontal plane. (See Figure P5.129.) It hits squarely an identical sphere $\mathcal{B}_2$ that is at rest. The coefficient of friction between a sphere and the plane is μ, and between spheres it is negligible. The impact is nearly elastic ($e \approx 1$).

a. Find v_{C_f} and ω_f of each sphere right after impact.
b. Find v_C of each sphere after it has started rolling uniformly.
c. Discuss the special case when $\mu = 0$.

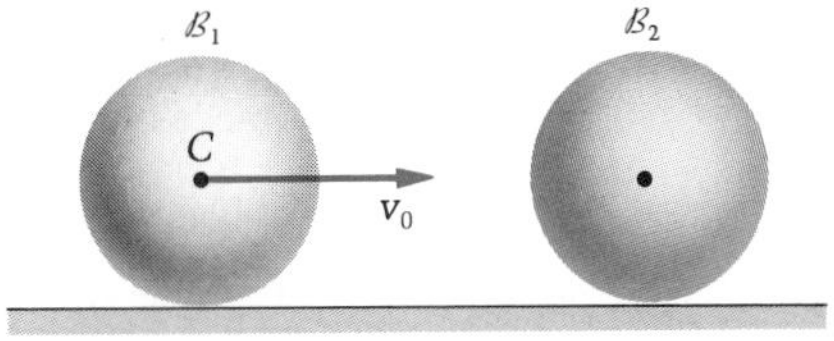

Figure P5.129

* **5.130** The 30-kg bent bar in Figure P5.130 falls from the dashed position onto the spinning cylinder, which was initially turning at 3000 rad/s ↺. If the bar does not bounce (coefficient of restitution is zero), find the stopping time for the cylinder following the impact.

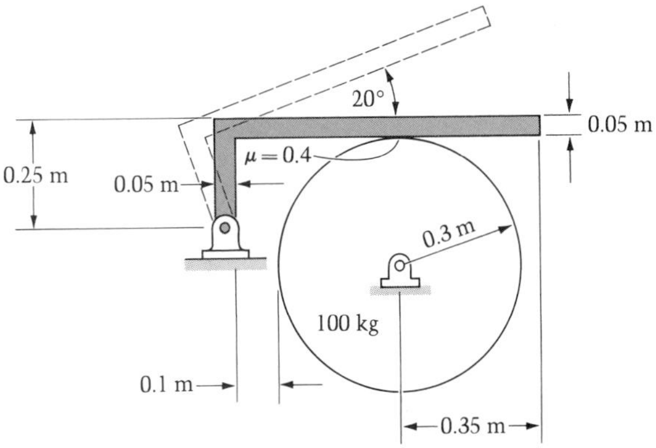

Figure P5.130

* **5.131** In Example 5.18 assume that the lower front striking corner of the vessel $\mathcal{B}$ rebounds back up the plane with velocity ev_{C_i}, where e is the coefficient of restitution ($0 < e \le 1$). Use the equations of impulse and momentum in the x and y directions, and the equation of angular impulse and angular momentum, to find the two mass-center velocity components and the angular velocity of $\mathcal{B}$ after impact. Compare the results of $\dot{x}_{C_f}$, $\dot{y}_{C_f}$, and $\dot{\theta}_f$ for $e = 1$ with those at $e = 0$. Show that no energy is lost when $e = 1$; that is, show

$$\frac{1}{2}mv_{C_i}^2 = \frac{1}{2}m(\dot{x}_{C_f}^2 + \dot{y}_{C_f}^2) + \frac{1}{2}I_C\dot{\theta}_f^2$$

* **5.132** *After* the impact in the preceding problem, show that the equations of motion of the vessel are

$$m\ddot{x}_C = -mg\sin\psi \quad (1)$$

$$m\ddot{y}_C = N - mg\cos\psi \quad (2)$$

$$I_C\ddot{\theta} = \frac{\ell}{2}N\sin\theta - rN\cos\theta \quad (3)$$

where N is the normal reaction at the corner Q (see Figure P5.132). Observe that until there is another impact, the mass center has a constant x component of acceleration. Use kinematics to prove that

$$\ddot{y}_C = (r\cos\theta - \frac{\ell}{2}\sin\theta)\ddot{\theta} - \dot{\theta}^2(r\sin\theta + \frac{\ell}{2}\cos\theta) \quad (4)$$

Use Equations (2) and (4) to eliminate N from (3), thus obtaining a single differential equation in θ governing the rotational motion of $\mathcal{B}$, and note its complexity.

5.133 Prove statement 1 near the end of Example 5.18 for the case when the plane is level ($\psi = 0$) and $e = 0$. *Hint*: Note carefully that the angle of the plane does not affect Equation (6), so you only need to alter the Δh in Equation (7) to obtain the new result.

5.134 Prove statement 2 near the end of Example 5.18. Again the plane is to be level in this problem, but now $e = 1$. *Hint*: The x_C-component of velocity is constant after impact, since with $\psi = 0$ all external forces (mg and N) are vertical. To find this velocity $\dot{x}_{C_f}$, use ω_f and the velocity of the striking corner Q just after impact (ω_f is the same as with $\psi = 3.719°$ in Problem 5.131 with $e = 1$; v_Q is v_{C_i} back to the left). Then use

$$W = \Delta T = \frac{1}{2}m\dot{x}_{C_f}^2 + \frac{1}{2}m\dot{y}_{C_f}^2 + \frac{1}{2}I_C\omega_f^2 - \frac{1}{2}mv_{C_i}^2$$

to show that C reaches the top with energy to spare.

* **5.135** Show that the cylinder in Figure P5.135, following release from rest, will reach the lower wall. Find the velocity with which C will rebound up the plane following impact, and determine the amount of energy lost. All data are shown on the figure.

* **5.136** A sphere rolling with speed v_C on a horizontal surface strikes an obstacle of height H. What is the largest value that H can have if the sphere is able to make it over the obstacle? Consider the coefficient of restitution to be zero during the sphere's impact with the corner point O. The answer will be a function of g, r, and v_C—in fact, H/r may be solved for as a function of the single nondimensional parameter gr/v_C^2. (See Figure P5.136.)

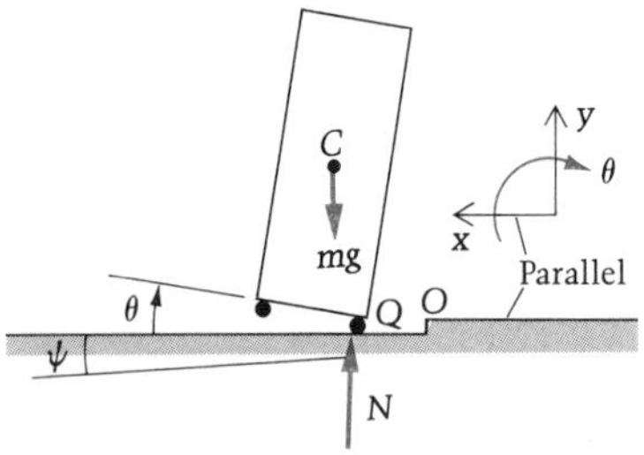

Figure P5.132

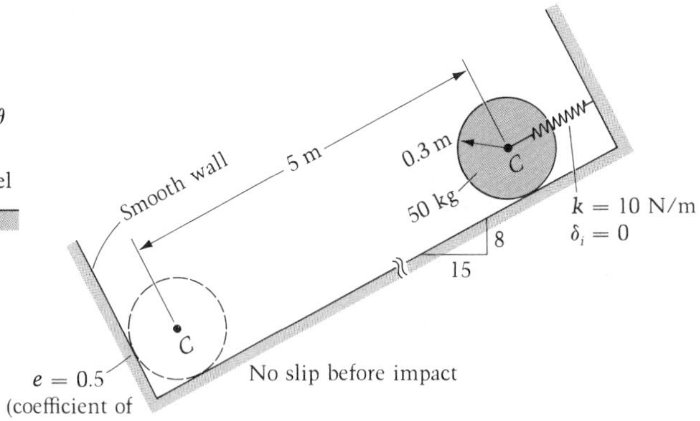

Figure P5.135

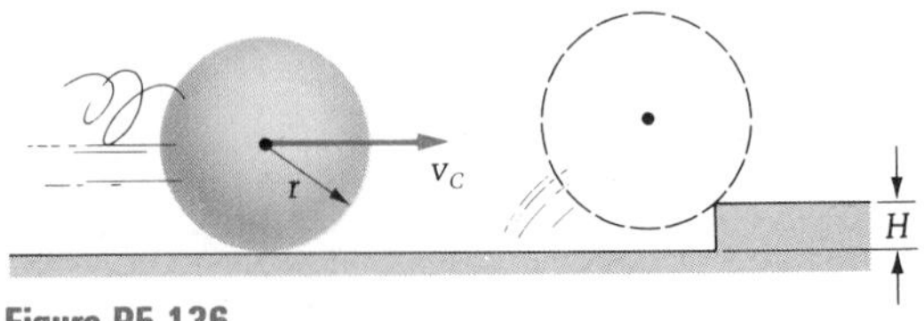

Figure P5.136

***** **5.137** Using a density of wood of 673 kg/m^3, find the mass center C of the baseball bat in Figure P5.137 and then determine its moment of inertia with respect to the z_0 axis, perpendicular to the axis of symmetry of the bat. Use the parallel-axis theorem to obtain I_{zz}^C, and find the bat's center of percussion if it is swinging about a fixed point O.

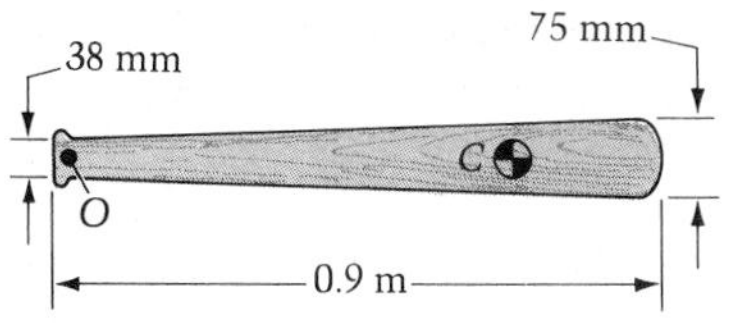

Figure P5.137

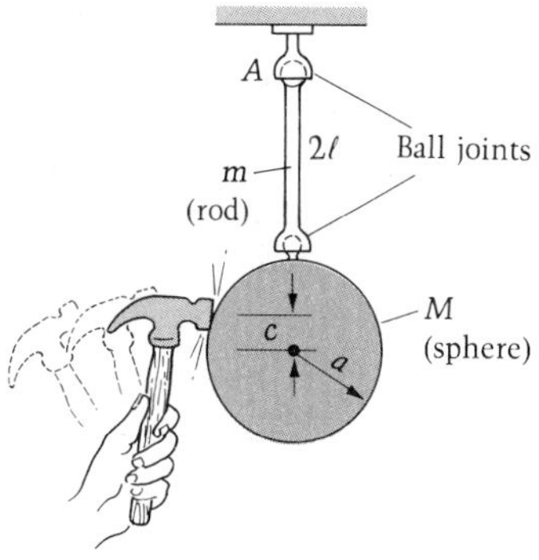

Figure P5.138

***** **5.138** The hammer in Figure P5.138 strikes the sphere and imparts a horizontal impulse I to it. Determine the initial angular velocity of the sphere.

5.139 Repeat the preceding problem but suppose that the sphere and rod are welded together to form one rigid body.

COMPUTER PROBLEM ▶ Chapter 5

***** **5.140** The system in Figure P5.140 is released from rest in the given position. With the help of a computer, generate data for a plot of the angle θ turned through by wheel $\mathcal{B}_1$ before first stopping, as a function of the mass ratio M/m. *Hint*: First show using $W = \Delta T$ that the equation governing θ is

$$\sin\theta = \frac{2m}{M}\theta$$

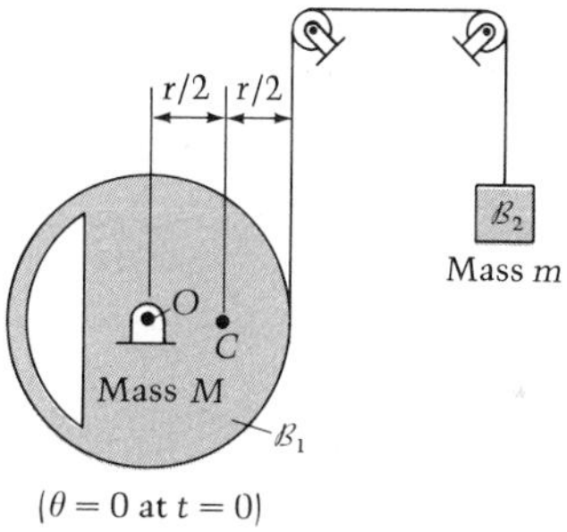

$(\theta = 0 \text{ at } t = 0)$

Figure P5.140

SUMMARY ▶ Chapter 5

For a rigid body in plane motion, the kinetic energy, T, can be expressed as

$$T = \tfrac{1}{2}mv_C^2 + \tfrac{1}{2}I_C\omega^2$$

or as

$$T = \tfrac{1}{2}I_{\text{①}}\omega^2$$

With W standing for the net work over a time interval (t_1, t_2) of all the *external* forces on the body, the principle of work and kinetic energy states

$$W = T_2 - T_1$$

or more compactly

$$W = \Delta T$$

The work done by a force, **F**, is defined as

$$\int_{t_1}^{t_2} \mathbf{F} \cdot \mathbf{v}\, dt$$

where $\mathbf{F} \cdot \mathbf{v}$ is called the power, or rate of work of force **F**, with **v** being the velocity of the material point being acted upon instantaneously by the force. This definition is necessary to accommodate the possibility that the force moves around on the body. When the force always acts on the same material point for which a position vector is **r**,

$$\int_{t_1}^{t_2} \mathbf{F} \cdot \mathbf{v}\, dt = \int_{t_1}^{t_2} \mathbf{F} \cdot \frac{d\mathbf{r}}{dt}\, dt$$
$$= \int_{\mathbf{r}_1}^{\mathbf{r}_2} \mathbf{F} \cdot d\mathbf{r}$$

The work done by a couple of moment **C** is

$$\int_{t_1}^{t_2} \mathbf{C} \cdot \omega\hat{\mathbf{k}}\, dt$$

which, with $\mathbf{C} = C\hat{\mathbf{k}}$, is

$$\int_{t_1}^{t_2} C\omega dt = \int_{t_1}^{t_2} C d\theta$$

For some special cases:

(a) **F** is constant:

$$W = \mathbf{F} \cdot (\mathbf{r}_2 - \mathbf{r}_1)$$

and for weight mg, with y being elevation,

$$W = -mg(y_2 - y_1)$$

(b) Spring forces (on two bodies):

$$W = -(k/2)(\delta_2^2 - \delta_1^2)$$

(c) Workless force:

$$W = \int_{t_1}^{t_2} \mathbf{F} \cdot \mathbf{v}\, dt = 0$$

if $\mathbf{F} \perp \mathbf{v}$ as with a normal force acting on a sliding body, or if $\mathbf{v} = \mathbf{0}$ at each instant as for the contact point of a rolling body.

(d) Constant couple:

$$W = C(\theta_2 - \theta_1)$$

A conservative force does work independent of path and can have associated with it a potential φ which we define so that its change is the negative of the work done by the force. Examples are weight, for which $\varphi = mgy$, and a linear spring, $\varphi = (k/2)\delta^2$. The minus sign is used for the convenience that follows when all forces acting on a body are conservative, so that

$$W = T_2 - T_1$$
$$-(\varphi_2 - \varphi_1) = T_2 - T_1$$

or

$$T_2 + \varphi_2 = T_1 + \varphi_1$$

which expresses the conservation of mechanical energy.

For a rigid body in plane motion, we found in Chapter 4 that

$$\mathbf{H}_C = I_{xz}^C\,\omega\hat{\mathbf{i}} + I_{yz}^C\,\omega\hat{\mathbf{j}} + I_{zz}^C\,\omega\hat{\mathbf{k}},$$

and similarly for a pivot. Concerning ourselves only with the case when the products of inertia vanish,

$$\mathbf{H}_C = I_{zz}^C\,\omega\hat{\mathbf{k}}$$

As long as a body is behaving rigidly before and after an interval of time of interest, the principle of angular impulse and angular momentum from Chapter 2 gives

$$\int_{t_1}^{t_2} \Sigma M_C\, dt = (I_C\omega)_2 - (I_C\omega)_1$$

even though it may be that $I_C(t_2) \neq I_C(t_1)$ as in the example of the spinning ice skater. The above is, of course, paired with the principle of linear impulse and momentum,

$$\int_{t_1}^{t_2} \Sigma\mathbf{F}\, dt = m\mathbf{v}_C\,(t_2) - m\mathbf{v}_C\,(t_1)$$

to effect solution of collision problems.

REVIEW QUESTIONS ▶ Chapter 5

True or False?

These questions all refer to rigid bodies in plane motion.

1. If you raise a 2-lb object 3 ft from rest and stop it there, gravity has done −6 ft-lb of work and you have done +6 ft-lb on the object.
2. The work of a constant couple $C\hat{\mathbf{k}}$ on body $\mathcal{B}$ is always $C\hat{\mathbf{k}} \cdot \theta\hat{\mathbf{k}} = C\theta$, where θ is the angle through which the body turns.
3. The work done by a linear spring is always $k(\delta_i^2 - \delta_f^2)/2$, where δ_i and δ_f are the amounts of initial and final stretch. (If negative, they represent compression.)
4. There are actually three separate work and kinetic energy principles; two of the equations add to give the third.
5. The principles of work and kinetic energy, and (linear and angular) impulse and momentum, result from general integrations of the equations of motion, and thus they are free of accelerations.

6. Not all forces acting on a body have to do non-zero work on it in general.
7. The friction force beneath a rolling wheel does work on it if the surface of contact is curved and fixed.
8. The normal force exerted on a rolling wheel by a surface, whether fixed or in motion, never does work on the wheel.
9. The principle $\int \Sigma \mathbf{F}\, dt = \int \dot{\mathbf{L}}\, dt = m\mathbf{v}_{C_f} - m\mathbf{v}_{C_i}$ is valid for deformable bodies.
10. The principle $\int \Sigma M_C\, dt = I_{zz}^C \omega_f - I_{zz}^C \omega_i$ is valid for deformable bodies.
11. Any problem that can be solved by $W = \Delta T$ can likewise be solved by using "kinetic + potential energy = constant."
12. The formula $T = \frac{1}{2} I_{①} \omega^2$ gives *all* the kinetic energy of the rigid body in plane motion, assuming the body is not translating.

Answers: 1. T 2. T 3. T 4. T 5. T 6. T 7. F 8. F 9. T 10. F 11. F 12. T

6 KINEMATICS OF A RIGID BODY IN THREE-DIMENSIONAL MOTION

6.1 Introduction

In this chapter we study the kinematics of a rigid body in general motion — we now do for general motion what we did in Chapter 3 for plane motion. There we found that at any instant the velocities of different points are linked together because of the rigidity of the body, and the connecting link is the angular velocity of the body. We also found there that it is angular velocity that links the derivatives of the *same* vector relative to two different frames of reference. We now wish to remove the plane-motion restriction.

The principal difficulty encountered in the study of general motion of a rigid body is that the angular velocity $\boldsymbol{\omega}$ does not always take the $\dot{\theta}\hat{\mathbf{k}}$ form of plane motion. The fact that $\boldsymbol{\omega}$ can be changing in direction as the body moves causes it to be difficult to visualize. An efficient way to deal with this abstract concept is to start with derivatives of the same vector in different frames of reference (rigid bodies). Angular velocities will naturally arise out of connecting these derivatives, and properties of relative angular velocities amongst several bodies, almost self-evident for the case of plane motion, will surface for the general case.

Application of the derivative/angular-velocity relationship to position vectors, in two frames, of a point leads us to the velocities of the point as observed in those frames, connected, in part, by the relative angular velocity of the frames. Subsequent mathematical analysis leads to: a relationship between accelerations of a point in two frames; a relationship between velocities, relative to a frame, of two points fixed in the same rigid body; and a relationship between accelerations, relative to a frame, of two such body-fixed points.

We close this chapter with development of methods to describe the orientation of a rigid body (relative, of course, to some reference frame). This was easy to do in plane motion — an angle θ is all that was required. In the general case we shall see the need for three angles, called Euler angles in one popular scheme.

The reflective reader will notice that the sequence of coverages in this chapter is almost precisely the reverse of its counterpart in Chapter 3, which is after all just a special case. There we were able to capitalize on an ease of visualization not available to us here.

6.2 Relation Between Derivatives/The Angular Velocity Vector

In this section we consider the relationship between the derivatives of a vector taken in two different frames. In the process we shall arrive at a concise and useful definition of angular velocity. The reader is strongly encouraged to persevere until this section and the next are fully understood. Even though the angular velocity vector in three dimensions is a difficult subject at first, it *must* be comprehended before we can consider the kinematics and kinetics of general rigid-body motion. The angular velocity vector is the key to the subject. It will either make life easier for students of three-dimensional motion (if they work hard at understanding it) or much more difficult (if they do not).

Let $\mathbf{Q}$ be an arbitrary vector. We may express $\mathbf{Q}$ in terms of its components (Q_x, Q_y, Q_z) associated with directions fixed in a frame $\mathcal{B}$ by

$$\begin{aligned}\mathbf{Q} &= (\mathbf{Q}\cdot\hat{\mathbf{i}})\hat{\mathbf{i}} + (\mathbf{Q}\cdot\hat{\mathbf{j}})\hat{\mathbf{j}} + (\mathbf{Q}\cdot\hat{\mathbf{k}})\hat{\mathbf{k}} \\ &= Q_x\hat{\mathbf{i}} + Q_y\hat{\mathbf{j}} + Q_z\hat{\mathbf{k}} \end{aligned} \tag{6.1}$$

in which the unit vectors ($\hat{\mathbf{i}}$, $\hat{\mathbf{j}}$, $\hat{\mathbf{k}}$) are parallel at all times to the respective axes of a Cartesian coordinate system fixed in $\mathcal{B}$. Now consider another reference frame $\mathcal{A}$, in which we wish to differentiate vector $\mathbf{Q}$ (see Figure 6.1). As an example, we may wish to find the velocity of a point in frame $\mathcal{A}$ even though the point's location may be defined in $\mathcal{B}$ (say by the vector $\mathbf{Q}$). In this case, part of the solution will require that we be able to differentiate $\mathbf{Q}$ in $\mathcal{A}$ even though it is expressed in terms of its components in $\mathcal{B}$.

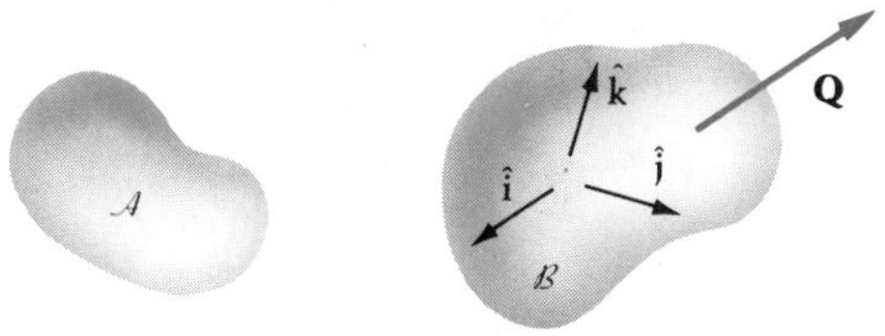

Figure 6.1 Vector Q and frames $\mathcal{A}$ and $\mathcal{B}$.

Therefore it is now time to learn how to relate derivatives of a vector taken in two different frames. We emphasize at the outset that these vectors are completely arbitrary — they need not even be related to dynamics! Nor does the derivative have to be with respect to time, although this is the independent variable of interest to us in dynamics and thus the one we shall use in the development to follow.

Letting ${}^{\mathcal{A}}\dot{\mathbf{Q}}$ represent (see Equation 1.8) the derivative of $\mathbf{Q}$ with respect to time in $\mathcal{A}$, we have

$${}^{\mathcal{A}}\dot{\mathbf{Q}} = \dot{Q}_x\hat{\mathbf{i}} + \dot{Q}_y\hat{\mathbf{j}} + \dot{Q}_z\hat{\mathbf{k}} + Q_x\,{}^{\mathcal{A}}\dot{\hat{\mathbf{i}}} + Q_y\,{}^{\mathcal{A}}\dot{\hat{\mathbf{j}}} + Q_z\,{}^{\mathcal{A}}\dot{\hat{\mathbf{k}}} \tag{6.2}$$

Recognizing the first three terms on the right of Equation (6.2) as the derivative of $\mathbf{Q}$ in $\mathcal{B}$, we have

$${}^{\mathcal{A}}\dot{\mathbf{Q}} = {}^{\mathcal{B}}\dot{\mathbf{Q}} + (Q_x\,{}^{\mathcal{A}}\dot{\hat{\mathbf{i}}} + Q_y\,{}^{\mathcal{A}}\dot{\hat{\mathbf{j}}} + Q_z\,{}^{\mathcal{A}}\dot{\hat{\mathbf{k}}}) \tag{6.3}$$

Clearly the last three (parenthesized) terms in Equation (6.3) represent a vector depending upon both $\mathbf{Q}$ *and* the change of orientation of frame $\mathcal{B}$ with respect to $\mathcal{A}$. We now proceed to obtain a useful and compact expression for this vector; in the process, the angular velocity vector will arise.

Since

$$\hat{\mathbf{i}}\cdot\hat{\mathbf{i}} = \hat{\mathbf{j}}\cdot\hat{\mathbf{j}} = \hat{\mathbf{k}}\cdot\hat{\mathbf{k}} = 1 \tag{6.4}$$

it follows that

$$\hat{\mathbf{i}}\cdot{}^{\mathcal{A}}\dot{\hat{\mathbf{i}}} = 0 = \hat{\mathbf{j}}\cdot{}^{\mathcal{A}}\dot{\hat{\mathbf{j}}} = \hat{\mathbf{k}}\cdot{}^{\mathcal{A}}\dot{\hat{\mathbf{k}}} \tag{6.5}$$

so that the three derivatives of the unit vectors in Equation (6.3) are each perpendicular to the respective unit vectors themselves.*

> **Question 6.1** Will this be true for *any* vector of constant magnitude (not necessarily a unit vector)?

This means that there are three vectors $\boldsymbol{\alpha}$, $\boldsymbol{\beta}$, and $\boldsymbol{\gamma}$ for which

$$\begin{aligned} {}^{\mathcal{A}}\dot{\hat{\mathbf{i}}} &= \boldsymbol{\alpha} \times \hat{\mathbf{i}} \\ {}^{\mathcal{A}}\dot{\hat{\mathbf{j}}} &= \boldsymbol{\beta} \times \hat{\mathbf{j}} \\ {}^{\mathcal{A}}\dot{\hat{\mathbf{k}}} &= \boldsymbol{\gamma} \times \hat{\mathbf{k}} \end{aligned} \tag{6.6}$$

The cross products ensure that $\hat{\mathbf{i}}$, $\hat{\mathbf{j}}$, and $\hat{\mathbf{k}}$ are each perpendicular to their derivatives ($\hat{\mathbf{i}} \perp {}^{\mathcal{A}}\dot{\hat{\mathbf{i}}}$ and so on) and the magnitudes of $\boldsymbol{\alpha}$, $\boldsymbol{\beta}$, and $\boldsymbol{\gamma}$ give to ${}^{\mathcal{A}}\dot{\hat{\mathbf{i}}}$, ${}^{\mathcal{A}}\dot{\hat{\mathbf{j}}}$, and ${}^{\mathcal{A}}\dot{\hat{\mathbf{k}}}$ their correct magnitudes.

In terms of their components in $\mathcal{B}$, we can write $\boldsymbol{\alpha}$, $\boldsymbol{\beta}$, and $\boldsymbol{\gamma}$ as

$$\begin{aligned} \boldsymbol{\alpha} &= \alpha_x\hat{\mathbf{i}} + \alpha_y\hat{\mathbf{j}} + \alpha_z\hat{\mathbf{k}} \\ \boldsymbol{\beta} &= \beta_x\hat{\mathbf{i}} + \beta_y\hat{\mathbf{j}} + \beta_z\hat{\mathbf{k}} \\ \boldsymbol{\gamma} &= \gamma_x\hat{\mathbf{i}} + \gamma_y\hat{\mathbf{j}} + \gamma_z\hat{\mathbf{k}} \end{aligned} \tag{6.7}$$

Substituting these component expressions into Equations (6.6) results in

$$\begin{aligned} {}^{\mathcal{A}}\dot{\hat{\mathbf{i}}} &= \alpha_z\hat{\mathbf{j}} - \alpha_y\hat{\mathbf{k}} \\ {}^{\mathcal{A}}\dot{\hat{\mathbf{j}}} &= \beta_x\hat{\mathbf{k}} - \beta_z\hat{\mathbf{i}} \\ {}^{\mathcal{A}}\dot{\hat{\mathbf{k}}} &= \gamma_y\hat{\mathbf{i}} - \gamma_x\hat{\mathbf{j}} \end{aligned} \tag{6.8}$$

and we see that α_x, β_y, and γ_z, at this point, remain arbitrary.

> **Question 6.2** Why do they remain arbitrary?

Here we are seeking to relate the components of the vectors $\boldsymbol{\alpha}$, $\boldsymbol{\beta}$, and $\boldsymbol{\gamma}$ in the hope of finding a way to express the last three terms of Equation (6.3). To this end we note that, for all time t,

$$\hat{\mathbf{i}} \cdot \hat{\mathbf{j}} = \hat{\mathbf{j}} \cdot \hat{\mathbf{k}} = \hat{\mathbf{k}} \cdot \hat{\mathbf{i}} = 0$$

from the first of which, differentiation yields

$${}^{\mathcal{A}}\dot{\hat{\mathbf{i}}} \cdot \hat{\mathbf{j}} + {}^{\mathcal{A}}\dot{\hat{\mathbf{j}}} \cdot \hat{\mathbf{i}} = 0 \tag{6.9}$$

---

* This assumes that the unit vectors are not constant in frame $\mathcal{A}$. If two of them are constant in $\mathcal{A}$, then all three are and the angular velocity vanishes, if only *one* is constant in $\mathcal{A}$, we have a simple special case to be considered later.

Answer 6.1 Sure, as we have seen in Section 1.6.

Answer 6.2 Since $\alpha_x\hat{\mathbf{i}} \times \hat{\mathbf{i}} = \mathbf{0}$, then α_x can be anything and not affect the first of Equations (6.6).

Substitution of the first two equations of (6.6) into (6.9) yields

$$(\boldsymbol{\alpha} \times \hat{\mathbf{i}}) \cdot \hat{\mathbf{j}} + (\boldsymbol{\beta} \times \hat{\mathbf{j}}) \cdot \hat{\mathbf{i}} = 0 \tag{6.10}$$

Interchanging the dot and cross in each term (which leaves the scalar triple product unchanged) results in

$$\boldsymbol{\alpha} \cdot \hat{\mathbf{k}} - \boldsymbol{\beta} \cdot \hat{\mathbf{k}} = 0 \tag{6.11}$$

so that

$$\alpha_z = \beta_z \tag{6.12}$$

Similarly from

$$\hat{\mathbf{j}} \cdot \hat{\mathbf{k}} = 0 \qquad \text{and} \qquad \hat{\mathbf{k}} \cdot \hat{\mathbf{i}} = 0 \tag{6.13}$$

we respectively obtain (as the student should verify)

$$\beta_x = \gamma_x \qquad \text{and} \qquad \gamma_y = \alpha_y \tag{6.14}$$

The only components not involved in Equations (6.12) and (6.14) are α_x, β_y, and γ_z, which were arbitrary. If we now select them as follows,

$$\begin{aligned} \alpha_x &= \beta_x = \gamma_x \\ \beta_y &= \gamma_y = \alpha_y \\ \gamma_z &= \alpha_z = \beta_z \end{aligned} \tag{6.15}$$

then *all three vectors are identical,* and we call the resulting common vector $\boldsymbol{\omega}_{\mathcal{B}/\mathcal{A}}$:

$$\boldsymbol{\alpha} = \boldsymbol{\beta} = \boldsymbol{\gamma} = \boldsymbol{\omega}_{\mathcal{B}/\mathcal{A}} \tag{6.16}$$

If we now dot the three equations (6.8) respectively with $\hat{\mathbf{j}}$, $\hat{\mathbf{k}}$, and $\hat{\mathbf{i}}$, we get the three components of $\boldsymbol{\omega}_{\mathcal{B}/\mathcal{A}}$:

$$\begin{aligned} {}^{\mathcal{A}}\dot{\hat{\mathbf{i}}} \cdot \hat{\mathbf{j}} &= \alpha_z = \omega_{\mathcal{B}/\mathcal{A}_z} \\ {}^{\mathcal{A}}\dot{\hat{\mathbf{j}}} \cdot \hat{\mathbf{k}} &= \beta_x = \omega_{\mathcal{B}/\mathcal{A}_x} \\ {}^{\mathcal{A}}\dot{\hat{\mathbf{k}}} \cdot \hat{\mathbf{i}} &= \gamma_y = \omega_{\mathcal{B}/\mathcal{A}_y} \end{aligned} \tag{6.17}$$

Thus the vector $\boldsymbol{\omega}_{\mathcal{B}/\mathcal{A}}$ may be expressed as*

$$\boldsymbol{\omega}_{\mathcal{B}/\mathcal{A}} = ({}^{\mathcal{A}}\dot{\hat{\mathbf{j}}} \cdot \hat{\mathbf{k}})\hat{\mathbf{i}} + ({}^{\mathcal{A}}\dot{\hat{\mathbf{k}}} \cdot \hat{\mathbf{i}})\hat{\mathbf{j}} + ({}^{\mathcal{A}}\dot{\hat{\mathbf{i}}} \cdot \hat{\mathbf{j}})\hat{\mathbf{k}} \tag{6.18}$$

We call the vector $\boldsymbol{\omega}_{\mathcal{B}/\mathcal{A}}$ defined by Equation (6.18) the **angular velocity of frame $\mathcal{B}$ with respect to frame $\mathcal{A}$**, or more briefly, the **angular velocity of $\mathcal{B}$ in $\mathcal{A}$**. It is clear that the angular velocity vector depends intimately on the way frame $\mathcal{B}$ is changing its orientation with respect to $\mathcal{A}$. In the next section we examine some special properties of this vector. We shall see that $\boldsymbol{\omega}_{\mathcal{B}/\mathcal{A}}$ is *unique,* which means that we lost no generality when we let $\alpha_x = \beta_x = \gamma_x$, $\beta_y = \gamma_y = \alpha_y$, and $\gamma_z = \alpha_z = \beta_z$ in our development above of angular velocity.

---

* This is the *definition* of angular velocity set forth by the dynamicist T. R. Kane. See his books *Dynamics: Theory and Applications* (New York: McGraw-Hill, 1985), p. 16 and *Spacecraft Dynamics* (New York: McGraw-Hill, 1983), p. 49.

6.3 Properties of Angular Velocity

The Derivative Formula

We now return to Equation (6.3). Substituting from Equations (6.6) for ${}^{\mathcal{A}}\dot{\hat{\mathbf{i}}}$, ${}^{\mathcal{A}}\dot{\hat{\mathbf{j}}}$, and ${}^{\mathcal{A}}\dot{\hat{\mathbf{k}}}$ and using Equation (6.16) to replace $\boldsymbol{\alpha}$, $\boldsymbol{\beta}$, and $\boldsymbol{\gamma}$ by $\boldsymbol{\omega}_{\mathcal{B}/\mathcal{A}}$, we obtain

$$ {}^{\mathcal{A}}\dot{\mathbf{Q}} = {}^{\mathcal{B}}\dot{\mathbf{Q}} + Q_x(\boldsymbol{\omega}_{\mathcal{B}/\mathcal{A}} \times \hat{\mathbf{i}}) + Q_y(\boldsymbol{\omega}_{\mathcal{B}/\mathcal{A}} \times \hat{\mathbf{j}}) + Q_z(\boldsymbol{\omega}_{\mathcal{B}/\mathcal{A}} \times \hat{\mathbf{k}}) \quad (6.19) $$

We call this the **derivative formula**, which may be expressed, using Equation (6.1), as

$$ {}^{\mathcal{A}}\dot{\mathbf{Q}} = {}^{\mathcal{B}}\dot{\mathbf{Q}} + \boldsymbol{\omega}_{\mathcal{B}/\mathcal{A}} \times \mathbf{Q} \quad (6.20) $$

Equation (6.20) will turn out to be of vital importance in this chapter and, moreover, to be equally invaluable in our later study of the kinetics of rigid bodies in general motion in Chapter 7. It permits us easily to calculate the derivative of a vector in one frame if it is expressed in terms of base vectors fixed in another; the only price we have to pay is to add the cross product $\boldsymbol{\omega}_{\mathcal{B}/\mathcal{A}} \times \mathbf{Q}$. Thus the first property of $\boldsymbol{\omega}_{\mathcal{B}/\mathcal{A}}$ is that it allows us to relate (by Equation 6.20) the derivatives of any vector in two different frames. We have already encountered this for the special case of plane motion in Section 3.7, where Equation (3.44) may be seen to be the plane motion counterpart of Equation (6.20).

Uniqueness of the Angular Velocity Vector

There remains the nagging question of whether there might be more than one vector satisfying Equation (6.20); remember that we arbitrarily selected the components α_x, β_y, and γ_z in the preceding section in order to make $\boldsymbol{\alpha} = \boldsymbol{\beta} = \boldsymbol{\gamma} = \boldsymbol{\omega}_{\mathcal{B}/\mathcal{A}}$. We now proceed to show that the angular velocity vector is indeed unique. We do so by postulating that *two* vectors $\boldsymbol{\omega}_{\mathcal{B}/\mathcal{A}_1}$ and $\boldsymbol{\omega}_{\mathcal{B}/\mathcal{A}_2}$ *both* satisfy Equation (6.20) and then showing that they are necessarily equal.* We have

$$ {}^{\mathcal{A}}\dot{\mathbf{Q}} = {}^{\mathcal{B}}\dot{\mathbf{Q}} + \boldsymbol{\omega}_{\mathcal{B}/\mathcal{A}_1} \times \mathbf{Q} \quad (6.21) $$

$$ {}^{\mathcal{A}}\dot{\mathbf{Q}} = {}^{\mathcal{B}}\dot{\mathbf{Q}} + \boldsymbol{\omega}_{\mathcal{B}/\mathcal{A}_2} \times \mathbf{Q} \quad (6.22) $$

so that, subtracting, we get

$$ (\boldsymbol{\omega}_{\mathcal{B}/\mathcal{A}_1} - \boldsymbol{\omega}_{\mathcal{B}/\mathcal{A}_2}) \times \mathbf{Q} = \mathbf{0} \quad (6.23) $$

Finally, since $\mathbf{Q}$ is arbitrary, the parenthesized expression of Equation (6.23) must vanish, and thus the angular velocity vector has been shown to be unique as the two $\boldsymbol{\omega}$'s are one and the same.

Nothing has yet been said about dynamics in this section or in the preceding one; thus it is clear that angular velocity is a far more general

* Let $\boldsymbol{\omega}_{\mathcal{B}/\mathcal{A}_1}$ be calculated with $\hat{\mathbf{i}}$, $\hat{\mathbf{j}}$, and $\hat{\mathbf{k}}$ as described in Section 6.2, for example, and let $\boldsymbol{\omega}_{\mathcal{B}/\mathcal{A}_2}$ be computed with another triad of unit vectors fixed in $\mathcal{B}$. The question of uniqueness is whether the resulting $\boldsymbol{\omega}_{\mathcal{B}/\mathcal{A}}$'s are the same.

vector than one that is simply useful in describing rotational motions of rigid bodies. We have seen that angular velocity is in fact the vector that may be used in relating the derivatives in two frames of any arbitrary vector. Furthermore, even though we have used time as the independent variable, these derivatives may be taken with respect to *any* scalar variable. Finally, we note from the defining equation (6.18) for $\boldsymbol{\omega}_{\mathcal{B}/\mathcal{A}}$ that angular velocity is a vector relating two *frames*; thus it is meaningless to talk about the angular velocity of a point.

Now let us consider several additional properties of $\boldsymbol{\omega}_{\mathcal{B}/\mathcal{A}}$ that will prove useful in what is to follow. First we note for emphasis that if two frames $\mathcal{A}$ and $\mathcal{B}$ maintain a constant orientation (even if they are each in motion in a third frame $\mathcal{C}$), then $\boldsymbol{\omega}_{\mathcal{B}/\mathcal{A}} \equiv \mathbf{0}$.* The proof is simply to observe that no unit vector fixed in direction in $\mathcal{B}$ can change with time in $\mathcal{A}$ if there is no change in orientation between $\mathcal{A}$ and $\mathcal{B}$. Thus from Equation (6.18), $\boldsymbol{\omega}_{\mathcal{B}/\mathcal{A}} \equiv \mathbf{0}$.

Next we shall prove that the angular velocity of $\mathcal{B}$ in $\mathcal{A}$ is the negative of the angular velocity of $\mathcal{A}$ in $\mathcal{B}$. If we add Equation (6.20) to the equation

$$ {}^{\mathcal{B}}\dot{\mathbf{Q}} = {}^{\mathcal{A}}\dot{\mathbf{Q}} + \boldsymbol{\omega}_{\mathcal{A}/\mathcal{B}} \times \mathbf{Q} \tag{6.24} $$

we obtain

$$ (\boldsymbol{\omega}_{\mathcal{B}/\mathcal{A}} + \boldsymbol{\omega}_{\mathcal{A}/\mathcal{B}}) \times \mathbf{Q} = \mathbf{0} \tag{6.25} $$

Again, since $\mathbf{Q}$ is arbitrary, we have the expected result:

$$ \boldsymbol{\omega}_{\mathcal{B}/\mathcal{A}} = -\boldsymbol{\omega}_{\mathcal{A}/\mathcal{B}} \tag{6.26} $$

The Addition Theorem

We now prove the **addition theorem**, which states that

$$ \boldsymbol{\omega}_{\mathcal{C}/\mathcal{A}} = \boldsymbol{\omega}_{\mathcal{C}/\mathcal{B}} + \boldsymbol{\omega}_{\mathcal{B}/\mathcal{A}} \tag{6.27} $$

For the proof, we know from the first property of $\boldsymbol{\omega}_{\mathcal{B}/\mathcal{A}}$ that

$$ {}^{\mathcal{A}}\dot{\mathbf{Q}} = {}^{\mathcal{C}}\dot{\mathbf{Q}} + \boldsymbol{\omega}_{\mathcal{C}/\mathcal{A}} \times \mathbf{Q} \tag{6.28} $$

$$ {}^{\mathcal{B}}\dot{\mathbf{Q}} = {}^{\mathcal{C}}\dot{\mathbf{Q}} + \boldsymbol{\omega}_{\mathcal{C}/\mathcal{B}} \times \mathbf{Q} \tag{6.29} $$

$$ {}^{\mathcal{A}}\dot{\mathbf{Q}} = {}^{\mathcal{B}}\dot{\mathbf{Q}} + \boldsymbol{\omega}_{\mathcal{B}/\mathcal{A}} \times \mathbf{Q} \tag{6.30} $$

Adding Equations (6.29) and (6.30) yields

$$ {}^{\mathcal{A}}\dot{\mathbf{Q}} = {}^{\mathcal{C}}\dot{\mathbf{Q}} + (\boldsymbol{\omega}_{\mathcal{C}/\mathcal{B}} + \boldsymbol{\omega}_{\mathcal{B}/\mathcal{A}}) \times \mathbf{Q} \tag{6.31} $$

and subtracting Equation (6.31) from (6.28) gives

$$ (\boldsymbol{\omega}_{\mathcal{C}/\mathcal{A}} - \boldsymbol{\omega}_{\mathcal{C}/\mathcal{B}} - \boldsymbol{\omega}_{\mathcal{B}/\mathcal{A}}) \times \mathbf{Q} = \mathbf{0} \tag{6.32} $$

---

* *Constant orientation* means that $\mathcal{A}$ and $\mathcal{B}$ move as if they were rigidly attached except for a possible translation of one with respect to the other.

Therefore, again since **Q** is arbitrary,

$$\boldsymbol{\omega}_{\mathcal{C}/\mathcal{A}} = \boldsymbol{\omega}_{\mathcal{C}/\mathcal{B}} + \boldsymbol{\omega}_{\mathcal{B}/\mathcal{A}} \tag{6.33}$$

and the theorem is proved. It may seem intuitively obvious to the reader that Equation (6.33) is true, but in the next section we show that such a relationship *does not exist* for angular acceleration!

The addition theorem is an extremely powerful result. With it we are able to build up the angular velocity, one pair of frames at a time, of a body turning in complicated ways relative to a reference frame. This theorem makes it possible for us to avoid using the definition (6.18), which has served us well but in practice is normally supplanted by the properties described in this section.

We recall for emphasis that if $\mathcal{A}$ and $\mathcal{B}$ both move in $\mathcal{C}$ and maintain constant orientation with respect to each other, then $\boldsymbol{\omega}_{\mathcal{B}/\mathcal{A}} = \mathbf{0}$. Thus, by the addition theorem,

$$\boldsymbol{\omega}_{\mathcal{B}/\mathcal{C}} = \boldsymbol{\omega}_{\mathcal{B}/\mathcal{A}} + \boldsymbol{\omega}_{\mathcal{A}/\mathcal{C}} = \boldsymbol{\omega}_{\mathcal{A}/\mathcal{C}}$$

so that, as expected, the angular velocities in $\mathcal{C}$ of two frames $\mathcal{A}$ and $\mathcal{B}$ maintaining constant orientation with each other are identical.

Conversely, if two frames $\mathcal{A}$ and $\mathcal{B}$ have equal angular velocities in $\mathcal{C}$, we may show that their relative orientation is constant. Using the addition theorem, $\boldsymbol{\omega}_{\mathcal{B}/\mathcal{A}} = \mathbf{0}$; thus with $\hat{\mathbf{i}}$, $\hat{\mathbf{j}}$, and $\hat{\mathbf{k}}$ still fixed in direction in $\mathcal{B}$, then from Equations (6.6) and (6.16)

$$
\begin{aligned}
{}^{\mathcal{A}}\dot{\hat{\mathbf{i}}} &= \boldsymbol{\omega}_{\mathcal{B}/\mathcal{A}} \times \hat{\mathbf{i}} = \mathbf{0} \\
{}^{\mathcal{A}}\dot{\hat{\mathbf{j}}} &= \boldsymbol{\omega}_{\mathcal{B}/\mathcal{A}} \times \hat{\mathbf{j}} = \mathbf{0} \\
{}^{\mathcal{A}}\dot{\hat{\mathbf{k}}} &= \boldsymbol{\omega}_{\mathcal{B}/\mathcal{A}} \times \hat{\mathbf{k}} = \mathbf{0}
\end{aligned}
$$

Therefore, $\hat{\mathbf{i}}$, $\hat{\mathbf{j}}$, and $\hat{\mathbf{k}}$ are constant in $\mathcal{A}$, so the orientation of $\mathcal{B}$ in $\mathcal{A}$ is constant.

Thus, for two frames $\mathcal{A}$ and $\mathcal{B}$, the descriptions "constant orientation" and "$\boldsymbol{\omega}_{\mathcal{B}/\mathcal{A}} \equiv \mathbf{0}$" are completely equivalent. Note also that the addition theorem can be extended to *any number of frames* by repetition of the following procedure, two frames at a time:

$$\boldsymbol{\omega}_{\mathcal{A}/\mathcal{D}} = \boldsymbol{\omega}_{\mathcal{A}/\mathcal{B}} + \boldsymbol{\omega}_{\mathcal{B}/\mathcal{D}} = \boldsymbol{\omega}_{\mathcal{A}/\mathcal{B}} + \boldsymbol{\omega}_{\mathcal{B}/\mathcal{C}} + \boldsymbol{\omega}_{\mathcal{C}/\mathcal{D}}$$

Simple Angular Velocity

Next we show that when there exists a unit vector $\hat{\mathbf{k}}$ whose time derivative in each of two frames $\mathcal{A}$ and $\mathcal{B}$ vanishes (that is, ${}^{\mathcal{A}}\dot{\hat{\mathbf{k}}} = {}^{\mathcal{B}}\dot{\hat{\mathbf{k}}} = \mathbf{0}$), then

$$\boldsymbol{\omega}_{\mathcal{B}/\mathcal{A}} = \dot{\theta}\hat{\mathbf{k}} \tag{6.34}$$

in which θ is the angle between a pair of directed line segments $\ell_{\mathcal{A}}$ and $\ell_{\mathcal{B}}$ fixed respectively in $\mathcal{A}$ and $\mathcal{B}$, each perpendicular to $\hat{\mathbf{k}}$. The angle is measured in a reference plane containing projections of the two lines intersecting at point P as shown in Figure 6.2. The sign of the angle θ is given by the right-hand rule: If the right thumb is placed in the positive $\hat{\mathbf{k}}$ direction at P, then the direction of positive θ is that of the right hand's fingers when they curl from $\ell_{\mathcal{A}}$ into $\ell_{\mathcal{B}}$ as shown.

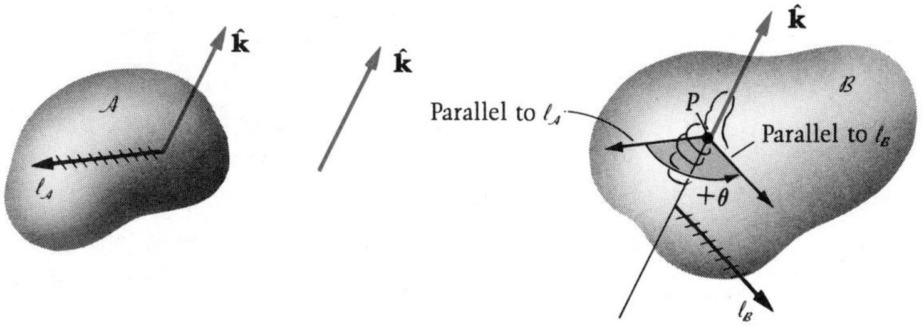

Figure 6.2 Simple angular velocity.

The type of rotational motion given by Equation (6.34) is called **simple angular velocity**. One case in which Equation (6.34) holds is that of plane motion; note, however, that there are more general cases of simple angular velocity in which the body $\mathcal{B}$ may also have a translational motion in $\mathcal{A}$ parallel to $\hat{\mathbf{k}}$ that would prevent the plane motion designation.

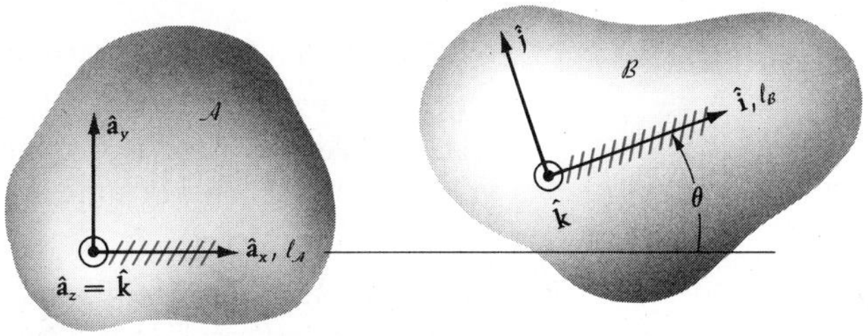

Figure 6.3 Unit vectors drawn in the reference plane for simple angular velocity.

To prove Equation (6.34), we make use of Figure 6.3. (The reference plane is the plane of the paper, and the unit vector that is constant in $\mathcal{A}$ and in $\mathcal{B}$ is $\hat{\mathbf{k}}$, perpendicular to the paper.) From this figure we may write

$$\begin{aligned} \hat{\mathbf{i}} &= \hat{\mathbf{a}}_x \cos\theta + \hat{\mathbf{a}}_y \sin\theta \\ \hat{\mathbf{j}} &= -\hat{\mathbf{a}}_x \sin\theta + \hat{\mathbf{a}}_y \cos\theta \\ \hat{\mathbf{k}} &= \hat{\mathbf{a}}_z \end{aligned} \tag{6.35}$$

Therefore, differentiating Equation (6.35), we get

$$\begin{aligned} {}^{\mathcal{A}}\dot{\hat{\mathbf{i}}} &= (-\hat{\mathbf{a}}_x \sin\theta + \hat{\mathbf{a}}_y \cos\theta)\dot{\theta} = \dot{\theta}\hat{\mathbf{j}} \\ {}^{\mathcal{A}}\dot{\hat{\mathbf{j}}} &= (-\hat{\mathbf{a}}_x \cos\theta - \hat{\mathbf{a}}_y \sin\theta)\dot{\theta} = -\dot{\theta}\hat{\mathbf{i}} \\ {}^{\mathcal{A}}\dot{\hat{\mathbf{k}}} &= \mathbf{0} \end{aligned} \tag{6.36}$$

and Equation (6.18) yields, upon direct substitution of Equation (6.36),

$$\boldsymbol{\omega}_{\mathcal{B}/\mathcal{A}} = 0\hat{\mathbf{i}} + 0\hat{\mathbf{j}} + \dot{\theta}\hat{\mathbf{k}} \tag{6.37}$$

which is the desired result.

One interesting final property of $\boldsymbol{\omega}_{\mathcal{B}/\mathcal{A}}$ is that its derivative is the same whether computed in $\mathcal{A}$ or in $\mathcal{B}$. Using Equation (6.20) and letting $\mathbf{Q}$ be $\boldsymbol{\omega}_{\mathcal{B}/\mathcal{A}}$ itself, we get

$$^{\mathcal{A}}\dot{\boldsymbol{\omega}}_{\mathcal{B}/\mathcal{A}} = {}^{\mathcal{B}}\dot{\boldsymbol{\omega}}_{\mathcal{B}/\mathcal{A}} + \boldsymbol{\omega}_{\mathcal{B}/\mathcal{A}} \times \boldsymbol{\omega}_{\mathcal{B}/\mathcal{A}} = {}^{\mathcal{B}}\dot{\boldsymbol{\omega}}_{\mathcal{B}/\mathcal{A}}$$

This result is not true for any other nonvanishing vector, unless it happens to be parallel to $\boldsymbol{\omega}_{\mathcal{B}/\mathcal{A}}$.

Summary of Properties of Angular Velocity

The properties of $\boldsymbol{\omega}_{\mathcal{B}/\mathcal{A}}$ that we have examined are summarized here:

1. It is a unique vector that satisfies

$$^{\mathcal{A}}\dot{\mathbf{Q}} = {}^{\mathcal{B}}\dot{\mathbf{Q}} + \boldsymbol{\omega}_{\mathcal{B}/\mathcal{A}} \times \mathbf{Q}$$

which is called "The Derivative Formula."

2. $\boldsymbol{\omega}_{\mathcal{B}/\mathcal{A}} = \mathbf{0}$ is synonymous with "the orientations of $\mathcal{B}$ and $\mathcal{A}$ do not change." And if $\mathcal{A}$ and $\mathcal{B}$ maintain constant orientation, their angular velocities in any third frame are equal.
3. $\boldsymbol{\omega}_{\mathcal{B}/\mathcal{A}} = -\boldsymbol{\omega}_{\mathcal{A}/\mathcal{B}}$.
4. "The Addition Theorem": $\boldsymbol{\omega}_{\mathcal{D}/\mathcal{A}} = \boldsymbol{\omega}_{\mathcal{D}/\mathcal{B}} + \boldsymbol{\omega}_{\mathcal{B}/\mathcal{A}} = (\boldsymbol{\omega}_{\mathcal{D}/\mathcal{C}} + \boldsymbol{\omega}_{\mathcal{C}/\mathcal{B}}) + \boldsymbol{\omega}_{\mathcal{B}/\mathcal{A}}$, which can be further extended to any number of frames.
5. "Simple Angular Velocity": If $\hat{\mathbf{k}}$ is constant in both $\mathcal{A}$ and $\mathcal{B}$, then

$$\boldsymbol{\omega}_{\mathcal{B}/\mathcal{A}} = \dot{\theta}\hat{\mathbf{k}}$$

where θ was defined earlier.

6. $^{\mathcal{A}}\dot{\boldsymbol{\omega}}_{\mathcal{B}/\mathcal{A}} = {}^{\mathcal{B}}\dot{\boldsymbol{\omega}}_{\mathcal{B}/\mathcal{A}}$.

In the following example we use the addition theorem to write an angular velocity, and then we express it in three different frames.

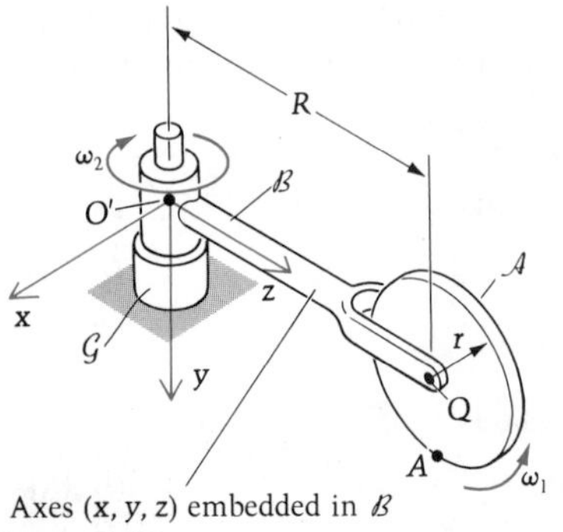

Axes (x, y, z) embedded in $\mathcal{B}$

Figure E6.1a

EXAMPLE 6.1

Body $\mathcal{B}$ in Figure E6.1a rotates in frame $\mathcal{G}$ about the vertical at constant angular speed ω_2; in $\mathcal{B}$, disk $\mathcal{A}$ rotates about its pinned axis at constant angular speed ω_1 relative to $\mathcal{B}$. (The directions of rotation are as shown. Determine the angular velocity of $\mathcal{A}$ in $\mathcal{G}$.

Solution

The coordinate axes shown are fixed in $\mathcal{B}$. By the addition theorem,*

$$\begin{aligned}\boldsymbol{\omega}_{\mathcal{A}/\mathcal{G}} &= \boldsymbol{\omega}_{\mathcal{A}/\mathcal{B}} + \boldsymbol{\omega}_{\mathcal{B}/\mathcal{G}} \\ &= \omega_1\hat{\mathbf{i}} + \omega_2\hat{\mathbf{j}} \qquad (1)\end{aligned}$$

We see from this answer that expressing $\boldsymbol{\omega}_{\mathcal{A}/\mathcal{G}}$ in terms of its components in the intermediate ("between" $\mathcal{A}$ and $\mathcal{G}$) frame $\mathcal{B}$ has yielded a neat, simple result. If we had chosen instead to write $\boldsymbol{\omega}_{\mathcal{A}/\mathcal{G}}$ in terms of its components in $\mathcal{G}$, then (see Figure E6.1b) with $\hat{\mathbf{I}}$, $\hat{\mathbf{J}}$, $\hat{\mathbf{K}}$ fixed in $\mathcal{G}$,

$$\hat{\mathbf{i}} = (\cos\omega_2 t)\hat{\mathbf{I}} - (\sin\omega_2 t)\hat{\mathbf{K}}$$

$$\hat{\mathbf{j}} = \hat{\mathbf{J}}$$

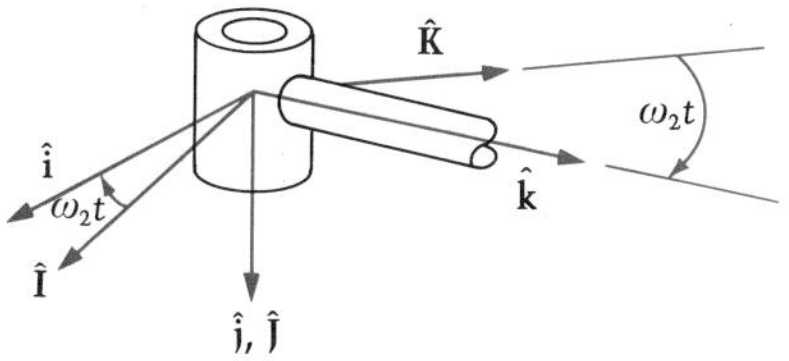

Figure E6.1b

so that, substituting into Equation (1),

$$\boldsymbol{\omega}_{\mathcal{A}/\mathcal{G}} = (\omega_1\cos\omega_2 t)\hat{\mathbf{I}} + \omega_2\hat{\mathbf{J}} - (\omega_1\sin\omega_2 t)\hat{\mathbf{K}}$$

And if we had written $\boldsymbol{\omega}_{\mathcal{A}/\mathcal{G}}$ in terms of its components along directions ($\hat{\mathbf{i}}_1$, $\hat{\mathbf{j}}_1$, $\hat{\mathbf{k}}_1$ in Figure E6.1c) fixed in $\mathcal{A}$, then with

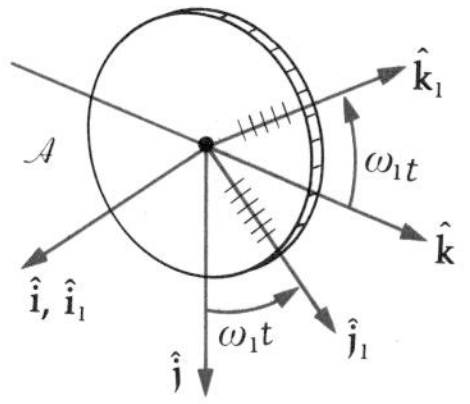

Figure E6.1c

$$\hat{\mathbf{i}} = \hat{\mathbf{i}}_1$$

$$\hat{\mathbf{j}} = (\cos\omega_1 t)\hat{\mathbf{j}}_1 - (\sin\omega_1 t)\hat{\mathbf{k}}_1$$

we obtain, again using Equation (1),

$$\boldsymbol{\omega}_{\mathcal{A}/\mathcal{G}} = \omega_1\hat{\mathbf{i}}_1 + (\omega_2\cos\omega_1 t)\hat{\mathbf{j}}_1 - (\omega_2\sin\omega_1 t)\hat{\mathbf{k}}_1$$

We see that expressing $\boldsymbol{\omega}_{\mathcal{A}/\mathcal{G}}$ in terms of its components in either $\mathcal{A}$ or $\mathcal{G}$ gives a lengthier expression than in $\mathcal{B}$; moreover, these expressions become even more complicated if ω_1 or ω_2 vary with time.

Question 6.3 Why?

The reader should note, however, that even though each of the three above representations of $\boldsymbol{\omega}_{\mathcal{A}/\mathcal{G}}$ appear to be different, they all yield the same vector.

---

* While the defining equation (6.18) is always available for directly computing the angular velocity, it is usually easier to build up the $\boldsymbol{\omega}$ vector by using the addition theorem.

Answer 6.3 Because then the angles (arguments of the sines and cosines) are not simply $\omega_1 t$ or $\omega_2 t$, but integrals of ω_1 or ω_2 with respect to time.

In the next example we illustrate the use of the "Derivative Formula" (Equation 6.20) three times.

EXAMPLE 6.2

Two children are playing in the park on a seesaw mounted on a merry-go-round as shown in Figure E6.2. The merry-go-round rotates about the ground-fixed (frame $\mathcal{G}$) vertical at $\omega_v = 3$ rad/sec, and at the instant shown the seesaw turns at $\omega_H = 2$ rad/sec relative to the merry-go-round $\mathcal{J}$. The vector from the girl to the boy is always $\mathbf{Q} = -10\,\hat{\mathbf{j}}$ ft, $(\hat{\mathbf{i}}, \hat{\mathbf{j}}, \hat{\mathbf{k}})$ being fixed in the seesaw board $\mathcal{B}$. Find ${}^{\mathcal{B}}\dot{\mathbf{Q}}$, ${}^{\mathcal{J}}\dot{\mathbf{Q}}$ and ${}^{\mathcal{G}}\dot{\mathbf{Q}}$ at the given instant.

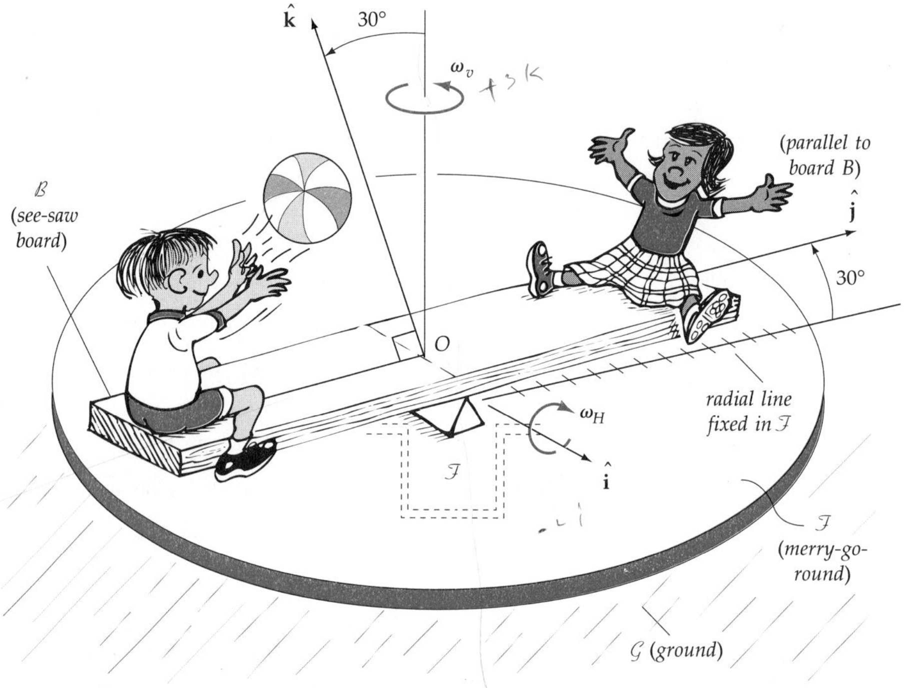

Figure E6.2

Solution

$\mathbf{Q} = -10\hat{\mathbf{j}}$ is constant relative to $\mathcal{B}$ so

$$ {}^{\mathcal{B}}\dot{\mathbf{Q}} = \mathbf{0} $$

To find ${}^{\mathcal{J}}\dot{\mathbf{Q}}$ we shall use the derivative formula:

$$ {}^{\mathcal{J}}\dot{\mathbf{Q}} = {}^{\mathcal{B}}\dot{\mathbf{Q}} + \boldsymbol{\omega}_{\mathcal{B}/\mathcal{J}} \times \mathbf{Q} $$

where

$$ \boldsymbol{\omega}_{\mathcal{B}/\mathcal{J}} = \omega_H(-\hat{\mathbf{i}}) = -2\hat{\mathbf{i}} \text{ rad/sec,} $$

so

$$
\begin{aligned}
{}^{\mathcal{I}}\dot{\mathbf{Q}} &= \mathbf{0} + (-2\hat{\mathbf{i}}) \times (-10\hat{\mathbf{j}}) \\
&= 20\hat{\mathbf{k}} \text{ ft/sec}
\end{aligned}
$$

One way to find ${}^{\mathcal{G}}\dot{\mathbf{Q}}$ is to use

$$
{}^{\mathcal{G}}\dot{\mathbf{Q}} = {}^{\mathcal{I}}\dot{\mathbf{Q}} + \boldsymbol{\omega}_{\mathcal{I}/\mathcal{G}} \times \mathbf{Q}
$$

where

$$
\begin{aligned}
\boldsymbol{\omega}_{\mathcal{I}/\mathcal{G}} &= \omega_v(\sin 30°\hat{\mathbf{j}} + \cos 30°\hat{\mathbf{k}}) \\
&= 3(0.500\hat{\mathbf{j}} + 0.866\hat{\mathbf{k}}) \\
&= 1.50\hat{\mathbf{j}} + 2.60\hat{\mathbf{k}}
\end{aligned}
$$

so

$$
\begin{aligned}
{}^{\mathcal{G}}\dot{\mathbf{Q}} &= 20\hat{\mathbf{k}} + (1.50\hat{\mathbf{j}} + 2.60\hat{\mathbf{k}}) \times (-10\hat{\mathbf{j}}) \\
&= 20\hat{\mathbf{k}} + 26\hat{\mathbf{i}} \text{ ft/sec}
\end{aligned}
$$

Had we not desired to obtain ${}^{\mathcal{I}}\dot{\mathbf{Q}}$ we might have used

$$
{}^{\mathcal{G}}\dot{\mathbf{Q}} = {}^{\mathcal{B}}\dot{\mathbf{Q}} + \boldsymbol{\omega}_{\mathcal{B}/\mathcal{G}} \times \mathbf{Q}
$$

where, by the addition theorem,

$$
\begin{aligned}
\boldsymbol{\omega}_{\mathcal{B}/\mathcal{G}} &= \boldsymbol{\omega}_{\mathcal{B}/\mathcal{I}} + \boldsymbol{\omega}_{\mathcal{I}/\mathcal{G}} \\
&= -2\hat{\mathbf{i}} + (1.50\hat{\mathbf{j}} + 2.60\hat{\mathbf{k}})
\end{aligned}
$$

so that

$$
\begin{aligned}
{}^{\mathcal{G}}\dot{\mathbf{Q}} &= \mathbf{0} + (-2\hat{\mathbf{i}} + 1.50\hat{\mathbf{j}} + 2.60\hat{\mathbf{k}}) \times (-10\hat{\mathbf{j}}) \\
&= 20\hat{\mathbf{k}} + 26\hat{\mathbf{i}} \text{ ft/sec}
\end{aligned}
$$

as before.

We now present an extended practical example of the use of the $\boldsymbol{\omega}$ properties. In this example three separate bodies are in motion in a reference frame, and their angular velocities are related by using the simple angular velocity and addition theorem properties.

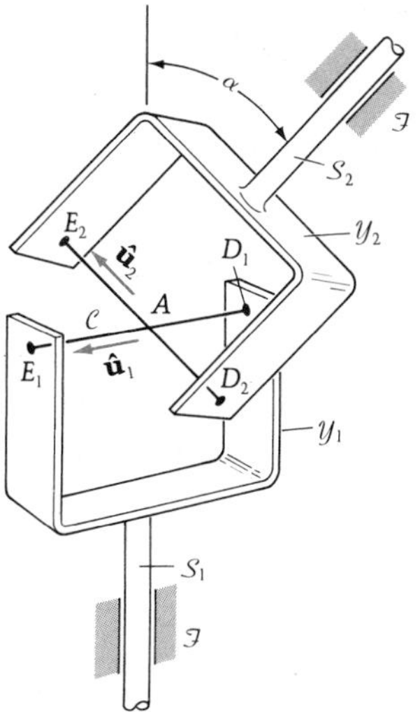

Figure E6.3a

Example 6.3

The *Hooke's joint,* or universal joint, is a device used to transmit power between two shafts that are not collinear. Figure E6.3a shows a Hooke's joint in which the shafts $\mathcal{S}_1$ and $\mathcal{S}_2$ are out of alignment by the angle α.

Each shaft is mounted in a bearing fixed to the reference frame $\mathcal{I}$. The shafts, whose axes intersect at point A, are rigidly attached to the yokes $\mathcal{Y}_1$ and $\mathcal{Y}_2$. A rigid cross $\mathcal{C}$ is the connecting body between the yokes. One leg of the cross (indicated by the unit vector $\hat{\mathbf{u}}_1$) turns in bearings fixed in $\mathcal{Y}_1$ at D_1 and E_1, while the other leg (unit vector $\hat{\mathbf{u}}_2$) turns in bearings fixed in $\mathcal{Y}_2$ at D_2 and E_2. The arms of cross $\mathcal{C}$ are identical; they form a right angle with each other, and each is perpendicular to its respective shaft.

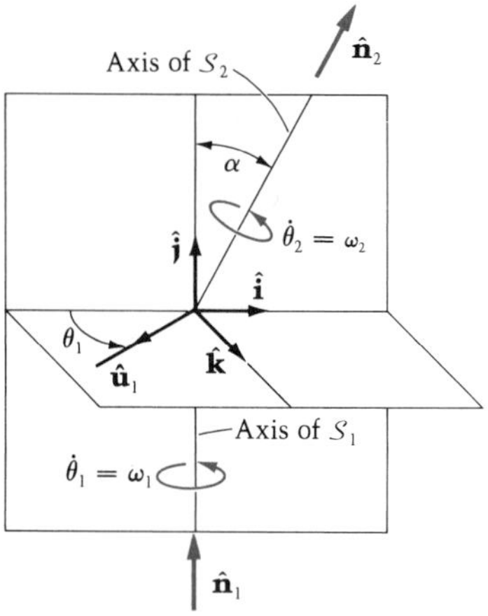

Figure E6.3b

Figure E6.3b shows that θ_1 measures the angular position of $\mathcal{S}_1$ in $\mathcal{I}$. If $\mathcal{S}_1$ (considered to be the drive shaft) has angular velocity $\boldsymbol{\omega}_{\mathcal{S}_1/\mathcal{I}} = \omega_1\hat{\mathbf{n}}_1$ and the resulting angular velocity of $\mathcal{S}_2$ is $\boldsymbol{\omega}_{\mathcal{S}_2/\mathcal{I}} = \omega_2\hat{\mathbf{n}}_2$, find the ratio of ω_2 to ω_1 in terms of θ_1 and α, and plot ω_2/ω_1 versus θ_1 for $\alpha = 0, 20, 40, 60$, and $80°$. Letting θ_2 be the rotation angle of $\mathcal{S}_2$, further investigate θ_2 versus θ_1 for the same five α values.

Question 6.4 In Figure E6.3b note that $\hat{\mathbf{n}}_2$ has no $\hat{\mathbf{k}}$ component. Why does this not represent a loss of generality?

Solution

Using the addition theorem, we may relate the angular velocities of the four rigid bodies $\mathcal{B}_2$ (shaft $\mathcal{S}_2$ plus its yoke $\mathcal{Y}_2$), $\mathcal{C}$, $\mathcal{B}_1$ (shaft $\mathcal{S}_1$ plus its yoke $\mathcal{Y}_1$), and $\mathcal{I}$:

$$\boldsymbol{\omega}_{\mathcal{B}_2/\mathcal{I}} = \boldsymbol{\omega}_{\mathcal{B}_2/\mathcal{C}} + \boldsymbol{\omega}_{\mathcal{C}/\mathcal{B}_1} + \boldsymbol{\omega}_{\mathcal{B}_1/\mathcal{I}} \tag{1}$$

Since $\mathcal{B}_1$ and $\mathcal{B}_2$ both have simple angular velocity in $\mathcal{I}$, we may write

$$\boldsymbol{\omega}_{\mathcal{B}_1/\mathcal{I}} = \omega_1\hat{\mathbf{n}}_1 \text{ or } \dot{\theta}_1\hat{\mathbf{n}}_1 \qquad \boldsymbol{\omega}_{\mathcal{B}_2/\mathcal{I}} = \omega_2\hat{\mathbf{n}}_2 \text{ or } \dot{\theta}_2\hat{\mathbf{n}}_2 \tag{2}$$

We also know from Figure E6.3a that the cross $\mathcal{C}$ has a simple angular velocity in *each* of $\mathcal{B}_1$ and $\mathcal{B}_2$. For example, the only motion that $\mathcal{C}$ can have with respect to $\mathcal{B}_1$ is a rotation about D_1E_1, the line fixed in *both* bodies. The same is true for the motion of $\mathcal{C}$ in $\mathcal{B}_2$. Thus

$$\boldsymbol{\omega}_{\mathcal{C}/\mathcal{B}_1} = \omega_{\mathcal{C}/\mathcal{B}_1}\hat{\mathbf{u}}_1 \qquad \boldsymbol{\omega}_{\mathcal{C}/\mathcal{B}_2} = -\boldsymbol{\omega}_{\mathcal{B}_2/\mathcal{C}} = -\omega_{\mathcal{B}_2/\mathcal{C}}\hat{\mathbf{u}}_2 \tag{3}$$

in which $\omega_{\mathcal{C}/\mathcal{B}_1}$ and $\omega_{\mathcal{B}_2/\mathcal{C}}$ are the unknown magnitudes of the respective vectors.

Next we must express all of $\hat{\mathbf{n}}_1$, $\hat{\mathbf{n}}_2$, $\hat{\mathbf{u}}_1$, and $\hat{\mathbf{u}}_2$ in terms of a common set of unit vectors. Then we shall be able to obtain three scalar equations from (1) and hence solve for ω_2 in terms of ω_1. From Figure E6.3b, three of the unit vectors are obvious:

$$\begin{aligned} \hat{\mathbf{n}}_1 &= \hat{\mathbf{j}} \\ \hat{\mathbf{n}}_2 &= \sin\alpha\hat{\mathbf{i}} + \cos\alpha\hat{\mathbf{j}} \\ \hat{\mathbf{u}}_1 &= -\cos\theta_1\hat{\mathbf{i}} + \sin\theta_1\hat{\mathbf{k}} \end{aligned} \tag{4}$$

To obtain $\hat{\mathbf{u}}_2$, we note that it is perpendicular to both $\hat{\mathbf{n}}_2$ and $\hat{\mathbf{u}}_1$. Crossing $\hat{\mathbf{u}}_1$ into $\hat{\mathbf{n}}_2$ then gives the assigned direction of $\hat{\mathbf{u}}_2$ (note that $\hat{\mathbf{n}}_2 \times \hat{\mathbf{u}}_1$ is opposite!); $\hat{\mathbf{u}}_1 \times \hat{\mathbf{n}}_2$ is not generally a unit vector, however, so to get $\hat{\mathbf{u}}_2$ we divide this vector by its magnitude:

$$\hat{\mathbf{u}}_2 = \frac{\hat{\mathbf{u}}_1 \times \hat{\mathbf{n}}_2}{|\hat{\mathbf{u}}_1 \times \hat{\mathbf{n}}_2|} = \frac{-\cos\alpha\sin\theta_1\hat{\mathbf{i}} + \sin\alpha\sin\theta_1\hat{\mathbf{j}} - \cos\alpha\cos\theta_1\hat{\mathbf{k}}}{\sqrt{\cos^2\alpha + \sin^2\alpha\sin^2\theta_1}} \tag{5}$$

Substituting Equations (4) and (5) for the four unit vectors into Equations (2) and (3), and then substituting the resulting angular velocity expressions into Equation (1), we get a vector equation that has the following three scalar component equations:

$$\hat{\mathbf{i}} \text{ coefficients:} \qquad \begin{aligned} \omega_2\sin\alpha = {} & R\omega_{\mathcal{B}_2/\mathcal{C}}(-\cos\alpha\sin\theta_1) \\ & + \omega_{\mathcal{C}/\mathcal{B}_1}(-\cos\theta_1) \end{aligned} \tag{6}$$

Answer 6.4 The (xy) plane of the paper can be chosen to be the plane containing $\hat{\mathbf{n}}_1$ and $\hat{\mathbf{n}}_2$ without loss of generality.

$\hat{\mathbf{j}}$ coefficients: $\omega_2 \cos \alpha = R\omega_{\mathcal{B}_2/\mathcal{C}}(\sin \alpha \sin \theta_1) + \omega_1$ (7)

$\hat{\mathbf{k}}$ coefficients: $0 = R\omega_{\mathcal{B}_2/\mathcal{C}}(-\cos \alpha \cos \theta_1) + \omega_{\mathcal{C}/\mathcal{B}_1}(\sin \theta_1)$ (8)

in which $R = 1/\sqrt{\cos^2 \alpha + \sin^2 \alpha \sin^2 \theta_1}$. Eliminating $\omega_{\mathcal{C}/\mathcal{B}_1}$ between (6) and (8) gives

$$\omega_{\mathcal{B}_2/\mathcal{C}} = \frac{-\omega_2 \tan \alpha \sin \theta_1}{R} \tag{9}$$

and substitution of (9) into (7) yields

$$\omega_2 = \left(\frac{\cos \alpha}{\cos^2 \alpha + \sin^2 \alpha \sin^2 \theta_1}\right) \omega_1$$

so that

$$\frac{\omega_2}{\omega_1} = \frac{\cos \alpha}{1 - \sin^2 \alpha \cos^2 \theta_1} \tag{10}$$

A plot of this expression (Figure E6.3c) shows the manner in which ω_2 changes over a quarter-turn of $\mathcal{S}_1$ in space. Note that since $\cos \theta_1$ is squared, the curves reflect around the vertical line at $\theta_1 = 90°$ for $90° \le \theta_1 \le 180°$; between $180°$ and $360°$ we again have a mirror image, this time of the curves between $0°$ and $180°$. Note that for large misalignment angles α, shaft $\mathcal{S}_2$ must turn very rapidly at and near $\theta_1 = 0$; in fact, when $\alpha = 90°$ the bodies reach a configuration in which they cannot turn at all. This is called *gimbal lock.* Note further that a misalignment of as much as $10°$ results in an output speed variation (ω_2) over a revolution of only about 3 percent.

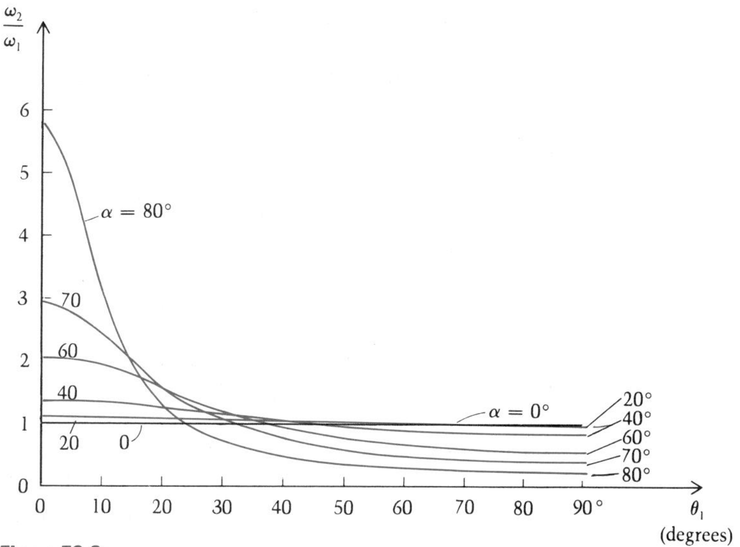

Figure E6.3c

Next we examine the angles of rotation θ_1 (of $\mathcal{S}_1$) and θ_2 (of $\mathcal{S}_2$). Since $\mathcal{S}_1$ and $\mathcal{S}_2$ both have simple angular velocity in $\mathcal{I}$, we have $\omega_1 = \dot{\theta}_1$ and $\omega_2 = \dot{\theta}_2$, so that (from Equation 10):

$$\dot{\theta}_2 = \frac{\dot{\theta}_1 \cos \alpha}{1 - \sin^2 \alpha \cos^2 \theta_1} \tag{11}$$

Integrating (11) gives

$$\theta_2 + \text{constant} = \int \frac{\cos\alpha\, d\theta_1}{1 - \sin^2\alpha\cos^2\theta_1}$$

$$= \tan^{-1}\left(\frac{\tan\theta_1}{\cos\alpha}\right) \quad (12)$$

The constant of integration is zero if we define $\theta_2 = 0$ when $\theta_1 = 0$. Then θ_2 may be plotted as a function of θ_1 for the same representative values of α (see Figure E6.3d). If $\alpha = 0$, then $\theta_1 \equiv \theta_2$ and the curve is a 45° line. If there is

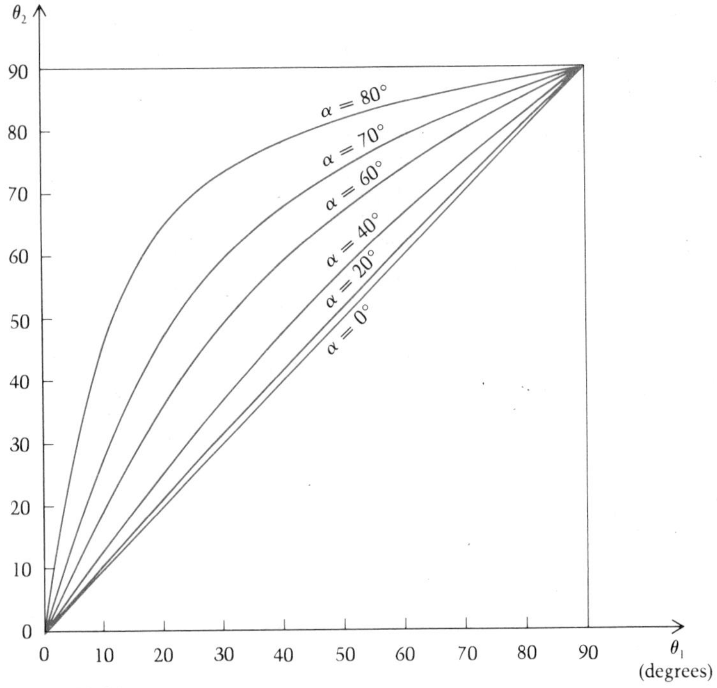

Figure E6.3d

misalignment ($\alpha \neq 0$), then note from the curves that for $0 \leq \theta_1 \leq 90°$, shaft $\mathcal{S}_2$ is always turned *more* than $\mathcal{S}_1$. Then $\mathcal{S}_1$ catches up at $\theta = 90°$, and from $\theta_1 = 90°$ to 180° the angle θ_2 of $\mathcal{S}_2$ lags *behind* θ_1.

Question 6.5 Explain why this is so by using Equation (12).

From 180° to 360°, the cycle repeats and everything returns to the same starting position ($\theta_1 = \theta_2 = 360°$) at the same time.

Answer 6.5 For θ_1 in the second quadrant, and with $0 < \cos\alpha \leq 1$, we see from Equation (12) that $\tan\theta_2$ is a more negative number than $\tan\theta_1$ (unless $\alpha = 0$, in which case $\theta_2 = \theta_1$). Therefore, θ_1 and θ_2 are angles between 90° and 180°, and $\theta_2 < \theta_1$.

PROBLEMS ▶ Section 6.3

(See also the Project Problem 6.93 at the end of this chapter.)

6.1 Verify, in Example 1.1, that ${}^{\mathcal{I}}\dot{\mathbf{A}}$ is indeed $\boldsymbol{\omega}_{\mathcal{B}/\mathcal{I}} \times \mathbf{A}$, where $\boldsymbol{\omega}_{\mathcal{B}/\mathcal{I}} = \dot{\theta}(-\hat{\mathbf{k}})$. (See Figure P6.1.)

6.2 The angular velocities of $\mathcal{A}$ and $\mathcal{B}$ in a reference frame $\mathcal{I}$ are, respectively, $10\hat{\mathbf{n}}_1$ rad/sec and $7\hat{\mathbf{n}}_2$ rad/sec. Find the angular velocity of $\mathcal{B}$ in $\mathcal{A}$ expressed in terms of $\hat{\mathbf{i}}$ and $\hat{\mathbf{j}}$. (See Figure P6.2.)

6.3 A vector **v** is given as a function of time t by $\mathbf{v} = t^3\hat{\mathbf{i}} + t^2\hat{\mathbf{j}} + t\hat{\mathbf{k}}$ m/s, where ($\hat{\mathbf{i}}$, $\hat{\mathbf{j}}$, $\hat{\mathbf{k}}$) are unit vectors whose directions are fixed in a frame $\mathcal{H}$. The angular velocity of $\mathcal{H}$ in frame $\mathcal{R}$ is $\boldsymbol{\omega}_{\mathcal{H}/\mathcal{R}} = t\hat{\mathbf{i}} + t^2\hat{\mathbf{j}} + t^3\hat{\mathbf{k}}$ rad/s. Find the derivative of **v** in frame $\mathcal{R}$, that is, ${}^{\mathcal{R}}\dot{\mathbf{v}}$: (a) as a function of t; (b) at $t = 1$ s; (c) at $t = 2$ s.

6.4 In the preceding problem, find ${}^{\mathcal{R}}\ddot{\mathbf{v}}$.

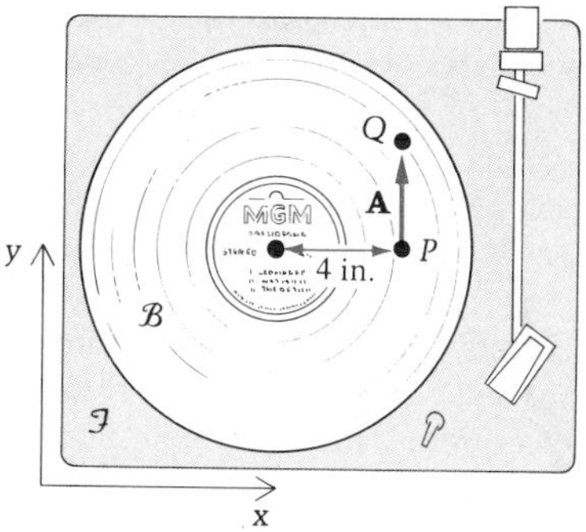

Figure P6.1

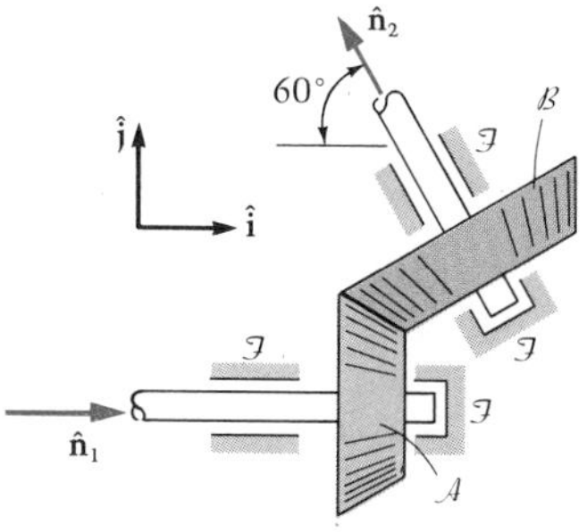

Figure P6.2

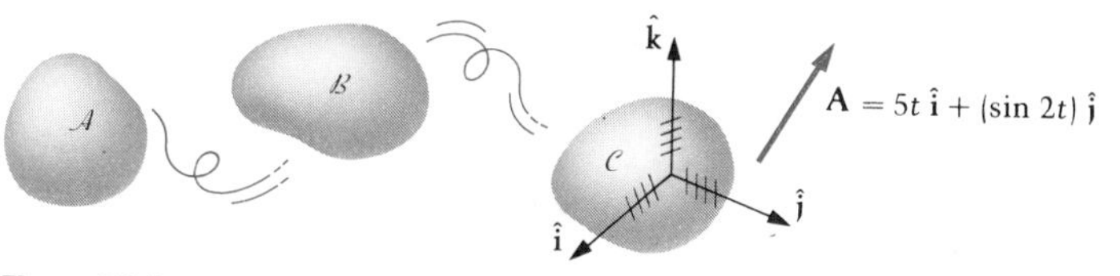

Figure P6.5

6.5 Note the three frames $\mathcal{A}$, $\mathcal{B}$, and $\mathcal{C}$, and the vector **A** defined in Figure P6.5 expressed in terms of its components in $\mathcal{C}$. Also, $\boldsymbol{\omega}_{\mathcal{C}/\mathcal{B}} = t^2\hat{\mathbf{i}} + t^3\hat{\mathbf{j}}$ and $\boldsymbol{\omega}_{\mathcal{B}/\mathcal{A}} = 5\hat{\mathbf{k}}$. Find ${}^{\mathcal{A}}\dot{\mathbf{A}}$ at $t = \pi/4$ sec.

6.6 Review Problem 1.155 in which the unit tangent, normal, and binormal of a curve in space $\mathcal{S}$ are defined. Let $\mathcal{B}$ be a frame moving relative to $\mathcal{S}$ in such a way that $\hat{\mathbf{e}}_t$, $\hat{\mathbf{e}}_n$, and $\hat{\mathbf{e}}_b$ are always fixed in $\mathcal{B}$. Use the definition (Equation (6.18)) of angular velocity to find the angular velocity of $\mathcal{B}$ in $\mathcal{S}$. Note that

$$\frac{d}{dt}(\ ) = \left[\frac{d}{ds}(\ )\right]\frac{ds}{dt} = \dot{s}\,\frac{d(\ )}{ds}$$

6.7 The antenna $\mathcal{A}$ in Figure P6.7 is oriented with the following three rotations:

1. Azimuth, about y fixed in $\mathcal{G}$, at the rate $\omega_{\mathcal{I}_1/\mathcal{G}} = 3t^2$ rad/sec
2. Elevation, about z_1 fixed in a first intermediate frame $\mathcal{I}_1$, at the constant rate $\omega_{\mathcal{I}_2/\mathcal{I}_1} = 1$ rad/sec
3. A polarization rotation about the antenna axis x_2 (fixed in both the second intermediate frame $\mathcal{I}_2$ and in $\mathcal{A}$) at $\omega_{\mathcal{A}/\mathcal{I}_2} = 4t$ rad/sec

If the structure is in the $\phi = 0$ position at $t = 0$, find $\boldsymbol{\omega}_{\mathcal{A}/\mathcal{G}}$ at the time $t = \pi/2$ sec. Use these unit vectors fixed in direction in $\mathcal{I}_1$: $\hat{\mathbf{j}}$ parallel to y, $\hat{\mathbf{k}}$ parallel to z_1, and $\hat{\mathbf{i}} = \hat{\mathbf{j}} \times \hat{\mathbf{k}}$.

6.8 Show that the output angle θ_2 of the Hooke's joint in Example 6.3 can be alternatively obtained by dotting $\hat{\mathbf{k}}$ with $\hat{\mathbf{u}}_2$.

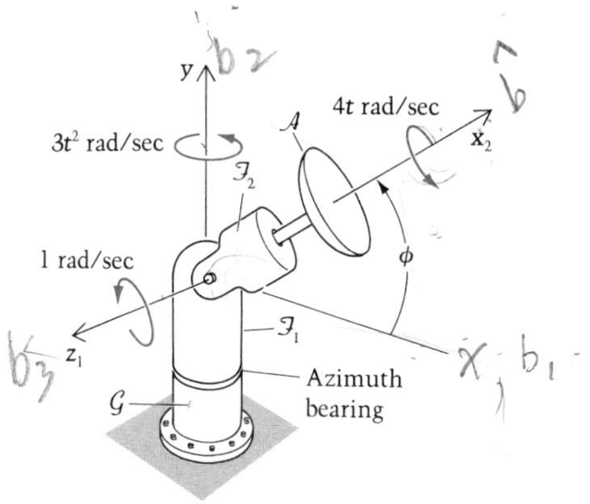

Figure P6.7

6.9 A device for simulating conditions in space allows rotations about orthogonal axes as shown in Figure P6.9. Determine the angular velocity in frame $\mathcal{G}$ of the capsule $\mathcal{C}$ containing the astronaut. Express the result in terms of unit vectors $(\hat{\mathbf{i}}, \hat{\mathbf{j}}, \hat{\mathbf{k}})$ fixed in the beam $\mathcal{B}$. Note that the ω_y rotation is about an axis fixed in $\mathcal{C}$ and in $\mathcal{A}$ but *not* in $\mathcal{B}$; this axis is parallel to y at $t = 0$, and ω_x is a constant.

** **6.10** The outer cone $\mathcal{B}$ in Figure P6.10 has the following prescribed motion with respect to the fixed inner cone $\mathcal{C}$:

1. The vertices remain together.

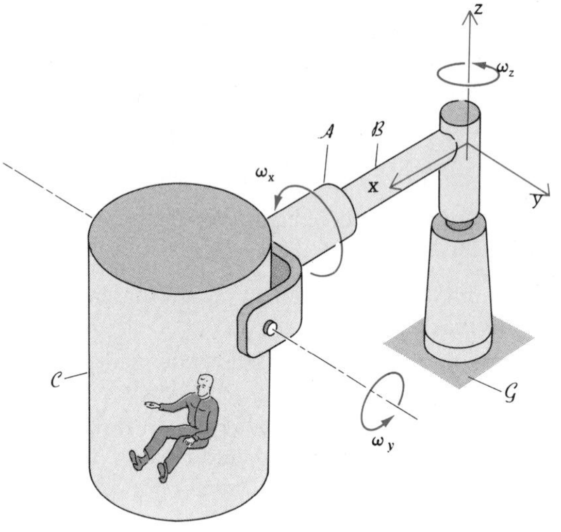

Figure P6.9

2. The line AB (a base radius fixed in $\mathcal{B}$) always lies in some vertical plane parallel to XY.
3. Cone $\mathcal{B}$ slides on $\mathcal{C}$; that is, there is always a line of contact between O and a point of the base circle of $\mathcal{C}$.
4. Point A of $\mathcal{B}$ revolves around the x axis in a vertical circle at constant speed $H\dot{\theta}$.

Use the addition theorem to show that the angular velocity of $\mathcal{B}$ in $\mathcal{C}$ is given by

$$\boldsymbol{\omega}_{\mathcal{B}/\mathcal{C}} = \frac{\dot{\theta}(\tan^2 \gamma \cos^2 \theta \hat{\mathbf{i}} - \tan \gamma \cos \theta \hat{\mathbf{j}} - \sin \theta \tan \gamma \hat{\mathbf{k}})}{1 + \tan^2 \gamma \cos^2 \theta}$$

6.11 In the preceding problem find $\boldsymbol{\omega}_{\mathcal{B}/\mathcal{C}}$ if the projection of AB into the YZ plane through A is always aligned with the radius. (See Figure P6.11.)

6.12 Find $\boldsymbol{\omega}_{\mathcal{B}/\mathcal{C}}$ in the preceding problem if the outer cone *rolls* on the fixed inner cone. (See Figure P6.12.)

A popular method of stabilizing shipboard antennas is by means of pendulous masses together with the gyroscopic effect of spinning flywheels. In Figure P6.13 the ship (frame $\mathcal{S}$) pitches (about x), rolls (about y), and yaws (about z) in the sea (frame $\mathcal{E}$). Frame $\mathcal{I}$, just above a Hooke's joint, is to form a stable platform on which the antenna can then be easily positioned in azimuth (angle A) and elevation (angle E). The frame $\mathcal{I}$ remains level by a "depitching" rotation P above the "derolling" rotation R. The following three problems are based on this system.

The INMARSAT communications satellite system required that shipboard antenna systems remain operational up to the following oscillatory limits:

Pitch:	±10° in 6 sec
Roll:	±30° in 8 sec
Yaw:	±8° in 50 sec

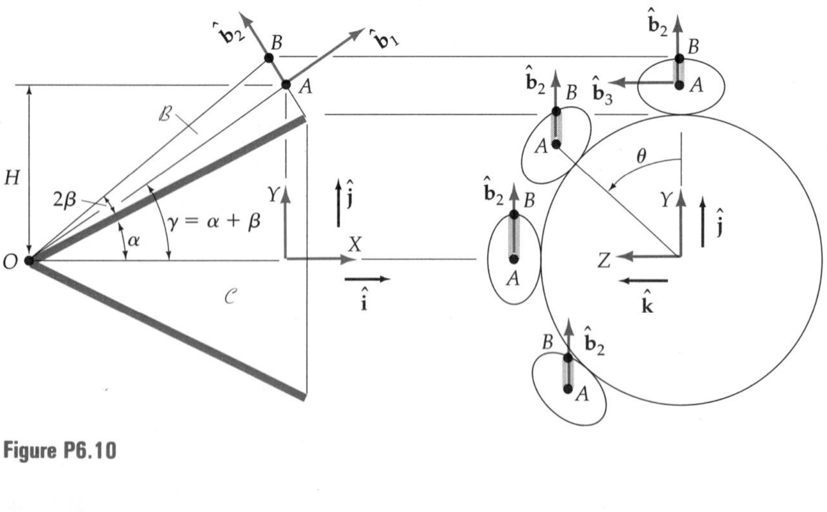

Figure P6.10

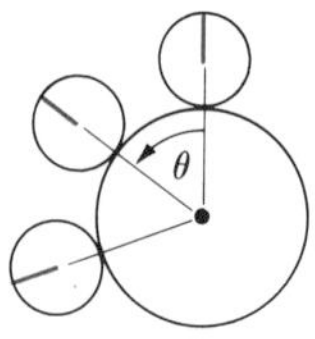

Figure P6.11

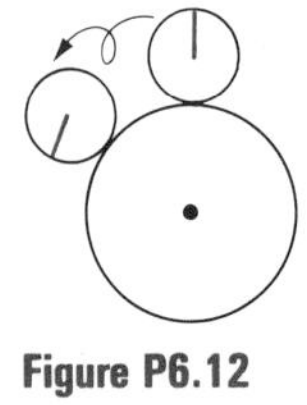

Figure P6.12

* **6.13** Assume sine waves for each of these three motions and assume yaw over roll over pitch — that is, the assumed order of ship rotations is (1) pitch, from frame $\mathcal{E}$ to an intermediate frame $\mathcal{I}_1$; and (2) roll, from $\mathcal{I}_1$ to a second

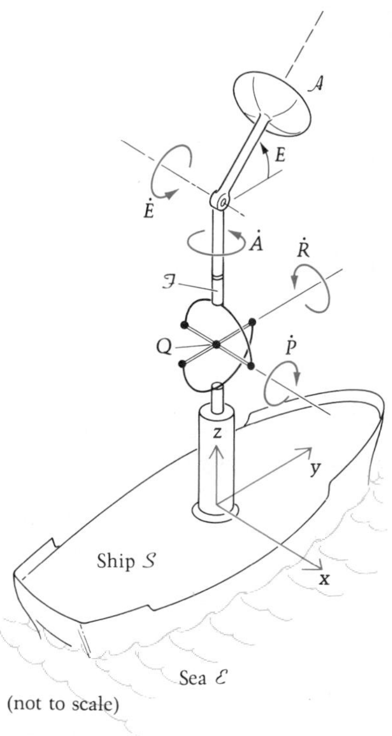

Figure P6.13

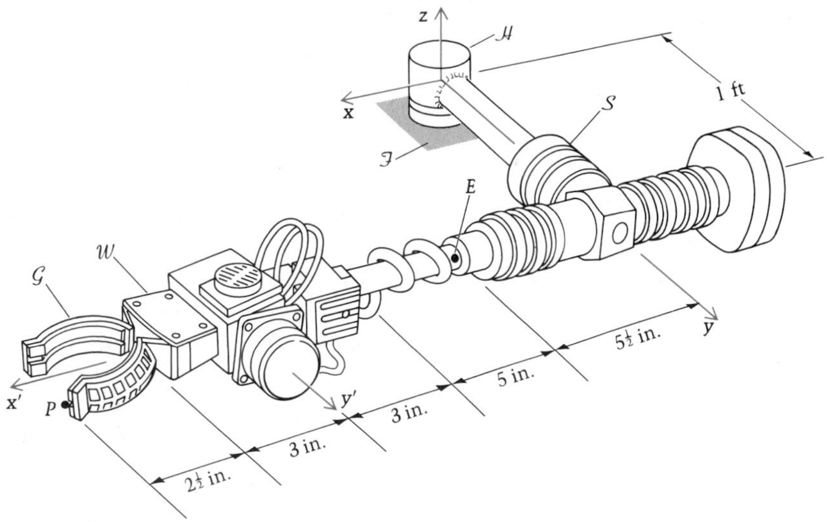

Figure P6.16(a)

intermediate frame $\mathcal{I}_2$; and (3) yaw, from $\mathcal{I}_2$ to frame $\mathcal{S}$. Write the angular velocity of the ship $\mathcal{S}$ in the sea (earth-fixed frame $\mathcal{E}$), expressed in the ship-fixed axes (x, y, z). *Hint*: For example, θ_{pitch} will be $10\pi/180 \sin 2\pi t/6$ rad.

6.14 Write the angular velocity of $\mathcal{I}$ in $\mathcal{S}$, expressed in the axes (x, y, z).

6.15 Write the angular velocity of $\mathcal{I}$ in $\mathcal{E}$, using the results of the preceding two problems together with the addition theorem.

* **6.16** A robot manufactured by the Heath Company has the mechanical arm shown in Figure P6.16(a) and accompanying photograph. Its shoulder $\mathcal{S}$ extends from the head $\mathcal{H}$, which can itself rotate 350° about z in 30 sec relative to the reference frame $\mathcal{I}$. The arm is able to travel 150° about axis y in 26 sec. (The axes (x, y, z) are fixed in the head $\mathcal{H}$.) The part of the arm to the left of point E is able to extend (and retract) up to 5 in. in 30 sec. The wrist $\mathcal{W}$ has two motions: It is able to pivot up to 180° about y' in 10 sec and to rotate (about x') up to 350° in 4 sec. Axes (x', y', z') are fixed in $\mathcal{W}$. Finally, the gripper $\mathcal{G}$ is able to open (and close) 3.5 in. in 3 sec, but is assumed here to be a closed circle with a 2.5-inch diameter. Approximate dimensions are shown in the figure.

For this problem, assume that all the robot's motions (except the gripper opening) are occurring simultaneously about positive axes with their respective average speeds. Find the angular velocity of the gripper $\mathcal{G}$, relative to $\mathcal{I}$ and

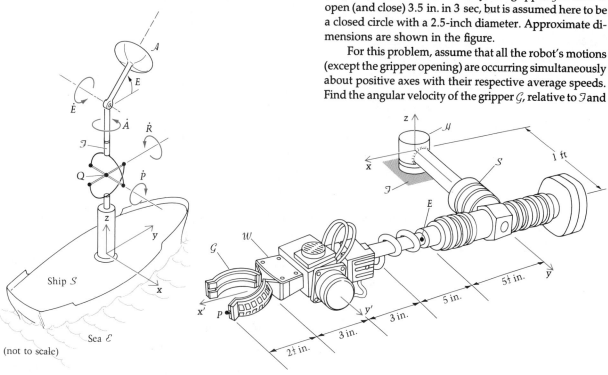

Figure P6.16(b) *(Courtesy of the Heath Company.)*

* Asterisks identify the more difficult problems.

expressed in terms of unit vectors in $\mathcal{H}$, at an instant when these two conditions hold:

1. The shoulder rotation angle θ is $-60°$:

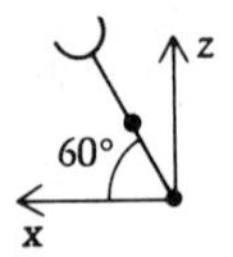

2. The wrist is pivoted 30°:

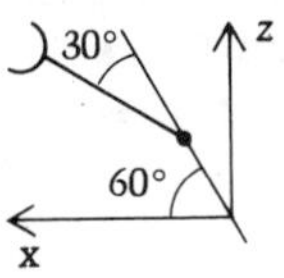

6.4 The Angular Acceleration Vector

In applying what we have learned about angular velocity to the kinematics of rigid bodies, we also need to understand its derivative. The **angular acceleration** of frame $\mathcal{B}$ relative to frame $\mathcal{A}$ is defined to be

$$\boldsymbol{\alpha}_{\mathcal{B}/\mathcal{A}} = {}^{\mathcal{A}}\dot{\boldsymbol{\omega}}_{\mathcal{B}/\mathcal{A}} \tag{6.38}$$

(Note from the last property of $\boldsymbol{\omega}_{\mathcal{B}/\mathcal{A}}$ in the previous section that the derivative could equally well be taken in $\mathcal{B}$, but generally not in any other frame.)

It is important to note that the addition theorem (Equation 6.27) *does not hold* for angular acceleration. Watch:

$$\begin{aligned}
\boldsymbol{\alpha}_{\mathcal{C}/\mathcal{A}} &= {}^{\mathcal{A}}\dot{\boldsymbol{\omega}}_{\mathcal{C}/\mathcal{A}} = {}^{\mathcal{A}}\dot{\overline{\boldsymbol{\omega}_{\mathcal{C}/\mathcal{B}} + \boldsymbol{\omega}_{\mathcal{B}/\mathcal{A}}}} \\
&= {}^{\mathcal{A}}\dot{\boldsymbol{\omega}}_{\mathcal{C}/\mathcal{B}} + {}^{\mathcal{A}}\dot{\boldsymbol{\omega}}_{\mathcal{B}/\mathcal{A}} \\
&= ({}^{\mathcal{B}}\dot{\boldsymbol{\omega}}_{\mathcal{C}/\mathcal{B}} + \boldsymbol{\omega}_{\mathcal{B}/\mathcal{A}} \times \boldsymbol{\omega}_{\mathcal{C}/\mathcal{B}}) + \boldsymbol{\alpha}_{\mathcal{B}/\mathcal{A}}
\end{aligned}$$

$$\boldsymbol{\alpha}_{\mathcal{C}/\mathcal{A}} = \boldsymbol{\alpha}_{\mathcal{C}/\mathcal{B}} + \boldsymbol{\alpha}_{\mathcal{B}/\mathcal{A}} + \boldsymbol{\omega}_{\mathcal{B}/\mathcal{A}} \times \boldsymbol{\omega}_{\mathcal{C}/\mathcal{B}} \tag{6.39}$$

We see that there is an extra term (the cross product of two angular velocity vectors) that prevents the simple theorem we have derived for $\boldsymbol{\omega}$'s from working for $\boldsymbol{\alpha}$'s. This term is sometimes called a *gyroscopic term;* note that it vanishes for plane motion, in which case we do have an addition theorem for the $\boldsymbol{\alpha}$'s (which are then of the form $\ddot{\theta}\hat{\mathbf{k}}$).

In each of the two examples to follow, the reader should notice how the various properties of $\boldsymbol{\omega}$ — simple angular velocity (Equation (6.34)), the addition theorem (6.27), and the derivative formula (6.20) — are used to great advantage.

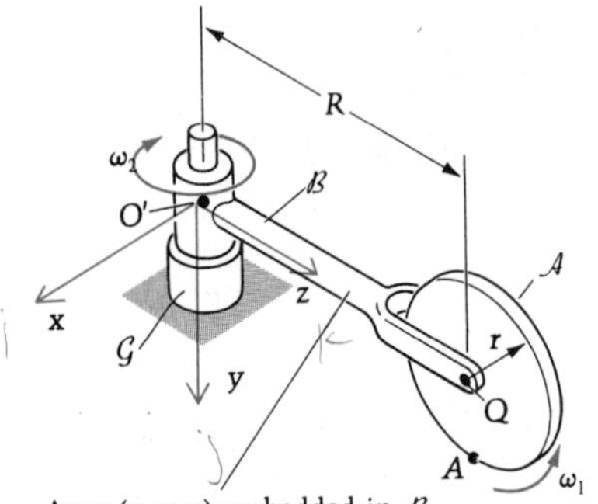

Axes (x, y, z) embedded in $\mathcal{B}$

Figure E6.4

EXAMPLE 6.4

Body $\mathcal{B}$ in Figure E6.4 rotates in frame $\mathcal{G}$ about the vertical at constant angular speed ω_2; in $\mathcal{B}$, disk $\mathcal{A}$ rotates about its pinned axis at constant angular speed ω_1. (The directions of rotation are as shown.) Determine the angular acceleration of $\mathcal{A}$ in $\mathcal{G}$.

Solution

As we saw in Example 6.1,

$$\boldsymbol{\omega}_{\mathcal{A}/\mathcal{G}} = \boldsymbol{\omega}_{\mathcal{A}/\mathcal{B}} + \boldsymbol{\omega}_{\mathcal{B}/\mathcal{G}} = \omega_1\hat{\mathbf{i}} + \omega_2\hat{\mathbf{j}}$$

Next, using Equation (6.20) and noting that $\boldsymbol{\omega}_{\mathcal{A}/\mathcal{G}}$ is expressed in terms of axes embedded in $\mathcal{B}$, we "move the derivative" using the derivative formula and obtain*

$$\begin{aligned}\boldsymbol{\alpha}_{\mathcal{A}/\mathcal{G}} &= {}^{\mathcal{G}}\dot{\boldsymbol{\omega}}_{\mathcal{A}/\mathcal{G}} \\ &= {}^{\mathcal{B}}\dot{\boldsymbol{\omega}}_{\mathcal{A}/\mathcal{G}} + \boldsymbol{\omega}_{\mathcal{B}/\mathcal{G}} \times \boldsymbol{\omega}_{\mathcal{A}/\mathcal{G}} \\ &= \mathbf{0} + \omega_2\hat{\mathbf{j}} \times (\omega_1\hat{\mathbf{i}} + \omega_2\hat{\mathbf{j}}) \\ &= -\omega_1\omega_2\hat{\mathbf{k}}\end{aligned}$$

Note that the same result is obtained by using Equation (6.39) with frames $\mathcal{C}$, $\mathcal{B}$, and $\mathcal{A}$ replaced by $\mathcal{A}$, $\mathcal{B}$, and $\mathcal{G}$, respectively.

Question 6.6 Why was ${}^{\mathcal{B}}\dot{\boldsymbol{\omega}}_{\mathcal{A}/\mathcal{G}} = \mathbf{0}$ in the above example?

EXAMPLE 6.5

Determine the angular acceleration of the cross $\mathcal{C}$ relative to frame $\mathcal{I}$ in Example 6.3, for the case $\dot{\theta}_1 =$ constant. Express the result in terms of unit vectors $\hat{\mathbf{n}}_2$, $\hat{\mathbf{u}}_2$, and $\hat{\mathbf{v}}_2 = \hat{\mathbf{n}}_2 \times \hat{\mathbf{u}}_2$ fixed in $\mathcal{B}_2$.

Solution

Using the definition of angular acceleration [Equation (6.38)], the addition theorem, and the derivative formula,

$$\begin{aligned}\boldsymbol{\alpha}_{\mathcal{C}/\mathcal{I}} &= {}^{\mathcal{I}}\dot{\boldsymbol{\omega}}_{\mathcal{C}/\mathcal{I}} \\ &= {}^{\mathcal{I}}\dot{\boldsymbol{\omega}}_{\mathcal{C}/\mathcal{B}_2} + {}^{\mathcal{I}}\dot{\boldsymbol{\omega}}_{\mathcal{B}_2/\mathcal{I}} \\ &= ({}^{\mathcal{B}_2}\dot{\boldsymbol{\omega}}_{\mathcal{C}/\mathcal{B}_2} + \boldsymbol{\omega}_{\mathcal{B}_2/\mathcal{I}} \times \boldsymbol{\omega}_{\mathcal{C}/\mathcal{B}_2}) + {}^{\mathcal{I}}\dot{\boldsymbol{\omega}}_{\mathcal{B}_2/\mathcal{I}} \qquad (1)\end{aligned}$$

Let us first concentrate on the first term on the right side of Equation (1). In Example 6.3 we had

$$\boldsymbol{\omega}_{\mathcal{C}/\mathcal{B}_2} = \dot{\theta}_2 T_\alpha s_1 \sqrt{1 - s_\alpha^2 c_1^2}\,\hat{\mathbf{u}}_2 \qquad (2)$$

where we are using the notation $s_\alpha = \sin\alpha$, $c_\alpha = \cos\alpha$, $T_\alpha = \tan\alpha$, $s_1 = \sin\theta_1$, and $c_1 = \cos\theta_1$. We also know from Example 6.3 that

$$\dot{\theta}_2 = \frac{\dot{\theta}_1 c_\alpha}{1 - s_\alpha^2 c_1^2} \qquad (3)$$

Substituting $\dot{\theta}_2$ from Equation (3) into (2) and differentiating the result in $\mathcal{B}_2$ (and noting $\hat{\mathbf{u}}_2$ is constant there) yields, after simplifying,

$${}^{\mathcal{B}_2}\dot{\boldsymbol{\omega}}_{\mathcal{C}/\mathcal{B}_2} = \frac{c_1 s_\alpha c_\alpha^2 \dot{\theta}_1^2}{(1 - s_\alpha^2 c_1^2)^{3/2}}\,\hat{\mathbf{u}}_2 \qquad (4)$$

* By "moving the derivative," we mean shifting it *from* a frame in which it is inconvenient to differentiate, *to* a frame in which we desire to differentiate.

Answer 6.6 ω_1 and ω_2 are constant scalars, and $\hat{\mathbf{i}}$ and $\hat{\mathbf{j}}$ are unit vectors fixed in direction in $\mathcal{B}$.

The second term in Equation (1) is

$$\dot{\theta}_2\hat{\mathbf{n}}_2 \times \dot{\theta}_2 T_\alpha s_1\sqrt{1 - s_\alpha^2 c_1^2}\,\hat{\mathbf{u}}_2$$

which upon simplification is

$$\frac{\dot{\theta}_1^2 s_\alpha c_\alpha s_1}{(1 - s_\alpha^2 c_1^2)^{3/2}}\,\hat{\mathbf{v}}_2 \tag{5}$$

The last term in Equation (1) is:

$$^{\mathcal{I}}\dot{\boldsymbol{\omega}}_{\mathcal{B}_2/\mathcal{I}} = \ddot{\theta}_2\hat{\mathbf{n}}_2 = \frac{-2\dot{\theta}_1^2 s_\alpha^2 c_\alpha s_1 c_1}{(1 - s_\alpha^2 c_1^2)^2}\,\hat{\mathbf{n}}_2 \tag{6}$$

where we have differentiated Equation (3).

The solution for $\boldsymbol{\alpha}_{\mathcal{C}/\mathcal{I}}$ is therefore the sum of the vectors in Equations (4), (5), and (6):

$$\boldsymbol{\alpha}_{\mathcal{C}/\mathcal{I}} = \left[c_\alpha c_1\hat{\mathbf{u}}_2 + s_1\hat{\mathbf{v}}_2 - \frac{2 s_\alpha s_1 c_1}{\sqrt{1 - s_\alpha^2 c_1^2}}\,\hat{\mathbf{n}}_2\right]\frac{s_\alpha c_\alpha \dot{\theta}_1^2}{(1 - s_\alpha^2 c_1^2)^{3/2}}$$

We note that up to this point in Chapter 6 we have not mentioned velocities or accelerations. As long as the angular velocities from one body to another are all simple, we can do a considerable amount of angular velocity and angular acceleration computation merely by using the definitions and properties of $\boldsymbol{\omega}$ and $\boldsymbol{\alpha}$.

PROBLEMS ▶ Section 6.4

6.17 We know that one of the properties of the angular velocity vector is that $^{\mathcal{I}}\dot{\boldsymbol{\omega}}_{\mathcal{B}/\mathcal{I}} = {}^{\mathcal{B}}\dot{\boldsymbol{\omega}}_{\mathcal{B}/\mathcal{I}}$. Show that this is *not* a property of the angular acceleration vector $\boldsymbol{\alpha}_{\mathcal{B}/\mathcal{I}}$.

6.18 In Problem 6.6, find the angular acceleration of $\mathcal{B}$ in $\mathcal{S}$.

6.19 The components of two angular velocity vectors are shown in the following table as functions of time. The orthogonal unit vectors $(\hat{\mathbf{i}}, \hat{\mathbf{j}}, \hat{\mathbf{k}})$ are fixed in direction in frame $\mathcal{B}$. Find the angular acceleration of $\mathcal{C}$ in $\mathcal{A}$: (a) as a function of time; (b) at $t = 0$ sec; (c) at $t = 0.5$ sec. (See Figure P6.19.)

	$\hat{\mathbf{i}}$	$\hat{\mathbf{j}}$	$\hat{\mathbf{k}}$
$\omega_{\mathcal{C}/\mathcal{B}}$	$4t^2$	$2t$	6
$\omega_{\mathcal{B}/\mathcal{A}}$	$\sin t$	$\cos t$	$7t$

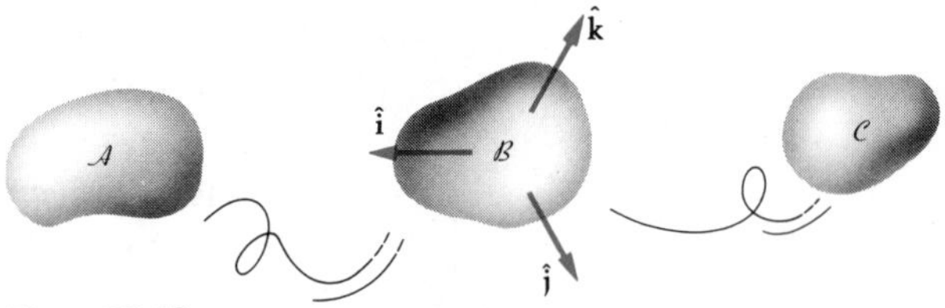

Figure P6.19

6.20 Let the angular velocity and angular acceleration vectors $\boldsymbol{\omega}_{\mathcal{C}/\mathcal{A}}$ and $\boldsymbol{\alpha}_{\mathcal{C}/\mathcal{A}}$ be expressed in terms of their components in a third frame $\mathcal{B}$:

$$\left.\begin{aligned}\boldsymbol{\omega}_{\mathcal{C}/\mathcal{A}} &= \omega_1\hat{\mathbf{i}} + \omega_2\hat{\mathbf{j}} + \omega_3\hat{\mathbf{k}}\\ \boldsymbol{\alpha}_{\mathcal{C}/\mathcal{A}} &= \alpha_1\hat{\mathbf{i}} + \alpha_2\hat{\mathbf{j}} + \alpha_3\hat{\mathbf{k}}\end{aligned}\right\}\quad \text{(The unit vectors are fixed in } \mathcal{B}\text{)}$$

Find the restriction on frame $\mathcal{B}$ for which $\alpha_i = \dot{\omega}_i$ $(i = 1, 2, 3)$.

6.21 The antenna $\mathcal{A}$ in Figure P6.21 (see Problem 6.7) is oriented with the following three rotations:

1. Azimuth, about y fixed in $\mathcal{G}$, at the rate $\omega_{\mathcal{I}_1/\mathcal{G}} = 3t^2$ rad/sec
2. Elevation, about z_1 fixed in a first intermediate frame $\mathcal{I}_1$, at the constant rate $\omega_{\mathcal{I}_2}/_{\mathcal{I}_1} = 1$ rad/sec

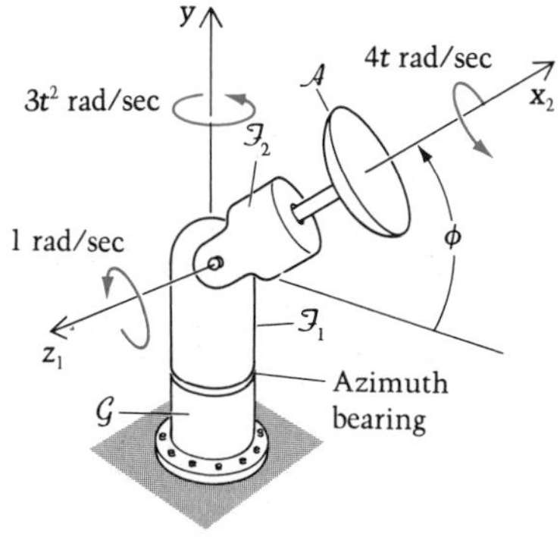

Figure P6.21

3. A polarization rotation about the antenna axis x_2 (fixed in both the second intermediate frame $\mathcal{I}_2$ and in $\mathcal{A}$) at $\omega_{\mathcal{A}/\mathcal{I}_2} = 4t$ rad/sec

If the structure is in the $\phi = 0$ position at $t = 0$, find $\boldsymbol{\alpha}_{\mathcal{A}/\mathcal{G}}$ at the time $t = \pi/2$ sec. Use unit vectors fixed in direction in $\mathcal{I}_1$.

6.22 In problem 6.9 find the angular acceleration of the capsule $\mathcal{C}$ in $\mathcal{G}$. Take ω_z and ω_x to be constants.

6.23 In Example 6.2 find the angular acceleration of the see-saw board $\mathcal{B}$ in the ground $\mathcal{G}$, if in addition to the given data $\dot{\omega}_V = 2$ rad/sec² and $\dot{\omega}_H = 1.5$ rad/sec².

6.24 See Figure P6.24. Axes x, y, and z are fixed in body $\mathcal{A}$, which rotates in $\mathcal{I}$ about the z axis with angular velocity $\omega_1\hat{\mathbf{k}}$. The arm $\mathcal{B}$, attached rigidly to $\mathcal{A}$, supports a bearing about which $\mathcal{C}$ turns with angular velocity $\omega_2\hat{\mathbf{j}}$ relative to $\mathcal{A}$. Finally, body $\mathcal{D}$ turns about the direction $\hat{\mathbf{u}}$ (which lies along axes of symmetry of both $\mathcal{C}$ and $\mathcal{D}$) with $\omega_3\hat{\mathbf{u}}$ relative to $\mathcal{C}$. If ω_1, ω_2, and ω_3 are all functions of time, find the angular acceleration of $\mathcal{D}$ in $\mathcal{I}$ at an instant when $\hat{\mathbf{u}}$ makes angles with x and z of 135° and 45°, respectively.

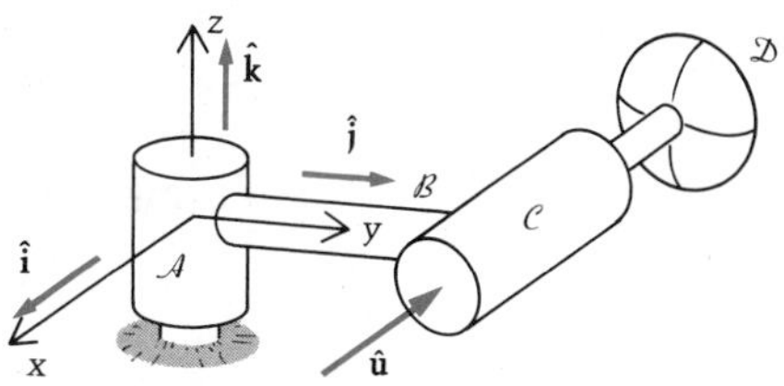

Figure P6.24

6.5 Velocity and Acceleration in Moving Frames of Reference

In certain practical situations a point is moving relative to *two* frames (or bodies) of interest. For example, a pin P may be sliding in a slot of a body $\mathcal{B}$ that is itself in motion in another frame $\mathcal{I}$ (see Figure 6.4). In problems such as these, we are often interested in the relationship between the velocities (and also the accelerations) of P in the two frames $\mathcal{B}$ and $\mathcal{I}$. We studied this problem in plane motion in Sections 3.7 and 3.8, and we now wish to expand the treatment to three dimensions.

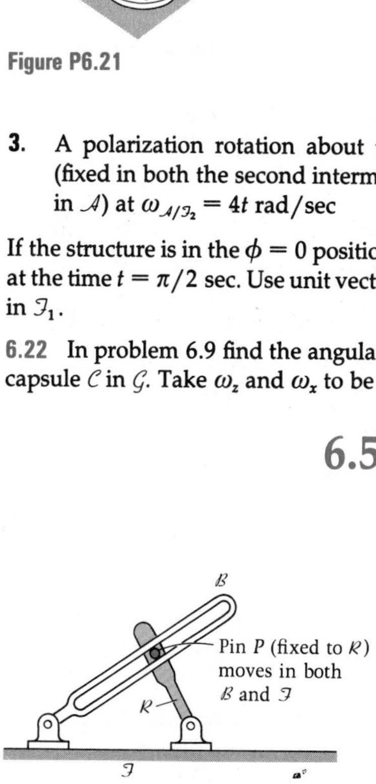

Figure 6.4 Example of a point moving relative to two frames.

The Velocity Relationship in Moving Frames

We shall arbitrarily choose $\mathcal{I}$ as a reference frame for the moving body $\mathcal{B}$, but we emphasize that both are frames and both are bodies; as long as they are considered rigid, the terms mean the same. Figure 6.5 shows the general picture.

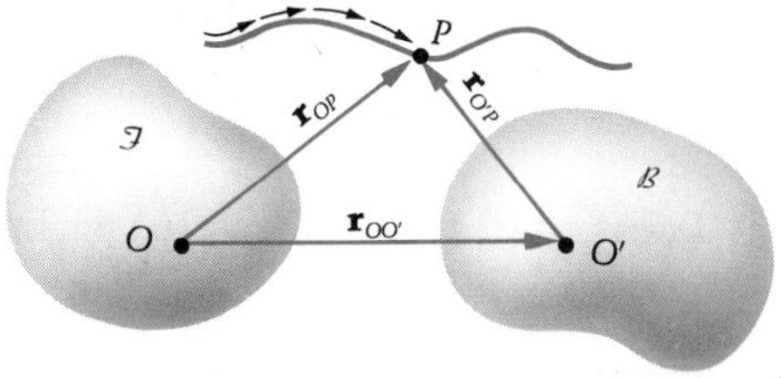

Figure 6.5 Point P moving with respect to frames $\mathcal{B}$ and $\mathcal{I}$.

Letting O and O' be fixed points of $\mathcal{I}$ and $\mathcal{B}$, respectively, we differentiate the connecting relation

$$\mathbf{r}_{OP} = \mathbf{r}_{OO'} + \mathbf{r}_{O'P} \tag{6.40}$$

in $\mathcal{I}$ and obtain*

$$^{\mathcal{I}}\dot{\mathbf{r}}_{OP} = {}^{\mathcal{I}}\dot{\mathbf{r}}_{OO'} + {}^{\mathcal{I}}\dot{\mathbf{r}}_{O'P} \tag{6.41}$$

Because O is fixed in $\mathcal{I}$, the first two of the three vectors in Equation (6.41) are the velocities in $\mathcal{I}$ of P and O':

$$\mathbf{v}_{P/\mathcal{I}} = \mathbf{v}_{O'/\mathcal{I}} + {}^{\mathcal{I}}\dot{\mathbf{r}}_{O'P} \tag{6.42}$$

The last vector in Equation (6.42) presents a problem. It is not the velocity of P in $\mathcal{I}$, because the point O' is not fixed in $\mathcal{I}$; nor is it the velocity of P in $\mathcal{B}$, because the derivative is not taken there. To overcome this dilemma, we shall rewrite the term by moving the derivative from $\mathcal{I}$ to $\mathcal{B}$ using Equation (6.20):

$$\mathbf{v}_{P/\mathcal{I}} = \mathbf{v}_{O'/\mathcal{I}} + ({}^{\mathcal{B}}\dot{\mathbf{r}}_{O'P} + \boldsymbol{\omega}_{\mathcal{B}/\mathcal{I}} \times \mathbf{r}_{O'P}) \tag{6.43}$$

$$= \mathbf{v}_{O'/\mathcal{I}} + \mathbf{v}_{P/\mathcal{B}} + \boldsymbol{\omega}_{\mathcal{B}/\mathcal{I}} \times \mathbf{r}_{O'P}$$

$$= \mathbf{v}_{P/\mathcal{B}} + (\mathbf{v}_{O'/\mathcal{I}} + \boldsymbol{\omega}_{\mathcal{B}/\mathcal{I}} \times \mathbf{r}_{O'P}) \tag{6.44}$$

We are now in a position to derive the equation relating the velocities of two points of the same rigid body from the (therefore more general) equation (6.44). Temporarily let P be a fixed point of $\mathcal{B}$; then $\mathbf{v}_{P/\mathcal{B}} = \mathbf{0}$ and

$$\mathbf{v}_{P/\mathcal{I}} = \mathbf{v}_{O'/\mathcal{I}} + \boldsymbol{\omega}_{\mathcal{B}/\mathcal{I}} \times \mathbf{r}_{O'P} \tag{6.45}$$

or

$$\mathbf{v}_P = \mathbf{v}_{O'} + \boldsymbol{\omega} \times \mathbf{r}_{O'P} \tag{6.46}$$

which is the same as the plane motion equation (3.5), except that now the $\mathbf{r}$ and $\mathbf{v}$ vectors may also have z components and the $\boldsymbol{\omega}$ vector can have x and y components in addition to z.

Returning to the general case in which P is not necessarily attached to either $\mathcal{B}$ or $\mathcal{I}$, let us denote the point of $\mathcal{B}$ (or $\mathcal{B}$ extended) coincident with P by $P_{\mathcal{B}}$. Then Equation (6.44) becomes

$$\mathbf{v}_{P/\mathcal{I}} = \mathbf{v}_{P/\mathcal{B}} + \mathbf{v}_{P_{\mathcal{B}}/\mathcal{I}} \tag{6.47}$$

In words, we may restate Equation (6.47) as follows:

$$\begin{bmatrix}\text{Velocity of}\\ P \text{ in } \mathcal{I}\end{bmatrix} = \begin{bmatrix}\text{Velocity of}\\ P \text{ in } \mathcal{B}\end{bmatrix} + \begin{bmatrix}\text{Velocity in } \mathcal{I} \text{ of}\\ \text{the fixed point of}\\ \mathcal{B} \text{ coincident with } P\end{bmatrix}$$

$$\mathbf{v}_{P/\mathcal{I}} \quad = \quad \mathbf{v}_{P/\mathcal{B}} \quad + \quad \mathbf{v}_{P_{\mathcal{B}}/\mathcal{I}}$$

Equation (6.47) has the virtue of compactness. However, it is a less con-

* The superscripts in Equation (6.41) are now necessary to denote the frame in which the derivative is taken.

venient form than is (6.44) for differentiation to produce a corresponding relationship of accelerations.

Question 6.7 Why?

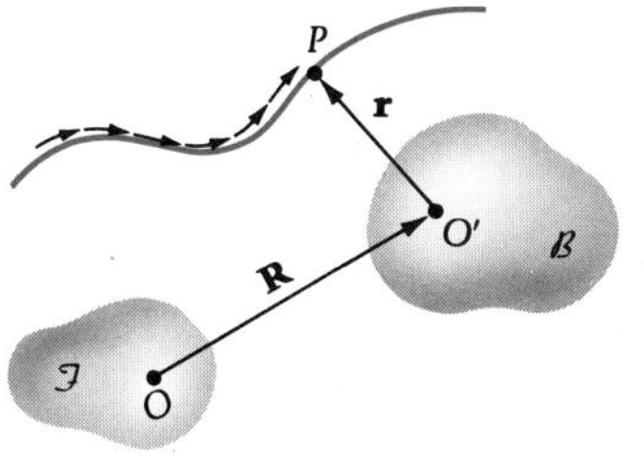

Figure 6.6

Another common alternative means of stating Equation (6.44) is to view body $\mathcal{B}$ as a moving frame — that is, as a body moving relative to another frame $\mathcal{J}$. (See Figure 6.6.) We may then rewrite Equation (6.44) as

$$\mathbf{v}_P = \dot{\mathbf{R}} + \mathbf{v}_{\text{rel}} + \boldsymbol{\omega} \times \mathbf{r} \tag{6.48}$$

in which

$$\begin{aligned} \mathbf{v}_P &= \text{velocity of } P \text{ in reference frame } \mathcal{J} \\ \dot{\mathbf{R}} &= \text{velocity of moving origin} = \mathbf{v}_{O'/\mathcal{J}} \\ \mathbf{v}_{\text{rel}} &= \text{velocity of } P \text{ in moving frame} = \mathbf{v}_{P/\mathcal{B}} \\ \boldsymbol{\omega} &= \text{angular velocity of moving frame} = \boldsymbol{\omega}_{\mathcal{B}/\mathcal{J}} \\ \mathbf{r} &= \text{position vector of } P \text{ in moving frame} = \mathbf{r}_{O'P} \end{aligned}$$

We now consider two examples involving the use of the velocity equation for moving frames (Equation 6.44, 6.47, or 6.48).

EXAMPLE 6.6

Find the velocity in $\mathcal{G}$ of point A at the bottom of the disk in Example 6.1. (See Figure E6.6.)

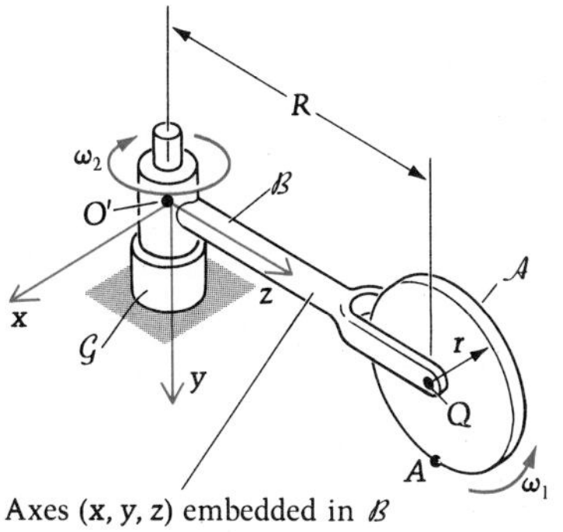

Axes (x, y, z) embedded in $\mathcal{B}$
Figure E6.6

Solution

We select $\mathcal{B}$ as the moving frame, and Equation (6.44) gives the following (where $\mathcal{G}$ is $\mathcal{J}$ and A is P):

$$\mathbf{v}_{A/\mathcal{G}} = \mathbf{v}_{A/\mathcal{B}} + \mathbf{v}_{O'/\mathcal{G}} + \boldsymbol{\omega}_{\mathcal{B}/\mathcal{G}} \times \mathbf{r}_{O'A}$$

In this case $\mathbf{v}_{O'/\mathcal{G}} = \mathbf{0}$; also $\boldsymbol{\omega}_{\mathcal{B}/\mathcal{G}} = \omega_2\hat{\mathbf{j}}$ from the previous example. The velocity of A in $\mathcal{B}$ is given by

$$\mathbf{v}_{A/\mathcal{B}} = \overset{\mathbf{0}}{\cancel{\mathbf{v}_{Q/\mathcal{B}}}} + \omega_1\hat{\mathbf{i}} \times r\hat{\mathbf{j}} = r\omega_1\hat{\mathbf{k}}$$

Therefore

$$\mathbf{v}_{A/\mathcal{G}} = r\omega_1\hat{\mathbf{k}} + \omega_2\hat{\mathbf{j}} \times (r\hat{\mathbf{j}} + R\hat{\mathbf{k}}) = R\omega_2\hat{\mathbf{i}} + r\omega_1\hat{\mathbf{k}}$$

Answer 6.7 If we differentiate Equation (6.47) in $\mathcal{J}$, we encounter the term ${}^{\mathcal{J}}\dot{\mathbf{v}}_{P_g/\mathcal{J}} (\neq \mathbf{a}_{P_g/\mathcal{J}})$, which is not recognizable as a standard kinematic quantity; this is because P_g denotes a *succession* of points of $\mathcal{B}$ which are at each instant coincident with P.

EXAMPLE 6.7

Crank $\mathcal{C}$ in Figure E6.7 rotates about axis z through point O. Its other end, Q, is attached to a ball and socket joint as shown. The ball forms the end of rod $\mathcal{B}$, which passes through a hole in the ceiling $\mathcal{J}$. find the velocity of point P of the bar, which is passing through the hole when $\theta = 90°$ as shown.

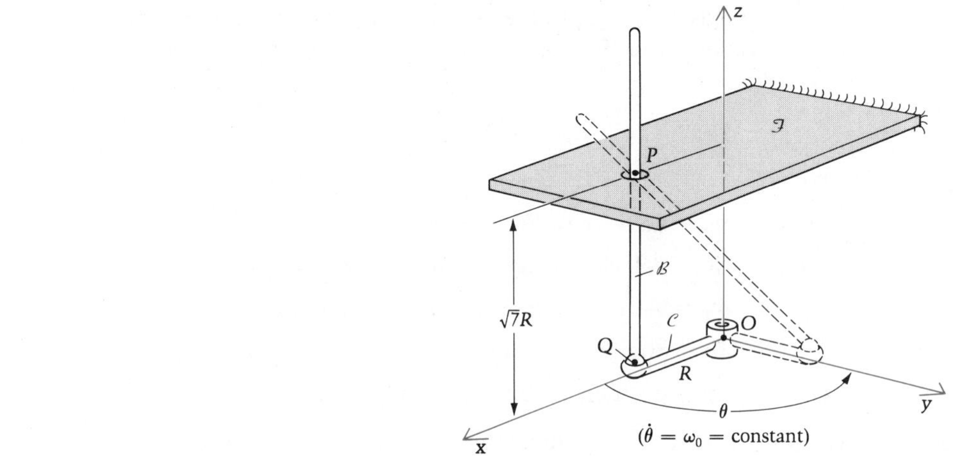

Figure E6.7

Solution

We denote by H the point of $\mathcal{J}$ at the center of the hole; then:

$$\overset{\mathbf{0}}{\mathbf{v}_{H/\mathcal{J}}} = \mathbf{v}_{H/\mathcal{B}} + \mathbf{v}_{P/\mathcal{J}} \tag{1}$$

in which P is the point of $\mathcal{B}$ coincident with point H. Knowing that the motion of H in $\mathcal{B}$ must be *along* the axis of $\mathcal{B}$, we have:

$$\mathbf{0} = v_{H/\mathcal{B}}\left(\frac{R\hat{\mathbf{i}} - R\hat{\mathbf{j}} + \sqrt{7}\,R\hat{\mathbf{k}}}{3R}\right) + \overbrace{(\mathbf{v}_{Q/\mathcal{J}} + \boldsymbol{\omega}_{\mathcal{B}/\mathcal{J}} \times \mathbf{r}_{QP})}^{\mathbf{v}_{P/\mathcal{J}}} \tag{2}$$

or

$$\mathbf{0} = v_{H/\mathcal{B}}\left(\frac{\hat{\mathbf{i}} - \hat{\mathbf{j}} + \sqrt{7}\hat{\mathbf{k}}}{3}\right)$$

$$+ [-R\omega_0\hat{\mathbf{i}} + (\omega_x\hat{\mathbf{i}} + \omega_y\hat{\mathbf{j}} + \omega_z\hat{\mathbf{k}}) \times (R\hat{\mathbf{i}} - R\hat{\mathbf{j}} + \sqrt{7}R\hat{\mathbf{k}})] \tag{3}$$

Collecting the coefficients of $\hat{\mathbf{i}}$, $\hat{\mathbf{j}}$, and $\hat{\mathbf{k}}$, respectively, we find

$$\frac{1}{3}v_{H/\mathcal{B}} + \sqrt{7}R\omega_y + R\omega_z = R\omega_O \tag{4}$$

$$-\frac{1}{3}v_{H/\mathcal{B}} + R\omega_z - \sqrt{7}R\omega_x = 0 \tag{5}$$

$$\frac{\sqrt{7}}{3}v_{H/\mathcal{B}} - R\omega_x - R\omega_y = 0 \tag{6}$$

These three equations (4–6) obviously cannot be solved for unique values of all four unknowns ω_x, ω_y, ω_z, $v_{H/\mathcal{B}}$. However, an answer for $v_{H/\mathcal{B}}$ is obtainable by subtracting Equation (5) from (4) and then adding $\sqrt{7}$ times Equation (6). The result is

$$v_{H/\mathcal{B}} = \frac{R\omega_0}{3}$$

Substituting this result back into Equations (4–6) gives three equations in ω_x, ω_y, and ω_z whose coefficient matrix is singular (has a zero determinant). Thus they cannot be solved for the angular velocity components. A more physical reason for this is that the component of $\boldsymbol{\omega}_{\mathcal{B}/\mathcal{I}}$ along bar $\mathcal{B}$ cannot affect the answer for $\mathbf{v}_{P/\mathcal{I}}$ because $\mathcal{B}$ can turn freely in its socket about its axis without altering $\mathbf{v}_{P/\mathcal{I}}$. Mathematically this is manifested by this "axial" component of $\boldsymbol{\omega}_{\mathcal{B}/\mathcal{I}}$ being parallel to $\mathbf{r}_{QP}$ and thus canceling out of Equation (2). These $\boldsymbol{\omega}$ components are not needed for a solution, however, because from Equation (1) we can obtain our desired result:

$$\mathbf{v}_{P/\mathcal{I}} = -\mathbf{v}_{H/\mathcal{B}} = \frac{-R\omega_0}{3}\left(\frac{\hat{\mathbf{i}} - \hat{\mathbf{j}} + \sqrt{7}\hat{\mathbf{k}}}{3}\right)$$

An added note is that although we cannot solve for the component of $\boldsymbol{\omega}_{\mathcal{B}/\mathcal{I}}$ along $\mathcal{B}$ with the given information, we *are* able to calculate the component of $\boldsymbol{\omega}_{\mathcal{B}/\mathcal{I}}$ normal to $\mathcal{B}$. It will be made up of values ω_x', ω_y', and ω_z' that enforce the relationship

$$\boldsymbol{\omega}' \cdot \mathbf{r}_{QP} = 0$$

That is,

$$R(\omega_x' - \omega_y' + \sqrt{7}\omega_z') = 0 \tag{7}$$

This equation states that these $\boldsymbol{\omega}'$ components form a vector normal to the line QP; again, this vector is the only part of $\boldsymbol{\omega}$ that can affect $v_{P/\mathcal{I}}$.

After adding primes to ω_x, ω_y, ω_z in Equations (4–6), the solution of Equations (4–7), as the reader may verify, is

$$\omega_x' = 0 \qquad \omega_y' = \frac{\sqrt{7}}{9}\omega_0 \qquad \omega_z' = \frac{\omega_0}{9} \qquad v_{H/\mathcal{B}} = \frac{R\omega_0}{3}$$

And we may now calculate the velocity of P from Equation (1) as before or from Equation (2) as follows:

$$\begin{aligned}
\mathbf{v}_{P/\mathcal{I}} &= \mathbf{v}_{Q/\mathcal{I}} + \boldsymbol{\omega}_{\mathcal{B}/\mathcal{I}} \times \mathbf{r}_{QP} \\
&= -R\omega_0\hat{\mathbf{i}} + \omega_0\left(\frac{\sqrt{7}\hat{\mathbf{j}} + \hat{\mathbf{k}}}{9}\right) \times R(\hat{\mathbf{i}} - \hat{\mathbf{j}} + \sqrt{7}\hat{\mathbf{k}}) \\
&= \frac{R\omega_0}{9}(-\hat{\mathbf{i}} + \hat{\mathbf{j}} - \sqrt{7}\hat{\mathbf{k}}) \qquad \text{(as before)}
\end{aligned}$$

The Acceleration Relationship in Moving Frames/Coriolis Acceleration

After these examples we are now ready to derive the corresponding relationship between the *accelerations* of P in two frames $\mathcal{B}$ and $\mathcal{I}$. Differentiating Equation (6.44) gives

$${}^{\mathcal{I}}\dot{\mathbf{v}}_{P/\mathcal{I}} = {}^{\mathcal{I}}\dot{\mathbf{v}}_{O'/\mathcal{I}} + {}^{\mathcal{I}}\dot{\boldsymbol{\omega}}_{\mathcal{B}/\mathcal{I}} \times \mathbf{r}_{O'P} + \boldsymbol{\omega}_{\mathcal{B}/\mathcal{I}} \times {}^{\mathcal{I}}\dot{\mathbf{r}}_{O'P} + {}^{\mathcal{I}}\dot{\mathbf{v}}_{P/\mathcal{B}} \tag{6.49}$$

or, again using Equation (6.20) (once in each of the last two terms), we get

$$\mathbf{a}_{P/\mathcal{I}} = \mathbf{a}_{O'/\mathcal{I}} + \boldsymbol{\alpha}_{\mathcal{B}/\mathcal{I}} \times \mathbf{r}_{O'P} + \boldsymbol{\omega}_{\mathcal{B}/\mathcal{I}} \times ({}^{\mathcal{B}}\dot{\mathbf{r}}_{O'P} + \boldsymbol{\omega}_{\mathcal{B}/\mathcal{I}} \times \mathbf{r}_{O'P}) + ({}^{\mathcal{B}}\dot{\mathbf{v}}_{P/\mathcal{B}} + \boldsymbol{\omega}_{\mathcal{B}/\mathcal{I}} \times {}^{\mathcal{B}}\dot{\mathbf{r}}_{O'P}) \quad (6.50)$$

Rearranging the terms, we have

$$\mathbf{a}_{P/\mathcal{I}} = \underbrace{{}^{\mathcal{B}}\dot{\mathbf{v}}_{P/\mathcal{B}}}_{\mathbf{a}_{P/\mathcal{B}}} + \underbrace{\mathbf{a}_{O'/\mathcal{I}} + \boldsymbol{\alpha}_{\mathcal{B}/\mathcal{I}} \times \mathbf{r}_{O'P} + \boldsymbol{\omega}_{\mathcal{B}/\mathcal{I}} \times (\boldsymbol{\omega}_{\mathcal{B}/\mathcal{I}} \times \mathbf{r}_{O'P})}_{\mathbf{a}_{P_{\mathcal{B}}/\mathcal{I}}} + 2\boldsymbol{\omega}_{\mathcal{B}/\mathcal{I}} \times {}^{\mathcal{B}}\dot{\mathbf{r}}_{O'P} \quad (6.51)$$

$$\mathbf{a}_{P/\mathcal{I}} = \mathbf{a}_{P/\mathcal{B}} + \mathbf{a}_{P_{\mathcal{B}}/\mathcal{I}} + 2\boldsymbol{\omega}_{\mathcal{B}/\mathcal{I}} \times \mathbf{v}_{P/\mathcal{B}} \quad (6.52)$$

The middle three terms on the right side of Equation (6.51) make up the acceleration of the point $P_{\mathcal{B}}$ of $\mathcal{B}$ (or $\mathcal{B}$ extended) coincident with P. (The proof is brief: If P is *fixed* to $\mathcal{B}$ at point $P_{\mathcal{B}}$, then the other two terms vanish since $\mathbf{r}_{O'P}$ becomes a constant vector in $\mathcal{B}$, and what remains is necessarily $\mathbf{a}_{P_{\mathcal{B}}/\mathcal{I}}$.) The term ${}^{\mathcal{B}}\dot{\mathbf{v}}_{P/\mathcal{B}}$ (which is ${}^{\mathcal{B}}\ddot{\mathbf{r}}_{O'P}$, with both derivatives taken in $\mathcal{B}$) is clearly the acceleration of P in $\mathcal{B}$. The last term, $2\boldsymbol{\omega}_{\mathcal{B}/\mathcal{I}} \times \mathbf{v}_{P/\mathcal{B}}$, is called the **Coriolis acceleration** of P. Note that due to the presence of the Coriolis acceleration it is *not true* that the acceleration of P in frame $\mathcal{I}$ is its acceleration in $\mathcal{B}$ plus the acceleration of the point of $\mathcal{B}$ with which it is coincident (as was in fact the case with the velocity of P). This result is interestingly analogous to the fact that the addition theorem for angular velocity is not true for angular acceleration.

As we did for the velocity equation, we now restate Equation (6.52) in words:

$$\begin{bmatrix}\text{Acceleration}\\ \text{of } P \text{ in } \mathcal{I}\end{bmatrix} = \begin{bmatrix}\text{Acceleration}\\ \text{of } P \text{ in } \mathcal{B}\end{bmatrix} + \begin{bmatrix}\text{Acceleration in } \mathcal{I} \text{ of}\\ \text{the fixed point of } \mathcal{B}\\ \text{coincident with } P\end{bmatrix} + \begin{bmatrix}\text{Coriolis}\\ \text{acceleration}\end{bmatrix}$$

$$\mathbf{a}_{P/\mathcal{I}} \quad = \quad \mathbf{a}_{P/\mathcal{B}} \quad + \quad \mathbf{a}_{P_{\mathcal{B}}/\mathcal{I}} \quad + \quad 2\boldsymbol{\omega}_{\mathcal{B}/\mathcal{I}} \times \mathbf{v}_{P/\mathcal{B}}$$

In the abbreviated notation of Equation (6.48) we have

$$\mathbf{a}_P = \ddot{\mathbf{R}} + \boldsymbol{\alpha} \times \mathbf{r} + \boldsymbol{\omega} \times (\boldsymbol{\omega} \times \mathbf{r}) + 2\boldsymbol{\omega} \times \mathbf{v}_{\text{rel}} + \mathbf{a}_{\text{rel}} \quad (6.53)$$

in which

$\mathbf{a}_P$ = acceleration of P in reference frame $\mathcal{I}$
$\ddot{\mathbf{R}}$ = acceleration of moving origin = $\mathbf{a}_{O'/\mathcal{I}}$
$\boldsymbol{\alpha}$ = angular acceleration of moving frame = $\boldsymbol{\alpha}_{\mathcal{B}/\mathcal{I}}$
$\mathbf{a}_{\text{rel}}$ = acceleration of P in moving frame = $\mathbf{a}_{P/\mathcal{B}}$

and in which all other terms in Equation (6.53) are defined directly after Equation (6.48).

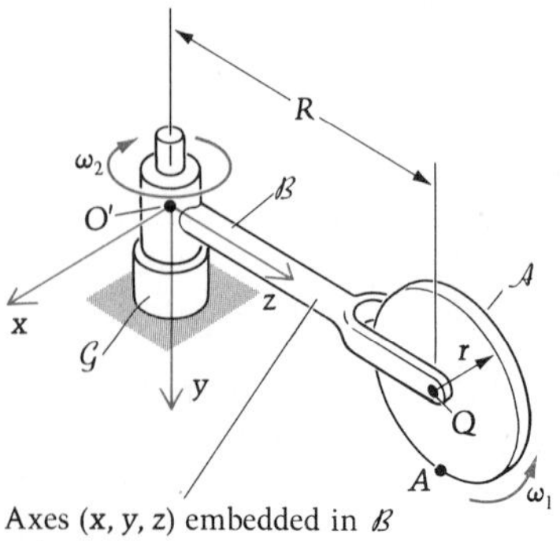

Axes (x, y, z) embedded in $\mathcal{B}$
Figure E6.8

EXAMPLE 6.8

Compute the acceleration in $\mathcal{G}$ of point A in Examples 6.1, 6.4, and 6.6 (see Figure E6.8).

Solution

We shall use Equation (6.51): $\mathcal{B}$ is again the moving frame; the reference frame $\mathcal{I}$ is $\mathcal{G}$, and the moving point P is A:

$$\mathbf{a}_{A/\mathcal{G}} = \mathbf{a}_{A/\mathcal{B}} + \mathbf{a}_{O'/\mathcal{G}} + \boldsymbol{\alpha}_{\mathcal{B}/\mathcal{G}} \times \mathbf{r}_{O'A} + \boldsymbol{\omega}_{\mathcal{B}/\mathcal{G}} \times (\boldsymbol{\omega}_{\mathcal{B}/\mathcal{G}} \times \mathbf{r}_{O'A}) + 2\boldsymbol{\omega}_{\mathcal{B}/\mathcal{G}} \times \mathbf{v}_{A/\mathcal{B}}$$

The various terms on the right side are calculated as follows:

1. $\mathbf{a}_{A/\mathcal{B}} = \overset{\mathbf{0}}{\cancel{\mathbf{a}_{Q/\mathcal{B}}}} + \overset{\mathbf{0}}{\cancel{\boldsymbol{\alpha}_{\mathcal{A}/\mathcal{B}}}} \times \mathbf{r}_{QA} - \underbrace{\omega_1^2 \mathbf{r}_{QA}}$

   Note that $\mathcal{A}$ is in plane motion relative to $\mathcal{B}$, so that this short form is acceptable.

   $= -r\omega_1^2\hat{\mathbf{j}}$ (A moves on a circle at constant speed in $\mathcal{B}$)
2. $\mathbf{a}_{O'/\mathcal{G}} = \mathbf{0}$. Note that O' is fixed in $\mathcal{G}$ in this example.
3. $\boldsymbol{\alpha}_{\mathcal{B}/\mathcal{G}} \times \mathbf{r}_{O'A} = \mathbf{0}$. Note that $\boldsymbol{\alpha}_{\mathcal{B}/\mathcal{G}} = {}^{\mathcal{G}}(d/dt)(\omega_2\hat{\mathbf{j}}) = \mathbf{0}$ since ω_2 is a constant, and $\hat{\mathbf{j}}$ does not change in direction in $\mathcal{G}$.
4. $\boldsymbol{\omega}_{\mathcal{B}/\mathcal{G}} \times (\boldsymbol{\omega}_{\mathcal{B}/\mathcal{G}} \times \mathbf{r}_{O'A}) = \omega_2\hat{\mathbf{j}} \times [\omega_2\hat{\mathbf{j}} \times (r\hat{\mathbf{j}} + R\hat{\mathbf{k}})] = -R\omega_2^2\hat{\mathbf{k}}$.
5. $2\boldsymbol{\omega}_{\mathcal{B}/\mathcal{G}} \times \mathbf{v}_{A/\mathcal{B}} = 2\omega_2\hat{\mathbf{j}} \times (r\omega_1\mathbf{k}) = 2r\omega_1\omega_2\hat{\mathbf{i}}$.

Thus the answer for the acceleration of point A is

$$\mathbf{a}_{A/\mathcal{G}} = 2r\omega_1\omega_2\hat{\mathbf{i}} - r\omega_1^2\hat{\mathbf{j}} - R\omega_2^2\hat{\mathbf{k}}$$

In Section 6.7, we shall rework the above example using another approach.

PROBLEMS ▶ Section 6.5

6.25 In Example 6.2, the children are each 5 ft from the fulcrum O. At a later time the girl slides a stone P along the see-saw board toward the boy. The velocity of the stone relative to the board is $-4\hat{\mathbf{j}}$ ft/sec at an instant when (a) the stone is 2 ft from the girl; (b) $\theta = 0$; (c) $\omega_V = 3$ rad/sec; and (d) $\omega_H = 2$ rad/sec. Find $\mathbf{v}_{P/\mathcal{I}}$ and $\mathbf{v}_{P/\mathcal{G}}$ at this instant.

6.26 In Problem 6.25, if at the given instant $\dot{\omega}_V = 0$, $\dot{\omega}_H = 1.5$ rad/sec^2, and the stone's acceleration relative to the board is $0.8\,\hat{\mathbf{j}}$ ft/sec^2, find $\mathbf{a}_{P/\mathcal{I}}$ and $\mathbf{a}_{P/\mathcal{G}}$.

6.27 The large disk $\mathcal{A}$ in Figure P6.27 rotates at 10 rad/sec counterclockwise (looking down on its horizontal surface). A small disk $\mathcal{B}$ rolls radially outward along a radius OD of $\mathcal{A}$. At the instant shown, the center C of $\mathcal{B}$ is 4 ft from the axis of rotation of $\mathcal{A}$, and this distance is increasing at the constant rate of 2 ft/sec. Determine the velocity and acceleration of point E, which is at the top of $\mathcal{B}$ at the given instant.

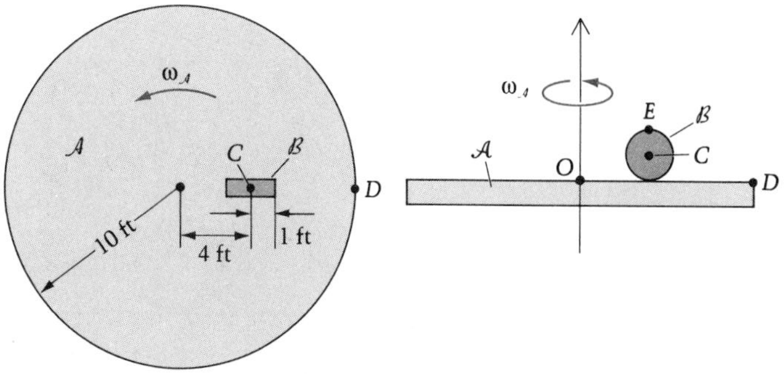

Figure P6.27

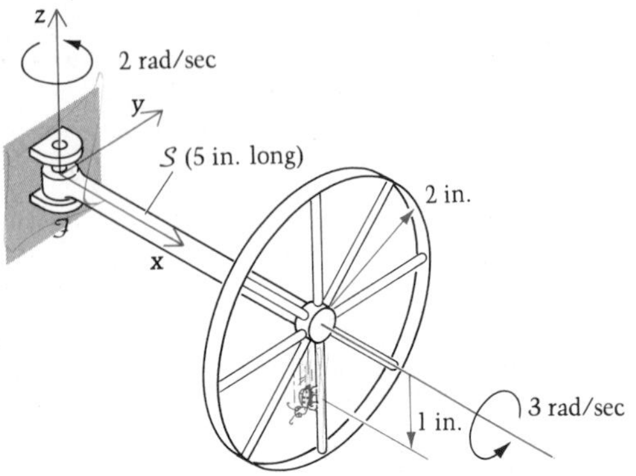

Figure P6.28

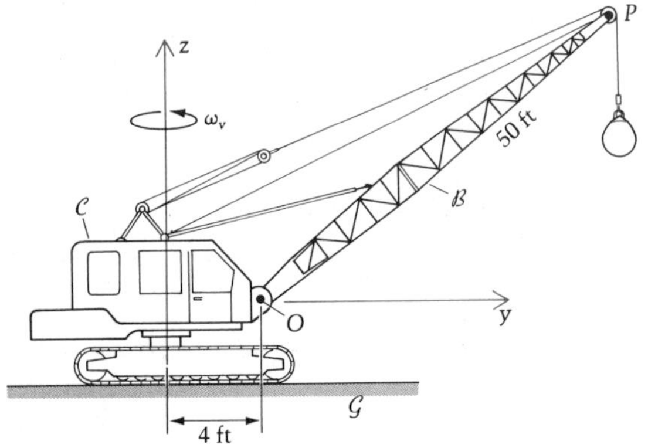

Figure P6.30

6.28 Shaft $\mathcal{S}$ in Figure P6.28 turns in the clevis $\mathcal{J}$ at 2 rad/sec in the direction shown. The wheel simultaneously rotates at 3 rad/sec about its axis as indicated. Both rates are constants. The bug is crawling outward on a spoke at 0.2 in./sec with an acceleration of 0.1 in./sec^2 both relative to the spoke. At the instant shown, find: (a) the angular velocity of the wheel; (b) the velocity of the bug.

6.29 Find the angular acceleration of the wheel and the acceleration of the bug in Problem 6.28.

* **6.30** The crane $\mathcal{C}$ in Figure P6.30 turns about the vertical at $\omega_v = 0.2$ rad/sec = constant, and simultaneously its boom $\mathcal{B}$ is being elevated at the increasing rate $\omega_H = 0.1t$ rad/sec. The (x, y, z) axes are fixed to the crane $\mathcal{C}$ at O, and the boom was along the y axis when t was zero. When the boom makes a 60° angle with the horizontal, find: (a) $\boldsymbol{\omega}_{\mathcal{B}/\mathcal{G}}$; (b) $\boldsymbol{\alpha}_{\mathcal{B}/\mathcal{G}}$, (c) $\mathbf{v}_{P/\mathcal{G}}$; (d) $\mathbf{a}_{P/\mathcal{G}}$.

6.31 A platform translates past a turntable at 6 mph. (See Figure P6.31.) People step onto the turntable and walk straight toward the center where they exit onto stairs. There is rolling contact between platform and turntable. Suppose the people walk at the constant rate of approximately 3 mph relative to the turntable. If it is desired that they do not experience more than 3 ft/sec^2 of lateral acceleration, find the required turntable radius.

6.32 The truck in Figure P6.32 moves to the left at a constant speed of 7.07 ft/sec. At the instant shown, the loading compartment has an angular speed $\dot{\theta} = \frac{1}{3}$ rad/sec and an angular acceleration of $\theta = -\frac{1}{15}$ rad/sec^2. The cylinder shown on the truck bed comes loose and rolls

Figure P6.31

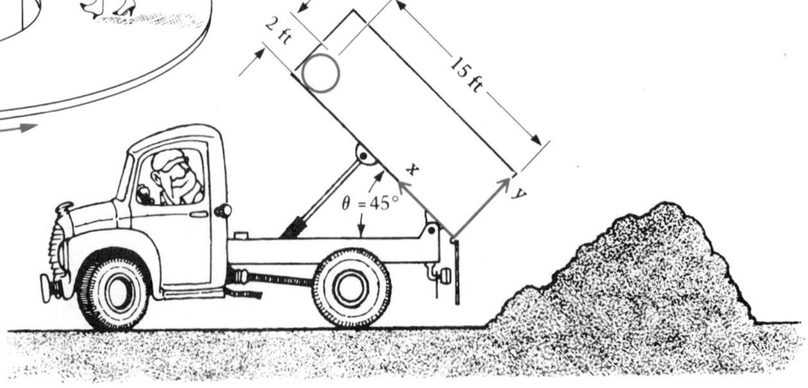

Figure P6.32

toward the ground at an angular velocity, relative to the compartment, of 1 ⟲ rad/sec at this instant; it is speeding up at a rate of $\frac{1}{2}$ rad/sec^2, also relative to the compartment. Find the velocity and acceleration of the center of the cylinder relative to the ground. Use the rotating reference frame shown.

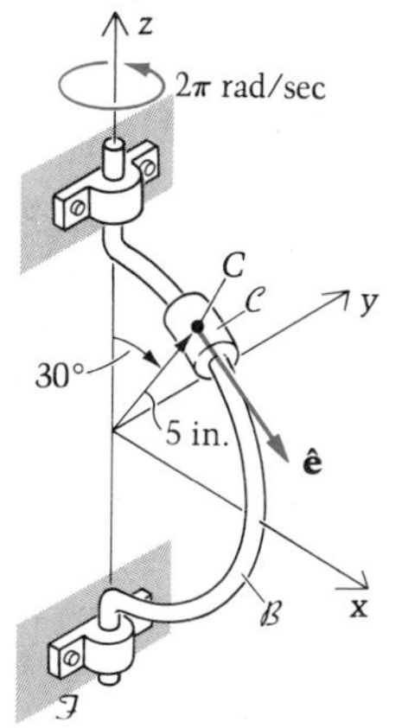

Figure P6.33

6.33 The bent bar $\mathcal{B}$ in Figure P6.33 revolves about the vertical at $2\pi\hat{\mathbf{k}}$ rad/sec. The center C of the collar $\mathcal{C}$ has velocity and acceleration relative to $\mathcal{B}$ of $20\hat{\mathbf{e}}$ in./sec and $-10\hat{\mathbf{e}}$ in./sec^2, respectively, where $\hat{\mathbf{e}}$ is in the direction of the velocity of C in $\mathcal{B}$. At the given instant, find the velocity and acceleration of C in the frame $\mathcal{I}$ in which $\mathcal{B}$ turns.

6.34 A centrifugal pump $\mathcal{P}$ turns at 500 rpm in $\mathcal{I}$, and the water particles have respective tangential velocity and acceleration components *relative to the blades* of 120 ft/sec and 80 ft/sec^2 outward when they reach the outermost point of their blades. (See Figure P6.34.) At the instant before exit determine the velocity and acceleration vectors of the water particle at P with respect to the ground $\mathcal{I}$.

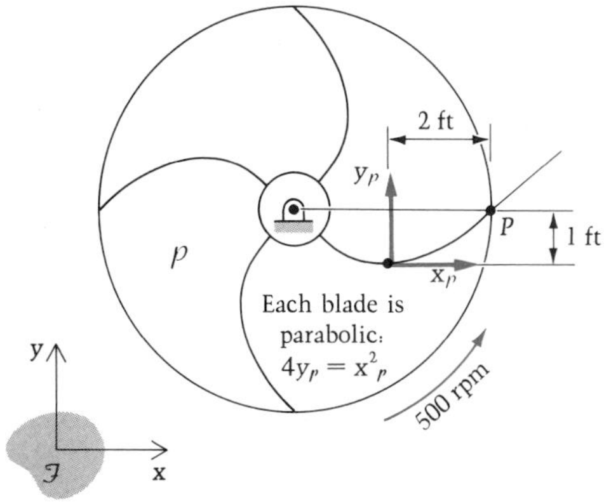

Figure P6.34

6.35 A man walks dizzily outward along a sine wave fixed to a merry go-round that is 40 ft in diameter and turns at 10 rpm. (See Figure P6.35.) If the man's speed relative to the turntable is a constant 2 ft/sec, what is the magnitude of his acceleration when he is 10 ft from the center?

6.36 Disk $\mathcal{D}$ rotates about its axis at the constant angular speed $\omega_0 = 0.1$ rad/s. The round wire is rigidly affixed to $\mathcal{D}$ at points A and B as shown in Figure P6.36. A bug crawls around the wire from A to B; its speed relative to the wire (initially zero) is always increasing at the constant rate of 0.001 m/s^2. Find the velocity and acceleration of the bug, relative to the reference frame $\mathcal{I}$ in which the disk turns, when it arrives at B.

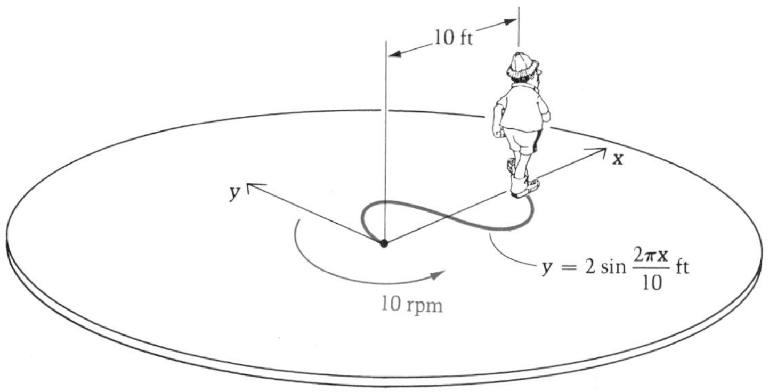

Figure P6.35

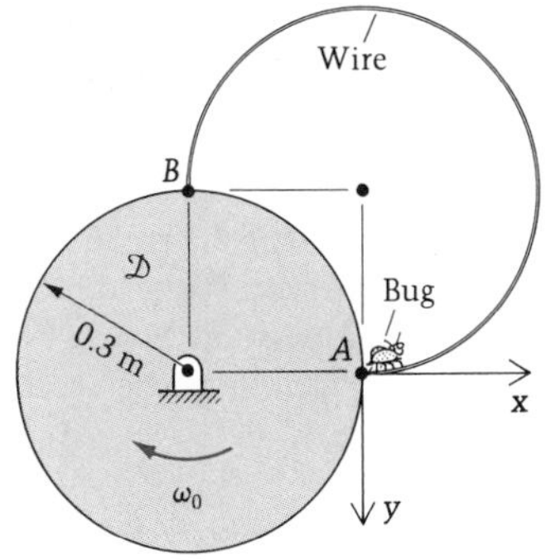

Figure P6.36

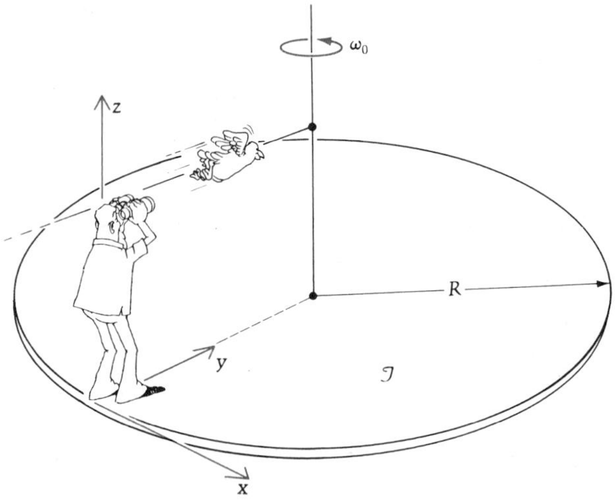

Figure P6.37

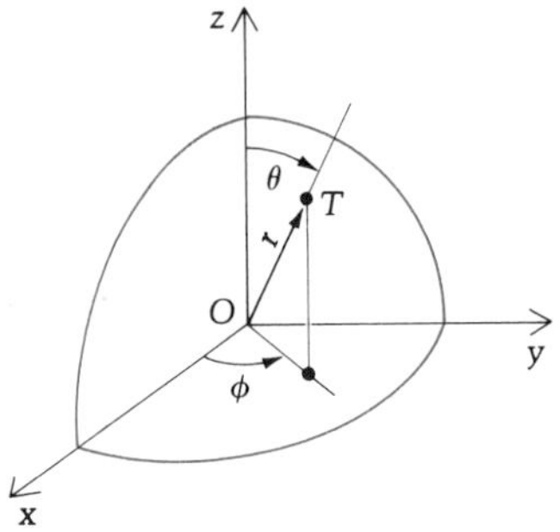

Figure P6.38

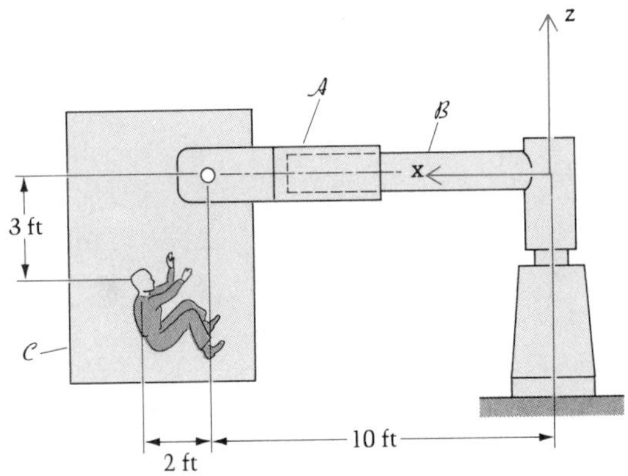

Figure P6.39

6.37 A bird flies horizontally past a man's head in a straight line at constant speed v_0 toward the axis of a turntable on which the man is standing. The axes xy have origin at the man's feet; the z axis is vertical and y always points toward the center of the turntable, which has radius R and angular velocity $\boldsymbol{\omega} = \omega_0\hat{\mathbf{k}}$. Derive $x(t)$ and $y(t)$ of the bird in terms of R, v_0, ω_0, and the time t. (See Figure P6.37.) Use Equation (6.47), integrate, then check your results by inspection.

* **6.38** In Figure P6.38 the axes x and y, and the origin O, are fixed on the deck of a ship. The ship $\mathcal{S}$ has an angular velocity relative to the earth $\mathcal{E}$ of

$$\boldsymbol{\omega}_{\mathcal{S}/\mathcal{E}} = \omega_r\hat{\mathbf{i}} + \omega_P\hat{\mathbf{j}}$$

where x and y are fore-and-aft and athwartships axes, respectively; thus ω_r is a rolling component and ω_P a pitching component of angular velocity. Point T is a target fixed relative to earth, such as a geosynchronous satellite. Find the angular rates $\dot{\theta}$ and $\dot{\phi}$, in terms of θ, ϕ, ω_r, and ω_P, that are required to track point T.

* **6.39** In Problems 6.9 and 6.22, let $\omega_z = 0.5$ rad/sec and $\omega_x = 0.7$ rad/sec in the directions indicated. With the dimensions given in Figure P6.39, find the maximum value of ω_y for which the acceleration magnitude of the astronaut's head will not exceed 5g at the given instant. Let $\omega_y =$ constant.

* **6.40** In Problem 6.16 find the velocity of point P at the tip of the gripper $\mathcal{G}$ at the instant given.

* **6.41** Prove that if two bodies are in rolling contact, the shaded arcs in Figure P6.41, representing the loci of former contact points, are equal in length. (Note from Problem 3.105 that the converse is not true.)

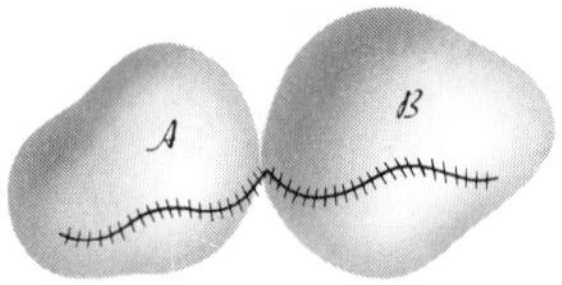

Figure P6.41

6.6 The Earth as a Moving Frame

In this section, we shall make use of Equation (6.53) to set up the differential equation governing the position of the mass center C of a body $\mathcal{B}$ that is in motion near the earth. This equation will allow us to measure the position of C with respect to a desired site O' at latitude λ, which is itself in motion as the earth turns on its axis from west to east. We assume that for describing certain motions near the earth, a frame $\mathcal{I}$ with origin at the earth's mass center O is "sufficiently fixed" to be justifiably called inertial. The frame $\mathcal{I}$ moves as does the (assumed rigid) earth, except that it does not share the earth's daily spin. Thus the site O' has an acceleration $\ddot{\mathbf{R}}$ in $\mathcal{I}$, directed toward the earth's north-south polar axis.

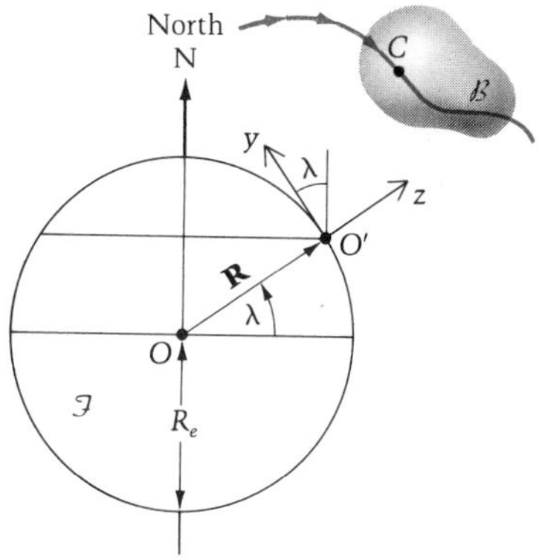

Figure 6.7

We set up the moving frame $\mathcal{F}$ as shown as Figure 6.7. The frame $\mathcal{F}$ is the turning earth, and the (x, y, z) axes are embedded in it at O' with x pointing east, y north, and z in the direction of local vertical. The acceleration of the mass center C of a body $\mathcal{B}$, moving near the earth and whose position is desired relative to O', is known from Equation (6.53) to be

$$\mathbf{a}_{C/\mathcal{I}} = \ddot{\mathbf{R}} + \boldsymbol{\alpha} \times \mathbf{r} + \boldsymbol{\omega} \times (\boldsymbol{\omega} \times \mathbf{r}) + 2\boldsymbol{\omega} \times \mathbf{v}_{\text{rel}} + \mathbf{a}_{\text{rel}}$$

where $\mathbf{r}$ is the position vector of C in $\mathcal{F}$, and $\mathbf{v}_{\text{rel}}$ and $\mathbf{a}_{\text{rel}}$ are the velocity and acceleration vectors of C in $\mathcal{F}$. Further, $\boldsymbol{\omega} = \boldsymbol{\omega}_{\mathcal{F}/\mathcal{I}}$ and $\boldsymbol{\alpha} = \boldsymbol{\alpha}_{\mathcal{F}/\mathcal{I}}$. We now use the mass-center equation of motion from Chapter 2:

$$\Sigma\mathbf{F} = m\mathbf{a}_{C/\mathcal{I}}$$

to obtain

$$\mathbf{F} - mg\hat{\mathbf{k}} = m[\ddot{\mathbf{R}} + \boldsymbol{\alpha} \times \mathbf{r} + \boldsymbol{\omega} \times (\boldsymbol{\omega} \times \mathbf{r}) + 2\boldsymbol{\omega} \times \mathbf{v}_{\text{rel}} + \mathbf{a}_{\text{rel}}] \quad (6.54)$$

where $\mathbf{F}$ represents all external forces on $\mathcal{B}$ besides gravity, which is written separately.

Question 6.8 This is a good place to ask: Why is Equation (6.54) restricted to bodies in motion near the earth?

We now proceed to compute the various terms in Equation (6.54). First we note that

$$\begin{aligned}
\mathbf{r} &= x\hat{\mathbf{i}} + y\hat{\mathbf{j}} + z\hat{\mathbf{k}} \\
\mathbf{v}_{\text{rel}} &= \dot{x}\hat{\mathbf{i}} + \dot{y}\hat{\mathbf{j}} + \dot{z}\hat{\mathbf{k}} \\
\mathbf{a}_{\text{rel}} &= \ddot{x}\hat{\mathbf{i}} + \ddot{y}\hat{\mathbf{j}} + \ddot{z}\hat{\mathbf{k}} \\
\boldsymbol{\omega} &= \omega_e(\cos\lambda\hat{\mathbf{j}} + \sin\lambda\hat{\mathbf{k}}) \qquad (\text{where } \omega_e = 2\pi \text{ rad/day} \\
&\qquad\qquad\qquad\qquad\qquad\qquad \approx 0.0000727 \text{ rad/sec}) \\
\boldsymbol{\alpha} &= \mathbf{0} \\
\mathbf{R} &= R_e\hat{\mathbf{k}}
\end{aligned}$$

Next we compute the acceleration $\ddot{\mathbf{R}}$ of the site (O'). We utilize Equation (6.53) again, this time with the "moving point" being O' and the origin in the moving frame being O:

Answer 6.8 It has been assumed that the strength of the gravitational field is constant, but of course gravity decreases as C moves farther and farther from the earth's surface.

$$\mathbf{a}_{O'} = \ddot{\mathbf{R}} = \cancelto{0}{\mathbf{a}_{O}} + \cancelto{0}{\boldsymbol{\alpha}} \times \mathbf{R} + \boldsymbol{\omega} \times (\boldsymbol{\omega} \times \mathbf{R}) + 2\boldsymbol{\omega} \times \mathbf{v}_{\text{rel}} + \mathbf{a}_{\text{rel}}$$

This time, however, note that $\mathbf{v}_{\text{rel}}$ and $\mathbf{a}_{\text{rel}}$ are zero; this follows from the fact that O' has *no* motion relative to the moving frame (the earth). Thus the only term surviving is:

$$\ddot{\mathbf{R}} = \boldsymbol{\omega} \times (\boldsymbol{\omega} \times \mathbf{R})$$

Therefore, from Equation (6.54),

$$\mathbf{F} - mg\hat{\mathbf{k}} = m\{\boldsymbol{\omega} \times [\boldsymbol{\omega} \times (\mathbf{R} + \mathbf{r})] + 2\boldsymbol{\omega} \times (\dot{x}\hat{\mathbf{i}} + \dot{y}\hat{\mathbf{i}} + \dot{z}\hat{\mathbf{k}}) + (\ddot{x}\hat{\mathbf{i}} + \ddot{y}\hat{\mathbf{j}} + \ddot{z}\hat{\mathbf{k}})\}$$

Neglecting $|\mathbf{r}|$ with respect to $|\mathbf{R}|$ and expressing $\mathbf{F}$ in terms of its components (F_x, F_y, F_z), we arrive at a set of differential equations governing the motion of C:

$$\ddot{x} = 2\omega_e(\dot{y} \sin \lambda - \dot{z} \cos \lambda) + \frac{F_x}{m}$$

$$\ddot{y} = -2\omega_e \dot{x} \sin \lambda - R_e\omega_e^2 \cos \lambda \sin \lambda + \frac{F_y}{m}$$

$$\ddot{z} = 2\omega_e \dot{x} \cos \lambda + R_e\omega_e^2 \cos^2\lambda - g + \frac{F_z}{m} \tag{6.55}$$

We complete this brief section with an example illustrating the use of Equations (6.55).

EXAMPLE 6.9

Due to the earth's rotation, the resultant force it exerts on a particle P at rest on its surface is not quite directed toward its center of mass. Use Equations (6.55) to find this deviation, assuming a spherical earth.

Solution

We have $\dot{x}$, $\dot{y}$, $\dot{z}$, $\ddot{x}$, $\ddot{y}$, and $\ddot{z}$ all zero, so that the equations of motion of P (which moves!) in the inertial frame $\mathcal{I}$ are:

$$F_x = 0$$

$$F_y = mR_e\omega_e^2 \cos \lambda \sin \lambda = k \sin \lambda$$

$$F_z = -mR_e\omega_e^2 \cos^2 \lambda + mg = mg - k \cos \lambda$$

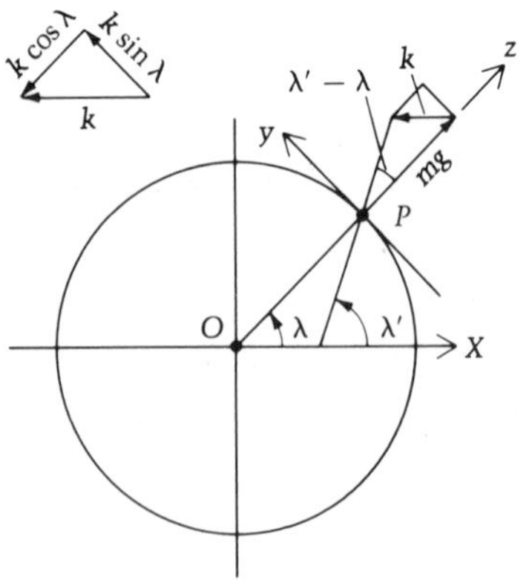

Figure E6.9a

Calling $mR_e\omega_e^2 \cos \lambda = k$, we see (Figure E6.9a) that the earth must push on P at a small angle ($\lambda' - \lambda$) with the *"geometric vertical"* in order for P to remain at rest in the "moving frame" $\mathcal{F}$, rigidly attached to the surface of the planet. It is the angle λ', *not* the angle λ, that defines *"local vertical"*; this is because λ' is the angle that a plumb bob string makes with the axis X in the equatorial plane.

Figure E6.9b shows that (with $\phi = \lambda' - \lambda$)

$$\tan \phi \approx \frac{k \sin \lambda}{mg - k \cos \lambda}$$

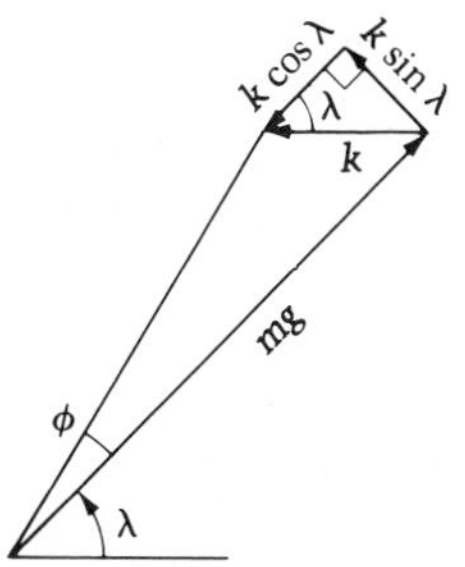

Figure E6.9b

We note that k, in comparison with mg, is quite small:

$$\frac{k}{mg} = \frac{R_e \omega_e^2 \cos \lambda}{g} \approx \frac{6370(0.0000727)^2 \cos \lambda}{(9.81/1000)}$$

$$= \frac{6370(1000)(0.0000727)^2 \cos \lambda}{9.81}$$

$$\approx 0.00343 \cos \lambda$$

This means that we can use small angle approximations on the angle ϕ:

$$\phi \approx \tan \phi \approx \frac{k \sin \lambda}{mg} = \frac{R_e \omega_e^2 \sin \lambda \cos \lambda}{g} = 0.00343 \sin \lambda \cos \lambda$$

so that the deviation of the local "plumb-line" vertical from the "geometric vertical" is

$$\lambda' - \lambda = \phi \approx 0.00343 \sin \lambda \cos \lambda$$

Of course there is *no* deviation in direction at the poles (where $\cos \lambda = 0$) or the equator (where $\sin \lambda = 0$). The maximum, at 45° latitude, is 0.0017 rad, or about 0.1°.

PROBLEMS ▸ Section 6.6

6.42 If it were possible for a train to travel continuously around the world on a meridional track as shown in Figure P6.42(a), one side of the track would wear out in time due to the Coriolis acceleration. Explain which side will wear out in each of the four numbered quadrants of the circular path. Figure P6.42(b) shows how the train's wheels rest on the track.

6.43 Explain in detail how the Coriolis acceleration is related to the deflection of the air rushing toward a low-pressure area, thereby forming a hurricane. (See Figure P6.43.) Do the problem for each hemisphere!

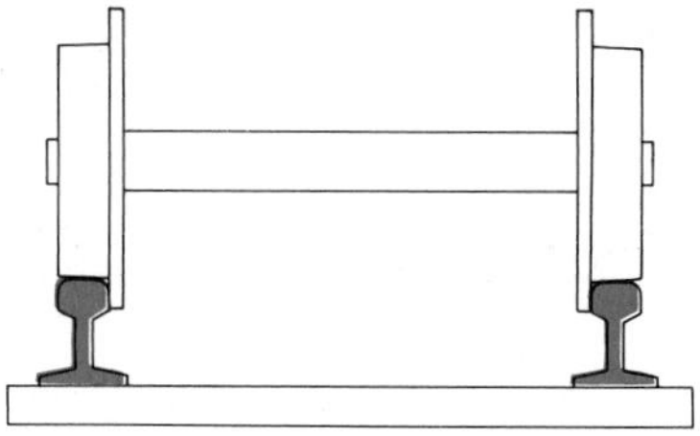

Figure P6.42(b)

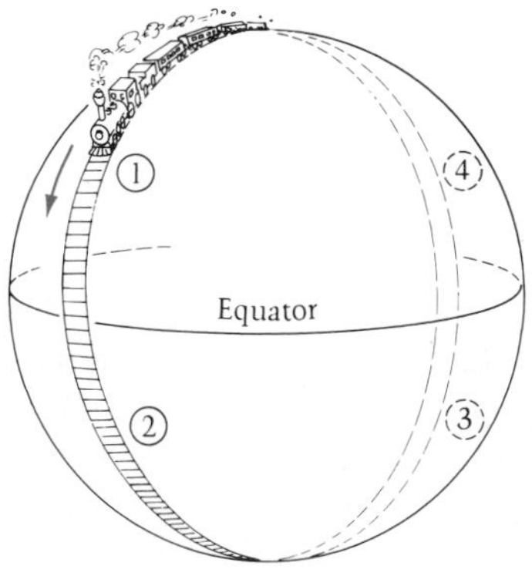

Figure P6.42(a)

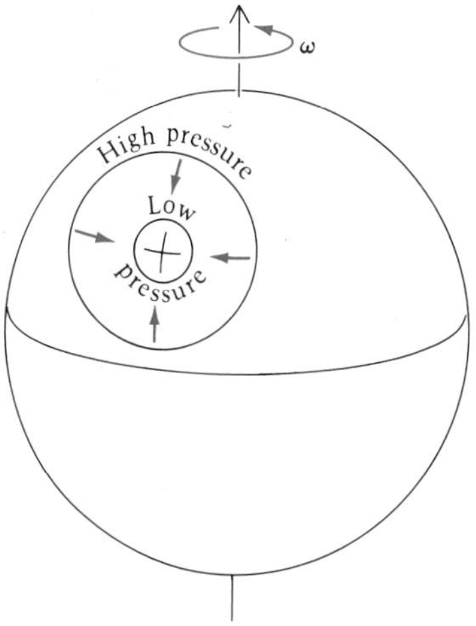

Figure P6.43

6.44 A car travels south along a meridian at a certain time; its speed relative to the earth is 60 mph, increasing at the rate of 2 ft/sec^2. (See Figure P6.44.) Find the acceleration of the car in a frame having origin always at the center of the earth and z axis along the polar axis of rotation, but not rotating about the axis with the earth.

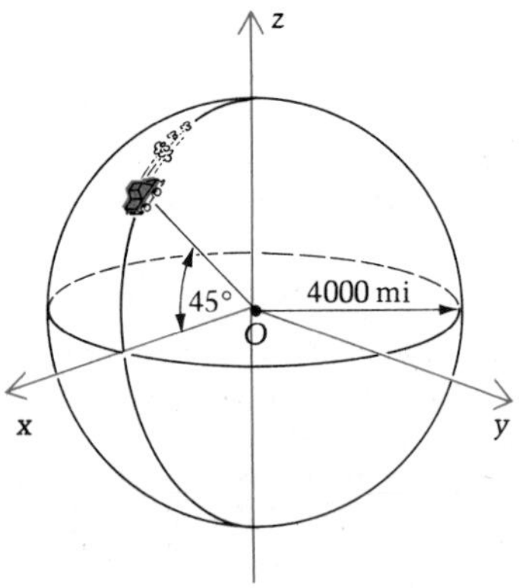

Figure P6.44

6.45 If in the preceding problem the car is traveling from west to east at 45°N latitude instead of along a meridian, find the car's acceleration in the same frame. (See Figure P6.45.)

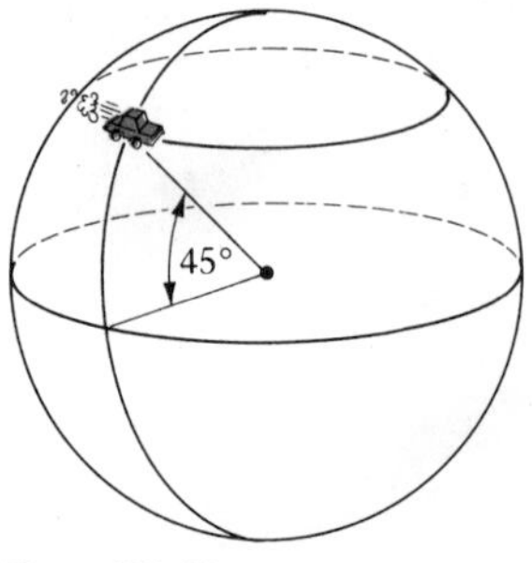

Figure P6.45

* **6.46** A projectile is fired from a site at latitude λ, with the initial velocity components (at time $t = 0$) $= \dot{x}_i, \dot{y}_i, \dot{z}_i$. Determine the maximum height reached by the projectile, neglecting air resistance and the $R_e\omega_e^2$ terms in Equations (6.55).

* **6.47** Refer to Example 6.9. Show that if the component equations of (6.54) are written in terms of axes (x', y', z') (see Figure P6.47) instead of (x, y, z), they will have the same form as do Equations (6.55) without their $R_e\omega_e^2$ terms, provided that

$$G = \sqrt{(g - R_e\omega_e^2 C_\lambda^2)^2 + (R_e\omega_e^2 C_\lambda S_\lambda)^2}$$

replaces g; λ' replaces λ; and y' and z', respectively, replace y and z.

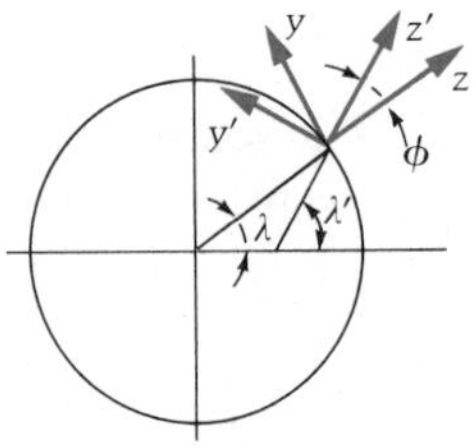

Figure P6.47

* **6.48** Using equations from the preceding problem, find the location at which a falling rock will strike the earth if dropped from rest on the z' axis from a height H. Neglect air resistance and assume the rock strikes the earth when $z' = 0$.

6.7 Velocity and Acceleration Equations for Two Points of the Same Rigid Body

We next apply the concepts of position, velocity, acceleration, angular velocity, and angular acceleration (developed in Chapter 1 and Sections 6.2 to 6.5) to the kinematics of a rigid body $\mathcal{B}$ in general motion in a frame $\mathcal{I}$. The equation relating the velocities in $\mathcal{I}$ of two points of a rigid body $\mathcal{B}$ to its angular velocity $\omega_{\mathcal{B}/\mathcal{I}}$ is a special case of Equation (6.44). We have seen in the text following that equation

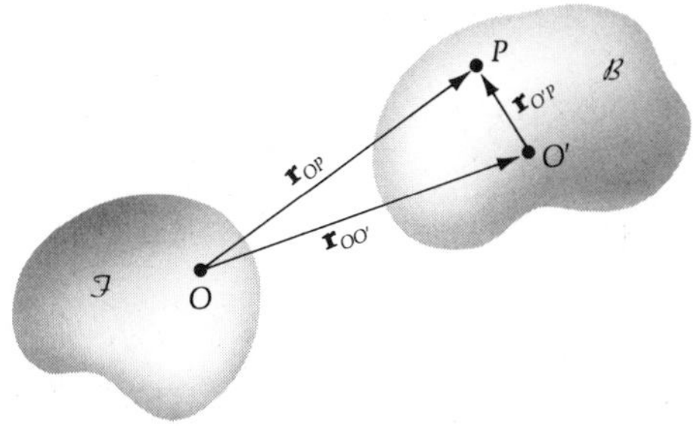

Figure 6.8 Points P and O' of a rigid body $\mathcal{B}$ in general motion.

that if P joins O' as a fixed point of $\mathcal{B}$ (see Figure 6.8), the relationship between the velocities of these two points is given by the equation

$$\mathbf{v}_{P/\mathcal{I}} = \mathbf{v}_{O'/\mathcal{I}} + \boldsymbol{\omega}_{\mathcal{B}/\mathcal{I}} \times \mathbf{r}_{O'P} \tag{6.56}$$

From the text following Equation (6.52), we also know the relationship between the accelerations of these two points of $\mathcal{B}$:

$$\mathbf{a}_{P/\mathcal{I}} = \mathbf{a}_{O'/\mathcal{I}} + \boldsymbol{\alpha}_{\mathcal{B}/\mathcal{I}} \times \mathbf{r}_{O'P} + \boldsymbol{\omega}_{\mathcal{B}/\mathcal{I}} \times (\boldsymbol{\omega}_{\mathcal{B}/\mathcal{I}} \times \mathbf{r}_{O'P}) \tag{6.57}$$

We emphasize that Equations (6.56) and (6.57) follow from the general equations (6.44) and (6.51) when and if $\mathbf{r}_{O'P}$ is a constant in $\mathcal{B}$; in that case, $\mathbf{v}_{P/\mathcal{B}} = {}^{\mathcal{B}}\dot{\mathbf{r}}_{O'P} = \mathbf{0}$ and $\mathbf{a}_{P/\mathcal{B}} = {}^{\mathcal{B}}\ddot{\mathbf{r}}_{O'P} = \mathbf{0}$. Furthermore, the two points P and O' in Equations (6.56) and (6.57) may be replaced by *any* pair of points fixed to $\mathcal{B}$, since being fixed to $\mathcal{B}$ is the only restriction on either of them. Speaking loosely, with Equations (6.56, 6.57) we are interested in *two points* on *one body*, whereas with Equations (6.44, 6.51) we were studying *one point* in motion relative to *two bodies*. We now consider examples involving the use of the two rigid-body equations (6.56) and (6.57).

EXAMPLE 6.10

Rework Examples 6.6 and 6.8 by treating point A as a point of body $\mathcal{A}$ instead of as a point moving with a known motion in $\mathcal{B}$.

Solution

From Equation (6.56), and Figure E6.10, we have

$$\mathbf{v}_{A/\mathcal{G}} = \mathbf{v}_{Q/\mathcal{G}} + \boldsymbol{\omega}_{\mathcal{A}/\mathcal{G}} \times \mathbf{r}_{QA}$$

Now recognizing that Q is also a point of $\mathcal{B}$, we obtain

$$\begin{aligned}\mathbf{v}_{A/\mathcal{G}} &= (\mathbf{v}_{O'/\mathcal{G}} + \boldsymbol{\omega}_{\mathcal{B}/\mathcal{G}} \times \mathbf{r}_{O'Q}) + \boldsymbol{\omega}_{\mathcal{A}/\mathcal{G}} \times \mathbf{r}_{QA}\\ &= \mathbf{0} + \omega_2\hat{\mathbf{j}} \times R\hat{\mathbf{k}} + (\omega_1\hat{\mathbf{i}} + \omega_2\hat{\mathbf{j}}) \times r\hat{\mathbf{j}}\\ &= R\omega_2\hat{\mathbf{i}} + r\omega_1\hat{\mathbf{k}} \quad \text{(as we obtained in Example 6.6)}\end{aligned}$$

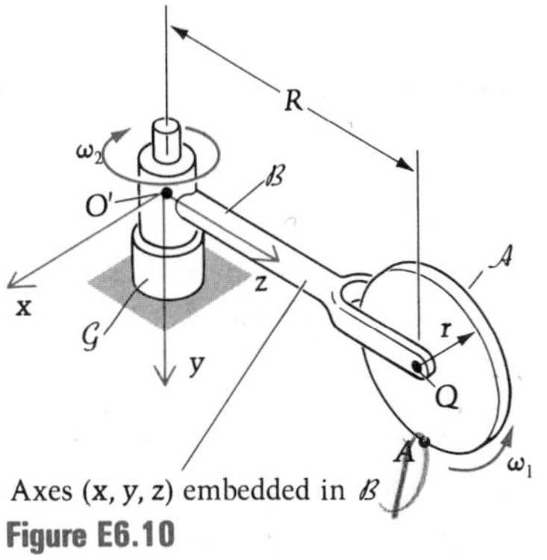

Axes (x, y, z) embedded in $\mathcal{B}$

Figure E6.10

Next we relate the accelerations of A and Q with Equation (6.57):

$$\mathbf{a}_{A/\mathcal{G}} = \mathbf{a}_{Q/\mathcal{G}} + \boldsymbol{\alpha}_{\mathcal{A}/\mathcal{G}} \times \mathbf{r}_{QA} + \boldsymbol{\omega}_{\mathcal{A}/\mathcal{G}} \times (\boldsymbol{\omega}_{\mathcal{A}/\mathcal{G}} \times \mathbf{r}_{QA})$$

The first term on the right side, using Q and O' as points of body $\mathcal{B}$, is

$$\mathbf{a}_{Q/\mathcal{G}} = \mathbf{a}_{O'/\mathcal{G}} + \boldsymbol{\alpha}_{\mathcal{B}/\mathcal{G}} \times \mathbf{r}_{O'Q} + \boldsymbol{\omega}_{\mathcal{B}/\mathcal{G}} \times (\boldsymbol{\omega}_{\mathcal{B}/\mathcal{G}} \times \mathbf{r}_{O'Q})$$
$$= 0 + 0 + \omega_2\hat{\mathbf{j}} \times (R\omega_2\hat{\mathbf{i}})$$
$$= -R\omega_2^2\hat{\mathbf{k}}$$

Thus

$$\mathbf{a}_{A/\mathcal{G}} = -R\omega_2^2\hat{\mathbf{k}} + (-\omega_1\omega_2\hat{\mathbf{k}}) \times (r\hat{\mathbf{j}}) + (\omega_1\hat{\mathbf{i}} + \omega_2\hat{\mathbf{j}}) \times (r\omega_1\hat{\mathbf{k}})$$

or

$$\mathbf{a}_{A/\mathcal{G}} = 2r\omega_1\omega_2\hat{\mathbf{i}} - r\omega_1^2\hat{\mathbf{j}} - R\omega_2^2\hat{\mathbf{k}}$$

which we previously obtained in Example 6.8 by another approach.

In the next example we will see that sometimes there is an indeterminate component of angular velocity.

EXAMPLE 6.11

Collars C_1 and C_2 in Figure E6.11a are attached at C_1 and C_2 to rod R by ball and socket joints. At the instant shown, C_2 is moving away from the origin at speed $\sqrt{13}$ cm/s. Find the velocity of C_1 at the same instant. Can the angular velocity of R in $\mathcal{J}$ be found?

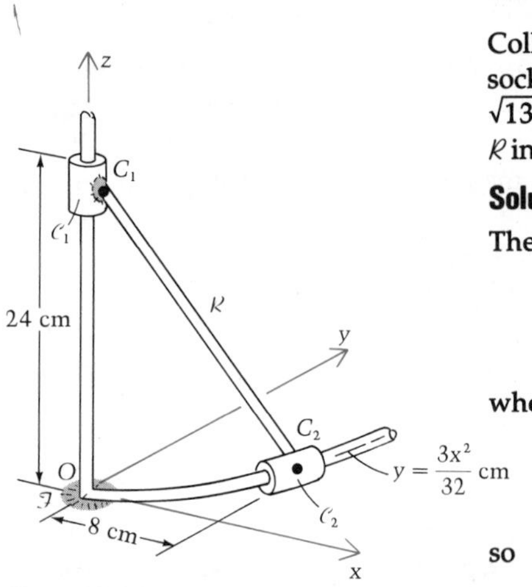

Solution

The velocity of C_2 is determined by (see Figure E6.11b):

$$\mathbf{v}_{C_2} = \sqrt{13}\hat{\mathbf{e}}_t = \sqrt{13}\left(\frac{2}{\sqrt{13}}\hat{\mathbf{i}} + \frac{3}{\sqrt{13}}\hat{\mathbf{j}}\right) = 2\hat{\mathbf{i}} + 3\hat{\mathbf{j}}$$

where

$$\frac{dy}{dx} = \frac{6x}{32}$$

so

$$\left.\frac{dy}{dx}\right|_{x=8} = \frac{3}{2}$$

Now we can calculate the velocities of C_1 and C_2 using Equation (6.56):

$$\begin{array}{cc} v_{C_1}\hat{\mathbf{k}} & (2\hat{\mathbf{i}} + 3\hat{\mathbf{j}}) \\ \mathbf{v}_{C_1} & = \mathbf{v}_{C_2} + \boldsymbol{\omega}_{R/\mathcal{J}} \times \mathbf{r}_{C_2C_1} \end{array} \qquad (1)$$

We note that the component of the angular velocity $\boldsymbol{\omega}_{R/\mathcal{J}}$ along the line C_1C_2 cannot be determined from the given information, because any value of it whatsoever will not affect Equation (1). However, dotting this equation with

$$\mathbf{r}_{C_2C_1} = -8\hat{\mathbf{i}} - 6\hat{\mathbf{j}} + 24\hat{\mathbf{k}} \text{ yields:}$$

$$24v_{C_1} = -8(2) - 6(3) + 0 = -34$$

$y = \dfrac{3x^2}{32}$ cm

Figure E6.11a

Figure E6.11b

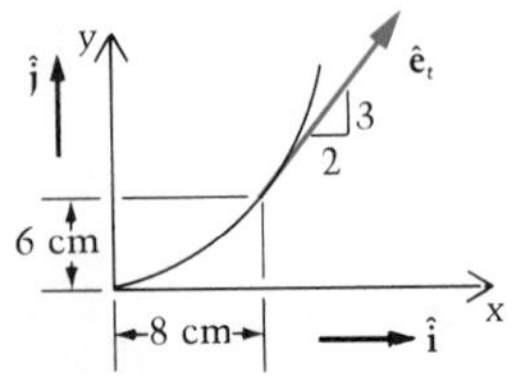

so that

$$v_{C_1} = -1.42$$

and

$$\mathbf{v}_{C_1} = -1.42\hat{\mathbf{k}} \text{ cm/s}$$

Example 6.12

The cone $\mathcal{B}_1$ in Figures E6.12a,b rolls on the floor in such a way that the center Q of the base of the cone travels on a horizontal circle at constant velocity $v_Q\hat{\mathbf{j}}$. Let $\mathcal{B}_2$ denote an intermediate frame ("between" cone $\mathcal{B}_1$ and the ground $\mathcal{B}_3$) in which $\hat{\mathbf{i}}$, $\hat{\mathbf{j}}$, and $\hat{\mathbf{k}}$ are fixed. The unit vector $\hat{\mathbf{i}}$ is always directed along OQ, and $\hat{\mathbf{j}}$ is normal to the plane of $\hat{\mathbf{i}}$ and the contact line, in a direction parallel to $\mathbf{v}_Q$; finally, $\hat{\mathbf{k}} = \hat{\mathbf{i}} \times \hat{\mathbf{j}}$. Find the angular velocity of $\mathcal{B}_1$ in $\mathcal{B}_3$.

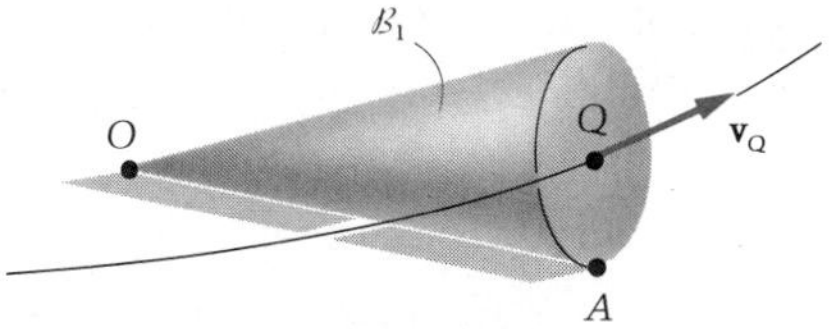

Figure E6.12a

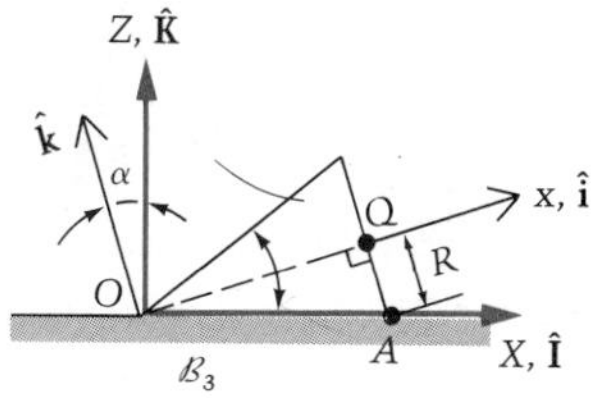

Figure E6.12b

Solution

We shall denote $\boldsymbol{\omega}_{\mathcal{B}_1/\mathcal{B}_3}$ by simply $\boldsymbol{\omega}_{1/3}$. Since $\mathbf{v}_A = \mathbf{v}_O = \mathbf{0}$, then their difference, $\boldsymbol{\omega}_{1/3} \times \mathbf{r}_{OA}$, must also vanish. This requires $\boldsymbol{\omega}_{1/3}$ to be parallel at all times to the line of contact of $\mathcal{B}_1$ with the ground. Thus

$$\boldsymbol{\omega}_{1/3} = \omega(-\cos\alpha\hat{\mathbf{i}} + \sin\alpha\hat{\mathbf{k}})$$

Next, using Equation (6.56),

$$\mathbf{v}_Q = \mathbf{v}_A + \boldsymbol{\omega}_{1/3} \times \mathbf{r}_{AQ}$$
$$v_Q\hat{\mathbf{j}} = \mathbf{0} + \omega(-\cos\alpha\hat{\mathbf{i}} + \sin\alpha\hat{\mathbf{k}}) \times R\hat{\mathbf{k}}$$

Therefore

$$v_Q = \omega R\cos\alpha$$

so that

$$\omega = v_Q/(R\cos\alpha)$$

Thus

$$\boldsymbol{\omega}_{1/3} = \frac{v_Q}{R}(-\hat{\mathbf{i}} + \tan\alpha\,\hat{\mathbf{k}})$$

We wish to make some further remarks about the preceding example. The addition theorem gives:

$$\omega_{1/3} = \omega_{1/2} + \omega_{2/3} \tag{1}$$

where $\omega_{2/3}$ is given by

$$\omega_{2/3} = \underbrace{\left[\frac{v_Q}{\text{horizontal projection of } OQ}\right]}_{(R/\tan\alpha)\cos\alpha} \underbrace{(\sin\alpha\hat{\mathbf{i}} + \cos\alpha\hat{\mathbf{k}})}_{\hat{\mathbf{K}}}$$

Substituting $\omega_{2/3}$ and $\omega_{1/3}$ into the addition theorem equation (1) above, we find

$$\omega_{1/2} = \frac{v_Q}{R\cos^2\alpha}(-\hat{\mathbf{i}})$$

Note the check on the direction $(-\hat{\mathbf{i}})$, since the only way $\mathcal{B}_1$ can move relative to $\mathcal{B}_2$ is to rotate around OQ.

Consider finally two ways of depicting the components of $\omega_{1/3}$. From Figure 6.9a, note that the vector sum of the two components of $\omega_{1/3}$ is parallel to the contact line since

$$\frac{\dfrac{v_Q}{R}\tan\alpha}{v_Q/R} = \tan\alpha$$

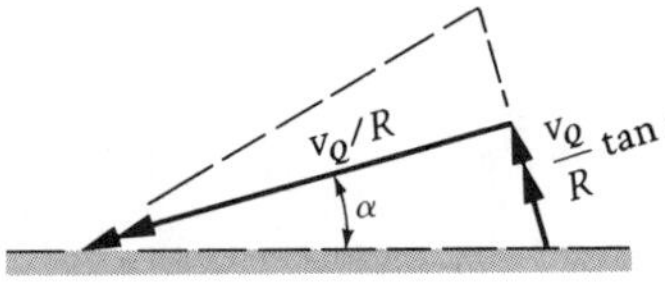

Figure 6.9a

Figure 6.9b illustrates the addition theorem. Note that the "direction check" again results in $\omega_{1/3}$ being along the contact line, this time because

$$\frac{\left(\dfrac{v_Q\tan\alpha}{R\cos\alpha}\right)}{\left(\dfrac{v_Q}{R\cos^2\alpha}\right)} = \sin\alpha$$

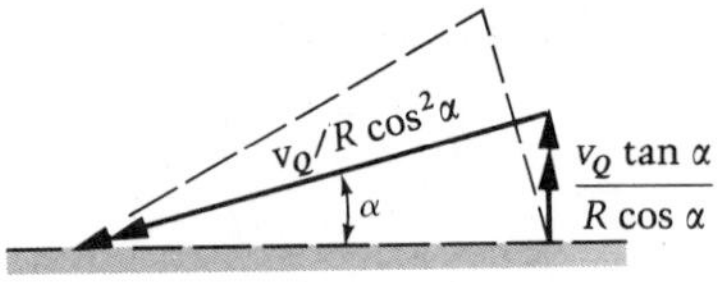

Figure 6.9b

EXAMPLE 6.13

In Example 6.12, find the angular acceleration of $\mathcal{B}_1$ in $\mathcal{B}_3$.

Solution

Differentiating $\omega_{1/3}$,

$$\alpha_{1/3} = {}^{\mathcal{B}_3}\dot{\omega}_{1/3} = {}^{\mathcal{B}_2}\dot{\omega}_{1/3} + \omega_{2/3} \times \omega_{1/3}$$

We had

$$\omega_{1/3} = \frac{v_Q}{R}(-\hat{\mathbf{i}} + \tan\alpha\hat{\mathbf{k}})$$

and

$$\omega_{2/3} = \frac{v_Q}{R}(\tan^2\alpha\hat{\mathbf{i}} + \tan\alpha\hat{\mathbf{k}})$$

Thus

$${}^{\mathcal{B}_2}\dot{\omega}_{1/3} = \mathbf{0}$$

and we obtain

$$\boldsymbol{\alpha}_{1/3} = \left(\frac{v_Q}{R}\right)^2 \begin{vmatrix} \hat{\mathbf{i}} & \hat{\mathbf{j}} & \hat{\mathbf{k}} \\ \tan^2\alpha & 0 & \tan\alpha \\ -1 & 0 & \tan\alpha \end{vmatrix}$$

$$= \frac{v_Q^2}{R^2}(-\hat{\mathbf{j}})(\tan^3\alpha + \tan\alpha)$$

or

$$\boldsymbol{\alpha}_{1/3} = \frac{\sin\alpha}{\cos^3\alpha}\frac{v_Q^2}{R^2}(-\hat{\mathbf{j}})$$

We see in the above example that acceleration equations need not be used to compute $\boldsymbol{\alpha}$ if $\boldsymbol{\omega}$ is known.

The final example in this section illustrates the workings of a complicated three-dimensional gear train.

Example 6.14

Very large alterations in speed along a given direction may be obtained by using the gear arrangement shown in the diagram. Gears $\mathcal{A}$, $\mathcal{B}$, and $\mathcal{D}$ all rotate about the x axis in $\mathcal{I}$, but $\mathcal{C}$ has a more complicated motion:

1. It rolls on the fixed (to $\mathcal{I}$) gear $\mathcal{G}$, currently contacting it at P.
2. It rolls on $\mathcal{A}$. (The contacting teeth are at A in Figure E6.14a.)
3. It rolls on $\mathcal{B}$ (at B in the figure).
4. It turns with respect to $\mathcal{D}$ about the line ℓ, which is fixed in both $\mathcal{C}$ and $\mathcal{D}$.

Considering $\mathcal{B}$ to be the driven gear, find the ratio of $\omega_{\mathcal{A}}$ to $\omega_{\mathcal{B}}$.

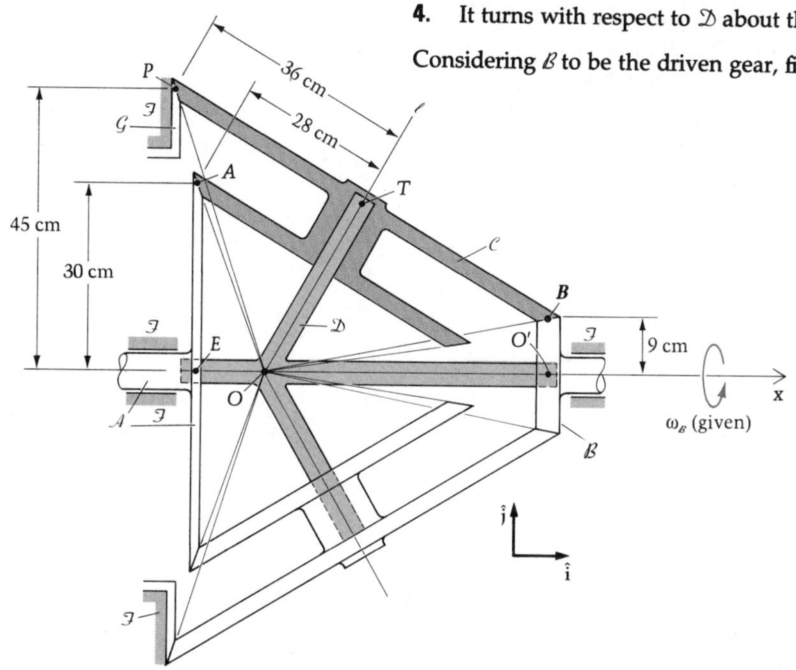

Figure E6.14a

Solution

The velocity of the contact points of $\mathcal{B}$ and $\mathcal{C}$ is, using body $\mathcal{B}$,

$$\mathbf{v}_B = \overset{0}{\cancel{\mathbf{v}_{O'}}} + \boldsymbol{\omega}_{\mathcal{B}/\mathcal{I}} \times \mathbf{r}_{O'B}$$

$$\mathbf{v}_B = \omega_{\mathcal{B}}\hat{\mathbf{i}} \times 9\hat{\mathbf{j}} = 9\omega_{\mathcal{B}}\hat{\mathbf{k}} \text{ cm/s} \tag{1}$$

in which we use just one subscript on ω when it is the angular velocity of a body with respect to $\mathcal{I}$.

Next we find another expression for $\mathbf{v}_B$, this time by relating the velocity of the tooth point of $\mathcal{C}$ at B to that of the point of $\mathcal{C}$ that contacts the reference frame ($\mathcal{G}$ is fixed to $\mathcal{I}$) at P.

$$\mathbf{v}_B = \overset{0}{\cancel{\mathbf{v}_P}} + \boldsymbol{\omega}_{\mathcal{C}} \times \mathbf{r}_{PB} \tag{2}$$

To obtain $\mathbf{r}_{PB}$, we use Figure E6.14b and see that

$$\phi = \cos^{-1}\left(\frac{36}{72}\right) = 60°$$

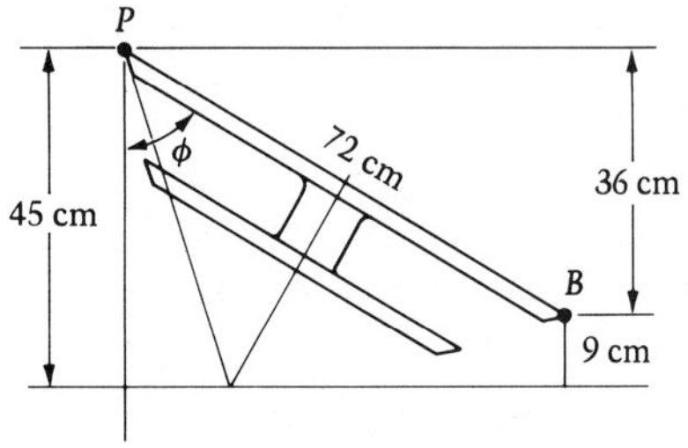

Figure E6.14b

Therefore

$$\mathbf{r}_{PB} = 72\left(\frac{\sqrt{3}}{2}\hat{\mathbf{i}} - \frac{1}{2}\hat{\mathbf{j}}\right) \text{ cm} \tag{3}$$

Also, by the addition theorem,

$$\boldsymbol{\omega}_{\mathcal{C}} = \boldsymbol{\omega}_{\mathcal{C}/\mathcal{I}} = \boldsymbol{\omega}_{\mathcal{C}/\mathcal{D}} + \boldsymbol{\omega}_{\mathcal{D}/\mathcal{I}}$$

$$= \omega_{\mathcal{C}/\mathcal{D}}\left(\frac{1}{2}\hat{\mathbf{i}} + \frac{\sqrt{3}}{2}\hat{\mathbf{j}}\right) + \omega_{\mathcal{D}}\hat{\mathbf{i}} \tag{4}$$

in which we have used the fact that we know the directions (but not the magnitudes yet) of $\boldsymbol{\omega}_{\mathcal{C}/\mathcal{D}}$ and $\boldsymbol{\omega}_{\mathcal{D}/\mathcal{I}}$.

Substituting Equations (3) and (4) into (2) and substituting the result into Equation (1) gives us

$$9\omega_{\mathcal{B}}\hat{\mathbf{k}} = \left[\omega_{\mathcal{C}/\mathcal{D}}\left(\frac{1}{2}\hat{\mathbf{i}} + \frac{\sqrt{3}}{2}\hat{\mathbf{j}}\right) + \omega_{\mathcal{D}}\hat{\mathbf{i}}\right] \times 72\left(\frac{\sqrt{3}}{2}\hat{\mathbf{i}} - \frac{1}{2}\hat{\mathbf{j}}\right)$$

$$= \left\{\omega_{\mathcal{C}/\mathcal{D}}\left[72\left(-\frac{1}{4} - \frac{3}{4}\right)\right] + \omega_{\mathcal{D}}\left[72\left(-\frac{1}{2}\right)\right]\right\}\hat{\mathbf{k}}$$

or, simplifying,

$$\omega_{\mathcal{B}} = -8\omega_{\mathcal{C}/\mathcal{D}} - 4\omega_{\mathcal{D}} \tag{5}$$

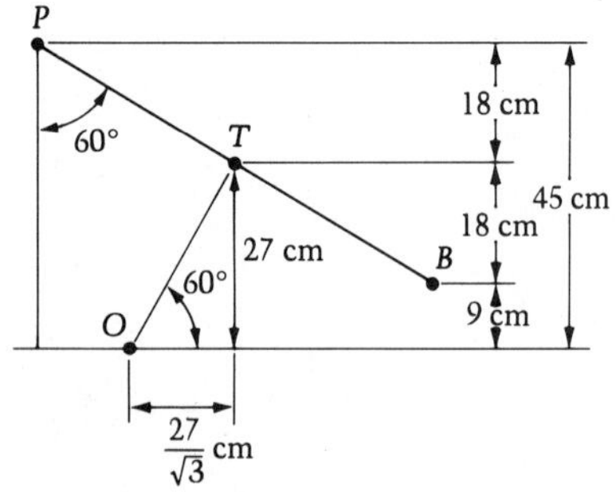

Figure E6.14c

To get another equation in these variables, we shall use the point T, which belongs to both $\mathcal{C}$ and $\mathcal{D}$ and is shown in Figure E6.14c along with some essential geometry. First, as a point of $\mathcal{D}$, we have

$$\mathbf{v}_T = \overset{0}{\cancel{\mathbf{v}_O}} + \boldsymbol{\omega}_{\mathcal{D}} \times \mathbf{r}_{OT}$$

$$= \omega_{\mathcal{D}}\hat{\mathbf{i}} \times \left(\frac{27}{\sqrt{3}}\hat{\mathbf{i}} + 27\hat{\mathbf{j}}\right)$$

$$= 27\omega_{\mathcal{D}}\hat{\mathbf{k}} \text{ cm/s} \tag{6}$$

Next, as a point of $\mathcal{C}$,

$$\mathbf{v}_T = \overset{0}{\cancel{\mathbf{v}_P}} + \boldsymbol{\omega}_{\mathcal{C}} \times \mathbf{r}_{PT} \tag{7}$$

The cross product in Equation (7) is one-half the cross product in Equation (2).

Question 6.9 Why is this so?

Therefore

$$\mathbf{v}_T = (-36\omega_{\mathcal{C}/\mathcal{D}} - 18\omega_{\mathcal{D}})\hat{\mathbf{k}} \tag{8}$$

Equating the right sides of Equations (6) and (8) gives

$$27\omega_{\mathcal{D}} = -36\omega_{\mathcal{C}/\mathcal{D}} - 18\omega_{\mathcal{D}}$$

or

$$5\omega_{\mathcal{D}} = -4\omega_{\mathcal{C}/\mathcal{D}} \tag{9}$$

Substituting Equation (9) into (5) leads to

$$\omega_{\mathcal{D}} = \frac{\omega_{\mathcal{B}}}{6} \text{ rad/s} \tag{10}$$

and

$$\omega_{\mathcal{C}/\mathcal{D}} = -\frac{5}{24}\,\omega_{\mathcal{B}} \text{ rad/s} \tag{11}$$

Substituting Equations (10) and (11) into (4) gives us $\boldsymbol{\omega}_{\mathcal{C}}$, which we shall need in the last step of the problem. The result is

$$\boldsymbol{\omega}_{\mathcal{C}} = \left(-\frac{5}{24}\,\omega_{\mathcal{B}}\right)\left(\frac{1}{2}\,\hat{\mathbf{i}} + \frac{\sqrt{3}}{2}\,\hat{\mathbf{j}}\right) + \left(\frac{\omega_{\mathcal{B}}}{6}\right)\hat{\mathbf{i}}$$

$$\boldsymbol{\omega}_{\mathcal{C}} = \frac{\omega_{\mathcal{B}}}{16}\,\hat{\mathbf{i}} - \frac{5\sqrt{3}}{48}\,\omega_{\mathcal{B}}\hat{\mathbf{j}} \tag{12}$$

We can now relate the velocities of the contacting points of bodies $\mathcal{C}$ and $\mathcal{A}$ at point A; first, using points A and E of $\mathcal{A}$, we get

$$\mathbf{v}_A = \overset{0}{\cancel{\mathbf{v}_E}} + \omega_{\mathcal{A}}\hat{\mathbf{i}} \times 30\hat{\mathbf{j}}$$

$$\mathbf{v}_A = 30\omega_{\mathcal{A}}\hat{\mathbf{k}} \text{ cm/s} \tag{13}$$

To get another expression for $\mathbf{v}_A$, we relate the velocities of the two points A and P on body $\mathcal{C}$:

$$\mathbf{v}_A = \overset{0}{\cancel{\mathbf{v}_P}} + \boldsymbol{\omega}_{\mathcal{C}} \times \mathbf{r}_{PA} \tag{14}$$

To obtain the position vector $\mathbf{r}_{PA}$, we use the geometry shown in Figure E6.14d. The distance x, needed in forming $\mathbf{r}_{PA}$, is equal to $(27 - d)/\sin 60°$:

$$x = \frac{27 - d}{\sqrt{3}/2} = \frac{27 - [30 - 28(\frac{1}{2})]}{\sqrt{3}/2} = \frac{22}{\sqrt{3}} \tag{15}$$

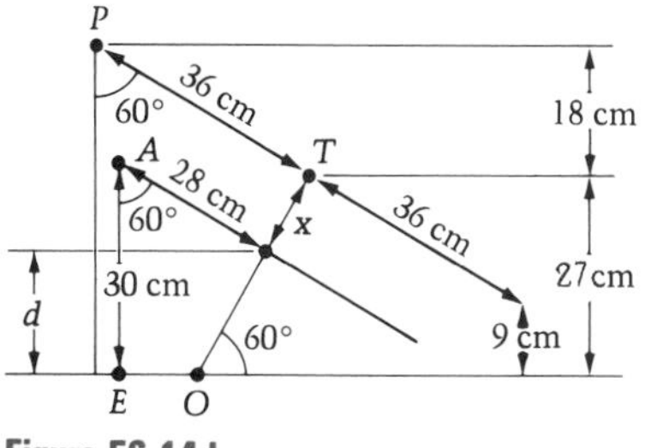

Figure E6.14d

Answer 6.9 Because $\mathbf{r}_{PT} = \mathbf{r}_{PB}/2$.

Therefore

$$\mathbf{r}_{PA} = (36 - 28)\left(\frac{\sqrt{3}}{2}\hat{\mathbf{i}} - \frac{1}{2}\hat{\mathbf{j}}\right) + \frac{22}{\sqrt{3}}\left(-\frac{1}{2}\hat{\mathbf{i}} - \frac{\sqrt{3}}{2}\hat{\mathbf{j}}\right)$$

$$= 0.577\hat{\mathbf{i}} - 15.0\hat{\mathbf{j}} \text{ cm} \tag{16}$$

Substituting Equations (12) and (16) into (14), we get

$$\mathbf{v}_A = \omega_{\mathcal{B}}\left(\frac{1}{16}\hat{\mathbf{i}} - \frac{5\sqrt{3}}{48}\hat{\mathbf{j}}\right) \times (0.577\hat{\mathbf{i}} - 15.0\hat{\mathbf{j}})$$

$$\mathbf{v}_A = -0.833\omega_{\mathcal{B}}\hat{\mathbf{k}} \text{ cm/s} \tag{17}$$

Equating the two expressions for $\mathbf{v}_A$ in Equations (13) and (17) gives

$$\omega_{\mathcal{A}} = -0.0278\omega_{\mathcal{B}} \text{ rad/s} \tag{18}$$

It is seen that the angular speed of gear $\mathcal{B}$ is 36 times that of $\mathcal{A}$, and in the opposite direction.

Question 6.10 Give an argument why $\mathcal{A}$ has to be turning in the opposite direction from that of $\mathcal{B}$. (*Hint:* Use the original figure and focus your attention on points P and O of $\mathcal{C}$.)

Does An Instantaneous Axis of Rotation Exist in General?

We recall that in Chapter 3 we were able to show that in plane motion a rigid body $\mathcal{B}$, except when its angular velocity vanishes, always has a point of zero velocity (the instantaneous center Ⓘ, and hence a line of points of zero velocity exists which we may call the instantaneous axis of rotation). We now show that in general (three-dimensional) motion, such an axis does not always exist. We start with an arbitrary point P with velocity $\mathbf{v}_P$, and sketch its velocity along with the angular velocity vector $\boldsymbol{\omega}_{\mathcal{B}/\mathcal{I}} = \boldsymbol{\omega}$ of $\mathcal{B}$ in the reference frame $\mathcal{I}$. (See Figure 6.10(a).)

Note that in Figure 6.10a there is a plane $\mathcal{P}$ defined by the vectors $\mathbf{v}_P$ and $\boldsymbol{\omega}$ drawn through P, unless the two vectors are parallel. If they are, then the motion of $\mathcal{B}$ in $\mathcal{I}$ is like that of a screwdriver — the body turns around a line that translates along its axis. The general case ($\mathbf{v}_P$ not parallel to $\boldsymbol{\omega}$) may also be reduced to a screwdriver motion as follows. First we replace $\mathbf{v}_P$ by its components parallel ($v_{P\parallel}$) and perpendicular ($v_{P\perp}$) to $\boldsymbol{\omega}$. (See Figure 6.10b.) Next we consider a plane $\mathcal{R}$ parallel to $\mathcal{P}$ and separated from $\mathcal{P}$ by the distance d as shown in Figure 6.10(b). Point Q is the projection of P into the plane $\mathcal{R}$, and we may write its velocity in

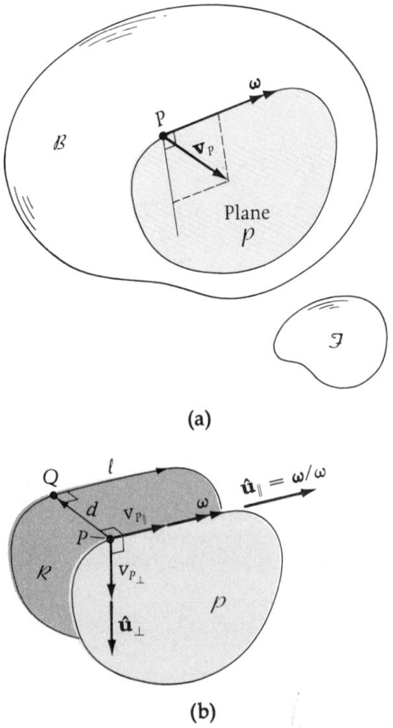

Figure 6.10

Answer 6.10 Each point of $\mathcal{C}$ (extended) which lies on line $\overline{PO}$ has zero velocity. Thus the velocity of B is its distance from $\overline{PO}$ times $|\boldsymbol{\omega}_{\mathcal{C}}|$, and the same is true for point A. The former is seen to be coming out of the paper, and the latter going into it, both about $\overline{PO}$. Therefore $\boldsymbol{\omega}_{\mathcal{A}}$, as determined by the direction of $\mathbf{v}_A$, is in the negative x direction, opposite to $\boldsymbol{\omega}_{\mathcal{B}}$.

of P by Equation (6.46):

$$\mathbf{v}_Q = \mathbf{v}_P + \boldsymbol{\omega} \times \mathbf{r}_{PQ}$$
$$= v_{P\|}\hat{\mathbf{u}}_{\|} + v_{P\perp}\hat{\mathbf{u}}_{\perp} + \boldsymbol{\omega} \times \underbrace{d(\hat{\mathbf{u}}_{\|} \times \hat{\mathbf{u}}_{\perp})}$$

Note from Figure 6.10b that this is the unit vector directed from P to Q.

The vector triple product is equal to

$$d\hat{\mathbf{u}}_{\|}(\boldsymbol{\omega} \cdot \hat{\mathbf{u}}_{\perp}) - d\hat{\mathbf{u}}_{\perp}(\boldsymbol{\omega} \cdot \hat{\mathbf{u}}_{\|}) = 0 - d\hat{\mathbf{u}}_{\perp}(\omega)$$

so that

$$\mathbf{v}_Q = v_{P\|}\hat{\mathbf{u}}_{\|} + \hat{\mathbf{u}}_{\perp}(v_{P\perp} - d\omega)$$

Therefore if d is chosen equal to $v_{P\perp}/\omega$, the line ℓ will be a "screwdriver line" and the motion of $\mathcal{B}$ will be as shown in Figure 6.10(c).

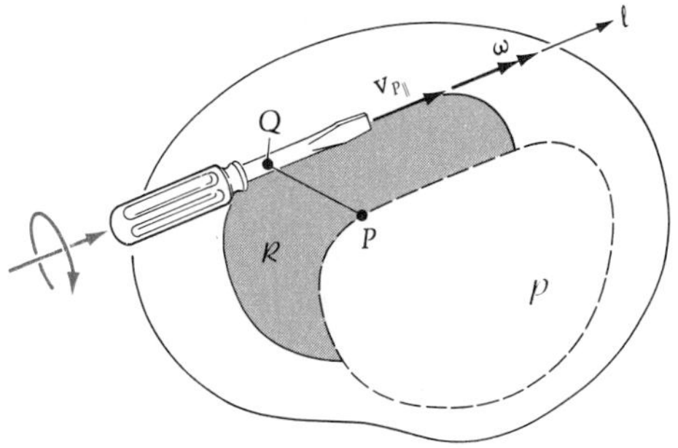

Figure 6.10(c)

All points on ℓ have velocities *along* ℓ at the given instant, while those off the line have the same velocity component parallel to ℓ; in addition, they rotate around it. This is the simplest reduction possible for the motion of $\mathcal{B}$, and it is clear that unless $v_{P\|} = 0$, *no* points can have zero velocity. Thus in three dimensions we are no longer assured of having an instantaneous axis as we were in dealing with plane motion.

There are, however, special cases in three dimensions in which an instantaneous axis exists; a good example is a cone rolling on a plane (Figure 6.11). Note that in this case the *entire line* of contact is at each

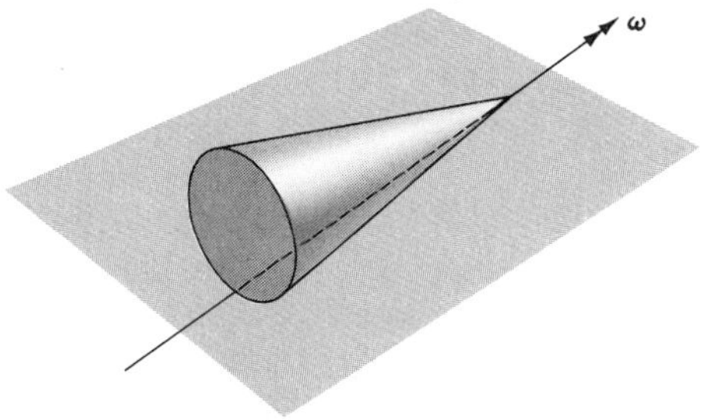

Figure 6.11

instant at rest on the plane. Since the angular velocity of the cone is parallel to this line, *all* the contact points have $v_{P\|} = 0$, which we have shown must be true if the instantaneous axis is to exist in three dimensions.

PROBLEMS ▶ Section 6.7

6.49 A youngster finds an old wagon wheel $\mathcal{W}$ and pushes it around with the center C moving at constant speed in a horizontal circle. (See Figure P6.49(a).) If the end O of the axle stays fixed while C returns to its starting point in T seconds, compute the angular velocity vector of the wheel $\boldsymbol{\omega}_{\mathcal{W}/\mathcal{J}}$, where $\mathcal{J}$ is the ground frame (Figure P6.49(b)). Give the result in terms of b, β, and T.

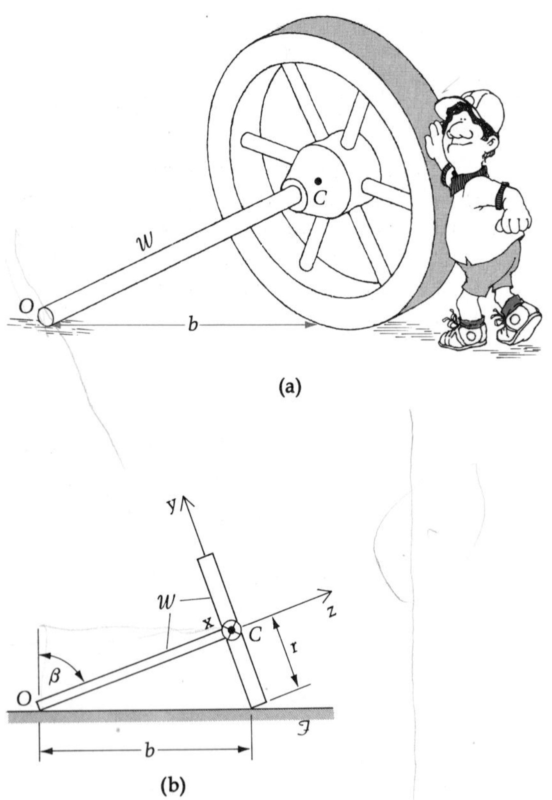

Figure P6.49

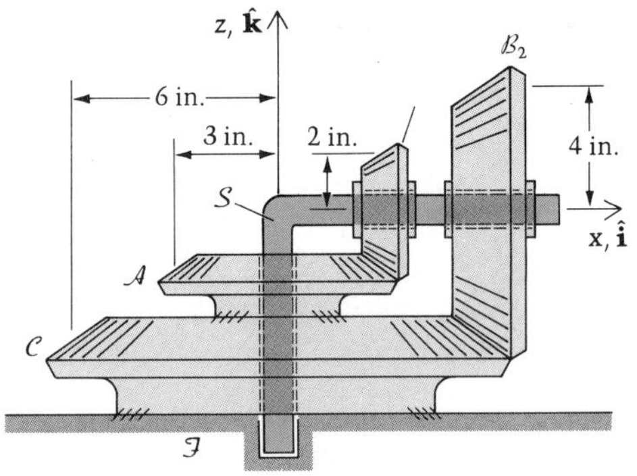

Figure P6.50

6.50 The bevel gears $\mathcal{B}_1$ and $\mathcal{B}_2$ in Figure P6.50 support the turning shaft $\mathcal{S}$, whose angular velocity is given by $\boldsymbol{\omega}_{\mathcal{S}/\mathcal{J}} = 30\hat{\mathbf{k}}$ rad/sec, in which $\hat{\mathbf{k}}$ is parallel to the z direction in both frames $\mathcal{J}$ and $\mathcal{S}$. (Gears $\mathcal{A}$ and $\mathcal{C}$ are seen to be part of $\mathcal{J}$.) Find $\boldsymbol{\omega}_{\mathcal{B}_1/\mathcal{J}}$ and $\boldsymbol{\omega}_{\mathcal{B}_2/\mathcal{J}}$.

6.51 In the preceding problem, find $\boldsymbol{\alpha}_{\mathcal{B}_1/\mathcal{J}}$ and $\boldsymbol{\alpha}_{\mathcal{B}_2/\mathcal{J}}$.

6.52 Find the angular acceleration of the wagon wheel of Problem 6.49.

6.53 Using the data of Problems 6.25 and 6.26, find the velocity and acceleration in $\mathcal{G}$ of the point Q at the end of board $\mathcal{B}$ (beneath the little girl) having position vector $\mathbf{r}_{OQ} = 5\hat{\mathbf{j}}$ ft. Do this using Equations (6.56, 6.57) of this section, then check your answers using Equations (6.44, 6.51).

6.54 The angular velocity of a rigid body $\mathcal{B}$, in motion in frame $\mathcal{J}$, is $\boldsymbol{\omega} = 3\hat{\mathbf{i}} + 2\hat{\mathbf{j}}$ rad/sec. If possible, locate (from P) a point Q of $\mathcal{B}$ with zero velocity when:

a. $\mathbf{v}_{P/\mathcal{J}} = 5\hat{\mathbf{i}}$ in./sec

b. $\mathbf{v}_{P/\mathcal{J}} = 6\hat{\mathbf{i}} - 9\hat{\mathbf{j}} + 3\hat{\mathbf{k}}$ in./sec

In both cases explain why Q exists or not in light of the discussion about points of zero velocity at the end of the preceding section.

6.55 Disk $\mathcal{D}$ in Figure P6.55 spins relative to the bent shaft $\mathcal{B}$ at constant angular speed Ω_1 rad/sec; $\mathcal{B}$ rotates in the reference frame $\mathcal{J}$ at the constant rate Ω_2 rad/sec. (The directions are indicated in the figures.) Using the rigid-body equations (6.56 and 6.57), find $\mathbf{v}_{Q/\mathcal{J}}$ and $\mathbf{a}_{Q/\mathcal{J}}$ for point Q on the periphery of the disk. Express the result in terms of components along the (x, y, z) axes, which are fixed in $\mathcal{B}$.

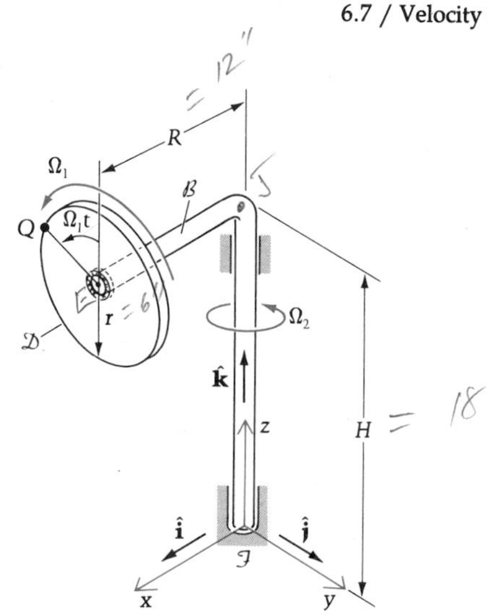

Figure P6.55

6.56 Rework the preceding problem, this time using the moving-frame concept of Section 6.5. Let bar $\mathcal{B}$ be the frame in motion with respect to $\mathcal{I}$, and let Q be the point moving relative to both $\mathcal{B}$ and $\mathcal{I}$.

6.57 A wheel of radius r turns on an axle that rotates with angular velocity $\omega_1\hat{\mathbf{k}}$ about a vertical axis (z) fixed relative to ground (Figure P6.57). If the wheel rolls on the horizontal plane and ω_1 is constant, find:

- **a.** The angular velocity and angular acceleration of the wheel relative to the ground
- **b.** The acceleration, relative to the ground, of the point on the wheel in contact with the horizontal plane.

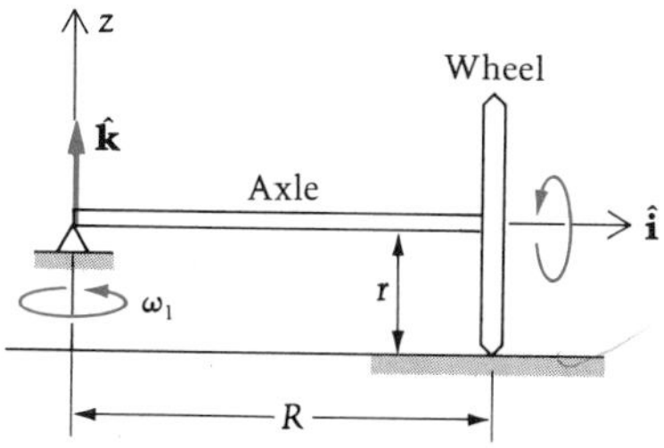

Figure P6.57

6.58 Rework Example 6.14, but this time suppose that the radius of gear $\mathcal{A}$ is 44 cm instead of 30 cm (and that its center is at the same point of $\mathcal{I}$). Explain why gears $\mathcal{A}$ and $\mathcal{B}$ are now moving in the same direction.

6.59 The bevel gear $\mathcal{A}$ in Figure P6.59 is fixed to a reference frame in which the mating gear $\mathcal{B}$ moves. The axis OC of $\mathcal{B}$ turns about the z axis at the constant rate $\Omega = 0.2$ rad/sec, and the angle OQC is 30°. Find the angular velocity of $\mathcal{B}$ in $\mathcal{A}$.

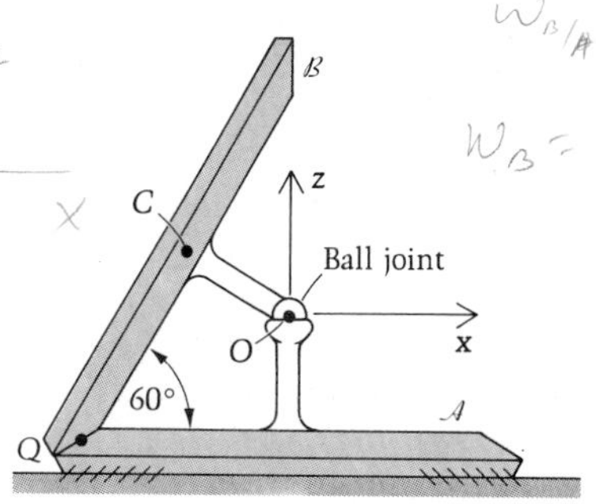

Figure P6.59

6.60 In the preceding problem, find the angular acceleration of $\mathcal{B}$ in $\mathcal{A}$ for the same defined motion.

6.61 A differential friction gear can be made with either bevel gears, as shown at the top of Figure P6.61, or friction disks, as shown at the bottom. In each case, body $\mathcal{D}$ rolls on $\mathcal{A}$ and $\mathcal{B}$ and may turn without resistance on the crank arm $\mathcal{C}$. Find $\boldsymbol{\omega}_{\mathcal{C}/\mathcal{I}}$, $\boldsymbol{\omega}_{\mathcal{D}/\mathcal{C}}$, and the velocities of points A and B of $\mathcal{D}$.

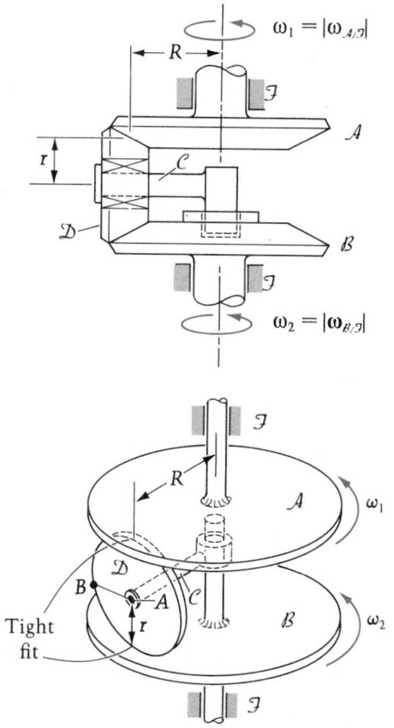

Figure P6.61

6.62 The uniform, solid, right circular cone $\mathcal{C}$ in Figure P6.62 rolls on the horizontal plane $\mathcal{I}$. Let $\mathcal{P}$ represent a frame in which the vertical axis z and the cone's axis ℓ are fixed. (Hence $\mathcal{P}$ has a simple angular velocity in $\mathcal{I}$ about the vertical.) Show that the cone can roll so that $|\boldsymbol{\omega}_{\mathcal{C}/\mathcal{I}}| = \omega$ and $|\boldsymbol{\omega}_{\mathcal{P}/\mathcal{I}}| = n$ are constants and

$$\omega \sin \alpha = n \cos \alpha$$

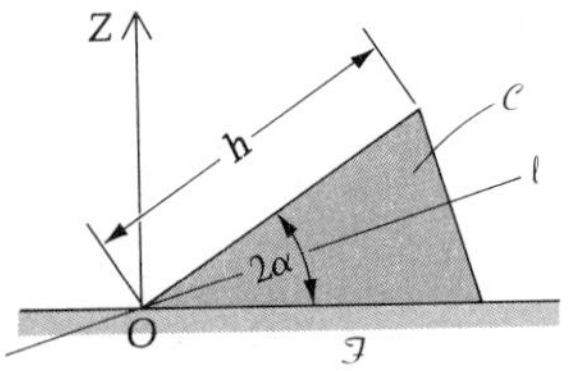

Figure P6.62

6.63 The center C of the bevel gear $\mathcal{C}$ in Figure P6.63 rotates in a horizontal circle at a constant speed of 40 mm/s (clockwise when viewed from above). The mating gear $\mathcal{D}$ is fixed to the reference frame $\mathcal{I}$; shaft $\mathcal{S}$, rigidly attached to $\mathcal{C}$, is connected to $\mathcal{I}$ through a ball and socket joint at O. Find the angular velocity vector of $\mathcal{C}$ in $\mathcal{I}$.

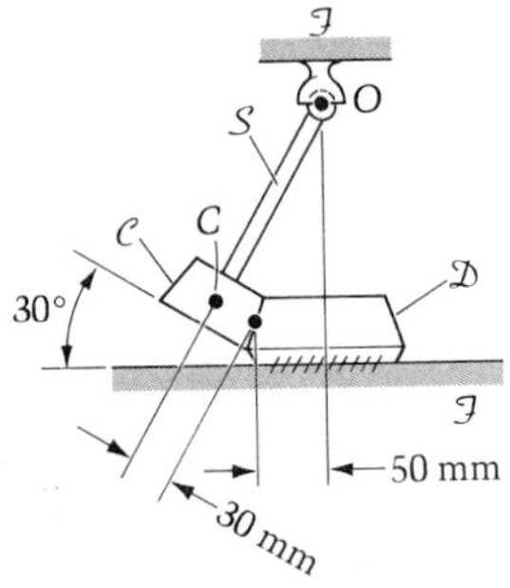

Figure P6.63

6.64 Find $\boldsymbol{\alpha}_{\mathcal{C}/\mathcal{I}}$ for the gear in Problem 6.63.

* **6.65** Plate $\mathcal{P}$ in Figure P6.65 has the following motion:

1. Corner A moves on the x axis.
2. Corner B moves on the y axis with constant velocity $6\hat{\mathbf{j}}$ in./sec.
3. Some point of the top edge of $\mathcal{P}$ (point Q at the instant shown) is always in contact with the z axis.

Find the angular velocity vector of the plate when $x_A = 3$ in.

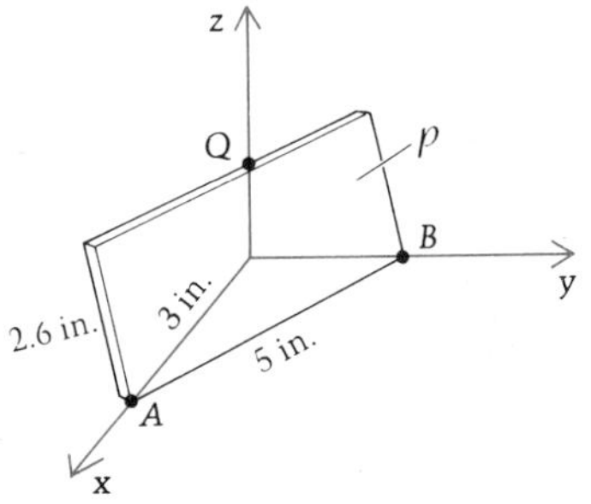

Figure P6.65

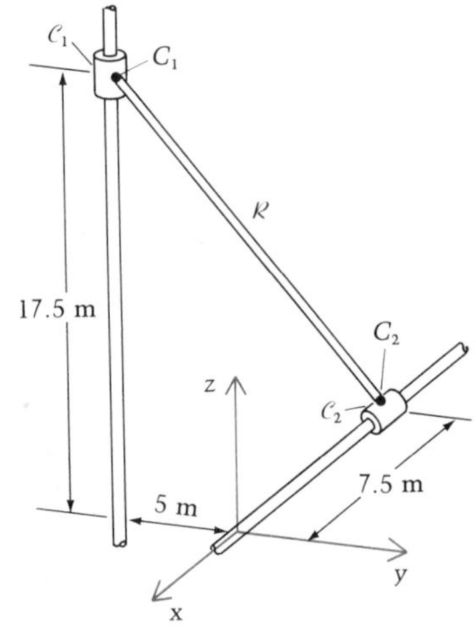

Figure P6.69

6.66 In Example 6.14 find the radius of gear $\mathcal{A}$ for which it will remain stationary in $\mathcal{I}$ as $\mathcal{B}$, $\mathcal{C}$, and $\mathcal{D}$ turn.

6.67 In Example 6.14 label the radius of gear $\mathcal{A}$ as H and call the radius of the 28-cm gear R. Show that the relationship between $\omega_{\mathcal{A}}$ and $\omega_{\mathcal{B}}$ is given by $96\omega_{\mathcal{A}}H = (16H - 20R)\omega_{\mathcal{B}}$. (The center of $\mathcal{A}$ is fixed in $\mathcal{I}$.)

6.68 If the speed of point C_2 is constant in Example 6.11, find the acceleration of C_1 at the instant given.

6.69 Collars $\mathcal{C}_1$ and $\mathcal{C}_2$ in Figure P6.69 are attached at C_1 and C_2 to rod $\mathcal{R}$ by ball and socket joints. Point C_2 has a motion along the x axis given by $x_2 = -0.012t^3$ m. Find the velocity of C_1 as a function of time.

* **6.70** Cone $\mathcal{C}_1$ rolls on cone $\mathcal{C}_2$ so that its axis of symmetry (x) moves in a horizontal plane through O, turning about z at rate Ω_2 rad/sec. (See Figure P6.70.) Cone $\mathcal{C}_2$ is rotating about ($-z$) at Ω_1 rad/sec. Find, with respect to the frame $\mathcal{I}$ in which $\mathcal{C}_2$ turns, the angular velocity and angular acceleration of $\mathcal{C}_1$, and the acceleration of point A. (Axes (x, y, z) are fixed in frame $\mathcal{B}$, which turns so that x is always along the axis of symmetry of $\mathcal{C}_1$ and z is always vertical. Also, Ω_1 and Ω_2 are constants.)

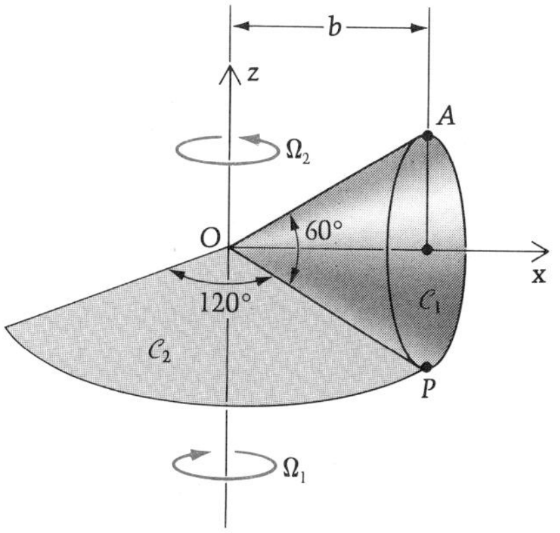

Figure P6.70

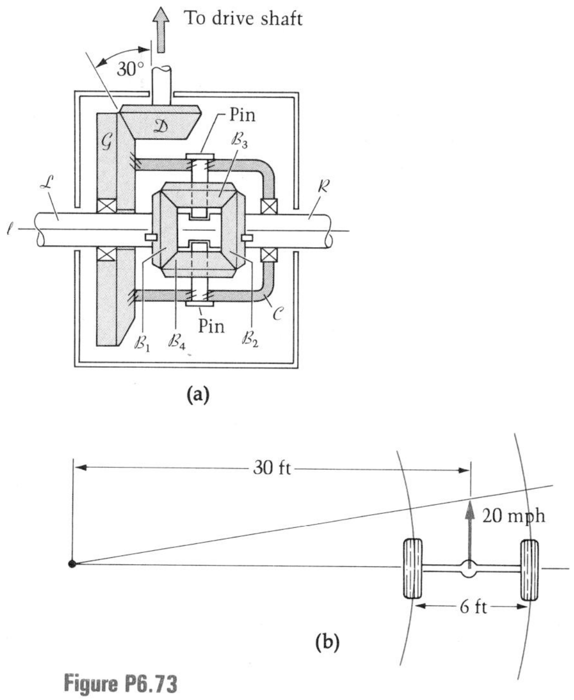

Figure P6.73

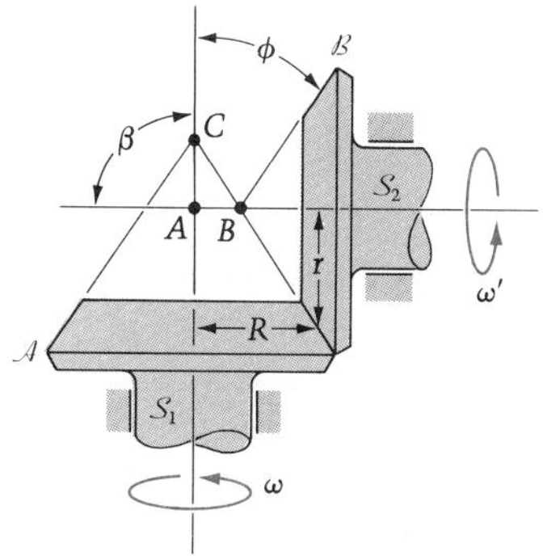

Figure P6.71

6.71 The two shafts $\mathcal{S}_1$ and $\mathcal{S}_2$ are fixed to bevel gears $\mathcal{A}$ and $\mathcal{B}$ as shown in Figure P6.71. (a) Prove that if the velocities of each pair of contacting points are to match along the line of contact of the gears, then points A, B, and C must coincide. (b) Let A, B, and C coincide and find the ratio of ω to ω'.

6.72 (a) In the preceding problem show that if $0 < \beta < 90°$, the result is still true about A, B, and C coinciding (b) Find ω/ω' for this case.

6.73 Depicted in Figure P6.73(a) are the main features of an automobile differential. The left and right axles, $\mathcal{L}$ and $\mathcal{R}$, are keyed to the bevel gears $\mathcal{B}_1$ and $\mathcal{B}_2$. Gear $\mathcal{G}$ is fixed to the case $\mathcal{C}_1$ and the combination is free to turn in bearings around line ℓ. Gear $\mathcal{G}$ meshes with gear $\mathcal{D}$ attached to the car's drive shaft. As the casing turns about the common axis ℓ of $\mathcal{L}$ and $\mathcal{R}$, its pins bear against the other two bevel gears within $\mathcal{C}$, which are $\mathcal{B}_3$ and $\mathcal{B}_4$. (Observe that these two gears do not turn about their axes at all on a straight road.)

Suppose the car makes a 30-ft turn at a speed of 20 mph (Figure P6.73(b)). If the tire radius is 14 in., find the angular velocities of $\mathcal{L}$ and $\mathcal{R}$ and use them to compute the angular velocity of $\mathcal{D}$. In the process note that the differential allows the driven wheels to turn at different angular speeds.

It is assumed in the following three problems that not all points of body $\mathcal{B}$ have zero acceleration.

6.74 Show that for a rigid body $\mathcal{B}$ in general motion, there is a point Q of zero acceleration if $\boldsymbol{\omega} \neq \mathbf{0}, \boldsymbol{\alpha} \neq \mathbf{0}$, and $\boldsymbol{\alpha}$ is not parallel to $\boldsymbol{\omega}$. *Hint:* Let P be an arbitrary point, and let $\boldsymbol{\omega} = \omega\hat{\mathbf{i}}$, $\boldsymbol{\alpha} = \alpha_1\hat{\mathbf{i}} + \alpha_2\hat{\mathbf{j}}$, and $\mathbf{a}_P = a_{P_1}\hat{\mathbf{i}} + a_{P_2}\hat{\mathbf{j}} + a_{P_3}\hat{\mathbf{k}}$, noting that there is no loss in generality in these assumptions. Set $\mathbf{a}_Q = \mathbf{0} = \mathbf{a}_P + \boldsymbol{\alpha} \times \mathbf{r}_{PQ} + \boldsymbol{\omega} \times (\boldsymbol{\omega} \times \mathbf{r}_{PQ})$, set $\mathbf{r}_{PQ} = x\hat{\mathbf{i}} + y\hat{\mathbf{j}} + z\hat{\mathbf{k}}$, and solve for x, y, and z.

6.75 (a) Following up the previous problem, show that if $\boldsymbol{\omega} = \mathbf{0}$ and $\boldsymbol{\alpha} \neq \mathbf{0}$ at a given instant, then at this instant there is a point Q of zero acceleration if and only if the accelerations of all points of B are perpendicular to $\boldsymbol{\alpha}$. (b) Investigate the case $\boldsymbol{\omega} \neq \mathbf{0}$ and $\boldsymbol{\alpha} = \mathbf{0}$.

6.76 Here is another follow-up on Problem 6.74: Show that at any instant when the two vectors $\boldsymbol{\omega}$ and $\boldsymbol{\alpha}$ are parallel, there is a point of $\mathcal{B}$ with zero acceleration if and only if the accelerations of all its points are perpendicular to $\boldsymbol{\omega}$ and $\boldsymbol{\alpha}$.

* **6.77** Find the angular acceleration of the plate of Problem 6.65 at the same instant of time.

6.8 Describing the Orientation of a Rigid Body

In the case of plane motion of a rigid body $\mathcal{B}$, if we know the angular velocity

$$\omega = \omega\hat{\mathbf{k}} = \dot{\theta}\hat{\mathbf{k}}$$

as a function of time, we may clearly integrate to find the orientation of $\mathcal{B}$ at any time t:

$$\int_0^t \dot{\theta}(\xi)\,d\xi = \theta(t) - \theta(0)$$

Thus in plane motion we may completely specify the position of $\mathcal{B}$ by giving the xy coordinates of a point (usually the mass center C is chosen to be the point) and the orientation angle θ.

The Eulerian Angles

A major difference between planar and general motion is that the angles which yield a body's orientation in space in three-dimensional motion are *not* the integrals by simple quadratures of the angular velocity components of the body. In fact, finding (in general) the orientation of a body $\mathcal{B}$ in closed form, given $\omega(t)$ and the orientation of $\mathcal{B}$ at $t = 0$, is an unsolved fundamental problem in rigid-body kinematics. We now introduce the **Eulerian angles** in order to show the difficulty of determining a body's orientation in space when the motion is nonplanar.

We begin with the body $\mathcal{B}$ oriented so that the body-fixed axes (x, y, z) initially coincide respectively with axes (X, Y, Z) embedded in the reference frame $\mathcal{I}$. Let $(\hat{\mathbf{i}}, \hat{\mathbf{j}}, \hat{\mathbf{k}})$ and $(\hat{\mathbf{I}}, \hat{\mathbf{J}}, \hat{\mathbf{K}})$ be sets of unit vectors respectively parallel to (x, y, z) and (X, Y, Z). Three successive rotations about specific axes will now be described that will orient $\mathcal{B}$ in $\mathcal{I}$. (See Figure 6.12(a))

In Figure 6.12(b) the first rotation is through the angle ϕ about the Z axis. Let the new positions of (x, y, z) after this first rotation be denoted by (x_1, y_1, z_1) as shown; these positions are embedded in an intermediate frame $\mathcal{I}_1$. Note that axes Z and z_1 are identical and that $\mathcal{I}_1$ has the simple angular velocity $\dot{\phi}\hat{\mathbf{K}}$ in $\mathcal{I}$. Note also from Figure 6.12 that $\hat{\mathbf{n}}_{11}$, $\hat{\mathbf{n}}_{12}$, and $\hat{\mathbf{n}}_{13}$ are unit vectors that are respectively and always parallel to x_1, y_1, z_1. Next (Figure 6.12(c)) a rotation through the angle θ about axis y_1 moves the body axes into the coordinate directions (x_2, y_2, z_2) of a second intermediate frame $\mathcal{I}_2$ having unit vectors $(\hat{\mathbf{n}}_{21}, \hat{\mathbf{n}}_{22}, \hat{\mathbf{n}}_{23})$. A final rotation, this time of amount ψ about z_2 (Figure 6.12(d)), turns the body axes into their final positions in $\mathcal{B}$, indicated by (x, y, z).

It is clear that we may use the addition theorem to express the angular velocity of $\mathcal{B}$ in $\mathcal{I}$ as follows:

$$\omega_{\mathcal{B}/\mathcal{I}} = \omega_{\mathcal{B}/\mathcal{I}_2} + \omega_{\mathcal{I}_2/\mathcal{I}_1} + \omega_{\mathcal{I}_1/\mathcal{I}}$$

$$\omega_{\mathcal{B}/\mathcal{I}} = \dot{\psi}\hat{\mathbf{k}} + \dot{\theta}\hat{\mathbf{n}}_{12} + \dot{\phi}\hat{\mathbf{K}} \qquad (6.58)$$

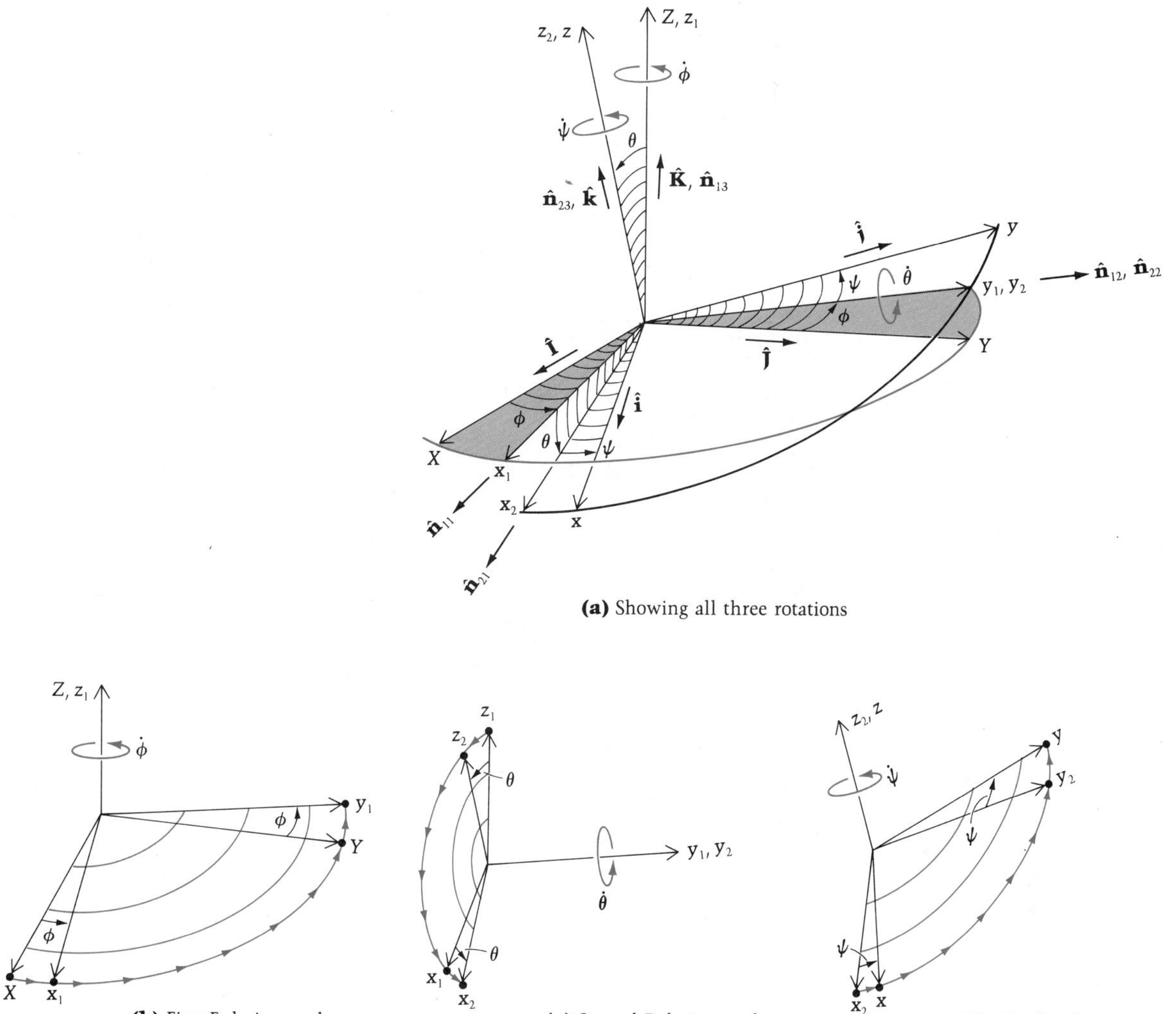

(a) Showing all three rotations

(b) First Eulerian angle rotation (ϕ), about Z

(c) Second Eulerian angle rotation (θ), about y_1

(d) Third Eulerian angle rotation (ψ), about z_2

Figure 6.12 Eulerian angles.

in which we again remark that $\hat{\mathbf{n}}_{ij}$ is a unit vector in the jth coordinate direction of frame $\mathcal{F}_i (i = 1, 2)$.

If the preceding expression for $\boldsymbol{\omega}_{\mathcal{B}/\mathcal{I}}$ is to be functionally useful, its three terms should all be expressed in the *same frame* — that is, in terms of components associated with the directions of base vectors all fixed in direction in a common frame. Now in practice, as we shall see, they are sometimes expressed by components associated with body-fixed directions, at other times with space-fixed directions, and at still other times with directions fixed in intermediate frames such as $\mathcal{F}_1$ or $\mathcal{F}_2$. For example, let us write $\boldsymbol{\omega}_{\mathcal{B}/\mathcal{I}}$ in body components. This means, looking at Equation (6.58), that we must express $\hat{\mathbf{n}}_{12}$ and $\hat{\mathbf{K}}$ in terms of $\hat{\mathbf{i}}$, $\hat{\mathbf{j}}$, and $\hat{\mathbf{k}}$.

We see from Figure 6.12 that

$$\hat{\mathbf{n}}_{12} = \sin\psi\hat{\mathbf{i}} + \cos\psi\hat{\mathbf{j}} \tag{6.59}$$

and that

$$\begin{aligned}\hat{\mathbf{K}} &= -\sin\theta\hat{\mathbf{n}}_{21} + \cos\theta\hat{\mathbf{k}} \\ &= -\sin\theta(\cos\psi\hat{\mathbf{i}} - \sin\psi\hat{\mathbf{j}}) + \cos\theta\hat{\mathbf{k}}\end{aligned}$$

or

$$\hat{\mathbf{K}} = -\mathrm{s}_\theta \mathrm{c}_\psi \hat{\mathbf{i}} + \mathrm{s}_\theta \mathrm{s}_\psi \hat{\mathbf{j}} + \mathrm{c}_\theta \hat{\mathbf{k}} \tag{6.60}$$

where $\mathrm{s}_\theta = \sin\theta$, $c_\psi = \cos\psi$, and so forth. Substituting these expressions into Equation (6.58) then gives $\boldsymbol{\omega}_{\mathcal{B}/\mathcal{I}}$ in body components:

$$\boldsymbol{\omega}_{\mathcal{B}/\mathcal{I}} = (\mathrm{s}_\psi\dot{\theta} - \mathrm{s}_\theta \mathrm{c}_\psi\dot{\phi})\hat{\mathbf{i}} + (\mathrm{c}_\psi\dot{\theta} + \mathrm{s}_\theta \mathrm{s}_\psi\dot{\phi})\hat{\mathbf{j}} + (\dot{\psi} + \mathrm{c}_\theta\dot{\phi})\hat{\mathbf{k}} \tag{6.61}$$

Alternatively, we may express $\boldsymbol{\omega}_{\mathcal{B}/\mathcal{I}}$ in terms of its components in frame $\mathcal{I}$ by writing $\hat{\mathbf{k}}$ and $\hat{\mathbf{n}}_{12}$ in terms of $\hat{\mathbf{I}}$, $\hat{\mathbf{J}}$, and $\hat{\mathbf{K}}$. Again referring to the figure, we see that

$$\hat{\mathbf{n}}_{12} = -\mathrm{s}_\phi\hat{\mathbf{I}} + \mathrm{c}_\phi\hat{\mathbf{J}} \tag{6.62}$$

and

$$\begin{aligned}\hat{\mathbf{k}} &= \sin\theta\hat{\mathbf{n}}_{11} + \cos\theta\hat{\mathbf{K}} \\ &= \sin\theta(\cos\phi\hat{\mathbf{I}} + \sin\phi\hat{\mathbf{J}}) + \cos\theta\hat{\mathbf{K}}\end{aligned}$$

or

$$\hat{\mathbf{k}} = \mathrm{s}_\theta \mathrm{c}_\phi\hat{\mathbf{I}} + \mathrm{s}_\theta \mathrm{s}_\phi\hat{\mathbf{J}} + \mathrm{c}_\theta\hat{\mathbf{K}} \tag{6.63}$$

so that, substituting into Equation (6.58), we have $\boldsymbol{\omega}_{\mathcal{B}/\mathcal{I}}$ written in $\mathcal{I}$:

$$\boldsymbol{\omega}_{\mathcal{B}/\mathcal{I}} = (-\mathrm{s}_\phi\dot{\theta} + \mathrm{s}_\theta \mathrm{c}_\phi\dot{\psi})\hat{\mathbf{I}} + (\mathrm{c}_\phi\dot{\theta} + \mathrm{s}_\theta \mathrm{s}_\phi\dot{\psi})\hat{\mathbf{J}} + (\dot{\phi} + \mathrm{c}_\theta\dot{\psi})\hat{\mathbf{K}} \tag{6.64}$$

Finally, we notice that expressing $\boldsymbol{\omega}_{\mathcal{B}/\mathcal{I}}$ in the *intermediate* frames is easier still. To write it in terms of its components in $\mathcal{I}_1$, we note that

$$\hat{\mathbf{K}} \equiv \hat{\mathbf{n}}_{13} \tag{6.65}$$

and

$$\hat{\mathbf{k}} = \mathrm{c}_\theta\hat{\mathbf{n}}_{13} + \mathrm{s}_\theta\hat{\mathbf{n}}_{11} \tag{6.66}$$

so that from Equation (6.58) we get

$$\boldsymbol{\omega}_{\mathcal{B}/\mathcal{I}} = \mathrm{s}_\theta\dot{\psi}\hat{\mathbf{n}}_{11} + \dot{\theta}\hat{\mathbf{n}}_{12} + (\mathrm{c}_\theta\dot{\psi} + \dot{\phi})\hat{\mathbf{n}}_{13} \tag{6.67}$$

All these expressions for $\boldsymbol{\omega}_{\mathcal{B}/\mathcal{I}}$ appear different, but of course they all represent the *same vector* written in different frames. The components may vary from frame to frame, but the vector is the same. In the first exercise at the end of this section, the reader will be asked to write $\boldsymbol{\omega}_{\mathcal{B}/\mathcal{I}}$ in yet another frame, $\mathcal{I}_2$.

The angles (ϕ, θ, ψ) are known as the Eulerian angles. They represent one way of orientating a rigid body in space. Unfortunately the Eulerian angles (ϕ, θ, ψ) do not carry the same symbol from one book to the next; worse still, the order and even the directions of the rotations vary from writer to writer. Obviously, then, it is important to choose a set to work with and then be consistent.

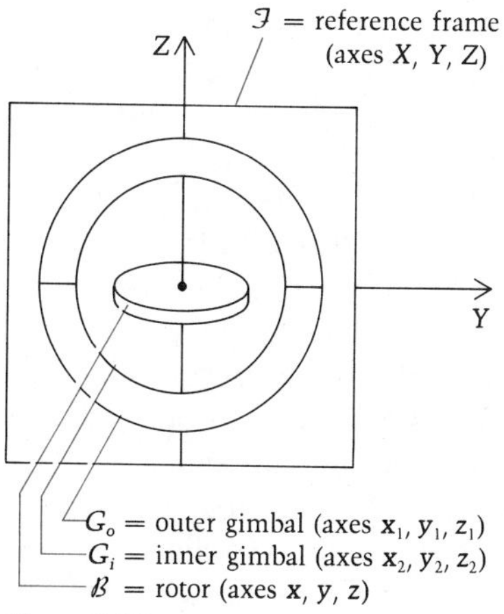

Figure 6.13

A physical feel for the Eulerian angles may be gained by considering a gyroscope $\mathcal{B}$ spinning in a Cardan suspension as shown in Figure 6.13. In this system the Eulerian angles (ϕ, θ, ψ) may be used as follows to pinpoint the orientation of the rotor $\mathcal{B}$ in space:

1. First Rotation: With respect to frame $\mathcal{I}$, we rotate the plane of the outer gimbal G_o (frame $\mathcal{I}_1$ in the earlier theory) about axis Z through angle ϕ. Axis X is turned into x_1 and axis Y into y_1; frames (bodies) G_i and $\mathcal{B}$ are not shown yet, but they move rigidly with G_o in this first rotation. (See Figure 6.14.)

2. Second Rotation: Next we turn the plane of the inner gimbal $G_i(\mathcal{I}_2)$ about axis y_1, through angle θ, thereby tilting G_i with respect to G_o. Axis z_1 is thereby rotated into z_2 and axis x_1 into x_2. Body $\mathcal{B}$ (not shown in Figure 6.15) goes along for the ride.

3. Third Rotation: The third and last rotation turns the rotor $\mathcal{B}$ about axis z_2 through angle ψ. (See Figure 6.16.) This allows $\mathcal{B}$ to spin relative to G_i. Axis x_2 is turned into x and axis y_2 into y.

Through these three Eulerian angle rotations, the body ($\mathcal{B}$) can be positioned in any desired orientation in space ($\mathcal{I}$). We are now able to see the difficulty of solving for the orientation of a rigid body in general motion. If we write $\omega_{\mathcal{B}/\mathcal{I}}$ as $\omega_1\hat{\mathbf{i}} + \omega_2\hat{\mathbf{j}} + \omega_3\hat{\mathbf{k}}$, then Equation (6.61) is equivalent to

$$\begin{aligned} s_\psi\dot{\theta} - s_\theta c_\psi\dot{\phi} &= \omega_1 \\ c_\psi\dot{\theta} + s_\theta s_\psi\dot{\phi} &= \omega_2 \\ \dot{\psi} + c_\theta\dot{\phi} &= \omega_3 \end{aligned} \tag{6.68}$$

Solving for the rates of change of the Eulerian angles gives

$$\begin{aligned} \dot{\theta} &= \omega_1 s_\psi + \omega_2 c_\psi \\ \dot{\phi} &= \frac{\omega_2 s_\psi - \omega_1 c_\psi}{s_\theta} \\ \dot{\psi} &= \frac{\omega_3 s_\theta - \omega_2 c_\theta s_\psi + \omega_1 c_\theta c_\psi}{s_\theta} \end{aligned} \tag{6.69}$$

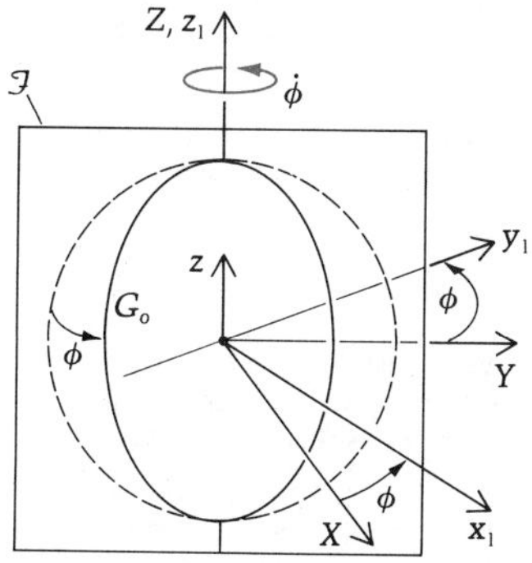

Figure 6.14 First rotation.

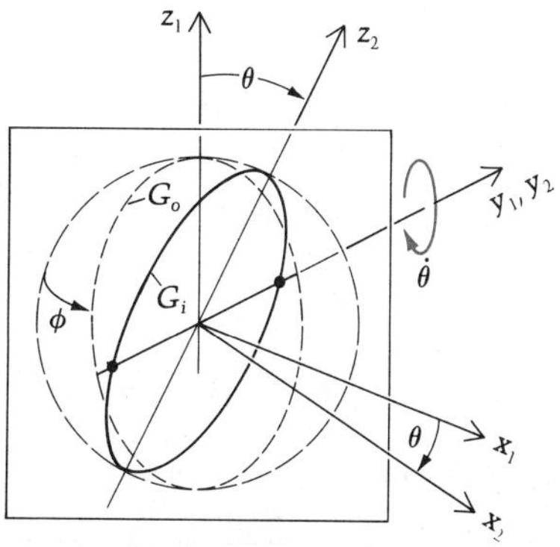

Figure 6.15 Second rotation.

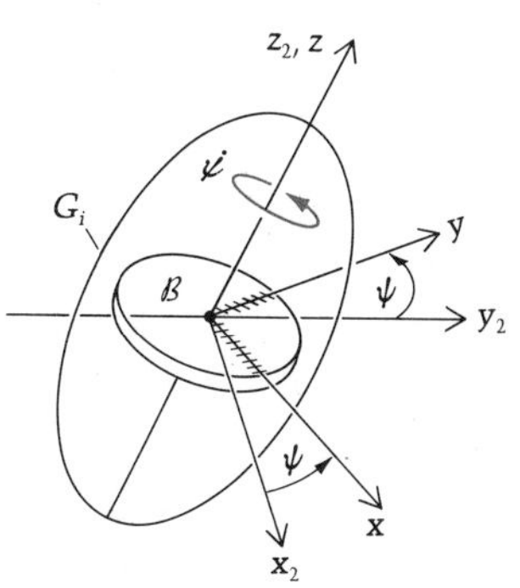

Figure 6.16 Third rotation.

We see from these equations that even if we knew the ω components as functions of time in closed form,* it would still be a formidable task to integrate Equations (6.69) analytically to obtain the Eulerian angles and thus to know the body's orientation in space. This is usually not even possible, and resort is made to computers that can numerically carry out integrations with a step-by-step scheme such as Runge-Kutta.

Incidentally, the sin θ denominators in Equations (6.69) present serious obstacles in the dynamics of space vehicles; whenever θ is zero or a multiple of π, the equations develop a singularity. Sophisticated programming or, in some cases, completely different mathematical schemes for orienting the body are required to overcome such difficulties.

We mention that use of the preceding set of Eulerian angles as defined requires that we maintain the *order* of rotation. To illustrate the importance of rotation order, we remark that if this book is rotated through two $\pi/2$ rotations about the space axes Y and Z, in opposite orders as suggested by Figure 6.17, it will end up in a different position. We should point out, however, that there are ways of setting up the axes and angles which make a body's final orientation independent of order. For instance, just as in our Eulerian angle development, let Z be fixed in $\mathcal{I}$, let z be fixed in $\mathcal{B}$, and let y be always perpendicular to both Z and z. But now restrict Z and z to be nonparallel. (Let z, the axis of $\mathcal{B}$, lie along X initially, for example.) In this case, the angles (ϕ, θ, ψ) as defined earlier may be performed in any of the six possible orders and the body's resulting orientation in $\mathcal{I}$ will be the same each time!†

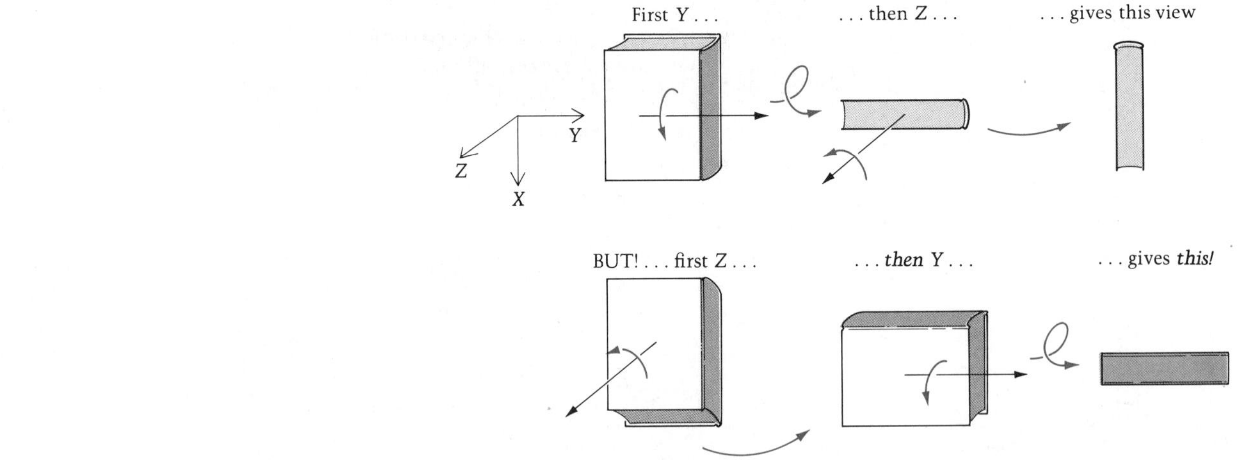

Figure 6.17 Finite rotations.

* The equations governing these three components of angular velocity are the Euler equations of rigid-body kinetics. They themselves are also nonlinear and unsolvable in closed form in general three-dimensional motion except for a few special cases. We shall be studying these equations in Chapter 7.

† Also, subsequent reorientations will likewise be order-independent; see "Successive Finite Rotations," by T. R. Kane and D. A. Levinson, *Journal of Applied Mechanics*, Dec. 1978, Vol. 45, pp. 945–946.

An alternative set of three rotations has become popular in the literature in recent years:

1. Rotate through θ_1 about X.
2. Then rotate through θ_2 about y_1.
3. Then rotate through θ_3 about z_2.

This sequence results in angular velocity components in $\mathcal{B}$ — that is, in the body-axis system (x, y, z) — of

$$\begin{aligned}\omega_1 &= \dot{\theta}_1 \cos\theta_2 \cos\theta_3 + \dot{\theta}_2 \sin\theta_3 \\ \omega_2 &= -\dot{\theta}_1 \cos\theta_2 \sin\theta_3 + \dot{\theta}_2 \cos\theta_3 \\ \omega_3 &= \dot{\theta}_1 \sin\theta_2 + \dot{\theta}_3\end{aligned} \tag{6.70}$$

Although (6.70) is not made up of classic Eulerian angles, the result is an equally valid set of relations between the $\omega_{\mathcal{B}/\mathcal{I}}$ components and the angles θ_i of rotation. The order and axes of rotations of the θ_i in this system are quite simple to remember.

There are alternatives to using the Eulerian angles or the angles in Equation (6.70) to orient a body in space. One such alternative is to use the quaternions of Hamilton, which are free of the singular points caused by zero denominators at certain values of the Eulerian angles. (For this reason, quaternions were used in the Skylab Orbital Assembly's attitude control system.) Another approach to the orientation of a body in space is to determine the direction cosines of a unit vector, fixed in direction in space $\mathcal{I}$, with respect to a set of axes fixed in the body $\mathcal{B}$. Let this unit vector (call it $\hat{\mathbf{u}}$) have direction cosines (p, q, r) with respect to axes (x, y, z) fixed in $\mathcal{B}$. Further, let $(\hat{\mathbf{i}}, \hat{\mathbf{j}}, \hat{\mathbf{k}})$ be unit vectors, always respectively parallel to (x, y, z). Then

$$\hat{\mathbf{u}} = p\hat{\mathbf{i}} + q\hat{\mathbf{j}} + r\hat{\mathbf{k}}$$

and, differentiating in $\mathcal{I}$, we obtain

$$\begin{aligned}{}^{\mathcal{I}}\dot{\hat{\mathbf{u}}} = \mathbf{0} &= {}^{\mathcal{B}}\dot{\hat{\mathbf{u}}} + \boldsymbol{\omega}_{\mathcal{B}/\mathcal{I}} \times \hat{\mathbf{u}} \\ &= (\dot{p}\hat{\mathbf{i}} + \dot{q}\hat{\mathbf{j}} + \dot{r}\hat{\mathbf{k}}) + (\omega_x\hat{\mathbf{i}} + \omega_y\hat{\mathbf{j}} + \omega_z\hat{\mathbf{k}}) \times (p\hat{\mathbf{i}} + q\hat{\mathbf{j}} + r\hat{\mathbf{k}}) \\ &= (\dot{p} + \omega_y r - \omega_z q)\hat{\mathbf{i}} + (\dot{q} + \omega_z p - \omega_x r)\hat{\mathbf{j}} + (\dot{r} + \omega_x q - \omega_y p)\hat{\mathbf{k}}\end{aligned}$$

This vector equation has the following scalar component equations:

$$\begin{aligned}\dot{p} &= \omega_z q - \omega_y r \\ \dot{q} &= \omega_x r - \omega_z p \\ \dot{r} &= \omega_y p - \omega_x q\end{aligned} \tag{6.71}$$

If the $\boldsymbol{\omega}$ components are either prescribed or else found from kinetics equations (to be studied in Chapter 7), then Equations (6.71) may be solved for the direction cosines, thereby orienting $\mathcal{B}$ in $\mathcal{I}$. Equations (6.71) are known as the Poisson equations. They are alternatives to equations such as (6.68) and (6.70).

PROBLEMS ▶ Section 6.8

6.78 Using the Eulerian angles (ϕ, θ, ψ) discussed in this section, express $\omega_{\mathcal{B}/\mathcal{I}}$ in (terms of its components in) the frame $\mathcal{I}_2$.

6.79 Show that the magnitudes of $\omega_{\mathcal{B}/\mathcal{I}}$ are all the same as expressed (a) in $\mathcal{B}$ in Equation (6.61); (b) in $\mathcal{I}$ in Equation (6.64); and (c) in $\mathcal{I}_1$ in Equation (6.67).

6.80 Derive Equations (6.70).

6.81 Write $\omega_{\mathcal{B}/\mathcal{I}}$ in $\mathcal{I}$ by using the successive rotations θ_1, θ_2, and θ_3 that resulted in Equation (6.70) when expressed in $\mathcal{B}$.

6.82 Euler's theorem for finite rotations is stated as follows: The most general rotation of a rigid body $\mathcal{B}$ with respect to a point A is equivalent to a rotation about some axis through A. Prove the theorem. *Hint*: Let A be considered fixed in the reference frame in which $\mathcal{B}$ moves. (See Figure P6.82.) Let point P be at P_1 prior to the rotation and at P_2 afterward; assume the same for point Q (Q_1 before, Q_2 after). Bisect angle P_1AP_2 with a plane normal to the plane of the angle. Do the same for Q_1AQ_2 and consider the intersection of the two planes.

6.83 Chasle's theorem states: The most general displacement of a rigid body is equivalent to the translation of some point A followed by a rotation about an axis through A. Show that this result follows immediately from the previous problem.

6.84 The circular drum of radius R in Figure P6.84 is pivoted to a support at O, where O is a distance $R/2$ from the center C of the drum. A weight $\mathcal{W}$ (particle) hangs from a cord wrapped around the drum. The drum is slowly rotated $\pi/2$ rad clockwise about O. Find the displacement of $\mathcal{W}$. *Hint*: Use the result of the preceding problem, with C being the point A and with a $\pi/2$ ↻ rotation following the translation. Add the displacements of $\mathcal{W}$ during each part.

6.85 In Figure P6.85, the sphere $\mathcal{S}$ rolls on the plane, and its angular velocity in the reference frame $\mathcal{I}$, in which (x, y, z) are fixed, is given by Equation (6.64). Noting that $\mathbf{v}_C = \dot{x}\hat{\mathbf{i}} + \dot{y}\hat{\mathbf{j}}$, write the constraint equations (the "no slip" conditions) relating $\dot{x}$ and $\dot{y}$ to the Eulerian angles ϕ, θ, ψ and their derivatives.

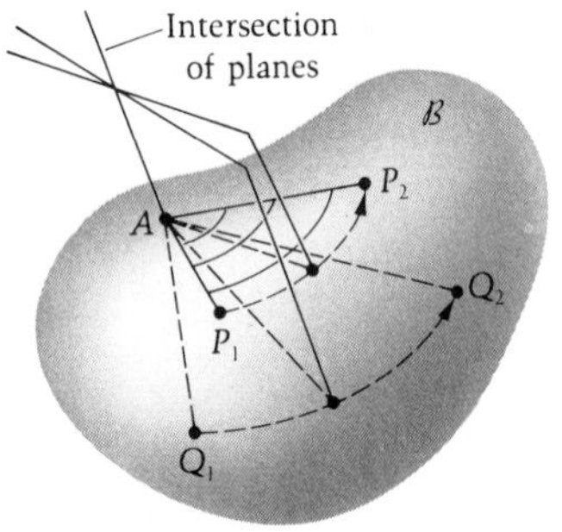

Figure P6.82

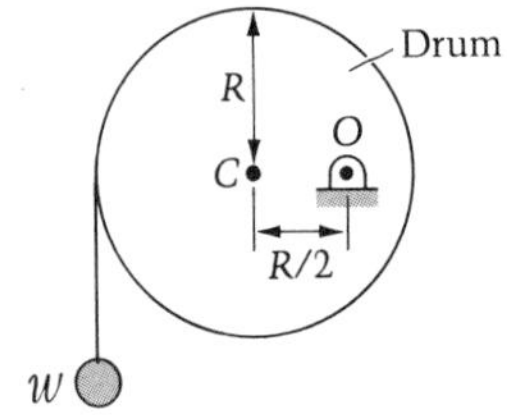

Figure P6.84

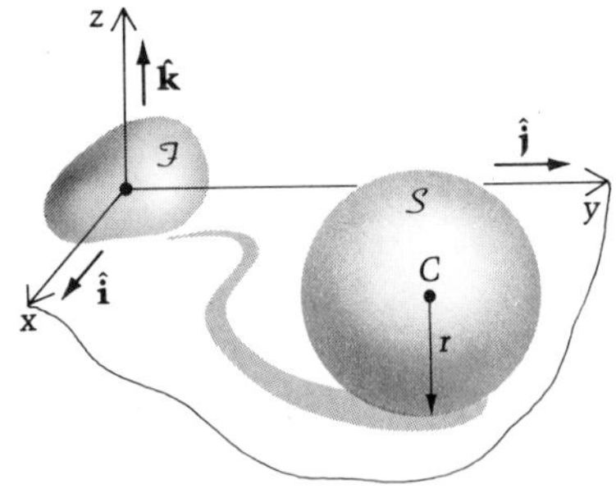

Figure P6.85

6.9 Rotation Matrices

Suppose that a vector $\mathbf{Q}$ is written in a frame $\mathcal{I}$ and that its components are the elements of a column matrix $\{Q\}_{\mathcal{I}}$. It is possible to develop a set of 3 × 3 matrices $[T_x]$, $[T_y]$, and $[T_z]$ each of which, when postmultiplied by $\{Q\}_{\mathcal{I}}$, gives the components of $\mathbf{Q}$ in a new frame rotated about an x, y, or z axis, respectively, of $\mathcal{I}$. These matrices are handy work-savers. For example, they reduce the work involved, in going from Equation (6.58) to (6.61) or (6.64), to a pair of simple matrix multiplications.

We shall develop $[T_z]$ and then state the results for $[T_x]$ and $[T_y]$, which are derived similarly. Let $Q = Q_x\hat{\mathbf{i}} + Q_y\hat{\mathbf{j}} + Q_z\hat{\mathbf{k}}$, in which ($\hat{\mathbf{i}}$, $\hat{\mathbf{j}}$, $\hat{\mathbf{k}}$) are a triad of unit vectors having fixed directions along axes (x, y, z) of frame $\mathcal{I}$. Suppose further that $\mathcal{B}$ is a frame whose orientation may be obtained from that of $\mathcal{I}$ by a rotation through the angle θ_z about z. The rotated axes, which were aligned with (x, y, z) prior to the rotation, will be denoted (x_1, y_1, z_1) with associated unit vectors $(\hat{\mathbf{i}}_1, \hat{\mathbf{j}}_1, \hat{\mathbf{k}}_1)$.

We note that in the "new" frame $\mathcal{B}$, we may write $\mathbf{Q} = Q_{x_1}\hat{\mathbf{i}}_1 + Q_{y_1}\hat{\mathbf{j}}_1 + Q_{z_1}\hat{\mathbf{k}}_1$, where $Q_{z_1} = Q_z$ and $\hat{\mathbf{k}}_1 = \hat{\mathbf{k}}$ since the rotation is about this axis, common to both frames. To get Q_{x_1} and Q_{y_1} in terms of Q_x, Q_y, and θ_z, we use (see Figure 6.18):

$$\begin{aligned} Q_{x_1} &= \mathbf{Q} \cdot \hat{\mathbf{i}}_1 = (Q_x\hat{\mathbf{i}} + Q_y\hat{\mathbf{j}} + Q_z\hat{\mathbf{k}}) \cdot \hat{\mathbf{i}}_1 \\ &= Q_x(\hat{\mathbf{i}} \cdot \hat{\mathbf{i}}_1) + Q_y(\hat{\mathbf{j}} \cdot \hat{\mathbf{i}}_1) + Q_z \cancelto{0}{(\hat{\mathbf{k}} \cdot \hat{\mathbf{i}}_1)} \\ &= Q_x \cos\theta_z + Q_y \sin\theta_z \end{aligned} \tag{6.72}$$

and similarly,

$$\begin{aligned} Q_{y_1} &= \mathbf{Q} \cdot \hat{\mathbf{j}}_1 = Q_x(\hat{\mathbf{i}} \cdot \hat{\mathbf{j}}_1) + Q_y(\hat{\mathbf{j}} \cdot \hat{\mathbf{j}}_1) + Q_z \cancelto{0}{(\hat{\mathbf{k}} \cdot \hat{\mathbf{j}}_1)} \\ &= Q_x(-\sin\theta_z) + Q_y \cos\theta_z \end{aligned} \tag{6.73}$$

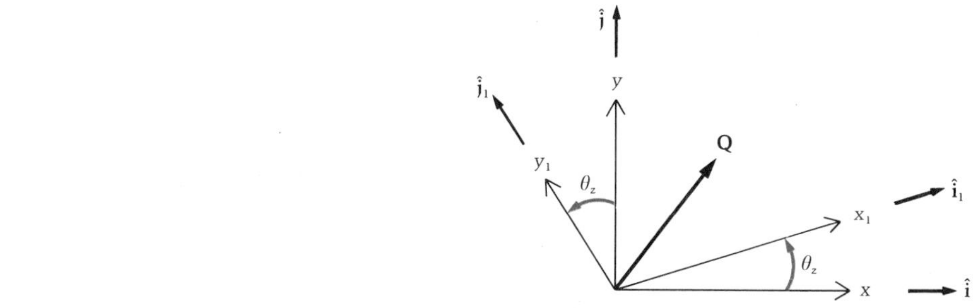

Figure 6.18

Together with $Q_{z_1} = Q_z$, the Equations (6.72, 6.73) give the components of $\mathbf{Q}$ in the rotated frame $\mathcal{B}$, in terms of its components Q_x, Q_y, Q_z) back in $\mathcal{I}$. Now we are ready to observe that if the matrix $[T_z]$ is defined as

$$[T_z] = \begin{bmatrix} \cos\theta_z & \sin\theta_z & 0 \\ -\sin\theta_z & \cos\theta_z & 0 \\ 0 & 0 & 1 \end{bmatrix}$$

then the same results for the components of $\mathbf{Q}$ in the rotated frame $\mathcal{B}$ are obtained from the matrix product $[T_z]\{Q\}_{\mathcal{I}}$:

$$\begin{aligned} \{Q\}_{\mathcal{B}} = [T_z]\{Q\}_{\mathcal{I}} &= \begin{bmatrix} \cos\theta_z & \sin\theta_z & 0 \\ -\sin\theta_z & \cos\theta_z & 0 \\ 0 & 0 & 1 \end{bmatrix} \begin{Bmatrix} Q_x \\ Q_y \\ Q_z \end{Bmatrix} \\ &= \begin{Bmatrix} Q_x \cos\theta_z + Q_y \sin\theta_z \\ -Q_x \sin\theta_z + Q_y \cos\theta_z \\ Q_z \end{Bmatrix} \end{aligned}$$

The rotation matrices for rotations of θ_x about x, and θ_y about y, are respectively given by $[T_x]$ and $[T_y]$:

$$[T_x] = \begin{bmatrix} 1 & 0 & 0 \\ 0 & \cos\theta_x & \sin\theta_x \\ 0 & -\sin\theta_x & \cos\theta_x \end{bmatrix} \qquad [T_y] = \begin{bmatrix} \cos\theta_y & 0 & -\sin\theta_y \\ 0 & 1 & 0 \\ \sin\theta_y & 0 & \cos\theta_y \end{bmatrix}$$

The student may wish to verify one or both of these matrices, as we did for $[T_z]$. Note the change in the "sign of sine" in the $[T_y]$ matrix. Moreover, if we must turn about an axis through a negative angle, we need only change the signs of both sine terms; this follows from the fact that $\cos(-\theta) = \cos\theta$, while $\sin(-\theta) = -\sin\theta$. We now consider examples of the use of the rotation matrices. We shall use some shorthand common in the literature of kinematics: s_θ for $\sin\theta$, c_ϕ for $\cos\phi$, etc.

EXAMPLE 6.15

Use rotation matrices to obtain the components of $\omega_{\mathcal{B}/\mathcal{I}}$ in body coordinates, given its representation (Equation 6.64) in the reference or space frame $\mathcal{I}$.

Solution

We premultiply $\omega_{\mathcal{B}/\mathcal{I}}$, expressed in $\mathcal{I}$ in matrix form, with rotation matrices of ϕ about the 3-axis, then θ about the new 2-axis, and then ψ about the new and final 3-axis:

$$\{\omega_{\mathcal{B}/\mathcal{I}}\}_{\mathcal{B}} = \underset{(\text{angle } \psi)}{[T_z]} \quad \underset{(\text{angle } \theta)}{[T_y]} \quad \underset{(\text{angle } \phi)}{[T_z]} \quad \{\omega_{\mathcal{B}/\mathcal{I}}\}_{\mathcal{I}}$$

$$= \begin{bmatrix} c_\psi & s_\psi & 0 \\ -s_\psi & c_\psi & 0 \\ 0 & 0 & 1 \end{bmatrix} \begin{bmatrix} c_\theta & 0 & -s_\theta \\ 0 & 1 & 0 \\ s_\theta & 0 & c_\theta \end{bmatrix} \underbrace{\begin{bmatrix} c_\phi & s_\phi & 0 \\ -s_\phi & c_\phi & 0 \\ 0 & 0 & 1 \end{bmatrix} \overbrace{\begin{Bmatrix} s_\theta c_\phi \dot{\psi} - s_\phi \dot{\theta} \\ c_\phi \dot{\theta} + s_\theta s_\phi \dot{\psi} \\ \dot{\phi} + c_\theta \dot{\psi} \end{Bmatrix}}^{\text{components of } \omega_{\mathcal{B}/\mathcal{I}} \text{ in } \mathcal{I}}}_{\text{This gives the components of } \omega_{\mathcal{B}/\mathcal{I}} \text{ in } \mathcal{I}_1}$$

$$= \begin{bmatrix} c_\psi & s_\psi & 0 \\ -s_\psi & c_\psi & 0 \\ 0 & 0 & 1 \end{bmatrix} \underbrace{\begin{bmatrix} c_\theta & 0 & -s_\theta \\ 0 & 1 & 0 \\ s_\theta & 0 & c_\theta \end{bmatrix} \begin{Bmatrix} s_\theta \dot{\psi} \\ \dot{\theta} \\ \dot{\phi} + c_\theta \dot{\psi} \end{Bmatrix}}_{\text{This yields the components of } \omega_{\mathcal{B}/\mathcal{I}} \text{ in } \mathcal{I}_2}$$

$$= \begin{bmatrix} c_\psi & s_\psi & 0 \\ -s_\psi & c_\psi & 0 \\ 0 & 0 & 1 \end{bmatrix} \begin{Bmatrix} -s_\theta \dot{\phi} \\ \dot{\theta} \\ \dot{\psi} + c_\theta \dot{\phi} \end{Bmatrix}$$

Finally, we obtain the components of $\omega_{\mathcal{B}/\mathcal{I}}$ in $\mathcal{B}$:

$$= \begin{Bmatrix} -s_\theta c_\psi \dot{\phi} + s_\psi \dot{\theta} \\ s_\psi s_\theta \dot{\phi} + c_\psi \dot{\theta} \\ \dot{\psi} + c_\theta \dot{\phi} \end{Bmatrix}$$

Comparing the elements of this matrix with the components in Equation (6.61), we see that rotation matrices indeed furnish us with a rapid means of "converting" a vector from one frame to another. Note also that the bracket in the second line above contains $\boldsymbol{\omega}_{\mathcal{B}/\mathcal{I}}$ expressed in $\mathcal{I}_1$, previously derived as Equation (6.67). The bracket in the third line gives the components of $\boldsymbol{\omega}_{\mathcal{B}/\mathcal{I}}$ in $\mathcal{I}_2$.

The next example illustrates a use that was made of rotation matrices by one of the authors in the development of earth stations.

Example 6.16

Using rotation matrices, compute the angles A (azimuth) and E (elevation) through which an antenna must respectively turn about (a) the negative of local vertical and then (b) the new, rotated position of the elevation axis in order to sight a satellite in geosynchronous orbit (see Problems 2.64,65). Angles A and E are called *look angles,* and an antenna that performs azimuth followed by elevation in this manner is said to have "el over az" positioning. The azimuth angle A is a function of the local latitude λ and the relative west longitude δ of the satellite; the elevation angle E depends additionally on R_e/R, the ratio of the earth and orbit radii. (See Figure E6.16a.)

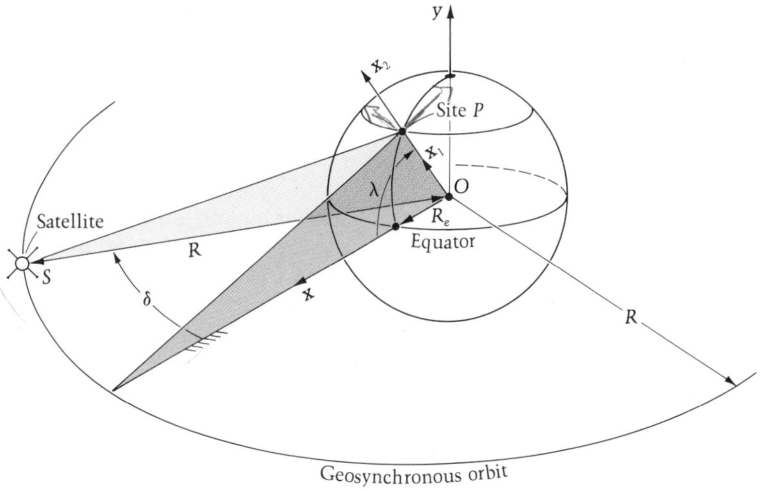

Figure E6.16a

Solution

The frame $\mathcal{I}(x, y, z)$ has origin at the center of the earth as shown; the xy plane contains the site P and its meridian. The coordinates of the satellite in this frame are seen to be given by the position vector

$$\mathbf{r}_{OS} = R \cos \delta \hat{\mathbf{i}} + R \sin \delta \hat{\mathbf{k}}$$

First we rotate the frame $\mathcal{I}$ through the latitude angle λ about z in order to line up the new axis x_1 with the local vertical at the site P. We call the resulting rotated frame $\mathcal{I}_1$ and obtain the following for the new components of $\mathbf{r}_{OS}$ ($c_\lambda = \cos\lambda$, $s_\delta = \sin\delta$, and so forth):

$$\{r_{OS}\}_{\mathcal{I}_1} = \underset{(\text{angle } \lambda)}{[T_z]} \{r_{OS}\}_{\mathcal{I}}$$

$$\{r_{OS}\}_{\mathcal{I}_1} = \begin{bmatrix} c_\lambda & s_\lambda & 0 \\ -s_\lambda & c_\lambda & 0 \\ 0 & 0 & 1 \end{bmatrix} \begin{Bmatrix} Rc_\delta \\ 0 \\ Rs_\delta \end{Bmatrix} = \begin{Bmatrix} Rc_\lambda c_\delta \\ -Rs_\lambda c_\delta \\ Rs_\delta \end{Bmatrix}$$

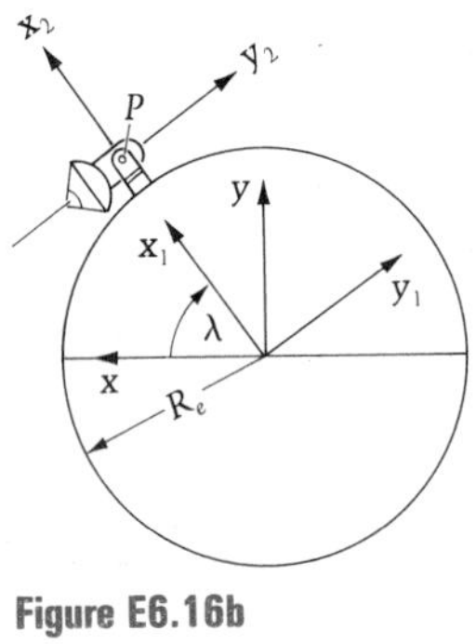

Figure E6.16b

Next we translate the axes to the site, as shown in Figure E6.16b. The only component of $\{r_{OS}\}_{\mathcal{I}_1}$ that changes is x_1, and we see by inspection that

$$\{r_{PS}\}_{\mathcal{I}_2} = \begin{Bmatrix} Rc_\lambda c_\delta - R_e \\ -Rs_\lambda c_\delta \\ Rs_\delta \end{Bmatrix}$$

Note that we must subtract the earth radius R_e from x_1 to get the proper x_2 coordinate of the satellite.

The second rotation is the azimuth rotation about $-x_2$; if we call the rotated frame $\mathcal{I}_3$, the coordinates (x_3, y_3, z_3) of S in this frame are given by

$$\{r_{PS}\}_{\mathcal{I}_3} = \underset{(\text{angle } -A)}{[T_x]} \{r_{PS}\}_{\mathcal{I}_2}$$

$$= \begin{bmatrix} 1 & 0 & 0 \\ 0 & c_A & -s_A \\ 0 & s_A & c_A \end{bmatrix} \begin{Bmatrix} Rc_\lambda c_\delta - R_e \\ -Rs_\lambda c_\delta \\ Rs_\delta \end{Bmatrix}$$

$$= \begin{Bmatrix} Rc_\lambda c_\delta - R_e \\ -Rs_\lambda c_\delta c_A - Rs_\delta s_A \\ -Rs_\lambda c_\delta s_A + Rs_\delta c_A \end{Bmatrix}$$

Here we take an important step. We want the z_3 component of $\mathbf{r}_{PS}$ to be zero because we wish to rotate next in elevation about z_3 and end up with the "bore-sight" (axis) of the antenna aiming at the satellite. Thus angle A (see Figure E6.16c) is determined by setting the third element of the preceding matrix to zero:

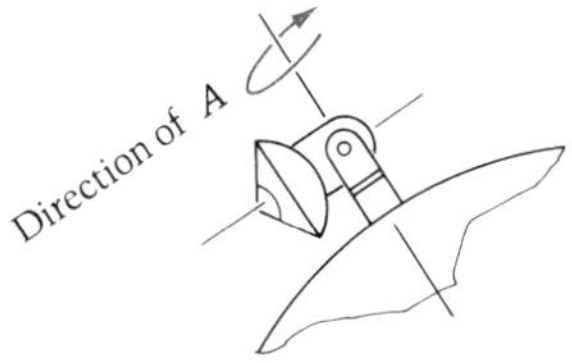

Figure E6.16c

$$-Rs_\lambda c_\delta s_A + Rs_\delta c_A = 0$$

$$\tan A = \tan\delta \csc\lambda$$

$$A = \tan^{-1}(\tan\delta \csc\lambda) \tag{1}$$

Finally we rotate through angle E about the z_3 axis:

$$\{r_{PS}\}_{\mathcal{I}_4} = \underset{(\text{angle } E)}{[T_z]} \{r_{PS}\}_{\mathcal{I}_3}$$

$$= \begin{bmatrix} c_E & s_E & 0 \\ -s_E & c_E & 0 \\ 0 & 0 & 1 \end{bmatrix} \begin{Bmatrix} Rc_\lambda c_\delta - R_e \\ -Rs_\lambda c_\delta c_A - Rs_\delta s_A \\ 0 \end{Bmatrix}$$

$$= \begin{Bmatrix} c_E(Rc_\lambda c_\delta - R_e) - s_E(Rs_\lambda c_\delta c_A + Rs_\delta s_A) \\ -s_E(Rc_\lambda c_\delta - R_e) - c_E(Rs_\lambda c_\delta c_A + Rs_\delta s_A) \\ 0 \end{Bmatrix}$$

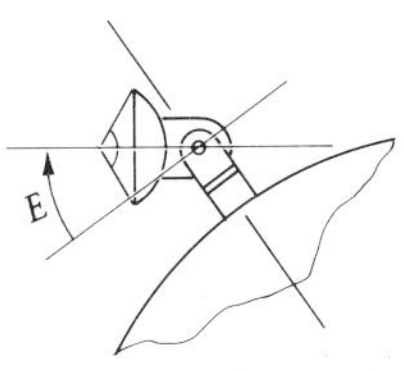

Figure E6.16d

Now we come to the condition that will allow us to determine the value of angle E (see Figure E6.16d): We wish the antenna to aim directly at the satellite. Since the antenna boresight is now in the $-y_3$ direction, we wish the elevation rotation to stop when the x_3 coordinate is zero:

$$c_E(Rc_\lambda c_\delta - R_e) - s_E(Rs_\lambda c_\delta c_A + Rs_\delta s_A) = 0$$

$$E = \tan^{-1}\left(\frac{Rc_\lambda c_\delta - R_e}{Rs_\lambda c_\delta c_A + Rs_\delta s_A}\right) \tag{2}$$

If $r = R_e / R(\approx 1/6.61)$, then (2) becomes

$$E = \tan^{-1}\left(\frac{c_\lambda c_\delta - r}{s_\lambda c_\delta c_A + s_\delta s_A}\right)$$

in which the azimuth angle A is given by Equation (1), so that

$$E = \tan^{-1}\left(\frac{c_\lambda c_\delta - r}{\sqrt{1 - c_\delta^2 c_\lambda^2}}\right) \tag{3}$$

There is a single circle in the sky in which geosynchronous satellites can exist. This circle, which was examined in Problem 2.64, has rapidly become very crowded, however. As of the summer of 1994, there were nearly 500 satellites in geosynchronous orbit. Six years earlier, there were just over 100, and in early 1982, only around 30.

PROBLEMS ▶ Section 6.9

6.86 For the United States the eastern and western limits of usable satellite positions in the geosynchronous arc (see Problem 2.64) are about 70° and 143°W longitude, respectively. Find the ranges in azimuth and elevation that are required if the antenna in Example 6.16 is to sweep from the eastern to the western limit for a site at: (a) 34°N latitude and 84°W longitude; (b) your home town. (Select a city in the contiguous United States if you are from another country.)

6.87 Use Equations (6.70) together with rotation matrices to compute the angular velocity components in space-fixed axes.

6.88 An antenna $\mathcal{P}$ has three rotational degrees of freedom (see Figure P6.88):

1. Azimuth angle A about local vertical z,
2. Elevation angle E about an axis originally parallel to x,
3. Polarization angle P about the axis of symmetry of the dish (originally parallel to y).

Use the addition theorem together with rotation matrices to calculate $\boldsymbol{\omega}_{\mathcal{P}/\mathcal{I}}$ in terms of its components in $\mathcal{I}$.

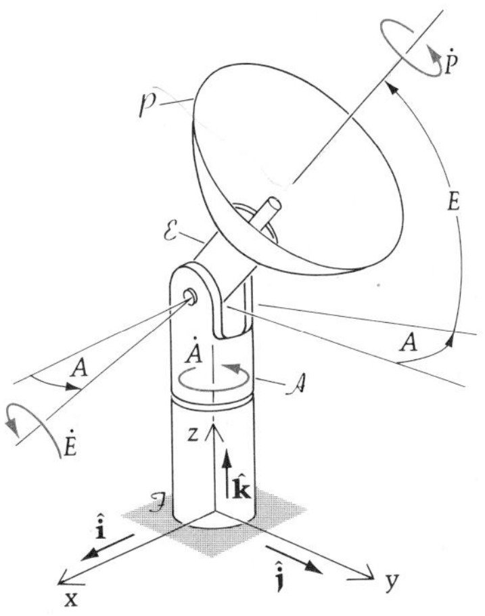

Figure P6.88

6.89 In the preceding problem calculate $\boldsymbol{\omega}_{\mathcal{P}/\mathcal{I}}$ in terms of its components in $\mathcal{E}$.

* **6.90** Plot the elevation angle E versus the satellite angle δ (see Example 6.16) for the following values of λ (on the same graph): $\lambda = 0°$, $20°$, $40°$, $60°$, and $80°$. What do the crossings of the δ axis of these curves physically represent?

* **6.91** Calculate the look angles (see Example 6.16) for the case in which the elevation rotation is performed *prior* to azimuth ("az over el" positioning).

6.92 It takes six *orbital parameters* to establish the location of a planet with respect to a frame fixed in space. (See Figure P6.92.) To find its orbital path, we first turn through the angle Ω in the ecliptic plane (the plane containing the path of the sun as we see it from earth) to the *ascending node* (the intersection of the ecliptic plane with the planet's path when going north). Next we turn in inclination through the angle i about x_1' to obtain the tilt of the planet's plane. (Thus earth's inclination is defined as zero.) Finally, the angle θ_o locates the perihelion of the planet's orbit. This is the closest point of the orbit to the sun's center S. Two other quantities give the orbit's shape, and a sixth one locates the planet in its orbit with respect to the perihelion. If $\mathbf{v} = (v_x, v_y, v_z)$ is a vector defined in the space frame $\mathcal{I}$, use rotation matrices to obtain the components of $\mathbf{v}$ in the frame $\mathcal{B}(x, y, z)$ located as shown in the orbital path at P, in terms of Ω, i, and θ_o.

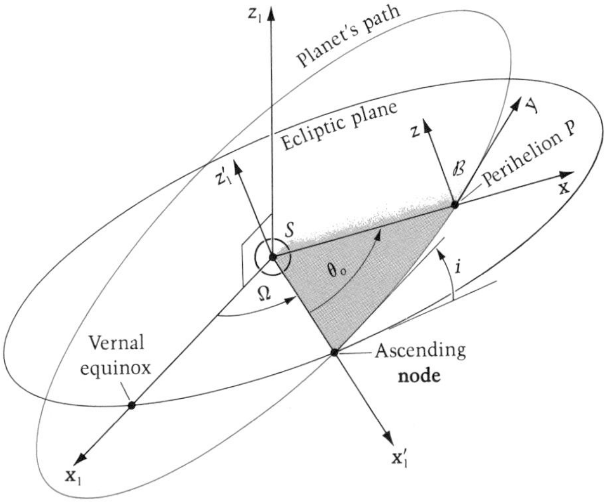

Figure P6.92

PROJECT PROBLEM ▶ Chapter 6

6.93 After reading Example 6.3, construct a simple model that illustrates the workings of a universal joint (see Figure P6.93 for some examples).

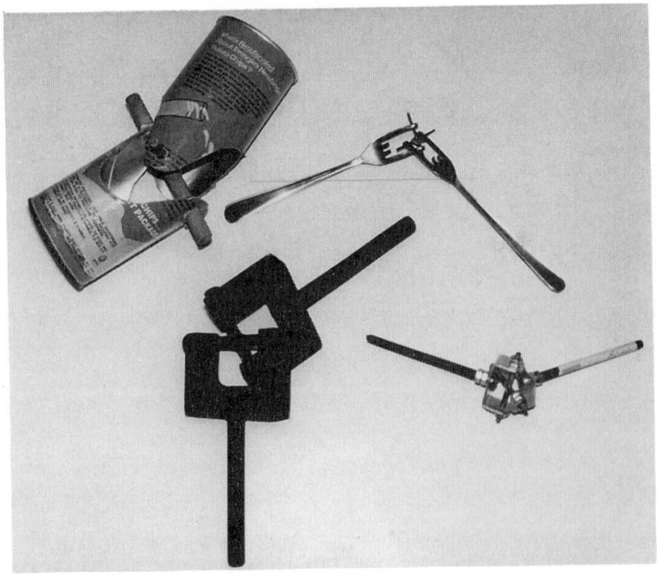

Figure P6.93

SUMMARY ▶ Chapter 6

The key concept of this chapter is angular velocity. If $\mathcal{A}$ and $\mathcal{B}$ are two rigid bodies (or frames of reference), the angular velocity of $\mathcal{B}$ relative to $\mathcal{A}$ (or of $\mathcal{B}$ *in* $\mathcal{A}$), $\boldsymbol{\omega}_{\mathcal{B}/\mathcal{A}}$, is the unique vector that connects the derivatives relative to $\mathcal{A}$ and $\mathcal{B}$ of any vector, say **Q**, by

$$^{\mathcal{A}}\dot{\mathbf{Q}} = {}^{\mathcal{B}}\dot{\mathbf{Q}} + \boldsymbol{\omega}_{\mathcal{B}/\mathcal{A}} \times \mathbf{Q}$$

The angular velocity describes the rate at which the orientation of one body changes relative to another. Among the important properties are:

$$\boldsymbol{\omega}_{\mathcal{B}/\mathcal{A}} = -\boldsymbol{\omega}_{\mathcal{A}/\mathcal{B}}$$

and the addition theorem (a chain rule):

$$\boldsymbol{\omega}_{\mathcal{C}/\mathcal{A}} = \boldsymbol{\omega}_{\mathcal{C}/\mathcal{B}} + \boldsymbol{\omega}_{\mathcal{B}/\mathcal{A}}$$

which can be extended by repetition to

$$\boldsymbol{\omega}_{\mathcal{D}/\mathcal{A}} = \boldsymbol{\omega}_{\mathcal{D}/\mathcal{C}} + \boldsymbol{\omega}_{\mathcal{C}/\mathcal{B}} + \boldsymbol{\omega}_{\mathcal{B}/\mathcal{A}}$$

and in fact to any number of bodies.

This rather abstract-seeming concept of angular velocity reduces to the familiar form of Chapter 3 when we have "plane-motion" situations in which a direction is fixed in both of two bodies. So if $\hat{\mathbf{k}}$ is the constant vector (denotes a fixed direction) in both of the bodies $\mathcal{A}$ and $\mathcal{B}$, we have

$$\boldsymbol{\omega}_{\mathcal{B}/\mathcal{A}} = \dot{\theta}\hat{\mathbf{k}}$$

This equation, along with the addition theorem, gives us an important tool to deal with systems of bodies that appear complex but where each body moves relative to a neighbor by rotating about an axis fixed in the neighbor.

Angular acceleration is the derivative of angular velocity, and the derivative may be calculated in either of the two bodies (frames) involved.

$$\boldsymbol{\alpha}_{\mathcal{B}/\mathcal{A}} = {}^{\mathcal{A}}\dot{\boldsymbol{\omega}}_{\mathcal{B}/\mathcal{A}} = {}^{\mathcal{B}}\dot{\boldsymbol{\omega}}_{\mathcal{B}/\mathcal{A}}$$

Angular accelerations cannot be added with the simplicity of an addition theorem such as (6.27), but instead obey the formula:

$$\boldsymbol{\alpha}_{\mathcal{C}/\mathcal{A}} = \boldsymbol{\alpha}_{\mathcal{C}/\mathcal{B}} + \boldsymbol{\alpha}_{\mathcal{B}/\mathcal{A}} + \boldsymbol{\omega}_{\mathcal{B}/\mathcal{A}} \times \boldsymbol{\omega}_{\mathcal{C}/\mathcal{B}}$$

Velocities of a point P relative to two frames $\mathcal{B}$ and $\mathcal{I}$ are related by

$$\mathbf{v}_{P/\mathcal{I}} = \mathbf{v}_{P/\mathcal{B}} + \mathbf{v}_{O'/\mathcal{I}} + \boldsymbol{\omega}_{\mathcal{B}/\mathcal{I}} \times \mathbf{r}_{O'P}$$

where O' is a point fixed in $\mathcal{B}$. Similarly for accelerations, we have

$$\mathbf{a}_{P/\mathcal{I}} = \mathbf{a}_{P/\mathcal{B}} + \mathbf{a}_{O'/\mathcal{I}} + \boldsymbol{\alpha}_{\mathcal{B}/\mathcal{I}} \times \mathbf{r}_{O'P} + \boldsymbol{\omega}_{\mathcal{B}/\mathcal{I}} \times (\boldsymbol{\omega}_{\mathcal{B}/\mathcal{I}} \times \mathbf{r}_{O'P}) + 2\boldsymbol{\omega}_{\mathcal{B}/\mathcal{I}} \times \mathbf{v}_{P/\mathcal{B}}$$

By letting P be fixed in $\mathcal{B}$ (as is O') we can deduce from the two previous equations a pair of expressions relating the velocities and then the accelerations (relative to a frame $\mathcal{I}$) of two points such as P and O', both fixed in $\mathcal{B}$:

$$\mathbf{v}_{P/\mathcal{I}} = \mathbf{v}_{O'/\mathcal{I}} + \boldsymbol{\omega}_{\mathcal{B}/\mathcal{I}} \times \mathbf{r}_{O'P}$$
$$\mathbf{a}_{P/\mathcal{I}} = \mathbf{a}_{O'/\mathcal{I}} + \boldsymbol{\alpha}_{\mathcal{B}/\mathcal{I}} \times \mathbf{r}_{O'P} + \boldsymbol{\omega}_{\mathcal{B}/\mathcal{I}} \times (\boldsymbol{\omega}_{\mathcal{B}/\mathcal{I}} \times \mathbf{r}_{O'P})$$

These are the counterparts in three dimensions of Equations (3.8) and (3.19).

Finally, we have discussed a system that is often used to describe the orientation of a body relative to a frame of reference — Euler angles. Angular velocity in terms of rates of change of Euler angles has been developed.

REVIEW QUESTIONS ▶ Chapter 6

True or False?

1. The angular velocity of body $\mathcal{B}$ in frame $\mathcal{I}$ depends only upon the changes in orientation of $\mathcal{B}$ with respect to $\mathcal{I}$.
2. The addition theorem for angular velocity applies equally well to angular acceleration.
3. The formula relating the velocities of two points of a rigid body in plane motion, $\mathbf{v}_B = \mathbf{v}_A + \boldsymbol{\omega} \times \mathbf{r}_{AB}$, applies to three-dimensional problems provided that $\boldsymbol{\omega}$, $\mathbf{r}_{AB}$, and the $\mathbf{v}$'s become three-dimensional vectors.
4. The formula relating the accelerations of two points of a rigid body in plane motion, $\mathbf{a}_B = \mathbf{a}_A + \boldsymbol{\alpha} \times \mathbf{r}_{AB} - \omega^2 \mathbf{r}_{AB}$, applies to three-dimensional problems provided that $\boldsymbol{\alpha}$, $\mathbf{r}_{AB}$, $\boldsymbol{\omega}$, and the $\mathbf{a}$'s become three-dimensional vectors.
5. The equation $\omega_z = \dot{\theta}$ in plane motion extends to three similar linear equations in general motion for determining the orientation angles.
6. For a point P to have a nonvanishing Coriolis acceleration, there must be both a relative velocity of P with respect to the "moving frame" and an angular velocity of the moving frame relative to the reference frame.
7. If we premultiply a vector $\{v\}$ by a rotation matrix $[T]$, the 3×1 vector we get contains the "new" components of $\mathbf{v}$ in the rotated frame.
8. The Eulerian angles are used to orient a body in three-dimensional space.
9. The Eulerian angles are three rotations ϕ, θ, and ψ about axes which were originally distinct and orthogonal.
10. It is possible, for any moving point P, to choose a moving frame such that the Coriolis acceleration of P vanishes identically.
11. The angular velocity vector is used to relate the derivatives of a vector in two frames.
12. If one yoke of a misaligned universal (Hooke's) joint turns at constant angular speed, so does the other.

13. In general motion of a rigid body $\mathcal{B}$, as long as $\boldsymbol{\omega} \neq \mathbf{0}$ there is a point of zero velocity of $\mathcal{B}$ or $\mathcal{B}$-extended.

14. The order of rotations is important in orienting a body if the Eulerian angles are used as defined in this chapter (in conjunction with the Cardan suspension of the gyroscope).

Answers: 1. T 2.F 3. T 4. F 5. F 6. T 7. T 8. T 9. F 10. T 11. T 12. F 13. F 14. T

7 KINETICS OF A RIGID BODY IN GENERAL MOTION

7.1 Introduction

In Chapter 2 we found that, relative to an inertial frame of reference, motion of a body is governed by

$$\Sigma \mathbf{F} = \frac{d\mathbf{L}}{dt} = m\mathbf{a}_C \tag{7.1}$$

and a moment equation

$$\Sigma \mathbf{M}_C = \frac{d\mathbf{H}_C}{dt} \tag{7.2}$$

or

$$\Sigma \mathbf{M}_O = \frac{d\mathbf{H}_O}{dt} \tag{7.3}$$

with O being fixed in the inertial frame. These general equations were specialized to *plane* motion of a rigid body $\mathcal{B}$ in Chapter 4 and will now be used to study the *general* motion of $\mathcal{B}$ in three dimensions.

As we indicated in Chapter 2, the first of the two vector equations given above describes the mass center motion of *any* system.* It is applicable, for example, to rigid or deformable solids, systems of small masses, liquids, and gases. For a body in general (three-dimensional) motion, Equation (7.1) now possesses three nontrivial scalar component equations whose solutions allow us to locate the mass center C.

We note that, as was the case with plane motion, the mass center can move independently of the body's changing orientation (provided that the external forces do not themselves depend on the body's angular motion, which is frequently the case). We saw such an example in Section 6.6 when we examined the motion of (a particle or) the mass center of a body near the rotating earth. We emphasize that such a simple and natural extension from two to three dimensions will *not* occur with the *orientation* (or angular) motion of $\mathcal{B}$, as we shall see in Section 7.5. The reason is that $\dot{\mathbf{H}}_C$ in Equation (7.2) *cannot* be written as the sum of three terms of the form $I_{zz}^C \ddot{\theta}\hat{\mathbf{k}}$.

In the remainder of the chapter we shall first develop the expression for the moment of momentum of a rigid body $\mathcal{B}$ in general motion. This will then lead us into a study of the inertia properties of $\mathcal{B}$. Having partially examined the concept of inertia in Chapter 4, we shall extend this study to include transformations at a point as well as principal moments and axes of inertia. Then and only then shall we be fully prepared to derive the Euler equations that govern the rotational motion of a rigid body in general motion. We shall also examine, as we did for the plane

* Excluding throughout, of course, relativistic effects occurring when velocities are not small compared to the speed of light.

motion in Chapter 5, some special integrals of the equations of motion, which are known as the principles of impulse and momentum, angular impulse and angular momentum, and work and kinetic energy.

7.2 Moment of Momentum (Angular Momentum) in Three Dimensions

We saw in Chapter 4 that when it is reasonable to treat a body $\mathcal{B}$ as rigid, the equations of motion of $\mathcal{B}$ are greatly simplified. The mass center becomes fixed in the body, and the moment of momentum is expressible in terms of the angular velocity and the inertia properties of $\mathcal{B}$—hence the other name of moment of momentum: **angular momentum**. We shall proceed now to study the angular momentum $\mathbf{H}_P$ of $\mathcal{B}$ about a point P in general (three-dimensional) motion. We shall see that the equations that result are much more complicated than their plane motion counterparts.

Let us begin by introducing a system of rectangular axes (x, y, z) which have their origin at P. The angular velocity of $\mathcal{B}$ in reference frame $\mathcal{I}$ may then be expressed in terms of its components along these axes by

$$\boldsymbol{\omega}_{\mathcal{B}/\mathcal{I}} = \boldsymbol{\omega} = (\boldsymbol{\omega} \cdot \hat{\mathbf{i}})\hat{\mathbf{i}} + (\boldsymbol{\omega} \cdot \hat{\mathbf{j}})\hat{\mathbf{j}} + (\boldsymbol{\omega} \cdot \hat{\mathbf{k}})\hat{\mathbf{k}}$$
$$= \omega_x\hat{\mathbf{i}} + \omega_y\hat{\mathbf{j}} + \omega_z\hat{\mathbf{k}} \tag{7.4}$$

The location, relative to P, of a typical point in the body is given by

$$\mathbf{r} = x\hat{\mathbf{i}} + y\hat{\mathbf{j}} + z\hat{\mathbf{k}} \tag{7.5}$$

The moment of momentum of $\mathcal{B}$ relative to P is now defined (see Sections 2.6 and 4.3) to be

$$\boxed{\mathbf{H}_P = \int \mathbf{r} \times \mathbf{v} \, dm} \tag{7.6}$$

Moment of Momentum of $\mathcal{B}$ relative to P

in which $\mathbf{v}$, the velocity of the mass element dm, is not the derivative of $\mathbf{r}$ but rather of the position vector to the element from a point fixed in the reference frame $\mathcal{I}$, as shown in Figure 7.1.

$$V = V_P + \omega \times r_{P\,dm}$$

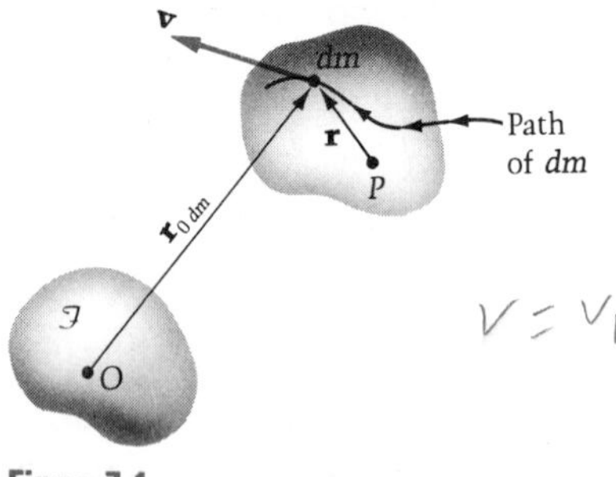

Figure 7.1

Since $\mathcal{B}$ is a rigid body, we know from Equation (6.56) that $\mathbf{v} = \mathbf{v}_P + \boldsymbol{\omega} \times \mathbf{r}$, and we may substitute this expression into Equation (7.6) to obtain

$$\mathbf{H}_P = \int \mathbf{r} \times \mathbf{v}_P \, dm + \int \mathbf{r} \times (\boldsymbol{\omega} \times \mathbf{r}) \, dm$$
$$= \left(\int \mathbf{r} \, dm \right) \times \mathbf{v}_P + \int \mathbf{r} \times (\boldsymbol{\omega} \times \mathbf{r}) \, dm$$

Using the mass center definition, the integral in the first term on the right-hand side is $m\mathbf{r}_{PC}$. Therefore

$$H_p = \left(\int r\, dm\right) \times V_p + \int r \times (\omega \times r)\, dm$$

$$m\, r_{pc}$$

$$\mathbf{H}_P = m\mathbf{r}_{PC} \times \mathbf{v}_P + \int \mathbf{r} \times (\omega \times \mathbf{r})\, dm \tag{7.7}$$

In the cases where either (a) P is chosen to be the mass center C, or (b) $\mathbf{v}_P = \mathbf{0}$, or (c) $\mathbf{r}_{PC} \parallel \mathbf{v}_P$, the first term on the right side of Equation (7.7) vanishes. For these cases,

$$\mathbf{H}_P = \int \mathbf{r} \times (\omega \times \mathbf{r})\, dm \tag{7.8}$$

Substituting ω and $\mathbf{r}$ from Equations (7.4) and (7.5) and using the identity

$$\mathbf{A} \times (\mathbf{B} \times \mathbf{C}) \equiv (\mathbf{A} \cdot \mathbf{C})\mathbf{B} - (\mathbf{A} \cdot \mathbf{B})\mathbf{C} \tag{7.9}$$

we obtain the result

$$\begin{aligned}
\mathbf{H}_P = & \left[\omega_x \int (y^2 + z^2)\, dm - \omega_y \int xy\, dm \quad - \omega_z \int xz\, dm\right]\hat{\mathbf{i}} \\
& + \left[-\omega_x \int xy\, dm + \omega_y \int (x^2 + z^2)\, dm - \omega_z \int yz\, dm\right]\hat{\mathbf{j}} \\
& + \left[-\omega_x \int xz\, dm - \omega_y \int yz\, dm \quad + \omega_z \int (x^2 + y^2)\, dm\right]\hat{\mathbf{k}}
\end{aligned} \tag{7.10}$$

Question 7.1 Why may the ω components be brought outside the various integrals in Equation (7.10)?

Once we recognize the inertia properties (see Sections 4.3 and 4.4), this angular momentum expression becomes, for the case when P is C,

$$\begin{aligned}
\mathbf{H}_C = & (I_{xx}^C \omega_x + I_{xy}^C \omega_y + I_{xz}^C \omega_z)\hat{\mathbf{i}} \\
& + (I_{xy}^C \omega_x + I_{yy}^C \omega_y + I_{yz}^C \omega_z)\hat{\mathbf{j}} \\
& + (I_{xz}^C \omega_x + I_{yz}^C \omega_y + I_{zz}^C \omega_z)\hat{\mathbf{k}}
\end{aligned} \tag{7.11}$$

The form of the equation is identical if point P is not C, but rather either $\mathbf{v}_P = \mathbf{0}$ or $\mathbf{r}_{PC} \parallel \mathbf{v}_P$; the only difference is that the inertia properties are calculated with respect to axes at P instead of at the mass center:

angular Momentum
when P is not C

$$\begin{aligned}
\mathbf{H}_P = & (I_{xx}^P \omega_x + I_{xy}^P \omega_y + I_{xz}^P \omega_z)\hat{\mathbf{i}} \\
& + (I_{xy}^P \omega_x + I_{yy}^P \omega_y + I_{yz}^P \omega_z)\hat{\mathbf{j}} \\
& + (I_{xz}^P \omega_x + I_{yz}^P \omega_y + I_{zz}^P \omega_z)\hat{\mathbf{k}}
\end{aligned} \tag{7.12}$$

Both of these forms for the angular momentum vector (Equations 7.11 and 7.12) will prove important to us in the sections to follow.

Answer 7.1 As we have seen in Chapter 6, $\omega_{\mathcal{B}/\mathcal{J}}$ depends only on how a set of unit vectors, locked into $\mathcal{B}$, change their directions in $\mathcal{J}$. The angular velocity is a constant with regard to integration at a particular instant over the body's volume.

PROBLEMS ▶ Section 7.2

7.1 Find the angular momentum vector $\mathbf{H}_O$ of the wagon wheel of Problem 6.49.

7.2 Find the angular momentum vector of the disk $\mathcal{B}$ in Problem 6.27 about (a) C and (b) O.

7.3 Find the angular momentum vector for the bent bar of Example 4.16 about the mass center, if it turns about the z axis at angular speed ω.

7.4 A thin homogeneous disk $\mathcal{D}$ of mass M and radius r rotates with constant angular speed ω_2 about the shaft $\mathcal{S}$ (Figure P7.4). This shaft is cantilevered from the vertical shaft $\mathcal{R}$ and rotates with constant angular speed ω_1 about the axis of $\mathcal{R}$. Find the angular momentum of the disk about point Q, and show the direction of the vector in a sketch.

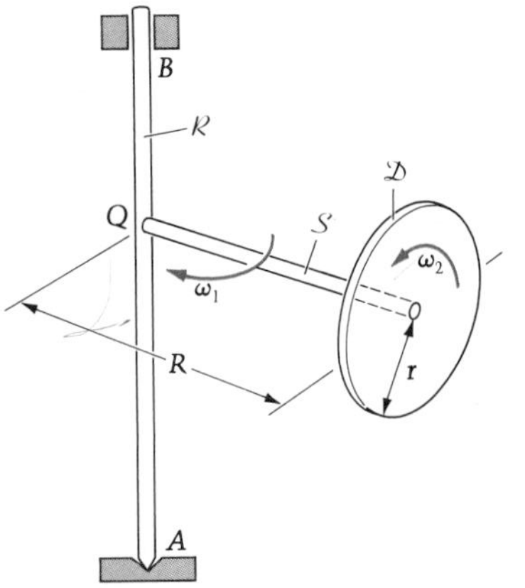

Figure P7.4

7.5 Depicted in Figure P7.5 is a grinder in a grinding mill that is composed of three main parts:

1. The vertical shaft $\mathcal{S}$, which rotates at constant angular speed Ω
2. The slanted shaft $\mathcal{B}$ of length ℓ, which is pinned to $\mathcal{S}$ and turns with it
3. The grinder $\mathcal{D}$ of radius r, turning in bearings at C about $\mathcal{B}$, and rolling on the inner surface of $\mathcal{I}$.

As shaft $\mathcal{S}$ gets up to speed Ω, body $\mathcal{B}$ swings outward, and then the angle ϕ remains constant during operation. Treat the grinder $\mathcal{D}$ as a disk and find its angular momentum vector $\mathbf{H}_C$ in convenient coordinates. (A suggested set is shown.)

7.6 In the preceding problem note that point O is a fixed point of all three bodies $\mathcal{S}$, $\mathcal{B}$, and $\mathcal{D}$ extended. Compute the angular momentum $\mathbf{H}_O$ of $\mathcal{D}$, and verify that $\mathbf{H}_O = \mathbf{H}_C + \mathbf{r}_{OC} \times \mathbf{L}$.

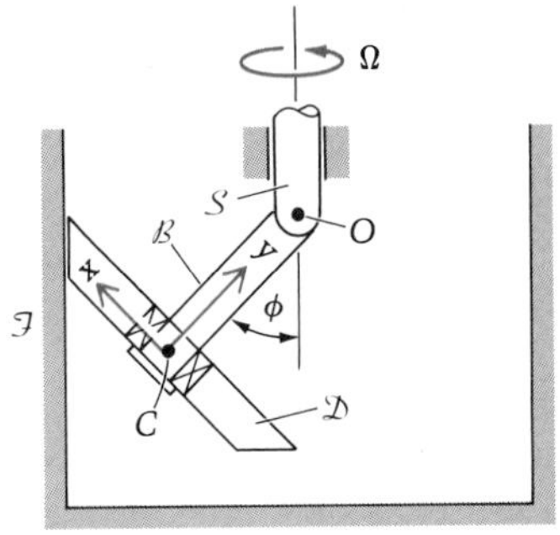

Figure P7.5

7.3 Transformations of Inertia Properties

Sometimes we need the moments and products of inertia at points other than the mass center C of a rigid body $\mathcal{B}$. These properties can be found without further integration by using the parallel-axis theorems, which were derived in Chapter 4. These are restated below, where $(\bar{x}, \bar{y}, \bar{z})$ are the coordinates of the mass center C relative to axes at P. For the moments of inertia at P:

$$I_{xx}^P = I_{xx}^C + m(\bar{y}^2 + \bar{z}^2) \quad (7.13a)$$

$$I_{yy}^P = I_{yy}^C + m(\bar{z}^2 + \bar{x}^2) \quad (7.13b)$$

$$I_{zz}^P = I_{zz}^C + m(\bar{x}^2 + \bar{y}^2) \quad (7.13c)$$

And for the products of inertia at P:

$$I_{xy}^{P} = I_{xy}^{C} - m\bar{x}\bar{y} \tag{7.14a}$$

$$I_{yz}^{P} = I_{yz}^{C} - m\bar{y}\bar{z} \tag{7.14b}$$

$$I_{zx}^{P} = I_{zx}^{C} - m\bar{z}\bar{x} \tag{7.14c}$$

EXAMPLE 7.1

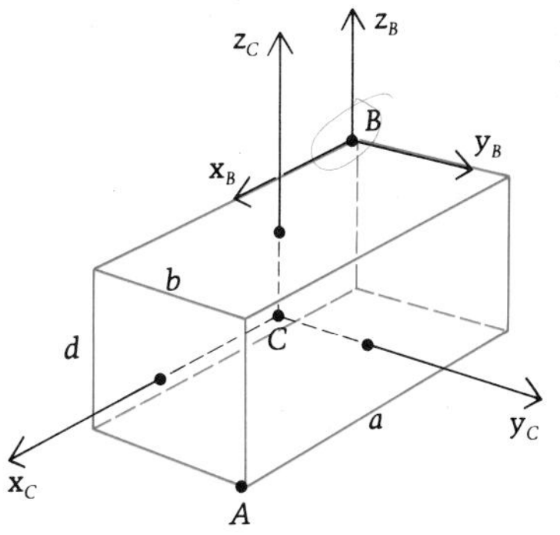

Figure E7.1

As a review example, compute the inertia properties at the corner B of the uniform rectangular solid of mass m shown in Figure E7.1.

Solution

For the moments of inertia we obtain

$$I_{xx}^{B} = I_{xx}^{C} + m(\bar{y}^2 + \bar{z}^2) = \frac{m}{12}(b^2 + d^2) + m\left[\left(\frac{b}{2}\right)^2 + \left(\frac{d}{2}\right)^2\right] = \frac{m}{3}(b^2 + d^2)$$

Note that the distance between x_B and x_C is $\sqrt{(b/2)^2 + (d/2)^2}$. In the same way,

$$I_{yy}^{B} = \frac{m}{3}(d^2 + a^2) \quad \text{and} \quad I_{zz}^{B} = \frac{m}{3}(a^2 + b^2)$$

For the products of inertia,

$$I_{xy}^{B} = I_{xy}^{C} - m\bar{x}\bar{y} = 0 - m\left(\frac{a}{2}\right)\left(\frac{b}{2}\right) = \frac{-mab}{4}$$

$$I_{yz}^{B} = I_{yz}^{C} - m\bar{y}\bar{z} = 0 - m\left(\frac{b}{2}\right)\left(-\frac{d}{2}\right) = \frac{mbd}{4}$$

$$I_{xz}^{B} = I_{xz}^{C} - m\bar{x}\bar{z} = 0 - m\left(\frac{a}{2}\right)\left(-\frac{d}{2}\right) = \frac{mad}{4}$$

Transformation of Inertia Properties at a Point

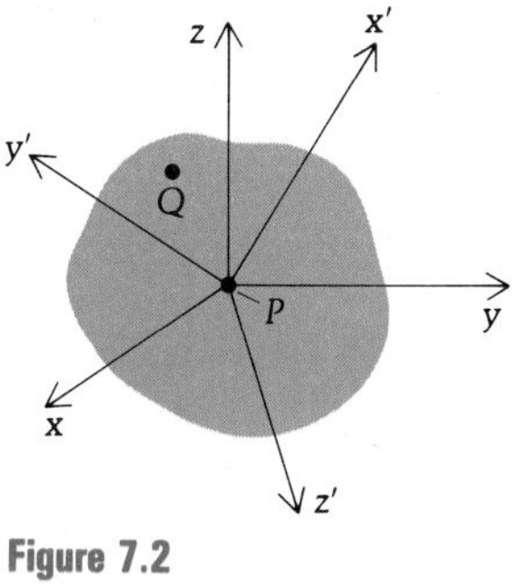

Figure 7.2

We now consider a second and equally important transformation, which will demonstrate that if we know the moments and products of inertia associated with a set of orthogonal axes through a point P, we can easily compute the moments and products of inertia associated with any *other* set of axes having the same origin. Consider two sets of axes with a common origin at P. (See Figure 7.2.) Let ℓ_x, ℓ_y, ℓ_z be the direction cosines of x' relative to x, y, and z, respectively. Then the rectangular coordinate x' of a point Q in the body is related to the rectangular coordinates x, y, z by

$$x' = \mathbf{r}_{PQ} \cdot \underbrace{(\ell_x\hat{\mathbf{i}} + \ell_y\hat{\mathbf{j}} + \ell_z\hat{\mathbf{k}})}_{\text{unit vector along } x' \text{ axis}}$$

$$= (x\hat{\mathbf{i}} + y\hat{\mathbf{j}} + z\hat{\mathbf{k}}) \cdot (\ell_x\hat{\mathbf{i}} + \ell_y\hat{\mathbf{j}} + \ell_z\hat{\mathbf{k}})$$

$$x' = x\ell_x + y\ell_y + z\ell_z \tag{7.15}$$

We seek a formula for $I^P_{x'x'}$ in terms of the inertia properties written with respect to the (x, y, z) axes. The definition of $I^P_{x'x'}$ is

$$I^P_{x'x'} = \int (y'^2 + z'^2)\, dm \tag{7.16}$$

Since $x'^2 + y'^2 + z'^2 = x^2 + y^2 + z^2$ (each is the square of the length of $\mathbf{r}_{PQ}$), we may add and subtract x'^2 to produce this quantity in Equation (7.16):

$$\begin{aligned} I^P_{x'x'} &= \int [(x'^2 + y'^2 + z'^2) - x'^2]\, dm \\ &= \int [(x^2 + y^2 + z^2) - x'^2]\, dm \end{aligned} \tag{7.17}$$

Substituting x' from Equation (7.15) into (7.17) gives

$$I^P_{x'x'} = \int [x^2 + y^2 + z^2 - (x\ell_x + y\ell_y + z\ell_z)^2]\, dm$$

Expanding the trinomial and rearranging, we get

$$\begin{aligned} I^P_{x'x'} = \int [&(1 - \ell_x^2)x^2 + (1 - \ell_y^2)\, y^2 + (1 - \ell_z^2)z^2 \\ &- 2xy\ell_x\ell_y - 2xz\ell_x\ell_z - 2yz\ell_y\ell_z]\, dm \end{aligned} \tag{7.18}$$

Since the ℓ's are the direction cosines of the vector in the direction of the x' axis, we know that

$$\ell_x^2 + \ell_y^2 + \ell_z^2 = 1$$

Using this relation in the first three terms of the integrand in Equation (7.18) gives

$$\begin{aligned} I^P_{x'x'} = \int [&(\ell_y^2 + \ell_z^2)\, x^2 + (\ell_x^2 + \ell_z^2)\, y^2 + (\ell_x^2 + \ell_y^2)\, z^2 \\ &- 2xy\ell_x\ell_y - 2xz\ell_x\ell_z - 2yz\ell_y\ell_z)]\, dm \end{aligned}$$

Rearranging, we have

$$\begin{aligned} I^P_{x'x'} = \ell_x^2 \int (y^2 + z^2)\, dm + \ell_y^2 \int (x^2 + z^2)\, dm + \ell_z^2 \int (x^2 + y^2)\, dm \\ + 2\ell_x\ell_y\left(-\int xy\, dm\right) + 2\ell_x\ell_z\left(-\int zx\, dm\right) + 2\ell_y\ell_z\left(-\int yz\, dm\right) \end{aligned} \tag{7.19}$$

Recognizing the six integrals in Equation (7.19) as the inertia properties associated with the (x, y, z) directions at P, we arrive at our goal:

$$I^P_{x'x'} = \ell_x^2 I^P_{xx} + \ell_y^2 I^P_{yy} + \ell_z^2 I^P_{zz} + 2\ell_x\ell_y I^P_{xy} + 2\ell_x\ell_z I^P_{xz} + 2\ell_y\ell_z I^P_{yz} \tag{7.20}$$

This formula allows us to compute the moment of inertia of the mass of $\mathcal{B}$ about any line through P if we know the properties at P for any set of orthogonal axes. We now illustrate its use with an example.

EXAMPLE 7.2

Compute the moment of inertia about the diagonal BA of the rectangular solid of Example 7.1.

Solution

We define the axis x' to emanate from B, pointing toward A as in Figure E7.2. The inertia properties at B were computed in the prior example. The direction cosines

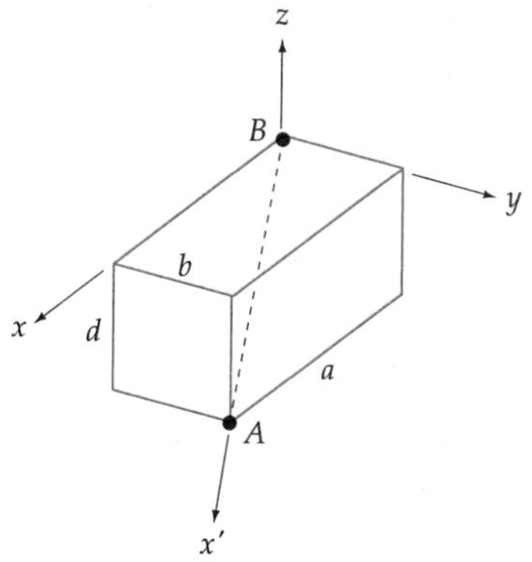

Figure E7.2

of x' are seen by inspection to be

$$\ell_x = \frac{a}{H} \qquad \ell_y = \frac{b}{H} \qquad \ell_z = \frac{-d}{H}$$

in which

$$H = \sqrt{a^2 + b^2 + d^2} = |\mathbf{r}_{BA}|$$

Substituting the ℓ_i's and the inertia properties at B into Equation (7.20) gives

$$I_{x'x'}^B = \frac{m}{3H^2}[(b^2+d^2)a^2 + (d^2+a^2)b^2 + (a^2+b^2)d^2]$$
$$+ \frac{2m}{H^2}\left[(ab)\left(\frac{-ab}{4}\right) + (-ad)\left(\frac{ad}{4}\right) + (-bd)\left(\frac{bd}{4}\right)\right]$$
$$= \frac{m(a^2b^2 + b^2d^2 + d^2a^2)}{6(a^2+b^2+d^2)}$$

In the preceding example we observe that the line BA also passes through the mass center C. The moment of inertia about line BA is of course the same no matter which point on the line one uses to make the calculation. (It would actually be easier to compute at C in this case, because the ℓ_i's are the same while the products of inertia vanish!)

A result similar to Equation (7.20) for products of inertia will now be derived. Let n_x, n_y, and n_z be the direction cosines of axis y'. Note that the rectangular coordinate y' may then be written, in the same way as Equation (7.15), as follows:

$$y' = xn_x + yn_y + zn_z \tag{7.21}$$

By definition,

$$I_{x'y'}^P = -\int x'y'\,dm$$
$$= -\int (x\ell_x + y\ell_y + z\ell_z)(xn_x + yn_y + zn_z)\,dm$$

Therefore, expanding and recognizing the product of inertia integrals,

$$I_{x'y'}^P = \int -(x^2\ell_x n_x + y^2\ell_y n_y + z^2\ell_z n_z)\,dm + (\ell_x n_y + \ell_y n_x)I_{xy}^P$$
$$+ (\ell_x n_z + \ell_z n_x)I_{xz}^P + (\ell_y n_z + \ell_z n_y)I_{yz}^P \tag{7.22}$$

Since (ℓ_x, ℓ_y, ℓ_z) and (n_x, n_y, n_z) are components of unit vectors along the mutually perpendicular axes x' and y', we may dot these vectors together and obtain

$$\ell_x n_x + \ell_y n_y + \ell_z n_z = 0$$

The integrand in Equation (7.22) may therefore be written as

$$-(x^2\ell_x n_x + y^2\ell_y n_y + z^2\ell_z n_z)$$
$$= x^2(\ell_y n_y + \ell_z n_z) + y^2(\ell_x n_x + \ell_z n_z) + z^2(\ell_x n_x + \ell_y n_y)$$
$$= \ell_x n_x(y^2+z^2) + \ell_y n_y(x^2+z^2) + \ell_z n_z(x^2+y^2) \tag{7.23}$$

Substituting Equation (7.23) into (7.22) then yields the desired transformation equation for the products of inertia:

$$I^P_{x'y'} = \ell_x n_x I^P_{xx} + \ell_y n_y I^P_{yy} + \ell_z n_z I^P_{zz} + (\ell_x n_y + \ell_y n_x) I^P_{xy} + (\ell_x n_z + \ell_z n_x) I^P_{xz} + (\ell_y n_z + \ell_z n_y) I^P_{yz} \quad (7.24)$$

EXAMPLE 7.3

In Examples 7.1 and 7.2 let the solid be a cube ($a = b = d$) and let y' be defined as follows (see Figure E7.3):

1. y' is perpendicular to x'.
2. y' is the same plane as z and x'.

Find $I^B_{x'y'}$.

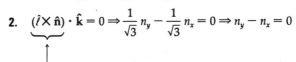

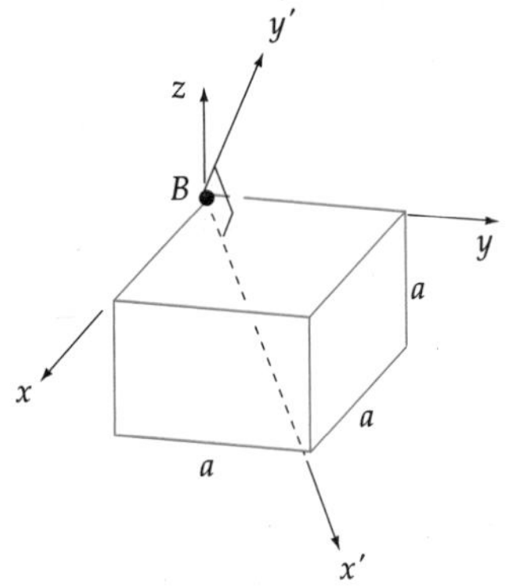

Figure E7.3

Solution

From the equations in Example 7.2 we have $\hat{\ell} = (\ell_x, \ell_y, \ell_z) = (1/\sqrt{3}, 1/\sqrt{3}, -1/\sqrt{3})$ for the unit vector along x'; we now force the components of $\hat{\mathbf{n}}$ — that is, (n_x, n_y, n_z) — to be such that conditions 1 and 2 are satisfied:

1. $\hat{\ell} \perp \hat{\mathbf{n}} \Rightarrow \frac{1}{\sqrt{3}} n_x + \frac{1}{\sqrt{3}} n_y - \frac{1}{\sqrt{3}} n_z = 0 \Rightarrow n_x + n_y - n_z = 0$
2. $(\hat{\ell} \times \hat{\mathbf{n}}) \cdot \hat{\mathbf{k}} = 0 \Rightarrow \frac{1}{\sqrt{3}} n_y - \frac{1}{\sqrt{3}} n_x = 0 \Rightarrow n_y - n_x = 0$

(a vector perpendicular to the plane of x' and y')

These two equations give $n_y = n_x$ and $n_z = 2n_x$. We must also ensure that $\hat{\mathbf{n}}$ is a unit vector:

$$1 = n_x^2 + n_y^2 + n_z^2 = n_x^2(1 + 1 + 4) = 6n_x^2$$

Thus $n_x = 1/\sqrt{6}$, so

$$n_x = \frac{1}{\sqrt{6}} \qquad n_y = \frac{1}{\sqrt{6}} \qquad n_z = \frac{2}{\sqrt{6}}$$

Substituting the components of $\hat{\ell}$ and $\hat{\mathbf{n}}$ and the inertia properties at B (letting the general point P in (7.24) be B in this problem) into Equation (7.24) gives

$$I^B_{x'y'} = \frac{2ma^2}{3}\left[\frac{1}{\sqrt{3}}\frac{1}{\sqrt{6}} + \frac{1}{\sqrt{3}}\frac{1}{\sqrt{6}} - \frac{1}{\sqrt{3}}\left(\frac{2}{\sqrt{6}}\right)\right] + \frac{ma^2}{4}\left[-\left(\frac{1}{\sqrt{3}}\frac{1}{\sqrt{6}} + \frac{1}{\sqrt{3}}\frac{1}{\sqrt{6}}\right) + \left(\frac{1}{\sqrt{3}}\frac{2}{\sqrt{6}} - \frac{1}{\sqrt{3}}\frac{1}{\sqrt{6}}\right) + \left(\frac{1}{\sqrt{3}}\frac{2}{\sqrt{6}} - \frac{1}{\sqrt{3}}\frac{1}{\sqrt{6}}\right)\right] = 0$$

We note that in Example 7.3 the zero result is not obvious at this point in our study. While it is true that, for this case of $a = b = d$, the $x'y'$

plane is a plane of symmetry, this guarantees (see Section 4.4) that $I^B_{x'z'}$ and $I^B_{y'z'}$ are zero but not necessarily $I^B_{x'y'}$. Note also that there are two directions (180° apart) for y' that both satisfy conditions 1 and 2 in the preceding example.

> **Question 7.2** Where in the solution did we choose one of these directions? (And does it matter?)

We close this section by noting that Equations (7.20) and (7.24) are the transformation equations satisfied by a symmetric second-order tensor; thus the inertia properties do indeed form such a tensor. We also note from these two equations that only if the products of inertia are defined with the minus sign (see Equations 4.2) do we get the correct tensor transformation equations.

Answer 7.2 When we said that $n_x = +\sqrt{1/6}$, that is, took the positive square root, we chose y' to be in the direction making an acute angle with x_B; had we chosen $-\sqrt{1/6}$, we would have gotten the opposite direction for y'. And had $I_{x'y'}$ been nonzero, the sign of the answer would have been opposite also.

PROBLEMS ▶ Section 7.3

7.7 The three homogeneous rods in Figure P7.7 are welded together at O to form a rigid body. Find the mass moments and products of inertia at point Q with respect to axes there that are parallel to x, y, and z.

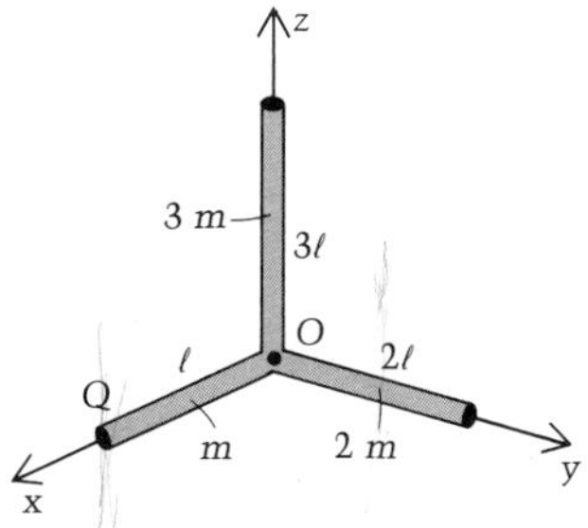

Figure P7.7

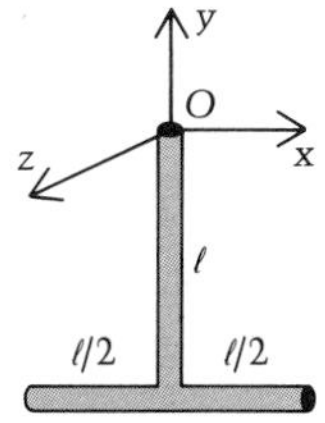

Figure P7.8

7.8 Find the mass moments and products of inertia of the body in Figure P7.8, with respect to a set of axes through the point P at $\left(\frac{\ell}{2}, \frac{\ell}{2}, \frac{\ell}{2}\right)$, respectively, parallel to x, y, and z. Each of the two perpendicular rods of the "T" has mass m and length ℓ.

7.9 In the preceding problem, find the moment of inertia about the line OP.

7.10 Compute the moment of inertia with respect to line AB for the bent bar in Figure P7.10. The bar lies in a plane and has mass 4 m.

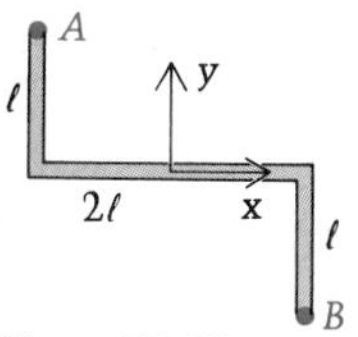

Figure P7.10

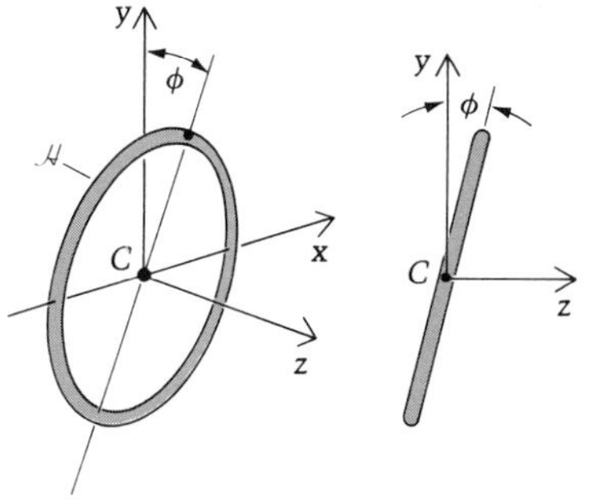

Figure P7.11

7.11 Find the product of inertia I^C_{yz} for the hoop $\mathscr{H}$ of mass m and radius R in Figure P7.11. The plane of $\mathscr{H}$ is misaligned with the xy plane by angle ϕ.

7.12 Compute the moment of inertia about line BA in Example 7.2 by using Equation (7.20) at C instead of B. Show that if $a = b = d$, the answer becomes equal to I_{xx}^C (which equals $I_{yy}^C = I_{zz}^C$ in this case).

7.13 The centroidal moments of inertia for the solid ellipsoid $\mathcal{E}$ in Figure P7.13 are

$$I_{xx}^C = \frac{m}{5}(b^2 + c^2)$$

$$I_{yy}^C = \frac{m}{5}(a^2 + c^2)$$

$$I_{zz}^C = \frac{m}{5}(a^2 + b^2)$$

Further, the mass of $\mathcal{E}$ is $(4\pi/3)\,\rho abc$, where ρ is the mass density. Find the moment of inertia of the mass of $\mathcal{E}$ about the line making equal angles with x, y, and z.

7.14 Show that the sum of any two of I_{xx}^P, I_{yy}^P, and I_{zz}^P always exceeds the third.

7.15 Part of a special-purpose, dual-driven antenna system consists of an octagonal rotator as shown in Figure P7.15. Each of the eight equal sections is a square steel tube with the indicated dimensions and thickness $\frac{1}{8}$ in. Find the moment of inertia of the rotator about the axis of rotation. (Consider each section to have squared-off ends at the average 18-in. length and ignore the small overlaps. Use a density of 15.2 slug/ft³.)

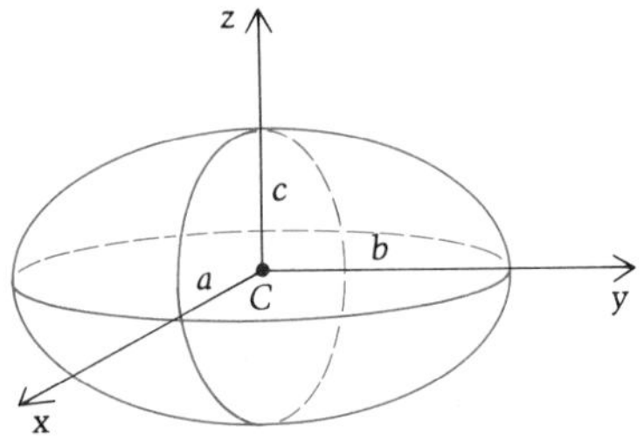

Figure P7.13

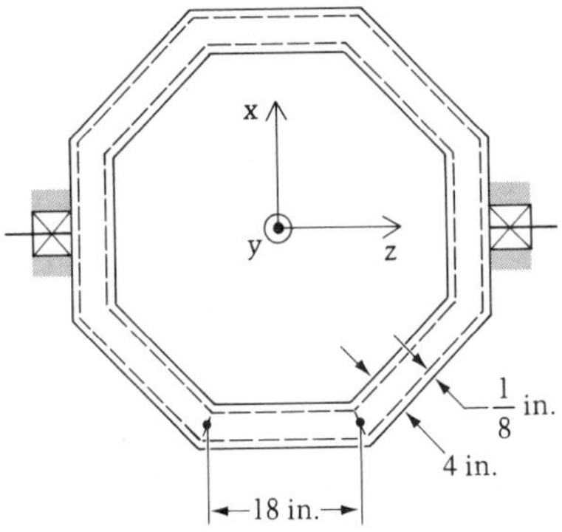

Figure P7.15

7.16 Find the inertia properties at O for the body shown in Figure P7.16, which is composed of a rod and ring that have equal cross sections and densities. The rod is perpendicular to the plane of the ring.

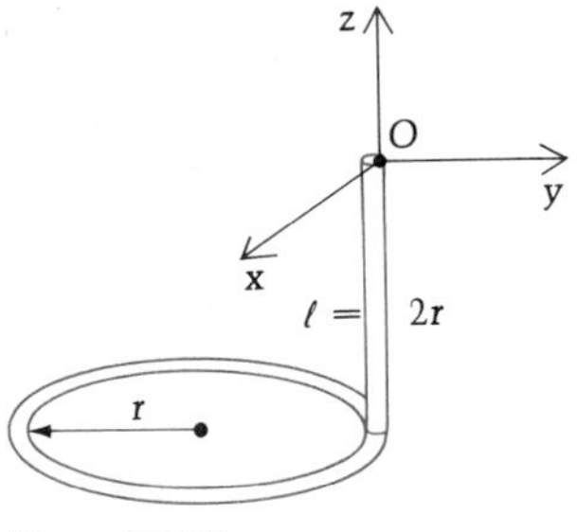

Figure P7.16

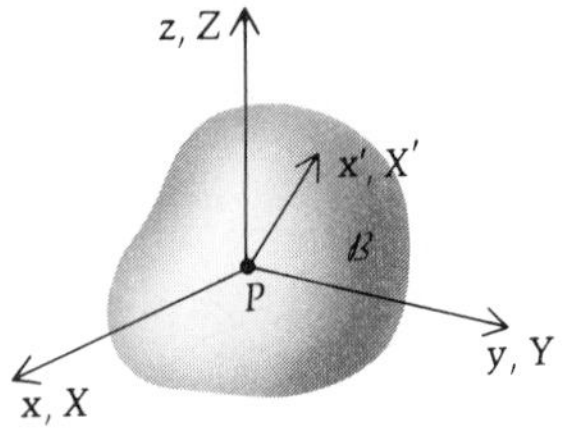

Figure P7.18

* **7.17** Find $I_{x'y'}^B$ for the problem of Example 7.3 if $b = d = a/2$. The axes have their origin at B just as in the example.

* **7.18** For the rigid body $\mathcal{B}$ in Figure P7.18 let the inertia properties at P be known ($I_{xx}^P, \ldots, I_{yz}^P$). Further, let there be measured along the same axes the quantities

$$X = \frac{\ell_x}{\sqrt{I_{x'x'}^P}}$$

$$Y = \frac{\ell_y}{\sqrt{I_{x'x'}^P}}$$

$$Z = \frac{\ell_z}{\sqrt{I_{x'x'}^P}}$$

where $I_{x'x'}^P$ and (ℓ_x, ℓ_y, ℓ_z) are as defined in Section 7.3. Show that Equation (7.20) then implies

$$I_{xx}^P X^2 + I_{yy}^P Y^2 + I_{zz}^P Z^2 + 2XYI_{xy}^P + 2XZI_{xz}^P + 2YZI_{yz}^P = 1$$

This is the equation of an ellipsoid centered at P. Developed by Cauchy in 1827, it is called the *ellipsoid of inertia.* Show that the moment of inertia about any line x' through P equals the reciprocal of the square of the distance from P to the point where X' intersects the ellipsoid.

* Asterisks identify the more difficult problems.

7.19 In the preceding problem, if the products of inertia vanish, the equation of the ellipsoid of inertia written in terms of the resulting moments of inertia (I_1^P, I_2^P, I_3^P) is

$$I_1^P X^2 + I_2^P Y^2 + I_3^P Z^2 = 1$$

Show that not all ellipsoids of the form $ax^2 + by^2 + cz^2 = 1$ can be ellipsoids of inertia. *Hint:* The sum of any two moments of inertia must always exceed the third, as stated in Problem 7.14.

** **7.20** Calculate the inertia properties at O of the three circular fan blades connected by light rods in Figure P7.20. The blades are tilted 30° with respect to the axes OC_1, OC_2, and OC_3; the shaded halves of each are behind the plane of the drawing and the unshaded halves are in front of it.

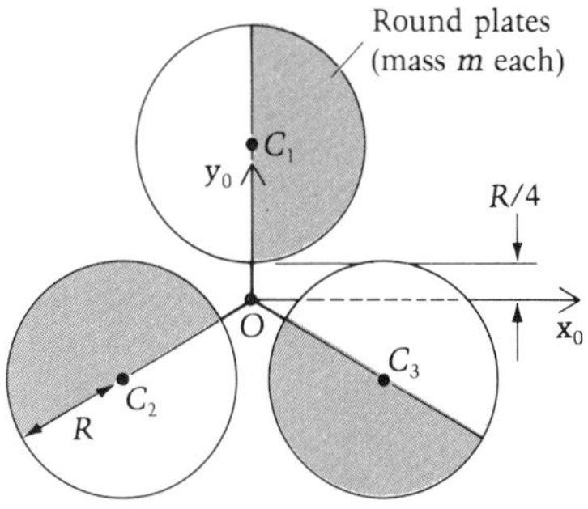

Figure P7.20

7.4 Principal Axes and Principal Moments of Inertia

In this section we demonstrate a particularly useful way of describing the inertia characteristics of a rigid body $\mathcal{B}$. It happens that, at any point P of $\mathcal{B}$, it is always possible to find a set of rectangular axes so that the products of inertia at P with respect to these axes all vanish. These axes are called the **principal axes of inertia** at P, and the moments of inertia with respect to them are known as the **principal moments of inertia** for the point.

Specifically, an axis x is a principal axis at P if $I_{x\beta}^P = 0$, where β is any axis through P that is perpendicular to x. (See Figure 7.3.) We can show that if $I_{xy}^P = I_{xz}^P = 0$, in which (x, y, z) form a triad of rectangular axes at P, then x is a principal axis at P. For the proof, we shall show that $I_{xy}^P = 0 = I_{xz}^P$ implies $I_{x\beta}^P = 0$, where β is the arbitrary axis through P normal to x. Using Equation (7.24) we may write

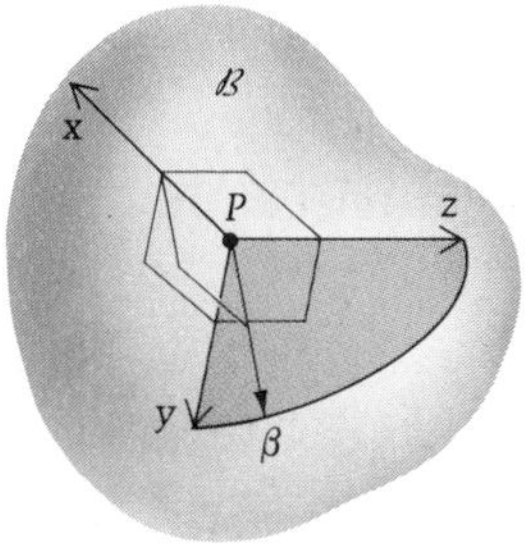

Figure 7.3

$$I_{x\beta}^P = \ell_x n_x I_{xx}^P + \ell_y n_y I_{yy}^P + \ell_z n_z I_{zz}^P + (\ell_x n_y + \ell_y n_x) I_{xy}^P + (\ell_x n_z + \ell_z n_x) I_{xz}^P + (\ell_y n_z + \ell_z n_y) I_{yz}^P$$

In this equation $(\ell_x, \ell_y, \ell_z) = (1, 0, 0)$ are the direction cosines of x (the first subscript in $I_{x\beta}$) with respect to x, y, and z, respectively, while (n_x, n_y, n_z) are the direction cosines of β (the second subscript in $I_{x\beta}$), also with respect to x, y, and z. Since $\beta \perp x$, we have $(n_x, n_y, n_z) = (0, n_y, n_z)$. Substituting these ℓ's and n's, we get

$$I_{x\beta}^P = 0 + 0 + 0 + (\ell_x n_y + 0) I_{xy}^P + (\ell_x n_z + 0) I_{xz}^P + 0$$

But $I_{xy}^P = I_{xz}^P = 0$, so that $I_{x\beta}^P = 0$ and we see that all it takes to make an axis, such as x, a principal axis at a point is to show that $I_{xy}^P = 0$ and $I_{xz}^P = 0$, where (x, y, z) form an orthogonal triad at P. We shall need this result in what is to follow.

Principal Axes at C

We now proceed to a computational procedure for finding the principal axes and moments of inertia. We shall do the derivation when P is C and then later explain how it applies equally well to *all* points of $\mathcal{B}$.

For the moment let (x, y, z) be centered at C and at some instant let x be parallel to the angular velocity $\boldsymbol{\omega}_{\mathcal{B}/\mathcal{I}} = \boldsymbol{\omega}$ of $\mathcal{B}$ in a reference frame $\mathcal{I}$. Then $\boldsymbol{\omega} = \omega\hat{\mathbf{i}}$ and Equation (7.11) gives, for the angular momentum of $\mathcal{B}$ about C, the simplified expression

$$\mathbf{H}_C = I_{xx}^C\omega\hat{\mathbf{i}} + I_{xy}^C\omega\hat{\mathbf{j}} + I_{xz}^C\omega\hat{\mathbf{k}} \tag{7.25}$$

We note from Equation (7.25) that $\mathbf{H}_C$ is parallel to $\boldsymbol{\omega}$ if and only if $I_{xy}^C = I_{xz}^C = 0$ — that is, if x is a principal axis at C. Thus we see, as did Leonhard Euler himself in the middle of the eighteenth century, that the principal axes have the property that when the angular velocity lies along one of them, so will the angular momentum. Euler was seeking an axis through C for which, when $\mathcal{B}$ was set spinning about it, the motion would continue about this axis without any need for external moments to maintain it. Note further that when $\mathbf{H}_C \parallel \boldsymbol{\omega}$ ("$\mathbf{H}_C$ is parallel to $\boldsymbol{\omega}$"), the proportionality constant is necessarily the moment of inertia I about their common axis.

Now we are ready for the big step. We let $\hat{\mathbf{n}} = n_x\hat{\mathbf{i}} + n_y\hat{\mathbf{j}} + n_z\hat{\mathbf{k}}$ be a unit vector and we seek the direction of $\hat{\mathbf{n}}$ that will ensure its being a principal axis. In other words, we want to find the values of the direction cosines of $\hat{\mathbf{n}}$ (n_x, n_y, n_z) such that if $\boldsymbol{\omega} = \omega\hat{\mathbf{n}}$, then $\mathbf{H} = I\boldsymbol{\omega}$. Writing $\boldsymbol{\omega}$ in component form gives

$$\boldsymbol{\omega} = \omega n_x\hat{\mathbf{i}} + \omega n_y\hat{\mathbf{j}} + \omega n_z\hat{\mathbf{k}} \tag{7.26}$$

Substituting from Equation (7.11) for $\mathbf{H}_C$, the vector relation $\mathbf{H}_C = I\boldsymbol{\omega}$ then gives the following three scalar component equations:

$$\begin{aligned} I_{xx}^C\omega_x + I_{xy}^C\omega_y + I_{xz}^C\omega_z &= I\omega n_x \\ I_{xy}^C\omega_x + I_{yy}^C\omega_y + I_{yz}^C\omega_z &= I\omega n_y \\ I_{xz}^C\omega_x + I_{yz}^C\omega_y + I_{zz}^C\omega_z &= I\omega n_z \end{aligned} \tag{7.27}$$

If we divide all three of Equations (7.27) by ω and note that since $\boldsymbol{\omega} = \omega\hat{\mathbf{n}}$ we have $n_x = \omega_x/\omega$, $n_y = \omega_y/\omega$, and $n_z = \omega_z/\omega$, then we may rearrange the equations as follows:

$$\begin{aligned} (I_{xx}^C - I)n_x + I_{xy}^C n_y + I_{xz}^C n_z &= 0 \\ I_{xy}^C n_x + (I_{yy}^C - I)n_y + I_{yz}^C n_z &= 0 \\ I_{xz}^C n_x + I_{yz}^C n_y + (I_{zz}^C - I)n_z &= 0 \end{aligned} \tag{7.28}$$

We now have a set of three equations that are algebraic, linear, and homogeneous in the three variables n_x, n_y, n_z. Such a system is known to have a nontrivial solution if and only if the determinant of the coefficients of the variables is zero.* In this case, we may drop the "nontrivial" adjective because the trivial solution ($n_x = n_y = n_z = 0$) fails to satisfy the side condition

$$n_x^2 + n_y^2 + n_z^2 = 1 \tag{7.29}$$

which must always be true for the direction cosines of a vector.

Calculation of Principal Moments of Inertia

Setting the determinant of the coefficients in (7.28) equal to zero will lead first to the special values of I for which the three equations have a solution. Each special value I is called an *eigenvalue,* or *characteristic value,* and will be a principal moment of inertia; the corresponding $\hat{\mathbf{n}}$ (with components n_x, n_y, n_z) is called the *eigenvector* associated with this eigenvalue I. The unit vector $\hat{\mathbf{n}}$ points in the direction of a principal axis of inertia at C. The determinant is equated to zero below:

$$\begin{vmatrix} I_{xx}^C - I & I_{xy}^C & I_{xz}^C \\ I_{xy}^C & I_{yy}^C - I & I_{yz}^C \\ I_{xz}^C & I_{yz}^C & I_{zz}^C - I \end{vmatrix} = 0 \tag{7.30}$$

If we expand this characteristic determinant, we clearly get a cubic polynomial in I:

$$I^3 + a_1 I^2 + a_2 I + a_3 = 0 \tag{7.31}$$

The a_i are of course functions of the inertia properties. Now we know from algebra that if polynomials with real coefficients have any complex roots, they must occur in conjugate pairs. Thus the polynomial derived above has at this point at least one real root I_1. (It is positive by the definition of the quantity "moment of inertia" that it represents.) We now are guaranteed at least one principal moment of inertia and corresponding principal axis of inertia.

In order to show that there are two others, we next reorient our orthogonal triad of reference axes so that one of them (x) coincides with the already identified principal axis; this then allows us to write $I_{xy}^C = 0$, $I_{xz}^C = 0$, and $I_{xx}^C = I_1$, where y and z are now a new pair of axes normal to our new (principal) x axis. Equations (7.28) now appear as

$$\begin{aligned} (I_1 - I)n_x &+ 0n_y &+ 0n_z &= 0 \\ 0n_x &+ (I_{yy}^C - I)n_y &+ I_{yz}^C n_z &= 0 \\ 0n_x &+ I_{yz}^C n_y &+ (I_{zz}^C - I)n_z &= 0 \end{aligned} \tag{7.32}$$

and the determinantal equation becomes:

$$\begin{vmatrix} I_1 - I & 0 & 0 \\ 0 & I_{yy}^C - I & I_{yz}^C \\ 0 & I_{yz}^C & I_{zz}^C - I \end{vmatrix} = 0 \tag{7.33}$$

---

* Cramer's rule clearly gives $n_x = n_y = n_z = 0$ as the only solution if the equations are independent, in which case the determinant D of the coefficients is not zero. If $D = 0$, then Cramer's rule yields the indeterminate form $0/0$ for the n's and there is a chance for other solutions, as the equations are then dependent.

This time the resulting cubic becomes factorable. Expanding the determinant, we have

$$(I_1 - I)[(I_{yy}^C - I)(I_{zz}^C - I) - I_{yz}^{C^2}] = 0 \qquad (7.34)$$

The principal moments of inertia at C are the roots of Equation (7.34). The first root (the one we already knew) is reaffirmed by setting the first factor to zero:

$$I = I_1 \qquad (= I_{xx}^C)$$

The others will be seen to come from equating the second factor to zero:

$$I^2 - (I_{yy}^C + I_{zz}^C)I + (I_{yy}^C I_{zz}^C - I_{yz}^{C^2}) = 0 \qquad (7.35)$$

This is, of course, a quadratic equation in I. Recalling that the two roots to

$$aI^2 + bI + c = 0$$

are

$$\frac{-b \pm \sqrt{b^2 - 4ac}}{2a}$$

we see that we shall have two (more) real roots I_2 and I_3 if the discriminant is positive or zero:

$$\begin{aligned} b^2 - 4ac &= (I_{yy}^C + I_{zz}^C)^2 - 4(1)(I_{yy}^C I_{zz}^C - I_{yz}^{C^2}) \\ &= I_{yy}^{C^2} - 2I_{yy}^C I_{zz}^C + I_{zz}^{C^2} + 4I_{yz}^{C^2} \\ &= (I_{yy}^C - I_{zz}^C)^2 + 4I_{yz}^{C^2} \geq 0 \end{aligned} \qquad (7.36)$$

Therefore all three roots of the characteristic cubic equation are real (and positive), and so we always have three principal moments of inertia at C, each with its own corresponding principal axis.*

Calculation of Principal Directions

We mention at this point the procedure for obtaining the principal direction, given by n_x, n_y, and n_z, for *each* of the principal moments of inertia (I_1, I_2, or I_3). Equations (7.28), being dependent, may not be solved for the three components of each $\hat{\mathbf{n}}$ in themselves; however, together with the identity $n_x^2 + n_y^2 + n_z^2 = 1$, a solution may be found. The idea is to solve for, say, n_y and n_z in terms of n_x from two of Equations (7.28); then we substitute into Equation (7.29) and solve for n_x. Either sign may be used in taking the final square root, because there are obviously two legitimate sets of direction cosines. These two sets are negatives of each other, and each yields the correct principal axis. In Figure 7.4, either $\hat{\mathbf{n}}$ or $-\hat{\mathbf{n}}$ defines a principal axis through C. The principal axis is an undirected line.

In the first of two examples to follow, we shall again see (as in the preceding discussion) that if at least two of the products of inertia vanish,

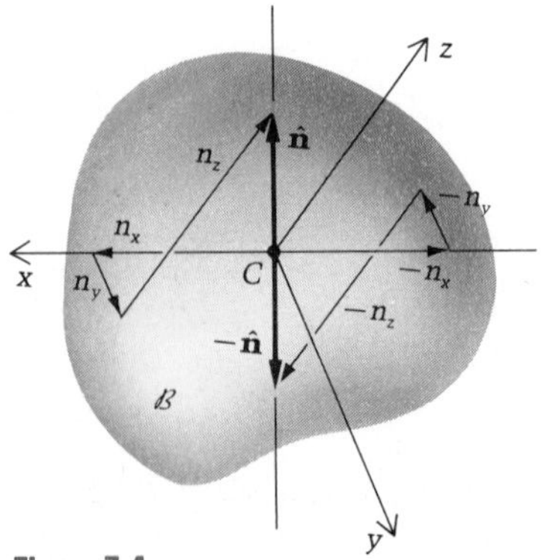

Figure 7.4

* This was proved in 1755 for the first time by Segner, a contemporary of Euler. Segner also showed that the principal axes (for distinct principal moments of inertia) are orthogonal.

then the cubic equation (7.31) becomes factorable. In this case, we do not have to solve it numerically.

EXAMPLE 7.4

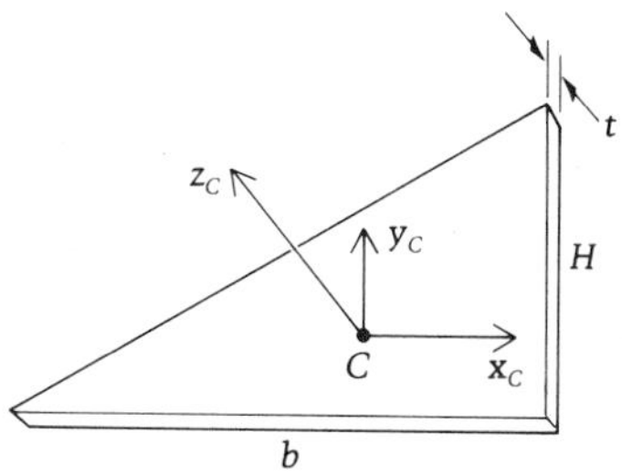

Figure E7.4a

The inertia properties of a right triangular plate (see Figure E7.4a) are

$$I_{xx}^C = \frac{mH^2}{18} \qquad I_{xy}^C = \frac{-mbH}{36}$$

$$I_{yy}^C = \frac{mb^2}{18} \qquad I_{xz}^C = 0$$

$$I_{zz}^C = \frac{m(H^2+b^2)}{18} \qquad I_{yz}^C = 0$$

Find the principal moments of inertia of the plate. Then find their associated principal axes when $b = H$.

Solution

Equations (7.28), which lead to the principal moments and axes of inertia, become, for the plate,

$$\left(\frac{mH^2}{18} - I\right) n_x - \frac{mbH}{36} n_y + 0n_z = 0 \tag{1}$$

$$\frac{-mbH}{36} n_x + \left(\frac{mb^2}{18} - I\right) n_y + 0n_z = 0 \tag{2}$$

$$0n_x + 0n_y + \left(\frac{m(b^2+H^2)}{18} - I\right) n_z = 0 \tag{3}$$

The algebra is simplified by dividing by $mH^2/36$ and defining

$$B = \frac{b}{H} \quad \text{and} \quad I^* = \frac{I}{(mH^2/36)} \tag{4}$$

This gives, in terms of the nondimensional parameters B and I^*,

$$(2 - I^*)n_x + (-B)n_y + 0n_z = 0 \tag{5}$$

$$(-B)n_x + (2B^2 - I^*)n_y + 0n_z = 0 \tag{6}$$

$$0n_x + 0n_y + [2(B^2+1) - I^*]n_z = 0 \tag{7}$$

Therefore the determinantal equation becomes

$$\begin{vmatrix} 2 - I^* & -B & 0 \\ -B & 2B^2 - I^* & 0 \\ 0 & 0 & 2(B^2+1) - I^* \end{vmatrix} = 0 \tag{8}$$

Expanding across the third row (or down the third column), we get

$$[2(B^2+1) - I^*][(2 - I^*)(2B^2 - I^*) - B^2] = 0 \tag{9}$$

Therefore one of the brackets must vanish and the roots come from

$$I^* = 2(B^2+1) \tag{10}$$

and

$$I^{*2} - 2I^*(B^2 + 1) + 3B^2 = 0 \tag{11}$$

Equation (10) gives, using Equations (4),

$$I_3 = \frac{m(b^2 + H^2)}{18} \tag{12}$$

Equation (11) gives, by the quadratic formula,

$$I^*_{1,2} = \frac{2(B^2 + 1) \pm \sqrt{4(B^2 + 1)^2 - 12B^2}}{2} \tag{13}$$

Thus

$$I_{1,2} = \frac{mH^2}{36}\left(\frac{b^2}{H^2} + 1 \pm \sqrt{\frac{b^4}{H^4} - \frac{b^2}{H^2} + 1}\right)$$

$$= \frac{m(b^2 + H^2)}{36} \pm \frac{m}{36}\sqrt{b^4 - b^2H^2 + H^4} \tag{14}$$

The three principal moments of inertia of the plate are given by Equations (12) and (14). In the case when the right triangle is isosceles ($b = H$), we have $B = 1$ so that, from Equation (13),

$$I^*_{1,2} = 2 \pm \sqrt{4 - 3} = 2 \pm 1$$

or

$$I^*_1 = 3 \qquad \text{and} \qquad I^*_2 = 1 \tag{15}$$

Also, from Equation (10),

$$I^*_3 = 2(1^2 + 1) = 4$$

Changing back to dimensional inertias by Equation (4), we see that for a right isosceles triangular plate,

$$I_1 = \frac{mH^2}{12} \qquad I_2 = \frac{mH^2}{36} \qquad I_3 = \frac{mH^2}{9} \tag{16}$$

We shall now determine the principal axis associated with each of these principal moments of inertia. We first substitute $I^*_1 = 3$ (with $B = 1$) in each of Equations (5) to (7) and get

$$\begin{aligned} -n_x - n_y &= 0 \\ -n_x - n_y &= 0 \\ n_z &= 0 \end{aligned} \tag{17}$$

The third of these equations says that the principal axis for I_1 is in the plane of the plate (xy); the other two equations both give

$$n_x = -n_y \tag{18}$$

Substituting this result into

$$n_x^2 + n_y^2 + n_z^2 = 1 \tag{19}$$

gives

$$(-n_y)^2 + n_y^2 = 1 \tag{20}$$

$$n_y^2 = \frac{1}{2}$$

$$n_y = \pm\frac{1}{\sqrt{2}} \tag{21}$$

Thus, by Equation (18),

$$n_x = \mp\frac{1}{\sqrt{2}} \tag{22}$$

so that either

$$(n_x, n_y, n_z) = \left(-\frac{1}{\sqrt{2}}, \frac{1}{\sqrt{2}}, 0\right) \tag{23}$$

or

$$(n_x, n_y, n_z) = \left(\frac{1}{\sqrt{2}}, -\frac{1}{\sqrt{2}}, 0\right) \tag{24}$$

The lines ℓ_1 defined by these two sets of direction cosines are shown in Figure E7.4b and Figure E7.4c. It is seen that the two preceding results represent the *same line;* the positive directions are opposite but unimportant. The inertia value, being the integral of $r^2\,dm$, is independent of the directivity of the line.

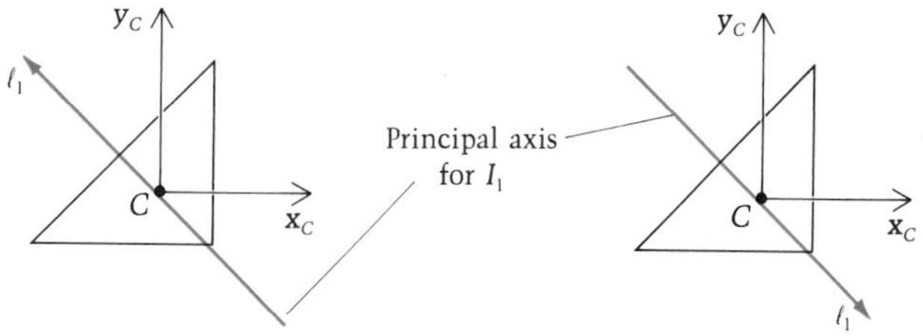

(a) Line of Eq. (7.12)

Figure E7.4b

(b) Line of Eq. (7.13)

Figure E7.4c

For I_2^*, Equations (5) to (7) become

$$\begin{aligned} n_x - n_y &= 0 \\ -n_x + n_y &= 0 \\ n_z &= 0 \end{aligned} \tag{25}$$

This time $n_x = n_y$ with n_z again equaling zero, so this principal axis makes equal angles with x_C and y_C. (See Figure E7.4d and Figure E7.4e.) Note from the pre-

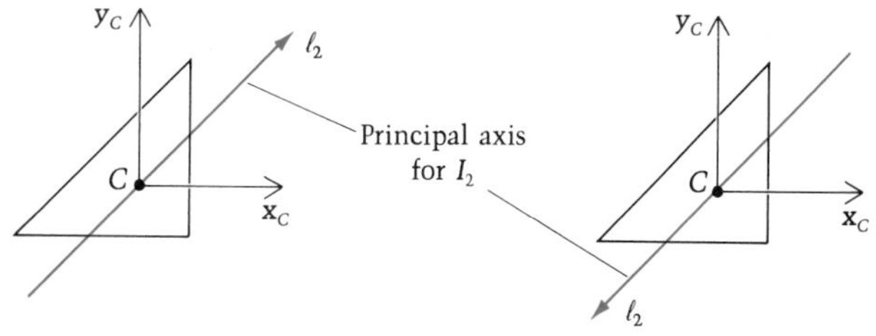

$n_x = n_y = 1/\sqrt{2}$

Figure E7.4d

$n_x = n_y = -1/\sqrt{2}$

Figure E7.4e

ceding diagrams that there is much more inertia about the principal axis for I_1 than there is for I_2; the mass is more closely clustered about ℓ_2 than it is about ℓ_1.

The third principal axis is found from Equations (5) to (7) when, for $B = 1$,

$$I^* = I_3^* = 2(B^2 + 1) = 4 \tag{26}$$

These equations are

$$\begin{aligned} -2n_x - n_y &= 0 \\ -n_x - 2n_y &= 0 \\ [2(2) - 4]n_z &= 0 \end{aligned} \tag{27}$$

The first two of these equations have the solution $n_x = n_y = 0$. The third leaves n_z indeterminate. But from Equation (19) we have $n_z = 1$ or -1. Thus the principal axis for I_3 is the line normal to the plate at C. This will in fact always be true: When a body is a *plate* (that is, flat with negligible thickness compared to its other dimensions), the moment of inertia about the axis normal to the plate at any point is principal for that point. It is even true that it is the sum of the other two and therefore the largest.

Principal Axes at Any Point

We now wish to remark that a set of principal axes exists at *every* point of $\mathcal{B}$, not just at the mass center C. To show this we first recall Equation (7.7), which with Equation (7.12) gives the angular momentum $\mathbf{H}_P$ about any point P of body $\mathcal{B}$:

$$\begin{aligned} \mathbf{H}_P = m\mathbf{r}_{PC} \times \mathbf{v}_P &+ [I_{xx}^P\omega_x + I_{xy}^P\omega_y + I_{xz}^P\omega_z]\hat{\mathbf{i}} \\ &+ [I_{xy}^P\omega_x + I_{yy}^P\omega_y + I_{yz}^P\omega_z]\hat{\mathbf{j}} \\ &+ [I_{xz}^P\omega_x + I_{yz}^P\omega_y + I_{zz}^P\omega_z]\hat{\mathbf{k}} \end{aligned}$$

Whenever the cross-product term in $\mathbf{H}_P$ vanishes, the terms that remain are identical to those of Equation (7.11) if P replaces C. Therefore, for cases in which $\mathbf{r}_{PC} \times \mathbf{v}_P = \mathbf{0}$, we need only recall our arguments made for C and we shall be led, through an identical procedure, to the principal axes and moments of inertia for *any* point P of $\mathcal{B}$.

Therefore we only need to imagine that at some instant our point P of interest has either $\mathbf{v}_P = \mathbf{0}$ or $\mathbf{v}_P \parallel \mathbf{r}_{PC}$. In either case, $\mathbf{r}_{PC} \times \mathbf{v}_P = \mathbf{0}$ and $\mathbf{H}_P$ then has the form of Equation (7.12). Retracing our steps from that point, we arrive at the three axes (through P this time) for which all three products of inertia are zero; these are the same three axes through P for which $\mathbf{H}_P \parallel \boldsymbol{\omega}$ whenever $\boldsymbol{\omega}$ is aligned with one of them and $\mathbf{v}_P$ either vanishes or is parallel to $\mathbf{r}_{PC}$. Of course, all references to the motion conditions that led to the determinantal equation are again lost (as they were for C), so that the principal moments and axes of inertia depend only on the body's mass distribution.

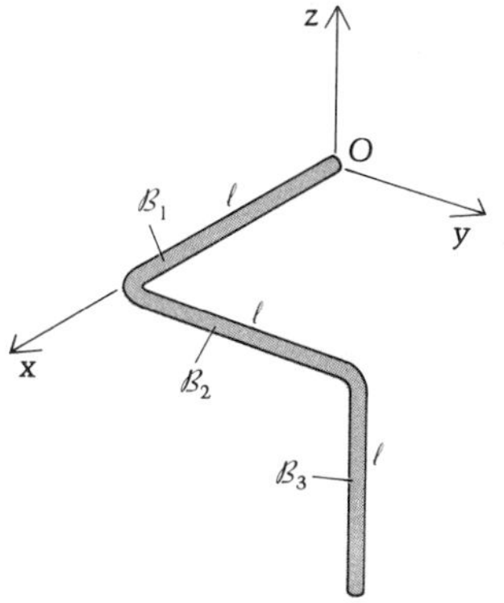

Figure E7.5a

EXAMPLE 7.5

Find the principal moments of inertia at O and the directions of their associated principal axes for the body shown in Figure E7.5a. It is made up of three rigid, identical slender rods welded together at right angles to form a single rigid body $\mathcal{B}$.

Solution

Using the moments of inertia shown for one rod in Figure E7.5b, plus the parallel-axis (transfer) theorem, the six inertia properties are calculated below. The reader should verify each of the entries.

$$I_{xx}^{O} = \underbrace{\frac{m\ell^2}{3}}_{\mathcal{B}_2} + \underbrace{\frac{m\ell^2}{12} + m\left[\ell^2 + \left(\frac{\ell}{2}\right)^2\right]}_{\mathcal{B}_3} = m\ell^2\left(\frac{4+1+15}{12}\right) = \frac{10}{6}m\ell^2$$

$$I_{yy}^{O} = \underbrace{\frac{m\ell^2}{3}}_{\mathcal{B}_1} + \underbrace{m\ell^2}_{\mathcal{B}_2} + \underbrace{\frac{m\ell^2}{12} + \frac{5}{4}m\ell^2}_{\mathcal{B}_3} = m\ell^2\left(\frac{4+12+1+15}{12}\right) = \frac{16}{6}m\ell^2$$

$$I_{zz}^{O} = \underbrace{\frac{m\ell^2}{3}}_{\mathcal{B}_1} + \underbrace{\frac{m\ell^2}{12} + \frac{5}{4}m\ell^2}_{\mathcal{B}_2} + \underbrace{2m\ell^2}_{\mathcal{B}_3} = m\ell^2\left(\frac{4+1+15+24}{12}\right) = \frac{22}{6}m\ell^2$$

$$I_{xy}^{O} = \underbrace{-m\ell\frac{\ell}{2}}_{\mathcal{B}_2} \underbrace{- m\ell^2}_{\mathcal{B}_3} = -\frac{3}{2}m\ell^2 = -\frac{9}{6}m\ell^2$$

$$I_{yz}^{O} = \underbrace{-\left(-m\frac{\ell^2}{2}\right)}_{\mathcal{B}_3} = +\frac{m\ell^2}{2} + \frac{3m\ell}{6}$$

$$I_{xz}^{O} = \underbrace{-\left(-m\frac{\ell^2}{2}\right)}_{\mathcal{B}_3} = +\frac{m\ell^2}{2} = \frac{3m\ell^2}{6}$$

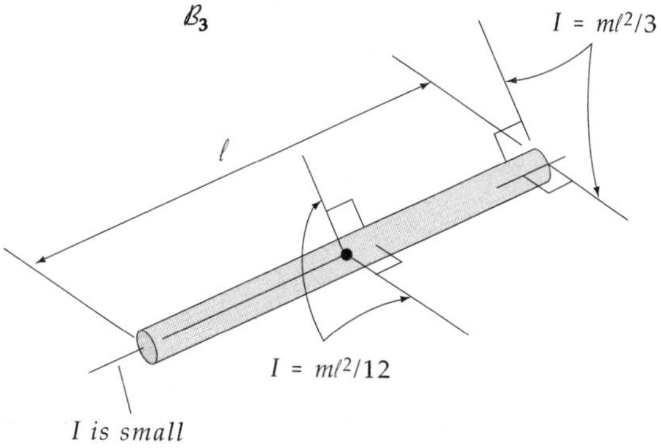

Figure E7.5b

For this problem, then, Equations (7.28) may be expressed as

$$\left(10 - \frac{6I}{m\ell^2}\right) n_x - 9n_y + 3n_z = 0$$

$$-9n_x + \left(16 - \frac{6I}{m\ell^2}\right) n_y + 3n_z = 0$$

$$3n_x + 3n_y + \left(22 - \frac{6I}{m\ell^2}\right) n_z = 0 \qquad (1)$$

in which we have multiplied the three equations by $(6/m\ell^2)$; we may replace $6I/m\ell^2$ by $\mathcal{I}$ and write the determinant as

$$\begin{vmatrix} 10 - \mathcal{I} & -9 & 3 \\ -9 & 16 - \mathcal{I} & 3 \\ 3 & 3 & 22 - \mathcal{I} \end{vmatrix} = 0$$

Expanding gives the characteristic cubic equation:

$$f(\mathcal{I}) = -\mathcal{I}^3 + 48\mathcal{I}^2 - 633\mathcal{I} + 1342 = 0$$

If a computer or programmable calculator is not available,* we can always solve a cubic by trial and error rather quickly. Noting that $f(\mathcal{I})$ is 1342 at $\mathcal{I} = 0$ and is negative at $\mathcal{I} = 3$, for example, on a calculator we may proceed and within a few minutes obtain the root between these values.† The procedure is as follows:

$\mathcal{I}$	$f(\mathcal{I})$	
3	−206	(thus a root is probably just past $\mathcal{I} = 2.5$)
2	260	
2.6	3.10	(still positive)
2.61	−0.93	(so it is >2.6 and <2.61, closer to the latter)
2.608	−0.123	(back up slightly!)
2.607	0.2802	(so it is about two-thirds of the way from 2.607 to 2.608)
2.6076	0.03835	(so just a little farther)
2.6077	−0.00195	(the root is close to this number!)
2.60769	0.00208	(should be halfway between the last one and this one . . .)
2.607695	0.00006	(now double-check)
2.607696	−0.00034	(so $\mathcal{I}_1$ = 2.607695 to seven figures)

Next we use synthetic division to obtain the reduced quadratic:

$$\begin{array}{r|rrrr} & -1 & 48 & -633 & 1342 \\ 2.607695 & & -2.607695 & 118.369287 & -1341.999938 \\ \hline & -1 & 45.392305 & -514.630713 & 0.000062 \approx 0 \end{array}$$

$$-\mathcal{I}^2 + 45.392305\mathcal{I} - 514.630713 = 0$$

* See Appendix B for a numerical solution to this problem using the Newton-Raphson method.

† We abandon our three-digit consistency in numerical analyses like this one in order to illustrate the speed of convergence.

Using the quadratic formula, we obtain

$$\mathcal{I}_2 = 22.000002$$

$$\mathcal{I}_3 = 23.392303$$

The value of $\mathcal{I}_2$ strongly hints that 22 might be a rational root. Synthetic division shows that it is, and refined values from the reduced quadratic are then

$$\mathcal{I}_1 = 2.607695$$

$$\mathcal{I}_2 = 22$$

$$\mathcal{I}_3 = 23.392305$$

Since $\mathcal{I} = 6I/m\ell^2$, our dimensional principal moments of inertia are, to six significant figures,

$$I_1 = 0.434623m\ell^2$$

$$I_2 = 3.66667m\ell^2$$

$$I_3 = 3.89872m\ell^2$$

We next illustrate the computation of the direction cosines, which locate for us the principal axes of inertia. We find them from Equations (1), for which $\mathcal{I}_1$, $\mathcal{I}_2$, and $\mathcal{I}_3$ are the only special values (eigenvalues) of $\mathcal{I}$ for which these equations have a solution. First we seek the principal axis associated with $\mathcal{I}_1 = 2.607695$. The first of Equations (1) becomes

$$7.392305n_x - 9n_y + 3n_z = 0$$

Solving for n_z in terms of n_x and n_y and substituting the result into the second of Equations (1) gives

$$n_y = 0.732051n_x$$

Thus

$$n_z = -0.267950n_x$$

Substituting these expressions for n_y and n_z into $n_x^2 + n_y^2 + n_z^2 = 1$ yields

$$\text{Unit vector } \hat{\mathbf{n}}_1 \begin{cases} n_x = 0.788675 \\ n_y = 0.577350 \\ n_z = -0.211325 \end{cases}$$

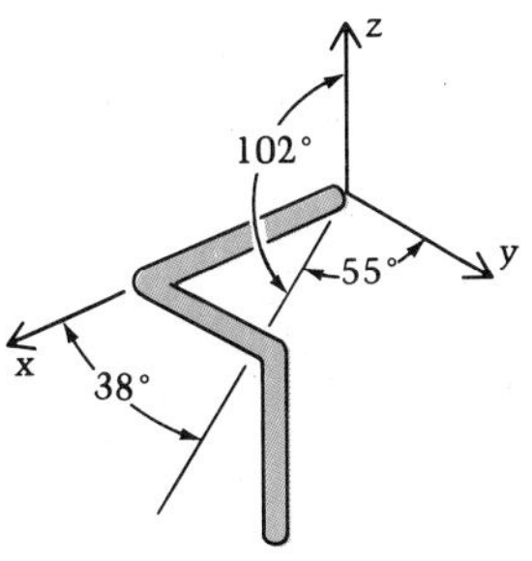

Figure E7.5c

Thus the angles that the principal axis of minimum moment of inertia makes with x, y, and z are, respectively, 37.94°, 54.74°, and 102.20°. This axis has to be the one to which, loosely speaking, the mass finds itself closest. Examination of Figure E7.5c at the left, together with these angles, shows that this makes sense.

Next we may follow the same procedure for the principal axis of maximum moment of inertia ($I_3 = \mathcal{I}_3 m\ell^2/6 = 3.898718m\ell^2$). The results are, as the reader may verify,

$$\text{Unit vector } \hat{\mathbf{n}}_3 \begin{cases} n_x = 0.211325 & (\theta_x = \cos^{-1} n_x = 77.80°) \\ n_y = -0.577350 & (\theta_y = 125.26°) \\ n_z = -0.788675 & (\theta_z = 142.06°) \end{cases}$$

And for the intermediate moment of inertia ($I_2 = \mathcal{I}_2 m\ell^2/6 = 3.666667\ m\ell^2$), the principal axis is defined by

$$\text{Unit vector } \hat{\mathbf{n}}_2 \begin{cases} n_x = \ \ \ 0.577350 & (\theta_x = 54.74°) \\ n_y = -0.577350 & (\theta_y = 125.26°) \\ n_z = \ \ \ 0.577350 & (\theta_z = 54.74°) \end{cases}$$

Orthogonality of Principal Axes

Note in the preceding example that $\hat{\mathbf{n}}_1 \cdot \hat{\mathbf{n}}_2 = \hat{\mathbf{n}}_2 \cdot \hat{\mathbf{n}}_3 = \hat{\mathbf{n}}_3 \cdot \hat{\mathbf{n}}_1 = 0$ and that $\hat{\mathbf{n}}_1 \times \hat{\mathbf{n}}_2 = \hat{\mathbf{n}}_3$.* These are very good checks on the solution since the principal axes, when the principal moments of inertia are distinct ($I_1 \neq I_2 \neq I_3 \neq I_1$), are orthogonal. To prove this in general, let $I_1 \neq I_2$ or I_3, and let x lie along the axis of I_1. Then from the first of Equations (7.32), which is

$$(I_1 - I)n_x = 0$$

we see that if I is either I_2 or I_3, then $n_x = 0$; that is, the cosine of the angle between x and the corresponding principal axis has to be zero. This means that x is perpendicular to the other two principal axes. In turn, these two axes are normal to each other; this follows from reorienting the axes once more so that x still lies along the axis of I_1 but now y lies along the axis of I_2. This time I_{yz}^C is also zero, so the new second equation becomes

$$(I_2 - I)n_y = 0$$

This shows that for $I = I_3 (I_2 \neq I_3)$, the value of n_y for the third principal axis vanishes and it is then normal not only to x (which it still is, since we have only rotated it about x) but also to y. Thus the three principal axes are orthogonal if the principal moments of inertia are all different. We now turn our attention to what happens if they are not.

Equal Moments of Inertia

A commonly occurring case is for two of the principal moments of inertia at a point P to be equal to each other but different from the third. When this happens, we can show that *every* line through P in the plane of the two axes (call them x and y) with equal moments of inertia (say $I_1 = I_2$) is a principal axis having this same value for its moment of inertia.

To do this, we first show that if $I_1 = I_2 \neq I_3$, then the axis associated with I_3 (call it z) is perpendicular to those (x, y) of I_1 and I_2. The third equation of (7.32) gives

$$(I_3 - I)n_z = 0$$

Thus when I is I_1 or I_2, then $n_z = 0$. Hence the angle between z and x (and

* This could have been $\hat{\mathbf{n}}_1 \times \hat{\mathbf{n}}_2 = -\hat{\mathbf{n}}_3$—equally correct!

between z and y) is also 90°. It does not follow in like manner from the equations, however, that axes x and y are perpendicular. To handle this case, we begin by showing that if the axes of the other two principal moments of inertia ($I_1 = I_2$) are *not* presumed to be orthogonal (say they are q and x in Figure 7.5 with $I^P_{qq} = I_1 = I^P_{xx}$), then all is well because I_{yy} is *also* equal to I_1.

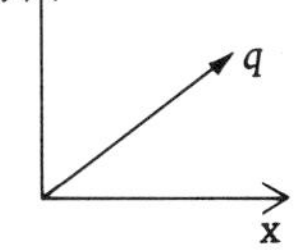

Figure 7.5

To prove this, we use Equation (7.20):

$$I^P_{qq} = I^P_{xx}\ell_x^2 + I^P_{yy}\ell_y^2 + I^P_{zz}\overset{0}{\cancel{\ell_z^2}} + \overbrace{0 + 0 + 0}^{x \text{ and } z \text{ are principal!}}$$

$$I_1 = I_1\ell_x^2 + I^P_{yy}\ell_y^2$$

$$\underbrace{I_1(1 - \ell_x^2)}_{\ell_y^2} = I^P_{yy}\ell_y^2$$

Thus we see that

$$I_{yy} = I_1$$

Next we let $\hat{\ell}$ be a unit vector in the direction of an arbitrary axis x' in the plane of the perpendicular axes x and y. Then using Equation (7.20) again gives

$$I^P_{x'x'} = \ell_x^2 I_1 + \ell_y^2 \overset{=I_1}{\cancel{I_2}} + \overset{0}{\cancel{\ell_z^2}} I_3 + \underbrace{0 + 0 + 0}_{I^P_{xy} = I^P_{xz} = I^P_{yz} = 0}$$

$$= I_1(\ell_x^2 + \ell_y^2) = I_1$$

And equation (7.24) yields (with y' perpendicular to x' and lying in the plane of x', x, and y as in Figure 7.6:

$$I^P_{x'y'} = \ell_x n_x I_1 + \ell_y n_y \overset{=I_1}{\cancel{I_2}} + \overset{0}{\cancel{\ell_z}}\, \overset{0}{\cancel{n_z}}\, I_3 + 0 + 0 + 0$$

$$= I_1(\ell_x n_x + \ell_y n_y)$$

$$= I_1(\hat{\ell} \cdot \hat{\mathbf{n}}) = 0$$

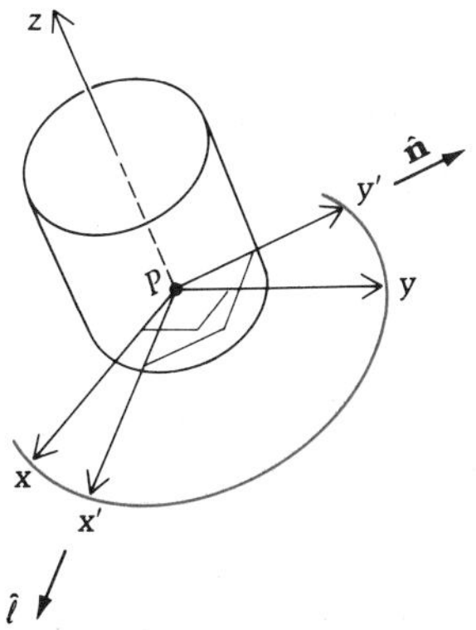

Figure 7.6

Noting that z is perpendicular to x' and that $I^P_{x'z} = 0$ since z is principal, we then have the result that x' is principal with the same moment of inertia as x and y, and this is what we wanted to prove. Note then, for a body with an axis of symmetry, for example, that any axis which passes through and is normal to the symmetry axis of $\mathcal{B}$ is always principal; this is true *even if the axis is not fixed in the body,* which will prove useful to us later.

Finally, if all three principal axes (x, y, z) through P have the same corresponding principal moment of inertia I_1, then *every* axis through P is principal with principal moment of inertia I_1. Let $\hat{\ell}$ be the unit vector along an arbitrary axis x' through P in this case of three equal principal moments of inertia. Also let y' and z' complete an orthogonal triad with x', with $\hat{\mathbf{m}}$ and $\hat{\mathbf{n}}$ being unit vectors in the respective y' and z' directions. Thus (ℓ_x, ℓ_y, ℓ_z), (m_x, m_y, m_z), and (n_x, n_y, n_z) are the respective sets of direction cosines of x', y', and z' with respect to (x, y, z). Equation (7.24) then gives

$$I^P_{x'y'} = (\ell_x m_x + \ell_y m_y + \ell_z m_z)I_1 + \underbrace{0 + 0 + 0}_{\substack{I^P_{xy} = I^P_{yz} = I^P_{xz} = 0 \text{ since} \\ (x,\, y,\, z) \text{ are principal axes}}}$$

$$= 0 \quad (\text{since } \hat{\ell} \perp \hat{\mathbf{m}})$$

In the same way, $I^P_{x'z'} = 0$ since $\hat{\ell} \perp \hat{\mathbf{n}}$ as well. Thus the arbitrary axis x' through P is principal, and Equation (7.20) shows that its moment of inertia is also I_1:

$$I^P_{x'x'} = \underbrace{(\ell_x^2 + \ell_y^2 + \ell_z^2)}_{= 1 \text{ (since } \hat{\ell} \text{ is a unit vector)}} I_1 = I_1$$

Examples of the preceding results regarding two and three equal principal moments of inertia are given in the table on the next page.

Maximum and Minimum Moments of Inertia

An important property of principal moments of inertia is that the largest and smallest of these are the largest and smallest moments of inertia associated with *any* axis through the point in question. To show that this is true, let $I_1 \le I_2 \le I_3$ be the principal moments of inertia at P and let the corresponding principal axes be x, y, and z. The moment of inertia about some other axis, x', is given by Equation (7.20):

$$I^P_{x'x'} = I_1\ell_x^2 + I_2\ell_y^2 + I_3\ell_z^2$$

or

$$\frac{I^P_{x'x'}}{I_1} = \ell_x^2 + \left(\frac{I_2}{I_1}\right)\ell_y^2 + \left(\frac{I_3}{I_1}\right)\ell_z^2 \ge 1$$

since

$$\frac{I_3}{I_1} \ge \frac{I_2}{I_1} \ge 1 \qquad \text{and} \qquad \ell_x^2 + \ell_y^2 + \ell_z^2 = 1$$

Thus $I^P{}_{x'x'} \ge I_1$, so that *no* line through P has a smaller moment of inertia than the smallest *principal* moment of inertia. By a similar argument $I^P_{x'x'}/I_3 \le 1$, which demonstrates that $I^P_{x'x'} \le I_3$ so that no line through P has a larger associated moment of inertia than the largest *principal* moment of inertia. Thus the largest and smallest moments of inertia at a point P are found among the principal moments of inertia for P. It may now be shown quite easily that the smallest moment of inertia at the mass center is the minimum I for *any* line through *any point* of $\mathcal{B}$ or $\mathcal{B}$ extended.

Two Equal I's

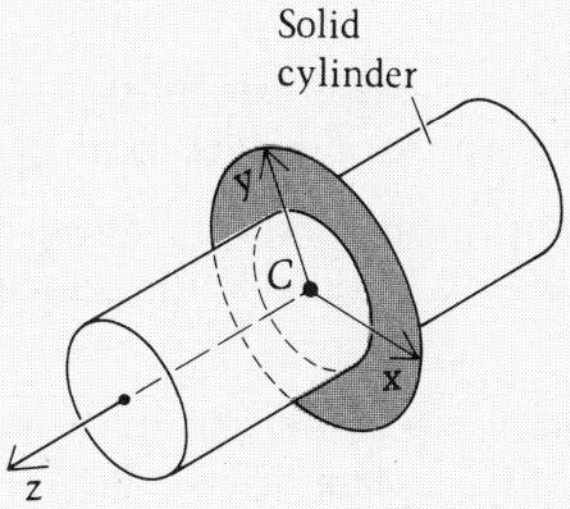

$$I_{xx}^{C} = I_{yy}^{C} = \frac{mR^2}{4} + \frac{m\ell^2}{12}$$

All axes through C in the shaded plane have this same inertia and are principal. Note that $I_{zz}^{C} = mR^2/2$ and is generally not equal to I_{xx} and I_{yy}; if, however, $\ell = \sqrt{3}R$, then *every* axis through C is principal with the same principal moment of inertia!

Three Equal I's

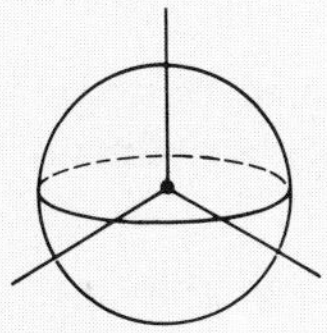

Solid sphere:

$$I = \frac{2}{5}mR^2$$

in any direction through C.

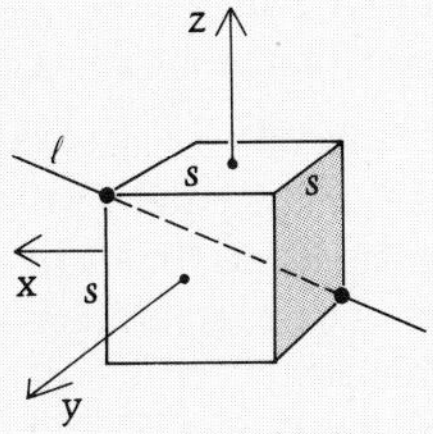

Solid cube:

$$I_x = I_y = I_z = \frac{m}{12}(s^2 + s^2) = \frac{ms^2}{6}$$

in any direction through C. Therefore if line ℓ is a diagonal of the cube, then $I_{\ell\ell} = ms^2/6$ even though it would be formidable to obtain this result by integration. Note that the direction cosines of ℓ are $(\ell_x, \ell_y, \ell_z) = (1/\sqrt{3}, 1/\sqrt{3}, 1/\sqrt{3})$ since it makes equal angles with x, y, and z. Thus Equation (7.20) gives

$$I_{\ell\ell}^{C} = \frac{1}{3}I_x + \frac{1}{3}I_y + \frac{1}{3}I_z + 0 + 0 + 0$$

$$= \left(\frac{1}{3} + \frac{1}{3} + \frac{1}{3}\right)\frac{ms^2}{6} = \frac{ms^2}{6}$$

Question 7.3 Write a one-sentence proof of this statement by using the preceding results together with the parallel-axis theorem.

Answer 7.3 The moment of inertia about any line ℓ through any point P other than C is larger than the moment of inertia about the line through C parallel to ℓ, by the transfer term md^2, and thus the smallest I at C is the smallest of all.

PROBLEMS ▶ Section 7.4

7.21 Find the principal axes and associated principal moments of inertia at O for the semicircular plate of mass m and radius R shown in Figure P7.21.

7.22 Find the principal axes and associated principal moments of inertia for the planar wire shown in Figure P7.22, at the mass center.

7.23 Find the vector from O to the mass center of the bent bar in Example 7.5. Observe that it does not lie along any of the three principal axes at O.

7.24 Use the definitions of the moments of inertia to prove that if a body lies essentially in the xy plane (that is, it has very small dimensions normal to it), then $I_{zz}^P \approx I_{xx}^P + I_{yy}^P$, where P is *any point* in the plane, and that z is a principal axis at P.

7.25 Show that if an axis through the mass center C of body $\mathcal{B}$ is principal at C, then it is a principal axis for *every point* on that axis. *Hint*: Use the transfer theorem for products of inertia, together with the orthogonality of principal axes. (See Figure P7.25.) Transfer I_{yz}^C and I_{xz}^C to P!

7.26 Show that if a principal axis for a point (such as P in the preceding problem) passes through C, then it is also principal for C. (Same hint!)

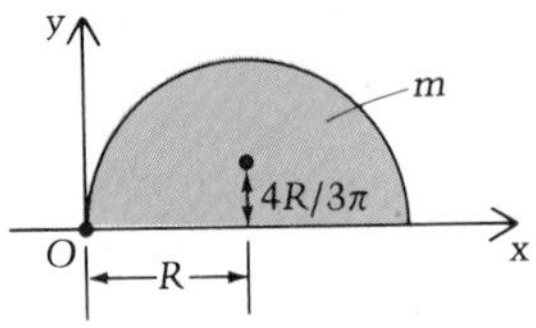

Figure P7.21

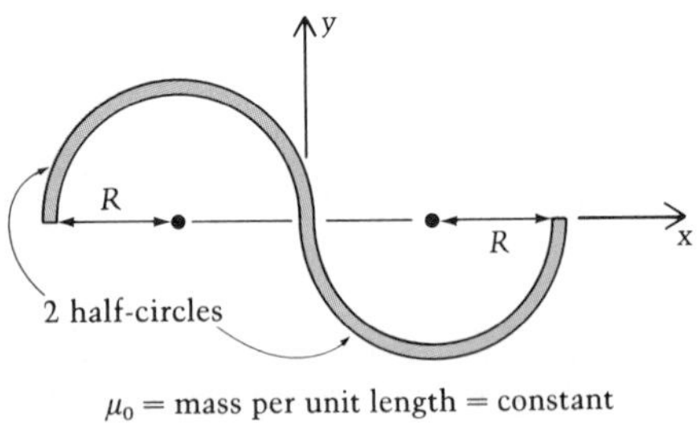

Figure P7.22

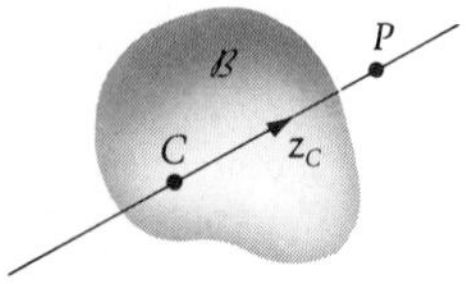

Figure P7.25

7.27 Show that if a line is a principal axis for two of its points, then it is a principal axis for the mass center.

7.28 Show that the three principal axes for any point lying on a principal axis for C are parallel to the principal axes for C.

7.29 Find the principal moments of inertia and their associated axes at O for the thin plate (Figure P7.29) in terms of its density ρ and thickness t.

* **7.30** Find the moments of inertia I_{xx}^P, I_{yy}^P, and I_{zz}^P for the body depicted in Example 4.16. Then find the principal moments of inertia and corresponding principal axes at P.

* **7.31** In Problem 7.16 extend the problem and find the principal moments of inertia, and their principal axes, at point O.

* **7.32** Calculate the principal moments of inertia at O, and the direction cosines of their respective principal axes, for body $\mathcal{B}$ in Figure P7.32. It is made up of three bent bars welded together; all legs are either along, or parallel to, the coordinate axes.

* **7.33** Find the principal moments of inertia and related principal axes at the origin for the body in Figure P7.33.

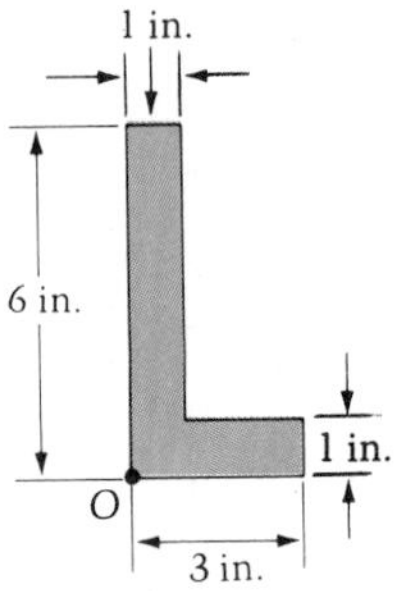

Figure P7.29

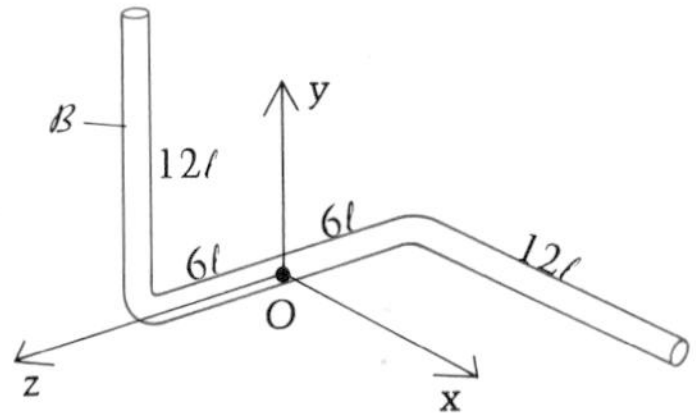

Figure P7.32

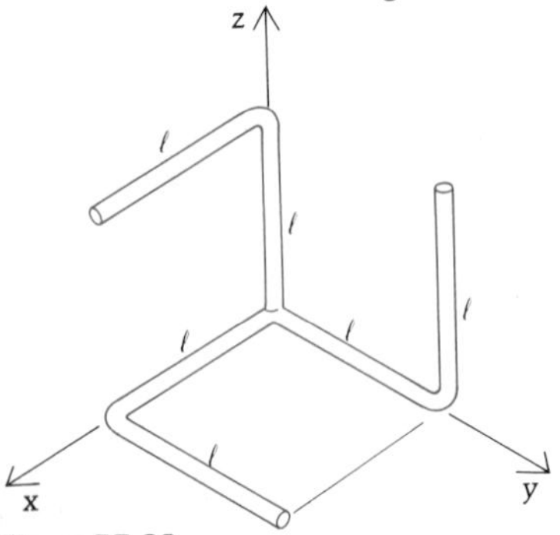

Figure P7.33

* **7.34** In Example 7.5 find the principal moments and axes of inertia at the mass center C.

* **7.35** Find the principal moments of inertia and their associated principal axes for the plate of Example 7.4 when $b = 2H$.

7.36 Find the principal moments and axes of inertia if the body of Example 7.4 has depth L instead of being a thin plate. (See Figure P7.36.) *Hint*: The xy plane is still one of symmetry, so $I_{xz}^C = I_{yz}^C = 0$ again!

* **7.37** For the homogeneous rectangular solid shown in Figure P7.37, find the smallest of the three angles between line AB and the principal axes of inertia at A.

* **7.38** At the origin, find the principal moments of inertia and associated principal axes for a body consisting of three square plates welded along their edges as shown in Figure P7.38. (The axis of the smallest value of I should be the one that the mass lies closest to in an overall sense. Make this rough check on your solution.) Mass $= 3m$, side $= a$.

* **7.39** Four slender bars, each of mass m and length ℓ, are welded together to form the body shown in Figure P7.39. Find: (a) the inertia properties at the mass center C; (b) the principal axes and principal moments of inertia at C.

* **7.40** In Figure P7.40, the axis of symmetry, y_C, of the disk is parallel to y; the plane of the disk is parallel to xz. Find the principal moments of inertia of $\mathcal{D}$ at the origin O, and for the smallest one, determine the angles that its associated principal axis forms with x, y, and z.

** **7.41** Figure P7.41 shows part of a space station being constructed in orbit. Find the principal axes and moments of inertia at C_3. The modules have 33-ft diameters, but due to the material within they are not hollow. For the purposes of this problem, treat each as a uniform hollow shell with a radius of gyration about its axis of 12 ft.

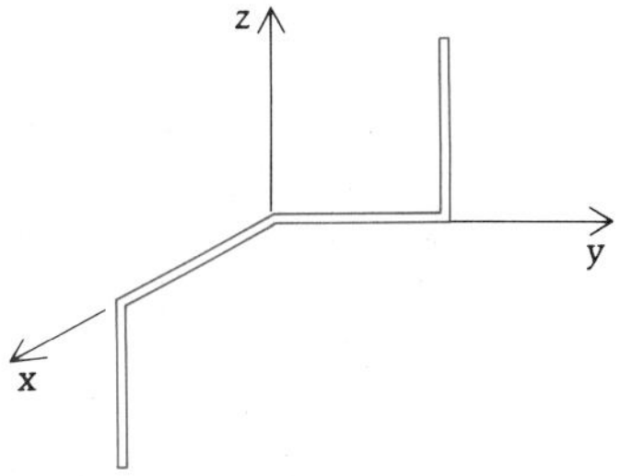

Figure P7.39

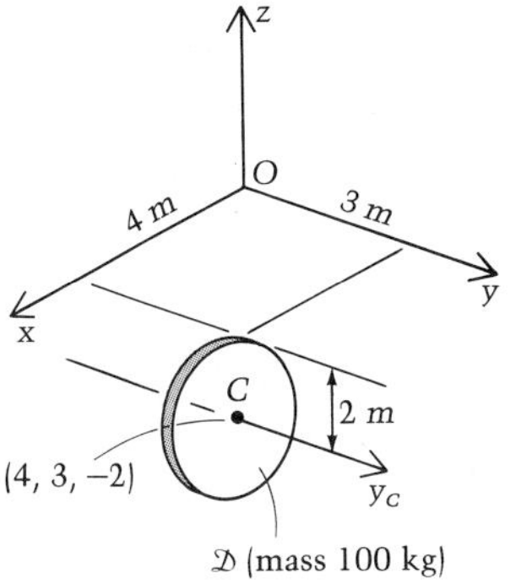

Figure P7.40

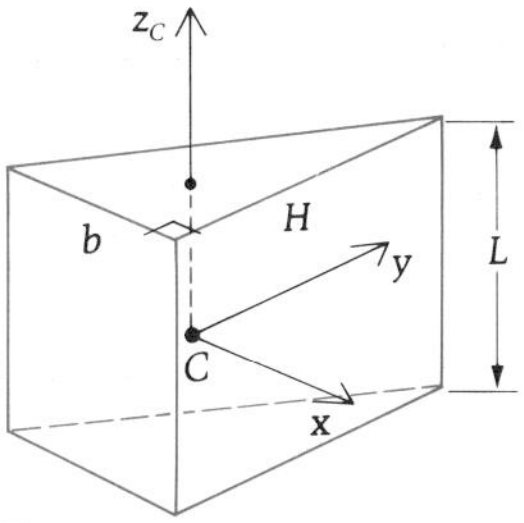

Figure P7.36

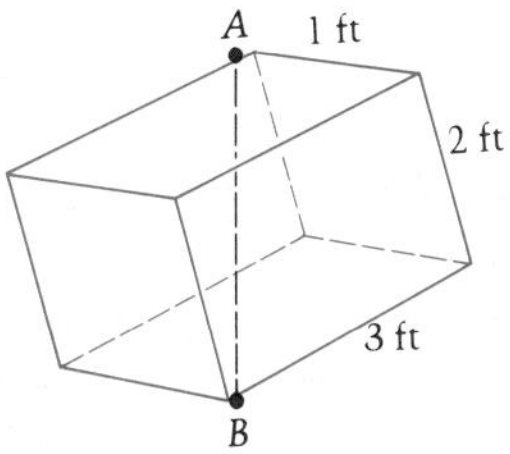

Figure P7.37

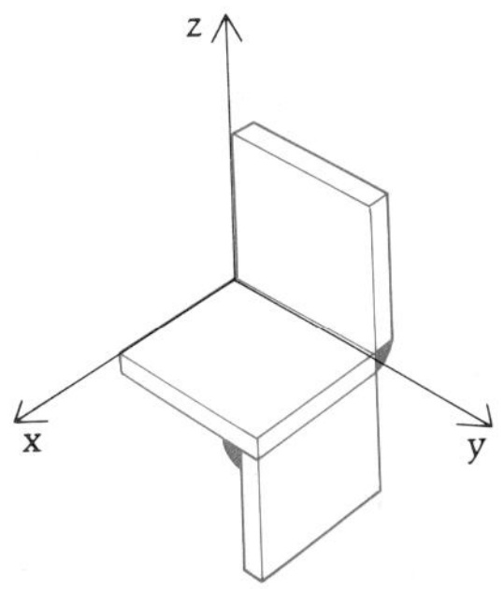

Figure P7.38

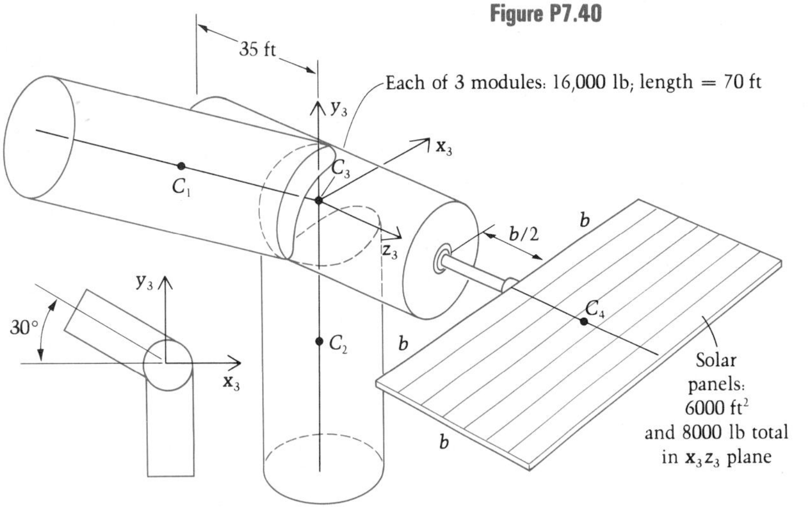

Figure P7.41

7.5 The Moment Equation Governing Rotational Motion

The Euler Equations

In this section we derive the three differential equations governing the angular motion of a rigid body $\mathcal{B}$. Their solution, which is difficult to obtain in closed form in most cases, yields the three components of the angular velocity of $\mathcal{B}$ in an inertial frame $\mathcal{I}$. We begin with Equation (7.2), noting that the time derivative is taken in $\mathcal{I}$. However, the angular momentum $\mathbf{H}_C$ was most conveniently expressed "in"* body $\mathcal{B}$ in Equation (7.11). Thus we shall use Equation (6.20) to move the derivative in (7.2) from $\mathcal{I}$ to $\mathcal{B}$:

$$\Sigma \mathbf{M}_C = {}^{\mathcal{I}}\dot{\mathbf{H}}_C = {}^{\mathcal{B}}\dot{\mathbf{H}}_C + \boldsymbol{\omega}_{\mathcal{B}/\mathcal{I}} \times \mathbf{H}_C \tag{7.37}$$

We now fix the axes (x, y, z) to body $\mathcal{B}$ so that, relative to $\mathcal{B}$, the inertia properties are constant. Using Equation (7.11), the first term on the right side of (7.37) is

$$\begin{aligned} {}^{\mathcal{B}}\dot{\mathbf{H}}_C = {} & (I_{xx}^C \dot{\omega}_x + I_{xy}^C \dot{\omega}_y + I_{xz}^C \dot{\omega}_z)\hat{\mathbf{i}} \\ & + (I_{xy}^C \dot{\omega}_x + I_{yy}^C \dot{\omega}_y + I_{yz}^C \dot{\omega}_z)\hat{\mathbf{j}} \\ & + (I_{xz}^C \dot{\omega}_x + I_{yz}^C \dot{\omega}_y + I_{zz}^C \dot{\omega}_z)\hat{\mathbf{k}} \end{aligned} \tag{7.38}$$

where the unit vectors $\hat{\mathbf{i}}$, $\hat{\mathbf{j}}$, and $\hat{\mathbf{k}}$ are respectively parallel to x, y, and z, and therefore now are fixed in direction in $\mathcal{B}$. The second term in Equation (7.37), after computing the cross product, is

$$\begin{aligned} \boldsymbol{\omega}_{\mathcal{B}/\mathcal{I}} \times \mathbf{H}_C = {} & [(I_{zz}^C - I_{yy}^C)\omega_y\omega_z + I_{yz}^C(\omega_y^2 - \omega_z^2) + \omega_x(\omega_y I_{xz}^C - \omega_z I_{xy}^C)]\hat{\mathbf{i}} \\ & + [(I_{xx}^C - I_{zz}^C)\omega_z\omega_x + I_{xz}^C(\omega_z^2 - \omega_x^2) + \omega_y(\omega_z I_{xy}^C - \omega_x I_{yz}^C)]\hat{\mathbf{j}} \\ & + [(I_{yy}^C - I_{xx}^C)\omega_x\omega_y + I_{xy}^C(\omega_x^2 - \omega_y^2) + \omega_z(\omega_x I_{yz}^C - \omega_y I_{xz}^C)]\hat{\mathbf{k}} \end{aligned} \tag{7.39}$$

The sum of Equations (7.38) and (7.39) yields the right side of Equation (7.37), which in turn equals the moment about the mass center C of all the external forces and couples acting on $\mathcal{B}$.

It is clear that this equation is extremely lengthy and complicated. If we select the body-fixed axes (x, y, z) to be the *principal* axes through C, however, then all product-of-inertia terms vanish and we obtain

$${}^{\mathcal{B}}\dot{\mathbf{H}}_C = I_{xx}^C \dot{\omega}_x \hat{\mathbf{i}} + I_{yy}^C \dot{\omega}_y \hat{\mathbf{j}} + I_{zz}^C \dot{\omega}_z \hat{\mathbf{k}}$$

and

$$\begin{aligned} \boldsymbol{\omega}_{\mathcal{B}/\mathcal{I}} \times \mathbf{H}_C = {} & (I_{zz}^C - I_{yy}^C)\omega_y\omega_z\hat{\mathbf{i}} + (I_{xx}^C - I_{zz}^C)\omega_z\omega_x\hat{\mathbf{j}} \\ & + (I_{yy}^C - I_{xx}^C)\omega_x\omega_y\hat{\mathbf{k}} \end{aligned}$$

so that, substituting into Equation (7.37) and equating the respective

* "Expressing a vector in a frame" simply means the vector is expressed in terms of unit vectors fixed in that frame.

coefficients of $\hat{\mathbf{i}}$, $\hat{\mathbf{j}}$, and $\hat{\mathbf{k}}$, we obtain the **Euler equations:**

$$\begin{aligned} \Sigma M_{Cx} &= I_{xx}^C \dot{\omega}_x - (I_{yy}^C - I_{zz}^C)\omega_y\omega_z \\ \Sigma M_{Cy} &= I_{yy}^C \dot{\omega}_y - (I_{zz}^C - I_{xx}^C)\omega_z\omega_x \\ \Sigma M_{Cz} &= I_{zz}^C \dot{\omega}_z - (I_{xx}^C - I_{yy}^C)\omega_x\omega_y \end{aligned} \tag{7.40}$$

We note that the Euler equations are nonlinear in the $\boldsymbol{\omega}$ components and that the plane-motion equation $\Sigma M_{Cz} = I_{zz}^C \dot{\omega}$ does not extend simply to general motion.

It is very important to realize that, if a body has a pivot (permanently fixed point), equations analogous to the preceding pertain. And in fact these are merely what is obtained by substituting O (the pivot) for C in all equations from (7.37) through (7.40).

We can use Equations (7.40) to make an important observation about the special case of "torque-free" motion, meaning $\Sigma \mathbf{M}_C = \mathbf{0}$. Suppose at an instant $\omega_x \neq 0$, $\omega_y = \omega_z = 0$, where x, y, and z are principal axes. Then, with $\Sigma \mathbf{M}_C = \mathbf{0}$, Equations (7.40) tell us that $\dot{\omega}_x = \dot{\omega}_y = \dot{\omega}_z = 0$; that is, if the body were initially to be spun about a principal axis it would continue to spin about that axis and at constant rate. Conversely, if we seek conditions for which $\dot{\omega}_x = \dot{\omega}_y = \dot{\omega}_z = 0$, we find from Equations (7.40) that two of ω_x, ω_y and ω_z must vanish. Thus the spin will persist in the absence of external moment if and only if the axis of initial spin is a principal axis. This investigation is what led Euler to discover the principal-axis concept in 1750.

Question 7.4 Can a similar conclusion be drawn for spinning about an axis through a pivot?

Use of Non-Principal Axes

Sometimes other forms of moment equations are more advantageous to apply in particular problems than Equations (7.40). Firstly, we may find it convenient to use reference axes that are body-fixed but not principal. Often it is less trouble to simply deal with nonzero products of inertia and axes that are convenient to the body (or its angular velocity) than to compute principal directions and associated inertia properties. The component equations are formed by combining Equations (7.37)–(7.39). When specialized to (x, y) plane motion, we recover the following equations developed in Chapter 4:

$$\begin{aligned} \Sigma M_{Cx} &= I_{xz}^C \dot{\omega}_z - I_{yz}^C \omega_z^2 \\ \Sigma M_{Cy} &= I_{yz}^C \dot{\omega}_z + I_{xz}^C \omega_z^2 \\ \Sigma M_{Cz} &= I_{zz}^C \dot{\omega}_z \end{aligned} \tag{7.41}$$

Answer 7.4 Yes, provided there is no net moment about the pivot.

Use of an Intermediate Frame

Secondly, we often find it convenient to express external moments and/or angular momentum in terms of components associated with directions fixed neither in the body nor in the inertial frame. That is, we may choose to involve an intermediate frame, say $\mathcal{J}$, and use

$$\Sigma \mathbf{M}_C = {}^{\mathcal{I}}\dot{\mathbf{H}}_C = {}^{\mathcal{J}}\dot{\mathbf{H}}_C + \boldsymbol{\omega}_{\mathcal{J}/\mathcal{I}} \times \mathbf{H}_C \qquad (7.42)$$

or, when there is a pivot O,

$$\Sigma \mathbf{M}_O = {}^{\mathcal{I}}\dot{\mathbf{H}}_O = {}^{\mathcal{J}}\dot{\mathbf{H}}_O + \boldsymbol{\omega}_{\mathcal{J}/\mathcal{I}} \times \mathbf{H}_O \qquad (7.43)$$

This approach is particularly useful when, usually because of symmetries, moments and products of inertia of the body remain constant relative to axes fixed in the intermediate frame.

We close this section with five examples, the first employing Euler's equations to study torque-free motion when the initial spin is not about a principal axis. In the second we use body-fixed axes that are not principal in a practical problem of a satellite dish antenna. The final three examples illustrate the use of intermediate frames of reference.

EXAMPLE 7.6

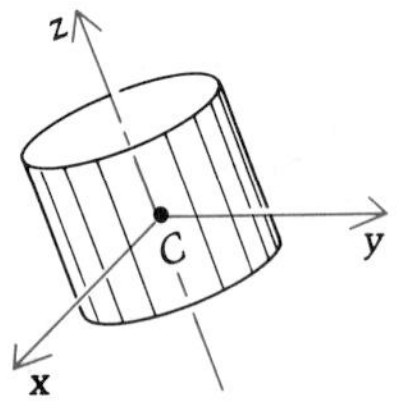

Figure E7.6

A satellite* is moving through deep space far from the influence of atmospheric drag and gravity. (See Figure E7.6.) If the z axis is one of symmetry and if at some instant called $t = 0$ we have $\boldsymbol{\omega} = (\omega_{x_i}, \omega_{y_i}, \omega_{z_i})$ along the body-fixed mass-center axes, find $\boldsymbol{\omega}(t)$. Assume the satellite to be a rigid body.

Solution

The Euler equations, if $I_{xx}^C = I_{yy}^C = I$ and $I_{zz}^C = J$, are

$$I\dot{\omega}_x - (I - J)\omega_y\omega_z = 0 \qquad (1)$$

$$I\dot{\omega}_y - (J - I)\omega_z\omega_x = 0 \qquad (2)$$

$$J\dot{\omega}_z - (I - I)\omega_x\omega_y = 0 \qquad (3)$$

in which the moment components are zero in the absence of external forces and couples. Equation (3) gives

$$\omega_z = \text{constant} = \omega_{z_i} \qquad (4)$$

so that Equations (1) and (2) become linear and are:

$$I\dot{\omega}_x - (I - J)\omega_y\omega_{z_i} = 0 \qquad (5)$$

$$I\dot{\omega}_y - (J - I)\omega_{z_i}\omega_x = 0 \qquad (6)$$

Differentiating Equation (6) and solving for $\dot{\omega}_x$, we get

$$\dot{\omega}_x = \frac{I\ddot{\omega}_y}{(J - I)\omega_{z_i}} \qquad (7)$$

* Such as the *Voyager* spacecraft, which left our solar system in 1983.

This expression may be substituted into (5) to yield an equation free of ω_x:

$$\ddot{\omega}_y + \left(\frac{(J-I)\omega_{z_i}}{I}\right)^2 \omega_y = 0$$

or

$$\ddot{\omega}_y + p^2\omega_y = 0$$

in which $p = (J - I)\omega_{z_i}/I$. The solution to this equation is harmonic:

$$\omega_y = A \cos pt + B \sin pt$$

Since $\omega_y = \omega_{y_i}$ at $t = 0$, we see that $A = \omega_{y_i}$. Finally, Equation (6) gives

$$\omega_x = \frac{\dot{\omega}_y}{p} = \frac{-\omega_{y_i}p \sin pt + Bp \cos pt}{p}$$

or

$$\omega_x = -\omega_{y_i} \sin pt + B \cos pt$$

The initial condition for ω_x gives us

$$\omega_{x_i} = 0 + B \Rightarrow B = \omega_{x_i}$$

so that the other two components (besides $\omega_z = \omega_{z_i}$) of $\boldsymbol{\omega}(t)$ are

$$\omega_x = \omega_{x_i} \cos pt - \omega_{y_i} \sin pt$$
$$\omega_y = \omega_{y_i} \cos pt + \omega_{x_i} \sin pt$$

Example 7.7

For reasons of interference with other bodies, an antenna was recently designed and built with an offset axis as shown in Figure E7.7a and Figure E7.7b. The antenna is composed of a 12-ft, 1200-lb parabolic reflector $\mathcal{R}$, a counterweight $\mathcal{W}$, a reflector support structure $\mathcal{S}$, and a positioner. The positioner consists of (1) a pedestal $\mathcal{P}$ that is fixed to the (inertial) reference frame; (2) an azimuth bearing at

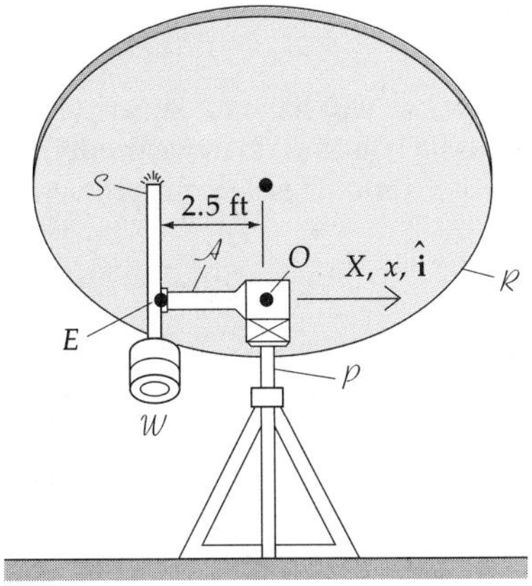

Figure E7.7a Back View

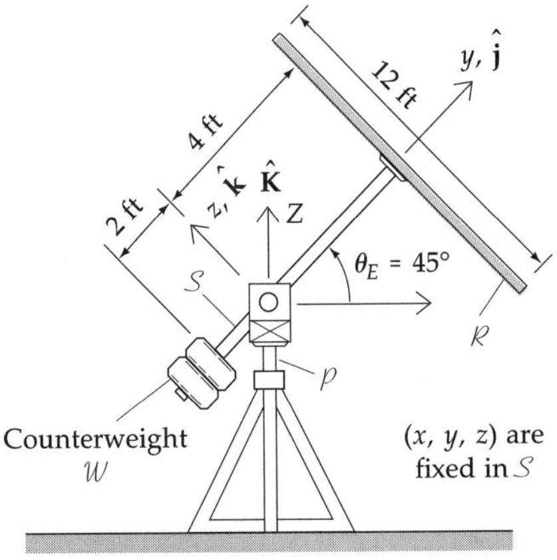

Figure E7.7b Side View

O and ring gear by means of which the arm $\mathcal{A}$ is made to rotate about the vertical; and (3) an elevation torque motor at E that rotates the support structure $\mathcal{S}$ with respect to the arm $\mathcal{A}$. Rotations of the reflector consist of an azimuth angle θ_A about Z, and an elevation angle θ_E about x.

a. Find the weight W of the counterweight $\mathcal{W}$.

b. Find the inertia properties at point O of the (assumed rigid) body $\mathcal{B}$ composed of $\mathcal{R} + \mathcal{W} + \mathcal{S}$.

c. Write the equations of rotational motion of $\mathcal{B}$.

d. Determine the net moments that must be exerted about axes through O for the position shown if first (in units of radians and seconds) $\dot{\theta}_A = \pi/6$, $\ddot{\theta}_E = \pi/6$, $\dot{\theta}_E = \ddot{\theta}_A = 0$; and then $\dot{\theta}_E = \pi/6$, $\ddot{\theta}_A = \pi/6$, $\dot{\theta}_A = \ddot{\theta}_E = 0$.

Solution

Let us model the antenna as follows. The reflector is treated as a thin disk, the counterweight as a point mass, and the support structure as being rigid but light. Axes (X, Y, Z) are fixed in the inertial frame $\mathcal{I}$ while (x, y, z) are attached to $\mathcal{B}$.

In part (a) the purpose of the counterweight is to place the mass center of $\mathcal{B}$ on its elevation axis. Thus

$$1200(4) = W(2)$$

$$W = 2400 \text{ lb}$$

For part (b) we generate the inertia properties of $\mathcal{B}$ about the point O. (See the table on the next page.)

Therefore the *inertia matrix* may be written as

$$\begin{bmatrix} 1230 & -373 & 0 \\ -373 & 1140 & 0 \\ 0 & 0 & 1700 \end{bmatrix} \quad \text{slug-ft}^2 \tag{1}$$

For part (c), one way to proceed is to compute the principal axes and moments of inertia from this matrix and then use the Euler equations (7.40). This would be a very unwise approach in this case, however. Not only is it tedious to locate the principal axes, but following this we would have to break up the angular velocity into its components along these directions; we would then obtain not-so-useful moment components about axes skewed with respect to the rotation axes. It is much simpler to use the second of Euler's laws:*

$$\Sigma \mathbf{M}_O = {}^{\mathcal{I}}\dot{\mathbf{H}}_O$$

The only prices we will have to pay are (1) to retain and deal with the nonzero product of inertia I_{xy}^O and (2) to move the derivative from frame $\mathcal{I}$ to $\mathcal{B}$. The angular momentum of $\mathcal{B}$ about O is, for our problem,

$$\begin{aligned} \mathbf{H}_O = {} & (I_{xx}^O \omega_x + I_{xy}^O \omega_y + \overset{0}{\cancel{I_{xz}^O}} \omega_z)\hat{\mathbf{i}} \\ & + (I_{yx}^O \omega_x + I_{yy}^O \omega_y + \overset{0}{\cancel{I_{yz}^O}} \omega_z)\hat{\mathbf{j}} \\ & + (\overset{0}{\cancel{I_{zx}^O}} \omega_x + \overset{0}{\cancel{I_{zy}^O}} \omega_y + I_{zz}^O \omega_z)\hat{\mathbf{k}} \end{aligned} \tag{2}$$

* All that is required of O for this equation to be valid is that it be a point of the inertial frame; in what follows, however, it also needs to be, and is, a pivot of $\mathcal{B}$.

	Reflector	+	Counterweight	=	Total
I_{xx}^{O}	$= m\left(\frac{r^2}{4} + d^2\right)$	+	$\frac{2400}{32.2} d^2$		
	$= \frac{1200}{32.2}\left(\frac{6^2}{4} + 4^2\right)$	+	$\frac{2400}{32.2} 2^2$		
	+ 932	+	298		= 1230 slug-ft²
I_{yy}^{O}	$= m\frac{r^2}{2}$	+	md^2		
	$= \frac{1200}{32.2}\frac{6^2}{2}$	+	$\frac{2400}{32.2} 2.5^2$		
	= 671	+	466		= 1140 slug-ft²
I_{zz}^{O}	$= m\left(\frac{r^2}{4} + d^2\right)$	+	md^2		
	$= \frac{1200}{32.2}\left(\frac{6^2}{4} + 4^2\right)$	+	$\frac{2400}{32.2}(2^2 + 2.5^2)$		
	= 932	+	764		= 1700 slug-ft²
I_{xy}^{O}	= 0*	+	$(-m\overline{x}\overline{y})$		
	= 0	+	$-\frac{2400}{32.2}(2.5)(2)$		
	= 0	+	−373		= −373 slug-ft²
I_{yz}^{O}	= 0	+	$(-m\overline{y}\overline{z})$		
	= 0	+	$-\frac{2400}{32.2}(2)(0)$		
	= 0	+	0	=	0
I_{xz}^{O}	= 0	+	$(-m\overline{x}\overline{z})$		
	= 0	+	$-\frac{2400}{32.2}(2.5)(0)$		
	= 0	+	0	=	0

* The y axis through O is an axis of symmetry of $\mathcal{R}$; hence $I_{xy}^{O} = 0 = I_{yz}^{O}$. And the third product of inertia of $\mathcal{R}$, I_{xz}^{O}, vanishes because zy is a plane of symmetry for body $\mathcal{R}$.

The angular velocity of $\mathcal{B}$ in frame $\mathcal{I}$ is found by the addition theorem. The reflector and its supporting structure rotate in elevation with a simple angular velocity $\dot{\theta}_E\hat{\mathbf{i}}$ with respect to the housing $\mathcal{A}$; likewise, $\mathcal{A}$ rotates in azimuth with a simple angular velocity $\dot{\theta}_A\hat{\mathbf{K}}$ with respect to the pedestal (which is rigidly fixed to the reference frame $\mathcal{I}$). Therefore

$$\begin{aligned}\boldsymbol{\omega}_{\mathcal{B}/\mathcal{I}} &= \boldsymbol{\omega}_{\mathcal{B}/\mathcal{A}} + \boldsymbol{\omega}_{\mathcal{A}/\mathcal{I}} \\ &= \omega_{\mathcal{B}/\mathcal{A}}\hat{\mathbf{i}} + \omega_{\mathcal{A}/\mathcal{I}}\hat{\mathbf{K}} \\ &= \dot{\theta}_E\hat{\mathbf{i}} + \dot{\theta}_A(\sin\theta_E\hat{\mathbf{j}} + \cos\theta_E\hat{\mathbf{k}}) \qquad (3)\end{aligned}$$

Substituting the ω components into Equation (2) gives the angular momentum of $\mathcal{B}$ in $\mathcal{I}$, expressed in (terms of its components in) $\mathcal{B}$:

$$\begin{aligned}\mathbf{H}_O &= (I_{xx}^O\dot{\theta}_E + I_{xy}^O\dot{\theta}_A \sin \theta_E)\hat{\mathbf{i}} \\ &\quad + (I_{xy}^O\dot{\theta}_E + I_{yy}^O\dot{\theta}_A \sin \theta_E)\hat{\mathbf{j}} + I_{zz}^O\dot{\theta}_A \cos \theta_E\hat{\mathbf{k}}\end{aligned} \tag{4}$$

Next we use Equation (6.20) to differentiate $\mathbf{H}_O$ (note that point O is fixed in $\mathcal{B}$ and $\mathcal{I}$):

$$\Sigma \mathbf{M}_O = {}^{\mathcal{I}}\dot{\mathbf{H}}_O = {}^{\mathcal{B}}\dot{\mathbf{H}}_O + \boldsymbol{\omega}_{\mathcal{B}/\mathcal{I}} \times \mathbf{H}_O \tag{5}$$

Taking the derivative and performing the cross product, we find that the three component equations are as follows. (Note that the inertia properties do not change in $\mathcal{B}$.)

$$\begin{aligned}\Sigma M_{Ox} &= I_{xx}^O\ddot{\theta}_E + I_{xy}^O(\ddot{\theta}_A \sin \theta_E + \dot{\theta}_A\dot{\theta}_E \cos \theta_E) + \dot{\theta}_A \sin \theta_E(I_{zz}^O\dot{\theta}_A \cos \theta_E) \\ &\quad - \dot{\theta}_A \cos \theta_E(I_{xy}^O\dot{\theta}_E + I_{yy}^O\dot{\theta}_A \sin \theta_E) \\ &= I_{xx}^O\ddot{\theta}_E + I_{yy}^O(-\dot{\theta}_A^2 \sin \theta_E \cos \theta_E) + I_{zz}^O(\dot{\theta}_A^2 \sin \theta_E \cos \theta_E) \\ &\quad + I_{xy}^O(\ddot{\theta}_A \sin \theta_E)\end{aligned} \tag{6a}$$

$$\begin{aligned}\Sigma M_{Oy} &= I_{xy}^O\ddot{\theta}_E + I_{yy}^O(\ddot{\theta}_A \sin \theta_E + \dot{\theta}_A\dot{\theta}_E \cos \theta_E) + \dot{\theta}_A \cos \theta_E(I_{xx}^O\dot{\theta}_E + I_{xy}^O\dot{\theta}_A \sin \theta_E) \\ &\quad - \dot{\theta}_E(I_{zz}^O\dot{\theta}_A \cos \theta_E)\end{aligned} \tag{6b}$$

$$\begin{aligned}\Sigma M_{Oz} &= I_{zz}^O(\ddot{\theta}_A \cos \theta_E - \dot{\theta}_A\dot{\theta}_E \sin \theta_E) + \dot{\theta}_E(I_{xy}^O\dot{\theta}_E + I_{yy}^O\dot{\theta}_A \sin \theta_E) \\ &\quad - \dot{\theta}_A \sin \theta_E(I_{xx}^O\dot{\theta}_E + I_{xy}^O\dot{\theta}_A \sin \theta_E)\end{aligned} \tag{6c}$$

As an indication of the increased difficulty of three-dimensional dynamics problems, note that *all four* of the nonvanishing inertia properties contribute to *each component* of the external moment acting on $\mathcal{B}$ at O!

In the indicated position, $\theta_E = 45°$. Therefore

$$\Sigma M_{Ox} = I_{xx}^O\ddot{\theta}_E + I_{yy}^O\left(\frac{-\dot{\theta}_A^2}{2}\right) + I_{zz}^O\left(\frac{\dot{\theta}_A^2}{2}\right) + I_{xy}^O\left(\frac{\ddot{\theta}_A}{\sqrt{2}}\right) \tag{7a}$$

$$\begin{aligned}\Sigma M_{Oy} &= I_{xx}^O\left(\frac{\dot{\theta}_A\dot{\theta}_E}{\sqrt{2}}\right) + I_{yy}^O\left(\frac{\ddot{\theta}_A + \dot{\theta}_A\dot{\theta}_E}{\sqrt{2}}\right) + I_{zz}^O\left(\frac{-\dot{\theta}_A\dot{\theta}_E}{\sqrt{2}}\right) \\ &\quad + I_{xy}^O\left(\ddot{\theta}_E + \frac{\dot{\theta}_A^2}{2}\right)\end{aligned} \tag{7b}$$

$$\begin{aligned}\Sigma M_{Oz} &= I_{xx}^O\left(\frac{-\dot{\theta}_A\dot{\theta}_E}{\sqrt{2}}\right) + I_{yy}^O\left(\frac{\dot{\theta}_A\dot{\theta}_E}{\sqrt{2}}\right) + I_{zz}^O\left(\frac{\ddot{\theta}_A - \dot{\theta}_A\dot{\theta}_E}{\sqrt{2}}\right) \\ &\quad + I_{xy}^O\left(\dot{\theta}_E^2 - \frac{\dot{\theta}_A^2}{2}\right)\end{aligned} \tag{7c}$$

In part (d), for the case specified, $\ddot{\theta}_E = \pi/6$ while $\ddot{\theta}_A = 0$. Also, $\dot{\theta}_A = \pi/6$ while $\dot{\theta}_E = 0$. This case physically corresponds to the antenna, at 45° elevation, swinging around the vertical at 30°/sec and suddenly sensing an object traveling toward zenith; the controls activate a motor whose torque produces an angular acceleration that will send the antenna upward in elevation. The angular accelerations are large because the need is to get there quickly.

Substituting these values of $\dot{\theta}_E$, $\ddot{\theta}_E$, $\dot{\theta}_A$, and $\ddot{\theta}_A$ into Equations (7), along with the inertia values, gives our answer:

$$\Sigma M_{Ox} = 1230\left(\frac{\pi}{6}\right) + (-1140 + 1700)\frac{(\pi/6)^2}{2} + (-373)(0)$$

$$= 721 \text{ lb-ft}$$

$$\Sigma M_{Oy} = 1230(0) + 1140(0) + 1700(0) + (-373)\left[\frac{\pi}{6} + \frac{(\pi/6)^2}{2}\right]$$

$$= -246 \text{ lb-ft}$$

$$\Sigma M_{Oz} = 1230(0) + 1140(0) + 1700(0) + (-373)\left(\frac{-(\pi/6)^2}{2}\right)$$

$$= 51 \text{ lb-ft} \tag{8}$$

These are the moments exerted by $\mathcal{P}$ onto $\mathcal{A}$, excluding that required to balance the dead weight of the antenna. In the inertial reference frame, the moments are

$$\Sigma M_{OX} = \Sigma M_{Ox} = 721 \text{ lb-ft}$$
$$\Sigma M_{OY} = \Sigma M_{Oy} \cos\theta_E + \Sigma M_{Oz}(-\sin\theta_E) = -210 \text{ lb-ft}$$
$$\Sigma M_{OZ} = \Sigma M_{Oy} \sin\theta_E + \Sigma M_{Oz}(\cos\theta_E) = -138 \text{ lb-ft} \tag{9}$$

In the opposite case when the antenna is tracking in elevation, say $\dot{\theta}_E = \pi/6$ rad/sec, and receives a sudden command resulting in $\ddot{\theta}_A = \pi/6$ rad/sec^2 at $\theta_E = 45°$, the moment components become (here $\ddot{\theta}_E = 0 = \dot{\theta}_A$):

$$\Sigma M_{Ox} = 0 + 0 + 0 + (-373)\frac{\pi}{6}\frac{1}{\sqrt{2}}$$

$$= -138 \text{ lb-ft}$$

$$\Sigma M_{Oy} = 0 + 1140\left(\frac{\pi/6 + 0}{\sqrt{2}}\right) + 0 + 0$$

$$= 422 \text{ lb-ft}$$

$$\Sigma M_{Oz} = 0 + 0 + 1700\left(\frac{\pi/6 - 0}{\sqrt{2}}\right) + (-373)\left[\left(\frac{\pi}{6}\right)^2 - 0\right]$$

$$= 629 - 102$$

$$= 527 \text{ lb-ft} \tag{10}$$

Once again we see the considerable effect of the product of inertia term.

The negatives of the X, Y, Z components respectively bend, bend, and twist the pedestal and are considerations in its design; far larger and more important moments, however, arise from the wind blowing against the "dish" and from gravity. There are also forces exerted on $\mathcal{A}$ at O due to gravity and the mass center acceleration.

EXAMPLE 7.8

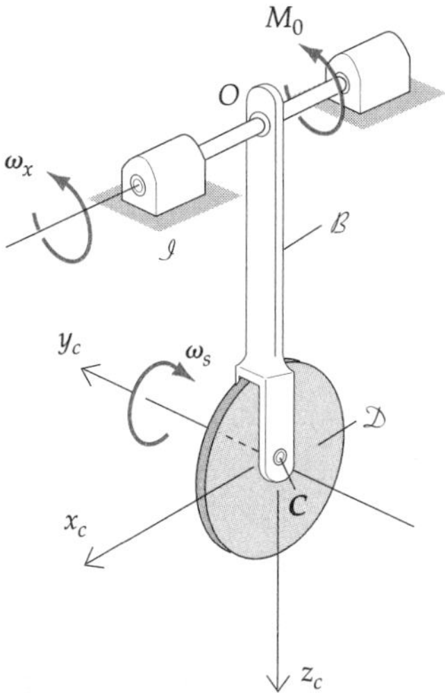

Figure E7.8a

A symmetric wheel $\mathcal{D}$ spins at angular speed ω_s about its axis, which is a line fixed in body $\mathcal{B}$ as well as in $\mathcal{D}$. (See Figure E7.8a.) The moments of inertia of $\mathcal{D}$ at C are $I_{yy}^C = J$ and $I_{xx}^C = I_{zz}^C = I$. Body $\mathcal{B}$ has negligible mass and rotates at angular speed

ω_x about the x axis at O. If a moment M_0 is applied to $\mathcal{B}$ parallel to the x_C axis, find, at the instant shown, the rates of change of ω_s and ω_x and the force and moment exerted on $\mathcal{D}$ at C by the pin. Neglect friction.

Solution

Let us denote the inertial reference frame (in which $\mathcal{B}$ moves) by $\mathcal{I}$. The addition theorem for angular velocity then gives, for the wheel,

$$\begin{aligned}\boldsymbol{\omega}_{\mathcal{D}/\mathcal{I}} &= \boldsymbol{\omega}_{\mathcal{D}/\mathcal{B}} + \boldsymbol{\omega}_{\mathcal{B}/\mathcal{I}} \\ &= \omega_s\hat{\mathbf{j}} + \omega_x\hat{\mathbf{i}}\end{aligned}$$

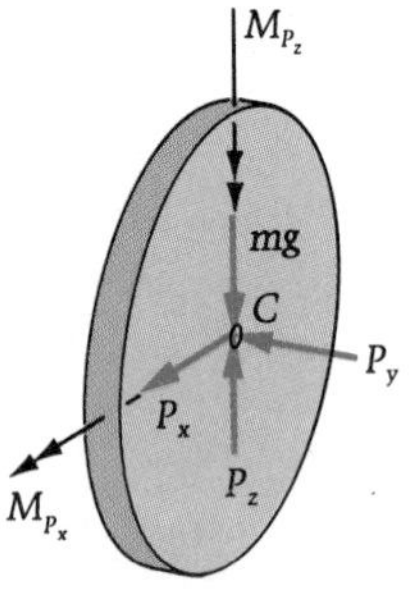

Figure E7.8b

where the axes (x_C, y_C, z_C) are fixed in $\mathcal{B}$ and their associated unit vectors $(\hat{\mathbf{i}}, \hat{\mathbf{j}}, \hat{\mathbf{k}})$ are fixed in direction in $\mathcal{B}$. Note that we may use the equation

$$\mathbf{H}_C = I_{xx}^C\omega_x\hat{\mathbf{i}} + I_{yy}^C\omega_y\hat{\mathbf{j}} + I_{zz}^C\omega_z\hat{\mathbf{k}}$$

for the angular momentum of $\mathcal{D}$ because even though (x_C, y_C, z_C) are not fixed in body $\mathcal{D}$, they are nonetheless permanently principal. Therefore

$$\mathbf{H}_C = I\omega_x\hat{\mathbf{i}} + J\omega_s\hat{\mathbf{j}}$$

The second law of Euler then yields

$$\Sigma\mathbf{M}_C = {}^{\mathcal{I}}\dot{\mathbf{H}}_C = {}^{\mathcal{B}}\dot{\mathbf{H}}_C + \boldsymbol{\omega}_{\mathcal{B}/\mathcal{I}} \times \mathbf{H}_C$$

where for convenience we differentiate $\mathbf{H}_C$ in frame $\mathcal{B}$ since the vector has been written in terms of its components there. Continuing,

$$\begin{aligned}\Sigma\mathbf{M}_C &= I\dot{\omega}_x\hat{\mathbf{i}} + J\dot{\omega}_s\hat{\mathbf{j}} + (\omega_x\hat{\mathbf{i}}) \times (I\omega_x\hat{\mathbf{i}} + J\omega_s\hat{\mathbf{j}}) \\ &= I\dot{\omega}_x\hat{\mathbf{i}} + J\dot{\omega}_s\hat{\mathbf{j}} + J\omega_x\omega_s\hat{\mathbf{k}}\end{aligned}$$

From Figure E7.8b, a free-body diagram of $\mathcal{D}$, we get the components of $\Sigma\mathbf{M}_C$ so that

$$M_{P_x}\hat{\mathbf{i}} + M_{P_z}\hat{\mathbf{k}} = I\dot{\omega}_x\hat{\mathbf{i}} + J\dot{\omega}_s\hat{\mathbf{j}} + J\omega_x\omega_s\hat{\mathbf{k}}$$

Thus we see that

$$M_{P_x} = I\dot{\omega}_x \tag{1}$$

$$0 = J\dot{\omega}_s \Rightarrow \omega_s = \text{constant} \tag{2}$$

$$M_{P_z} = J\omega_x\omega_s \tag{3}$$

In addition, $\Sigma\mathbf{F} = m\mathbf{a}_C$ for the disk yields

$$P_x\hat{\mathbf{i}} + P_y\hat{\mathbf{j}} - P_z\hat{\mathbf{k}} + mg\hat{\mathbf{k}} = m(-\ell\dot{\omega}_x\hat{\mathbf{j}} - \ell\omega_x^2\hat{\mathbf{k}})$$

so that

$$P_x = 0 \tag{4}$$

$$P_y = -m\ell\dot{\omega}_x \tag{5}$$

$$P_z = mg + m\ell\omega_x^2 \tag{6}$$

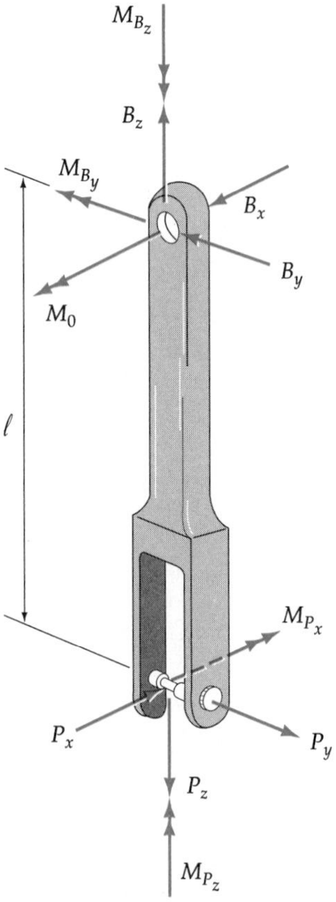

Figure E7.8c

Turning now to Figure E7.8c, a free-body diagram of the *light* body $\mathcal{B}$, we have $\Sigma M_{Ox} \approx 0$ so that

$$\ell P_y + M_0 - M_{P_x} = 0 \tag{7}$$

Combining Equations (1), (5), and (7),

$$\dot{\omega}_x = \frac{M_0}{I + m\ell^2} \tag{8}$$

and

$$P_y = \frac{-m\ell M_0}{I + m\ell^2} \tag{9}$$

and

$$M_{P_x} = \frac{IM_0}{I + m\ell^2} \tag{10}$$

The results (4), (6), (8), (9), and (10) are recognizable as what would have been obtained had the disk been frozen in its bearings, in which case $\mathcal{B}$ and $\mathcal{D}$ would constitute a single rigid body in plane motion.

Equation (3), however, does not follow intuitively from the study of plane motion. The term $J\omega_x\omega_s$ is sometimes called a gyroscopic moment, and the equation says that a moment of this magnitude must act on $\mathcal{D}$ about z_C if the given motion is to occur. Note that in this case the body $\mathcal{D}$ is *not allowed* to turn about z_C as it spins (about y_C). If it were, say by means of a bearing between C and O, then the moment component M_{P_z} would become zero, and a third ω component (about z_C) would appear. Note also from Figure E7.8c that the gyroscopic moment *twists* the shaft of $\mathcal{B}$. This, for example, is a consideration in the retraction of the wheels of some airplanes.

Example 7.9

The thin disk $\mathcal{D}$ turns on the light arm $\mathcal{B}$ by way of a smooth bearing that keeps the axes of $\mathcal{D}$ and $\mathcal{B}$ aligned. The arm is hinged to a shaft driven at constant angular speed Ω by a motor. The system is set up so that the arm is horizontal when the disk contacts the ground as shown in Figure E7.9a. Assuming the disk, of mass m, to roll on the ground, find the forces exerted by the ground on the disk, the forces and moments exerted by the arm on the disk, and the forces and moments exerted by the shaft on the arm.

Solution

We shall begin the analysis by applying to the disk the equations of motion:

$$\Sigma \mathbf{F} = m\mathbf{a}_C$$

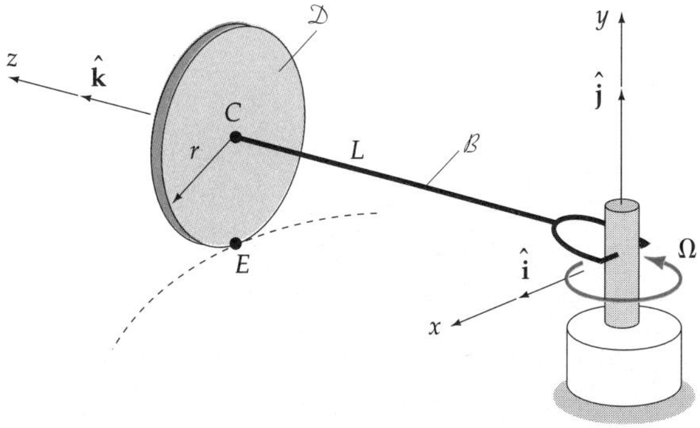

Figure E7.9a

and

$$\Sigma \mathbf{M}_C = {}^{\mathcal{I}}\dot{\mathbf{H}}_C$$

Because C is on the axis of the arm which turns at constant rate Ω, we know that

$$\mathbf{a}_C = -L\Omega^2\hat{\mathbf{k}}$$

where we are using the axes and unit vectors shown in Figure E7.9a and these are *fixed in the arm* $\mathcal{B}$. To obtain an expression for $\mathbf{H}_C$ we need to determine $\boldsymbol{\omega}_{\mathcal{D}/\mathcal{I}}$. So, using the addition theorem,

$$\begin{aligned}\boldsymbol{\omega}_{\mathcal{D}/\mathcal{I}} &= \boldsymbol{\omega}_{\mathcal{D}/\mathcal{B}} + \boldsymbol{\omega}_{\mathcal{B}/\mathcal{I}} \\ &= \omega\hat{\mathbf{k}} + \Omega\hat{\mathbf{j}}\end{aligned}$$

where the only motion $\mathcal{D}$ can have in $\mathcal{B}$ is to turn (in simple angular velocity) about their common axis.

Because the disk rolls, its point E in contact with the ground has zero velocity; using Equation (6.56), we obtain

$$\begin{aligned}\mathbf{v}_E &= \mathbf{v}_C + \boldsymbol{\omega}_{\mathcal{D}/\mathcal{I}} \times \mathbf{r}_{CE} \\ \mathbf{0} &= \mathbf{v}_C + (\omega\hat{\mathbf{k}} + \Omega\hat{\mathbf{j}}) \times (-r\hat{\mathbf{j}}) \\ \mathbf{0} &= L\Omega\hat{\mathbf{i}} + r\omega\hat{\mathbf{i}}\end{aligned}$$

or

$$\omega = -\frac{L}{r}\Omega$$

and

$$\boldsymbol{\omega}_{\mathcal{D}/\mathcal{I}} = -\frac{L}{r}\Omega\hat{\mathbf{k}} + \Omega\hat{\mathbf{j}}$$

Therefore, since we assume the disk to be relatively thin so that to good approximation

$$I_{xx}^C = I_{yy}^C = \frac{mr^2}{4},$$

and since $I_{zz}^C = \dfrac{mr^2}{2}$, we have

$$\mathbf{H}_C = \frac{mr^2}{4}\Omega\hat{\mathbf{j}} - \frac{mrL}{2}\Omega\hat{\mathbf{k}}$$

Note that since Ω is constant,

$${}^{\mathcal{B}}\dot{\mathbf{H}}_C = \mathbf{0}$$

and so

$$\begin{aligned}{}^{\mathcal{I}}\dot{\mathbf{H}}_C &= \overset{\mathbf{0}}{\cancel{{}^{\mathcal{B}}\dot{\mathbf{H}}_C}} + \boldsymbol{\omega}_{\mathcal{B}/\mathcal{I}} \times \mathbf{H}_C \\ &= \mathbf{0} + \Omega\hat{\mathbf{j}} \times \left(\frac{mr^2}{4}\Omega\hat{\mathbf{j}} - \frac{mrL}{2}\Omega\hat{\mathbf{k}}\right) \\ &= -\frac{mrL}{2}\Omega^2\hat{\mathbf{i}}\end{aligned}$$

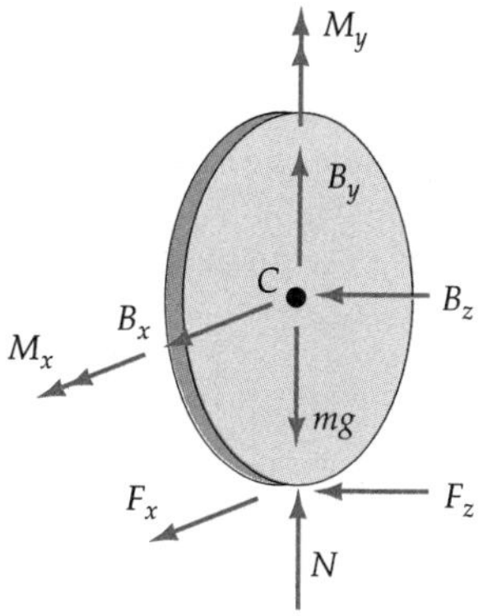

Figure E7.9b

Now using Figure E7.9b, the free-body diagram of the disk, the equation

$$\Sigma \mathbf{F} = m\mathbf{a}_C$$

yields

$$B_x + F_x = 0 \tag{1}$$

$$N + B_y - mg = 0 \tag{2}$$

$$B_z + F_z = -mL\Omega^2 \tag{3}$$

and the equation

$$\Sigma \mathbf{M}_C = {}^{\mathcal{I}}\dot{\mathbf{H}}_C$$

likewise yields

$$M_x - rF_z = -\frac{mrL}{2}\Omega^2 \tag{4}$$

$$M_y = 0 \tag{5}$$

$$rF_x = 0 \tag{6}$$

Using Figure E7.9c, the free-body diagram of the massless bar, we have

$$\Sigma M_{Ax} = 0$$

$$LB_y - M_x = 0 \tag{7}$$

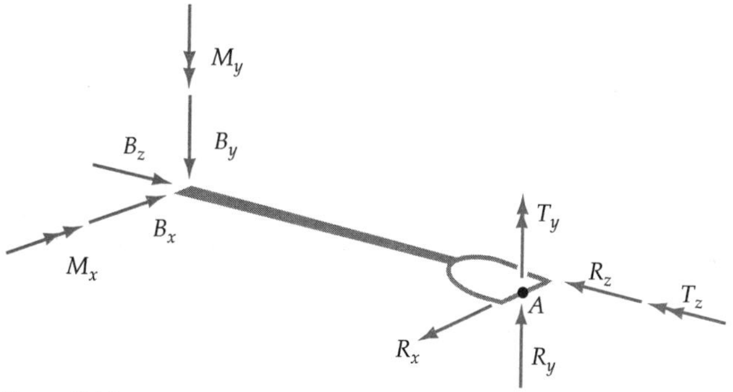

Figure E7.9c

We shall delay using $\Sigma M_{Ay} = 0$, $\Sigma M_{Az} = 0$ and $\Sigma \mathbf{F} = \mathbf{0}$, which will ultimately yield R_x, R_y, R_z, T_y and T_z, and focus on Equations (1)–(7) from which we obtain

$$M_y = F_x = B_x = 0$$

$$M_x = \frac{-mLr}{2}\Omega^2 + rF_z$$

$$B_y = \frac{-mr\Omega^2}{2} + \frac{r}{L}F_z$$

$$B_z = -mL\Omega^2 - F_z$$

$$N = mg + \frac{mr}{2}\Omega^2 - \frac{r}{L}F_z$$

which is as far as we can go because this problem is actually "dynamically" indeterminate. However, if we assume friction to be small (just enough to produce rolling as this system is slowly brought up to speed) so that $F_z \approx 0$, then we have M_x, B_y, B_z and N uniquely determined. With that condition we may now use the remaining "equilibrium" equations for the light rod to obtain

$$R_x = 0$$

$$R_y = \frac{-mr}{2}\Omega^2$$

$$R_z = -mL\Omega^2$$

$$T_y = T_z = 0$$

Let us make a couple of observations: first, note that $T_y = 0$ means that the motor doesn't have to supply any driving torque in order to maintain the constant Ω — remember we have smooth bearings; and secondly note that if Ω is large, so is N:

$$N = mg + \frac{mr\Omega^2}{2}$$

and that extra part of N, over and above the weight mg, is often said to be due to gyroscopic action.

EXAMPLE 7.10

It is possible for the homogeneous cone $\mathcal{B}_1$ of base radius R in the Figure E7.10a to roll steadily around on a flat horizontal table $\mathcal{B}_3$ in such a way that its unconstrained vertex O remains fixed and the center point Q of its base travels on a horizontal circle at constant speed. Let this motion be begun by forces which are then released. Assume that there is sufficient friction between the cone and the table to prevent slipping. Besides having enough friction, there is yet another special condition that must be satisfied in order that the motion occur. Find the friction and normal force resultants, their lines of action, and the special condition. Refer to Example 6.12 for some related kinematics.

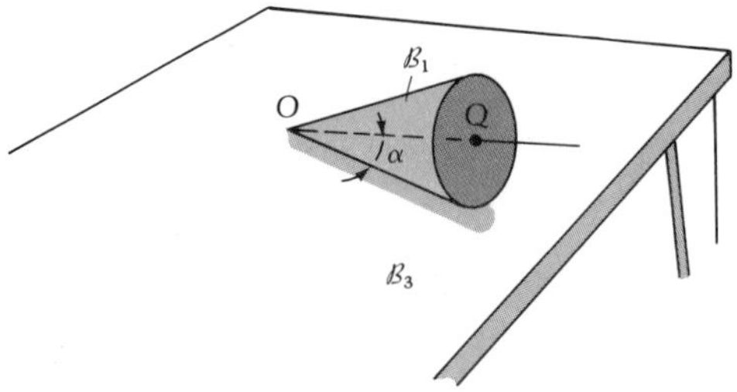

Figure E7.10a

Solution

Since the mass center C moves on a horizontal circle at the constant speed $\frac{3}{4}v_Q$ (see Figure E7.10b), the friction force f is given by:

$$\overset{+}{\longleftarrow} \qquad \Sigma F_x = ma_{CX} = \frac{m\dot{s}_C^2}{\rho_C}$$

or

$$f = \frac{m\left(\frac{3}{4}v_Q\right)^2}{\left(\frac{3}{4}H\cos\alpha\right)} = \frac{3}{4}\frac{mv_Q^2\sin\alpha}{R\cos^2\alpha}$$

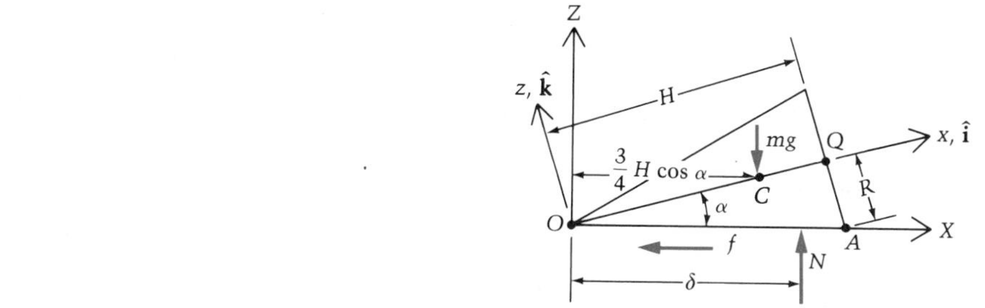

Figure E7.10b

since $H = R/\tan\alpha$. The normal force N is simply mg since $a_{CZ} = 0$. There is no friction force component normal to the plane of the paper, since a_{CY} also vanishes. The directed line of action of f is as shown above, along AO. For the line of action of N, we use the moment equation of motion; note that $\mathcal{B}_3$ is our inertial frame:

$$\Sigma\mathbf{M}_O = {}^{\mathcal{B}_3}\dot{\mathbf{H}}_O = {}^{\mathcal{B}_2}\dot{\mathbf{H}}_O + \boldsymbol{\omega}_{2/3} \times \mathbf{H}_O$$

where frame $\mathcal{B}_2$ contains the Z axis and the axis of the cone. Now the angular momentum $\mathbf{H}_O$ is, since $(\hat{\mathbf{i}}, \hat{\mathbf{j}}, \hat{\mathbf{k}})$ are in principal directions at the fixed point O:

$$\mathbf{H}_O = J\omega_x\hat{\mathbf{i}} + I\omega_y\hat{\mathbf{j}} + I\omega_z\hat{\mathbf{k}}$$

In the preceding equation, the $\boldsymbol{\omega}$-components are those of the body $\mathcal{B}_1$ in $\mathcal{B}_3$. From the previous examples, $\omega_y = 0$ and, substituting for ω_x, ω_z, J, and I,

$$\mathbf{H}_O = \frac{mR^2v_Q}{R}\left[\frac{-3\hat{\mathbf{i}}}{10} + \left(\frac{3}{20} + \frac{3H^2}{5R^2}\right)\tan\alpha\hat{\mathbf{k}}\right]$$

$$= \frac{3mR^2v_Q}{20R}[-2\hat{\mathbf{i}} + (1 + 4/\tan^2\alpha)\tan\alpha\hat{\mathbf{k}}]$$

Noting that ${}^{\mathcal{B}_2}\dot{\mathbf{H}}_O = \mathbf{0}$, we find

$$\Sigma\mathbf{M}_O = \frac{v_Q\sin\alpha}{R\cos^2\alpha}\frac{3mR^2v_Q}{20R}\begin{vmatrix} \hat{\mathbf{i}} & \hat{\mathbf{j}} & \hat{\mathbf{k}} \\ \sin\alpha & 0 & \cos\alpha \\ -2 & 0 & \left(1 + \dfrac{4}{\tan^2\alpha}\right)\tan\alpha \end{vmatrix}$$

$$= \frac{3mv_Q^2\sin\alpha}{20\cos^3\alpha}(1 + 5\cos^2\alpha)(-\hat{\mathbf{j}})$$

and from the free-body diagram, using $N = mg$,

$$-mg\,\delta + mg\frac{3R\cos\alpha}{4\tan\alpha} = \frac{-3mv_Q^2\sin\alpha}{20\cos^3\alpha}(1 + 5\cos^2\alpha)$$

or

$$\delta = \frac{3R \cos \alpha}{4 \tan \alpha} + \frac{3v_Q^2 \sin \alpha}{20g \cos^3 \alpha}(1 + 5 \cos^2 \alpha)$$

which gives the line of action of the normal force resultant.

The "special condition" mentioned in the problem statement is that the normal force must intersect the plane at a point of physical contact with the cone, i.e.:

$$\delta \leq R/\sin \alpha$$

Thus

$$\frac{3R \cos^2 \alpha}{4 \sin \alpha} + \frac{3v_Q^2 \sin \alpha}{20g \cos^3 \alpha}(1 + 5 \cos^2 \alpha) \leq \frac{R}{\sin \alpha}$$

which when simplified is

$$3 \frac{v_Q^2}{gR} \sin^2 \alpha (1 + 5 \cos^2 \alpha) \leq 5 \cos^3 \alpha (4 - 3 \cos^2 \alpha)$$

Holding α constant, we see that what happens if v_Q is too large (or g or R too small) is that the normal force needs to act beyond the point A. Since that cannot physically happen, the specified motion will not occur but will give way to another, that of tipping outward.

PROBLEMS ▶ Section 7.5

7.42 The rod $\mathcal{L}$ is rigidly attached to shaft $\mathcal{S}$ which is free to turn in the two bearings as indicated in Figure P7.42. The y and y' axes point into the page at C. Show that the moment with respect to C that must be supplied by the bearings to shaft $\mathcal{S}$ to sustain the motion must have the components:

$$\Sigma M_{Cx'} = -\frac{m\ell^2}{12}(\sin \beta \cos \beta)\alpha$$

$$\Sigma M_{Cy'} = -\frac{m\ell^2}{12}(\sin \beta \cos \beta)\omega^2$$

Do this in two ways: (1) Use the Euler equations (7.40) with the principal axes (x, y, z); (2) use the Equations (7.41) with the axes (x', y', z') in the figure.

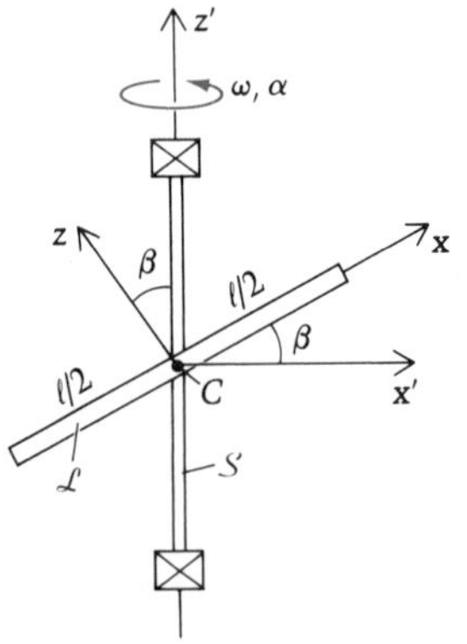

Figure P7.42

7.43 Find the reactions exerted by the bearings on the shaft to which a 30-kg thin plate $\mathcal{P}$ is welded. (See Figure P7.43.) The assembly is turning at the constant angular speed of 30 rad/s. Work the problem by using

$$\Sigma \mathbf{M}_C = {}^{\mathcal{I}}\dot{\mathbf{H}}_C = {}^{\mathcal{P}}\dot{\mathbf{H}}_C + \boldsymbol{\omega}_{\mathcal{P}/\mathcal{I}} \times \mathbf{H}_C$$

that is, by expressing $\mathbf{H}_C$ using principal directions in $\mathcal{P}$ (which omits the need for computing the nonzero product of inertia I_{xz}).

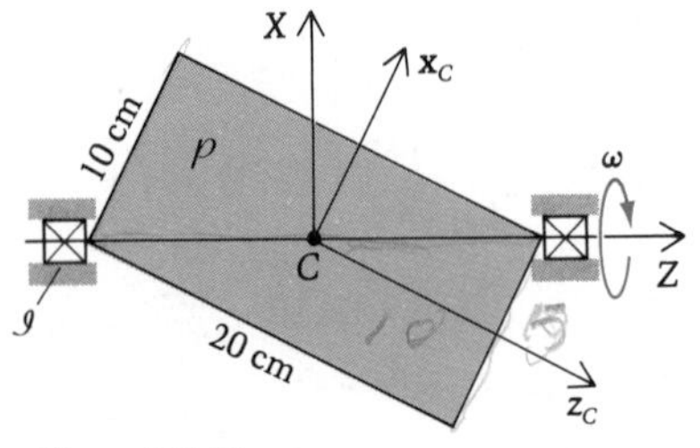

Figure P7.43

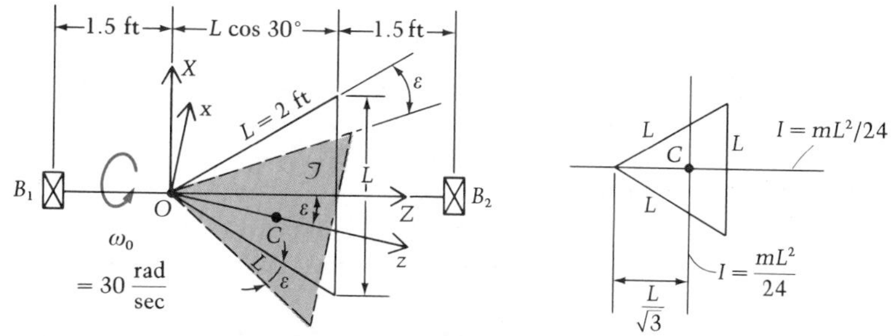

Figure P7.45

7.44 Rework and check the results of the preceding problem by using Equations (7.24) to calculate I_{xz}; then use Equations (7.41) to obtain the bearing reactions.

7.45 The equilateral triangular plate $\mathcal{T}$ was to be mounted onto the rotating shaft as shown in Figure P7.45 in the solid figure. The installation resulted in the misalignment angle ϵ (see the dashed position). The plate has mass 2 slugs, and the shaft is light. Determine the dynamic bearing reactions at B_1 and B_2 caused by the misalignment, in terms of ϵ, by: (a) using principal axes; (b) without using principal axes (i.e., calculate and utilize the products of inertia).

7.46 A slender rod $\mathcal{A}$ and a small ball each weigh 0.3 lb. (See Figure P7.46.) The bodies rotate about the vertical along with the slender shaft $\mathcal{S}$ and are supported by a smooth step or thrust bearing at D and by the cord $\mathcal{C}$. Find the tension in the cord if the angular speed of the system is 20 rad/sec.

7.47 The disk (mass m, radius r) in Figure P7.47 is rigidly attached to the shaft, and the assembly is spun up to angular speed Ω about the z_C axis. Determine the bearing reactions at A and B in terms of m, r, g, β, Ω, and L.

7.48 A bicycle wheel weighing 5 lb and having a 14-in. radius is misaligned by 1° with the vertical. The top of the wheel tilts toward the right with the bike moving forward. If the bike is driven along a straight path at 15 mph, find the moment $\Sigma \mathbf{M}_C$ exerted on the wheel. Use the result of Problem 7.11, neglecting the spokes and hub.

7.49 In the preceding problem, suppose instead that the bicycle is in a 50-ft-radius turn to the (a) right and (b) left. Determine the new values of the moment exerted on the misaligned wheel. Neglect the lean angle.

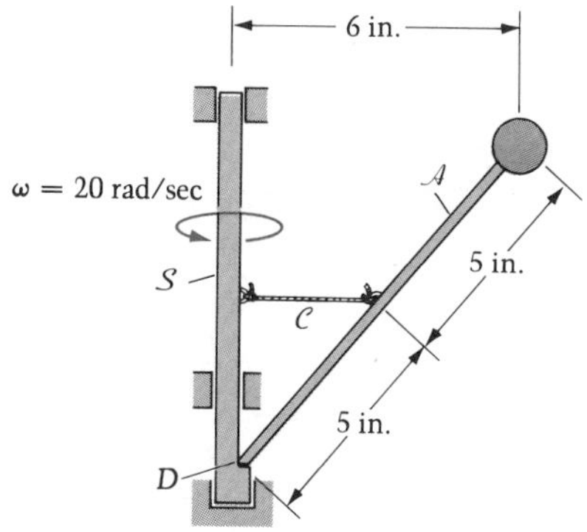

Figure P7.46

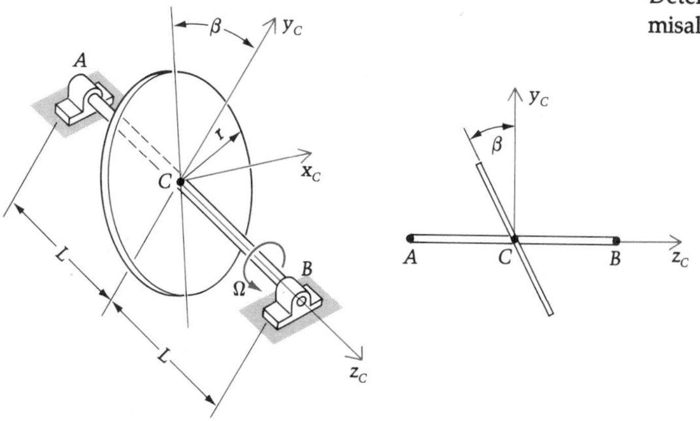

Figure P7.47

7.50 Writing Euler's laws as $\Sigma\mathbf{F} + (-m\mathbf{a}_C) = \mathbf{0}$ and $\Sigma\mathbf{M}_C + (-\dot{\mathbf{H}}_C) = \mathbf{0}$ results in what is known as the *reversed effective force* ($-m\mathbf{a}_C$) and the *inertia torque* ($-\dot{\mathbf{H}}_C$). If these quantities are added to the free-body diagram, the object may be treated as though it were in equilibrium. For a car at the instant of overturning while traveling at speed v on a curve of radius R and bank angle θ, such a diagram would appear as shown in Figure P7.50.

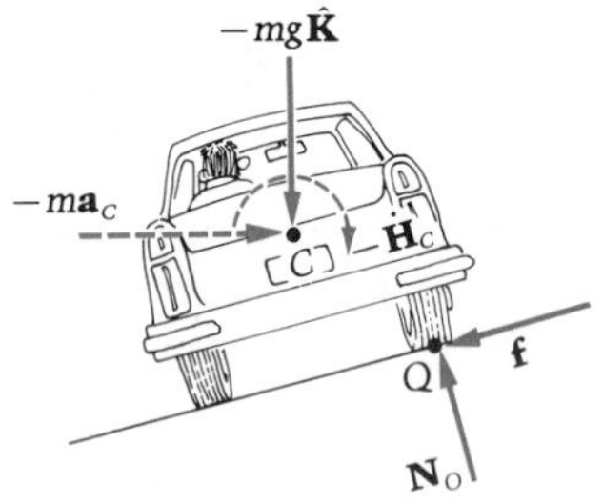

Figure P7.50

Calculate the moment about the point Q and, noting that $\Sigma\mathbf{M}_Q = \mathbf{0}$ for this free-body diagram, compare the effects of the reversed effective force (or inertia force) $-m\mathbf{a}_C$ with that of the inertia torque $-\dot{\mathbf{H}}_C$ on the overturning tendency. What factors make the car more likely to turn over?

7.51 Suppose that the components of the mass center velocity $\mathbf{v}_C$ are written in body $\mathcal{B}$ instead of in an inertial frame $\mathcal{I}$. Use the property (6.20) of $\boldsymbol{\omega}_{\mathcal{B}/\mathcal{I}}$ to derive the scalar equations of motion of the mass center from Euler's first law: $\Sigma\mathbf{F} = {}^{\mathcal{I}}\dot{\mathbf{L}}$.

7.52 Show that if a rigid body $\mathcal{B}$ undergoing torque-free motion in an inertial frame $\mathcal{I}$ has three equal principal moments of inertia at its mass center, then its angular velocity is constant in $\mathcal{I}$.

7.53 In Example 7.8 define the (x, y, z) axes at O as principal for $\mathcal{B}$, with associated respective moments of inertia $\bar{I}$, $\bar{J}$, and $\bar{K}$. Rework the problem without assuming that $\mathcal{B}$ has negligible mass. The two sets of axes are respectively parallel prior to the application of M_O.

7.54 Show that if the solution $\omega(t)$ of Example 7.6 is projected into the xy plane, the tip of the projection vector travels on a circle of radius $\sqrt{\omega_{x_i}^2 + \omega_{y_i}^2}$ at the frequency $(J - I)\omega_{z_i}/I$.

7.55 A result of the earth's bulge is that $I/J = 0.997$. Use this result to compute the period of one revolution of the earth's angular velocity vector (North Pole!) about its axis of symmetry. (The answer, obtained by Euler in 1752, is about 4 months less than the actual period first observed by S. Chandler in 1891. The difference is attributed to the nonrigidity of the earth. Although energy dissipation should damp out this "wobble," in fact it does not. The ongoing cause of the wobble is an unsolved problem in geodynamics at this time. See *Science News*, 24 October 1981.)

7.56 A screwdriver-like motion between the planar case and general (three-dimensional) motion is defined as follows. All points of the body $\mathcal{B}$ have, at any time, identical z components of velocity in a reference frame $\mathcal{I}$. The unit vector $\hat{\mathbf{k}}$ of this $\dot{z}\hat{\mathbf{k}}$ component is constant in both $\mathcal{B}$ and $\mathcal{I}$, though $\dot{z}$ can vary with time. Thus the angular velocity vector is still expressible as $\boldsymbol{\omega} = \dot{\theta}\hat{\mathbf{k}}$. Derive a moment equation for $\mathcal{B}$ that is valid for this motion.

7.57 The disk in Figure P7.57 is spinning about the light axle, and the axle is precessing at the constant rate of $32\hat{\mathbf{i}}$ rad/s. If the axle is observed to remain horizontal, find the magnitude and direction of the spin of the disk.

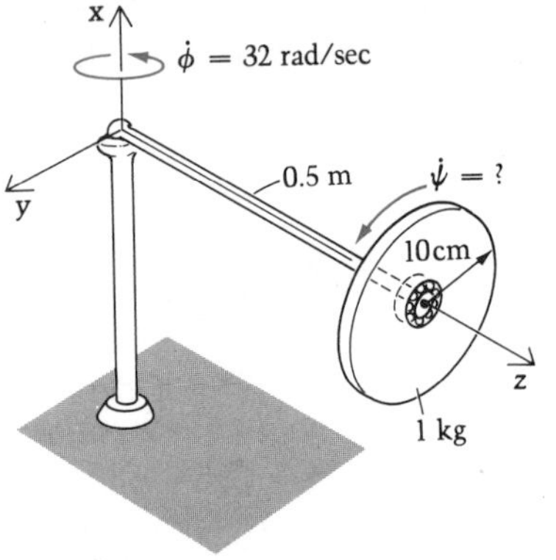

Figure P7.57

7.58 A single-engine aircraft has a four-bladed propeller weighing 128 lb with a radius of gyration about its center of mass of 3 ft. It rotates counterclockwise at 2000 rpm when viewed from the rear. Find the gyroscopic moment on the propeller shaft when the plane is at the bottom of a vertical loop of 2000-ft radius with a speed of 500 mph. In which direction will the tail of the plane tend to move because of this moment?

7.59 The blades of a fan turn at 1750 rpm, and the fan oscillates about the vertical axis z (Figure P7.59(a)) at the rate of one cycle every 10 sec. Assuming that the fan travels at the constant angular velocity of 0.2 rad/sec except when it is reversing direction (Figure P7.59(b)), find the moment exerted by the base on the arm section $\mathcal{A}$ at the $\frac{1}{4}$-cycle point due to gyroscopic action. For the calculation (*only!*) consider the blades (Figure P7.59(c)) to be 4-in.-diameter circular aluminum plates all in the same plane and $\frac{1}{32}$ in. thick. Use a density of 0.1 lb/in.^{3}.

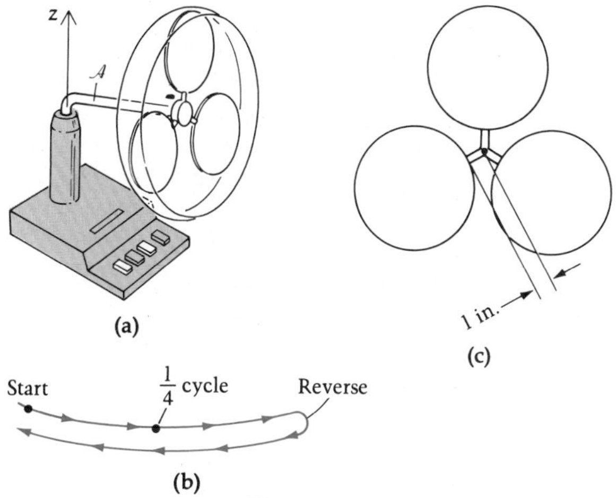

Figure P7.59

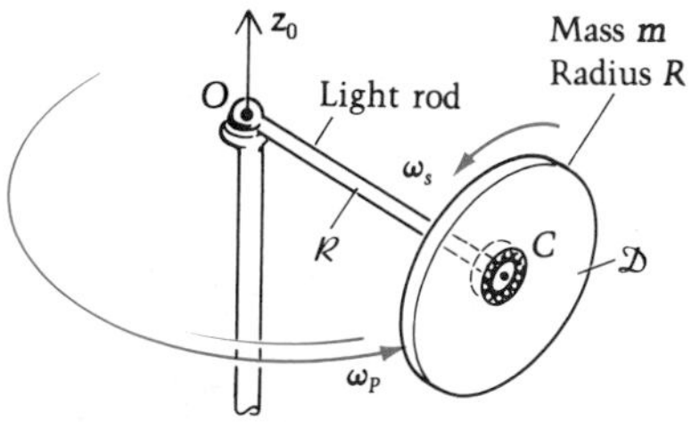

Figure P7.60

7.60 Disk $\mathcal{D}$ in Figure P7.60 turns in bearings at C at angular rate ω_s about the light rod $\mathcal{R}$, and both precess about axis z_O at angular rate ω_P as shown. Show with a free-body diagram how it is possible for the mass center C to remain in a horizontal plane. Then find the reactions exerted onto $\mathcal{R}$ by the socket at O. Is there any difference in the solution if $\mathcal{D}$ and $\mathcal{R}$ are rigidly connected?

7.61 In the grinding mill of Problem 7.5, suppose that the wall is absent. (See Figure P7.61.) Find, for a given Ω (constant angular speed of $\mathcal{S}$), the angle ϕ that the axis of the grinder $\mathcal{D}$ will make with the vertical. Observe that with the wall present and ϕ fixed, larger speeds than this Ω will allow the grinder to work. In particular, show that the following set of parameters is satisfactory: $r = 2.5$ ft, $\ell = 6$ ft, $\Omega = 2\pi$ rad/sec, and $\phi = 60°$. Neglect the mass of body $\mathcal{B}$ in comparison with the heavy grinding disk $\mathcal{D}$.

7.62 Find the grinding force N produced at the wall of the grinding mill of Problems 7.5 and 7.61 for the given parameters.

7.63 Find the magnitude and direction of the force and/or couple exerted on disk $\mathcal{D}$ by the shaft $\mathcal{S}$ in Problem 7.4.

7.64 A disk $\mathcal{D}$ rolls around in a circle with its plane vertical and its center traveling at constant speed v_C. Find the tension in the string, and the friction force exerted on $\mathcal{D}$ by the floor. (See Figure P7.64.)

7.65 There is a relationship among v_C, g, r, R, and θ such that the disk can roll around in a circle as shown in Figures P7.65(a) and (b), with v_C and θ remaining constant. Find this relationship.

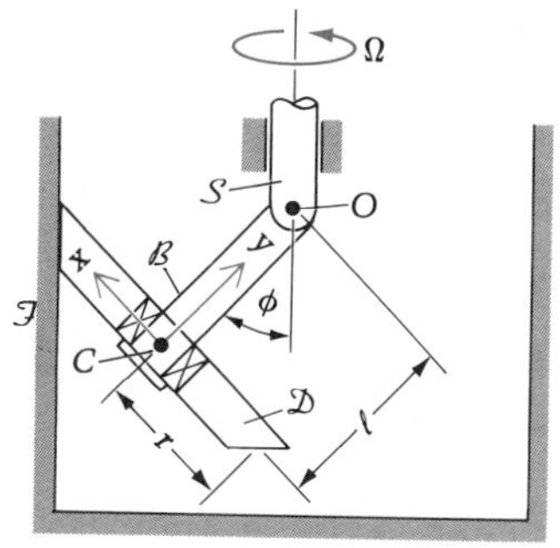

Figure P7.61

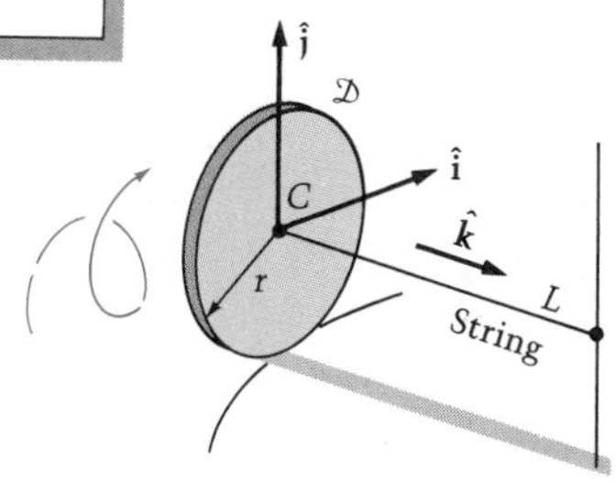

Figure P7.64

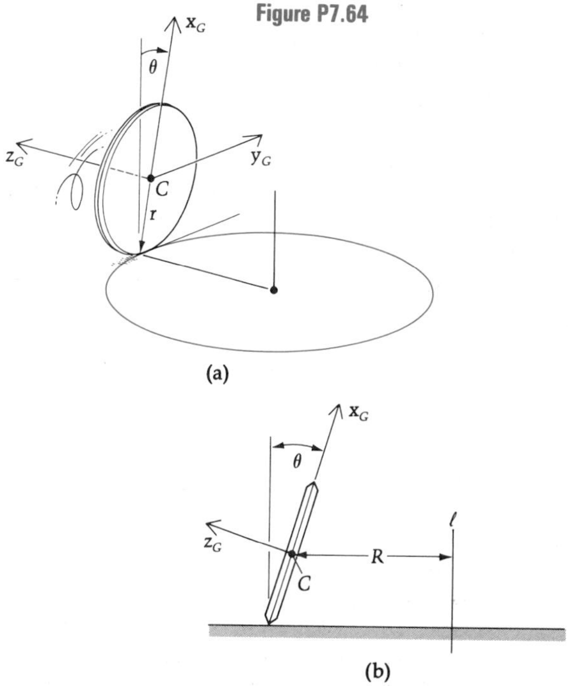

Figure P7.65

7.66 Obtain the results of Example 7.9 by using the Euler equations (7.42). *Hint*: This time the axes are body-fixed, and the Ω part of $\boldsymbol{\omega}$ changes direction in $\mathcal{D}$; therefore the components ω_x and ω_y have derivatives that were formerly zero in $\mathcal{B}$. The Ω component of $\boldsymbol{\omega}_{\mathcal{D}/\mathcal{I}}$ is

$$\Omega\,(\cos\theta_r\hat{\mathbf{j}} - \sin\theta_r\hat{\mathbf{i}})$$

where θ_r is the angle of roll as shown in Figure P7.66. Differentiate this expression, substitute $\dot{\theta}_r = v_C/r$, and *then* take $\theta_r = 0$. Finally, go to Equations (7.42) and substitute your results.

7.67 A ship's turbine has a mass of 2500 kg and a radius of gyration about its axis (y_c in Figure P7.67) of 0.45 m. It is mounted on bearings as indicated and turns at 5000 rpm clockwise when viewed from the stern (rear) of the boat.

a. If the ship is in a steady turn to the right of radius 500 m and is traveling at 15 knots, what are the reactions exerted on the shaft by the bearings? (1 knot = 1.15 mph = 1.85 km/hr)

b. If the ship on a straight course in rough seas pitches sinusoidally at $\pm 12°$ amplitude with a 6-s period, what are the maximum bearing reactions then?

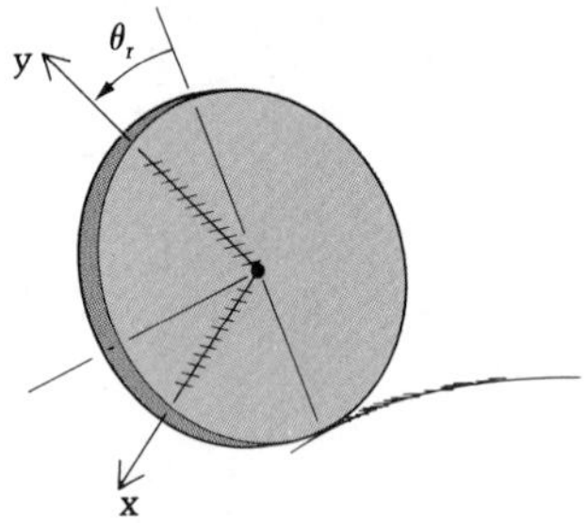

Figure P7.66

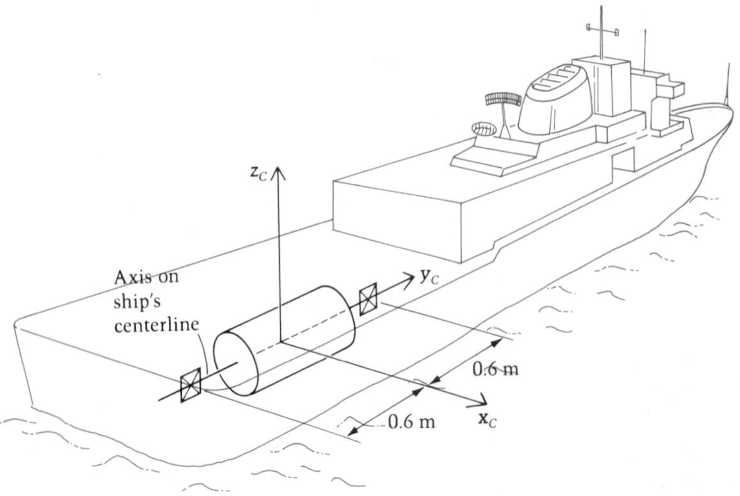

Figure P7.67

7.68 A heavy disk $\mathcal{D}$ of mass m and radius r spins at the angular rate ω_3 $(=|\boldsymbol{\omega}_{\mathcal{D}/\mathcal{B}}|)$ with respect to the rigid, but light, bent bar $\mathcal{B}$. (See Figure P7.68.) Body $\mathcal{B}$ turns at rate ω_2 $(=|\boldsymbol{\omega}_{\mathcal{B}/\mathcal{G}}|)$ about a vertical axis through O, a point of both $\mathcal{B}$ and the inertial frame $\mathcal{G}$. Find the force and couple

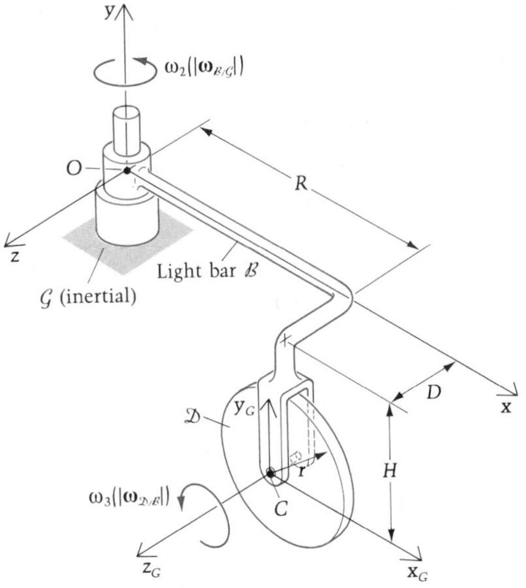

Figure P7.68

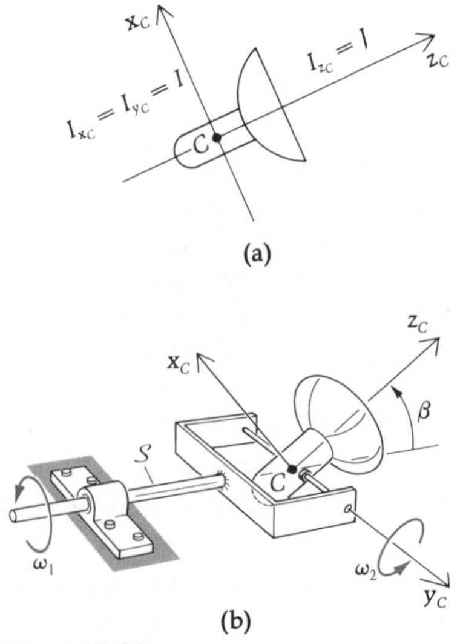

Figure P7.69

that must be acting on $\mathcal{B}$ at O to produce a motion of the system for which ω_2 and ω_3 are constants. Both sets of axes in the figure are fixed in $\mathcal{B}$, and note that (x_C, y_C, z_C) are always principal axes for $\mathcal{D}$ at C even though they are not fixed in $\mathcal{D}$.

7.69 Compute the moment M applied to the shaft $\mathcal{S}$ in Figures P7.69(a) and (b), on the preceding page, as a function of the angle $\mathcal{B}$ if ω_1 and ω_2 $(=\dot{\beta})$ are constants.

7.70 A bike rider enters a turn of radius R at a constant speed of v_C. (See Figure P7.70.) Other quantities are defined below:

r = radius of wheel

d = distance between axle and C

I_1, I_2 = principal moments of inertia of entire bike plus rider with respect to $\hat{\mathbf{n}}_1$ and $\hat{\mathbf{n}}_2$ directions through C

i = moment of inertia, with respect to $\hat{\mathbf{n}}_2$ direction, of one wheel about its axis of symmetry

m = total mass

ϕ = angle shown

Solve for the resultant force $\Sigma\mathbf{F}$ and moment $\Sigma\mathbf{M}_C$ in terms of these quantities. Compare the effects of the *D'Alembert force* $(-m\mathbf{a}_C)$ and the *inertia torque* $(-\dot{\mathbf{H}}_C)$ in righting the bike when ϕ is small. Neglect the products of inertia.

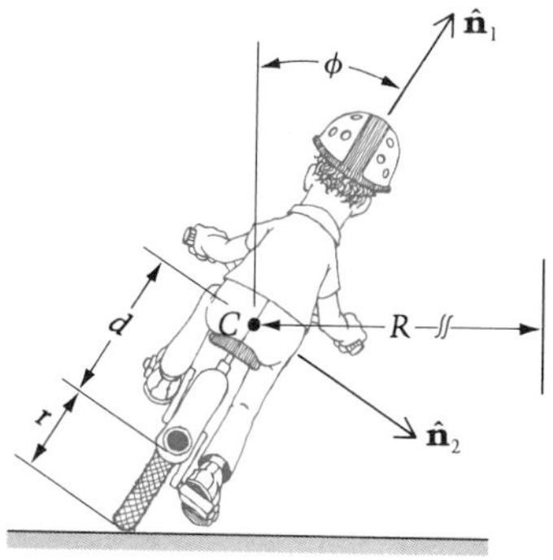

Figure P7.70

* **7.71** Disk $\mathcal{D}$ in Figure P7.71 turns in bearings around rod $\mathcal{R}$ as it rolls on the ground. Find all the forces acting on $\mathcal{D}$, and contrast the solution with that of Example 7.9. The rod is free to slide along the vertical post about which it turns at constant rate.

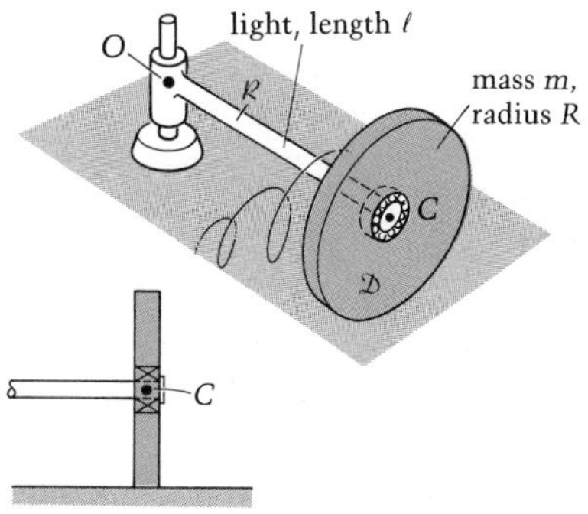

Figure P7.71

7.72 In Problem 6.24, if ω_1, ω_2, and ω_3 are constants, find the resultant moment exerted on $\mathcal{D}$ at its mass center, when $\hat{\mathbf{u}}$ is pointing straight up. The body $\mathcal{D}$ is symmetrical, with centroidal principal moments of inertia J along $\hat{\mathbf{u}}$, and I normal to $\hat{\mathbf{u}}$.

* **7.73** The body $\mathcal{B}$ in Figure P7.73 is an ellipsoid of revolution with mass = 1 slug, and semiminor and semimajor axis lengths a and $2a$, with a = 1 ft. Thus

$$I_{xx}^C = \frac{m}{5}(a^2 + a^2) = 0.4 \text{ slug-ft}^2$$

and

$$I_{yy}^C = I_{zz}^C = \frac{m}{5}[a^2 + (2a)^2] = 1.0 \text{ sl-ft}^2$$

The shaded light frame $\mathcal{A}$ is driven around the fixed post $\mathcal{P}$, with angular velocity $\omega_2\hat{\mathbf{j}}$, by a motor torque T_2 applied at P. Another motor (neither is shown) between $\mathcal{A}$ and $\mathcal{B}$ applies a torque $T_1\hat{\mathbf{i}}$ which causes the body $\mathcal{B}$ to spin in the frame. The axes and unit vectors shown are fixed in $\mathcal{A}$. During an interval of motion, $\omega_1 = 3t^2$ rad/sec and $\omega_2 = 2t$ rad/sec. Find all forces and couples applied onto $\mathcal{A}$ at P when t = 1 sec. The distance from P to the x axis is 2 ft.

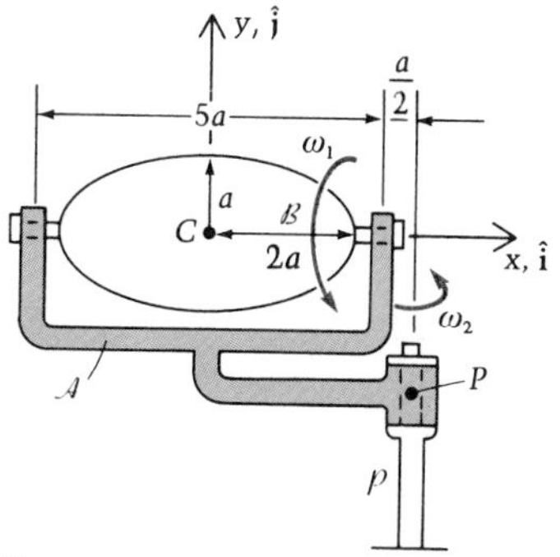

Figure P7.73

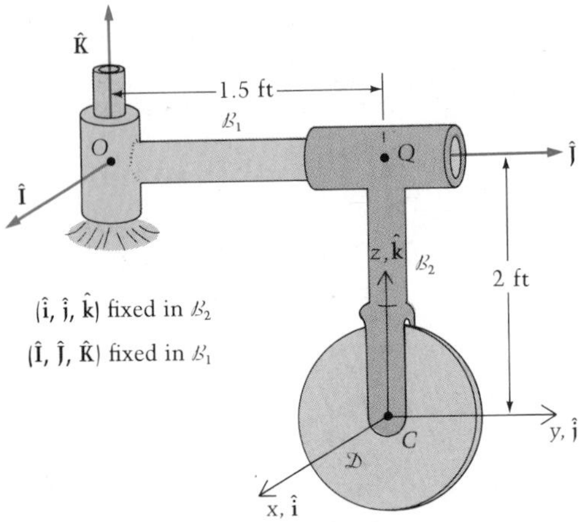

Figure P7.74

7.74 By means of appropriately mounted control motors (not shown), the disk $\mathcal{D}$ in Figure P7.74 is made to turn about its axis of symmetry relative to the arm $\mathcal{B}_2$; $\mathcal{B}_2$ is made to turn relative to arm $\mathcal{B}_1$ about the axis of $\mathcal{B}_1$; and $\mathcal{B}_1$ is made to turn with respect to the ground (inertial frame) $\mathcal{I}$ about the vertical through point O. At the given instant, the angular velocity of $\mathcal{D}$ in $\mathcal{I}$, expressed in terms of unit vectors $(\hat{\mathbf{i}}, \hat{\mathbf{j}}, \hat{\mathbf{k}})$ fixed in $\mathcal{B}_2$, is $\omega_1\hat{\mathbf{i}} + \omega_2\hat{\mathbf{j}} + \omega_3\hat{\mathbf{k}} = \hat{\mathbf{i}} + 2\hat{\mathbf{j}} + 3\hat{\mathbf{k}}$ rad/sec. In addition, at this instant we are given $\dot{\omega}_1 = 4$, $\dot{\omega}_2 = 5$, and $\dot{\omega}_3 = 6$ rad/sec². For the disk, $m = 10$ slugs, $I_{xx}^C = 1.4$ slug-ft², and $I_{yy}^C = 0.7$ slug-ft². Find all forces and couples that are acting on $\mathcal{D}$ at C at the given instant.

* **7.75** In the preceding problem, use $\omega_1, \omega_2, \omega_3$, and their derivatives to compute the right-hand sides of the Euler equations (7.40). Explain why these results are not the components of the moments of external forces acting on $\mathcal{D}$ at C.

* **7.76** In the preceding two problems, at the given instant find $\dot{\theta}_i$ and $\ddot{\theta}_i$, for $i = 1$, 2, and 3, where θ_1 is the angle of rotation of $\mathcal{D}$ with respect to $\mathcal{B}_2$, θ_2 is the angle of rotation of $\mathcal{B}_2$ with respect to $\mathcal{B}_1$, and θ_3 is the angle of rotation of $\mathcal{B}_1$ with respect to $\mathcal{I}$.

7.6 Gyroscopes

We now return our attention to the gyroscope whose orientation was examined in Section 6.8. First we shall derive the equations of rotational motion of such a gyroscope $\mathcal{G}$.* We begin by expressing its angular velocity $\boldsymbol{\omega}_{\mathcal{G}/\mathcal{I}}$ in the frame $\mathcal{F}_2$; see Figures 7.7, 7.8 and 7.9, repeated from Section 6.8.

$$\boldsymbol{\omega}_{\mathcal{G}/\mathcal{I}} = -\dot{\phi}\sin\theta\hat{\mathbf{i}}_2 + \dot{\theta}\hat{\mathbf{j}}_2 + (\dot{\psi} + \dot{\phi}\cos\theta)\hat{\mathbf{k}}_2 \tag{7.44}$$

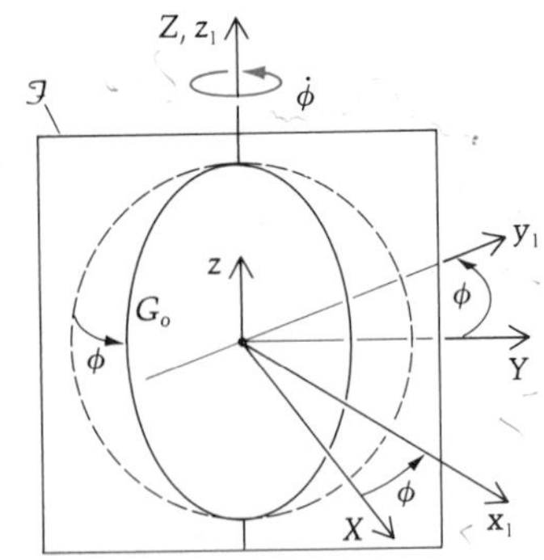

Figure 7.7 First rotation.

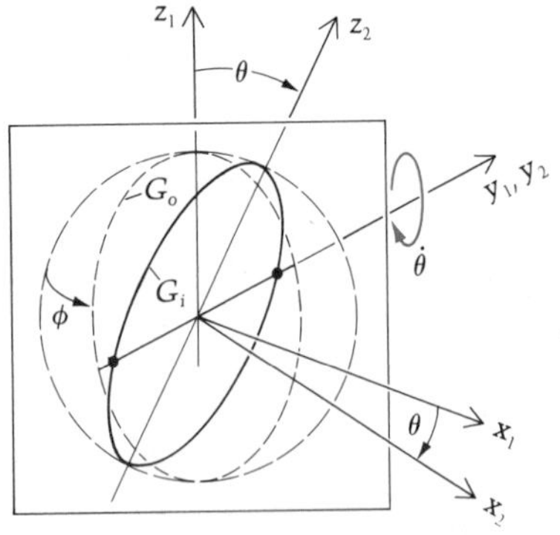

Figure 7.8 Second rotation.

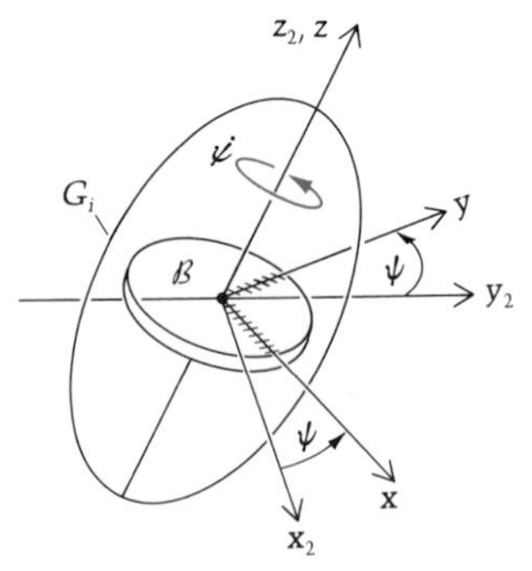

Figure 7.9 Third rotation.

* We are taking $\mathcal{G}$ to be the rotor and are considering it as heavy with respect to the inner and outer gimbals, whose mass we then neglect. We also assume $\mathcal{G}$ to be symmetric about its axis. Of course, a gyroscope does not have to possess *any* gimbals; the earth is a massive gyro, as will be seen in an example to follow.

The axes (x_2, y_2, z_2) of the inner gimbal are not fixed in $\mathcal{G}$ because of its spin $\dot{\psi}$ — *but they are nonetheless permanently principal.* This important fact allows us to write the angular momentum of the gyroscope in the frame $\mathcal{I}_2$ as

$$\mathbf{H}_C = -I\dot{\phi}\sin\theta\hat{\mathbf{i}}_2 + I\dot{\theta}\hat{\mathbf{j}}_2 + J(\dot{\psi} + \dot{\phi}\cos\theta)\hat{\mathbf{k}}_2 \tag{7.45}$$

Euler's second law, together with Property (6.20) of the angular velocity vector, then gives the equations of motion of the gyroscope as follows:

$$\begin{aligned}\Sigma\mathbf{M}_C &= {}^{\mathcal{I}}\dot{\mathbf{H}}_C = {}^{\mathcal{I}_2}\dot{\mathbf{H}}_C + \boldsymbol{\omega}_{\mathcal{I}_2/\mathcal{I}} \times \mathbf{H}_C \\ &= [-I(\ddot{\phi}\sin\theta + \dot{\phi}\dot{\theta}\cos\theta) + \dot{\theta}\dot{\phi}\cos\theta(J - I) + \dot{\theta}\dot{\psi}J]\hat{\mathbf{i}}_2 \\ &\quad + [I\ddot{\theta} + (J - I)\dot{\phi}^2\sin\theta\cos\theta + J\dot{\phi}\dot{\psi}\sin\theta]\hat{\mathbf{j}}_2 \\ &\quad + \left[J\frac{d}{dt}(\dot{\psi} + \dot{\phi}\cos\theta)\right]\hat{\mathbf{k}}_2\end{aligned} \tag{7.46}$$

In the preceding calculations we have used the addition theorem to observe the following:

$$\begin{aligned}\boldsymbol{\omega}_{\mathcal{G}/\mathcal{I}} &= \boldsymbol{\omega}_{\mathcal{G}/\mathcal{I}_2} + \boldsymbol{\omega}_{\mathcal{I}_2/\mathcal{I}} \\ &= \dot{\psi}\hat{\mathbf{k}}_2 + \boldsymbol{\omega}_{\mathcal{I}_2/\mathcal{I}}\end{aligned} \tag{7.47}$$

so that $\boldsymbol{\omega}_{\mathcal{I}_2/\mathcal{I}}$ is the same vector as in Equation (7.44) if $\dot{\psi}$ is omitted. The equations of motion of $\mathcal{G}$ are therefore

$$\begin{aligned}\Sigma M_{Cx_2} &= -I(\ddot{\phi}\sin\theta + 2\dot{\phi}\dot{\theta}\cos\theta) + J\dot{\theta}(\dot{\psi} + \dot{\phi}\cos\theta) \\ \Sigma M_{Cy_2} &= I(\ddot{\theta} - \dot{\phi}^2\sin\theta\cos\theta) + J\dot{\phi}\sin\theta(\dot{\psi} + \dot{\phi}\cos\theta) \\ \Sigma M_{Cz} &= J\frac{d}{dt}(\dot{\psi} + \dot{\phi}\cos\theta) = J\frac{d\omega_z}{dt}\end{aligned} \tag{7.48}$$

where ω_z is the component of $\boldsymbol{\omega}_{\mathcal{G}/\mathcal{I}}$ about the spin axis of symmetry of the gyroscope. Note that it is made up of part of the precession speed as well as all of the spin.

The gyroscope equations are seen to be nonlinear, including not only products of the angles' derivatives but also trigonometric functions of them. Their general solution is an unsolved problem; however, there are two special solutions that are quite worthy of study. The first of these is steady precession; the second is torque-free motion. We shall have a look at each in turn.

Steady Precession

Steady precession is defined by the nutation angle θ, the precession speed $\dot{\phi}$, and the spin speed $\dot{\psi}$ each being constant throughout the motion. Let us call these constants θ_0, $\dot{\phi}_0$, and $\dot{\psi}_0$, and substitute them into Equation (7.48) to obtain:

$$\begin{aligned}\Sigma M_{Cx_2} &= 0 \\ \Sigma M_{Cy_2} &= -I\dot{\phi}_0^2\sin\theta_0\cos\theta_0 + J\dot{\phi}_0\sin\theta_0(\dot{\psi}_0 + \dot{\phi}_0\cos\theta_0) \\ \Sigma M_{Cz} &= 0\end{aligned} \tag{7.49}$$

We see that only a moment about the y_2 axis is needed to sustain steady precession. Also $\Sigma M_{Cz} = 0 = J(d\omega_z/dt)$ means that ω_z is a constant.*

In the case in which $\theta = 90°$, we have the precession and spin axes orthogonal; the situation is shown in Figure 7.10. If the gyro is spinning and the torque is applied, there will simultaneously occur a precession that tends to turn the spin vector toward the torque vector. This is sometimes called the law of gyroscopic precession. The torque in this case, from Equation (7.49), is

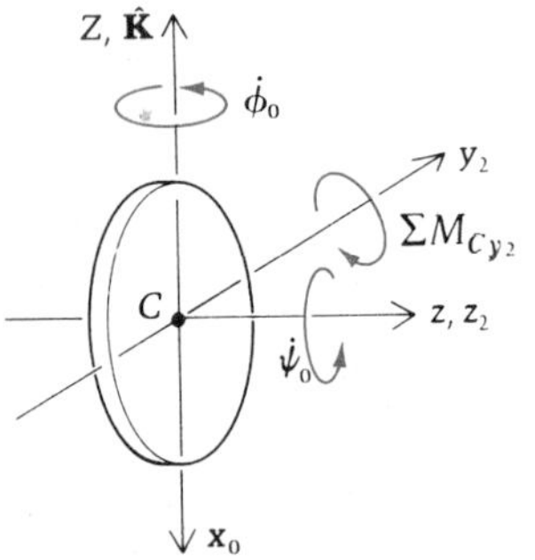

Figure 7.10

$$\Sigma M_{Cy_2} = J\dot{\psi}_0\dot{\phi}_0 \tag{7.50}$$

and it is seen to be the product of the spin momentum $J\dot{\psi}_0$ and the precessional angular speed $\dot{\phi}_0$.

We turn now to an illustration of the law of gyroscopic precession. We have just seen that when a freely spinning body is torqued about an axis normal to the spin axis, it precesses about a third axis that forms an orthogonal triad with the spin and torque vectors. The direction of the precession is such that it turns the spin vector toward the torque vector. This law of gyroscopic precession is responsible for the lunisolar precession of the equinoxes.

What is the lunisolar precession? Because of the billions of years of gravitational pull from the sun and moon, the earth is slightly bulged instead of round. It is in fact about 27 miles shorter across the poles than it is across the equator. This bulge, plus the fact that its axis is tilted $23\frac{1}{2}°$ to the ecliptic, causes the sun (and moon) to torque the earth in addition to the gravity pull, as shown in Figure 7.11.

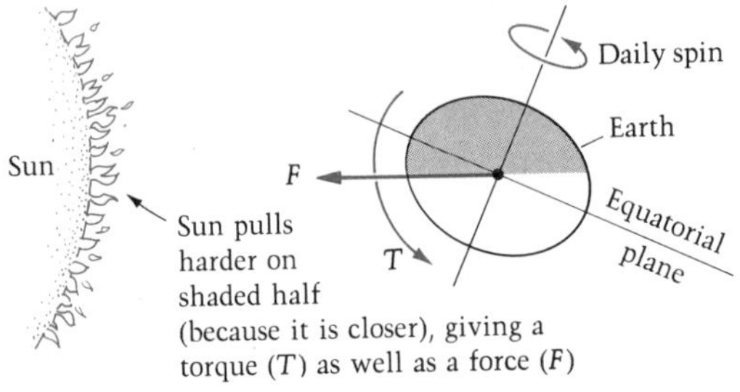

Figure 7.11

We see, then, that we live on the surface of a spinning gyroscope that is constantly being acted on by an external torque, and a precession is thus ongoing. This precession, shown in Figure 7.12, turns the spin axis of the earth out of the plane of the paper toward the torque vector.† This motion results in counterclockwise movement, on the celestial sphere, of the celestial pole to which the earth's rotation axis is directed. (This point

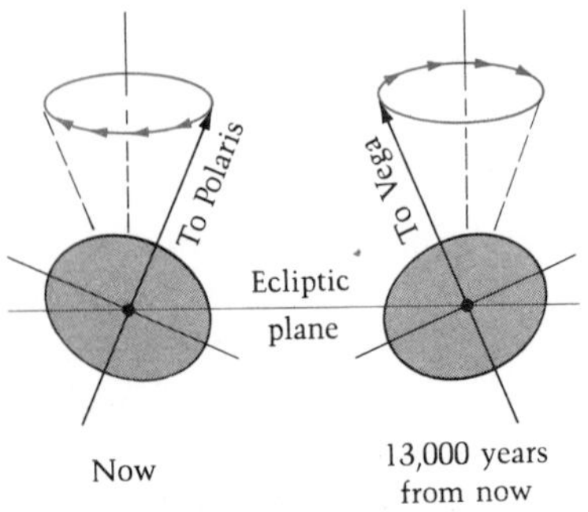

Figure 7.12

* Note that z_2 and z are the same axis.

† The spin axis of the earth is aligned with its angular velocity vector ω. The point where the ω vector, placed at C, cuts the surface of the earth is the real meaning of the North Pole. The North Pole wanders about the geometric pole (on the symmetry axis) as time passes; it has remained within a few feet of it in this century.

is currently close to Polaris, the North Star.) The period of this rotation is about 26,000 years, and it is interesting that the moon's effect is 2.2 times that of the sun's because it is much closer.

Torque-Free Motion

We now take up the other example of a solution to the gyroscope equations: the case of **torque-free motion**. "Torque-free" means that $\Sigma \mathbf{M}_C$ vanishes, so that $\mathbf{H}_C$ is a constant. This follows from Euler's second law:

$$\Sigma \mathbf{M}_C = \mathbf{0} = {}^{\mathcal{I}}\dot{\mathbf{H}}_C \Rightarrow \mathbf{H}_C = \text{constant vector in the inertial frame } \mathcal{I}$$

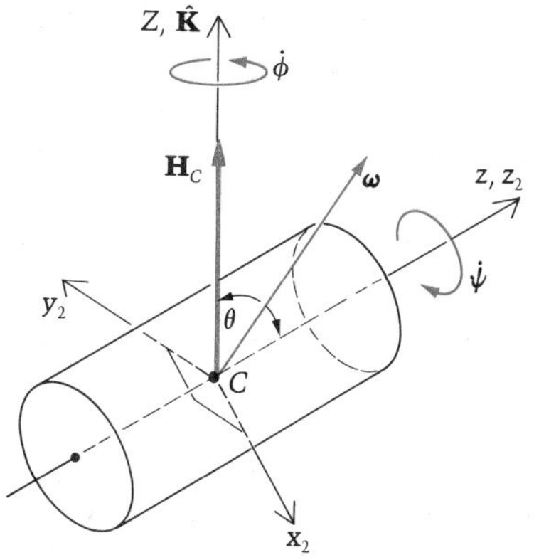

Figure 7.13

We shall conveniently let the direction of the Z axis (see Figure 7.13), which is arbitrary, coincide with the constant direction of $\mathbf{H}_C$. Then Z becomes the precession axis of the motion, and the (x_2, y_2, z_2) axes appear as shown in Figure 7.13. it is seen that since

$$\begin{aligned} \mathbf{H}_C &= H_{C_{x_2}}\hat{\mathbf{i}}_2 + H_{C_{y_2}}\hat{\mathbf{j}}_2 + H_{C_{z_2}}\hat{\mathbf{k}}_2 = \text{constant} \\ &= -I\dot{\phi}\sin\theta\,\hat{\mathbf{i}}_2 + I\dot{\theta}\hat{\mathbf{j}}_2 + J(\dot{\psi} + \dot{\phi}\cos\theta)\hat{\mathbf{k}}_2 \end{aligned} \qquad (7.51)$$

and since $\mathbf{H}_C$ is seen always to lie in the x_2z_2 plane, then $H_{C_{y_2}}$ must vanish:

$$H_{C_{y_2}} = 0 = I\dot{\theta} \Rightarrow \theta = \text{constant} \qquad (7.52)$$

Note that $\boldsymbol{\omega}$ lies in the x_2z_2 plane along with $\mathbf{H}_C$, since its y_2 component, $\dot{\theta}$, vanishes. Furthermore, we see that

$$\begin{aligned} H_{C_{x_2}} &= -I\dot{\phi}\sin\theta \quad \text{and} \quad & H_{C_{z_2}} &= J(\dot{\psi} + \dot{\phi}\cos\theta) \\ &= I\omega_{x_2}{}^* & &= J\omega_{z_2} \end{aligned} \qquad (7.53)$$

But also, as we can see from Figure 7.13,

$$H_{C_{x_2}} = -H_C\sin\theta \quad \text{and} \quad H_{C_{z_2}} = H_C\cos\theta \qquad (7.54)$$

Therefore, equating the first of Equations (7.53) and (7.54) for $H_{C_{x_2}}$, we obtain

$$\omega_{x_2} = \frac{-H_C\sin\theta}{I} = \text{constant} \quad \text{(Since } H_C,\ \theta, \text{ and } I \text{ are constants)} \qquad (7.55)$$

$$= -\dot{\phi}\sin\theta \qquad (7.56)$$

and we see that

$$\dot{\phi} = \frac{H_C}{I} \quad \text{(a constant)} \qquad (7.57)$$

Similarly, equating the two preceding values of $H_{C_{z_2}}$ gives

$$\omega_{z_2} = \frac{H_C\cos\theta}{J} = \text{constant} \qquad (7.58)$$

$$= \dot{\psi} + \dot{\phi}\cos\theta \qquad (7.59)$$

---

* Note again that the (x_2, y_2, z_2) axes are permanently principal, even though x_2 and y_2 are not body-fixed, and this lets us write $\mathbf{H}_C$ in terms of the $I\omega$'s along these axes.

so that

$$\dot{\psi} = \frac{H_C \cos\theta}{J} - \left(\frac{H_C}{I}\right)\cos\theta = H_C \cos\theta \left(\frac{I-J}{IJ}\right)$$
$$= \text{constant} \tag{7.60}$$

Therefore all conditions are satisfied for the torque-free body to be in a state of steady precession about the z axis fixed in $\mathcal{I}$!

Dividing Equation (7.55) by (7.58) leads to

$$\frac{\omega_{x_2}}{\omega_{z_2}} = -\frac{J}{I}\tan\theta \tag{7.61}$$

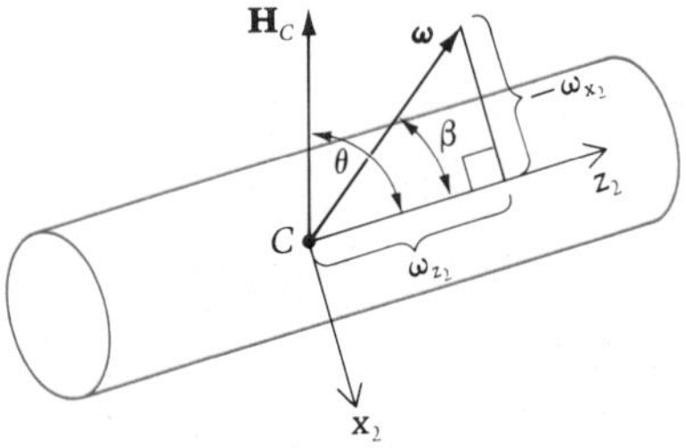

Figure 7.14

and Figure 7.14 shows that

$$\frac{-\omega_{x_2}}{\omega_{z_2}} = \tan\beta \tag{7.62}$$

where β is the angle between z_2 and ω. Therefore

$$\tan\beta = \frac{J}{I}\tan\theta$$

and we see that the answer to whether β is larger or smaller than θ depends on the ratio of J to I. If $J < I$, as in the elongated shape in Figure 7.14, then $\beta < \theta$ and the angular velocity vector lies inside of $\mathbf{H}_C$ and z_2, making a constant angle with each. Two cones may be imagined — one fixed to the body, the other in space ($\mathcal{I}$). The body cone is seen to roll on the fixed space cone as its spin and precession vectorially add to the vector $\boldsymbol{\omega}$, which changes only in direction.

This precession (Figure 7.15) is called *direct* because $\dot{\phi}$ and $\dot{\psi}$ have the same counterclockwise sense when observed from the $\boldsymbol{\omega}$ vector outside the cones. If, however, $J > I$, then $\beta > \theta$ and $\boldsymbol{\omega}$ lies *outside* the angle

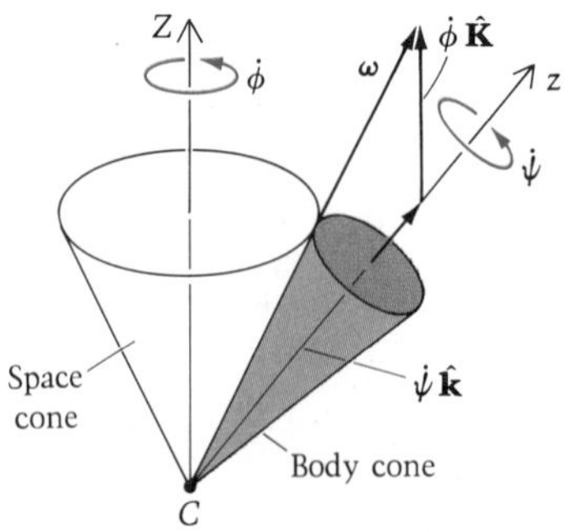

Figure 7.15

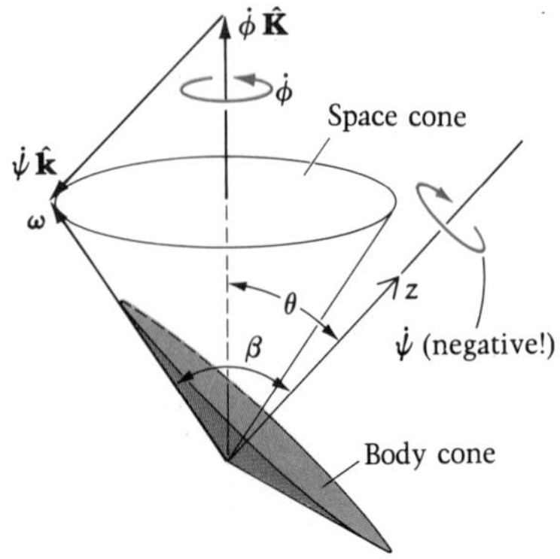

Figure 7.16

ZCz. This situation is harder to depict, but just as important. This time the body cone rolls around the outside of the enclosed, fixed space cone (Figure 7.16), and the two rotations $\dot{\phi}$ and $\dot{\psi}$ have opposite senses. This precession is called *retrograde*.

A final note on the theory of the torque-free body: If the constant value of θ is either 0 or 90°, there is *no* precession and the gyroscope is simply in a state of pure rotation, planar motion:

$\theta = 0$:

Here $z = Z$, so that $\omega_{x_2} = 0$ and $\omega_{z_2} = H_C/J$. If $\theta = 0°$, the body simply spins about its axis. In this case, the rates $\dot{\phi}$ and $\dot{\psi}$ cannot be distinguished.

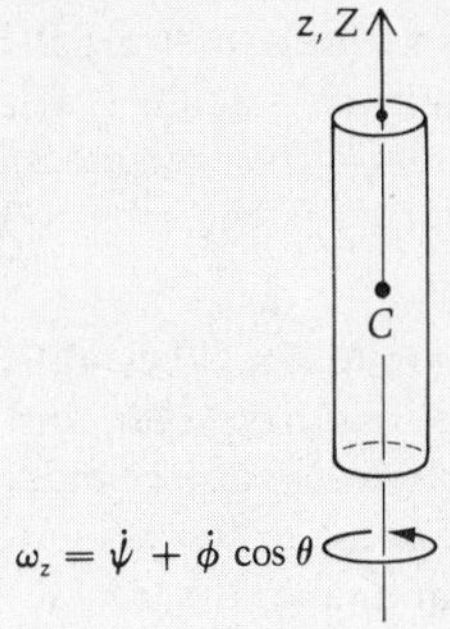

Question 7.5. Why can we not use Equations (7.57) and (7.60) to get $\dot{\phi}$ and $\dot{\psi}$ in this case?

$\theta = 90°$:

Here we have $\omega_{x_2} = -H_C/I$ and $\omega_{z_2} = 0$. Thus:

$$\dot{\phi} = \frac{H_C}{I}$$

$$\dot{\psi} = 0$$

If $\theta = 90°$, the body spins about a transverse axis without precessing.

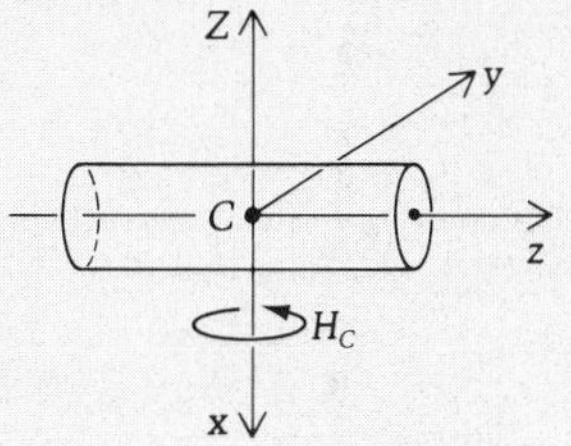

It requires a great many terms to describe the complex motion of the earth. We have seen one example of this in the lunisolar precession caused by the gravity torque exerted on the earth by the sun and moon. This motion is analogous to a differential equation's particular solution, which it has whenever the equation has a nonzero right-hand side. The complementary, or homogeneous, solution is analogous to the torque-free part of the solution to the earth's rotational motion. This part, called the free precession of the earth, is in fact retrograde. Both the space and body cones are very thin as the $\boldsymbol{\omega}$, $\mathbf{H}_C$, and z axes are all quite close together; each lies about $23\frac{1}{2}°$ off the normal to the ecliptic plane.

Answer 7.5 In deriving (7.57), if $\theta = 0$ we have divided both sides of an equation by zero. This result is then used in getting $\dot{\psi}$ in (7.60).

PROBLEMS ▶ Section 7.6

7.77 Find the angular acceleration in $\mathcal{I}$ of the gyroscope for the case of steady precession.

7.78 The spinning top (Figure P7.78) is another example of a gyroscope. Show that if the top's peg is not moving across the floor, the condition for steady precession is given by

$$mgd = J\dot{\psi}\dot{\phi} + (J - I)\dot{\phi}^2 \cos\theta$$

7.79 A top steadily precesses about the fixed direction Z at 60 rpm. (See Figure P7.79.) Treating the top as a cone of radius 1.2 in. and height 2.0 in., find the rate of spin $\dot{\psi}$ of the top about its axis of symmetry.

7.80 Cone $\mathcal{C}$ in Figure P7.80 has radius 0.2 m and height 0.5 m. It is precessing about the vertical axis through the ball joint, in the direction shown, at the rate of $\dot{\phi} = 0.5$ rad/s. If the angle θ is observed to be 20° and unchanging, what must be the rate of spin $\dot{\psi}$ of the cone?

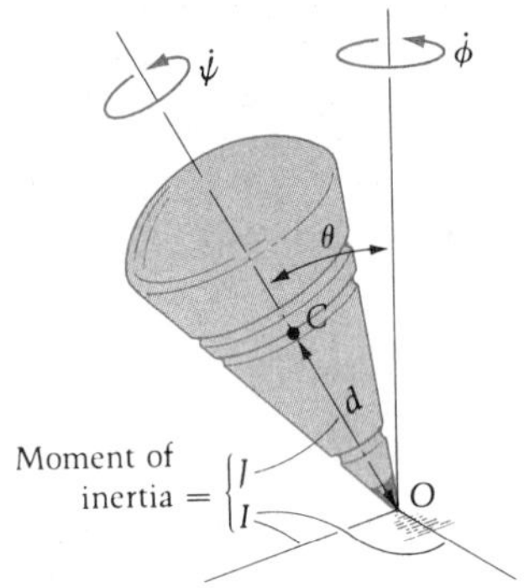

Figure P7.78

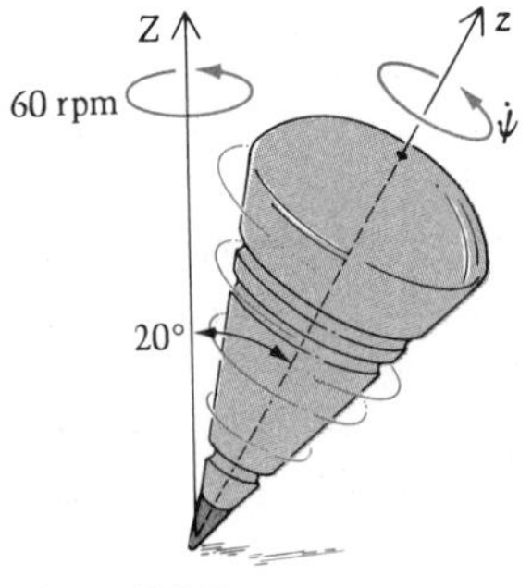

Figure P7.79

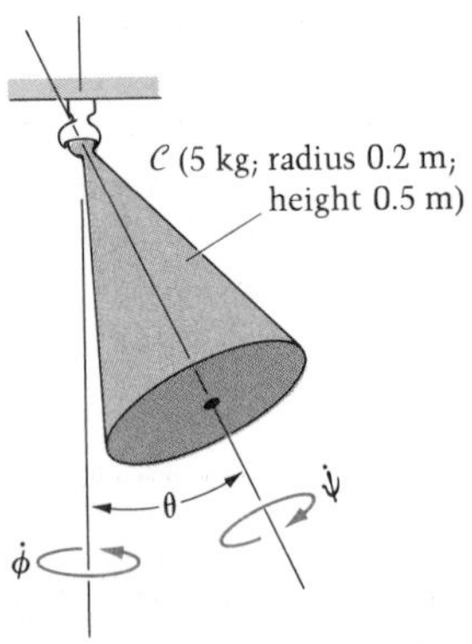

Figure P7.80

7.81 In the preceding problem, suppose that $\dot{\psi}$ is given to be 400 rad/s in the same direction as given in the figure and that the cone's height H is not given. Find the value of H for which this steady precession will occur.

7.82 Using the fact that the sum of any two moments of inertia at a point is always larger than the third (Problem 7.14), show that for a torque-free axisymmetric body undergoing retrograde precession, $\dot{\phi} \geq 2|\dot{\psi}|$ and that the z axis of the body is always outside the space cone.

7.83 The graph in Figure P7.83a depicts the stability of symmetrical satellites spinning about the axis z_C normal to the orbital plane. The abscissa is the ratio of I_{z_C} to the moment of inertia I_ℓ about any lateral axis (they are all the same for what is called a "symmetrical" satellite—it need not be *physically* symmetric about z_C). The ordinate is the ratio of the spin speed ω_s (about z_C) *in* the orbit to the orbital angular speed ω_0.

a. For a satellite equivalent to four solid cylinders each of mass m, radius R, and height $3R$, find I_{z_C} and I_ℓ. The distance from C to any cylinder's center is $2R$, and the connecting cross is light. The cylinders' axes are normal to the orbital plane. (See Figure 7.83b.)

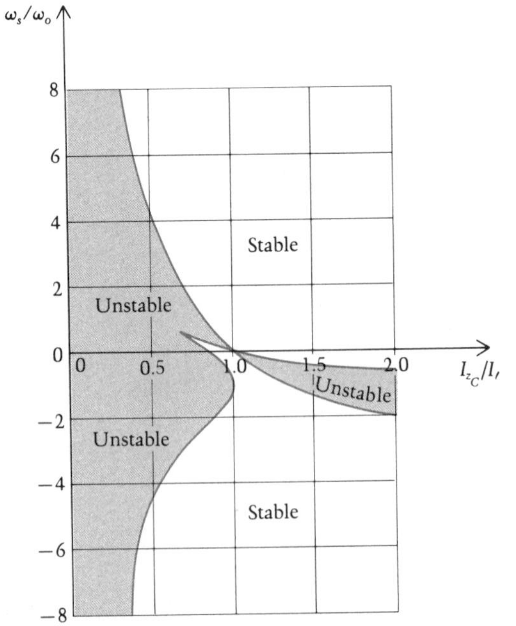

Figure P7.83a

b. Determine whether the station is stable for the following cases:

i. The station's orientation is fixed in inertial space.

ii. The station travels around the earth as the moon does.

iii. The station has twice the angular velocity of the orbiting frame.

iv. The same as (iii), but the spin is opposite in direction to the orbital angular speed.

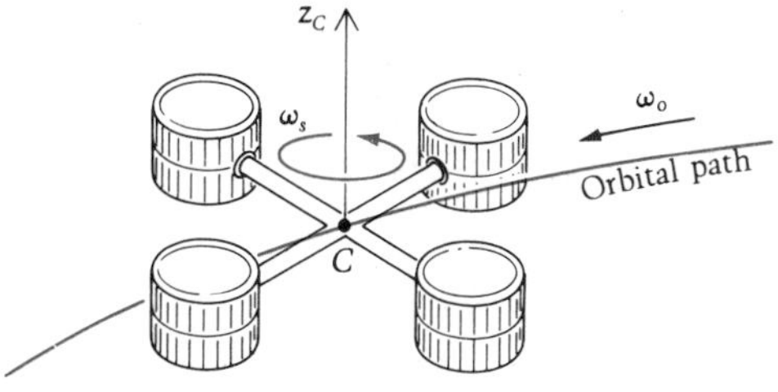

Figure P7.83b

The following five problems are advanced looks at statics of rigid bodies that depend on our study of dynamics.

7.84 It is possible for a spinning top to "sleep," meaning that its axis remains vertical and its peg stationary as it spins on a floor. (See Figure P7.84.) In the absence of a friction couple about the axis of the top, note that the spin speed ω_z is constant and that the equations of motion then reduce to $\Sigma \mathbf{M}_C = \mathbf{0}$ and $\Sigma \mathbf{F} = \mathbf{0}$. It thus follows that

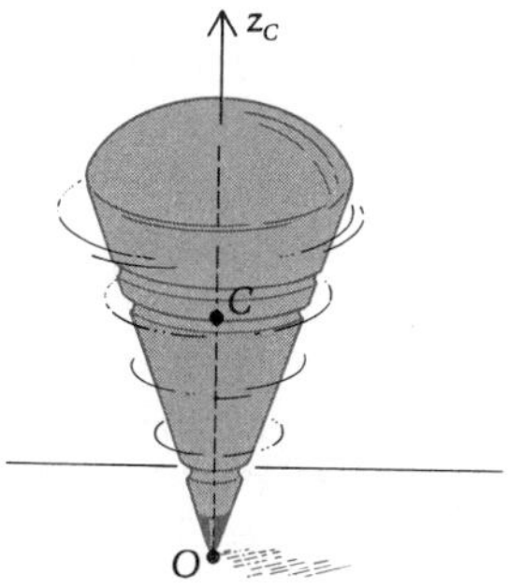

Figure P7.84

$\Sigma \mathbf{M}_O$ is also zero. None of the particles of the top not on the z_C axis are in equilibrium, however, because they all have (inward) accelerations $r\omega^2$. A body is in equilibrium if and only if all its particles are in equilibrium, so the sleeping top cannot be in equilibrium. Explain this statement in light of $\Sigma \mathbf{F} = \mathbf{0}$ and $\Sigma \mathbf{M}_O = \mathbf{0}$, which were the equilibrium equations for a body in statics. *Hint:* If O is a fixed point of rigid body $\mathcal{B}$ in an inertial frame $\mathcal{I}$, then

$$\Sigma \mathbf{M}_O = {}^{\mathcal{I}}\dot{\mathbf{H}}_O = {}^{\mathcal{I}}\frac{d}{dt}(\mathbf{H}_C + \mathbf{r}_{OC} \times \mathbf{L})$$

$$= {}^{\mathcal{B}}\dot{\mathbf{H}}_C + \boldsymbol{\omega}_{\mathcal{B}/\mathcal{I}} \times \mathbf{H}_C + \mathbf{r}_{OC} \times \underbrace{m\mathbf{a}_C}_{\Sigma \mathbf{F}}$$

Therefore show that just because $\Sigma \mathbf{F} = \mathbf{0}$ and $\Sigma \mathbf{M}_O = \mathbf{0}$, $\boldsymbol{\omega}_{\mathcal{B}/\mathcal{I}}$ need not be zero. Use the top as a counterexample and explain why the first two terms on the right side of the preceding equation vanish. Thus $\Sigma \mathbf{F} = \Sigma \mathbf{M}_O = \mathbf{0}$ are necessary but not *sufficient* conditions for equilibrium of a rigid body.

7.85 In the sleeping top counterexample of Problem 7.84, the terms ${}^{\mathcal{B}}\dot{\mathbf{H}}_C$ and $\boldsymbol{\omega}_{\mathcal{B}/\mathcal{I}} \times \mathbf{H}_C$ both vanish independently. Show that there are more complicated counterexamples in which $\boldsymbol{\omega}_{\mathcal{B}/\mathcal{I}}$ is not constant in direction in $\mathcal{B}$ and $\mathcal{I}$ and in which the two terms *add* to zero. *Hint:* $\Sigma \mathbf{M}_O - \mathbf{r}_{OC} \times \Sigma \mathbf{F} = \Sigma \mathbf{M}_C$. What is $\Sigma \mathbf{M}_C$ for the torque-free body?

7.86 Show that if $\boldsymbol{\omega}_{\mathcal{B}/\mathcal{I}} = \mathbf{0}$ at all times, then so is $\Sigma \mathbf{M}_C$. Is the converse true?

7.87 If a frame $\mathcal{B}$ is moving relative to an inertial frame $\mathcal{I}$, it can be shown that $\mathcal{B}$ is also an inertial frame if and only if $\boldsymbol{\omega}_{\mathcal{B}/\mathcal{I}} = \mathbf{0}$ at all times *and* the acceleration in $\mathcal{I}$ of at least one point of $\mathcal{B}$ is zero at all times. Use this theorem to show that if a rigid body $\mathcal{B}$ is in equilibrium in an inertial frame $\mathcal{I}$, then $\mathcal{B}$ is *itself* an inertial frame. Is the converse true?

7.88 Show that a rigid body $\mathcal{B}$ is in equilibrium in an inertial frame $\mathcal{I}$ if and only if (a) at least one point of $\mathcal{B}$ is fixed in $\mathcal{I}$ and (b) $\boldsymbol{\omega}_{\mathcal{B}/\mathcal{I}} = \mathbf{0}$ at all times. What is the minimum number of constraints on $\mathcal{B}$ that will satisfy (a) and (b)? Describe one set of physical constraints that will assure equilibrium.

7.7 Impulse and Momentum

As we did in Chapter 5 for the case of plane motion, we could apply the principles of impulse and momentum and those of angular impulse and angular momentum to the three-dimensional motion of a rigid body $\mathcal{B}$. As we saw in Section 5.3, however, these applications are really nothing more than time integrations of the equations of motion.

There is one type of problem, however, in which these two principles furnish us with a means of solution—problems involving impact. Some three-dimensional aspects are sufficiently different from the planar case to warrant an example. But first we use the integrals of the Euler laws to derive the needed relations:

$$\int_{t_i}^{t_f} \Sigma \mathbf{F}\, dt = \mathbf{L}_f - \mathbf{L}_i \tag{7.63}$$

$$\int_{t_i}^{t_f} \Sigma \mathbf{M}_C\, dt = \mathbf{H}_{C_f} - \mathbf{H}_{C_i} \tag{7.64}$$

An alternative to the rotational equation (7.64) is to integrate the equally general equation

$$\Sigma \mathbf{M}_O = \dot{\mathbf{H}}_O \tag{7.65}$$

where O is now a fixed point of the inertial frame $\mathcal{I}$:

$$\int_{t_i}^{t_f} \Sigma \mathbf{M}_O\, dt = \mathbf{H}_{O_f} - \mathbf{H}_{O_i} = (\mathbf{H}_{C_f} - \mathbf{H}_{C_i}) + (\mathbf{r}_{OC} \times m\mathbf{v}_C)\Big|_i^f \tag{7.66}$$

To use either Equation (7.64) or (7.66) in an impact situation, we use Equation (7.11) for the body's angular momentum *before the deformation starts* (at t_i) and then again *after it ends* (at t_f). The following example illustrates the procedure.

EXAMPLE 7.11

The bent bar $\mathcal{B}$ of Example 4.16 is dropped from a height H and strikes a rigid, smooth surface on one end of $\mathcal{B}$ as shown in Figure E7.11. If the coefficient of restitution is e, find the angular velocity of $\mathcal{B}$, as well as the velocity of C, just after the collision.

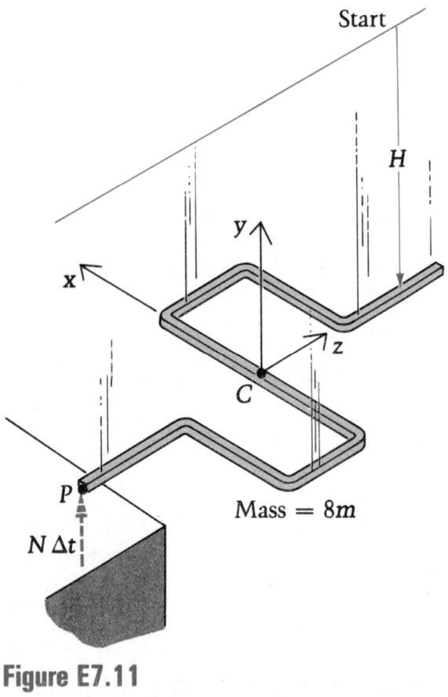

Figure E7.11

Solution

Using the y component equation of (7.63) yields

$$N\,\Delta t = 8m\dot{y}_{C_f} - 8m(-\sqrt{2gH}) \tag{1}$$

in which the impulse of the gravity force is neglected as small in comparison with the impulsive upward force exerted by the surface over the short time interval Δt.

Next we write the component equations of (7.64); we first need the inertia properties of the body, which can be computed to be

$$I_{xx}^C = \frac{22}{3} m\ell^2 \qquad I_{xy}^C = 0$$

$$I_{yy}^C = \frac{32}{3} m\ell^2 \qquad I_{yz}^C = 0$$

$$I_{zz}^C = \frac{10}{3} m\ell^2 \qquad I_{zx}^C = -2m\ell^2 \tag{2}$$

We then obtain, from Equation (7.64),

$$2\ell N\,\Delta t\hat{\mathbf{i}} = m\ell^2\left[\frac{22}{3}\,\omega_x + 0\omega_y - 2\omega_z\right]\hat{\mathbf{i}}$$
$$+\, m\ell^2\left[0\omega_x + \frac{32}{3}\,\omega_y + 0\omega_z\right]\hat{\mathbf{j}}$$
$$+\, m\ell^2\left[-2\omega_x + 0\omega_y + \frac{10}{3}\,\omega_z\right]\hat{\mathbf{k}} \tag{3}$$

in which the initial angular velocity components vanish and the desired final components are $(\omega_x, \omega_y, \omega_z)$. The component equations of (3) are

$$\frac{22}{3}\,\omega_x - 2\omega_z = \frac{+2N\,\Delta t}{m\ell} \tag{4}$$

$$\frac{32}{3}\,\omega_y = 0 \tag{5}$$

$$-2\omega_x + \frac{10}{3}\,\omega_z = 0 \tag{6}$$

At this point we have four equations in the five unknowns $\dot{y}_C, \omega_x, \omega_y, \omega_z$, and the impulse $N\,\Delta t$. We get a fifth equation from the definition of the coefficient of restitution together with the y component of the rigid-body velocity relationship between P and C:

$$e = \frac{\dot{y}_{P_f} - 0}{0 - (-\sqrt{2gH})} \Rightarrow \dot{y}_{P_f} = \sqrt{2gH}\,e \tag{7}$$

and

$$\mathbf{v}_C = \mathbf{v}_P + \boldsymbol{\omega} \times \underbrace{\mathbf{r}_{PC}}_{2\ell\hat{\mathbf{k}}} \tag{8}$$

which has the y-component equation

$$\dot{y}_{C_f} = \dot{y}_{P_f} - 2\ell\omega_x \tag{9}$$

Using Equation (7), we obtain

$$\dot{y}_{Cf} = \sqrt{2gH}\,e - 2\ell\omega_x \tag{10}$$

The solution to the five equations (1, 4, 5, 6, 10) is

$$\omega_x = \frac{60(1+e)\sqrt{2gH}}{143\ell} \qquad \dot{y}_{C_f} = \frac{(23e - 120)\sqrt{2gH}}{143}$$
$$\omega_y = 0 \tag{11}$$
$$\omega_z = \frac{36(1+e)\sqrt{2gH}}{143\ell} \qquad N\,\Delta t = \frac{184m(1+e)\sqrt{2gH}}{143}$$

Returning to Equation (8), we find that the x and z components of $\mathbf{v}_{C_f}$ vanish:

$$\begin{aligned} x \text{ components} &\Rightarrow \dot{x}_{C_f} = 0 + 2\omega_y\ell = 0 \\ z \text{ components} &\Rightarrow \dot{z}_{C_f} = 0 + \quad 0 \quad = 0 \end{aligned} \tag{12}$$

The results in Equations (12) are obvious, since if there is no friction at the point of contact there can be no impulsive forces in the horizontal plane to change the momentum (from zero) in the x or z directions.

It is seen that the single nonzero product of inertia causes a coupling between ω_x and ω_z (see Equations (4) and (6)), which prevents ω_z from vanishing—even though the only moment component with respect to C is about the x axis!

We shall now see with another example the advantages of Equation (7.66), which may be used to eliminate undesired forces from moment equations, just as was done in our study of statics.

EXAMPLE 7.12

Rework the preceding example by using Equation (7.66) instead of the combination of Equations (7.63) and (7.64). Find the value of ω after impact.

Solution

Equation (7.66) allows us to eliminate the impulse $N\,\Delta t$ by summing moments about the point (P') of impact:

$$\int_{t_i}^{t_f} \Sigma \mathbf{M}_{P'}\,dt = 0 = (\mathbf{H}_{C_f} - \overset{0}{\cancel{\mathbf{H}_{C_i}}}) + (\mathbf{r}_{P'C} \times m\mathbf{v}_C)\Big|_i^f$$

$$= \left[\left(\frac{22}{3}\,\omega_x - 2\omega_z\right) m\ell^2\hat{\mathbf{i}} + \frac{32}{3}\,\omega_y m\ell^2\hat{\mathbf{j}} + \left(-2\omega_x + \frac{10}{3}\,\omega_z\right) m\ell^2\hat{\mathbf{k}}\right]$$

$$+ 2\ell\mathbf{k} \times [8m\dot{y}_{C_f}\hat{\mathbf{j}} - 8m\sqrt{2gH}(-\hat{\mathbf{j}})]$$

We still have to use the coefficient of restitution and relate $\mathbf{v}_P$ and $\mathbf{v}_C$ exactly as before; making this substitution for $\dot{y}_{C_f}$ leads to the following three scalar component equations:

$$\frac{118}{3}\,\omega_x - 2\omega_z = \frac{16\sqrt{2gH}}{\ell}\,(1+e)$$

$$\frac{32}{3}\,\omega_y = 0$$

$$-2\omega_x + \frac{10}{3}\,\omega_z = 0$$

These equations, of course, have the same solution as $(\omega_x, \omega_y, \omega_z)$ in the preceding example.

PROBLEMS ▶ Section 7.7

7.89 Bend a coat hanger or pipe cleaner into the shape of the bent bar of Example 7.11. Drop it onto the edge of a table as in the example and observe that the angular velocity direction following impact agrees with the results of the example.

*** 7.90** The equilateral triangular dinner bell in Figure P7.90 is struck with a horizontal force in the y direction that imparts an impulse $F\,\Delta t\hat{\mathbf{j}}$ to the bell. Find the angular velocity of the bell immediately after the blow is struck. Is the answer the same if the bell is an equilateral triangular *plate* of the same mass? Why or why not?

*** 7.91** In the preceding problem, suppose the hammer is replaced by a bullet of mass m and speed v_b that rebounds straight back with a coefficient of restitution $e = 0.1$. Determine the resulting angular velocity of the bell.

7.92 Repeat Problem 7.90, but this time suppose the bell hangs from a string instead of from a ball and socket joint.

7.93 The bent bar $\mathcal{B}$ of Figure P7.93 has the inertia properties listed below. It is in motion in an inertial frame $\mathcal{I}$, and at a certain instant has angular velocity $\boldsymbol{\omega}_{\mathcal{B}/\mathcal{I}} = \omega(4\hat{\mathbf{i}} + 2\hat{\mathbf{j}} + 7\hat{\mathbf{k}})$ rad/sec. Use the angular impulse and angular momentum principle to answer the following question: Is it possible to strike $\mathcal{B}$ at point Q with an impulse $\mathbf{F} = F_x\,\Delta t\hat{\mathbf{i}} + F_y\,\Delta t\hat{\mathbf{j}} + F_z\,\Delta t\hat{\mathbf{k}}$ that reduces $\boldsymbol{\omega}_{\mathcal{B}/\mathcal{I}}$ to zero after the impulse? If so, find the components of the impulse in terms of m, ℓ, ω, and Δt. If not, show why not.

$$I^C_{xx} = \frac{22}{3}m\ell^2 \qquad I^C_{zz} = \frac{10}{3}m\ell^2$$

$$I^C_{yy} = \frac{32}{3}m\ell^2 \qquad I^C_{xz} = -2\,m\ell^2$$

$$I^C_{yz} = I^C_{xy} = 0$$

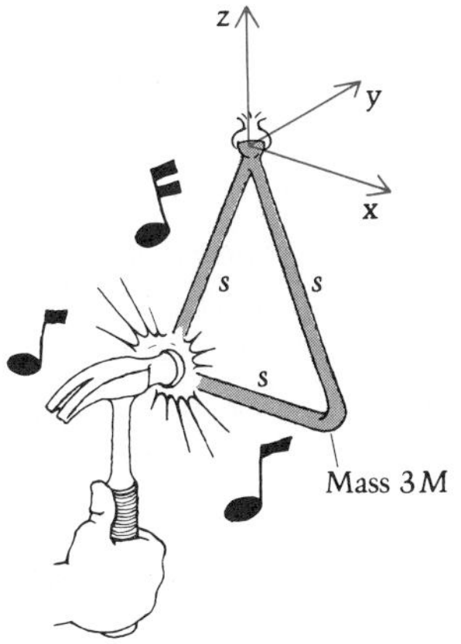

Figure P7.90

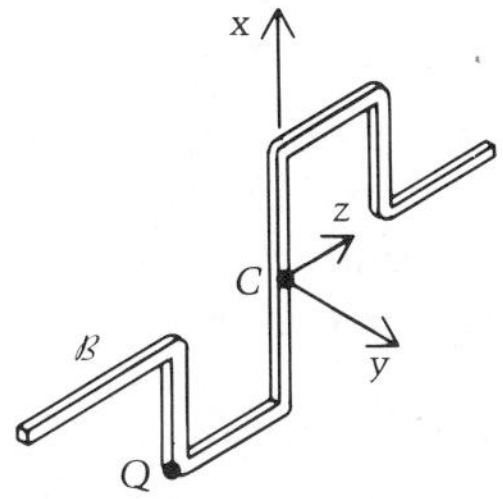

Figure P7.93

*** 7.94** A diver $\mathcal{D}$ leaves a diving board in a straight, symmetric position with angular velocity and angular momentum vectors each in the x direction as indicated in Figure P7.94a. Since $\Sigma\mathbf{M}_C$ is zero, there will be no change in the angular momentum $\mathbf{H}_C$ in the inertial frame (the swimming pool) as long as the diver is in the air. Therefore, as long as he remains in the straight position, his constant angular momentum is expressed by

$$I^C_{xx}\omega_x + \overset{0}{\cancel{I^C_{xy}}}\,\overset{0}{\cancel{\omega_y}} + \overset{0}{\cancel{I^C_{xz}}}\,\overset{0}{\cancel{\omega_z}} = H_{C_1} = \text{constant} \qquad (1)$$

$$\overset{0}{\cancel{I^C_{yx}}}\,\omega_x + I^C_{yy}\overset{0}{\cancel{\omega_y}} + \overset{0}{\cancel{I^C_{yz}}}\,\overset{0}{\cancel{\omega_z}} = 0 \qquad (2)$$

$$\overset{0}{\cancel{I^C_{zx}}}\,\omega_x + \overset{0}{\cancel{I^C_{zy}}}\,\overset{0}{\cancel{\omega_y}} + I^C_{zz}\overset{0}{\cancel{\omega_z}} = 0 \qquad (3)$$

where we assume the body to be sufficiently internally symmetric so that the products of inertia all vanish. Now suppose the diver instantaneously moves his arms as shown in Figure P7.94b to initiate a twist. Following the maneuver, he may again be treated as a rigid body and we may use the same body-fixed axes as before. (Note that the mass center changes very little.)

a. From Figure P7.94c argue that the indicated changes in the products of inertia occur. (Only the shaded arms contribute to the products of inertia.) Argue also that I^C_{yz} is smaller than I^C_{xz} and also less than $|I^C_{xy}|$. Observe that all three

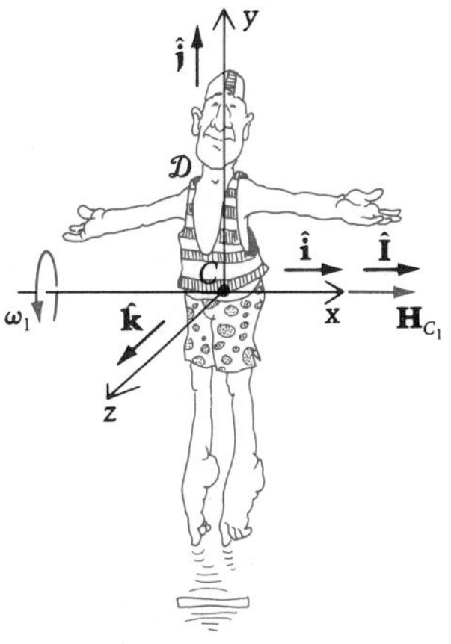

Figure P7.94a

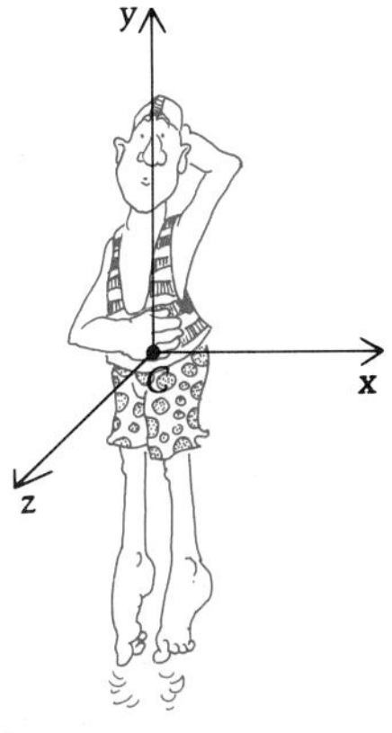

Figure P7.94b

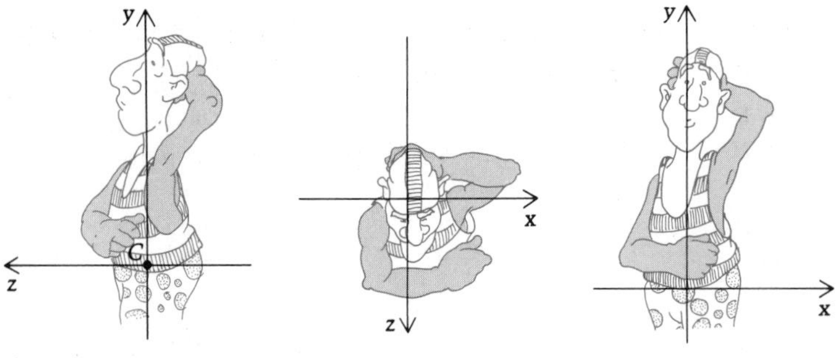

$I_{yz}^C > 0$ $I_{xz}^C > 0$ $I_{xy}^C < 0$

Figure P7.94c

products of inertia are small compared with the three moments of inertia and that $I_{yy}^C < I_{xx}^C < I_{zz}^C$, with I_{yy}^C being very much smaller than the other two moments of inertia. Note that (x, y, z) are no longer principal, but this does not matter since we are not making use of principal axes here.

As the diver's body begins to twist and turn, the right sides of Equations (1) to (3) will change and *none* of the quantities on the left will remain zero. But the right sides will constitute the components *in the body frame* $\mathcal{D}$ of the vector $\mathbf{H}_C$, which will still vectorially add to $H_{C_1}\hat{\mathbf{I}}$, where $\hat{\mathbf{I}}$ is the original direction in $\mathcal{I}$ of $\mathbf{H}_C$ after the diver leaves the diving board (to the right in the first sketch).

b. After the rapid twist maneuver, but *before* the diver begins to twist, his axes are still instantaneously aligned with those of the frame $\mathcal{I}$. Use Equations (1) to (3), with the right sides $(H_{C_1}, 0, 0)$ and the now nonzero products of inertia, to show that:
   i. There will be a small (compared to the original ω_x) angular velocity developed about the $-z_C$ direction (negative ω_z).
   ii. There will be an angular velocity of twist developed about y_C (positive ω_y).
   iii. There will be an increase in the somersaulting angular velocity component ω_x.

In arguing statements (i) to (iii), assume nothing about the ω's following the maneuver except that ω_x is still in the same direction as before.

7.8 Work and Kinetic Energy

A special integral of the equations of motion of a rigid body $\mathcal{B}$ yields a relationship between the work of the external forces (and/or couples) and the change in the kinetic energy of $\mathcal{B}$. To develop this relation, we must first explore expressions for the kinetic energy of the rigid body. **Kinetic energy** is usually denoted by the letter T and is defined by (see Section 5.2)

$$T = \frac{1}{2}\int \mathbf{v} \cdot \mathbf{v}\, dm \tag{7.67}$$

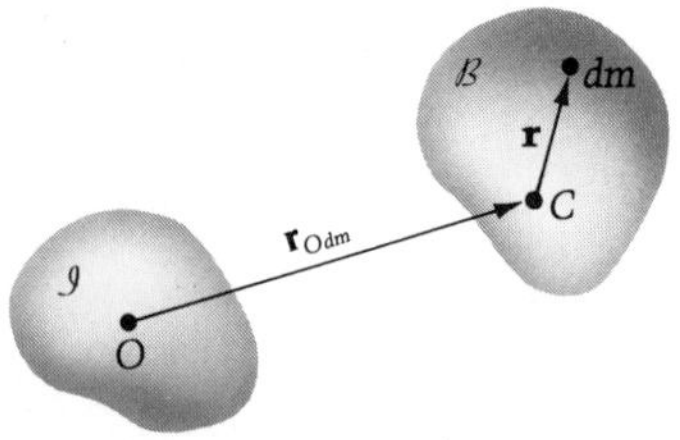

Figure 7.17

in which $\mathbf{v}$ is the derivative of the position vector from O (fixed point in the inertial frame $\mathcal{I}$ in Figure 7.17) to the differential mass element dm. In this section all time derivatives, velocities, and angular velocities are taken in $\mathcal{I}$ unless otherwise specified.

Since $\mathcal{B}$ is a rigid body, we may relate $\mathbf{v}$ to the velocity $\mathbf{v}_C$ of the mass center C of $\mathcal{B}$:

$$\mathbf{v} = \mathbf{v}_C + \boldsymbol{\omega} \times \mathbf{r} \tag{7.68}$$

in which ω is $\omega_{\mathcal{B}/\mathcal{I}}$ and $\mathbf{r}$ is the position vector from C to dm as shown in Figure 7.17. Substituting Equation (7.68) into (7.67), we get

$$T = \frac{1}{2}\mathbf{v}_C \cdot \mathbf{v}_C \int_{\mathcal{B}} dm + \frac{1}{2}\int_{\mathcal{B}} (\omega \times \mathbf{r}) \cdot (\omega \times \mathbf{r})\, dm + \mathbf{v}_C \cdot \left[\omega \times \int_{\mathcal{B}} \mathbf{r}\, dm\right] \tag{7.69}$$

where $\mathbf{v}_C$ and ω do not vary over the body's volume and can thus be taken outside the integrals. The integral in the last term is zero by virtue of the definition of the mass center:

$$\int_{\mathcal{B}} \mathbf{r}\, dm = m\mathbf{r}_{CC} = \mathbf{0} \tag{7.70}$$

The integral in the first term on the right side of Equation (7.69) is of course the mass m of $\mathcal{B}$. The integrand of the remaining term may be simplified by the vector identity:*

$$(\mathbf{A} \times \mathbf{B}) \cdot (\mathbf{C} \times \mathbf{D}) = \mathbf{A} \cdot [\mathbf{B} \times (\mathbf{C} \times \mathbf{D})] \tag{7.71}$$

Therefore Equation (7.69) becomes

$$T = \frac{m}{2}(\mathbf{v}_C \cdot \mathbf{v}_C) + \frac{1}{2}\omega \cdot \int [\mathbf{r} \times (\omega \times \mathbf{r})]\, dm \tag{7.72}$$

As we have already seen in Section 7.2, the integral in Equation (7.72) is the angular momentum (moment of momentum) of the body with respect to C, and thus we can write

$$T = \overbrace{\frac{m}{2}\mathbf{v}_C \cdot \mathbf{v}_C}^{T_v} + \overbrace{\frac{1}{2}\omega \cdot \mathbf{H}_C}^{T_\omega} \tag{7.73}$$

It is seen that the kinetic energy can be represented as the sum of two terms:

1. A part $T_v = (m/2)\mathbf{v}_C \cdot \mathbf{v}_C$ that the body possesses if its mass center is in motion.
2. A part $T_\omega = \frac{1}{2}\omega \cdot \mathbf{H}_C$ that is due to the difference between the velocities of the points of $\mathcal{B}$ and the velocity of its mass center.

The term T_ω can be interpreted quite simply if at an instant we let $\omega = \omega\hat{\mathbf{i}}$—that is, if we align the reference axis x with the angular velocity vector at that instant. In this case, using Equations (7.11), we obtain

$$\mathbf{H}_C = I_{xx}^C \omega\hat{\mathbf{i}} + I_{xy}^C \omega\hat{\mathbf{j}} + I_{xz}^C \omega\hat{\mathbf{k}} \tag{7.74}$$

* Which is nothing more than interchanging the dot and cross of the scalar triple product $(\mathbf{A} \times \mathbf{B}) \cdot \mathbf{E}$, where $\mathbf{E}$ is the vector $\mathbf{C} \times \mathbf{D}$.

so that

$$\frac{1}{2}\,\boldsymbol{\omega}\cdot\mathbf{H}_C = \frac{1}{2}\,I_{xx}^{C}\omega^2 \tag{7.75}$$

This means that the "rotational part" of T is *instantaneously* of the same form as it was for the plane case in Chapter 4. The difference, of course, is that the direction of the angular velocity vector $\boldsymbol{\omega}$ changes in the general (three-dimensional) case.

Suppose the body $\mathcal{B}$ has a point P with zero velocity. (This is not always the case in general motion as we have already seen in Chapter 6.) Then if $\mathbf{v}$ in Equation (7.67) is replaced by $\mathbf{v}_P + \boldsymbol{\omega} \times \mathbf{r}' = \boldsymbol{\omega} \times \mathbf{r}'$, where $\mathbf{r}'$ extends from P to the mass element dm, we obtain

$$T = \int (\boldsymbol{\omega} \times \mathbf{r}') \cdot (\boldsymbol{\omega} \times \mathbf{r}')\, dm \tag{7.76}$$

The identical steps that produced the second term of Equation (7.73) from the middle term of (7.69) then give

$$T = \frac{1}{2}\,\boldsymbol{\omega}\cdot\mathbf{H}_P \tag{7.77}$$

and the two terms of Equation (7.73) have collapsed into one if $\mathbf{H}$ is expressed relative to a point of zero velocity instead of C.

In Sections 2.4 and 5.2 we demonstrated one work and kinetic energy principle that remains true for the general case. This result came from integrating $\Sigma\mathbf{F} = m\mathbf{a}_C$:

$$\int_{t_1}^{t_2} \Sigma\mathbf{F}\cdot\mathbf{v}_C\, dt = \frac{1}{2}\,m|\mathbf{v}_C(t_2)|^2 - \frac{1}{2}\,m|\mathbf{v}_C(t_1)|^2$$

$$= \frac{1}{2}\,m(v_{C_2}^2 - v_{C_1}^2) \tag{7.78}$$

A second principle will now be deduced from the moment equation*

$$\Sigma\mathbf{M}_C = \dot{\mathbf{H}}_C \tag{7.79}$$

but first we need to prove the non-obvious result that:

$$\dot{\boldsymbol{\omega}}\cdot\mathbf{H}_C = \boldsymbol{\omega}\cdot\dot{\mathbf{H}}_C$$

To do this, we first recall that

$$\mathbf{H}_C = \int [\mathbf{r} \times (\boldsymbol{\omega} \times \mathbf{r})]\, dm \tag{7.80}$$

If ${}^{\mathcal{B}}\dot{\mathbf{H}}_C$ is the derivative of $\mathbf{H}_C$ taken in the body $\mathcal{B}$, then the derivative relative to the inertial frame can be written

$$\dot{\mathbf{H}}_C = {}^{\mathcal{B}}\dot{\mathbf{H}}_C + \boldsymbol{\omega} \times \mathbf{H}_C \tag{7.81}$$

---

* Derivatives such as $\dot{\boldsymbol{\omega}}$ are taken in the inertial frame $\mathcal{I}$ in this section unless the letter $\mathcal{B}$ appears to the left of the dot, in which case the derivative is taken in the body.

Dotting ω with both sides of Equation (7.81) shows that

$$\omega \cdot \dot{\mathbf{H}}_C = \omega \cdot {}^{\mathcal{B}}\dot{\mathbf{H}}_C \tag{7.82}$$

and since $\mathbf{r}$ is constant in time relative to body $\mathcal{B}$, we can differentiate Equation (7.80) there and obtain

$${}^{\mathcal{B}}\dot{\mathbf{H}}_C = \int \mathbf{r} \times (\dot{\omega} \times \mathbf{r})\, dm \tag{7.83}$$

In Equation (7.83) we have used the property of ω that its derivatives in $\mathcal{I}$ and $\mathcal{B}$ are the same; that is,

$${}^{\mathcal{I}}\dot{\omega} = {}^{\mathcal{I}}\dot{\omega}_{\mathcal{B}/\mathcal{I}} = {}^{\mathcal{B}}\dot{\omega}_{\mathcal{B}/\mathcal{I}} + \omega_{\mathcal{B}/\mathcal{I}} \times \omega_{\mathcal{B}/\mathcal{I}} = {}^{\mathcal{B}}\dot{\omega}_{\mathcal{B}/\mathcal{I}} = {}^{\mathcal{B}}\dot{\omega}$$

Substituting Equation (7.83) into (7.82) then gives

$$\begin{aligned}
\omega \cdot \dot{\mathbf{H}}_C &= \omega \cdot \int \mathbf{r} \times (\dot{\omega} \times \mathbf{r})\, dm = \int \omega \cdot [\mathbf{r} \times (\dot{\omega} \times \mathbf{r})]\, dm \\
&= \int (\omega \times \mathbf{r}) \cdot (\dot{\omega} \times \mathbf{r})\, dm = \int [(\dot{\omega} \times \mathbf{r}) \cdot (\omega \times \mathbf{r})]\, dm \\
&= \dot{\omega} \cdot \int \mathbf{r} \times (\omega \times \mathbf{r})\, dm
\end{aligned}$$

Hence

$$\omega \cdot \dot{\mathbf{H}}_C = \dot{\omega} \cdot \mathbf{H}_C \tag{7.84}$$

We are now in a position to observe that

$$\omega \cdot \Sigma \mathbf{M}_C = \omega \cdot \dot{\mathbf{H}}_C = \frac{d}{dt}\left(\frac{\omega \cdot \mathbf{H}_C}{2}\right) \tag{7.85}$$

Integrating Equation (7.85), we have

$$\int_{t_1}^{t_2} \Sigma \mathbf{M}_C \cdot \omega\, dt = \frac{1}{2}\, \omega(t_2) \cdot \mathbf{H}_C(t_2) - \frac{1}{2}\, \omega(t_1) \cdot \mathbf{H}_C(t_1) \tag{7.86}$$

Note that the right sides of Equations (7.78) and (7.86) each represents the change, occurring in the time interval $t_1 \le t \le t_2$, of part of the kinetic energy of the body. The left sides of these equations are usually called a form of *work*.

While the relationships between work and kinetic energy that have been developed are important, another relationship that combines them is often more useful. We can differentiate Equation (7.73) and get

$$\frac{dT}{dt} = m\mathbf{v}_C \cdot \mathbf{a}_C + \frac{1}{2}\, \dot{\omega} \cdot \mathbf{H}_C + \frac{1}{2}\, \omega \cdot \dot{\mathbf{H}}_C$$

Using Euler's laws and Equation (7.84), this may be put into the form

$$\frac{dT}{dt} = \Sigma \mathbf{F} \cdot \mathbf{v}_C + \Sigma \mathbf{M}_C \cdot \omega \tag{7.87}$$

If we now let $\mathbf{F}_1, \mathbf{F}_2, \ldots$ represent the external forces acting on the body, and $\mathbf{C}_1, \mathbf{C}_2, \ldots$ represent the moments of the external couples, then

$$\Sigma\mathbf{F} = \mathbf{F}_1 + \mathbf{F}_2 + \cdots \tag{7.88a}$$

$$\Sigma\mathbf{M}_C = \mathbf{r}_1 \times \mathbf{F}_1 + \mathbf{r}_2 \times \mathbf{F}_2 + \cdots + \mathbf{C}_1 + \mathbf{C}_2 + \cdots \tag{7.88b}$$

where $P_1, P_2, \ldots$ are the points of $\mathcal{B}$ where $\mathbf{F}_1, \mathbf{F}_2, \ldots$ are respectively applied and where $\mathbf{r}_1 = \mathbf{r}_{CP_1}$, $\mathbf{r}_2 = \mathbf{r}_{CP_2}$, and so forth, as shown in Figure 7.18. Recall from statics that a couple has the same moment about any point in space, so that the $\mathbf{C}_i$'s are simply added into the moment equation (7.88b).

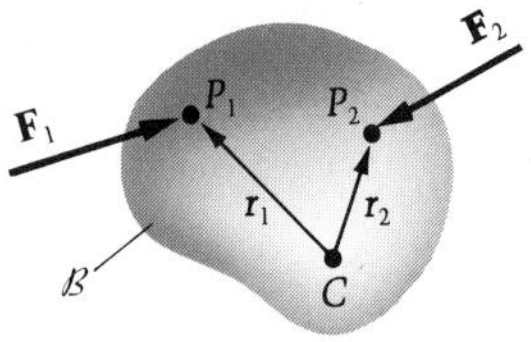

Figure 7.18

Substituting Equation (7.88) into (7.87), we obtain

$$\frac{dT}{dt} = \mathbf{F}_1 \cdot \mathbf{v}_C + \mathbf{F}_2 \cdot \mathbf{v}_C + \cdots + \boldsymbol{\omega} \cdot (\mathbf{r}_1 \times \mathbf{F}_1)$$
$$+ \boldsymbol{\omega} \cdot (\mathbf{r} \times \mathbf{F}_2) + \cdots + \boldsymbol{\omega} \cdot \mathbf{C}_1 + \boldsymbol{\omega} \cdot \mathbf{C}_2 + \cdots \tag{7.89}$$

However,

$$\boldsymbol{\omega} \cdot (\mathbf{r}_i \times \mathbf{F}_i) = \mathbf{F}_i \cdot (\boldsymbol{\omega} \times \mathbf{r}_i) \qquad (i = 1, 2, \ldots)$$

so that

$$\frac{dT}{dt} = \mathbf{F}_1 \cdot (\mathbf{v}_C + \boldsymbol{\omega} \times \mathbf{r}_1) + \mathbf{F}_2 \cdot (\mathbf{v}_C + \boldsymbol{\omega} \times \mathbf{r}_2) + \cdots$$
$$+ \boldsymbol{\omega} \cdot \mathbf{C}_1 + \boldsymbol{\omega} \cdot \mathbf{C}_2 + \cdots \tag{7.90}$$

We note that $\mathbf{v}_C + \boldsymbol{\omega} \times \mathbf{r}_1$ is the velocity $\mathbf{v}_1$ of point P_1, the point of application of $\mathbf{F}_1$. Therefore

$$\frac{dT}{dt} = \dot{T} = \Sigma\mathbf{F}_i \cdot \mathbf{v}_i + \boldsymbol{\omega} \cdot \Sigma\mathbf{C}_i \tag{7.91}$$

Equation (7.91) leads us to define the **power**, or **rate of work**, as follows:

$$\text{Power (rate of work) of a force } (\mathbf{F}_1) = \mathbf{F}_1 \cdot \mathbf{v}_1$$
$$\text{Power (rate of work) of a couple } (\mathbf{C}_1) = \boldsymbol{\omega} \cdot \mathbf{C}_1 \tag{7.92}$$

Therefore

$$\dot{T} = \text{rate of work of external forces and couples}$$

Integrating Equation (7.91), we get

$$\int_{t_1}^{t_2} (\text{rate of work})\, dt = T(t_2) - T(t_1) = \Delta T \tag{7.93}$$

The integral on the left side of Equation (7.93) is called the **work** done on $\mathcal{B}$ between t_1 and t_2 by the external forces and couples. Hence

$$\text{Work} = \int (\Sigma\mathbf{F}_i \cdot \mathbf{v}_i + \boldsymbol{\omega} \cdot \Sigma\mathbf{C}_i)\, dt = \Delta T \tag{7.94}$$

That is, the work done on $\mathcal{B}$ equals its change in kinetic energy. It is left as an exercise for the reader to show that Equation (7.94) is in fact the sum of the two "subequations" (7.78) and (7.86).

EXAMPLE 7.13

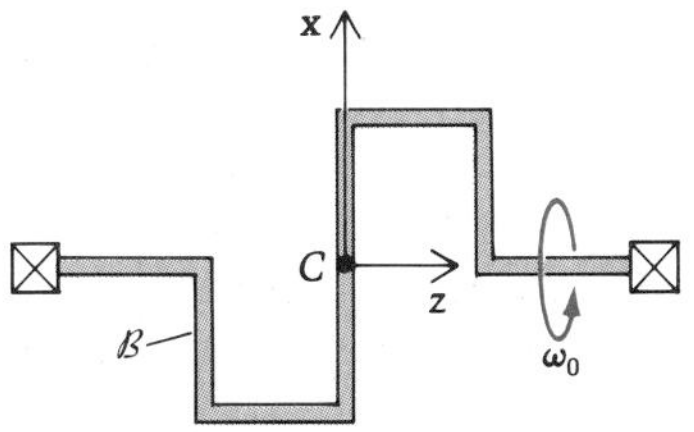

Figure E7.13

Find the work done on the bent bar of Example 7.11 by a motor that brings it up to speed ω_0 from rest. (See Figure E7.13.)

Solution

The mass center C does not move, so that Equation (7.73) gives, in this case,

$$T_f = \frac{1}{2}\,\boldsymbol{\omega}\cdot\mathbf{H}_C \tag{1}$$

Since $\boldsymbol{\omega}$ has only a $\hat{\mathbf{k}}$ component, $\omega_0\hat{\mathbf{k}}$, we may substitute Equation (7.11) into (1) and get

$$T_f = \frac{1}{2}\,\omega_0(I_{xz}^C\overset{0}{\cancel{\omega_x}} + I_{yz}^C\overset{0}{\cancel{\omega_y}} + I_{zz}^C\overset{\omega_0}{\cancel{\omega_z}}) \tag{2}$$

We note that even though I_{xz}^C is not zero, it has no effect on the kinetic energy of $\mathcal{B}$ since it is multiplied by ω_x, which is forced to vanish by the bearings aligned with z.

Thus the work done by the motor on $\mathcal{B}$ is given simply by Equation (7.94):

$$\begin{aligned} W = \Delta T &= T_f - \overset{0}{\cancel{T_i}} \\ &= \frac{1}{2}\,I_{zz}^C\omega_0^2 \\ &= \frac{5}{3}\,m\ell^2\omega_0^2 \end{aligned} \tag{3}$$

where $I_{zz}^C = (10/3)m\ell^2$ from Example 7.11. The motor would, of course, have to do additional work besides that given by (3) to overcome its own armature inertia, bearing and belt friction, and air resistance.

We now consider an example in three dimensions in which the products of inertia do play a role in the kinetic energy calculation.

EXAMPLE 7.14

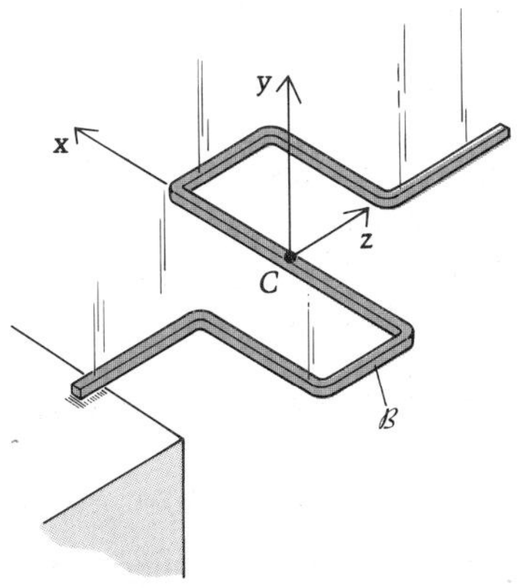

Figure E7.14a

Find the kinetic energy lost by the bent bar of Example 7.10 when it strikes the table top as shown in Figure E7.14a.

Solution

During the impact with the table top, the bodies do not behave rigidly. The kinetic energy lost by bar $\mathcal{B}$ is transformed into noise, heat, vibration, and both elastic and permanent deformation. In Example 7.11 we found $\mathbf{v}_C$ and $\boldsymbol{\omega}_i$ just before and after impact; we now use these vectors to find the kinetic energy lost by $\mathcal{B}$. Just after impact we have

$$T_f = \frac{1}{2}\,mv_{C_f}^2 + \frac{1}{2}\,\boldsymbol{\omega}_f\cdot\mathbf{H}_{C_f}$$

The term $\boldsymbol{\omega}_f \cdot \mathbf{H}_{C_f}$ can be written just after impact, using Equation (7.11), as follows. (Note that ω_y and two of the products of inertia are zero here.)

$$\begin{aligned}\omega_f \cdot \mathbf{H}_{C_f} &= \omega_x(I_{xx}^C\omega_x + \overset{0}{\cancel{I_{xy}^C}}\overset{0}{\cancel{\omega_y}} + I_{xz}^C\omega_z)\\ &\quad + \omega_y(\overset{0}{\cancel{I_{yx}^C}}\omega_x + I_{yy}^C\overset{0}{\cancel{\omega_y}} + \overset{0}{\cancel{I_{yz}^C}}\omega_z)\\ &\quad + \omega_z(I_{zx}^C\omega_x + \overset{0}{\cancel{I_{zy}^C}}\overset{0}{\cancel{\omega_y}} + I_{zz}^C\omega_z)\\ &= I_{xx}^C\omega_x^2 + 2I_{xz}\omega_x\omega_z + I_{zz}^C\omega_z^2\end{aligned}$$

Using this result and Equations (11) and (12) from Example 7.14, we obtain

$$\begin{aligned}T_f = \frac{1}{2}(8m)\left[\frac{(23e-120)\sqrt{2gh}}{143}\right]^2 + \frac{1}{2}\Bigg\{\left[\frac{60(1+e)\sqrt{2gh}}{143\ell}\right]^2\frac{22}{3}m\ell^2\\ + 2\left[\frac{60(1+e)\sqrt{2gh}}{143\ell}\right]\left[\frac{36(1+e)\sqrt{2gh}}{143\ell}\right](-2m\ell^2)\\ + \left[\frac{36(1+e)\sqrt{2gh}}{143\ell}\right]^2\frac{10}{3}m\ell^2\Bigg\}\end{aligned}$$

which, after simplification, equals

$$T_f = mgh(1.29e^2 + 6.71)$$

The initial kinetic energy (just prior to the collision) was

$$T_i = \frac{1}{2}(8m)(\sqrt{2gh})^2 = 8mgh$$

Thus the change in kinetic energy of the bent bar is given by

$$\Delta T = T_f - T_i = mgh(1.29e^2 - 1.29)$$

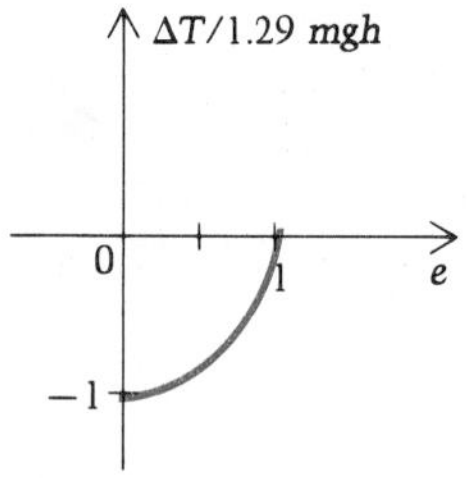

Figure E7.14b

We see that if $e = 1$ (purely elastic collision), no loss in kinetic energy occurs and hence no work is done in changing T. The energy lost is seen in Figure E7.14b to vary quadratically, with a maximum percentage loss (when $e = 0$) of

$$\frac{1.29mgh}{8mgh}\cdot 100 = 16.1\%$$

in this case. Because the point of striking is the end of the bar, 83.9 percent of the kinetic energy is retained. If the mass center of the bar were the point that struck the table, however, *all* the kinetic energy would have been lost if $e = 0$.

PROBLEMS ▶ Section7.8

7.95 Find the kinetic energy of disk $\mathcal{B}$ in Problem 6.27.

7.96 The center of mass C of a gyroscope $\mathcal{G}$ is fixed. Show that the kinetic energy of $\mathcal{G}$ is

$$\tfrac{1}{2}A(\dot\theta^2 + \dot\phi^2\sin^2\theta) + \tfrac{1}{2}C(\dot\phi\cos\theta + \dot\psi)^2$$

where ϕ, θ, ψ are the Eulerian angles and A, A, C are the principal moments of inertia of $\mathcal{G}$ at C.

7.97 Find the kinetic energy of the wagon wheel in Problem 6.49 and use it to deduce the work done by the boy in getting it up to its final speed from rest.

7.98 A disk $\mathcal{D}$ of mass 10 kg and radius 25 cm is welded at a 45° angle to a vertical shaft $\mathcal{S}$. (See Figure P7.98.) The shaft is then spun up from rest to a constant angular speed $\omega_f = 10$ rad/s.

a. How much work is done in bringing the assembly up to speed?

b. Find the force and couple system acting on the plate at C after it is turning at the constant speed ω_f.

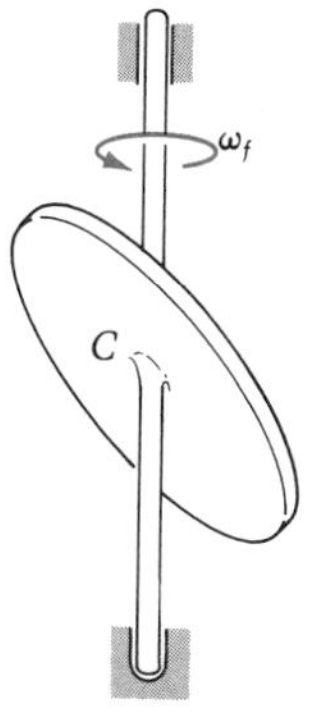

Figure P7.98

7.99 A thin rectangular plate (Figure P7.99) is brought up from rest to speed ω_0 about a horizontal axis Y.

a. Find the work that is done.

b. If two concentrated masses of $m/2$ each are added on the x_C axis, one on each side of the mass center, find their distances d from the mass center that will eliminate the bearing reactions.

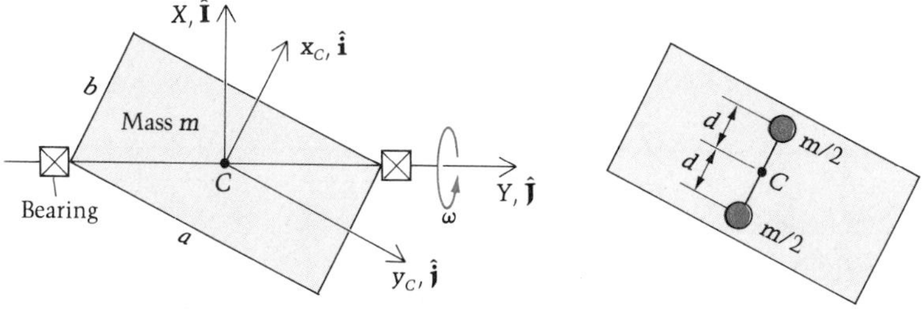

Figure P7.99

*** 7.100** The rigid body in Figure P7.100 consists of a disk $\mathcal{D}$ and rod $\mathcal{R}$, welded together perpendicularly as shown in the figure. If the body is spun up to angular speed ω_o about the z axis, how much work was done on it (excluding the overcoming of frictional resistance)?

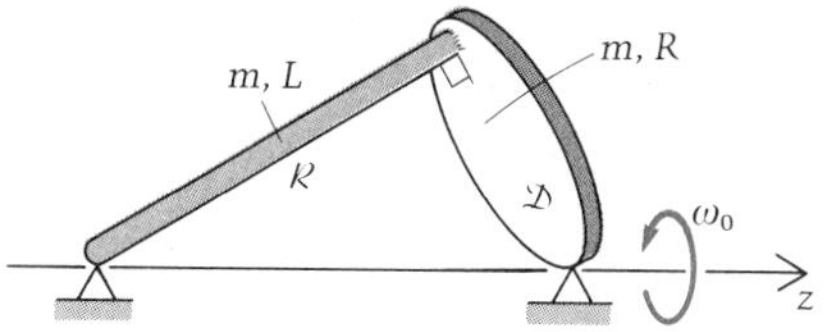

Figure P7.100

*** 7.101** Figure P7.101 shows a thin homogeneous triangular plate of mass m, base a, and height $2a$. It is welded to a light axle that can turn freely in bearings at A and B. Given:

$$I_{xx}^A = \frac{2ma^2}{3} \qquad I_{yy}^A = \frac{ma^2}{6} \qquad I_{xz}^A = I_{yz}^A = 0$$

$$I_{zz}^A = \frac{5ma^2}{6} \qquad I_{xy}^A = \frac{-ma^2}{6}$$

a. If the plate is turning at constant angular speed ω, find the torque that must be applied to the axle, and find the dynamic bearing reactions.

b. Find the principal axes at A and the principal moments of inertia there. Draw the axes on a sketch.

c. If possible, give the radius of a hole that, when drilled at C, will eliminate the bearing reactions. Give the answer in terms of m and ρt (density times thickness) of the plate.

d. Find the work done in bringing the plate up to speed ω from rest.

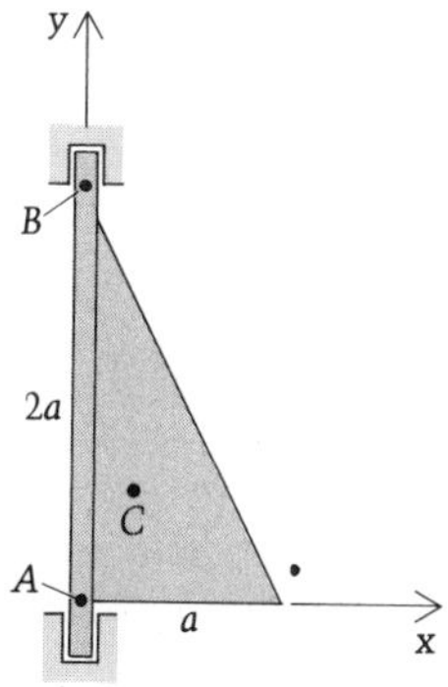

Figure P7.101

* **7.102** A thin equilateral triangular plate $\mathcal{P}$ of side s is welded to the vertical shaft at A in Figure P7.102. The shaft is brought up to speed ω_0 from rest by a motor.

a. How much work is done in bringing the system up to speed?

b. Find the force and couple system acting on the plate at A after it is turning at the speed ω_0 and the motor is turned off.

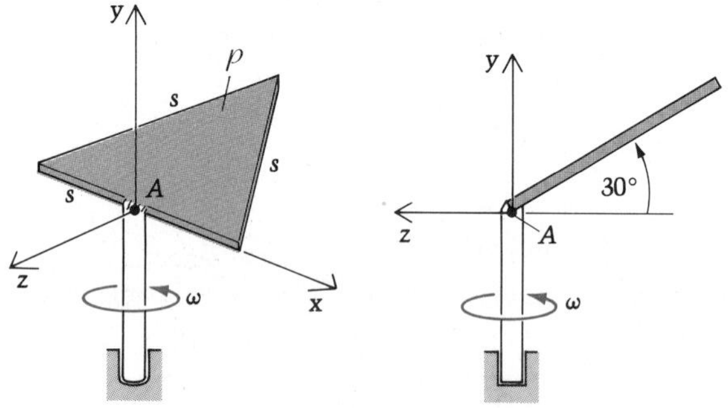

Figure P7.102

7.103 Two concentrated masses $m_1 = 10$ kg and $m_2 = 20$ kg are connected by a 15-kg slender rod m_3 of length 1.5 m. As shown in Figure P7.103, $(\hat{\mathbf{I}}, \hat{\mathbf{J}}, \hat{\mathbf{K}})$ are unit vectors fixed in direction in the inertial frame $\mathcal{I}$ and $(\hat{\mathbf{i}}, \hat{\mathbf{j}}, \hat{\mathbf{k}})$ are parallel to principal axes fixed at C in the combined body. At two times t_1 and t_2, the velocities of C and the angular velocities of the combined body are

$$\mathbf{v}_C(t_1) = \hat{\mathbf{I}} + 2\hat{\mathbf{J}} \text{ m/s} \qquad \boldsymbol{\omega}(t_1) = \hat{\mathbf{i}} + 2\hat{\mathbf{j}} - 4\hat{\mathbf{k}} \text{ rad/s}$$
$$\mathbf{v}_C(t_2) = 3\hat{\mathbf{J}} - 4\hat{\mathbf{K}} \text{ m/s} \qquad \boldsymbol{\omega}(t_2) = 3\hat{\mathbf{j}} - \hat{\mathbf{k}} \text{ rad/s}$$

Find the total work done on the system between t_1 and t_2.

7.104 Find the kinetic energy of the grinder in Problem 7.62. Is this equal to the work done by a motor on $\mathcal{S}$ which brings the system up to speed? (Neglect the masses of $\mathcal{S}$ and $\mathcal{B}$.)

* **7.105** A ring is welded to a rod at a point A as shown in Figure P7.105. The cross sections and densities of the rod and ring are the same. The combined body is released with a gentle nudge with end B of the rod connected to the smooth plane by a ball joint and with point A at its highest point as shown. At the instant when A reaches its lowest point, find the relationship between the horizontal and vertical angular velocity components of the body.

7.106 If in the preceding problem the plane is rough enough to prevent slipping, find the magnitude of the angular velocity when A reaches the floor.

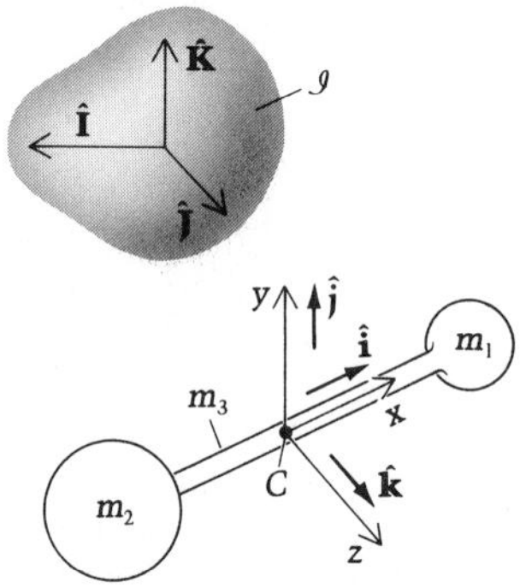

Figure P7.103

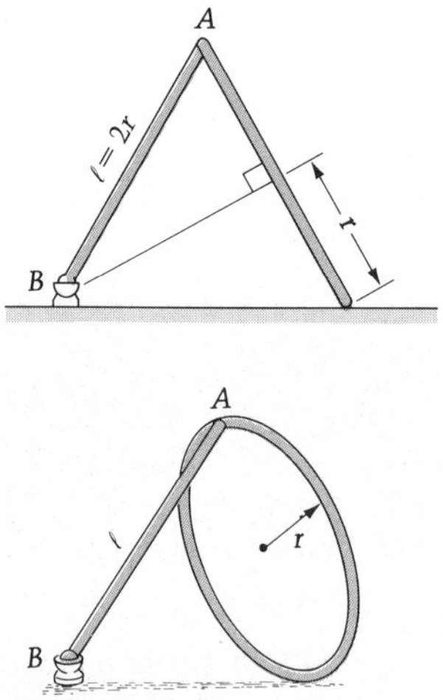

Figure P7.105

COMPUTER PROBLEM ▸ Chapter 7

* **7.107** Use a computer to generate data for a plot of maximum values of $V_Q^2/(gR)$ versus α in Example 7.10, for $0 < \alpha \leqq \pi/2$.

SUMMARY ▶ Chapter 7

In this chapter we have developed expressions for angular momentum of a rigid body in general three-dimensional motion. With respect to the mass center, it is

$$\begin{aligned}\mathbf{H}_C = &(I^C_{xx}\omega_x + I^C_{xy}\omega_y + I^C_{xz}\omega_z)\hat{\mathbf{i}} \\ &+ (I^C_{xy}\omega_x + I^C_{yy}\omega_y + I^C_{yz}\omega_z)\hat{\mathbf{j}} \\ &+ (I^C_{xz}\omega_x + I^C_{yz}\omega_y + I^C_{zz}\omega_z)\hat{\mathbf{k}}\end{aligned}$$

And if P is the location of a point of the body with zero velocity,

$$\begin{aligned}\mathbf{H}_P = &(I^P_{xx}\omega_x + I^P_{xy}\omega_y + I^P_{xz}\omega_z)\hat{\mathbf{i}} \\ &+ (I^P_{xy}\omega_x + I^P_{yy}\omega_y + I^P_{yz}\omega_z)\hat{\mathbf{j}} \\ &+ (I^P_{xz}\omega_x + I^P_{yz}\omega_y + I^P_{zz}\omega_z)\hat{\mathbf{k}}\end{aligned}$$

Transformation properties of moments and products of inertia include the parallel-axis theorems

$$\begin{aligned}I^P_{xx} &= I^C_{xx} + m(\bar{y}^2 + \bar{z}^2) \\ I^P_{yy} &= I^C_{yy} + m(\bar{z}^2 + \bar{x}^2) \\ I^P_{zz} &= I^C_{zz} + m(\bar{x}^2 + \bar{y}^2) \\ I^P_{xy} &= I^C_{xy} - m\bar{x}\bar{y} \\ I^P_{yz} &= I^C_{yz} - m\bar{y}\bar{z} \\ I^P_{zx} &= I^C_{zx} - m\bar{z}\bar{x}\end{aligned}$$

together with formulas for obtaining the moments and products of inertia associated with axes through a point when those properties are known for other axes through the point:

$$\begin{aligned}I^P_{x'x'} &= \ell_x^2 I^P_{xx} + \ell_y^2 I^P_{yy} + \ell_z^2 I^P_{zz} + 2\ell_x\ell_y I^P_{xy} + 2\ell_x\ell_z I^P_{xz} + 2\ell_y\ell_z I^P_{yz} \\ I^P_{x'y'} &= \ell_x n_x I^P_{xx} + \ell_y n_y I^P_{yy} + \ell_z n_z I^P_{zz} + (\ell_x n_y + \ell_y n_x) I^P_{xy} \\ &\quad + (\ell_x n_z + \ell_z n_x) I^P_{xz} + (\ell_y n_z + \ell_z n_y) I^P_{yz}\end{aligned}$$

In these two equations ℓ_x, ℓ_y, ℓ_z and n_x, n_y, n_z are respectively the direction cosines of x' and y' relative to axes x, y, and z.

Principal axes of inertia are very important and have the key property that were a body to rotate about a principal axis at a point P, then the angular momentum with respect to P would be in the same direction as the angular velocity, or

$$\mathbf{H}_P = I\boldsymbol{\omega}$$

where I is the moment of inertia about the principal axis, and is called a principal moment of inertia.

All products of inertia associated with a principal axis vanish, and at any point there are three mutually perpendicular principal axes. The largest and smallest of the principal moments of inertia are the largest and smallest of all the moments of inertia about axes through the point.

Some important special cases are:

1. If P lies in a plane of symmetry of the body, then the axis through P and perpendicular to the plane is a principal axis.
2. If P lies on an axis of symmetry of the body, then that axis and every line through P and perpendicular to it is a principal axis. Furthermore, the moments of inertia about these transverse axes through a given point are all the same.
3. If P is a point of spherical symmetry, e.g., the center of a uniform sphere, then every line through P is a principal axis and all of the corresponding principal moments of inertia are equal.

The most convenient form of Euler's second law, $\Sigma \mathbf{M}_C = \dot{\mathbf{H}}_C$, to use in a particular problem is often dependent on the problem. When body-fixed principal axes are used for reference, then we have what are usually referred to as the Euler equations:

$$\Sigma M_{C_x} = I^C_{xx}\,\dot{\omega}_x - (I^C_{yy} - I^C_{zz})\,\omega_y\omega_z$$
$$\Sigma M_{C_y} = I^C_{yy}\,\dot{\omega}_y - (I^C_{zz} - I^C_{xx})\,\omega_z\omega_x$$
$$\Sigma M_{C_z} = I^C_{zz}\,\dot{\omega}_z - (I^C_{xx} - I^C_{yy})\,\omega_x\omega_y$$

However, it is very often more convenient to express the angular momentum in terms of its components parallel to reference axes associated with some intermediate frame of reference, say $\mathcal{I}$, which is neither the body itself nor the inertial frame $\mathcal{J}$, so that

$$\Sigma \mathbf{M}_C = {}^{\mathcal{I}}\dot{\mathbf{H}}_C + \boldsymbol{\omega}_{\mathcal{I}/\mathcal{J}} \times \mathbf{H}_C$$

Just as in the case of plane motion (Chapter 5), the work of external forces equals the change in kinetic energy for rigid bodies in general motion. In three-dimensional motion the kinetic energy, T, can be written in general as

$$T = \frac{m}{2}\,\mathbf{v}_C \cdot \mathbf{v}_C + \frac{1}{2}\,\boldsymbol{\omega} \cdot \mathbf{H}_C$$

The second term may be compactly written as

$$\frac{1}{2}\,I\omega^2$$

where I is the moment of inertia about the axis, through C, that is instantaneously aligned with $\boldsymbol{\omega}$.

REVIEW QUESTIONS ▶ Chapter 7

True or False?

1. Products of inertia associated with principal axes always vanish, but only at the mass center.
2. If the principal moments of inertia at a point are distinct, then the principal axes of inertia associated with them are orthogonal.

3. The maximum moment of inertia about any line through point P of rigid body $\mathcal{B}$ is the largest principal moment of inertia at P.
4. General motion is a much more difficult subject than plane motion. A major reason for this is that neither the kinematics nor kinetics differential equations governing the orientation motion of the body are linear.
5. If we solve the Euler equations (7.40), we immediately know the orientation of the rigid body in space.
6. The sun and the moon exert gravity torques on the earth, and they cause the axis of our planet to precess.
7. If at a certain instant the moment of inertia of the mass of body $\mathcal{B}$ about an axis through C parallel to the angular velocity vector is I, then the kinetic energy of $\mathcal{B}$ at that instant is $\frac{1}{2}mv_C^2 + \frac{1}{2}I\omega^2$.
8. The earth's lunisolar precession is the result of *both* the bulge at the equator *and* the tilt of the axis.
9. The kinetic energy lost during a collision of two bodies does not depend on the angular velocities of the bodies prior to impact.
10. The work-energy and impulse-momentum principles are general integrals of the equations of motion for a rigid body.
11. Sometimes it is better to use the products of inertia in $\Sigma \mathbf{M}_C = {}^{\mathcal{I}}\dot{\mathbf{H}}_C$ than to take the time to compute principal moments and axes of inertia so as to be able to utilize Euler's equations (7.40).
12. In steady precession with the nutation angle θ equaling $90°$, the spin vector always precesses away from the torque vector.

Answers: 1. F 2. T 3. T 4. T 5. F 6. T 7. T 8. T 9. F 10. T 11. T 12. F

8 Special Topics

8.1 Introduction

In this chapter, we examine three subjects which are of considerable practical importance in Dynamics. In the first of these special topics, we introduce the reader to the subject of vibrations, limiting the presentation to a single degree of freedom (in which the oscillatory motion can be described by just one coordinate).

The second special topic deals with problems in which mass is continuously leaving and / or entering a region of space known as a control volume. A rocket is a good example: as the fuel is burned and the combustion products are ejected from a control volume enveloping the rocket, its momentum changes and it is propelled through the atmosphere. Euler's laws still apply, though the resulting equations are a bit more complicated than they were for the "constant mass" particles and bodies of earlier chapters.

The final topic in the chapter is central force motion, the most common example of which is that of orbits—such as the motions of planets around the sun, and of the moon and of man-made satellites around the earth.

The topics of Sections 8.2, 3, and 4 all stand alone, and can be read and understood after the reader has mastered Chapters 1 and 2, except for some of the problems in Section 8.2 in which the moment equation for rigid bodies in plane motion from Chapter 4 is also needed.

8.2 Introduction to Vibrations

Vibration is a term used to describe oscillatory motions of a body or system of bodies. These motions may be caused by isolated disturbances as when the wheel of an automobile strikes a bump or by fluctuating forces as in the case of the fuselage panels in an airplane vibrating in response to engine noise. Similarly, the oscillatory ground motions resulting from an earthquake cause vibrations of buildings. In each of these cases the undesirable motion may cause discomfort to occupants; moreover, the oscillating stresses induced within the body may lead to a fatigue failure of the structure, vehicle, or machine.

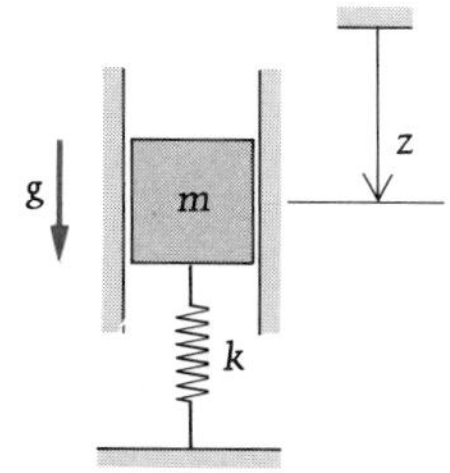

Figure 8.1

Free Vibration

For perhaps the simplest example of a mechanical oscillator consider the rigid block and linear spring shown in Figure 8.1. The block is constrained to translate vertically; thus a single parameter (scalar) is sufficient to establish position and hence the system is called a **single-degree-of-freedom system**. We choose z to be the parameter and let $z = 0$ correspond to the configuration in which the spring is neither stretched nor compressed.

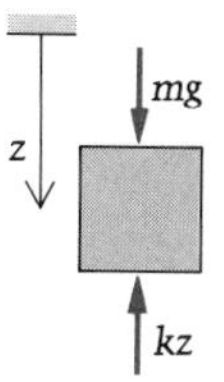

Figure 8.2

Using a free-body diagram of the block in an arbitrary position (Figure 8.2), Euler's first law yields

$$m\ddot{z} = mg - kz$$

or

$$m\ddot{z} + kz = mg \tag{8.1}$$

which is a second-order linear differential equation with constant coefficients describing the motion of the block. The fact that the differential equation is nonhomogeneous (the right-hand side is not zero) is a consequence of our choice of datum for the displacement parameter z. For if we make the substitution $y = z - mg/k$, the governing equation (8.1) becomes

$$m\ddot{y} + ky = 0 \tag{8.2}$$

which is a *homogeneous* differential equation. It is not a coincidence that this occurs when the displacement variable is chosen so that it vanishes when the block is in the equilibrium configuration—that is, when the spring is compressed mg/k.

Motion described by an equation such as (8.2) is called a **free vibration** since there is no external force (external, that is, to the spring-mass system) stimulating it.

Rewriting Equation (8.2), we obtain

$$\ddot{y} + \frac{k}{m} y = 0$$

or, defining $\omega_n = \sqrt{k/m}$,

$$\ddot{y} + \omega_n^2 y = 0 \tag{8.3}$$

which has as its general solution

$$y = A \sin \omega_n t + B \cos \omega_n t \tag{8.4}$$

or

$$y = C \sin(\omega_n t + \varphi) \tag{8.5}$$

where

$$C = \sqrt{A^2 + B^2} \qquad \text{and} \qquad \tan \varphi = \frac{B}{A}$$

Whether expressed in the form of (8.4) or (8.5), y is called a **simple harmonic** function of time, ω_n is called the **natural circular frequency**, C is called the **amplitude** of the displacement y, and φ is said to be the **phase angle** by which y *leads* the reference function, $\sin \omega_n t$. The simple harmonic function is *periodic* and its **period** is $\tau_n = 2\pi/\omega_n$. Another quantity called **frequency** is $f_n = 1/\tau_n = \omega_n/2\pi$, which gives the number of cycles in a unit of time. When the unit of time is the second, the unit for f_n is the hertz (Hz); 1 Hz is 1 cycle per second.

The constants A and B in (8.4), or equivalently C and φ in (8.5), are determined from initial conditions of position and velocity. Thus if

$$y(0) = y_0$$

and

$$\dot{y}(0) = v_0$$

then

$$B = y_0$$

and

$$A = \frac{v_0}{\omega_n}$$

Now let us investigate what might seem an entirely different situation—that of a rigid body constrained to rotate about a fixed horizontal axis (through O as in Figure 8.3). Since the only kinematic freedom the body has is that of rotation, a single angle is sufficient to describe a configuration of the body. Let the angle be θ as shown, where we note that when $\theta = 0$ the mass center C is located directly below the pivot O.

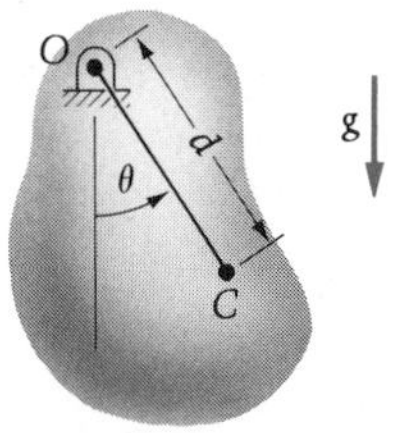

Figure 8.3

Neglecting any friction at the axis of rotation, the free-body diagram appropriate to an arbitrary instant during the motion is shown in Figure 8.4. Summing moments about the axis of rotation, we get

$$-mg\, d \sin\theta = I_0 \ddot{\theta} \tag{8.6}$$

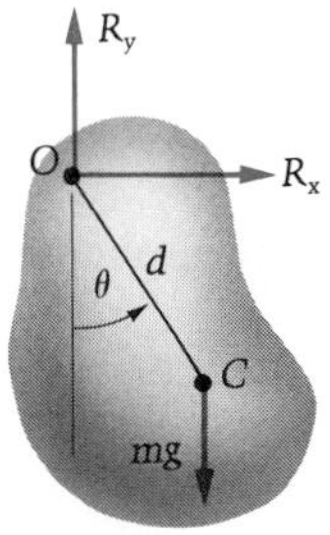

Figure 8.4

where I_0 is the mass moment of inertia about the axis of rotation. Equation (8.6) is a nonlinear differential equation because $\sin\theta$ is a nonlinear function of θ, but if we restrict our attention to sufficiently small angles so that $\sin\theta \approx \theta$, Equation (8.6) becomes

$$I_0 \ddot{\theta} + (mg\, d)\theta = 0 \tag{8.7}$$

That is, θ is a simple harmonic function:

$$\theta = A \sin \omega_n t + B \cos \omega_n t$$

where now

$$\omega_n^2 = \frac{mg\, d}{I_0}$$

The two preceding examples have an important feature in common: Motion near the equilibrium configuration is governed by a homogeneous, second-order, linear differential equation with constant coefficients, and in each case the motion is simple harmonic. A point of difference is that in the block-spring case the gravitational field plays no role other than establishing the equilibrium configuration; in particular the natural frequency does not depend on the strength (g) of the field. In the second case where the body is basically behaving as a pendulum, the gravitational field provides the "restoring action" and the natural frequency is proportional to $\sqrt{g}$.

EXAMPLE 8.1

Find the natural frequency of small oscillations about the equilibrium position of a uniform ball (sphere) rolling on a cylindrical surface.

Solution

Let m be the mass of the ball, let R be the radius of the path of its center, and let θ be the polar coordinate angle locating the center as shown in Figure E8.1a. Thus

$$\mathbf{a}_C = -R\dot{\theta}^2\hat{\mathbf{e}}_R + R\ddot{\theta}\hat{\mathbf{e}}_\theta$$

and the angular acceleration of the ball is $\boldsymbol{\alpha} = -(R\ddot{\theta}/r)\hat{\mathbf{k}}$ because of the no-slip condition. We shall now use $\mathbf{a}_C$ and $\boldsymbol{\alpha}$ in the equations of motion:

$$\Sigma\mathbf{F} = m\mathbf{a}_C \tag{1}$$

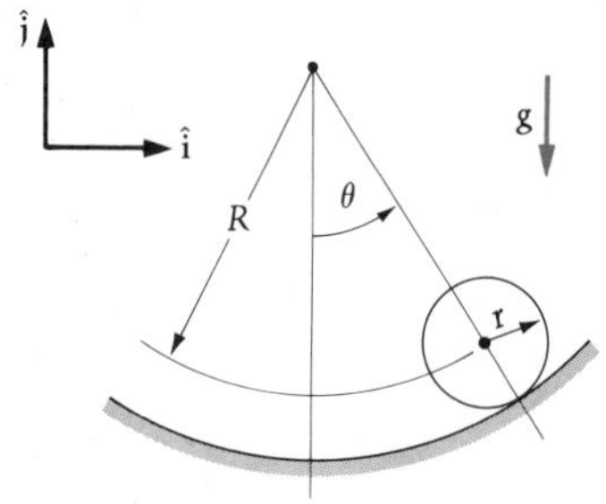

Figure E8.1a

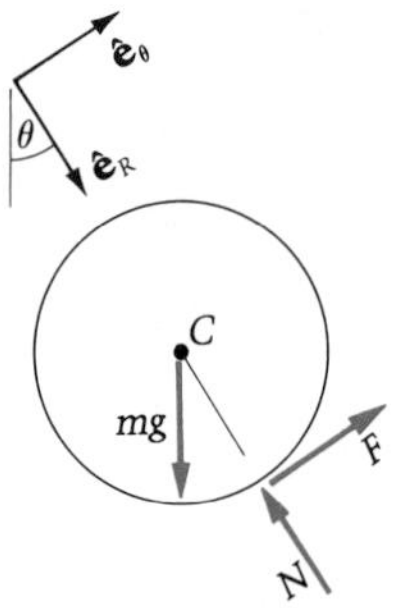

Figure E8.1b

Hence, from the free-body diagram shown in Figure E8.1b, the $\hat{\mathbf{e}}_R$ and $\hat{\mathbf{e}}_\theta$ component equations of (1) are:

$$F - mg\sin\theta = mR\ddot{\theta} \tag{2}$$

and

$$N - mg\cos\theta = mR\dot{\theta}^2 \tag{3}$$

Also, from summing moments about C, we have

$$Fr = \left(\frac{2}{5}mr^2\right)\left(-\frac{R\ddot{\theta}}{r}\right)$$

or

$$F = -\frac{2}{5}mR\ddot{\theta} \tag{4}$$

Eliminating the friction force F between Equations (2) and (4), we obtain the differential equation

$$\frac{7}{5}mR\ddot{\theta} + mg\sin\theta = 0$$

For small θ so that $\sin\theta \approx \theta$,

$$\frac{7}{5}R\ddot{\theta} + g\theta = 0$$

from which we see that

$$\omega_n^2 = \frac{5g}{7R}$$

or

$$\omega_n = 0.845\sqrt{\frac{g}{R}}$$

Damped Vibration

The simple harmonic motion in our examples of free vibration has a feature that conflicts with our experience in the real world; that is, the motion calculated persists forever unabated. Intuition would suggest decaying oscillations and finally the body coming to rest. Of course the problem here is that we have not incorporated any mechanism for energy dissipation in the analytical model. To do that, we shall return to the simple block-spring system and introduce a new element: a viscous damper (Figure 8.5). The *rate* of extension of this element is proportional to the force applied, through a damping constant c, so that the force is c times the rate of extension.

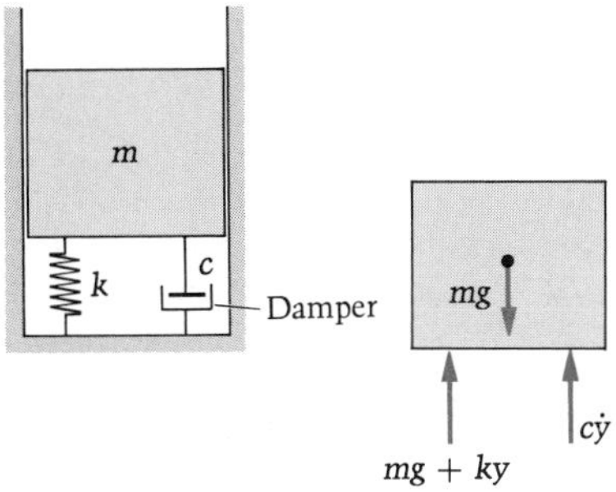

Figure 8.5

Referring to the free-body diagram in Figure 8.5 and letting $y = 0$ designate the equilibrium position as before, we have

$$m\ddot{y} = mg - (mg + ky) - c\dot{y}$$

or

$$m\ddot{y} + c\dot{y} + ky = 0 \tag{8.8}$$

The appearance of the $c\dot{y}$ term in (8.8) has a profound effect on the solution to the differential equation and hence on the motion being described. Solutions to (8.8) may be found from

$$y = Ae^{rt} \tag{8.9}$$

where A is an arbitrary constant and r is a characteristic parameter. Substituting (8.9) into (8.8), we obtain

$$(mr^2 + cr + k)Ae^{rt} = 0 \tag{8.10}$$

which is satisfied nontrivially (i.e., for $A \neq 0$) with

$$mr^2 + cr + k = 0 \tag{8.11}$$

This characteristic equation has two roots given by

$$r = -\frac{c}{2m} \pm \sqrt{\left(\frac{c}{2m}\right)^2 - \frac{k}{m}} \tag{8.12}$$

Except for the case in which $(c/2m)^2 = k/m$, the roots are distinct; if we call them r_1 and r_2, then the general solution to (8.8) is

$$y = A_1e^{r_1t} + A_2e^{r_2t}$$

In the exceptional case $(c/2m)^2 = k/m$, there is only the one repeated root $r = -c/2m$, but direct substitution will verify that there is a solution to (8.8) of the form $te^{-(c/2m)t}$ so that the general solution in that case is

$$y = A_1 e^{-(c/2m)t} + A_2 t e^{-(c/2m)t} \tag{8.13}$$

With initial conditions

$$y(0) = y_0$$

and

$$\dot{y}(0) = v_0$$

we find that

$$A_1 = y_0$$

and

$$A_2 = v_0 + \left(\frac{c}{2m}\right) y_0$$

Since

$$\left(\frac{c}{2m}\right)^2 = \frac{k}{m}$$
$$= \omega_n^2$$

the solution is

$$y = e^{-\omega_n t}[y_0 + (v_0 + \omega_n y_0)t] \tag{8.14}$$

Displacements given by (8.14) are plotted in Figure 8.6 for several representative sets of initial conditions (positive y_0 but positive and negative v_0). Two features of the motion are apparent:

1. $y \to 0$ (the equilibrium position) as $t \to \infty$.
2. The motion is not oscillatory; the equilibrium position is "overshot" at most once and only then when the initial speed is sufficiently large and in the direction opposite to that of the initial displacement.

In the case we have just studied, the damping is called **critical damping**, because it separates two quite different mathematical solu-

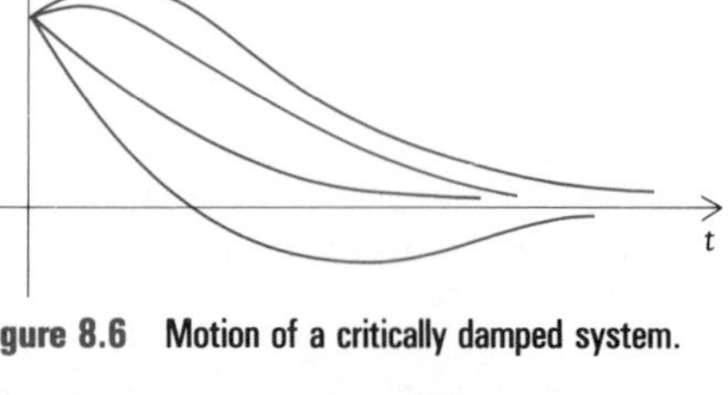

Figure 8.6 Motion of a critically damped system.

tions: For greater damping the roots of the characteristic equation (8.11) are both real and negative, and for small damping the roots are complex conjugates. If we let the critical damping be denoted by c_{crit} then we have seen that

$$\begin{aligned} c_{crit} &= 2\sqrt{km} \\ &= 2m\omega_n \end{aligned} \tag{8.15}$$

Now let us consider the case for which $c > c_{crit}$; the mechanical system is then said to be **overdamped** or the damping is said to be **supercritical**. In this case the roots given by (8.12) are both real and negative since $(c/2m)^2 > k/m$; if we call these roots $-a_1$ and $-a_2$, with $a_2 > a_1 > 0$, then the general solution to the differential equation of motion is

$$y = A_1 e^{-a_1 t} + A_2 e^{-a_2 t} \tag{8.16}$$

The motion described here is in no way qualitatively different from that for the case of critical damping, which we have just discussed. For a given set of initial conditions, Equation (8.16) yields a slower approach to $y = 0$ than does (8.13). That is, the overdamped motion is more "sluggish" than the critically damped motion as we would anticipate because of the greater damping.

Finally we consider the case in which the system is said to be **underdamped** or **subcritically** damped; that is, $c < c_{crit}$. The roots given by (8.12) are the complex conjugates

$$-\frac{c}{2m} \pm i\sqrt{\frac{k}{m} - \left(\frac{c}{2m}\right)^2}$$

where $i = \sqrt{-1}$. It is possible to express the general solution to the governing differential equation as

$$y = e^{-(c/2m)t}(A_1 \sin \omega_d t + A_2 \cos \omega_d t) \tag{8.17}$$

where $\omega_d = \sqrt{k/m - (c/2m)^2}$. A typical displacement history corresponding to (8.17) is shown in Figure 8.7. We note that, just as in the preceding cases, $y \to 0$ as $t \to \infty$; however, here the motion is oscillatory. We see that the simple harmonic motion obtained for the model without

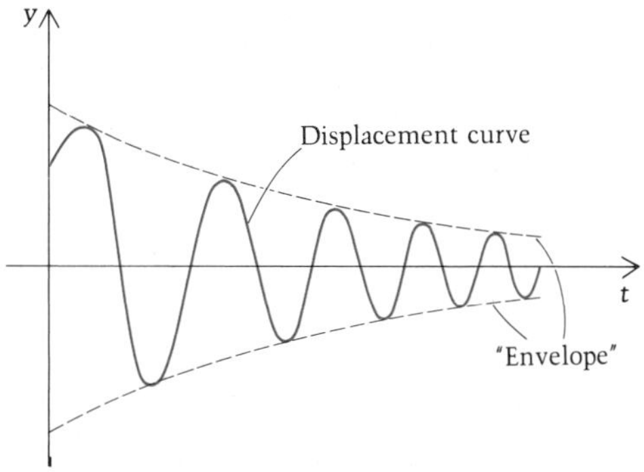

Figure 8.7 Motion of an underdamped system.

damping is given by (8.17) with $c = 0$. Moreover, we see that with light damping (small c) the analytical model that does not include damping adequately describes the motion during the first several oscillations. It is this case—subcritical damping—that is of greatest practical importance in studies of vibration.

EXAMPLE 8.2

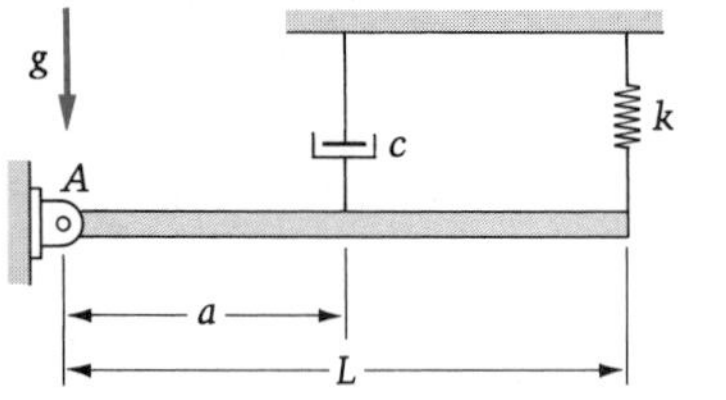

Figure E8.2a

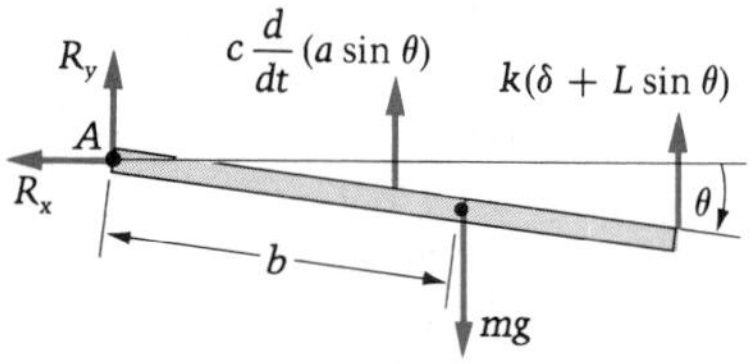

Figure E8.2b

Find the damping constant c that gives critical damping of the rigid bar executing motions near the equilibrium position shown in Figure E8.2a.

Solution

We are going to restrict our attention to small angles θ, and thus we may ignore any tilting of the damper or the spring. However, it may help us develop the equation of motion in an orderly way if we assume that the upper ends of the spring and damper slide along so that each remains vertical as the bar rotates through the angle θ. Without further restriction, if we sum moments about A with the help of the FBD in Figure E8.2b, we obtain, with I_A the moment of inertia of the mass of the bar about the axis of rotation at A,

$$I_A\ddot{\theta} = mg(b \cos \theta) - \left[c \frac{d}{dt}(a \sin \theta)\right](a \cos \theta) - k(\delta + L \sin \theta)(L \cos \theta) \quad (1)$$

where δ is the spring stretch at equilibrium. Thus for small θ (that is, $\sin \theta \approx \theta$, $\cos \theta \approx 1$) we linearize Equation (1) and obtain

$$I_A\ddot{\theta} = mgb - ca^2\dot{\theta} - kL\delta - kL^2\theta$$

Of course, $\theta = 0$ is the equilibrium configuration so that

$$mgb = kL\delta$$

The linear governing differential equation is then

$$I_A\ddot{\theta} + ca^2\dot{\theta} + kL^2\theta = 0$$

and for critical damping we get, associating the coefficients of $\theta, \dot{\theta}, \ddot{\theta}$ with those of $y, \dot{y}, \ddot{y}$ in Equation (8.8),

$$c_{\text{crit}}a^2 = 2\sqrt{(kL^2)I_A}$$

or

$$c_{\text{crit}} = 2\frac{L}{a^2}\sqrt{kI_A}$$

Any c less than this critical value will result in oscillations of decreasing amplitude.

Forced Vibration

Fluctuating external forces may have destructive effects on mechanical systems; this is perhaps the primary motivation for studying mechanical vibration. It is common for the external loading to be a periodic function of time, in which case the loading may be expressed as a series of simple

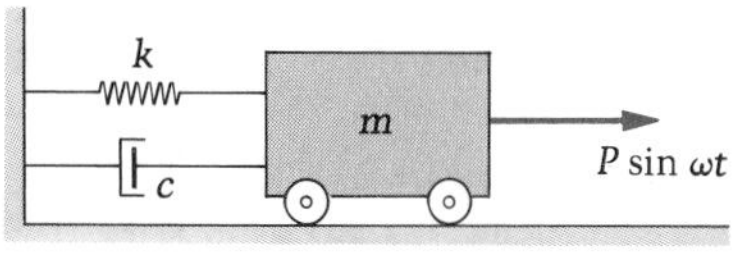

Figure 8.8

harmonic functions (Fourier series). Consequently it is instructive to consider the case in which the loading is simple-harmonic. For the mass-spring-damper system shown in Figure 8.8, the differential equation of motion is

$$m\ddot{x} + c\dot{x} + kx = P \sin \omega t \tag{8.18}$$

The general solution is composed of two parts: a *particular* solution (anything that satisfies the differential equation) and what is called the *complementary* solution (the general solution to the homogeneous differential equation). A particular solution of the form $x = X \sin(\omega t - \varphi)$ may be found. If we substitute this expression in (8.18) we obtain

$$-m\omega^2 X \sin(\omega t - \varphi) + c\omega X \cos(\omega t - \varphi) + kX \sin(\omega t - \varphi) = P \sin \omega t$$

or

$$(k - m\omega^2)X(\sin \omega t \cos \varphi - \sin \varphi \cos \omega t) + c\omega X(\cos \omega t \cos \varphi + \sin \omega t \sin \varphi) = P \sin \omega t$$

or

$$[(k - m\omega^2) \cos \varphi + c\omega \sin \varphi]X \sin \omega t - [(k - m\omega^2) \sin \varphi - c\omega \cos \varphi]X \cos \omega t = P \sin \omega t$$

For this to be satisfied at every instant of time,

$$[(k - m\omega^2) \cos \varphi + c\omega \sin \varphi]X = P \tag{8.19}$$

and

$$-c\omega \cos \varphi + (k - m\omega^2) \sin \varphi = 0 \tag{8.20}$$

From (8.20) we get

$$\tan \varphi = \frac{c\omega}{k - m\omega^2} \tag{8.21}$$

so that

$$\sin \varphi = \frac{c\omega}{\sqrt{(k - m\omega^2)^2 + (c\omega)^2}}$$

and

$$\cos \varphi = \frac{k - m\omega^2}{\sqrt{(k - m\omega^2)^2 + (c\omega)^2}}$$

Substituting these expressions for $\sin \varphi$ and $\cos \varphi$ into (8.19), we obtain

$$\left[\frac{(k - m\omega^2)^2}{\sqrt{(k - m\omega^2)^2 + (c\omega)^2}} + \frac{(c\omega)^2}{\sqrt{(k - m\omega^2)^2 + (c\omega)^2}}\right] X = P$$

so that

$$X = \frac{P}{\sqrt{(k - m\omega^2)^2 + (c\omega)^2}} \tag{8.22}$$

We may now write the complete solution to the differential equation (8.18):

$$x = x_c(t) + X\sin(\omega t - \varphi) \tag{8.23}$$

where x_c is the complementary solution and is one of the three cases enumerated in the preceding section. That is, the form of x_c depends on whether the system is overdamped, critically damped, or underdamped. However, in each of these cases the negative exponent causes the function to approach zero as time becomes large. Thus for large time x_c tends to zero and $x(t)$ tends to the particular solution. For this reason the simple-harmonic particular solution is called the **steady-state displacement**, since it represents the long-term behavior of the system.

We note that the steady-state motion is a simple harmonic function having amplitude X and lagging the excitation (force) function by the phase angle φ. We may put these in a convenient form by dividing numerator and denominator of (8.21) and (8.22) by k, so that

$$\tan\varphi = \frac{c\omega/k}{1 - \omega^2/\omega_n^2} \qquad \left(\text{where } \omega_n^2 = \frac{k}{m}\right) \tag{8.24}$$

and

$$X = \frac{P/k}{\sqrt{(1 - \omega^2/\omega_n^2)^2 + (c\omega/k)^2}} \tag{8.25}$$

Investigating the dimensionless quantity $c\omega/k$, we find

$$\frac{c\omega}{k} = \frac{c\omega}{m\omega_n^2} = \frac{2c}{2m\omega_n}\left(\frac{\omega}{\omega_n}\right)$$

But we know that $2m\omega_n = c_{crit}$, the critical damping, so if we let ζ be the damping ratio (c/c_{crit}),

$$\frac{c\omega}{k} = 2\zeta\frac{\omega}{\omega_n} \tag{8.26}$$

and

$$\tan\varphi = \frac{2\zeta(\omega/\omega_n)}{1 - \omega^2/\omega_n^2} \tag{8.27}$$

and

$$X = \frac{P/k}{\sqrt{(1 - \omega^2/\omega_n^2)^2 + [2\zeta(\omega/\omega_n)]^2}} \tag{8.28}$$

The phase angle φ and the dimensionless displacement amplitude kX/P are plotted against the frequency ratio ω/ω_n in Figures 8.9 and 8.10, respectively, for various values of the damping ratio ζ. We see that, with small damping, large amplitudes of displacement occur when the excitation frequency ω is near the natural frequency ω_n. This phenomenon is called **resonance**, and the desire to avoid it has led to the development of methods for estimating natural frequencies of mechanical sys-

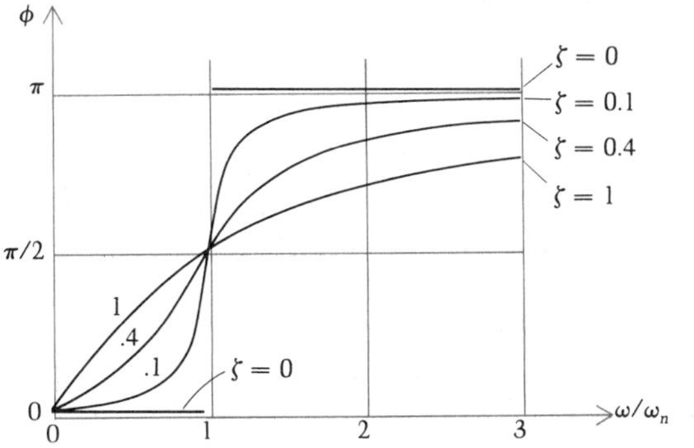

Figure 8.9

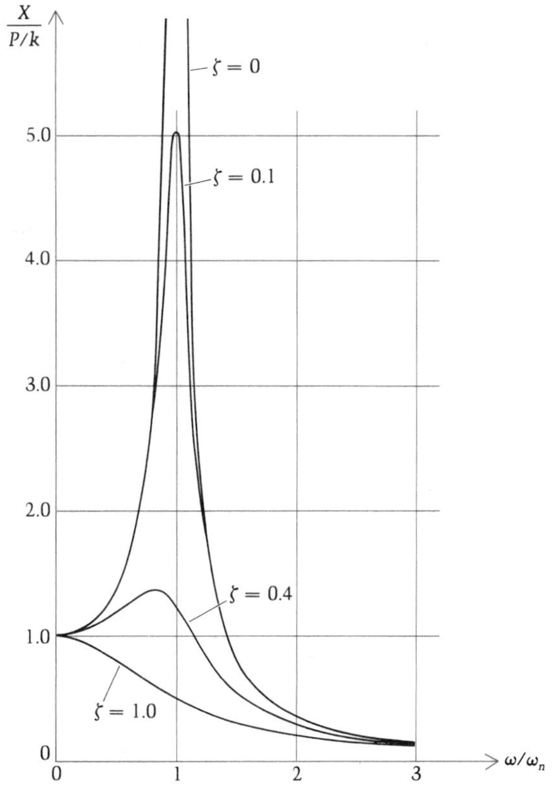

Figure 8.10

tems. Note that the steady-state response curves are insensitive to the damping for sufficiently small damping (say $\zeta < 0.1$) provided that we are not in the near vicinity of $\omega / \omega_n = 1$. This is an important observation because often in engineering practice we have reason to believe that the damping is small but we do not have accurate quantitative information about it.

We close this section by discussing the usual source of a simple harmonic external loading—an imbalance in a piece of rotating machinery. Let the machine be made up of two parts. The first, of mass m_1, is a rigid body constrained to rotate about an axis fixed in the second body (mass m_2), which translates relative to the inertial frame of reference. Let the mass center of the rotating body lie off the axis of rotation a distance e and let the body rotate at constant angular speed ω. (See Figure 8.11(a).)

Referring to the free-body diagram in Figure 8.11(b), we have

$$-kx - c\dot{x} = m_2\ddot{x} + m_1 \frac{d^2}{dt^2}(x + e \sin \omega t)$$

$$= (m_1 + m_2)\ddot{x} - m_1 e\omega^2 \sin \omega t$$

If we denote the total mass of the machine by m, then $m = m_1 + m_2$ and

$$m\ddot{x} + c\dot{x} + kx = m_1 e\omega^2 \sin \omega t$$

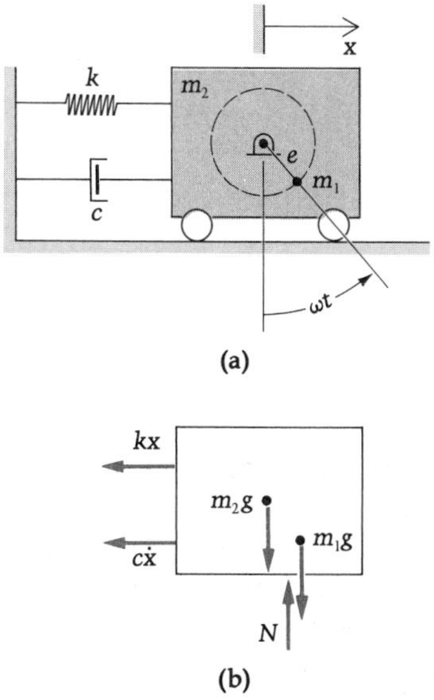

Figure 8.11

Thus the amplitude of the apparent "external" sinusoidal loading is $m_1 e\omega^2$ and its frequency is the angular speed of the rotating element.

EXAMPLE 8.3

A piece of machinery weighing 200 lb has a rotating element with imbalance ($m_1 e$ times the acceleration of gravity, which is $32.2 \times 12 = 386$ in./sec^2) 5 lb-in. and an operating speed of 1200 rpm. There are four springs, each of stiffness 1500 lb/in., supporting the machine whose frame is constrained to translate vertically. The damping ratio is $\zeta = 0.3$. Find the steady-state displacement of the frame.

Solution

The effective spring stiffness is

$$k = 4(1500) = 6000 \text{ lb/in.}$$

so that

$$\omega_n = \sqrt{\frac{6000}{200/386}}$$

$$= 108 \text{ rad/sec}$$

$$\omega = \frac{1200}{60}(2\pi)$$

$$= 126 \text{ rad/sec}$$

The effective external force amplitude is

$$P = m_1 e\omega^2 = \left(\frac{5}{386}\right)(126)^2 = 206 \text{ lb}$$

From Equation (8.28) we get

$$X = \frac{P/k}{\sqrt{(1 - \omega^2/\omega_n^2)^2 + (2\zeta\omega/\omega_n)^2}}$$

$$= \frac{206/6000}{\sqrt{[1 - (126/108)^2]^2 + [2(0.3)(126/108)]^2}}$$

$$= 0.0436 \text{ in.}$$

The phase angle φ is given by

$$\tan\varphi = \frac{2\zeta\omega/\omega_n}{1 - \omega^2/\omega_n^2}$$

$$= \frac{2(0.3)(126/108)}{1 - (126/108)^2} = -1.94$$

so that $\varphi = 205$ rad (117°).

EXAMPLE 8.4

The machine of Example 8.3 (weight = 200 lb, imbalance = 5 lb-in., operating speed = 1200 rpm) is to be supported by springs with negligible damping. If the machine were bolted directly to the floor, the amplitude of force transmitted to the floor would be

$$(m_1 e)\omega^2 = 206 \text{ lb}$$

What should the stiffness of the support system be so that the amplitude of the force transmitted to the floor is less than 20 lb?

Solution

The force exerted on the floor is transmitted through the supporting springs and is of amplitude kX, where X is the amplitude of displacement of the machine. From Equation (8.28) we have

$$kX = \frac{m_1 e\omega^2}{\sqrt{(1-\omega^2/\omega_n^2)^2 + (2\zeta\omega/\omega_n)^2}}$$

or with negligible damping (i.e., $\zeta \approx 0$)

$$kX = \frac{m_1 e\omega^2}{\sqrt{(1-\omega^2/\omega_n^2)^2}} = \frac{m_1 e\omega^2}{|1-\omega^2/\omega_n^2|}$$

Thus for

$$\frac{1}{|1-\omega^2/\omega_n^2|} = \frac{kX}{m_1 e\omega^2} < \frac{20}{206} = 0.0971$$

it is clear that $1-\omega^2/\omega_n^2$ is negative. Note that only when $\omega^2/\omega_n^2 > 2$ is

$$\frac{1}{|1-\omega^2/\omega_n^2|} < 1$$

Therefore we inquire into the condition for which

$$-\frac{1}{1-\omega^2/\omega_n^2} < -0.0971$$

or

$$\frac{\omega^2}{\omega_n^2} > 1 + \frac{1}{0.0971} = 11.3$$

or

$$\omega_n^2 < \frac{(126)^2}{11.3} = 1400$$

since $\omega = 126$ rad/sec. But

$$k = m\omega_n^2 < \frac{200}{386}(1400) = 725 \text{ lb/in.}$$

Thus to satisfy the given conditions the support stiffness must be *less than* 725 lb/in.

If the only springs available give a greater stiffness, the problem may be solved by increasing the mass; particularly we might mount the machine on a block of material, say concrete, and then support the machine and block by springs. For example, if the only springs available were those of Example 8.3 for which $k = 6000$ lb/in., then we need m to be *at least* that given by

$$m = \frac{k}{\omega_n^2} = \frac{6000}{1400}$$

$$= 4.29 \text{ lb-sec}^2/\text{in.}$$

for which the weight is

$$(4.29)(386) = 1660 \text{ lb}$$

Therefore we need a slab or block weighing

$$1660 - 200 = 1460 \text{ lb}$$

EXAMPLE 8.5

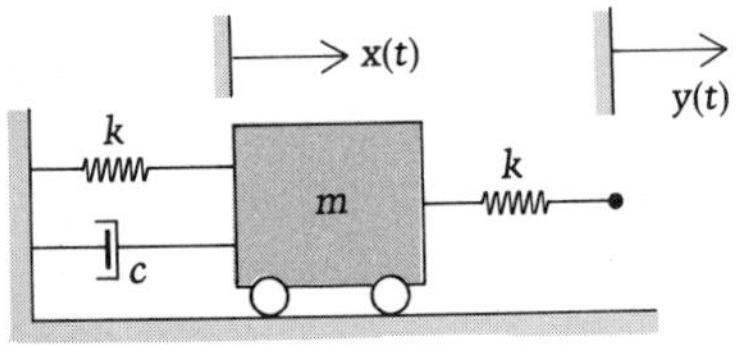

Figure E8.5

Find the steady-state displacement $x(t)$ of the mass in Figure E8.5 if $y(t) = 0.1 \cos 120t$ inch, where t is in seconds, $m = 0.01$ lb-sec²/in., $k = 100$ lb/in., and $c = 2$ lb-sec/in. In particular: (a) What is the amplitude of $x(t)$? (b) What is the angle by which $x(t)$ leads or lags $y(t)$?

Solution

The differential equation of motion of the mass is seen to be

$$m\ddot{x} = -kx - c\dot{x} - k(x - y)$$

or

$$m\ddot{x} + c\dot{x} + 2kx = kY \cos \omega t$$

where $Y = 0.1$ in. and $\omega = 120$ rad/sec. Using Equation (8.18), we see that kY is playing the same role as the oscillating force P, so that the steady-state amplitude is

$$X = \frac{kY}{\sqrt{(2k - m\omega^2)^2 + (c\omega)^2}}$$

$$= \frac{100(0.1)}{\sqrt{[200 - 0.01(120)^2]^2 + [2(120)]^2}}$$

or

$$X = 0.0406 \text{ in.}$$

The phase angle is

$$\varphi = \tan^{-1}\left(\frac{c\omega}{2k - m\omega^2}\right)$$

$$= \tan^{-1}\left[\frac{2(120)}{200 - 0.01(120^2)}\right]$$

$$= 76.9° \text{ or } 1.34 \text{ rad} \qquad \text{(lagging)}$$

Thus the steady-state motion is

$$x_{ss} = 0.0406 \cos(120t - 1.34) \text{ in.}$$

PROBLEMS ▶ Section 8.2

8.1 Find the frequency of small vibrations of the round wheel $\mathcal{C}$ as it rolls back and forth on the cylindrical surface in Figure P8.1. The radius of gyration of $\mathcal{C}$ with respect to the axis through C normal to the plane of the figure is k_C. Verify the result of Example 8.1 with your answer.

8.2–8.4 Find the equations of motion and periods of vibration of the systems shown in Figure P8.2 to P8.4. In each case, neglect the mass of the rigid bar to which the ball (particle) is attached.

8.5 The cylinder in Figure P8.5 is in equilibrium in the position shown. For no slipping, find the natural frequency of free vibration about this equilibrium position.

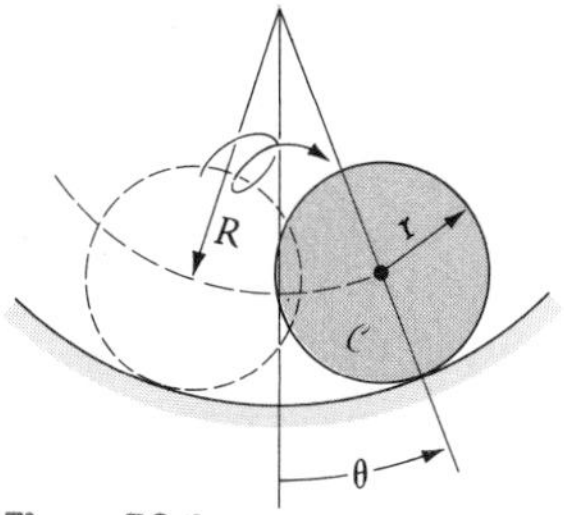

Figure P8.1

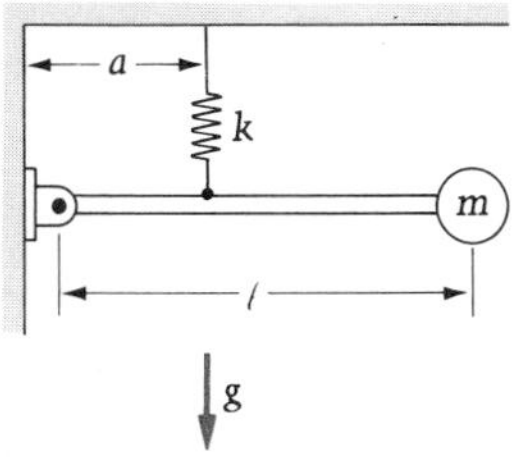

Figure P8.2

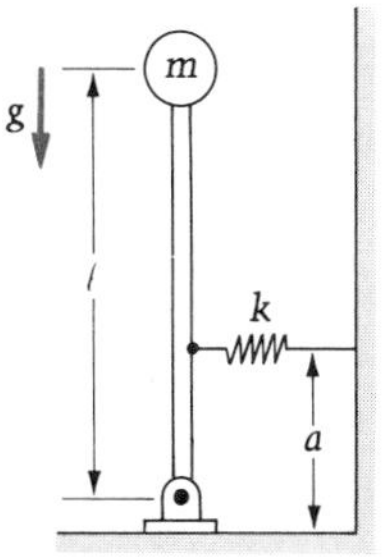

Figure P8.3

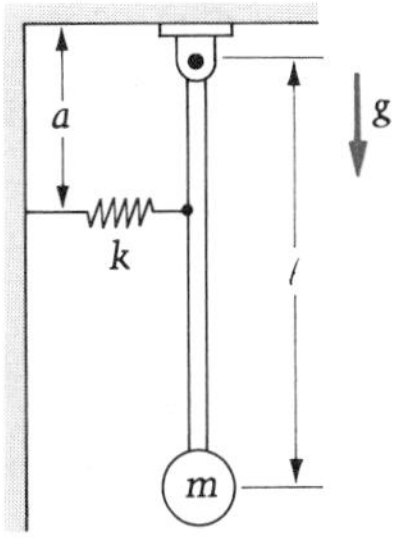

Figure P8.4

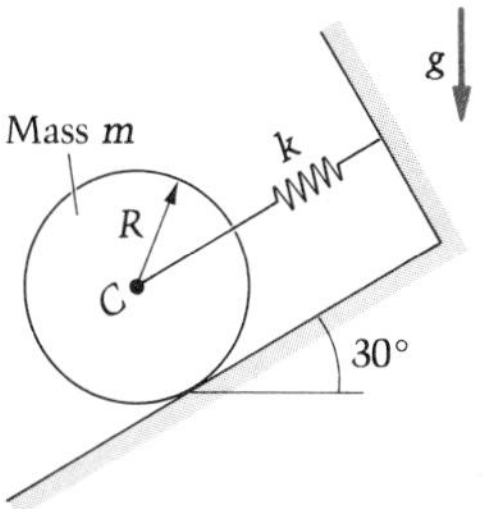

Figure P8.5

8.6 A uniform cylinder of mass m and radius R is floating in water. (See Figure P8.6.) The cylinder has a spring of modulus k attached to its top center point. If the specific weight of the water is γ, find the frequency of the vertical bobbing motion of the cylinder. *Hint*: The upward (buoyant) force on the bottom of the cylinder equals the weight of water displaced at any time (Archimedes' principle).

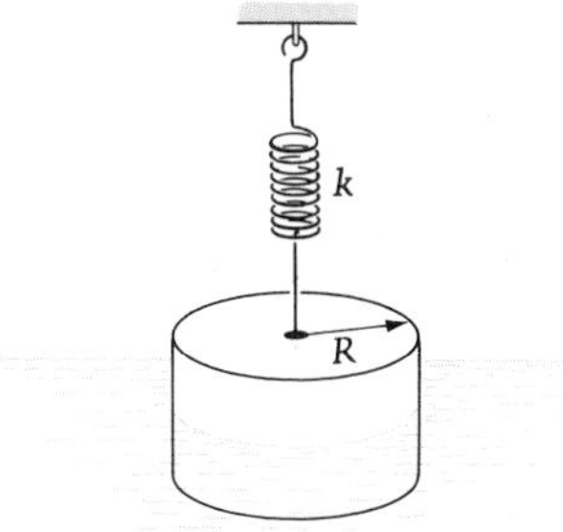

Figure P8.6

8.7 It is possible to determine experimentally the moments of inertia of large objects, such as the rocket shown in Figure P8.7. If the rocket is turned through a slight angle about z_C and released, for example, it oscillates with a period of 2.8 sec. Find the radius of gyration k_{z_C}.

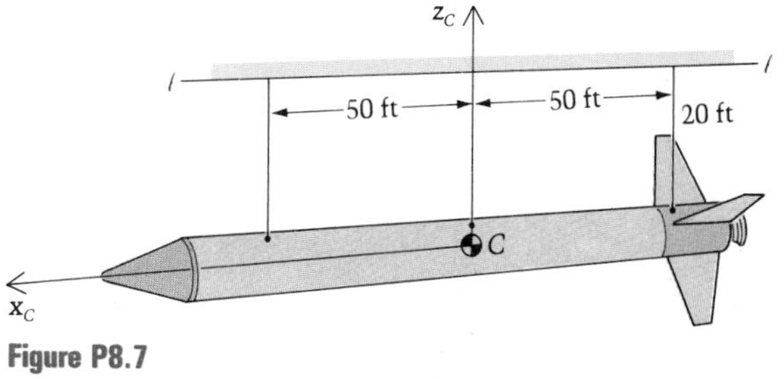

Figure P8.7

8.8 In the preceding problem, when the rocket is caused to swing with small angles about axis $\ell\ell$ as shown, the period is observed to be 8 sec. Find from this information the value of k_{x_C}.

8.9 Prove statements (1) and (2) on page 522.

* **8.10** Find the frequency of small amplitude oscillations of the uniform half-cylinder near the equilibrium position shown in Figure P8.10. Assume that the cylinder rolls on the horizontal plane.

8.11 A particle of mass m is attached to a light, taut string. The string is under tension, T, sufficiently large that the string is, for all practical purposes, straight when the system is in equilibrium as shown in Figure P8.11. Find the natural frequency of small transverse oscillations of the particle.

* **8.12** The masses in Figure P8.12 are connected by an inextensible string. Find the frequency of small oscillations if mass m is lowered slightly and released.

* **8.13** The solid homogeneous cylinder in Figure P8.13 weighs 200 lb and rolls on the horizontal plane. When the cylinder is at rest, the springs are each stretched 2 ft. The modulus of each spring is 15 lb / ft. The mass center C is given an initial velocity of $\frac{1}{2}$ ft / sec to the right.

- **a.** How far to the right will C go?
- **b.** How long will it take to get there?
- **c.** How long will it take to go halfway to the extreme position?

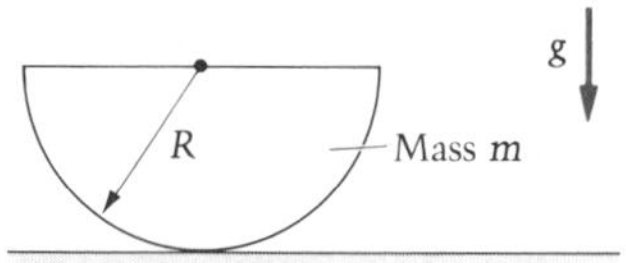

Figure P8.10

Figure P8.11

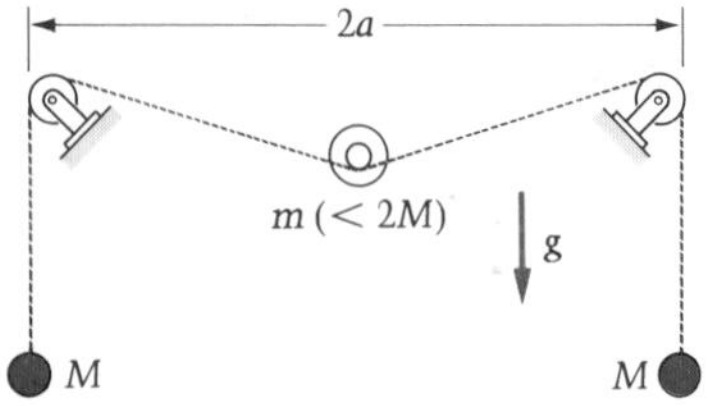

Figure P8.12

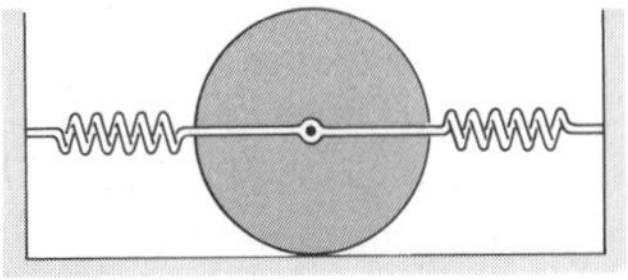

Figure P8.13

8.14 A sack of cement of mass m is to be dropped on the center of a simply supported beam as shown in Figure P8.14. Assume that the mass of the beam may be neglected, so that it may be treated as a simple linear spring of stiffness k. Estimate the maximum deflection at the center of the beam.

* **8.15** A particle P of mass m moves on a rough, horizontal rail with friction coefficient μ. (See Figure P8.15.) It is attached to a fixed point on the rail by a linear spring of modulus k. The initial stretch of the spring is $7\,\mu g m / k$. Describe the subsequent motion if it is known that the particle starts from rest. Show that the mass stops for good when $t = 3\pi / \sqrt{k/m}$. *Hint*: The differential equation doesn't have quite the form found in the text; also, every time the particle reverses direction, so does the friction force—thus the equation needs rewriting with each stop.

8.16 A spring with modulus 120 lb / in. supports a 200-lb block. (See Figure P8.16.) The block is fastened to the spring. A 400-lb downward force is applied to the top of the block at $t = 0$ when the block is at rest. Find the maximum deflection of the spring in the ensuing time.

* **8.17** A block weighing 1 lb is dropped from height $H = 0.1$ in. (See Figure P8.17.) If $k = 2.5$ lb / in., find the time interval for which the ends of the springs are in contact with the ground.

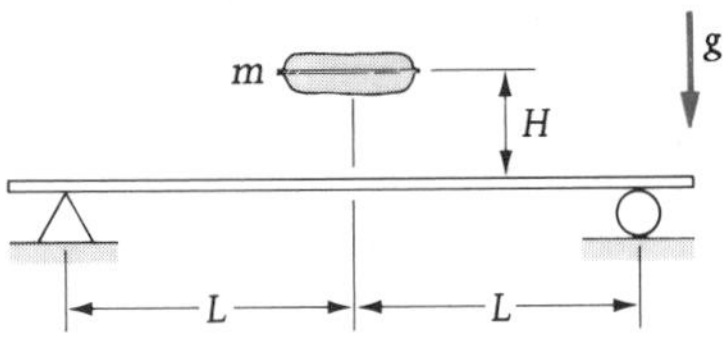

Figure P8.14

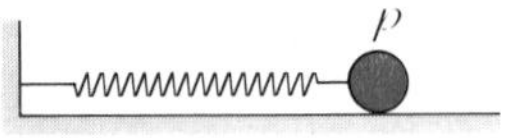

Figure P8.15

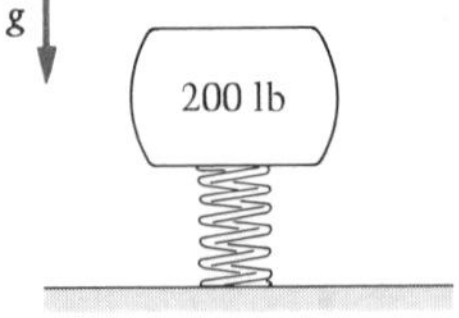

Figure P8.16

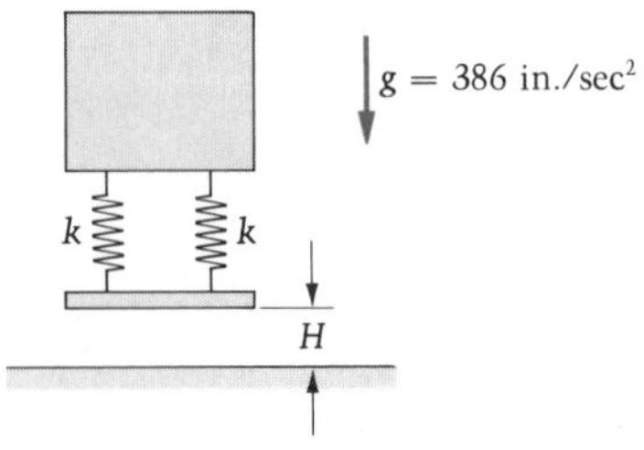

Figure P8.17

8.18 Assume that the slender rigid bar $\mathcal{B}$ in Figure P8.18 undergoes only small angles of rotation. Find the angle of rotation $\theta(t)$ if the bar is in equilibrium prior to $t = 0$, at which time the constant force P begins to traverse the bar at constant speed v.

• **8.19** Refer to the preceding problem: (a) Find the work done by P in traversing the bar $\mathcal{B}$; (b) show that this work equals the change in mechanical energy (which is the kinetic energy of $\mathcal{B}$ plus the potential energy stored in the spring).

• **8.20** The turntable in Figure P8.20 rotates in a horizontal plane at a *constant* angular speed ω. The particle P (mass $= m$) moves in the frictionless slot and is attached to the spring (modulus k, free length ℓ) as shown.

a. Derive the differential equation describing the motion $y(t)$ of the particle relative to the slot.

b. What is the extension of the spring such that P does not accelerate relative to the slot?

c. Suppose the motion is initiated with the spring unstretched and the particle at rest relative to the slot. Find the ensuing motion $y(t)$.

8.21 Find the value of c to give critical damping of the pendulum in Figure P8.21. Neglect the mass of the rigid bar to which the particle of mass m is attached.

8.22 If $k = 100$ lb/in. and the mass of the uniform, slender, rigid bar in Figure P8.22 is 0.03 lb-sec²/in., what damping constant c results in critical damping?

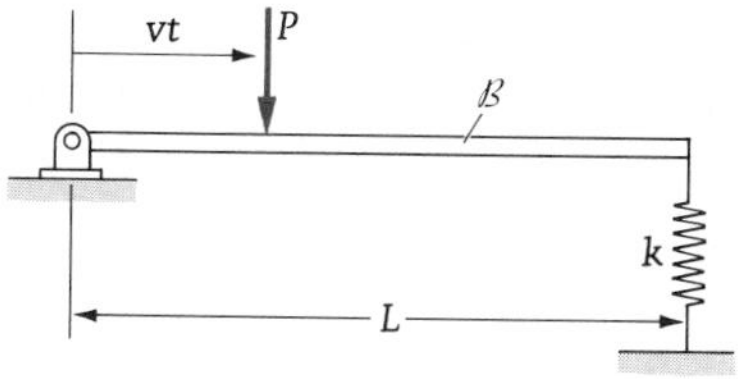

Figure P8.18

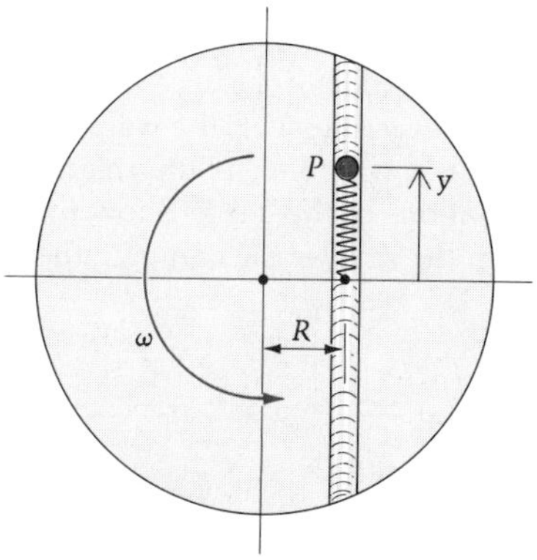

Figure P8.20

8.23 If $k = 100$ lb/in. and the mass of the uniform, slender, rigid bar in Figure P8.23 is 0.03 lb-sec²/in., what damping modulus c results in critical damping? Compare with the c from Problem 8.22. For this damping, find $\theta(t)$ if the bar is turned through a small angle θ_0 and then released from rest. If the dashpot were removed, what would be the period of free vibration?

8.24 A cannon weighing 1200 lb shoots a 100-lb cannonball at a velocity of 600 ft/sec. (See Figure P8.24.) It then immediately comes into contact with a spring of stiffness 149 lb/ft and a dashpot that is set up to critically damp the system. Assuming that there is no friction between the wheels and the plane, find the displacement toward the wall after $\frac{1}{2}$ sec has elapsed.

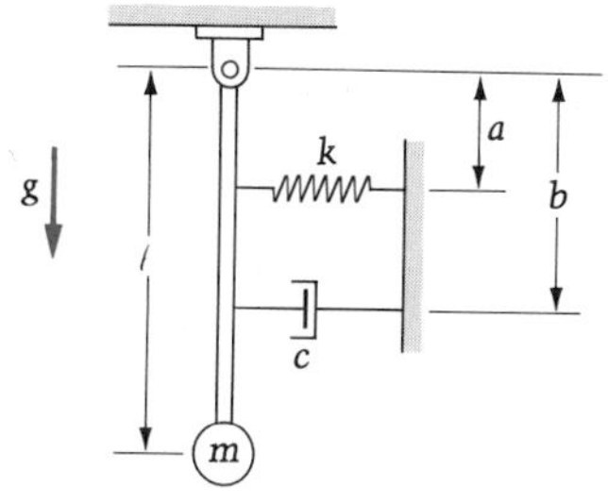

Figure P8.21

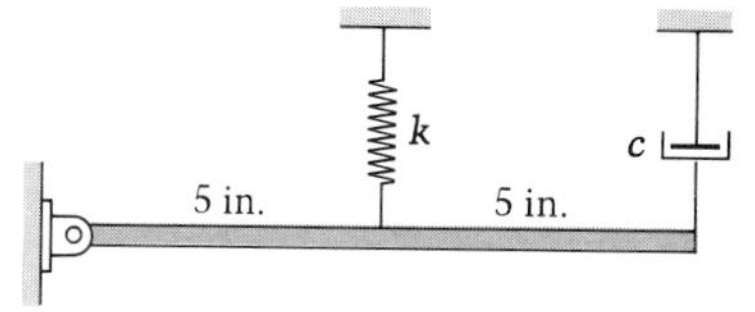

Figure P8.22

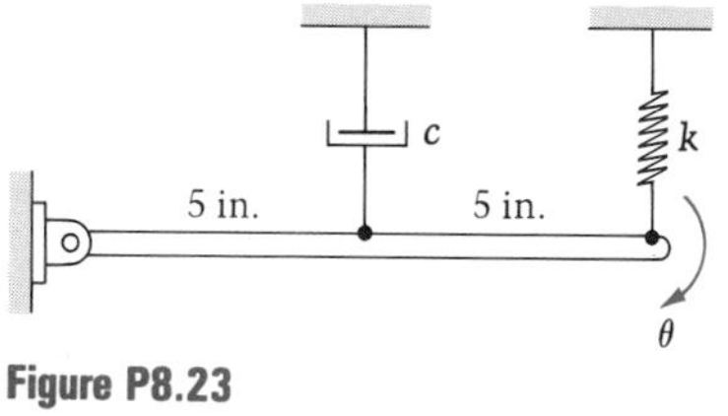

Figure P8.23

Figure P8.24

8.25 Consider free oscillations of a subcritically damped oscillator. Do local maxima in the response occur periodically?

8.26 In Figure P8.26, find the steady-state displacement $x(t)$ if $y(t) = 0.1 \sin 100t$ inch, where t is in seconds, $m = 0.01$ lb-sec^2/in., $k = 100$ lb/in., and $c = 2$ lb-sec/in. In particular:

a. What is the amplitude of $x(t)$?

b. What is the angle by which $x(t)$ leads or lags $y(t)$?

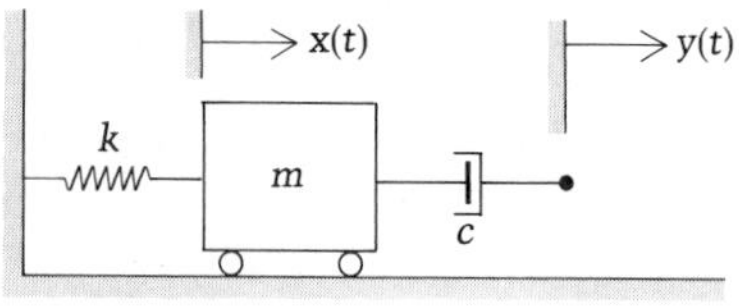

Figure P8.26

8.27 In Figure P8.27 find the steady-state displacement $x(t)$ if $y(t) = 0.2 \sin 90t$ inch, where t is in seconds, $m = 0.01$ lb-sec^2/in., $k = 50$ lb/in., and $c = 1$ lb-sec/in. In particular:

a. What is the amplitude of $x(t)$?

b. What is the phase angle by which $x(t)$ leads or lags $y(t)$?

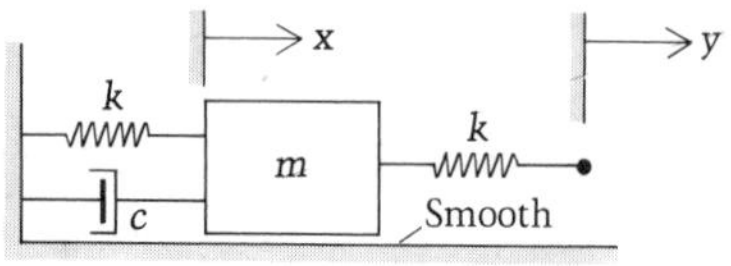

Figure P8.27

8.28 The cart in Figure P8.28 is at rest prior to $t = 0$, at which time the right end of the spring is given the motion $y = vt$, where v is a constant. Find $x(t)$.

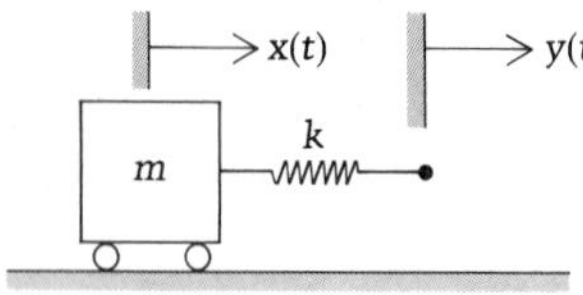

Figure P8.28

8.29 The block in Figure P8.29 is at rest in equilibrium prior to the application of the constant force $P = 50$ lb at $t = 0$. If $k = 100$ lb/in., $m = 0.01$ lb-sec^2/in., and the system is critically damped, find $x(t)$.

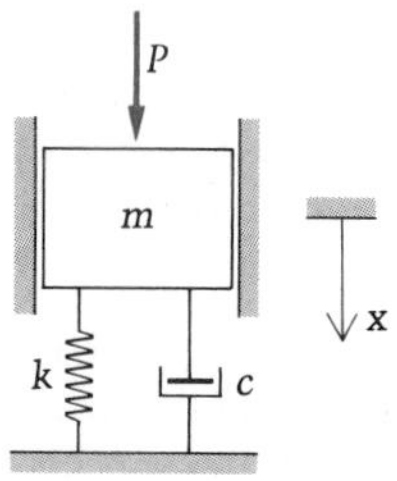

Figure P8.29

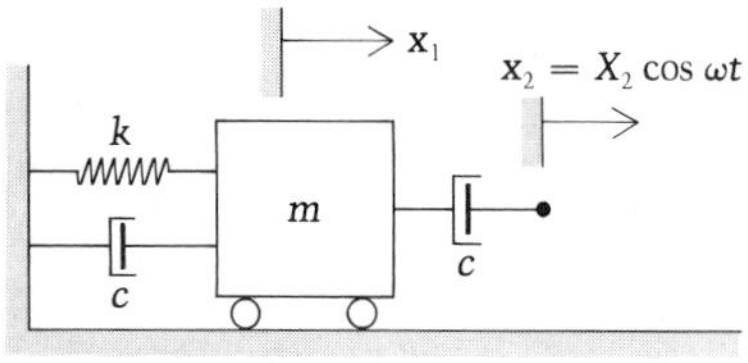

Figure P8.30

* **8.30** In Figure P8.30 find the response $x_1(t)$ for the initial conditions $x_1(0) = \dot{x}_1(0) = 0$ if

$$k = 100 \text{ lb/in.}$$
$$m = 0.01 \text{ lb-sec}^2\text{/in.}$$
$$c = 1.0 \text{ lb-sec/in.}$$
$$X_2 = 0.05 \text{ in.}$$
$$\omega = 100 \text{ rad/sec}$$

* **8.31** Repeat the preceding problem if (a) $c = 2.0$ lb-sec/in.; (b) $c = 0.5$ lb-sec/in.

8.32 Optical equipment is mounted on a table whose four legs are pneumatic springs. If the table and equipment together weigh 700 lb, what should be the stiffness of each spring so that the amplitude of steady, simple-harmonic, vertical displacement of the table will not be greater than 5 percent of a corresponding motion of the floor? The forcing frequency is 30 rad/sec. Neglect damping in your calculations.

* **8.33** The block of mass m in Figure P8.33 is mounted through springs k and damper c on a vibrating floor. Derive an expression for the steady-state acceleration of the block (whose motion is vertical translation). Show that the amplitude of the acceleration is less than that of the floor, regardless of the value of c, provided that $\omega > \sqrt{2}\omega_n$, where ω_n is the frequency of free undamped vibrations of the block. Show further that if $\omega > \sqrt{2}\omega_n$, then the smaller the damping the better the isolation.

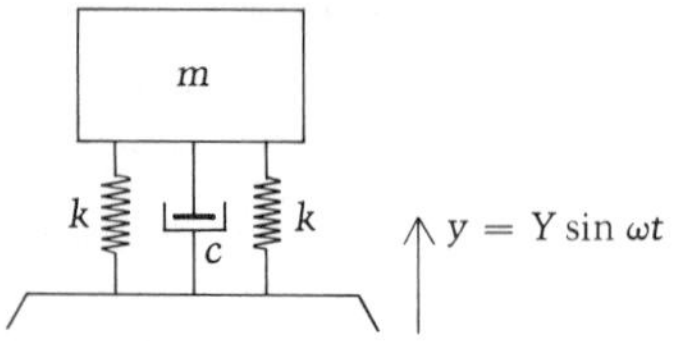

Figure P8.33

8.3 Euler's Laws for a Control Volume

Euler's laws describe the relationship between external forces and the motion of any body whether it be a solid, liquid, or gas. Sometimes, however, it is desirable to focus attention on some region of space (control volume) through which material may flow rather than on the fixed collection of particles that constitute a body. Examples of this sort are abundant in the field of fluid mechanics and include the important problem of describing and analyzing rocket-powered flight. Our purpose in this section is to discuss the forms taken by Euler's laws when the focus of attention is the control volume rather than the body.

We take as self-evident what might be called the "law of accumulation, production, and transport" — that is, the rate of accumulation of something within a region of space is equal to the rate of its production within the region plus the rate at which it is transported into the region.* Thus, for example, the rate of accumulation of peaches in Georgia equals the rate of production of peaches in the state plus the net rate at which they are shipped in. This idea can be applied in mechanics whenever we are dealing with a quantity whose measure for a body is the sum of the measures for the particles making up the body. Thus we can apply this principle to things such as mass, momentum, moment of momentum, and kinetic energy.

Suppose that at an instant a closed region V (control volume) contains material (particles) making up body $\mathcal{B}$. Let $m_{\mathcal{B}}$ denote the mass of body $\mathcal{B}$ and m_V denote the mass associated with V (that is, the mass of whatever particles happen to be in V at some time). Instantaneously $m_V = m_{\mathcal{B}}$, but because some of the material of $\mathcal{B}$ is flowing out of V and some other material is flowing in, $\dot{m}_V \neq \dot{m}_{\mathcal{B}}$. In fact by the accumulation principle stated above

$$\dot{m}_V = \dot{m}_{\mathcal{B}} + (\text{net rate of mass flow into } V) \tag{8.29}$$

since clearly $\dot{m}_V$ represents the rate of buildup (accumulation) of mass in V and since $\dot{m}_{\mathcal{B}}$, the rate of change of mass of the material instantaneously within V, represents the production term. Of course a body, being a specific collection of particles, has constant mass; thus $\dot{m}_{\mathcal{B}} = 0$ and (8.29) becomes $\dot{m}_V =$ (rate of mass flow into V), which is often called the **continuity equation**.

For momentum **L**, the statement corresponding to Equation (8.29) is

$$\dot{\mathbf{L}}_V = \dot{\mathbf{L}}_{\mathcal{B}} + (\textbf{net rate of flow of momentum into } V) \tag{8.30}$$

But Euler's first law applies to a body (such as $\mathcal{B}$) so that $\Sigma\mathbf{F} = \dot{\mathbf{L}}_{\mathcal{B}}$, where $\Sigma\mathbf{F}$ is the resultant of the external forces on $\mathcal{B}$ — or, in other words, the resultant of the external forces acting on the material instantaneously in V. Thus Equation (8.30) becomes

$$\dot{\mathbf{L}}_V = \Sigma\mathbf{F} + (\textbf{net rate of flow of momentum into } V) \tag{8.31}$$

* The mathematical statement of this is known as the Reynolds Transport Theorem.

which is the **control-volume form of Euler's first law.** The momentum flow rate on the right of (8.31) is calculated by summing up (or integrating) the momentum flow rates across infinitesimal elements of the boundary of V, where the momentum flow rate per unit of boundary area is the product of the mass flow rate per unit of area and the instantaneous velocity of the material as it crosses the boundary.

A similar derivation produces a **control-volume form of Euler's second law,** for which the result is

$$(\dot{\mathbf{H}}_O)_V = \Sigma \mathbf{M}_O + \begin{matrix}\textbf{(net rate of flow into } V \textbf{ of moment}\\ \textbf{of momentum with respect to } O\textbf{)}\end{matrix} \tag{8.32}$$

where O is a point fixed in the inertial frame of reference.

It is important to realize that nothing in our derivations here has restricted the control volume except that it be a closed region in space. It may be moving relative to the frame of reference in almost any imaginable way, and it may be changing in shape or volume with time. We conclude this section with examples of two of the most common applications of Equation (8.31).

EXAMPLE 8.6

A fluid undergoes steady flow in a pipeline and encounters a bend at which the cross-sectional area of pipe changes from A_1 to A_2. At inlet 1 the density is ρ_1 and the velocity (approximately uniform over the cross section) is $v_1\hat{\mathbf{i}}$. At outlet 2 the density is ρ_2. Find the resultant force exerted on the pipe bend by the fluid. (See Figure E8.6a.)

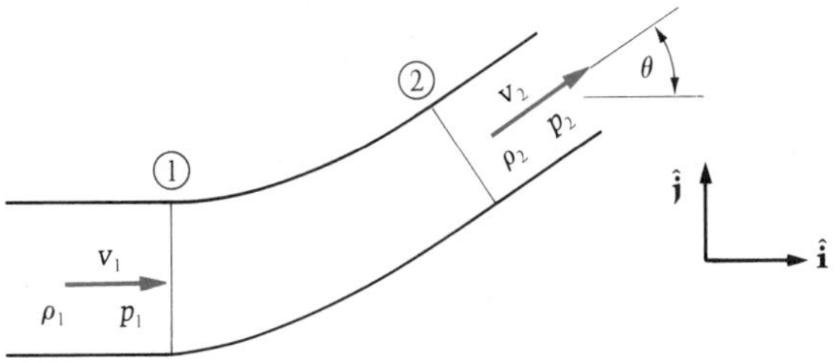

Figure E8.6a

Solution

Let the velocity of flow at the outlet be given by $v_2(\cos\theta\hat{\mathbf{i}} + \sin\theta\hat{\mathbf{j}})$. Then for steady flow the rate of mass flow at the inlet section is the same as that at the outlet section:

$$\rho_1 A_1 v_1 = \rho_2 A_2 v_2$$

so that

$$v_2 = \frac{\rho_1 A_1}{\rho_2 A_2} v_1$$

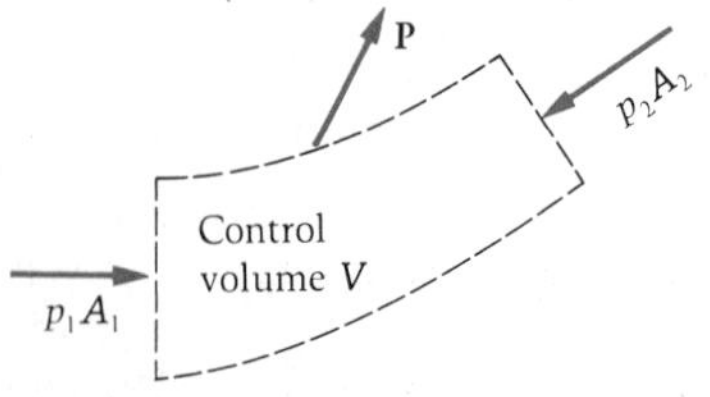

Figure E8.6b

Let the control volume be the region bounded by the inner surface of the pipe bend and the inlet and outlet cross sections. (See Figure 8.6b.) A consequence of the condition of steady flow is that within the control volume the distributions of velocity and density are independent of time. Thus the total momentum associated with V is a constant and

$$\frac{d\mathbf{L}_V}{dt} = 0$$

But

$$\frac{d\mathbf{L}_V}{dt} = \Sigma\mathbf{F} + (\text{net rate of momentum flow into } V)$$

Therefore

$$\begin{aligned}
\Sigma\mathbf{F} &= (\text{net rate of momentum flow out of } V) \\
&= \rho_2 A_2 v_2 (v_2 \cos\theta\hat{\mathbf{i}} + v_2 \sin\theta\hat{\mathbf{j}}) - \rho_1 A_1 v_1 (v_1 \hat{\mathbf{i}}) \\
&= \rho_1 A_1 v_1 (v_2 \cos\theta\hat{\mathbf{i}} + v_2 \sin\theta\hat{\mathbf{j}} - v_1 \hat{\mathbf{i}}) \\
&= \rho_1 A_1 v_1 \left[\left(\frac{\rho_1 A_1}{\rho_2 A_2} v_1 \cos\theta - v_1\right)\hat{\mathbf{i}} + \left(\frac{\rho_1 A_1}{\rho_2 A_2} v_1 \sin\theta\right)\hat{\mathbf{j}}\right] \\
&= \rho_1 A_1 v_1^2 \left[\left(\frac{\rho_1 A_1}{\rho_2 A_2} \cos\theta - 1\right)\hat{\mathbf{i}} + \frac{\rho_1 A_1}{\rho_2 A_2} \sin\theta\hat{\mathbf{j}}\right]
\end{aligned}$$

And if $\mathbf{P}$ is the force exerted on the fluid by the bend, then

$$\Sigma\mathbf{F} = \mathbf{P} + p_1 A_1 \hat{\mathbf{i}} + p_2 A_2 (-\cos\theta\hat{\mathbf{i}} - \sin\theta\hat{\mathbf{j}})$$

where p_1 and p_2 are the inlet and outlet fluid pressures respectively. Therefore

$$\mathbf{P} = \Sigma\mathbf{F} - p_1 A_1 \hat{\mathbf{i}} + p_2 A_2 (\cos\theta\hat{\mathbf{i}} + \sin\theta\hat{\mathbf{j}})$$

and the force exerted on the bend by the fluid is $-\mathbf{P}$ with

$$\begin{aligned}
-\mathbf{P} = &\left[p_1 A_1 - p_2 A_2 \cos\theta + \rho_1 A_1 v_1^2 \left(1 - \frac{\rho_1 A_1}{\rho_2 A_2} \cos\theta\right)\right]\hat{\mathbf{i}} \\
&- \left[p_2 A_2 \sin\theta + \rho_1 A_1 v_1^2 \left(\frac{\rho_1 A_1}{\rho_2 A_2}\right) \sin\theta\right]\hat{\mathbf{j}}
\end{aligned}$$

EXAMPLE 8.7

To illustrate how the control volume form of Euler's first law is used to describe the motion of a rocket vehicle, consider such a vehicle climbing in a vertical rectilinear flight. Let $v\hat{\mathbf{j}}$ be the velocity of the vehicle from which combustion products are being expelled at velocity $-v_e\hat{\mathbf{j}}$ relative to the rocket. Further let $M(t)$ be the mass at time t of the vehicle and its contents, let μ be the rate of mass flow of the ejected gases, and let p be the gas pressure at the nozzle exit of cross section A.

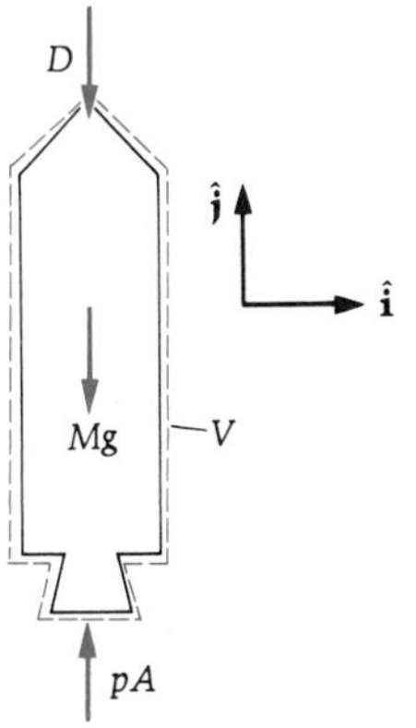

Figure E8.7 Free-body diagram of rocket.

Solution

Force D in the free-body diagram (Figure E8.7), representing the drag or resistance to motion, is the resultant of (1) all the shear stresses acting on the surface of the vehicle and (2) all the pressure on the surface. Thus $(pA - D)\hat{\mathbf{j}}$ represents the resultant of all the surface-distributed forces on the control volume. The two terms have been separated so that we may point out that the force pA remains even after the rocket has cleared the atmosphere. That is, p is pressure exerted by the gas particles *about to pass across* the nozzle exit plane* on the particles that have *just passed across,* and vice-versa.

If we let the control volume V surround the vehicle, Equation (8.31) becomes

$$\frac{d}{dt}\mathbf{L}_V = (pA - D)\hat{\mathbf{j}} - Mg\hat{\mathbf{j}} - \mu(v - v_e)\hat{\mathbf{j}}$$

At this point we must approximate $\mathbf{L}_V$ by $Mv\hat{\mathbf{j}}$; this is an approximation because some of the products of combustion inside the rocket are of course moving relative to the vehicle. Therefore

$$\frac{d}{dt}(Mv\hat{\mathbf{j}}) = (pA - D)\hat{\mathbf{j}} - Mg\hat{\mathbf{j}} - \mu(v - v_e)\hat{\mathbf{j}}$$

or

$$v\frac{dM}{dt} + M\frac{dv}{dt} = pA - D - Mg - \mu v + \mu v_e$$

But of course

$$\frac{dM}{dt} = -\mu$$

so that

$$M\frac{dv}{dt} = pA - D - Mg + \mu v_e$$

which is of the form of force = mass × acceleration, where one of the "forces" is the "thrust" μv_e.

* This term may be neglected if exhaust gases have expanded to atmospheric pressure or nearly so.

PROBLEMS ▶ Section 8.3

8.34 Let dm_i/dt and dm_o/dt be the respective rates at which mass enters and leaves a system. Show that Equation (8.31) may be expressed in terms of these rates as

$$\Sigma\mathbf{F} = m\frac{dv}{dt} + \left(\frac{dm_i}{dt} - \frac{dm_o}{dt}\right)\mathbf{v} + \frac{dm_o}{dt}\mathbf{v}_o - \frac{dm_i}{dt}\mathbf{v}_i$$

where $m\mathbf{v} = \mathbf{L}_V$ and it is assumed that all the incoming particles have a common velocity $\mathbf{v}_i$ (in an inertial frame) and that all the exiting particles have a common velocity $\mathbf{v}_o$.

8.35 Liquid of specific weight w flows out of a hole in the side of a tank in a jet of cross section A. If the velocity of the jet is $\mathbf{v}$, determine the force exerted on the tank by the supporting structure that holds the tank at rest. Note that the pressure in the jet will be atmospheric pressure.

8.36 A child aims a garden hose at the back of a friend. (See Figure P8.36.) If the water (specific weight 62.4 lb/ft^3) stream has a diameter of $\frac{1}{4}$ in. and a speed of 50 ft/sec, estimate the force exerted on the "target" if: (a) he is stationary; (b) he is running away from the stream at a speed of 10 ft/sec. Assume the flow in contact with the

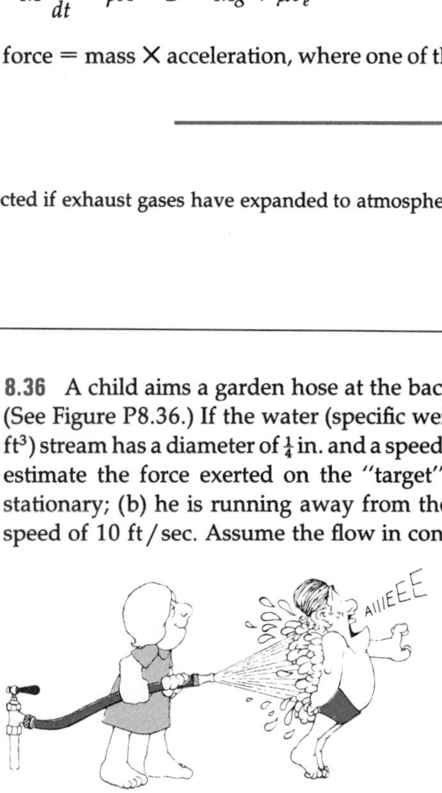

Figure P8.36

boy's back to be vertical relative to him; that is, neglect any splashback.

8.37 A steady jet of liquid is directed against a smooth rigid surface and the jet splits as shown in Figure P8.37. Assume that each fluid particle moves in a plane parallel to that of the figure and ignore gravity. Ignoring gravity and friction, it can be shown that the particle speed after the split is still v as depicted. Estimate the fraction of the flow rate occurring in each of the upper and lower branches. *Hint*: Use the fact that no external force tangent to the surface acts on the liquid.

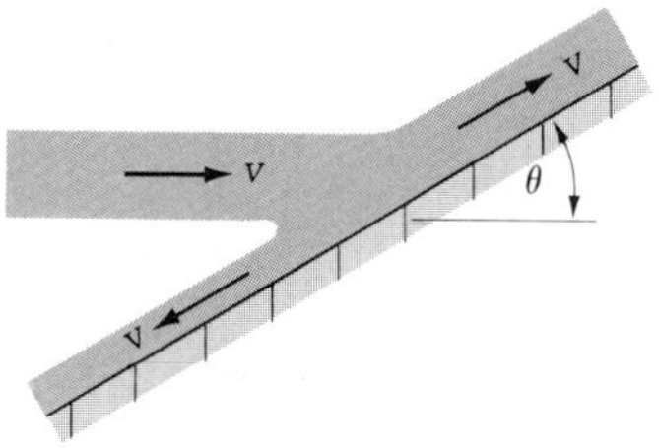

Figure P8.37

8.38 Air flows into the intake of a jet engine at mass flow rate q (slug / sec or kg / s). If v is the speed of the airplane flying through still air and u is the speed of engine exhaust relative to the plane, derive an expression for the force (thrust) of the flowing fluid on the engine. Neglect the fact that the rate of exhaust is slightly greater than q because of the addition of fuel in the engine.

8.39 Revise the analysis of the preceding problem to account for the mass of fuel injected into the engine. Let f be the mass flow rate of the fuel, and assume that the fuel is injected with no velocity relative to the engine housing.

8.40 In a quarry, rocks slide onto a conveyor belt at the constant mass flow rate k, and at speed v_{rel} relative to the ground. (See Figure P8.40.) The belt is driven by a motor (with torque M applied to the drum on the right) at constant speed v_B. Find the power that the motor must deliver, neglecting friction in the shaft bearings and assuming the belt does not slip. *Hint:* Use the control volume indicated by the dashed lines to compute the difference in belt tensions, neglecting any sag of the belt due to the weight of the rocks.

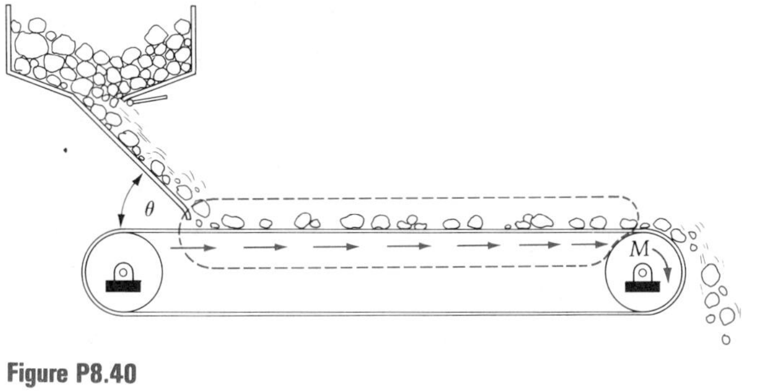

Figure P8.40

8.41 Sand is being dumped on a flatcar of mass M at the constant mass flow rate of q. (See Figure P8.41.) The car is being pulled by a constant force P, and friction is negligible. The car was at rest at $t = 0$. Determine the car's acceleration as a function of P, M, q, and t.

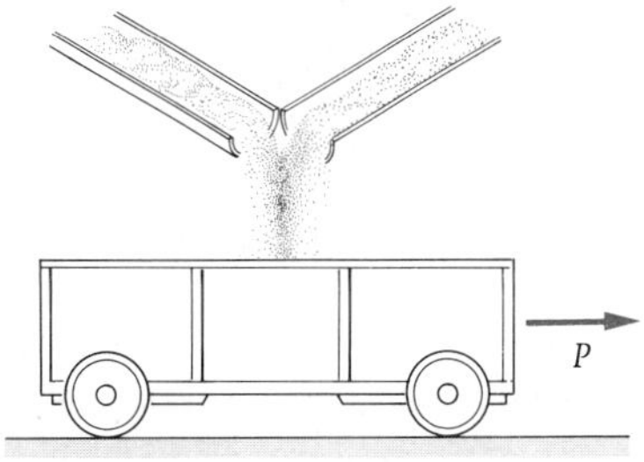

Figure P8.41

8.42 The pressure in a 90° bend of a water pipe is 2 psi (gauge). The inside diameter of the pipe is 6 in. and water flows steadily at the rate of 1.5 ft^3 / sec. Find the magnitude of the force exerted on the bend by the fluid. The specific weight of water is 62.4 lb / ft^3.

8.43 The reducing section in Figure P8.43 connects a 36-in. inside-diameter pipe to a 24-in. inside-diameter pipe. Water enters the reducer at 10 ft / sec and 5 psi (gauge) and leaves at 2 psi. Find the force exerted on the reducer by the steadily flowing water.

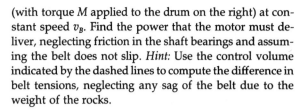

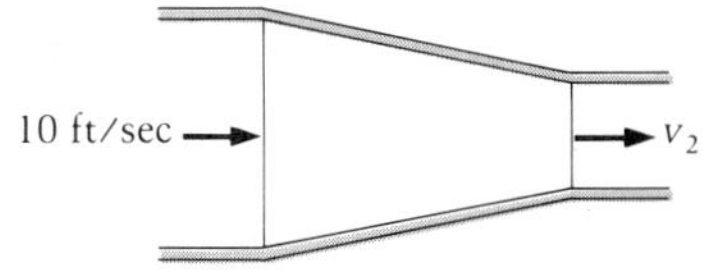

Figure P8.43

8.44 If the plane of Figure P8.44 is horizontal, find the force and moment at O that will allow the body $\mathcal{B}$ to remain in equilibrium when the open stream of water impinges steadily on it as shown. The stream's velocity is 60 ft/sec, and its constant area is 12 in.^{2}; its specific weight is 62.4 lb/ft^3.

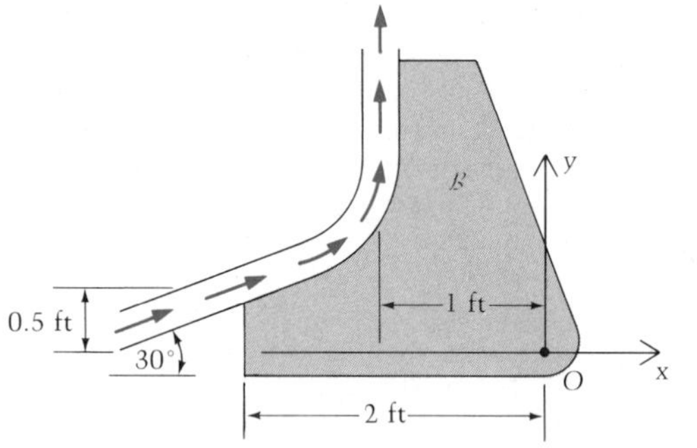

Figure P8.44

8.45 A coal truck weighs 5 tons when empty. It is pushed under a loading chute by a constant force of 500 lb. The chute, inclined at 60° as shown in Figure P8.45, delivers 100 lb of coal per second to the truck at a velocity of 30 ft/sec. When the truck contains 10 tons of coal, its velocity is 10 ft/sec to the right. (a) What is its acceleration at this instant? The wheels are light and all horizontal frictional forces may be taken as included in the 500-lb force. (b) What is the horizontal component of force on the truck from the coal at this instant?

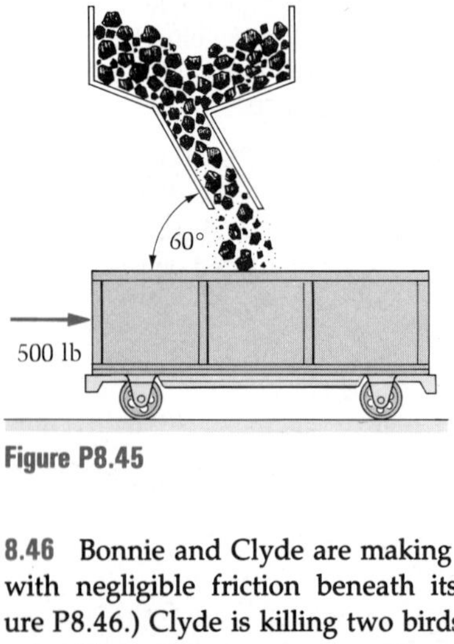

Figure P8.45

8.46 Bonnie and Clyde are making a getaway in a cart with negligible friction beneath its wheels. (See Figure P8.46.) Clyde is killing two birds with one stone by using his machine gun to propel the car as well as to ward off pursuers. He fires 500 rounds (shots) per minute with each bullet weighing 1 oz and exiting the muzzle with a speed relative to the car of 2500 ft/sec. The bullets originally comprised 2 percent of an initial total mass of $m_0 = 20$ slugs. If the system starts from rest at $t = 0$, find: (a) the maximum speed of Bonnie and Clyde; (b) how long it takes to attain this speed.

Figure P8.46

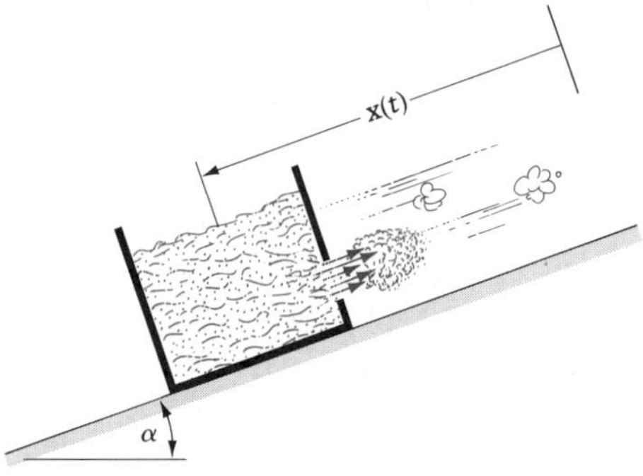

Figure P8.47

8.47 A black box with an initial mass of m_0 (of which 10 percent is box and 90 percent is fuel) is released from rest on the inclined plane in Figure P8.47. The coefficient of friction is μ between the box and the plane.

a. Show that with $\tan \alpha > \mu$, the box will begin to slide down the plane.
b. Assume now that $\tan \alpha > \mu$ and that a mechanism in the box is able to sense its velocity and eject particles rearward (up the plane) at a constant mass flow rate of k_0, and at a relative velocity always equal to the negative of the velocity of the box. Find the velocity of the box at the time t_f when the last of the fuel leaves.
c. Show that the box is going 5.5 times faster at $t = t_f$ than it would have gone if no fuel had been ejected.

8.48 Santa Claus weighs 450 lb and drops down a 20-ft chimney (Figure P8.48). He gains mass in the form of ashes and soot at the rate of 3 slugs / sec from a *very* dirty chimney.

a. Find Santa's velocity as a function of time. (Neglect friction.)

b. Calculate the velocity v_b and the time t_b at which he would hit bottom *without* adding mass and then compare v_b with his "ashes and soot velocity" at the same t_b.

Figure P8.48

8.49 A small rocket is fired vertically upward. Air resistance is neglected. Show that for the rocket to have constant acceleration upward, its mass m must vary with time t according to the equation

$$\frac{dm}{dt} = -\frac{a + g}{u} m$$

where a is the acceleration of the rocket and u is the velocity of the escaping gas relative to the rocket.

8.50 The end of a chain of length L and weight per unit length w, which is piled on a platform, is lifted vertically by a variable force P so that it has a constant velocity v. (See Figure P8.50.) Find P as a function of x. *Hint*: Choose a control-volume boundary so that material crosses the boundary (with negligible velocity) just *before* it is acted on by the moving material already in the control volume. That is, there is no force transmitted across the boundary of the control volume. The solution will be an approximation to reality because of assuming arbitrarily small individual links; but the more links having the common velocity of the fully engaged links within the volume, the better the approximation will be.

8.51 Solve the final equation of Example 8.7 for the velocity $v\hat{\mathbf{j}}$ of the rocket as a function of time in the case in which the pressure force pA and the drag D are negligible, the initial mass of the rocket is m_0, and the gravitational acceleration g, the rate μ, and the relative velocity v_e are all constants. Initially, the rocket is at rest.

8.52 Repeat the preceding problem, but this time include a drag force of $-kv$, where k is a positive constant.

8.53 (a) Extending Problem 8.51, find the height of the rocket as a function of time. (b) If the fraction of m_0 which is fuel is f, find the rocket's "burnout" velocity and position when all the fuel is spent.

8.54 Spherical raindrops produced by condensation are precipitated form a cloud when their radius is a. They fall freely from rest, and their radii increase by accretion of moisture at a uniform rate k. Find the velocity of a raindrop at time t, and show that the distance fallen in that time is

$$\frac{gt^2}{8}\left(\frac{2a + kt}{a + kt}\right)^2$$

* **8.55** A chain of length L weighing γ per unit length begins to fall through a hole in a ceiling. (See Figure P8.55.) Referring to the hint in Problem 8.50:

a. Find $v(x)$ if $v = 0$ when x and t are zero.

b. Show that the acceleration of the end of the falling chain is the constant $g/3$.

c. Show that when the last link has left the ceiling, the chain has lost more potential energy than it has gained in kinetic energy, the difference being $\gamma L^2/6$. Give the reason for this loss.

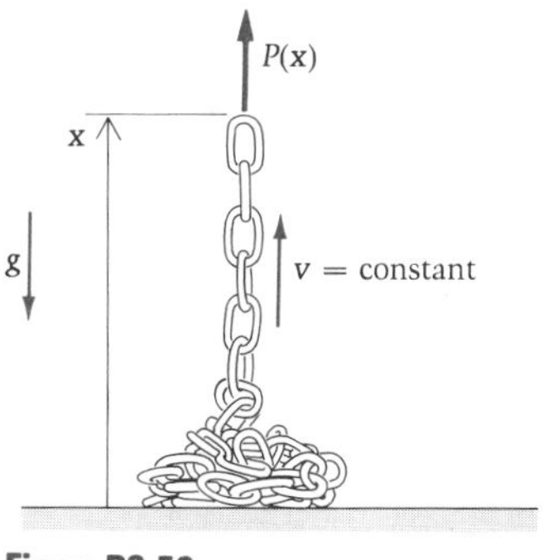

Figure P8.50

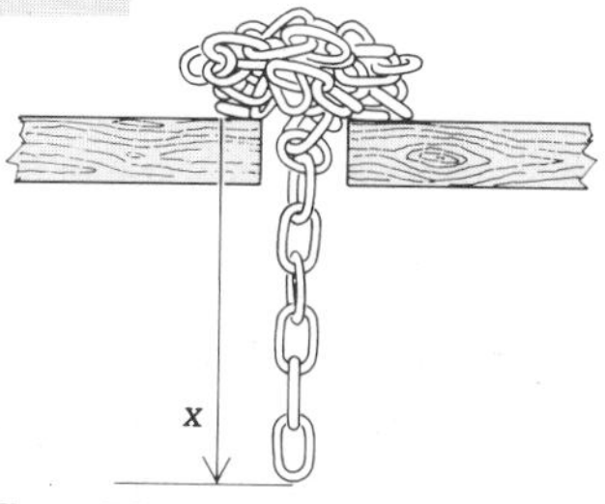

Figure P8.55

8.56 The machine gun in Figure P8.56 has mass M exclusive of its bullets, which have mass M' in total. The bullets are fired at the mass rate of K_0 "slugs" per second, with velocity u_0 relative to the ground. If the coefficient of friction between the gun's frame and the ground is μ, find the velocity of the gun at the instant the last bullet is fired.

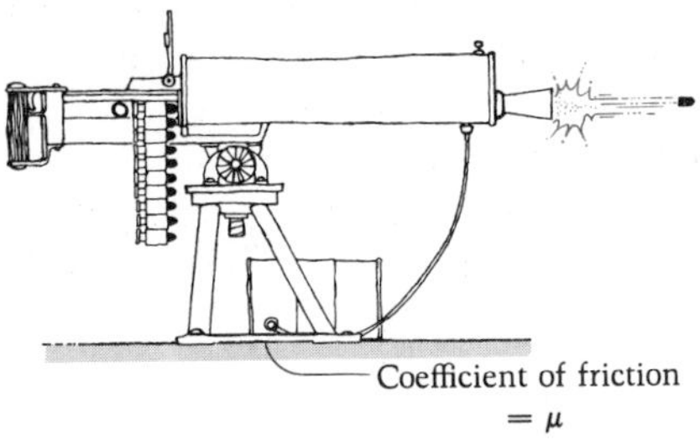

Figure P8.56

* **8.57** A particle of mass m, initially at rest, is projected with velocity $\mathbf{v}_0$ at an angle α to the horizontal and moves under gravity. (See Figure P8.57.) During its flight, it gains mass at the uniform rate k. If air resistance is neglected, show that its equation of motion is

$$(m + kt)\ddot{\mathbf{r}} + k\dot{\mathbf{r}} = (m + kt)g\hat{\mathbf{k}}$$

and that the equation of its path is

$$\mathbf{r} = \frac{m^2}{4k^2}\left[\left(1 + \frac{kt}{m}\right)^2 - 1 - 2\log\left(1 + \frac{kt}{m}\right)\right]g\hat{\mathbf{k}}$$
$$+ \frac{m}{k}\log\left(1 + \frac{kt}{m}\right)\mathbf{v}_0$$

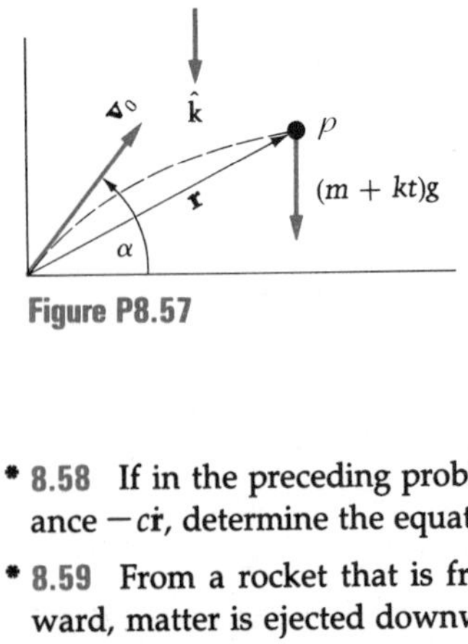

Figure P8.57

* **8.58** If in the preceding problem the air offers a resistance $-c\dot{\mathbf{r}}$, determine the equation of the path.

* **8.59** From a rocket that is free to move vertically upward, matter is ejected downward with a constant relative velocity gT at a constant rate $2M/T$. Initially the rocket is at rest and has mass $2M$, half of which is available for ejection. Neglecting air resistance and variations in the gravitational attraction, (a) show that the greatest upward speed is attained when the mass of the rocket is reduced to M, and determine this speed. (b) Show also that the rocket rises to a height

$$\tfrac{1}{2}gT^2(1 - \ln 2)^2$$

8.60 With the same notation and conditions as in Problem 8.34, show that Equation (8.32) may be written as

$$\Sigma\mathbf{M}_O = (\dot{\mathbf{H}}_O)_V + \frac{dm_o}{dt}(\mathbf{r}_o \times \mathbf{v}_o) - \frac{dm_i}{dt}(\mathbf{r}_i \times \mathbf{v}_i)$$

where $\mathbf{r}_i$ and $\mathbf{r}_o$ are position vectors for the mass centers of the incoming and exiting particles.

* **8.61** A pinwheel of radius a, which can turn freely about a horizontal axis, is initially of mass M and moment of inertia I about its center. A charge is spread along the rim and ignited at time $t = 0$. While the charge is burning, the rim of the wheel loses mass at a constant rate m_1 mass units per second, and at the rim a mass m_2 of gas is taken up per second from the atmosphere, which is at rest. The total mass $m_1 + m_2$ is discharged per second tangentially from the rim, with velocity v relative to the rim. Prove that if θ is the angle through which the wheel has turned after t sec, then

$$\theta = \frac{v}{a(\mu - \lambda)}[\mu t - 1 + (1 - \lambda t)^{\mu/\lambda}]$$

where

$$\lambda = \frac{m_1 a^2}{I} \qquad \mu = \frac{(m_1 + m_2)a^2}{I}$$

* **8.62** A wheel of radius a starts from rest and fires out matter at a uniform rate from all points on the rim (Figure P8.62). The matter leaves tangentially with relative speed v and at such a rate that the mass decreases at the rim by m mass units per second. Show that the angle θ turned through by the wheel is given by

$$\theta = \frac{vI_0}{ma^3}\left[\left(1 - \frac{ma^2t}{I_0}\right)\ln\left(1 - \frac{ma^2t}{I_0}\right) + \frac{ma^2t}{I_0}\right]$$

in which I_0 is the initial moment of inertia of the wheel about its axis.

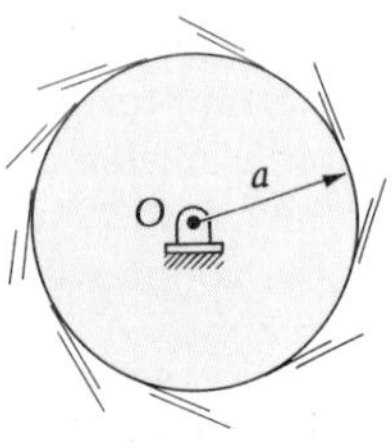

Figure P8.62

8.4 Central Force Motion

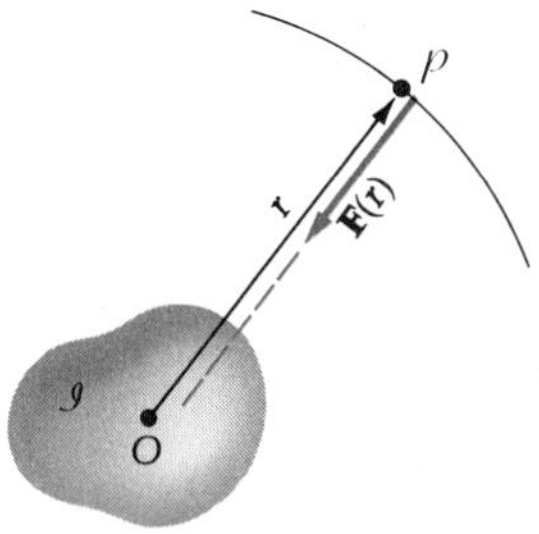

Figure 8.12

In Chapter 2 we defined a **central force** acting on a particle $\mathcal{P}$ as one which (1) always passes through a certain point O fixed in the inertial reference frame $\mathcal{I}$ and (2) depends only on the distance r between O and $\mathcal{P}$. (See Figure 8.12.) In this section we are going to treat the central force in more detail. We shall go as far as we can without specializing $\mathbf{F}(r)$ — that is, without saying how $\mathbf{F}$ depends on r. In the second part of the section we shall study the most important of central forces: gravitational attraction.

If the central force $\mathbf{F}$ is the only force acting on the particle, then $\mathbf{F} = m\mathbf{a}$; and since the central force always passes through point O, $\mathbf{r} \times \mathbf{F}$ is identically zero. These two facts allow us to write:

$$\mathbf{r} \times \mathbf{F} = \mathbf{r} \times m\mathbf{a} = \mathbf{0}$$

or, since $\dot{\mathbf{r}} = \mathbf{v}$,

$$\frac{d}{dt}(\mathbf{r} \times m\mathbf{v}) = \mathbf{0}$$

Therefore for a particle acted on only by a central force,

$$\mathbf{r} \times \mathbf{v} = \text{constant vector in } \mathcal{I} = \mathbf{h}_O \tag{8.33}$$

Dotting this equation with $\mathbf{r}$, we find, since $\mathbf{r} \times \mathbf{v}$ is perpendicular to $\mathbf{r}$,

$$\mathbf{r} \cdot (\mathbf{r} \times \mathbf{v}) = 0 = \mathbf{r} \cdot \mathbf{h}_O$$

and we see that $\mathbf{r}$ is always perpendicular to a vector that is constant in $\mathcal{I}$; therefore $\mathcal{P}$ moves in a plane in $\mathcal{I}$. Using polar coordinates to then describe the motion of $\mathcal{P}$ in this plane, the governing equations are:

$$\Sigma F_r = -F(r) = m(\ddot{r} - r\dot{\theta}^2) \tag{8.34}$$

and

$$\Sigma F_\theta = 0 = m(r\ddot{\theta} + 2\dot{r}\dot{\theta}) = \frac{m}{r}\frac{d}{dt}(r^2\dot{\theta}) \tag{8.35}$$

From Equation (8.35) we see immediately that

$$r^2\dot{\theta} = \text{constant} = h_O \tag{8.36}$$

where h_O is the magnitude of the constant vector $\mathbf{h}_O$ of Equation (8.33) because, expressing $\mathbf{r}$ and $\mathbf{v}$ in polar coordinates, we find

$$\mathbf{h}_O = \mathbf{r} \times \mathbf{v} = \text{constant} = r\hat{\mathbf{e}}_r \times (\dot{r}\hat{\mathbf{e}}_r + r\dot{\theta}\hat{\mathbf{e}}_\theta) = r^2\dot{\theta}\hat{\mathbf{k}}$$

so that

$$|\mathbf{h}_O| = h_O = r^2\dot{\theta} = \text{constant} \tag{8.37}$$

Equation (8.37) is a statement of the conservation of the moment of momentum, or angular momentum of $\mathcal{P}$; the constant h_O is the magnitude of the angular momentum $\mathbf{H}_O$ of $\mathcal{P}$ divided by its mass m. Thus we shall call h_O the angular momentum (magnitude) per unit mass.

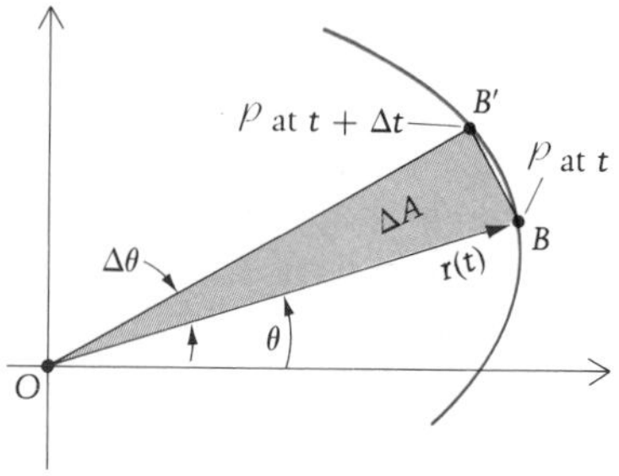

Figure 8.13

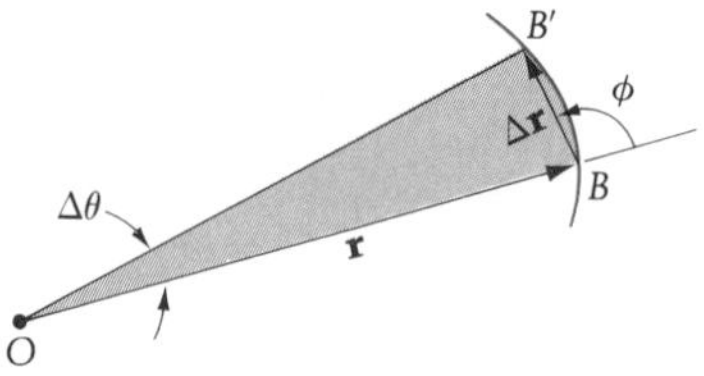

Figure 8.14

We can use the previous pair of results to show that the second of Kepler's three laws of planetary motion is in fact valid for any central force. This law states that the radius vector from the sun to a planet sweeps out equal areas in equal time intervals. From Figure 8.13 the incremental planar area ΔA swept out by ℘ between θ (at t) and $\theta + \Delta\theta$ (at $t + \Delta t$) is approximately given by the area of the triangle OBB'* (see Figure 8.14):

$$\begin{aligned}\Delta A &\approx \tfrac{1}{2} \times \text{base} \times \text{height} \\ &\approx \tfrac{1}{2} r(\Delta r \sin \phi) \approx \tfrac{1}{2} |\mathbf{r} \times \Delta\mathbf{r}|\end{aligned}$$

Dividing by the time increment Δt and taking the limit as $\Delta t \to 0$, we have

$$\lim \frac{\Delta A}{\Delta t} = \frac{dA}{dt} = \frac{1}{2} \lim_{\Delta t \to 0} \left| \mathbf{r} \times \frac{\Delta \mathbf{r}}{\Delta t} \right| = \frac{1}{2} \overbrace{|\mathbf{r} \times \mathbf{v}|}^{\mathbf{h}_O}$$

or

$$\frac{dA}{dt} = \frac{h_O}{2} \qquad \text{(a constant that is } r^2\dot{\theta}/2\text{)} \tag{8.38}$$

Thus the rate of sweeping out area is a constant. This is why a satellite or a planet in elliptical orbit (Figure 8.15) has to travel faster when it is near the *perigee* than the *apogee*—the same area must be swept out in the same period of time.† We emphasize again that this result is valid for *all* central force trajectories, not just elliptical orbits and not just if the central force is gravity.

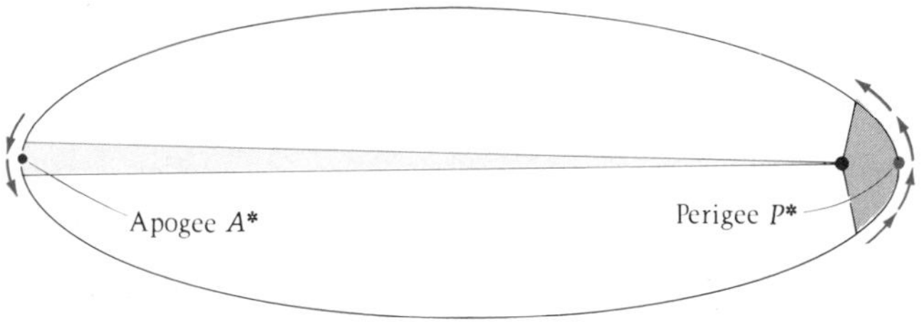

Figure 8.15

Next we focus our attention on the most important central force: gravitational attraction. If G is the universal gravitation constant and M and m are the masses of what we are considering to be the attracting and attracted bodies,‡ then the central force acting on m for this case is

---

* "Approximately" because the area between arc and chord is outside the triangle.

† We use *perigee* and *apogee* in a general sense; technically these words refer to the nearest and farthest points, respectively, for the moon and artificial satellites. For the orbits of planets, the proper terms are *perihelion* and *aphelion.*

‡ Actually, of course, both are attracting and both are attracted — each to the other! The constant GM is, for the sun, 4.68×10^{21} ft^3/sec^2.

$$F(r) = \frac{GMm}{r^2} \tag{8.39}$$

and Equation (8.34) becomes

$$m(\ddot{r} - r\dot{\theta}^2) = -\frac{GMm}{r^2} \tag{8.40}$$

Canceling m and inserting h_O for $r^2\dot{\theta}$ gives

$$\ddot{r} - \frac{h_O^2}{r^3} = -\frac{GM}{r^2} \tag{8.41}$$

Multiplying Equation (8.41) by $\dot{r}$ will allow us to integrate it:

$$\dot{r}\ddot{r} - h_O^2 r^{-3}\dot{r} = -GMr^{-2}\dot{r} \tag{8.42}$$

Integrating, we get

$$\frac{\dot{r}^2}{2} + h_O^2\frac{r^{-2}}{2} = \frac{-GMr^{-1}}{-1} + C_1 \tag{8.43}$$

If we multiply Equation (8.43) by m and replace h_O by $r^2\dot{\theta}$, we see that

$$\frac{m}{2}[\dot{r}^2 + (r\dot{\theta})^2] - \frac{GMm}{r} = C_1 m \tag{8.44}$$

and the left side of Equation (8.44) is seen to be the total energy of $\mathcal{P}$, kinetic plus potential. Thus we shall replace C_1 by E, the energy of $\mathcal{P}$ per unit mass, and obtain

$$\dot{r}^2 + h_O^2 r^{-2} = +2GMr^{-1} + 2E \tag{8.45}$$

This equation will be helpful to us later. But now we are interested in studying the trajectory of particle $\mathcal{P}$— that is, in finding r as a function of θ. By the chain rule,

$$\frac{dr}{dt} = \frac{dr}{d\theta}\frac{d\theta}{dt} = \dot{\theta}\frac{dr}{d\theta}$$

and since $\dot{\theta} = h_O/r^2$ from Equation (8.36),

$$\frac{dr}{dt} = \frac{h_O}{r^2}\frac{dr}{d\theta} \tag{8.46}$$

We need the second derivative of r in Equation (8.41), so we apply the chain rule once more:

$$\begin{aligned}\frac{d^2r}{dt^2} &= \left[\frac{d}{d\theta}\left(\frac{h_O}{r^2}\frac{dr}{d\theta}\right)\right]\frac{d\theta}{dt} \\ &= \frac{h_O}{r^2}\left[-\frac{2h_O}{r^3}\left(\frac{dr}{d\theta}\right)^2 + \frac{h_O}{r^2}\frac{d^2r}{d\theta^2}\right] \\ &= \frac{h_O^2}{r^4}\frac{d^2r}{d\theta^2} - \frac{2h_O^2}{r^5}\left(\frac{dr}{d\theta}\right)^2 = \frac{h_O^2}{r^2}\frac{d}{d\theta}\left(\frac{1}{r^2}\frac{dr}{d\theta}\right)\end{aligned} \tag{8.47}$$

Substituting into Equation (8.41), we get

$$\frac{h_O^2}{r^2}\frac{d}{d\theta}\left(\frac{1}{r^2}\frac{dr}{d\theta}\right)-\frac{h_O^2}{r^3}=-\frac{GM}{r^2}$$

or

$$\frac{d}{d\theta}\left(\frac{1}{r^2}\frac{dr}{d\theta}\right)-\frac{1}{r}=-\frac{GM}{h_O^2} \tag{8.48}$$

The following simple change of variables will make the solution to this differential equation immediately recognizable:

$$u=\frac{1}{r} \tag{8.49}$$

Substituting Equation (8.49) into (8.48) along with

$$\frac{dr}{d\theta}=\frac{dr}{du}\frac{du}{d\theta}=\frac{-1}{u^2}\frac{du}{d\theta} \tag{8.50}$$

gives

$$\frac{d}{d\theta}\left[u^2\left(\frac{-1}{u^2}\frac{du}{d\theta}\right)\right]-u=\frac{-GM}{h_O^2} \tag{8.51}$$

or

$$\frac{d^2u}{d\theta^2}+u=\frac{GM}{h_O^2} \tag{8.52}$$

The solution to Equation (8.52), from elementary differential equations, consists of a homogeneous (or complementary) part plus a particular part:

$$u=\overbrace{u_H}^{}+\overbrace{u_P}^{}$$
$$=\underbrace{A_1\cos\theta+B_1\sin\theta}+\underbrace{\frac{GM}{h_O^2}} \tag{8.53}$$

Switching variables back from u to r by Equation (8.49), we obtain

$$r=\frac{h_O^2/GM}{1+(h_O^2/GM)(A_1\cos\theta+B_1\sin\theta)} \tag{8.54}$$

This solution for $r(\theta)$ is the equation of a **conic**; it can be put into a more recognizable form after a brief review of conic sections. For every point P on a conic, the ratio of the distances from P to a fixed point (O: the focus) and to a fixed line (ℓ: the directrix) is a constant called the **eccentricity** of the conic:

$$e=\frac{OP}{L} \tag{8.55}$$

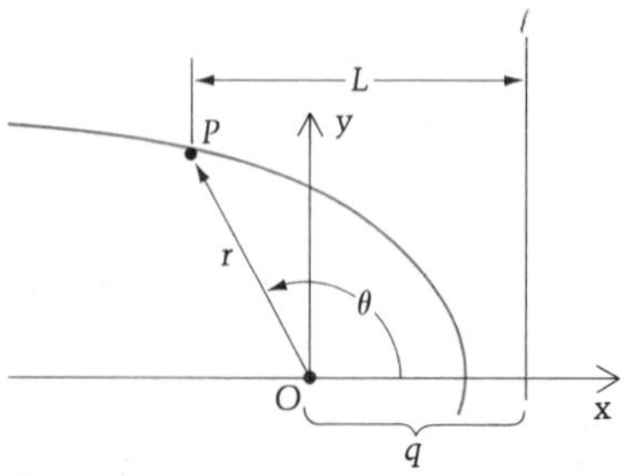

Figure 8.16

Therefore, in terms of the parameters in Figure 8.16,

$$e=\frac{r}{q-r\cos\theta} \tag{8.56}$$

or, solving for r,

$$r = \frac{eq}{1 + e \cos \theta} \tag{8.57}$$

The conic specified by Equation (8.57) is a:

$$\begin{aligned} &\text{Hyperbola if } |e| > 1 \\ &\text{Parabola if } e = 1 \\ &\text{Ellipse if } -1 < e < 1 \end{aligned} \tag{8.58}$$

The ellipse becomes a circle if $e = 0$. It is an ellipse with perigee (closest point to O) at $\theta = 0$ if $0 < e < 1$ and an ellipse with apogee (farthest point from O) at $\theta = 0$ if $-1 < e < 0$; this latter type is called a subcircular ellipse.

Returning to our solution (8.54) for $r(\theta)$, it is customary to select one of the constants A_1 and B_1 so that, as suggested by Figure 8.16, $dr/d\theta = 0$ when $\theta = 0$. This condition easily gives $B_1 = 0$, as the reader may wish to demonstrate using calculus. The result simply means that we are measuring θ from the perigee of the conic. At this point we should compare Equations (8.57) and (8.54) with $B_1 = 0$:

$$r = \frac{h_O^2/GM}{1 + (h_O^2/GM)A_1 \cos \theta} \tag{8.59}$$

and

$$r = \frac{eq}{1 + e \cos \theta} \tag{8.60}$$

By direct comparison of these two expressions for r, we see that

$$A_1 = \frac{eGM}{h_O^2} \qquad \text{and} \qquad eq = \frac{h_O^2}{GM}$$

It is more customary, however, to express the constant A_1 (as well as the eccentricity) in terms of the energy E of the orbit. To do this, Equations (8.45) and (8.46) give

$$\frac{h_O^2}{r^4}\left(\frac{dr}{d\theta}\right)^2 + \frac{h_O^2}{r^2} - \frac{2GM}{r} = 2E \tag{8.61}$$

At the point r_P where $\theta = 0$ and $dr/d\theta = 0$, we see that

$$\frac{h_O^2}{r_P^2} - \frac{2GM}{r_P} = 2E \tag{8.62}$$

Thus not all of h_O, r_P, and E are independent. We shall eliminate r_P. Multiplying Equation (8.62) by r_P^2, we get

$$2Er_P^2 + 2GMr_P - h_O^2 = 0 \tag{8.63}$$

Solving via the quadratic formula, we have

$$r_P = \frac{-2GM + \sqrt{4G^2M^2 + 8Eh_O^2}}{4E} \tag{8.64}$$

in which we use the plus sign since we need the smaller root for closed conics ($E < 0$). The positive sign also ensures a positive r_P for open conics ($E > 0$).

Returning to our solution (8.59), when $\theta = 0$ then

$$r_P = \frac{h_O^2/GM}{1 + (h_O^2/GM)A_1} \tag{8.65}$$

Equating the two expressions for r_P, Equations (8.64) and (8.65), we can solve for A_1.

We see by comparing Equations (8.59) and (8.60) that the eccentricity e of our conic will be $(h_O^2/GM)A_1$. Equating the right sides of Equations (8.64) and (8.65) and solving for this quantity, we get

$$\frac{h_O^2}{GM}A_1 = e = \sqrt{1 + \frac{2Eh_O^2}{G^2M^2}} \tag{8.66}$$

Therefore

$$r = \frac{h_O^2/GM}{1 + \sqrt{1 + (2Eh_O^2/G^2M^2)} \cos\theta} \tag{8.67}$$

which expresses r as a function of θ, the constant GM, the energy E, and the angular momentum per unit mass h_O. Note that by again comparing Equations (8.59) and (8.60) we can obtain the distance q between the focus O and the directrix ℓ:

$$\frac{h_O^2}{GM} = eq \Rightarrow q = \frac{h_O^2/GM}{\sqrt{1 + 2Eh_O^2/G^2M^2}} \tag{8.68}$$

The first of **Kepler's three laws of planetary motion** states that the planets travel in elliptical orbits with the sun at one focus.* These ellipses are very nearly circular for most of the planets; the eccentricity of earth is $e = 0.017$. To obtain the third of Kepler's laws, we return once more to our equations and obtain for elliptic orbits, from (8.67), the distance r when $\theta = 90°$:

$$r_{90} = \frac{h_O^2/GM}{1+0} = \ell \tag{8.69}$$

This distance, the *semilatus rectum,* may be used to express the distance r_A between the focus O and apogee A^*, and the distance r_P between O and the perigee P^*. (See Figure 8.17.) At apogee, $\theta = \pi$ and Equations (8.59), (8.60), and (8.69) give

$$r_A = \frac{h_O^2/GM}{1-e} = \frac{\ell}{1-e} \tag{8.70}$$

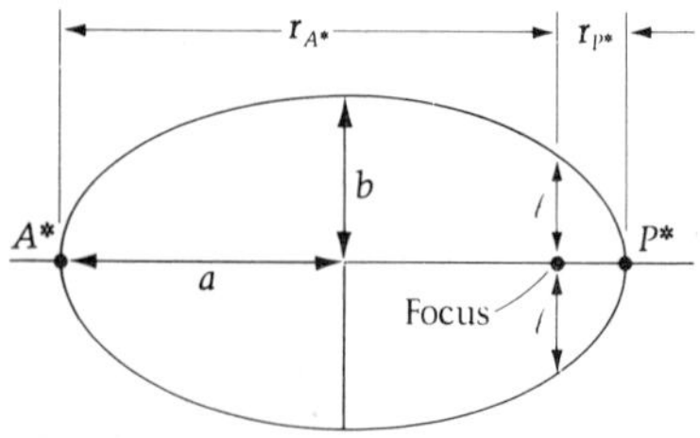

Figure 8.17

* Kepler's laws, based on his astronomical observations and set forth in 1609 and 1619, were studied by Newton before the Englishman published the *Principia,* which contained his own laws of motion.

and at perigee ($\theta = 0$),

$$r_{P^*} = \frac{\ell}{1+e} \tag{8.71}$$

The semimajor axis length of the ellipse is

$$a = \frac{r_{A^*} + r_{P^*}}{2} = \frac{\ell}{1-e^2} \tag{8.72}$$

and the semiminor axis length is, from analytic geometry,

$$b = a\sqrt{1-e^2} = \frac{\ell}{\sqrt{1-e^2}} \tag{8.73}$$

An ellipse has area

$$A_T = \pi ab = \pi\left(\frac{\ell}{1-e^2}\right)\left(\frac{\ell}{\sqrt{1-e^2}}\right)$$

or

$$A_T = \frac{\pi\ell^2}{(1-e^2)^{3/2}} = \pi a^2\sqrt{1-e^2} \tag{8.74}$$

With these results in hand, we shall now prove Kepler's third law. Since dA/dt is constant,

$$\frac{dA}{dt} = \frac{h_O}{2} \Rightarrow A = \frac{h_O t}{2} \tag{8.75}$$

where we take $A = 0$ when $t = 0$, say at the perigee. Over one orbit we have, with T being the orbit period,

$$\begin{aligned} A_T &= \text{area of ellipse} \\ &= \pi a^2\sqrt{1-e^2} = \frac{h_O T}{2} \end{aligned} \tag{8.76}$$

Since (from Equations (8.69) and (8.72))

$$h_O = \sqrt{GM\ell} = \sqrt{GMa(1-e^2)} \tag{8.77}$$

we obtain the following from Equation (8.76):

$$\pi a^2\sqrt{1-e^2} = \frac{\sqrt{GMa(1-e^2)}}{2}\,T \tag{8.78}$$

so that

$$T = \frac{2\pi a^{3/2}}{\sqrt{GM}}$$

or

$$T^2 = \frac{4\pi^2}{GM}\,a^3 \tag{8.79}$$

Equation (8.79) states the third of Kepler's laws: The squares of the planets' orbital periods are proportional to the cubes of the semimajor axes of their orbits.

EXAMPLE 8.8

Calculate the semimajor axis length of an earth satellite with a period of 90 min.

Solution

We can solve this problem using Kepler's third law. The weight of a particle (mass m) on the earth's (mass M) surface is both mg and GMm/r_e^2; thus

$$mg = \frac{GMm}{r_e^2} \Rightarrow GM = gr_e^2$$

and we see that the product of the unwieldy constants G and M is

$$GM = gr_e^2 = \frac{32.2}{5280}(3960)^2 = 95{,}600 \text{ mi}^3/\text{sec}^2$$

Therefore

$$T^2 = (90 \times 60)^2 = \frac{4\pi^2}{95{,}600}a^3$$

$$a = 4130 \text{ mi.}$$

which is about 170 mi above the earth's surface.

We shall present one more example on elliptical orbits under gravity, but first we need two equations relating the velocities v_1 and v_2 at any two points P_1 and P_2 with radii r_1 and r_2 on the orbit. The first of these comes from Equation (8.37), which states that $r^2\dot{\theta} = \text{constant}$. From Figure 8.18, since the velocity **v** is always tangent to the path, we see that if ϕ is the angle between **r** and **v**, then in cylindrical coordinates

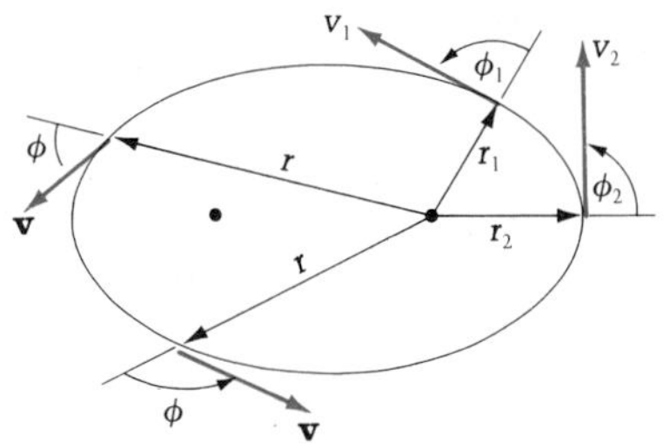

Figure 8.18

$$v \sin\phi = (\text{transverse component of } \mathbf{v}) = r\dot{\theta}$$

so that

$$r(r\dot{\theta}) = rv\sin\phi = \text{constant}$$

or, for two points P_1 and P_2 on the orbital path,

$$r_1 v_1 \sin\phi_1 = r_2 v_2 \sin\phi_2 \tag{8.80}$$

Note that at apogee and perigee, $\phi = 90°$. Thus letting P_1 and P_2 be these two points, we get from Equation (8.80)

$$r_{A\bullet} v_{A\bullet} = r_{p\bullet} v_{p\bullet} \tag{8.81}$$

and the two velocities are inversely proportional to the radii, with v being faster at perigee as we have already seen from Kepler's second law.

The other equation relating v_1 and v_2 comes from the potential for gravity, which from Equation (2.27) and Example 8.8 is

$$\varphi = -\frac{gr_e^2 m}{r} = -\frac{GMm}{r}$$

Using conservation of energy between P_1 and P_2,

$$T_1 + \varphi_1 = T_2 + \varphi_2$$

$$\frac{mv_1^2}{2} - \frac{GMm}{r_1} = \frac{mv_2^2}{2} - \frac{GMm}{r_2}$$

$$v_2^2 - v_1^2 = 2GM\left(\frac{1}{r_2} - \frac{1}{r_1}\right) \tag{8.82}$$

If we let point P_2 represent the perigee P^*, as suggested in Figure 8.18, then Equation (8.80) becomes

$$v_2 = v_{P*} = \frac{r_1 v_1 \sin\phi_1}{r_{P*}} \tag{8.83}$$

where $\sin\phi_2 = \sin 90° = 1$. Now if r_1, v_1, and ϕ_1 are initial (launch) values of r, v, and ϕ, then we may consider these as given quantities. Substituting Equation (8.83) into (8.82), we can obtain an equation for the perigee radius $r_2(=r_{P*})$:

$$\frac{r_1^2 v_1^2 \sin^2\phi_1}{r_{P*}^2} - v_1^2 = 2GM\left(\frac{1}{r_{P*}} - \frac{1}{r_1}\right)$$

Multiplying through by $-r_{P*}^2/(r_1^2 v_1^2)$ and rearranging, we get

$$\left(\frac{r_{P*}}{r_1}\right)^2\left(1 - \frac{2GM}{r_1 v_1^2}\right) + \left(\frac{r_{P*}}{r_1}\right)\frac{2GM}{r_1 v_1^2} - \sin^2\phi_1 = 0 \tag{8.84}$$

We see that Equation (8.84) is simply a quadratic equation in the ratio (r_{P*}/r_1) and that $[2GM/(r_1 v_1^2)]$ is a nondimensional parameter of the orbit. We now illustrate the use of this important equation in an example.

Example 8.9

A satellite is put into an orbit with the following launch parameters: $H = 1000$ mi, $v_1 = 17{,}000$ mph, and $\phi = 100°$ (see Figure E8.9). Find the apogee and perigee radii of the resulting orbit.

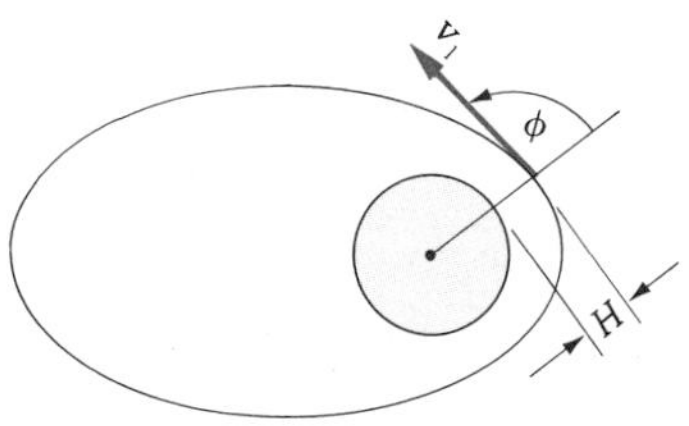

Figure E8.9

Solution

We need GM in mi^3/hr^2; therefore

$$GM = gr_e^2 = 32.2(3960)^2 \text{ ft-mi}^2/\text{sec}^2$$

$$= 32.2(3960)^2 \times \frac{1}{5280} \times 3600^2 \text{ mi}^3/\text{hr}^2$$

$$= 124 \times 10^{10} \text{ mi}^3/\text{hr}^2$$

The parameter $2GM/(r_1 v_1^2)$ in Equation (8.84) is therefore

$$\frac{2GM}{r_1 v_1^2} = \frac{2(124 \times 10^{10})}{(1000 + 3960)17{,}000^2} = 1.73$$

Equation (8.84) becomes

$$\left(\frac{r_{P^\bullet}}{r_1}\right)^2(-0.73) + 1.73\left(\frac{r_{P^\bullet}}{r_1}\right) - 0.970 = 0$$

The quadratic formula gives

$$\left(\frac{r_{P^\bullet}}{r_1}\right)_{1,2} = \frac{-1.73 \pm \sqrt{1.73^2 - 4(-0.73)(-0.970)}}{2(-0.73)}$$

$$= 0.911 \text{ and } 1.46$$

Therefore

$$r_{P_1} = r_{P^\bullet} = 0.911(4960) = 4520 \text{ mi}$$

The other root corresponds to the apogee. (Since the starting condition of sin ϕ = 1 is the same for apogee and perigee, both answers are produced by the quadratic formula!)

$$r_{P_2} = r_{A^\bullet} = 1.46(4960) = 7240 \text{ mi}$$

The altitudes are

$$\text{Perigee height} = 4520 - 3960 = 560 \text{ mi}$$

$$\text{Apogee height} = 7240 - 3960 = 3280 \text{ mi}$$

To pin down the orbit in space, we need to know the angle to the perigee point from the launch point and also the orbit's eccentricity. Problems 8.81 and 8.82 will be concerned with finding these two quantities given initial values of r, v, and ϕ.

PROBLEMS ▶ Section 8.4

8.63 Show that a satellite in orbit has a period T given by

$$T = \frac{2\pi ab}{r_{A^\bullet}v_{A^\bullet}(\text{or } r_{P^\bullet}v_{P^\bullet})}$$

8.64 Show that, for a body in elliptical orbit (Figure P8.64), $b = \sqrt{r_{A^\bullet} r_{P^\bullet}}$.

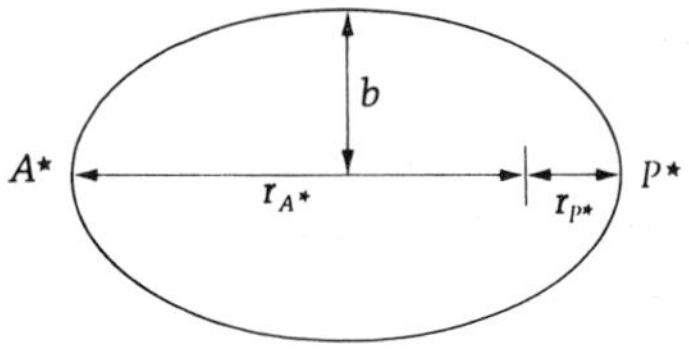

Figure P8.64

8.65 Halley's comet's latest return to Earth was in 1986. The comet orbits the sun in an elongated ellipse every 74 to 79 years; the period varies due to perturbations in its orbit caused by the four largest (Jovian) planets. (Its passages have been recorded since 240 B.C.!) What is the approximate semimajor axis length of Halley's comet? (Use 76 years as the period.)

8.66 Find the minimum period of a satellite in circular orbit about the earth. Upon what assumption is your answer based?

8.67 Repeat the preceding problem if the satellite orbits the moon. Assume

$$g_{\text{moon}} = \tfrac{1}{6}g_{\text{earth}}$$

$$r_{\text{moon}} = 0.27r_{\text{earth}}$$

8.68 Show that for circular orbits around an attracting body of mass M, $rv^2 = GM$. Then use the 93×10^6 mi average orbital radius of earth, and the fact that its orbit is nearly circular, to find the constant GM_s for the sun as attracting center (heliocentric system).

8.69 The first artificial satellite to orbit the earth was the Russians' Sputnik I. Following insertion into orbit it had a period of 96.2 min. Find the semimajor axis length. If the initial eccentricity was 0.0517, find the maximum and minimum distances from earth following its injection into orbit.

8.70 Show that if a satellite is in a circular orbit at radius r around a planet of mass M, the velocity to which it must increase to *escape* the planet's gravitational attraction is given by

$$v_{escape} = \sqrt{\frac{2GM}{r}}$$

Find v_{escape} if "to which" is replaced by "by which."

8.71 Show that if the launch velocity in Example 8.9 is 15,000 mi/hr, the satellite will fail to orbit the earth.

8.72 Using Equation (8.54), show that $B_1 = 0$ follows from the condition $dr/d\theta = 0$ when $\theta = 0$.

* **8.73** Prove that Equation (8.66) follows from (8.64) and (8.65).

8.74 Find the form of the central force $\mathbf{F}(r)$ for which all circular orbits of a particle about an attracting center O have the same angular momentum (and the same rate of sweeping out area).

8.75 Show that, in terms of the radius r_{P^*} and speed v_{P^*} at perigee, the energy and eccentricity of the orbit may be expressed as

$$\frac{Er_{P^*}}{GM} = \frac{r_{P^*}v_{P^*}^2}{2GM} - 1$$

and

$$e = \frac{r_{P^*}v_{P^*}^2}{2GM} - 1$$

8.76 Use Equations (8.81) and (8.82) to show that the velocities at apogee and perigee, in terms of the known radii r_{A^*} and r_{P^*}, are:

$$v_{A^*} = \sqrt{\frac{2GMr_{P^*}}{r_{A^*}(r_{A^*} + r_{P^*})}}$$

and

$$v_{P^*} = \sqrt{\frac{2GMr_{A^*}}{r_{P^*}(r_{A^*} + r_{P^*})}}$$

* **8.77** A rocket is in a 200-mi-high circular parking orbit above a planet. What velocity boost at point P will result in the new, elliptical orbit shown in Figure P8.77? *Hint:* Use the results of Problem 8.75.

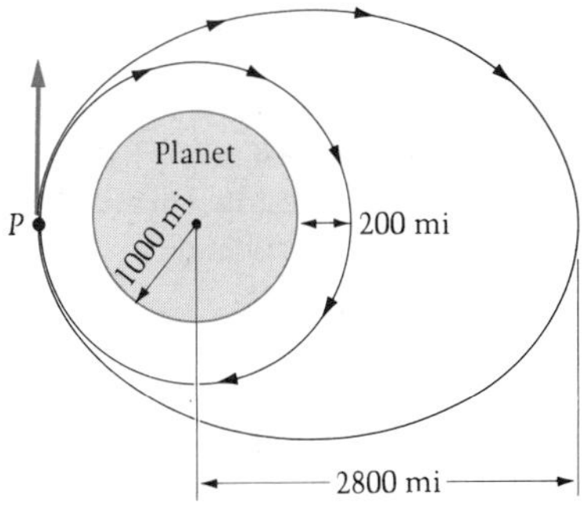

Figure P8.77

* **8.78** A satellite is in a circular orbit of radius R_1. (See Figure P8.78.) Find the (negative) velocity increment that will send the satellite to position A, at radius R_2 ($<R_1$), 180° away. Then find the second negative velocity increment, this time applied at A, that will put the satellite in a circular orbit of radius R_2. *Hint*: Use Problem 8.76.

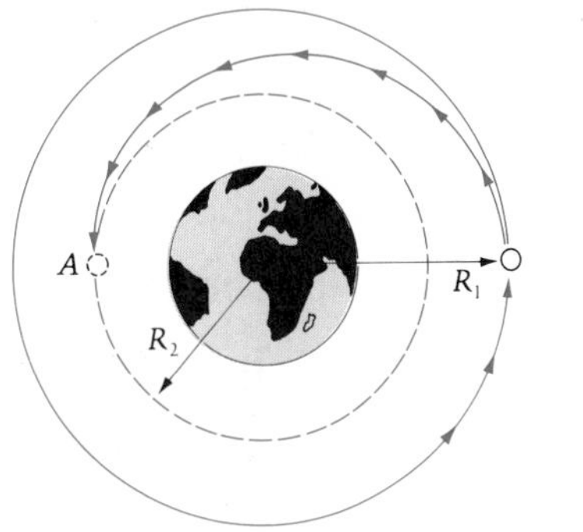

Figure P8.78

8.79 A satellite has $r_{A^*} = 8000$ mi and $r_{P^*} = 5000$ mi. If it was launched with a velocity of 15,000 mi/hr, what was its launch radius? What was the angle ϕ between $\mathbf{r}$ and $\mathbf{v}$ at launch? *Hint*: Use Problem 8.76.

8.80 What is the period of the satellite in the preceding problem?

* **8.81** Show that if the launch parameters, r_1, v_1, and ϕ_1 are known, then the angle θ from perigee to the launch point of a satellite in orbit is given by

$$\tan\theta = \frac{r_1v_1^2}{GM}\sin\phi_1\cos\phi_1 \Bigg/ \left[\frac{r_1v_1^2}{GM}\sin^2\phi_1 - 1\right]$$

* **8.82** Show that if the launch parameters r_1, v_1, and ϕ_1 are known, then the eccentricity of the resulting conic is:

$$e = \sqrt{\left(\frac{r_1v_1^2}{GM} - 1\right)^2 \sin^2\phi_1 + \cos^2\phi_1}$$

* **8.83** A large meteorite approaches the earth. (See Figure P8.83.) Measurements indicate that at a given time it has a speed of 8000 mph at a radius of 100,000 mi. Will it orbit the earth? If so, what is the period? If not, what is the maximum velocity v that would have resulted in an orbit? *Hint*: Use the result of the preceding problem.

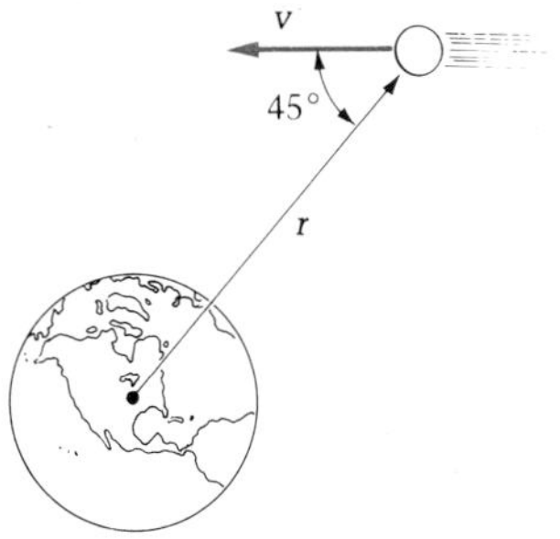

Figure P8.83

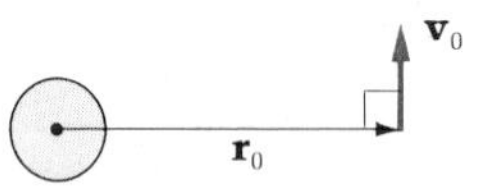

Figure P8.84

8.84 Classify the various orbits according to values of the dimensionless parameter $GM/(r_0 v_0^2)$ for a satellite launched with the conditions of Figure P8.84.

* **8.85** Find the kinetic energy increase needed to move a satellite from radius R to $nR(n > 1)$ in circular orbits. *Hint*: Use Problem 8.76.

* **8.86** A particle of mass m moves in the xy plane under the influence of an attractive central force that is proportional to its distance from the origin ($F(r) = kr$). It has the same initial conditions as Problem 8.84. Find the largest and smallest values of r in the ensuing motion.

8.87 A satellite has apogee and perigee points 1000 and 180 mi, respectively, above the earth's surface. Compute the satellite's period.

8.88 In the preceding problem find the speeds of the satellite at perigee and at apogee. *Hint*: Use Problem 8.76.

REVIEW QUESTIONS ▶ Chapter 8

True or False?

1. Frequencies of vibration of periodic motion are associated only with small amplitudes.
2. Natural frequencies of vibration are associated only with translational motion.
3. Natural frequencies of vibration of bodies in a gravitational field do not have to depend on "g."
4. Free vibrations will always decrease in time due to "real world damping."
5. There are three types of damped vibrations, and the subcritical case has the greatest practical importance.
6. In forced vibration, the steady-state part of the response dies out due to damping inherent in the physical system.
7. In general, the rate of accumulation of momentum within a region of space equals the rate at which it is transported into that region.

In using the control volume form of Euler's Laws, the control volume:

8. has to be fixed in the inertial frame of reference;
9. may change in shape or volume with time;
10. has to be a closed region of space.
11. In a one-dimensional control-volume problem, Euler's first law becomes, in general,

$$\Sigma F = \frac{d}{dt}(mv_C) = \dot{m}v_C + m\dot{v}_C$$

12. The law of accumulation, production, and transport applies only to scalar quantities.

13. A central force depends only on the distance *r* between the attracted and attracting particles.

14. There are central forces besides gravity.

15. In every central force problem the angular momentum about the attracting point *O* is a constant.

16. All three of Kepler's laws apply to motions of a particle under the influence of any type of central force.

17. All central force problems result in paths which are conics.

18. In a gravitational central force problem, the type of conic is determined by the eccentricity.

Answers

1. F **2.** F **3.** T **4.** T **5.** T **6.** F **7.** F **8.** F **9.** T **10.** T **11.** F **12.** F **13.** T **14.** T **15.** T **16.** F **17.** F **18.** T

Appendices Contents

A UNITS

The numerical value assigned to a physical entity expresses the relationship of that entity to certain standards of measurement called **units**. There is currently an international set of standards called the International System (SI) of Units, a descendant of the meter-kilogram-second (mks) metric system. In the SI system the unit of time is the **second** (s), the unit of length is the **meter** (m), and the unit of mass is the **kilogram** (kg). These independent (or *basic*) units are defined by physical entities or phenomena. The second is defined by the period of a radiation occurring in atomic physics; the meter is defined by the wavelength of a different radiation; the kilogram is defined to be the mass of a certain body of material stored in France. Any other SI units we shall need are *derived* from these three basic units. For instance, the unit of force, the **newton** (N), is a derived quantity in the SI system, as we shall see.

Until recently almost all engineers in the United States used the system (sometimes called British gravitational or U.S.) in which the basic units are the second (sec) for time, the **foot** (ft) for length,* and the **pound** (lb) for force. The pound is the weight, at a standard gravitational condition (location), of a certain body of material stored in the United States. In this system the unit of mass, the **slug**, is a derived quantity. It is a source of some confusion that sometimes there is used a unit of mass called the pound (the mass whose weight is one pound of force at standard gravitational conditions); also, particularly in Europe, the term *kilogram* is also sometimes used for a unit of force.† Grocery shoppers in the United States are exposed to this confusion by the fact that packages are marked by weight (or is it mass?) both in pounds and in kilograms. Throughout this text, without exception, *the pound is a unit of force* and *the kilogram is a unit of mass.*

The United States is currently in the painful process of gradual changeover to the metric system of units after more than 200 years of attachment to the U.S. system. The new engineers who begin practicing their profession in the 1990s will doubtless encounter both systems, and thus it is crucial to master both (including *thinking* in terms of the units of either) and to be able to convert from one to the other. The units mentioned here are summarized in Table A.1 for the SI and the U.S. systems.

---

* Sometimes, particularly in the field of mechanical vibrations, the inch is used as the unit of length; in that case the unit of mass is 1 lb-sec^2/in., which equals 12 slugs.

† A kilogram *was* a force unit in one of two mks systems, compounding the misunderstanding.

Table A.1

Quantity	SI (Standard International or "Metric") Unit	U.S. Unit
force	newton (N)	pound (lb)
mass	kilogram (kg)	slug
length	meter (m)	foot (ft)
time	second (s)	second (sec)

We now examine how the newton of force is derived in SI units and the slug of mass is derived in U.S. units. Let the dimensions of the four basic dimensional quantities be labeled as F (force), M (mass), L (length), and T (time). From the first law of motion (discussed in detail in Chapter 2), $\mathbf{F} = m\mathbf{a}$, we observe that the four basic units are always related as follows:

$$F = \frac{ML}{T^2}$$

This means, of course, that we may select three of the units as basic and derive the fourth. Two ways in which this has been done are the *gravitational* and the *absolute* systems. The former describes the U.S. system; the latter describes SI. (See Table A.2.) Therefore, in U.S. units the mass of an object weighing W lb is $W/32.2$ slugs. Similarly, in SI units the weight of an object having a mass of M kg is $9.81M$ newtons.

Table A.2

Gravitational System	Absolute System
The basic units are force, length, and time, and mass is derived: $M = \frac{FT^2}{L}$ This system has traditionally been more popular with engineers. As an example, in the U.S. system of units the pound, foot, and second are basic. Thus the mass unit, the slug, is derived: $1 \text{ slug} = 1 \frac{\text{lb-sec}^2}{\text{ft}}$ This is summed up by: A slug is the quantity of mass that will be accelerated at 1 ft/sec^2 when acted upon by a force of 1 lb.	The basic units are mass, length, and time, and force is derived: $F = \frac{ML}{T^2}$ This system has traditionally been more popular with physicists. As an example, in the SI (metric) system of units the kilogram, meter, and second are basic. Thus the force unit, the newton, is derived: $1 \text{ newton} = 1 \frac{\text{kg} \cdot \text{m}}{\text{s}^2}$ This is summed up by: A newton is the amount of force that will accelerate a mass of 1 kg at 1 m/s^2.

In the SI system the unit of moment of force is the newton · meter (N · m); in the U.S. system it is the pound-foot (lb-ft). Work and energy have this same dimension; the U.S. unit is the ft-lb whereas the SI unit is the joule (J), which equals 1 N · m. In the SI system the unit of power is called the watt (W) and equals one joule per second (J/s); in the U.S. system it is the ft-lb/sec. The unit of pressure or stress in the SI system is called the pascal (Pa) and equals 1 N/m^2; in the U.S. system it is the lb/ft^2, although often the inch is used as the unit of length so that the unit of pressure is the $lb/in.^2$ (or psi). In both systems the unit of frequency is called the hertz (Hz), which is one cycle per second. Other units of interest in dynamics include those in Table A.3.

Table A.3

Quantity	SI Unit	U.S. Unit
velocity	m/s	ft/sec
angular velocity	rad/s	rad/sec
acceleration	m/s^2	ft/sec^2
angular acceleration	rad/s^2	rad/sec^2
mass moment of inertia	$kg \cdot m^2$	$slug\text{-}ft^2$
momentum	$kg \cdot m/s$	slug-ft/sec
moment of momentum	$kg \cdot m^2/s$	$slug\text{-}ft^2/sec$
impulse	$N \cdot s(= kg \cdot m/s)$	lb-sec
angular impulse	$N \cdot m \cdot s(= kg \cdot m^2/s)$	lb-ft-sec
mass density	kg/m^3	$slug/ft^3$
specific weight	N/m^3	lb/ft^3

Moreover, in the SI system there are standard prefixes to indicate multiplication by powers of 10. For example, kilo (k) is used to indicate multiplication by 1000, or 10^3; thus 5 kilonewtons, written 5 kN, stands for 5×10^3 N. Other prefixes that commonly appear in engineering are shown in Table A.4. We reemphasize that for the foreseeable future American engineers will find it desirable to know both the U.S. and SI systems well; for that reason we have used both sets of units in examples and problems throughout this book.

Table A.4

tera	T	10^{12}	centi	c	10^{-2}
giga	G	10^9	milli	m	10^{-3}
mega	M	10^6	micro	μ	10^{-6}
kilo	k	10^3	nano	n	10^{-9}
hecto	h	10^2	pico	p	10^{-12}
deka	da	10^1	femto	f	10^{-15}
deci	d	10^{-1}	atto	a	10^{-18}

We turn now to the question of unit conversion. The conversion of units is quickly and efficiently accomplished by multiplying by equivalent fractions until the desired units are achieved. Suppose we wish to know how many newton-meters ($N \cdot m$) of torque are equivalent to 1 lb-ft. Since we know there to be 3.281 ft per meter and 4.448 N per pound,

$$1 \text{ lb-ft} = 1 \cancel{\text{lb}}\text{-}\cancel{\text{ft}} \left(\frac{1 \text{ m}}{3.281 \cancel{\text{ft}}}\right)\left(\frac{4.448 \text{ N}}{1 \cancel{\text{lb}}}\right) = 1.356 \text{ N} \cdot \text{m}$$

Note that if the undesired unit (such as lb in this example) does not cancel, the conversion fraction is upside-down!

For a second example, let us find how many slugs of mass there are in a kilogram:

$$1 \text{ kg} = 1\,\frac{\cancel{N}\cdot s^2}{\cancel{m}}\cdot\left(\frac{1\text{ lb}}{4.448\ \cancel{N}}\right)\cdot\left(\frac{1\ \cancel{m}}{3.281\text{ ft}}\right) = \frac{1}{14.59}\,\frac{\text{lb-sec}^2}{\text{ft}} = 0.06852\text{ slug}$$

Inversely, 1 slug = 14.59 kg. A set of conversion factors to use in going back and forth between SI and U.S. units is given in Table A.5.*

Table A.5

To Convert From	To	Multiply By	Reciprocal (to Get from SI to U.S. Units)
Length, area, volume			
foot (ft)	meter (m)	0.30480	3.2808
inch (in.)	m	0.025400	39.370
statute mile (mi)	m	1609.3	6.2137×10^{-4}
foot² (ft²)	meter² (m²)	0.092903	10.764
inch² (in.²)	m²	6.4516×10^{-4}	1550.0
foot³ (ft³)	meter³ (m³)	0.028317	35.315
inch³ (in.³)	m³	1.6387×10^{-5}	61024
Velocity			
feet/second (ft/sec)	meter/second (m/s)	0.30480	3.2808
feet/minute (ft/min)	m/s	0.0050800	196.85
knot (nautical mi/hr)	m/s	0.51444	1.9438
mile/hour (mi/hr)	m/s	0.44704	2.2369
mile/hour (mi/hr)	kilometer/hour (km/h)	1.6093	0.62137
Acceleration			
feet/second² (ft/sec²)	meter/second² (m/s²)	0.30480	3.2808
inch/second² (in./sec²)	m/s²	0.025400	39.370
Mass			
pound-mass (lbm)	kilogram (kg)	0.45359	2.20462
slug (lb-sec²/ft)	kg	14.594	0.068522
Force			
pound (lb) or pound-force (lbf)	newton (N)	4.4482	0.22481
Density			
pound-mass/inch³ (lbm/in.³)	kg/m³	2.7680×10^{4}	3.6127×10^{-5}
pound-mass/foot³ (lbm/ft³)	kg/m³	16.018	0.062428
slug/foot³ (slug/ft³)	kg/m³	515.38	0.0019403
Energy, work, or moment of force			
foot-pound or pound-foot (ft-lb) (lb-ft)	joule (J) or newton · meter (N · m)	1.3558	0.73757
Power			
foot-pound/minute (ft-lb/min)	watt (W)	0.022597	44.254
horsepower (hp) (550 ft-lb/sec)	W	745.70	0.0013410
Stress, pressure			
pound/inch² (lb/in.² or psi)	N/m² (or Pa)	6894.8	1.4504×10^{-4}
pound/foot² (lb/ft²)	N/m² (or Pa)	47.880	0.020886
Mass moment of inertia			
slug-foot² (slug-ft² or lb-ft-sec²)	kg · m²	1.3558	0.73756

* Rounded to the five digits cited. Note, for example, that 1 ft = 0.30480 m, so that

$$(\text{Number of feet}) \times \left(\frac{0.30480\text{ m}}{1\text{ ft}}\right) = \text{number of meters}$$

Table A.5 Continued

To Convert From	To	Multiply By	Reciprocal (to Get from SI to U.S. Units)
Momentum (or linear momentum)			
slug-foot/second (slug-ft/sec)	kg · m/s	4.4482	0.22481
Impulse (or linear impulse)			
pound-second (lb-sec)	N · s (or kg · m/s)	4.4482	0.22481
Moment of momentum (or angular momentum)			
slug-foot^2/second (slug-ft^2/sec)	kg · m^2/s	1.3558	0.73756
Angular impulse			
pound-foot-second (lb-ft-sec)	N · m · s (or kg · m^2/s)	1.3558	0.73756

Note that the units for time (s or sec), angular velocity (rad/s or 1/s), and angular acceleration (rad/s^2 or 1/s^2) are the same for the two systems. To five digits, the acceleration of gravity at sea level is 32.174 ft/s^2 in the U.S. system and 9.8067 m/s^2 in SI units.

We wish to remind the reader of the care that must be exercised in numerical calculations involving different units. For example, if two lengths are to be summed in which one length is 2 ft and the other is 6 in., the simple sum of these measures, $2 + 6 = 8$, does not of course provide a measure of the desired length. It is also true that we may not add or equate the numerical measures of different types of entities; thus it makes no sense to attempt to add a mass to a length. These are said to have different dimensions. A dimension is the name assigned to the *kind* of measurement standard involved as contrasted with the choice of a particular measurement standard (unit). In science and engineering we attempt to develop equations expressing the relationships among various physical entities in a physical phenomenon. We express these equations in symbolic form so that they are valid regardless of the choice of a system of units, but nonetheless they must be *dimensionally consistent.* In the following equation, for example, we may check that the units on the left and right sides agree; r is a radial distance, P is a force, and dots denote time derivatives:

$$P - mg\cos\theta = m(\ddot{r} - r\dot{\theta}^2)$$

Dimensions of . . .	. . . are
P	F
$mg\cos\theta$	$M\left(\frac{L}{T^2}\right)(1) = F$
$m\ddot{r}$	$M\frac{L}{T^2} = F$
$-mr\dot{\theta}^2$	$ML\left(\frac{1}{T}\right)^2 = F$

Therefore the units of (every term in) the equation are those of force. If such a check is made prior to the substitution of numerical values, much time can be saved if an error has been made.

PROBLEMS ▸ Appendix A

A.1 Find the units of the universal gravitational constant G, defined by

$$F = \frac{GMm}{r^2}$$

in (a) the SI system and (b) the U.S. system.

A.2 Find the weight in pounds of 1 kg of mass.

A.3 Find the weight in newtons of 1 slug of mass.

A.4 One pound-mass (lbm) is the mass of a substance that is acted on by 1 lb of gravitational force at sea level. Find the relationship between (a) 1 lbm and 1 slug; (b) 1 lbm and 1 kg.

A.5 The momentum of a body is the product of its mass m and the velocity v_C of its mass center. A child throws an 8-oz ball into the air with an initial speed of 20 mph. Find the magnitude of the momentum of the ball in (a) slug-ft/sec; (b) kg · m/s.

A.6 Is the following equation dimensionally correct?

$$\int_0^5 Fv\,dt = \frac{mv^2}{2} + ma \qquad (v = \text{velocity}; \; a = \text{acceleration})$$

A.7 The equation for the distance r_s from the center of the earth to the geosynchronous satellite orbit is

$$gr_e^2 = r_s^3\omega^2 \qquad (\omega = \text{angular speed of earth}; \; r_e = \text{earth radius})$$

a. Show that the equation is dimensionally correct.

b. Use the equation to find the ratio of the orbit radius to earth radius.

A.8 The universal gravitational constant is $G = 6.67 \times 10^{-11}$ N · m²/kg². Express G in units of lb-ft²/slug².

B Examples of Numerical Analysis/The Newton-Raphson Method

There are a few places in this book where equations arise whose solutions are not easily found by elementary algebra; they are either polynomials of degree higher than 2 or else transcendental equations. In this appendix we explain in brief the fundamental idea behind the Newton-Raphson numerical method for solving such equations. We shall first do this while applying the method to the solution for one of the roots of a cubic polynomial equation that occurs in Chapter 7.

To solve the cubic equation of Example 7.5,

$$f(\mathcal{J}) = -\mathcal{J}^3 + 48\mathcal{J}^2 - 633\mathcal{J} + 1342 = 0$$

we could, alternatively, use the Newton-Raphson algorithm. This procedure finds a root of the equation $f(\mathcal{J}) = 0$ (it need not be a polynomial equation, however) by using the slope of the curve. The algorithm, found in more detail in any book on numerical analysis, works as follows. If $\mathcal{J}_{1_0}$ is an initial estimate of a root $\mathcal{J}_1$, then a better approximation is

$$\mathcal{J}_{1_1} = \mathcal{J}_{1_0} - \frac{f(\mathcal{J}_{1_0})}{f'(\mathcal{J}_{1_0})}$$

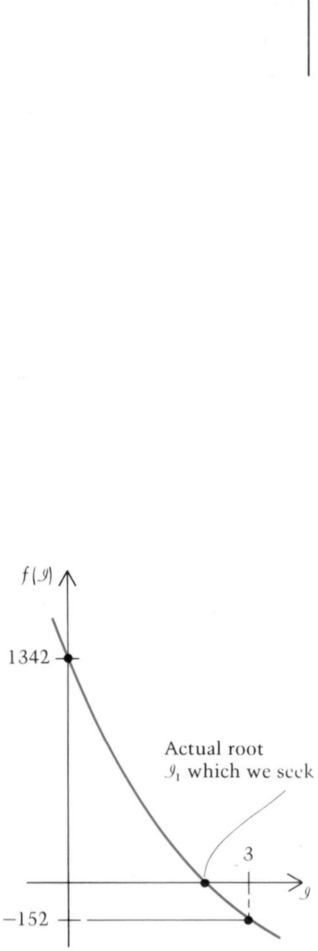

Figure B.1

Figures B.1 and B.2 indicate what is happening. The quantity $f(\mathcal{J}_{1_0})/f'(\mathcal{J}_{1_0})$ causes a backup in the $\mathcal{J}_1$ approximation — in our case from the *initial* value of 3 to the *improved* estimate $\mathcal{J}_{1_1}$:

$$\begin{aligned}\mathcal{J}_{1_1} = 3 - \frac{f(3)}{f'(3)} &= 3 - \frac{-152}{-372}\\ &= 3 - 0.408602150\\ &= 2.591397850\end{aligned}$$

where

$$f'(\mathcal{J}) = -3\mathcal{J}^2 + 96\mathcal{J} - 633$$

so that $f'(3) = -372$. Repeating the algorithm, we get

$$\begin{aligned}\mathcal{J}_{1_2} &= \mathcal{J}_{1_1} - \frac{f(\mathcal{J}_{1_1})}{f'(\mathcal{J}_{1_1})}\\ &= 2.591397850 - \frac{6.579491260}{-404.3718348}\\ &= 2.591397850 + 0.016270894\\ &= 2.607668744\end{aligned}$$

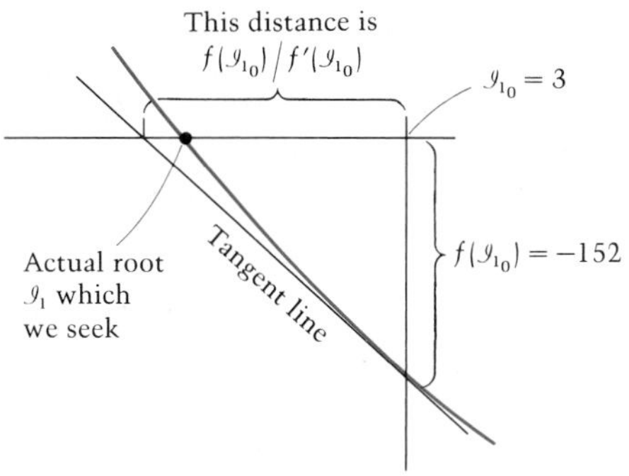

Figure B.2

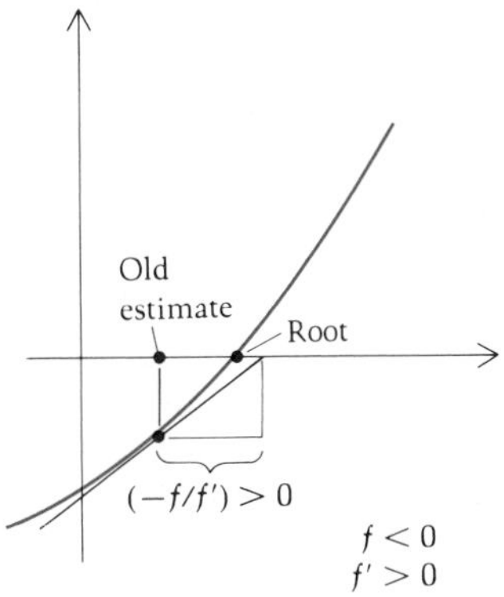

Figure B.3

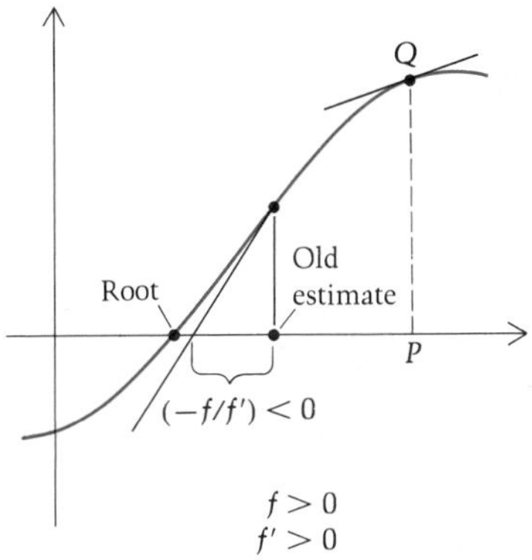

Figure B.4

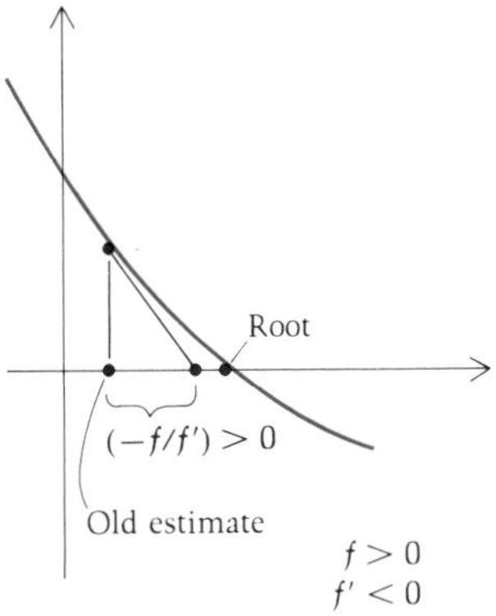

Figure B.5

And one more time:

$$\vartheta_{1_3} = \vartheta_{1_2} - \frac{f(\vartheta_{1_2})}{f'(\vartheta_{1_2})}$$

$$= 2.607668744 - \frac{0.010645000}{-403.0636094}$$

$$= 2.607668744 + 0.000026410$$
$$= 2.607695154$$

This algorithm is easily programmed on a computer. After doing this, the results (with the same initial guess $\vartheta_{1_0} = 3$) are:

$$\vartheta_{1_0} = 3$$
$$\vartheta_{1_1} = 2.591397850$$
$$\vartheta_{1_2} = 2.607668744$$
$$\vartheta_{1_3} = 2.607695154$$
$$\vartheta_{1_4} = 2.607695156$$
$$\left.\begin{aligned} \vartheta_{1_5} &= 2.607695153 \\ \vartheta_{1_6} &= 2.607695153 \\ \vartheta_{1_7} &= 2.607695153 \end{aligned}\right\} \quad \text{convergence!}$$

which is in agreement with the results in Example 7.5.

Incidentally, note from Figures B.3 to B.5 that adding $(-f/f')$ to form the new estimate works equally well for the three other sign combinations of f and f'. Note also that if the estimate is *too far* from the root, such as P in Figure B.4, the procedure might not converge; the tangent at Q in this case would send us far from the desired root.

We next consider the equation from Problem 5.140 when $M = 4m$:

$$f(\theta) = \sin\theta - \frac{\theta}{2} = 0 \tag{B.1}$$

with the derivative of f being

$$f'(\theta) = \cos\theta - \frac{1}{2}$$

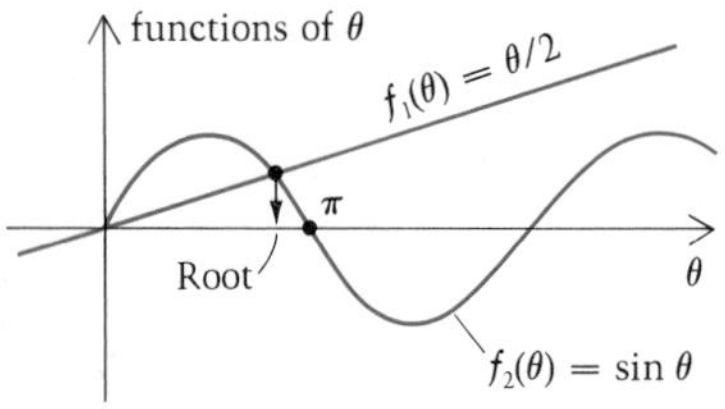

Figure B.6

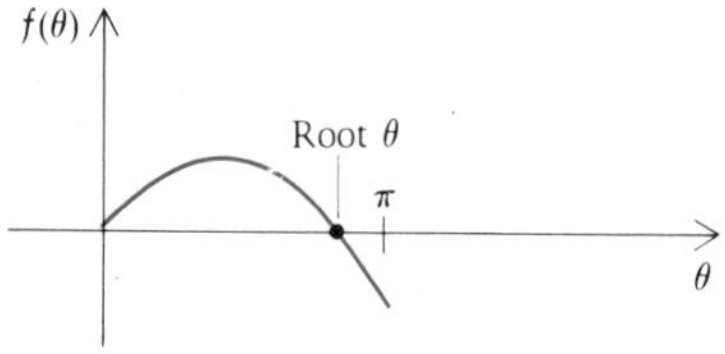

Figure B.7

There is but one root of Equation (B.1) for $\theta > 0$, as can be seen from Figure B.6, which shows the two functions making up $f(\theta)$. To find this root, we can use Newton-Raphson as previously described. Figure B.7 suggests that π might serve as a good first guess at the root. A Newton-Raphson program shows that it is, and yields the answer below very quickly:

$$\theta_0 = 3.141592654$$

$$\theta_1 = 2.094395103$$

$$\theta_2 = 1.913222955$$

$$\theta_3 = 1.895671752$$

$$\theta_4 = 1.895494285$$

$$\left.\begin{aligned}\theta_5 &= 1.895494267\\ \theta_6 &= 1.895494267\\ \theta_7 &= 1.895494267\end{aligned}\right\} \quad \text{convergence!}$$

The last example in this appendix will be to solve the equation

$$\cos\left(\frac{\pi}{4} - q\right) = 0.373q$$

from Example 2.6. We write this equation as

$$f(q) = \cos\left(\frac{\pi}{4} - q\right) - 0.373q = 0$$

with

$$f'(q) = \sin\left(\frac{\pi}{4} - q\right) - 0.373$$

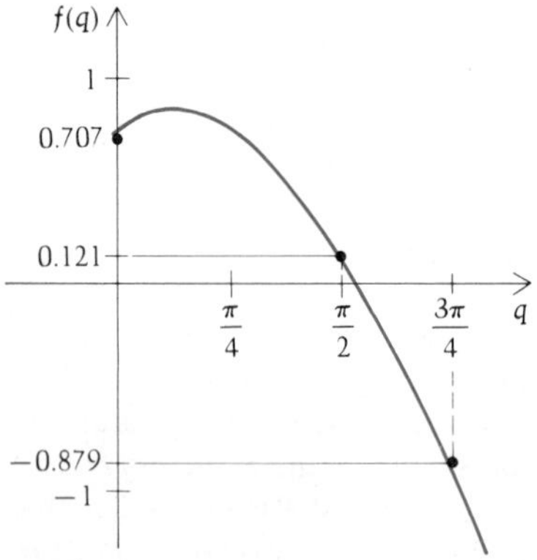

Figure B.8

The rough plot in Figure B.8 shows a few points which indicate that $\pi/2$ is fairly close to the root. Here are the results of a program, which uses the Newton-Raphson method as in the first two examples, to narrow down on the root quickly and accurately:

$$q_0 = 1.570796327$$

$$q_1 = 1.683007224$$

$$q_2 = 1.679300543$$

$$\left.\begin{aligned}q_3 &= 1.679296821\\ q_4 &= 1.679296821\\ q_5 &= 1.679296821\end{aligned}\right\} \quad \text{convergence!}$$

C

Moments of Inertia of Masses (See also Section 4.3)

	Object	Mass Center Coordinates and Volume V	Moments of Inertia About Indicated Axes
slender rod	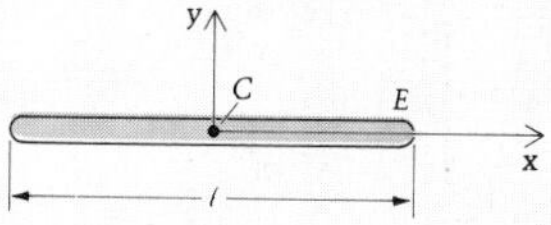	$(0, 0, 0)$ $V = A\ell$ (A = area of cross section)	$I^C_{xx} \approx 0,\ I^C_{yy} = I^C_{zz} \approx \dfrac{m\ell^2}{12}$ (Note: $I^E_{yy} = I^E_{zz} \approx m\ell^2/3$.)
slender circular rod	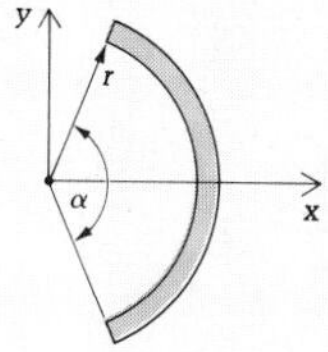	$\left(\dfrac{r \sin(\alpha/2)}{\alpha/2}, 0, 0\right)$ $V = Ar\alpha$ (A = area of cross section)	$I_{xx} = \dfrac{mr^2}{2}\left(1 - \dfrac{\sin\alpha}{\alpha}\right)$ $I_{yy} = \dfrac{mr^2}{2}\left(1 + \dfrac{\sin\alpha}{\alpha}\right)$ $I_{zz} = mr^2$ (Note special cases for $\alpha = \pi$ and 2π.)
bent slender rod	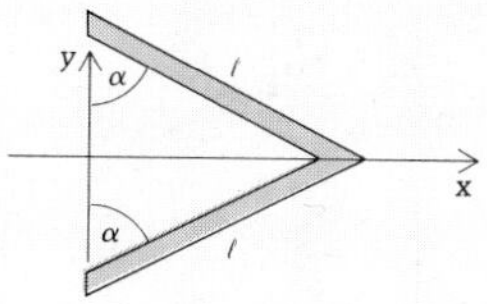	$\left(\dfrac{\ell \sin\alpha}{2}, 0, 0\right)$ $V = 2A\ell$ (A = area of cross section)	$I_{xx} = \dfrac{m\ell^2 \cos^2\alpha}{3}$ $I_{yy} = \dfrac{m\ell^2 \sin^2\alpha}{3}$ $I_{zz} = \dfrac{m\ell^2}{3}$

	Object	Mass Center Coordinates and Volume V	Moments of Inertia About Indicated Axes
rectangular solid	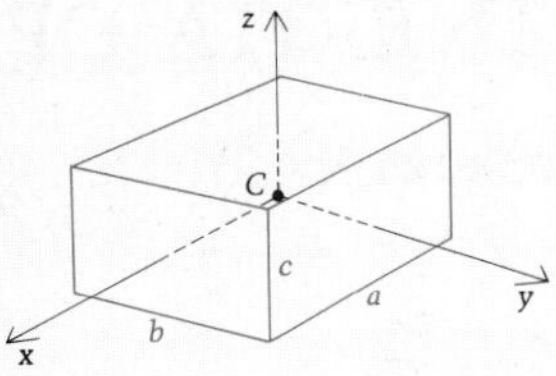	(0, 0, 0) $V = abc$	$I_{xx} = \frac{m}{12}(b^2 + c^2)$ $I_{yy} = \frac{m}{12}(a^2 + c^2)$ $I_{zz} = \frac{m}{12}(a^2 + b^2)$ (Note: If it is a rectangular plate, the thickness dimension is neglected.)
hollow cylinder	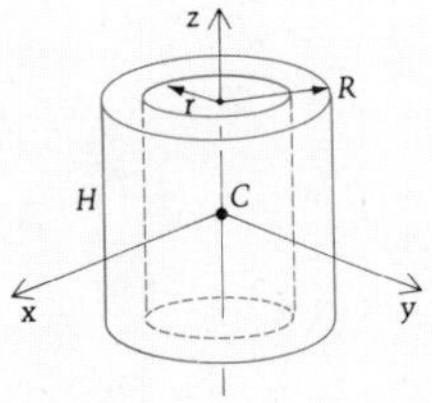	(0, 0, 0) $V = \pi(R^2 - r^2)H$	$I_{xx} = I_{yy} = \frac{m}{4}\left(R^2 + r^2 + \frac{H^2}{3}\right)$ $I_{zz} = \frac{m}{2}(R^2 + r^2)$ (Note special cases when $r = 0$, $r \approx R$, and $H \ll R$, below.) For the case $R \ll H$, see the slender rod above.
Four Special Cases 1. If $r = 0$: solid cylinder	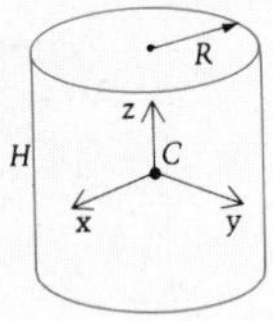	(0, 0, 0) $V = \pi R^2 H$	$I_{xx} = I_{yy} = \frac{mR^2}{4} + \frac{mH^2}{12}$ $I_{zz} = \frac{mR^2}{2}$
2. If $r \approx R$: cylindrical shell	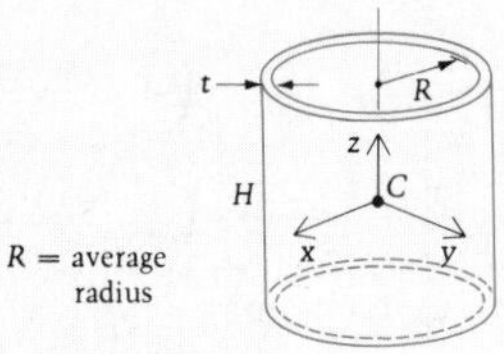	(0, 0, 0) $V = 2\pi RtH$	$I_{xx} = I_{yy} \approx \frac{mR^2}{2} + \frac{mH^2}{12}$ $I_{zz} \approx mR^2$
3. If $H \approx r$: annular disk	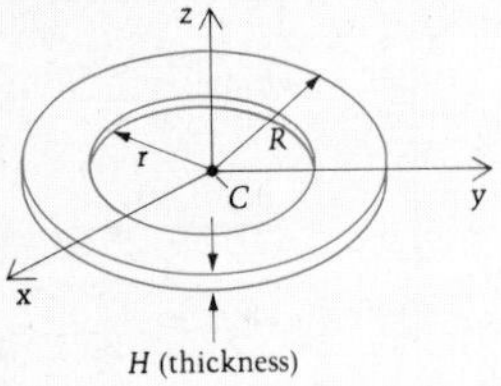	(0, 0, 0) $V = r(R^2 - r^2)H$	$I_{xx} = I_{yy} \approx \frac{m}{4}(R^2 + r^2)$ $I_{zz} = \frac{m(R^2 + r^2)}{2}$

	Object	Mass Center Coordinates and Volume V	Moments of Inertia About Indicated Axes
4. If $H \ll r \approx R$: thin ring with rectangular cross section (area)	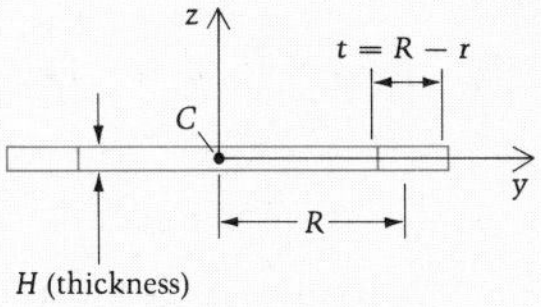	(0, 0, 0) $V = 2\pi RtH$ $= 2\pi R(A)$ $(A = tH$ = area of cross section)	$I_{xx} = I_{yy} \approx \frac{mR^2}{2}$ $I_{zz} \approx mR^2$ (Note: These apply for *any* cross-sectional shape as long as the ring is thin.)
thin right triangular plate	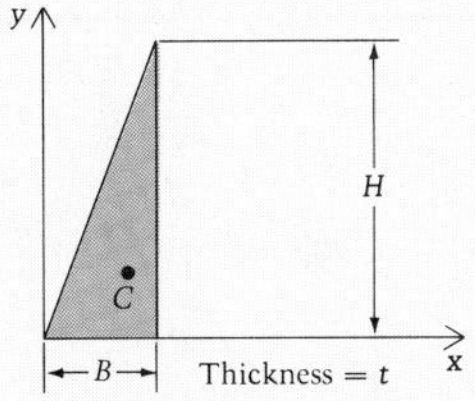	(We are interested here in properties for any base axis, such as x in the figure.) $\left(\frac{2B}{3}, \frac{H}{3}, 0\right)$ $V = \frac{1}{2}BHt$	$I_{xx} = \frac{mH^2}{6}$ $I_{yy} = \frac{mB^2}{2}$ $I_{zz} = \frac{mH^2}{6} + \frac{mB^2}{2}$ $I_{xy} = -\frac{mBH}{4}$
thin elliptical plate	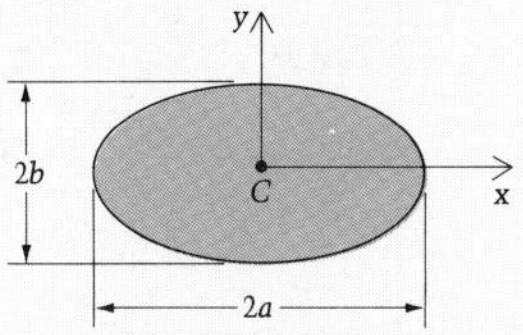 Thickness = t in z direction	(0, 0, 0) $V = \pi abt$	$I_{xx} = \frac{mb^2}{4}$ $I_{yy} = \frac{ma^2}{4}$ $I_{zz} = \frac{m(a^2 + b^2)}{4}$
thin paraboloidal plate	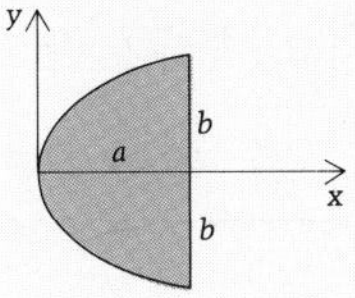 Thickness = t in z direction	$\left(\frac{3a}{5}, 0, 0\right)$ $V = \frac{4}{3}abt$	$I_{xx} = \frac{mb^2}{5}$ $I_{yy} = \frac{3ma^2}{7}$ $I_{zz} = \frac{m(15a^2 + 7b^2)}{35}$
thin circular sector plate	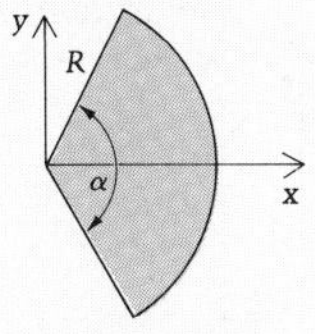 Thickness = t	$\left(\frac{2R \sin(\alpha/2)}{3\alpha/2}, 0, 0\right)$ $V = \frac{\alpha R^2 t}{2}$	$I_{xx} = \frac{mR^2}{4}\left(1 - \frac{\sin\alpha}{\alpha}\right)$ $I_{yy} = \frac{mR^2}{4}\left(1 + \frac{\sin\alpha}{\alpha}\right)$ $I_{zz} = \frac{mR^2}{2}$
Two Special Cases 1. If $\alpha = \pi$: semicircular plate	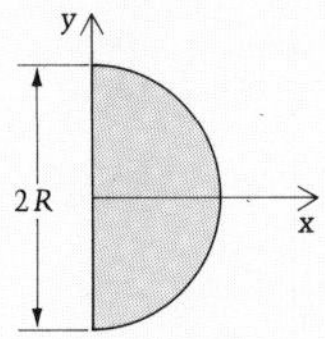 Thickness = t	$\left(\frac{4R}{3\pi}, 0, 0\right)$ $V = \frac{\pi R^2 t}{2}$	$I_{xx} = I_{yy} \approx \frac{mR^2}{4}$ $I_{zz} = \frac{mR^2}{2}$

	Object	Mass Center Coordinates and Volume V	Moments of Inertia About Indicated Axes
2. If $\alpha = 2\pi$: circular plate	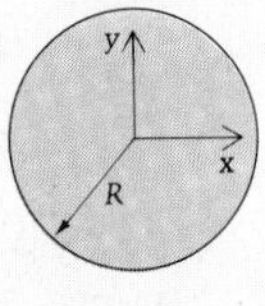 Thickness = t	$(0, 0, 0)$ $V = \pi R^2 t$	$I_{xx} = I_{yy} \approx \dfrac{mR^2}{4}$ $I_{zz} = \dfrac{mR^2}{2}$ (Note: The results appear to be the same as those of the semicircular plate, but the masses differ by a factor of 2.)
thin circular segment plate	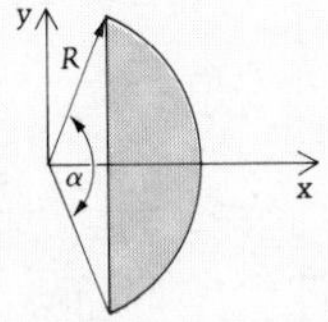	$\left(\dfrac{4R \sin^3 (\alpha/2)}{3(\alpha - \sin \alpha)}\right)$ $V = \dfrac{R^2 t}{2}(\alpha - \sin \alpha)$	$I_{xx} = \dfrac{mR^2}{12}(3 - k)$ $I_{yy} = \dfrac{mR^2}{4}(1 + k)$ $I_{zz} = \dfrac{mR^2}{6}(3 + k)$ where $k = \dfrac{(1 - \cos \alpha) \sin \alpha}{\alpha - \sin \alpha}$ (Note special cases for $\alpha = \pi, 2\pi$.)
rectangular tetrahedron	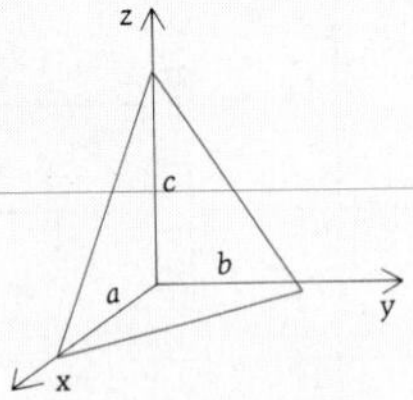	$\left(\dfrac{a}{4}, \dfrac{b}{4}, \dfrac{c}{4}\right)$ $V = \dfrac{abc}{6}$	$I_{xx} = \dfrac{m}{10}(b^2 + c^2)$ $I_{yy} = \dfrac{m}{10}(a^2 + c^2)$ $I_{zz} = \dfrac{m}{10}(a^2 + b^2)$
hollow sphere	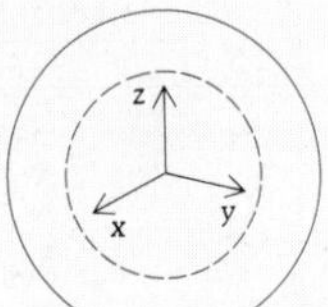	$(0, 0, 0)$ $V = \frac{4}{3}\pi(R^3 - r^3)$	$I_{xx} = I_{yy} = I_{zz} = \dfrac{2}{5} m \left(\dfrac{R^5 - r^5}{R^3 - r^3}\right)$ (Note: If $r = 0$, get $\frac{2}{5}mR^2$, and if $r \approx R$ (spherical shell), get $\frac{2}{3}mR^2$ with $V \approx 4\pi R^2 t$, where $t = R - r =$ thickness of shell. For the shell result, $R - r$ divides both numerator and denominator evenly.)
solid ellipsoid	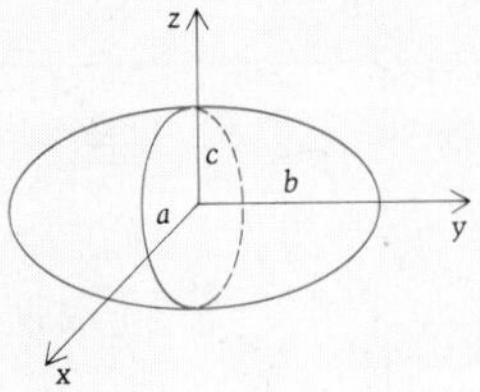	$(0, 0, 0)$ $V = \frac{4}{3}\pi abc$	$I_{xx} = \dfrac{m}{5}(b^2 + c^2)$ $I_{yy} = \dfrac{m}{5}(a^2 + c^2)$ $I_{zz} = \dfrac{m}{5}(a^2 + b^2)$

	Object	Mass Center Coordinates and Volume V	Moments of Inertia About Indicated Axes
solid spherical cap	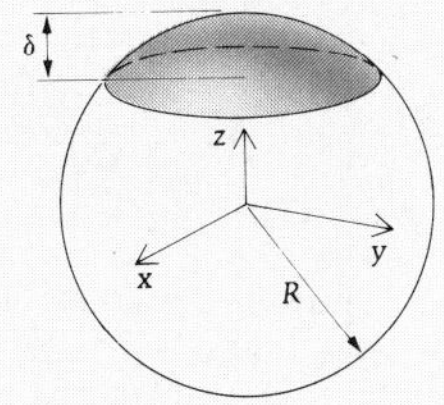	$\left(0, 0, \frac{3(2R-\delta)^2}{4(3R-\delta)}\right)$ $V = \frac{\pi}{3}\delta^2(3R-\delta)$	$I_{xx} = I_{yy}$ $= \frac{m}{2}\left[2R^2 - \frac{3(10R^2-\delta^2)\delta}{5(3R-\delta)} + \frac{3\delta^2}{2}\right]$ $I_{zz} = \frac{m\delta}{10}\left[\frac{20R^2 - 15R\delta + 3\delta^2}{3R-\delta}\right]$ (Note: If $\delta = R$, we have a hemisphere and $I_{xx} = I_{yy} = I_{zz} = \frac{2}{5}mR^2$, with $V = \frac{2}{3}\pi R^3$ and $\bar{z} = \frac{3}{8}R$.)
paraboloid of revolution	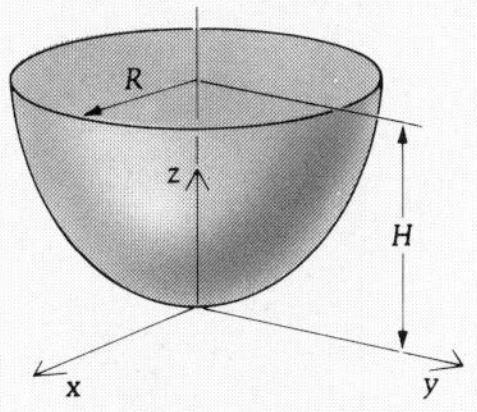	$(0, 0, \frac{2}{3}H)$ $V = \frac{\pi R^2 H}{2}$	$I_{xx} = I_{yy} = \frac{m}{6}(R^2 + 3H^2)$ $I_{zz} = \frac{mR^2}{3}$
elliptic paraboloid	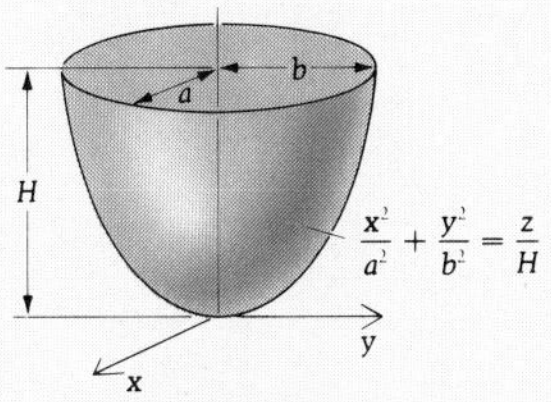	$\left(0, 0, \frac{2H}{3}\right)$ $V = \frac{\pi abH}{2}$	$I_{xx} = \frac{m}{6}(b^2 + 3H^2)$ $I_{yy} = \frac{m}{6}(a^2 + 3H^2)$ $I_{zz} = \frac{m}{6}(a^2 + b^2)$
solid cone	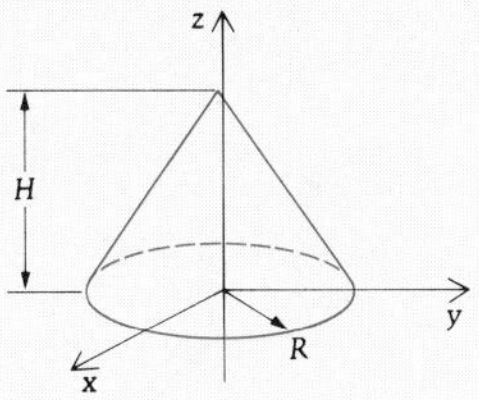	$\left(0, 0, \frac{H}{4}\right)$ $V = \frac{\pi R^2 H}{3}$	$I_{xx} = I_{yy} = \frac{m}{20}(3R^2 + 2H^2)$ $I_{zz} = \frac{3}{10}mR^2$
solid right rectangular prism	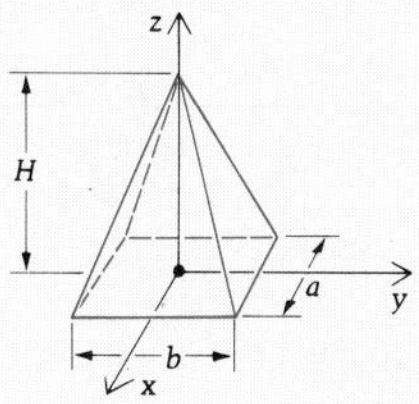	$\left(0, 0, \frac{H}{4}\right)$ $V = \frac{abH}{3}$	$I_{xx} = \frac{m}{80}(4b^2 + 8H^2)$ $I_{yy} = \frac{m}{80}(4a^2 + 8H^2)$ $I_{zz} = \frac{m}{20}(a^2 + b^2)$

	Object	Mass Center Coordinates and Volume V	Moments of Inertia About Indicated Axes
solid toroid	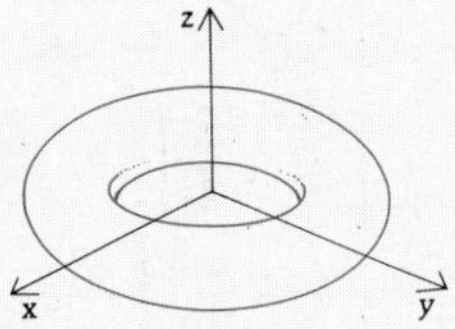	$(0, 0, 0)$ $V = 2\pi^2 R r^2$	$I_{xx} = I_{yy} = \frac{m}{8}(4R^2 + 5r^2)$ $I_{zz} = \frac{m}{4}(4R^2 + 3r^2)$ (Note: If $R \gg r$, we have a hoop for which $I_{xx} = I_{yy} = mR^2/2$ and $I_{zz} = mR^2$.)
frustum of cone	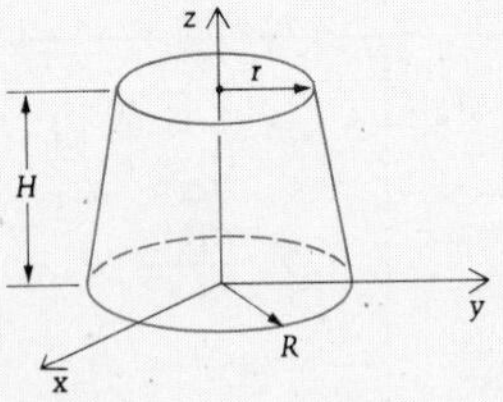	$\left(0, 0, \frac{H(R^2 + 2Rr + 3r^2)}{4(R^2 + Rr + r^2)}\right)$ $V = \frac{\pi H}{3}(R^2 + Rr + r^2)$	$I_{xx} = I_{yy} = \frac{m}{20}\left[3(R^2 + r^2) + \frac{(2R^2 + 6Rr + 12r^2)H^2 - 3r^2R^2}{R^2 + Rr + r^2}\right]$ $I_{zz} = \frac{3m(R^5 - r^5)}{10(R^3 - r^3)}$

D ANSWERS TO ODD-NUMBERED PROBLEMS

In the solutions to problems in Chapters 1–5, unless identified otherwise below, $\hat{\mathbf{i}}$, $\hat{\mathbf{j}}$, and $\hat{\mathbf{k}}$ are unit vectors in the respective directions →, ↑, and out of the page. In Chapters 6–8, the unit vectors are respectively parallel to axes defined in the problems.

CHAPTER 1

1.1 $18\hat{\mathbf{j}} - 8\hat{\mathbf{k}}$ kg · m/s²
1.3 $11.2\hat{\mathbf{i}} + 20\hat{\mathbf{j}} - 30\hat{\mathbf{k}}$ slug-ft/sec²
1.5 $5\hat{\mathbf{i}} + 0.889\hat{\mathbf{j}} - 0.222\hat{\mathbf{k}}$ slug-ft/sec²
1.7 $0.00420\hat{\mathbf{i}} + 29.7\hat{\mathbf{j}}$ kg · m/s²
1.9 $6\hat{\mathbf{i}} + 117\hat{\mathbf{j}} - 84\hat{\mathbf{k}}$ N · s
1.11 $-57.2\hat{\mathbf{i}} + 210\hat{\mathbf{j}} - 195\hat{\mathbf{k}}$ lb-sec
1.13 $52.5\hat{\mathbf{i}} - 3.60\hat{\mathbf{j}} + 0.630\hat{\mathbf{k}}$ lb-sec
1.15 $-0.0183\hat{\mathbf{i}} + 213\hat{\mathbf{j}}$ N · s
1.17 Answer given in problem.
1.19 $-1.64\hat{\mathbf{i}} + 12.9\hat{\mathbf{j}}$ ft/sec² **1.21** $\hat{\mathbf{i}} - (\pi/2)\hat{\mathbf{j}}$ ft/sec²
1.23 $120\hat{\mathbf{i}} + 3\hat{\mathbf{j}} - 152\hat{\mathbf{k}}$ m; 194 m
1.25 $\dfrac{-80}{\pi}\hat{\mathbf{i}} + \dfrac{80}{\pi}\hat{\mathbf{j}}$ m; 36.0 m
1.27 $6.08\hat{\mathbf{i}} - 1.11\hat{\mathbf{k}}$ m; 6.18 m **1.29** 387 ft
1.31 $32.3\hat{\mathbf{i}}$ ft/sec²
1.33 15 sec to *return* (21 seconds total elapsed time)
1.35 $v_P = -10t + 150$

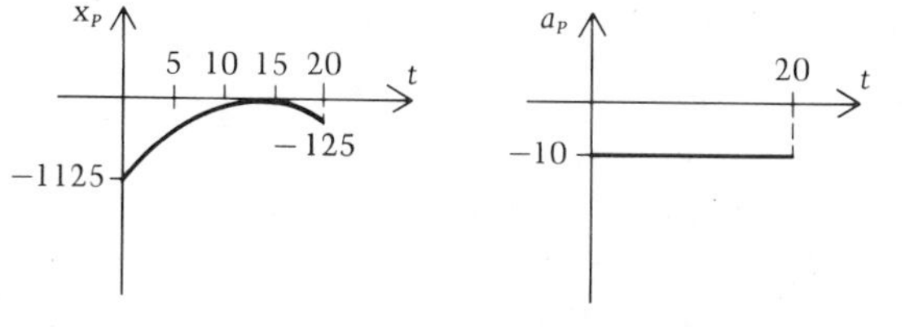

$x_P = -5t^2 + 150t - 1125$ $a_P = -10$

1.37 $v_P = t^2/2$, $a_P = t$, $x_P = \dfrac{t^3}{6} + 10$

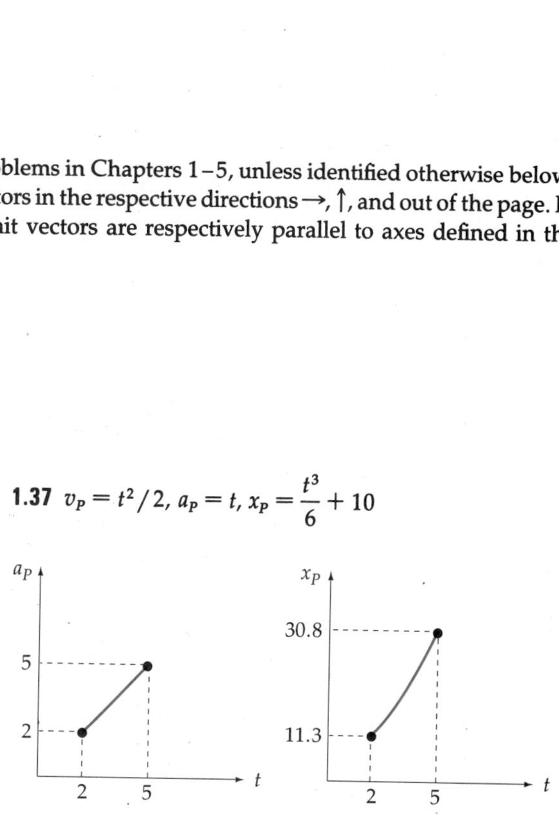

1.39 (a) 2.57 s; (b) 11.7 m/s; (c) 15.0 m
1.41 2 → ft/sec **1.43** 512 ft
1.45 (a) $230\hat{\mathbf{i}}$ m; (b) 234 m **1.47** 8.46 m
1.49 0.469 s **1.51** $\dot{x} = 1/(5t + 1.67)$ m/s
1.53 52.6 sec
1.55 (a) $\mathscr{B}$ wins by 198 ft; (b) 660 ft; (c) 2.4 miles
1.57 $\mathbf{v}_A = 10 \downarrow$ m/s and $\mathbf{v}_B = 20 \uparrow$ m/s (at $t = 2$ s)
1.59 11.3 m **1.61** 6.21 ↑ m/s
1.63 $1.73\left(\dfrac{5\hat{\mathbf{i}} + 12\hat{\mathbf{j}}}{13}\right)$ m/s **1.65** $(8a_0/15)\hat{\mathbf{j}}$
1.67 $20\pi\hat{\mathbf{j}}$; $-20\pi\hat{\mathbf{i}}$; $20\pi\hat{\mathbf{i}}$, $20\pi\hat{\mathbf{j}}$ m/s
1.69 $\mathbf{v}_P = -10 \sin 5t\hat{\mathbf{i}} + 10 \cos 5t\hat{\mathbf{j}}$ m/s;
$\mathbf{a}_P = -50 \cos 5t\hat{\mathbf{i}} - 50 \sin 5t\hat{\mathbf{j}}$ m/s²
1.71 $3\hat{\mathbf{i}} + 6\hat{\mathbf{j}}$ ft/sec; $18\hat{\mathbf{j}}$ ft/sec²
1.73 (a) $1.5\hat{\mathbf{i}} + 11.8\hat{\mathbf{j}}$ m/s²; (b) 0 and 2.52 s

1.75 (a) $\mathbf{v}_P = -6\pi \sin \frac{\pi t}{2}\hat{\mathbf{i}} + 4\pi \cos \frac{\pi t}{2}\hat{\mathbf{j}}$ m/s;

$\mathbf{a}_P = -3\pi^2 \cos \frac{\pi t}{2}\hat{\mathbf{i}} - 2\pi^2 \sin \frac{\pi t}{2}\hat{\mathbf{j}}$ m/s²

(b) $\mathbf{r}_{OP} = 12\hat{\mathbf{i}}$ m; $\mathbf{v}_P = 4\pi\hat{\mathbf{j}}$ m/s; $\mathbf{a}_P = -3\pi^2\hat{\mathbf{i}}$ m/s²
(c) $(x/12)^2 + (y/8)^2 = 1$ (an ellipse) **1.77** kr
1.79 $y = \frac{4}{3}x - 11$, a straight line
1.81 $\frac{(x-2)^2}{3^2} + \frac{y^2}{4^2} = 1$, an ellipse
1.83 $3\hat{\mathbf{i}} + 0.0968\hat{\mathbf{j}}$ in./sec
1.85 $3\hat{\mathbf{i}} - 0.00312\hat{\mathbf{j}}$ in./sec **1.87** $-0.00937\hat{\mathbf{j}}$ in./sec²
1.89 $0.000604\hat{\mathbf{j}}$ in./s² **1.91** 14.9 m at 0.690 s
1.93 $0.671\hat{\mathbf{i}} + 2.30\hat{\mathbf{j}} + 0.500\hat{\mathbf{k}}$ m/s
1.95 (a) 7 m/s; (b) $4.46 \times 10^5\hat{\mathbf{k}}$ m/s²; (c) 3 m/s
1.97 For x within the intervals (290, 1200) ft and (1800, 2700) ft
1.99 0.000219 ft/sec²; at $x = 0$, 1500 and 3000 ft
1.101 $(\dot{\theta}, \ddot{\theta}) = (1.29$ rad/sec, -3.44 rad/sec²$)$ or $(-1.29$ rad/sec, 3.44 rad/sec²$)$
1.103 $2\sqrt{2}\hat{\mathbf{j}}$ ft/sec; $-12\sqrt{2}\hat{\mathbf{i}} + 6\sqrt{2}\hat{\mathbf{j}}$ ft/sec²
1.105 In the order $\theta = 0, \pi/2, \pi$, and $-\pi/2$: $\mathbf{v}_P = 2aK\hat{\mathbf{j}}, -aK(\hat{\mathbf{i}} + \hat{\mathbf{j}}), 0$, and $aK(\hat{\mathbf{i}} - \hat{\mathbf{j}})$. Note respectively that $\hat{\mathbf{e}}_r = \hat{\mathbf{i}}, \hat{\mathbf{j}}, -\hat{\mathbf{i}}, -\hat{\mathbf{j}}$ and $\hat{\mathbf{e}}_\theta = \hat{\mathbf{j}}, -\hat{\mathbf{i}}, -\hat{\mathbf{j}}, \hat{\mathbf{i}}$.
1.107 $\mathbf{v}_P = \frac{9}{\sqrt{2}}\hat{\mathbf{e}}_r + \frac{3}{\sqrt{2}}\hat{\mathbf{e}}_\theta$; $\mathbf{a}_P = \frac{18}{\sqrt{2}}(\hat{\mathbf{e}}_r + \hat{\mathbf{e}}_\theta)$
1.109 (a) 0.577 ft/sec; (b) $-0.136\hat{\mathbf{i}} - 0.385\hat{\mathbf{j}}$ ft/sec²
1.111 (a) $-535\hat{\mathbf{e}}_r + 560\hat{\mathbf{e}}_\theta$ m/s² (at $\theta = 114°$, $t = 0.446$ s) (b) $-1380\hat{\mathbf{e}}_r + 160\hat{\mathbf{e}}_\theta$ m/s²
1.113 (a) yes; (b) yes
1.115 $\frac{2.93x}{\sqrt{361 + x^2}}$ ↑ ft/sec; measuring r from pulley to bumper, $\dot{r}$ is the velocity of the shingles; it is also the component of $\mathbf{v}_A$ along the rope.
1.117 First,

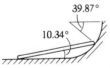

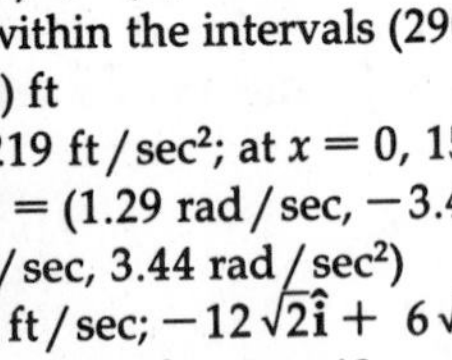

Then, $\mathbf{v}_B = 0.640$ ∠50.1° ft/sec.
1.119 $1.65\hat{\mathbf{i}} + 1.13\hat{\mathbf{j}} + 0.600\hat{\mathbf{k}}$ m; $-0.0226\hat{\mathbf{i}} + 0.0330\hat{\mathbf{j}} + 0.0200\hat{\mathbf{k}}$ m/s
1.121 $R\hat{\mathbf{e}}_r + H\hat{\mathbf{k}}$; $k_1\hat{\mathbf{e}}_r + \frac{2k_2R^2}{k_1}\hat{\mathbf{e}}_\theta + \frac{2Hk_1}{R}\hat{\mathbf{k}}$;

$\frac{-4k_2^2R^3}{k_1^2}\hat{\mathbf{e}}_r + 6Rk_2\hat{\mathbf{e}}_\theta + \frac{2Hk_1^2}{R^2}\hat{\mathbf{k}}$
1.123 largest: 73.3 ft/sec, at the top; smallest: 1.17 ft/sec, at the bottom.
1.125 Answer given in problem.
1.127 Answer given in problem.
1.129 $6.5t^2 + C$, where C is a constant of integration
1.131 16 ft **1.133** -34.6 m/s²; 0.2 m
1.135 4 m/s²; 1.11 m **1.137** 0.471 ft
1.139 $(-\dot{x}_0^2/a)\hat{\mathbf{j}}$ **1.141** $\frac{-aK^2}{\sqrt{2}}\hat{\mathbf{e}}_t + \frac{3aK^2}{\sqrt{2}}\hat{\mathbf{e}}_n$; $\frac{2\sqrt{2}}{3}a$
1.143 $(x, y) = (2.88, -0.584)$ m
1.145 $-1.85\hat{\mathbf{i}} - 1.91\hat{\mathbf{j}} - 3.41\hat{\mathbf{k}}$ m; 28.9 m/s²
1.147 $11.3\hat{\mathbf{i}} - 6.24\hat{\mathbf{k}}$ m
1.149 Use $\mathbf{v} = \dot{x}\hat{\mathbf{i}} + \dot{y}\hat{\mathbf{j}}$ and $\mathbf{a} = \ddot{x}\hat{\mathbf{i}} + \ddot{y}\hat{\mathbf{j}}$ for the proof; 9680 ft
1.151 $a_t = \frac{1}{Ka}$; $a_n = \frac{1}{a}$ m/s²

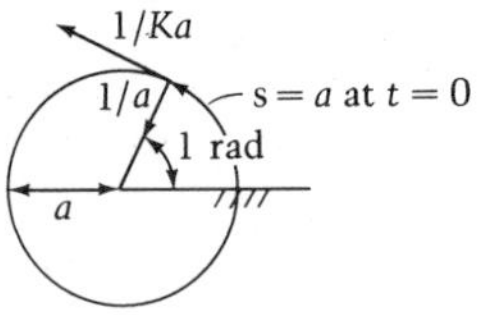

1.153 $a_t = 0$; $a_n = 0.708$ ft/sec²; $\ddot{x} = 0.0637$ and $\ddot{y} = -0.701$ with $\sqrt{\ddot{x}^2 + \ddot{y}^2} = 0.704$ ft/sec²
1.155 Answers given in problem.
1.157 $y_M = 2D/3$ at $t = 2D/(3V_0)$

CHAPTER 2

2.1 Answer given in problem.
2.3 With $x \rightarrow$ and $y \uparrow$ in center of sphere, $(\bar{x}, \bar{y}) = (-1.55, -0.20)$ m **2.5** $\left(-\frac{\ell}{5}, \frac{\ell}{5}, -\frac{\ell}{2}\right)$
2.7 $1.34R$ **2.9** $\sqrt{3}R$ **2.11** By inspection, $g \downarrow$
2.13 Home run, clears fence by 2 ft
2.15 88.1 ft/sec at $\theta = 14.9°$; 0.705 sec
2.17 3 ft **2.19** 63.8 ft/sec **2.21** 35.2 ft/sec
2.23 $\phi = \tan^{-1}\left[\frac{H}{d} + \frac{g(D-d)^2}{2dv_0^2}\right]$; time $= \frac{D-d}{v_0}$;

$$v_i = \sqrt{\left(\frac{dv_0}{D-d}\right)^2 + \left(\frac{Hv_0}{D-d}\right)^2 + gH + \frac{g^2(D-d)^2}{4v_0^2}}$$

2.25 43.4° **2.27** 540 m
2.29 5.45 m/s²; Apollo: 1.94 m/s² **2.31** W
2.33 time $= v_i/(\mu g)$; distance $= v_i^2/(2\mu g)$
2.35 225 ft **2.37** 3.35 → ft/sec²
2.39 No slip, and the force is 40 → N
2.41 164 N **2.43** $\mu < 0.10$ **2.45** 43.7 lb
2.47 Answer given in problem. **2.49** 0.788 sec
2.51 (a) 13.1 m (b) smaller because now μN resists the motion of $\mathcal{A}$. **2.53** 0.032 lb **2.55** 136 N
2.57 $\mathbf{a}_{\mathcal{A}} = 0.0204g \downarrow$, $\mathbf{a}_{\mathcal{B}} = 0.224g \uparrow$, $\mathbf{a}_{\mathcal{C}} = 0.184g \downarrow$; $\mathcal{C}$; 0.941 s ($g = 9.81$ m/s²)
2.59 Answer given in problem.
2.61 (a) a (vertical) component of the string tension; (b) $\sqrt{(mg)^2 + \left(\frac{mv_0^2}{R}\right)^2}$;

(c) another component of the string tension, this one in the $\hat{\mathbf{e}}_t$ direction.
2.63 170 miles **2.65** Answer given in problem.
2.67 3.13 rad/sec **2.69** $\rho = a^2/b$; 52.0 mph, so yes.
2.71 0.5, at $\theta = 0$
2.73 $\ddot{\mathbf{r}} = 1$ ft/sec^2; $\mathbf{F} = 0.00207\hat{\mathbf{e}}_\theta + 0.0500\hat{\mathbf{k}}$ lb
2.75 $m\ell\omega_0^2 \cos\phi = (m + M)g$ **2.77** $mr_0^4\,\dot{\theta}_0^2/(r_0 - v_C t)^3$
2.79 Answer given in problem.
2.81 1.85 sec; 24.6 ↑ ft/sec **2.83** $\sqrt{2gR}$
2.85 $\dfrac{v_0 m}{K}(1 - e^{-Kt/m})$ **2.87** 1.16 m/D
2.89 191 ↓ ft/sec, compared to 200 ↓ ft/sec without air resistance. **2.91** 15.9 ↓ ft/sec **2.93** 0.736 sec
2.95 $m\ddot{x} = -kx + mg$, x measured down from point of contact; $x(0) = 0 = \dot{x}(0)$; $2mg$; $\pi\sqrt{m/k}$
2.97 Answer given in problem.
2.99 Answer given in problem. (Set $N = 0$ to find the leaving point.) **2.101** 76,300 lb/ft
2.103 20.7 ft/sec down the plane.
2.105 (a) 12 lb (b) 2.84 ft/sec **2.107** 72 lb/ft
2.109 2.56 ← m/s **2.111** 4.0 ft **2.113** $5d/4$
2.115 1020 ↓ N; 9.80 ↑ m/s^2 **2.117** $\sqrt{2gR}$
2.119 $\sqrt{2K/(3mS)}$ **2.121** W and ΔT are each 400 J.
2.123 $-0.849\, mgr_e$
2.125 (a) 4.78 ft (b) $\mathcal{A}$ will start back upward.
2.127 $v_A = 5.3$ ↓ ft/sec; $v_B = 10.6$ ↑ ft/sec
2.129 Distance between them is $(\ell_u - 0.02)$ m, where ℓ_u = unstretched length. They are 0.22 m closer together. Final spring force = 1 lb (compressive).
2.131 1.41 miles/sec **2.133** 7.45 sec
2.135 36.2 → ft/sec **2.137** 7.86 sec; 491 ft/sec
2.139 0.400 → ft/sec **2.141** $v/3$ →
2.143 $\dfrac{m^2U}{(M + 2m)(M + m)}$ → **2.145** e^4H
2.147 Answer given in problem. (If $e > 1$, there would be an energy *gain*!) **2.149** 0.446
2.151 $\mathbf{v}_A = 2\hat{\mathbf{i}} - 1.53\hat{\mathbf{j}}$ ft/sec; $\mathbf{v}_B = -4\hat{\mathbf{i}} + 3.65\hat{\mathbf{j}}$ ft/sec; Impulse on $\mathcal{B}$ = 0.077 ↑ lb-sec = −(Impulse on $\mathcal{A}$)
2.153 With 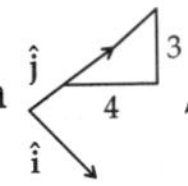

, $\mathbf{v}_A = -0.479\hat{\mathbf{i}} - 2.49\hat{\mathbf{j}}$ ft/sec;
$\mathbf{v}_B = 4.69\hat{\mathbf{j}}$ ft/sec; impulse on $B = 0.113\hat{\mathbf{j}}$ lb-sec
2.155 2.00 ft **2.157** 0.997 ft
2.159 $\dfrac{W_1 + W_2}{k} + \sqrt{\dfrac{W_1^2(W_1 + W_2) + 2HkW_1^2}{k^2(W_1 + W_2)}}$
2.161 $\mathbf{v}_A = 2.97$ → m/s; $\mathbf{v}_B = 1.48$ ← m/s
2.163 $3\beta gy$ **2.165** Answers given in problem.
2.167 60 ↻ lb-ft
2.169 The derivation of Equation (2.36) nowhere requires that the point be the mass center.
2.171 Answer given in problem.
2.173 11.5 ft upward **2.175** $d/4$ upward
2.177 Answer given in problem.
2.179 Answer given in problem.
2.181 $H_1 = 57.0$ in.; $H_2 = 56.5$ in.

CHAPTER 3

3.1 a,c,d,e **3.3** $0.1\hat{\mathbf{i}} + 0.1\hat{\mathbf{j}}$ m/s
3.5 0; $-6.67\hat{\mathbf{k}}$ rad/sec
3.7 $\omega_1 = 2$ ↺ rad/s; $\omega_2 = 0.640$ ↺ rad/s
3.9 $\omega_1 = 2$ ↻ rad/s; $\omega_2 = 2$ ↻ rad/s
3.11 $\omega_2 = 0$; $\omega_3 = -2\hat{\mathbf{k}}$ rad/sec
3.13 $0.058\hat{\mathbf{j}}$ m/s; 0.385 ↺ rad/s
3.15 $0.3\hat{\mathbf{i}} - 1.3\hat{\mathbf{j}}$ m/s **3.17** $0.272\hat{\mathbf{i}}$ m/s
3.19 1.08 → m/s
3.21 $-1.6\hat{\mathbf{i}} - 1.2\hat{\mathbf{j}}$ m/s; $-6\hat{\mathbf{k}}$ rad/s; $4\hat{\mathbf{k}}$ rad/s
3.23 $1.73\left(\dfrac{5\hat{\mathbf{i}} + 12\hat{\mathbf{j}}}{13}\right)$ m/s
3.25 (a) **0**, 0.688 ↻ rad/sec; **0**, **0**
3.27 The plots can be constructed from the answer to 3.26, which is $[(r\cos\theta)/\sqrt{\ell^2 - r^2\sin^2\theta}\,]\,\dot{\theta}$
3.29 $\dfrac{-12\cos\theta\,\hat{\mathbf{j}}}{\sqrt{169 - 120\sin\theta}}$ m/s
3.31 In each case, Ⓘ is at the intersection of the radial line OA and the normal to the slot at B; $\mathbf{v}_B = \mathbf{0}$ when $\overline{OAB}$ and $\overline{AOB}$ are straight lines.
3.33 See 3.7. **3.35** $2\hat{\mathbf{j}}$ m/s; $\frac{2}{3}\hat{\mathbf{k}}$ rad/s **3.37** See 3.11.
3.39 0.129 ↻ rad/s; 0.129 ↺ rad/s
3.41 See Example 3.5. **3.43** $\omega_2 = 0.2$ ↺ rad/sec $= \omega_3$
3.45 $-0.389\hat{\mathbf{k}}$ rad/sec **3.47** 2.11 → ft/sec; 0.201 ft above P;

$\mathbf{v}_Q = 60.7$ 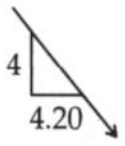

ft/sec;

$\mathbf{v}_S = 60.7$ 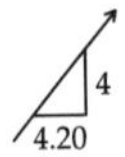

ft/sec;

$\mathbf{v}_R = 85.9$ → ft/sec
3.49 0.0400 ↻ rad/s; **0** **3.51** 14 ↻ rad/sec
3.53 8.95 ↺ rad/sec^2 **3.55** $-29\hat{\mathbf{i}} - 24.4\hat{\mathbf{j}}$ ft/sec^2
3.57 $\boldsymbol{\alpha}_2 = 0.0128\hat{\mathbf{k}}$ rad/sec^2; $\boldsymbol{\alpha}_3 = -0.055\hat{\mathbf{k}}$ rad/sec^2
3.59 3.91 ↺ rad/sec^2
3.61 84.5 → in./sec; 38.1 ← in./sec^2; 565 ← in./sec^2
3.63 $-30\hat{\mathbf{i}} - 30\hat{\mathbf{j}}$ in./sec^2 **3.65** $\ddot{\varphi} = \dfrac{r\dot{\theta}^2 \sin\theta(r^2 - \ell^2)}{(\ell^2 - r^2\sin^2\theta)^{3/2}}$
3.67 $6.26\hat{\mathbf{i}} + 0.320\hat{\mathbf{j}}$ m/s^2; 0.320 ↺ rad/s^2
3.69 $-0.0938\hat{\mathbf{i}} - 0.225\hat{\mathbf{j}}$ m/s^2
3.71 $-18\hat{\mathbf{i}} - 24\hat{\mathbf{j}}$ in./sec^2
3.73 0.0735 ↑ m/s^2; 0.0172 ↻ rad/s^2
3.75 0.779 ↻ rad/sec^2; 6.56 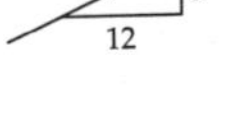

in./sec^2

3.77 $\frac{\omega_o}{K}(e^{Kt}-1)$; $\omega_o e^{Kt}$; $\omega_o K e^{Kt}$ **3.79** $(v_0^2/2)\hat{\mathbf{j}}$ ft/sec²
3.81 (a) 4 ↻ rad/sec²; (b) $\theta = 45°$; $\overline{PA} = \sqrt{2}/2$ ft
3.83 $x = \dfrac{a_{P_x}\omega^2 - a_{P_y}\alpha}{\omega^4+\alpha^2}$; $y = \dfrac{a_{P_x}\alpha - a_{P_y}\omega^2}{\omega^4+\alpha^2}$
3.85 (a) 1.4 ft above P; (b) $-0.27\hat{\mathbf{i}} + 1.13\hat{\mathbf{j}}$ ft/sec
3.87 $\mathbf{v}_A = 117 \leftarrow$ in./sec; $\mathbf{v}_B = \mathbf{0}$; $\mathbf{v}_D = 2820 \rightarrow$ in./sec; $\mathbf{v}_E = 2930 \rightarrow$ in./sec; A is going backwards (to the left) since it's below Ⓘ, and the wheel turns ↻.
3.89 $\mathbf{v}_C = 1\hat{\mathbf{i}}$ m/s; $\mathbf{v}_B = 1\hat{\mathbf{i}} + 4\hat{\mathbf{j}}$ m/s
(Note: radius superfluous.)
3.91 $\dfrac{\omega}{\dot{\phi}} = \dfrac{R+r}{r}$
3.93 In order, $2v_0\hat{\mathbf{i}}$; $1.71v_0\hat{\mathbf{i}} - 0.707v_0\hat{\mathbf{j}}$; $v_0\hat{\mathbf{i}} - v_0\hat{\mathbf{j}}$; $\mathbf{0}$; $v_0\hat{\mathbf{i}} + v_0\hat{\mathbf{j}}$
3.95 $\mathbf{v}_A = 0.1\hat{\mathbf{i}} + 0.1\hat{\mathbf{j}}$ m/s; $\mathbf{v}_B = 0.2\hat{\mathbf{i}}$ m/s;
$\mathbf{v}_D = 0.1\hat{\mathbf{i}} - 0.1\hat{\mathbf{j}}$ m/s; $\mathbf{v}_E = \mathbf{0}$; (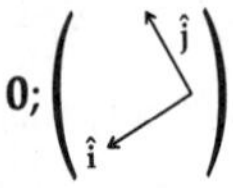)

3.97 $\omega_3 = 0.113$ ↻ rad/sec; $\omega_2 = 0.653$ ↻ rad/sec
3.99 $\omega_2 = 7.5$ ↻ rad/sec; $\omega_3 = 0$
3.101 34 ↻ rad/s; $680\hat{\mathbf{i}}$ cm/s, the speed of the tooth point of $\mathcal{B}_4$ in contact with $\mathcal{B}_3$ **3.103** 0; 80 → m/s²
3.105 $\dot{x}_C$ = any positive constant k; $\dot{\theta} = k/R$
(Directions are → and ↺.)
3.107 $2a_0\hat{\mathbf{i}}$; $(1.71a_0 - 0.707v_0^2/R)\hat{\mathbf{i}} + (0.707a_0 + 0.707v_0^2/R)\hat{\mathbf{j}}$; $(a_0 - v_0^2/R)\hat{\mathbf{i}} + a_0\hat{\mathbf{j}}$; 0; $(a_0 + v_0^2/R)\hat{\mathbf{i}} - a_0\hat{\mathbf{j}}$ **3.109** -35.4 ft/sec²
3.111 A: $9890\hat{\mathbf{j}}$; B: $9130\hat{\mathbf{j}}$; C: $\mathbf{0}$; D: $-9130\hat{\mathbf{j}}$; E: $-9890\hat{\mathbf{j}}$, all in ft/sec². **3.113** (a) 4.69 sec (b) 3.5 revs
3.115 $0.347\hat{\mathbf{i}} + 0.0198\hat{\mathbf{j}}$ m/s²
3.117 $-0.889\hat{\mathbf{i}} - 5.41\hat{\mathbf{j}}$ m/s²
3.119 $\alpha_3 = 0.0889\omega_0^2$ ↻; $\alpha_2 = 0.178\,\omega_0^2$ ↻
3.121 $-400\hat{\mathbf{i}}$ ft/sec²
3.123 14.7 ↻ rad/s²; $-11.9\hat{\mathbf{j}}$ m/s²
3.125 $-\dfrac{v_0^2}{R-r}\hat{\mathbf{i}} - \dfrac{v_0^2}{r}\hat{\mathbf{j}}$
3.127 (a) $-6\hat{\mathbf{i}}$ and $-6\hat{\mathbf{i}} - 6\hat{\mathbf{j}}$ cm/s;
(b) $4\hat{\mathbf{i}} - 3\hat{\mathbf{j}}$ and $22\hat{\mathbf{i}} + \hat{\mathbf{j}}$ cm/s²; -16.3 cm/s²
3.129 2.70 ↺ rad/sec²; 0.216 ↻ rad/sec²
3.131 Let x and y respectively be directed down and toward the plane, with origin at the center of the disk. Then the point has $(x, y) = (4.80, 3.60)$ ft.
3.133 $-43.3\hat{\mathbf{i}}$ in./sec²
3.135 Let $\hat{\mathbf{i}}$ be from the center of $\mathcal{B}_2$ along $\mathcal{B}_1$, and $\hat{\mathbf{k}}$ be out of the page. Answers are then $-24\pi^2\hat{\mathbf{i}} + 3\pi\hat{\mathbf{j}}$ in./sec² ($\mathcal{B}_2$), and $138\pi^2\hat{\mathbf{i}} + 3\pi\hat{\mathbf{j}}$ in./sec² ($\mathcal{B}_3$).
3.137 21.6 m/s² (It is the highest point of $\mathcal{P}$.)
3.139 (a) Answer given in problem.
(b) Curve is concave downward.
(c) First is $a\alpha\hat{\mathbf{i}} + a\alpha\hat{\mathbf{i}} - \omega^2 a\hat{\mathbf{j}}$, and second is $(2a\alpha)\hat{\mathbf{i}} - [(2a\omega)^2/(4a)]\hat{\mathbf{j}}$, the same.
3.141 Answer given in problem. **3.143** 0.25 ↻ rad/s
3.145 0.781 ↻ rad/sec **3.147** 0.120 ↺ rad/s
3.149 0.592 ↻ rad/s; $11.6\hat{\mathbf{i}} - 4.85\hat{\mathbf{j}}$ cm/s
3.151 $(D\dot{\theta}\sin\theta/\cos^2\theta)(\cos\theta\hat{\mathbf{i}} + \sin\theta\hat{\mathbf{j}})$
3.153 Answer given in problem.
3.155 0.0334 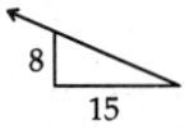 m/s; 0.428 ↻ rad/s

3.157 Answer given in problem.
3.159 $26\pi\sin\pi t \downarrow$; $26\pi^2\cos\pi t \downarrow$
3.161 $-0.165\hat{\mathbf{i}}$ m/s²; 0.170 ↺ rad/s²
3.163 $0.186\hat{\mathbf{k}}$ rad/sec²
3.165 $\mathbf{v}_{A/\mathcal{A}} = 0.240\hat{\mathbf{i}} + 0.180\hat{\mathbf{j}}$ ft/sec;
$\mathbf{v}_{A/\mathcal{B}} = -0.480\hat{\mathbf{i}} + 0.640\hat{\mathbf{j}}$ ft/sec;
$\mathbf{a}_{A/\mathcal{A}} = 0.512\hat{\mathbf{i}} + 0.384\hat{\mathbf{j}}$ ft/sec²;
$\mathbf{a}_{A/\mathcal{B}} = -0.180\hat{\mathbf{i}} + 0.240\hat{\mathbf{j}}$ ft/sec²
3.167 $-8.51\hat{\mathbf{k}}$ rad/s²; $-118\hat{\mathbf{i}} + 49.2\hat{\mathbf{j}}$ cm/sec²
3.169 $r\omega_0^2$ 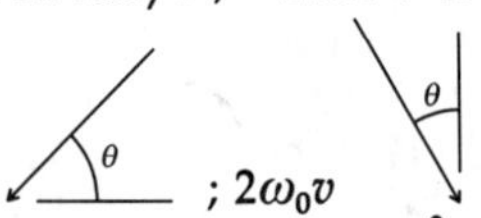 ; $2\omega_0 v$

3.171 (a) 103 ↻ rad/s²; (b) $2.25\hat{\mathbf{i}} + 1.62\hat{\mathbf{j}}$ m/s²
3.173 9.11 ft
3.175 $\theta = 70.7°$ for the maximum piston speed

CHAPTER 4

4.1 $\frac{3}{4}mg$
4.3 (a) 12.9 → ft/sec²; (b) left: 40 ↑ lb; right: 60 ↑ lb **4.5** $\dfrac{b}{2} \le H \le \dfrac{3}{2}b$ **4.7** 8.05 ft/sec²
4.9 (a) 108 N; (b) 64.8 N **4.11** $8g/15 \rightarrow$
4.13 $10\mu/(1+9\mu)$ **4.15** $-12.9 \le a \le 8.05$ ft/sec²
4.17 $(m+M)g/4$ **4.19** $g/5$ up the plane
4.21 (a) $(b-d)g\mu/(b-\mu H)$
(b) $d = \mu H$ gives $\ddot{x}_{C_{MAX}} = \mu g$
4.23 time $= (a+b+\mu H)v/(\mu a g)$;
distance $= (a+b+\mu H)v^2/(2\mu g a)$
4.25 $0.847 \le \ddot{x}_C \le 33.9$ ft/sec²
4.27 16π slug-ft² **4.29** $\pi R^2 L^3(17\rho_1 + 35\rho_0)/60$
4.31 (a) $\dfrac{mH^2}{6} + \dfrac{mt^2}{12}$; (b) $\dfrac{mB^2}{2} + \dfrac{mt^2}{12}$;
(c) $\dfrac{mH^2}{6} + \dfrac{mB^2}{2}$; (d) $-\dfrac{mBH}{4}$;
(e) 0, 0; neglect t^2 terms in (a) and (b).
4.33 Answer given in problem. (I_{zz}^O exceeds $I_{xx}^O + I_{yy}^O$ by $mt^2/6$.) **4.35** $mR^2\left(\dfrac{1}{4} + \dfrac{\sin\alpha}{4\alpha} - \dfrac{16}{9\alpha^2}\sin^2\dfrac{\alpha}{2}\right)$
4.37 3619 kg · m²; 1660 kg · m²
4.39 $0.145\,mR^2 + 0.0481\,mH^2$

4.41 From the corner, 1.56 ft ← and 0.563 ft ↑; $I_{z_C} = 4.45$ slug-ft². **4.43** 1.00 slug-ft²
4.45 3.44 slug-ft² **4.47** $0.259\, mR^2$
4.49 37.7 kg · m² **4.51** 29.3 kg · m²
4.53 2.69, 0.0200, and 2.71 kg · m², respectively.
4.55 Answer given in problem
4.57 $M = md/k_G$; $D = k_G$ **4.59** 23.1 kg · m²
4.61 Answer given in problem; only (b) starts without approximation. **4.63** $2g/3$; $5g/7$; $g/2$ ↓
4.65 (a) Wally, Sally, Carolyn, Harry;
(b) 1.41K, 1.67K, 1.73K, 2K where $K = \sqrt{\dfrac{D}{g \sin \beta}}$
4.67 $0.200\, mg \sin \beta$ (compression)
4.69 (a) $0.428\, \mu mg$ (b) $1.86\, \mu g\hat{\mathbf{i}}$; $(0.360\, \mu g/r)\hat{\mathbf{k}}$
4.71 1/2 **4.73** $\dfrac{0.714\, M_O}{mr} \rightarrow$ **4.75** 0.577; 1.02 s
4.77 (a) 4.00 → m/s² (b) 0.23 m
4.79 (a) 14.4 in. wrapped; 0.064 **4.81** 300 ft
4.83 0.450 ↺ rad/sec² **4.85** 78.6 m
4.87 (a) 14.6°; (b) 1.38 s **4.89** 0.753 slug-ft²; 38.3 → ft
4.91 $2.88\hat{\mathbf{k}}$ rad/s²; 0.180 m to the right

4.93 (a)
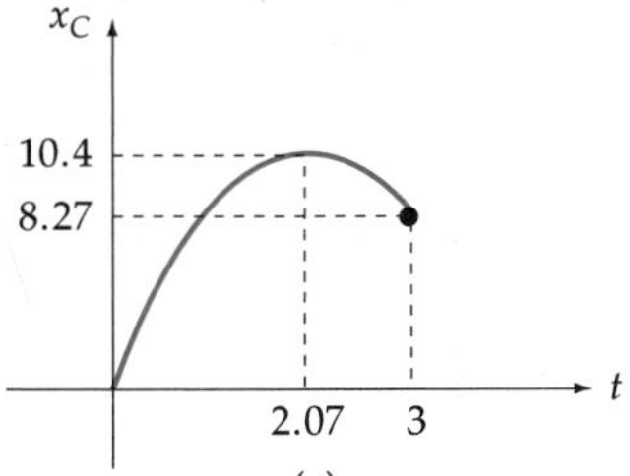

(a) ;

(b)
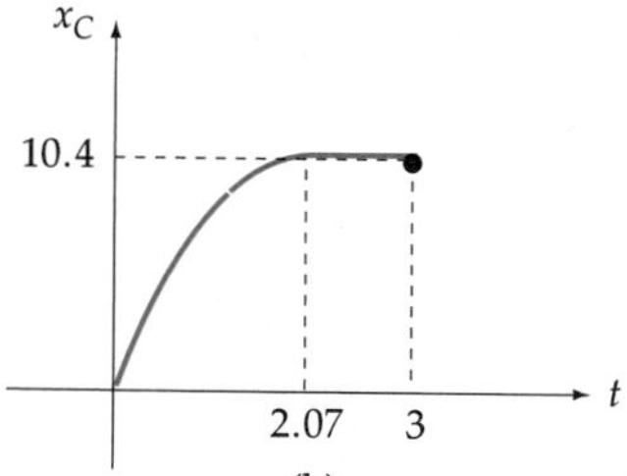

(b) ;

(c)
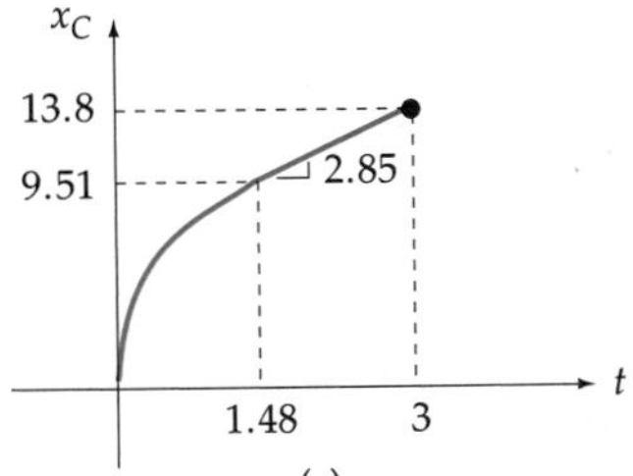

(c)

4.95 (a) Answer given in problem.
(b) 0.676 → m/s², 1.69 ↻ rad/s²
4.97 21.8 ↻ rad/sec²; Yes, because the geometry doesn't change. **4.99** 5.50 kg · m²; 4.04 → m/s²
4.101 Largest for P_E, and is 4.14 → m
4.103 2.09 s (rolls off left corner)
4.105 (a) 13.3 ↺ rad/sec² (b) 1.11 ↻ rad/sec² (c) 0.0197
4.107 28.2 sec **4.109** 440 m **4.111** 7.16 ← ft/sec²
4.113 (a) $\dfrac{F}{M + 3mn/8}$ (b) $\dfrac{Ft^2}{2M + 3mn/4}$
4.115 4.47 sec **4.117** (a) No (b) 0.585 → ft/sec²
4.119 177 lb up the plane **4.121** 2.10 s
4.123 straight down; $mg/4$ ↓
4.125 (a) $3\sqrt{3}/7$ (b) For $\mu \geq \mu_{\text{min}}$, $\dfrac{3\sqrt{3}}{16} g\hat{\mathbf{i}} - \dfrac{9}{16} g\hat{\mathbf{j}}$;
for $\mu < \mu_{\text{min}}$, $\dfrac{g}{13 - 3\sqrt{3}\,\mu}\,[4\mu\hat{\mathbf{i}} - (9 - 3\sqrt{3}\,\mu)\hat{\mathbf{j}}]$
4.127 $mg/4$ **4.129** 43.3 lb **4.131** $-0.400\, g\hat{\mathbf{i}}$; $0.308\, mg$
4.133 Answers given in problem.
4.135 $\mathbf{a}_C = 3g/4$ ↓; $\alpha = 3g/(2s)$ ↻
4.137 $\mathbf{a}_C = g$ ↓; $\alpha = 6g/s$ ↻
4.139 $\mathbf{a}_C = 2g/3$ ↓; $\alpha = 2g/s$ ↻ **4.141** 6.38 lb
4.143 On the section to the left of the cut, $V = 3WL/64$ ↓ and $M = 9WL^2/256$ ↺.
4.145 At O, $\dfrac{2}{7} mg$ ↑; at A, $\dfrac{mg}{14}$ ↑
4.147 $3T_0/[(2R^2)(9M + 2m)]$ ↺
4.149 Answer given in problem. **4.151** 1.5 ↻ rad/s²
4.153 $\alpha = \dfrac{3}{8}\dfrac{g}{\ell}$ ↻ **4.155** 17.4 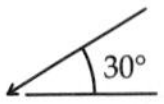

ft/sec²
4.157 For rolling on fixed surface, $\mathbf{a}_Q$ is normal to the surface, hence toward geometric center of round body; thus $\mathbf{r}_{QC} \times \mathbf{a}_Q = \mathbf{0}$ since geometric center is mass center.
4.159 (a) 4 m/s² (b) 11.8 ↺ rad/s²; no
4.161 0.400 ↻ rad/s² **4.163** 55.8 ↑ N
4.165 (a) 30°: 0.805 ↻ rad/sec²; 60°: 2.28 ↻ rad/sec²; 90°: 2.82 ↻ rad/sec² (b) 163°
4.167 10.8 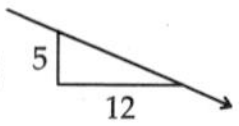

N;

497 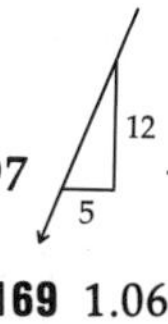

+ 34.0 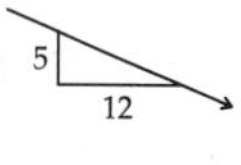

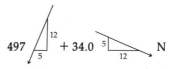

N

4.169 $1.06\, g/L$ ↻ **4.171** OB: $0.405\, mg$; OA: $0.595\, mg$
4.173 7.85 ↻ rad/s² **4.175** 17.3 ↓ ft/sec²; 14.9 ↑ lb
4.177 (a) 0.188 (b) 53.1°
4.179 counterclockwise, 1.48 sec
4.181 (a) $mg\sqrt{\cos^2\theta + (10 \sin \theta)^2}/4$ (b) $5mg/2$

4.183 $M\ell\omega^2 \sin\theta\hat{\mathbf{i}} - Mg\hat{\mathbf{j}}$;
$(Mg\ell \sin\theta + M\ell^2\omega^2 \sin\theta\cos\theta)$ ↻
4.185 (a) $\frac{m\delta\omega^2}{2}\hat{\mathbf{i}} - \frac{m\delta\alpha}{2}\hat{\mathbf{j}}$ each,
with $(\hat{\mathbf{i}}, \hat{\mathbf{j}})$ along (x, y) in figure; (b) $0.258\, r$
4.187 $\left(m_p + m_e + \frac{m_{\mathcal{A}}}{2}\right) g \downarrow$ each,
plus $\left[\frac{m_e\ell_{\mathcal{A}}\ell_e}{2} + m_p(\ell_{\mathcal{A}} + \ell_p)\ell_e\right]\frac{\omega^2}{2d}$,
up on left bearing and down on right bearing.
4.189 $\frac{13\, mr^2\omega_0^2 \sin\psi\cos\psi}{96\ell}$, down on left and up on right, onto shaft and turning with it.
4.191 By parallel-axis theorem, $I_{xz}^D = 0 \Rightarrow I_{xz}^C = 0$ since C is on z-axis. Thus using the theorem again, $I_{xz}^Q = 0$. Same arguments for I_{yz}.
4.193 In $\mathcal{A}$, $(x, y) = (-56/3, -21)$ in.; in $\mathcal{B}$, $(x, y) = (-70/3, -35)$ in. **4.195** 24.0 rad/sec

CHAPTER 5

5.1 25 ft-lb **5.3** 100 J **5.5** 140 N/m
5.7 3.75 ↻ rad/sec
5.9 $\sqrt{\frac{3g\sin\theta_0}{L}}$ (↻ for $\mathcal{A}$ and ↺ for $\mathcal{B}$)
5.11 103,000 lb/ft **5.13** 4.18 ↺ rad/s
5.15 2.95 ↺ rad/s **5.17** 2.00 ↺ rad/s
5.19 12.1 ↻ rad/s **5.21** (a) 2440 N/m
5.23 (a) 0.0794 m; (b) back to the starting point
5.25 4.36 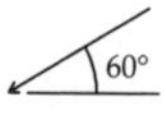 m/s **5.27** $\sqrt{\frac{18ag}{7}} \leftarrow$

5.29 (a) 30 lb; (b) 26.7 in./sec
5.31 (a) 1.56 ↺ rad/sec; (b) 1.40 ↺ rad/sec
5.33 $0.0931 \leftarrow$ m/s
5.35 (a) 7.19 ↻ rad/s; (b) 7.53 ↻ rad/s; (c) 8.58 ↻ rad/s
5.37 $\sqrt{24\pi gr\sin\beta/13}$  **5.39** 35 lb/ft; yes

5.41 4.43 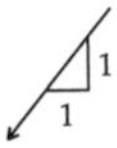 ft/sec

5.43 8 in.; the two points are the intersections of the perimeter of $\mathcal{B}_2$ (in the starting position) with a circle of radius 12 in. and center at Q (in the final position)
5.45 $\frac{L\omega}{2}\sqrt{1 - \left(\frac{2H}{L}\right)^2} \downarrow$
5.47 $4\sqrt{2}R\sqrt{\frac{g(1-\cos\theta)}{R(9\pi - 16)}}$ 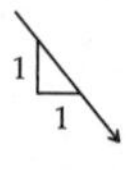

5.49 4.91 ↺ rad/sec; $\frac{1}{3}$ ft **5.51** 3.82 m
5.53 2.11 ↻ rad/sec
5.55 For $T \uparrow$, $v_C = \sqrt{\frac{2RrTx_C}{I_C + mR^2}}$ (moves to the left);
For $T \rightarrow$, $v_C = \sqrt{\frac{2R(R-r)Tx_C}{I_C + mR^2}}$ (moves to the right)
5.57 (a) 4/17 m; (b) 50 N *increase*, $\mu_{\min} = 0.220$
5.59 It starts out to the right, the spring goes slack, and then it leaves on the right. (It would need one more foot of plane to stay on.)
5.61 $\sqrt{6\sqrt{2}g\ell} \downarrow$ **5.63** 1.33 ↻ rad/s
5.65 0.106 m to the left; 0.000428 **5.67** $41.5\hat{\mathbf{j}}$ lb
5.69 (a) $\frac{3g\cos\theta_o}{2L}$ ↻; $W\sin\theta_o$ 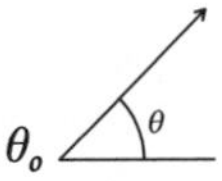 $+ \frac{W\cos\theta}{4}$ 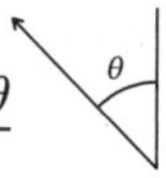;

(b) $\frac{3W\sin\theta_o}{2} \leftarrow + \frac{W}{4} \uparrow$
5.71 $\tan^{-1}[\mu/(1 + 36k^2)]$; 5.27°
5.73 Answer given in problem.
5.75 5.66 ↑ m/s **5.77** $2gt_0/(R\pi)$
5.79 (a) $0.00240 \rightarrow$ m/s; (b) $0.0889 \rightarrow$ m/s
5.81 0.366 sec
5.83 (a) 1.88 ↓ m/s; (b) 4.51 ↑ m/s **5.85** See 4.166(c). **5.87** 4.02 sec **5.89** 1.38 sec
5.91 1.19 ↻ rad/sec **5.93** $\frac{5\, gt\sin\beta}{7}$
5.95 $32.2 \rightarrow$ ft/sec **5.97** (a) $2.82 \leftarrow$ m/s; (b) zero
5.99 $2d/9$ upward
5.101 $\Delta t = \frac{I_1 I_2(\omega_2 - \omega_1)}{\mu m_1 gR(I_1 + I_2)}$; $\omega_f = \frac{I_1\omega_1 + I_2\omega_2}{I_1 + I_2}$;
$\Delta T = \frac{I_1 I_2(\omega_1 - \omega_2)^2}{2(I_1 + I_2)}$; ω_f and ΔT are the same
5.103 (a) no; (b) yes, if $R^2H \le dfk^2$
5.105 $6v_0 m/[\ell(4m + M)]$, counterclockwise looking down
5.107 (a) $\sqrt{1.26g/r + 0.399\, v_0^2/r^2}$; (b) 36.8%
5.109 $3I^2/(2M)$ **5.111** $0.526\frac{m_2 L}{m_1 h}\sqrt{gL}$
5.113 0.159 sec **5.115** $-0.25v_0\hat{\mathbf{i}} + 0.75v_0\hat{\mathbf{j}}$
5.117 0.8 ft; $19.0(10^6)$ slug-ft²; 1.05% difference
5.119 $\Delta T = \frac{35}{72} mv_0^2$, or 72.9% loss
5.121 Answer given in problem. **5.123** $\frac{3M^2\ell}{2\mu(M + 3m)^2}$
5.125 $e = 0.8$; $\mu = 0.32$
5.127 (a) 1.66 rad/sec for both; (b) 17%
5.129 (a) $\mathbf{v}_{G_{\mathcal{A}}} = 0$; $\mathbf{v}_{G_{\mathcal{B}}} = v_0 \rightarrow$; $\boldsymbol{\omega}_{\mathcal{A}} = v_0/r$ ↻; $\boldsymbol{\omega}_{\mathcal{B}} = \mathbf{0}$;
(b) $\mathbf{v}_{G_{\mathcal{A}}} = 2v_0/7 \rightarrow$; $\mathbf{v}_{G_{\mathcal{B}}} = 5v_0/7 \rightarrow$;
(c) If $\mu = 0$, final motion is given by (a).

5.131 $\omega_f = \dfrac{3\ell v_{C_i}(1+e)}{2\ell^2 + 9r^2}$ clockwise;
$\mathbf{v}_{C_f} = (-\ell\omega_f/2 + ev_{C_i})\hat{\mathbf{i}} + r\omega_f\hat{\mathbf{j}}$
(see Example 5.18 for $\hat{\mathbf{i}}, \hat{\mathbf{j}}$); $\omega_f(e=0) = \frac{1}{2}\omega_f(e=1)$;
$\dot{x}_{C_f}(e=0) = 1.91\dot{x}_{C_f}(e=1)$; $\dot{y}_{C_f}(e=0) = \frac{1}{2}\dot{y}_{C_f}(e=1)$
5.133 Answer given in problem.

5.135 $v_{C_{\text{rebound}}} = 2.63$ 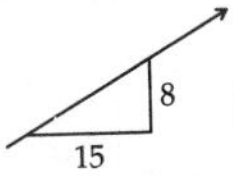 m/s; $\Delta T = 512$ ft-lb

5.137 0.545 m from left end; 0.562 kg-m²;
0.0957 kg-m²; 0.657 m from left end

5.139 $\dfrac{F\,\Delta t(2\ell + a - c)}{\dfrac{4m\ell^2}{3} + M\left[\dfrac{2a^2}{5} + (a+2\ell)^2\right]}$ ↺

CHAPTER 6

6.1 Answer given in problem.
6.3 (a) $(-t^5 + t^3 + 3t^2)\hat{\mathbf{i}} + (t^6 - t^2 + 2t)\hat{\mathbf{j}} + (-t^5 + t^3 + 1)\hat{\mathbf{k}}$ m/s²;
(b) $3\hat{\mathbf{i}} + 2\hat{\mathbf{j}} + \hat{\mathbf{k}}$ m/s²; (c) $-12\hat{\mathbf{i}} + 64\hat{\mathbf{j}} - 23\hat{\mathbf{k}}$ m/s²
6.5 $19.6\hat{\mathbf{j}} - 1.29\hat{\mathbf{k}}$ **6.7** $\boldsymbol{\omega}_{\mathcal{A}/\mathcal{G}} = 13.7\hat{\mathbf{j}}_1 + \hat{\mathbf{k}}_1$
6.9 $\omega_x\hat{\mathbf{i}} + \omega_y\cos\omega_x t\,\hat{\mathbf{j}} + (\omega_z + \omega_y \sin\omega_x t)\hat{\mathbf{k}}$ **6.11** $\dot{\theta}\hat{\mathbf{i}}$
6.13 The solution is Equation (6.70) with
$\theta_1 = \theta_{\text{pitch}} = \dfrac{\pi}{18}\sin\dfrac{2\pi t}{6}$; $\theta_2 = \theta_{\text{roll}} = \dfrac{\pi}{6}\sin\dfrac{2\pi t}{8}$;
and $\theta_3 = \theta_{\text{yaw}} = \dfrac{\pi}{22.5}\sin\dfrac{2\pi t}{50}$.
6.15 To the components in 6.13, respectively, add $-\dot{P}\cos R$, $-\dot{R}$, and $-\dot{P}\sin R$.
6.17 ${}^{\mathcal{I}}\dot{\boldsymbol{\alpha}}_{\mathcal{B}/\mathcal{I}} = {}^{\mathcal{B}}\dot{\boldsymbol{\alpha}}_{\mathcal{B}/\mathcal{I}} + \boldsymbol{\omega}_{\mathcal{B}/\mathcal{I}} \times \boldsymbol{\alpha}_{\mathcal{B}/\mathcal{I}}$, and the cross-product is not generally zero this time.
6.19 (a) $(7\cos t + 8t - 14t^2)\hat{\mathbf{i}} + (-7\sin t + 2 + 28t^3)\hat{\mathbf{j}} + (2t\sin t - 4t^2\cos t + 7)\hat{\mathbf{k}}$; (b) $7\hat{\mathbf{i}} + 2\hat{\mathbf{j}} + 7\hat{\mathbf{k}}$;
(c) $6.64\hat{\mathbf{i}} + 2.14\hat{\mathbf{j}} + 6.60\hat{\mathbf{k}}$ rad/sec²
6.21 $\boldsymbol{\alpha}_{\mathcal{A}/\mathcal{G}} = 1.12\hat{\mathbf{i}}_1 + 13.4\hat{\mathbf{j}}_1$
6.23 $-1.5\hat{\mathbf{i}} - 4.20\hat{\mathbf{j}} + 4.73\hat{\mathbf{k}}$ rad/sec²
6.25 $\mathbf{v}_{P/\mathcal{I}} = -4\hat{\mathbf{j}} - 6\hat{\mathbf{k}}$ ft/sec;
$\mathbf{v}_{P/\mathcal{G}} = -9\hat{\mathbf{i}} - 4\hat{\mathbf{j}} - 6\hat{\mathbf{k}}$ ft/sec
6.27 $4\hat{\mathbf{i}} + 40\hat{\mathbf{j}}$ ft/sec; $-400\hat{\mathbf{i}} + 80\hat{\mathbf{j}} - 4\hat{\mathbf{k}}$ ft/sec²
6.29 $6\hat{\mathbf{j}}$ rad/sec²; $-32.0\hat{\mathbf{i}} + 1.20\hat{\mathbf{j}} + 8.90\hat{\mathbf{k}}$ in./sec²
6.31 36.5 ft
6.33 $17.3\hat{\mathbf{i}} + 15.7\hat{\mathbf{j}} - 10\hat{\mathbf{k}}$ in./sec;
$-147\hat{\mathbf{i}} + 218\hat{\mathbf{j}} - 64.3\hat{\mathbf{k}}$ in./sec²
6.35 $\mathbf{a} = -14.3\hat{\mathbf{i}} + 2.61\hat{\mathbf{j}}$ ft/sec²;
magnitude = 14.5 ft/sec²
6.37 $x = (v_o t - R)\sin\omega_o t$; $y = (v_o t - R)\cos\omega_o t + R$
6.39 6.60 rad/sec **6.41** Answer given in problem.

6.43

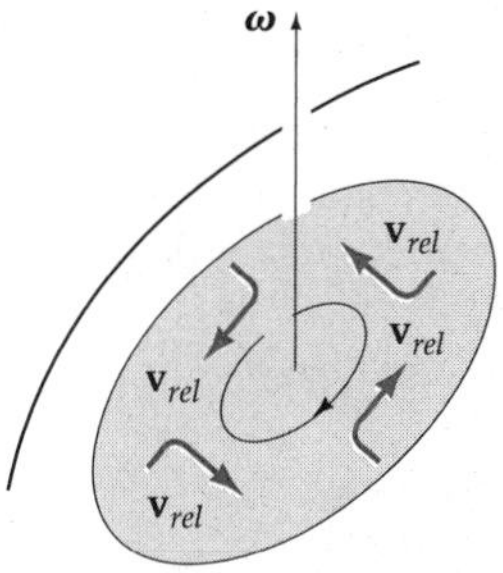

"$-2m\boldsymbol{\omega} \times \mathbf{v}_{rel}$" is a "force" that will change the particles' velocity directions relative to the earth so as to produce ccw rotation in the northern hemisphere. The effect is opposite in the southern hemisphere.

6.45 $-0.0923\hat{\mathbf{i}}$ ft/sec² **6.47** Answer given in problem.
6.49 $\dfrac{2\pi}{T}\left(\sin\beta\hat{\mathbf{j}} - \dfrac{\sin^2\beta}{\cos\beta}\hat{\mathbf{k}}\right)$
6.51 $-1350\hat{\mathbf{j}}$ rad/sec² (each)
6.53 $\mathbf{v}_{P/\mathcal{G}} = -15\hat{\mathbf{i}} - 10\hat{\mathbf{k}}$ ft/sec;
$\mathbf{a}_{P/\mathcal{G}} = -65\hat{\mathbf{j}} - 7.5\hat{\mathbf{k}}$ ft/sec²
6.55 $\mathbf{v}_Q = r\Omega_2 s_1\hat{\mathbf{i}} + (R\Omega_2 - r\Omega_1 C_1)\hat{\mathbf{j}} - r\Omega_1 s_1\hat{\mathbf{k}}$;
$\mathbf{a}_Q = (-R\Omega_2^2 + 2r\Omega_1\Omega_2 C_1)\hat{\mathbf{i}} + (r\Omega_1^2 s_1 + r\Omega_2^2 s_1)\hat{\mathbf{j}} - r\Omega_1^2 C_1\hat{\mathbf{k}}$, where $s_1 = \sin\Omega_1 t$ and $C_1 = \cos\Omega_1 t$
6.57 (a) $\boldsymbol{\omega} = -(R\omega_1/r)\hat{\mathbf{i}} + \omega_1\hat{\mathbf{k}}$; $\boldsymbol{\alpha} = -(R\omega_1^2/r)\hat{\mathbf{j}}$
(b) $R\omega_1^2\hat{\mathbf{i}} + (R^2\omega_1^2/r)\hat{\mathbf{k}}$
6.59 $0.2\left(\dfrac{\sqrt{3}}{2}\hat{\mathbf{i}} + \dfrac{1}{2}\hat{\mathbf{k}}\right)$ rad/sec
6.61 $\dfrac{\omega_1+\omega_2}{2}\hat{\mathbf{k}}$; $\dfrac{R(\omega_1-\omega_2)}{2r}\hat{\mathbf{j}}$; $\dfrac{R(\omega_1+\omega_2)}{2}\hat{\mathbf{i}}$;
$\dfrac{R(\omega_1+\omega_2)}{2}\hat{\mathbf{i}} - \dfrac{r(\omega_1+\omega_2)}{2}\hat{\mathbf{j}} - \dfrac{R(\omega_2-\omega_1)}{2}\hat{\mathbf{k}}$
6.63 1.36 rad/s, directed from O through the line of contact between $\mathcal{C}$ and $\mathcal{D}$.
6.65 $-1.68\hat{\mathbf{i}} + 2.24\hat{\mathbf{j}} - 2\hat{\mathbf{k}}$ rad/sec
6.67 Answer given in problem.
Also, $R = \dfrac{H}{2} + 13$ so right side is $(6H - 260)\omega_{\mathcal{B}}$
6.69 $-0.000432t^5/\sqrt{363 - 0.000144t^6}\,\hat{\mathbf{k}}$ m/s
6.71 (a) Answer given in problem; (b) r/R
6.73 $\boldsymbol{\omega}_{\mathcal{L}} = -22.6\hat{\mathbf{i}}$ rad/sec; $\boldsymbol{\omega}_{\mathcal{R}} = -27.7\hat{\mathbf{i}}$ rad/sec;
$\boldsymbol{\omega}_{\mathcal{D}} = -43.6\hat{\mathbf{j}}$ rad/sec with 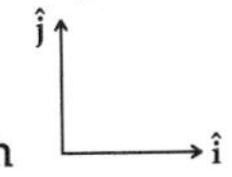

6.75 (a) Answer given in problem.
(b) Answer same if $\boldsymbol{\alpha}$ is replaced by $\boldsymbol{\omega}$.
6.77 $-11.8\hat{\mathbf{i}} + 25.0\hat{\mathbf{j}} - 5.33\hat{\mathbf{k}}$ rad/sec²
6.79 $|\boldsymbol{\omega}_{\mathcal{B}/\mathcal{I}}| = \sqrt{\dot{\phi}^2 + \dot{\theta}^2 + \dot{\psi}^2 + 2\dot{\phi}\dot{\psi}\cos\theta}$
6.81 Components are:
$(\dot{\theta}_1 + s_2\dot{\theta}_3,\ c_1\dot{\theta}_2 - s_1 c_2\dot{\theta}_3,\ s_1\dot{\theta}_2 + c_1 c_2\dot{\theta}_3)$

6.83 Let point A be displaced from its original to its final position. Then, using Euler's Theorem, all other points of the body may be placed in their final positions via a single rotation about an axis through A.
6.85 $\dot{x} = r(\dot{\psi} s_\phi s_\theta + \dot{\theta} c_\phi)$; $\dot{y} = -r(\dot{\psi} c_\phi s_\theta - \dot{\theta} s_\phi)$
6.87 Components are:
$(\dot{\theta}_1 + \dot{\theta}_3 s_2, \dot{\theta}_2 c_1 - \dot{\theta}_3 s_1 c_2, \dot{\theta}_2 s_1 + \dot{\theta}_3 c_1 c_2)$,
the same as Problem 6.73.
6.89 $\dot{E}\hat{\mathbf{n}}_{21} + (S_E \dot{A} + \dot{P})\hat{\mathbf{n}}_{22} + \dot{A} C_E \hat{\mathbf{n}}_{23}$
6.91 $E = \tan^{-1}[(c_\lambda c_\delta - r)/(s_\lambda c_\delta)]$;
$A = \tan^{-1}[s_\delta / \sqrt{c_\delta^2 - 2rc_\lambda c_\delta + r^2}]$, where $r = R_e/R$
6.93 Be sure the cross is rigid and planar!

CHAPTER 7

7.1 With $\hat{\mathbf{i}}$ along the axle from O through the wheel center, and $\hat{\mathbf{k}}$ out of the page,

$$\mathbf{H}_O = \frac{2\pi m b^2}{T}\left[-\cos\beta \sin^2\beta\,\hat{\mathbf{i}} + \left(1 - \frac{\cos^2\beta}{2}\right)\sin\beta\,\hat{\mathbf{j}}\right]$$

7.3 $m\ell^2\omega_0[2\hat{\mathbf{i}} + (10/3)\hat{\mathbf{k}}]$
7.5 With $\hat{\mathbf{i}}$ and $\hat{\mathbf{j}}$ parallel to x and y of the figure,

$$\mathbf{H}_C = \frac{mr^2}{4}\Omega \sin\phi\left(\hat{\mathbf{i}} - \frac{2\ell}{r}\hat{\mathbf{j}}\right)$$

7.7 $I_{xx}^Q = 11.7\,m\ell^2$; $I_{yy}^Q = 14.3m\ell^2$; $I_{zz}^Q = 8.00m\ell^2$;
$I_{xy}^Q = 2.00m\ell^2$; $I_{xz}^Q = 4.50m\ell^2$; $I_{yz}^Q = 0$ **7.9** $0.944m\ell^2$
7.11 $\dfrac{-mR^2}{2}\sin\phi\cos\phi$ **7.13** $2m(a^2 + b^2 + c^2)/15$
7.15 4.33 slug-ft² **7.17** $-0.0186ma^2$
7.19 If $c > a + b$, the ellipsoid $ax^2 + by^2 + cz^2 = 1$ cannot be an ellipsoid of inertia, for then it would represent a body having one moment of inertia > sum of other two, a physical impossibility.
7.21 $I_1 = 1.5mR^2$, d.c.'s (0,0,1);
$I_2 = 1.41mR^2$, $(-0.345, 0.939, 0)$;
$I_3 = 0.0942mR^2$, $(0.939, 0.345, 0)$
7.23 $\frac{5}{6}\ell\hat{\mathbf{i}} + \frac{1}{2}\ell\hat{\mathbf{j}} - \frac{1}{6}\ell\hat{\mathbf{k}}$; the direction is different.
7.25 $I_{xz}^P = \overset{0}{\cancel{I_{xz}^C}} - m\overset{0}{\cancel{a}}c = 0$ and $I_{yz}^P = \overset{0}{\cancel{I_{xz}^C}} - m\overset{0}{\cancel{b}}c = 0$,
so that z is a principal axis for every point on that axis

7.27 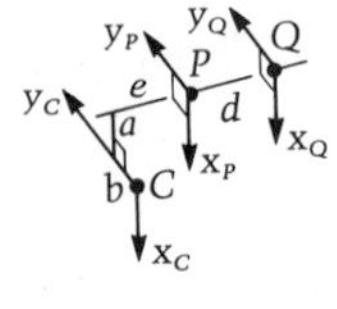

$$I_{xz}^C = \overset{0}{\cancel{I_{xz}^P}} + m(-a)e = mae$$
$$I_{xz}^C = \overset{0}{\cancel{I_{xz}^Q}} + m(-a)(e + d) = ma(e + d)$$

$\therefore a = 0$. From I_{yz} equations, get $b = 0$, so C is on the z axis and by Problem 7.26, z is a principal axis for C.
7.29 $I_1 = 83.4\rho t$, d.c.'s (0, 0, 1);
$I_2 = 74.6\rho t$, $(0.985, -0.170, 0)$;
$I_3 = 8.80\rho t$, $(0.170, 0.985, 0)$
7.31 $I_1 = 37.2\rho A r^3$, d.c.'s (1, 0, 0);
$I_2 = 6.18\rho A r^3$, $(0, 0.454, 0.891)$;
$I_3 = 37.3\rho A r^3$, $(0, -0.891, 0.454)$
7.33 $I_1 = \dfrac{7}{3}m\ell^2$ with $\hat{\mathbf{n}}_1 = \left(\dfrac{1}{\sqrt{3}}, \dfrac{1}{\sqrt{3}}, \dfrac{1}{\sqrt{3}}\right)$;
$I_2 = I_3 = \dfrac{23}{6}m\ell^2$ with $\hat{\mathbf{n}}$ in any direction normal to $\hat{\mathbf{n}}_1$
7.35 $I_1 = 0.278mH^2$, d.c.'s (0, 0, 1);
$I_2 = 0.239mH^2$, $(0.290, -0.957, 0)$;
$I_3 = 0.0387mH^2$, $(0.957, 0.290, 0)$
7.37 3.58°, working with six digits and rounding at the end
7.39 (a)
$$\begin{bmatrix} 1.44m\ell^2 & 0.563m\ell^2 & 0.500m\ell^2 \\ & 1.44\ m\ell^2 & -0.500m\ell^2 \\ & \text{Symmetric} & 1.54\ m\ell^2 \end{bmatrix};$$
(b) $I_1 = 2.00m\ell^2$, d.c.'s $(0.707, 0.707, 0)$;
$I_2 = 1.99m\ell^2$, $(0.379, -0.379, 0.844)$;
$I_3 = 0.427m\ell^2$, $(0.597, -0.597, -0.535)$
7.41 $I_1 = 34.3 \times 10^5$ slug-ft², $\hat{\mathbf{n}} = (0, 0, 1)$
$I_2 = 33.7 \times 10^5$ slug-ft², $\hat{\mathbf{n}} = (0.566, -0.824, 0)$
$I_3 = 47.5 \times 10^5$ slug-ft², $\hat{\mathbf{n}} = (0.824, 0.566, 0)$
Note: There is a precision problem here because (1) I_{xx} and I_{yy} are so much larger than I_{xy}, and (2) I_{xx} and I_{yy} are nearly equal.
7.43 Left: 121 ↓ N; right: 121 ↑ N
7.45 At B_1: $(1419 - 507\cos\epsilon)\sin\epsilon$ ↑;
at B_2: $(659 + 507\cos\epsilon)\sin\epsilon$ ↑
7.47 At A: $\dfrac{mr^2\Omega^2 \sin 2\beta}{16L}$ ↓;
at B: same magnitude but ↑
7.49 (a) 2.41 ⮍ lb-ft; (b) 1.09 ⮌ lb-ft
7.51 $\Sigma F_x = m\dot{v}_{Cx} + \omega_y m v_{Cz} - \omega_z m v_{Cy}$;
$\Sigma F_y = m\dot{v}_{Cy} + \omega_z m v_{Cx} - \omega_x m v_{Cz}$;
$\Sigma F_z = m\dot{v}_{Cz} + \omega_x m v_{Cy} - \omega_y m v_{Cx}$;
7.53 $\dot{\omega}_s = 0$, $\dot{\omega}_x = \dfrac{M_o}{I + \bar{I} + m\ell^2}$,

$$\text{Force} = \left(0, \frac{-m\ell M_o}{I + \bar{I} + m\ell^2}, mg + m\ell\omega_x^2\right),$$
$$\text{Couple} = \left(\frac{IM_o}{I + \bar{I} + m\ell^2}, 0, J\omega_x\omega_s\right)$$

7.55 332 days
7.57 $\dot{\psi} = 30.7$ rad/s, about the $+z$ axis
7.59 0.123 lb-in.
7.61 $\phi = \cos^{-1}\{g\ell/[\Omega^2(\ell^2 - r^2/4)]\}$; for $\phi = 60°$, $\Omega_{\min} = 3.35$ rad/sec $< 2\pi$ rad/sec
7.63 With x along $\mathcal{S}$ from Q and y upward,

$$\mathbf{F} = (-mR\omega_1^2, mg, 0) \text{ and } \mathbf{M}_C = \left(0, 0, \frac{mr^2}{2}\omega_1\omega_2\right)$$

7.65 $\sin\theta - \dfrac{v_C^2}{gR}\left(\dfrac{3}{2}\cos\theta + \dfrac{r\sin\theta\cos\theta}{4R}\right) = 0$

7.67 (a) stern bearing: $149\hat{\mathbf{i}} + R_y\hat{\mathbf{j}} + 8850\hat{\mathbf{k}}$ N; bow bearing: $149\hat{\mathbf{i}} - R_y\hat{\mathbf{j}} + 15650\hat{\mathbf{k}}$ N; (b) stern: $48460\hat{\mathbf{i}} + R'_y\hat{\mathbf{j}} + 12250\hat{\mathbf{k}}$ N; bow: $-48460\hat{\mathbf{i}} - R'_y\hat{\mathbf{j}} + 12250\hat{\mathbf{k}}$ N (R_y and R'_y are indeterminate).

7.69

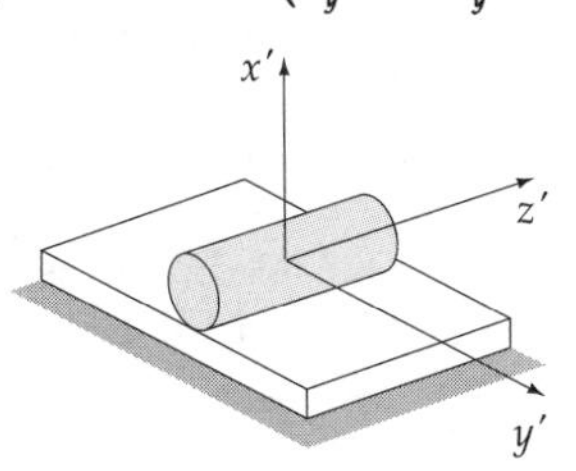

$M_{x'} = \omega_1\dot{\beta}\,(2IC_\beta^2 - J\cos 2\beta)$;
$M_{y'} = (J - I)\omega_1^2 S_\beta C_\beta$;
$M_{z'} = 2(J - I)\omega_1\dot{\beta}S_\beta C_\beta$

7.71 With $\hat{\mathbf{i}}$ from O toward C, and $\hat{\mathbf{j}}$ in the direction of $\mathbf{v}_c$, $\mathbf{F}$ on $\mathcal{D} = \dfrac{-mv_C^2}{R}\hat{\mathbf{i}}$ by the bearing at C, $-mg\hat{\mathbf{k}}$ from gravity, and $mg\hat{\mathbf{k}}$ from the ground; and $\mathbf{C}$ on $\mathcal{D}$ by bearing $= \dfrac{-2IR\Omega^2}{r}\hat{\mathbf{j}}$.
The difference is that in this problem the normal force from the ground is just the weight.

7.73 $12\hat{\mathbf{i}} + 32.2\hat{\mathbf{j}} + 6\hat{\mathbf{k}}$ lb; $14.4\hat{\mathbf{i}} + 20\hat{\mathbf{j}} - 123\hat{\mathbf{k}}$ lb-ft

7.75 RHS of x-eqn = 5.6 (the same);
RHS of y-eqn $= 5.6 \neq 7.7$; RHS of z-eqn $= 2.8 \neq 1.4$;
Axes used for Euler Equations must be *body*-fixed, not just permanently principal.

7.77 $(-\ddot{\phi}s_\theta - \dot{\phi}\dot{\theta}C_\theta + \dot{\psi}\dot{\theta})\hat{\mathbf{i}}_2 + (\ddot{\theta} + \dot{\phi}\dot{\psi} - s_\theta)\hat{\mathbf{j}}_2 + (\ddot{\psi} + \ddot{\phi}C_\theta - \dot{\theta}\dot{\phi}s_\theta)\hat{\mathbf{k}}_2$ **7.79** $\dot{\psi} = 244$ rad/sec

7.81 0.328 m (Also, 51.9 m is a solution!)

7.83 $I_{zz}^C = 18mR^2$; $I_t = 12mR^2$; ratio = 1.5;
(i) is unstable and the other three are stable.

7.85 For a torque-free body $\mathcal{B}$ in general motion in an inertial frame $\mathcal{I}$, with $\boldsymbol{\omega}_{\mathcal{B}/\mathcal{I}}$ not parallel to $\mathbf{H}_C$, we have $\Sigma M_C = \mathbf{0} = {}^{\mathcal{B}}\dot{\mathbf{H}}_C + \boldsymbol{\omega}_{\mathcal{B}/\mathcal{I}} \times \mathbf{H}_C$, and the two terms *add* to zero.

7.87 If $\mathcal{B}$ is in equilibrium in $\mathcal{I}$, all its points are stationary there; thus $\mathbf{a} = \mathbf{0}$ for all these points, and also $\boldsymbol{\omega}_{\mathcal{B}/\mathcal{I}} = \mathbf{0}$. Hence $\mathcal{B}$ is an inertial frame. But if $\mathcal{B}$ is an inertial frame, it can at most translate at constant velocity with respect to another inertial frame $\mathcal{I}$. Thus it need not be stationary in $\mathcal{I}$, i.e., need not be in equilibrium in $\mathcal{I}$ even though none of its points accelerates in $\mathcal{I}$!

7.89 Answer agrees with Example 7.11.

7.91 $\dfrac{11.4\,v_b m}{s(15M + 24m)}\hat{\mathbf{i}} - \dfrac{33v_b m}{s(15M + 24m)}\hat{\mathbf{k}}$

7.93 No. Two different results are obtained for F_y.

7.95 $815m$ ft-lb, where m is the mass in slugs

7.97 $W = T_f = \dfrac{4\pi^2 m(b^2 - r^2)(b^2 - 0.75r^2)}{b^2T^2}$

7.99 (a) $\dfrac{a^2b^2m\omega^2}{12(a^2 + b^2)}$; (b) $\sqrt{\dfrac{a^2 - b^2}{12}}$

7.101 (a) Torque = 0, At B: $\dfrac{ma\omega^2}{12} \leftarrow$, At A: $\dfrac{ma\omega^2}{4} \leftarrow$;
(b) $I_1 = \dfrac{5}{6}ma^2$, z axis;
$I_2 = 0.116ma^2$, d.c.'s (0.290, 0.957, 0);
$I_3 = 0.717ma^2$, d.c.'s (0.957, −0.290, 0);
(c) no hole is physically possible; (d) $ma^2\omega^2/12$

7.103 358 J

7.105 $0.339\,\dfrac{g}{r} = \omega_x^2 - 1.33\omega_x\omega_y + 2.03\,\omega_y^2$

7.107 Some check values: $\left(\alpha, \dfrac{v_Q^2}{gR}\right) = (1°, 913)$, (45°, 0.842), and (80°, 0.031)

CHAPTER 8

8.1 $\omega_n = \sqrt{\dfrac{gr^2/R}{k_C^2 + r^2}}$; for $k_C = \sqrt{\dfrac{2}{5}}\,r$, $\omega_n = \sqrt{\dfrac{5g}{7R}}$

8.3 $m\ell^2\ddot{\theta} + ka^2\theta - mg\ell\theta = 0$; $2\pi\sqrt{\dfrac{m\ell^2}{ka^2 - mg\ell}}$

8.5 $\sqrt{2k/(12m\pi^2)}$ **8.7** 40.0 ft

8.9 Answer given in problem. **8.11** $\omega_n = 3\sqrt{\dfrac{T}{2m\ell}}$

8.13 (a) 0.278 ft (b) 0.876 sec (c) 0.291 sec

8.15 (1) moves to left with $x = \dfrac{6\mu g m}{k}\left(1 - \cos\sqrt{\dfrac{k}{m}}\,t\right)$
(2) moves to right with $x = \dfrac{4\mu g m}{k}\left(1 - \cos\sqrt{\dfrac{k}{m}}\,t\right)$
(3) moves to left with $x = \dfrac{2\mu g m}{k}\left(1 - \cos\sqrt{\dfrac{k}{m}}\,t\right)$ and stops for good $\mu gm/k$ to left of unstretched position. Time in *each* interval is $\pi\sqrt{m/k}$, and total distance traveled $= 24\,\mu gm/k$. **8.17** 0.107 sec

8.19 (a) Work $= \displaystyle\int_0^{L/v} Pvt[\dot{\theta}(t)]\,dt =$
$(P^2/k)\left[\dfrac{1}{2} - \dfrac{v}{\omega L}\sin\dfrac{\omega L}{v} - \dfrac{v^2}{\omega^2L^2}\cos\dfrac{\omega L}{v} + \dfrac{v^2}{\omega^2L^2}\right]$,
(b) which agrees with $T + \phi$.

8.21 $\dfrac{2\ell}{b^2}\sqrt{(mg\ell + ka^2)m}$

8.23 8 lb-sec/in.; 8 times as great;
$\theta_o e^{-100t}(1 + 100t)$; 0.0628 sec

8.25 Yes, they do. (Start with $x = Ae^{-\zeta\omega_n t}\sin(\omega_d t - \varphi)$ and investigate when $\dot{x} = 0$!)

8.27 (a) 0.109 in.;
(b) 1.36 rad or 78.1°, with x lagging y
8.29 $x(t) = 0.5[1 - (1 + 100t)e^{-100t}]$ in.
8.31 (a) $x = 0.025 \cos 100t - 0.0269e^{-26.8t} + 0.0019e^{-373t}$ in. (b) $x = 0.025 \cos 100t - e^{-50t} \cdot (0.0144 \sin 86.6t + 0.025 \cos 86.6t)$ in.
8.33 $-Y\omega^2\sqrt{\dfrac{(c\omega)^2 + (2k)^2}{(2k - m\omega^2)^2 + (c\omega)^2}} \sin(\omega t - \phi)$,
where $\phi = \tan^{-1}\left[\dfrac{cm\omega^3}{2k(2k - m\omega^2) + (c\omega)^2}\right]$;
the radicand is < 1 if $\omega > \sqrt{2}\,\omega_n$, and for these ω's, the radicand is smaller for smaller values of c.
8.35 $wA|\mathbf{v}|\mathbf{v}/g - \mathbf{W}$,
where $\mathbf{W}$ is the weight of tank plus fluid.
8.37 $Q_u/Q = (1 + \cos\theta)/2$; $Q_l/Q = (1 - \cos\theta)/2$
8.39 $q(u - v) + fu$ **8.41** $MP/(M + qt)^2$
8.43 $2480\hat{\mathbf{i}}$ lb **8.45** (a) 0.553 ft/sec²; (b) 328 ← lb
8.47 (a, c) Answer given in problem;
(b) $4.95 \dfrac{m_o g}{k_o}(\sin\alpha - \mu\cos\alpha)$
8.49 Answer given in problem.
8.51 $v(t) = -gt + v_e \ln[m_0/(m_0 - \mu t)]$, valid until fuel gone
8.53 $x(t) = -gt^2/2 + v_e\left[\left(\dfrac{m_0}{\mu} - t\right)\ln\left(1 - \dfrac{\mu t}{m_0}\right) + t\right]$;

$$v_{\text{burnout}} = \frac{-gfm_0}{\mu} - v_e \ln(1 - f);$$

$$x_{\text{burnout}} = \frac{-gf^2m_0^2}{2\mu^2} + \frac{v_e m_0}{\mu}[(1 - f)\ln(1 - f) + f]$$

8.55 (a) $v = \sqrt{2gx/3}$; (b) $a = g/3$;
(c) Mechanical energy is lost (to heat, deformation, vibration, etc.) as the links suddenly join the falling part of the chain. **8.57** Answers given in problem.
8.59 (a) $0.193gT$. (b) Answer given in problem.
8.61 Answer given in problem.
8.63 Answer given in problem. (Use Kepler's laws!)
8.65 1.67×10^9 mi
8.67 107 min ("No air resistance" needn't be assumed this time!) **8.69** 4320 mi; 583 mi and 137 mi
8.71 r_{p*} is 3620 mi, and this is $< r_{\text{earth}}$
8.73 Answer given in problem. (Isolate the radical and square both sides.) **8.75** Answer given in problem.
8.77 $0.00530\sqrt{GM}$ mi/hr **8.79** 5960 mi; 77.5°
8.81 Answer given in problem.
8.83 No; 4980 mph **8.85** $GMm(n - 1)/(2Rn)$
8.87 104 min

APPENDIX A

A.1 (a) lb-ft²/slug²; (b) N · m²/kg² **A.3** 143 N
A.5 (a) 0.456 slug-ft/sec; (b) 2.03 kg · m/s
A.7 (a) $\dfrac{L}{T^2}L^2 = L^3\left(\dfrac{1}{T}\right)^2$;
(b) 6.61, using $g = 32.17$ ft/sec²,
$\omega = 2\pi\left(1 + \dfrac{1}{365}\right)$ rad/day, and $r_e = 3960$ mi

E ADDITIONAL MODEL-BASED PROBLEMS FOR ENGINEERING MECHANICS

KINEMATICS OF ROLLING

Draw radial lines on the ends of a large and a small cylinder or tube, measure their outer diameters, and lay them on a table as shown in (a) below. Predict the number of times the small cylinder will turn in space if it rolls completely around the large cylinder. Hold the large cylinder still and slowly and carefully turn the small cylinder clockwise, maintaining contact and not allowing them to slip. If slipping is a problem, increase the friction between the surfaces by placing a rubber band or strip of tape around the small cylinder. Record the number of times the small cylinder turns in space before it contacts the starting point on the large cylinder. Did you correctly predict this number? Dividing the circumferences to predict the number of revolutions is incorrect! Place a small cylinder inside a large tube as shown in (b). Carefully rotate the small cylinder, not allowing it to slip, on the inside of the tube and measure the number of turns before it returns to the starting point. Use the appropriate kinematic relation, the inside diameter of the large tube, and the outside diameter of the small cylinder to calculate the same result.

REFERENCE — *Dynamics,* Section 3.6.

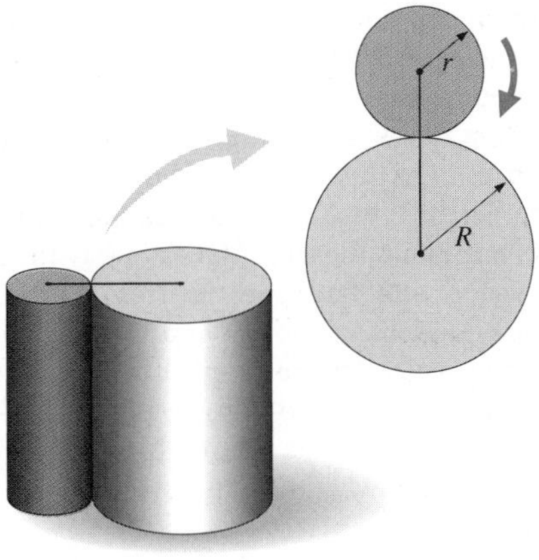

(a)

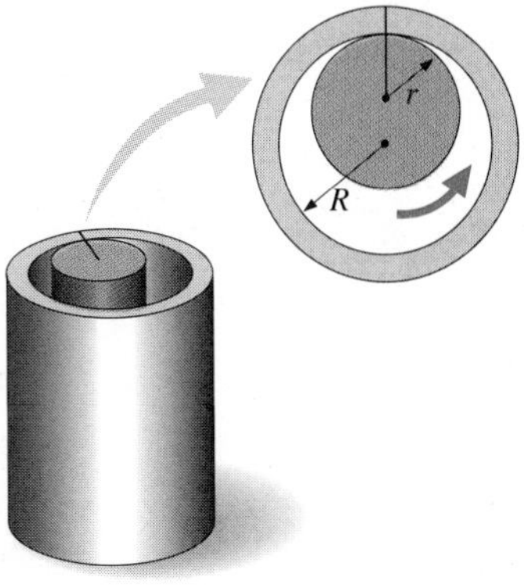

(b)

IMPACT AND TIPPING OF A SLIDING BLOCK

Install a small lip or bump on an inclined straight board and place a rectangular block on the board. Hold the board at an angle $\theta > \tan^{-1}\mu_s$ and release the block from rest behind the lip, as depicted in (a) below. Observe the motion of the block as it slides toward and impacts the lip. Release the block from different positions until you find the length L for which the block slides to the lip and barely tips over after impact, as in (b). Use the equations of motion and the impulse–momentum and work–energy principles to calculate length L in terms of μ_k, θ, and the dimensions of the block. Compare the experimental and theoretical results and explain any differences. Repeat this demonstration for several different values of θ and sizes of blocks. Graph the predicted and actual values of L and θ.

REFERENCES — *Dynamics,* Sections 5.2 and 5.3.

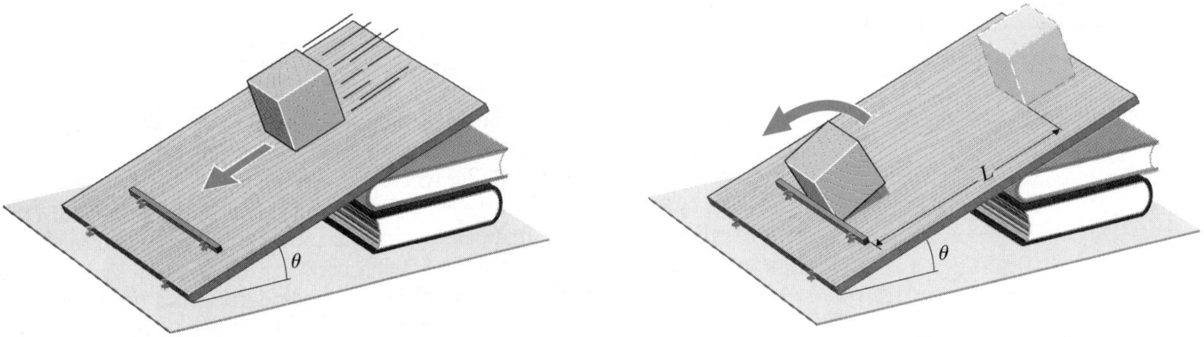

(a) (b)

ROLLING A CYLINDER OVER A STEP

Select a cylinder of radius R and install a step of height $H \cong 0.1R$ on an inclined straight board (a). Hold the board at some convenient angle θ between 10° and 20°. Release the cylinder from rest behind the step as depicted in (b) below. Observe the motion of the cylinder as it rolls toward and impacts the step. Release the cylinder from different positions until you find the length L for which the cylinder just rolls over the step after impact. Repeat the experiment for several values of step height $H \leq R$ and measure the minimum distance L the cylinder must roll to barely climb the step. (An easy way to vary the height is to build the steps from layers of stiff cardboard cut from the backing of a note pad.)

Using the equations of motion and the principles of impulse–momentum and work–energy, calculate the length L in terms of the ratio H/R and the slope angle θ. Plot the predicted and measured values of L versus H/R for different slopes and compare the results. Show that your results are independent of the friction coefficient and weight of the cylinder. Do your results depend upon the coefficient of restitution?

REFERENCES — *Dynamics,* Sections 5.2 and 5.3.

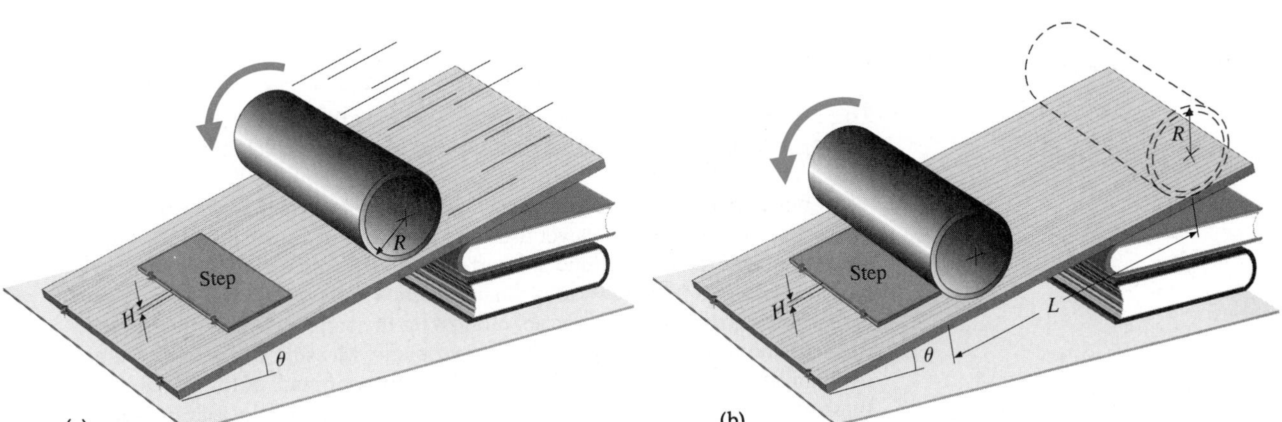

(a) (b)

MOTION, MOMENTUM, AND ENERGY OF A SIMPLE PENDULUM

Build a simple pendulum of length L using string and a small sphere as shown in (a) below. Release the pendulum from rest at angle θ_0 and measure its period of vibration. Repeat the experiment for different lengths and release angles. Compare your values with those predicted from the linear theory of vibrations. Indicate where the actual behavior of your pendulum begins to deviate from the theoretical. Explain any differences.

Place a small peg a distance d below the point of attachment of the pendulum (b). Predict what will happen if the mass is released from rest at $\theta_0 = 90°$. Calculate the minimum distance d_{min} for which the ball will make at least one **complete** revolution around the peg. Determine d_{min} experimentally and explain any difference in the two values. Repeat the calculation and experiment for different release angles. Do you find better agreement between the values?

REFERENCES — *Dynamics,* Sections 2.4 and 8.2.

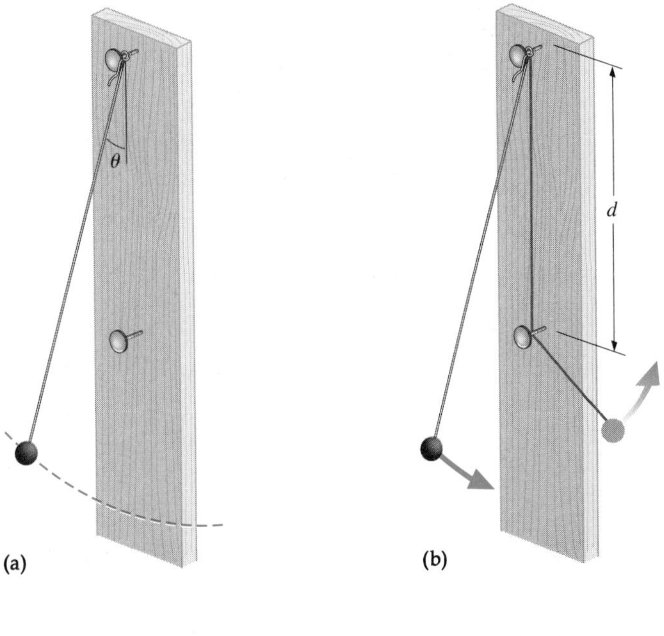

(a) (b)

USING VIBRATIONS TO DETERMINE MOMENT OF INERTIA

This experiment explores a method for determining the mass moment of inertia of a rigid body about an axis of rotation and about a parallel axis through the body's center of mass.

Construct a triangle from a piece of stiff wire (steel coathanger wire is suitable). The sides should be several inches long and of unequal length. Pass a small-diameter rod or thin rigid support through the frame and hold it horizontally, as shown in (a) below (an ordinary notched key works well as a support). One of the corners of the frame should rest on the support. Allow the frame to oscillate about the fixed support in its own plane. Measure the period of these oscillations and the mass and location of the center of mass of the frame. From these values you can compute the moment of inertia I_0 about the axis of rotation. You can also calculate I_0 from the definition of mass moment of inertia if you know the density of the material and the geometry of the frame. Compare the experimental and theoretical values of I_0.

Devise a scheme to determine the mass moment of inertia I_C about the centroidal axis perpendicular to the plane of a steel wire coathanger, as shown in (b). *Hints*: Measure the periods of oscillation about axes through points A and B and the distance AB. Do not attempt to locate the center of mass C. Use the parallel axis theorem. Generalize this approach for irregular rigid bodies with a plane of symmetry.

REFERENCES — *Dynamics,* Sections 4.4, 8.2, and Appendix C.

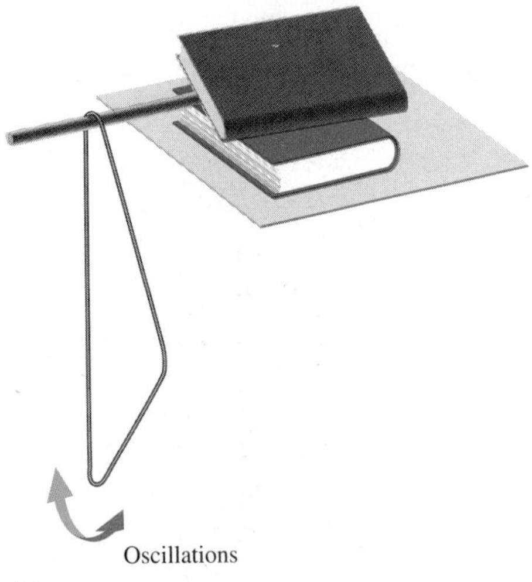

(a)

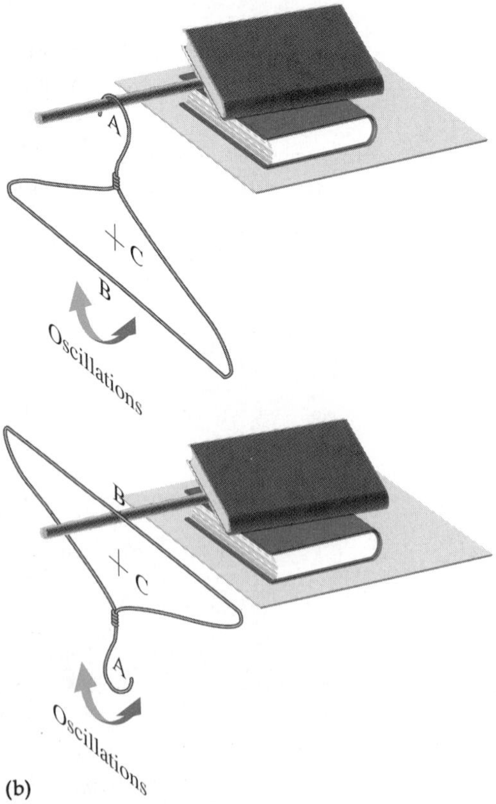

(b)

INDEX